COHERENCE AND QUANTUM OPTICS IV

COHERENCE AND QUANTUM OPTICS IV

Proceedings of the Fourth Rochester Conference on
Coherence and Quantum Optics held at the
University of Rochester, June 8-10, 1977

Edited by

Leonard Mandel and Emil Wolf

Department of Physics and Astronomy
The University of Rochester
Rochester, New York

PLENUM PRESS · NEW YORK AND LONDON

Library of Congress Cataloging in Publication Data

Rochester Conference on Coherence and Quantum Optics, 4th, 1977.
  Coherence and quantum optics IV.

  Includes bibliographical references and index.
  1. Coherence (Optics) – Congresses. 2. Quantum optics – Congresses. I. Mandel,
Leonard. II. Wolf, Emil. III. Title.
QC476.C6R63   1977   535'.2   78-15470
ISBN 0-306-40038-3

PREFACE

This volume contains most of the papers that were presented at
the Fourth Rochester Conference on Coherence and Quantum Optics,
held at the University of Rochester during June 8-10, 1977. The Conference was attended by about 300 registrants, plus a substantial
number of other physicists who participated in some sessions on an
informal basis. Altogether 98 papers were presented in 20 sessions.
One session, on resonance fluorescence, was held jointly with the
International Conference on Multiphoton Processes, that took place
on the Campus during part of the same period.

The Conference was organized by a committee consisting of

N. Bloembergen	(Harvard University)
J.H. Eberly	(University of Rochester)
P. Franken	(University of Arizona)
E.L. Hahn	(University of California-Berkeley)
H. Haken	(University of Stuttgart, Germany)
M. Lax	(City College of CUNY)
A.L. Schawlow	(Stanford University)
C.R. Stroud, Jr.	(University of Rochester)
L. Mandel	(University of Rochester) }joint secretaries
E. Wolf	(University of Rochester)

It was sponsored in part by the Air Force Office of Scientific Research, and we are grateful to them for their support. The Office
of Special Events of the University of Rochester provided valuable
assistance with the organization, as in previous years.

As a glance at the contents of the book will show, the field of
quantum optics is not only alive and well, but it is capable of
spawning new topics from time to time. As new experimental and theoretical techniques are developed, new subjects move to the center of
the stage. For example, the development of the tunable dye laser is
undoubtedly responsible for the renewed interest in resonance fluorescence to which two sessions were devoted, whereas this subject
heading did not appear at all in previous conferences of the series.
New work on superradiance within the last year or two enlivened this
topic and resulted in some spirited controversies. In addition, the
new and potentially important subject of optical bistability grew out
of the fusion of resonance fluorescence with superradiance. Similarly,

v

the well established field of radiative energy transfer received new
interest and emphasis as a result of recent developments in optical
coherence theory. One of the highlights of the Conference occurred
during the June 9 session on quantum electrodynamics and alternatives,
when W.E. Lamb, Jr. declared the winner of the celebrated bet between
P. Franken and E.T. Jaynes made on the occasion of the Second Rochester Conference in the series.

Several secretaries in our Department helped with the typing
of the manuscripts for publication, and we appreciate their assistance. The formidable task of organizing all the editorial and typing work and of keeping track of all the authors' corrections and
the correspondence with them was handled by Mrs. Ruth Andrus, who
also typed a large proportion of the manuscript herself. The success of the project is largely due to her; any shortcomings are ours.

L. Mandel, E. Wolf
Department of Physics and Astronomy
University of Rochester
Rochester, New York 14627

May 1978

CONTENTS

CONTENTS   ix

Two-Photon Resonances

Radiative Transfer and Coherence

QED and Alternatives

Field Correlations and Their Measurements

Superradiance II

CONTENTS xiii

Page

Participants at the
Fourth Rochester Conference on Coherence & Quantum Optics
June 8–10, 1977

RECENT DEVELOPMENTS IN FOUR-WAVE LIGHT MIXING AND TWO-PHOTON

ABSORPTION SPECTROSCOPY

N. Bloembergen

Harvard University, Cambridge, Massachusetts

When three tunable laser beams, at frequencies ω_0, ω_1 and ω_2, respectively, are incident on an atomic vapor, liquid or crystal, a third order polarization at the combination frequency $\omega_3 = \omega_0 + \omega_1 - \omega_2$ is created. The intensity generated at ω_3 reveals resonant interference terms of the material system at $\omega_0 + \omega_1$, $\omega_0 - \omega_2$ and $\omega_1 - \omega_2$, respectively. Such interference terms permit the relative non-linearities of various two-photon absorption and Raman resonant excitations to be calibrated.

Particular attention will be given to interference between imaginary parts of the third order susceptibility and to one-photon resonant contributions. These effects depend sensitively on the sign of $\omega_1 - \omega_2$, whether coherent stokes and antistokes processes are involved.

1

SECOND-ORDER SUM- AND DIFFERENCE-FREQUENCY GENERATION VIA

QUADRUPOLE TRANSITIONS IN ATOMIC VAPORS[*]

Donald Bethune, Robert W. Smith, and Y. R. Shen

University of California, Berkeley, California

Atomic vapors have played an important role in recent work on
optical mixing. They have been used to generate tunable IR radiation, tunable vacuum UV, and coherent radiation with wavelengths as
short as 380 Å. Until recently, the schemes used all involved susceptibilities derived by keeping only the dipole interaction of
light with the medium. Recently we reported second order sum-frequency generation in Na vapor.[1] The dipole susceptibility for
second order mixing vanishes for a vapor, and a more general type
of susceptibility must be considered to describe this process.[2]
For example, our Na sum-frequency generation can be described as
phase matched emission from oscillating atomic quadrupole moments
driven at frequency $\omega_3 = \omega_1 + \omega_2$ by two applied fields, as shown
schematically in Fig. 1. The quadrupole moment density is given by:

$$\underline{\underline{Q}} = \underline{\underline{\chi}}^{(Q)} : \underline{E}_1 \underline{E}_2$$
$$= \chi^{(Q)} \left[\underline{E}_1 \underline{E}_2 + \underline{E}_2 \underline{E}_1 - \frac{2}{3} \underline{\underline{I}} (\underline{E}_1 \cdot \underline{E}_2) \right] . \tag{1}$$

The second form shows that the spherical symmetry of the atoms allows us to define a *scalar* susceptibility, and to derive the quadrupole geometry directly from the incident field vectors. The
quadrupole radiation with $\underline{k}_3 = n_3\omega_3/c \; \hat{e}_3$ is equivalent to that radiated by dipole of strength

$$\underline{P}_{\text{eff}} = -i\underline{k}_3 \cdot \underline{\underline{Q}}$$

$$= -i\chi^{(Q)} \left[(\underline{k}_3 \cdot \underline{E}_1)\underline{E}_2 + (\underline{k}_3 \cdot \underline{E}_2)\underline{E}_1 - \frac{2}{3}(\underline{E}_1 \cdot \underline{E}_2)\underline{k}_3 \right] . \tag{2}$$

The dot products in Eq.(2) show immediately that a non-collinear
geometry must be used, and can also be used to show that the sum-

3

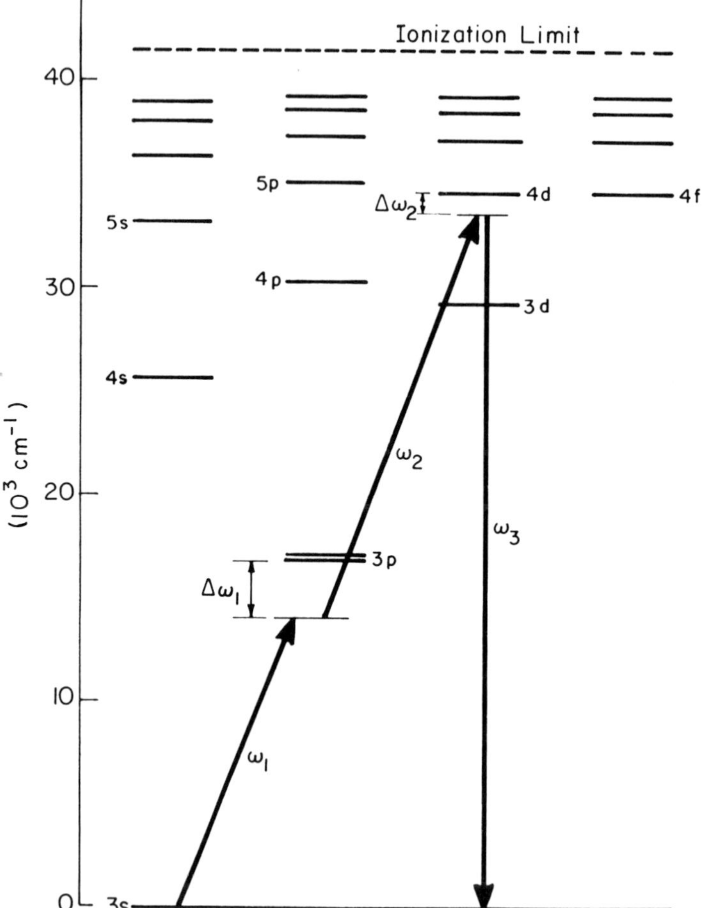

Figure 1: Partial level diagram of the sodium atom. Quadrupole sum-frequency generation with $\omega_3 = \omega_2 + \omega_1$ is shown schematically.

frequency is maximum for orthogonally polarized beams. The output power can be explicitly calculated. The quadrupole susceptibility is given by

$$\chi^{(Q)} = \frac{Ne^3}{\hbar^2}\sum_{rst}\left[\frac{\langle r|yz/2|s\rangle\langle s|y|t\rangle\langle t|z|r\rangle}{(\omega_{sr}-\omega_1-\omega_2-i\gamma)(\omega_{tr}-\omega_1)} + (1\leftrightarrow2)\right] , \qquad (3)$$

which can be evaluated using tabulated matrix elements. We find the output power can be expressed as

$$\mathcal{P}(\omega_3) = \frac{4\pi^3\omega_3^4}{c^5} |\chi^{(Q)}|^2 \, \mathcal{P}(\omega_1)\mathcal{P}(\omega_2) \, e^{-\Delta k_z^2/2\sigma_k^2} \quad , \tag{4}$$

where $\Delta k_z = |k_3 - k_1 - k_2|$ is the wave vector mismatch. For example, for Na with $N = 10^{16}$ cm^{-3}, $\gamma_{Laser} = .25$ cm^{-1} and $\omega_{3p_{1/2}} - \omega_1 = 10$ cm^{-1}, $\chi^{(Q)} = 3 \times 10^{-14}$ esu, which gives an effective dipole susceptibility $|k_3\chi^{(Q)}| \sim 10^{-9}$ esu, as large as $\chi^{(2)}$ for quartz. Equation (4) then gives $\mathcal{P}(\omega_3) = \mathcal{P}(\omega_1)\mathcal{P}(\omega_2)/(11.7 \times 10^6)$ watts.

The experimental arrangement used to observe quadrupole sumfrequency generation is shown in Fig. 2. Generally the characteristics of the observed output agree with those predicted by Eq. (4). In particular, the output shows the sharp two-photon resonance expected from the expression for $\chi^{(Q)}$ when $\omega_1 + \omega_2 = \omega_{4d}$ (Fig. 3), and also shows the expected variation with $\underline{k}$ vector mismatch, which is adjusted by changing the beam intersection angle (Fig. 4).

At high densities and high input powers, however, deviations from the simple theory occur, and the output is less than one would expect from Eq. (4). First, with increasing densities, the linear absorption associated with the D-line resonances causes the output

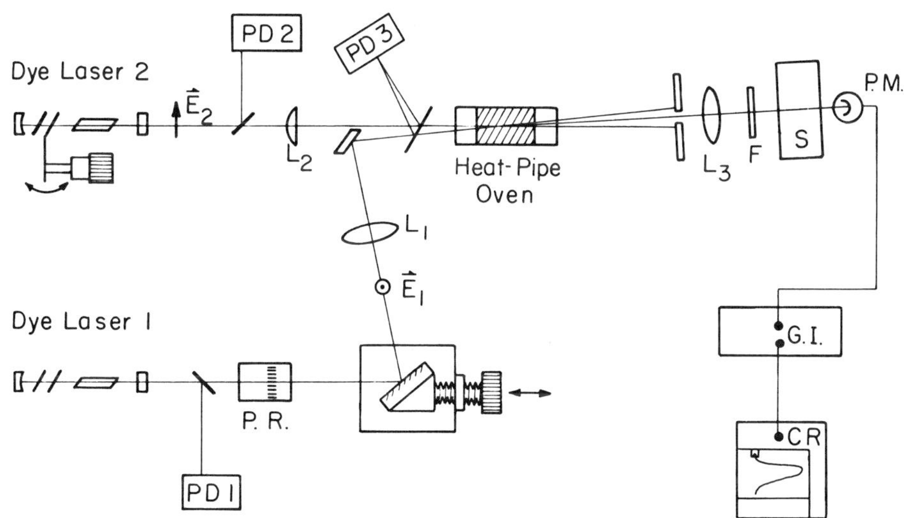

Figure 2: Experimental set-up. PD1, PD2, PD3- monitor photodetectors; L_1, L_2 - 40 cm and 50 cm lenses; L_3 - 10 cm quartz lens; PR - polarization rotator; F- Corning 7-54 filter and pyrex attenuators; S -25 cm spectrometer; PM- RCA 4837; G. I. - gated integrator; CR - chart recorder.

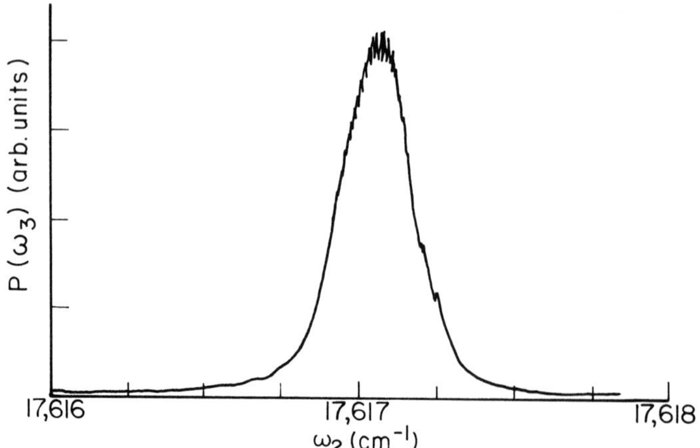

Figure 3: Sum-frequency output $P(\omega_3)$ as a function of ω_2 showing the sharp resonance at $\omega_1 + \omega_2 = \omega_{4d} = 34548.8$ cm^{-1}. $P_1 = 2W$; $P_2 = 25W$; $\theta = 47.9$ mrad; $N = 1.6 \times 10^{16}$ cm^{-3}.

to fall sharply as shown in Fig. 5. Since we rely on the dispersion of the D-lines for phase matching and resonance enhancement of $\chi^{(Q)}$, we cannot tune very far from the D-lines, and the density must therefore be limited to 10^{17} cm^{-3} or less. In addition, Na$_2$ absorp-

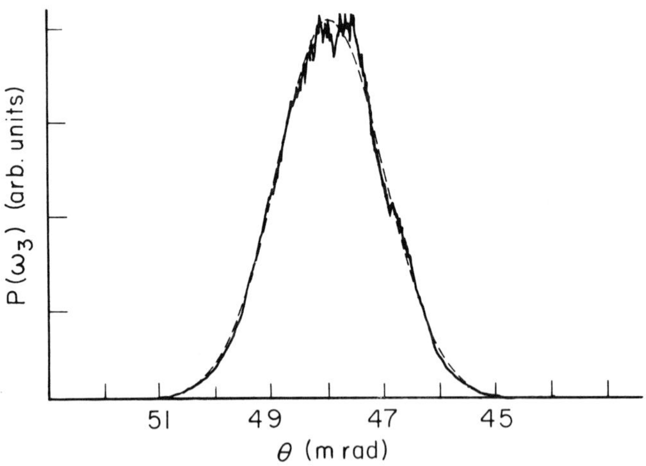

Figure 4: Phase-matching curve $P(\omega_3)$ versus θ. $P_1 = 2W$; $P_2 = 25W$; $\Delta\omega_1 \equiv \omega_{3p\frac{1}{2}} - \omega_1 = 25.6$ cm^{-1}; $\omega_1 + \omega_2 = \omega_{4d}$; $N = 1.6 \times 10^{16}$ cm^{-3}. The dashed curve is a theoretical curve calculated from Eq. (4).

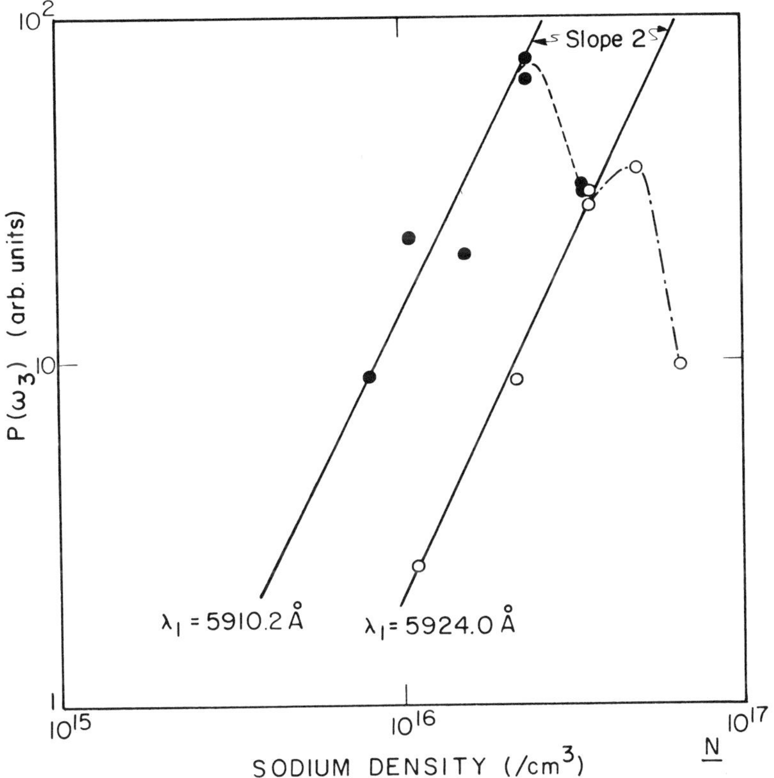

Figure 5: Phase-matched sum-frequency output $P(\omega_3)$ as a function of sodium density N at $\Delta\omega_1$ = 40.8 cm^{-1} and 80.4 cm^{-1}. The other parameters fixed in the experiment are $P_1 \simeq 2W$, $P_2 \simeq 20W$, and $\omega_1 + \omega_2 = \omega_{4d}$.

tion increases strongly at densities above 10^{17}.

A second limiting factor is two-photon saturation. Since the output power is proportional to the square of the population difference between the 3s and 4d states, we have

$$P(\omega_3) \propto [1/W_{tp}T]^2 ,\tag{5}$$

where W_{tp} is the two-photon excitation rate from 3s to 4d and T is an effective upper-state relaxation time. For Gaussian beams the spatial variation of the intensities must be considered, and in this case

$$\mathcal{P}(\omega_3) \propto \frac{I_1 I_2}{|\Delta\omega_2|^2} \int_{-\infty}^{\infty} \frac{dy}{e^{y^2} + \frac{wI_1I_2}{|\Delta\omega_2|^2}} \quad , \tag{6}$$

where w is a constant related to the two-photon transition probability and $\Delta\omega_2 = \omega_{4d} - \omega_1 - \omega_2 - i\gamma$. Eq. (6) predicts both saturation of the output with increasing $I_1 \cdot I_2$, and broadening of the two-photon resonance curve. In Fig. 6 we compare these predicted variations with the experimentally observed saturation and broadening. The constant, w, was chosen to give the best fit to the saturation data. The fit value, $w = 1.55 \times 10^{-3}$ is in satisfactory agreement with the theoretical value $w = 1.2 \times 10^{-3}$. Saturation of the signal on two-photon resonance becomes significant for $I_1 I_2 \approx 10^{11} w^2/ cm^4$, for an intermediate state detuning of 41.2 cm.

Finally, at high intensities, third order nonlinearities and population redistribution can cause changes in the index of refraction of the vapor which defocus the pump beams and, more importantly, destroy the phase matching of the output radiation. By measuring the defocusing of the beam nearest the D lines, we deduce that induced changes of the refractive index for that beam may be on the order of 10^{-5}. Both theoretical estimates and experiment show that index changes of that magnitude are sufficient to reduce the output substantially below the value expected if no induced index changes were present. A refractive index change of this magnitude is produced in Na by an intensity as low as 1 MW/cm^2 at 40 cm^{-1} detuning from the D lines and a pressure of 2 torr, by saturation of the dispersion. It can also be produced by fractional ionization (due to resonant multiphoton ionization) of only one percent. Such index changes are the strongest factors limiting conversion efficiency in our sum-frequency generation experiments.[3]

Besides frequency mixing, quadrupole sum-frequency generation also allows us to use a novel and accurate technique to measure quadrupole transition moments of atoms and molecules.[4] The technique uses the interference of quadrupole sum-frequency generation and DC field-induced third order sum-frequency generation to allow direct comparison of the quadrupole moment with products of dipole moments. Since the dependence of the two processes on the beam geometry and intensities, phase matching, and atomic density are all common, only the matrix elements themselves are compared. Thus, we expect the technique to be capable of high accuracy.

To describe the interference theoretically, the total effective polarization radiating in the direction $\underline{k}_3$ is written:

$$\underline{P} = [-i\underline{k}_3 \cdot \underline{\chi}^{(Q)} + \underline{\chi}^{(3)} \cdot \underline{E}_0] : \underline{E}_1 \underline{E}_2 , \tag{7}$$

where E_0 is the DC field. This polarization is proportional to:

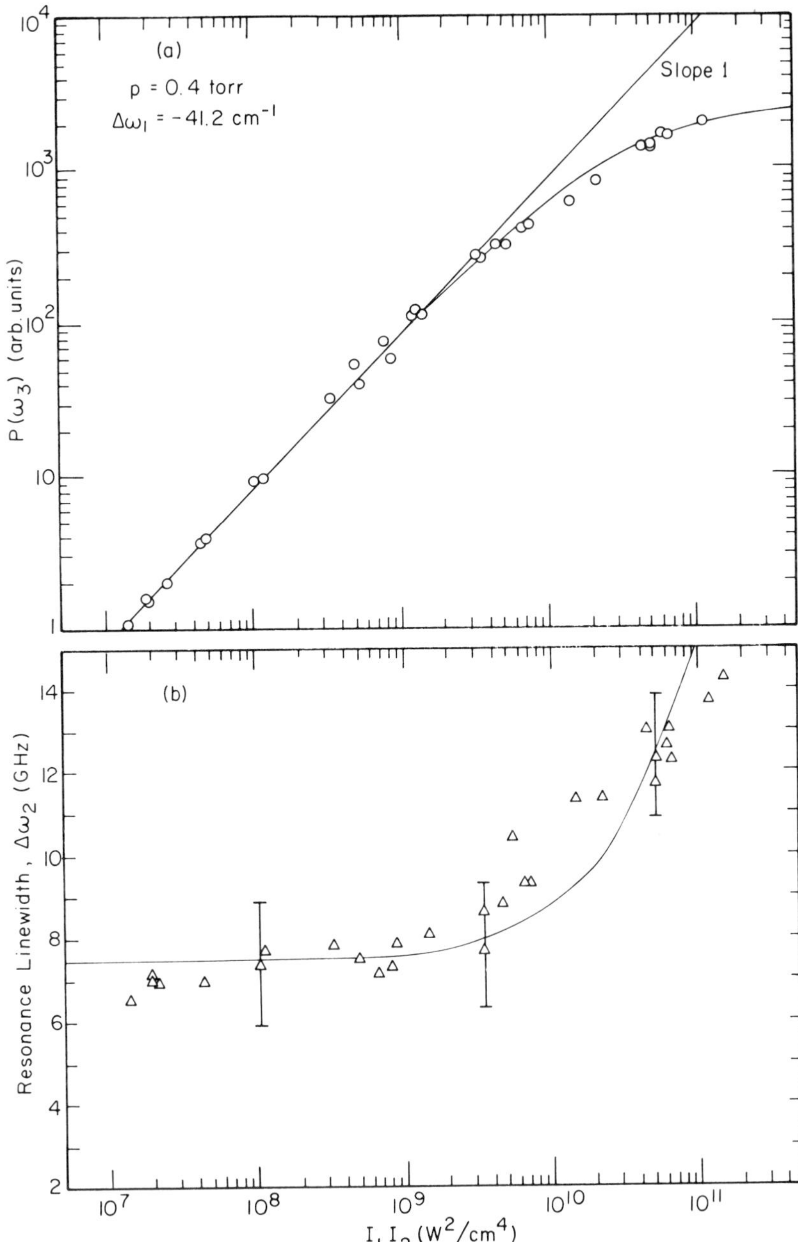

Fig. 6: (a) Sum-frequency power $P(\omega_3)$ and (b) resonance linewidth $\Delta\omega_2$ as functions of the product of the input intensities $I_1 I_2$. The solid curves are both derived from Eq.(6) for the effects of two-photon saturation with w = 1.55×10^{-3}.

$$\underline{P} \propto [-i\underline{k}_3 \cdot \underline{\underline{M}}_Q + \underline{\underline{M}}_D \cdot \underline{E}_0] \ , \tag{8}$$

where:

$$\underline{\underline{M}}_Q = <a|\underline{r}\ \underline{r}/2|b>$$

$$\underline{\underline{M}}_D = e\sum_c [\frac{-<a|\underline{r}|c><c|\underline{r}|b>}{\hbar(\omega_3 - \omega_c)} + \frac{<c|\underline{r}|b><a|\underline{r}|c>}{\hbar\omega_c}] \ .$$

When E_0 is adjusted and both light fields are linearly polarized, the output is *circularly* polarized when:

$$E_0 = \pm k_3 M_{Qzz}\tan\theta_2/M_{Dzz} \ , \tag{9}$$

where θ_2 is the angle between $\underline{k}_2$ and $\underline{k}_3$. When the ω_2 input beam is circularly polarized, the component of the output parallel to $\underline{k}_1 \times \underline{k}_2$ has a minimum at a field

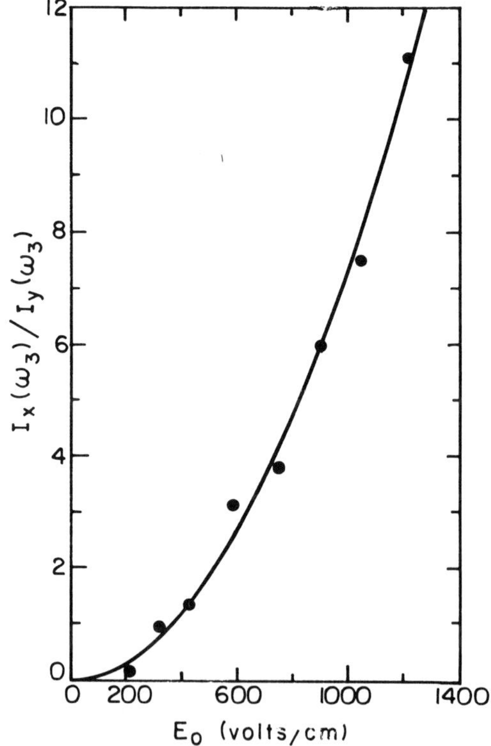

Figure 7: $I_x(\omega_3)/I_y(\omega_3)$ as a function of the applied DC field E_0 for linear polarized input beams. $I_x(\omega_3)$ and $I_y(\omega_3)$ are phase-matched sum-frequency signals polarized along $\hat{x}$ and $\hat{y}$, respectively. The solid curve is a theoretical curve obtained from Eq.(7) to fit the data points.

$$E_0 = 3/4(k_3 M_{Qzz} \sin\theta_2/M_{Dzz}). \tag{10}$$

Either condition allows us to determine (M_Q/M_D), including sign, if we know the electric field strength. To demonstrate this technique, we have measured the 3s-4d quadrupole transition moment of the sodium atom. $\underline{E}_0$ was applied with a pair of flat plate electrodes, heated to prevent Na condensation. The results for linearly polarized input light are shown in Fig. 7, while those for one input beam circularly polarized are shown in Fig. 8. Using tabulated values for the dipole matrix elements, we derive the values $<3s|zz/2|4d> = +2.24a_0^2$ and $<3s|zz/2|4d> = +2.27a_0^2$,respectively, from the two measurements. Laser fluctuations limited our accuracy in this experiment to about $\pm20\%$.

We have so far considered the sum-frequency generation process in which mixing of two pump beams at ω_1 and ω_2 induces a quadrupole polarization at $\omega_3 = \omega_1 + \omega_2$. Another interesting possibility involves the creation of an electric *dipole* polarization by mixing of two pump fields, one of which is coupled to a quadrupole transition. Because the transition is so weak, a laser can be tuned directly to

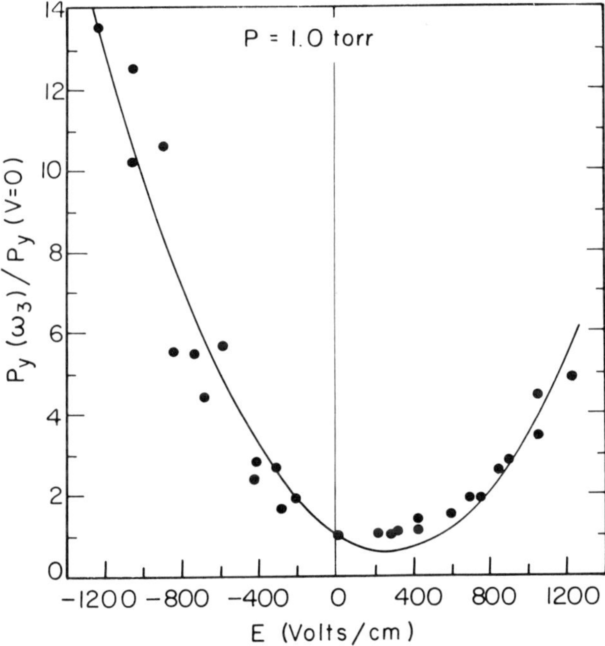

Figure 8: $I_y(\omega_3)/I_y(\omega_3, E_0 = 0)$ as a function of the applied DC field E_0 for $\underline{E}(\omega_1)$ linearly polarized along y and $\underline{E}(\omega_2)$ left circularly polarized.

the quadrupole transition to achieve maximum resonant enhancement. Such a process can be described by the *difference* frequency polarization

$$\underline{P}(\omega_1 - \omega_2) = i\chi^{(Q_1)} [\underline{E}_1 (\underline{k}_1 \cdot \underline{E}_2^*) + \underline{k}_1 (\underline{E}_1 \cdot \underline{E}_2^*)] \ , \tag{11}$$

where

$$\chi^{(Q_1)} = \frac{Ne^3}{\hbar^2} \sum_{rst} \frac{\langle r|z|s\rangle\langle s|y|t\rangle\langle t|zy/2|r\rangle}{(\omega_{sr} - \omega_2)(\omega_{tr} - \omega_1)} \ .$$

As an example, if we tune two lasers near the 6s-6d quadrupole and 6s-7p dipole transitions of cesium vapor, we should obtain infrared difference frequency generation in the range 14-15μ near the 7p-6d dipole allowed transition of the atoms. For $N = 10^{17} cm^{-3}$, $\omega_{7p_{3/2}} - \omega_2 = 2 cm^{-1}$ and $\omega_{6d_{5/2}} - \omega_1 = .25$ cm^{-1} we find $\chi^{(Q_1)} = 2.1 \times 10^{-12} esu$ and $\mathscr{P}(\omega_3) = \mathscr{P}(\omega_1)\,\mathscr{P}(\omega_2)/15\mu$W.

The inclusion of higher multipole terms in the interaction of light with matter allows second order nonlinear processes to be observed in isotropic media. These processes may be useful both for generating new wavelengths of radiation, and for measuring atomic or molecular quadrupole transition moments, which are difficult to measure using other techniques. We may thus observe and exploit weak transitions, which until recently, were routinely neglected.

*
This work was done with support from the U.S. Energy Research and Development Administration.

References

1. D.S. Bethune, R.W. Smith and Y.R. Shen, Phys. Rev. Letts. *37*, 431 (1976).
2. P.S. Pershan, Phys. Rev. *130*, 919 (1963); E. Adler, Phys. Rev. *134*, A728 (1964).
3. Further details on these measurements to be published in Phys. Rev. A.
4. D.S. Bethune, R.W. Smith and Y.R. Shen, Phys. Rev. Letts. *38*, 647 (1977).

FINE STRUCTURE OF THIRD HARMONIC GENERATION

M. L. Ter-Mikaelian

Armenian Academy of Science, Ashtarak, USSR

Various types of coherent effects present when atoms interact
with intense resonant field are considered. Relaxation and collisions
are disregarded. Three- and four-level systems are investigated.
As an example, the fine structure in third harmonic generation is
calculated.

RABI FLOPPING AND COHERENT TRANSIENTS IN A JOSEPHSON JUNCTION

A. DiRienzo[*], D. Rogovin[*] and M. Scully

University of Arizona, Tucson, Arizona

I.   INTRODUCTION

Recently, studies involving coherent transient phenomena in
Josephson devices have been discussed from a quantum optical per-
spective [1,2]. The motivation for these investigations has been
the close analogy between two-level quantum optical systems and
Josephson devices [3]. In particular, the angular momentum formulation of an n-atom system according to Dicke [4] has a close analogy
to the corresponding problem of n-Cooper pairs interacting with the
radiation field in a Josephson tunnel junction. In this article,
we briefly review the fundamental concepts of superconductivity and
superconducting tunnel junctions, and emphasize the relation between
Josephson junctions and superradiant physics. We then sketch the
analysis as applied to a tunnel junction undergoing dynamical transients first from a quantum optical and then from the common phenomenological point of view. We find that the present results differ
from those obtained in the usual phenomenological fashion.

II.   REVIEW OF CONCEPTS AND NOTATION

A. Superconductivity

We begin with a brief review of several concepts underlying
superconductivity. As Cooper [5] showed in 1956, it is electron
pairs that are all-important in superconductivity. An attractive
interaction between these electrons comes about as a result of the
attractive electron-phonon interaction. The electrons tend to

"pair up" in states of the appropriate momenta $\underline{p}$ and $-\underline{p}$. This is physically reasonable if we think of the attractive electron interaction as binding the electron pairs in an s state similar to positronium. For if we were to solve the hydrogenic atom in a basis of plane waves its bound states would be described in terms of a continuous scattering from one (paired $\underline{p}$ and $-\underline{p}$) plane wave to another. Bardeen, Cooper and Schrieffer [6] then used these concepts to derive a successful microscopic theory of superconductivity in 1957. They gave, at that time, a state which emphasized the central role played by the pair concept

$$|\psi>_{BCS} = \prod_k |\psi_{pair}>_k \ .$$

$|\psi_{pair}>_k$ is a state describing a pair of electrons with momenta $\underline{k}$ and $-\underline{k}$. Each of these pair states, in the BCS ground state, is a coherent superposition of occupied and unoccupied electron pair states

$$|\psi_{pair}>_k = U_k |empty>_k + V_k |full>_k \ . \tag{1}$$

For our purposes, we can take $U_k = V_k = 1/\sqrt{2}$ for all states within a Debye energy of the Fermi surface (Figure 1).

We now choose to write things in terms of Anderson's pseudospin formulation [7]. Anderson's notation is based on an analogy between two-level systems and the occupation or non-occupation of a pair-state. In this framework

$$|empty>_k = \begin{pmatrix} 1 \\ 0 \end{pmatrix}_k \ ,$$

$$|full>_k = \begin{pmatrix} 0 \\ 1 \end{pmatrix}_k \ , \tag{2}$$

and the following operators are defined

$$\sigma_k^+ = \begin{pmatrix} 0 & 1 \\ 0 & 0 \end{pmatrix}_k \qquad \text{— annihilation operator for an electron pair of momentum k,} \tag{3a}$$

$$\sigma_k^- = \begin{pmatrix} 0 & 0 \\ 1 & 0 \end{pmatrix}_k \qquad \text{— creation operator for an electron pair of momentum k,} \tag{3b}$$

$$\sigma_{kz} = \frac{1}{2} \begin{pmatrix} 1 & 0 \\ 0 & -1 \end{pmatrix}_k \quad \text{— acts like a number operator.} \tag{3c}$$

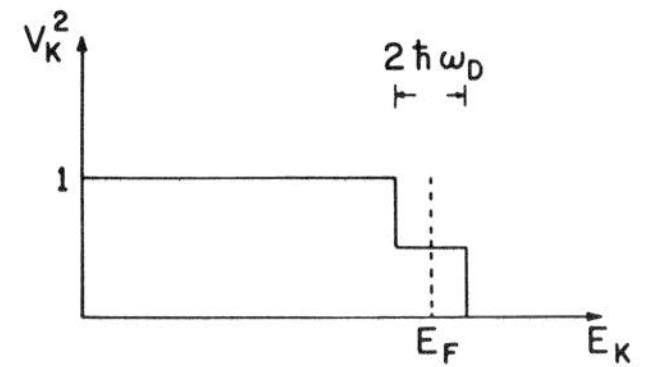

Figure 1. V_k^2 is the probability of occupation, and m is the number of pairs within $\pm\hbar\omega_D$ of the Fermi energy.

These then allow us to write at zero temperature,

$$|\psi\rangle_{BCS} = \prod_k (U_k + V_k\sigma_k^-)\begin{pmatrix}1\\0\end{pmatrix}_k \quad . \tag{4}$$

B. Josephson Junctions

In 1962, Josephson predicted [8] that when two superconductors are brought close to one another, coherent tunneling of electron pairs can take place. The above pseudospin notation lends itself remarkably well to the study of such devices. The problem here is to describe the tunneling of Cooper pairs from one side of the junction to the other (Figure 2). Mathematically we can describe tunneling from left to right by an operator [9]

$$\hat{S}^- = \frac{\hat{S}_L^- \hat{S}_R^+}{\sqrt{2}\,m} \quad ,$$

where

$$\hat{S}_L^- = \sum_k \sigma_k^+ \quad \text{— destroys electron pairs on the left,} \tag{5a}$$

$$\hat{S}_R^+ = \sum_q \sigma_q^- \quad \text{— creates electron pairs on the right,} \tag{5b}$$

and m = number of pairs within $\hbar\omega_{Debye}$ of ε_{Fermi}, (see Fig. 1).

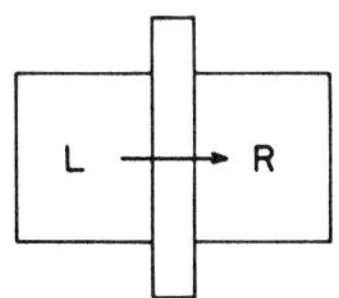

Figure 2. Two superconductors separated by a thin oxide layer with coherent tunneling of electron pairs.

Similarly, we can describe the inverse process by

$$\hat{S}^{+} = \frac{\hat{S}^{-}_{R}\,\hat{S}^{+}_{L}}{\sqrt{2}\,m} \quad .$$

Finally, a third operator $\hat{S}_{z} = 1/2(\hat{S}_{zL} - \hat{S}_{zR}) = -1/2(\Sigma\sigma_{kz} - \Sigma\sigma_{qz})$ can also be defined. This operator is a direct measure of the charge imbalance across the junction. These three operators: $\hat{S}^{\pm}$, $\hat{S}_{z}$, the pair transfer operators and pair charge difference operator, respectively, form a vector, the components of which satisfy approximate angular momentum commutation rules.

It is straightforward now to calculate the dc Josephson current. We write the tunneling Hamiltonian as[10]

$$H_{T} = \sqrt{2}\,mg'(\hat{S}^{+} + \hat{S}^{-}) \quad .$$

Next, we define a current operator

$$J = -2e\dot{\hat{S}}_{z} \quad ,$$

which may be written as $J = -2e(i/\hbar)[H_{T},S_{z}]$. Using the above form of H_{T} and angular momentum commutation relations, we find

$$J = \frac{2\sqrt{2}\,emg'}{i\,\hbar}\,(\hat{S}^{-} - \hat{S}^{+}) \quad ,$$

which leads us to the expression $J = j_{1}\sin\phi$, where we have used Eqs.(11) and factored $m\sin\theta$ into j_{1}. The mean value of J then gives the famous Josephson equation.

III.   SUPERRADIANCE AND THE JOSEPHSON JUNCTION

Before going into the details of our Josephson junction calculation, we show that the Hamiltonian describing Dicke superradiance processes is, in form, very close to that which we will use in describing the dynamics of our Josephson junction[11].

We saw in the last section how the BCS state could be written in the language of two-level systems.

$$|\psi\rangle_{BCS} = \prod_{k}[U_{k}\begin{pmatrix}1\\0\end{pmatrix}_{k} + V_{k}\begin{pmatrix}0\\1\end{pmatrix}_{k}] \quad .$$

It now becomes clear that there is a striking similarity between this state and the Dicke superradiant state [4]

$$|\psi_n> = \prod_{i=1}^{n} [U_i \begin{pmatrix} 1 \\ 0 \end{pmatrix}_i + V_i \begin{pmatrix} 0 \\ 1 \end{pmatrix}_i] \; . \tag{6}$$

Here $\begin{pmatrix} 1 \\ 0 \end{pmatrix}_i$ and $\begin{pmatrix} 0 \\ 1 \end{pmatrix}_i$ are, respectively, the upper and lower energy states of an atom. This analogy can be carried further.

The Hamiltonian used by Dicke to describe the interaction of a many-atom system with a radiation field is

$$H_D = \hbar\omega_o \hat{R}_3 - \underline{A} \cdot (\underline{e}_1 \hat{R}_1 + \underline{e}_2 \hat{R}_2) \; . \tag{7}$$

The operator R_k is defined as

$$\hat{R}_k = \sum_{j=1}^{n} R_{jk} \; , \qquad k = 1,2,3 \tag{8}$$

where, apart from a factor of $\hbar$, R_{jk} is the k-th component of the spin-1/2 angular momentum operator. Note that $\underline{R}$ satisfies angular momentum commutation relations. Also $\hbar\omega_o$ is the atomic excitation energy, while $\underline{A}$ is the applied radiation field.

This Hamiltonian can be put into the form

$$H_D = \hbar\omega_o \hat{R}_3 + \tilde{g}[A^+ \hat{R}^- + A^- \hat{R}^+] \; , \tag{9}$$

where $A^{\pm} = (1/2)A_o e^{\pm i\omega t}$. Here the first term represents the internal energies of the n atoms while the second is the interaction energy between the field and the atomic system.

In the next section we will describe the transient behavior of a Josephson junction in the presence of an external radiation field. The Hamiltonian used to describe the dynamics of such a system is

$$H = \frac{(2e)^2}{2C} \hat{S}_z^2 + i\hbar g(A^+ \hat{S}^- - A^- \hat{S}^+) \; , \tag{10}$$

as discussed in Figure 3.

$A^+ = (1/2)A_o e^{+i\omega t}$ is the negative frequency part of the applied field, while C is the capacitance of the junction. The first term, in this case, represents the electrostatic energy of the pairs due to an applied dc voltage, while the second describes the interaction of the tunneling pairs with the applied electromagnetic field.

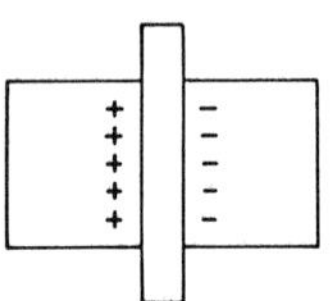

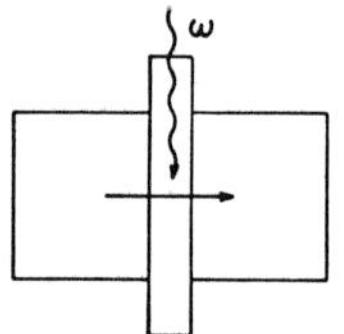

Figure 3a. When a voltage is applied across the junction, it acts as a capacitor with charges piled up on each side of the oxide layer

$$H_C = \frac{Q^2}{2C} = ((2e)^2 S_z^2 / 2C) \ .$$

Figure 3b. As pairs tunnel through the oxide layer they interact with the electromagnetic field via their current

$$H_I = \frac{1}{c} \int d^3r \underline{A}(r) \cdot \underline{j}(r)$$

$$= i\hbar g(A^+ S^- - A^- S^+) \ .$$

Note that Eqs. (9) and (10) differ in their dependence on the z-component of the angular momentum operator.

IV. JUNCTION DYNAMICS

A. Present Approach

We consider the following situation. A Josephson junction is initially held in series with a battery and a resistance. In addition, we subject the junction to an external radiation field of frequency $\omega \equiv \omega_0 = 2eV_0/\hbar$ (Figure 4). With the system prepared in this manner, we open the switch at t = 0 and observe how the junction evolves via pair tunneling. Throughout the experiment $T \ll T_c$ so that quasi-particle tunneling processes are negligible. We can also ignore spontaneous emission by choosing the external field to be sufficiently strong [14].

The dynamics of the junction are governed by the Hamiltonian in Eq.(10). In light of the analogy between n two-level atoms and a Josephson junction [3], it is apparent that not only the phase difference ϕ between the two sides of the junction, but also the Bloch angle θ, of the vector $\underline{S}$, is needed to describe the state of the system. We incorporate these variables into our formalism by noting that $\underline{S}^2 = S_z^2 + (S^+S^- + S^-S^+)/2$ is conserved, and setting

$$S_z = m \cos \theta \ , \tag{11a}$$

$$S^{\pm} = me^{\pm i\phi} \sin \theta \ . \tag{11b}$$

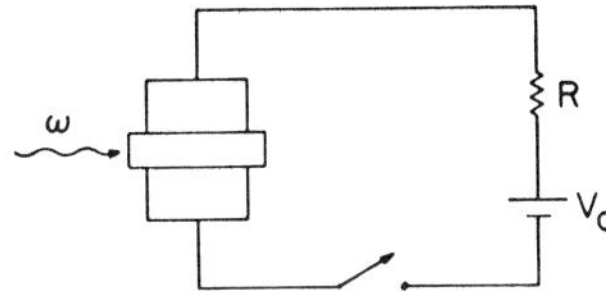

Figure 4. A Josephson junction emersed in a radiation field, while initially having an applied voltage across it.

Using H, and the commutation rules for angular momentum the equations of motion for S_z and S^+ become

$$\dot{S}_z = -g(S^+A^- + S^-A^+) \quad , \tag{12a}$$

$$\dot{S}^+ = i\,\frac{(2e)^2}{\hbar C}\,S_z S^+ + 2gA^+ S_z \quad . \tag{12b}$$

Equations (12) determine the transient response of the junction to the incident radiation. Inserting Eqs.(11) into Eqs.(12), we find the following equations of motion for the Bloch angle θ and the detuning frequency $\dot{\Phi} \equiv \dot{\phi} - \omega_0$

$$\dot{\theta} = \Omega_R \cos \Phi \, , \tag{13a}$$

$$\dot{\phi} = \frac{(2e)^2}{\hbar C}\,m \cos \theta - \Omega_R \left[\frac{\cos\theta}{\sin\theta} \right] \sin \Phi \, . \tag{13b}$$

Here, $\Omega_R = gA_0$ is the Rabi flopping frequency. The first term on the right-hand side of Eq.(13b) corresponds to the electrostatic contribution to the Josephson frequency and the second arises from the interaction of the tunneling Cooper pairs with the applied electromagnetic field.

In order to facilitate the comparison given in Section IV-B, we now obtain the equation for $\ddot{\Phi}$. From Eq.(13b)

$$\dot{\Phi} = -\omega_0 + \frac{(2e)^2}{\hbar C}\,m \cos \theta - \Omega_R \left(\frac{\cos\theta}{\sin\theta}\right) \sin\Phi \quad .$$

By substituting $J_0 = m \cos \theta_0$ and letting $\omega_m = \dfrac{\omega_0 m}{J_0} = \dfrac{(2e)^2}{\hbar C}\,m$

$$\dot{\Phi} = \omega_m (\cos\theta - \cos\theta_0) - \Omega_R \left(\frac{\cos\theta}{\sin\theta}\right) \sin\Phi \quad .$$

Differentiating and using Eq.(13a) gives the final result

$$\ddot{\Phi} = -\omega_m \Omega_R \, F(\theta) \cos\Phi + G(\theta,\Phi) \, , \tag{14}$$

where

$$F(\theta) = [\sin\theta + \frac{\cos\theta}{\sin\theta} (\cos\theta - \cos\theta_o)] \quad , \tag{15a}$$

$$G(\theta,\Phi) = \frac{\Omega^2_R}{\sin^2\theta} (1 + \cos^2\theta) \cos\Phi \sin\Phi \quad . \tag{15b}$$

B. Phenomenological Approach

Now consider the phenomenological approach to this problem as typified by Harrison [12]. Here, we study the junction shown in Figure 5. The junction, denoted by a cross, has parameters J_1 and capacitance C, while the battery has voltage $\sqrt{B}$ across it. The standard equations governing the dynamics of this circuit are

$$\hbar \frac{d\phi}{dt} = 2eV \quad ,$$

$$J = -J_1 \sin\phi - C \frac{dV}{dt} \quad ,$$

$$V = V_B + RJ \quad .$$

By eliminating J and V we obtain

$$\ddot{\phi} = \frac{-2eJ_1}{\hbar C} \sin\phi + \frac{2eV_B}{\hbar RC} - \frac{1}{RC} \dot{\phi} \quad . \tag{16}$$

If we now open the switch at t = 0, equivalently letting $R \to \infty$ in Eq.(16), the situation will correspond to the one we have treated in Section IV-A. Equation (16) becomes

$$\ddot{\phi} = \frac{-2eJ_1}{\hbar C} \sin\phi \quad . \tag{17}$$

We now consider the application of this expression to our problem.

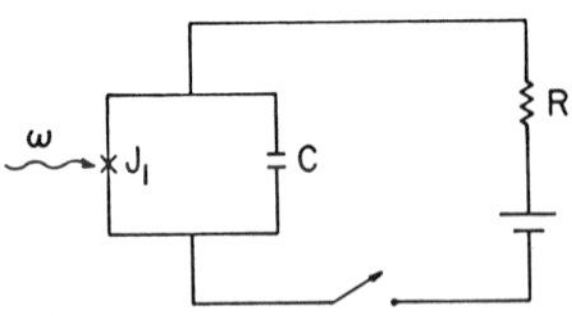

Figure 5. A Josephson junction, designated by a cross, in a simple circuit. The junction capacitance is drawn in parallel to it. Here, again, the junction is emersed in a radiation field while initially having an applied voltage across. See Ref. 12.

We can represent the effect of opening the switch in the above treatment by defining

$$\tilde{V}(t) = V_O + \nu(t) \quad,$$

where $\nu(t)$ represents the change in the measured voltage. This allows us to write an expression for ϕ as a function of time directly [13]

$$\phi(t) = \phi_O + \frac{2e}{\hbar} \int^t \tilde{V}(t')dt' - \frac{2e}{\hbar} \int^t dt' \int d\ell E(\ell,t') \quad,$$

$$\phi(t) = \phi_O + \frac{2e}{\hbar} \int^t \nu(t')dt' + \omega_O t + \frac{2e\ell}{\hbar} A_O \cos(\omega t) \quad.$$

Here $E(\ell,t)$ is the external field of frequency $\omega \equiv \omega_O = (2eV_O)/\hbar$.

Physically we identify $\phi_O + (2e/\hbar)\int^t \nu(t')dt'$ as Φ of our work.

$$\phi(t) = \Phi + \omega_O t + \frac{2e\ell}{\hbar} A_O \cos(\omega t) \tag{18a}$$

$$\sin\phi(t) = \sin[(\Phi + \omega_O t) + \frac{2e\ell}{\hbar} A_O \cos(\omega t)] \quad.$$

By noting that Φ is slowly varying and A_O is small, and using the rotating wave approximation we get

$$\sin\phi \cong \frac{+e\ell}{\hbar} A_O \cos\Phi - \frac{2e\ell}{\hbar} A_O \frac{\sin(2\omega t)}{2} \sin\Phi \quad. \tag{18b}$$

Substituting Eq.(18a) into the left-hand side of Eq.(17) and Eq.(18b) into the right-hand side, we get

$$\ddot{\Phi} - \left(\frac{2e\ell A_O}{\hbar}\right)\omega^2 \cos\omega t = -\left(\frac{2eJ_1}{\hbar C}\right)[\frac{+e\ell}{\hbar} A_O \cos\Phi - \frac{e\ell}{\hbar} A_O \sin(2\omega t)\sin\Phi] \quad.$$

Keeping only low frequency terms we obtain

$$\ddot{\Phi} = -\left(\frac{2eJ_1}{\hbar C}\right) \left(\frac{e\ell A_O}{\hbar}\right) \cos\Phi \quad. \tag{19}$$

Finally, we identify the coefficient of $\cos\Phi$ by matching the DC currents predicted by the phenomenological approach to that of our work,

(a) Phenomenological approach

$$J = -J_1 \sin(\Phi + \omega_o t + \frac{2e\ell}{\hbar} A_o \cos\omega_o t) \ ,$$

$$J \cong \frac{-eA_o \ell}{\hbar} \, J_1 \cos\Phi \ .$$

(b) Our work:

$$J = -2e\langle \dot{S}_z \rangle \ ,$$

$$J = 2egmA_o \sin\theta_o \cos\Phi \ .$$

These imply $g \sin\theta_o = + \ell J_1 / 2m\hbar$ [15].

Substituting this into Eq.(19), we obtain

$$\ddot{\Phi} = \frac{-(2e)^2 mgA_o}{\hbar C} \sin\theta_o \cos\Phi \ ,$$

$$\ddot{\Phi} = -\omega_m \Omega_R \sin\theta_o \cos\Phi \ . \tag{20}$$

This then, is the phenomenological result for the special case we are considering and in terms of our notation.

Comparing Eq.(20) to Eq.(14) we see that there is an additional dynamical variable θ in the equation of motion arrived at via the present quantum optical approach. Details of these calculations and implications of the differences between the two results will follow in a future publication.

*Supported by the Office of Naval Research

References

1.   Proc. IX Intern. Conf. on Quantum Electronics, Amsterdam, June 1976.
2.   ONR Research Symposium on Transient Effects in Josephson Junctions, Point Loma, California, March 1977; see also Ref. (9).
3.   D. Rogovin and M. Scully, Phys. Rept. *25C*, 175 (1976).
4.   R. Dicke, Phys. Rev. *93*, 99 (1954).

5. L.N. Cooper, Phys. Rev. *104*, 1189 (1956).
6. J. Bardeen, L. Cooper, J. Schrieffer, Phys. Rev. *108*, 1175 (1957).
7. P.W. Anderson, Phys. Rev. *110*, 827 (1958); *112*, 1900 (1958).
8. B. Josephson, Phys. Letters *1*, 251 (1962).
9. R. Bonifacio, D. Rogovin and M. Scully, Opt. Comm., to be published.
10. P. Lee and M. Scully, Phys. Rev. B*3*, 769 (1971).
11. D. Rogovin, Phys. Rev. B*12*, 130 (1976).
12. W.A. Harrison, *Solid State Theory*, (McGraw-Hill, Hightstown, N.J., 1970).
13. For more details on this type of calculation see D. Rogovin, M. Scully, and P. Lee, Prog. in Quantum Electronics *2*, 238 (1973).
14. Strictly speaking V_0 as it appears in this paper is the measured dc voltage. The electrostatic voltage will differ slightly from the measured voltage as discussed in detail in Ref. 13. This difference is of secondary importance to our main arguments about the variable θ. Consequently, we will often not distinguish between these two voltages.
15. The symbol e used by Harrison is e>o, whereas ours is e<o.

HIGHER ORDER COHERENT RAMAN SCATTERING

S. Chandra, A. Compaan and E. Wiener-Avnear*

Kansas State University, Manhattan, Kansas

There has recently been much interest in the third order non-linear light mixing, especially the process of coherent anti-Stokes (Stokes) Raman scattering - the CARS (CSRS) effect[1-4]. The resonance generation of the Raman frequency $\omega_{as}^{(1)}$ ($\omega_s^{(1)}$) when the frequency difference of the two laser beams $|\omega_1-\omega_2|$ is close to a Raman active vibrational mode ω_R, is characterized by the energy and wave vector conservation ($\omega_1>\omega_2$),

$$\omega_{as}^{(1)} = 2\omega_1 - \omega_2 \qquad \underline{k}_{as}^{(1)} = 2\underline{k}_1 - \underline{k}_2 \qquad \text{anti-Stokes}$$

$$\omega_s^{(1)} = 2\omega_2 - \omega_1 \qquad \underline{k}_s^{(1)} = 2\underline{k}_2 - \underline{k}_1 \qquad \text{Stokes}$$

where $\underline{k}_1$, $\underline{k}_2$ are the wave vectors of the two incident beams, $\underline{k}_{as}^{(1)}$ and $\underline{k}_s^{(1)}$ of the generated anti-Stokes and Stokes radiation. The dispersive lineshape of the observed CARS (CSRS) signal is attributed [2-3] to the interference between the following contributions to the third order nonlinear susceptibility $\chi^{(3)}$: a nonresonance term $\chi_{NR}^{(3)}$, due mostly to a two-photon absorption and a resonant contribution, $\chi_{RR}^{(3)}$. For certain mixing media, a lower cascaded process involving $\chi^{(2)}$, the nonlinear second order susceptibility, should be considered[5-7]. When high power dye lasers are used for excitation, the CARS signal may exceed 10^{-3} of the original laser power[8]. In this case, higher order light mixing effects can occur, as the first order coherent Raman beam itself acts as the driving field. The second order CARS effect was also observed by Chabay et al.[9].

In the present paper a detailed study of the dispersion curve

---

*On leave from the Ben-Gurion University, Beer Sheva, Israel.

of the higher order coherent Raman scattering has been carried out near the Raman active vibrational mode, ω_R = 992 cm^{-1}, of benzene. Especially the frequency dependence of the second order process was compared with the first order data. The possible mechanisms for the second order process and the dispersion expected in the most important higher order nonlinear susceptibilities are discussed. Several experimental parameters have been manipulated in order to distinguish between the different possible mechanisms which can produce the second order CARS signal. In addition, we were able for the first time to produce the third order CARS and CSRS in benzene and up to the fourth order scattering in CS$_2$. A partial discussion of the characteristic lineshapes and intensity dependence of these higher orders will be presented.

In general, the higher order coherent Raman scattering processes are determined by the energy conservation of the excitations and no net energy transfer to the medium. The n'th order CARS (CSRS) will generate photons at the following frequencies:

$$\omega_{as}^{(n)} = (n + 1)\,\omega_1 - n\omega_2 , \tag{1a}$$

$$\omega_{s}^{(n)} = (n + 1)\,\omega_2 - n\omega_1 . \tag{1b}$$

The $\underline{k}$ matching condition is essential for the enhancement of the coherent Raman signal and can be expressed as:

$$\underline{k}_{as}^{(n)} = (n + 1)\underline{k}_1 - n\underline{k}_2 , \tag{2a}$$

$$\underline{k}_{s}^{(n)} = (n + 1)\underline{k}_2 - n\underline{k}_1 . \tag{2b}$$

Consequently, the various orders are spatially separated. The $\underline{k}$ matching condition for the second order CARS process and the spatial separation of the various orders are demonstrated in Fig. 1. By adjusting the angle, θ, between the two incident laser beams to account for the dispersion in the linear index of refraction, the four-wave mixing will occur along the overlapping interaction length, and the scattering angles $\theta_1 \approx 2\theta$, $\theta_2 \approx 3\theta$ will be used to collect the signal of the first and second order, respectively.

EXPERIMENTAL

The experimental setup was similar to that used before [8] and consists of two independently tunable dye lasers, pumped simultaneously with a UV 1000 Molectron nitrogen laser (1MW peak power, 15 pps, 7 ns pulse width). The separation between the two parallel beams could be adjusted before focusing into the sample, using a 12 cm achromatic focusing lens. Appropriate spatial filtering enables selecting the various CARS and CSRS higher order signals. The selected component (Fig. 1) was directed into a low resolution spec-

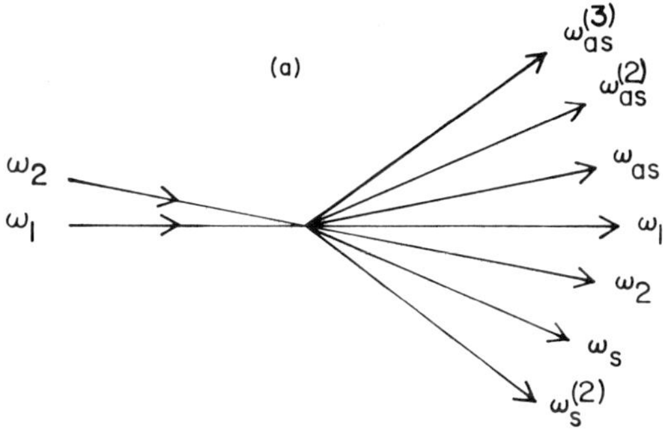

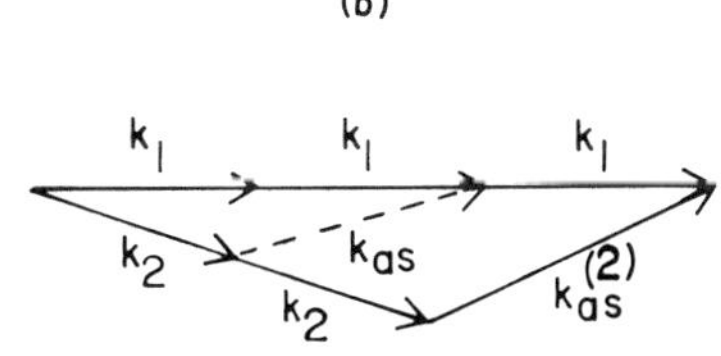

Figure 1: (a) The angular distribution of the CARS and CSRS higher order effects. (b) k phase matching for the second order CARS effect.

trometer and a D. C. averaging system was used to monitor the signal. A set of optical filters further improved the signal-to-noise ratio to better than 10^7:1 in a typical experiment. This enables one to monitor the first, second and the resonant peak of the third order CARS signal of the benzene 992 cm^{-1} Raman line. The scan over the Raman resonance was performed by tuning one dye laser and simultaneously driving the output spectrometer to meet condition (1a). For several cases the discrimination of the spatial and chromatic filtering was enough to obtain the necessary signal-to-noise ratio and the spectrometer was not needed. The average dye laser resolution and monochromaticity was evaluated using a Fabry-Perot interferometer to be 0.2 A°; thus, no deconvolution corrections were applied in the analysis of the 2.2 cm^{-1} wide benzene 992 cm^{-1} line. Detailed analysis of the 656 cm^{-1} CS$_2$ CARS resonance, which was observed up to the fourth order, requires such corrections, since the Raman line-width is 0.6 cm^{-1}. However, in this paper only the benzene data are discussed.

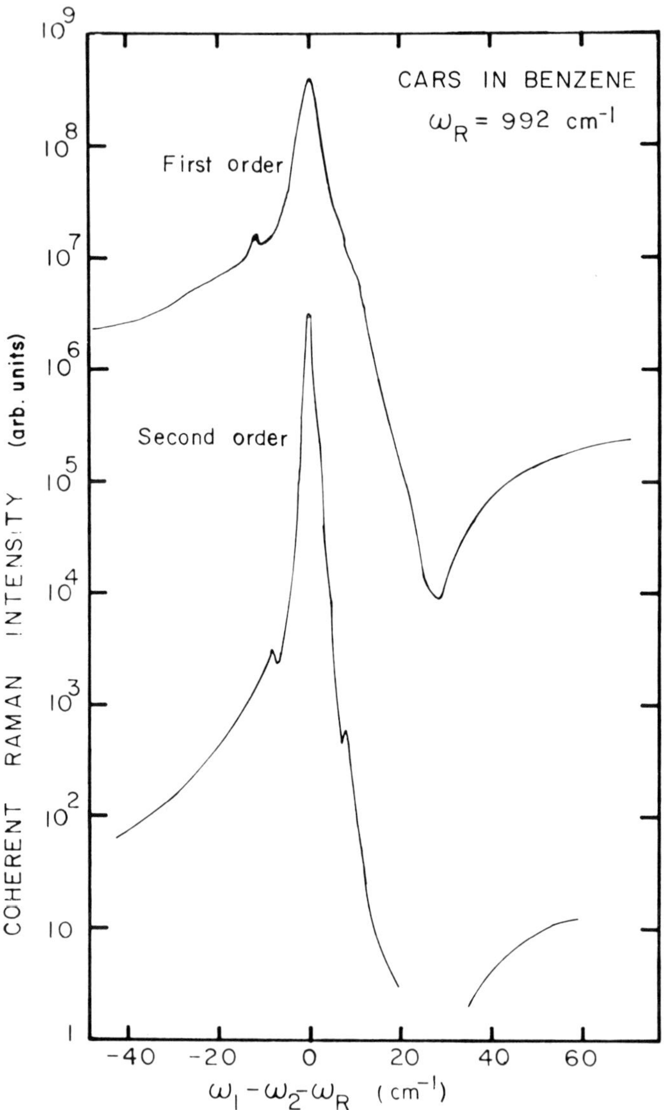

Figure 2: Observed intensity traces of the first and second order CARS signals for benzene at $\omega_1 - \omega_2 \approx \omega_R = 992$ cm^{-1}. The polarizations of the incident beams were mostly parallel to each other and no analyzer was used for detecting the CARS signals.

RESULTS

The frequency dependence of the first and second order coherent anti-Stokes spectra for the benzene 992 cm^{-1} resonance are compared

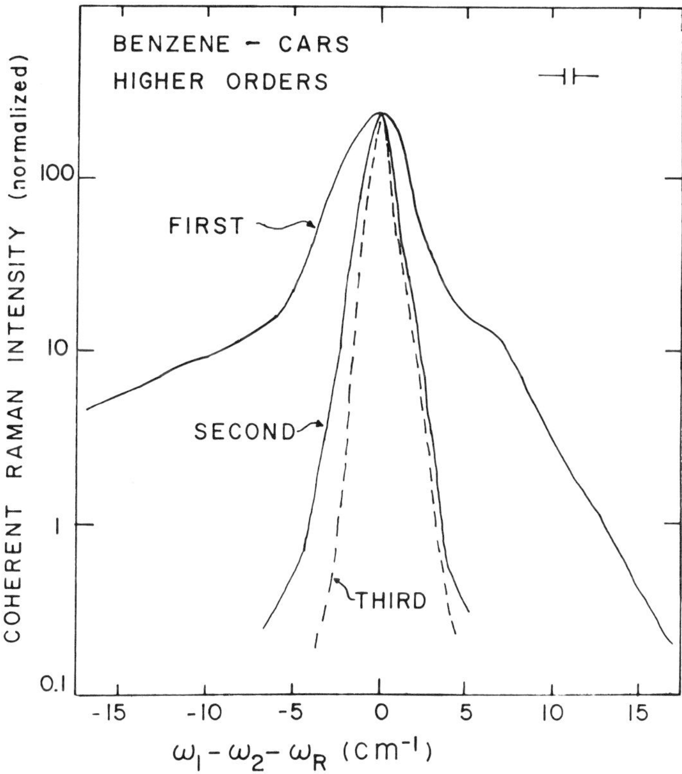

Figure 3: The normalized dispersion curves of the first, second and third order CARS of benzene near the Raman resonance, $\omega_1 - \omega_2 \approx \omega_R = 992$ cm^{-1}. Experimental conditions are similar to those used in the previous figure.

in Fig. 2. The characteristic constructive and destructive inter-ferences in the lineshapes are observed for both curves at the same frequencies; however, a narrowing of the resonance and higher dynamic range in the second order data should be noted. The dispersion of the third order CARS signal could be monitored only very near the resonant frequency (Fig. 3) as the peak intensity was reduced further by about three orders of magnitude from the second order. The narrowing of the resonant signals on increasing order is demonstrated in Fig. 3, where the peak intensities of all the three orders were arbitrarily normalized.

The polarized spontaneous Raman spectrum in benzene was measured in 90° scattering configuration and resulted in a linewidth of 2.2 cm^{-1} (FWHM) for the 992 cm^{-1} line. Other satellites with much lower Raman scattering efficiencies were detected in the spontaneous scat-tering. The 983.3 cm^{-1} line is attributed to the 6% abundance of

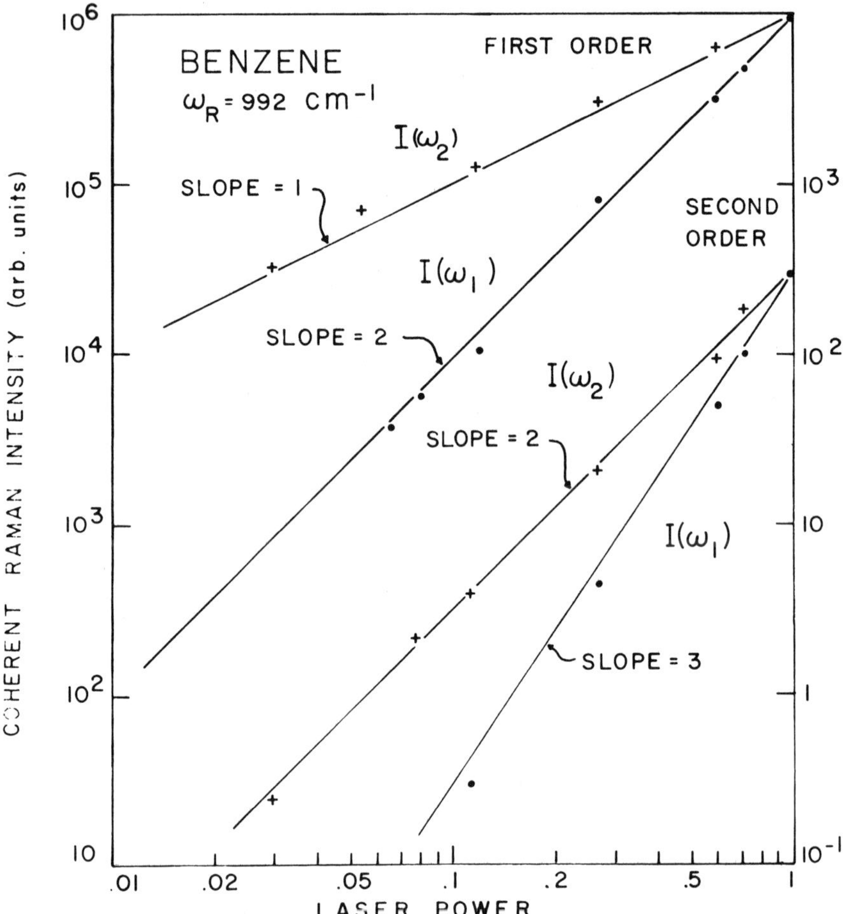

Figure 4: The dependence of the first and second order CARS intensities at resonance $\omega_1 - \omega_2 = \omega_R$ on the incident intensities of the two lasers ω_1 and ω_2.

C^{14} and the recorded linewidth was 1.3 cm^{-1}.   Other spontaneous Raman lines at 979 cm^{-1} and 998 cm^{-1} with about 3% and 6% intensity of the main line, respectively, and similar linewidths were also detected.   All these Raman-allowed lines appear at the corresponding frequencies as small features in the CARS first and second order spectra (Fig. 2).

In order to further examine the characteristics of the higher order process and to insure against leaking due to the more powerful low order coherent scattering process, the power dependence of the second order CARS signal was measured, utilizing calibrated neutral density filters in front of the ω_1 and ω_2 laser beams. The results are demonstrated in Fig. 4 and the slopes were found to correspond to the following relation:

$$I^{(2)}(\omega_1 + 2\omega_R) \propto I^3(\omega_1)I^2(\omega_2) \tag{3}$$

within the experimental error. The power dependence was further compared with the first order signal, which shows only a second power dependence with $I(\omega_1)$ and a linear relation with $I(\omega_2)$. This has ruled out the possibility of a stimulated Raman process, which is a threshold phenomenon. Also, no Terhune rings could be observed and the second order had a crescent shaped characteristic angular distribution, similar to the first order signal. However, narrower and more elongated shapes were visually observed and photographed for the second and third order processes.

Another evidence for the mechanism of the second order coherent process is gained from measuring the signal power dependence on concentration in mixtures. The power of the first and second order CARS signals for the 992 cm^{-1} benzene line was measured in mixtures diluted with cyclohexane, which has a Raman-active mode at 1034 cm^{-1} (FWHM = 12 cm^{-1}) and its intensity is lower by a factor of 20 than the benzene line. The plot of the first and second order outputs as function of the volumetric benzene concentration is given in Fig. 5 down to 10% dilution. A quadratic dependence of the second order power on the benzene concentration was observed for the out-coming signal of the 1st order, whereas a fourth power dependence was recorded for the second order CARS process. The possible effect of dilution on the nonresonant terms may require some corrections at low concentrations and may explain the observed small deviations.

Generally,   the CARS and CSRS spectra of benzene show almost the same results for the present excitation laser lines, $\omega_1 \cong 17190$ cm^{-1}, $\omega_2 \cong 16200$ cm^{-1}; thus, only the anti-Stokes results will be discussed. The CSRS effect can be accounted for by the exchange, $\omega_1 \leftrightarrow \omega_2$.

DISCUSSION

A general theory[1,6] has recently been developed for explaining the characteristic interference lineshape of the CARS (CSRS) first order dispersion curve. The first order coherent Raman scattering in liquids is determined by the appropriate third order nonlinear susceptibility $\chi^{(3)}$. The diagram 6a(I), which describes the first step of the cascaded process to be discussed below, presents the dominating resonance perturbation factor near the Raman resonance at $\omega_1 - \omega_2 \sim \omega_R$. The interference of the Raman resonant third order light mixing term, $\chi_{RR}^{(3)}$ with the nonresonant contribution $\chi_{NR}^{(3)}$, due mostly to a two photon absorption process not shown in Fig. 6, determines the dispersive characteristic lineshape of the coherent Raman

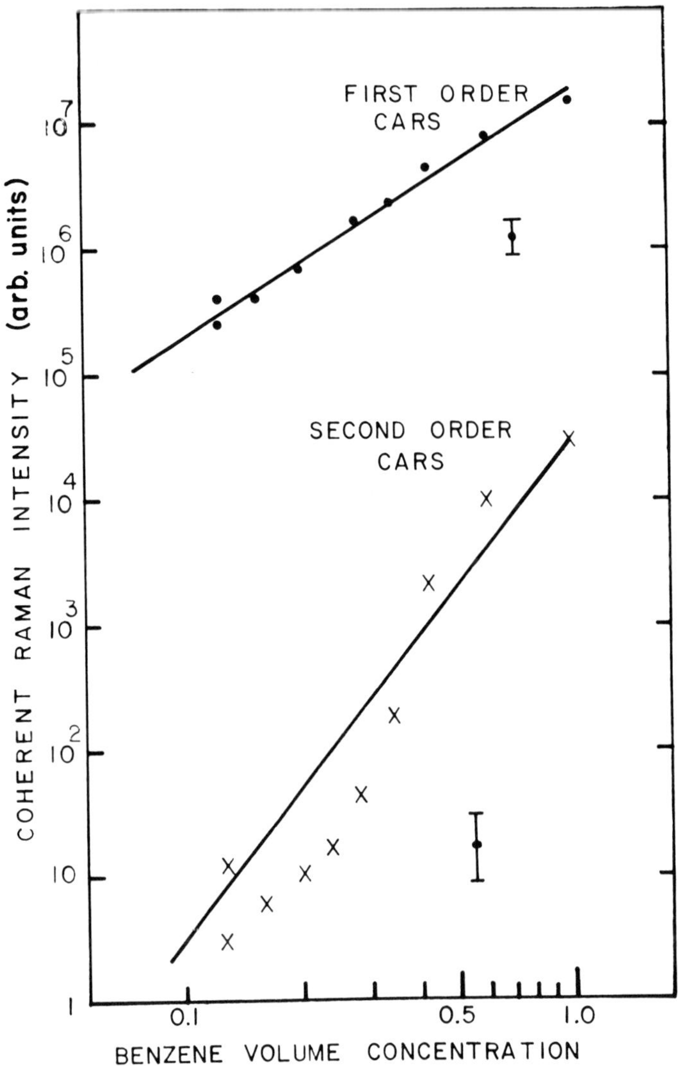

Figure 5: The concentration dependence of the benzene first and second order CARS intensities for benzene-cyclohexane solutions. The experimental errors for both measurements are also presented. The predicted slopes of two and four are plotted for the first and second order CARS processes, respectively.

scattering [2, 3, 6]. The observed intensity is proportional to

$$\left|\chi^{(3)}(-\omega_{as},\omega_1,\omega_1,-\omega_2)\right|^2 = \left|\chi_{NR}^{(3)} + \chi_{RR}^{(3)}\right|^2 = \left|\chi_{NR}^{(3)} + \sum_R \frac{A_R}{\omega_R-(\omega_1-\omega_2)+i\Gamma_R}\right|^2 \quad (4)$$

where all the participating Raman-allowed modes near the main resonance line have to be included in the summation and the phenomenological line-widths, Γ_R, are introduced. An excellent agreement was
obtained with the observed first order CARS spectrum in benzene, [2,
3, 8] utilizing the spontaneous Raman data mentioned above, and fitting only a proportionality factor.

The coherent Raman second order scattering near the $\omega_1 - \omega_2 = \omega_R$
resonance can be produced either by a six-wave mixing, which depends
on the appropriate $\chi^{(5)}$ $(-\omega_{as}^{(2)}, \omega_1, \omega_1, \omega_1, -\omega_2, -\omega_2)$ nonlinear susceptibility or a cascaded first order CARS process involving $\chi^{(3)}$
$(-\omega_{as}^{(1)}, \omega_1, \omega_1, -\omega_2)$ combined with $\chi^{(3)} (-\omega_{as}^{(2)}, \omega_1, \omega_{as}^{(1)}, -\omega_2)$. Thus,
the induced polarization can be described as:

$$P(-\omega_{as}^{(2)}, \omega_1, \omega_1, \omega_1, -\omega_2, -\omega_2) = \chi^{(5)}(-\omega_{as}^{(2)}, \omega_1, \omega_1, \omega_1, -\omega_2 -\omega_2) E_1 E_1 E_1 E_2 E_2$$

$$+ \chi^{(3)}(-\omega_{as}^{(2)}, \omega_1, \omega_{as}^{(1)}, -\omega_2) \times \chi^{(3)}(-\omega_{as}^{(1)}, \omega_1, \omega_1, -\omega_2) E_1 E_1 E_1 E_2 E_2 \quad . (5)$$

The main energy diagrams and perturbation processes are demonstrated
in Figs. 6a, b. Both processes will meet the necessary $\underline{k}$ matching
condition for transparent media with constant $dn/d\omega$ as $\underline{is}$ demonstrated in Fig. 1.

In general, $\chi^{(5)}$ represents 60 independent terms due to the
various time ordering of the six interactions. However, only the
most significant Raman term enhanced by a double resonace process
will be considered (Fig. 6b). This term is given by

$$\chi_{RR}^{(5)} \propto \sum_{\substack{R,a \\ b,c}} \frac{M}{(\omega_c - 3\omega_1 + 2\omega_2)(\omega_{2R} - 2\omega_1 + 2\omega_2)(\omega_b - 2\omega_1 + \omega_2)(\omega_R - \omega_1 + \omega_2)(\omega_a - \omega_1)} , \quad (6)$$

where the M is a product of the appropriate matrix elements of the
six-wave interaction. Assuming that the dipole-allowed electronic
states a, b, c are far off-resonance, the intensity of the second
order CARS signal, due to these doubly resonant terms in the fifth
order susceptibility, is proportional to:

$$|\chi^{(5)}|^2 = \left| \chi_{NR}^{(5)} + \sum_R \frac{B_R}{[\omega_R - (\omega_1 - \omega_2) + i\Gamma_R][\omega_{2R} - 2(\omega_1 - \omega_2) + i\Gamma_{2R}]} \right|^2 , \quad (7)$$

where all the nonresonant terms are included in $\chi_{NR}^{(5)}$, and the phenomenological Raman linewidths Γ_R and Γ_{2R} are added. If $\omega_{2R} = 2\omega_R$ a
symmetric line will result, which is in contrast to the present ob-
servation of the second order dispersion curve (Fig. 2). Fluoresence
measurements [10] on dilute benzene solutions in cyclohexane at 77°K
have indicated $\omega_{2R} - 2\omega_R = -3$ cm^{-1}, thus the anharmonic nature of the
benzene vibration can not explain the present observations on the
basis of a $\chi^{(5)}$ double resonance.

The observed second order interference lineshape may, however,
be explained as a cascaded process (Fig. 6a) involving the third

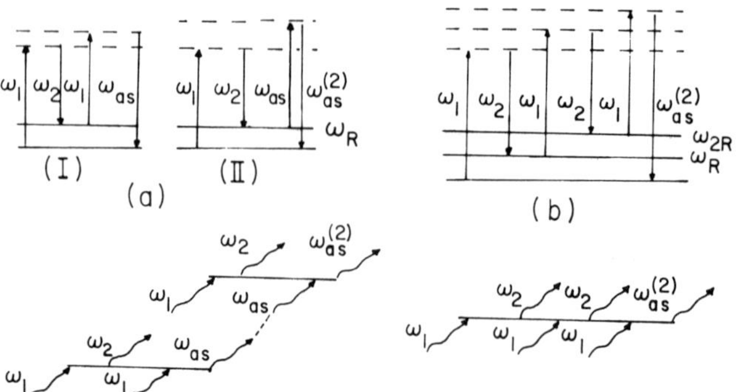

Figure 6: Energy level and perturbation diagrams describing two processes, which can contribute to the second order coherent anti-Stokes Raman scattering:
(a) a cascaded process involving $\chi^{(3)}$ twice.
(b) a double resonance process contributing to $\chi^{(5)}$.

order susceptibility $\chi^{(3)}$ twice. When all photon frequencies are far from the electronic dipole allowed states, then the two $\chi^{(3)}$ terms in Eq. (5) are equal [11] and the intensity of the second order CARS signal is proportional to:

$$\left| \chi^{(3)} \right|^4 = \left| \chi_{NR}^{(3)} + \sum_R \frac{A_R}{\omega_R - (\omega_1 - \omega_2) + i\Gamma_R} \right|^4 \quad . \tag{8}$$

The theoretical fit based on Eq. (8), utilizing the same spontaneous Raman parameters which were used previously to fit the first order dispersion curve [8] and fitting only a scaling factor, have produced the dispersion curve in Fig. 7 for the second order CARS effect in benzene. Generally, the theoretical curve fits the observed spectrum (Fig. 2) within the experimental error. through all the studied range, except for the high frequency wing above 30 cm^{-1}. We believe that this deviation may be due to additional small contributions of the other processes (e.g. $\chi^{(5)}$), which were not included in the above treatment of the cascaded effect. The narrowing of the higher order resonances and the higher dynamic range are obviously explained. Additional support for the cascaded process is gained from the concentration measurement (Fig. 5). Since the susceptibility χ is proportional to the concentration of active molecules, the cascaded process yields a fourth power dependence on concentration whereas the $\chi^{(5)}$ term would produce a quadratic dependence.

Generally a cascaded mechanism, involving $\chi^{(3)}$ twice, could be

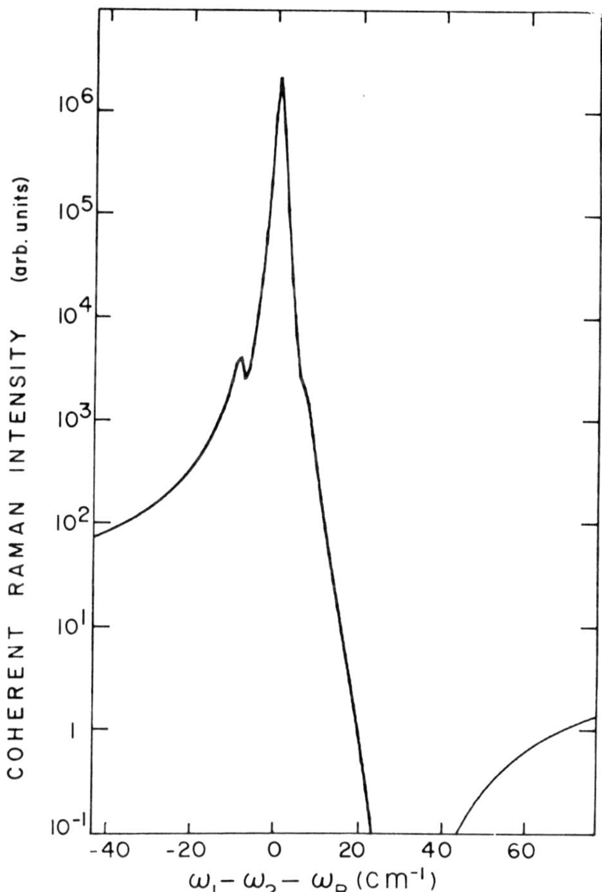

$$\omega_1 - \omega_2 - \omega_R \ (cm^{-1})$$

Figure 7: The theoretical dispersion curve predicted for the second order CARS in benzene by the cascaded process. Except for a normalization factor the parameters are the same as those used for fitting the first order dispersion curve in benzene (Ref. 8).

operative either as a local *microscopic* process, relating the fields by an internal constitutive relationship, or as a *macroscopic* process, involving retarded fields from different molecules along the overlapping region. The cross section for the latter process will be highly dependent on the interaction length of the signals, which provides a method to distinguish between the two alternatives.

Observation of the second and third order CARS in CS_2 near $\omega_R = 656 \ cm^{-1}$ indicate interference structures similar to the benzene data. However, with a spontaneous Raman linewidth of $0.3 \ cm^{-1}$ in CS_2, the laser resolution set a practical lower limit to the observed linewidth of the higher order signals. Nevertheless the narrowing of the linewidth in higher order coherent Raman resonances seems to be a general feature of these higher order effects (see also Fig. 3). As almost the same factorial reduction of the intensity was observed on shifting from second to third order as from first to second, it is suggested that similar cascaded mechanisms are also responsible for the third and higher order coherent Raman scattering processes.

References

1.   N. Bloembergen, *Nonlinear Optics*, (Benjamin, New York, 1965).
2.   M.D. Levenson and N. Bloembergen, J. Chem. Phys. *60*, 1323 (1974) and Phys. Rev. B *10*, 4447 (1974).
3.   H. Lotem, R.T. Lynch and N. Bloembergen, Phys. Rev. A *14*, 1748 (1976).
4.   W.M. Tolles, J.W. Nibler, J.R. McDonald and A.B. Harvey, Appl. Spectroscopy (in press).
5.   C. Flytzanis, in *Quantum Electronics*, Vol. 1, ed. H. Rabin and C.L. Tang (Academic Press, New York, 1975).
6.   C. Flytzanis and N. Bloembergen, Proc. Quant. Electr. *4*, 271 (1976).
7.   S.D. Kramer, F.G. Parson and N. Bloembergen, Phys. Rev. B *9*, 1853 (1974).
8.   S. Chandra, A. Compaan and E. Wiener-Avnear, Bull. Amer. Phys. Soc. *22*, 315 (1977); and to be published.
9.   I. Chabay, G.K. Klauminzer and B.S. Hudson, Appl. Phys. Lett. *28*, 27 (1976).
10. J.D. Spangler and N.G. Kilmer, J. Chem. Phys. *48*, 698 (1968).
11. E. Wiener-Avnear, S. Chandra and A. Compaan (to be published).

MEMORY FUNCTION METHODS FOR QUANTUM SYSTEMS IN CONTACT WITH
RESERVOIRS[*]

A. Zardecki

Université Laval, Québec, P.Q., Canada

1. INTRODUCTION

In recent years increasing use has been made of methods of
quantum statistics and stochastic processes in a variety of problems
in quantum optics. A comprehensive list of references can be found
in the review articles of Agarwal [1] and Haken [2], see also a
recent work of Gronchi and Lugiato [3]. In many cases of practical
interest one is concerned with the dynamics of an open systems S
moving irreversibly under the influence of a reservoir R. Ultimately, the properties of S are inferred through the elimination of
the R-variables. This is usually accomplished by two complementary
approaches:

(i) The elimination procedure in the Schrödinger picture leads
to an equation for a reduced density operator (master equation).

(ii) The elimination procedure in the Heisenberg picture leads
to a generalized, including memory effects, Langevin equation.

While the former approach is now widely used, the merits of
the latter one have not yet sufficiently been recognized, and most
of the work is limited to a Markoffian approximation to the dynamics
of S. A step forward was made by Bose and Ng who derived the equations of motion of the correlation functions for the problems of
two-level atoms [4] and the Josephson radiation field [5]. Their
treatment, based on the Zwanzig-Mori projection formalism, applies
to the total system S + R specified by a Hamiltonian accounting for
the coupling between S and R.

The purpose of the present study is to set up the equations
of motion for the operators of S, or for their correlation functions, in the case where the influence of R may be treated in a
Markoffian approximation. In Sec. 2, we consider a simple system
S in contact with a reservoir. In Sec. 3, we analyse the case of
S being a subsystem of a larger system S + T where the coupled
variables S and T are each in a Markoffian contact with their own
reservoirs. After elimination of the T variables, nonmarkoffian
effects for S are then found.

2.   OPEN SYSTEM S AS SUBSYSTEM OF S + R

We consider the equation of motion

$$i\dot{W}(t) = LW(t) \tag{1}$$

for the density operator of an open system S, where the Liouville
operator L consists of a coherent and a dissipative part

$$LX \equiv L_c X + iL_d X \equiv [H,\ X] + iL_d X \ . \tag{2}$$

In (2), H denotes the internal Hamiltonian of S and L_d describes
the influence of a reservoir R. We assume the reservoir correlation time to be small on a time scale in which S loses an
appreciable portion of energy. This amounts to the Markoff
approximation for L_d, and (1) becomes the master equation in the
Schrödinger-Markoff picture [6].

With respect to the scalar product in the Liouville space

$$(X|Y) = \mathrm{Tr}(X^\dagger Y) \ , \tag{3}$$

we may introduce the adjoint operator $L^\dagger$, which governs the
motion of an operator A of the system S

$$\dot{A}(t) = iL^\dagger A(t) \ . \tag{4}$$

Explicit examples of L and $L^\dagger$ are contained in the work of Weidlich
[7].

In order to derive a generalized Langevin equation for A(t),
we start by defining a projection operator P onto A = A(t = 0).
This is given as

$$PX = A(AW_o|A)^{-1} (AW_o|X) \ , \tag{5}$$

where $W_0 = W(t = 0)$.

If we decompose $A(t)$ into a secular part

$$PA(t) = \Xi(t)A \tag{6}$$

and a nonsecular part

$$(1 - P)\, A(t) = A'(t) \quad, \tag{7}$$

then, by acting on (4) with the operator $Q = 1 - P$, we get

$$A'(t) = \int_0^t \Xi(t - s)\, f(s)\, ds \quad, \tag{8}$$

where the random force $f(t)$ is given by

$$f(t) = i \exp(itQL^\dagger)\, QL^\dagger A \quad. \tag{9}$$

We introduce the frequency function ω as

$$\omega = (AW_0|A)^{-1} \, (AW_0|L^\dagger A) \quad. \tag{10}$$

It is seen that

$$\dot{A} \equiv \dot{A}(t = 0) = i\omega A + Q\dot{A} \tag{11}$$

and that the function $\Xi(t) = (AW_0|A)^{-1} \, (AW_0|A(t))$ satisfies the equation

$$\frac{d}{dt}\, \Xi(t) = i\omega\, \Xi(t) - \int_0^t \Xi(t - s)\, \psi(s)\, ds \quad, \tag{12}$$

where the memory function $\psi(t)$ is defined as

$$\psi(t) = (AW_0|A)^{-1} \, (AW_0|L^\dagger \exp[itQL^\dagger]QL^\dagger A) \quad. \tag{13}$$

The equation of motion (4) can now be transformed into the generalized Langevin equation

$$\frac{d}{dt}A(t) - i\omega A(t) + \int_0^t \psi(t - s)\, A(s)\, ds = f(t) \quad. \tag{14}$$

Since the random force given by (9) is orthogonal to A, i.e.

$(A|f(t)) = 0$, it follows that the correlations function

$$C_A(t) = (AW_o|A(t))$$ (15)

obeys

$$\frac{d}{dt}C_A(t) - i\omega C_A(t) + \int_o^t \psi(t-s)\, C_A(s)\, ds = 0 \; ,$$ (16)

which is Eq. (14) with the inhomogeneous term set equal zero. Equations (14) and (16) constitute the main results of this section. Their derivation follows closely the original procedure used by Mori [8] but, with the projection operator (5) not necessarily referring to equilibrium average, a relation between the random force and the memory function takes a new form. This relation

$$\psi(t) = (AW_o|A)^{-1} (iAW_o|L^\dagger f(t))$$ (17)

is to be considered as a manifestation of the fluctuation - dissipation theorem for open systems [9].

In practice, a representation of the operators in terms of the field or atomic coherent states provides a convenient framework for the evaluation of the memory function. In fact, the action of the operator L can be represented through a differential operator

$$LW = \int d^2z \; | \; z > < z \; | \; \mathcal{L} \, w(z) \; ,$$ (18)

where $| \, z >$ is a generic symbol denoting the field or atomic coherent state, and $w(z)$ is the function specifying the P - representation of the density operator W. Similarly, if $X(z) = < z \, | \, X \, | \, z >$ defines the Q - representation of an operator X, then

$$< z \; | \; L^\dagger X \; | \; z > = \mathcal{L}^\dagger X \, (z) \; ,$$ (19)

where $\mathcal{L}^\dagger$ is a differential operator formally adjoint to $\mathcal{L}$. Apart from the normalization factor $(AW_o|A)$, the memory function now becomes

$$\psi(t) = \int d^2z \; w(z) \; A^*(z) \; \mathcal{L}^\dagger < z \; | \; \exp(itQL^\dagger)QL^\dagger A \; | \; z >. $$ (20)

Suppose we know the biorthogonal set of eigenfunctions, $u_n(z)$ and $v_n(z)$, of L and $L^\dagger$

$$\mathcal{L}\, u_n(z) = \lambda_n u_n(z)$$

$$\mathcal{L}^\dagger v_n(z) = \lambda_n^* v_n(z) \ ,$$

$$(21)$$

the λ_n's denoting the corresponding eigenvalues. Then (20) can
be determined by a successive evaluation of the expression

$$< z \mid QL^\dagger X \mid z > = \sum_n \lambda_n^* \left\{ v_n(z) - A(z)\, \frac{<AW_o \mid v_n>}{(AW_o \mid A)} \right\} x_n \ , \quad (22)$$

where the x_n's are the expansion coefficients of $X(z)$ in terms of
$v_n(z)$, i.e.

$$x_n = \int d^2z \ u_n^*(z) \ X(z) \tag{23}$$

and

$$< AW_o \mid v_n > = \int d^2z \ w(z) \ A^*(z) v_n(z) \ . \tag{24}$$

To illustrate those general results, we consider a damped harmonic
oscillator of frequency ω and the damping constant κ. The
Liouville operator has the form

$$LW = \omega[a^\dagger a,\, W] + i\kappa \left\{ (\bar{n}+1)[a,\, Wa^\dagger] + \bar{n}[a^\dagger,\, Wa] + \text{H.c.} \right\}, \tag{25}$$

where $[a, a^\dagger] = 1$, while the adjoint operator $L^\dagger$ is

$$L^\dagger X = \omega[a^\dagger a,\, X] - i\kappa \left\{ (\bar{n}+1)[a^\dagger, X]\, a + \bar{n}[a, X]\, a + \text{H.c.} \right\}. \tag{26}$$

In the coherent state representation, the desired eigenfunctions are well-known [10]. Written in terms of the associated
Laguerre polynomials $L_m^{(n)}$, they are

$$u_{mn}(z) = N_{mn} \ z^n L_m^{(n)}(\mid z \mid^2/\bar{n}) \ \exp(-\mid z \mid^2/\bar{n})$$

$$(27)$$

$$v_{mn}(z) = z^n \ L_m^{(n)} (\mid z \mid^2/\bar{n})$$

where N_{mn} is a normalization constant. The eigenvalues of $\mathcal{L}$ are

$$\lambda_{mn} = -n\omega - i(2m + \mid n \mid)\kappa \ . \tag{28}$$

Consider first $A_1 = a^n$. We assume that at $t = 0$ the system is in
the state of thermal equilibrium. Then the Gaussian weight func-

tion $w(z)$ simply becomes $u_{oo}(z)$. With the help of (20), (22) and (27), we readily get

$$\omega_1 = -n\omega + i \, | \, n \, | \, \kappa$$

$$\psi_1(t) = 0 \quad , \tag{29}$$

Therefore the memoryless equation for the correlation function of A_1 is

$$\frac{d}{dt} < (a^\dagger)^n a^n (t) >_{av} + (in\omega + | \, n \, |\kappa) < (a^\dagger)^n a^n (t) >_{av} = 0 \tag{30}$$

This describes Markoffian spontaneous damped fluctuations. On the other hand, the thermal equilibrium fluctuations of the photon number operator $A_2 = a^\dagger a$ do contain the memory effects. In this case, we have

$$\omega_2 = i\gamma(\bar{n} + 1)/\bar{n}$$

$$\psi_2(t) = -\gamma^2 \exp(-\gamma t) \, (\bar{n} + 1)/\bar{n} \quad , \tag{31}$$

with $\gamma = \kappa\bar{n}/(\bar{n} + 1/2)$. Since the correlation function $<a^\dagger aa^\dagger(t) a(t)>_{av}$ differs from $<(a^\dagger)^2 a^2(t)>_{av}$, the memory can be thought of to be contained in the commutation relations for $a^\dagger$ and a. As the correlation function deviates from the normal form, the memory effects become more and more pronounced. By virtue of (16) and (31), the photon number correlation function is

$$<a^\dagger aa^\dagger(t) a (t) >_{av} = \bar{n}[(\bar{n} + 1) \exp(-2\kappa t) + \bar{n}] \quad , \tag{32}$$

The same result can also be obtained by the density matrix methods [6].

3.   OPEN SYSTEM S AS A SUBSYSTEM OF A LARGER OPEN SYSTEM S + T

We extend our former considerations to the case where S is a subsystem of a larger system S + T, the total system S + T being open as a result of the Markoffian contact of each part S and T with their own reservoirs. The Liouville operator has now the form

$$L = L_S + L_T + L_{ST} + i\Lambda_S + i\Lambda_T \quad , \tag{33}$$

where Λ_S and Λ_T are the dissipative parts, and where L_{ST} accounts for the coupling between S and T. The problem of motion of the observables of the relevant part S is solved by introducing two kinds of the projection operators. First

$$X_S \equiv P_1 X = (Tr_T W)^{-1} Tr_T (WX) \qquad (34)$$

reduces the dynamical variables to those of S. In (34), W is the density operator of S + T at t = 0. By acting with P_1 on the operator equation of motion of the form (4), but with $L^\dagger$ being the adjoint of (33), we obtain an equation for A_S in a standard way. It reads

$$\frac{d}{dt} A_S(t) = iL_{ef}^\dagger A_S(t) + \int_o^t K(t - s) A_S(s) \, ds \quad , \qquad (35)$$

with $L_{ef}^\dagger = P_1 L^\dagger P_1$, and

$$K(t) = -P_1 L^\dagger \exp[it(1 - P_1)L^\dagger](1 - P_1) L^\dagger P_1 \quad . \qquad (36)$$

For the sake of simplicity, we have assumed that $(1 - P_1) A(0) = 0$, rendering thus equation (35) homogeneous.

Second, with $W_S = Tr_T W$, and $(X|Y) = Tr_S (X^\dagger Y)$, the Zwanzig-Mori projection operator

$$P_2 X_S(t) = A_S (A_S W_S | A_S)^{-1} (A_S W_S | X_S(t)) \qquad (37)$$

separates the secular part of $X_S(t)$. When applying to (35) the procedure sketched in Sec. 2, the generalized Langevin equation obeyed by A_S is found again to have the form (14). However, the frequency and memory functions are now replaced by

$$\omega = \omega_r + i\omega_i$$

$$\omega_r = (A_S W_S | A_S)^{-1} (A_S W_A L_{ef}^\dagger | A_S) \qquad (38)$$

$$\omega_i = (A_S W_S | A_S)^{-1} \lim_{t \to o} \int_o^t ds (A_S W_S | K(t - s) A_S(s))$$

and

$$\psi(t) = (A_S W_S | A_S)^{-1} \int_o^t ds \left(A_S W_S | [iL_{ef}^\dagger \, \delta(s) + K(s)] \, f(t - s) \right) \quad . \qquad (39)$$

The random force $f(t)$ in (39) is given as an inverse Laplace transform through the formula

$$f(t) = \frac{1}{2\pi i} \int_{\varepsilon - i\infty}^{\varepsilon + i\infty} e^{zt} \frac{(1 - P_2)[iL_{ef}^{\dagger} + \hat{K}(z)] A_S}{z - (1 - P_2)[iL_{ef}^{\dagger} + \hat{K}(z)]} dz \quad , \qquad (40)$$

$\hat{K}(z)$ being the Laplace transform of (36). Equations (35)-(40) have formally the same form as the Kawasaki equations [11] even though they describe different systems. While S + T is closed in Kawasaki's treatment, in our case it remains open, which is accounted for by the Λ_S- and Λ_T- terms in (33).

Equation (14) with ω and ψ given by (38)-(40) describes evidently nonmarkoffian effects, reflected by the structure of the memory function (39). On a time scale much larger than the characteristic time of the T-variables, a short time approximation to the kernal function will be adequate. Setting $K(t) = K_T \delta(t)$, (40) becomes

$$f(t) = i \exp[it(1 - P_2) L_T^{\dagger}](1 - P_2) L_T^{\dagger} A_p \qquad (41)$$

with $L_T^{\dagger} = L_{ef}^{\dagger} - iK_T$.

The memory function (39) acquires then the form (13) with $L^{\dagger}$ replaced by $L_T^{\dagger}$.

As an application of this formalism, consider the superradiant emission from a pencil shaped volume. In the interaction picture, the equation of motion for the density operator is

$$i\dot{W}(t) = (L_{af} + i\Lambda_f) W \quad , \qquad (42)$$

where a and f refer to the coupled atomic and field variables, respectively. In (42)

$$L_{af} X = [H_{af}, X] = g[aR^{+} + a^{\dagger}R^{-}, X] \quad , \qquad (43)$$

where g is the coupling constant, a and $a^{\dagger}$ are the photon operators introduced earlier in (25), while R^{+} and R^{-} are the dipole moment operators. The dissipative term Λ_f is given by the second term of the right-hand side of (25). With those definitions, (42) corresponds to the Bonifacio, Schwendimann and Haake (BSH) theory [12] with a nonzero temperature reservoir. Let

$$W = W_f \, W_e = \mid 0 >_f \, <_f 0 \mid W_a \quad,$$

then the first projection operator $P_1 X = {}_f< 0 \mid X \mid 0 >_f$ does not alter the atomic variables built up of R^+ and R^-. It can readily be verified that

$$K(t) \, X = -\exp(-t\kappa) P_1 L_{af} L_{af} X$$

$$= \kappa \, \exp(-t\kappa) \, \Lambda_S^\dagger \, X \quad, \tag{44}$$

where

$$\Lambda_S^\dagger \, X = \frac{g^2}{\kappa} \, \{ (\bar{n} + 1) [R^+, X] R^- + \bar{n}[R^-, X] R^+ + H.c.\} \, . \tag{45}$$

For an operator X_a pertaining to the atomic subsystem, i.e. $X_a = P_1 X_a$, Eqs. (35) and (44) yield

$$\frac{d}{dt} \, X_a = \kappa \int_0^t e^{-s\kappa} \, \Lambda_S^\dagger \, X_a(t - s) \, ds \tag{46}$$

In the adiabatic regime, (46) is approximated by its Markoffian version

$$\frac{d}{dt} \, X_a = \Lambda_S^\dagger X_a \quad, \tag{47}$$

which is the adjoint to the BSH superradiance equation.

With the aid of the now familiar transformations, (47) can be converted to the form (14). In the atomic coherent state representation, an asymptotic approximation to $\Lambda_S^\dagger$ is easily obtained for large values of the ratio of the cooperative number to the mean number $\bar{n}$ of thermal photons. This is similar to the result of Glauber and Haake [13] for Λ_S. The memory function can then explicitly be evaluated by using, as discussed in Sec. 2, the expansion into a biorthogonal set of functions.

4.   CONCLUSION

In relation to the master equation methods, the memory function formalism provides a complementary approach to quantum statistical problems. Conceived initially as a formalism describing the frequency and damping of collective modes in a system fluctuating from thermal equilibrium, it has now been extended to a larger class of systems. For a simple system in Markoffian

contact with a reservoir, the memory effects are manifested only
through the quantum fluctuations, as the example of the photon
number operator shows. In compound systems discussed in Sec. 3,
there are inherent nonmarkoffian effects due to the interaction
between the subsystems of an open system.

* Research supported by the National Research Council of Canada.

References

1.   G.S. Agarwal, in *Progress in Optics*, Vol. XI, ed. E. Wolf
     (North-Holland, Amsterdam, 1973) p. 3.
2.   H. Haken, Rev. Mod. Phys. *47*, 67 (1975).
3.   M. Gronchi and L.A. Lugiato, Phys. Rev. A*13*, 830 (1976).
4.   S.K. Bose and D.T. Ng, Phys. Rev. A*11*, 1001 (1975).
5.   S.K. Bose and D.T. Ng, Phys. Rev. A*13*, 1548 (1976).
6.   W.H. Louisell, *Quantum Statistical Properties of Radiation*
     (John Wiley, New York, 1973) chapters 6 and 7.
7.   W. Weidlich, Z. Physik, *241*, 325 (1971).
8.   H. Mori, Prog. Theoret. Phys. (Kyoto) *33*, 423 (1965).
9.   W. Weidlich, Z. Physik, *248*, 234 (1971).
10. R. Bonifacio and F. Haake, Z. Physik, *200*, 526 (1967).
11. T. Kawasaki, Progr. Theoret. Phys. *49*, 2151 (1973).
12. R. Bonifacio, P. Schwendimann and F. Haake, Phys. Rev. A*4*,
     302 (1971).
13. R.J. Glauber and F. Haake, Phys. Rev. A*13*, 357 (1976).

THE LASER - TRAILBLAZER OF SYNERGETICS

Hermann Haken

Universität Stuttgart, Stuttgart, West Germany

In my paper I want to discuss how a thorough theoretical study of the laser process has led us to understand basic mechanisms of self-organization in physics, chemistry, biology and other sciences. A typical laser [1] consists of a rod of active material (compare fig. 1) with two mirrors at its endfaces. The laser-active atoms of the material are pumped energetically from the outside and emit laser light through one of the mirrors. The two mirrors form a Perot-Fabry-interferometer which allows only for those modes, whose wavelengths obey the relation (fig. 2)

$$n \cdot \lambda/2 = L$$

(n is an integer and L is the distance between the mirrors).

Since the emission line of the laser atoms has a certain profile we should expect that the emitted laser light corresponds in its frequency distribution to the corresponding profile (r.h.s. of fig. 2). I will show, however, that this is not true. The laser

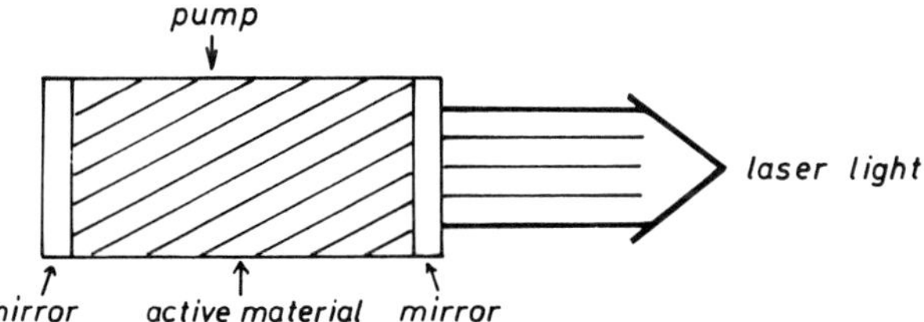

Fig. 1. Typical set-up of a laser.

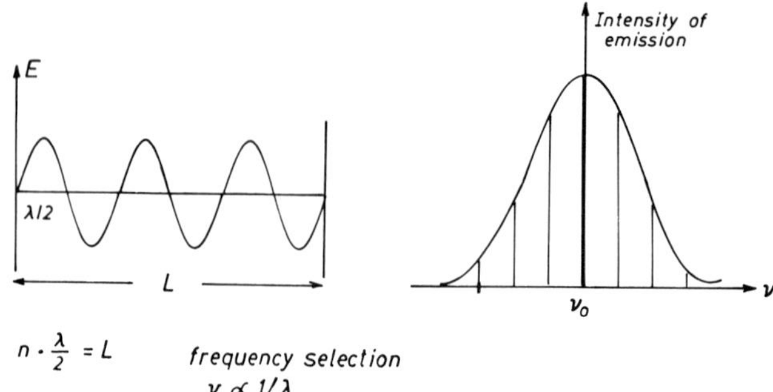

Fig. 2. The mirrors act as Perot-Fabry-interferometer (left), line
profile of spontaneous emission (right).

selects only one or very few modes. When we pump a laser only
weakly, it acts as a lamp emitting short wave tracks with random
phases. Beyond a certain pump power the laser emits a coherent
wave (fig. 3).

This indicates that the internal states of the laser atoms
have changed drastically. Let us visualize the atoms in the frame
of Bohr's model. In a lamp, the electrons jump from an outer orbit
to an inner orbit at random, whereas in the laser the electronic
transitions are well correlated. The change of the properties of
light below and above the laser threshold shows up dramatically in
macroscopic properties of laser light (fig. 4). While below that
threshold the emitted light intensity increases only slowly with
increasing pump power, above the laser threshold the intensity
increases very rapidly. Apparently the efficiency has changed
drastically. Furthermore when we plot the relative intensity or

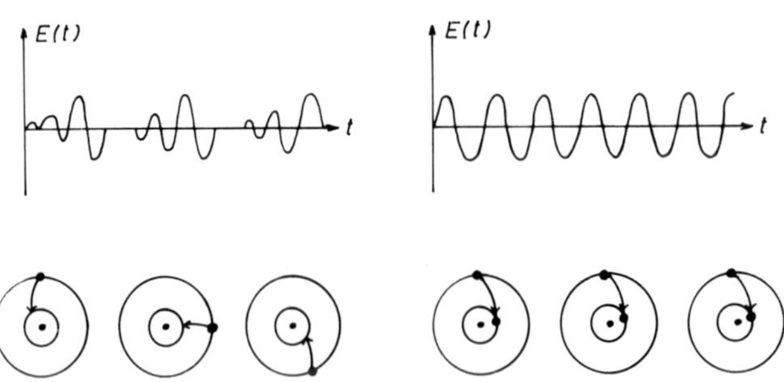

Fig. 3. Electric field strength of light as function of time.
Left: lamp, right: laser.

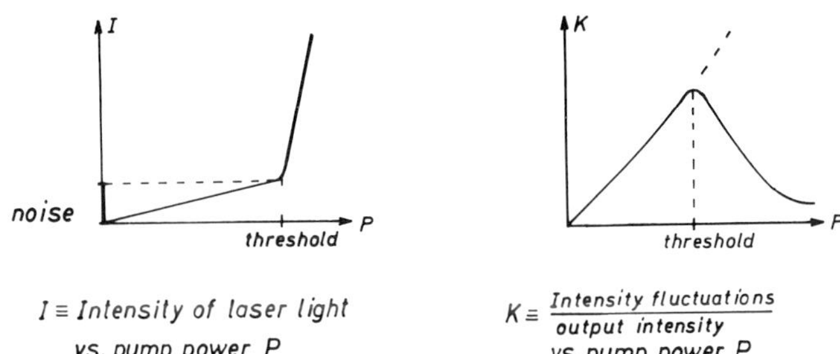

Fig. 4. Left: Light intensity vs. pump power; right: Intensity fluctuations over output intensity vs. pump power.

amplitude fluctuations against pump power above threshold the relative fluctuations are more and more reduced as had been predicted in my 1964 paper [2] and verified experimentally by a number of authors [2a]. Quite obviously there exists a certain threshold at which the macroscopic properties of laser light change drastically.

In nature there are numerous other examples known, either since a long time, or very recently, in which the macroscopic pattern of systems changes when certain external parameters are altered. In the Bénard problem [3] (or convection instability) a fluid layer is heated from below so that a temperature gradient is established between its lower and upper surface (fig. 5). When the

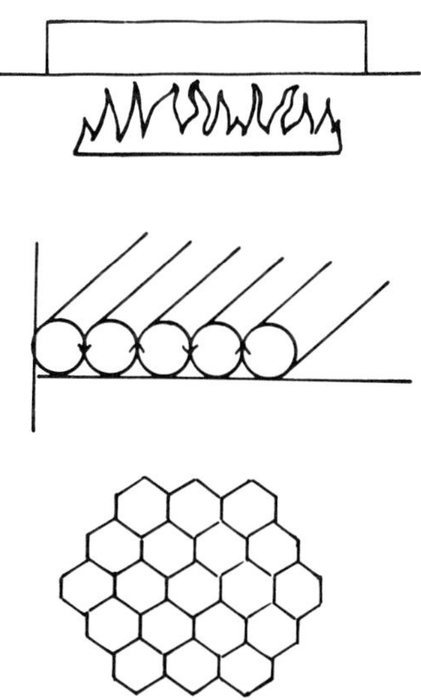

Fig. 5. Convection instability: fluid at rest, motion of fluid with roll pattern, hexagonal cells of motion (seen from above).

gradient is small, heat is transported by conduction and no macroscopic pattern occurs. Beyond a certain critical value of the
gradient a macroscopic motion sets in either in form of rolls or
hexagonal cells.

A seemingly completely different example is provided by
chemical reactions, for instance the Belousov-Zhabotinsky reaction.[4] It produces spatial structures, for instance a layer
structure (fig. 6) or oscillations, or chemical waves (fig. 7) [5].
Exactly the same wave patterns are produced by slime mold, a certain kind of mushrooms, when its cells are aggregating (fig. 8) [6].

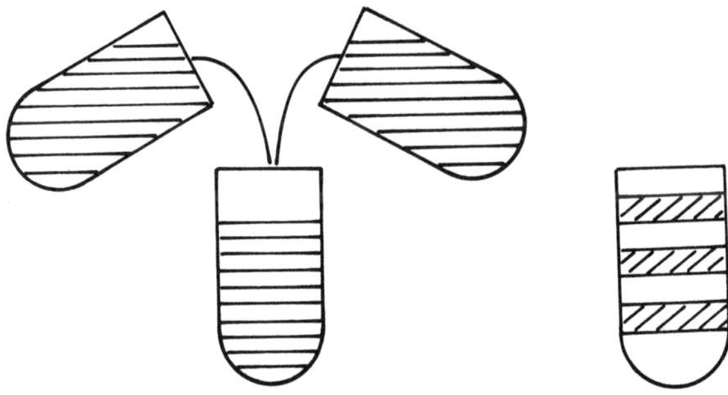

Fig. 6. Chemical reaction producing layer structure (red-blue-
red etc.)

Fig. 7. Spirals of chemical
activity in a shallow dish.
Wherever two waves collide head
on, both vanish. Photograph
taken by A. T. Winfree with
Polaroid Sx 70.

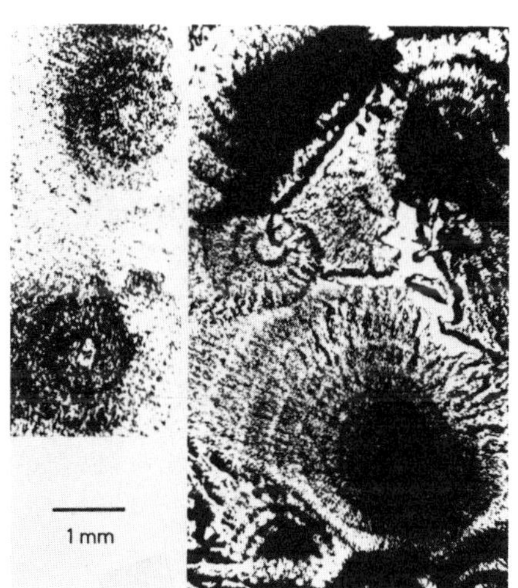

Fig. 8. Wave pattern of chemotactic activity in dense cell layers of slime mold (after Gerisch et al.).

Biologists have observed this phenomenon since some time and suggested that it may serve as a model for selforganization of organisms.

If we change the external parameters (pump power of the laser, temperature gradients in fluids, etc.) still more, new kinds of patterns may occur. For instance in the laser, pulses occur above a second threshold (fig. 9). A similar phenomenon is the Gunn effect of semiconductors.

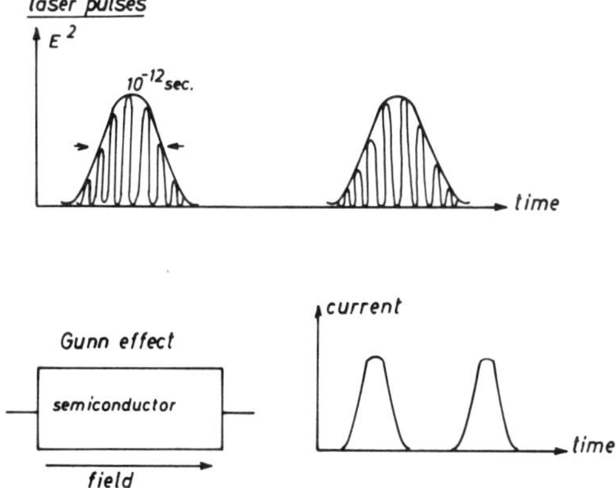

Fig. 9. Above: Laser pulses; below left: Setup of Gunn oscillator; right: Pulses of electrical current.

Considering these examples (and many others) we observe the
following: all these systems are composed of very many subsystems.
Under certain conditions they exhibit order phenomena on a macroscopic scale. This brings us to the definition of synergetics.
It is a new field of interdisciplinary research which analyzes
and compares ordering phenomena and tries to find the underlying
common principles governing these cooperative effects.[7] This
field comprises both disciplines where physical conceptions (e.g.
dissipation [8]) apply and nonphysical disciplines (for instance
ecology or sociology). I now want to show how the laser helps us
to find out and exemplify common principles of how such macroscopic
patterns are established. To do this in a mathematically more
explicit manner let us consider the laser equations.

I decompose the electric field strength $E(x,t)$ into a spatial
mode, sin kx, a rapidly oscillating factor $e^{i\omega t}$, where ω is the
circular transition frequency of the atoms, and a slowly varying
amplitude $E(t)$,

$$E(x,t) = E(t) \, e^{i\omega t} \sin kx + c.c. \tag{1}$$

As is shown in laser theory [1] this amplitude $E(t)$ obeys the
equation

$$\dot{E}(t) = - \kappa E(t) + \Sigma_\mu g_\mu p_\mu(t) + F(t) \, , \tag{2}$$

The first term on the r.h.s. of (2) describes the decay of the
field due to losses, for instance by the mirrors. The second term
describes the generation of the field by oscillating dipole moments,
where p_μ is the dipole moment of the atom with label μ, and g_μ are
certain coupling coefficients. F is a force describing fluctuations
which are always present because of a fluctuation-dissipation
theorem. In turn the dipole moments p_μ also change during the
laser process which is described by the equation

$$\dot{p}_\mu(t) = -\gamma p_\mu(t) + E(t)D_\mu h_\mu + F_\mu(t) \, . \tag{3}$$

Due to interactions with its surroundings, the dipole moment decays.
This is described by the first term. The second term on the right-
hand side describes how the dipole moment is driven by the electric
field strength E. h_μ are again coupling constants. D_μ represents
the difference of the occupation numbers of the upper and lower
levels of the atom μ. Now let us assume that the decay constant γ
is much larger than κ. We therefore expect that E decays slowly.
However, since E drives p_μ according to eq. (3) the temporal derivation of p_μ is much smaller than γp_μ. This means that we can neglect
$\dot{p}_\mu$ and thus express the dipole moment p_μ by the instantaneous value
of $E(t): p_\mu(t) \sim E(t)$. In the following we shall describe this

situation by saying: *The dipole moment is slaved by the field strength.*

In a similar manner one can show that also the atomic inversion D_μ is slaved by the field strength. This allows us to express p_μ and D_μ by E which leads to the equation[9]

$$\dot{E} = (-\kappa + G)E - C|E|^2 E + F_{tot} .\qquad (4)$$

G is a constant depending on the power with which the laser is pumped. For a lamp, $G<\kappa$ but for a laser, $G>\kappa$. We have discussed this equation many times. It can best be interpreted when we identify E with the coordinate q of a particle. To this end we replace E in Eq. (4) by q and add an acceleration term $m\ddot{q}$ with a very small mass. Furthermore we note that the first terms on the r.h.s. of eq. (4) can be derived from a fixed potential V(q). Thus we are led to the equation (compare fig. 10)[2]

$$m\ddot{q} + \dot{q} = -\partial V/\partial q + F_{tot}$$

We have formerly shown that this allows us to put the laser threshold in analogy with a second order phase transition of a system in thermodynamic equilibrium [10]. Indeed, nearly all phenomena known from second order phase transitions show up, such as symmetry breaking, critical slowing down, critical fluctuations, etc. What is important in our present context is this: the laser consists of very many atoms. Thus, it has very many degrees of freedom. However, as it is apparent from eq. (4), the essential behavior of

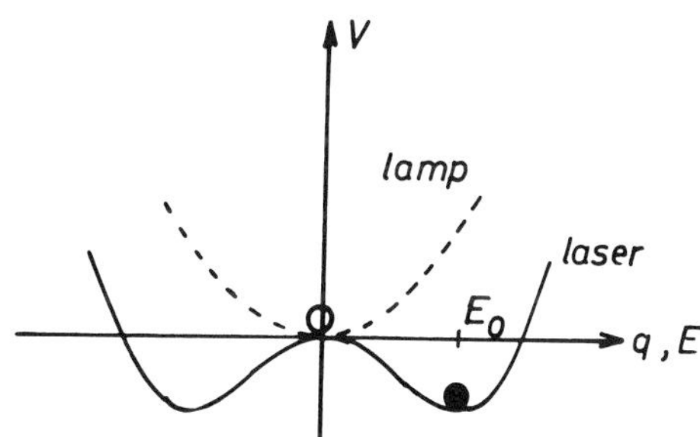

Fig. 10.   Potential of fictitious particle.

the laser is described by a single variable E, i.e. by a single
degree of freedom. Due to the "slaving principle" we have achieved
an enormous reduction of the number of degrees of freedom. Only
one degree of freedom shows up macroscopically in the laser light
field, and apparently is able to describe the state of the total
system. Since E describes the (macroscopic) order and simultaneously
gives orders to ("slaves") the atoms, we call E an order parameter.
Because below threshold ($G<\kappa$) E is zero (aside from fluctuations)
it cannot give orders and the dipole moments are just driven by
fluctuations incorporated in the fluctuating forces F_μ.

When we try to apply these ideas to systems other than the
laser, a problem seems to arise. When we consider, for instance,
a fluid layer, all its particles have the same relaxation times.
Therefore the "slaving principle" seems to fail. However, we can
overcome this difficulty.[12] Let us assume that we describe the
system by variables q_μ which obey equations of the type

$$\dot{q}_\mu = N_\mu(q_1,\ldots,q_n;\alpha) \tag{5}$$

where N_μ are in general nonlinear functions of the coordinates q_μ ,
and α is an external parameter (explicit examples for (5) are our
laser equations above). When we deal with continuously extended
systems, q_μ will depend on the space coordinate x and N_μ may
contain differential operators acting on $q_\mu(x)$.

Let us assume that we have found stationary solutions q_μ of (5)
which we call q^o_μ. To check the stability of that solution we make
as usual the hypothesis

$$q_\mu(t) = q^o_\mu + u_\mu(t) \tag{6}$$

and linearize (5) with respect to u_μ. For the example of two
variables this leads us to two equations of the form

$$\dot{u}_1 = a\,u_1 + b\,u_2 \ , $$
$$ \tag{6a}$$
$$\dot{u}_2 = c\,u_1 + d\,u_2 \ , $$

and to corresponding equations in the case of many variables. Equations (6a) can be solved by the hypothesis

$$u_\mu(t) = e^{\lambda t}v_\mu \ , \tag{7}$$

For $\lambda<0$, we obtain a damped solution while for $\lambda>0$ an undamped
solution results and the stationary solution is unstable. We
represent $q_\mu(t)$ in (6) as a superposition of solutions v_μ i.e. in

our example as

$$q_\mu(t) = q^o_{\ \mu} + \xi_\mu \, \nu_\mu^{\ (u)} + \xi_s \, \nu_\mu^{\ (s)} \tag{8}$$

(and correspondingly in the case of higher numbers of variables).
In continuously extended media ν_μ may have the form of waves, for
instance

$$\nu_\mu \equiv \nu_\mu(x) = a_\mu \, \sin kx \ , \tag{9}$$

where a_μ are certain constant coefficients and k wavevectors. One
can then, after some manipulations, show that the amplitudes obey
the equations

$$\dot{\xi}_u = \lambda_u \xi_u - \xi_u \xi_s \tag{10}$$

$$\dot{\xi}_s = \lambda_s \xi_s + \xi_u^2 \ , \tag{11}$$

where we have presented only a special case to show the general
argument. We assume that $\lambda_u \geq 0$. That means ξ_u is the slowly
varying coordinate and may thus serve as the order parameter.
Because in our model we assume λ_s negative and finite, we conclude
that ξ_s is slaved by ξ_u

$$\xi_s \sim \xi_u^2 \ , \tag{12}$$

i.e. the stable modes are slaved by unstable modes. In our present
example, inserting (12) into (10) gives rise to an equation of
exactly the same type as the laser equation (4). We have seen in
many examples of real systems that this classification holds so that
we have now a means to translate what we have learned from the
laser to other systems. Again the behavior of the whole system is
governed by one or very few order parameters. We now turn to the
interesting case when several order parameters are present. We
find two main patterns, namely competition or cooperation.[13]
Competition is found in lasers or in autocatalytic chemical processes
or in the multiplication of species. Denoting the number of different kinds of photons by n_j (or molecules or species), typical
equations for these order parameters are

$$\dot{n}_j = G_j n_j - \kappa_j n_j \ . \tag{13}$$

The first term on the r.h.s. represents the gain, the second the
losses. In the case of species (or laser photons) the gain is,
for instance, proportional to the amount of food (excited atoms)

available. When several kinds of species (photons) are present
the gain is lowered so that G_j has for instance the form

$$G_j = g_j(G_o - \Sigma_k \ldots n_k) \, , \tag{14}$$

Then the following situation arises: first only a few individuals
of the different species are present. Then their numbers grow
leading to segregation. But finally only one species, which is
the fittest, survives (fig. 11).

An example for cooperation is provided by certain mode coupling
effects in lasers. A still more drastic example is that of Bénard
cells where the velocity field in the vertical direction can be
represented by

$$q = \xi_1 e^{ik_1 x} + \xi_2 e^{ik_2 x} + \xi_3 e^{ik_3 x} + \text{c.c.} \, , \tag{15}$$

where the amplitudes ξ_1, ξ_2, ξ_3 serve as order parameters. The
corresponding equations show that the stable configuration is
given by $\xi_1 = \xi_2 = \xi_3 = \xi$, i.e. the three modes stabilize each
other giving rise to hexagon formation [11] (fig. 12). More
recently it has turned out that the same general principles are
valid for pulse instabilities.[14] The evolving pulse amplitude
q with the wave vector k_u and the frequency ω_u can be represented
as a superposition

$$q(x,t) = \xi_u \sin(k_u x - \omega_u t) + \xi_s \sin[2(k_u x - \omega_u t) + \phi] + \ldots , \tag{16}$$

where ξ_u serves as order parameter slaving the amplitudes of all
higher harmonics. ξ_u obeys again a very simple equation for
instance of the same type as eq. (4). Also other types of potentials indicating a first order transition may occur here.

Fig. 11. Development of competing species. From left to right:
initial state; medium state: segregation; final state: survival
of the fittest.

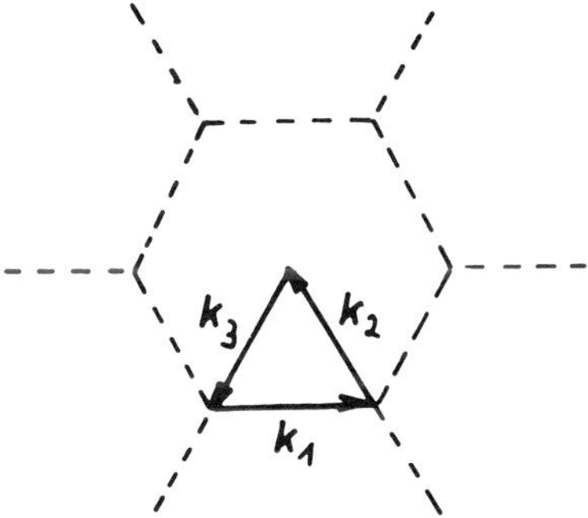

Fig. 12. Construction of hexagon of Benard cell out of three fundamental waves.

Some time ago Lorenz [15] developed a model of turbulence including 3 variables. We have found that when we identify these variables with the electric field strength, the atomic polarization and the atomic inversion of the laser, the Lorenz equations are equivalent to those of a single mode laser.[16] Though Lorenz' equations are completely deterministic their solutions show a completely irregular behavior as if caused by a stochastic process (fig. 13). The laser when pumped high enough is the first realistic system obeying Lorenz' equations. In this paper we have not yet discussed the role of fluctuations. They play a decisive role in driving systems from one state to another. They thus allow the

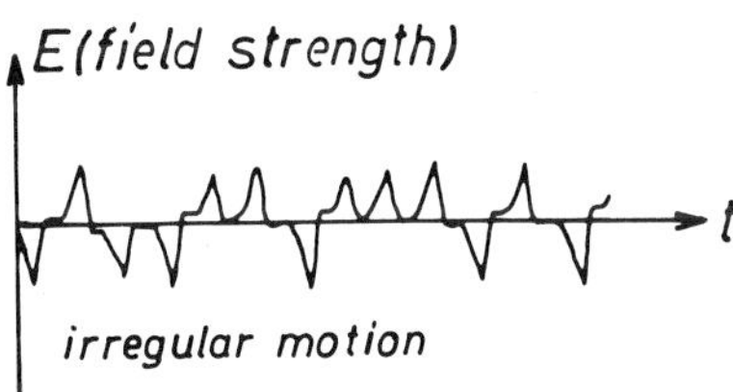

Fig. 13. Irregular development of a variable, after Lorenz, J. Atmos. Sci. *20*, 130 (1963).

systems to explore more stable or more "successful" configurations or functionings. The actually realized order parameters and thus the resulting macroscopic patterns stem from the interplay between deterministic forces (the potential of fig. 10, for example) and the fluctuations.

In conclusion, let us describe the general mechanism through which macroscopic patters (spatial, temporal, spatiotemporal, or functional structures) evolve (scheme). We start with a certain structure for given external parameters. When we change one (or several) external parameter this structure may become unstable and a new structure evolves. At the instability point we can distinguish between unstable and stable modes. Since the unstable modes slave the stable modes we eliminate the latter. Therefore in many cases near instability points an enormous reduction of the number of degrees of freedom is possible. The resulting equations for the unstable modes, called order parameters, can be grouped into few classes. This means that quite different systems show equivalent groups of phenomena. These classes are thus very analogous to the universality classes known from conventional phase transition theory. The competition or cooperation of the order parameters decides upon the new evolving structure. When we change external parameters further the structure can become unstable and we may pass through several instabilities leading to more and more structures. While, for instance, in fluid dynamics it seems that there is a number of finite instabilities until turbulence is reached, we observe that in living systems higher and higher order is achieved. How nature avoids "turbulent" states of living organisms is still one of its great mysteries.

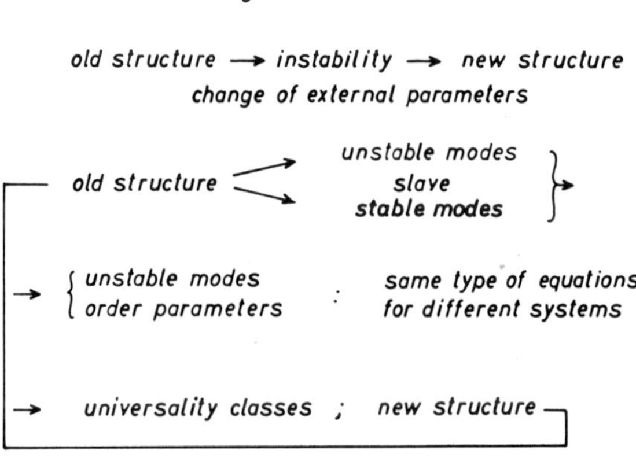

Scheme

References

1.   H. Haken, in *Encyclopedia of Physics*, S. Flügge, ed. (Springer
     Verlag, Berlin, 1970); M. Sargent, III, M.O. Scully and W.E.
     Lamb, *Laser Physics* (Addison-Wesley, Reading, Mass., 1974);
     an excellent collection of papers in *Selected Papers on Coher-
     ence and Fluctuations of Light*, Vols. I and II, L. Mandel and
     E. Wolf, eds. (Dover Publ. Inc., New York, 1970).
2.   H. Haken, Z. Physik *181*, 96 (1964).
2a. See for instance F. T. Arecchi, V. DeGiorgio in *Laser Handbook*,
     F. T. Arecchi and E. O. Schulz-Dubois, eds. (North Holland,
     Amsterdam, 1972); E. R. Pike in *Quantum Optics*, S. M. Kay and
     A. Maitland, eds. (Academic Press, London, New York, 1970).
3.   H. Bénard, Revue générale des sciences pures et appliquées *11*,
     1261-71 and 1309-28 (1900), Annales de Chimie et de Physique
     23, 62-144 (1901).
4.   B. P. Belusov, Sb. ref. redak. med. Moscow (1959).
5.   A. T. Winfree, Scient. Amer. *230*, 83 (1974).
6.   G. Gerisch, B. Hess, Progr. Mat. Ac. Sci. *71*, 2118 (1974);
     G. Gerisch, Naturwiss. *58*, 430 (1971).
7.   H. Haken and R. Graham, Umschau *6*, 191 (1971); *Synergetics*,
     H. Haken, ed. (Teubner, Stuttgart, 1973); *Cooperative Effects*,
     H. Haken, ed. (North Holland, Amsterdam, 1974); *Synergetics.
     An Introduction. Nonequilibrium Phase Transitions and Self-
     Organization in Physics, Chemistry and Biology*, H. Haken,
     (Springer, Berlin, Heidelberg, New York, 1977). These books
     contain many further references. While stability (or insta-
     bility) criteria (see in particular [8]) are important to
     find "instabilities," synergetics goes far beyond that, for
     instance, treating the question of what happens at the insta-
     bility point (where also fluctuations are to be considered)
     and which new structures evolve.
8.   P. Glansdorff and I. Prigogine, *Thermodynamic Theory of
     Structure, Stability and Fluctuations* (Wiley Intersc., New
     York, 1971).
9.   This "adiabatic elimination" procedure together with a self-
     consistency requirement for E (also in the multimode case)
     is the basic method of solution in semiclassical laser theory.
     H. Haken, talk given at the International Conference on Optical
     Pumping, Heidelberg, 1962; H. Haken and H. Sauermann, Z.
     Physik *173*, 261 (1963); *176*, 47 (1963); W. E. Lamb, Phys. Rev.
     134, 1429 (1964); some historical aspects of semiclassical
     laser theory stressing its independent development by Haken
     and by Lamb are described by M. Sargent III, Opt. Comm. (1974).
10. H. Haken, in *Festkorperprobleme* X, O. Madelung, ed. (Pergamon,
     Vieweg, 1970); R. Graham and H. Haken, Z. Physik *213*, 420
     (1968); *237*, 31 (1970); V. DeGiorgio and M. O. Scully, Phys
     Rev. A*2*, 1170 (1970); for references to phase transition
     analogies in other fields see [11].

11. H. Haken, Rev. Mod. Phys. 47, 67 (1975).
12. H. Haken, Z. Physik B21, 105 (1975); B22, 69 (1975); B23, 388 (1976).
13. H. Haken, in *Statistical Physics* (Proc. of the IUPAP Conf. on Stat. Mechanics, Budapest, 1975) L. Pàl and P. Szepfalusky, eds.
14. H. Haken and H. Ohno, Opt. Comm. 16, 205 (1976); H. Ohno and H. Haken, Phys. Lett. 59A, 4, 261 (1976).
15. E.N. Lorenz, J. Atmos. Sci. 20, 130 (1963).
16. H. Haken, Phys. Lett. 53A, 77 (1975).

PHOTON STATISTICS, INSTABILITIES AND PHASE TRANSITIONS IN DYE LASERS

Charles R. Willis

Boston University, Boston, Massachusetts

1.   INTRODUCTION

Recently Ray Schaefer and I [1] showed that the behavior of a
dye laser could be considerably different than the usual laser behavior. Subsequently we indicated that the anomalous behavior of the
dye laser could be interpreted in terms of a first order phase transition [2], instead of the conventional second order [3-5] laser
phase transition analogy. In the present paper I want to develop
the laser phase transition analogy further, in order to understand
better the fundamentally different behavior possible in dye lasers,
to develop further the differences between the second order and
first order laser phase transitions analogies and to examine the
role of fluctuations in phase transitions.

Previously, [1,2] we demonstrated that of the many dimensionless
parameters possible in a dye laser, two, which we denoted by α and β
and which are defined below, were particularly important. We want
to call attention to the parameter β because as we will show,
the parameter β plays a crucial role in determining the type of
phase transition the laser undergoes and as far as we can tell the
value of the parameter β has not been determined experimentally.

It is useful to have a mean field treatment of the dye laser
phase transition, even though [1] contains a complete quantum statistical treatment of a dye laser model, because the presence of
triplet levels in dye molecules considerably increases the complexity
of the laser model. At the same time we find the presence of triplets enhances the fluctuations and we find that in order to understand phenomena like hysteresis and bistability [6] we have to go
beyond a mean field treatment and include the fluctuations that are

contained in [1]. When we apply the dynamical correspondence
between the density matrix of [1] and the associated classical
function, developed by Lax and Louisell [7], the resultant equation
is a special case of the generalized Fokker-Planck approach [8].

One of the ways of thinking of a dye laser is that the triplets behave as a saturable absorber. Consequently the theories
developed by Kazantsev and Surdutovich [9] and Salomaa [10] apply,
with the only modifications concerning the fact that the triplets
(saturable absorbers) are in the same molecule and are created by
crossover from the singlet levels. The article by Scott et al.
[11] also called attention to the fact that the saturable absorber
problem could be treated as a first order transition.

2.   DYE LASER MODEL

We will start with a brief description of our model for a dye
laser. For a more thorough discussion see [1] and the references
contained therein. A dye molecule is composed of singlet and triplet levels with the triplet levels playing the role of a loss mechanism. The presence of the triplets requires that an adequate model
for a dye laser requires the inclusion of six states, four states
of which interact directly with the laser radiation and two states
of which interact with the pump as depicted in Fig. 1. The bottom
of the S_0 level is the ground state and the top of S_1 is the upper
pump level, where S_0 is the lowest energy singlet band, S_1 is the
first excited singlet state band, T_1 is the lowest energy triplet
band and T_2 is the first excited state triplet band. The bottom
of the S_1-level (state 3) is a single level with width $\nu_{S,u}$ (the
u refers to upper and S refers to singlet), the bottom of the T_1-
level (state 5) is a single level with width $\nu_{T,\ell}$ (the ℓ refers to
lower and the T to triplet), the top of the S_0 level (state 2) is
a level with width $\nu_{S,\ell}$ and the top of the T_2 level (state 6) is a
level with width $\nu_{T,u}$. These four levels are directly involved in
transition processes with the laser radiation. The pump pumps the
atom to state 4, after which there is a fast nonradiative decay to
state 3, then the dye molecule is induced to emit laser radiation,
with the molecule ending up in one of the states of the S_0-band,
after which there is a rapid nonradiative decay to the singlet
ground state 1. Thus, the singlet levels are equivalent to a conventional four-level laser system. However, the lower triplet band
T generally has energy levels lower than the lowest S_1 levels and
some of the dye molecules decay nonradiatively from the bottom of
the S, to the T-band, at a rate denoted by K_{ST}. The decay to the
triplet level causes losses in two ways, which we collectively term
"triplet-state losses." One loss is the depletion of the upper
singlet population available for lasing. The other loss is an

absorption of laser radiation in a transition to an upper triplet band T_2. The triplet state losses are dependent on the intensity of the laser radiation indirectly through K_{ST} and directly through the absorption of the laser radiation. Consequently, the net effect of the triplets on the singlet laser levels is to appear as a laser intensity-dependent loss mechanism, which is responsible for the first order phase transition behavior.

There are a minimum of ten inverse lifetimes and transition rates needed to describe dye laser behavior. However in [1] we found and discussed two dimensionless ratios that played a crucial role in determining dye laser behavior. The first which we denoted by α was defined by $\alpha \equiv (TK_{ST})/(S\nu_{T,\ell})$, where S is the stimulated singlet emission rate and T is the stimulated triplet absorption rate. In [1] we showed that a minimum condition necessary for laser emission is $\alpha < 1$. The physical meaning of the condition $\alpha < 1$ is that the probability of the emission from the singlet level in the time of the inverse crossover rate $S(K_{ST})^{-1}$ is greater than the probability of the absorption from the triplet level T in the life-time of the triplet level $T(\nu_{T,\ell})^{-1}$. For small values of α triplet losses are relatively small; as α increases the minimum pumping needed to obtain laser oscillation (threshold) increases as α increases. For $\alpha > 1$, triplet losses dominate and continuous operation cannot take place. The second important dimensionless parameter we found and discussed in [1] is the fractional return parameter

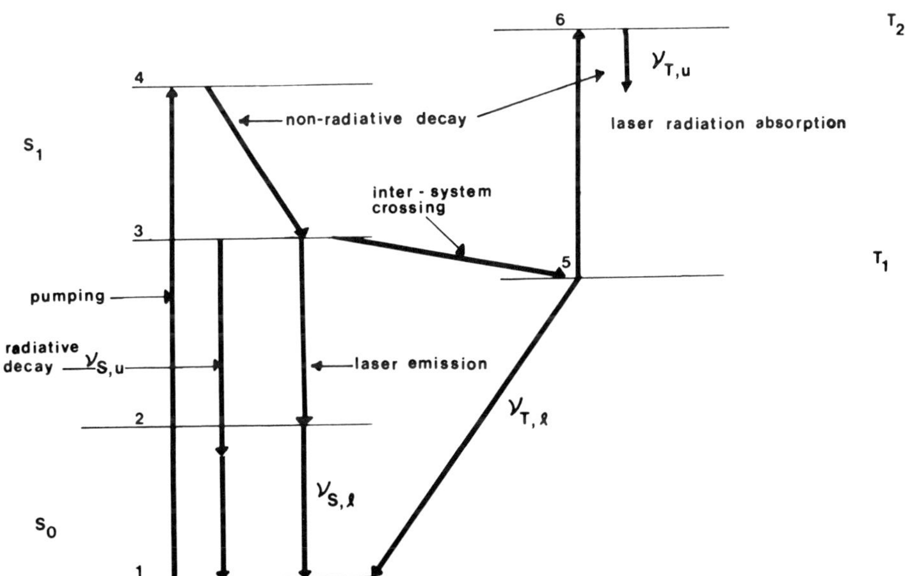

Figure 1. Organic dye laser as a six-state system.

$\beta(0 \leq \beta < 1)$, defined as the fraction of excited triplets that de-excite by returning to the triplet ground state (level 5 of Fig. 1). The condition $\beta < 1$ implies that a fraction $(1-\beta)$ of dye molecules excited to the T_2 level de-excite by nonradiative or radiative processes that take the dye molecule directly to some singlet level. If $\beta = 1$, then the triplet loss rate becomes intensity-independent and the dye laser becomes a conventional two-state laser, with the triplets playing the role of a conventional intensity-independent loss rate.

In order to provide an intuitive basis for understanding of the role of the β parameter and the role of the triplets generally in the first order phase transition behavior of the dye laser, we discuss some of the crucial details that appear in the derivation of the equations of motion for a dye laser presented in [1]. The equation of motion for the diagonal matrix elements of the radiation density operator, $R(n)$, of a single-mode dye laser (Eq. (4.3) of [1]) is

$$\dot{R}(n) + \nu_R[nR(n) - (n+1)R(n+1)] = -NS\Delta\{(n+1)D^S(n)\} + NT\Delta\{(n+1)D^T(n+1)\},$$

$$(2.1)$$

where

$$\Delta\{(n+1)D^S(n)\} \equiv (n+1)D^S(n) - nD^S(n-1)$$

$$\Delta\{(n+1)D^T(n)\} \equiv (n+1)D^T(n) - nD^T(n-1) ,$$

and $D^S(n) \equiv D_n^{S,u}$ is the correlated density matrix element for a single dye molecule in the upper singlet laser state when there are n photons present in the radiation mode, and $D^T(n) \equiv D_n^{T,\ell}$ is the correlated density matrix element for a single dye molecule in the lower triplet state of T, when there are n photons present in the radiation mode. In Eq. (2.1) we have omitted purely for convenience of exposition the small terms corresponding to the lower singlet state and the upper triplet state, as is usually done in laser theories.

The steady state solution for $D^T(n)$ given by Eq. (5.1) in [1] can be written in the following form:

$$D^T(n) = (K_{ST}/\nu_{T,\ell}) \; f(n)D^S(n) , \qquad (2.2)$$

where

$$f(n) = [1 + (n/n^T)(1-\beta)(1 + 2nT/\nu_{T,u})^{-1}]^{-1} , \qquad (2.3)$$

where $n^T \equiv \nu_{T,\ell}(2T)^{-1}$ is the triplet saturation photon number. When we substitute Eq. (2.2) into Eq. (2.1) we obtain

$$\dot{R}(n) + \nu_R[nR(n) - (n+1)R(n+1)] = -NS\Delta\{(n+1)[1-\alpha f(n)]D^S(n)\}. \qquad (2.4)$$

When $\beta = 1$, $f(n) = 1$ independent of n and Eq. (2.3) reduces to a conventional two-level laser, where the net effect of the triplets is to raise the threshold from N_T to $N_T[1+(K_{ST}/\nu_{T,\ell})](1-\alpha)^{-1}$, as we showed in Eq.(5.16) of [1].

When $\beta \neq 1$, $f(n)$ depends on n and we have the possibility of changing the usual second order laser transition into a first order transition if $f(n)$ has the proper dependence on n. There is no requirement that $<n>$ be small compared with n^T or $n^S \equiv \nu_{S,u}(2S)^{-1}$. In fact, we believe that in many dye lasers $<n>>n^T$ and some $<n>><n^S>$. However, for this paper we will consider $<n><<n^T,n^S$ because the inequalities are sometimes satisfied in practice and it is mathematically more convenient to deal with a cubic equation in n rather than an infinite series. The qualitative phenomenon of the Van der Waals type of behavior as we shall show is the same in both cases; the only difference is quantitative.

We can understand how the triplets change the stability character of the dye laser by expanding the right-hand side of Eq.(2.4) to second order in n, i.e.,

$$(n+1)[1-\alpha f(n)]D^S(n) \approx (n+1)[1-\alpha - \frac{n}{2}\{(n^S)^{-1}-\alpha(1-\beta)(n^T)^{-1}\}+\ldots]R(n),$$

$$(2.5)$$

where we used $D^S(n) \approx [1+(n/n^S)]^{-1}R(n)$ from Eq.(5.a) of [1]. From Eq. (2.5) we see if $n_e^T \equiv n^T(1-\beta)^{-1}>\alpha n^S$ then the coefficient of n^2 is negative as in conventional laser theory and we have a second order phase transition. For $n^S>n_e^T$ the coefficient of n^2 in Eq.(2.5) is positive which we will show below leads to a first order phase transition, i.e. Van der Waals behavior. At $\alpha n^S = n_e^T$ we have the critical point of the first order phase transition. We introduce a third dimensionless parameter, the relative saturability $r \equiv (n^S/n_e^T) \equiv (n^S/n^T)(1-\beta)$, which determines the type of phase transition the dye laser undergoes. As we will see in the next section, the important parameters are α and r. The parameter β will appear only in the definition of r, and the crucial point is that β must not equal one for the dye to display anomalous behavior. Experimentally we don't know the value of β; it certainly is not equal to one, but in some cases $(\beta-1)$ may be so small that $r \to 0$ and then the dye laser will behave conventionally.

3. DYE LASER FOKKER-PLANCK EQUATION IN THE $P(\alpha)$ REPRESENTATION

In Section 1 we derived and analyzed the equation of motion for the radiation density operator quantum mechanically. However, as DeGiorgio and Scully [3] showed, it is very useful to analyze mean field and phase transition behavior in terms of the c-number function $P(\alpha)$ where [12]

$$R = \int d^2\alpha \, P(\alpha,\alpha^*,t)|\alpha><\alpha| \quad . \tag{3.1}$$

In order to treat first order phase transition behavior it is more convenient to express the generalized Fokker-Planck for P in terms of polar coordinates $|\alpha|$ and ψ instead of the coordinates α and α^*. However, for problems in which an external polarization is on the system, the α,α^* representation is more useful. Furthermore it is easier to treat the analogy of the second order laser phase transition with the Landau theory in the α,α^* representation. Consequently in this section we will use the α,α^* representation to treat the critical point of the first order phase transition of the dye laser, deferring until the next section an analysis of how the critical point of the dye laser first order phase transition arises, in analogy to the phenomena of critcal opalescence at the critical point of the Van der Waals first order gas-liquid phase transition.

When Eq.(3.1) is substituted in the equation of motion of R for a dye laser Eqs. (4.3) and (5.1) of [1] we find the c-number function P satisfies to a good approximation the following Fokker-Planck equation,

$$\frac{\partial P}{\partial t} = -\frac{\partial}{\partial \alpha}\left\{\frac{\alpha}{2}((A'-C')-B'\alpha^*\alpha-\frac{3D'}{4}(\alpha^*\alpha)^2)P\right\}$$

$$-\frac{\partial}{\partial \alpha^*}\left\{\frac{\alpha^*}{2}((A'-C')-B'\alpha\alpha^*-\frac{3D'}{4}(\alpha\alpha^*)^2)P\right\} + A'\frac{\partial^2 P}{\partial\alpha\partial\alpha^*} , \qquad (3.2)$$

where we have retained not only the first nonlinear term proportional to B' but also the next highest order nonlinearity proportional to D', because, as contrasted with the usual laser theory where B' is positive and much greater than D', in dye lasers B' can be small, zero and even negative. Consequently, we will find the stability role played by B' > 0 in the usual laser theory is played by the condition D' > 0 in dye lasers. In Eq.(3.2), A' is the unsaturated gain in the medium, B' is the lowest order nonlinearity in the presence of triplets, C' is the loss parameter and D' is the next order nonlinear term of the saturation. If the triplets were absent, A', B', and C' would reduce to A, B and C of DeGiorgio and Scully [3] and D' would be unnecessary. The parameters A', B', and D' are proportional to the inversion σ so we have A' = a'σ, B' = b'σ, D' = d'σ and C' = a'σ_T, where σ_T is the "threshold" in the presence of the triplets, i.e. $N_T(1+(K_{ST}/\nu_{T,\ell}))(1-\alpha)^{-1}$, and N_T is the inversion required for laser action in the singlets in the absence of the triplet levels.

Before studying the steady state solution Eq.(3.2), which contains all the steady state information about the dye laser including fluctuations, we first investigate the mean field treatment of the critical point. We obtain the mean field results in the following manner. We first find the equations of motion for the average value <E> of the dimensionless electric field operator $E \equiv \frac{1}{2}(a_{op} + a_{op}^+)$ (where a_{op} and a_{op}^+ are the usual photon annihilation and creation operators), which is obtained from Eq.(3.2), namely,

$$\langle\dot{E}\rangle = \frac{1}{2}(A'-C')\langle E\rangle - \frac{1}{2}B'\langle E^3\rangle - (3/8)D'\langle E^5\rangle \quad . \tag{3.3}$$

Next we set $\langle\dot{E}\rangle$ equal to zero and neglect fluctuations, by replacing $\langle E^3\rangle$ by $\langle E\rangle^3$, and $\langle E^5\rangle$ by $\langle E\rangle^5$ which yields

$$(A'-C')\langle E\rangle - B'\langle E\rangle^3 - \frac{3}{4}D'\langle E\rangle^5 = 0 \quad . \tag{3.4}$$

We show in the next section that the mean field behavior contained in Eq. (3.4) is the same as the mean field behavior of the Van der Waals equation. Therefore, the dye laser instability is not analogous to the order-disorder transition of the usual laser theory. However, there is a critical point of the first order phase transition analogous to the phenomena of critical opalescence which is a second order transition. This critical point occurs at B'=0 as we now demonstrate. The solution of Eq. (3.4) for B'=0 is

$$\langle E\rangle = 0 \qquad\qquad \text{if } \sigma-\sigma_T < 0, \quad \text{(below threshold)}, \tag{3.5a}$$

$$\langle E\rangle = \left\{\frac{a'}{d'}\left(\frac{\sigma-\sigma_T}{\sigma}\right)\right\}^{\frac{1}{4}} \quad \text{if } \sigma-\sigma_T > 0. \quad \text{(above threshold)}. \tag{3.5b}$$

Consequently the critical point exponent is $(1/4)$ instead of $(1/2)$ which means the photon number $\langle n\rangle$ $(\langle n\rangle = \langle E\rangle^2)$ increases as $((\sigma-\sigma_T)/\sigma)^{\frac{1}{2}}$ instead of linearly as in conventional laser theory.

An external classical field resonant with the laser introduced into the cavity induces an addition polarization denoted by S, which couples to the laser photons and adds an inhomogeneous source term to Eq. (3.4) in the following manner [3]:

$$(A'-C')\langle E\rangle - B'\langle E\rangle^3 - \frac{3}{4}D'\langle E\rangle^5 + 2S = 0 \quad . \tag{3.6}$$

The critical point of the first order transition is at B'=0 and the critical inversion curve (A'=C') is

$$\langle E\rangle = \left(\frac{8S}{3D'}\right)^{1/5} = \left(\frac{8S}{3\sigma_T d'}\right)^{1/5} \quad , \tag{3.7}$$

which is in contrast to the usual laser case where D' is negligible and B > 0, so that $\langle E\rangle = (2S/b\sigma_T)^{\frac{1}{3}}$. As DeGiorgio and Scully [3] showed, the analogue of the zero field magnetic susceptibility $\chi = (\partial\langle M\rangle/\partial H)_{H=0}$ is the reciprocal polarizability ξ of the laser $(\partial\langle E\rangle/\partial S)_{S=0}$. Consequently, when we differentiate Eq. (3.6) with respect to S, setting B' = 0, then set S = 0, we obtain

$$\xi = \begin{cases} [\frac{1}{2}a'(\sigma_T - \sigma)]^{-1} & \text{for } \sigma < \sigma_T \\[2mm] [2a'(\sigma - \sigma_T)]^{-1} & \text{for } \sigma > \sigma_T \quad , \end{cases} \tag{3.8}$$

which has the same critical exponent as the conventional theory and

differs only in the factor 2 in place of 1 in the second line of
Eq.(3.8). Equation (3.8) is an example of how the critical point
behavior depends on $(\sigma_T/\sigma)-1$ in the same way that the magnetic analogy depends on $1-(T_c/T)$. Consequently, apart from an unimportant
change in sign, σ corresponds to T, in the same manner as in the
usual laser theory because of the critical point behavior.

We can go beyond mean-field theory and include fluctuations by
solving Eq.(3.2) exactly for the steady state solution obtained by
setting $\dot{P}=0$. The solution is

$$P_s = Z \exp\left\{(\tfrac{1}{4}(A'-C')|\alpha|^2 - \tfrac{1}{8}B'|\alpha|^4 - \tfrac{1}{16}D'|\alpha|^6)/\tfrac{1}{4}A'\right\} \qquad (3.9a)$$

$$\equiv Z \exp\left\{-\tfrac{\bar{a}}{2}|\alpha|^2 - \tfrac{\bar{b}}{2}|\alpha|^4 - \tfrac{\bar{c}}{6}|\alpha|^6\right\} \equiv Z \exp\{-\Phi\} \quad , \qquad (3.9b)$$

where Z is the partition function, $\bar{a} \equiv 2[(\sigma_T/\sigma)-1]$ and

$$\bar{b} \equiv \frac{b'}{a'} = \frac{b^S}{a^S}(1-\alpha(1+r)) \equiv (n^S)^{-1}(1-\alpha(1+r)) \quad , \qquad (3.10a)$$

$$\bar{c} \equiv \frac{3}{2}(\frac{d'}{a'}) = \left(\frac{b^S}{a^S}\right)^2\{\alpha(1+r(1+r))-1\} \equiv (n^S)^{-2}\{\alpha(1+r(r+1))-1\}. \qquad (3.10b)$$

The symbols a^S and b^S refer to the unsaturated gain and the saturation parameters per particle of the singlets respectively, and are
the same as the a and b of DeGiorgio and Scully. The "potential" Φ
for the dye laser, Eq.(3.9b), is the same Φ we had obtained previously [2] without going through the intermediary step of the α
representation.

When we calculate the steady state value of $\langle\alpha\rangle$ from P in Eq.
(3.9) we find $\langle\alpha\rangle$ vanishes in the steady state. If we calculate
$\langle|\alpha|^2\rangle$ from Eq.(3.9) we find the behavior is smooth at threshold
(even at B'=0); thus there is no non-analyticity and the fluctuations
ultimately due to spontaneous emission wipe out all the singular
behavior. The limiting procedures under which the singular behavior
can be reinstated involves complicated and subtle questions about
which variables of the system and reservoirs should be allowed to
approach infinity and in which order. These questions have been most
carefully investigated by Graham [13]. It is beyond the scope of
this paper to enter into a discussion of these points, except to
point out in most laser situations we can obtain the full steady
state distribution function, and the mean-field results constitute
the "most probable" values of the potential which appears in the
steady state distribution function as in Eq.(3.9). For finite systems, at least, the fluctuations are large compared with the mean
values at the instability points, and the "most probable" values lose
their usefulness. In the next section we will see that the mean-
field treatment of the dye lasers leads to different nonanalyticities

than the usual laser theory, and in the following section we will
see the effect the inclusion of the fluctuations has on these new
singularities.

4. DIMENSIONLESS FOKKER-PLANCK EQUATION IN POLAR COORDINATES

In problems where one is not particularly interested in phase
information of the electric field, such as in the steady state Eq.
(3.9), it is convenient to transform Eq.(3.2) in the α representation
to amplitude and phase variables. The transformation has the added
advantage in our case of making the relationship with Van der Waals
behavior more transparent. Introducing the variables $\alpha = \xi^{\frac{1}{2}}e^{i\psi}$ and
$\tau = \frac{1}{4}A't$ into Eq.(3.2) we obtain

$$\frac{\partial P}{\partial \tau} = \frac{\partial}{\partial \xi}\left(\xi\frac{\partial \Phi}{\partial \xi} + \xi\frac{\partial}{\partial \xi}\right)P + \frac{1}{4\xi}\frac{\partial^2}{\partial \psi^2}P \quad , \tag{4.1}$$

where Φ is the potential in Eq.(3.9b). Throughout this paper we are
considering exact resonance for ease in presentation. In the case
of off-resonance there would be changes in the coefficients and there
would be a linear first derivative term in ψ in Eq.(4.1). Actually
Kazantsez and Surdutovich [9] have derived an equation essentially
the same as Eq.(4.1) in treating the saturable absorber problem.
They obtained their equation from a previous important fundamental
work of theirs [14] on the laser.

When we examine the coefficients of Φ which appear in Eq.(3.10a,
b) we find the photon number ξ scales with n^S, so we make a final
transformation of Eq.(4.1) to scaled variables $\eta \equiv \xi(n^S)^{-1}$, and the
variables a, b, and c

$$\frac{dP}{d\tau} = -\frac{\partial}{\partial \eta}\left(A(\eta) - \frac{\eta}{n^S}\frac{\partial}{\partial \eta}\right)P(\eta) \quad , \tag{4.2a}$$

where

$$A(\eta) = -\eta\frac{\partial}{\partial \eta}\Phi = -(a + b\eta + c\eta^2)\eta \tag{4.2b}$$

$$\Phi(\eta) = a\eta + \frac{b}{2}\eta^2 + \frac{c}{3}\eta^3 \tag{4.2c}$$

and

$$a \equiv \left(\frac{\sigma_T}{\sigma} - 1\right) \quad ; \quad b \equiv (1 - \alpha(1+r)) \quad ;$$

$$c \equiv \frac{1}{2}\{\alpha(1 + r(1+r)) - 1\} = \frac{1}{2}(\alpha r^2 - b) \quad , \tag{4.2d}$$

where for the purposes of the present paper we have dropped the
dependence on the phase ψ.

The steady state solution of Eq.(4.2a) is

$$P(\eta) = Z \exp(-n^S \Phi(\eta)) \quad . \tag{4.3}$$

Before studying the full statistical distribution Eq.(4.3), which includes fluctuations, we obtain a mean field treatment in the new variables by neglecting fluctuations. There are two ways of going over to a mean field approach. One way is to calcualte $\langle\eta\rangle$ from Eq.(4.2a), neglect correlations (by replacing $\langle\eta^2\rangle$ by $\langle\eta\rangle^2$ etc.) and obtain information in the same manner as we did in Section 3. The second way we can neglect fluctuations is to replace $\Phi(\eta)$ in Eq.(4.3) by $\Phi(\langle\eta\rangle)$; then $\Phi(\langle\eta\rangle)$ becomes the appropriate "free energy" approximating the correct "free energy" $\sim\ln\int P_S(\eta)d\eta$ which contains contributions from fluctuations. This mean-field replacement is valid as long as the fluctuations are small compared with mean values, i.e. away from points of instability. In the laser problem both in critical phenomena and in first order phase transitions the fluctuations wipe out the singluar behavior and replace it by a smooth but rapidly changing behavior. In this paper we find it useful to use both approaches to the mean-field problem, starting with the free energy approach.

The properties of $\Phi(\langle\eta\rangle)$ and $\Phi(\eta)$ are determined by three parameters a, b, and c given in Eqs.(4.2c) and (4.2d). The extrema of Eq.(4.2c) are

$$\eta_\pm \equiv \langle\eta\rangle_\pm = \frac{-b \pm (b^2-4ac)^{\frac{1}{2}}}{2c} \tag{4.4a}$$

with

$$\Phi''(\eta) = 2c\eta + b \ ,$$

$$\Phi''(\eta_\pm) = \pm(b^2-4ac)^{\frac{1}{2}} \ , \tag{4.4b}$$

where the prime indicates differentiation with respect to the argument. In all our cases $c > 0$ so the qualitative behavior is determined by a and b. If $b > 0$, we have the conventional theory; however, for $b < 0$, $a > 0$ and $b^2 > 4ac$ we have a maximum at η_-, a minimum at η_+, and the origin is the absolute minimum when $\Phi(\eta) > 0$ for all η. In the α-representation the origin is a minimum for $a > 0$ and a maximum for $a < 0$. In the ξ or η coordinates $\Phi'(\eta)$ does not vanish at $\eta=0$, except for the one value $a=0$. However, recall the force in the η variables is not $\Phi'(\eta)$ but $\eta\Phi'(\eta)$, so the force vanishes at $\eta=0$. The zeros of $\Phi(\eta)$ are $\langle\eta\rangle = 0$, and $\eta_\pm^0 \equiv \{-3/2)b\pm[(9/4)b^2-12ac]^{\frac{1}{2}}\}(2c)^{-1}$. The height of the barrier defined as $h = \Phi(\eta_-)-\Phi(\eta_+)$ is equal to $(b^2-4ac)^{\frac{3}{2}}(6c^2)^{-1}$. The value of η at which the minimum of $\Phi(\eta)$ crosses the η axis is $\eta^*=(3|b|/4c)$. In Figs. (2a-2d) we show how the usual Van der Waals-like loop appears for $a > 0$ and $b < 0$. For fixed

b we can consider the sequence of diagrams as taking the limit as a
decreases. Qualitatively, for $\Phi(\eta_+) > 0$, the origin remains the
absolute minimum as in Fig. 2a. In Fig 2b the minimum at η_+ has the
same value as the minimum at $\langle\eta\rangle = 0$. As a decreases further the
minimum at η_+ becomes the absolute minimum and the origin becomes
the relative minimum until finally the barrier disappears entirely
and the origin becomes maximum at a=0, as in Fig. 2d. In going from
Fig. 2c to Fig. 2d the point η_1 moves to $\eta=0$ and the point η_2 moves
to $(3|b|/2c)$ at a=0. When a becomes negative, i.e. for $\sigma > \sigma_T$, the
well becomes deeper and moves toward increasing values of η qualitatively in the same manner as a conventional laser above the threshold, but in a quantitatively different manner.

We now want to investigate the physical consequences of the
characteristics of Fig. 2. When $\Phi(\eta_+) > \Phi(\eta=0)$ then η_+ is a relative
minimum and consequently represents a metastable state, with the
stable state at $\langle\eta\rangle=0$ as in Fig. 2a. At $\eta=\eta^*$, as in Fig. 2b, the
free energy of the two states $\eta=0$ and $\eta=\eta^*$ are equal and we have the
coexistence of the two states or "phases". As the variable a further
decreases the state at $\eta=\eta_+$ becomes the absolute minimum and $\eta=0$
becomes the metastable state, as in Fig. 2c. Finally, when a becomes
zero, $\langle\eta\rangle=0$ becomes a maximum and the only minimum of the free energy
is at η_+. What we have just described is a phase transition from
the situation where the absolute minimum is $\langle\eta\rangle=0$, representing incoherent photons, with a metastable state at η_+ representing coherent
photons. As a decreases, i.e. approaches the conventional threshold
from below, we come to the phase transition when the free energy of
the incoherent phase $\Phi(\eta=0)$ equals the free energy of the coherent
phase $\Phi(\eta_+)$; then as a decreases further the coherent phase becomes

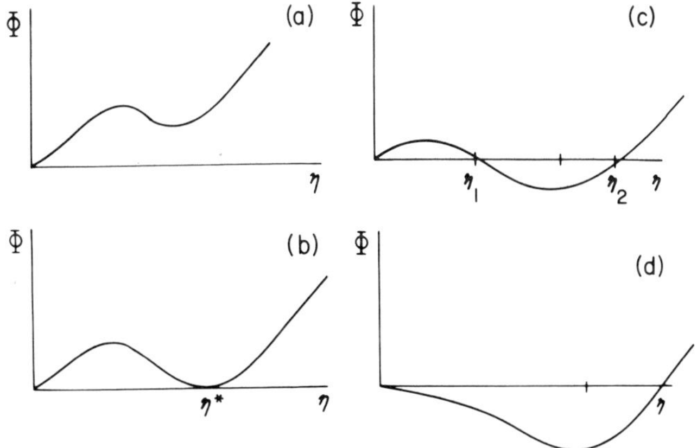

Figure 2. The "free energy" $\Phi(\eta)$ for a dye laser versus the scaled
number of photons η, $\eta \equiv (n/n^S)$. The pump parameter $a \equiv [(\sigma_T/\sigma)-1]$
decreases in going from (a) to (d), is positive in (a)-(c), and
negative in (d).

the absolute minimum of the free energy. In our problem of the
single mode laser there is no space dependence, so the free energy
is not a functional of spatial gradients and consequently there are
no spatial phase separations. The two phases incoherent photons
and coherent photon coexist throughout the same volume.

The description in the previous paragraph is analogous to the
usual discussion of an equilibrium first order phase transition,
where there is a free energy for each of two phases. For some range
of parameters one phase is lower in free energy than the second
phase. At some point the free energies become equal where the two
phases coexist, as in Fig. 2c, then, as the parameters change further, the second phase becomes the absolute minimum of the free
energy. The free energy is continuous but usually the first derivatives of the free energy are discontinuous, as in the present case.

In the above argument it has not been necessary to specify a
thermodynamic analogue variable for $\langle\eta\rangle$. However, we find it convenient to consider $\langle\eta\rangle$ analogous to the thermodynamic variable density. The selection of the minimum of the free energy for the state
of the system avoids the question of metastable states (which in the
thermodynamic analogy are the phenomena of superheating and supercooling) by ignoring them. We will consider the question of metastable states and hysteresis in Section 5.

The existence of the first-order transition behavior requires
$a = [(\sigma_T/\sigma)-1]$ to be greater than zero, that is, the inversion has to
be less than $\sigma_T = N_T(1-\alpha)^{-1}[1+(\kappa_{ST}/\nu_{t,\ell})]$, but a is greater than N_T,
the inversion in the absence of the triplets. Consequently, the
first order transition behavior takes place in a region where there
would be laser behavior if the triplets were absent, but below the
threshold for conventional second order laser behavior when the
triplets are present.

The alternate approach to mean field theory is to first calculate $\langle\dot{\eta}\rangle$ from Eq.(4.2a), obtaining

$$\frac{d\langle\eta\rangle}{dt} = -\langle(a+b\eta+c\eta^2)\eta\rangle \overset{m.f.}{\rightarrow} -(a\langle\eta\rangle+b\langle\eta\rangle^2+c\langle\eta\rangle^3) , \qquad (4.5)$$

where the mean field approximation follows when we neglect correlations. In Fig. 3 we have plotted the right-hand side of Eq.(4.5)
versus -a. The mean field theory in the form of the right-hand side
of Eq.(4.5) is a standard example of a hard oscillation in the theory
of bifurcations of nonlinear oscillations [15].

The value of $\underline{a}$ is the value of a where there is the first non-
zero real root for $\langle\eta\rangle$ from the equation $\langle\dot{\eta}\rangle = 0$, and corresponds to
the value $(b^2-4ac)^{\frac{1}{2}} = 0$, or $\underline{a} = b^2(4c)^{-1}$ and $\langle\eta\rangle = (|b|/2c)$. An alter-
nate way to obtain the same point is to set $\langle\eta\rangle = 0$ in Eq.(4.5) and

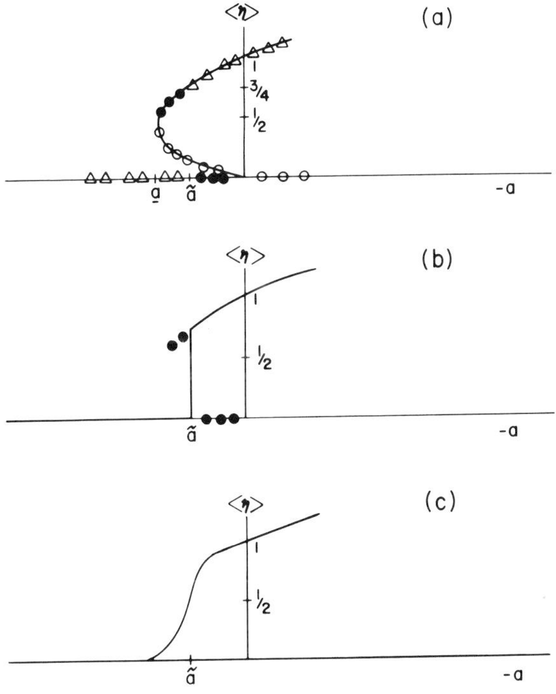

Figure 3. (a) A plot of the most probable number of photons versus
the negative of the pump parameter $-a \equiv [1-(\sigma_T/\sigma)]$. The triangles
represent stable states, solid circles metastable states and open
circles unstable states. (b) A Maxwell construction that picks out
the absolute minimum state of $\Phi(\langle\eta\rangle)$. The solid circles are metastable states. (c) The effect of fluctuations on the average value
of η.

differentiate $\langle\eta\rangle$ with respect to a, and thus obtain

$$\frac{d\langle\eta\rangle}{da} = \frac{-\langle\eta\rangle}{a + 2b\langle\eta\rangle + 3c\langle\eta\rangle^2} \quad . \tag{4.6}$$

The general "susceptibility" $d\langle\eta\rangle/da$ diverges for the values of η
which make the denominator of Eq.(4.6) vanish, i.e. $\eta=(-b/3c)\{1\pm$
$[1-(3ca/b^2)]^{\frac{1}{2}}\}$. When we substitute $\eta = (|b|/2c)$ in the positive
root we obtain $\underline{a}$. Figure 3 contains the information we previously
obtained from the mean field "free energy." Starting with $\sigma<\sigma_T$ to
the left-hand side of Fig.3 we find the branch $\langle\eta\rangle=0$ is the stable
stationary state denoted by crosses; at $a=\underline{a}$ two new roots appear.
The upper branch denoted by solid circles is metastable, the lower
branch denoted by open circles is unstable. At $a = \tilde{a} \equiv (3b^2/16c)$ the
upper branch becomes the stable state, corresponding to $\eta=\eta^*$ becoming

the absolute minimum, and the $\langle\eta\rangle=0$ branch changes from stable to
metastable. At a=0 the $\langle\eta\rangle=0$ branch becomes unstable and the upper
branch becomes the only stable solution. The $\langle\eta\rangle=0$ branch is the
incoherent phase and the upper stable branch is the coherent phase.
In Fig. 3b we have plotted the stable states alone in order to show
the close analogy to the Maxwell construction in Van der Waals type
theories.

In the mean field theory $d\langle\eta\rangle/da$ is infinite. However, if
fluctuations are included, the discontinuity disappears, as we show
in Fig. 3c. However, by means of the relation

$$\frac{1}{n^S} \cdot \frac{d\langle\eta\rangle}{da} = \langle\eta^2\rangle - \langle\eta\rangle^2 \quad , \tag{4.7}$$

we see that the large value of $d\langle\eta\rangle/da$ at $\tilde{a}$, corresponds to large
but finite fluctuations. Note however, these fluctuations are not
critical point fluctuations as long as $|b| \neq 0$, because they do not
occur at a critical point but at an ordinary first order phase
transition point and represent large fluctuations in a generalized
"compressibility."

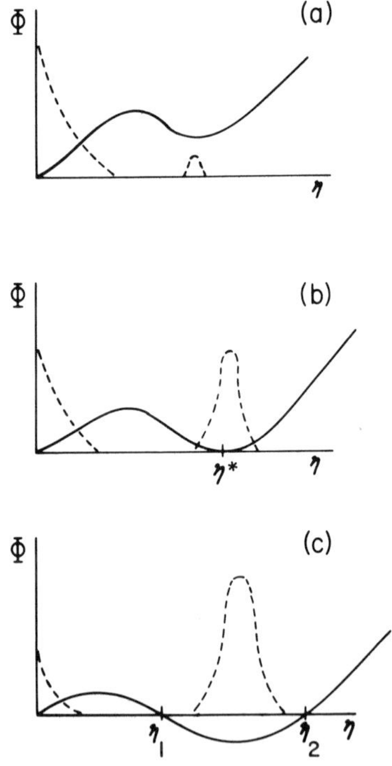

Figure 4. Same as Fig. 2(a)-2(c), with the addition of the steady
state distribution $P_S(\eta)$ represented by dashed lines.

We observe there is a complete analogy of the critical point
of the first order laser transition and the gas-liquid critical
point where the large scale fluctuations lend to critical opalescence.
As Kazantsev and Surdutovich [9] showed, for dye lasers it is possible to calculate explicitly all the moments of the distribution
function at b=0. In particular the root mean square photon number
in the threshold region is larger for dye lasers.

5. FLUCTUATIONS, HYSTERESIS AND BISTABILITY

When fluctuations become qualitatively important, as at points
of instability, mean field theory breaks down and for steady state
phenomena we must use the exact steady state solution Eq.(4.3) for
P_S. From our investigation of the behavior of $\Phi(\eta)$ in Section 4 we
are able to determine the qualitative behavior of $P_S(\eta)$ rather
easily for most ranges of the parameters a, b, and c. The quantitative analysis of the phenomena discussed in this section will be
presented elsewhere [16]. Except for a rather small range of parameters the behavior of $P_S(\eta)$ is depicted in Fig. 4. The small range
of parameters is that where there is no clear cut distinction between
the two main parts of the distribution function in narrow cross-over
regions. In Fig. 4a, where the incoherent minimum of Φ is much
lower than the relative minimum at η_+, the distribution function is
dominated by the incoherent "thermal" distribution $\exp\text{-}(n/n^S)$, with
a small coherent contribution represented by a gaussian centered on
η_+. As a decreases toward zero the relative contribution of the
incoherent photons decreases and the contribution of the coherent
photons increases until their contributions are comparable, as shown
in Fig. 4b. At this point in equilibrium statistical mechanics we
would have a first order phase transition, with the two phases
coexisting with their free energies equal. In thermodynamics the
realized state is the absolute minimum of the free energy. The
question of whether or not the relative minimum at η_+ is observed
before coexistence is a dynamical question we will discuss later in
this section. As a decreases further the coherent photon state
becomes the absolute minimum and the incoherent photons become the
metastable state, as shown in Fig. 4c.

The thermodynamic theory of phase transitions is based on having
an infinite time for the fluctuations to cause all of the system to
end up in the state of absolute minimum of the free energy. In
practice we observe phenomena like superheating, supercooling and
hysteresis, which are a consequence of the fact that the times involved in the experiment may be long compared with the times required
for the fluctuations to find the absolute minimum.

Consider the system in a coherent state, in a situation depicted
in Fig. 4c, with no incoherent photons and suppose we increase a

until we arrive at Fig. 4b. Will the incoherent phase appear and
grow? The answer will be yes if the relaxation time for surmounting
the potential barrier $\{\Phi(\eta_-)-\Phi(\eta_+)\}$ is short compared with the time
the parameter a changes. This problem has been studied in the
similar problem of saturable absorbers in Refs. 9 and 10. These
authors base their work on Kramer's [17] treatment of the escape of
particles over potential barriers. Here we have space to mention
only the qualitative [16] results. The dominant point is the relaxation time from one well to another and is determined by two factors.
First the relaxation time is proportional to the exponential of the
height of the barrier, i.e. $\exp\{n^S[\Phi(\eta_-)-\Phi(\eta_+)]\}$ in the case of
relaxation from the coherent to the incoherent phase. The exponential is a very large number unless $(b^2-4ac)^{\frac{3}{2}}$ is a very small number.
The second factor determining the relaxation time is a prefactor
determined by the inverse of the curvature at η_+ and η_-.

Consequently the relaxation times are usually very long and we
expect to observe hysteresis. For example, we could start in the
coherent state in Fig. 4c, go all the way to Fig. 4a without the
incoherent phase appearing. Eventually the barrier will become
sufficiently small that the fluctuations will cause some of the
system to make a transition. Conversely, if the system starts out
with incoherent photons and proceeds toward Fig. 4c, coherent oscillation will not appear until the relaxation time proportional to
$\exp\{-n^S\Phi(\eta_-)\}$ is shorter than the time scale of the change of a.
Eventually, as a gets close to zero, the barrier will become small
enough for fluctuations to surmount it easily and coherent oscillations will appear and we may expect large fluctuations in dye lasers
near threshold.

The cause of the fluctuations in lasers is not the temperature
of a reservoir as in thermodynamics but the incoherent spontaneous
emission of the laser molecules. Consequently, questions of metastability, bistability, and hysteresis are determined by the competition between the nonlinear restoring force and the diffusion
determined by the spontaneous emission of the laser molecules.

We have found the mean field treatment is useful in indicating
some of the behavior of the dye laser. However, the behavior of a
dye laser near points of instability and in regions of metastability
is determined by fluctuations and we have to go beyond mean field
theory to the exact steady state with fluctuations. The result of
fluctuations is to cause the behavior of $\langle\eta\rangle$, for example, in the
steady state to be perfectly smooth in going from Fig. 4a to 4d,
including going through the phase transition. The behavior of $\langle\eta\rangle$
and the higher moments is perfectly smooth and analytical but rapidly
changing at points of instability. As we have seen, because of the
presence of metastable states it is necessary to go beyond the
stationary state and to include dynamics to determine whether or not
hysteresis, bistability and other phenomena actually occur.

6. CONCLUSIONS

We have already seen in [1] that the anomalously large fluctuations of a dye laser in the threshold region require changes in
the definition of laser threshold. Besides the practical importance
of the change in laser threshold, we have the possibilities by
varying the dye molecule parameters of converting the first order
phase transition into a second order phase transition. It is also
possible as we have shown by varying the dye molecule parameters to
observe a critical point of the laser transition. If experiments
could be performed accurately in the transition region they could
be used to obtain worthwhile information about dye molecule parameters.

The only direct experimental evidence related to the dye laser
phase transition phenomena is the work of Abate and coworkers [18],
who did find excess fluctuations but did not determine if the excess
fluctuations were intrinsic or instrumental. One of the problems
is the parameter β is not known experimentally. The main purpose
of the present work is to encourage experimental investigation into
the phenomena described in this paper.

There are significant differences between dye lasers and gas-
liquid systems for example (lasers do not undergo phase separation),
which make it unlikely that the understanding of the dye laser
transition will directly increase our understanding of the gas-liquid
phase transition. However, it is possible that a more thorough
understanding of the simpler metastable states and their dynamics
in dye lasers could be useful in understanding the more complicated
metastable state behavior in more complex systems.

There have been two approaches to the dye laser that are related
to the material presented in this paper. Independent of our work,
Baczynski et al. [19] derived a semiclassical model without fluctuations, which they subsequently generalized to include fluctuations
[20] by adding a stochastic noise source in the spirit of Ref. [8].
Recently [21] they have found the nature of the triplet losses
determine the order of the dye laser phase transition in a manner
that is related to our analysis of the order of the phase transition
represented by the parameter β.

References

1.   R.B. Schaefer and C.R. Willis, Phys. Rev. A*13*, 1874 (1976),
     hereafter referred to as [1].
2.   R.B. Schaefer and C.R. Willis, Phys. Letters *58A*, 53 (1976).
3.   V. DeGiorgio and Marlan O. Scully, Phys. Rev. A2, 1170 (1970).
4.   R. Graham and H. Haken, Z. Physik *237*, 31 (1970).

5. V. DeGiorgio, Physics Today *29*, 42 (1976).

6. R. Landauer, J. Appl. Phys. *33*, 2209 (1962).

7. M. Lax and W.H. Louisell, IEEE J. Quantum Electron. *3*, 67 (1967).

8. H. Haken, Rev. Mod. Phys. *47*, 67 (1975).

9. A.P. Kazantsev and G.L. Surdutovich, Zh. eksp. teor. Fiz. *58*, 245, (1970) [Sov. Phys. - JETP *31*, 133 (1970)].

10. R. Salomaa, J. Phys. A*7*, 1094 (1974).

11. J.F. Scott, M. Sargent III, and C.D. Cantrell, Optics Comm. *15*, 13 (1975).

12. R.J. Glauber in *Quantum Optics and Electronics*, edited by C. DeWitt, A. Blandin and C. Cohen-Tannoudji (Gordon and Breach, New York, 1965).

13. R. Graham, in *Fluctuations, Instabilities and Phase Transitions*, edited by T. Riste (Plenum Press, New York, 1975), as well as the earlier papers of R. Graham cited therein.

14. A.P. Kazantsev and G.L. Surdutovich, Zh. eksp. teor. Fiz. *56*, 2001 (1969) [Sov. Phys. - JETP *29*, 1075 (1969)].

15. See, for example, N. Minorsky, *Nonlinear Oscillations* (Van Nostrand, New York, 1962).

16. C.R. Willis, to be published.

17. H.A. Kramers, Physica *7*, 284 (1940).

18. J.A. Abate, H.J. Kimble, and L. Mandel, Phys. Rev. A*14*, 788 (1976).

19. A. Baczynski, A. Kossakowski and T. Marszalek, Zeits. Physik B *23*, 205 (1976).

20. S.T. Dembinski and A. Kossakowski, Zeits Physik B*24*, 141 (1976).

21. A. Baczynski, A. Kossakowski and T. Marszalek, Zeits. Physik B *26*, 93 (1977).

LASER AS CATASTROPHE[*]

R. Gilmore

University of South Florida, Tampa, Fla.

L.M. Narducci

Drexel University, Philadelphia, Pa.

1. ELEMENTARY CATASTROPHES

Many physical systems can be described by a potential $V=V(x_1,x_2,\ldots,x_n)$. The local properties of the system are then characterized by the local properties of the potential. If the gradient ∇V is nonvanishing at a value $\underline{x} = \underline{x}^0$ of the state variables $\underline{x}$, the implicit function theorem[1] guarantees that there is a smooth change of coordinates $x_i \rightarrow y_i(\underline{x})$ so that in the coordinate system $\underline{y}$, the potential assumes a canonical form

$$V(y_1,\ldots,y_n) = y_1 \quad . \tag{1}$$

Under equilibrium or steady state conditions $\nabla V = 0$ and the implicit function theorem is not applicable. However, if at the equilibrium or 'critical' point the matrix of mixed second partial derivatives (or 'Hessian') $\partial^2 V/\partial x_i \partial x_j = V_{ij}$ is nonsingular, a theorem of Morse[2] guarantees that the potential can be brought locally to a canonical quadratic form under a smooth change of coordinates

$$V(y_1,\ldots,y_n) = \sum_{j=1}^{n} \lambda_j y_j^2 \quad . \tag{2a}$$

If the eigenvalues λ_j of V_{ij} are absorbed into the new coordinates, V assumes the canonical Morse form

$$V(y_1,\ldots,y_n) = V_i^n = +\tilde{y}_1^2 + \ldots + \tilde{y}_i^2 - \tilde{y}_{i+1}^2 - \ldots - \tilde{y}_n^2 \tag{2b}$$

The form (2b) is called the Morse i-saddle.

If the potential depends on one or more external control parameters c_1, c_2,..,c_k as well as the state variables x_1, x_2,...,x_n, then the eigenvalues λ_1, λ_2,...,λ_n of the Hessian depend on the control parameters $\underline{c}$. When one or more eigenvalues vanish, the Morse theorem is not applicable. However, a theorem of Thom [3] guarantees that the potential can be brought locally to the following canonical form under a smooth change of coordinates

$$V(y_1,\ldots,y_n) = \text{Cat }(j) + V_i^{n-j} \ . \tag{3}$$

Here V_i^{n-j} is a Morse i-saddle in the coordinates $y_{j+1},\ldots,y_n$, and Cat (j) is a 'catastrophe' function in the state variables $y_1,\ldots,$ y_j.

A catastrophe function is a sum of two terms. One is a function with vanishing Hessian. The other plays the role of a "perturbation expansion" (universal unfolding) around the first. That is, every function, or family of functions, 'near' the first term can be obtained by a smooth change of variables in the second term. 'Near' is to be interpreted in a topological sense, the topology is that of a finite dimensional linear vector space, where the vector components are the coefficients of the potential in a Taylor series expansion about the non-Morse critical point, truncated after a suitably high order.

Three important classes of elementary catastrophes are ($y_1 = x$, $y_2 = y$)[4]

$$A_k: \quad x^{k+1} \quad + \quad \sum_{j=1}^{k-1} C_j x^j \quad ,$$

$$D_k: \quad x^2 y + y^{k-1} + \sum_{j=1}^{k-3} C_j y^j + C_{k-2} x + C_{k-1} x^2 \ ,$$

$$E_6: \quad x^3 + y^4 + \sum_{j=1}^{2} C_j y^j + \sum_{j=3}^{5} C_j xy^{j-3} \ ,$$

$$E_7: \quad x^3 + xy^3 + \sum_{j=1}^{4} C_j y^j + \sum_{j=5}^{6} C_j xy^{j-5} \ ,$$

$$E_8: \quad x^3 + y^5 + \sum_{j=1}^{3} C_j y^j + \sum_{j=4}^{7} C_j xy^{j-4} \ . \tag{4}$$

Thom's theorem asserts that when k, the number of control parameters, is less than or equal to 5 [4], the most general form a potential $V(x_1,\ldots,x_n; c_1,\ldots,c_k)$ can assume in the neighborhood of a non-Morse critical point is given by (3), where Cat (j) is A_{k+1}, D_{k+1}, or E_{k+1}.

The importance of this theorem is that it provides a short list of canonical forms [3,4] into which an arbitrary potential depending on fewer than six external controls can be transformed. It is then sufficient to catalog [5] the geometric properties of these families of canonical forms. For example, the A_3 or 'cusp' catastrophe has the form

$$V(x;\ A,B) = \frac{1}{4}\ x^4 - \frac{1}{2}\ Ax^2 - Bx\ . \tag{5}$$

The stationary points of this potential are determined from the equation $\nabla V = 0$

$$x^3 - Ax - B = 0\ \ . \tag{5'}$$

The two dimensional surface (5') is shown embedded in the three dimensional $x - A - B$ space in Fig. 1. Outside the cusp-shaped region in the control plane $A - B$ the equilibrium manifold is single valued. Inside, it is triple valued, with the upper and lower sheets corresponding to local minima, the middle, to an unstable local maximum. As the control parameters A, B are changed, the equilibrium value of the state variable x moves on the equilibrium surface (5').

Three conventions have been used to describe at what point the system state jumps from one locally stable equilibrium to another:
1. *Delay Convention.* The system remains in a locally stable equilibrium as long as possible. A jump occurs from a metastable to a stable equilibrium only when the local maximum separating the stable and metastable states coalesces with, and annihilates the metastable equilibrium.
2. *Maxwell Convention.* The system moves immediately into the state which provides a global minimum value for the potential.
3. *Maxwell Construction.* This convention is described in Refs. [6,7].

For the canonical potential (5), there is no difference between conventions 2 and 3. The appropriate criterion for deciding which (if any) convention to adopt is the ratio of the mean square fluctuations to the height of the potential barrier separating the local minima. Below we adopt the delay convention. A more complete treatment of catastrophe theory is given in Refs. [4,8-11].

2. LANGEVIN EQUATIONS

The Hamiltonian we study is

$$H = H(\alpha) + H_B\ \ . \tag{6}$$

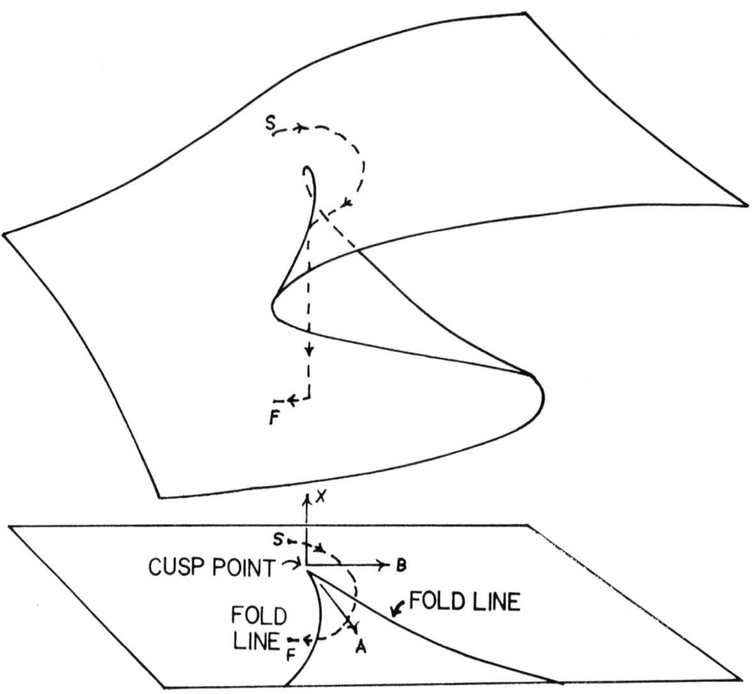

Fig. 1   The cusp manifold (5') is a pleated two-dimensional surface
in the three-dimensional state variable (x) - control parameter
(A-B) space.

Here H(α) is the Dicke Hamiltonian[12] including also the interaction of the N identical two-level atoms with a classical near-resonant external field α

$$H(\alpha) = \hbar\omega a^+ a + \sum_{j=1}^{N} \frac{\varepsilon}{2} \sigma_j^z + \frac{\lambda}{\sqrt{N}} \sum_{j=1}^{N} (a^\dagger + \alpha^*) \sigma_j^- + (a + \alpha) \sigma_j^+ \; . \quad (7)$$

The Hamiltonian H_B describes the interaction of the field mode and
each individual atom with external heat baths.

We study the Hamiltonian H(α) rather than the usual Dicke
Hamiltonian H(α=0) for two reasons:
(1) Physical reason: the external field is a useful tool for altering experimental conditions;
(2) Mathematical reason: the Hamiltonian H(α) is a 'universal unfolding' of the Dicke Hamiltonian.

The Langevin equations[13] for the operators a, σ_j^- and σ_j^z are easily derived. The equations derived using $H(\alpha)$ only are

$$i\hbar \frac{d}{dt} a = \hbar\omega a + \lambda\sqrt{N}\,\sigma^- \ ,$$

$$i\hbar \frac{d}{dt} \sigma^- = \varepsilon\,\sigma^- - \frac{\lambda}{\sqrt{N}}\,(a{+}\alpha)\sigma^z \ ,$$

$$i\hbar \frac{d}{dt} \sigma^z = 2\frac{\lambda}{\sqrt{N}}\,(a{+}\alpha)^\dagger\,\sigma^- - 2\frac{\lambda}{\sqrt{N}}\,(a{+}\alpha)\sigma^+ \ . \tag{8}$$

The effect of H_B on these equations is described below.

3.   SIMPLIFYING ASSUMPTIONS

One effect of coupling the interacting N atom-field mode system to external heat baths is to introduce stochastic driving terms $\Gamma_a(t),\Gamma_\sigma(t),\Gamma_3(t)$ on the right-hand sides of Eqs.(8). This problem can be eliminated by taking the expectation values (8), since $\langle\Gamma(t)\rangle = 0$ for the stochastic terms.

A second effect of the bath is to force regression of fluctuations back to equilibrium values with typical exponential relaxation constants. This is simply taken into account by the ansätz

$$\frac{d}{dt}\langle a\rangle \to (\frac{d}{dt} + \gamma_a)\,(\langle a\rangle - \langle a\rangle_e) \ ,$$

$$\frac{d}{dt}\langle\sigma^-\rangle \to (\frac{d}{dt} + \gamma_\sigma)\,(\langle\sigma^-\rangle - \langle\sigma^-\rangle_e) \ ,$$

$$\frac{d}{dt}\langle\sigma^z\rangle \to (\frac{d}{dt} + \gamma_3)\,(\langle\sigma^z\rangle - \langle\sigma^z\rangle_e) \ . \tag{9}$$

Here $\langle a\rangle_e$, $\langle\sigma^-\rangle_e$, $\langle\sigma^z\rangle_e$ are the equilibrium expectation values of the corresponding operators.

The equations obtained from (8) using the first two approximations above are not closed. For example, $d\langle\sigma^-\rangle/dt$ depends on $\langle a\sigma^z\rangle$. Closure can be forced by assuming factorization of operator products. This can be carried out in a selfconsistent way,[14] and amounts to a coupled mean field theory for the coupled system. This is equivalent to the assumption that the system density operator factors into the direct product of two reduced density operators, one for the field subsystem, the other for the atomic subsystem.

It is useful to remove the fast time dependence from $\langle a\rangle$, $\langle\sigma^-\rangle$ by making the ansätz $\langle a\rangle = \langle\tilde{a}\rangle e^{-i\omega t}$, $\langle\sigma^-\rangle = \langle\tilde{\sigma}^-\rangle e^{-i\varepsilon t/\hbar}$, and similarly for their equilibrium expectation values.

With these four simplifying assumptions the Langevin equations reduce to the system of coupled nonlinear equations

$$i\hbar \left(\frac{d}{dt} + \gamma_a\right) \langle a \rangle = \hbar\omega \langle a \rangle_e + \lambda\sqrt{N} \langle \sigma^- \rangle \quad ,$$

$$i\hbar \left(\frac{d}{dt} + \gamma_\sigma\right) \langle \sigma^- \rangle = \varepsilon \langle \sigma^- \rangle_e - \frac{\lambda}{\sqrt{N}} (\langle a \rangle + \alpha) \langle \sigma^z \rangle \quad ,$$

$$i\hbar \left(\frac{d}{dt} + \gamma_3\right) (\langle \sigma^z \rangle - \langle \sigma^z \rangle_e) =$$

$$2\frac{\lambda}{\sqrt{N}} (\langle a \rangle + \alpha) \langle \sigma^- \rangle^* - 2\frac{\lambda}{\sqrt{N}} (\langle a \rangle + \alpha)^* \langle \sigma^- \rangle \quad .$$

$$(10)$$

For convenience, all tildes have been removed from (10).

4. BOUNDARY CONDITIONS

The system of equations (10) has been studied subject to two different sets of boundary conditions: equilibrium and nonequilibrium steady states. Under both sets of conditions all time derivatives vanish.

Under nonequilibrium steady state boundary conditions (10) describes a model for a single mode CW laser operation. The equilibrium value $\langle \sigma^z \rangle_e$ is the population inversion produced by an external pumping mechanism. For simplicity it is generally assumed that $\langle a \rangle_e = \langle \sigma^- \rangle_e = 0$.

Under equilibrium boundary conditions the steady state and equilibrium expectation values are identical, and $\langle \sigma^z \rangle_e$ is determined in a selfconsistent way from the order parameters $\langle a \rangle$, $\langle a^\dagger \rangle$ by

$$\langle \sigma^z \rangle_e = - \frac{\varepsilon}{2\theta} \tanh \beta\theta \quad ,$$

$$\theta^2 = (\varepsilon/2)^2 + \lambda^2 |\langle a \rangle + \alpha|^2/N \quad . \tag{11}$$

5. NONEQUILIBRIUM STATIONARY MANIFOLD

The equation of state for stationary nonequilibrium states is obtained from (10) by eliminating the steady state expectation values $\langle \sigma^- \rangle_s$, $\langle \sigma^z \rangle_s$. The resulting equation has the form(5'), with

$$x = \left(\langle a \rangle_s + \frac{2\alpha}{3}\right)/\sqrt{N} \quad ,$$

$$A = \frac{\alpha^2}{3} + \frac{\gamma_3}{4\gamma_a} \left[\langle \sigma^z \rangle_e - \frac{\gamma_a \gamma_\sigma}{(\lambda/\hbar)^2}\right] \quad ,$$

$$B = \frac{2}{27} \alpha^3 + \frac{\alpha}{3} \frac{\gamma_3}{4\gamma_a} [<\sigma^z>_e - \frac{\gamma_a\gamma_\sigma}{(\lambda/\hbar)^2}] + \frac{\gamma_\sigma\gamma_3}{4(\lambda/\hbar)^2} \alpha$$

In the absence of external fields, B=0. Increasing the population inversion makes A less negative. If $(\lambda/\hbar)^2 > \gamma_a\gamma_\sigma$, the system state passes through the cusp point (x,A,B) = (0,0,0), (Fig. 1), for sufficiently strong pumping. For A>0, the system state moves to the upper or lower sheet in a second order phase transition of Ginzburg-Landau type.

Trajectories in control parameter space obtained by increasing α from 0 when the pump rate is held below threshold are shown in Fig. 2. For $Q \equiv (\lambda/\hbar)^2(-<\sigma^z>_e)/8\gamma_a\gamma_\sigma < 1$, the trajectory passes to the right of the cusp point and the system state passes continuously onto the upper sheet. For Q > 1, the control parameter trajectory passes to the left of the cusp point. The system state moves continuously onto the lower sheet until the right-hand fold line is crossed (delay convention). The system state then jumps discontinuously onto the upper sheet. As a function of decreasing external field amplitude, hysteresis is observed, with the jump occurring at the left-hand fold line. The second order phase transition at Q→1+ is a limit of lines of first order phase transitions exhibiting hysteresis.

The phenomenon described above is called optical bistability. It has been observed[15] and treated theoretically.[16] For the experimental data exhibited in Fig. 1a of Ref. 15, the corresponding value of Q is Q = 1.65.

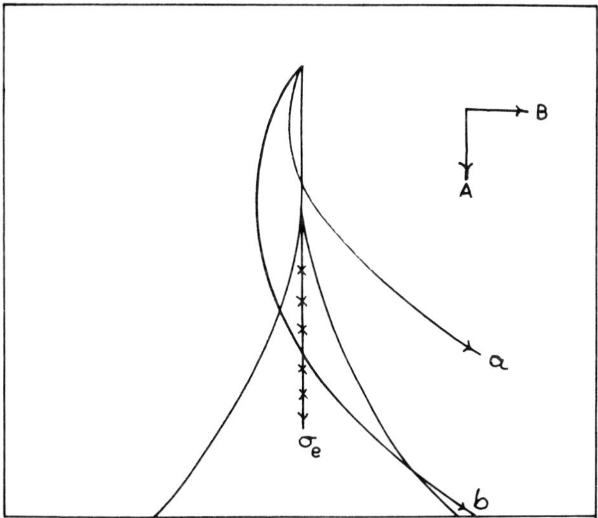

Fig. 2 Control parameter trajectories for increasing external field α when the pump parameter is below threshold. a,Q<1; b,Q>1.

6.   EQUILIBRIUM STATIONARY MANIFOLD

The equation of state for the equilibrium configurations of (6) and (7) are obtained from (10) in a similar manner, using the constitutive relation (11):

$$\hbar\omega\langle a\rangle_e - \frac{\lambda^2}{2\theta}\ (\langle a\rangle_e + \alpha)\ \tanh\ \beta\theta = 0\ \ . \tag{12}$$

The surface described by (12) is similar (math: "diffeomorphic") to the cusp manifold, and may be brought into the canonical form (5') by expanding in powers of $|\langle a\rangle + \alpha|/\sqrt{N}$ and truncating after third order. The truncation is justifiable because the function x-tanhx is 3-determinate.[9] Then (12) assumes the form (5') with

$$x = (\langle a\rangle_e + \alpha)/\sqrt{N}\ \ ,$$

$$A = -(\epsilon\hbar\omega - \lambda^2\ \tanh\ \beta\epsilon/2)/C\ \ ,$$

$$B = \epsilon\hbar\omega\alpha/C\sqrt{N}\ \ , \tag{13}$$

where C is a positive function of T, $0 \leq T < \infty$.

In the absence of external fields, B=0 and the system state passes through the cusp point (x,A,B) = (0,0,0) at sufficiently low temperature provided $\lambda^2 > \epsilon\hbar\omega$. For lower temperatures, the system state moves to the upper or lower sheet in a second order phase transition of Ginzburg-Landau type.[17]

For $\alpha > 0$ and constant, the control parameter trajectory passes to the right of the cusp and the system state moves continuously onto the upper sheet as a function of decreasing temperature. The second order phase transition (bifurcation) is destroyed by the presence of an arbitrarily small classical external field.[18,19]

If the temperature is held below the critical temperature and if α can be changed from positive to sufficiently large negative values, a first order phase transition will occur as the left-hand fold line is crossed (Fig. 3). This first order phase transition is also accompanied by hysteresis.

7.   REDUCED DENSITY OPERATORS

The unnormalized reduced density operator for the single mode field subsystem can be expressed as the exponential of a linear superposition of the photon operators

$$\rho_F = EXP(Ma^\dagger a + Ra^\dagger + La)\ . \tag{14}$$

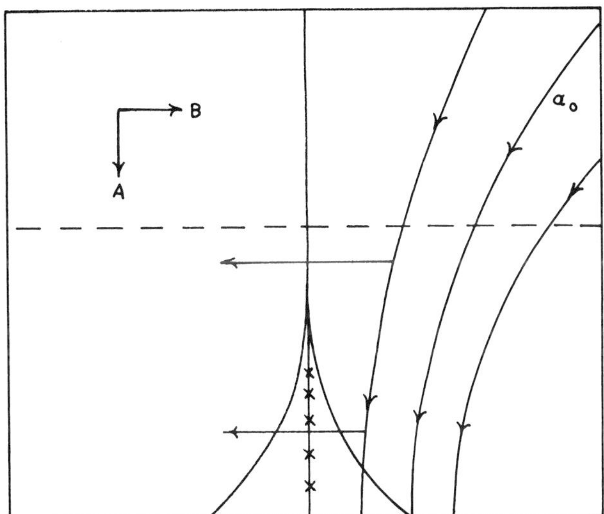

Fig. 3 Control parameter trajectories under equilibrium conditions.
Vertical motion corresponds to decreasing temperature, horizontal
motion to field reversal.

The parameters M,R,L can in turn be expressed in terms of operator
expectation values:

$$\langle a \rangle = -R/M$$

$$\langle a^\dagger \rangle = -L/M$$

$$\langle a^\dagger a \rangle - \langle a^\dagger \rangle \langle a \rangle = (e^{-M}-1)^{-1} \quad . \tag{15}$$

As a result

$$\rho_F = [\rho_F(\text{geometry})]^{M(\text{physics})} \quad ,$$

$$\rho_F(\text{geometry}) = \text{EXP}\{a^\dagger a - \langle a \rangle (a^\dagger + a)\} \quad , \tag{16}$$

and M(physics) is determined from the correlation function in (15).
In deriving (16) we have made the simplifying assumption that $\langle a \rangle$
is real.

The result (16) describes the factorization of the reduced
field density operator into a geometric part and a physical part.
The geometric part is an operator which depends on the system state,
represented by a point x on the cusp catastrophe manifold through
(12) or (13), so that

$$\langle a \rangle = (x - k\alpha)\ \sqrt{N}\ , \tag{17}$$

where $k = 2/3$ for nonequilibrium steady state boundary conditions and $k = 1$ for equilibrium conditions.

The unnormalized reduced atomic density operators for each individual atomic subsystem may be treated similarly. The single atom density operator may be written

$$\rho_A = \frac{1}{2}\, I_2 + \langle\sigma^z\rangle\, \frac{1}{2}\, \sigma^z + \langle\sigma^+\rangle\, \sigma^- + \langle\sigma^-\rangle\, \sigma^+ \ .$$

This can be written in exponential form analogous to (16)

$$\rho_A = [\rho_A(\text{geometry})]^{M'(\text{physics})}\ ,$$

$$\rho_A(\text{geometry}) = \text{EXP}\{\langle\sigma^z\rangle\, \frac{1}{2}\, \sigma^z + \langle\sigma^+\rangle\, \sigma^- + \langle\sigma\rangle\, \sigma^+\}\ ,$$

$$M'(\text{physics}) = \left|\sigma\right|^{-1} \text{Tanh}^{-1} 2\left|\sigma\right|\ ,$$

$$\left|\sigma\right|^2 = (\langle\tfrac{1}{2}\,\sigma^z\rangle)^2 + \langle\sigma^+\rangle\, \langle\sigma^-\rangle\ .$$

The relation between the expectation values $\langle\sigma^\pm\rangle$, $\langle\sigma^z\rangle$ and the co-ordinate x on the cusp manifold may be determined through the time independent form of Eqs.(10).

8.   DISCUSSION

We have identified the equation of state of the model (6,7) subject to either equilibrium or to nonequilibrium steady state boundary conditions with the cusp catastrophe manifold. This allows a 1-1 correspondence to be made between the equilibrium and non-equilibrium properties of this model Hamiltonian. In particular, the reduced field and atomic subsystem density operators factor into a geometric operator defined by a point on the cusp manifold, and a number characterizing noise.[20]

The existence of a second order phase transition of Ginzburg-Landau type [(5) with B=0)], requires that the most general family of perturbed potentials have the form of the cusp catastrophe [(5), Fig. 1)]. This in turn suggests that all possible physical phenomena produced by arbitrary perturbations can be studied by including only one other control parameter in the original system. It is for the reason we have investigated the Hamiltonian H(α) in (6). The presence of a second order phase transition forces a pleat into the stationary state manifold. This pleat, in turn, guarantees the presence of first order phase transitions, provided one "steers" along the external controls in an appropriate way.

*This paper is based in part on a longer contribution submitted to The Physical Review.

References

1. L.H. Loomis, S. Sternberg, *Advanced Calculus* (Addison-Wesley, Reading, Mass., 1968).
2. M. Morse, *The Calculus of Variations in the Large* (American Mathematical Society, New York, 1934).
3. R. Thom, Topology *8*, 313 (1969).
4. V.I. Arnol'd, Russian Math. Surveys *30* (5) 1 (1975).
5. A.E.R. Woodcock, T.M. Poston, *A Geometrical Study of the Elementary Catastrophes*, Springer Lecture Notes 373 (Springer-Verlag, Berlin, 1974).
6. D.H. Fowler, in *Towards a Theoretical Biology*, ed. C.H. Waddington (Aldine, Chicago, 1972) p. 1.
7. K. Huang, *Statistical Mechanics* (Wiley, New York, 1963).
8. E.C. Zeeman, Scientific American *234* (4) 65 (1976).
9. T. Poston and I.N. Stewart, *Taylor Expansions and Catastrophes* (Pitman, London, 1976).
10. E.C. Zeeman, *Readings in Catastrophe Theory* (Addison-Wesley, Reading, Mass., 1977).
11. T. Poston and I.N. Stewart, *Catastrophe Theory and its Applications* (Pitman, London, 1977).
12. R.H. Dicke, Phys. Rev. *93*, 99 (1954).
13. H. Haken, Rev. Mod. Phys. *47*, 67 (1975).
14. R. Gilmore and C.M. Bowden, Phys. Rev. A *13*, 1898 (1976).
15. H.M. Gibbs, S.L. McCall, T.N.C. Venkatesan, Phys. Rev. Lett. *36*, 1135 (1976).
16. R. Bonifacio, L.A. Lugiato, Opt. Commun. *19*, 172 (1976).
17. K. Hepp, E.H. Lieb, Ann. Phys. (N.Y.) *76*, 360 (1973).
18. J.P. Provost, F. Rocca, G. Vallee, M. Sirugue, Physica *85A*, 202 (1976).
19. R. Gilmore, Phys. Lett. *A55*, 459 (1976).
20. R. Gilmore, L.M. Narducci, Phys. Rev. A (to be published).

DRESSED-ATOM APPROACH TO RESONANCE FLUORESCENCE[*]

C. Cohen-Tannoudji

École Normale Supérieure, Paris, France

S. Reynaud

Collège de France, Paris, France

Resonance fluorescence, i.e., absorption and re-emission of
resonance radiation by free atoms is theoretically described as
spontaneous emission from the total system atom + impinging photons
interacting together (dressed-atom). One shows how such an approach
gives a simple quantitative understanding of the behaviour of an
atom irradiated by an intense resonant laser beam.

The positions of the various components of the fluorescence
spectrum are simply given by the allowed Bohr frequencies of the
dressed atom. The master equation describing spontaneous emission
from such a system is easily solved at the limit of high intensities
(well resolved lines in the fluorescence spectrum). Simple analytical expressions, taking into account the effect of cascades, are
derived for the widths of the components. The intensities of the
lines are explicitly related to the populations of the dressed-
atom energy levels and to transition rates towards lower levels.

The now well known predictions concerning two-level systems
are easily derived from such an approach, which proves its equivalence with other methods using a c-number description of the
applied laser field. One discusses the advantages of the dressed-
atom approach for dealing with multilevel systems.

Another interesting problem is the absorption spectrum of a
second weak laser beam probing the saturated atomic transition.
The intensities of the various components of such an absorption
spectrum appear now as products of transition rates by differences
of population (competition between absorption and stimulated emission processes). Important differences between fluorescence and

absorption spectra receive a simple explanation. As observed in
very recent experiments, the central component vanishes at high
intensities of the saturating laser beam; when this laser is
slightly off resonance, half of the lateral components are amplifying, half are absorbing.

*This paper was presented at a joint session of the 4'th Rochester
Conference on Coherence and Quantum Optics and of the International
Conference on Multiphoton Processes. A fuller account is published
in *Multiphoton Processes*, ed. J.H. Eberly and P. Lambropoulos (John
Wiley, New York, 1978) p. 103.

RESONANCE FLUORESCENCE UNDER FINITE BANDWIDTH EXCITATION[*][†]

L. Mandel and H. J. Kimble

University of Rochester, Rochester, New York

With the development of increasingly precise experimental
methods for studying the phenomenon of resonance fluorescence from
atoms, as in atomic beam experiments, the need arises for a more
realistic description of the excitation field in the theory. We
have generalized our earlier quantum field treatment of the problem of resonance fluorescence from a two-level atom,[1,2] to take
account of the finite bandwidth of the exciting laser field. In
the analysis we take the state of this field to be a statistical
mixture of coherent states, in which the phase performs a random
walk in time, and we ultimately average over the ensemble of
phases.[3] We investigate the properties of the fluorescent
light, and find some new features, that are absent under monochromatic excitation, and appear not to have been encountered in
previous treatments of the problem.[4,5] In particular, we show
that the spectral density should become asymmetric under off-
resonance, non-monochromatic excitation.

In an earlier treatment of the problem of resonance fluorescence under monochromatic excitation,[2] we investigated the time
development of the mean light intensity

$$\langle \hat{E}_i^{(-)}(\underline{r},\tau)\hat{E}_i^{(+)}(\underline{r},\tau)\rangle \equiv I(\underline{r},\tau),$$

the two-time intensity correlation function

$$\langle \hat{E}_i^{(-)}(\underline{r},t)\hat{E}_j^{(-)}(\underline{r},t+\tau)\hat{E}_j^{(+)}(\underline{r},t+\tau)\hat{E}_i^{(+)}(\underline{r},t)\rangle \equiv \Gamma^{(2,2)}(\underline{r},t,\tau),$$

the two-time amplitude correlation

$$\langle \hat{E}_i^{(-)}(\underline{r},t)\hat{E}_i^{(+)}(r,t+\tau)\rangle \equiv \Gamma^{(1,1)}(\underline{r},t,\tau)$$

and its Fourier transform - the spectral density. We showed that all three functions of τ, which we here represent collectively by $F(\tau)$, are expressible as solutions of a Volterra integral equation

$$F(\tau) = y(\tau) + \int_0^\tau d\tau' \, K(\tau-\tau') \, F(\tau'), \tag{1}$$

in which the kernel $K(\tau)$ is real and is the same for all three functions, whereas the inhomogeneous term $y(\tau)$ is different in every case. The solutions $F(\tau)$ in all three cases were expressible as combinations of exponential functions of τ of the form $\exp(p\tau)$, where the p's are roots of the cubic equation

$$p^3 + 4\beta p^2 + (5\beta^2+\beta^2\theta^2+\Omega^2)p + (2\beta^2+2\beta^2\theta^2+\Omega^2)\beta = 0 \; . \tag{2}$$

Here β is a decay constant equal to half the Einstein A-coefficient, θ is a dimensionless measure of the detuning of the exciting field in units of β, and Ω is the atomic Rabi frequency.

When the effects of finite bandwidth excitation are included in the analysis, we find that the three functions of $F(\tau)$ continue to obey integral equations of the form of Eq. (1), but that the kernels $K(\tau)$ are not always real and are not always the same in every case. When $F(\tau)$ represents the expectation value of the light intensity $I(r,\tau)$ or the two-time intensity correlation $\Gamma^{(2,2)}(\underline{r},t,\tau)$, the kernels $K(\tau)$ are still real and still identical, although they are more complicated than before. The solutions for $F(\tau)$ are again exponential functions of the form $\exp(p\tau)$, but the p's are the roots of the more complicated cubic equation

$$p^3 + (4\beta+2\lambda)p^2 + [(\beta+\lambda)(5\beta+\lambda) + \beta^2\theta^2 + \Omega^2] \, p + 2\beta(\beta+\lambda)^2$$

$$+ 2\beta^3\theta^2 + (\beta+\lambda)\Omega^2 = 0 \; , \tag{3}$$

in which λ is a measure of the effective bandwidth of the exciting field. These complications are reflected in changes in the shape of the functions $F(\tau)$, although the changes are not profound so long as $\lambda \lesssim \beta$.

However, when $F(\tau)$ is identified as the second order correlation function $\Gamma^{(1,1)}(r,t,\tau)$, the corresponding integral kernel $K(\tau)$ becomes complex in general, and differs significantly from the other integral kernels, although the three become identical in the limit $\lambda\to 0$. Moreover, the cubic equation for the corresponding p's also becomes complex in general, and takes the form

$$p^3 + (4\beta+5\lambda)p^2 + [5\beta^2+\beta^2\theta^2+\Omega^2+\lambda(14\beta+4\lambda-4i\beta\theta)]p$$

$$+ 2\beta^3 + \beta\Omega^2 + 2\beta^3\theta^2 + \lambda(9\beta^2+2\Omega^2+4\lambda\beta+\beta^2\theta^2-8i\beta^2\theta-4i\beta\lambda\theta) = 0.$$

$$(4)$$

The complexity of the kernel $K(\tau)$ and of the equation for p is reflected in an asymmetry of the spectrum of the fluorescent light when the atom is excited off-resonance ($\theta\neq0$). We show by numerical examples, that the effect on the spectrum can be quite significant even when $\lambda\lesssim\beta$, and that very marked asymmetries may appear when $\lambda>\beta$. Moreover, the A.C. Stark effect may become a very small effect even in a strong field for large detuning, when the bandwidth λ is only moderately large compared with the decay constant β. The effect of finite bandwidth excitation on the second order correlation and on the spectrum is therefore much more profound than on the other features of the fluorescence. It is possible that these spectral asymmetries played a role in some recent measurements of resonance fluorescence.[6]

*This work was supported by the National Science Foundation.

†This paper was presented at a joint session of the 4'th Rochester Conference on Coherence and Quantum Optics and of the International Conference on Multiphoton Processes. A fuller account is published in *Multiphoton Processes*, ed. J.H. Eberly and P. Lambropoulos (John Wiley, New York, 1978) p. 119.

References

1. H.J. Kimble and L. Mandel, Phys. Rev. Lett. *34*, 1485 (1975).
2. H.J. Kimble and L. Mandel, Phys. Rev. A *13*, 2123 (1976).
3. H.J. Kimble and L. Mandel, Phys. Rev. A *15*, 689 (1977).
4. J.H. Eberly, Phys. Rev. Lett. *37*, 1387 (1976).
5. G.S. Agarwal, Phys. Rev. Lett. *37*, 1383 (1976).
6. R.E. Grove, F.Y. Wu and S. Ezekiel, Phys. Rev. A *15*, 327 (1977).

ATOMIC FLUORESCENCE WITH MONOCHROMATIC EXCITATION[*]

H. Walther

Universität München, Garching, Germany

The investigation of the resonance fluorescence induced by monochromatic laser radiation has received large attention in recent years. A review of level crossing experiments performed under those conditions has been given. Furthermore experiments in which the frequency distribution of the resonance radiation was analysed using a Fabry-Perot interferometer have been discussed as well as correlation experiments now under way.

[*]This paper was presented at a joint session of the 4'th Rochester Conference on Coherence and Quantum Optics and of the International Conference on Multiphoton Processes. A fuller account is published in *Multiphoton Processes*, ed. J.H. Eberly and P. Lambropoulos (John Wiley, New York, 1978) p. 129.

TWO-LEVEL ATOMS IN AN INTENSE MONOCHROMATIC FIELD: A REVIEW OF

RECENT EXPERIMENTAL INVESTIGATIONS[*]

Shaoul Ezekiel

Massachusetts Institute of Technology, Cambridge, MA

The effect of an intense monochromatic field on the behavior
of a two-level atom with natural damping has recently been studied
experimentally using stable tunable lasers and atomic beams. A
number of researchers have measured the spectrum of the emission,
and more recently, the absorption of the strongly driven atoms for
both on-resonance and off-resonance excitation. In addition, the
energy levels of the atoms have been carefully probed to examine in
some detail, the ac Stark effect induced by a strong resonant and
nearly resonant field. The paper will cover a brief review of the
measurements that have been performed so far by the various groups,
a comparison of the experimental results with theoretical predictions, as well as a discussion of the experimental problems in
performing such measurements.

[*] This paper was presented at a joint session of the 4'th Rochester
Conference on Coherence and Quantum Optics and of the International
Conference on Multiphoton Processes. A fuller account is published
in *Multiphoton Processes*, ed. J.H. Eberly and P. Lambropoulos (John
Wiley, New York, 1978) p. 145.

QUANTUM MECHANICAL RESONANT LIGHT SCATTERING[*]

B. R. Mollow

University of Massachusetts, Boston, Mass.

The problem of resonant light scattering was first treated
correctly [1] by an adaptation of a Markoff method borrowed from
quantum statistical mechanics [2]. The accuracy of the method,
which relied upon an assumption of statistical factorization of
atomic and emission field variables, was subsequently confirmed
by theoretical analyses carried out both in the Schrödinger picture,
where solutions for the complete correlated atom-field state vector
were obtained [3,4], and the Heisenberg picture, where methods based
on operator radiation reaction theory [5,6] yielded the same results
[7-9]. In other analyses the incident field, which had been represented by a c-number in Ref. 1, was treated quantum mechanically
[10,11,30]. Subsequently, the use of a c-number applied field
was shown quite generally to be an exact quantum mechanical method
[3]. Experimental studies, beginning with Schuda, Stroud, and
Hercher's important experiment [12] have with the more recent and
precise measurements of Ezekiel and coworkers [13] and Walther
and coworkers [14] amply confirmed the predictions of what is now
the generally accepted theory [1].

A central point from which the many formulations of resonant
light scattering can be understood is specified by the relation
which gives the value of the photon absorption term in the
Schrödinger Hamiltonian, [3]

$$\vec{\mu}_{10} \cdot \vec{E}_R^{(+)} (r = 0)|t> = i\hbar(\tfrac{1}{2}\kappa + i\delta\omega)\, a_{10}|t> , \qquad (1)$$

where $\vec{\mu}_{10}$ is the dipole matrix element connecting the excited
atomic state $|1>_A$ with the ground state $|0>_A$, κ is the Einstein
A-coefficient for the transition, $\delta\omega$ is the Lamb shift (hereafter

ignored), a_{10} is the atomic lowering operator, $|t\rangle$ is the joint correlated atom-field pure state vector, and $\vec{E}_R^{(+)}(\vec{r})$ is the positive frequency part of the quantum mechanical radiation field operator. The total field is

$$\vec{E}(\vec{r},t) = \vec{E}_R(\vec{r}) + \vec{E}_c(\vec{r},t), \tag{2}$$

where $\vec{E}_c$ is the c-number eigenvalue of the coherent state which represents the incident field.

The relation (1), like the corresponding relation in the Heisenberg picture [5,6]

$$\vec{\mu}_{10} \cdot \vec{E}^{(+)}(\vec{r} = 0, t) = i\hbar(\tfrac{1}{2}\kappa + i\delta\omega)a_{10}(t) + \vec{\mu}_{10} \cdot \vec{E}_{FREE}^{(+)}(r = 0, t)$$

$$\tag{3}$$

is a radiation reaction theory statement, which expresses the radiated field at the position of the atom in terms of atomic variables evaluated exactly at the time in question. It is thus a kind of Markoff statement, albeit one which applies to the pure state of a correlated system. Its validity is based on the simple fact that any photons emitted at prior times will have traveled too far from the atom to affect it any longer, and its validity is assured in the rotating wave approximation as long as the state of the atom does not change appreciably within an optical period.

It follows directly from (1) that the state vector for the atom-field system obeys the time-development equation

$$i\hbar\frac{d}{dt}|t\rangle = [\tilde{H}_{OA} + H_{OF} + \tilde{H}_I(t)]|t\rangle , \tag{4a}$$

where the modified atomic Hamiltonian $\tilde{H}_{OA}$ has the imaginary term $-\tfrac{1}{2}i\hbar\kappa$ added to the excited state energy,

$$\tilde{H}_{OA} = \hbar(\omega_{10} - \tfrac{1}{2}i\kappa)|1\rangle_A \, _A\langle 1| , \tag{4b}$$

H_{OF} is the familiar free field Hamiltonian, and in the interaction Hamiltonian $\tilde{H}_I(t)$ the photon absorption term is absent, having been incorporated into $\tilde{H}_{OA}$ in the form of the imaginary term, so that

$$\tilde{H}_I(t) = -\vec{\mu}_{10}^* \cdot [\vec{E}_R^{(-)}(0) + \vec{E}_c^{(-)}(0,t)]a_{10} - a_{10}^\dagger \vec{\mu}_{10} \cdot \vec{E}_c^{(+)}(0,t). \tag{4c}$$

It is important to understand that it is the state of the joint atom-field system that is governed by (4), not the state of the atom alone. Purely atomic wave functions governed by the non-Hermitian Hamiltonian $\tilde{H}_{OA}$ will decay to zero rather than maintain unit probability. If the atom is initially in its ground state, for example, and the incident field is $\vec{E}_c(0,t)$ = $\vec{E}_0\cos\omega t$, the (rotating wave picture) atomic state amplitudes at time t are

$$\psi_0(t) = (\cos\tfrac{1}{2}\bar{\Omega}t - i\frac{\delta}{\bar{\Omega}}\sin\tfrac{1}{2}\bar{\Omega}t)e^{\tfrac{1}{2}i\delta t} \quad ,$$

$$(5)$$

$$\psi_1(t) = i\frac{\Omega}{\bar{\Omega}}\sin\tfrac{1}{2}\bar{\Omega}t \ e^{\tfrac{1}{2}i\delta t} \quad ,$$

where $\Omega = |\vec{\mu}_{10} \cdot \vec{E}_0/\hbar|$, and δ and $\bar{\Omega}$ are the complex quantities

$$\delta = \omega - \omega_{10} + \tfrac{1}{2}i\kappa \quad ,$$

$$(6)$$

$$\bar{\Omega} = (\Omega^2 + \delta^2)^{\tfrac{1}{2}} \quad .$$

The amplitudes given by (5) accurately describe the projection of the joint atom-field state vector onto the vacuum state of the field (assuming the field is in the vacuum state initially), and their decay to zero with increasing t simply describes the diminishing probability that there are still no photons present at time t. It is an easy matter to show that total probability is conserved for the system as a whole (allowing for arbitrary numbers of radiated photons), even though the Hamiltonian in (4a) is non-Hermitian [3].

The reduced atomic density matrix is defined as a sum (trace) over all field states, and hence, unlike the vacuum amplitudes in (5), conserves probability. The increase in the probability of finding the atom in its ground state as the result of photon emission, in particular, is a direct consequence of the fundamental condition (1), involving no *ad hoc* steps. Indeed, the optical Bloch equations in their familiar form follow immediately from Eqs. (1) and (4). [3] The same relations can be shown [3] to prove the validity of the quantum fluctuation-regression theorem [2] as it applies to the radiatively damped system under discussion, and hence to prove the validity of the results found in Ref. 1 for the frequency spectrum of the scattered field.

In the Heisenberg picture, the optical Bloch equations follow directly from Eq. (3).[5,15] The fluctuation-regression theorem, on the other hand, requires in addition to Eq. (3), the commutation relation [15]

$$[\vec{E}_{FREE}(r = 0, t), a_{10}(t')] = 0 \qquad \text{for } t > t' . \tag{7}$$

The proof of this relation, which is given in Ref. 7, is easily understood in terms of configuration-space concepts. If the initial time is taken as -T in the distant past, the free field at the position of the atom (r = 0) at time t originates from initial inward traveling waves, evaluated on the surface of a sphere of radius c(T + t). The full Heisenberg atomic operator $a_{10}(t')$, on the other hand, will involve the free field, but only that part of it which has had a chance to reach the atom before time t'. This part originates from the portion of the initial free field which is located (initially) within a sphere of radius c(T + t'), smaller than the radius of the aforementioned sphere, for t' < t. Thus for t' < t the free field which can affect the atom before time t' is spatially distinct from (and hence commuting with) the field which reaches the atom at time t.

If it is desired to evaluate contributions made by terms representing a fixed number n of photons (rather than the full, n-summed solutions), this can be done directly with the aid of Eqs. (4). The n-photon subspace of the full state-space of the atom-field system may be described by joint atom-field amplitudes $\beta^{(n)}_{\mu\,k_n \ldots k_1}(t)$, where $\mu = 0,1$ is the atomic state index and $k_n, \ldots k_1$ are the momenta of the emitted photons. These amplitudes evolve in time in an extremely simple way, the amplitude $\beta^{(n)}_{\mu k}(t)$ being generated continuously by photon emission from the amplitude $\beta^{(n-1)}_{\mu k}(t)$, and decaying continuously as the result of the imaginary term in the excited state energy, which of course represents the effect of further photon emission.

By an appropriate sum over photon momenta, one may construct the projection $\rho^{(n)}_{\mu\nu}(t)$ of the reduced atomic density matrix onto the subspace representing a fixed total number n of photons. This has been done by Oliver et al, [10] and by the author in Ref. 3, Sec. VII. The probability $P^{(n)}(t) = \rho^{(n)}_{00}(t) + \rho^{(n)}_{11}(t)$ of finding exactly n photons in the field at time t is found to be well approximated for large n and t by the Poisson formula [10] with mean photon number $\kappa\bar{\rho}_{11}t$. The projected atomic density matrix is well approximated, in the same limit, as $P^{(n)}(t)$ times the steady state solution $\bar{\rho}_{\mu\nu}$ to the optical Bloch equations for the full atomic density matrix,

$$\rho^{(n)}_{\mu\nu}(t) = \bar{\rho}_{\mu\nu}P^{(n)}(t) . \tag{8}$$

This relation describes an approximate statistical independence between atomic and field variables, which simply reflects the fact that most of the photons which have been radiated by the atom are far from it, and uncorrelated with it.

One can also evaluate the n-photon contribution to the emission spectrum, i.e., the joint probability $N_k^{(n)}(t)$ of finding one photon in mode k and n photons in the remaining modes of the field at time t. [The only published evaluation of this quantity, unlike the simpler quantity $\rho_{\mu\nu}^{(n)}(t)$, appears to be the one in Ref. 3. A careful distinction must be drawn between a specific n-photon contribution, on the one hand (whether evaluated for short or for long times), and the full n-summed solution as evaluated in the initial transient regime, on the other hand. The latter solution, which has direct experimental significance, has been evaluated by Renaud et al. [16] Smithers and Freedhoff [17], however, have considered specific n-photon terms in an approximation in which they appear not as time-dependent functions, but instead as quantities evaluated at the times at which they are appreciable. For large n their results are in agreement with those found by other methods.] Here too one finds for large n and t an approximate correspondence with the full n-summed result, i.e., [3]

$$N_k^{(n)}(t) = W_k t P^{(n)}(t) \ , \tag{9}$$

where W_k is the steady-state spectrum evaluated in Ref. 1. A more detailed analysis (in Ref. 3, Sec. VII) shows the n-photon spectrum $N_k^{(n)}(t)$ to have *time-dependent widths*, which for large n and t are approximately equal to the constant steady state widths during the limited time interval within which $P^{(n)}(t)$ is appreciable.

The configuration space wave functions

$$\varepsilon_\mu^{(n)}(r_n,\ldots r_1,t) \equiv {}_A\langle\mu|{}_F\langle VAC|E_R^{(+)}(r_n)\ldots E_R^{(+)}(r_1)|t\rangle$$

$$\equiv \prod_{j=1}^{n}\left[\frac{-(\vec{\mu}_{10}\times\hat{r}_j)\times\hat{r}_j}{|\vec{\mu}_{10}|r_j}\left(\frac{3\hbar\omega}{16\pi}\right)^{\frac{1}{2}}e^{i\omega(r_j/c-t)}\right]\phi_\mu^{(n)}(r_n,\ldots r_1,t) \tag{10}$$

can be evaluated by an extremely simple and direct method, based upon little more than the fundamental relation (1) and the solutions (5) for the vacuum amplitudes. Equation (1) is easily shown to imply that the functions $\phi^{(n)}$ in (10) obey the boundary conditions at $r_n = 0$

$$\phi_0^{(n)}(0, r_{n-1}, \ldots r_1, t) = \sqrt{\frac{\kappa}{c}}\, \phi_1^{(n-1)}(r_{n-1}, \ldots r_1, t)$$

$$\phi_1^{(n)}(0, r_{n-1}, \ldots r_1, t) = 0 , \tag{11}$$

which describe the system immediately after the n'th photon has been emitted as being in its ground state (so that $\phi_1^{(n)} = 0$), and as originating directly from the pre-emission excited state amplitude $\phi_1^{(n-1)}$. As time goes on, all of the photons travel away from the atom at speed c. The wave functions $\phi^{(n)}$ for $r_n \neq 0$ can therefore be obtained simply by translating the arguments of the functions on the left sides of (11) by the amount r_n (and the time argument by r_n/c), and multiplying the result by the atomic amplitude $\psi_\mu(r_n/c)$ which describes the evolution of the atom during the time r_n/c it takes the n'th photon to travel from the atom to the distance r_n from it. One finds then

$$\phi_\mu^{(n)}(r_n, \ldots r_1, t) =$$

$$\sqrt{\kappa/c}\; \psi_\mu\!\left(\frac{r_n}{c}\right) \phi_1^{(n-1)}(r_{n-1} - r_n, \ldots r_1 - r_n, t - \frac{r_n}{c}) . \tag{12}$$

By iterating this relation n times one then finds

$$\phi_\mu^{(n)}(r_n, \ldots r_1, t) = \left(\frac{\kappa}{c}\right)^{\frac{1}{2}n} \psi_\mu\!\left(\frac{r_n}{c}\right) \psi_1\!\left(\frac{r_{n-1} - r_n}{c}\right) \ldots \psi_1\!\left(\frac{r_1 - r_2}{c}\right) \psi_1\!\left(t - \frac{r_1}{c}\right),$$

where

$$r_n \leq \ldots \leq r_1 \leq ct \tag{13}$$

which together with Eqs. (5) agrees with the solution found by a more complicated method in Eq. (7.16a) of Ref. 3.

Equation (13) represents a complete solution for the wave functions of the joint atom-field system in terms of the purely atomic amplitudes $\psi_\mu(t)$, as given by (5), for finding the driven damped atom in the state $|\mu>_A$ at time t if it is in its ground state $|0>_A$ at $t = 0$. This solution is easily understood if one thinks of the arguments r_j of the wave function as representing the positions of photons emitted at the specified instants

$$t_j = t - r_j/c , \tag{14}$$

where $0 \leq t_1 \leq \ldots \leq t_n \leq t$. The factor $\psi_1(t - r_1/c) = \psi_1(t_1)$ then describes the evolution of the atom between the time the field is turned on ($t = 0$) and the time t_1 at which the first photon is emitted from the atomic excited state, with amplitude $\psi_1(t_1)$. The intermediate factors similarly describe the evolution of the atom between subsequent emissions, while the factor $\psi_\mu(r_n/c) = \psi_\mu(t - t_n)$ describes the evolution of the atom between the time t_n of the last emission and the later time t at which the evaluation of the wave function $\phi_\mu^{(n)}$ is made.

A noteworthy feature of the solution for the configuration space wave function, as emphasized in Ref. 3, is that it vanishes when any two spatial arguments are equal, "and hence that one cannot find two scattered photons at the same point in space (or, more generally, at equal distances from the atom) at the same time." [18] This feature of the solution is seen in the light of the above discussion as a consequence of the identity $\psi_1(0) = 0$, which simply means that it takes the atom at least some time, following one emission, to be raised back up to its excited state by the applied field. Ultimately the relation in question is a consequence of Eq. (1), which directly implies that there cannot be two photons present at the position of the atom at the same time,

$$[E_R^{(+)}(r = 0)]^2 |t\rangle = 0 \ . \tag{15}$$

The same relation at other spatial points then follows from the outward propagation of the emitted field.

The intensity autocorrelation function which exhibits the "anti-bunching" effect under discussion has been evaluated by Carmichael and Walls [19] and discussed further by others [8,20].

The possibility of thinking of the emission of photons as taking place at specified instants is ultimately a consequence of the (near) resonant character of the process under discussion, which gives to an intermediate state in a complicated sequence a real (as opposed to a virtual) character which it would not have if the incident field were far off resonance. A picture of the emission process as characterized by classical emission-instants, to be averaged over in the evaluation of expectation values, is constructed in Sec. VIII of Ref. 3; the function $\psi_\mu(t|t_n...t_1)$ described there is in fact identical under the substitution (14) to the configuration space wave functions evaluated in (13).

The picture under discussion leads to a decomposition [in Eqs. (8.9) of Ref. 3] of the full spectral emission rate W_k into the sum of two terms

$$W_k = W_k^{(1)} + \Delta W_k \ , \tag{16}$$

of which $W_k^{(1)}$ is a species of preparation-time-averaged one-photon
approximation, and ΔW_k is a correction term involving coherent
multiphoton effects. This decomposition is identical to the one
found by Swain [4], which in turn describes a spectrum identical
to the one found in Ref. 1. The one photon approximation $W_k^{(1)}$,
as well as an analogous one-collision approximation developed in
Sec. IX of Ref. 3, have been useful in analyzing a number of
approximations to the full emission spectrum [21-24].

A generalization of the methods which describe the radiatively
damped spectrum appropriate for the treatment of other relaxation
mechanisms (notably collisional ones) has been undertaken by the
author, first in Ref. 25, where the Karplus-Schwinger [26] (instant-
aneously thermalizing) model is adopted, and subsequently with
arbitrary ratios of elastic to inelastic collision rates, i.e.,
of transverse to longitudinal relaxation rates [27,28]. The
latter results have been found to be in reasonable agreement with
recent measurements of the collisional redistribution function [29].

When more than two levels are present (particularly in cases
of degeneracy or near degeneracy), simple generalizations of two-
level methods are available. Cohen-Tannoudji [30] has employed
the "dressed atom" picture to obtain a set of simple rules which
describe the radiatively damped spectrum when the spectral lines
are well separated. These rules have been re-derived in the
formalism in which the incident field is described by a c-number
[28] and generalized to include collisional effects [28,31].

Modifications of the emission-spectral analyses have yielded
predictions about the absorption, amplification, and coupled propagation of weak probe fields in the presence of a strong near
resonant "pump" field [27,32]. A recent experiment has confirmed
the prediction of amplification without population inversion [33].

When many atoms are present, generalizations of one-atom
methods are straightforward, though the calculations become quite
complicated. The two and three atom cases have been treated in
detail by Agarwal et al [34], while McCall [32] and Bonifacio and
Lugiato [35] have made an interesting prediction of bistability in
the many atom case.

For the case of weak incident fields, the contributions made
by the one- and two-photon processes to the scattering spectrum
in the n-atom case have relatively simple forms when all of the
(identical) atoms are located at the same point in space. The
one-photon spectrum is a delta-function at the incident field
frequency, and has the intensity

$$I_{n\ atom}^{(1\ photon)} = \frac{\frac{1}{4}n^2\Omega^2}{\Delta^2 + \frac{1}{4}n^2\kappa^2} \quad, \tag{17}$$

where $\Delta = \omega - \omega_{10}$. This intensity is thus governed by the increased width $\frac{1}{2}n\kappa$, and stands in the ratio

$$\frac{I_{n\ atom}^{(1\ photon)}}{I_{1\ atom}^{(1\ photon)}} = \frac{n^2(\Delta^2 + \frac{1}{4}\kappa^2)}{\Delta^2 + \frac{1}{4}n^2\kappa^2} \tag{18}$$

to the corresponding one-atom result.

The two-photon emission spectral density consists of two symmetrical terms, one centered at the frequency $\omega_{10} = \omega - \Delta$ and the other at the frequency $\omega + \Delta$. [1,3] In the n-atom case, these spectral terms are found to have the (equal) widths $\frac{1}{2}n\kappa$, again broadened from the one-atom width $\frac{1}{2}\kappa$. The spectrally integrated intensity of the two-photon spectrum is found to be

$$I_{n\ atom}^{(2\ photon)} = \frac{n\Omega^4\Delta^2/8}{(\Delta^2 + \frac{1}{4}n^2\kappa^2)^2[\Delta^2 + \frac{1}{4}(n-1)^2\kappa^2]} \quad, \tag{19}$$

and hence to stand in the ratio

$$\frac{I_{n\ atom}^{(2\ photon)}}{I_{1\ atom}^{(2\ photon)}} = \frac{n\Delta^2(\Delta^2 + \frac{1}{4}\kappa^2)^2}{(\Delta^2 + \frac{1}{4}n^2\kappa^2)^2[\Delta^2 + \frac{1}{4}(n-1)^2\kappa^2]} \tag{20}$$

to the corresponding one-atom intensity.

For $n > 1$, the two-photon intensity vanishes at zero detuning ($\Delta = 0$), while at large detuning it has a value n times as great as the corresponding one-atom intensity. The one-photon intensities, by contrast, are at zero detuning equal in the n-atom and one-atom cases, and stand in the ratio $n^2 : 1$ at large detuning. The ratio of two-photon to one-photon intensities, which is the same as the ratio of incoherent to coherent emission in the presently considered weak field limit, stands relative to the same quantity in the one-atom case as

$$\frac{I_{n\ atom}^{(2\ photon)}/I_{n\ atom}^{(1\ photon)}}{I_{1\ atom}^{(2\ photon)}/I_{1\ atom}^{(1\ photon)}} = \frac{\Delta^2(\Delta^2 + \frac{1}{4}\kappa^2)}{n(\Delta^2 + \frac{1}{4}n^2\kappa^2)[\Delta^2 + \frac{1}{4}(n-1)^2\kappa^2]} \quad, \tag{21}$$

which approaches zero as $n \to \infty$.

The vanishing in the weak field case of the incoherent intensity relative to the coherent intensity with increasing n (found also in Ref. 34 for n = 2,3) is understandable in view of the fact that the weakly excited n-atom system in the limit $n \to \infty$ resembles a harmonic oscillator, and hence exhibits no incoherent effects. This is apparent also in the solution for the two-photon configuration-space wave function, which has the value

$$\varepsilon^{(2)}(r_2,r_1,t) =$$

$$\varepsilon^{(1)}(r_2,t)\varepsilon^{(1)}(r_1,t)\ \{1 - \frac{\Delta}{n\delta_{n-1}}\ \exp[(i\Delta-\tfrac{1}{2}n\kappa)|r_2-r_1|/c]\}, \quad (22)$$

where $\varepsilon^{(1)}(r,t)$ is the one-photon wave function and $\delta_{n-1} = \Delta + \tfrac{1}{2}i(n-1)\kappa$. The intensity autocorrelation function at equal spatial points thus has the value

$$\frac{|\varepsilon^{(2)}(r,r)|^2}{|\varepsilon^{(1)}(r)|^4} = \frac{(n-1)^2(\Delta^2 + \tfrac{1}{4}n^2\kappa^2)}{n^2[\Delta^2 + \tfrac{1}{4}(n-1)^2\kappa^2]}\ , \quad (23)$$

and vanishes only in the one-atom case. It is clear from (23) that the photon anti-bunching effect disappears in the limit $n \to \infty$, as one should expect it to do in the harmonic oscillator limit.

* Supported by the National Science Foundation.

References

1. B. R. Mollow, Phys. Rev. *188*, 1969 (1969).
2. M. Lax, Phys. Rev. *172*, 350 (1968), and references cited therein.
3. B. R. Mollow, Phys. Rev. A*12*, 1919 (1975).
4. S. Swain, J. Phys. B*8*, L437 (1975).
5. J. R. Ackerhalt, P. L. Knight, and J. H. Eberly, Phys. Rev. Lett. *30*, 456 (1973); J. R. Ackerhalt and J. H. Eberly, Phys. Rev. D*10*, 3350 (1974).
6. R. Saunders, R. K. Bullough and F. Ahmad, J. Phys. A*8*, 759 (1975).
7. B. R. Mollow, J. Phys. A*8*, L130 (1975).
8. H. J. Kimble and L. Mandel, Phys. Rev. A*13*, 2123 (1976).
9. B. Renaud, R. M. Whitley, and C. R. Stroud, Jr., J. Phys. B*9*, L19 (1976).
10. G. Oliver, E. Ressayre, and A. Tallet, Nuovo Cimento Lett. *2*, 777 (1971).
11. H. J. Carmichael and D. F. Walls, J. Phys. B*8*, L77 (1975); J. Phys. B*9*, 1199 (1976).
12. F. Schuda, C. R. Stroud, Jr., and M. Hercher, J. Phys. B*7*, L198 (1974).

13. F. Y. Wu, R. E. Grove, and S. Ezekiel, Phys. Rev. Lett. *35*, 1426 (1975); R. E. Grove, F. Y. Wu, and S. Ezekiel, Phys. Rev. A*15*, 227 (1977).

14. H. Walther, in *Proceedings of the Second Laser Spectroscopy Conference*, Megeve, France, 1975 (Springer, Berlin, 1975); W. Hartig, W. Rasmussen, R. Schieder, and H. Walther, Z. Phys. A*278*, 205 (1976).

15. S. S. Hassan and R. K. Bullough, J. Phys. B*8*, L147 (1975).

16. B. Renaud, R. M. Whitley, and C. R. Stroud, Jr., J. Phys. B*10*, 19 (1977). See also H. J. Kimble and L. Mandel, Opt. Commun. *14*, 167 (1975) for the time development of the light intensity.

17. M. E. Smithers and H. S. Freedhoff, J. Phys. B*8*, 2911 (1975).

18. Ref. 3, immediately following Eq. (4.16). Note also the manifest vanishing of Eq. (7.16a) for equal spatial arguments.

19. H. J. Carmichael and D. F. Walls, J. Phys. B*9*, L43 (1976).

20. G. S. Agarwal, Phys. Rev. A*15*, 814 (1977).

21. C. R. Stroud, Jr., Phys. Rev. A*3*, 1044 (1971); in *Coherence and Quantum Optics*, edited by L. Mandel and E. Wolf (Plenum, N. Y., 1973) p.537.

22. G. E. Notkin, S. G. Rautian, and A. A. Feoktistov, Zh. Eksp. Teor. Fiz. *52*, 1673 (1967)[Sov. Phys.-JETP *25*, 1112 (1967)].

23. M. Newstein, Phys. Rev. *167*, 89 (1968).

24. R. Gush and H. P. Gush, Phys. Rev. A*6*, 129 (1972).

25. B. R. Mollow, Phys. Rev. A*2*, 76 (1970).

26. R. Karplus and J. Schwinger, Phys. Rev. *73*, 1020 (1948).

27. B. R. Mollow, Phys. Rev. A*5*, 2217 (1972); Phys. Rev. A*8*, 1949 (1973).

28. B. R. Mollow, Phys. Rev. A*15*, 1023 (1977).

29. J. L. Carlsten, A. Szöke, and M. G. Raymer, Phys. Rev. A*15*, 1029 (1977).

30. C. Cohen-Tannoudji, in Ref. 14, p. 324; *Frontiers in Laser Spectroscopy*, Les Houches Summer School, 1975, edited by R. Balian et al. (North-Holland, Amsterdam, to be published); C. Cohen-Tannoudji and S. Reynaud, J. Phys. B*10*, 345 (1977).

31. E. Courtens and A. Szöke, Phys. Rev. A*15*, 1588 (1977).

32. S. L. McCall, Phys. Rev. A*9*, 1515 (1974).

33. F. Y. Wu, S. Ezekiel, M. Ducloy, and B. R. Mollow, Phys. Rev. Lett. *38*, 1077 (1977).

34. G. S. Agarwal, A. C. Brown, L. M. Narducci, and G. Vetri, Phys. Rev. A*15*, 1613 (1977).

35. R. Bonifacio and L. A. Lugiato, Opt. Commun. *19*, 172 (1976).

RESONANCE FLUORESCENCE IN 3-LEVEL SYSTEMS*

C.R. Stroud, Jr. and H.R. Gray

University of Rochester, Rochester, New York

Two types of double optical resonance are described theoretically, and studied experimentally in sodium. One is a ladder-like set of states, and the other has two lower levels each coupled to a single upper level. The atomic steady state populations and the spectrum of the scattered light are determined as a function of the various atomic and laser parameters. It is assumed that the only damping is due to spontaneous emission.

In the ladder-type transition it is found that when the laser resonant with the lower transition is intense, the intermediate level is effectively split by the Autler-Townes effect. When the Rabi frequencies of the two transitions are equal this splitting is replaced by power broadening. When the lower laser is tuned off-resonance the 2-photon resonance becomes quite sharp allowing high resolution study of upper state splittings. The excited state population is maximized by a proper choice of field parameters and may appreciably exceed the rate equation predictions. In addition to these populations, the spectrum of the scattered light is studied; it may contain up to seven components at each transition frequency. This system is studied experimentally in the

$$(3S_{1/2}, \ F = 2, \ m_f = 2) \rightarrow (3P_{3/2}, \ F = 3, \ m_f = 3) \rightarrow$$

$$(4D_{5/2}, \ F = 4, \ m_f = 3)$$

system and the results are compared with theory.

The second type of system is studied in a similar way and compared with the

$$(3S_{1/2}, \ F = 1,2) \rightarrow (3P_{1/2}, \ F = 2)$$

transitions in sodium. This system is particularly interesting because it illustrates the difficulty of obtaining efficient pumping of all the population out of a degenerate or nearly degenerate ground state. An effect which might be called coherent trapping is described which places limits on the efficiency with which such population can be extracted.

*Supported in part by Office of Naval Research under Contract No. N0014-68-A-0091.

THERMODYNAMICS OF A NON-EQUILIBRIUM QUANTUM SYSTEM - RESONANCE
FLUORESCENCE

D. F. Walls

University of Waikato, Hamilton, New Zealand

1. INTRODUCTION

The problem of resonance fluorescence from a two-level atom
is the subject of intense current research activity [1]. One
aspect that has so far received little comment is that it provides
a most interesting example of nonequilibrium thermodynamics in a
quantum system. Conceptually the situation is simple. The two-
level atom is subject to two interactions (a) a coherent interaction with a driving field, (b) an incoherent interaction with a
thermal reservoir (spontaneous emission into the radiation field).
If the driving field were absent the atom would merely relax into
equilibrium with the reservoir and the usual Wigner Weisskopf
theory for a damped atom would apply. The effect of the driving
field is to maintain the atom away from equilibrium in a new non-
equilibrium steady state.

While for equilibrium systems we have the principle of detailed balance and Boltzmann's H theorem, general properties of non-
equilibrium steady states are not at all well known. We test the
quantum formulation of detailed balance for two nonequilibrium
steady states, the laser and resonance fluorescence. We find that
in general the principle of detailed balance is not obeyed; however,
in the special case of a strong driving field a two-level atom
undergoing resonance fluorescence is found to satisfy detailed
balance. We consider several definitions of entropy and test
their application for nonequilibrium steady states. We find that
in the nonequilibrium steady state where detailed balance is
satisfied a generalised H theorem results for several definitions
of generalised entropy. Thus the property of detailed balance,

indicating the microscopic mechanism maintaining the steady state
is the same as that for equilibrium, enables similar macroscopic
properties to equilibrium to be derived.

2. DETAILED BALANCE IN OPEN QUANTUM MARKOFFIAN SYSTEMS

The principle of detailed balance is a property of the microscopic dynamics of systems in equilibrium. Well understood in a
classical context, its formulation in the quantum mechanical regime
has only recently been accomplished [2,3].

We begin by restating the conditions for detailed balance for
a classical system described by the Pauli master equation

$$\frac{\partial P_n}{\partial t} = \sum \gamma_{mn} P_m - \sum \gamma_{nm} P_n \, , \tag{2.1}$$

where $P_n(t)$ is the probability of the system being in state n
at time t and γ_{mn} is the transition probability from state m to n.

In an equilibrium steady state the principle of detailed
balance states that

$$\gamma_{mn} P_n^{ss} = \gamma_{nm} P_n^{ss} \, . \tag{2.2}$$

In a quantum mechanical situation however one has an equation of
motion for the density operator ρ of the open system

$$\frac{\partial \rho}{\partial t} = - i \mathcal{L} \rho \, , \tag{2.3}$$

where $\mathcal{L}$ is the Liouvillian operator. Now if the matrix elements
of ρ are taken in the energy representation one obtains the equation

$$\frac{\partial \rho_{nm}}{\partial t} = \sum \gamma_{n'm'}^{nm} \, \rho_{n'm'} \, , \tag{2.4}$$

where

$$\gamma_{n'm'}^{nm} = \langle n | (\mathcal{L} | n' \rangle \langle m' |) | m \rangle \, .$$

What are the detailed balance conditions for this master equation?
One can see that in general there is no equivalent condition to

the above Pauli condition. In cases where the diagonal matrix
elements and off-diagonal matrix elements are uncoupled, the Pauli
condition may be applied to the diagonal elements alone with
$P_n = \langle n|\rho|n\rangle$. A formulation of detailed balance for an open
quantum Markoffian system in terms of two time averages was formulated by Agarwal [2] and shown to be a direct consequence of micro-
reversibility by Carmichael and Walls [3].

This statement of detailed balance reads

$$\langle S^{(\alpha)}(\tau)S^{(\beta)}(0)\rangle = \langle \tilde{S}^{(\beta)}(\tau)\tilde{S}^{(\alpha)\dagger}(0)\rangle \; , \qquad (2.5)$$

where $S^{(\alpha)}$, $S^{(\beta)}$ are two operators of the open system and $\tilde{S}$ represents the time reversed operator.

In terms of the energy representation this condition takes
the form [3]

$$\sum_k \gamma_{mk}^{n'm'} \rho_{nk}^{ss} = \sum_k \gamma_{n'k}^{mn} \rho_{m'k}^{ss} \; , \qquad (2.6)$$

where the energy states are time reversal invariant. Though the
principle of detailed balance is guaranteed to hold in equilibrium
systems, it is not a general property of nonequilibrium systems.
We shall test the detailed balance condition for two examples of
nonequilibrium steady states.

(a)   Single Mode Laser

The single mode laser may be described by the Scully-Lamb
master equation for the density operator of the field mode [4]

$$\frac{\partial \rho_{nn'}}{\partial t} = -[(n+1)\,R_{nn'} + (n'+1)\,R_{n'n}^{*}]\,\rho_{nn'}$$

$$+ \left(R_{n-1,n'-1} + R_{n'-1,n-1}^{*}\right)(nn')^{\frac{1}{2}}\rho_{n-1,n'-1} \qquad (2.7)$$

$$- \frac{1}{2}C(n+n')\rho_{nn'} + C[(n+1)(n'+1)]^{\frac{1}{2}}\rho_{n+1,n'+1} \; ,$$

where $R_{nn'}$ is defined in terms of laser parameters. The steady
state solution of this equation is

$$\rho_{nn'}^{ss} = \delta_{nn'}\,N\prod_{k=0}^{n}\frac{A}{C}\left(1 + \frac{kB}{A}\right)^{-1} \; , \qquad (2.8)$$

where A, B, C and N are constants.

To test if the laser satisfies detailed balance in the steady state we test eq. (2.6) which reads [3]

$$A\left(1 + (n+1)\frac{B}{A}\right) = A\left[1 + \left(\frac{n+n'}{2} + 1\right)\frac{B}{A} + (n-n')^2\frac{DB}{A}\right] , \qquad (2.9)$$

This is clearly satisfied only for the diagonal elements n = n', a circumstance which corresponds simply to the Pauli conditions. Thus the full detailed balance conditions are not satisfied by the laser.

(b)   Resonance Fluorescence

We shall now test the conditions of detailed balance in the resonance fluorescence problem. The density operator for a two-level atom subject to a resonant driving field, obeys the following master equation in the Markoffian approximation [1]

$$\frac{\partial}{\partial t}\begin{pmatrix}\rho_{22}\\[4pt]\rho_{11}\\[4pt]\rho_{21}\\[4pt]\rho_{12}\end{pmatrix} = \begin{pmatrix}-\gamma & 0 & \frac{i\Omega}{2} & \frac{-i\Omega}{2}\\[6pt]\gamma & 0 & \frac{-i\Omega}{2} & \frac{i\Omega}{2}\\[6pt]\frac{i\Omega}{2} & \frac{-i\Omega}{2} & \frac{-\gamma}{2} & 0\\[6pt]\frac{-i\Omega}{2} & \frac{i\Omega}{2} & 0 & \frac{-\gamma}{2}\end{pmatrix}\begin{pmatrix}\rho_{22}\\[4pt]\rho_{11}\\[4pt]\rho_{21}\\[4pt]\rho_{12}\end{pmatrix} \qquad (2.10)$$

where the driving field is assumed to be in a coherent state $|\alpha\rangle$. $\gamma/2$ is the natural linewidth of the atom and $\Omega = 2|\alpha|K$ is the Rabi frequency (K is the atomic dipole matrix element). The matrix elements are taken in the energy eigenstates of the free atom $|2\rangle$ and $|1\rangle$.

The steady state solution to this equation is

$$\rho_{22} = 1 - \rho_{11} = \frac{\Omega^2}{\gamma^2+2\Omega^2}$$

$$\rho_{12} = \rho_{21}^* = \frac{i\Omega\gamma}{\gamma^2+2\Omega^2} . \qquad (2.11)$$

For this system the detailed balance conditions given by eq. (2.6) reduce to six independent conditions.

For a general steady state three conditions are always satisfied, one trivially, the other two being

$$\frac{i\Omega}{2} (\rho_{21} - \rho_{12}) = \gamma\rho_{22} \ ,$$

$$\frac{i\Omega}{2} (\rho_{11} - \rho_{22}) = \frac{\gamma}{2} \rho_{12} \ . \tag{2.12}$$

These are precisely the conditions for the semiclassical equations to achieve a steady state.

The remaining three conditions are not in general satisfied. However if $\rho_{12} = \rho_{21} = 0$ these remaining conditions are satisfied. Thus detailed balance holds for equilibrium ($\Omega = 0$) and for strong driving fields ($\Omega \gg \gamma$). In the strong field limit care must be taken with one of the conditions, which reads

$$-i\Omega\rho_{12} = 0 \ . \tag{2.13}$$

However the limit of $-i\Omega\rho_{12}$ for $\Omega \gg \gamma$ is $\gamma/2$.

However in the limit $\Omega \gg \gamma$ the system may be described by the equation

$$\frac{\partial}{\partial t}
\begin{pmatrix} \rho_{++} + \rho_{--} \\ \rho_{++} - \rho_{--} \\ \rho_{+-} \\ \rho_{-+} \end{pmatrix}
=
\begin{pmatrix}
0 & 0 & 0 & 0 \\
0 & \dfrac{-\gamma}{2} & 0 & 0 \\
0 & 0 & -\left(\dfrac{3\gamma}{4} + i\Omega\right) & 0 \\
0 & 0 & 0 & -\left(\dfrac{3\gamma}{4} - i\Omega\right)
\end{pmatrix}
\begin{pmatrix} \rho_{++} + \rho_{--} \\ \rho_{++} - \rho_{--} \\ \rho_{+-} \\ \rho_{-+} \end{pmatrix} \ ,$$

$$\tag{2.14}$$

The solution to this equation corresponds to those of the full master equation in the limit $\Omega \gg \gamma$. It is readily apparent that this equation satisfies detailed balance in the steady state. Thus in this sense we may deduce that an atom undergoing resonance fluorescence satisfies detailed balance in the limit $\Omega \gg \gamma$.

Thus we have a far-from-equilibrium steady state which is
described by a set of equations which satisfy detailed balance.
In physical terms the microscopic dynamics which maintain the
steady state are the same as for equilibrium. Thus one is motivated
to investigate the use of equilibrium techniques to determine some
macroscopic properties of the steady state.

Similar ideas relating to the validity of detailed balance and
the existence of a thermodynamic potential were conjectured by
Landauer [5] and discussed in detail by Graham and Haken [6] with
particular application to the laser.

3.   ENTROPY PRODUCTION IN RESONANCE FLUORESCENCE

While for equilibrium systems we have Boltzmann's H theorem,
for nonequilibrium systems no such general theorem exists. We
shall consider several definitions of entropy and investigate their
behaviour far from equilibrium.

(a) Gibbs Formula

We begin with the generalization of Boltzmann's definition
of entropy to quantum systems as given by the Gibbs formula

$$S = -Tr(\rho\ln\rho) \,. \tag{3.1}$$

For a two-level system on diagonalising the product $\rho\ln\rho$
and taking the trace the entropy may be written as

$$-S = \lambda_+\ln\lambda_+ + \lambda_-\ln\lambda_- \,, \tag{3.2}$$

where

$$\lambda_\pm = \frac{1}{2}\{1 \pm [1-4(\rho_{22} - \rho_{22}^2 - |\rho_{21}|^2)]^{\frac{1}{2}}\} \,. \tag{3.3}$$

For the atom initially in its ground state the solutions of Eq. (2.10)
for the matrix elements are

$$\rho_{22}(t) = 1 - \rho_{11}(t) =$$

$$\frac{\Omega^2}{\gamma^2+2\Omega^2} \left[1 - \exp\left(\frac{-3\gamma}{4} t\right)\left(\cosh Kt + \frac{3\gamma}{4K} \sinh Kt\right)\right] \,,$$

$$\rho_{12}(t) = \rho_{21}^{*}(t) =$$

$$i\Omega\left\{\frac{\gamma}{\gamma^2+2\Omega^2}\left[1 - \exp\left(\frac{-3\gamma}{4}\,t\right)\left(\cosh Kt + \frac{3\gamma}{4K}\,\sinh Kt\right)\right]\right.$$

$$\left.+ \frac{1}{2}\,\exp\left(\frac{-3\gamma}{4}\,t\right)\frac{\sinh Kt}{K}\right\} \quad , \qquad (3.4)$$

where

$$K = \left[\left(\frac{\gamma}{4}\right)^2 - \Omega^2\right]^{\frac{1}{2}} .$$

In general S is a function of γ and Ω. Below the threshold for the spectrum splitting into three peaks $\Omega < \gamma/4$, the solutions are a continuation of the equilibrium branch and S is a monotonically increasing function.

Above threshold $\Omega > \gamma/4$, S exhibits oscillations and no general H theorem applies. However in the strong coupling limit $\Omega \gg \gamma$, the atom saturates and the solutions for the matrix elements simplify to

$$\rho_{22}(t) = \frac{1}{2}\left[1 + \exp\left(\frac{-3\gamma}{4}\,t\right)\cos\Omega t\right]$$

$$(3.5)$$

$$\rho_{21}(t) = \frac{-i}{2}\,\exp\left(\frac{-3\gamma}{4}\,t\right)\sin\Omega t$$

In this limit S is a monotonically increasing function of t and is a function only of γ, the external parameter, and has no dependence on the internal parameter Ω.

The entropy production is given by

$$\frac{\partial S}{\partial t} = \frac{3\gamma}{8}\,\exp\left(\frac{-3\gamma}{4}\,t\right)\ln\left[\frac{1 + \exp\left(\frac{-3\gamma}{4}\,t\right)}{1 - \exp\left(\frac{-3\gamma}{4}\,t\right)}\right] \quad , \qquad (3.6)$$

It is clear that $\partial S/\partial t > 0$, thus a generalised H theorem (H = -S) holds in this nonequilibrium steady state which possesses detailed balance. The conditions for a Liapounov function are satisfied; thus the stability of the steady state is guaranteed.

We conclude that the property of detailed balance, indicating the underlying microscopic structure of the steady state is the same as for equilibrium, enables the extension of certain equilibrium properties to nonequilibrium steady states to be made.

(b) Schlogl's K Function

As a generalisation to quantum systems of a Liapounov function introduced by Bergmann and Lebowitz [7] for classical systems Schlogl [8] has suggested the function

$$K(\rho,\rho_{ss}) = Tr(\rho\ell n\rho) + Tr(\rho\ell n\rho_{ss}^{-1}) = -S + Tr(\rho\ell n\rho_{ss}^{-1}). \quad (3.7)$$

However this function will only be a Liapounov function for quantum systems with diagonal density operators. For resonance fluorescence with $\Omega > \gamma/4$ this K function will oscillate and thus in general not satisfy the conditions for a Liapounov function. However in the limit of the saturated atom where detailed balance is obeyed we have

$$\frac{\partial K}{\partial t} = -\frac{\partial S}{\partial t} \qquad\qquad (3.8)$$

where $\partial S/\partial t$ is given by eqs. (3.6). Thus in this limit K monotonically increases from 0 to ∞ and clearly satisfies the requirements for a Liapounov function.

(c) Generalised Entropy of Prigogine, George, Henin and Rosenfeld

A theory of generalised entropy developed by Prigogine et al [9] suggests a definition for entropy given by

$$S = -\frac{1}{2}\ell n\beta, \qquad\qquad (3.9)$$

where

$$\beta = \frac{1}{2} Tr(\rho^{\dagger}\rho),$$

For the two-level atom we have

$$S = -\frac{1}{2}\ell n \left[\frac{1}{4}(1 + d^2) + |\rho_{12}|^2\right], \qquad\qquad (3.10)$$

where $d = \rho_{22} - \rho_{11}$.

In general this yields a function of Ω and γ which exhibits oscillations for $\Omega > \gamma/4$. However in the limit of the saturated atom the expression for S is a monotonically increasing function of t and is a function of the external parameter γ only.

The entropy production in this limit is given by

$$\frac{\partial S}{\partial t} = \frac{3\gamma}{4} \; \frac{\exp\left(\frac{-3\gamma}{2} t\right)}{1+\exp\left(\frac{-3\gamma}{2} t\right)} \; . \tag{3.11}$$

We see that $\partial S/\partial t \geqslant 0$. Thus a generalised H theorem will hold for this definition of entropy (H = -S) and either H or β may be used as a Liapounov function.

Further insight may be gained by considering an approximate expression for $\partial S/\partial t$. It has been shown [10] that close to the steady state we may write

$$\dot{S} = \frac{\partial S}{\partial t} = - \frac{1}{2} \, \mathrm{Tr}(\dot{\rho}^{+}\rho) \; . \tag{3.12}$$

Using this expression for $\dot{S}$ we obtain

$$\dot{S} = \dot{S}_D + \dot{S}_{OD} \; , \tag{3.13}$$

where $\dot{S}_D$ and $\dot{S}_{OD}$ refer to the terms involving diagonal and off-diagonal elements of ρ respectively,

$$\dot{S}_D = - \frac{d\dot{d}}{2}$$

$$\dot{S}_{OD} = -2\dot{\rho}_{12}\rho_{12}^{*} \; . \tag{3.14}$$

In the limit where detailed balance holds

$$\dot{S}_D = \frac{\Omega}{4} \sin 2\Omega t \, \exp\left(\frac{-3\gamma}{2} t\right) + \frac{3\gamma}{8} \cos^2\Omega t \, \exp\left(\frac{-3\gamma}{2} t\right)$$

$$\dot{S}_{OD} = \frac{-\Omega}{4} \sin 2\Omega t \, \exp\left(\frac{-3\gamma}{2} t\right) + \frac{3\gamma}{8} \sin^2\Omega t \, \exp\left(\frac{-3\gamma}{2} t\right) \; , \tag{3.15}$$

We see that the second terms of both these expressions are always positive. The first terms however are oscillatory and may assume negative values. These oscillation terms are precisely out of phase. That is, an increase in entropy through the diagonal elements is exactly balanced by a decrease in entropy through the off-diagonal elements. Thus the rate of increase in the entropy

$$\dot{S} = \dot{S}_D + \dot{S}_{OD} = \frac{3\gamma}{8} \exp\left(- \frac{3\gamma}{2} t\right) \tag{3.16}$$

is a quantity which always is positive and is a function of the bath parameter γ only.

4. SUMMARY

A two-level atom undergoing resonance fluorescence may be described by a set of equations which obey detailed balance in the saturation limit. In this limit a generalised H theorem is shown to hold for this non-equilibrium steady state for several definitions of generalised entropy.

Acknowledgment

The author is grateful to Dr. H. Carmichael for several clarifying discussions.

References

1. The reader is referred to the articles on resonance fluorescence in this volume by C. Cohen-Tannoudji and S. Reynaud, p. 93; L. Mandel and H.J. Kimble, p. 95; H. Walther, p. 99; S. Ezekiel, p. 101; B.R. Mollow, p. 103; C.R. Stroud, Jr. and H.R. Gray, p. 115; W.A. McClean and S. Swain, p. 127; and the references contained therein.
2. G.S. Agarwal, Z. Physik *258*, 409 (1973).
3. H.J. Carmichael and D.F. Walls, Z. Physik *B23*, 299 (1976).
4. M.O. Scully and W.E. Lamb, Phys. Rev. *159*, 208 (1967).
5. R. Landauer, IBM Research RC 2960 (1970).
6. R. Graham and H. Haken, Z. Physik *243*, 289 (1971); ibid. *245*, 141 (1971).
7. P.G. Bergman and J.L. Lebowitz, Annals of Physics *1*, 1 (1957).
8. F. Schlogl, Z. Physik *249*, 1 (1971).
9. I. Prigogine, C. George, F. Henin and L. Rosenfeld, Chemica Scripta *4*, 5 (1973).
10. A.R. Bulsara and W.C. Schieve, J. Chem. Phys. *65*, 2532 (1976).

THE THEORY OF THE OPTICAL AUTLER-TOWNES EFFECT IN SODIUM INCLUDING

DEGENERACY AND DETUNING

S. Swain and W. A. McClean

The Queen's University of Belfast

Belfast, Northern Ireland

1. INTRODUCTION

Consider an atomic system illuminated by an intense monochromatic laser field tuned to resonance between two isolated atomic levels. According to the Autler-Townes theory (1955) each of the levels will be split into a doublet, the magnitude of the splitting being proportional to the square root of the intensity of the laser field. If one of the levels is connected to a third level by a second weak laser beam, the doublet structure can be probed by monitoring the population of the third level as a function of the frequency of the weak (probe) laser. The theory of this three-level model has been discussed by several authors, two recent examples being Feneuille and Schweighofer, (1975) and Whitley and Stroud (1976).

The observation of the Autler-Townes effect in the optical region by such methods has been reported recently by Schabert, Keil and Toschek (1975), Picqué and Pinard (1976) and Delsart and Keller (1976), and in the infra-red region by Cahuzac and Vetter (1976). Two very recent experiments are those of Moody and Lambropoulos (1977) and Bjorkholm and Liao (1977). Here we are mainly concerned with the experiments of Picqué and Pinard, which were conducted on an atomic beam of sodium, but, as we shall see, the analysis of this experiment is complicated by the hyperfine structure and magnetic degeneracy of the sodium atomic levels. (See Fig. 1). In these experiments a powerful pump laser (intensity I_1, frequency Ω_1) was tuned to the $3^2S_{1/2} F = 2 \leftrightarrow 3^2P_{3/2} F = 3$ transition (resonant at frequency ω_1) of the sodium atoms in an atomic beam, and a weak probe laser (intensity I_2, frequency Ω_2) tunable

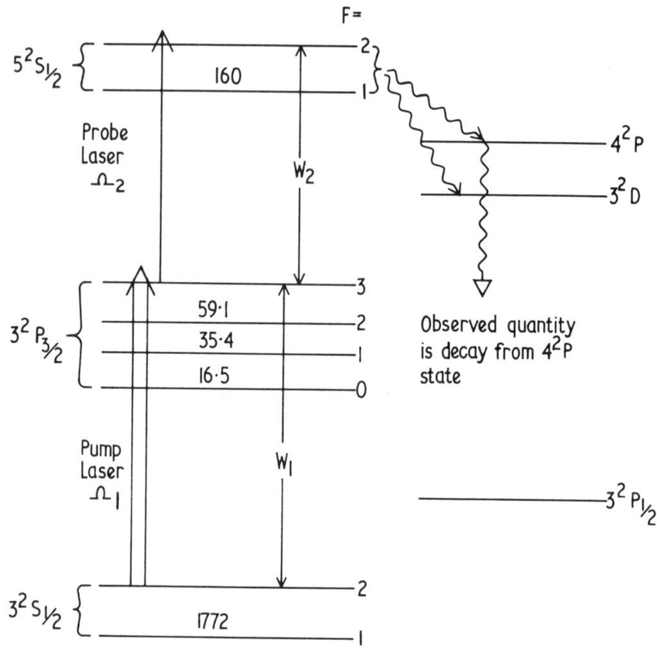

Fig. 1 The energy levels of sodium considered at various stages
in this calculation. The hyperfine level separations are given
in MHz, and the hyperfine structures of the 4^2P, 3^2D, 4^2S and
$3^2P_{\frac{1}{2}}$ levels are not shown.

in the neighbourhood of the $3^2P_{3/2}$ $F = 3 \leftrightarrow 5\,^2S_{\frac{1}{2}}$ $F = 2$ transition
(resonant at frequency ω_2) was used to populate the $5^2S_{\frac{1}{2}}$ levels.
These levels decay spontaneously to the 4P and 3D levels, and the
quantity observed was the intensity of the spontaneous U.V. decay
from the 4P levels as a function of the detuning $\delta_2 \equiv \Omega_2 - \omega_2$ of
the probe laser. The spectra observed for pump laser intensities
$I_1 = 0.9$ W cm^{-2} and 0.3 W cm^{-2} were doublets, but with the high
frequency peak larger and considerably broader than the low
frequency peak. For an intensity $I_1 = 0.09$ W cm^{-2}, only a single
peak (with some structure) was observed. The peak separations of
the doublets agreed well with the three-level theory.

It is convenient to characterize the strength of the coupling
of the pump field to the atom by the Rabi-frequency,

$$V_1 = \frac{1}{2}\hbar^{-1} E_1 \; \langle 3^2S_{1/2}F = 2 | \underline{D} \cdot \underline{\hat{e}}_1 | 3^2P_{3/2}F = 3\rangle$$

rather than by the intensity I_1. Note that $I_1 \propto V_1^2$ and that in the three-level model the separation of the peaks of the doublet spectrum is equal to just $2V_1$. For the various values of the intensity I_1 of the pump laser, we assume that the intensity I_2 of the probe laser remains at one tenth of the particular value $L_1 = 0.9$ cm^{-2}, which corresponds to $V_1 = 45$ MHz in the experiments of Picqué and Pinard.

We have undertaken a numerical analysis of the situation; a preliminary account, dealing only with the situation in which the pump laser is tuned to exact resonance, has already been published (McClean and Swain, 1977). An alternative approach has been developed by Higgins, Hassan and Bullough (1977). In all the work which follows we have assumed that the polarizations of the probe and intense laser are linear and parallel to each other. A further simplification is that we have ignored all effects due to the distribution of velocities of the sodium atoms; this is justified as the experiments were carried out on a well collimated atomic beam.

2. THE CASE OF RESONANT PUMPING

Before describing the results for off-resonant tuning of the pump laser we summarize the main conclusions to be drawn from the resonant case: (1) If we treat the system as a simple, three-level system, i.e. if we take into account only the transitions $3^2S_{1/2}F = 2 \leftrightarrow 3^2P_{3/2}$ $F = 3$ and $3^2P_{3/2}$ $F = 3 \leftrightarrow 5^2S_{1/2}F = 2$, then theory predicts a symmetric doublet whose separation coincides very well with that observed in Picqué and Pinard's experiments. We would expect the three-level theory to be reasonable if the Rabi frequency V_1 of the pump laser beam is much smaller than the separation of the $3^2P_{3/2}F=3$ level from the nearest hyperfine level; however for sodium the separation of the F=3 and F=2 sublevels of the $3^2P_{3/2}$ state is only 59 MHz, which is comparable to the typical values of V_1 we consider (e.g. $V_1 = 45$ MHz). In fact, for $V_1 \lesssim 10$ MHz, the doublet structure disappears, so that for sodium one can never operate comfortably in the region $V_1 \ll 59$ MHz and still observe the Autler Townes splitting. Hence the three-level model is inadequate for sodium. (2) Turning the system into a four-level model by taking into account the transition $3^2S_{1/2}F = 2 \leftrightarrow 3^2P_{3/2}$ $F = 2$ changes the situation drastically; one now obtains a three peaked spectrum with two of the peaks on the high frequency detuning side, so that there is a minimum where in the three-level model a maximum occurred. (See Fig. 2). Thus taking into account an additional transition *worsens*, rather than improves the agreement between theory and experiment.

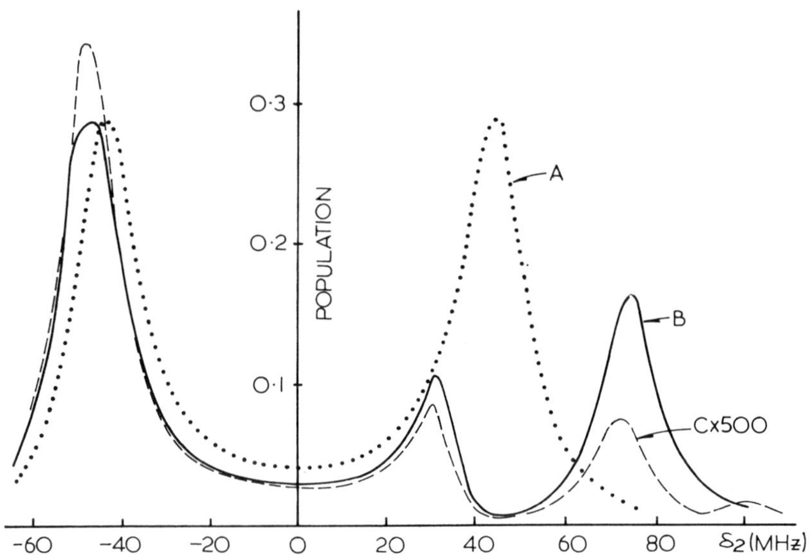

Fig. 2 Population of the $5^2S_{1/2}F = 2$ level for $V_1 = 45$ MHz as a function of the probe field detuning $\delta_2 = \Omega_2 - \omega_2$ for A — three-level
model, B — four-level model and C — eight-level model.

    To show that it is necessary to take into account the transition
$3^2S_{1/2}F = 2 \leftrightarrow 3^2P_{3/2}F = 2$ consider only the three levels, a, b, and c
with corresponding energies E_a, E_b and E_c which are connected by the
pump laser, where $a \equiv 3^2S_{1/2}F = 2$, $b \equiv 3^2P_{3/2}F = 2$, $c \equiv 3^2P_{3/2}F = 3$.
One can write down the density matrix equations as previously out-
lined, introducing spontaneous decay constants γ_{ij} in the usual
way, and solve for the steady state density matrix elements ρ_{bb}
and ρ_{cc}. We assume that the pump laser is resonant with the $a \leftrightarrow c$
transition, and for simplicity take $\gamma_{aa} = 0$, $\gamma_{bb} = \gamma_{cc} = \gamma$, and
$\gamma_{ij} = \frac{1}{2}(\gamma_{ii} + \gamma_{jj})$ for $i \neq j$. Consider first the case where the
frequency separation ω_{cb} of the levels c and b is much larger than
the other parameters, i.e. $\omega_{cb} \gg \gamma, V_{ab}, V_{ac}$ where V_{ab}, V_{ac} are the
Rabi frequencies for the transitions $a \leftrightarrow b, a \leftrightarrow c$, respectively.
We find

$$\frac{\rho_{bb}}{\rho_{cc}} = \frac{V_{ab}^2(\frac{1}{4}\gamma^2 + 2V_{ac}^2)}{V_{ac}^2 \omega_{cb}^2} \ll 1 \quad .$$

On the other hand, assuming the values $\omega_{cb} = 59$ MHz, $\gamma = 10$ MHz

V_{ac} = 45 MHz, V_{ab} = 26 MHz, relevant to the Picqué-Pinard experiments
we find ρ_{cc} = 0.304 and ρ_{bb} = 0.147. Thus the population of the
intermediate level is a significant fraction of the population of
the resonant level. (3) Taking into account all the remaining
transitions, i.e. allowing for transitions between all the eight
hyperfine levels shown in Fig. 1, produces quantitative but not
very great qualitative changes in the spectrum. Note that, because
of optical pumping into the ground state, the population of the
$5^2 S_{\frac{1}{2}} F$ = 2 level is of the order 6 x 10^{-4}, as compared with about
0.3 in the four-level case (for V_1 = 45 MHz). Even taking into
account eight levels, we are still no nearer restoring the agreement
between theory and experiment. (4) Finally, we allowed for the
Zeeman degeneracy of all the eight hyperfine levels shown in Fig. 1.
The results are shown in Fig. 3 for pump laser fields with
V_1 = 1, $1/\sqrt{3}$, $1/\sqrt{10}$ and $1/\sqrt{30}$ times 45 MHz. The most striking
feature is that the spectra are not very different from those of
the three-level model. This is particularly true at the lower
pump laser fields, $V_1 \lesssim$ 45 MHz. At the largest value shown,
V_1 = 45 MHz, the simple doublet structure is beginning to break up.
For greater intensities (not shown) the high frequency peak breaks
into two, three, or more peaks as V_1 is increased, whilst the low
frequency peak remains a single, smooth peak. For values of
$V_1 \lesssim$ 45 MHz, the separation of the doublet is approximately (but
only approximately) equal to $2V_1$, as Table 1 shows.

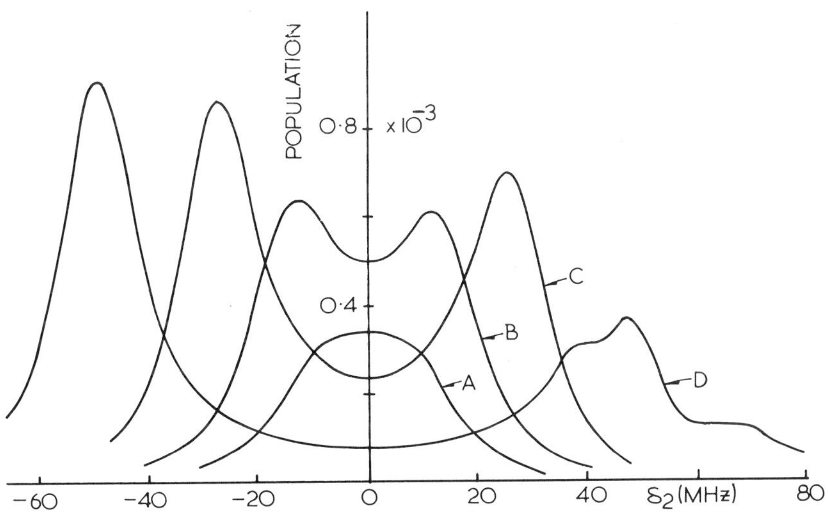

Fig. 3 Population of the $5^2 S_{\frac{1}{2}} F$ = 2 level against δ_2, taking into
account all eight hyperfine levels and their degeneracy, with
V_1 = $1/\sqrt{30}$, $1/\sqrt{10}$, $1/\sqrt{3}$, and 1 times 45 MHz as shown in A, B, C and
D, respectively.

Table 1. Comparison of the three-level model peak separation $(2V_1)$ and the peak separation of the degenerate eight-level model shown in Fig. 3. All units are MHz. The first and last columns should be compared.

$2V_1$	Position of high frequency peak	Position of low frequency peak	Separation
$\dfrac{2 \times 45}{\sqrt{10}} = 28.4$	12	12	24
$\dfrac{2 \times 45}{\sqrt{3}} = 52.0$	25	27.5	52.5
$\dfrac{2 \times 45}{\sqrt{2}} = 63.6$	31	35	66
$\dfrac{2 \times 45}{1} = 90$	47	50	97

Note that the peaks are not only asymmetric in height but that they are also asymmetrically displaced about the origin. For $V_1 < 45$ MHz the peaks of the degenerate eight-level spectra are only slightly broader than those in the three-level non-degenerate model.

We thus have the somewhat embarrassing situation that the much more sophisticated degenerate eight-level model produces spectra not greatly different from those of the simple three-level model, and that neither model reproduces all the features of the experimental spectra. In Fig. 3, the low frequency peak is the largest, whereas in the experiments the high frequency peak is the largest. Also, the high frequency peaks in Fig. 3 begin to show broadening due to degeneracy only at values of $V_1 = 45$ MHz, whereas in the experiment such effects are pronounced for $V_1 = 45$ MHz, and $V_1 = 1/\sqrt{3} \times 45$ MHz.

However, we have assumed for purposes of comparison that the polarizations π_1 and π_2 of the two lasers in Picqué and Pinard's experiments are parallel, although this is not explicitly stated in their paper. It is evident that if the polarizations were not

parallel, we would obtain very different theoretical spectra from
those shown here (Delsart and Keller, 1976). Also, as we shall see,
the spectra are modified considerably by even quite small detunings
of the pump laser.

3. THE CASE OF OFF-RESONANT PUMPING

Now we consider the effect of detuning of the pump laser from
exact resonance with the $3^2S_{1/2}F = 2 \leftrightarrow 3^2P_{3/2}F = 3$ transition. In
Figs. 4, 5 and 6 we show the spectra for detunings $\delta_1 \equiv \Omega_1 - \omega_1 = \pm 10$ MHz
for pump fields V_1 of $1/\sqrt{2}$, 1 and $\sqrt{2}$ times 45 MHz respectively.

It is evident that the spectrum is very sensitive to detunings
of this order; in particular, the relative heights of the two peaks
change very rapidly with the detuning. It seems clear that for a
sufficiently large negative detuning a theoretical spectrum with a
similar asymmetry to the experimentally observed ones would be
obtained. In Figs. 5 and 6 it is also apparent that if there were
some additional line broadening mechanism present, such as would
occur in practice using lasers of finite linewidths etc., the
structure in the right-hand peaks would be smeared out, producing a
right-hand peak which is considerably broader than the left-hand
peak. This feature is consistent with the experimental results.

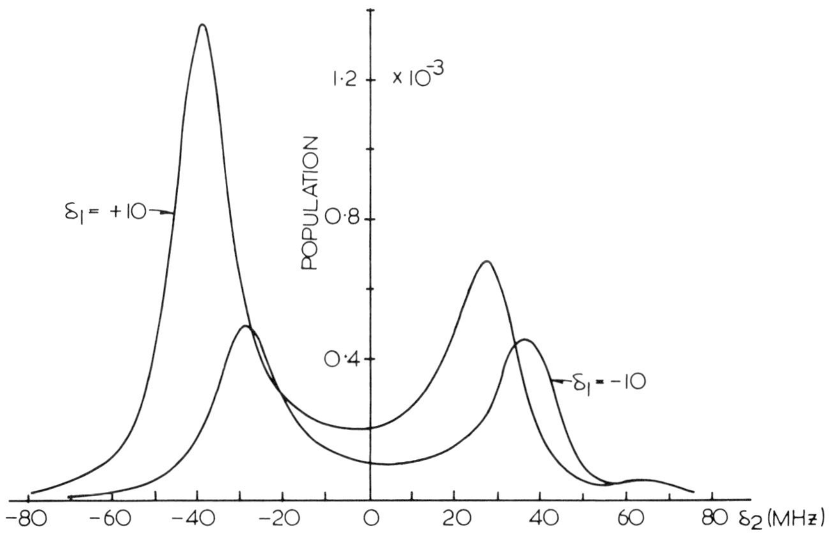

Fig. 4 Population of the $5^2S_{1/2}F = 2$ level against δ_2 for $V_1 = 45/\sqrt{2}$ MHz
for detunings $\delta_1 = \pm 10$ MHz.

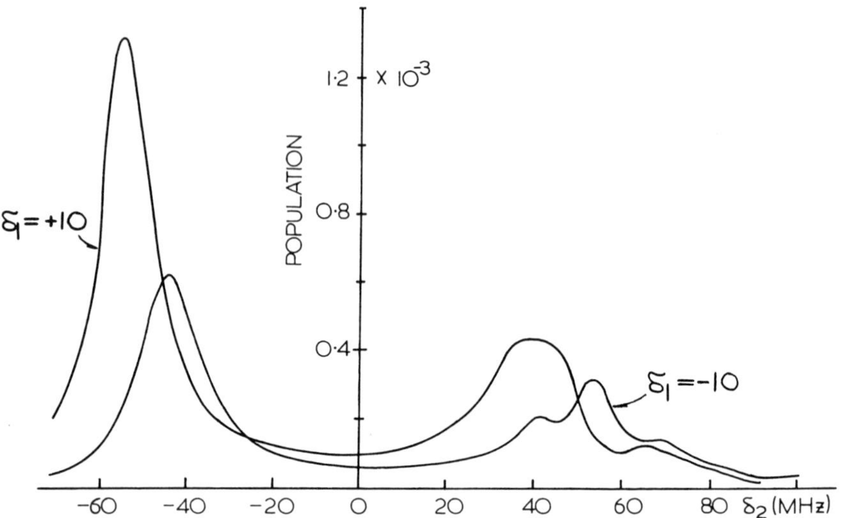

Fig. 5. Population of the $5^2 S_{1/2} F = 2$ level against δ_2 for V_1 = 45 MHz
for detunings δ_1 = ±10 MHz.

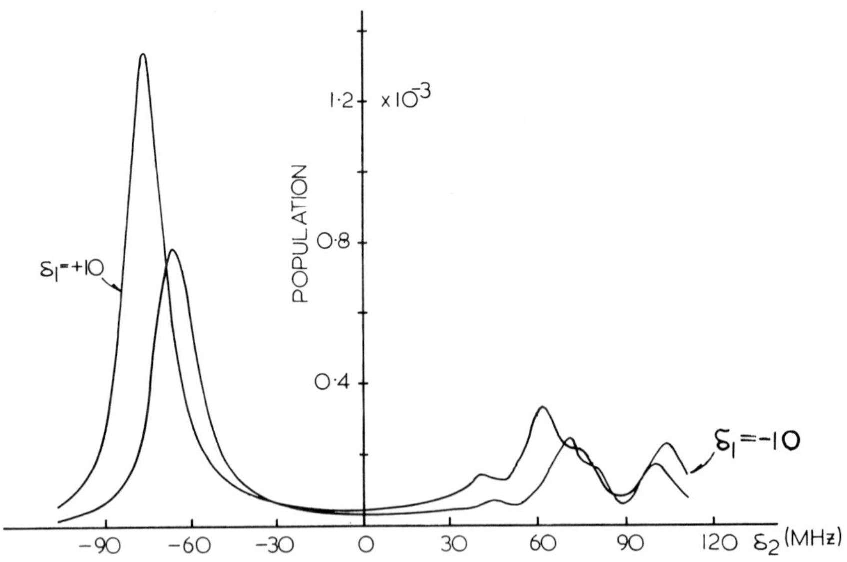

Fig. 6 Population of the $5^2 S_{1/2} F = 2$ level against δ_2 for
V_1 = $45\sqrt{2}$ MHz for detunings δ_1 = ±10 MHz.

According to the simple three-level theory of Feneuille and
Schweighofer (1975), for such small values of the detuning δ_1 the
spectra should merely be shifted by $-\frac{1}{2}\delta_1$, the shape being essentially
unaltered. Whilst the shift is indeed of this order, the relative
heights of the main peaks change drastically, and the structure of
the right-hand peak also changes qualitatively.

We have also calculated the spectra to be expected when the
pump laser is tuned to resonance with the F = 2 and F = 1 hyperfine
levels of the $3^2P_{3/2}$ state. The corresponding detunings are
δ_1 = -59.1 and δ_1 = -94.5 MHz, respectively.

In Fig. 7 we give the theoretical spectra for the situation
where the pump laser is exactly resonant with the $3^2S_{1/2}F = 2 \leftrightarrow 3^2P_{3/2}F=2$
hyperfine transition for pump laser fields of $V_1 = 1/\sqrt{2}$, 1 and $\sqrt{2}$
times 45 MHz. It is evident that the structure is most clearly
resolved for a field of V_1 = 45 MHz, and that for $V_1 = 45/\sqrt{2}$ and
$45\times\sqrt{2}$ MHz the spectrum is basically a doublet. In this case the
high frequency peak is the larger. In practice the spectra would
be smeared out by additional line broadening effects not considered
here.

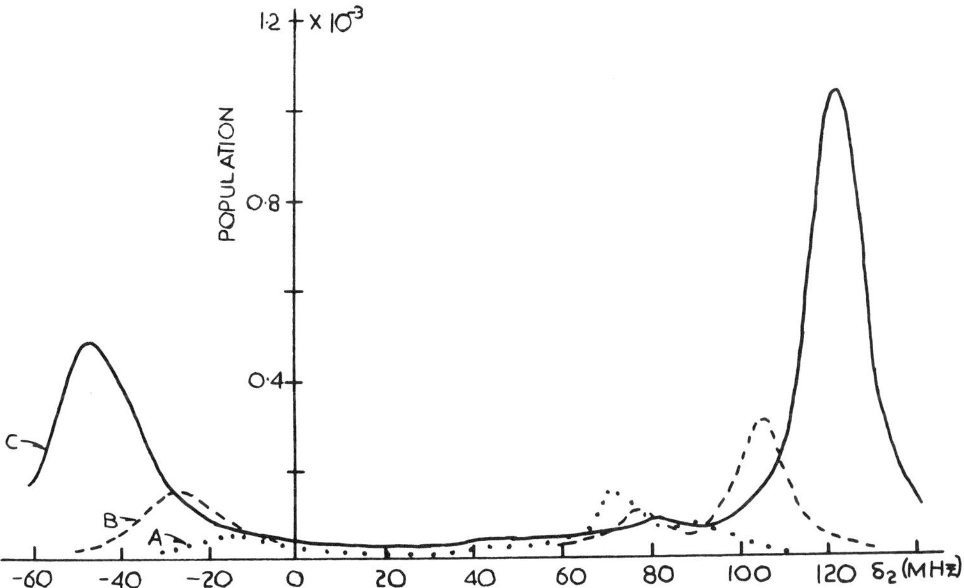

Fig. 7 Population of the $5^2S_{1/2}F$ = 2 level against δ_2 for the pump
laser tuned to the $3^2S_{1/2}F$ = 2 $\leftrightarrow$ $3^2P_{3/2}F$ = 2 transition, i.e.
δ_1 = -59.1 MHz, with $V_1 = 1/\sqrt{2}$, 1, $\sqrt{2}$ times 45 MHz as shown in
A, B and C, respectively.

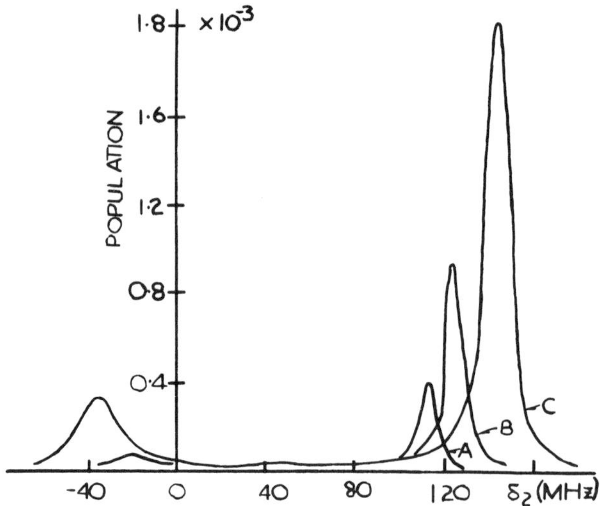

Fig. 8 Population of the $5^2S_{\frac{1}{2}}F = 2$ level against δ_2 for the pump
laser tuned to the $3^2S_{\frac{1}{2}}F = 2 \leftrightarrow 3^2P_{3/2}F = 1$ transition, i.e.
$\delta_1 = -94.5$ MHz, with $V_1 = 1/\sqrt{2}$, 1, $\sqrt{2}$ times 45 MHz as shown in A,
B and C, respectively.

In Fig. 8 we show the theoretical spectrum for the case where
the intense laser is tuned to resonance with the $3^2S_{\frac{1}{2}}F = 2 \leftrightarrow 3^2P_{3/2}F = 1$
transition. The significant features of this spectra are that for
low powers the spectrum is practically a single peak, and even for
higher powers the low frequency component is much less intense than
the high frequency component. All peaks are smooth and show no
structure. We note that in none of the cases we have considered in
this paper does the low frequency peak show structure.

Next we consider the situation where the pump and probe field
interactions are equal and strong ($V_1 = V_2 = 45$ MHz) for $\delta_1 = 0$
and $\delta_1 = +50$ MHz. In both cases we obtain single smooth peaks.
The results are rather similar to the absorption spectra reported
by Whitley and Stroud (1976) for the three-level problem with pump
and probe field both strong except that in our case the half-width
increased as the pump laser was detuned whereas in Whitley and
Stroud's case it decreased.

Finally, we consider the case $V_1 = 450$ MHz, a value for the
Rabi frequency which considerably exceeds the frequency separation,
111 MHz, of the $F = 3$ and $F = 1$ levels of the $3^2P_{3/2}$ state. In this

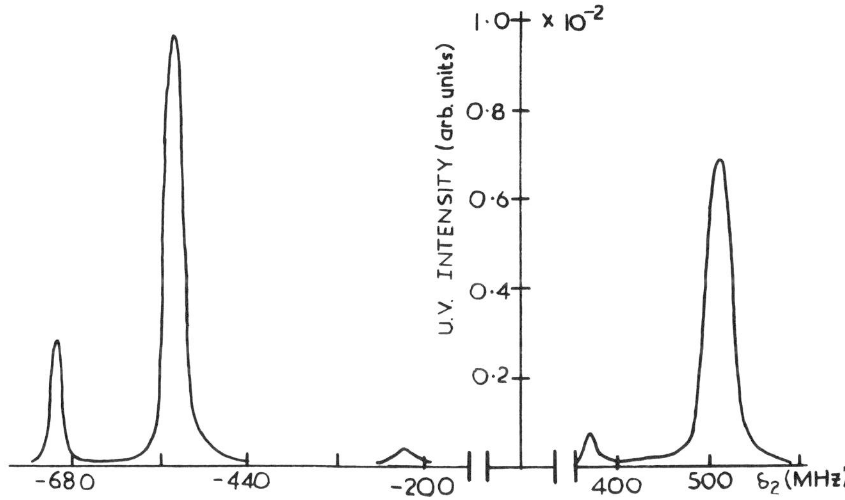

Fig. 9   U.V. emission against δ_2 for V_1 = 450 MHz and δ_1 = 0.

case, the population of the $5^2S_{1/2}F = 2$ level is no longer a good approximation to the observable, as the population of the $5^2S_{1/2}F = 1$ level becomes comparable, and it is necessary to calculate directly the UV emission from the 4^2P state. As shown in Fig. 9, the spectrum again closely resembles that of the three-level model (cf Moody and Lambropoulos, 1977; Bjorkholm and Liao, 1977). The two small peaks, separated from the large ones by about 160 MHz, are contributions to the U.V. emission from the $5^2S_{1/2}F = 1$ state.

SUMMARY

We have investigated the theory of the optical Autler-Townes effect in a sodium atomic beam using lasers with parallel linear polarizations. Whereas the simplest three-level model gives reasonable agreement with the experiments of Picqué and Pinard, four-, five-, ... and eight-level models do not. To achieve partial restoration of the situation we find it essential to allow for the magnetic degeneracy of the hyperfine levels, when we obtain an asymmetric doublet, with the high frequency peak considerably broadened relative to the low frequency one. The latter feature is in accord with experiment. However, the asymmetry of the theoretical spectrum is in the opposite sense to that observed experimentally. It is pointed out that detuning of the pump laser and non-alignment of the laser polarizations can have strong effects on the shape of the spectrum. We have also reported the spectra obtained when the pump is tuned to the F = 2 and F = 1 hyperfine levels of the $3^2P_{3/2}$ state.

Our results indicate that for relatively low laser powers, i.e. for values of the Rabi frequency $V_1 \lesssim 45$ MHz, the predictions of the degenerate eight-level model are very close to those of the non-degenerate three-level model. If experiments could be performed with lasers for which $V_1 \gtrsim \sqrt{2} \times 45$ MHz, significant deviations from the two peaked spectra could be expected. This work suggests (although of course it does not prove) that in related experiments on sodium, such as resonance fluorescence, reasonable agreement with theories which ignore hyperfine structure and degeneracy may be expected if the laser powers are not too large. This is consistent with the experimental results.

References

Autler, S.H. and Townes, C.H., Phys. Rev. *100*, 703 (1955).
Bjorkholm, J.E. and Liao, P.F., Opt. Comm. *21*, 132 (1977).
Cahuzac, P.H. and Vetter, R., Phys. Rev. A *14*, 220 (1976).
Delsart, C. and Keller, J-C., J. Phys. B: Atom. Molec. Phys. *9*, 2769 (1976).
Feneuille, S. and Schweighofer, M.G., J. Physique *36*, 781 (1975).
Higgins, R.B., Hassan, S.S. and Bullough, R.K., to be published(1977).
Picqué, J.L. and Pinard, J., J. Phys. B: Atom. Molec. Phys. *9*, L77 (1976).
McClean, W.A. and Swain, S., J. Phys. B: Atom. Molec. Phys. *10*, L143 (1977).
Moody, S.E. and Lambropoulos, M., Phys. Rev. A *15*, 1497 (1977).
Schabert, A., Keil,R. and Toschek, P.E., Opt. Comm. *13*, 265 (1975).
Whitley, R.M. and Stroud, C.R. Jr., Phys. Rev. A *14*, 1498 (1976).

TRANSIENT QUANTUM PHENOMENA IN THE OPTICAL REGION

Richard G. Brewer

IBM Research Laboratories, San Jose, Calif.

The class of optical coherence effects which are now accessible by laser techniques permit new ways of studying dynamic interactions in atoms, molecules, and solids. Systems which are prepared by laser light may exhibit photon echoes, free induction decay (FID), coherent Raman beats and a variety of other optical transient effects which closely parallel the transients of pulsed NMR. Time-dependent interactions involving population and dipole dephasing can be examined in a selective manner and without the influence of Doppler broadening in the case of a gas, or crystalline strain broadening in the case of a solid.

In the infrared, the Stark switching technique[1-3] has been used to monitor the dephasing effects of elastic or inelastic collisions of vibrationally excited molecules. Most recently, the optical analog of pulse Fourier transform NMR spectroscopy has been demonstrated.[4] By employing suitable coherent optical transient effects in a sample of $^{13}CH_3F$, we are able to resolve Doppler-free spectra in a set of closely spaced Stark split lines. The decay characteristics for each transition are obtained also and reveal the separate contributions of elastic and inelastic molecular collisions. Experiments are performed by interfacing the Stark switching apparatus with a computer which Fourier analyzes coherent transient signals to yield a many-line spectrum. By varying the pulse delay time, as in an echo experiment, the decay behavior for each line can be mapped in a three-dimensional diagram of signal amplitude versus frequency and elapsed time. Elastic collisions are due to small angle scattering and in an echo experiment produce a characteristic non-exponential decay, which also follows from a solution of the Boltzmann transport equation. By probing different parts of the Doppler lineshape, the dependence of collision cross-section on

molecular velocity can be found, and with the aid of scattering
theory, the force laws that operate for elastic or inelastic collisions can be deduced.

In the visible region, the laser frequency switching technique
[5] permits similar studies of electronically excited atoms, molecules, and solids. The dephasing time of coherently prepared Pr^{3+}
impurity ions in a LaF_3 crystal has been measured by optical free-
induction decay.[6] Above 4°K, phonon processes severely shorten
the decay time. Below 4°K, we observe a temperature-independent
decay time of 0.38 μsec. This is surprisingly short considering
that there are no contributions from first-order hyperfine, phonon,
or Pr^{3+}-Pr^{3+} interactions.

References

1.   R.G. Brewer and R.L. Shoemaker, Phys. Rev. Lett. *27*, 631
     (1971).
2.   R.G. Brewer, in *Very High Resolution Spectroscopy*, ed. R.A.
     Smith (Academic Press, London, 1976) p. 127.
3.   R.G. Brewer, in *Proceedings of the International School of
     Physics "Enrico Fermi"*, Varenna, Italy, 1975, ed. N. Bloembergen
     (North-Holland, Amsterdam, 1977) p. 87.
4.   S.B. Grossman, A. Schenzle and R.G. Brewer, Phys. Rev. Lett.
     38, 275 (1977).
5.   R.G. Brewer and A.Z. Genack, Phys. Rev. Lett. *36*, 959 (1976).
6.   A.Z. Genack, R.M. Macfarlane and R.G. Brewer, Phys. Rev. Lett.
     37, 1078 (1976).

PICOSECOND DEPHASING OF COHERENTLY EXCITED VIBRATIONAL LEVELS IN
LIQUID N_2/AR MIXTURES

J.D.W. van Voorst, H.M.M. Hesp, J. Langelaar

and D. Bebelaar

University of Amsterdam, Amsterdam, The Netherlands

The molecular vibration of N_2 is coherently excited by a
single high power picosecond laser pulse via stimulated Raman scattering. With a second picosecond probe pulse the dephasing time is
obtained from the intensity of the stimulated anti-Stokes Raman
scattering as a function of the time delay between both laser
pulses.

In mixtures of liquid N_2 with Ar the surroundings of the excited N_2 molecules can be changed by varying the mole fraction of
Ar in the mixture. In this way the influence of intermolecular
interactions on the dephasing of vibrationally excited N_2 molecules
can be studied.

The measured dephasing times are in good agreement with Raman
linewidth measurements. It turns out that the dephasing time is
strongly dependent on the mole fraction of Ar and varies from 90 ps
in pure N_2 to 35 ps in almost pure Ar.

The correlation time of the modulation is calculated to be
2,1 ps and 4,4 ps in pure N_2 and pure Ar, respectively. These
rather long correlation times suggest that the dephasing results
from the average force field determined by the number of nearest
neighbour molecules, for which it is known from molecular dynamics
calculations that the correlations of their motions is in the order
of 10^{-12} -10^{-11} seconds.

STRONG DEPARTURES FROM UNIFORM PLANE WAVE PULSE PROPAGATION AS A
RESULT OF COHERENT TRANSVERSE EFFECTS

F.P. Mattar[*]

University of Rochester, Rochester, New York

M.C. Newstein

Polytechnic Institute of New York, Farmingdale, New York

P.E. Serafim

Naval Research Laboratories, Washington, D.C.

H.M. Gibbs[**]

Bell Telephone Laboratories, Murray Hill, New Jersey

B. Bölger

Philips Research Laboratories, Eindhoven, The Netherlands

G. Forster and P.E. Toschek[***]

University of Heidelberg, Heidelberg, Germany

INTRODUCTION

Analytic and numerical treatments of the coupled Maxwell-
Bloch[1] equations, including transverse and time-dependent phase
terms, predict a new self-focusing (SF) effect that does not vanish
on resonance and requires coherent pulse-matter interaction.[2] This
new SF falls in the category of transient SF problems associated with
the inertial response of coherent nonlinear active media.[3] The
evolution of this new SF effect (for a given pulse shape and beam
profile), its strength and location, the sharpness of its threshold,
and the spacing between multiple foci depend on the Fresnel number,
absorption coefficient, relaxation times, laser-absorber frequency
mismatch, and input on-axis electric field "area".[4] The effect
also occurs in the presence of inhomogeneous broadening.[5] Recently

two independent experiments demonstrating coherent self-focusing
were reported with as much as a factor of 2 increase in fluency on-
resonance and 4.5 off-resonance. The first of these[6] was made in
inhomogeneously broadened Na; SF and temporal reshaping were also
studied as a function of laser detuning. The observation of SF on-
resonance for coherent pulses (but not for cw light[7]) clearly
illustrates that coherent transient SF is different from previous
SF involving resonant interactions and investigated either in the
rate-equation[8] or in the adiabatic-following[9] approximations.
The second experiment[10] involved the investigation of the on-
resonance coherent SF in quasi-nondegenerate inhomogeneously broadened Ne and its dependence on input pulse on-axis "area" and on
pulse duration.

In summary, coherent transient SF has been demonstrated. It
greatly alters the transverse and temporal evolution of optical
pulses propagating in thick resonant absorbers when sufficient diffraction (transverse coupling) is present. Larger beam diameters
with small irregularities (ripples) will often break up into small-
scale self-focusing filaments.[11] Coherent transient self-focusing
may explain previously observed, but not understood, transverse effects
in SIT experiments.[12-14]

EQUATIONS OF MOTION

In the slowly varying envelope approximation the dimensionless
field-matter equations are:

$$-i \, F \, \nabla_\rho^2 \, \xi + \frac{\partial \xi}{\partial \eta} = p \quad , \tag{1}$$

$$\partial p / \partial \tau = \xi W - [i\Delta\Omega + 1/\tau_2] \, p \quad , \tag{2}$$

and

$$\partial W / \partial \tau = \tfrac{1}{2}(\xi^* p + \xi p^*) - (W - W_o)/\tau_1 \quad , \tag{3}$$

where $\xi = (2\mu \, \tau_p/\hbar)\xi'$, $p = (2/\mu)p'$, $E = \mathrm{Re} \, \xi' \, \exp(i[\omega t - (\omega n/c)z])$
and $P = \mathrm{Re} \, i \, p' \, \exp(i[\omega t - (\omega n/c)z])$.

The complex field amplitude ξ, the complex polarization density p, and the energy stored per atom W, are functions of the transverse coordinate $\rho = r/r_p$, the longitudinal coordinate $\eta = \alpha_{eff} \, z$,
and the retarded time $\tau = (t - zn/c)/\tau_p$. The time scale is normalized
to a characteristic time τ_p of the input pulse, and the transverse
dimension scales to a characteristic spatial width r_p of the input
pulse. The longitudinal distance is normalized to the effective
absorption length $(\alpha_{eff})^{-1}$, where[15] $\alpha_{eff} = [8\pi\omega\mu^2 N\tau_p/n \, \hbar c]$. In this

expression ω is the angular carrier frequency of the optical pulse, μ is the dipole moment of the resonant transition, N is the number density of resonant molecules, and n is the index of refraction of the background material. The dimensionless quantities $\Delta\Omega = (\omega-\omega_0)\tau_p$, $\tau_1 = T_1/\tau_p$, and $\tau_2 = T_2/\tau_p$ measure the offset of the optical carrier frequency ω from the central frequency of the molecular resonance ω_0, the thermal relaxation time T_1, and the coherence relaxation time T_2, respectively. The dimensionless parameter F is given by $F = (\alpha_{eff})^{-1}\lambda/4\pi\ r_p^2$. The reciprocal of F is the Fresnel number associated with an aperture radius r_p and a propagation distance $(\alpha_{eff})^{-1}$. The magnitude of the parameter F determines whether it is possible to divide up the transverse dependence of the field into "pencils" (one pencil for each beam radius ρ) which may be treated in the plane-wave approximation; i.e., the parameter F plays an important role in the transverse propagation behavior.[16] If F is small, each pencil may be treated independently and the situation is equivalent to the geometrical optics approximation. The beam distortion is solely due to self-induced nonlinear interaction with the active medium. On the other hand, if F is larger, diffraction will strongly couple the different pencils preventing this separation across the beam; we are essentially dealing with the wave-optics limit where the nonlinear interaction can be neglected. Clearly, the behavior of the beam varies smoothly between these two limits.

From the field-matter relations (1-3) and using the polar representation of the complex envelope, $\xi = A\ \exp[-i\phi]$, one finds that the transverse energy flow is given by $J_T = 2F\ i\ A^2\ \partial\phi/\partial\rho$. Thus, the existence of transverse energy is clearly associated with the radial variation of the phase of the complex field amplitude ξ. When J_T is negative (i.e., $\partial\phi/\partial\rho < 0$), self-induced focusing dominates diffraction spreading.

Since the inverse Fresnel number per characteristic length F, the "area" of the input pulse on axis, the relaxation times and the off-line center frequency shift are the pertinent parameters describing the temporal and transverse evolution of these coherent pulses in the absorbing media, we have studied the dependence of the SF characteristics on these parameters by numerical integration of Eqs. (1) to (3). The location of the focusing, the sharpness of its threshold, as well as the appearance of the multiple foci on-axis along the direction of propagation were studied. Furthermore, it was verified that coherent SF predicted in sharp-line remains in a broad-line atomic system characterized by T_2^*.[5] The purpose was to see ahead of time if the proposed experiment in Na[6] would work.

This SF is distinct from previous SF phenomena.[3] SF involving resonant interactions has been investigated in rate-equation[8] or adiabatic-following[9] approximations in which the medium's response

to the field is describable as a simple intensity-dependent dispersive susceptibility. For SIT the response is a more general functional of the field.[1]   SIT off-resonance and defocusing would occur
for a non-uniform input through the imaginary part of the p term in
Eq.(1) even if diffraction were absent. This off-resonance SIT self-
focusing and defocusing is evident in the Na data.

SIMPLE PHYSICAL PICTURES

There are at least two ways of visualizing the coherent transient on-resonance self-focusing. The first is to see that phase
variations can lead to a frequency offset, so that index arguments
used for off-resonance self-focusing apply. By Eq.(1) the in-phase
component of the polarization leads to phase changes with propagation. If the input is uniform transversely and on resonance (hence
also unchirped), the in-phase component of p always vanishes, even
when integrated over a symmetric inhomogeneous linewidth, so no
intensity dependent phase changes or focusing occur. If the input
is non-uniform transversely, the in-phase p is initially zero, but
phase variations begin immediately through the ordinary diffraction
term. These phase changes lead to diffractive defocusing; but they
also lead to chirps which appear as a frequency offset. This allows
for an intensity dependent in-phase p which can result in focusing.

The second argument is less mathematical. It is well known
that the more intense the large-area coherent pulse the faster it
travels (see Ref. 1 and Fig. 1). Therefore, after a few absorption
lengths, the most intense central part of the beam is ahead of the
larger radii portions. The surface of peak intensity appears much
like a shock wave propagating through the sample. Diffraction from
the trailing edges of the beam experiences strong absorption except
along this shock front, which may provide net amplification from
atoms excited by faster, smaller diameter, portions of the beam.
Such amplified diffraction is directed inward as needed for focusing
(see Figs. 2 through 5). Perturbation and numerical analyses substantiate this physical picture.

PERTURBATION APPROACH TO THE INITIAL BEHAVIOR

In the following example, a comparison of the distinct temporal
evolution of two separate pencils will be made for short propagation
distances. One of these is on-axis, where the intensity is at maximum, and the second is just off-axis where the intensity is smaller.
The nonlinear polarization that results will make the group velocity
of the pulse peak at the center pencil exceed the corresponding off-
axis group velocity. This is sketched in Fig. 6. For the particular
instant of time τ_0, the off-axis field is larger than the on-axis
field.

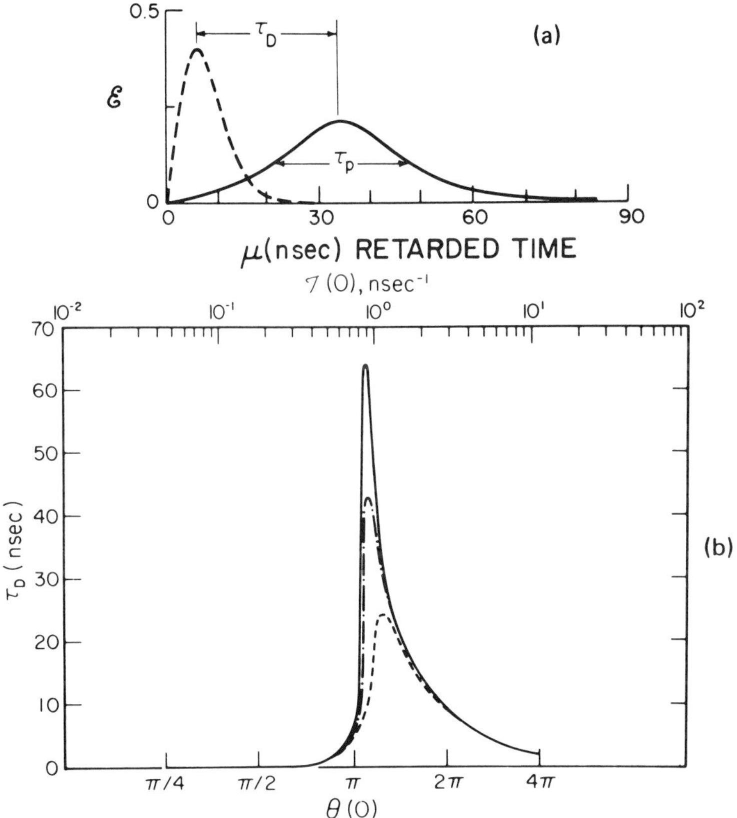

Fig. 1 (a) Illustration of pulse delay. (b) Delay time τ_D versus input area; corresponding value of input energy is given at the top scale. The solid curve is for the case $T_2 > \tau_p$, $T_1 > \tau_p$, no loss, and zero detuning. The dot-dash curve is for $T_2 = 150$ nsec, and the dashed curve is for $T_2 = 50$ nsec. Other parameters are the same as for the solid curve.

Associated with this relative motion between adjacent pencils, there is a variation in the sign for the on-axis transverse coupling (transverse Laplacian) term. At the input plane its contribution is negative. As the pulse propagates along η, at a later instant of time τ_0 the transverse coupling term eventually vanishes. Still further away, its contribution for lagging times τ_0 becomes positive. Thus, the sign of the transverse Laplacian is a function of time differing at the leading and lagging portions of the pulse. Along with these changes in amplitude, the phase also changes. These amplitude A and phase ϕ delays lead to buckling (larger phase accumulation on-axis). Transverse variation of the phase will induce a dynamic transverse energy current, defined by $J_T = 2\mathcal{F}iA^2 \, \partial\phi/\partial\rho$, flowing inwards at some times and outwards at others.

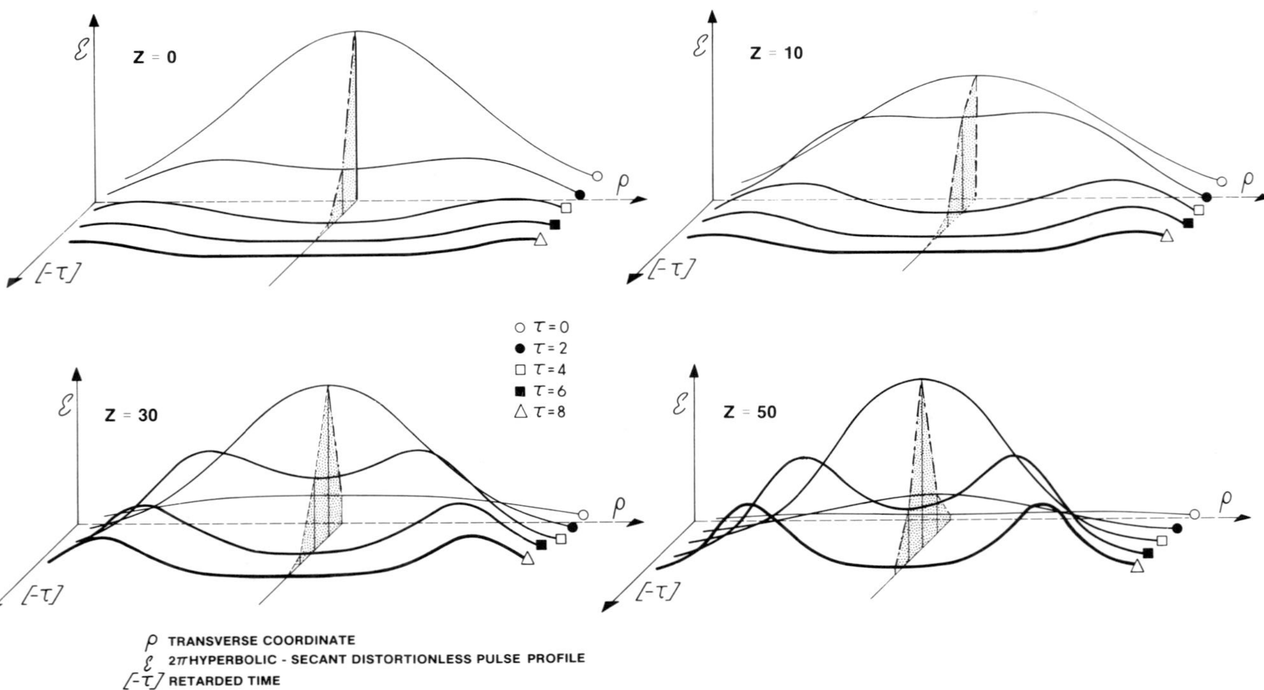

Fig. 2   Isometric profile plots of a family of 2π hyperbolic-secant pulses with radially-dependent pulse-width at both the input plane and after a distance of propagation z. This graph illustrates the relative motion between adjacent pencils. At the new location z, the profile will develop a hole near the axis for certain instants of the retarded time.

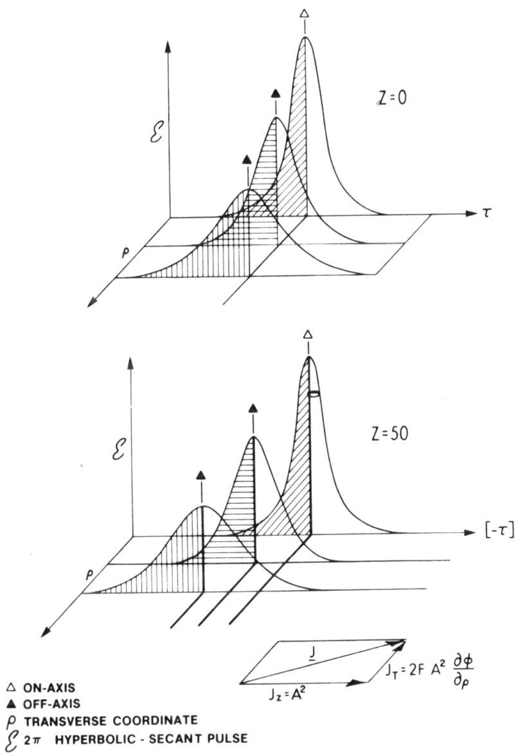

Fig. 3 Isometric time evolution for three distinct radii representing three 2π hyperbolic-secant pulses (with radially dependent pulse-width) after a propagation distance z. This plot illustrates the relative motion.

Fig. 4 Isometric time evolution for three distinct radii representing three 2π hyperbolic-secant pulses (with radially dependent pulse-width) after a propagation distance z. This plot illustrates the relative motion as well as the boosting operation that the light diffracted from the tail of the pulse on the rim of the beam experiences while flowing toward the axis.

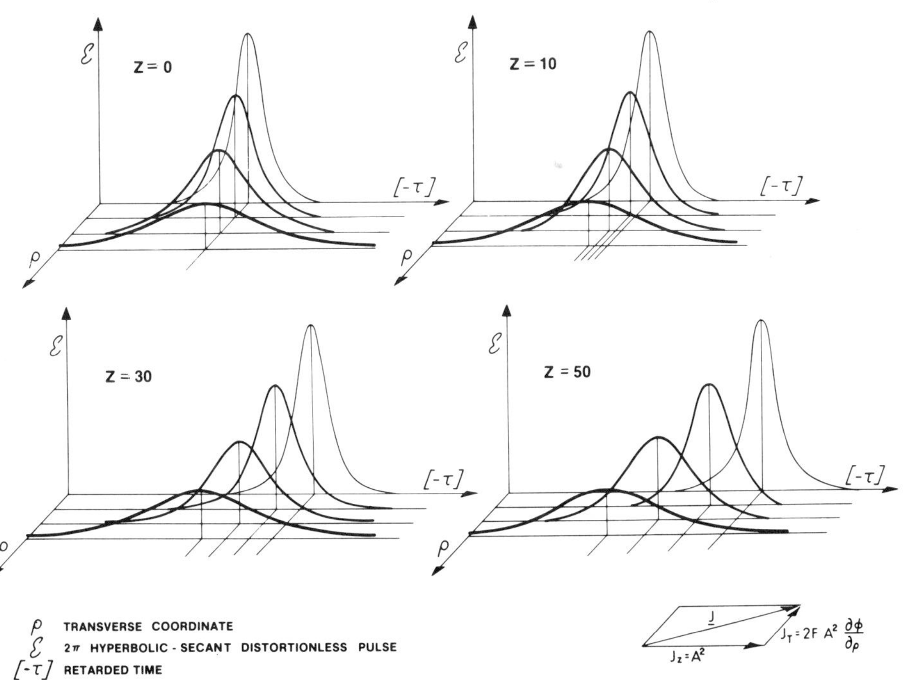

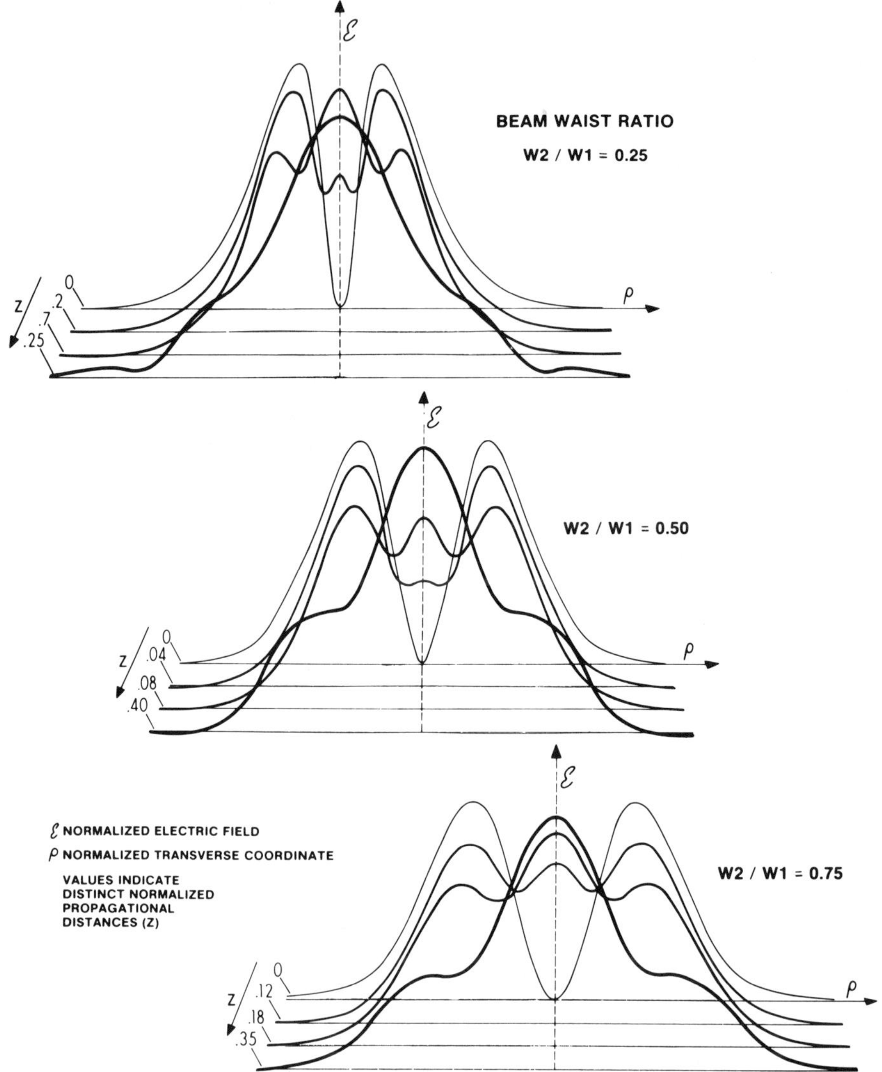

Fig. 5 Effects of linear diffraction on the propagation of an intensity profile with a hole near the axis. This input profile is obtained by the subtraction of two Gaussians with different beam width. The propagation follows the analytical work of Kogelnik and Li.[21]

Expanding the relevant variables in powers of the suitably small quantity F, and keeping only the first order terms, we obtain

$$\xi = \xi_0 + F\,\xi_1 = \xi_0 + F(\xi_{1r} + i\,\xi_{1i}) \quad ,$$

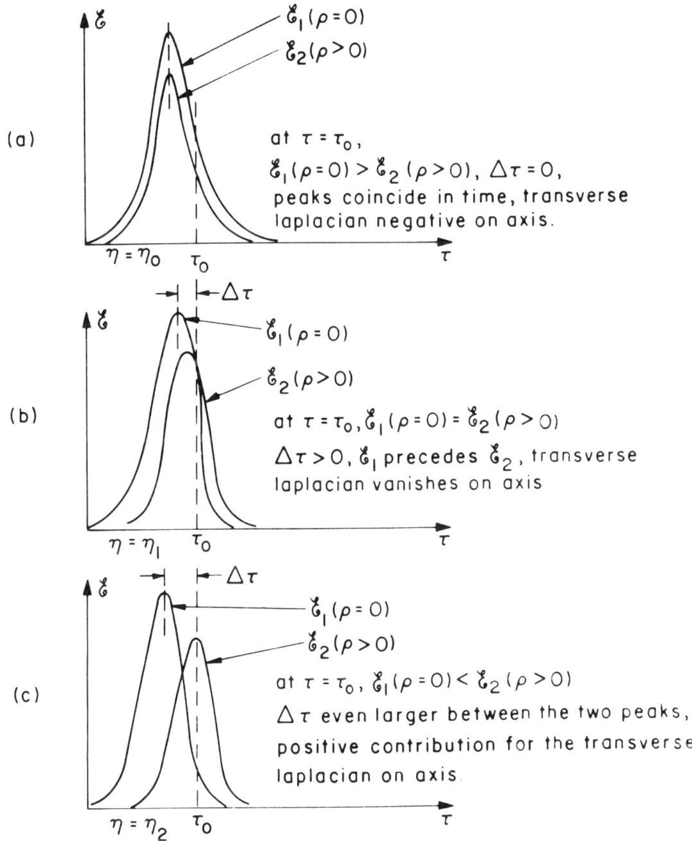

Fig. 6 Relative motion among adjacent pencils propagating coherently in a nonlinear resonant absorber.

$$p = p_0 + F\,p_1 = p_0 + F(p_{1r} + p_{1i}) \quad ,$$

$$W = W_0 + F\,W_1 = W_0 + F\,W_{1r} \quad ,$$

where ξ_0 and p_0 are the (real) solution of the uniform plane-wave
propagation problem. At the input plane $\xi_1 = 0$, $p_1 = 0$, and $W_1 = 0$;
as the pulse propagates, the imaginary part of the three-dimensional
field grows. The rate of growth of ξ_{1i} and its sign depend on the
radial variation of ξ_0. From the known uniform plane-wave properties these transverse variations can be calculated assuming each
pencil propagates independently of the other. It turns out that
the quantities ξ_{1r}, p_{1r} and W_{1r} remain zero for all η. From this,
one can calculate ξ_{1i} and hence obtain the phase

$$\phi = \arctan\,(F\xi_{1i}/\xi_0) \quad .$$

The validity of this procedure is limited to the range of propagation distances where one-dimensional pulses do not differ significantly from their three-dimensional counterparts. A significant
prediction of the perturbation theory is the development of a substantial focusing curvature of the wavefront in the tail of the
pulse. This occurs well within the reshaping region before any substantial focusing begins to occur. The fact that the phase is smaller on the wings of the beam than on the axis is indicative of an inward radial flow of energy current. The results of the perturbation
theory using the plane-wave solution as the unperturbed behavior are
in good agreement with the corresponding three-dimensional results
obtained from rigorous numerical calculations.[18] This is illustrated in Fig. 7 which refers to a situation well within the reshaping region (substantial focusing has not yet occurred). Nevertheless, there is already substantial phase variation on the tail
of the pulse. This is indicated by the curves labeled $\phi(3D;\rho=0)$ and
$\xi(3D;\rho=\Delta\rho)$.

If one assumes that for small propagating distances the induced
polarization is driven by the field at the input plane, an analytic

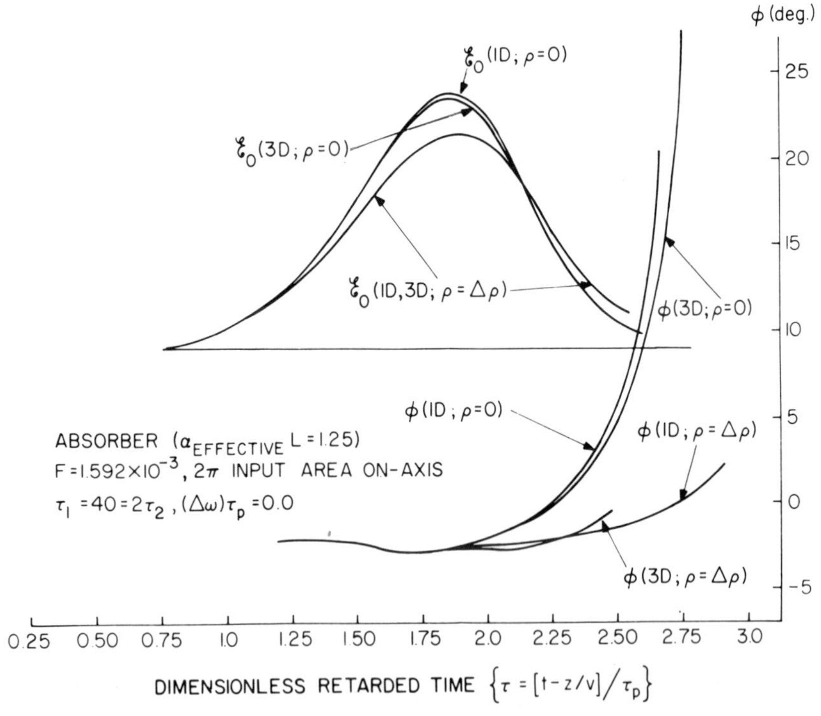

Fig. 7 Comparison of perturbation theory results concerning the
development of longitudinal and transverse phase variations with
the 3-D numerical results.

expression for the one-dimensional field is obtained from the energy equation $\partial\eta\,\xi_0^2 = 2\,\xi_0\,p_0$ as follows:

$$\xi_0 \cong [\xi_a - 2\eta\,\sin\theta_a]^{\frac{1}{2}} \quad ,$$

where ξ_a and θ_a are the field and its time-integrated area at the aperture (input plane). For large value of τ, ξ_a tends to zero causing ξ_0 to be bound and finite. From the previous pencils approach, one obtains an analytic expression for the phase that emerges due to transverse coupling, as well as for J_T, the transverse energy current. The resulting expression[18] for J_T is a power series in η with coefficients of alternate sign. For small enough η (where higher order terms of η can be neglected), the energy current is positive indicating an outward transverse energy flow in agreement with the spreading due to a near-field diffraction effect. As η increases the second term must be considered also. Since its contribution is a negative quantity, J_T eventually changes sign and begins flowing inwards towards the axis forming a converging lens. The latter may counteract and overcome the diffraction giving rise to the coherent self-focusing. If η increases further, the third term contribution becomes important; its effect tends to reduce the inward transverse energy flow due to the second term. The third term may be interpreted as the manifestation of additional diffraction due to the narrowing of the optical beam which is the result of propagating through the nonlinear converging lens-like resonant absorber.

The three-dimensional numerical calculations[18] substantiate the physical picture based on time changes in the phase. In Fig. 8 the field amplitude is plotted isometrically versus the retarded time for three stages of the propagation process: (a) the reshaping region; (b) the build-up region; and (c) the focal region. The transverse energy current is plotted isometrically versus the retarded time for the same three distances in Fig. 8(d), (e) and (f). In each case the plots are given for several values of the transverse coordinate ρ. Positive values of the transverse energy flow correspond to outward flow and negative values correspond to inward flow. Figure 8 clearly illustrates the following features of the self-focusing process. In the earliest stages of the propagation (Figs. (a) and (d)) the near axis energy current is outward for most of the pulse time, but becomes inward (self-focusing) toward the rear (~ 2.4). For this value of τ the field amplitude (Fig. 8(a)) is already past its peak and has a small value. As we proceed into the reshaping region (Figs. (b) and (e)) the near axis peak amplitude moves back in time (corresponding to the fact that the group velocity is less than c/n) while the temporal location of the change from focusing to defocusing energy flow remains the same. This leads to a large increase in the value of the transverse energy flow. In Figs. (c) and (f) (in the focal plane) the peak amplitude occurs at $\tau = 2.4$. The energy current

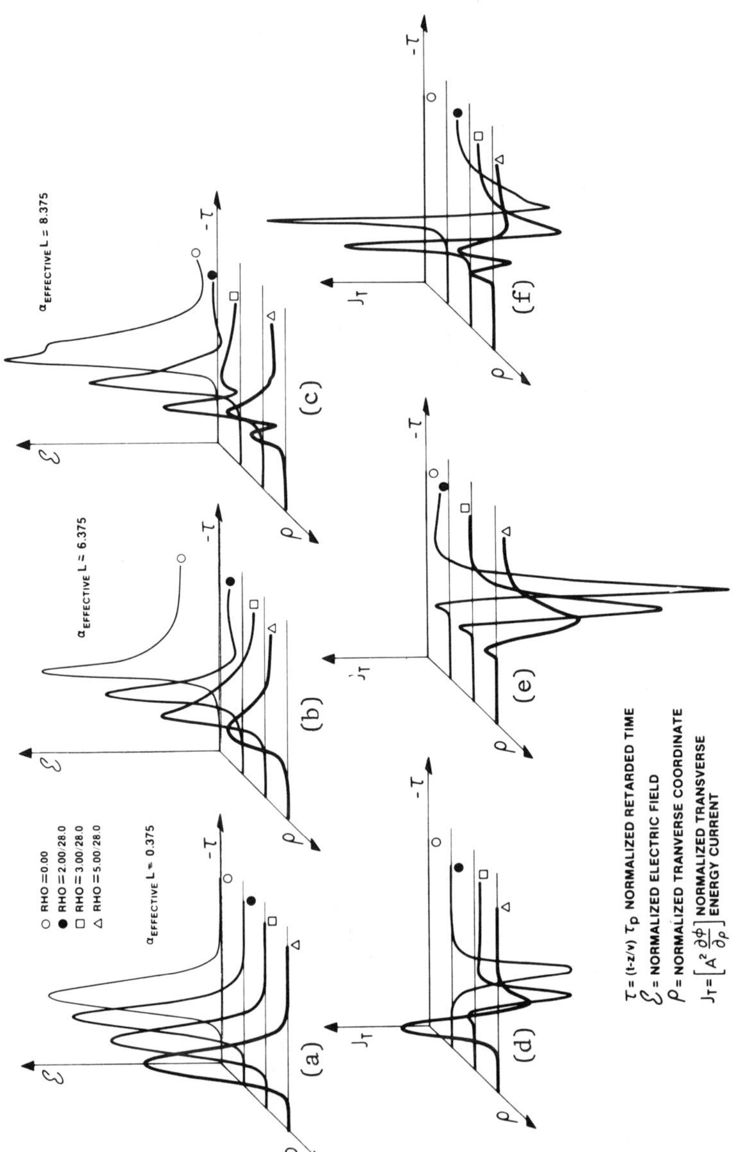

Fig. 8   The field amplitude (a,b,c) and the transverse energy current (d,e,f) for several radii versus the retarded time for three stages of the propagation: the re-shaping region, the build-up region and the focal region (isometric display as a function of the transverse coordinate).

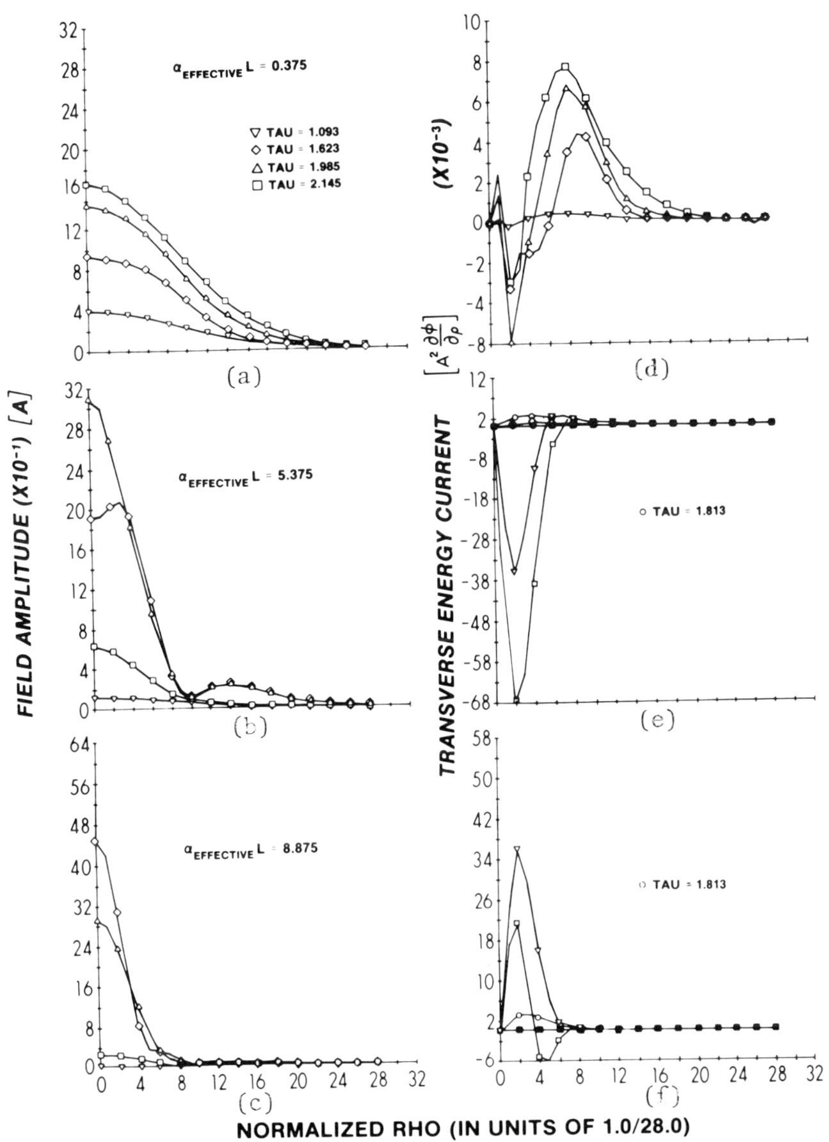

Fig. 9 The profile of the field amplitude (a,b,c) and the profile of the transverse energy current (d,e,f) for several earlier (small) instants of time for three stages of the propagation: the reshaping region, the build-up region and the focal region, as a function of earlier retarded time.

flow in the earlier stages of the pulse is now outgoing corresponding to power which has already focused and is now diverging. The results of the earlier stage (a),(d) are in quantitative agreement
with the analytic predictions of the perturbation theory presented
earlier. From the isometric representation, the relative motion between neighboring pencils is clearly evident as well as the occurence of temporal distortion due to the dynamic self-action previously described.

In graphs (a),(b) and (c) in Fig. 9 and in Fig. 10 the profile
of the field amplitude is plotted for several instants of time in
the leading part of the pulse or in the lagging parts of the pulse,
respectively, for the same three stages of the propagation process.
In the plots of Fig. 9(a),(b),(c) one can notice the distinct deviation of the profile from the input Gaussian shape. The beam splits
into more than one lobe, indicating the different concentration of
energy in more than one ring around the axis outwardly due to diffraction. The quantitative agreement of the hole formation and its
filling up (see Fig. 5) with the simplified physical picture predictions presented earlier is quite evident. Graphs (d),(e) and (f) in
Fig. 9 and in Fig. 10 display the transverse energy current for the
earlier and later instants of time, respectively, at the main stages
of the propagation process. In the earlier stage of propagation,
the energy current flows outwardly according to traditional diffraction spreading (see Fig. 11). Once a sufficient relative motion between neighboring pencils arises, a burn hole in the profile appears;
by diffraction radial energy flows in both directions, outwards and
inwards. Due to the interaction with the active medium, the inward
transverse energy current flows in a quiet substantial manner leading to a build-up of field energy on-axis, which is the new coherent
self-focusing phenomenon. Immediately after the focal plane, both
inward and outward transverse energy flow occur within the same
pulse at different slices of time; the outward flow occurs in the
leading portion resulting from the diffraction of the focused beam,
while the inward flow occurs later and prepares the second (but
weaker) focusing.

This same pattern of pulse compression and beam narrowing was
clearly observed in the first experiment conducted in Na, illustra-
ting SIT-SF with transverse energy flow. The drastic changes in
spatial and temporal profiles are shown in Fig. 12 for the maximum
off-resonance (a and a1) and on-resonance (b and b1) coherent self-
focusing. The significant contribution of the transverse effects in
SIT experiments is clearly displayed in Fig. 13 where we contrast the
temporal behavior for both uniform-plane wave SIT and non-uniform SIT.

EXPERIMENTAL VERIFICATION

This SIT near-resonance SF and defocusing was first observed at
Philips[6] on inhomogeneously broadened Na. An increase in fluency

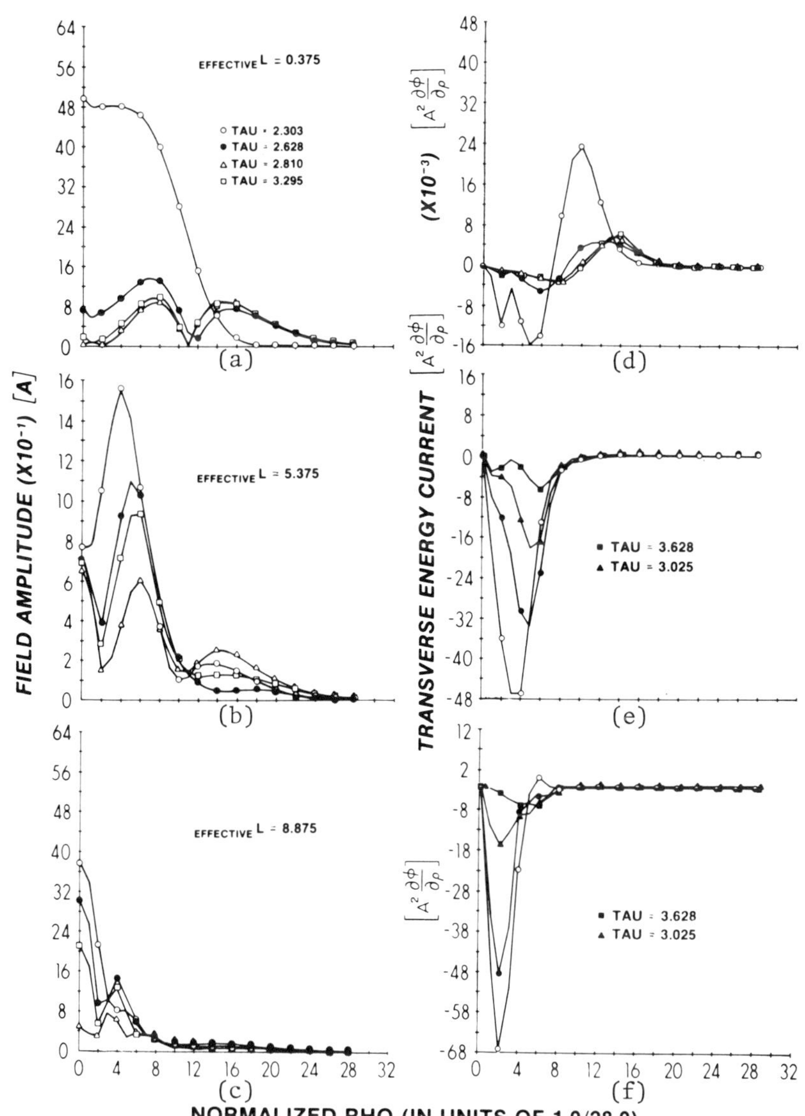

Fig. 10 The profile of the field amplitude (a,b,c) and the profile
of the transverse energy current (d,e,f) for several later (subsequent) instants of time for the three stages of the propagation:
the reshaping region, the build-up region and the focal region as
a function of subsequent retarded times.

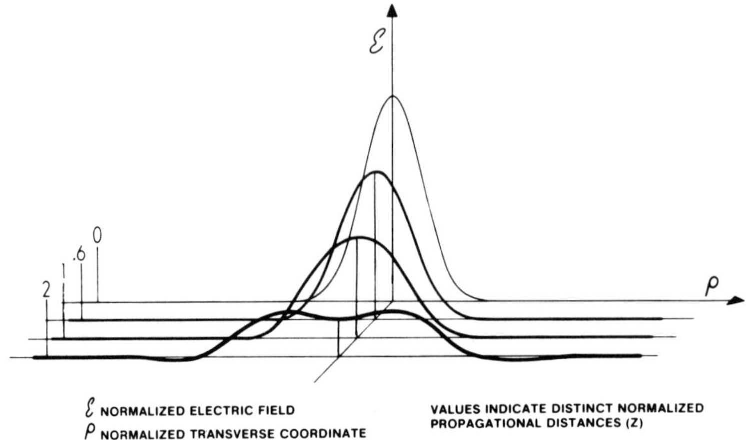

Fig. 11 Spreading effects of linear diffraction on the propagation of a gaussian beam profile.

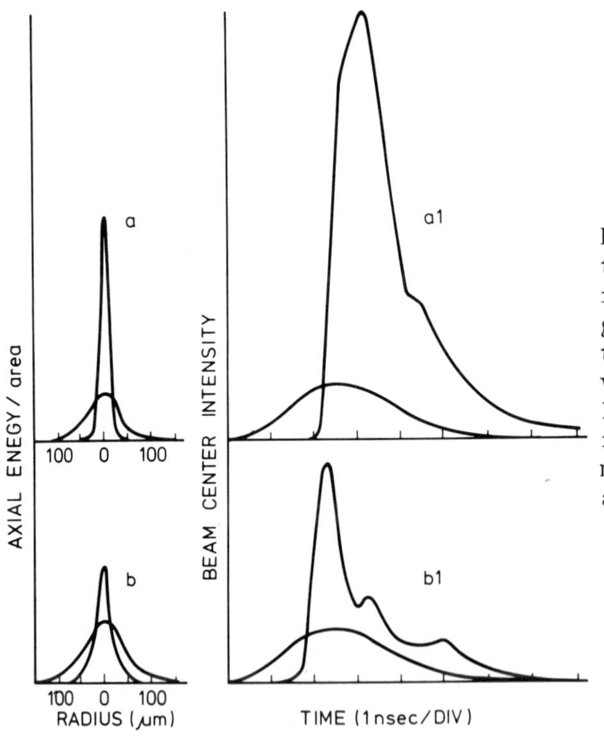

Fig. 12 Changes in spatial and temporal profiles for pulses undergoing self-induced-transparency with transverse energy flow, that lead to coherent self-focusing, a and a1, maximum off-resonance SF; b and b1, on-resonance.

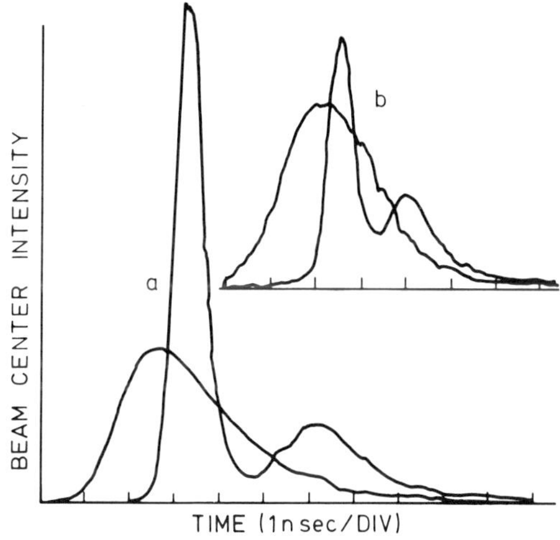

Fig. 13 Temporal behavior of pulses. Break-up of a 4π input pulse is shown under (a) self-focusing and (b) uniform plane-wave conditions in Na. In (a) the integrated output is 34% larger than the input.

Fig. 14 Cross-section of beam at cell exit in the first experiment: curve a, without Na, and curve b, with Na on-resonance, with input area of 3π to 4π and magnetic field of 3.5 kG.

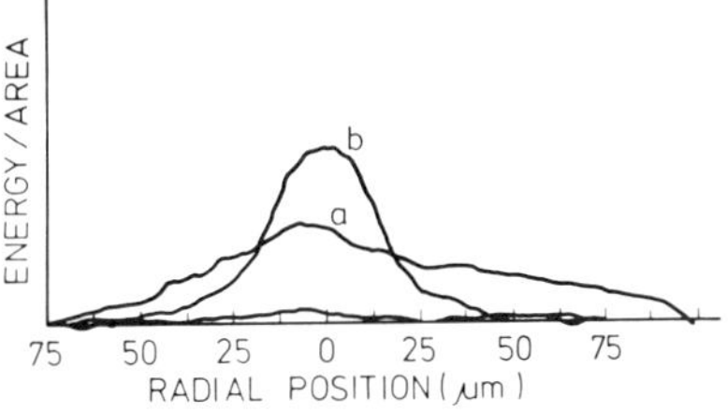

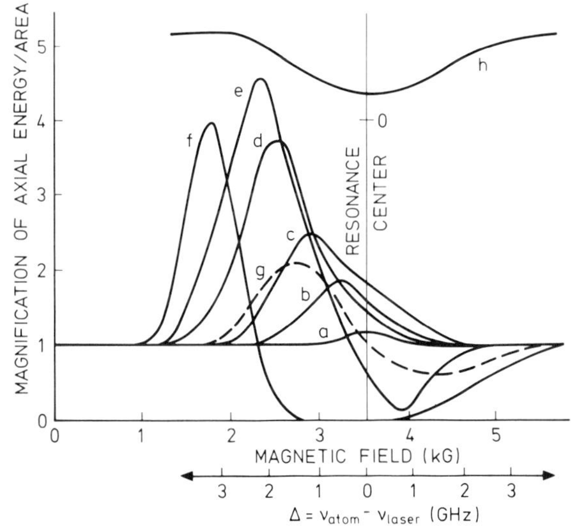

Fig. 15 Experimental energy-density magnification in Na as a function of detuning. Curves a to f are for 2ns, 5 pulses of 125 μm diam. in an 11mm cell. The absorption increases from curve a to curve f; curve g is self-focusing of cw light; curve h shows the atomic absorption.

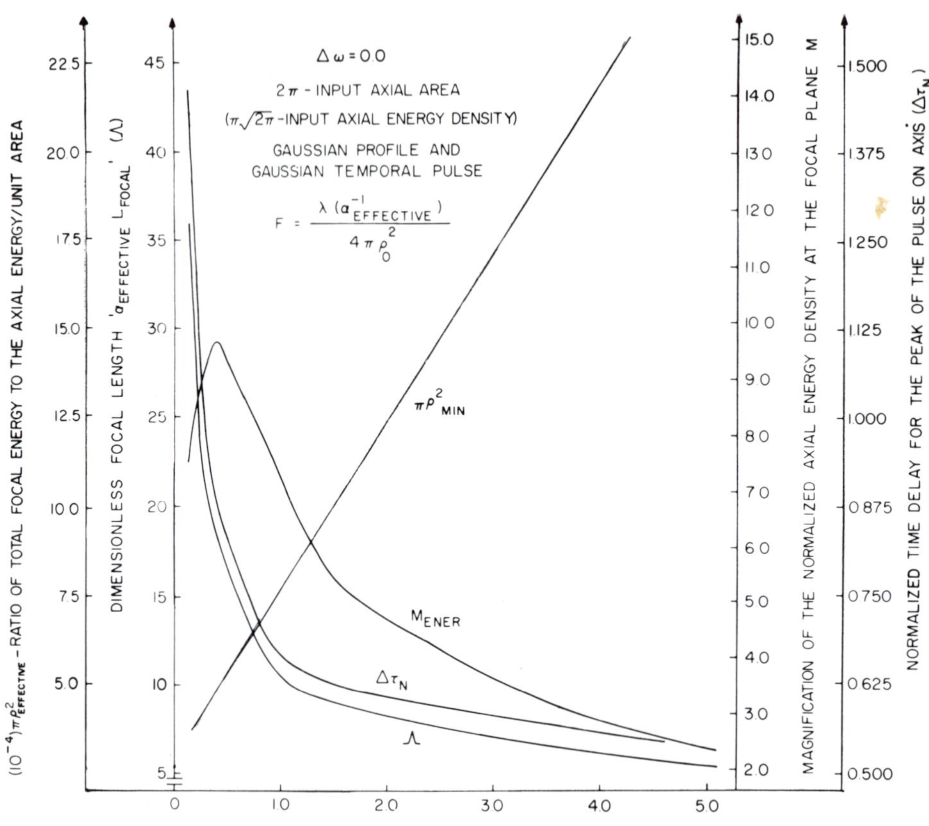

Fig. 16 Principal characteristics of the focal plane as a function of the parameter F: the dimensionless focal length ($\alpha_{eff}\cdot$ L(focal))= Λ(focal); the ratio M of the axial energy per unit area at the focal plane to that at the input plane; the time delay at the focal plane of the peak of the pulse on axis; the ratio of the total field energy to the axial energy per unit area $\pi\rho^2_{eff}$.

(energy per unit area) at the center of the beam was used as definition of focusing as shown in Fig. 14. The experimental dependence of coherent transient SF upon absorption and magnetic detuning in Na is shown in Fig. 15. CW light has no transient SF and focuses for $\Delta\omega < 0$ and defocuses for $\Delta\omega > 0$ (Fig. 15, curve g.[7]) For 2ns pulses and $\alpha L < 3$, slight SF occurs peaked on-resonance (Fig. 15, curve a). At higher absorption the maximum SF occurs for $\Delta\omega < 0$ but SF is still seen on-resonance (Fig. 15, curves a-d). Only for very high absorption ($\alpha L \approx 20$) is there no magnification on-resonance (Fig. 15, curves e and f); presumably absorption destroys the pulse after it

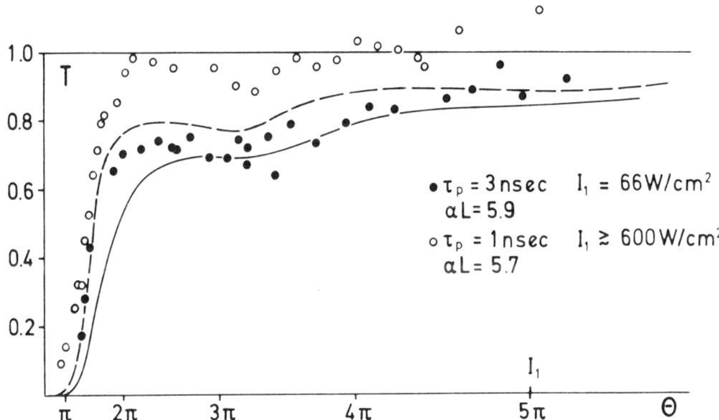

Fig. 17 Pulse energy transmission (output per input in N vs. squared pulse area for 3ns (full dots) and 1ns (open dots) pulses. Curves are the corresponding plane-wave computer simulations.

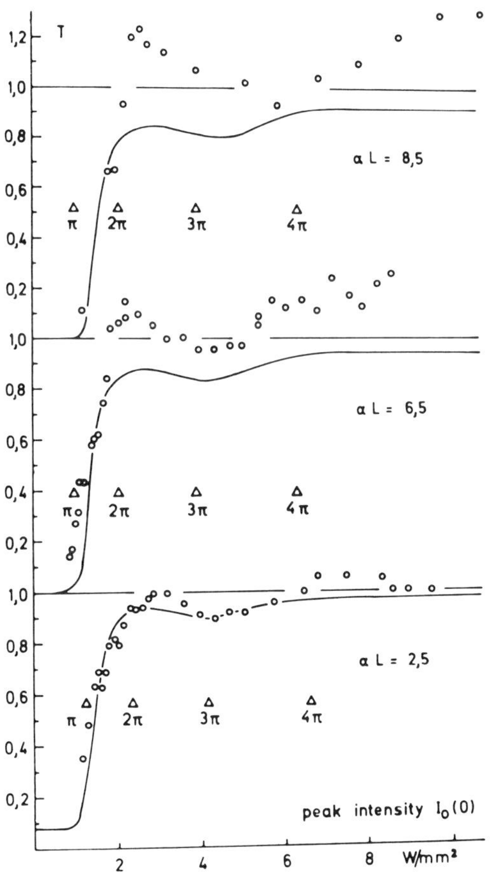

Fig. 18 Energy-density magnification vs. peak intensity for 0.8ns pulses, showing increase of self-focusing with Ne absorption αL. Curves are plane wave computer simulations with $T_2' = 10$ns.

passes its focus. The observation of focusing on-resonance for
coherent pulses and not for cw light clearly illustrates that coherent transient self-focusing is different from previous self-
focusing.[3,7,9] This dependency on the frequency mismatch agrees
perfectly with the trend predicted theoretically.[4,5]

The experimental F value of 10^{-2} corresponds to a Fresnel number   of about 3, i.e. diffraction is strong enough to provide
radial communications, but weak enough that the induced nonlinear
lensing effect overcomes the spreading associated with diffraction.
Numerical simulations predict, for sharpline atomic systems, maxi-
mum magnification of almost 10 for $F = 4\times10^{-4}$ and $\alpha L \sim 18$(~ 44).[4]
Furthermore, if one adjusts in the numerical simulation for sharp-
line two-level atomic system, the absorption in such a manner that
it is equal to the effective absorption that one gets in the broad-
line case (which is the situation in most of the experiments), one
finds that on-resonance coherent SF does indeed occur at an αL of
7.5 with magnification of about 2.45 (see Fig. 16). *This is an
astonishingly good semi-quantitative agreement between theory and
experiment.*

Attempts to observe on-resonance magnifications greater than 2
with large-diameter beams resulted in hot spots in the output for
$\alpha L < 15$. These probably result from focusing of smaller diameter
regions with $F \sim 10^{-2}$ initiated by input phase and intensity variations (small ripples) differing from the theoretically assumed
perfect Gaussian input. Consequently, self-focusing may be unavoid-
ably important in the propagation of coherent optical pulses through
thick absorbers ($\alpha L \geq 15$) even when large-diameter beams (large)
are used on-resonance.

The second experiment was conducted in Heidelberg by Toschek[10]
et al. They investigated the propagation of linearly polarized
1.15 μm light pulses in an absorbing inhomogeneously broadened quasi-
non-degenerate neon discharge ($\alpha L < 9$) and observed both SIT and coherent SF. The transmission T (or magnification) is determined by
graphic integration of about 40 superimposed pulses. Particular
care is taken to produce a perfect Gaussian beam with less than 3%
variation of diameter along the absorber to avoid small-scale self-
focusing.[11] Comparison with numerical simulations of uniform plane
wave Maxwell-Schrödinger equations with $T_1 = 33$ns and $T_2' = 10$ns
calibrates the squared pulse area θ^2, in agreement with power measurements. Data and plane-wave simulations are compared in Fig. 17.
For $\tau_p = 3$ns, experimental T's slightly exceed those from uniform-
plane-wave theory; for 1ns pulses the discrepancy is even larger.
This increase in SF with increased T_2'/τ_p agrees with 3-dimensional
calculations[4] and emphasizes the coherent nature of the effect.
Magnifications up to 40% (Fig. 18) and more recently up to 60% were
seen.[20] Note that the experimental transmissions increase and the

uniform plane-wave simulations decrease with increasing αL.

Data and theory agree that coherent SF can be as large on-resonance as off-resonance; it can occur for a wide range of input areas above π, and is most effective when the relaxation times are long compared to the pulse length. There is no doubt that the recently experimentally observed SF [6,10,19,20] is the new mechanism recently predicted numerically.[2,4,5] Coherent transient SF may also explain previously not understood observed transverse effects.[12-14]

*All the computations performed from June 1973 to January 1977 were jointly supported by F.P. Mattar, the joint services Electronic Program, the Office of Naval Research, the International Division of Mobil Oil Corporation, and the University of Montreal. From February 1976, the Laboratory for Laser Energetics of the University of Rochester extended its support in the form of a post-doctoral fellowship until April 1977 when one of the authors (F.P.M.) transferred to the Department of Physics and Astronomy at the same University.

The theoretical work reported here is part of a dissertation submitted by F.P. Mattar in December 1975 as a partial fulfillment for the Ph.D. degree to Professor M.C. Newstein at the Department of Electrical Engineering and Electrophysics, Polytechnic Institute of New York, Farmingdale, New York 11735.

**Resident visitor 1975-76 at Philips Research Laboratory, Eindhoven, The Netherlands.

***Work supported by the Deutsche FORSCHUNGSGEMEINSCHAFT.

References

1. S.L. McCall and E.L. Hahn, Phys. Rev. Lett. *18*, 308 (1967) and Phys. Rev. *183*, 457 (1969).
2. N. Wright and M.C. Newstein, Opt. Commun. *1*, 8 (1973); M.C. Newstein and F.P. Mattar, J. Opt. Soc. Am. *65*, 1181 (1975) and Opt. Commun. *18*, 70 (1976).
3. J. Marburger, in *Progress in Quantum Electronics*, Vol. 4, ed. J.H. Sanders and S. Stenholm (Pergamon, New York, 1975) p.35
4. F.P. Mattar and M.C. Newstein, IEEE J. Quant. Electr. *13*, 507 (1977); in *Cooperative Effects in Matter and Radiation*, ed. C.M. Bowden, D.W. Howgate and H.R. Robl (Plenum Press, New York, 1977) p.139; and in *Recent Advances in Optical Physics*, ed. B. Havelka and J. Blabla (Society of Czechoslovak Mathematicians and Physicists, Prague, 1976) p.299.
5. F.P. Mattar and M.C. Newstein (unpublished calculations).

6.   H.M. Gibbs, B. Bölger and L. Baede, Opt. Commun. *18*, 199 (1976).

7.   J.E. Bjorkholm and A. Ashkin, Phys. Rev. Lett. *32*, 28 (1974).

8.   A. Javan and P.L. Kelley, IEEE J. Quant. Electr. *2*, 470 (1966);
     J.A. Fleck, Jr. and R.L. Carman, Appl. Phys. Lett. *22*, 546 (1973).

9.   D. Grischkowsky, Phys. Rev. Lett. *24*, 866 (1970); D. Grischkowsky and J. Armstrong, in *Coherence and Quantum Optics,* ed. L. Mandel and E. Wolf (Plenum Press, New York, 1973) p. 829.

10. W. Krieger, G. Gaida and P.E. Toschek, Z. Physik B *25*, 297 (1976).

11. V.I. Bestalov and V.I. Talanov, JETP Lett. *3*, 307 (1966); B.R. Suydam, in *Laser Induced Damage in Optical Material* (N.B.S. Special Publication 387, 1973) p.42; IEEE J. Quant. Electr. *10*, 837 (1974) and *11*, 225 (1975).

12. Perhaps the reason this effect has not been seen clearly before is that it becomes significant only after the reshaping region, i.e. $\alpha L < 5$, on which most SIT experiments have concentrated. Much higher absorption ($\alpha L \sim 25$) was used by McCall and Hahn in the first SIT experiment; they reported hot spots in the output (also seen in the Na experiment for large input diameters) which they attributed to transverse instabilities. Zembrod and Gruhl (Ref. 13) used $\alpha L \sim 11$ and $F \sim 10^{-2}$, so self-focusing should be beginning; their Fig. 1c has an output about 14% higher than the input and most of their data have higher transmissions than expected by uniform plane-wave simulations. Rhodes and Szöke (Ref. 14) also reported transverse effects seen in SF_6 with $\alpha l \sim 20$, which may have resulted from self-focusing.

13. A. Zembrod and T. Gruhl, Phys. Rev. Lett. *27*, 287 (1971).

14. C.K. Rhodes and A. Szöke, Phys. Rev. *184*, 25 (1969); C.K. Rhodes, A. Szöke and A. Javan, Phys. Rev. Lett. *21*, 1151 (1968); F.A. Hopf, C.K. Rhodes and A. Szöke, Phys. Rev. B *1*, 2833 (1970).

15. The concept of "effective absorption coefficient" has been introduced by Gibbs and Slusher in their discussion of coherent propagation in the regime where the inhomogeneous width $(T_2^*)^{-1}$ was greater than the homogeneous width $(T_2)^{-1}$ but less than the inverse pulse duration $(\tau_p)^{-1}$. Under these conditions, the pulse behavior corresponds to that in medium with a homogeneously broadened line. [See for example, H.M. Gibbs and R.E. Slusher, Appl. Phys. Lett. *18*, 505 (1971); Phys. Rev. A *5*, 1634 (1972); Phys. Rev. A *6*, 2326 (1972).]

16. S.L. McCall, Phys. Rev. A *9*, 1515 (1974).

17. F.A. Hopf and M.O. Scully, Phys. Rev. B *1*, 50 (1970).

18. For details see F.P. Mattar, Ph.D. Thesis, *Transverse Effects Associated with the Propagation of Coherent Optical Pulses in Resonant Absorbing Media* (Polytechnic Institute of New York, December 1975)(distributed by University Microfilms, Ann Arbor, Mich., U.S.A.).

19. H.M. Gibbs, B. Bölger, F.P. Mattar, M.C. Newstein, G. Forster and P.E. Toschek, Phys. Rev. Lett. *37*, 1743 (1976).

20. F.P. Mattar, G. Forster and P.E. Toschek, Spring meeting of the German Physical Society, Mayence, West Germany, 1977.

21. H. Kogelnik and T. Li, Appl. Opt. *5*, 1550 (1966).

PHOTON-ECHO ZEEMAN QUANTUM BEATS IN CESIUM HYPERFINE LEVELS*

T. Baer and I.D. Abella

University of Chicago, Chicago, Illinois

1. INTRODUCTION

Photon echo quantum beats were first reported by Lambert et al. [1] who observed echo modulations as a function of excitation pulse separation in ruby at low temperature. The modulations in the echo signal arise when a set of nondegenerate levels in the excited or ground state simultaneously participates in the echo process giving rise to interference terms in the coherent oscillating dipole moment. In ruby the quantum beats were attributed to Cr-Al superhyperfine interactions [2]. A semi-classical analysis of the general problem of photon echo quantum beats was given by Lambert, Compaan, and Abella [3], using methods of Gordon et al. [4]. Subsequent theoretical work was doen by Schenzle et al. [5] while observations of echo quantum beats have been reported in NH_2D [6] and in cesium vapor [7].

A variety of systems which exhibit quantum beats in fluorescence have been studied over the past twenty years and this topic has been discussed from several perspectives [8]. For example, Haroche et al. [9] have observed beats in cesium fluorescence following pulsed laser excitation. Recently, quantum beats in superfluorescence have been reported [10] in cesium vapor with a good signal-to-noise ratio.

In this paper we report the first observation of Zeeman hyperfine photon echo quantum beats in an atomic vapor. The multilevel system in this work is cesium vapor in a weak magnetic field. We apply the theoretical approach of Lambert et al. [3] to the specific case of the ground and excited states split by an external magnetic field.

2. EXPERIMENTAL DETAILS

In our initial work in cesium [7], we generated echoes on transitions between the ground state hyperfine levels $6S_{1/2}$ (F = 3,4) and the second excited P doublet, $7P_{3/2, 1/2}$ corresponding to 4555 Å and 4591 Å, respectively. The measured echo lifetime of 150 nsec is sufficiently long to permit convenient observation of modulations. Bolger and Diels have reported [11] echoes on the $6S_{1/2}$ - $6P_{3/2}$ transition at 8521 Å with a lifetime of 28 nsec.

The present work is confined to the $6S_{1/2}$ to $7P_{1/2}$ transition. Figure 1 shows the experimental arrangement. A pulsed nitrogen laser, N.R.G. Inc. model .5-5-150 is used to pump a dye laser operating with 7-diethyamino-4-methylcoumarin in ethanol. The laser pulse is ∿5 nsec in duration, and the tuning element is a grating with intracavity etalon, resulting in a linewidth of 3 GHz, which resolved the ground state hyperfine splitting of 9.2 GHz. A White cell delay line provides the second pulse.

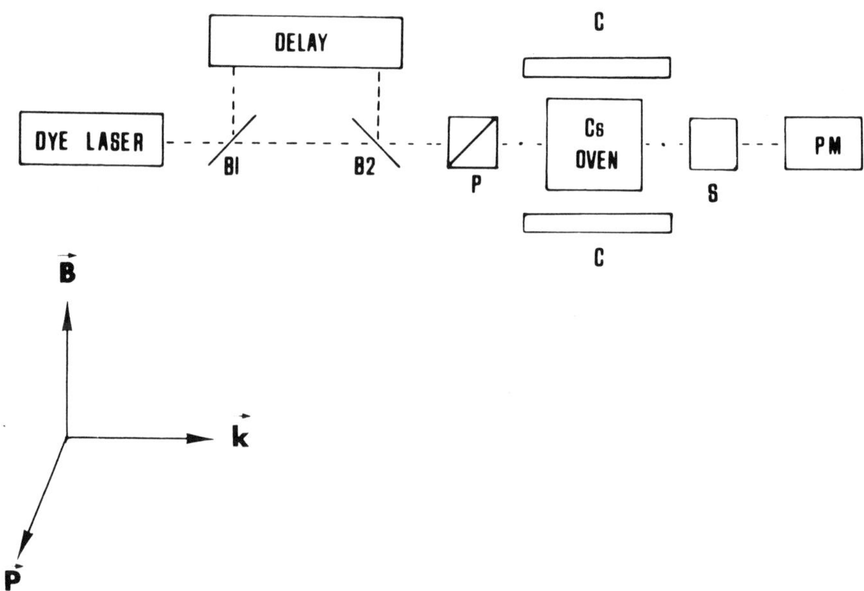

Fig. 1. Schematic diagram of the apparatus used in the Zeeman quantum beats experiments. B_1 and B_2 are beam splitters, P is a linear polarizer, S is a Kerr-cell shutter, C indicates Helmholtz coil, and PM is a photomultiplier. The coordinate system indicates the relative orientation of the excitation pulse polarization (P) and propagation (k) direction with respect to the magnetic field (B).

The cesium is contained in an outgassed glass cell which was sealed following distillation at a background pressure of 10^{-6} Torr. The cell is heated to about 80°C, which corresponds to a vapor pressure of 5×10^{-4} Torr. or 3×10^{12} atoms/cm^3. The optical path length in the cell is 2 cm. The pulses are incident collinear on the cell, with their planes of polarization parallel. The component of the echo signal which is polarized parallel to the excitation pulse polarization is transmitted by a Kerr-cell electro-optical switch. The Kerr-cell is required to reduce scattered light at the photomultiplier. It should be emphasized that the echo signal measured in this work and the excitation pulses all have the same polarization direction.

A magnetic field of up to 50 G can be applied with a pair of Helmholtz coils. The magnetic field direction is perpendicular to both the polarization and propagation vectors of the incident light. No attempt was made to compensate for the earth's magnetic field.

The signals were fed to a Chronetics 116 linear gate which was gated to pass the echo only. The echo pulse was stretched, the pulse height was digitized, and was averaged. Each data point represents about 100 echo pulses.

In our experiments we measure the echo intensity as a function of magnetic field at a fixed delay between the incident excitation pulses. This experiment is quite similar in form to those performed by Schenck et al. [12] in their study of time resolved fluorescence of Ca and Ba vapors in a magnetic field. However, the modulations in the echo experiments will be different from those observed in fluorescence.

3. PHYSICAL DISCUSSION

Figure 2 contains a level diagram indicating the hyperfine structure of the $6S_{\frac{1}{2}}$ and $7P_{\frac{1}{2}}$ states of Cs^{133}. This isotope of cesium has a nuclear spin of 7/2. If we apply a magnetic field, both upper and lower states split into their $2F + 1$ components, as illustrated.

We now consider applying a photon echo excitation sequence of two resonant pulses separated by a time delay τ. The pulse widths Δt_1, Δt_2 are assumed to be short enough so that all allowed transitions from the ground state to the excited state are in resonance. That is, Δt_1, $\Delta t_2 \ll 1/\Delta\omega$, where $\Delta\omega$ is a typical ground or excited state splitting.

To illustrate the process responsible for the echo beats we imagine an atom which is initially in a particular ground state

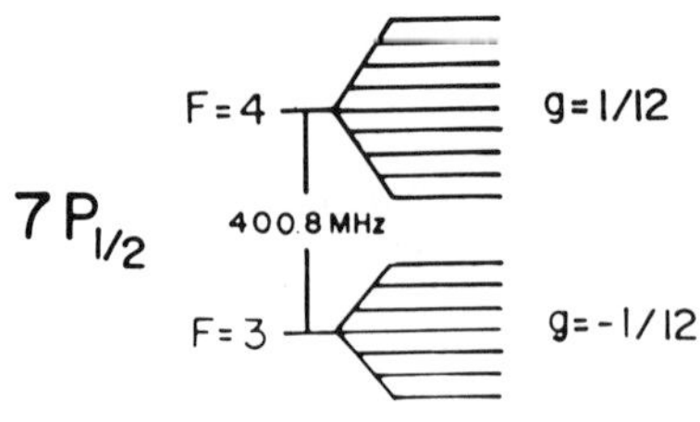

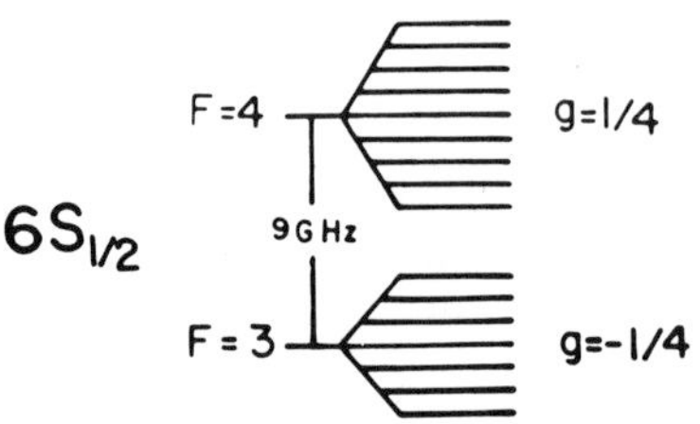

Fig. 2. Schematic diagram of the $6S_{1/2}$ and $7P_{1/2}$ hyperfine level structure of cesium showing zero field splittings and the theoretical g factors for the different F levels. The splittings between the different m levels occur when a magnetic field is applied. (The intervals are not drawn to scale.)

$|6S_{1/2}, F = 4, m_1\rangle$. We assume the first pulse couples the ground state to an excited state $|7P_{1/2}, F = 4, m_2\rangle$ with the same total angular momentum. The second pulse introduces another set of excited and ground state levels by coupling this initial excited state $|7P_{1/2}, F = 4, m_2\rangle$ to a ground state with a different magnetic quantum number $|6S_{1/2}, F = 4, m_3\rangle$. The second pulse further couples this new ground state $|6S_{1/2}, F = 4, m_3\rangle$ to a different excited state $|7P_{1/2}, F = 4, m_4\rangle$. We thus have a pair of two level systems: one defined by the first pulse interaction with the atom, the other determined by the 'remixing' caused by the second pulse. The frequency of the photon echo quantum beat is determined by the energy *difference* between the *pair* of two level systems mixed by the first and second pulses, i.e. $(\omega_{m_2} - \omega_{m_1}) - (\omega_{m_4} - \omega_{m_3})$ where $\hbar\omega_{m_1}$ is the energy of the state $|6S_{1/2}, F = 4, m_1\rangle$ etc.

If the first and second pulses couple the same levels in the ground and excited states ($m_1 = m_3$ and $m_2 = m_4$ in the above example) then there will be no quantum beats in the echo signal since

$$(\omega_{m_2} - \omega_{m_1}) - (\omega_{m_4} - \omega_{m_3}) = 0 .$$

The atoms in a cesium vapor sample are more accurately described as being initially in a state which is a superposition of all the ground level m states. The interaction of the two-pulse sequence with such an initial state is of course much more complicated than the previous example. Nevertheless, the expression for the photon echo intensity contains terms which can be identified with the processes described above for each possible coupling of

ground to excited state. Those terms which represent different
levels mixed by the second pulse give rise to modulations in the
echo signal. Those terms which represent the same levels mixed by
the two pulses do not give rise to modulations.

Chow et al. [13] have discussed the existence of photon echo
quantum beats due to interference effects arising in situations
where there are many atoms in a volume λ^3. These beats arise from
interference between different atoms in the sample. We have performed our experiments using number densities less than one atom
per wavelength cubed and the analysis of Chow et al. does not apply
to the experiment reported here.

4. THEORETICAL TREATMENT

Lambert et al. [3], using a density matrix approach, derived
an expression for the time evolution of the atomic polarization in
a multilevel system, assuming two-pulse resonant excitation. The
square of the atomic polarization evaluated at the time of the echo
pulse is proportional to the echo intensity. Using this result, we
calculate the echo intensity as a function of magnetic field for an
atom whose hyperfine levels are split by the Zeeman effect. In our
experiments the applied field is too weak to uncouple nuclear and
electronic angular momenta.

We assume that all levels in the ground state and the excited
state (denoted by a and b, respectively) can be characterized by
the total angular momentum quantum number F and its z-axis projection m. We take the z-axis to be parallel to the magnetic field
direction. Two resonant pulses of the form $\underline{E}(t) = \underline{\varepsilon}(t) \cos (\omega t)$
are applied, separated by a time delay τ. The duration of the pulses
is short enough so that all allowed transitions from the ground to
the excited state are in resonance with the applied field. The
excitation pulses are polarized in the x-direction and propagate in
the y-direction as shown in Fig. 1. With these assumptions Eq. (5)
of Lambert et al. becomes

$$\langle \underline{P}(2\tau) \rangle = \frac{1}{2} \sum \exp \{-i(\Delta\omega_{m_1 m_2} - \Delta\omega_{m_3 m_4})\tau\}$$

$$\times [\langle aF_1 m_1 | \underline{P} | bF_2 m_2 \rangle \langle bF_2 m_2 | \sin (A_2 P_x) | aF_3 m_3 \rangle$$

$$\times \langle aF_3 m_3 | \sin (2A_1 P_x) | bF_4 m_4 \rangle \langle bF_4 m_4 | \sin(A_2 P_x) | aF_1 m_1 \rangle$$

$$+ \text{C.C.}] \qquad\qquad (1)$$

where

$$A_1 = \hbar^{-1} \int_0^{\Delta t_1} \underline{\varepsilon}_1(t)\,dt \;, \qquad A_2 = \hbar^{-1} \int_\tau^{\tau+\Delta t_2} \underline{\varepsilon}_2(t)\,dt \;,$$

$$\Delta\omega_{m_1 m_2} = \omega_{F_2 m_2}^b - \omega_{F_1 m_1}^a - \omega \;.$$

The summation is over all allowed values of m_1 through m_4 and F_1 through F_4 and $\hbar\,\omega_{F_1 m_1}^a$ is the energy of the level $|aF_1 m_1\rangle$.

The physical interpretation we made in section 2 is suggested by the structure of Eq. (1). The term $\langle aF_3 m_3 | \sin(2A_1 P_x) | bF_4 m_4 \rangle$ indicates that the first pulse couples the ground state $\langle aF_3 m_3 |$ to the excited state $|bF_4 m_4\rangle$. The second pulse introduces a new pair of levels $|aF_1 m_1\rangle$ and $|bF_2 m_2\rangle$, as described in the last section. The echo modulation is equal to the difference between the frequencies corresponding to each two-level system. If the two pulses mix the same levels, then no echo modulations will occur since the argument of the exponential $\omega_{m_1 m_2} - \omega_{m_3 m_4}$ will be zero.

Equation (1) can be solved easily when F_a and F_b are small (i.e. $F_a=0$, $F_b=1$ or $F_a=\frac{1}{2}$, $F_b=\frac{1}{2}$). In cesium $F_a = 3$ or 4 depending on which ground state is in resonance with the laser while $F_b = 3$ *and* 4 since both excited states participate in the echo process. The solution to Eq. (1) is much more complicated in this case.

The computation simplifies considerably while still yielding accurate results if we expand the sine matrix elements in Eq. (1) and consider only the first order terms. That is, we expand

$$\langle bF_2 m_2 | \sin(A_2 P_x) | aF_3 m_3 \rangle = A_2 \langle bF_2 m_2 | P_x | aF_3 m_3 \rangle$$

$$- A_2^3 \langle bF_2 m_2 | P_x^3 | aF_3 m_3 \rangle / 6 + \ldots \qquad (2)$$

If the products $A_2 \langle bF_2 m_2 | P_x | aF_3 m_3 \rangle$ are small for all values of F_3, m_3, F_2, and m_2, then the first order terms will dominate. In our experiments we estimate the $A_1 \langle bF_2 m_2 | P_x | aF_3 m_3 \rangle$ product is $\sim.1$.

Since we measure the intensity of the echo signal polarized parallel to the excitation pulse polarization, we consider only $\langle P_x(2\tau) \rangle$. Equation (1) thus simplifies to

$$\langle P_x(2\tau) \rangle = A_1 A_2^2 \sum \exp\{-i(\Delta\omega_{m_1 m_2} - \Delta\omega_{m_3 m_4})\tau\}$$

$$\times \left[\langle aF_1 m_1 | P_x | bF_2 m_2 \rangle \langle bF_2 m_2 | P_x | aF_3 m_3 \rangle \right.$$

$$\times \left. \langle aF_3 m_3 | P_x | bF_4 m_4 \rangle \langle bF_4 m_4 | P_x | aF_1 m_1 \rangle + \text{C.C.} \right] \;, \qquad (3)$$

where now the summation is restricted by dipole selection rules so that: $m_1 = m_2 \pm 1$, $m_2 = m_3 \pm 1$, etc. These restrictions greatly decrease the number of terms in the summation. The matrix elements in Eq. (3) can be reduced to forms demonstrating explicitly their F and m dependence [14]. For given values of m_1 to m_4 and F_1 to F_4, the coefficient multiplying the exponential can be determined. The results are then summed for all allowed values of m_1 to m_4 and F_1 to F_4. We find for the F=4 ground state of cesium that Eq. (3) becomes

$$<P_x> = <P>^4 A_1 A_2^2 \{588 \cos (\Omega_2 \tau/6) + 924 \cos (\Omega_2 \tau/2)$$

$$+ 1304 + \cos(\Omega_1 \tau) [206 \cos(\Omega_2 \tau/2) + 394 \cos(\Omega_2 \tau/3)$$

$$+ 524 \cos(\Omega_2 \tau/6) + 364 - 180 \cos(2\Omega_2\tau/3) - 126 \cos(5\Omega_2 \tau/6)$$

$$- 56 \cos(\Omega_2 \tau)]\} \quad , \tag{4}$$

where <P> is the reduced matrix element for the $6S_{1/2}$ to $7P_{1/2}$ transition. $\Omega_1/2\pi$ = 400.8 MHz and $\Omega_2 = \mu_0 B/\hbar$. We have assumed that the level spacings vary linearly in the field. The term $\cos(\Omega_1\tau)$ arises from the zero field hyperfine splitting of the excited state (see Figure 2). The coefficients of the cosine terms are a measure of the total line strength of transitions with a particular magnetic field dependence.

Figure 3 shows some experimental data, below which is the beat spectrum given by Eq. (4). The time delay τ between the two

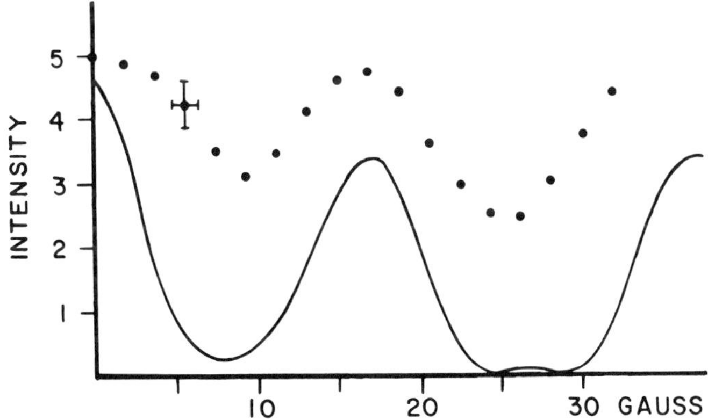

Fig. 3. Echo intensity versus magnetic field for the $6S_{1/2}$ (F=4) ground state of cesium. The black dots are experimental points with the typical error indicated. The solid curve is a graph of $|<P_x(2\tau)>|^2$ calculated from Eq. (4) with $\cos(\Omega_1\tau)$ = -.6. The delay between excitation pulses was approximately 80 nsec.

excitation pulses is ∿80 nsec. The frequency of the dominant beat and the position of maxima appear to be predicted by the theory in the linear approximation. Since the modulations in the experimental data match the theoretical curves we conclude that the data is consistent with the values of $g_a = \pm 1/4$, $g_b = \pm 1/12$.

The experimental data in Fig. 3 indicates that the echo modulations are superimposed on a large DC signal. The DC signal is probably due to inhomogeneities in the magnetic field. With sufficiently homogeneous fields the modulations in the echo signal should approach 100% as indicated by the theoretical curves.

5. CONCLUSIONS

The theory of Lambert et al. predicts the observed quantum beats in the photon echo intensity versus magnetic field graph of the $6S_{\frac{1}{2}}$ (F=4) to $7P_{\frac{1}{2}}$ (F=3,4) transition in cesium. Other relative orientations of the excitation pulse polarization direction and the magnetic field lead to a rotation of the photon echo polarization direction. This effect is also well described by applying the theory of Lambert et al. [3] and will be discussed elsewhere [15].

Acknowledgement

We thank Steven Aoki for valuable discussions and a critical reading of the manuscript.

*Work supported in part by the National Science Foundation, DMR-73-02450-A01.

References

1.   L.Q. Lambert, A. Compaan and I.D. Abella, Phys. Letters *30A*, 153 (1969).
2.   I.D. Abella, A. Compaan and L.Q. Lambert, in *Laser Spectroscopy*, ed. R.G. Brewer and A. Mooradian (Plenum Press, New York, 1974) p. 457; L.Q. Lambert, Phys. Rev. *B7*, 1834 (1973); D. Grischkowsky and S.R. Hartmann, Phys. Rev. *B2*, 60 (1970).
3.   L.Q. Lambert, A. Compaan and I.D. Abella, Phys. Rev. *A4*, 2022 (1971).
4.   J.P. Gordon, C.H. Wang, C.K.N. Patel, R.E. Slusher and W.J. Tomlinson, Phys. Rev. *179*, 294 (1969).

5. A. Schenzle, S. Grossman, R.G. Brewer, Phys. Rev. *A13*, 1891
    (1976).
6. R.L. Shoemaker and F.A. Hopf, Phys. Rev. Letters *33*, 1527 (1974).
7. T. Baer and I.D. Abella, Phys. Letters *59A*, 371 (1976).
8. U. Fano and J. Macek, Rev. Mod. Phys. *45*, 553 (1973) and refer-
    ences therein.
9. S. Haroche, J.A. Paisner, and A.L. Schawlow, Phys. Rev. Letters
    30, 948 (1973).
10. Q.H.F. Vrehen, H.M.J. Hikspoors and H.M. Gibbs, Phys. Rev.
    Letters *38*, 764 (1977).
11. B. Bolger and J.C. Diels, Physics Letters *28A*, 401 (1968).
12. P. Schenck, R.C. Hilborn and H. Metcalf, Phys. Rev. Letters *31*,
    189 (1973).
13. W.W. Chow, M.O. Scully and J.O. Stoner, Jr., Phys. Rev. *A11*,
    1380 (1975).
14. E.U. Condon and G.H. Shortley, *The Theory of Atomic Spectra*
    (Cambridge University Press, 1935), Chap. 3.
15. T. Baer and I.D. Abella, Phys. Rev. *A16*, 2093 (1977).

OBSERVATION OF A RAMAN ECHO

P. Hu, S. Geschwind and T. M. Jedju

Bell Laboratories, Murray Hill, N. J.

The microscopic decay time T_2 of induced coherence in a two-level system is often an important parameter in characterizing interactions between the two-level systems or between them and their surroundings. This has been traditionally studied by spin echo [1] and photon echo experiments.[2] In this powerful class of experiments, radiation which is resonant with the two-level system is required to induce coherence between the levels. It would be advantageous, however, in those many cases where resonant radiation is not readily available or its use inconvenient, to be able to induce the coherence in the system by a nonresonant method. It has been shown in a variety of experiments in past years that coherence can be produced by two-wave light mixing or by stimulated Raman scattering.[3]

However, reported measurements of dephasing times by these techniques have generally involved only a measurement of the free induction decay or the *macroscopic* dephasing time. This is the Fourier transform of the linewidth and corresponds to the linewidth observed in spontaneous Raman scattering. To measure the microscopic dephasing time or irreversible loss of phase memory of the constituent homogeneous packets of an inhomogeneous line would require an echo experiment which has been dubbed a Raman echo.[4] We recently reported the first such observation of a two-pulse Raman echo [5] and present here further new details including the observation of a three-pulse stimulated Raman echo.

The Raman echo reported here involves scattering between the spin levels of bound donor states in CdS. Spin flip Raman scattering was first suggested by Yafet [6] following a suggestion by

Wolff on Raman scattering from Landau levels from carriers in a
semiconductor.[7] Spontaneous [8] and stimulated [9] spin-flip
scattering were first observed in InSb and later in CdS.[10,11]
For two time reversed states $|a>$ and $|b>$ (see Fig. 1a) in cubic
symmetry, an effective spin flip Raman Hamiltonian was first given
by Yafet [6,12] as

$$\mathcal{H}^{(2)}_{sp} = \frac{1}{2}\,\alpha\underline{\sigma}\,\cdot\,(\underline{E}_L{}^{\times}\underline{E}_R)e^{i[(\omega_L-\omega_R)t\,-\,(\underline{k}_L-\underline{k}_R)\,\cdot\,\underline{r}]} + c.c. \qquad (1)$$

where c.c. denotes a complex conjugate. $\underline{E}_L$ and $\underline{E}_R$ are the laser and
Raman fields, respectively, with frequencies ω_L and ω_R and wave
vectors $\underline{k}_L$ and $\underline{k}_R$; $\underline{\sigma}$ are the Pauli matrices; $\underline{r}$ is the positron vector
of the electron center; α is related to the spontaneous differential
Raman cross section $d\sigma/d\Omega$, by $d\sigma/d\Omega = 4|\alpha|^2(\omega_L+\omega_{ba})^3\omega_L/c^4$. α becomes
very large when $\hbar\omega_L$ is near resonant to an excited state, which is
an exciton bound to the neutral donor in our case.[10] The form of
Eq. 1 is such that $\alpha\underline{E}_L{}^{\times}\underline{E}_R$ acts as an effective R.F. magnetic
field [13] of frequency $\omega_L-\omega_R = \omega_{ba}$ which therefore generates a
coherent transverse magnetization $\propto \sigma_T$, which may be calculated
using the standard expressions of magnetic resonance. σ_T is given
by

$$\sigma_T = \sin\theta\cos[(\underline{k}_L-\underline{k}_R)\,\cdot\,\underline{r}], \qquad (2)$$

where θ is the "tipping" angle given by

$$\theta = \frac{1}{2\hbar}\frac{c^2}{\omega_L{}^2}\,(\frac{d\sigma}{d\Omega})^{1/2}\,E_R E_L\,\Delta t, \qquad (3)$$

where Δt is the duration of the light pulse and $(\omega_L\pm\omega_{ba}) \sim \omega_L$. The
coherence, or σ_T, which is thus established, may now be probed by
Stokes or anti-Stokes coherent forward scattering with another laser
beam at a different frequency ω_p. The fraction of the probe beam
power P_0 converted to Stokes or anti-Stokes sideband power P_S is
given by

$$\frac{P_S}{P_0} = \frac{1}{16}(\frac{d\sigma}{d\Omega})_{\omega_p}\frac{\lambda^2\eta^2L^2}{\varepsilon}\,(\sin^2\theta)I. \qquad (4)$$

λ is the free space wavelength; ε the dielectric constant; L the
sample length along the beam; and η is the impurity concentration
times the appropriate spin Boltzmann factor which is 0.45 in our
case. I is a phase matching factor given by

$$I = \frac{2(1-\cos\Delta\underline{k}\,\cdot\,\underline{L})}{(\Delta\underline{k}\,\cdot\,\underline{L})^2}, \qquad (5)$$

where $\Delta\underline{k} = [(\underline{k}_L-\underline{k}_R) - (\underline{k}_L^P-\underline{k}_R^P)]$ and where k_L^P and k_R^P are the wave vectors of the probe beam and its associated Raman sideband, respectively. I is of the order unity in our case. If the probe beam is set in time immediately after pulse I in Fig. 1, it will measure the free induction decay. To observe a Raman echo a second light pulse labeled pulse II in Fig. 1b is applied a time τ after pulse I whose effect is to rephase the spins at time 2τ, and this rephasing or echo is now observed by scattering of the probe beam by σ_T at time 2τ. In addition to its dependence on τ which provides a direct determination of T_2, the strength of the echo depends upon the strength of the excitation pulses and is given by the familiar expression

$$\sigma_T^{\,2}(t=2\tau) \propto \sin^2\theta_I \sin^4(\frac{\theta_{II}}{2}), \qquad (6)$$

where θ_I and θ_{II} are the tipping angles produced by pulses I and II, respectively.

In our initial experimental arrangement,[5] coherence was established in the spin system by a 5 ns dye laser pulse of 100 W peak power at 4905Å operating simultaneously in two longitudinal

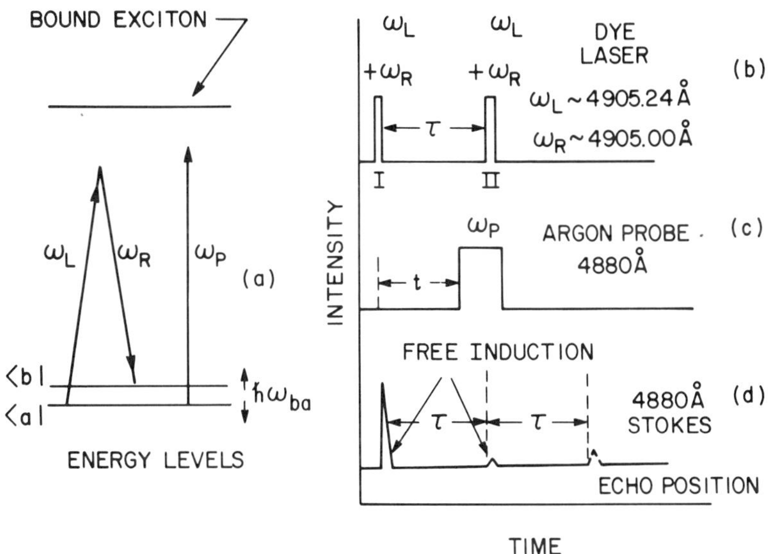

Fig. 1.   (a) Energy levels in a Raman echo experiment. The excited state here is an exciton bound to a neutral donor.
(b) The excitation pulse sequence of the dye laser.
(c) The Argon probe timing.
(d) Expected behavior of 4880Å Stokes free induction and echo signals.

modes ω_L and ω_R, each of spectral width 1.5 GHz, where $\omega_L - \omega_R$ = 32 GHz. The magnetic field which was perpendicular to the $\hat{c}$ axis of the crystal was adjusted to 12.75 kG to tune the donor spin splitting ω_{ab} to 32 GHz. The CdS samples had a donor concentration of $\sim 8 \times 10^{15}/cm^3$ and a beam path length of 0.5 mm. The exciting 4905Å dye laser beam and the 4880Å argon laser probe beam were made colinear and focused to a diameter of 100 μm in the crystal where they traveled along the $\hat{c}$ axis. The sample was cooled to 1.6 K. The probed signal (4880Å Stokes line) was observed in the *forward* direction through a triple pass Fabry-Perot interferometer in series with two 4880Å interference filters of 10Å bandwidth to block the exciting dye laser pulses.

When the probe pulse is set in the region of pulse I, the free induction signal is observed as shown in the inset of Fig. 2. The decay time is 3 ns which corresponds to an EPR linewidth of $\sim$22 G. The scattered Stokes and anti-Stokes peak intensities were of comparable magnitude and proportional to the intensity of the probing beam. The fractional probe sideband power vs. the intensity of the 4905Å excitation pulse is plotted in Fig. 2. From Eqs. 2-4, $P_S/P_0 \sim \sigma_T^2 \sim \sin^2\theta$ where θ is proportional to the intensity of

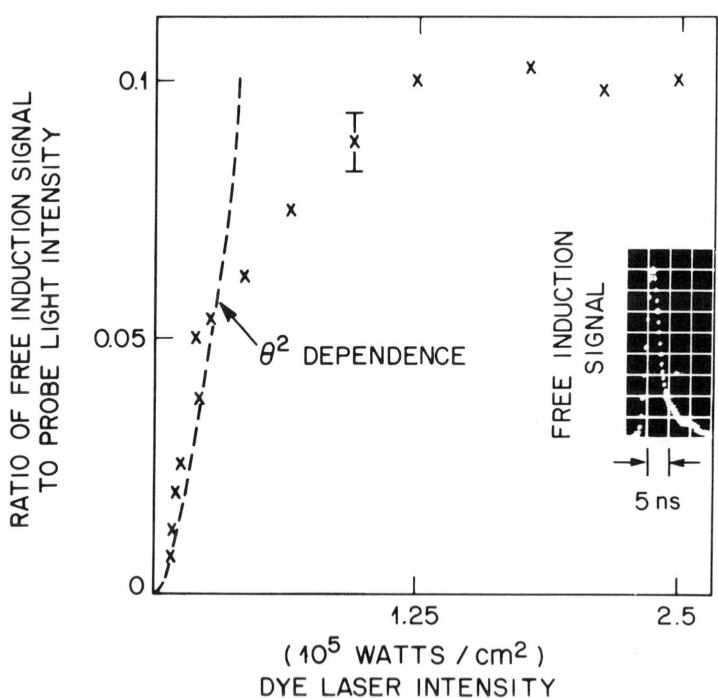

Fig. 2. Ratio of free induction signal to probing light intensity as a function of dye laser intensity. Dashed curve indicates the initial θ^2 dependence.

the exciting beam since $E_R \sim E_L$. This is seen to be the case at low exciting intensities where the signal is seen to have a θ^2 dependence. At higher intensities the signal saturates contrary to Eq. 3 which applies only to an ideally monochromatic pulse on resonance. When account is taken of the 1.5 GHz spectral width of the dye laser which is much larger than $1/\Delta t$, a maximum tipping angle, $\theta \sim \pi/2$, is predicted. The calculated maximum value of P_S/P_0 of 18% for $\theta \sim \pi/2$ is in reasonable agreement with the observed 10% conversion efficiency, given the uncertainty in several of the parameters.

To produce an echo, a rephasing pulse II is applied at a time τ after pulse I. This second pulse is derived from pulse I by means of an optical delay line and also made collinear with pulse I. When the probe pulse is set at time 2τ, the Raman echo signal is observed as shown in Fig. 3.

As is evident from our above discussion of our first successful Raman echo experiment, advantage was taken of two-wave mixing

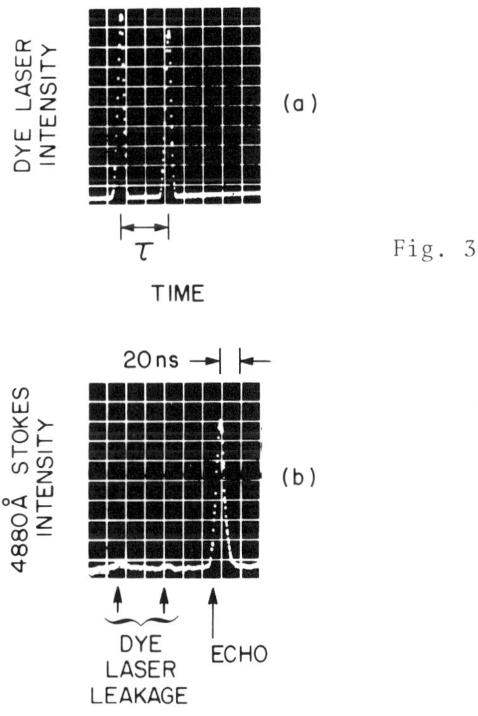

Fig. 3. a) Dye laser pulses separated by τ = 54 ns.
b) Dye laser leakage and the 4880Å Stokes echo signal at $t = 2\tau$.

by having our nitrogen laser pumped dye laser operated simulta-
neously in two longitudinal modes, ω_L and ω_R, with $\omega_L-\omega_R = \omega_{ba}$,
the electron spin Zeeman splitting. Obviously, it is more conven-
ient and desirable to obtain ω_R from ω_L directly by means of
stimulated Raman scattering. Another shortcoming in our first
experiment was the use of an optical delay line which only provided
discrete and very limited time separation between the two excitation pulses. This made the measurement of dephasing time unneces-
sarily complicated. A far more ideal arrangement is the use of a
single mode CW laser whose output is electronically gated to obtain
the necessary pulse sequences as has been done recently in the
photon echo and other optical coherent experiments.[14] An additional incentive for this new approach is the ability to generate
a three pulse stimulated echo and possibly other coherent phenomena
involving Raman processes. All of these advantages are being
realized in our latest experiments, in which the roles of the
4905Å dye laser and the 4880Å argon laser are *reversed*. Various
30 ns pulse sequences electronically gated from the cavity-pumped
Argon laser are used to create the spin coherence, while the dye
laser now operated in a single longitudinal mode and after large
attenuation is used to probe the coherence. The new pulse
sequence for the two-pulse echo experiment is shown schematically
in Fig. 4. Note that the 30 ns excitation pulses and, therefore,

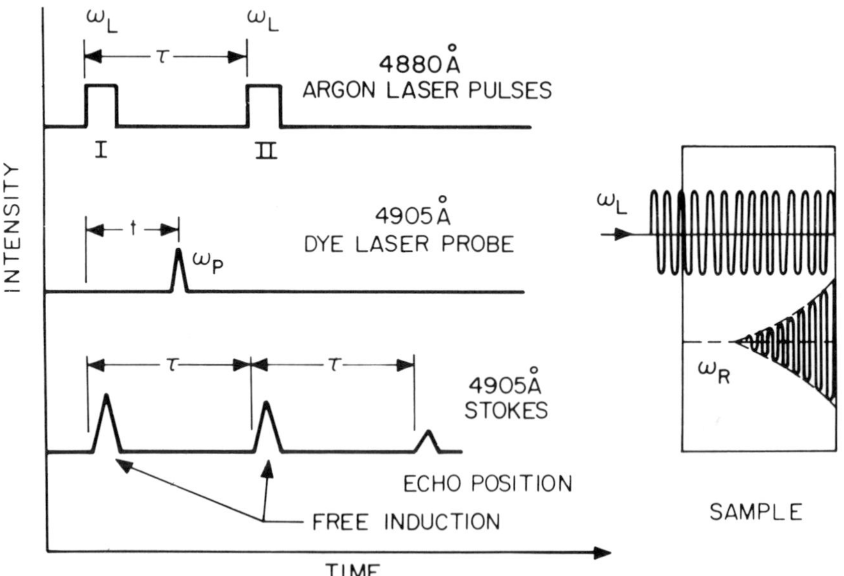

Fig. 4. Pulse sequence used in the new setup. Also shown is the
schematic diagram of the buildup of 4880Å Stokes inside
the sample.

the free induction and echo signals are now wider than the 5 ns
probing dye laser pulse. Furthermore, since ω_R is now generated by
stimulated Raman scattering in the crystal, as shown schematically
in Fig. 4, the resonance condition $\omega_L-\omega_R = \omega_{ba}$ is automatically
satisfied at any value of magnetic field. The echo has been observed over a continuous range of magnetic fields corresponding to
Zeeman splittings ranging from 10 GHz to 46 GHz. The low frequency
limit is set by the available Raman gain, while the upper frequency
limit is now restricted by the maximum field strength obtainable with
our conventional magnet. It is obvious that the frequency can be
extended to a few hundred GHz with a superconducting magnet.

The former optical delay line is now replaced in the new
arrangement by an electronic pulse sequence which drives the acousto-
optic coupler in our cavity-pumped Argon laser. A variety of pulse
sequences and variable pulse widths can now easily be obtained. The
dye laser whose role is now reversed to that of probe is coupled to
the Argon laser pulses by means of a beam splitter and propagated
collinearly along the c-axis of our sample as before. The samples
are now typically about 3 mm along the optical path and have a donor
(In) concentration ranging from 2×10^{16} to 4×10^{16}/cm^3 in order to
provide sufficient Raman gain. A 4905Å interference filter is now
placed in front of the RCA 8850 photomultiplier to block the 4880Å
exciting light. With an Argon laser power density of only the
order of 10^4 W/cm^2, the intensity of the stimulated Stokes light
produced by the single frequency argon pulse in the front end of
the n-type CdS crystal is sufficient to be used to produce a Raman
echo in the back half of the same crystal. When the dye laser
probe is set at $t=2\tau$, the echo signal can readily be observed on
the scope as shown in Fig. 5. The duty cycle of the experiment is
60 Hz, being limited by the maximum repetition rate of our nitrogen
laser.

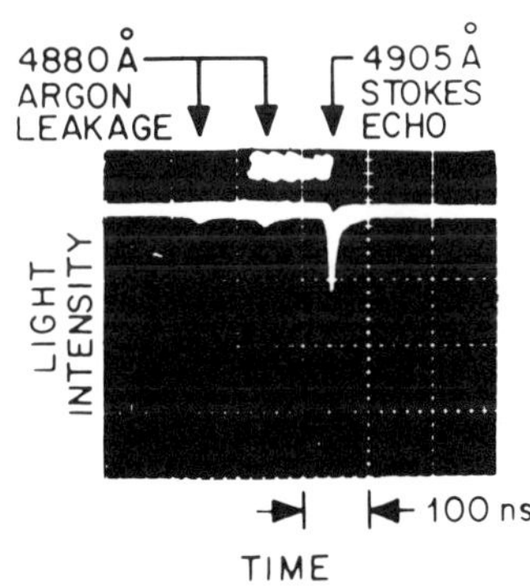

Fig. 5. Echo signal is
7.0 kG.

An automatic repetitive
electronic sweeping circuit scans
the time separation of the two
exciting pulses up to a few micro-
seconds and also provides a scanning triggering pulse for the boxcar which locks to the echo signals.
The output of the boxcar is then
fed into a multichannel analyzer
which is in turn synchronized to
the sweeping circuit. After
several sweeps, the averaged echo
signal as a function of τ can
then be displayed. A typical echo
decay at an external magnetic field
of 7.0 kG is shown in the inset of

Fig. 6. The data, together with another set of 17.4 kG are plotted
on a semi-log scale. The dephasing time T_2 for this particular
sample ($4 \times 10^{16}/cm^3$) is found to be about 200 ns, and is independent
of the magnetic field strength within experimental accuracy from
7 to 17.4 kG.

A stimulated echo [1] is generated by the application of three
excitation pulses I, II and III, separated by τ and T, respectively.

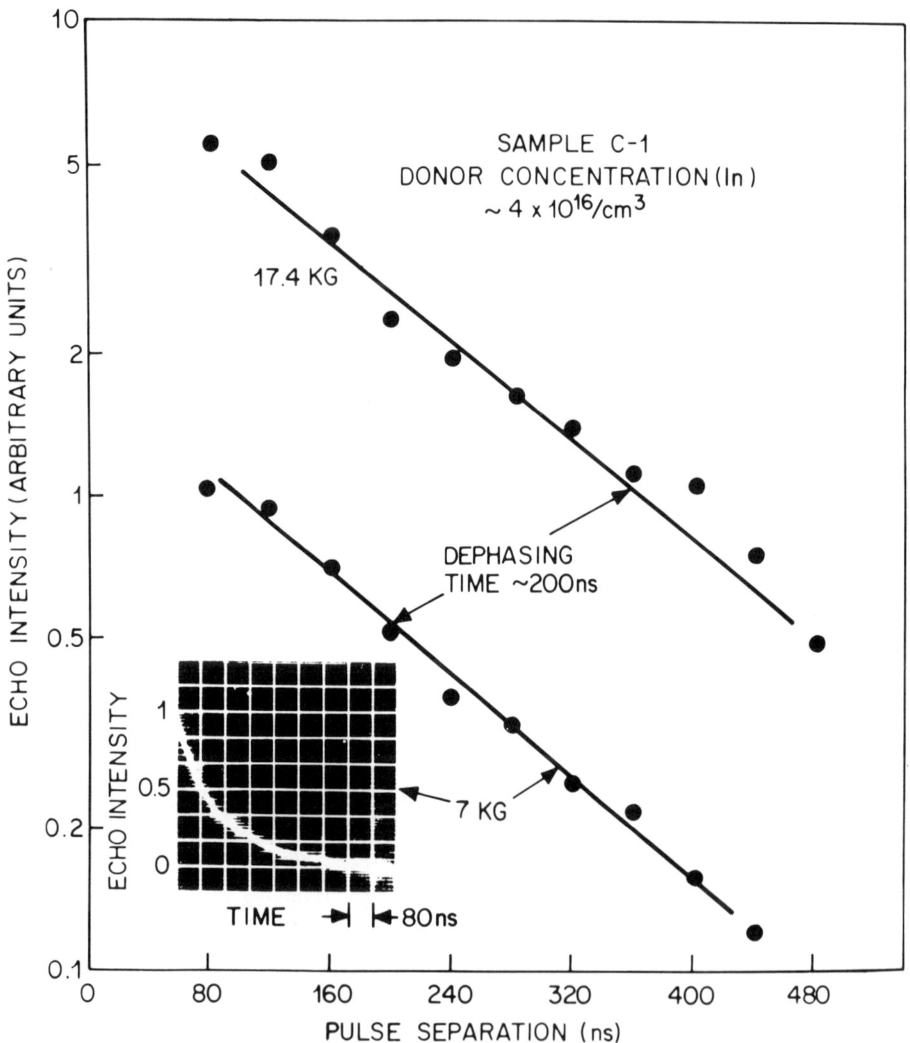

Fig. 6.   Echo signals as a function of pulse separation time τ on
         a semi-log plot for 7.0 and 17.4 kG. The inset is echo
         intensity vs. time on linear scale.

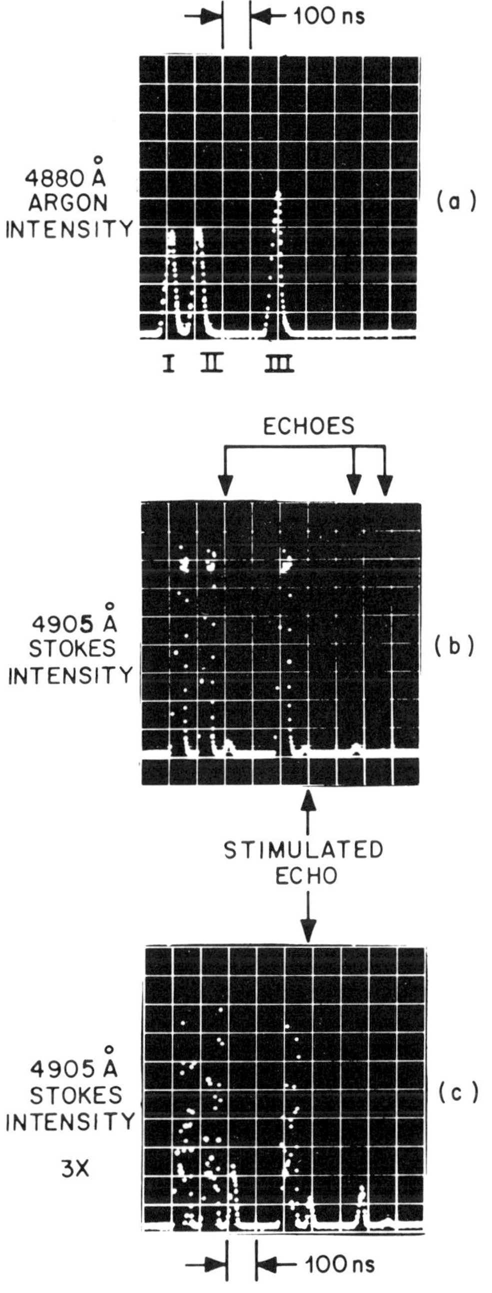

The echo is generated at t=τ after the third excitation pulses. The decay of the stimulated echo as function of T measures directly the spin lattice relaxation time T_1. In Fig. 7a, we show the three excitation argon laser pulses I, II and III. By scanning the 4905Å dye laser across the time scale range, and with the Fabry-Perot tuned to the Stokes of the 4905Å dye laser, we obtain Fig. 7b. Note the slight delay in the peak of the free induction decays relative to that of the argon laser pulses. Three two-pulse echoes are readily observed. The first, second and third echoes are generated by the set of pulses I, II; II, III and I, III, respectively. The stimulated echo appears at a time τ after the third pulse as is expected. This is the first observation of a stimulated Raman echo. Further details of T_1 and T_2 as a function of concentration and magnetic field will be published elsewhere.

In conclusion, we demonstrate the first observation of both a two-pulse Raman echo and a three pulse stimulated Raman echo in n-type CdS. The utility of the Raman echo is evident by the fact that this particular spin flip Raman echo has been studied over a *continuous* range of frequencies from

Fig. 7. a) 4880Å exciting laser pulses I, II and III.
   b) Free induction decay, two-pulse echoes and three-pulse stimulated echo.
   c) Same as (b) except 3 times gain on the vertical scale.

10 GHz to 46 GHz with the possibility of extension up to a few
hundred GHz, limited only by the available magnetic field strength.
In principle, Raman echo should be observable, with quite moderate
power, in a wide range of two-level systems not directly accessible
by the one-photon process. Furthermore, studies of the decay of
both two-pulse echo and three-pulse stimulated echo should provide
us with valuable information on the phase memory time and the
dynamics of a variety of systems.

We wish to thank G. Devlin for his advice and assistance on
many experimental details, and R. Romestain for many helpful
discussions.

References

1. E. L. Hahn, Phys. Rev. *80*, 580 (1950). For detailed dis-
    cussion see, for example, W. B. Mims, in *Electron Paramag-
    netic Resonance*, edited by S. Geschwind (Plenum, New York,
    1972).
2. N. A. Kurnit, I. D. Abella, and S. R. Hartmann, Phys. Rev.
    Lett. *13*, 567 (1964). See also, for example, J. P. Gordon,
    C. H. Wang, C. K. N. Patel, R. E. Slusher, and W. J. Tomlinson,
    Phys. Rev. *179*, 294 (1969).
3. J. A. Giordmaine and W. Kaiser, Phys. Rev. *144*, 676 (1966);
    for later references see A. Laubereau, G. Wochner, and
    W. Kaiser, Phys. Rev. A *13*, 2212 (1976). See, for examples,
    Richard G. Brewer and E. L. Hahn, Phys. Rev. A *8*, 464 (1973);
    M. M. T. Loy, Phys. Rev. Lett. *36*, 1454 (1976); V. T. Nguyen,
    E. G. Burkhardt, and P. W. Wolff, Opt. Commun. *16*, 145 (1976),
    and references therein.
4. S. R. Hartmann, IEEE J. Quantum Electron. *4*, 802 (1968);
    R. Weingarten, Ph.D. thesis, Columbia University, 1972
    (unpublished).
5. P. Hu, S. Geschwind, and T. M. Jedju, Phys. Rev. Lett. *37*,
    1357 (1976).
6. Y. Yafet, Phys. Rev. *152*, 858 (1966).
7. P. A. Wolff, Phys. Rev. Lett. *16*, 225 (1966).
8. R. E. Slusher, C. K. N. Patel, and P. A. Fleury, Phys. Rev.
    Lett. *18*, 77 (1967).
9. C. K. N. Patel and E. D. Shaw, Phys. Rev. Lett. *24*, 451 (1970);
    C. K. N. Patel, in *Laser Spectroscopy*, edited by R. G. Brewer
    and A. Mooradian (Plenum, New York, 1974).
10. D. G. Thomas and J. J. Hopfield, Phys. Rev. *175*, 1021 (1968).
11. J. F. Scott, T. C. Damen, and P. A. Fleury, Phys. Rev. B *6*,
    3856 (1972).
12. R. Romestain, S. Geschwind, G. E. Devlin, and P. A. Wolff,
    Phys. Rev. Lett. *33*, 10 (1974).

13. Terrence L. Brown and P. A. Wolff, Phys. Rev. Lett. *29*, 362
     (1972); V. T. Nguyen and T. J. Bridges, Phys. Rev. Lett. *29*,
     359 (1972).
14. R. G. Brewer, Physics Today *50*, May 1977.

PHOTON ECHO POLARIZATION MEASUREMENTS IN SF_6 AND SiF_4

W.M. Gutman and C.V. Heer

The Ohio State University, Columbus, Ohio

INTRODUCTION

The photon echo technique has become a valuable experimental
tool for studying molecular processes. The polarization of the
photon echo was first studied theoretically and experimentally by
Abella et al. [1]. They studied the photon echo stimulated in ruby
by exciting pulses from a ruby laser. Gordon et al. [2] studied
the polarization of the photon echo stimulated in gaseous SF_6 by
the $P(20)$ laser line of CO_2. They used their experimental results
and theoretical arguments to make inferences about the total angular momentum or the J-values of the levels which were involved in
the absorption of the radiation. They did not observe a component
of the echo polarization perpendicular to the polarization of the
second pulse. Some of their conclusions were in error. Alimpiev
and Karlov [3] studied the echo polarization with a linear-linear
excitation sequence, and with excitation by $P(16)$ were able to
observe a small echo polarization component perpendicular to the
second pulse polarization. Although their theoretical development
was applicable only to states with small J-values, they concluded
that $P(16)$ radiation was absorbed by a Q-branch transition in SF_6.
They did not observe the perpendicular component with $P(18)$ and
$P(20)$ excitation. Heer and Nordstrom [4] developed the photon echo
polarization theory for molecules with large total angular momentum
values for linear-linear, circular-linear, and linear-circular
excitation sequences. In their experimental work they observed a
large echo polarization perpendicular to the second pulse polarization with a $P(16)$ linear-linear or circular-linear excitation
sequence. From the details of the polarization behavior they concluded that $P(16)$ stimulated a Q-branch transition. The perpendicular polarization component was not observed for echoes stimulated

by other of the CO_2 laser lines. Observation of the perpendicular
component stimulated by P(14) was subsequently reported by us [5].
Initial polarization studies of the photon echo stimulated in SiF_4
by the P(30) and P(32) laser lines indicated that the perpendicular
polarization component was not observed [6].

Due to the lack of agreement with the large J-polarization
theory for the photon echo stimulated in SF_6 by the P(12) and P(18)
through P(24) CO_2 laser lines near 10.6μ and in SiF_4 by the P(30)
and P(32) CO_2 laser lines near 9.6μ, the previous apparatus [4] was
improved and the studies repeated. The results are summarized in
this paper and are in good agreement with the large J-polarization
theory [7]. Absorbing transitions can unambiguously be assigned
to the Q or P and R branch transitions by photon echo polarization
studies. Elastic cross-sections can be determined from the decay
in echo intensity with the interval between pulses. These results
are discussed in another paper [7].

REVIEW OF THEORY

A brief review of the theory of Heer and Nordstrom [4] of the
polarization of photon echoes is given. Two laser excitation
pulses are used and the electric dipole interaction with the laser
radiation is $V(t) = -P \cdot E(r_a, t)$. The time evolution of the molecule
at position r_a can be described by a density matrix $\sigma_a(t)$ and the
electric dipole moment of the radiating molecule by

$$P_a = \text{Tr } P_{op} \, \sigma_a(t) \; . \tag{1}$$

If the pulses have a simple "on" or "off" form and if relaxation
processes are ignored, the time evolution during a pulse is given
by a unitary operator $U(t'',t')$. For a first pulse with duration
from time t_0 to t_1 and a second pulse with duration from t_2 to t_3,
the density matrix describing the a'th molecule is

$$\sigma(t) = U_0(t,t_3)U_2(t_3,t_2)U_0(t_2,t_1)U_1(t_1,t_0)\sigma(t_0)U_1^\dagger(t_1,t_0)$$

$$\times \; U_0^\dagger(t_2,t_1)U_2^\dagger(t_3,t_2)U_0^\dagger(t,t_3) \tag{2}$$

and for convenience the subscript a is omitted. Denoting the upper
states by $J_a m_a$ and the lower energy states by $J_b m_b$, the general
matrix element between states m_a and m_b is written as

$$\sigma_{ab}(t) = (m_a|\sigma(t)|m_b) \tag{3}$$

and a sum over a,b implies the sum over the quantum numbers m_a and

m_b. An examination of Eq. (2) for $\sigma(t)$ indicates that

$$\sigma_{ab}(t) \propto \sigma_{b'a'}(t_2) = \sigma^*_{a'b'}(t_2) \; . \tag{4}$$

This dependence on the complex conjugate density matrix element at
an earlier time leads to the "reversal of runners in a race" at
time t_2 concept for nuclear spin echoes. This phase reversal in
the optical region can lead to the focusing of photon echoes [8],
deflection of photon echoes [1], and unusual properties in the
polarization of photon echoes [4-7,9].

Although a general solution is possible for arbitrary elliptic
polarization of the first and second pulses, this procedure is not
very useful for large J_a and J_b values where hundreds of states
are involved. Two special cases occur. The axis of quantization
can be chosen along the direction of linear polarization for a
linearly polarized pulse and along the direction of laser radiation
for a circularly polarized pulse. In each case the time evolution
which is described by the density matrix depends on only two states,
$m_a = m_b$ for linear polarization and $m_a = m_b \pm 1$ for circular polarization. Although the time evolution is simplified for each pulse,
the states for these two different axes of quantization must be
related by the rotation matrices. The problem can be reduced by
using the addition properties of the rotation matrices and the
relationships for the Clebsch-Gordon coefficients. For large values
of total angular momentum their theory yields the following results.
Let the stimulating laser radiation travel along the $\hat{Y}$ axis, let
$\hat{Z}$ be the linear polarization of the second pulse and $\hat{z}$ the linear
polarization of the first pulse. $\hat{z}$ makes angle β with $\hat{Z}$ as shown
in Fig. 1. A linearly polarized echo is formed with an unnormalized
electric field vector of

$$\underline{E} \cong \pm\hat{X}(\tfrac{1}{2}\sin\beta) + \hat{Z}\cos\beta \tag{5}$$

where the plus sign refers to $J_a = J_b$ or to a Q-branch transition
and the negative sign refers to the $J_a = J_b \pm 1$ or to the P-branch
and R-branch transitions. The photon echo is linearly polarized
at angle ϕ from the $\hat{Z}$ axis with $\tan\phi = \pm\tfrac{1}{2}\tan\beta$. For P-branch and
R-branch transitions ϕ rotates in the opposite sense of β and for
the Q-branch transitions ϕ rotates in the same sense as β. An
intensity ratio between the first and second pulses of 1:4 should
yield the best echoes.

If the first pulse is left circularly polarized and the second
pulse is linearly polarized along $\hat{Z}$, then the elliptical polarization of the photon electric field is given by the coefficient of
$\exp(-i\omega t)$ and the unnormalized echo polarization is

$$\underline{E} \cong \pm \frac{1}{2} i \, \hat{X} + \hat{Z} \; . \tag{6}$$

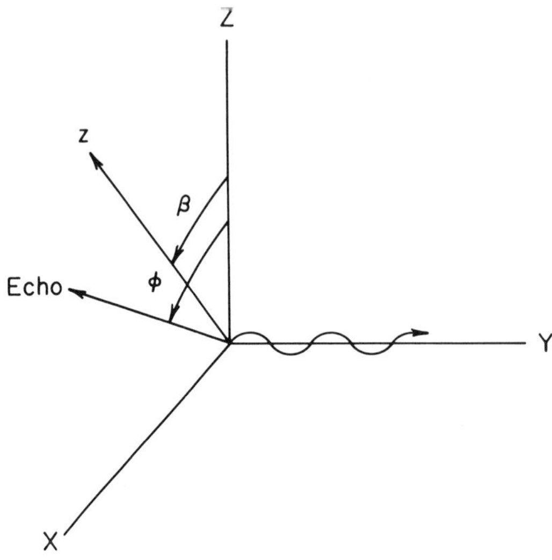

Fig. 1. The laser radiation propagates along the $\hat{Y}$ axis. The first pulse is linearly polarized along $\hat{z}$ at angle β with the linear polarization of the second pulse along $\hat{Z}$. ϕ is the angle between $\hat{Z}$ and the linear polarization of the photon echo.

The positive sign refers to the $J_a = J_b \pm 1$ or P-branch and R-branch transitions and the echo is left-elliptically polarized. The negative sign refers to the $J_a = J_b$ or Q-branch and the echo has right elliptic polarization. An intensity ratio for the Q-branch of 1:2 and for the P-branch and R-branch of 1:8 should give the best echoes.

If the first pulse is linear and the second pulse is left circular, then the echo is left circular. An intensity ratio of 1:8 for the Q-branch and 1:2 for the P-branch and R-branch should give the best echoes.

EXPERIMENTAL RESULTS

Photon echoes, which were stimulated in SF_6 by the P(12) through P(26) laser lines of the (00°1)-(10°0) vibrational band of CO_2 near 10.6μm and in SiF_4 by the P(28) through P(34) laser lines of the (00°1)-(02°0) vibrational band of CO_2 near 9.6μm, were observed. For comparison with the theory the polarization studies were made with an excitation pulse sequence which consisted of two linearly polarized pulses, a left circularly polarized pulse followed by a linearly polarized pulse, or a linearly polarized pulse followed by a left circularly polarized pulse. The experimental results are summarized.

Linear-Linear

Figure 1 shows the linear-linear pulse sequence. The linear polarization of the second pulse is along $\hat{Z}$ and the linear polarization of the first pulse is along $\hat{z}$ at angle β with $\hat{Z}$. The echo is analyzed with a linear polarizer and the intensities I_X, I_Z and $I(\alpha)$ are measured.

$\underline{SF_6}$: $\underline{P(12)}$ and $\underline{P(14)}$. With $P(12)$ and $P(14)$ laser excitation the echo is well described by $I_Z/I_0 = \cos^2\beta$ for both lines, and by $I_X/I_0 = 0.24 \sin^2\beta$ for $P(12)$ and $0.25 \sin^2\beta$ for $P(14)$. The echo is linearly polarized at angle ϕ and $\tan\phi = -\frac{1}{2} \tan\beta$ to a good approximation. The echo polarization rotates in the opposite sense as β. This result implies that $P(12)$ and $P(14)$ are absorbed by P-branch or R-branch transitions in SF_6. Using the results of conventional absorption spectroscopy [10], it may be determined that the transitions are R-branch, consistent with the conclusions of Houston and Steinfeld [11], and of McDowell et al [12] for $P(14)$.

$\underline{SF_6}$: $\underline{P(16)}$. With $P(16)$ $I_Z/I_0 = \cos^2\beta$ and $I_X/I_0 = 0.27 \sin^2\beta$ to a good approximation. The relationship between ϕ and β is well described by $\tan\phi = \frac{1}{2} \tan\beta$ and the echo polarization rotates in the same sense as β. This result indicates that $P(16)$ is absorbed by a Q-branch transition in SF_6, and this is consistent with the results of Houston and Steinfeld [11].

$\underline{SF_6}$: $\underline{P(18)}$, $\underline{P(20)}$, $\underline{P(22)}$, $\underline{P(24)}$, and $\underline{P(26)}$. Again the Z component of the echo follows $I_Z/I_0 = \cos^2\beta$. The X component follows $I_X/I_0 = A \sin^2\beta$, where $A = 0.06$ for $P(18)$, 0.10 for $P(20)$, 0.13 for $P(22)$ and 0.16 for $P(24)$. With $P(26)$, the echo size is too small for the polarization to be analyzed. The echo is linearly polarized and the linear polarization rotates in the opposite sense as β with all these lines. The dependence of ϕ on β is not well described by $\tan\phi = -\frac{1}{2} \tan\beta$. This reflects the fact that I_X is less than $\frac{1}{4} I_Z$. It is, nevertheless, reasonable to conclude that these lines are absorbed by P-branch or R-branch transitions. Using conventional spectroscopic results [10] allows the conclusion that the transitions are P-branch. This is consistent with the results of Houston and Steinfeld [11] for $P(18)$, $P(20)$ and $P(22)$, and of McDowell et al [12] for $P(18)$ and $P(20)$.

$\underline{SiF_4}$: $\underline{P(28)}$, $\underline{P(30)}$, $\underline{P(32)}$ and $\underline{P(34)}$. With $P(30)$, $P(32)$ and $P(34)$, the Z component of the echo follows $I_Z/I_0 = \cos^2\beta$, and the X component follows $I_X/I_0 = 0.25 \sin^2\beta$. The relationship between ϕ and β is well described by $\tan\phi = -\frac{1}{2} \tan\beta$, and the echo rotates in the opposite sense as β. This fact is consistent with the polarization theory if these lines are absorbed by a P-branch or R-branch transition in SiF_4. Conventional spectroscopic data [13] may be used to infer that the transitions are R-branch. With $P(28)$, the echo size is too small to analyze.

Left Circular-Linear

The first pulse is left circularly polarized and the second is linearly polarized along $\hat{Z}$. The echo is analyzed as a function of the angle α between the transmission axis of the analyzer and $\hat{Z}$.

$\underline{SF_6\colon\ P(12)\ \text{and}\ P(14)}$. For $P(12)$, $I/I_0 = 0.24\ \sin^2\alpha + \cos^2\alpha$ to a good approximation, and for $P(14)$, $I/I_0 = 0.25\ \sin^2\alpha + \cos^2\alpha$. The echo is left elliptically polarized. Further, the strongest echoes are obtained with a 1:8 pulse intensity ratio. These facts are consistent with predictions of the theory for absorption by an R-branch transition.

$\underline{SF_6\colon\ P(16)}$. With $P(16)$, $I/I_0 = 0.27\ \sin^2\alpha + \cos^2\alpha$, and a 1:2 pulse intensity ratio gives the best echoes. The echo is right elliptically polarized. This behavior is appropriate for a Q-branch absorbing transition.

$\underline{SF_6\colon\ P(18),\ P(20),\ P(22)\ \text{and}\ P(24)}$. For these lines, the echo intensity is well described by $I/I_0 = B\ \sin^2\alpha + \cos^2\alpha$, where $B = 0.08$ for $P(18)$, 0.23 for $P(20)$, 0.17 for $P(22)$, and 0.17 for $P(24)$. A 1:8 pulse intensity ratio gives best echoes. For $P(20)$, it may be determined that the echo is left elliptically polarized. For the other lines, the echo is too small for the sense of the elliptic polarization to be found.

$\underline{SiF_4\colon\ P(30),\ P(32),\ \text{and}\ P(34)}$. The relative echo intensity as a function of analyzer angle is given by $I/I_0 = 0.25\ \sin^2\alpha + \cos^2\alpha$. A 1:8 pulse intensity ratio gives the strongest echoes, and the echo polarization is left elliptic. This is appropriate for absorption by an R-branch transition.

Linear-Left Circular

$\underline{SF_6\colon\ P(12),\ P(14),\ P(16),\ P(18)\ \text{and}\ P(20)}$. With all of these lines, the echo is left circularly polarized. With $P(16)$, optimum echoes are obtained with a 1:8 pulse intensity ratio, while for the other lines, a 1:2 ratio is best.

$\underline{SiF_4\colon\ P(30),\ P(32)\ \text{and}\ P(34)}$. The echo is circularly polarized with excitation by $P(30)$, $P(32)$ and $P(34)$. The echoes are too small for the sense of the polarization to be determined. A 1:2 pulse intensity ratio provides optimum echoes.

Temperature Dependent Effects in SF_6

The sample cell used in the experiment was constructed so that
its temperature can be varied between room temperature and 180 K.
Several interesting effects are observed when the temperature of
the absorbing SF_6 is reduced. With P(12), P(14) and P(16), the
absolute echo size increases as the temperature is reduced. With
P(18), the temperature has little effect on the echo size. But
with P(20) through P(26), the echo size decreases as the temperature
is reduced. Echoes are no longer observed with P(24) or P(26)
below 204 K. This reduced echo size can probably be understood in
terms of the reduced contribution of hot bands to the SF_6 spectrum
at lower temperatures [11]. Absorption of these lines, and also
of P(12), decreases dramatically as the temperature is reduced.
The polarization properties of photon echoes stimulated by P(12)
through P(18) at reduced temperatures are in good agreement with
the theory for P-branch, Q-branch and R-branch transitions as
previously determined. For the other lines, the echoes are too
small to analyze.

A striking observation at reduced temperatures is the presence
of multiple photon echoes with P(14) and P(16). A **linear-linear P(16)**
excitation sequence with $\beta = 0°$ and separated in time by T produces
a very large primary echo, followed after time T by a second smaller
echo, followed again after time T by a third even smaller echo.
The intensity of the second echo is as large as 10% of the first.
The second echo is observed only for $T < 3\mu sec$, and the third only
for $T < 2\mu sec$. The primary echo is observed for $T < 10\mu sec$. The
second echo is also present with P(14), but it is smaller than with
P(16), and only observed with $T < 2\mu sec$.

CONCLUSION

In general, the experimental echo polarization results are in
agreement with the large angular momentum or large J-value theory.
Typical experimental results for the angle ϕ as a function of β
are shown in Fig. 2. Previously, the echo polarization component
along $\hat{X}$ had not been observed except for stimulation by P(14) and
P(16) in SF_6. Refinements in the experimental technique have
apparently made the $\hat{X}$ component of the echo observable with the
other lines. These refinements include a reduction of the excitation pulse duration by an improvement of the laser Q-switch, and
elimination of strain birefringence which was introduced by the
cell windows. It may be possible to obtain even better agreement
with the theory for stimulation of SF_6 by the lines P(18) through
P(26) by making further refinements in the experimental technique.

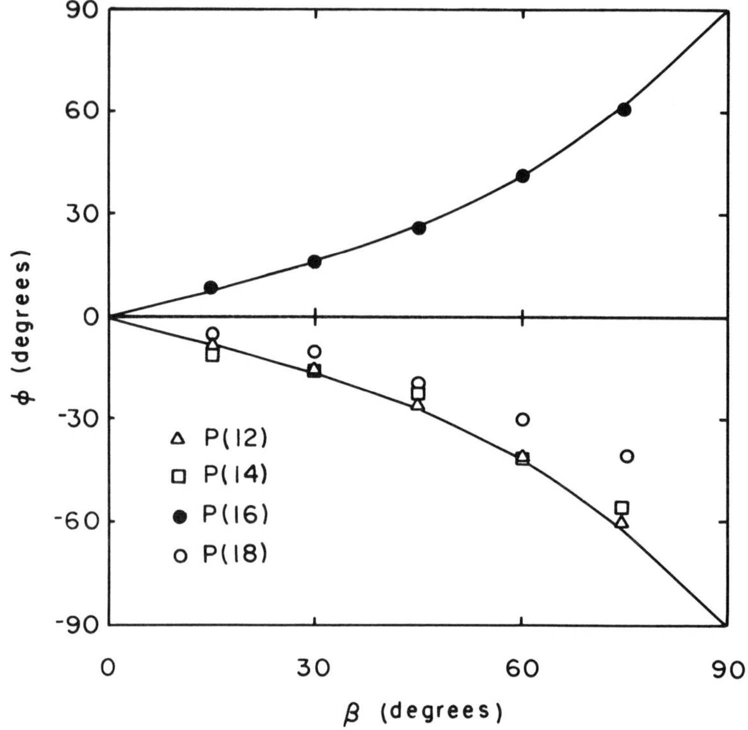

Fig. 2. The angle ϕ between the linear polarization of the echo and the $\hat{Z}$ axis is shown as a function of the angle β for stimulation of SF_6 by the P(12), P(14), P(16) and P(18) CO_2 laser lines. The solid curves are $\tan\phi = \pm\frac{1}{2}\tan\beta$. The upper curve applies to a Q-branch transition, and the lower curve applies to a P-branch or R-branch transition.

References

1.   I.D. Abella, N.A. Kurnit, and S.R. Hartmann, Phys. Rev. Lett. *13*, 567 (1964), and Phys. Rev. *141*, 391 (1966).
2.   J.P. Gordon, C.H. Wang, C.K.N. Patel, R.E. Slusher and W.J. Tomlinson, Phys. Rev. *179*, 294 (1969).
3.   S.S. Alimpiev and N.V. Karlov, Sov. Phys. JETP *36*, 255 (1973).
4.   C.V. Heer and R.J. Nordstrom, Phys. Rev. A *11*, 536 (1975).
5.   W.M. Gutman and C.V. Heer, Phys. Lett. *51A*, 437 (1975).
6.   R.J. Nordstrom, W.M. Gutman, and C.V. Heer, Phys. Lett. *50A*, 25 (1974).
7.   W.M. Gutman and C.V. Heer, Phys. Rev. A *16*, 659 (1977).

8.   C.V. Heer, Phys. Rev. A *7*, 1635 (1973).

9.   C.V. Heer and R.H. Kohl, Phys. Rev. A *1*, 693 (1970).

10. H. Brunet and M. Perez, J. Mol. Spectrosc. *29*, 472 (1969).

11. P.L. Houston and J.I. Steinfeld, J. Mol. Spectrosc. *54*, 335
     (1975).

12. R.S. McDowell, H.W. Galbraith, B.J. Krohn, and C.D. Cantrell,
     Optics Commun. *17*, 178 (1976).

13. E.A. Jones, J.S. Kirby-Smith, P.J.H. Woltz and A.H. Nielsen,
     J. Chem. Phys. *19*, 242 (1951).

THEORY OF ELECTROMAGNETIC TRANSIENTS IN SPATIALLY DISPERSIVE MEDIA
AND A NEW APPROACH TO PRECURSOR THEORY*

J. L. Birman, M. J. Frankel** and D. N. Pattanayak

City College, CUNY, New York, N. Y.

1. INTRODUCTION

The theoretical investigation of electromagnetic transients
in media has been an ongoing topic of interest since (at least)
the classic work of Sommerfeld and Brillouin[1] on precursor
theory. The prototypical problem is the analysis of the amplitude
$f(z,t)$ of the electromagnetic field at point z, time t in a material
dielectric when a wave with sharp front (for example, a sharply
truncated laser) impinges on the surface of a bounded medium.
Coming into play in this analysis are: the model of the medium,
(which in the classical theory is taken as a "Lorentz model" dielectric); the coupling of field and medium via some dielectric
response function; and the propagation of the field, described by
Maxwell equations. All these aspects of radiation and matter
theory can be explored by analysing the time evolution of the
amplitude $f(z,t)$, which in effect samples the dynamics of the
system in a time-resolved succession of stages, each such stage
referring to a different frequency regime.

In the present paper we shall present a brief review of the
results of the classical theory of transients (precursors, or
forerunners) due to Sommerfeld and Brillouin and their associates
in a frequency dispersive medium, and then discuss some new results
recently obtained by generalizing the Sommerfeld-Brillouin
analysis to spatially dispersive, or non-local media. Finally,
we shall report on a new approach to the theory of transients.

2.   THE CLASSICAL THEORY

To appreciate the types of physical processes which come into
play the discussion of Sommerfeld and Brillouin[1] is helpful and
even semi-quantitatively correct as verified by more exact mathematical analysis. Thus, the amplitude at (z,t) is zero until the
front arrives, which it does at time (z/c) since the effective
medium seen by the front of the wave is vacuum. The sharp front
(cut-off) of the laser means that in addition to the laser frequency ω_L, higher frequency and lower frequency sidebands are
contained within the incident field. Assume that the medium is a
Lorentz dielectric: an assembly of weakly interacting, classical,
polarizable, dipole oscillators in vacuum (Lorentz model), with
each oscillator having one "resonance" frequency ω_0, and that the
laser carrier frequency $\omega_L < \omega_0$. In this case one may take the
dielectric coefficient

$$\varepsilon(\omega) = 1 + \frac{4\pi\alpha_0\omega_0^2}{\omega_0^2 - \omega^2 - i\omega\Gamma} \qquad . \tag{2.1}$$

This expression will be used to describe a "classical" or "frequency
dispersive" *local* dielectric. Then the following stages emerge.
After the "dead time" a packet appears at (z,t) containing high
frequency components which travel with some average group velocity
v_{gH}; roughly speaking this packet peaks at time (z/v_{gH}). Physically
the "impulsive" or "instantaneous" response of the dipoles is
responsible for this part of the precursor pattern: damping or
restoring forces have not yet come into play. This is the Sommerfeld
Precursor (SP). In the work of Skrotskaya et al.[2] a Sommerfeld
Precursor regime is obtained following one of Sommerfeld's earlier
approaches[1] by taking the dispersion formula of the medium (for
$\omega/\omega_0 \gg 1$)

$$n^2(\omega) \cong 1 - \Omega^2/\omega^2 \tag{2.2}$$

with Ω^2 a plasma-like frequency ($\Omega^2 \equiv 4\pi\alpha_0\omega_0^2$ where α_0 is an
oscillator strength). This suggests in the time domain

$$\vec{E}(\vec{r},t) = \frac{1}{\alpha_0\omega_0^2} \frac{\partial^2\vec{P}}{\partial t^2}(\vec{r},t) , \tag{2.3}$$

which we note here for later comparison.

The next time regime corresponds to the arrival of a packet
containing low frequencies, the Brillouin Precursor (BP)
originating from the low frequency sidebands in the spectral

decomposition of the incident field. In this regime the oscillators respond to an effectively "stationary" electric field. A characteristic group velocity is v_{gL}, with peak of the packet arriving at $t_B \cong z/v_{gL}$. In the low frequency regime the dispersion formula may be written

$$n^2(\omega) \cong n_0 + n_1\omega^2 \qquad\qquad (2.4)$$

with $n_0 \equiv 1 + 4\pi\alpha_0/\omega_0^2$, $n_1 \equiv 4\pi\alpha_0/\omega_0^2$. Again this is suggestive (*vide infra*) in the time domain of the equation

$$\omega_0^2\vec{P}(\vec{r},t) = \alpha_0[\vec{E}(\vec{r},t) - \partial^2\vec{E}(\vec{r},t)/\partial t^2] \quad , \qquad\qquad (2.5)$$

or, to first approximation, neglecting the second term

$$\omega_0^2\vec{P}(\vec{r},t) \cong \alpha_0\vec{E}(\vec{r},t) \quad . \qquad\qquad (2.6)$$

This is a type of steady state behavior which is suggestive of the "stationary" approximation. However to properly take account of transient phenomena, the second term in (2.5) is clearly required.

Another manner of looking at the transient phenomena is to recognize that the constitutive relation, or linear response function, is time dependent, reflecting the time evolution of the response of the assembly of elementary dipoles to the passage of the time dependent electromagnetic field. This point of view is useful in our later consideration.

Thus, the physics of the classical situation can be epitomized in the four-step scenario as recorded by an observer at (z,t) in a bounded $(z{\geq}0)$ medium, with a pulse incident at $(z=0^-, t=0)$ as follows: (1) dead time (DT) $t{\leq} z/c$; (2) arrival of high frequency components in a Sommerfeld packet (SP) precursor at $t \sim z/v_{gH}$; (3) arrival of low frequency components in a Brillouin packet (BP) precursor at $t \sim z/v_{gL}$; (4) arrival of the laser at frequency ω_L for $t > z/v_g(\omega_L)$. If the laser frequency is in the region of the "stop" band of the oscillator $\omega_L \sim \omega_0$, then the "signal theory" comes into play: it requires a delicate analysis and is not discussed here (but see refs. 1 and 3).

3.   TRANSIENTS IN NON-LOCAL MEDIA[3]

In recent years the optics of non-local media has been actively investigated. Here we shall briefly review aspects of the transient optics of such media which have been examined by generalizing the classical Sommerfeld-Brillouin analysis. We consider a model bounded by a spatially dispersive medium with non-local susceptibility $\chi(\vec{r},\vec{r}';\omega)$ so that

$$\vec{P}(\vec{r},\omega) = \int_\Omega \chi(\vec{r},\vec{r}';\omega)\ \vec{E}(\vec{r}',\omega)d^3r' \ , \tag{3.1}$$

where $\vec{E}$ is the electric field, $\vec{P}$ is the polarization, and the inte-gration is over the bounded region Ω. We seek propagating electro-magnetic modes in such a medium; they satisfy a dispersion equation of form

$$k^2 = k_0^2\ \varepsilon_{EFF}(k,\omega) \ , \tag{3.2}$$

where k is the wave number of the propagating plane wave mode, $k_0 = \omega/c$ is the vacuum wave number and $\varepsilon_{EFF}(k,\omega)$ is some effective dielectric coefficient, which depends upon k in non-local media. In order to obtain physically meaningful results we need to specify the form of $\chi(\vec{r},\vec{r}',\omega)$ and ε_{EFF}: the physics, or microtheory, enters at that point. In the Appendix we give some details of a theory recently completed[3] which will apply to non-local phenomena in the exciton region of the optical spectrum, especially for Frenkel (tightly bound) excitons in insulators, and permits implementation of the analysis and comparison with experiments when they become available.

For local media ε does not depend on k, and one substitutes equation (2.1) into (3.2) to obtain propagating solutions.

Owing to k dependence of $\varepsilon_{EFF}(k,\omega)$ the order of the algebraic equation (3.2) is higher than for the local case, so there are more propagating solutions. Call the solutions k_j so that

$$k_j^2(\omega) = k_0^2\ \varepsilon_{EFF}(k_j,\omega) \ . \tag{3.3}$$

The solution of the electrodynamic problem in bounded non-local media requires boundary conditions in order to smoothly join ex-terior (incident plus reflected) to interior (propagating mode) fields. These boundary conditions include: *Maxwell b.c.* plus *additional b.c.*[4]. For an incident unit amplitude wave at $(z=0^+,$ $t=0)$ of form

$$f(t) = (\sin \omega_L t)\ \theta(t), \tag{3.4}$$

generalized Fresnel factors $A_j(\omega)$ can be determined, which give the relative amplitudes of the propagating modes. Then the generalization to non-local media of Sommerfeld's complex integral representation[3] gives for the amplitude of the field at (z,t)

$$f(z,t) = \frac{Re}{2\pi} \oint \sum_j \frac{A_j(\omega)\exp\, i[k_j(\omega)z-\omega t]}{\omega-\omega_L}\, d\omega \qquad . \qquad (3.5)$$

The contour is closed in the UHP for t<0 and in the LHP for t>0.

This simple integral representation contains the entire time development of the amplitude at (z,t). The strategy for analysing this integral is to locate the singularities of the integrand, and then use an asymptotic method, e.g. steepest descents to obtain f(z,t).

The details of this evaluation depend upon the form of $k_j(\omega)$, $A_j(\omega)$, which in turn depends upon $\varepsilon(k,\omega)$.

For a local medium $A_j \to (ck(\omega)-\omega)/(ck(\omega)+\omega)$; this is slowly varying in ω and is often taken as a constant, although even in the local case this is questionable[5]. The function $k(\omega)$ in the local case, as a function of complex ω, is analytic in UHP, and has four branch points in LHP connected by two branch lines: these line segments are at Im $\omega=-\Gamma/2$ (below the real axis) and are parallel to the real axis (schematically the situation below the real asix is like: •————• •————•). Finally there is the simple pole at $\omega=\omega_L$.

When the complex integral is asymptotically evaluated by steepest descents,[1] the different contributions give rise in succession to: dead time, Sommerfeld precursor, Brillouin precursor, and steady-state laser wave at ω_L. These contributions arise in time sequence because the *saddle point* depends on time and passes through the various singularities in order.

Now we turn to non-local media. Using the susceptibility given in the Appendix to this paper we find that $k_j(\omega)$ has a *pair* of singularities (branch points) in the UHP, but their total contribution to f(z,t) vanishes, preserving causality[6]. Also k(ω) has two branch points in the LHP, joined by an unbroken branch line parallel to the real axis. (Schematically the situation in LHP is like: •————•). Thus the "gap" in the branch line compared to the local case is absent. In fact the absence of a gap in the branch line corresponds to the absence of a "stop" gap for bulk propagation in non-local media owing to the extra modes which can propagate compared to the classical local case. The integrand has also the simple pole at $\omega=\omega_L$ and as usual $A_j(\omega)$ is taken slowly varying or constant about the saddle point frequency.

Asymptotic analysis by saddle point method of integral representation produces expressions which may be collected as:

$$f(z,t) \sim \sum_{\alpha} f_{\alpha}(z,t)\ \theta(t-t_{\alpha}),\ t < t_{sig}\ ,$$

$$\alpha = SP,\ BP,\ EP\ \ .$$

$$(3.6)$$

This expression represents three partially overlapping packets
which arrive prior to the laser wave (or signal) and after the
dead time. These precursors are the familiar Sommerfeld and
Brillouin precursors (SP and BP) plus a new high frequency precursor: the Exciton Precursor (EP), newly predicted for spatially
dispersive media. Details are given elsewhere[3] of the steps in the
analysis (cf. also the Appendix). The essential point here is the
successful generalization of the Sommerfeld integral expression and
asymptotic evaluations to non-local media. The last stage is the
arrival of the "steady" laser at ω_L.

4.   NEW APPROACH TO PRECURSOR THEORY

The traditional approach in the study of transients in material
media is to decompose the incident wavefront into its harmonics.
With the aid of dispersion theory the propagation of each harmonic
inside the medium is found and then the sum of them gives the field
in the time domain. This approach leads to evaluation of complicated
integrals in the frequency domain.

An alternative approach will be to formulate the problem in
the time domain entirely. This can be done following the formalisms
of time dependent scattering theory. We assume that the macroscopic
Maxwell equations are valid in the transient regime. Because the
Maxwell equations are differential equations we need to prescribe
proper initial and boundary conditions to obtain physical solutions
of the Maxwell equations. We extend the boundary value formulation
of time independent electromagnetic scattering theory due to
Pattanayak and Wolf[7] to the case in which the fields are not
monochromatic. According to this approach (the details of which
will be published elsewhere)[8], the physical solutions of the
Maxwell equations that satisfy the outgoing boundary conditions at
infinity and the time dependent Saltus conditions at the boundary
of the scattering medium are solutions to Maxwell equations subject
to the following time dependent boundary condition:

$$\vec{E}^I(\vec{r},t) + \frac{1}{4\pi} \int_S ds' \left\{ \left(\vec{n} \times \vec{E}\ (\vec{r}',t-R/c) + \frac{R}{c}\ \vec{n} \times \frac{\partial \vec{E}(\vec{r}',t-R/c)}{\partial t'} \right) \nabla' G_s(R) \right.$$

$$+ \left(1 + \frac{R}{c}\frac{\partial}{\partial t'}\right) \left(\vec{n} \cdot \vec{D}^{(-)}(\vec{r}',t-R/c) + 4\pi\ \sigma(\vec{r}',t-R/c)\right) \nabla' G_s(R)$$

$$+ \frac{1}{c}\left(\vec{n} \times \frac{\partial \vec{H}^{(-)}(\vec{r}',t-R/c)}{\partial t'} + 4\pi \frac{\partial \vec{K}(\vec{r}',t-R/c)}{\partial t'}\right) G_s(R) \left. \right\} = 0 \qquad (4.1)$$

for all points $\vec{r}$ inside the medium. $\vec{E}^I(\vec{r},t)$ denotes the incident field

$$R = |\vec{r} - \vec{r}'| \quad , \qquad\qquad\qquad\qquad\qquad (4.2)$$

$$G_s(R) = \frac{1}{R} \quad , \qquad\qquad\qquad\qquad\qquad\qquad (4.3)$$

where $\sigma(\vec{r},t)$ and $\vec{K}(\vec{r},t)$ are the surface charge and surface current densities, respectively. The superscripts on the field vectors denote their limiting values as one approaches the boundary from the interior.

Equation (4.1) relates the time dependent boundary values of the surface fields $\vec{n} \times \vec{E}$, $\vec{n} \times \vec{H}$ and $\vec{n} \cdot \vec{D}$ and the surface current and charge density $\vec{K}$ and σ with the incident field. The relationship is causal. Such a relationship in time independent electromagnetic scattering theory has been identified as generalized extinction theorem. We may therefore identify Eq. (4.1) as the time dependent extinction theorem. We shall not discuss the physical significance of this theorem here but only indicate briefly how this theorem can be used to obtain solutions in the transient regime.

5. USE OF TIME DEPENDENT EXTINCTION THEOREM

Let us consider a linear homogeneous, non-magnetic and isotropic ordinary dielectric medium. Inside this medium the electric field obeys the following wave equation

$$\nabla^2 \vec{E}(\vec{r},t) - \frac{1}{c^2}\frac{\partial^2 \vec{E}(\vec{r},t)}{\partial t^2} = \frac{4\pi}{c^2}\frac{\partial^2 \vec{P}(\vec{r},t)}{\partial t^2} \quad . \qquad\qquad (5.1)$$

For times very near the wavefront we have the approximate constitutive relation (2.3) which we rewrite here

$$\vec{E}(\vec{r},t) = \frac{1}{\alpha_o \omega_o^2}\frac{\partial^2 \vec{P}(\vec{r},t)}{\partial t^2} \qquad\qquad\qquad (5.2)$$

Eliminating $\partial^2 \vec{P}/\partial t^2$ between Eqs. (5.1) and (5.2) we obtain

$$\nabla^2 \vec{E}(\vec{r},t) - \kappa^2 \vec{E}(\vec{r},t) - \frac{1}{c^2}\frac{\partial^2 \vec{E}(\vec{r},t)}{\partial t^2} = 0 \quad , \tag{5.3}$$

where

$$\kappa = \left(\frac{4\pi\alpha_o \omega_o^2}{c^2}\right)^{\frac{1}{2}} . \tag{5.4}$$

In order to obtain a representation of the electric field inside the medium in terms of the boundary values we introduce the Green's function $g(\vec{r},\vec{r}',t,t')$ which satisfies the following equation

$$\left(\nabla^2 - \frac{1}{c^2}\frac{\partial^2}{\partial t^2} - \kappa^2\right) g(\vec{r},\vec{r}',t,t') = -4\pi\, \delta(\vec{r}-\vec{r}')\delta(t-t'). \tag{5.5}$$

The solution obeying the Sommerfeld radiation condition at infinity of Eq. (5.5) is known to be[9]

$$g(\vec{r},\vec{r}',t,t') = G(\vec{r},\vec{r}',t,t') - S(\vec{r},\vec{r}',t,t') \quad , \tag{5.6}$$

where

$$G(\vec{r},\vec{r}',t,t') = \frac{\delta(|\vec{r}-\vec{r}'|/c - (t-t'))}{|\vec{r}-\vec{r}'|} \quad , \tag{5.7}$$

$$S(\vec{r},\vec{r}',t,t') = \kappa\, U\left(t-(t' + \frac{|\vec{r}-\vec{r}'|}{c})\right) \frac{J_1\left(\kappa c\sqrt{(t-t')^2 - \frac{|\vec{r}-\vec{r}'|^2}{c^2}}\right)}{\left((t-t')^2 - \frac{|\vec{r}-\vec{r}'|^2}{c^2}\right)^{\frac{1}{2}}} \tag{5.8}$$

In Eq. (5.8) U is the Heaviside stepfunction and J_1 is the Bessel function of order one. Using Eqs. (5.3) - (5.8) and the extinction theorem (4.1) it can be shown after a lengthy calculation that the solution that satisfies the equation (5.3), subject to the time dependent extinction theorem (4.1), may be represented by the integral equation

$$\vec{E}(\vec{r},t) = \vec{E}^I(\vec{r},t) + \int_S ds' \left[\left\{(1+\frac{R}{c}\frac{\partial}{\partial t'}) (\vec{n}\cdot\vec{P}(\vec{r}',t'))\right\}\right.$$

$$\left.\times \nabla' G_S(R)\right] t'=(t-R/c) + \int_S ds' \int_0^{(t-R/c)} dt'\{\frac{\partial \vec{E}(\vec{r}',t')}{\partial n} S(\vec{r},\vec{r}',t,t')$$

$$- \frac{\partial S(\vec{r},\vec{r}',t,t')}{\partial n} \vec{E}(\vec{r}',t') \} - \frac{1}{c^2} \int_V d^3r' \{ \frac{\partial \vec{E}(\vec{r}',t')}{\partial t'}$$

$$\times S(\vec{r},\vec{r}',t,t') - \frac{\partial S(\vec{r},\vec{r}',t,t')}{\partial t'} \vec{E}(\vec{r}',t') \} \quad t'=(t-R/c) \quad . \quad (5.9)$$

The solution to Eq. (5.7) is very difficult to obtain because we
have posed the problem with a general boundary. However, from the
nature of the transient propagator $S(\vec{r},\vec{r}',t,t')$ it is clear that
the field at any point inside the medium consists, in addition to
the incident field, of transient fields whose time evaluation is
characterized by the propagator $S(\vec{r},\vec{r}',t,t')$. The causal behavior
of the field can be seen from (5.9), as the field inside the medium
is related to fields at time t' = t - R/C. A detailed account of
this will be given elsewhere.

Appendix

In order to implement and illustrate the theory described in
Section 3 of this paper we require a specific form for the new
non-local susceptibility. In Ref. 3 the analysis was carried out
using the form derived by Zeyher, Brenig and Birman[4] for the
linear dielectric response function of a medium near a single
exciton resonance including exciton center of mass motion
(Frenkel Exciton), and coherent exciton reflection at the z=0
boundary. This susceptibility is

$$\chi(\vec{r},\vec{r}',\omega) = \{\chi_0\delta(\vec{r}-\vec{r}') + \chi_1 G(\vec{r}-\vec{r}') - \chi_1 G(\xi)\theta(z)\theta(z')\} \quad (A.1)$$

with $\xi \equiv |\vec{r}_\perp - \vec{r}'_\perp + \hat{z}(z + z')|$ and $\theta(z)$ is the Heaviside function;
χ_0 and χ_1 are constants. The propagating plane wave solutions in
the volume satisfy the equation

$$k^2 = k_0^2 \, \varepsilon_{EFF}(\vec{k},\omega) \quad (A.2)$$

where k is the wave number of the mode, k_0 is the vacuum wave
number = ω/c, and

$$\varepsilon_{EFF}(k,\omega) = \left[1 + \frac{4\pi}{(2\pi)^{3/2}} \left(\chi_0 + \frac{4\pi\chi_1}{k^2 - \kappa^2(\omega)} \right) \right] \quad , \quad (A.3)$$

$$\kappa(\omega) = \left(\frac{(\omega_0^2 - \omega^2) + \omega^2\Gamma^2}{b^2} \right)^{1/4} \exp i\left(\frac{\theta}{2} + \pi \right) \quad , \quad (A.4)$$

$$\tan\theta = -\omega\Gamma/(\omega_0^2 - \omega^2); \quad -\pi < \theta < 0 \quad , \tag{A.5}$$

$$b \equiv h\omega_0/M^* \quad ,$$

where M^* is the total exciton mass. For normal incidence the additional boundary conditions (abc) are

$$\sum_{j=1}^{3} \vec{E}_j / \left(k_j^2 - \kappa^2(\omega) \right) \quad , \tag{A.6}$$

where $\vec{E}_j$ is the electric field, and k_j is the j'th wave number solution of the dispersion equation (A.2). There are three such k_j with which we are concerned: for $\omega > \omega_0$ the two solutions will be denoted $k^+(\omega)$ and $k^-(\omega)$ with (if $\omega \gg \omega_0$)

$$k^+(\omega) \approx \omega/\sqrt{b} + \frac{i\Gamma}{2\sqrt{b}} - \frac{\omega_0^2}{2b^{\frac{1}{2}}(\omega+ir)} \quad , \tag{A.7}$$

$$k^-(\omega) \sim \sqrt{\varepsilon_0}\,(\omega/c) - \left(\frac{4\pi\chi_1\hbar}{M^*\omega_0}\right)\frac{\omega_0^2}{2\sqrt{\varepsilon_0}c(\omega+i\Gamma)} \quad . \tag{A.8}$$

In the *local* case only $k^-(\omega)$ occurs: $k^+(\omega)$ corresponds to the additional wave and is the reason the abc are needed. Asymptotic evaluation of the integral by the saddle point method using (A.7) and (A.8) gives contributions of form

$$f_+(z,t) = \mathcal{A}_+(z,t)\cos\left\{\omega_s^+(t-t_+)+\pi/4\right\}\theta(t-t_+) \quad , \tag{A.9}$$

$$f_-(z,t) = \mathcal{A}_-(z,t)\cos\left\{\omega_s^-(t-t_-)+\pi/4\right\}\theta(t-t_-) \quad , \tag{A.10}$$

where $\mathcal{A}_\pm$ is a function of z, and a slowly varying function of $(t-t_\pm)$. The expressions (A.9) and (A.10) are, respectively, the new Exciton Precurson (EP) and the Sommerfeld Precursor (SP).

For low frequency $\omega \ll \omega_0$ one can use

$$k_+(\omega) \approx c_1\omega + c_2\omega^2 + c_3\omega^3 \quad , \tag{A.11}$$

where c_1, c_2, c_3 are constants which depend on the parameters $(b,\varepsilon_0,\alpha_0,\Gamma)$ and one can obtain a Brillouin Precursor of the form (A.9), making appropriate substitutions; or, taking account of the possibility of a divergence in the amplitude by expanding the phase to third order, one finds

$$f_{BP}(z,t) \cong \mathcal{A}'_{BP}(z,t)\,Ai\left(a(t-t_{BP})\right) \quad . \tag{A.12}$$

Assembling (A.9), (A.10) and (A.12) one obtains the expression cited in the text (3.6).

---

* Supported in part by the National Science Foundation, DMR76-20641, Army Research Office, DAHCO4-75-G-0052, and FRAP-CUNY 11453.

** Present address: Naval Surface Weapons Laboratory, Silver Springs, Maryland.

References

1.   See L. Brillouin *Wave Propagation and Group Velocity* (Academic Press, New York, 1960); papers of Sommerfeld and Brillouin are given there along with references. Also see *Transient Electromagnetic Fields*, ed. L.B. Felsen (Springer-Verlag, New York, 1976).
2.   E.G. Skrotskaya et al., Sov. Phys. JETP *29*, 123 (1969).
3.   M.J. Frankel and J.L. Birman, Phys. Rev. A *15*, 2000 (1977); J.L. Birman and M.J. Frankel, Opt. Comm. *3*, 303 (1975).
4.   See J.J. Sein, Ph.D. Thesis, New York University (1969); J.J. Sein and J.L. Birman, Phys. Rev. B *6*, 2482 (1972); G.S. Agarwal, D.N. Pattanayak, E. Wolf, Phys. Rev. Lett. *27*, 1022 (1971); Phys. Rev. B *10*, 1477 (1972); A.A. Maradudin and D.L. Mills, Phys. Rev. B *7*, 2787 (1973); R. Zeyher, J.L. Birman, W. Brenig, Phys. Rev. A *6*, 4613 and 4617 (1972).
5.   See E. Gitterman and M. Gitterman, Phys. Rev. A *13*, 763 (1976), who discuss the integral for the local case, extending Sommerfeld's work in several ways.
6.   See Ref. 3 and M.J. Frankel, Ph.D. Thesis, New York University (1975).
7.   D.N. Pattanayak and E. Wolf, Opt. Comm. *6*, 217 (1972).
8.   D.N. Pattanayak and J.L. Birman, to be published.
9.   P.M. Morse and H. Feshbach, *Methods of Theoretical Physics* (McGraw-Hill Book Co., New York, 1953) Vol. I, p. 856.

AN INTEGRAL EQUATION APPROACH TO THE THEORY OF OPTICAL RESONANCE[*]

Deva N. Pattanayak

City College, CUNY, New York, N.Y.

1. INTRODUCTION

In a bounded material medium, embedded in vacuuo, solutions
to Maxwell's equations can satisfy the outgoing boundary conditions
at infinity and the continuity conditions at the boundary even
without the presence of an incident electromagnetic field. Such
solutions are broadly classified as natural oscillations which can
further be divided into radiative modes and nonradiative modes
depending on whether the frequencies are complex or real. Surface
waves have complex wave vectors but real frequencies and hence
belong to the class of non-radiative natural modes. On the other
hand virtual, or resonance, modes have complex frequencies (and
hence, in general, have complex wave vectors) and belong to the
class of radiative natural modes. Clearly the frequencies of the
natural oscillations depend upon the geometry of the material medium
as much as upon the different physical parameters of the medium.
Natural oscillations satisfy, in addition to the normal dispersion
relations (the constraint due to the Maxwell wave equation), another
dispersion relation, i.e. the natural mode dispersion relation (also
known as the surface wave dispersion relation), a constraint necessary in order that the natural oscillations satisfy the Maxwell
continuity conditions at the boundary. In other words, the natural
mode dispersion relation is nothing but the eigenvalue equation
for the existence of these modes. Several investigations [1] have
been made to obtain dispersion relations for natural modes. Several
ways [2] of exciting these modes have also been developed. In this
work we shall formulate the problem of natural oscillations from
the point of view of an integral equation formalism. We will clas-
sify the natural modes appropriately and show that surface modes
and virtual modes belong to the class of natural modes in the sense
that they satisfy the same homogeneous integral equation. The

situation is analogous to the case of quantum mechanical potential
scattering theory [3] where the bound states and resonance states
also satisfy one homogeneous integral equation. We then develop an
optical resonance theory for a model medium analogous to the theory
of resonance in quantum mechanical potential scattering theory. Our
solution to the scattering integral equation follows closely a method
introduced by C. Miranda [4] to solve integral equations whose kernel
depends analytically on a parameter λ. The usual eigenfunction
expansion method is not very useful since the natural modes do not
form an orthogonal set. We introduce an auxiliary set of orthogonal
virtual modes by defining an eigenvalue problem with a perturbed
boundary condition. The procedure is very similar to the method of
Kapur and Peierls [5] in Nuclear Reaction Theory. We then express
the interior fields in terms of these orthogonal eigenmodes by using
the eigenfunction expansion method.

2. INTEGRAL EQUATION FORMULATION OF NATURAL OSCILLATIONS

Natural oscillations of a material medium are solutions
to Maxwell's equations which satisfy the outgoing boundary condition
at infinity. In the case of a finite medium with sharp boundary
separating the medium from vacuum, electromagnetic fields that
satisfy Maxwell's equations both inside and outside the medium are
subject to the Saltus conditions at every point on the surface.
Recent work [6] has shown that the Saltus conditions at the boundary
surface and the outgoing radiation condition at infinity implies
non-local boundary conditions on the electromagnetic fields at the
surface of the material medium. For our case, with no sources
present outside the material medium the non-local boundary condition
reduces to [7]

$$\int_{S^-} \left\{ ik \left(\underline{n}x\underline{H}(\underline{r}';\omega) + \frac{4\pi}{c}\,\underline{K}(\underline{r}';\omega) \right) \cdot \underline{\underline{G}}(\underline{r},\underline{r}';\omega) + \underline{n}x\underline{E}(\underline{r}';\omega) \right.$$

$$\left. \cdot \nabla x\underline{\underline{G}}(\underline{r},\underline{r}';\omega) \right\} ds' = 0 \quad , \tag{1}$$

and

$$\int_{S^-} \left\{ -ik \left(\underline{n}x\underline{E}(\underline{r}';\omega) \right) \cdot \underline{\underline{G}}(\underline{r},\underline{r}';\omega) + \left(\underline{n}x\underline{H}(\underline{r}';\omega) + \frac{4\pi}{c}\,\underline{K}(\underline{r}';\omega) \right) \right.$$

$$\left. \cdot \nabla x\underline{\underline{G}}(\underline{r},\underline{r}';\omega) \right\} ds' = 0 \quad . \tag{2}$$

where $\underline{\underline{G}}$ is the free space dyadic Green's function related to the
corresponding scalar Green's function G_0 as

$$\underline{\underline{G}}(\underline{r},\underline{r}';\omega) = (\underline{\underline{U}} + \frac{1}{k^2} \nabla\nabla) \, G_0(\underline{r},\underline{r}';\omega) \quad , \tag{3}$$

where

$$G_0(\underline{r},\underline{r}';\omega) = \frac{e^{ik|\underline{r}-\underline{r}'|}}{|\underline{r}-\underline{r}'|} \quad , \tag{3a}$$

$$k = \frac{\omega}{c} \quad . \tag{4}$$

Here $\underline{n}$ is the unit outward normal at the surface S bounding the volume V of the medium, $\underline{K}$ is the surface current, and the superscript (-) in S indicates that the field quantities appearing in the integrand must be taken in a limiting sense from inside the medium. $\underline{\underline{U}}$ is the unit dyadic. Equations (1) and (2) do not contain material parameters and are thus universal relations applicable to material medium obeying any constitutive relations. For the interior of the medium the electromagnetic fields satisfy the following set of coupled partial differential equations:

$$\nabla\times(\nabla\times\underline{E}) - k^2\underline{E} = 4\pi \, [\, \frac{ik}{c} \, \underline{J} + k^2\underline{P} + ik \, \nabla\times\underline{M}] \quad , \tag{5}$$

$$\nabla\times(\nabla\times\underline{H}) - k^2\underline{H} = 4\pi \, [\, \frac{1}{c} \, \nabla\times\underline{J} + k^2\underline{M} - ik \, \nabla\times\underline{P}] \quad . \tag{6}$$

Equations (5), (6), (1) and (2) thus define an eigenvalue problem for natural modes for a material medium of arbitrary shape, obeying any prescribed constitutive relations. It is obvious that the set of Eqs. (1), (2), (5) and (6) possesses solutions for some discrete set of frequencies which we will call the natural frequencies and the corresponding eigenfunctions, the natural oscillations. We have assumed a steady state situation and that the field vectors vary in time as $\underline{F}(r,\omega)e^{-i\omega t}$. It can be shown that the partial differential Eqs. (5) and (6) and the non-local boundary conditions (1) and (2) lead to the following set of coupled homogeneous integro-differential equations:

$$\underline{E}(\underline{r};\omega) = \nabla\Big(\nabla \cdot \int_V \{\underline{P}(\underline{r}';\omega) + \frac{i}{\omega} \, \underline{J}(\underline{r}';\omega)\}G_0(\underline{r},\underline{r}';\omega) \, d^3r'\Big)$$

$$+ \, ik \, \nabla\times \int_V \underline{M}(\underline{r}';\omega) \, G_0(\underline{r},\underline{r}';\omega) \, d^3r'$$

$$+ \, k^2 \int_V \Big(\underline{P}(\underline{r}';\omega) + \frac{i}{\omega} \, \underline{J}(\underline{r}';\omega)\Big) \, G_0(\underline{r},\underline{r}';\omega) \, d^3r'$$

$$+ \, \frac{i}{\omega} \, \nabla\times\Big(\nabla\times \int_S \underline{K}(\underline{r}';\omega) \, G_0(\underline{r},\underline{r}';\omega) \, ds'\Big) \, ; \tag{7}$$

$$\underline{H}(\underline{r};\omega) = \nabla\left(\nabla \cdot \int_V \underline{M}(\underline{r}';\omega)\, G_0(\underline{r},\underline{r}';\omega)\, d^3r'\right)$$

$$- ik\nabla x \int_V \left(\underline{P}(\underline{r}';\omega) + \frac{i}{\omega}\,\underline{J}(\underline{r}';\omega)\right) G_0(\underline{r},\underline{r}';\omega)\, d^3r'$$

$$+ k^2 \int_V \underline{M}(\underline{r}';\omega)\, G_0(\underline{r},\underline{r}';\omega)\, d^3r' \quad . \tag{8}$$

We note that the definition of natural modes leads to the coupled
integro-differential Eqs. (7) and (8) and, alternatively, we may
define the natural oscillations to be solutions of Eqs. (7) and (8).
Addition of an incident electric field to the r.h.s. of Eq. (7) and
an incident magnetic field to the r.h.s. of Eq. (8) would yield
scattering integral equations and therefore the natural modes are
solutions to the scattering problem with no incident field. In
order to develop the theory of natural oscillations in analogy with
the theory of resonance in nuclear reaction we will from now on
consider the simple case when the medium is a non-magnetic, linear
isotropic, homogeneous and local dielectric.

We take as an example: natural oscillations in a dielectric
medium. We consider a material medium whose constitutive relations
are

$$\underline{P}(\underline{r};\omega) = \chi(\omega)\, \underline{E}(\underline{r};\omega) \;, \tag{9}$$

$$\underline{M}(\underline{r};\omega) = 0 \;, \tag{10}$$

$$\underline{K}(\underline{r};\omega) = 0 \;. \tag{11}$$

Equations (7) and (8) then take the form

$$\underline{E}(\underline{r};\omega) = k^2\chi(\omega) \int_V G_0(\underline{r},\underline{r}';\omega)\, \underline{E}(\underline{r}';\omega)\, d^3r'$$

$$+ \chi(\omega)\, \nabla\left(\nabla \cdot \int_V G_0(\underline{r},\underline{r}';\omega)\, \underline{E}(\underline{r}';\omega)\, d^3r'\right) \;, \tag{12}$$

$$\underline{H}(\underline{r};\omega) = -ik\chi(\omega)\, \nabla x \int_V G_0(\underline{r},\underline{r}';\omega)\, \underline{E}(\underline{r}';\omega)\, d^3r'$$

$$= \frac{1}{ik}\, \nabla x \underline{E}(\underline{r};\omega) \;. \tag{13}$$

We see that for the case of no-retardation, i.e. $c \to \infty$, Eqs. (12) and
(13) reduce to

$$\underline{E}(\underline{r};\omega) = \chi(\omega)\, \nabla\left(\nabla \cdot \int_V \frac{\underline{E}(\underline{r}';\omega)}{|\underline{r} - \underline{r}'|}\, d^3r'\right) \;. \tag{14}$$

$$\underline{H}(\underline{r};\omega) = 0 \;. \tag{15}$$

From Eq. (14) we also have

$$\nabla \times \underline{E}(\underline{r};\omega) = 0 \ . \tag{16}$$

On taking the divergence of both sides of Eq. (14) we have for $4\pi\chi(\omega) \neq -1$,

$$\nabla \cdot \underline{E}(\underline{r};\omega) = 0 \ . \tag{17}$$

So for the case of no retardation the natural modes are solutions for which $\nabla \cdot E = 0$ and $\nabla \times E = 0$. This agrees with the result due to Fuchs and Kliewer [8]. On taking divergence of Eq.(12) we find that

$$\nabla \cdot \underline{E}(\underline{r};\omega) = 0 \ , \quad \text{for } 4\pi\chi(\omega) \neq -1 \ , \tag{18}$$

but $\nabla \times E \neq 0$. Thus, with retardation the natural modes are transverse in nature. We rewrite Eq. (12) in the tensorial notation as follows

$$E_i(\underline{r};\omega) = \int_V K_{ij}(\underline{r},\underline{r}';\omega) \ E_j(\underline{r}';\omega) \ d^3r', \tag{19}$$

where

$$K_{ij}(\underline{r},\underline{r}';\omega) = K_{ji}(\underline{r}',\underline{r};\omega) = \chi(\omega)(k^2\delta_{ij} + \nabla_i\nabla_j)G_0(\underline{r},\underline{r}';\omega) \ . \tag{20}$$

The corresponding scattering integral equation for electromagnetic scattering of an incident field $\underline{E}^{(0)}(\underline{r};\omega)$ is

$$E_i(\underline{r};\omega) = E_i^{(0)}(\underline{r};\omega) + \int_V K_{ij}(\underline{r},\underline{r}';\omega) \ E_j(\underline{r}';\omega) \ d^3r' \ . \tag{21}$$

According to our previous discussion the homogeneous integral equation is equivalent to the following eigenvalue problem

$$\nabla^2\underline{E}(\underline{r};\omega) + k^2\epsilon(\omega) \ \underline{E}(\underline{r},\omega) = 0 \ , \quad \epsilon(\omega) \equiv 1 + 4\pi\chi(\omega) \ , \tag{22}$$

where $\underline{E}(\underline{r},\omega)$ is subject to the boundary condition

$$\int_{S^-}\left\{ik\left(\underline{n}\times\underline{H}(\underline{r}';\omega)\right) \cdot \underline{\underline{G}}(\underline{r},\underline{r}';\omega) + \left(\underline{n}\times\underline{E}(\underline{r}';\omega)\right)\cdot \nabla\times\underline{\underline{G}}(\underline{r},\underline{r}';\omega)\right\}ds = 0 \tag{23}$$

for all $\underline{r}$ in V.

Although Eq. (22) has solutions for all values of ω, the boundary condition (23) restricts ω to the set of natural frequencies. We will now show that the eigenfrequencies corresponding to different eigenvalues ω_n are not orthogonal to each other. For this purpose we obtain from (22) the relation

$$\left(k_n^2 \ \epsilon(\omega_n) - k_m^2 \ \epsilon(\omega_m)\right) \int_V \underline{E}_n(\underline{r}';\omega_n) \cdot \underline{E}_m(\underline{r}';\omega_m) \ d^3r' =$$

$$= \int_{S^-} \left\{ \underline{E}_n(\underline{r}';\omega_n) \cdot \frac{\partial \underline{E}_m(\underline{r}';\omega_m)}{\partial n} - \frac{\partial \underline{E}_n(\underline{r}';\omega_n)}{\partial n} \cdot \underline{E}_m(\underline{r}';\omega_m) \right\} ds'. \quad (24)$$

For $m \neq n$, the r.h.s. is not equal to zero due to the nature of the boundary conditions (23). This can be seen explicitly for simple geometries of the surface such as half space, sphere, etc. Thus the natural modes do not form an orthogonal set. Let us briefly mention other properties of the natural modes. It can be shown by standard techniques that the homogeneous integral equation can have non-trivial solutions at real ω for which the scattering integral equation (21) has a solution, only if the following condition is met.

$$\int_V E_i^{(0)}(\underline{r};\omega) \, E_i^{(N)}(\underline{r};\omega) \, d^3r = 0 \quad , \quad (25)$$

where the superscript (N) denotes the natural oscillation modes. Such real ω solutions are known as non-radiative modes. When Eq. (25) is satisfied, the solution to the scattering problem is necessarily non-unique.

We may, however, have solutions to the homogeneous integral Eq.(19) at frequencies which are not real but complex. We must then restrict the imaginary part of the complex frequencies to take on only negative values in view of the $e^{-i\omega t}$ time dependence of the fields. We therefore require the condition that

$$\text{Im } \omega_n < 0 \quad . \quad (26)$$

Complex frequency solutions are also known as virtual modes or radiative modes. We conclude this section by noting that the incident field does not couple with non-radiative or real ω solutions (this can be seen from Eq. (25)). However, as Eq. (25) is not valid for virtual modes, these modes can couple with the incident field and we deal with this problem in the next section.

3.   SCATTERING VIA NATURAL OSCILLATIONS

From the discussion in the previous section it is clear that the incident radiation field can couple to the virtual modes which are natural oscillations with complex frequencies. The situation thus is similar to the theory of potential scattering resonances. Recently an integral equation approach to the theory of quantum mechanical potential scattering resonances has been developed by the author [9] which is based on an approach due to C. Miranda in solving integral equations whose kernel depends analytically on a parameter λ. We follow a similar approach here. We shall assume that the natural oscillations form a complete set and expand the electric field as follows

$$E_i(\underline{r};\omega) = E_i^{(0)}(\underline{r},\omega) + E_i^{(P)}(\underline{r};\omega) + \sum_m \frac{A_m E_i^{(N)}(\underline{r};\omega_m)}{\omega - \omega_m} \quad , \qquad (27)$$

where $E_i^{(P)}(\underline{r},\omega)$ is a particular solution which we take to be a regular function of ω, and $E_i^{(N)}(\underline{r},\omega_m)$ are the natural oscillation eigenfunctions corresponding to eigenvalue ω_m. The completeness of the natural oscillations have been proved by B.J. Hoenders [10] for the case of a spherical model dielectric medium. We choose the normalization such that

$$\int_V E_i^{(N)}(\underline{r},\omega_m) \; E_i^{(N)}(\underline{r};\omega_m) \; d^3r = 1 \; . \qquad (28)$$

We substitute Eq. (27) into Eq. (19) and expand the functions around ω_m in Taylor Series and equate the coefficients of $(\omega-\omega_m)$ to obtain the following results.

$$A_n = \frac{-\displaystyle\int_V E_i^{(0)}(\underline{r};\omega_n) \; E_i^{(N)}(\underline{r};\omega_n) \; d^3r}{\displaystyle\int_V\int_V \frac{\partial K_{ij}(\underline{r},\underline{r}';\omega_n)}{\partial\omega} E_i^{(N)}(\underline{r};\omega_n) \; E_j^{(N)}(\underline{r}';\omega_n) d^3r \; d^3r'} \quad , \qquad (29)$$

$$E_i^{(P)}(\underline{r};\omega_n) = R_i(\underline{r};\omega_n) + \int_V K_{ij}(\underline{r},\underline{r}';\omega_n) \; E_j^{(P)}(\underline{r}';\omega_n) \; d^3r', \qquad (30)$$

where

$$R_i(\underline{r};\omega_n) = A_n \int_V \frac{\partial K_{ij}(\underline{r},\underline{r}';\omega_n)}{\partial\omega} E_j^{(N)}(\underline{r}';\omega_n) \; d^3r'$$

$$+ \int_V K_{ij}(\underline{r},\underline{r}';\omega_n) \; E_j^{(0)}(\underline{r}';\omega_n) d^3r' - \sum_{m\neq n} \frac{A_m E_j^{(N)}(\underline{r}';\omega_m)}{(\omega_n - \omega_m)}$$

$$+ \sum_{m\neq n} \frac{A_m}{\omega_n - \omega_m} \int_V K_{ij}(\underline{r},\underline{r}';\omega_n) \; E_j^{(N)}(\underline{r}';\omega_m) \; d^3r' \; . \qquad (31)$$

Equation (30) may be solved by iteration and we note here that $E_i^{(P)}(\underline{r},\omega_n)$ is a complicated function of natural oscillation eigenfunctions other than ω_n. We shall not be interested in its evaluation here. The volume integral in Eq. (29) can be shown to be equivalent to a surface integral if we use Eq. (22) and the equation satisfied by $E_i^{(0)}(\underline{r};\omega)$, namely

$$(\nabla^2 + k^2)\underline{E}^{(0)}(\underline{r};\omega) = 0 \; ; \quad \nabla \cdot \underline{E}^{(0)}(\underline{r};\omega) = 0 \; . \tag{32}$$

where

$$k = \frac{\omega}{c} \; . \tag{33}$$

Then we have

$$A_n = \frac{i \int_S \{\underline{E}^{(N)}(\underline{r}';\omega_n) x \underline{H}^{(0)}(\underline{r}';\omega_n) + \underline{H}^{(N)}(\underline{r}';\omega_n) x \underline{E}^{(0)}(\underline{r}';\omega_n)\} \cdot d\underline{s}}{k_n(\varepsilon(\omega_n)-1) \int_V \int_V \frac{\partial K_{ij}(\underline{r},\underline{r}';\omega_n)}{\partial \omega} E_i^{(N)}(\underline{r}';\omega_n) E_j^{(N)}(\underline{r}';\omega_n) \, d^3r \, d^3r'} \; . \tag{34}$$

We have therefore expressed the field inside the medium in terms of
a scattering term $(E_i^{(P)}(\underline{r},\omega))$ and in terms of natural oscillations
$(E_i^{(N)}(\underline{r},\omega_n))$. For the case of non-radiative natural modes, i.e. for
surface waves, $A_n = 0$ and therefore the incident field does not
couple to the surface wave. We have discussed this point before.
For virtual modes Eq. (25) is no longer valid and Eq. (34) in conjunction with Eq. (27) couples the incident radiation to the virtual
modes. When the frequency of the incident field equals the real part
of the frequencies of the virtual modes, resonance occurs. Behavior
of the electromagnetic fields similar to this has been reported by
Fuchs, Kliewer and Padree [11] from a different approach. The fields
outside the medium can be written down from the knowledge of the
boundary values of the fields inside the medium [6] but we shall not
do so here.

4.   ORTHOGONAL EIGENMODES: EXPANSION OF ELECTRIC FIELD
        IN TERMS OF THESE EIGENFUNCTIONS

As we have discussed earlier the natural modes do not form an
orthogonal set. The situation is similar to the case of potential
scattering theory, where the resonance wavefunctions do not form an
orthogonal set. Kapur and Peierls [5] introduced an orthogonal set
of eigenmodes by perturbing the boundary condition and these modes
have been very useful in nuclear reaction theory. We now define a
set of eigenmodes which are solutions to the following eigenvalue
problem: the eigenmodes satisfy the differential equation

$$\nabla^2 \underline{E}(\underline{r};\tilde{\omega}_n) + \tilde{k}_n^2 \, \varepsilon(\tilde{\omega}_n) \, \underline{E}(\underline{r};\tilde{\omega}_n) = 0 \tag{35}$$

and are subject to the boundary condition

$$\int_{S^-} \left\{ i\tilde{k}_n \left(\underline{n} x \underline{H}(\underline{r}';\tilde{\omega}_n) \right) \cdot \underline{\underline{G}}(\underline{r},\underline{r}';\omega) + \left(\underline{n} x \underline{E}(\underline{r}';\tilde{\omega}_n) \right) \cdot \nabla x \underline{\underline{G}}(\underline{r},\underline{r}';\omega) \right\} ds' = 0.$$

(36)

Obviously the eigenmodes and eigenfrequencies will be a function of the arbitrary frequency ω. We may take this frequency ω to be the same as that of the incident electromagnetic field. It can now be shown that

$$\left(\tilde{k}_n^2 \varepsilon(\tilde{\omega}_n) - \tilde{k}_m^2 \varepsilon(\tilde{\omega}_m) \right) \int_V \underline{E}(\underline{r};\tilde{\omega}_n) \cdot \underline{E}(\underline{r};\tilde{\omega}_m) \, d^3r = 0 .$$

(37)

This is because both $\underline{E}(\underline{r};\tilde{\omega}_n)$ and $\underline{E}(\underline{r};\tilde{\omega}_m)$ satisfy the same boundary condition (36) at the surface of the scattering medium. Therefore we have

$$\int_V \underline{E}(\underline{r};\tilde{\omega}_n) \cdot \underline{E}(\underline{r};\tilde{\omega}_m) \, d^3r = N_n \, \delta_{nm} ,$$

(38)

where N_n is a normalization constant. If we write the field inside as

$$\underline{E}(\underline{r};\omega) = \underline{E}^{(0)}(\underline{r},\omega) + \underline{E}^{(S)}(\underline{r};\omega) ,$$

(39)

we find from the non-local boundary conditions of electromagnetic theory (Eq. 23) that $\underline{E}^{(S)}(\underline{r};\omega)$ satisfy the same boundary condition as the above set of eigenmodes. Therefore assuming these eigenmodes to form a complete set, we expand $\underline{E}^{(S)}(\underline{r};\omega)$ in terms of these eigenmodes to obtain

$$\underline{E}^{(S)}(\underline{r};\omega) = \sum_n d_n \underline{E}(\underline{r};\tilde{\omega}_n) ,$$

(40)

where

$$d_n = \frac{1}{N_n} \int_V \underline{E}^{(S)}(\underline{r};\omega) \cdot \underline{E}(\underline{r};\tilde{\omega}_n) \, d^3r .$$

(41)

We now convert the volume integral in (41) into a surface integral involving the boundary values of the eigenmodes $\underline{E}(\underline{r},\tilde{\omega}_n)$ by means of several vector manipulations to obtain

$$d_n = \frac{k^2 \left(\varepsilon(\omega) - 1 \right) \int_S i\tilde{k}_n \left(\underline{E}^{(0)}(\underline{r}';\omega) x \underline{H}(\underline{r}';\tilde{\omega}_n) + \underline{H}^{(0)}(\underline{r}';\omega) x \underline{E}(\underline{r}';\tilde{\omega}_n) \right) \cdot d\underline{S}}{N_n \left(k^2 - \tilde{k}_n^2 \varepsilon(\tilde{\omega}_n) \right) \left(k^2 \varepsilon(\omega) - \tilde{k}_n^2 \varepsilon(\tilde{\omega}_n) \right)}$$

(42)

Equation (40) in conjunction with Eq. (42) thus expresses the scattered field $\underline{E}^{(S)}(\underline{r};\omega)$ in terms of the boundary values of these orthogonal eigenmodes. Comparison with Eqs. (37) and (34) would

show that this eigenfunction expansion is much simpler than the
former; furthermore, the solution now depends only on the boundary
values of a set of eigenfunctions. Equations (40) and (42) may be
useful in boundary value problems in electromagnetic theory.
Resonance behavior is evident from the fact that d_n has a maximum
when $\omega = R_\ell \tilde{\omega}_n$.

CONCLUSION

We have presented the theory of natural modes in electromagnetic theory from an integral equation approach. This approach
leads to an expansion theorem for the fields inside the medium in
terms of the natural modes which are not orthogonal. An auxiliary
set is introduced which is an orthogonal set and the eigenfunction
expansion method leads to a simple expression for the interior
fields in terms of boundary values of these eigenmodes. Both the
expansions show resonance behavior in the sense that when the frequency of the incident electromagnetic field approaches the real
part of the complex eigenfrequencies of the eigenmodes, the interior
field exhibits a maximum in its amplitudes (A_n and d_n). The field
outside the scattering medium may be obtained in terms of the
boundary values of the interior field.

Acknowledgement

The author wishes to acknowledge the stimulating discussions
he had with Professor E. Wolf at the University of Toronto on the
subject of natural modes, and is indebted to Professor J.L. Birman
for discussions and critical reading of this manuscript.

* Work supported by National Science Foundation, DMR76-20641;
Army Research Office, DAHCO4-75-G-0052 and Faculty Research Award
Program, 11453.

References

1.   See for example the recent review articles by
     (a) E.N. Economu and K.L. Ngai, *Advances in Chemical Physics*
     (John Wiley and Sons, New York, 1974) Vol. XXVII, p. 265.
     (b) R. Fuchs and K.L. Kliewer, *Advances in Chemical Physics*
     (John Wiley and Sons, New York, 1974) Vol. XXVII, p. 355.
2.   A. Otto, Z. Phys. *216*, 398 (1968). See also D.L. Mills and
     E. Burstein, Reports on Progress in Physics *37*, 817 (1974)
     Chapter 10 and reference (1)(b) paragraph VI, p. 339.

3. D.N. Pattanayak and E. Wolf, Phys. Rev. D *13*, 913 (1976) and Phys. Rev. D *13*, 2287 (1976).
4. C. Miranda, Acta. Pontifica Acad. Scient. *3*, 1 (1939).
5. P.L. Kapur and R. Peierls, Proc. Roy. Soc. London *A166*, 277 (1938).
6. (a) D.N. Pattanayak and E. Wolf, Opt. Comm. *3*, 217 (1972);
   (b) E. Wolf in *Coherence and Quantum Optics*, ed. L. Mandel and E. Wolf (Plenum, New York, 1973) p. 339;
   (c) J.J. Sein, Opt. Commun. *2*, 170 (1970);
   (d) J.D. Goede and P. Mazur, Physica *58*, 805 (1965) and Alta Frequenza *38*, 348 (1968).
7. (a) D.N. Pattanayak, Ph.D. Thesis, The University of Rochester (1973), Section 2.7 (unpublished);
   (b) G.S. Agarwal, Phys. Rev. B*8*, 4768 (1973).
8. R. Fuchs and K.L. Kliewer, Phys. Rev. *140*, A2076 (1965).
9. D.N. Pattanayak, to be published.
10. B.J. Hoenders, this Volume, p. 221.
11. R. Fuchs, K.L. Kliewer and W.J. Pardee, Phys. Rev. *150*, 589 (1966).

ON THE DECOMPOSITION OF THE ELECTROMAGNETIC FIELD INTO ITS NATURAL
MODES

B.J. Hoenders

State University at Groningen, Groningen, The Netherlands

1. INTRODUCTION

Recently, amongst others, Wolf and Pattanayak [1] generalized
the classical Ewald-Oseen extinction theorem, which is usually
regarded to be only valid for molecular optics, to electromagnetic
fields in the presence of an arbitrary material medium. Their
theory leads to a general definition of the so-called natural modes
of the electromagnetic field.

It has been shown by Agarwal [2] that certain sets of modes
which have been known for a long time, like the natural modes in
Mie's theory, are consistent with this definition. The natural
modes can be defined as solutions of Maxwell's equations subject to
a non-local boundary condition, or, alternatively, as the solution
of a vector integral equation. Both definitions will be used. The
problem considered in this paper concerns the formulation of a
Sturm-Liouville (S-L) type of theory for these modes. Unfortunately
we cannot for such a theory rely on the Hilbert-Schmidt theory,
because the eigenvalue enters *non-linearly* in the vector integral
equation. This implies the not always recognized (Stratton [3])
non-hermiticity of the problems. From a physical point of view the
non-hermitian character of the problem is evident. The eigenfrequencies necessarily have a non-zero imaginary part, as pointed out
by Wolfsohn [4], because otherwise a once excited mode would radiate
an infinite amount of energy. However, this would mean for a medium
characterized by a real-valued index of refraction that the eigenvalues of the boundary value problem are complex-valued, which is
in contradiction to S-L theory.

It will be shown that the natural modes of the electromagnetic field associated with a medium characterized by a scalar complex index of refraction are complete, and an expansion theorem will be formulated. Using a procedure by Weatherburn [5] the vectorial character of the vector integral equation will be retained. Finally we will show that the natural modes connected with Mie scattering and potential scattering by a spherically symmetrical potential form an overcomplete set. However, it is conjectured that if we impose the condition that the modes must represent real-valued field vectors or, alternatively must represent outgoing waves, that the physically distinguishable subset of these modes is just complete.

2. CALCULATIONAL PROCEDURE

From Maxwell's equations for monochromatic fields,

$$\nabla \times \underline{H} = \underline{j} - ik\underline{D}, \quad \nabla \cdot \underline{D} = \rho, \quad \underline{H}(\underline{r},t) = \underline{H}(\underline{r})\exp(-i\omega t) \ ,$$

$$\nabla \times \underline{E} = ik\underline{B}, \quad \nabla \cdot \underline{B} = 0, \quad \underline{E}(\underline{r},t) = \underline{E}(\underline{r})\exp(-i\omega t), \quad \omega = ck \ , \qquad (2.1)$$

and the material equations

$$\underline{D} = \varepsilon\,\underline{E}, \quad \underline{B} = \mu\underline{H}, \quad \text{and} \quad \underline{j} = \sigma\underline{E} \ , \qquad (2.2)$$

where ε, μ and σ are constants, we derive the equations

$$\nabla \times \nabla \times \underline{E} - k^2\varepsilon\mu\underline{E} = ik\mu\underline{j} \ , \qquad (2.3)$$

$$\nabla \times \nabla \times \underline{H} - k^2\varepsilon\mu\underline{H} = \nabla\times\underline{j} \ . \qquad (2.4)$$

With the help of the dyadic Green's function

$$\underline{\underline{G}}(\underline{r},\underline{r}';k) = (\underline{\underline{e}} + k^{-2}\nabla\nabla)\,\frac{\exp\,ik|\underline{r}-\underline{r}'|}{|\underline{r}-\underline{r}'|} \ , \qquad (2.5)$$

satisfying $(\nabla \times \nabla \times - k^2)\underline{\underline{G}} = \underline{\underline{e}}\delta(\underline{r}-\underline{r}')$, where $\underline{\underline{e}}$ denotes the unit dyadic, and the vectorial form of Sommerfeld's radiation condition, Eq. (2.3) and the following tensor analogue of Green's theorem (Morse and Feshbach [6]) applied to every column vector of the dyadic $\underline{\underline{L}}$

$$\int_\tau \{\nabla\times\nabla\times\underline{P}\cdot\underline{\underline{L}} - \underline{P}\cdot\nabla\times\nabla\times\underline{\underline{L}}\} \ d\tau = \int_\sigma \left((\underline{n}\times\nabla\times\underline{P})\cdot\underline{\underline{L}} + (\underline{n}\times\underline{P})\cdot(\nabla\times\underline{\underline{L}})\right) \ d\sigma \ , \qquad (2.6)$$

for well-behaved vectors $\underline{P}$ and dyadics $\underline{\underline{L}}$ leads to:

$$\underline{E}(\underline{r}_<) = \int_\tau \{ik\mu \underline{j}(\underline{r}') + k^2(n^2-1)\underline{E}(\underline{r}')\} \cdot \underline{\underline{G}}(\underline{r}_<,\underline{r}';k)\,d\underline{r}' - \frac{1}{4\pi}\,\Sigma^{(-)}(\underline{r}_<),$$

$$(2.7)$$

where

$$\Sigma^{(-)}(\underline{r}_<) = \int_\sigma \{ik\,(\underline{n}\times\underline{B}(\underline{r}'))\cdot\underline{\underline{G}}(\underline{r}_<,\underline{r}';k)$$

$$+ (\underline{n}\times\underline{E}(\underline{r}'))\cdot(\nabla\times\underline{\underline{G}}(\underline{r}_<,\underline{r}';k))\}\,d\sigma\quad,\qquad(2.8)$$

and $n^2 = \varepsilon\mu$. (A dyadic product is defined by a contraction of the dyadic, Morse and Feshbach [6], §1.6.) The minus sign occuring on the left-hand side of Eq. (2.8) means that the values of $\underline{E}$ and $\underline{B}$ have to be taken at the inside of the surface σ, and $\underline{r}_<$ means that $\underline{r}$ is situated inside the domain τ. Considering a volume bounded by $\underline{\sigma}$ and a sphere with infinite radius, Eqs. (2.3) and (2.6) lead to (Wolf [1])

$$\Sigma^{(+)}(\underline{r}_<) = \underline{E}^{(i)}(\underline{r}_<)\quad,\qquad(2.9)$$

where $\underline{E}^{(i)}$ denotes the electric field vector of the incoming wave. The continuity and the saltus conditions of the electromagnetic field across the boundary σ yields:

$$\frac{1}{4\pi}\left[\Sigma^{(+)}(\underline{r}_<) - \Sigma^{(-)}(\underline{r}_<)\right] = -ik\int_\sigma (\underline{n}\times\underline{M})\cdot\underline{\underline{G}}(\underline{r}_<,\underline{r}';k)\,d\sigma,\qquad(2.10)$$

where

$$\underline{B} = \underline{H} + 4\pi\underline{M}\;.$$

Similar results can be obtained for the magnetic field vector, and, on using $\underline{j} = \sigma\underline{E}$, we finally obtain a vector integral equation:

$$\begin{pmatrix}\underline{E}(\underline{r}_<)\\[1em]\underline{H}(\underline{r}_<)\end{pmatrix}^T = \begin{pmatrix}\underline{E}^{(i)}(\underline{r}_<)\\[1em]\underline{H}^{(i)}(\underline{r}_<)\end{pmatrix}^T + \int_\tau \begin{pmatrix}\underline{E}(\underline{r}')\\[1em]\underline{H}(\underline{r}')\end{pmatrix}^T \cdot \underline{\underline{\Gamma}}_0(\underline{r}_<,\underline{r}';k)\,d\underline{r}'\;,$$

$$(2.11)$$

with

$$\underline{\underline{\Gamma}}_0(\underline{r}_<,\underline{r}';k) =$$

$$\begin{pmatrix}\underline{\underline{G}}(\underline{r}_<,\underline{r}';k)\left(k^2(n^2-1)+ik\mu\sigma\right) & -ik\tilde{\varepsilon}\,\underline{n}\times\underline{\underline{G}}(\underline{r}_<,\underline{r}';k)\,\delta\left(\underline{r}'-\underline{r}'(\sigma)\right)\\[1.5em]-ik\tilde{\mu}\,\underline{n}\times\underline{\underline{G}}(\underline{r}_<,\underline{r}';k)\,\delta\left(\underline{r}-\underline{r}'(\sigma)\right) & \underline{\underline{G}}(\underline{r}_<,\underline{r}';k)\left(k^2(n^2-1)+ik\mu\sigma\right)\end{pmatrix},$$

$$(2.12a)$$

$$\tilde{\varepsilon} = \varepsilon^{+} - \bar{\varepsilon} \, , \qquad \tilde{\mu} = \mu^{+} - \bar{\mu} \, , \tag{2.12b}$$

and the surface delta function satisfies

$$\int_{\tau} \delta\!\left(\underline{r}' - \underline{r}'(\sigma)\right) f(\underline{r}') \; d\underline{r}' = \int_{\sigma} f(\underline{r}') \; d\sigma \, . \tag{2.13}$$

The unique solution of Eq. (2.11) can be written as

$$\begin{pmatrix} \underline{E}(\underline{r}_{<}) \\[6pt] \underline{H}(\underline{r}_{<}) \end{pmatrix}^{T} = \begin{pmatrix} \underline{E}^{(i)}(\underline{r}_{<}) \\[6pt] \underline{H}^{(i)}(\underline{r}_{<}) \end{pmatrix}^{T} + \int_{\tau} \begin{pmatrix} \underline{E}^{(i)}(\underline{r}') \\[6pt] \underline{H}^{(i)}(\underline{r}') \end{pmatrix}^{T} \cdot \underline{\underline{\Gamma}}(\underline{r}_{<}, \underline{r}'; k) \; d\underline{r}' \, ,$$
$$\tag{2.14}$$

provided that $\underline{\underline{\Gamma}}$ is the (unique) solution of the dyadic Fredholm functional equation (Weatherburn [5])

$$\underline{\underline{\Gamma}}(\underline{r}_{<}, \underline{r}'; k) = \underline{\underline{\Gamma}}_{0}(\underline{r}_{<}, \underline{r}'; k) + \int_{\tau} \underline{\underline{\Gamma}}(\underline{r}_{<}, \underline{r}''; k) \cdot \underline{\underline{\Gamma}}_{0}(\underline{r}'', \underline{r}'; k) d\underline{r}'' \, , \tag{2.15a}$$

or

$$= \underline{\underline{\Gamma}}_{0}(\underline{r}_{<}, \underline{r}'; k) + \int_{\tau} \underline{\underline{\Gamma}}_{0}(\underline{r}_{<}, \underline{r}''; k) \cdot \underline{\underline{\Gamma}}(\underline{r}'', \underline{r}'; k) \; d\underline{r}'' \, . \tag{2.15b}$$

Consider the vector $\int_{\tau}\underline{\underline{\Gamma}}(\underline{r}_{<}, \underline{r}'; k) \cdot \underline{f}(\underline{r}') d\underline{r}'$ for large values of $|k|$, where $\underline{f}$ denotes an arbitrary vector whose components are of bounded variation. From a basic theorem of our theory,

$$\left| \int_{\tau} \underline{\underline{\Gamma}}(\underline{r}_{<}, \underline{r}'; k) \cdot \underline{f}(\underline{r}') d\underline{r}' - (n^{2}-1) \underline{f}(\underline{r}_{<}) \right| = 0\,(1) \, ,$$
$$\text{if } |k| \to \infty \, , \; 0 < \arg k < 2\pi \quad , \tag{2.16}$$

it follows that $\underline{\underline{\Gamma}}$ tends to a delta dyadic distribution. An outline of the proof will be given in the appendix. With the help of Eq. (2.16), Mittag-Leffler's theorem yields (Whittaker and Watson [7] evaluate the first integral of §7.4 by the theorem of residues and asymptotically on the contour):

$$\int_{\tau} \underline{\underline{\Gamma}}(\underline{r}_{<}, \underline{r}'; k) \cdot \underline{f}(\underline{r}') d\underline{r}' = \sum_{n} \int_{\tau} \frac{\underline{\underline{\Gamma}}_{n}(\underline{r}_{<}, \underline{r}')}{k - k_{n}} \cdot \underline{f}(\underline{r}') \; d\underline{r}'$$
$$+ (n^{2}-1) \underline{f}(\underline{r}_{<}) \, , \tag{2.17}$$

where the numbers k_{n} denote the singularities of the resolvent $\underline{\underline{\Gamma}}$ and we used the property that the left-hand side of Eq. (2.17) is uniformly bounded for a discrete set of values $|k_{n}|$ tending to infinity. The first term of the Laurent expansion of Eq. (2.15b) around the point $k = k_{n}$ and Eq. (2.17) with $\underline{f}(\underline{r}') = \underline{\delta}(\underline{r}' - \underline{r}_{i}')$, where the δ function has to be interpreted as a distribution, leads to

$$\underline{\underline{\Gamma}}_n(\underline{r}_<,\underline{r}_1') = \int_\tau \underline{\underline{\Gamma}}_0(\underline{r}_<,\underline{r}'';k_n) \cdot \underline{\underline{\Gamma}}_n(\underline{r}'',\underline{r}_1') \ d\underline{r}'' \ . \tag{2.18}$$

Equation (2.18) shows that the dyadics $\underline{\underline{\Gamma}}_n$ are the eigendyadics connected with the homogeneous vector equation (2.18). The dyadic Eq. (2.18) can be considered as a set of vector integral equations. Each vector integral equation can be written as a scalar equation (Fredholm [8]). Applying a method due to Titchmarsh [9] to each scalar equation we infer the existence of dyadics $\underline{\underline{a}}_j(\underline{r}')$ such that

$$\underline{\underline{\Gamma}}_n(\underline{r}_<,\underline{r}') = \sum_{j=1}^{p_n} \underline{\underline{\psi}}_{j,n}(\underline{r}_<) \cdot \underline{\underline{a}}_{j,n}(\underline{r}') \ . \tag{2.19}$$

Combination of Eqs. (2.16), (2.17) and (2.19) yields:

$$\int_\tau \underline{\underline{\Gamma}}(\underline{r}_<,\underline{r}';k) \cdot \underline{f}(\underline{r}')d\underline{r}' \ - \ (n^2-1)\underline{f}(\underline{r}_<)$$

$$= \sum_\ell \sum_{j=1}^{p_\ell} \frac{\underline{\underline{\psi}}_{j,\ell}(\underline{r}_<)}{k-k_n} \ \int_\tau \underline{\underline{a}}_{j,\ell}(\underline{r}') \cdot \underline{f}(\underline{r}')d\underline{r}' \ , \tag{2.20}$$

which shows the desired completeness of the eigendyadics $\underline{\underline{\psi}}_{j,\ell}$ for every function $\ell(\underline{r}_<)$ which can be expressed in terms of the function $\underline{f}(\underline{r}_<)$ by the left-hand side of Eq. (2.20). This is always possible if $k \neq k_n$. However, we would like to express the dyadics $\underline{\underline{a}}_{j,n}(\underline{r}')$ in terms of the eigendyadics $\underline{\underline{\psi}}_{j,n}(\underline{r}')$. Such a relation is implicitly contained in a formula conjectured by Miranda [10] for a similar scalar case. The author is deeply indebted to Dr. D.N. Pattanayak, who communicated a proof of Miranda's theorem to him. The core of this proof will be used for the derivation at hand. Let us insert Eq. (2.20) with $\underline{f}(\underline{r}')=\delta(\underline{r}'-\underline{r}_1')$ into Eq. (2.15a) and let us multiply from the right by $\underline{\underline{\Gamma}}_m(\underline{r}',\overline{f})$. Integrating over $\underline{r}'$, using Eq. (2.18) and the conjugo-symmetrical property of the kernel $\underline{\underline{\Gamma}}_0$, (see [5]), leads to:

$$\left(\underline{\underline{a}}_{m,j}(\underline{r})\right)_{\ell,k} = \frac{\left(\underline{\underline{\psi}}_{m,j}(\underline{r})\right)_{\ell,k}}{\left\{\int_\tau d\underline{r}' \int_\tau d\underline{r}'' \ \underline{\underline{\psi}}_{m,j}(\underline{r}'') \cdot \frac{\partial}{\partial k} \underline{\underline{\Gamma}}_0(\underline{r}'',\underline{r}';k)\Big|_{k=k_m} \cdot \underline{\underline{\psi}}_{m,j}(\underline{r}')\right\}_{\ell,k}} \tag{2.21}$$

where $(\ )_{\ell,k}$ denotes the ℓ,k component of the tensor within the brackets. The paper presented by Dr. D.N. Pattanayak [22] during this conference also contains the Eq. (2.21) as well as an application of this theory to the special problem of scattering by a non-magnetic, linear isotropic, homogeneous and local dielectric. In the following section we will show that for two special cases the natural modes form an overcomplete set.

3. THE OVERCOMPLETENESS OF THE SET OF NATURAL MODES CONNECTED WITH MIE SCATTERING AND WITH POTENTIAL SCATTERING

Consider a sphere of radius a, characterized by dielectric permeability ε_1, magnetic permeability μ_1 and conductivity σ_1, embedded in an infinite medium characterized by ε_2, μ_2, and σ_2, where ε_j, μ_j, and σ_j, (j=1,2) denote scalar constants. The so-called E-type field with $H_r=0$ can be derived from the Debye potential ψ_E satisfying the Helmholtz equation $(\nabla^2 + k_1^2)\psi_E = 0$, where $k_1^2 = \varepsilon_1\mu_1\omega^2 - i\sigma_1\omega$, inside the sphere and the same equation with $k_2^2 = \varepsilon_2\mu_2\omega^2 - i\sigma_2\omega$ outside the sphere, (Born and Wolf [11]). The incoming field can be derived from the potential $\psi_E^{(i)}$ and the potential $\psi_H^{(i)}$ corresponding to fields with $E_r=0$. The scattered field of the electric type is obtained from the potential $\psi_E^{(s)}$ and satisfies Sommerfeld's radiation condition at infinity. Therefore, from Green's theorem, applied to the potential $\psi_E^{(t)} = \psi_E^{(i)} + \psi_E^{(s)}$ within the domain bounded by the surface r=a of the sphere and a sphere with infinite radius, we obtain:

$$\psi_E^{(i)}(\underline{r}_<) = \int\limits_{r=a} \{\psi_E^{(t)}(\underline{r}')\frac{\partial}{\partial n} G(\underline{r}_<,\underline{r}';k)-G(\underline{r}_<,\underline{r}';k)\frac{\partial}{\partial n} \psi_E^{(t)}(\underline{r}')\}d\sigma \ ,$$

$$(3.1)$$

provided $\underline{r}_<$ is situated inside the sphere. The values of the potential and its normal derivative are taken at the exterior boundary of r=a and are, by the continuity requirements of the electromagnetic field, connected with the values inside r=a by (Stratton [3], §7.11, and §9.22; see also [11])

$$\frac{k_1}{\mu_1} \psi_E^{(-)} = \frac{k_2}{\mu_2} \psi_E^{(+)} \ ; \quad \frac{1}{k_1} \frac{\partial}{\partial r} (r\psi_E^{(-)}) = \frac{1}{k_2} \frac{\partial}{\partial r} (r\psi_E^{(+)})\Big|_{r=a} . \quad (3.2)$$

The natural modes are defined as the solutions of Helmholtz's equation inside the sphere subject to the condition of absence of an incoming wave (Wolf, Pattanayak [12]). The condition for the absence of an incoming wave yields the non-local boundary condition

$$\int\limits_{r=a} \{\psi_E^{(+)}(\underline{r}')\frac{\partial}{\partial n} G(\underline{r}_<,\underline{r}';k) - G(\underline{r}_<,\underline{r}';k)\frac{\partial}{\partial n} \psi_E^{(+)}(\underline{r}')\}d\sigma = 0 \ , \quad (3.3)$$

to be satisfied for all values of $\underline{r}_<$ situated within the sphere. Using the conditions (3.2) and the orthonormality properties of the spherical harmonics we obtain from Eq. (3.3), using the expansions

$$\psi_E^{(-)}(\underline{r}') = \sum_{\ell,m} a_{\ell,m} \, j_\ell(k_1 r) \, Y_\ell^m(\theta',\phi') , \qquad (3.4)$$

and

$$G(\underline{r}_<,\underline{r}';k) = \sum_{\ell,m} 2ik \, j_\ell(kr) h_\ell^{(1)}(kr') Y_\ell^m(\theta,\phi) Y_\ell^m(\theta',\phi'), \qquad (3.5)$$

the set of local boundary conditions is

$$0 = \alpha_\ell(k_1,k_2) = \mu_1 h_\ell^{(1)}(k_2 a) \frac{\partial}{\partial(k_1 a)} \left(k_1 a j(k_1 a)\right)$$

$$- \mu_2 \frac{k_1^2}{k_2^2} j_\ell(k_1 a) \frac{\partial}{\partial(k_2 a)} \{k_2 a h_\ell^{(1)}(k_2 a)\} , \qquad (3.6)$$

$$(\ell = 0,1,\ldots) .$$

These relations have been obtained by Debye [13] and Mie [14] in
an entirely different way, namely by considering the singularities
of the scattering coefficients occurring in the analysis of plane
wave scattering by a sphere. The natural modes of the electric-
type field are therefore the set of functions $j_\ell(k_{\ell,n} r) Y_\ell^m(\theta,\phi)$,
where the numbers $k_{\ell n}$ are the roots of the equations (3.6), which
reduce to

$$0 = \alpha_\ell(k_1,n_r k_1) = \mu_1 h_\ell^{(1)}(n_r k_1 a) \frac{\partial}{\partial(k_1 a)} \left(k_1 a j_\ell(k_1 a)\right)$$

$$- \mu_2 n_r^{-2} j_\ell(k_1 a) \frac{\partial}{\partial(n_r k_1 a)} \left(n_r k_1 a h_\ell^{(1)}(n_r k_1 a)\right) , \qquad (3.7)$$

$$(\ell = 0,1,\ldots)$$

where $n_r^2 = \varepsilon_2 \mu_2 (\varepsilon_1 \mu_1)^{-1}$ and σ_1 and σ_2 are zero. We will show that
for this case the radial parts of the modes are overcomplete.
From Eq. (3.7) we deduce that if $k_{\ell,n}$ is a root of Eq. (3.7) then
$-k_{\ell,n}^*$ is also a root, as is to be expected from S-matrix theory
(Nussenzweig [15]).

Consider the contour integral

$$I(c) = \frac{1}{2\pi i} \oint_{|k|=c} \frac{j_\ell(k_1 r) \exp(i n_r k_1 a)}{\alpha_\ell(k_1,n_r k_1)(k_1 - b)} \prod_{j=1}^{P} \frac{k_1 + k_{\ell,j}^*}{k_1 + k_{\ell,j}} \, dk_1 ,$$

$$(3.8)$$

with $b \neq k_{\ell,n}$ for large values of c, which are chosen in such a way that the contour passes between two singularities of the integrand of Eq. (3.8). From the asymptotic behaviour of the spherical Bessel and Hankel functions, $j_\ell(kr) \sim (kr)^{-1}\sin(kr)$, $h_\ell^{(1)}(kr) \sim (kr)^{-1}\exp(ikr)$, we deduce that

$$\mathop{\mathrm{Lim}}_{c\to\infty} I(c) = 0, \qquad \text{if } r < a . \qquad (3.9)$$

From the theorem of residues we obtain:

$$j_\ell(kb) = c(b) \left\{ \sum_n \frac{j_\ell(k_{\ell,n}r)\exp(in_r k_{\ell,n}a)}{\alpha_\ell'(k_{\ell n},n_r k_{\ell,n})(k_{\ell,n}-b)} \prod_{j=1}^{P} \frac{k_{\ell,n}+k_{\ell,j}^*}{k_{\ell,n}+k_{\ell,j}} \right. $$

$$\left. + (-1)^\ell \sum_{j=1}^{P} \frac{j_\ell(k_{\ell,n}r)(-k_{\ell,j}+k_{\ell,j}^*)}{\alpha_\ell(k_1,n_r k_1)(-k_\ell-b)} \prod_{j'=1,j'\neq j}^{P} \frac{-k_{\ell,j}+k_{\ell,j'}^*}{-k_{\ell,j}+k_{\ell,j'}} \right\} ,$$

$$(b \neq k_{\ell,n}) , \qquad (3.10a)$$

where

$$c(b) = \exp(-in_r ab)\alpha_\ell(b,n_r b) \left\{ \prod_{j=1}^{P} \frac{b+k_{\ell,j}^*}{b+k_{\ell,j}} \right\}^{-1} . \qquad (3.10b)$$

The summation over n has to be taken over all the numbers $k_{\ell,n}$ *with the exception of the roots* $-k_{\ell,j}^*$, j=1,...p. Consider the complete set of functions $\{j_\ell(b_{\ell,n}r)\}$ with $j_\ell(b_{\ell,n}a) = 0$ in the interval $0 \leq r \leq a$. Because the numbers $b_{\ell,n}$ are real and the numbers $k_{\ell,n}$ have a non-vanishing imaginary part, no $b_{\ell,n}$ is equal to a $k_{\ell,n}$. The overcompleteness is easily deduced from Eq. (3.10a) which shows that every member of the complete set $\{j_\ell(b_{\ell,n}r)\}$ can be written as a linear combination of natural modes with the exclusion of any arbitrary chosen finite subset. However, we conjecture that for physically admissible fields the set of distinguishable modes is just complete. In order to elucidate this conjecture we remark that the field vectors have to be real valued. The relation

$$\mathrm{Re}\{A_1 h_\ell^{(1)}(k_{\ell,n}r)\exp(-i\omega_{\ell,n}t) + A_2 h_\ell^{(1)}(-k_{\ell,n}^* r)\exp(i\omega_{\ell,n}^* t)\}$$

$$= \mathrm{Re}\{(A_1+(-1)^\ell A_2^*)h_\ell^{(1)}(k_{\ell,n}r)\exp(-i\omega_{\ell,n}t)\}, \qquad (\omega=ck) ,$$

$$(3.11)$$

with A_1 and A_2 arbitrary complex numbers shows that for a physically admissible field representing outgoing waves it is impossible to distinguish between a linear superposition of two modes with wavenumbers $k_{\ell,n}$ and $-k_{\ell,n}^*$ and the contribution of a single mode with wavenumber $k_{\ell,n}$. We can therefore dispose of one half of the set of modes $\{j_\ell(k_{\ell,n}r)\}$, for instance those modes for which $\mathrm{Re}(k_{s,n}) < 0$. Relying on results obtained by Paley and Wiener [16] it seems highly probable, in the opinion of the author, that the remaining set of modes is just complete.

A similar procedure, using the continuity of both the wavefunction and its normal derivative leads, in the case of potential scattering by a spherically symmetric potential $V(r)$ delimited by a sphere with radius a, to the set of local boundary conditions (Pattanayak, Wolf [12]):

$$\chi_\ell(a;k) \; \frac{\partial}{\partial(ka)} \; h_\ell^{(1)}(ka) \; - \; \frac{\partial}{\partial(ka)} \; \chi_\ell(a,k)\cdot h_\ell^{(1)}(ka) = 0 \;,$$

$$(\ell=0,1,2,\dots) \quad . \tag{3.12}$$

The condition with $\ell = 0$ was derived by Siegert [17], applying the continuity requirements directly to the explicit wavefunctions. The functions $\chi_\ell(r;k)$ are the regular solutions of the radial equation

$$\left[\frac{d^2}{dr^2} + \frac{2}{r}\frac{d}{dr} + k^2 - 2\frac{m}{\hbar^2}V(r) - \ell(\ell+1)\right]\chi_\ell(r;k) = 0 \;, \tag{3.13}$$

and $\phi_\ell(r;k) = r\chi_\ell(r;k) = 0\{\sin kr\}$, (Newton [18]). Therefore, the same analysis used previously shows that the set $\{\chi_\ell(r,k_{\ell,n})\}$, where the numbers $k_{\ell,n}$ are the roots of the Eq. (3.12) is overcomplete. However, we can only admit those values of $k_{\ell,n}$ which correspond to outgoing waves. Hence, $\mathrm{Re}\{k_{\ell,n}\} > 0$, and again it is conjectured that the set of modes satisfying this condition is just complete.

4. DISCUSSION

In this paper we developed a Sturm-Liouville (Schmidt-Hilbert) type theory for the natural modes of a medium characterized by a scalar complex index of refraction. These modes can be considered as the most natural basis for scattering problems, diagonalizing the resolvent, as in ordinary Hilbert-Schmidt theory. If our conjecture concerning the completeness of the physically acceptable modes is true, it seems also very probable that the conjecture put forward by Wolf and Pattanayak [1], [12], concerning the unique specification of the interior scattering problem by a non-local boundary condition is valid. An interesting application of this theory would be the calculation of the mode density inside an arbitrarily shaped medium with a scalar complex index of refraction as has been done by Weyl [19] for the case of a perfect conductor. This number would be fundamental for the theory of black body radiation and its corrections due to the finiteness of the cavity. (Baltes [20]).

APPENDIX

We will indicate how Eq. (2.16) can be derived for a scalar case. Consider the integral equation

$$\psi(\underline{r},k) = f(\underline{r}) + k^2 \int_\tau G(\underline{r},\underline{r}';k)V(\underline{r}')\psi(\underline{r}';k)d\underline{r}' . \tag{A.1}$$

If the function ψ, as a function of k is bounded as $|k| \to \infty$, (which has to be checked afterwards), we obtain from Eq. (A.1), using asymptotic techniques for $|k| \to \infty$, (Sirovich [21]),

$$\psi(\underline{r};k) \sim f(\underline{r}) + V(\underline{r})\psi(\underline{r};k) , \quad \text{if } 0 < \arg k < \pi , \tag{a}$$

$$\psi(\underline{r};k) \sim f(\underline{r}) + V(\underline{r})\psi(\underline{r};k) + \sum_{j=1,2} A_j k^{\alpha_1} G(\underline{r},\underline{r}_{m,j};k)V(\underline{r}_{m,j})\psi(\underline{r}'_{m,j};k) , \tag{b}$$

$$\psi(\underline{r}_{m,j};k) \sim f(\underline{r}_{m,j}) + A_4 k^{\alpha_1} G(\underline{r}_{m,j},\underline{r}'_{m,j};k)V(\underline{r}'_{m,j})\psi(\underline{r}'_{m,j})$$
$$+ A_5 k^{\alpha_2} V(\underline{r}_{m,j})\psi(\underline{r}_{m,j};k) , \tag{c}$$

$$\psi(\underline{r}'_{m,j};k) \sim f(\underline{r}_{m,j}) A_6 k^{\alpha_1} G(\underline{r}'_{m,j},\underline{r}_{m,j};k)V(\underline{r}_{m,j})\psi(\underline{r}_{m,j})$$
$$+ A_7 k^{\alpha_2} V(\underline{r}'_{m,j})\psi(\underline{r}'_{m,j};k), \tag{d}$$

$$\text{if } \pi < \arg k < 2\pi ,$$

where $\underline{r}_{m,j}, j=1,2$ denote points on the boundary of τ such that $|\underline{r}_\tau - \underline{r}_{m,j}|$ is maximal ($j=1$) or minimal ($j=2$). Analogously $\underline{r}'_{m,j}$ maximizes the distance $|\underline{r}_{m,j} - \underline{r}'_{m,j}|$, $j=1,2$. The numbers A_j, $j=1,\ldots 7$ are complex and α_j, $j=1,2$ are real. Equations (A.2c) and (A.2d) lead to $|\psi(\underline{r}_m;k)| = 0\{k^{-\alpha_1} \exp(-ik|\underline{r}_m - \underline{r}'_m|)\}$ which, in combination with Eqs. (A.2a) and (A.2b), yields:

$$\psi(\underline{r};k) \sim f(\underline{r}) + V(\underline{r})\psi(\underline{r};k) , \quad \text{if } |k| \to \infty , \quad 0 < \arg k < 2\pi. \tag{A.3}$$

However, because the solution of Eq. (A.1) can be written as

$$\psi(\underline{r};k) = f(\underline{r}) + \int_\tau \Gamma(\underline{r},\underline{r}';k)f(\underline{r}')d\underline{r}' , \tag{A.4}$$

combination of Eqs. (A.3) and (A.4) leads to

$$\Gamma(\underline{r},\underline{r}';k) \sim \left(1-V(r)\right)^{-1} V(r)\,\delta(\underline{r}-\underline{r}') \quad , \quad \text{if } |k| \to \infty, \quad 0 < \arg k < 2\pi.$$

Equation (2.16) can be derived in a similar manner.

ACKNOWLEDGEMENT

The author wishes to thank Dr. H.A. Ferwerda for a critical reading of this manuscript and Drs. A.M.J. Huiser for stimulating discussions.

References

1. E. Wolf in *Symposia Mathematica* (Academic Press, London and New York, 1976), vol. 18; see also D.N. Pattanayak, Ph.D. thesis, University of Rochester, Rochester, New York, 1973; Opt. Comm. *6*, 217 (1972).
2. G.S. Agarwal, Phys. Rev. *8*B, 4768 (1973).
3. J.A. Stratton, *Electromagnetic Theory* (McGraw-Hill, New York, 1941).
4. G. Wolfsohn, *Handbuch der Physik*, vol. 20 (H. Geiger and K. Scheel), (Springer, Berlin, 1928), chapter 7, §1 (265).
5. C.E. Weatherburn, Trans. Cambr. Phil. Soc. *22*, 133 (1916); see also the Quart. J. of Pure and Appl. Math. *46*, 334 (1915).
6. Ph. M. Morse and H. Feshbach, *Methods of Theoretical Physics* (McGraw-Hill, New York, 1953), vol. II, Eq. (13.1.7).
7. E.T. Whittaker and G.N. Watson, *A Course of Modern Analysis* (Cambridge University Press, 1963), §7.4.
8. E.J. Fredholm, Acta Math. *27*, 378 and 379 (1903).
9. E.C. Titchmarsh, *Eigenfunction Expansions*, vol. 2 (Clarendon Press, Oxford, 1970) §11.11.
10. C. Miranda, Act. Pontificia, Acad. Scient. *3*, 1, (1939).
11. M. Born and E. Wolf, *Principles of Optics* (Pergamon Press, Oxford, 1975), 5th edition, §13.5. See especially for the case μ_1 and $\mu_2 \neq 1$; T.J.I'A. Bromwich, Phil. Mag. *38*, 143 (1919).
12. D.N. Pattanayak and E. Wolf, Phys. Rev. D*13*, 913, 2287 (1976).
13. P. Debye, Ann. Physik *30*, 57 (1909).
14. C. Mie, Ann. Physik *25*, 377 (1900).
15. H.M. Nussenzweig, *Causality and Dispersion Relation* (Academic Press, New York, 1972), especially Eq. (2.3.13).
16. R.E.A.C. Paley and N. Wiener, *Am. Math. Soc. Coll. Publ.* (Am. Math. Soc. Proc. R.J., 1934), vol. 19, chapter 6.
17. A.J. Siegert, Phys. Rev. *56*, 750 (1939).
18. R.G. Newton, *Scattering Theory of Waves and Particles* (McGraw-Hill, New York, 1966) §12.2, Eq. (12.137).
19. H. Weyl, Journ. f. Math. *141*, 163 (1912) and *143*, 177 (1913); Math. Ann. *71*, 441 (1912); Rend. di Palermo *39*, 1 (1915).

20. H.P. Baltes, Appl. Phys. 12, 221 (1977), especially §2.4, §2.5
     and §2.6.
21. L. Sirovich, *Techniques of Asymptotic Analysis* (Springer, New
     York, 1971) §2.8 and §2.9.
22. D.N. Pattanayak, this Volume, p. 209.

THE ANALYTIC NATURE OF SCATTERED ELECTROMAGNETIC FIELDS AND THE

EFFECT OF COHERENCE ON PHASE RETRIEVAL

G. Ross and M.A. Fiddy

Queen Elizabeth College, London, U.K.

1. INTRODUCTION

At the Washington meeting of the American Physical Society in
1964, Wigner related that shortly after the invention of quantum
mechanics he asked Von Neumann if it were not strange that the
formalism failed to require analyticity. Von Neumann replied that
analytic functions constituted such a restricted and special class
of functions that there was no reason why they should be the only
ones admitted into physics. Wigner concluded with the remark that,
Von Neumann notwithstanding, the subsequent 35 years have shown
that, in a deep sense, physics is based on analytic functions.
G.F. Chew, who tells this story in "The Analytic S Matrix" (1966),
highlights the fact that, since this meeting, a further appreciation
of the relevance of analyticity to physical theories has taken place,
and he asks the fundamental question: does analyticity in itself
constitute a suitable a priori principle for the formulation of
theories?

The aim of this paper is to demonstrate that analyticity is of
fundamental importance in scattering theory. The use of this underlying analyticity in scattering phenomena will be illustrated by a
particular application, namely the retrieval of phase information
from intensity data. In the study of the interaction of radiation
with matter, only statistical information about the scattering
medium can be deduced from the intensity detected in the far field.
In order to obtain detailed information about the scatterer the
apparently lost phase of the scattered field is required. The treatment will indicate the significance of the coherence characteristics
of the field for this purpose.

2. THE ANALYTIC NATURE OF THE SCATTERED ELECTROMAGNETIC FIELD

When an electromagnetic field passes through an inhomogeneous
medium, part of it is scattered. In the Fraunhofer space, the
scattered field is related to the object wave by a simple Fourier
transform (Ross (1968)). The object wave is the wave developed in
the object as a result of the interaction between the incident wave
and the medium. Within the range of validity of the first Born
approximation, the relationship between the object wave and the
object may be expressed by a simple proportionality.

In any conceivable physical situation, all objects will be of
finite extent. As a result, both the scattered electromagnetic
field and the intensity in the Fraunhofer space will always display
remarkable properties. Let us discuss for simplicity, the scattering
process in one dimension only. The inhomgeneity distribution function is defined as the local deviation from the average of some
intensive parameter, such as the permittivity. The associated object
wave is denoted by $f(t)$, and the complex amplitude in the Fraunhofer
space, $F(x)$, is thus related to $f(t)$ by

$$F(x) = \int_a^b f(t) \, \exp(ixt) \, dt \tag{2.1}$$

where $|a| < b < \infty$.

Let us define by $F(z)$ the analytic continuation of $F(x)$ in the
complex plane, i.e.

$$F(z) = \int_b^a f(t) \, \exp(izt) \, dt \tag{2.2}$$

where $z = x + iy$. It is assumed that $f(t)$ is not zero almost everywhere in the neighbourhood of a and b. From a well known theorem
due to Paley and Wiener (Boas (1954)), $F(z)$ is an entire function,
i.e. analytic in the whold finite complex plane. Moreover, $F(z)$
belongs to a particular class of entire functions, namely those of
finite type and of order one. As a result of this $F(z)$ is said to
be of exponential type since its growth is bounded by that of a
simple exponential. In addition to this, $F(z) \in L^2(-\infty,\infty)$ on any
line parallel to the real axis.

Functions of exponential type are the simplest generalisation
of polynomials and in the same way are determined everywhere by
their zeros. This is expressed by the Hadamard product,

$$F(z) = \prod_{i=1}^{\infty} \left(1 - \frac{z}{z_i}\right) \quad .$$

The fact that $F(z) \in L^2(-\infty,\infty)$, means that the product is convergent and places a severe restriction on the distribution of the zeros of $F(z)$. The restriction consists of a regularity in the location of the zeros and an asymptotic tendency for them to lie on the real axis. One could imagine an equidistant array of real zeros corresponding to $f(t)$ being a constant, and the subsequent perturbation of the zeros about their lattice position as $f(t)$ becomes more and more structured. The lattice points about which the zeros are positioned are determined by the Nyquist frequency, i.e. $1/(b-a)$. On average, the density of zeros in the complex plane is given by $2(b-a)/\pi$ (Levin (1964)).

Asymptotically, the zeros tend to lie on the regular lattice points. In the region about the z origin, however, the location of the zeros about the lattice points is strongly determined by the behaviour of $f(t)$ near the ends of the interval. More exactly, the slower $f(t)$ decreases towards the end of the support, the larger the radius of the zero free circle about the z origin.

If we consider the intensity, it too is a function of exponential type being the Fourier transform of $\mathcal{F}(\tau)$, the convolution square of the object wave, where

$$\mathcal{F}(\tau) = \int_{-\left(\frac{T-|\tau|}{2}\right)}^{\left(\frac{T-|\tau|}{2}\right)} f\left(t-\frac{\tau}{2}\right)\, f\left(t+\frac{\tau}{2}\right)\, dt \qquad (2.3)$$

where $T = b-a$ and the intensity $I(z)$ is given by

$$I(z) = \int_{-T}^{T} \mathcal{F}(\tau)\, \exp(iz\tau)\, d\tau \quad . \qquad (2.4)$$

The intensity has the zeros of $F(z)$ as well as those of $F^*(z^*)$, which will be in the conjugate positions. Thus, $I(z)$ has its zeros symmetric about the real axis, (Schwartz reflection principle); their density is obviously $4T/\pi$.

In principle, the contribution from each zero extends over the entire complex plane. However, due to the inherent noise in every observation, the effective range of influence of each zero is finite. If the zeros are sufficiently near to the real axis, the observed intensity will clearly display a characteristic oscillatory profile with a frequency equal to that of the distribution of its zeros. This is the well known speckle phenomenon, (Fig. 1(a)). Should the zeros be far from the real axis, their effect on the

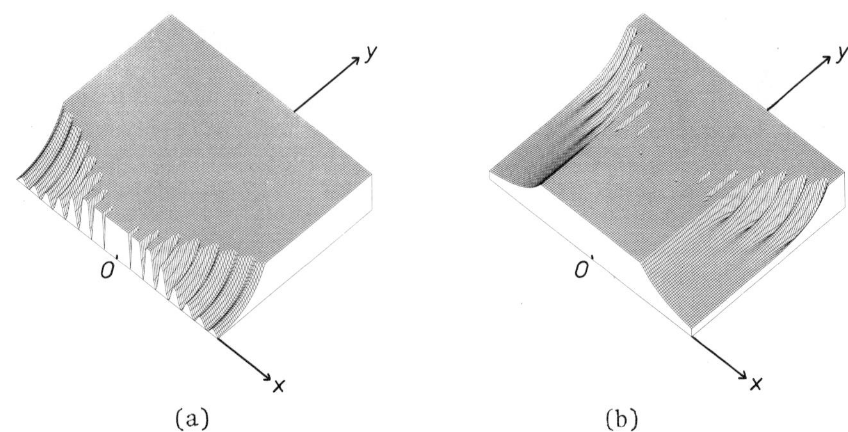

(a)   (b)

Figure 1. The effect on the real axis of complex zeros   (a) near
and (b) far from the real axis: (a) f(t) = 1 - $|t|$;
(b) f(t) = (1 - $|t|$)^3 ; $|t| \leq 1$.

intensity profile may be unnoticed, (Fig. 1(b)). We stress, however,
that provided the object is of finite extent, the regular distribution of zeros associated with the exponential type character of I(z),
is always present.

     In many situations of practical interest, the presence of a
high contrast speckle pattern is a bar to accurate measurements,
and hence it has been found to be desirable to eliminate it or, at
any rate, to render it insignificant. If f(t) happens to display
a suitable decreasing behaviour as t increases, then from the properties of the zeros outlined above, we can expect a zero free area
increasing in extent as T increases. On the real axis, this process
is seen as a reduction in the contrast of the speckle.

     If, however, f(t) does not decrease at a suitable rate, one can
still achieve a contrast reduction, i.e. smoothing of the intensity
profile. By using the same property, it is possible to construct
a modified function displaying the necessary conditions at the ends
of the interval, which produces a zero free area. The modified
function may be obtained by multiplying $\hat{f}(\tau)$ with a function, g(τ),
presenting an adequate rate of decrease. Such a spatial filtering
process is known in optics as apodisation, and may be achieved,
for example, when

$$g(\tau) = \frac{\sin(K\tau)}{\tau}$$

Experimentally, this is very convenient since, by the van Cittert-
Zernike theorem, g(τ), as defined here, is precisely the degree of

coherence for a one-dimension situation. Indeed, it has been observed that a reduction in the degree of coherence does reduce the
speckle contrast. From the point of view of the theory presented
here, this occurs because, as a result of apodisation, the zeros
are pushed further from the real axis. It is easy to recognise
that the apodisation process introduced here is completely equivalent to measuring the intensity with an integrating aperture. This
consists of convolving the intensity with a modulated rect function
of suitable width, i.e. a smoothing of the intensity.

Naturally, a wide range of apodising functions may be used,
such as a Gaussian. None of these $g(\tau)$ present the convenience and
practical availability of $\mathrm{sinc}(K\tau)$. Indeed, the spatial coherence
is one of the most easily adjustable parameters in any scattering
experiment.

3. THE PHASE PROBLEM IN SCATTERING

As an illustration of using the analytic nature of the scattered electromagnetic field and intensity, let us consider the
important problem of phase retrieval. It is well known (Ross et al
(1977)) that in order to obtain the local structure of any object
it is essential to determine the phase of the scattered field in
order to perform the inverse Fourier transform. The field is a
function of exponential type and thus it is possible, by considering
its logarithm, to write a Hilbert transform relating the measured
modulus to the phase:

$$\phi(x) = -\frac{x}{\pi} P \int_{-\infty}^{\infty} \frac{\log|F(x')|\,dx'}{x'(x'-x)} \tag{3.1}$$

where $F(x) = |F(x)| \exp(i\phi(x))$. The subtraction is introduced in
order to ensure the convergence of the integral (Burge et al (1976)).
Unfortunately, the phase obtained from this integral is only correct
if $\log F(z)$ has no branch points within the contour of integration,
i.e. $F(z)$ has no zeros in the upper half of the complex plane. From
the observable magnitude, i.e. the intensity, which displays both
the zeros of $F(z)$ and $F^*(z^*)$ it is not possible to establish when
the two sets of zeros are segregated into separate half-planes. The
exception would be when both half-planes are zero free and $I(z)$ has
only real zeros. However, the only class of non trivial $f(t)$'s
leading to this situation appears to be that of prolate spheroidal
functions. In general, it is necessary to remove the zeros from
one half-plane, or else establish the contribution to the phase from
those in the upper half-plane. The removal of the zeros can be
achieved by a holographic type experiment, i.e. by applying a suffi-

ciently strong reference wave which satisfies Rouché's theorem
(Burge et al (1976)). To obtain the contribution to the phase from
the zeros, one must locate them, a procedure requiring two measurements, which is tedious and susceptible to large errors. Disre-
garding the effect of zeros in the upper half plane, the phase calculated does not represent the actual phase but an associated phase.
This corresponds to a function known as the Hilbert function, having
the same modulus on the real axis, but a zero free half-plane. The
only way in which this can be achieved, also maintaining the density
of the zeros, is by the transfer of each zero to its conjugate position in the lower half-plane.

In practice, however, each zero has only a limited range of
influence due to inherent noise. Hence one need only create a
sufficiently large zero free area about the z origin, containing
the interval of the x axis where the intensity is measured. From
Section 2 it follows that this may be achieved by apodisation.

The reconstructed object from the apodised data is not identical to the original object because of the multiplication by $g(\tau)$.
The reconstruction will be, in general, slightly distorted, especially towards the ends of the support. If $g(\tau)$ is a characteristic
function, i.e. it is a convolution square, then it might become
possible to eliminate the distortion by dividing the reconstructed
object by the convolution square root. For example if $g(\tau)$ is the
degree of coherence then, since sinc(t) * sinc(t) = sinc(τ) one can
divide the reconstructed object by sinc(t). In principle, this
operation should remove the distortion.

Theoretically apodisation is useful only when $F(z)$ and $I(z)$
present complex zeros and should not be required when all their
zeros are real. Indeed, in this case one should be able to apply
directly the logarithmic Hilbert transform and thus obtain the
actual phase. However, a real zero on the real axis produces a
singularity in the logarithm and leads to numerical difficulties
in the computation. Hence even in the case of real zeros, although
apodisation is not necessary for theoretical reasons, it is required
for computational purposes.

Phase retrieval using apodisation is very simple, requiring
only data obtained in a single classical scattering experiment.
Once the computer program has been written, it becomes trivial to
operate. Apart from, or in addition to, the partial coherence,
one can use other forms of spatial filtering. Work is now in progress for establishing quantitative criteria for the optimum optical
data processing. At the moment, the activity still consists largely
of reconstructing simulated objects; only in this way, when the
original object is known, one can estimate the sucess of the method.
An example, shown in Figure 2, illustrates the main steps.

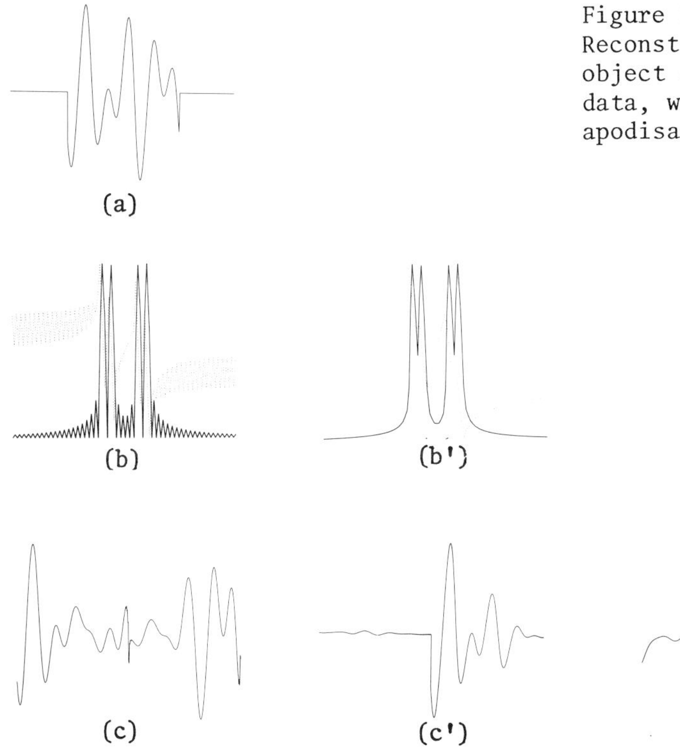

Figure 2.
Reconstruction of an
object from intensity
data, with and without
apodisation.

   Diagram (a) shows a simulated object and (b) its scattered
intensity. The dotted line is the calculated phase. In (c) the
reconstruction using this phase, i.e. the Fourier transform of the
Hilbert function is given; clearly it bears some resemblance to the
object but not a very close one. Diagram (b') shows the scattered
intensity following the multiplication of the convolution square of
the object with $sinc(K\tau)$. The reconstruction from this Hilbert
function, in (c'), exhibits the characteristic distortion at the
ends of the support. In (c'') following deapodisation, this distortion has been removed and a reconstruction closely resembling the
object is obtained.

4. CONCLUSIONS

   The phase problem in scattering phenomena illustrated here constitutes an important application of the concept of analyticity to
optical problems. For all phase problems, which occur in optics in
many contexts, a treatment conceptually identical to the one outlined

above could be applied whenever a finite Fourier transform can be
defined. Indeed, one of the first applications of analyticity for
phase retrieval in optics was to determine the energy spectrum of
light from measurements of the degree of coherence (Wolf (1962)).

A consistent application of the properties of analytic functions
to optics could be very rewarding. For example, the number of degrees of freedom of optical systems might acquire a new significance;
one is tempted to identify each zero of the entire function with a
degree of freedom. We feel that this is a worthwhile topic to pursue, with all its implications for super-resolution by analytic
continuation.

Another result of the analytic properties of a scattered field,
is the new meaning associated with the speckle phenomenon and related
effects. They appear as manifestations of the regular distribution
of zeros which is characteristic for functions of exponential type.
Knowledge of the properties of zeros of these functions could offer
additional means of influencing speckle patterns.

The use of analyticity is a direct extension of the now classical Fourier optics, formulated so far only for the real axis.
Fourier optics has proved very successful in developing the understanding of a wide range of processes. Since analyticity is implicit
whenever a finite Fourier transform may be defined, it is logical to
exploit this completely. The full weight of the properties of analytic functions will provide not only a very useful tool for solving
practical problems, but also, which is more important, a deeper insight into optical phenomena.

Acknowledgements

S.R.C. support for this work is gratefully acknowledged. The
authors thank M.W.L. Wheeler for computational help.

References

Boas, R.P., *Entire Functions* (Academic Press, New York, 1954).
Burge, R.E., Fiddy, M.A., Greenaway, A.H. and Ross, G., Proc. Roy.
    Soc. A *350*, 191 (1976).
Chew, G.F., *The Analytic S Matrix* (Benjamin, New York, 1966).
Levin, B. Ja., *Distribution of Zeros of Entire Functions* (Amer.
    Math. Soc., 1964).
Ross, G., Opt. Act. *15*, 451 (1968).
Ross, G., Fiddy, M.A., Nieto-Vesperinas, M. and Wheeler, M., Optik
    49, 71 (1977).
Wolf, E., Proc. Phys. Soc. *80*, 1269 (1962).

INFORMATION RATE FOR A SOURCE RELATIVE TO AN ABSOLUTE-MAGNITUDE
CRITERION

C. Bendjaballah

Université de Paris XI, Gif-Sur-Yvette, France

We present and discuss numerical calculations of the information
rate achievable for a source, relative to an absolute-magnitude criterion in a particular optical channel.

According to Shannon's theory [1] the rate distortion function
$R(v)$ for a source is defined as the minimum information rate necessary for transferring information from the given source to a receiver,
with mean distance value less than or equal to v. Then $R(v)$ represents a function of the statistical properties of the source and the
required distortion measure. Thus, for a communication channel, the
channel capacity must be greater than or equal to $R(v)$.

Let $P(x)$ be the probability density of the source and v the
evaluation determined by a distance function $\rho(x,y)$ which is assumed
continuous in both x (emitted symbol) and y (received symbol). The
most commonly used function $\rho(x,y)$ that is considered here depends
only on the difference between x and y

$$\rho(x,y) = |x - y| \tag{1}$$

Let us define the non-negative average distance measure by

$$v = \iint |x - y| \; P(x,y) \; dx \, dy \tag{2}$$

where $P(x,y)$ is the two-fold probability density of having the symbols
x and y simultaneously.

The corresponding information rate per symbol is

$$R = \iint P(x) \; Q(y|x) \; \text{Log} \frac{Q(y|x)}{Q(y)} \; dx \, dy \tag{3}$$

where $Q(y|x)$ is the conditional probability density, and

$$Q(y) = \int Q(y|x)\, P(x)\, dx \qquad (4)$$

is the a priori distribution. All the probability distributions are normalized to 1.

Then the rate of generating information for a given quality v of transmission, is the minimum over distributions $Q(y|x)$ of expression (3) subject to the condition that expression (2) has the given value v, for arbitrary ranges of x and y

$$R(v) = \min_{Q} (R) \; . \qquad (5)$$

In order to find an analytic expression for R(v) as parametrically defined by the system of equations (2), (3) and (5) for continuous-amplitude sources, it is necessary to solve the constrained minimization problem that is generally difficult.

We consider the particular communication channel modeled by Mandel [2] in which a light information source is modulated by a filter of continuously random variable transmittance x and the beam is detected by a photomultiplier giving a set of numbers $\{n\}$ in a time interval T. The signal which will be decoded by the receiver is determined by the conditional probability distribution $Q_{<n>}(n|x)$ which is the probability density that n photoelectrons are registered within a time interval T under the conditions that the symbol is emitted from the encoded source with a constant transmittance x. The transmittance is taken to be constant during the emission of the symbol. This communication channel having such a continuous source and a discrete receiver is therefore specified by a random representation of the continuous numbers x onto the set of integers $\{n\}$. The variable x is considered as a random variable in the range (0,1) with the probability $P_a(x)$.

The information rate per symbol has been shown by Mandel [2] to be

$$R = \sum_{n=0}^{\infty} \int_0^1 P_a(x)\, Q_{<n>}(n|x)\, \mathrm{Log} \frac{Q_{<n>}(n|x)}{\int_0^1 P_a(z)Q_{<n>}(n|z)\,dz}\, dx \; . \qquad (6)$$

It should be noted that this communication channel leads to a limited information rate.

The equivalent expression for v becomes

$$v = \sum_{n=0}^{\infty} \int_0^1 |<n>x-n|\, P_a(x)\, Q_{<n>}(n|x)\, dx \; . \qquad (7)$$

Note also that, using Eq. (4)

$$q_{<n>,a}(n) = \int_0^1 Q_{<n>}(n|x)\ P_a(x)\ dx\ . \tag{8}$$

The distribution $P_a(x)$ is taken to be of an **exponential** shape such that

$$\int_0^1 P_a(x)\ dx = 1\ .$$

By varying $<n>$ we describe a large variety of $Q_{<n>}(n|x)$ distributions. Now, R as given by Eq. (6), and v as given by Eq. (7), both considered as functions of $<n>$, will serve to determine the parametric form that we need to obtain $R(v)$.

Figure 1 displays the variation of $R(v)$ for a Poisson distribution $Q_{<n>}(n|x) = \exp(-<n>x)\cdot(<n>x)^n/n!$ for several different values of the parameter a. The corresponding values of v_{max} are in the range $(0.1, 0.458)$.

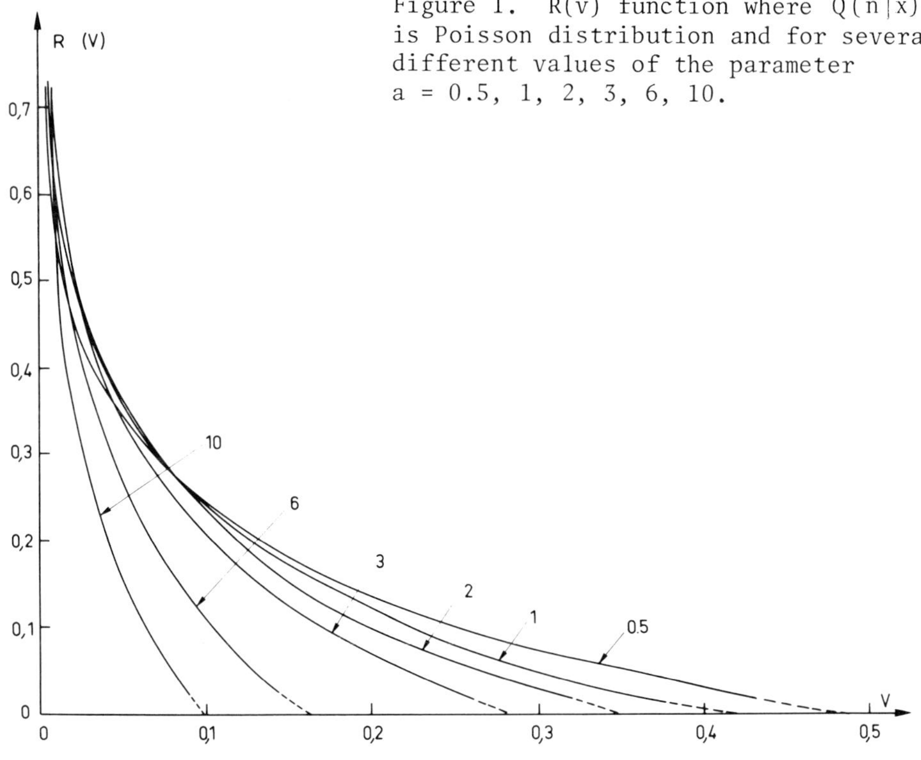

Figure 1. $R(v)$ function where $Q(n|x)$ is Poisson distribution and for several different values of the parameter $a = 0.5, 1, 2, 3, 6, 10.$

We also use Eqs. (6) and (7) to compute $R(v)$ for three distributions $Q_{<n>}(n|x)$ corresponding to different kinds of light signals. We assume that the information rate is still described by the same formula (6).

Figure 2 shows the variation of $R(v)$ for Poisson distribution, geometric distribution and distribution corresponding to the "real Gaussian" field [3].

The value of the parameter a is fixed at 2. This value corresponds to the optimum for which $R(a)$ is greatest for all values of $<n>$. The curves obtained show that $R(v)$ obtained from Poisson distribution is the upper limit to other distributions. We also note that although v_{max} is independent of Q, $R(v)$ depends significantly on it in the range $0 < v \leq 0.20$.

Let us restrict ourselves to the Poisson distribution and seek the $R_L(v)$ lower bound to $R(v)$, that will give some idea of how closely $R_L(v)$ approximates $R(v)$.

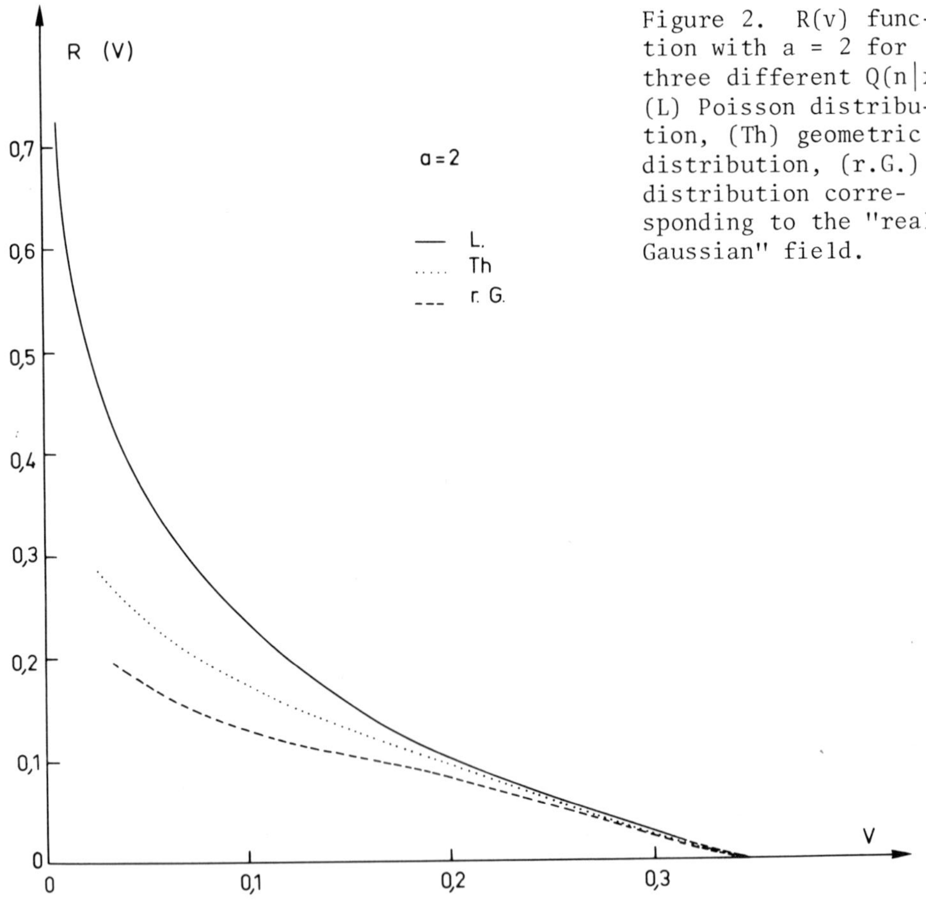

Figure 2. $R(v)$ function with a = 2 for three different $Q(n|x)$ (L) Poisson distribution, (Th) geometric distribution, (r.G.) distribution corresponding to the "real Gaussian" field.

We briefly recall some basic results of the method extensively discussed in Ref. [4], Chapter IV, in order to calculate $R_L(v)$. The equations for R_L and v in terms of s (Ref. [4] pp. 92-93) may be summarized in the following form

$$v(s) = \int dx\ \rho(x)\ g_s(x)\ , \tag{9}$$

where

$$g_s(x) = \frac{e^{-s\ \rho(x)}}{\int e^{-s\ \rho(z)}\ dz}\ (s > 0)\ , \tag{10}$$

and the parametric representation of the corresponding rate coordinate $R_L(v_s,s)$ can be written as follows

$$R_L(v_s,s) = h(P) - h(g_s) \tag{11}$$

with $h(f) = -\int f(t)\ \text{Log}\ f(t)\ dt$ and $R_L(v) \geq 0$. Note that for the continuous amplitude case $R_L(v) \to \infty$ as $v \to 0$ and $R_L(v) = 0$ for $v \geq v_{max}$. Another important result we shall use is the condition for $R_L(v)$ to coincide with $R(v)$, which reads

$$P(x) = \int dy\ q(y)\ g_s(x-y)\ . \tag{12}$$

We start by applying Eqs. (9)-(12) to the one-sided exponential profile of the source. The density is given by $P_a(x) = a\ e^{-ax}$ for $0 \leq x < \infty$. For simplicity, we set $\langle n \rangle = 1$. We easily find

$$v(s) = 1/s\ ;\quad h(g_s) = -\text{Log}(es)\ ;\quad h(P_a) = -\text{Log}(ea)$$

and

$$R_L(v) = -\text{Log}(av)\ . \tag{13}$$

Now, taking the Fourier transform of (4)

$$Q_s(\omega) = \mathcal{P}_a(\omega)/G_s(\omega) = \mathcal{P}_a(\omega)(1 + j\omega/s)\ .$$

Then the inverse transform, eliminating s, yields

$$q_s(n) = a\ e^{-an}(1-av)\ .$$

Hence $q_s(n) \geq 0$, provided that $0 < v \leq 1/a$. So $R_L(v) = 0$ for $v \geq v_{max}$ with $v_{max} = 1/a$.

Let us now consider the source density defined by $P_a(x) = a\ e^{-ax}/(1-e^{-a})$ with $0 \leq x \leq 1$. The same method of calculation for this particular case leads to

$$v(s) = \frac{1}{s} - \frac{e^{-s}}{1 - e^{-s}} \tag{14}$$

$$R_L(s) = \text{Log} \left[\frac{s(1-e^{-a})}{a(1-e^{-s})}\right] + \frac{s\,e^{-s}}{1-e^{-s}} - \frac{a\,e^{-a}}{1-e^{-a}} \quad . \tag{15}$$

For comparison with $q_{<n>,a}(n)$ given by (8)

$$q_{<n>,a}(n) = \frac{e^{-n\,\text{Log}\,[(1+\mu)/\mu]}}{1+\mu}\ \frac{1-e^{-a(1+\mu)}\sum_{m=0}^{n}[a(1+\mu)]^m/m!}{1-e^{-a}} \tag{16}$$

where $\mu = <n>/a$. We calculate $q_{s,a}(n)$ with Eq. (12)

$$q_{s,a}(n) = e^{-an}\ \frac{a(1-e^{-s})}{(1-e^{-a})(1-e^{-(a+s)})}\ \left(1 - \frac{a}{s}\right) \quad . \tag{17}$$

Hence $q(n) \geq 0$ for $0 < s \leq 1/a$ that involves $0 < v \leq v_{max}(a)$ with

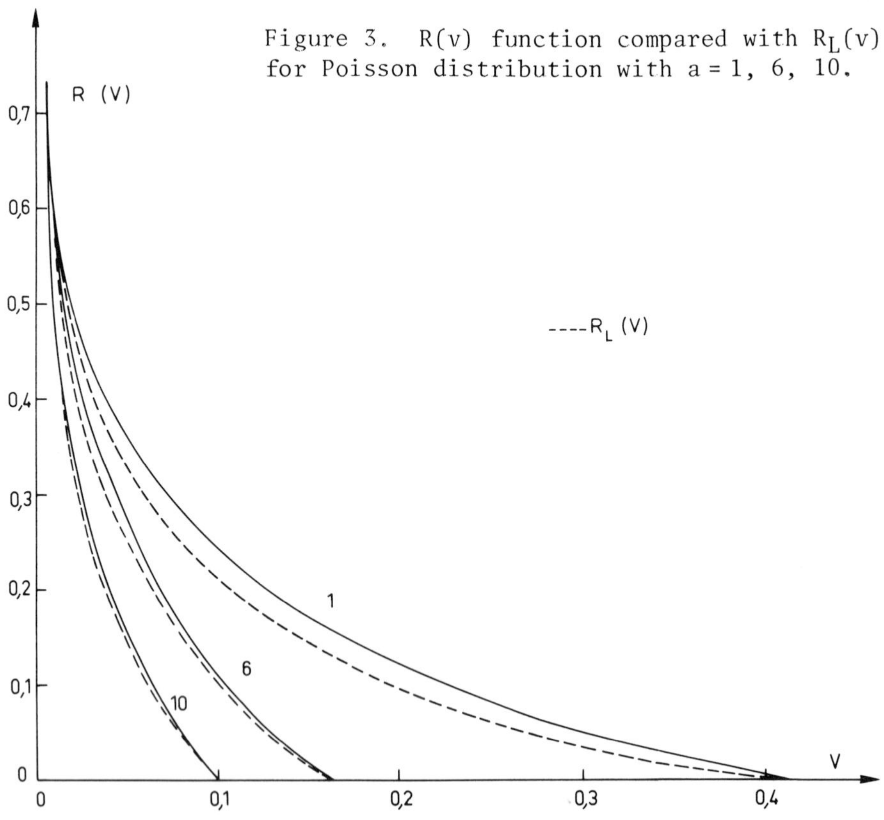

Figure 3. R(v) function compared with $R_L(v)$ for Poisson distribution with $a = 1,\ 6,\ 10$.

$$v_{max} = a - \frac{e^{-1/a}}{1 - e^{-1/a}} \ .$$

A closed expression for $R_L(v)$ is somewhat difficult to obtain. Expressions of q(n) given by Eqs. (16) and (17) are nearly equivalent and their variations display a similar behavior at large n.

The variations of R(v) and $R_L(v)$ given by (14) and (15) and fitted to R(v) at very low v, are plotted in Fig. 3 for a = 1, 6, 10.

We notice that these variations are very close for the values of a = 6, 10 far from the optimum (a = 1) for which the rate information is close to the channel capacity.

In conclusion, R(v) functions for different cases have been computed and compared with respect to the source properties. For a Poisson distribution the variations of R(v) and its lower bound have been compared.

References

1.   C.E. Shannon, Bell System Tech. Jour. *27*, 623 (1948), and Ref. in *Key Papers in the Development of Information Theory* (Part II), edited by D. Slepian (I.E.E.E. Press, 1973).
2.   R. Jodoin and L. Mandel, J. Opt. Soc. Am. *61*, 191 (1971).
3.   B. Picinbono and M. Rousseau, Phys. Rev. *A1*, 635 (1970).
4.   T. Berger, *Rate Distortion Theory*, A Mathematical Basis for Data Compression (Prentice-Hall, Inc., Englewood Cliffs, N.J., 1971).

COOPERATIVE EFFECTS IN OPTICAL BISTABILITY AND RESONANCE FLUORESCENCE

R. Bonifacio[+] and L.A. Lugiato

Università di Milano, Milano, Italy

1. INTRODUCTION

Cooperative phenomena in far-from-equilibrium open systems are presently the object of an ever increasing interest [1]. At a phenomenological level they are described by a set of nonlinear equations with suitable damping terms. Just the interplay of non-linearity and dissipation gives rise to a large variety of phenomena. Among them, one of the most interesting possibilities is the appearance of phase transitions in stationary nonequilibrium states.

It is already known that exactly solvable models are essential to get an insight into the physics of these phenomena, and to discover connections between different phenomena. A typical example is the one-mode laser model with distributed losses [2], which exhibits a second order phase transition [3]. In this paper we want to discuss a solvable model which shows a *first* order phase transition. It is a simple generalization of the one-mode laser model in the semiclassical approximation, which includes the effect of an external coherent field that continuously drives the system. The present model can be readily deduced from the Maxwell-Bloch equations, on taking into account the boundary conditions for the propagating field and performing the mean field approximation. It gives a unified description of optical bistability, resonance fluorescence, superfluorescence and allows a clear analysis of the role of atomic cooperation in all of them. The results that one finds using this model are given below.

1) We give a simple description of optical bistability [4]. This phenomenon, which has been recently experimentally observed [5], consists of the fact that the light transmitted by a filled optical

cavity varies discontinuously showing a hysteresis cycle (see Fig.
1). In other words, a Fabry-Perot filled with atoms behaves as an
optical transistor. As is well known, hysteresis is one of the main
characteristic features of first order phase transitions. Whereas
the theory of Ref. 4 gives only numerical results, the present
treatment is completely analytical. Furthermore, our model allows
a rigorous and complete stability analysis, which shows that there
is a critical slowing down in correspondence to the discontinuity
points $E_I^{(+)}$, $E_I^{(-)}$ of the hysteresis cycle (see Fig. 1). One
finds a direct connection between atomic cooperation and bistable
behaviour. In fact, the condition for the appearance of bistable
behaviour is $\gamma_R > 8\gamma_\perp$, where $\gamma_\perp$ is the incoherent transverse atomic
relaxation rate, while γ_R is the cooperative damping rate for pure
superfluorescence [6]. Since γ_R is proportional to the atomic

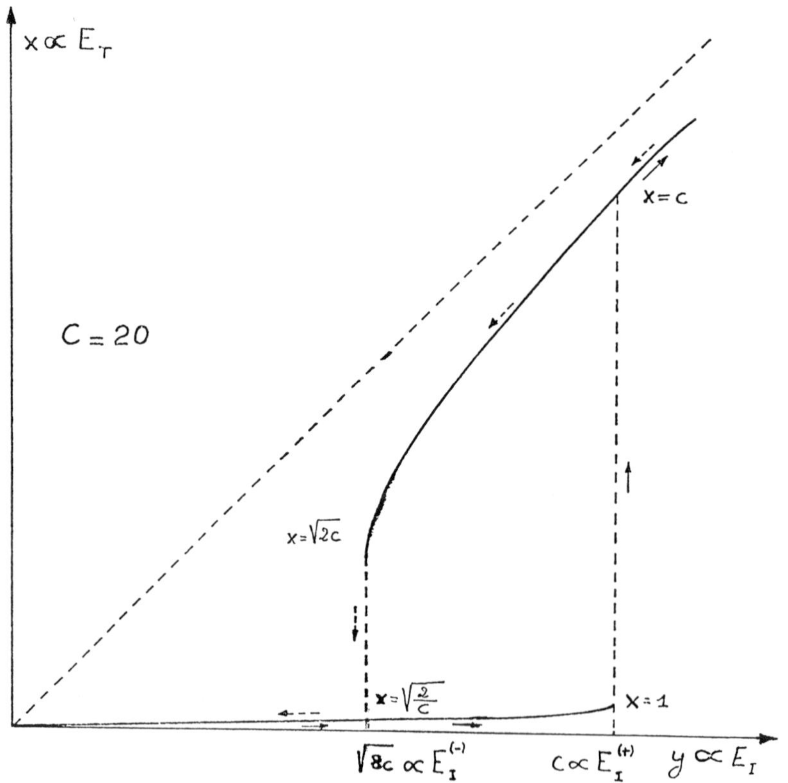

Fig. 1. Hysteresis cycle of the transmitted light. Full (dotted)
line arrows indicate the variations obtained on increasing (decreasing) the injected field E_I. The upper part of the plot corresponds to the one-atom stationary solution, the lower part corresponds to the cooperative stationary solution.

density, the bistable behaviour appears when the density is high
enough so that atomic cooperation can overcome the incoherent
single-atom processes. Actually condition $\gamma_R > 8\gamma_\perp$ fixes a critical
value of the density below which no bistable behaviour can arise.
So bistability is just the stationary counterpart of transient
cooperative spontaneous emission (superfluorescence), for a system
coherently and continuously excited. More specifically, the
analysis shows that in the bistable situation one of the two stable
steady states - termed "cooperative stationary state" (see the
lower branch in Fig. 1) - arises from atomic cooperation, whereas
in the other steady state - called "one-atom stationary state"
(see the upper branch in Fig. 1) - the system does not exhibit,
in general, a cooperative behaviour. The discontinuity points
$E_I^{(+)}$, $E_I^{(-)}$ correspond roughly [7] to the situations $\Omega_I = \gamma_R$ and
$\Omega_I = \sqrt{\gamma_R\gamma}$, where Ω_I is the Rabi frequency of the injected field
and γ is the natural linewidth of the atoms.

   2) The model describes the impact of atomic cooperation on
resonance fluorescence, showing that the usual one-atom theory of
resonance fluorescence [8] is valid only below the above mentioned
critical density of atoms. On the other hand, above the critical
density one finds a hysteresis cycle both in the *total fluorescence
light* and in the *spectrum* of the fluorescent light. The following
predictions should be capable of being experimentally checked.

   2a) When the system is in the one-atom stationary state, the
total fluorescence intensity is proportional to the number of atoms,
which is the normal situation. On the other hand, when the system
is in the cooperative stationary state the total fluorescence
intensity turns out to be inversely proportional to the number of
atoms. This amazing cooperative effect is just the opposite of
what occurs in superfluorescence, in which atomic cooperation gives
rise to a radiation output proportional to the square of the number
of atoms.

   2b) According to the one-atom theory of resonance fluorescence
[8] the spectrum becomes three-peaked (Dynamical Stark Shift, DSS)
in a continuous way as the intensity of the external field is
increased. In fact, the two sidebands gradually emerge from the
central line when $\Omega_I \sim \gamma$. This is a kind of second order phase
transition. In our theory, using the regression hypothesis we
tentatively associate the linewidth and the shift to the real and
imaginary parts of the damping constants given by the linear stability analysis. This shows that when the bistability condition
holds, the spectrum also varies discontinuously with the input
field. When the cavity damping constant k is much larger than γ,
the resulting picture is as follows: the spectrum remains one-
peaked whenever the system is in the cooperative steady state.
Hence no DSS appears when Ω_I becomes larger than γ, provided Ω_I

remains smaller than γ_R. Not only that, but one finds a line
narrowing as Ω_I approaches γ_R from below (i.e. approaches the discontinuity point $E_I^{(+)}$). This line narrowing is connected with
the critical slowing down mentioned above. On crossing the value
$\Omega_I = \gamma_R$ (i.e. jumping to the one-atom stationary state) one finds
an *abrupt* appearance of the three-peaked spectrum, with two already
largely separated sidebands.

A preliminary account of the present work has been given in
Ref. 9a; a more complete treatment is contained in Ref. 9b.

2. THE MODEL

A coherent monochromatic field E_I of frequency ω_0 is injected
into a pencil-shaped optical cavity of length L and volume V, such
that L is equal to an integer number of wavelengths. The cavity
has mirrors of reflectivity coefficient R, and contains N >> 1 two-
level atoms with transition frequency ω_0 (i.e. a homogeneously
broadened active medium, with no detuning from the incident field).
The incoherent atomic decay is characterized by the transverse and
longitudinal relaxation rates $\gamma_\perp = T_2^{-1}$, $\gamma_{\shortparallel} = T_1^{-1}$. Apart from the
external field E_1, no other mechanism to pump the atoms is con-
sidered in the present model. The atoms are assumed to be initially
in the ground state. Once it has entered the cavity, the injected
field induces a macroscopic polarization S and changes the population
difference $\Delta = (N_1 - N_2)/2$, where $N_1 (N_2)$ is the total population of
the lower (upper) level. The polarization in turn radiates a reaction
field which adds to the incident field giving rise to the total
internal field E. Obviously S and E both have the frequency of the
incident field. Our model equations are deduced in Ref. 9b from
the equations (Maxwell-Bloch type) which describe the interaction
of matter with two fields propagating in opposite directions in the
medium. The deduction is based on the proper boundary conditions
for the field [4] and on suitable approximations which hold when
the transmittivity coefficient T = I-R is small. The model consists
of three coupled time-evolution equations for the real quantities
S, Δ and E, which are precisely and respectively defined as the
space averages of the polarization, of the population difference
and of the field propagating in the direction of the incident field.
The equations are

$$\dot{S} = \frac{\mu}{\hbar} \, E\Delta - \gamma_\perp S \; , \tag{2.1a}$$

$$\dot{\Delta} = \frac{\mu}{h} \, ES - \gamma_{\shortparallel}(\Delta - N/2) \tag{2.1b}$$

$$\dot{E} = - \, gS - k(E - E_I/\sqrt{T}) \; , \tag{2.1c}$$

where g is the coupling constant $g = (4\pi\omega_0\mu)/3V$ (μ = modulus of
the atomic dipole moment times $\sqrt{3}$) and k is the cavity damping
constant $k = cT/2L$. In writing Eqs. (2.1) we have extracted a
rapidly varying factor exp $(-i\omega_0 t)$ from S, E and E_I. The field E_T
transmitted by the cavity is given by

$$E_T = \sqrt{T}\, E \quad , \tag{2.2}$$

while the field E_R reflected by the cavity is complementary to the
transmitted field:

$$(E_R)^2 = (E_I - E_T)^2 \quad . \tag{2.3}$$

The coherent terms in Eqs. (2.1) describe the interaction between
the field and the atoms in the dipole and rotating wave approximation, while the incoherent terms describe the atomic decay and the
escape of photons from the cavity (propagation). In particular,
the structure of the damping term in Eq. (2.1c) can be understood
as follows. When the cavity is empty of atomic material (μ=0) the
stationary solution of Eq. (2.1c) is $E = E_I/\sqrt{T}$, i.e. $E_T = E_I$. This
is the usual situation for a Fabry-Perot perfectly tuned to the
incident field. E_T is a fixed constant which we assume for
definiteness ≥ 0. Putting $E_I = 0$ and replacing -N/2 by a positive
inversion our Eqs. (2.1) reduce to the one-mode laser model in a
ring laser cavity in the semiclassical approximation [2]. Furthermore Eqs. (2.1) with $E_I = 0$ and R = 0 (cavity without mirrors)
reduce to the mean field model for superfluorescence, again in the
semiclassical approximation [6]. Hence Eqs. (2.1) generalize these
physical situations to include the effect of the injected field.
The reaction field is given by $E_{react} = E - E_I/\sqrt{T}$. Neglecting the
reaction field, i.e. putting $E = E_I/\sqrt{T}$, Eqs. (2.1a,b) become a
closed system of *linear* equations for S,Δ, which coincide with the
equations for resonance fluorescence (in the case of exact resonance
and for T = 1) studied in Ref. 9a. In Refs. 8 resonance fluorescence is described as the interaction of the external radiation
field with a single atom, assuming that the atoms evolve independently of one another, so that S and Δ are simply proportional to N.
On the other hand in the present paper we take into account the
cooperative effects arising from radiation reaction via the *nonlinear* terms in Eqs. (2.1).

   We mention finally that Eqs. (2.1) can be easily deduced from
a fully quantum mechanical model which again is a simple generalization of the one-mode laser model [9b]. This feature is very
important, since a rigorous discussion of the spectrum of the
fluorescent light requires a fully quantum mechanical treatment.

3. THE STATIONARY SOLUTIONS AND BISTABILITY

Let us now look at the steady-state solutions of Eqs. (2.1) (i.e. $\dot{S} = \dot{\Delta} = \dot{E} = 0$). To this extent, it is suitable to introduce the Rabi frequencies [10] Ω_T and Ω_I in the total internal field and in the incident field respectively:

$$\Omega_T = \frac{\mu E}{\hbar} = \frac{1}{\sqrt{T}} \frac{\mu E_T}{\hbar} \quad , \qquad \Omega_I = \frac{1}{\sqrt{T}} \frac{\mu E_I}{\hbar} \quad , \tag{3.1}$$

where we have used Eq. (2.2). Correspondingly, let us introduce the saturation parameters x and y in the total field and in the incident field respectively:

$$x = \Omega_T / \sqrt{\gamma_\perp \gamma_\parallel} \quad , \qquad y = \Omega_I / \sqrt{\gamma_\perp \gamma_\parallel} \quad . \tag{3.2}$$

Now in the stationary situation Eqs. (2.1a,b) lead to the following expressions for S, Δ in terms of x:

$$S = \frac{N}{2} \sqrt{\frac{\gamma_\parallel}{\gamma_\perp}} \frac{x}{1+x^2} \quad , \qquad \Delta = \frac{N}{2} \frac{1}{1+x^2} \quad , \tag{3.3a}$$

$$N_2 = \frac{N}{2} - \Delta = \frac{N}{2} \frac{x^2}{1+x^2} \quad . \tag{3.3b}$$

On the other hand inserting Eq. (3.3a) into Eq. (2.1c) with $\dot{E} = 0$ one gets an equation that for any given y (i.e. for any given E_I) fixes the stationary values of x (i.e. of E_T):

$$y = x + \frac{2Cx}{1+x^2} \quad , \tag{3.4}$$

where

$$C = \gamma_R / 2\gamma_\perp \quad , \qquad \gamma_R = (\mu g N)/(2K\hbar) = \frac{\gamma NL\lambda_0^2}{8\pi V(1-R)} \tag{3.5}$$

and λ_0 is the wavelength. We notice that for $R = 0$ γ_R is the cooperative decay rate of pure superfluorescence [6]. Hence C is the ratio of a cooperative to a non-cooperative decay rate. Taking into account the relation which links $\gamma_R, \gamma_\perp$ and the absorption coefficient α one has [11]

$$C = \alpha L/2(1-R) \quad . \tag{3.6}$$

In the one-atom theory (i.e. $E = E_I/\sqrt{T}$) one has by definition (cf. Eqs. (3.1) and (3.2)) $x = y$, which is the solution of Eq. (3.4) if

one neglects the term with C. In this case, we see from Eqs. (3.3)
that S, Δ, N_2 are proportional to N. Hence the term with C in
Eq. (3.4) describes atomic cooperation, as is apparent also from
the fact that C is proportional to the cooperative decay rate γ_R.
Since $\gamma_R \propto N$ the solutions of Eq. (3.4) depend in general on N, so
that when the term with C is important, S, Δ, N are no longer
extensive quantities. The physical interpretation of Eq. (3.4) is
clear. In fact, taking into account Eqs. (3.1) and (3.2) we see
that Eq. (3.4) can be re-expressed as follows:

$$E_I = E_T + \frac{2CE_T}{1+E_T^2/I_S} \quad , \qquad (3.7)$$

where $I_S = \mu^2/(T\hbar^2\gamma_\perp\gamma_\parallel)$ is the saturation intensity. Clearly the
cooperative term describes absorption from the atomic system. It
is a nonlinear absorption term which gives rise to all the interesting features of the present model [12], due to the typical
dependence of S on the saturation parameter x. In fact in the
saturation region x >> 1 the cooperative term is negligible, whereas
in the unsaturated region it is relevant and can dramatically change
the physical picture given by the single-atom theory.

To obtain the dependence of E_T on E_I one must invert the function y(x) given by Eq. (3.4). This amounts to solving a cubic
algebraic equation, so that E_T can be a multivalued function of E_I.
More specifically, the function y(x) has a qualitatively different
shape according to whether C > 4 or C < 4 (see Fig. 2). In fact
for C < 4, y is a monotonic function of x, so that the inverse
function x = x(y) is single-valued. So the transmitted light
increases monotonically and continuously with the incident light.
Also the other quantities which are continuous functions of x (as,
for example, the total fluorescence intensity I_F which is proportional to the population of the upper level N_2), vary continuously
with the incident field. Hence in this case the physical picture
is not qualitatively different from that given by the one-atom
theory. The only relevant difference is the presence of a region
in which the differential gain dE_T/dE_I exceeds unity, i.e. where
dy/dx < 1. This phenomenon has been experimentally checked in
Ref. 5 [13]. On the other hand, for C > 4 the function in Eq.(3.4)
has one maximum, x_M, and one minimum, x_m. Hence x is a multivalued
function of y. In fact Eq. (3.4) has one solution x_1 for $y < y_m$,
three solutions $x_1 < x_2 < x_3$ for $y_m < y < y_M$ and one solution x_3
for $y > y_M$. We stress that by exchanging ordinates with abscissae
in Fig. 2 for C > 4 one obtains a graph which closely resembles
the numerical plot of E_T versus E_I given in Ref. 4.

In the following, we shall consider the situation C >> 1 in
which the maximum and the minimum are well pronounced and separated.

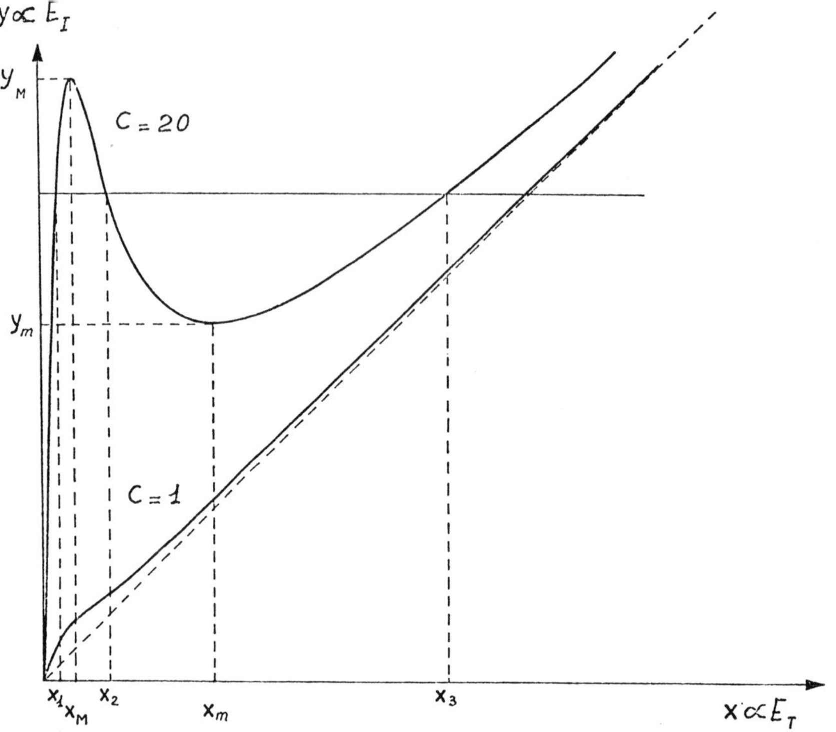

Fig. 2.   Plot of the function $y = x + 2Cx/(1+x^2)$ for $C = 1$ and $C = 20$. For $C \gg 1$ one has $x_M \simeq 1$, $y_M \simeq C$, $x_m \simeq \sqrt{2C}$, $y_m \simeq \sqrt{8C}$. Points x_1 and x_3 are stable, points x_2 are unstable.

In this situation, one has

$$x_M \simeq 1 \ , \qquad y_M \simeq C \ ,$$
$$x_m \simeq \sqrt{2C} \ , \qquad y_m \simeq \sqrt{8C} \ . \tag{3.8}$$

The stationary solution x_1 can be approximately calculated neglecting the term x in Eq. (3.4). This gives:

$$x_1 = \frac{y}{C + \sqrt{C^2 - y^2}} \tag{3.9}$$

For $y \ll y_m \simeq \sqrt{8C}$ one has the linear relation $x_1 \simeq y/2C \ll 1$. In particular for $y \simeq \sqrt{8C}$ $x_1 \simeq \sqrt{2/C}$. On the other hand, the stationary solution x_3 can be approximately calculated by neglecting 1 with respect to x^2 in the cooperative term of Eq. (3.4). One obtains

$$x_3 = \frac{y}{2}\left(1 + \sqrt{1 - \frac{8C}{y^2}}\right) \quad .\tag{3.10}$$

In particular for $y \gg y_M \simeq C$ one has the linear relation $x_3 \simeq y \gg 1$. The linear stability analysis of Eqs. (2.1) shows that points x_1 and x_3 are stable, whereas points x_2 are unstable. Since, for $y > C$, x_3 practically coincides with the one-atom solution, we shall call x_3 the "one-atom stationary state". On the other hand, point x_1 arises from atomic cooperation, so that we term it the "cooperative stationary state". So for $C > 4$ and $y_m < y < y_M$ the system is bistable, and it is easy to see how the hysteresis cycle of E_T versus E_I arises. In fact one must simply cancel from the plot of Fig. 2 ($C > 4$) the part with negative slope (i.e. the part corresponding to the unstable points x_2) and exchange the axes x and y (remembering that $E_I \propto y$, $E_T \propto x$). In this way one obtains Fig. 1. The transition points $y_m \simeq \sqrt{8C}$ and $y_M \simeq C$ correspond to $\Omega_I \simeq 2\sqrt{\gamma_R\gamma_\shortparallel}$ and $\Omega_I \simeq (\gamma_R/2)\sqrt{\gamma_\shortparallel/\gamma_\perp}$, respectively. Note that the jump is of order of the factor $C \gg 1$ at both transition points (see Fig. 1).

From Eqs. (3.3) we see clearly that one finds a hysteresis cycle in all other relevant physical quantities. In fact, using Eqs. (3.3) and Eq. (3.10) we see that when the system is in the one-atom stationary state x_3, which corresponds to practically complete saturation (i.e. $x_3 \gg 1$), one has

$$S_3 = \sqrt{\frac{\gamma_\shortparallel}{\gamma_\perp}}\,\frac{N}{y}\,\frac{1}{1 + \sqrt{1 - \frac{8C}{y^2}}} \quad , \qquad N_2^{(3)} \simeq \frac{N}{2} \quad .\tag{3.11}$$

In this regime S and N_2 are proportional to N, which is the normal situation. On the other hand, from Eqs. (3.3), (3.9) we see that when the system is in the cooperative stationary state x_1, one has

$$S_1 = \frac{Ny}{4C}\sqrt{\frac{\gamma_\shortparallel}{\gamma_\perp}} \quad , \qquad N_2^{(1)} = \frac{N}{4C^2}\,\frac{y^2}{1 + \sqrt{1 - \frac{y^2}{C^2}}} \quad .\tag{3.12}$$

Since $C \propto N$ in this regime S turns out to be independent of N and N is inversely proportional to N. This means that the total fluorescent intensity I_F, which is proportional to N_2, is inversely proportional to the number of atoms. This is a striking cooperative effect.

The hysteresis cycles of I_F versus E_I is shown in Fig. 3. Note that the jumps are at least as large as a factor 2.

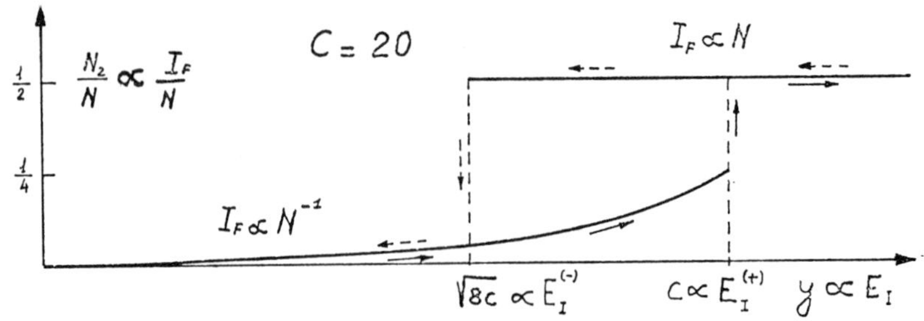

Fig. 3. Hysteresis cycles of the total fluorescence intensity.
Full (dotted) line arrows indicate the variation obtained on
increasing (decreasing) the input field E_I. The part of the plot
beginning from the origin corresponds to the cooperative stationary state x_1, while the other branch corresponds to the one-atom
stationary state x_3.

4. TRANSIENT BEHAVIOUR OF THE TRANSMITTED LIGHT

Let us consider a stable stationary state of our system. If
at a given time the state of the system differs slightly from this
stationary state, the approach to the steady situation is ruled by
the system of equations obtained by linearizing Eqs. (2.1). The
initial deviation from the stationary state can arise either from
a spontaneous fluctuation of the system or by an external perturbation. The latter possibility can be experimentally created as
follows.

Let us assume that the system is initially in a suitable steady
state corresponding to some value E_I of the external field. If E_I
is suddenly changed into $E_I + \delta E_I$ ($|\delta E_I| \ll E_I$), the system
approaches the new slightly different steady state corresponding
to $E_I + \delta E_I$. This approach is described by a solution of the
linearized equations (2.1) and can be experimentally observed by
looking at the transient behaviour of the transmitted field. The
solutions of the linearized system are linear combinations of three
exponentials $\exp(-\lambda_i t)$ (i = 1,2,3), where λ_i are the solutions of
the characteristic equation of the linearized system. In particular,
the approach to the steady state is ruled by the exponential $\exp(-\bar{\lambda} t)$
which decays the most slowly.

Let us now summarize the main results of the analysis of the
characteristic equation obtained in Ref. 9b. These results hold
for $C \gg 1$ and $K \gg \gamma_\perp, \gamma_\parallel$. The latter condition characterizes
superfluorescence rather than the usual laser amplifiers.

1) The approach to the cooperative stationary state is monotonic and is ruled by the damping constant

$$\bar{\lambda} = 2\gamma_\parallel \ \frac{\sqrt{1 - y^2/C^2}}{1 + \sqrt{1 - y^2/C^2}} \ . \tag{4.1}$$

Note that $\bar{\lambda} \to 0$ for $y \to C$. This means that when E_I is near $E_I^{(+)}$ the system approaches the stationary state x_1 slowly, and the more slowly the nearer E_I is to $E_I^{(+)}$. Hence one has a critical slowing down as E_I approaches $E_I^{(+)}$ from below; this is similar to what occurs in tunnel diodes [14].

2) For $E_I \gtrsim E_I^{(+)}$ the approach to the one-atom stationary state is oscillatory and is ruled by the complex damping constant

$$\bar{\lambda} = \frac{\gamma_\perp + \gamma_\parallel}{2} \pm i\Omega_I \ . \tag{4.2}$$

3) When E_I is only slightly larger than $E_I^{(-)}$ the approach to the one-atom stationary state is monotonic, and there is a critical slowing down corresponding to the transition point $E_I^{(-)}$ (i.e. $\bar{\lambda} \to 0$ for $E_I \to E_I^{(-)}$ from above).

5. SPECTRUM OF THE FLUORESCENT LIGHT

The description of the spectrum of the fluorescent light requires the analysis of the fluctuations (time correlation function) of the fluorescence field around the stable stationary states. As shown in Ref. 8a, the fluctuations of the fluorescence field are in turn simply related to the fluctuations of the polarization. Of course the study of these fluctuations requires a fully quantum mechanical analysis. However, on the basis of the regression hypothesis we can assume that the time dependence of the polarization fluctuations in the stationary state is well reproduced by the transient approach to the stationary state. Since we can analyze this approach by means of the linearized equations (2.1), we can get some information concerning the spectrum. We recall that the presence of oscillations in the time correlation function is a necessary condition for the existence of a DSS, which in fact appears when the shift becomes larger than the linewidth. Hence in this spirit we shall consider as a necessary condition for the appearance of a DSS the presence of oscillations in the approach to the stationary state. We shall tentatively estimate the linewidth of the incoherent part of the fluorescent light as $2\,\mathrm{Re}\bar{\lambda}$ and the shift as $|\mathrm{Im}\bar{\lambda}|$.

On the basis of Eqs. (4.1) and (4.2) we can now discuss the discontinuous change in the spectrum which occurs in crossing the transition value $y = C$ (i.e. $E_I = E_I^{(+)}$). Let us consider for simplicity the case $\gamma_\perp \simeq \gamma_{\shortparallel} \simeq \gamma$. From Eq. (4.1) we see that when the system is in the cooperative stationary state x_1 the spectrum is single-peaked because the approach to x_1 is monotonic. Furthermore, one has a line narrowing ($\lambda \to 0$) as y approaches C (i.e. $E_I \to E_I^{(+)}$) from below. On crossing the value $y = C$ one has an abrupt appearance of a large DSS, because from Eq. (4.2) the approach to the one-atom stationary state x_3 is oscillatory and the shift Ω_I is much larger than the linewidth γ (in fact $\Omega_I \simeq \gamma y \simeq \gamma C \gg \gamma$) [15]. Therefore the condition for the appearance of the DSS is $y > C$, which for $\gamma_\perp \simeq \gamma_{\shortparallel} \simeq \gamma$ means (neglecting factors 2) $\Omega_I > \gamma_R$.

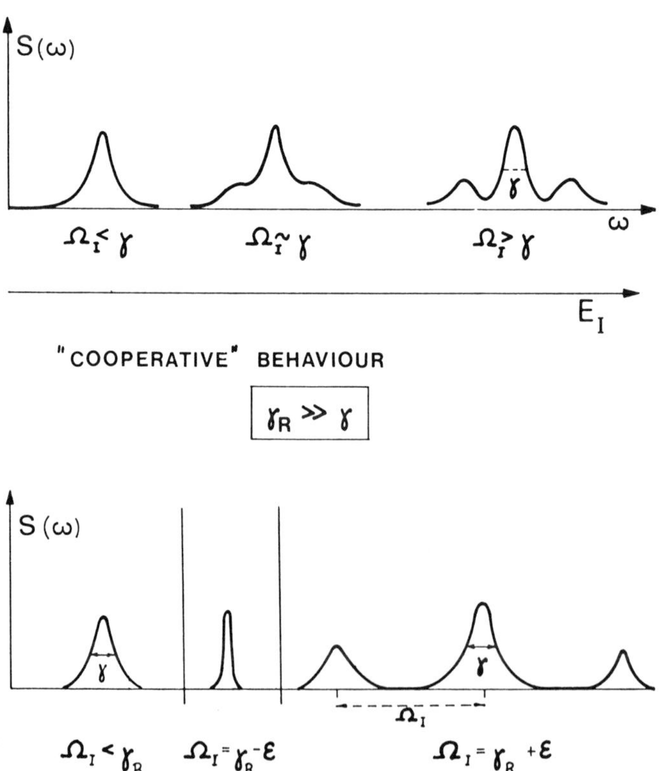

Fig. 4. Behaviour of the spectrum of the fluorescent light for increasing incident field E_I. For $\gamma_R < \gamma$ the DSS appears continuously for $\Omega_I \sim \gamma$; for $\gamma_R \gg \gamma$ it appears discontinuously for $\Omega_I = \gamma_R$.

The threshold $\Omega_I = \gamma_R$ can be well understood by referring to the
Rabi frequency Ω_T of the total internal field instead of the Rabi
frequency Ω_I of the incident field. In fact y > C corresponds to
x > 1, which, for $\gamma_\perp \simeq \gamma_\parallel \simeq \gamma$, means $\Omega_T > \gamma$. Now the atoms are
acted upon by the total internal field, which for y < C is not
intense enough to give rise to a DSS. Only on crossing the transition value $E_I^{(+)}$ the intensity of the internal field abruptly
increases, giving rise to a large DSS. Figure 4 compares the picture one gets below the critical density of atoms ($\gamma_R < \gamma$) with
the picture one obtains when atomic cooperation is dominant
($\gamma_R \gg \gamma$). In the latter case one has a kind of first order phase
transition in the spectrum at the transition value $E_I^{(+)}$ (i.e.
$\Omega_I = \gamma_R$).

Clearly all the statements of this section must be substan-
tiated by a complete quantum mechanical analysis, which should also
give the analytical shape of the spectrum of the fluorescent light.

† Also Istituto Nazionale di Ottica, Firenze, Italy

References

1.   See e.g. a) P. Glansdorff and I. Prigogine, *Thermodynamic
     Theory of Structure, Stability and Fluctuations* (Wiley, New
     York, 1971).
     b) *Synergetics*, ed. by H. Haken (Teubner, Stuttgart, 1973).
2.   a) H. Haken, *Handbuch der Physik*, vol. XXV/2C (Springer-Verlag,
     Berlin, 1970).
     b) M. Sargent III, M.O. Scully and W.E. Lamb, Jr., *Laser Physics*
     (Addison-Wesley, Mass., 1974).
3.   a) V. Degiorgio and M.O. Scully, Phys. Rev. A*2*, 1170 (1970).
     b) R. Graham and H. Haken, Z. Physik *213*, 420 (1968).
4.   S.L. McCall, Phys. Rev. A*9*, 1515 (1974).
5.   H.M. Gibbs, S.L. McCall and T.N.C. Venkatesan, Phys. Rev. Lett.
     36, 1135 (1976).
6.   R. Bonifacio and L.A. Lugiato, Phys. Rev. A*11*, 1507 (1975);
     12, 587 (1975).
7.   I.e. neglecting factors 2 and assuming that $\gamma_\perp \simeq \gamma_\parallel \simeq \gamma$, where
     $\gamma_\parallel$ is the longitudinal atomic relaxation rate.
8.   a) B.R. Mollow, Phys. Rev. *188*, 1969 (1969).
     b) G. Oliver, E. Ressayre and A. Tallet, Lett. Nuovo Cimento
     2, 777 (1971)
     c) H.S. Carmichael and D.F. Walls, J. Phys. B *9*, 1199 (1976).
9.   a) R. Bonifacio and L.A. Lugiato, Opt. Comm. *19*, 172 (1976).
     b) R. Bonifacio and L.A. Lugiato, Phys. Rev. March-April, 1978.

10. I.I. Rabi, Phys. Rev. *51*, 652 (1937).
11. R. Friedberg and S.R. Hartmann, Phys. Lett. *37*A, 285 (1971).
12. Other Optical Systems in which nonlinear absorption gives rise
     to bistability are i) the laser with saturable absorber, see
     J.F. Scott, M. Sargent III, C.D. Cantrell, Opt. Comm. *15*, 13
     (1975); ii) the dye laser, see S.T. Dembinski and A. Kossa-
     kowski, Z. Physik B*24*, 141 (1976) and references quoted therein.
13. Actually in the experiment of Ref. 5 nonlinear dispersion is
     dominant over nonlinear absorption. For the sake of simplicity,
     as in Ref. 4, in the present paper we consider the purely
     absorptive case.
14. R. Landauer and S.W.F. Woo, in Ref. 1 (b).
15. Note in Eq. (4.2) that the shift Ω_I coincides with that pre-
     dicted by the one-atom theory [9] in the high intensity limit.
     Furthermore, for $\gamma_{\parallel} = 2\gamma_{\perp}$ one sees that $2\,\mathrm{Re}\bar{\lambda}$ coincides with
     the width of the sidebands as predicted in Ref. 9.

THEORY OF FAR INFRARED SUPERFLUORESCENCE

R.K. Bullough, R. Saunders and C. Feuillade

*University of Manchester Institute of Science and
Technology, Manchester, U.K.*

1. INTRODUCTION

Different predictions have been made from semi-classical theories [1-3] and second quantised theories [4] concerning the emission
of superradiant pulses. This paper is concerned in showing that
these two different approaches to the problem, each of which is
necessarily approximate, are in fact wholly compatible with each
other. We have already discussed elsewhere [5] the problem of
drawing together the semi-classical and quantised theories. This
paper elaborates further on some points of that discussion and
reports a few more numerical results.

As we now know from the recent experiments [2,6-8] superradiant
emission from a system of inverted atoms or molecules n cm^{-3} is
characterised by the following: after complete inversion of each
atom of the population of density n at time t=0, the system emits
short, intense, spatially directed pulses of radiation of intensity
$\propto n^2$ at a time $t_D > 0$ where t_D, the delay time, $\propto n^{-1}$. It is useful
to make a distinction between superradiant emission from an *inverted*
population and the emission $\propto n^2$ from atomic samples phased by an
initial exciting pulse by calling the former "superfluorescence"
(SF). We concern ourselves solely with SF in this paper despite the
obvious interest of swept-gain systems [9] because we wish to compare current theory in detail with the precise experimental data on
SF now appearing [6].

In this paper we shall leave aside the subtlety of the directional property of SF emission [5] and limit ourselves almost entirely
to plane wave theory [5]. We investigate whether there are distinct
regimes of SF as predicted from mean field quantum theories [4],

263

whether the ideas of maximum co-operation time τ_C and maximum co-operation length $\ell_C = c\tau_C$, as introduced by Arecchi and Courtens [10], have significance, and go on to assess the measure of agreement between the results of current theory and the experimental data on HF [2] and Cs [6]. Our main results are to show that the mean field quantum theories [4] are contained within our semi-classical theory; that the plane wave semi-classical theory predicts two regimes of SF, not three [4]; that the trains of ringing pulses observed in the SF emission from HF [2] are not incompatible with the single pulse SF emission observed on Cs [6]; but that the theory in the rather simplified form in which we present it here is not yet good enough for an agreement *in detail* with the pulse profiles from Cs now obtained.

2. THE PURE SUPERFLUORESCENT REGIME

Pure SF in which the emitted intensity is single peaked and of sech^2 form is predicted by Bonifacio and Lugiato (BL) [4] and co-workers from the quantised theory in the regime in which the characteristic times T_2^*, T_2, $\tau_{SF}(=\tfrac{1}{2}\tau_R)$, τ_C and τ_E satisfy the inequalities T_2^*, $T_2 \gg \tau_{SF} \gg \tau_C \gg \tau_E$. We are concerned with a macroscopic rod-like geometry for the sample which has length L and area A. The times are defined as follows: T_2^* is the inhomogeneous broadening time which for simplicity we set to infinity here; T_2 is the usual T_2 including radiative damping and the shortening effects of collisions but again its effects are largely ignored here; τ_{SF} is the superfluorescent time defined for the rod by $\tau_{SF} \equiv (4\pi np^2 k_s L\hbar^{-1})^{-1}$ in which p is the dipole matrix element for a single non-degenerate 2-level atom transition (a model of 2-level atoms n cm^{-3} is assumed in this paper) and $k_s \equiv \omega_s c^{-1} \equiv 2\pi\lambda_s^{-1}$ where ω_s is the atomic transition frequency; $\tau_R = 2\tau_{SF}$ is the superradiant time used in our previous note [3]; $\tau_E (\equiv k^{-1}) = 2Lc^{-1}$ (k is the BL notation [4]) is the escape time; τ_C is the co-operation time mentioned above and here defined by $\tau_C^{-2} = 2\pi np^2\omega_s\hbar^{-1}$ so that $\tau_C^2 = \tau_{SF}\tau_E$. Both τ_E and τ_{SF} depend on the length of the sample, but τ_C does not.

BL [4] start from a quantised theory of the type we have analysed [5]. We show [3,5] that single- or two-mode quantised theories cannot be made internally consistent because of the spatially inhomogeneous terms arising in a finite length sample. These have the effect of generating a spatially dependent chirp. (Two modes need to be considered in a rod if this emits at each of its two ends). By considering the SF as an external process inducing internal damping in a large rod, BL ultimately find c-number equations of motion which in inhomogeneous form are

$$\pm c\,\frac{\partial \varepsilon}{\partial z}^{(\pm)} + \frac{\partial \varepsilon}{\partial t}^{(\pm)} + k\,\varepsilon^{(\pm)} = \tau_C^{-2}\,p^{(\pm)}\,, \qquad (2.1)$$

together with an associated set of Bloch equations. These have a
homogeneous limit for "one way going" waves (or indeed for both ways)

$$\sigma_{tt} + k\sigma_t - \tau_C^{-2} \sin \sigma = 0 \ . \tag{2.2}$$

No problem of consistency apparently arises.

The notation in (2.1) is that ε is a c-number field amplitude
and the $(\pm)$ describe waves traveling along z to the right $(+)$ and
left $(-)$ inside the rod of 2-level atoms; $P^{(\pm)}$ are the corresponding,
out-of-phase components of the atomic dipoles. Equation (2.2) fol-
lows from (2.1) by $\varepsilon^{(+)} = \sigma_t$ and $P^{(+)} = \sin\sigma$, a solution of the Bloch
equations for right going waves in the spatially independent limit
when the $\varepsilon^{(\pm)}$ fields do not interact with each other. BL give a
more general argument in this limit since the $\varepsilon^{(\pm)}$ fields are then
indistinguishable.

Equation (2.2) reduces to the pendulum equation $\sigma_{tt} = \tau_C^{-2}\sin \sigma$
when L is large so that $k\tau_C \ll 1$; it becomes a damped pendulum for
$k\tau_C \approx 1$; and, using $\tau_C^{-2} = kk^{-1}\tau_C^{-2} = k\tau_{SF}^{-1}$, is the equation

$$\sigma_t = \tau_{SF}^{-1} \sin \sigma \ , \tag{2.3a}$$

with solution

$$\sigma = \sin^{-1} \operatorname{sech} \ [\tau_{SF}^{-1}(t-t_D)] \ , \tag{2.3b}$$

$$\varepsilon^2 = \tau_{SF}^{-2} \operatorname{sech}^2 \ [\tau_{SF}^{-1}(t-t_D)] \tag{2.3c}$$

when n is small such that $\tau_{SF} \gg \tau_C \gg \tau_E$ and $k\tau_C (=\tau_C\tau_E^{-1}) \gg 1$.
This is the pure SF regime defined by BL [4] and the results (2.3)
agree in form with those for N_0 atoms on the same site (N_0 point
atom model [5]) and the single-mode model [5].

There is no ringing in the pure SF regime defined by $k\tau_C \gg 1$;
there is ringing of reducing amplitude in the "oscillatory SF regime"
defined by $k\tau_C \approx 1$; there is strong ringing in the "regime of steady
oscillation" defined by $k\tau_C \ll 1$. Three regimes are therefore predicted by BL [4]. It is easy to see that the regimes of pure and
oscillatory SF scale in time against τ_{SF}: thus $\varepsilon = \sigma_t = \tau_{SF}^{-1}\sigma_\eta$
(where $\eta \equiv t \ \tau_{SF}^{-1}$) and intensities go as $\tau_{SF}^{-2} \propto n^2$. In the regime
of steady oscillation, times scale against τ_C and intensities go as
$\tau_C^{-2} \propto n$.

In the following §3 we show in contrast that by considering a
rod of finite length L and following the fields and intensities

across the ends of the rod, the semi-classical Bloch-Maxwell theory
predicts only *two* regimes. We find (2.1) with the escape term in
$k \, \varepsilon^{(\pm)}$ missing. We find two regimes defined by $k\tau_C \ll 1$ and $k\tau_C \gtrsim 1$.
The former is indistinguishable from the regime of steady oscillation
predicted by BL; but the latter covers both the oscillatory and pure
SF regimes and is typified by n^2 intensities and at least one ringing
pulse even when $k\tau_C \gg 1$.

We show, as we have shown elsewhere [5], that the approximations
of the quantum theory are somewhat rougher than those implicit in the
semi-classical theory and neither theory in its simplest form predicts the single pulse emission observed from Cs. To simulate single
pulse emission we had invoked additional damping processes [5]. Some
data on this and on the effect of the initial condition on the level
of ringing are given again here. The nature of the regimes of SF,
and the significance of the several factors which control peak widths,
peak intensities, delays and levels of ringing finally emerged in
discussion, particularly with Rodolfo Bonifacio, at the Redstone
Arsenal Meeting [11]. Although this paper shows that the pure SF
regime as originally conceived is not a possibility we believe that
the earlier work [4] correctly stressed many of the factors important
to the understanding of the SF process.

The most striking additional results we have obtained since the
work of Ref. 5 are those on the effects of standing waves. Some of
these new results appear at the end of this paper. Standing wave
effects are apparently unimportant for HF. But for the rather different case of Cs we find that standing waves can reduce ringing
intensities by about 3% (from 15% to about 12%), whilst still lower
ringing intensities are not excluded. In consequence, the single
pulse emission actually observed in Cs becomes much better simulated.
Peak widths are in agreement with observations to a few % but are
2.4 times those expected for the pure SF regime of BL [4]. Standing
wave effects therefore seem to be a useful contributor to the single
pulse regime observed in Cs. The numerical results continue to show
spatial inhomogeneity in the fields, however, and our understanding
of this single pulse emission from Cs must therefore be rather different from that which is associated with the original discussions
[4] of the pure SF regime.

3. THE SEMI-CLASSICAL EQUATIONS AND THE TWO REGIMES OF SF

In terms of the ($\pm$) notation introduced for (2.1) the semi-
classical equations in simplest form prove to be

$$c^{-1} \varepsilon_t^{\pm} \pm \varepsilon_z^{\pm} = \alpha \, P^{\pm}$$

$$\varepsilon^{\pm} \{ c^{-1} \phi_t^{\,\pm} \pm \phi_z^{\,\pm} \} = \alpha \, Q^{\pm}$$

$$P_t^{\,\pm} = \phi_t^{\,\pm} \, Q^{\pm} + \varepsilon^{\pm} \, N$$

$$Q_t^{\,\pm} = -\phi_t^{\,\pm} \, P^{\pm}$$

$$N_t = -\{\varepsilon^+ P^+ + \varepsilon^- P^-\} \quad . \tag{3.1}$$

We use the notation $\varepsilon_t \equiv \partial\varepsilon/\partial t$ etc., we drop the parentheses on the $\pm$, and we define the coupling constant α (not the usual α!) by $\alpha = 2\pi n k_s p^2 \hbar^{-1}$ with $k_s \equiv \omega_s c^{-1}$.

An extended set of equations which includes standing wave effects is obtained by expressing the inversion density N (here scaled to a single atom by abstracting a factor n; see Ref. 5) in the form

$$N_o(z,t) + 2M_2 \sin 2k_s z + 2N_2 \cos 2k_s z + \ldots \tag{3.2a}$$

The dipoles, also scaled to a single atom, are expressed in the form

$$P_1^{\,+} \sin \theta_1^{\,+} + P_1^{\,-} \sin \theta_1^{\,-} + P_3^{\,+} \sin \theta_3^{\,+} + P_3^{\,-} \sin \theta_3^{\,-}$$

$$+ Q_1^{\,+} \cos \theta_1^{\,+} + Q_3^{\,+} \cos \theta_3^{\,+} + Q_1^{\,-} \cos \theta_1^{\,-} + Q_3^{\,-} \cos \theta_3^{\,-} + \ldots$$

$$\tag{3.2b}$$

in which $\theta_1^{\,\pm} = -\omega_s t \pm k_s z$ and $\theta_3^{\,\pm} = -\omega_s t \pm 3k_s z$.

Some data resulting from the numerical integration of the extension of equations (3.1) so obtained are given at the end of this paper and will be reported in greater detail elsewhere. These are the results on standing waves mentioned above.

Inhomogeneous broadening is excluded from both (3.1) and its extensions: it is heavy on computing time and results on this must be reported later. Homogeneous broadening replaces the last equation in (3.1) by

$$N_t = -\Gamma_o (1 + N) - \{\varepsilon^+ P^+ + \varepsilon^- P^-\} \quad , \tag{3.3}$$

in which Γ_0 is the A-coefficient. At t=0 there are no fields $\varepsilon^{\pm}$, N = +1 and (3.3) shows that $N = 2e^{-\Gamma_0 t} - 1$ for all t > 0! This result shows that in passing from the quantum theory the all important contribution of spontaneous emission in starting the development of intensity (not field) is inadequately treated. This very difficult problem must also be treated elsewhere. Here we follow others and take the initial condition

$$N(z,o) = 1-\delta \qquad\qquad (0 < \delta \ll 1) \quad ,$$

$$P^+(z,o) = P^-(z,o) = \{\frac{1}{2} [1 - (1-\delta)^2]\}^{\frac{1}{2}} \quad .$$

In so far as N is the third component of a Bloch vector, $N \sim \cos \sigma$ where σ is a tipping angle. In this picture $\delta \sim \frac{1}{2}\sigma_0^2$ where σ_0 is an initial tipping angle. Although σ_0 and δ are to be small we find [5] that they play an important role in controlling the level of ringing. We assume here that δ is a free parameter; but as it also controls the delay we have little actual freedom to vary it. The evidence has been [5] that δ is of order N_0^{-1} where N_0 is the total number of superfluorescing atoms: it has proved to be a very important parameter and now needs a more thorough theoretical justification. This is particularly so since we find that δ is more like $N_0^{-\frac{1}{2}}$ for agreement with the observations on Cs.

Other than $\varepsilon^{\pm}$, $P^{\pm}$, N and α, the new quantities $Q^{\pm}$ and $\phi^{\pm}$ in (3.1) are the oppositely directed in phase components $Q^{\pm}$ of the atomic dipoles and the slowly varying phase $\phi^{\pm}$. Equations (3.1) and their extension are in slowly varying phase and envelope approximation. The model we adopt includes δ as a parameter, sets $\Gamma_0 \equiv 0$, $T_2^* = T_2 = \infty$ and $\phi t^{\pm} = \phi z^{\pm} = 0$. The last choice eliminates any chirp whilst chirps certainly can arise in the extended standing wave theory. Nevertheless, and despite the fact that inconsistencies arise in the quantised theory [3,5] only at the level of the chirp, we do not examine the chirp in the semiclassical theory further here. Such analysis should be made sometime however.

So far we have integrated the following model equations numerically:

$$c^{-1} \varepsilon_t^{\pm} \pm \varepsilon_z^{\pm} = \alpha P^{\pm}$$

$$P_t^{\pm} = \varepsilon^{\pm} N$$

$$N_t = -(\varepsilon^+ P^+ + \varepsilon^- P^-) \qquad\qquad\qquad (3.4a)$$

with

$$N(z,o) = 1-\delta$$

$$P^{\pm}(z,o) = \{\tfrac{1}{2}\,[1 - (1-\delta)^2]\}^{\frac{1}{2}}$$

$$\varepsilon^{\pm}(z,o) = 0 \qquad\qquad\qquad\qquad\qquad\qquad (3.4b)$$

for all points z in $0 < z < L$; $\varepsilon^{\pm}(z,t)$ is continuous across $z = 0$ and $z = L$ and we follow fields both inside and outside $0 < z < L$ and compute intensities outside. The theory is a plane wave theory and a rough account [5] is taken of sideways loss in a rod-like geometry by additional damping.

Since $c\alpha \equiv \tau_C^{-2}$, equations (3.4a) are similar to BL's c-number equations (2.1); but they differ critically for regime analysis in that the escape term in k is missing. To fix ideas note that if the inversion breaks up into two-time evolving non-overlapping parts

$$N(z,t) = N^+(z,t) + N^-(z,t) \ , \qquad\qquad\qquad (3.5)$$

equations (3.4) break up into two independent sets of equations for the oppositely directed fields. The right going set is the sharp line limit of the SIT equations for fields traveling up z. Thus with $\varepsilon = \sigma_t$ and $N = 1-\delta$ at $t = 0$, $N = \cos\,\sigma$, $P = \sin\,\sigma$ and

$$c^{-1}\,\sigma_{tt} + \sigma_{zt} = \alpha\,\sin\,\sigma \ . \qquad\qquad\qquad (3.6)$$

This equation easily converts [5] to the Lorentz covariant form of the sine-Gordon equation with, however, the awkward condition that $N = 1-\delta$ at $t = 0$ and $N \to -1$ as $t \to +\infty$: this condition is outside the useful range of analytical methods for solving the sine-Gordon equation. We can analyse it here in the form (3.6): in this form it is apparently an inhomogeneous form of (2.2) with the term in $c\sigma_{zt}$ replacing the damping term $k\sigma_t$ there. If $c\sigma_z \approx k\sigma$ it would seem that (3.6) and (2.2) are closely equivalent. This ignores the fact that the replacement of $c\sigma_z$ by $k\sigma$ fundamentally changes the character of the equation.

Equation (2.2) is the homogeneous limit of (2.1) and so assumes that spatially homogeneous behaviour is a good approximation to the actual behaviour: the same limit also has the property of eliminating the term in $\varepsilon_z = \sigma_{tz}$ from (2.1) but the term in $k\sigma_t$ survives. Equation (3.6) is not spatially homogeneous and the numerical results based on (3.4) show that the assumption of inhomogeneity is not acceptable. Figure 1 shows numerical results for (3.4) for the data for HF gas [2] ($L = 100$ cm, $\lambda = 84\mu$m, $p = 6.7 \times 10^{-19}$ c.g.s. units): δ is chosen to have the large value of 10^{-3}.

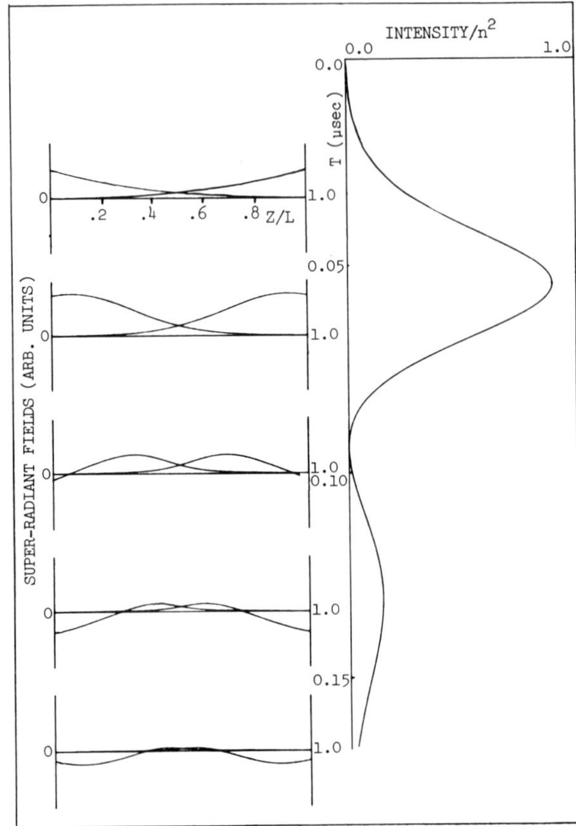

Figure 1.
τ_R = 5 nsec
τ_E = 6.7 nsec
τ_C = 4.1 nsec
κ_L = 0.0
δ = 10^{-3}
τ_W = 31.1 nsec
t_D = 54.4 nsec

For $\tau_R = 2\tau_{SF} \sim 5$ n sec (Fig. 1) (n = 9.4 x 10^{11} molecules cm^{-3}), intensities go $\propto \tau_R^{-2} (\propto n^2)$ and delays $t_D \sim 3\tau_R |\log_{10}\delta|$ (both results are obtained empirically by varying the parameters n and δ). The delays are too short for agreement with the observations [2] by a factor of about 12 but by varying δ towards $\delta \sim 10^{-12}$ this discrepancy is eliminated (see Ref. 5). The fields shown in the left-hand trace of Fig. 1 are manifestly inhomogeneous; ringing (seen in the output intensities in the right-hand trace) certainly occurs but the ($\pm$) fields overlap in the later ringing pulses. Under these conditions the approximation (3.4) leading to one way going fields is certainly acceptable at least for the leading radiated pulses. This justifies to that extent the one way approximation made by Feld and co-workers [1,2].

For $\tau_R \sim 5$ x 10^{-10} - 5 x 10^{-11}sec (n $\sim 10^{13}$ - 10^{14} molecules cm^{-3}, we find intensities $\propto \tau_R^{-1} (\propto n)$, delays and pulse widths $\propto \tau_R^{\frac{1}{2}}$ (for δ-fixed) and the $\pm$ fields interfere throughout the motion. Ringing is pronounced and the fields are again inhomogeneous. In Ref. 5 we compare the results for (3.6) ($\varepsilon = \sigma_t$ and one way going)

against (3.4), both ways going, for τ_R = 10 n sec.   There is some
reduction of ringing in the both ways case.

Figure 1 (τ_R = $2\tau_{SF}$ = 5 n sec, τ_C = 4.1 n sec, τ_E = 6.7 n sec)
is in the regime of oscillatory SF.   Figure 4 of Ref. 5 (τ_R = 100
n sec, τ_C = 18.2 n sec, τ_E = 6.7 n sec) is well into the regime of
pure SF.   Ringing persists and the general profile is very little
different from the case of τ_R = 50 n sec or indeed from Fig. 1.
This numerical work leads us to conclude that the semi-classical
equations (3.4) describe two, but only two, distinct regimes.

Since the one way going behaviour does not change qualitatively
the both ways going behaviour we can return to (3.6) and discuss it
further.   Equation (3.6) is spatially inhomogeneous and matched to
the fields outside $0 < z < L$.   Matching allows energy transfer, no
term $k\sigma_t$ is needed, and, by avoiding the homogeneous limit, $\varepsilon_z = \sigma_{tz}$
does not vanish.   We now examine the approximation $c\sigma_{tz} = k\sigma_t$, but
notice that if this approximation is assumed in (2.1) rather than
(2.2), the latter equation gains an additional term $k\sigma_t$.   The conceptual bases of (2.2) and (3.6) are therefore very different.

Figure 1 and other figures of Ref. 5 show that $\varepsilon_z \approx \varepsilon/2L$, so
that indeed $c\sigma_{tz} \approx k\sigma_t$.   But this is an order result.   The actual
replacement of σ_{tz} by $k\sigma_t$ changes the equations from wave equations
to ones of diffusive type.   It is the diffusive character which leads
to prediction of the pure SF regime.   To keep the wave-like character
of (3.6) we scale lengths z by 2L: we set $\zeta \equiv z/2L$ so that $0 < \zeta < \frac{1}{2}$
and (3.6) is

$$\sigma_{tt} + k\sigma_{t\zeta} - \tau_C^{-2} \sin \sigma = 0 \tag{3.7}$$

in that region.

There are now two cases: in terms of the "stretched" time
$\tau \equiv t\tau_C^{-1}$ equation (3.7) scales to

$$\sigma_{\tau\tau} + k\tau_C \sigma_{z\zeta} - \sin \sigma = 0 \ . \tag{3.8}$$

Since $\sigma_{\zeta\tau}$ is of the size of σ_τ from the numerical work it follows
that for $k\tau_C \ll 1$, the regime of steady oscillation, (3.8) is
governed by the pendulum equation $\sigma_{\tau\tau} - \sin \sigma = 0$.   Since $\varepsilon = \sigma_t$
= $\tau_C^{-1}\sigma_\tau$, $\varepsilon \propto n^{\frac{1}{2}}$ and $\varepsilon^2 \propto n$.   In the limit $k = c/2L \to 0$ this regime
is indistinguishable from the regime of steady oscillation predicted
by BL [4].   The output is a train of close to hyperbolic secant
squared intensities (BL [4]) with widths on the scale of τ_C.   (Figure 2 in Ref. 5 which is both ways going, shows that this limit is
still being approached but is not yet reached.)

Notice that τ_C is independent of L. In the oscillatory SF regime times scale instead against τ_{SF} and intensities go as $\tau_{SF}^{-2} \propto L^2$. It follows that with increasing L we can think of the SF pulses narrowing in width and increasing in peak intensity. But as L increases so that $k\tau_C \ll 1$ this gain of intensity with L is limited by the passage into the regime of steady oscillation. This occurs when $c\tau_C/2L < 1$ or where $\ell_C/2L < 1$ where ℓ_C is the maximum co-operation length of Arecchi and Courtens [10]. To this extent the semi-classical theory predicts the same co-operative limit (the limit of Ref. 10 has an extra factor $2\pi/3$ on τ_C^{-2}). However, a coherent behaviour is established in the regime of steady oscillation beyond that limit; the superfluorescent system does not break up into independently superradiating regions; and there is no apparent limit to the gain for fixed big enough L and increasing density n. In this respect homogeneously inverted SF has the same advantage as swept gain systems [9]. Only the proper passage from the quantum theory to the semi-classical theory could establish the independent spiking conjectured by Arecchi and Courtens [10] which could certainly limit the gain. Available experimental data may not support the conjecture: Gross et al. [7] see n dependent superradiant emission from the $5S_{\frac{1}{2}}$ - $4P_{\frac{3}{2}}$ lines in Na; but this may be evidence for the semi-classical steady oscillation regime or it may be the effect of steady pumping by the inverting pulse when t_D is approximately equal to the inversion time as Gross et al. themselves suggest. This would mean that the inversion process puts an important limitation on homogeneously inverted SF systems as sources of intense radiation.

For the oscillatory SF regime $k\tau_C \gg 1$ we scale t against τ_{SF}, and with $\eta \equiv t\tau_{SF}^{-1}$, (3.5) is

$$(k\tau_C)^{-2}\, \sigma_{\eta\eta} + \sigma_{\eta\zeta} - \sin \sigma = 0 \quad . \tag{3.9}$$

And since σ_η is of order 1 (see Fig. 1 and Figs. 3-5 of Ref. 5) we reach the sine-Gordon equation

$$\sigma_{\eta\zeta} - \sin \sigma = 0 \quad . \tag{3.10}$$

Fields are now proportional to $\tau_{SF}^{-1}\sigma_\eta$, intensities to $\tau_{SF}^{-2} \propto n^2L^2$, and widths to $\tau_{SF} \propto (nL)^{-1}$. However, there is no further regime where (3.10) is heavily damped. Solutions of (3.10) are always oscillatory [5] and ringing persists far into the SF regime as Fig. 4 of Ref. 5 shows.

Our conclusion from this work is that semi-classical theory predicts two and only two regimes. In order to simulate a single pulse regime like that observed by Gibbs and Vrehen [6], we replace (3.6) by

$$\sigma_{tt} + c\sigma_{tz} + \nu k\sigma_t = \tau_C^{-2} \sin \sigma \quad . \tag{3.11}$$

We have introduced [5] the damping term $\kappa\varepsilon = \kappa\sigma_t$ and set $\kappa L = 0.5\nu$ precisely, in order to introduce the reciprocal escape time $k = c/2L$. When $\nu = 1$ the diffraction term plays the role of BL's superfluorescence loss [4]. From (3.11) the regime of steady oscillation ($k\tau_C \ll 1$) is not changed. A transition to oscillatory SF occurs where $k\tau_C \sim 1$ and the n dependent intensities change to n^2 dependent; but this regime is governed now by the damped sine-Gordon equation

$$\sigma_{\eta\zeta} + \nu\sigma_\eta - \sin \sigma = 0 \quad . \tag{3.12}$$

For large ν (large diffraction loss), (3.12) describes pure SF behaviour with $\sigma_\eta = \nu^{-1}\sin \sigma$ and $\sigma = \sin^{-1}\mathrm{sech}[(t-t_D)/\nu\tau_{SF}]$ (compare (2.3)). Figures 6b and 6e of Ref. 5 calculated from (3.3) for both ways going fields with $\kappa\varepsilon^{\pm}$ added to the Maxwell equations for $\varepsilon^{\pm}$, show that ringing is reduced as κL increases from 0.5 to 3.5. The data are for HF with $\tau_R = 5$n sec ($\tau_{SF} = 2.5$n sec) and the case $\kappa L = 0$ appears in Fig. 1 (with, however, different scaling).

The behaviour changes little well inside the pure SF regime. The reader is referred to Ref. 5 for details. The pulse (2.3c) has width $3.52\tau_R$. Ratios given in Ref. 5 for $(\tau_W/3.52\tau_R)$ (τ_W is the calculated width) are 1.60, 1.75 and 1.97 for $\tau_R = 50$ n sec and $\kappa L = 0.0$, 0.5 and 1.5. Thus though single pulse emission is predicted through diffraction loss damping, widths are larger than would be expected in a pure SF regime. Ratios $(\tau_W/3.52\tau_R) \sim 2.5$ are observed from Cs [6].

We turn briefly to the delay. This is influenced by κL (t_D changes from 531 n sec for $\kappa L = 0$ and $\tau_R = 50$ n sec to 703 n sec for $\kappa L = 1.5$ and $\tau_R = 50$ n sec; see Ref. 5, Figs. 5, 7a and 7b). In Figs. 9a,b of Ref. 5 we have HF data, $\tau_R = 5$ n sec, $\delta = 10^{-12}$ and $\kappa L = 0$ (Fig. 9a) and $\kappa L = 3.5$ (Fig. 9b). Fields are going both ways. The delays and widths prove to be: $\kappa L = 0$, $t_D = 399$ n sec, $\tau_W = 81$ n sec; $\kappa L = 3.5$, $t_D = 543$ n sec, $\tau_W = 89$ n sec. Note first that the dramatic change in δ changes $t_D \sim 3/2\ \tau_R|\ln\delta|$ to about $3\tau_R|\ln\delta|$ for $\kappa L = 0$. Then this is substantially changed by changing κL to 3.5. The profile in Fig. 9b of Ref. 5, ($\kappa L = 3.5$), and the delay and width agree moderately well with the observations on HF (Ref. 2, Fig. 2c: $\tau_W = 120$ n sec there, however). We find no simple formula for t_D ($t_D = \tau_R \ln N_0^{\frac{1}{2}}$ is not a good expression, for example [4]); but $\delta \approx N_0^{-1}$ seems a fairly good empirical choice for this numerical work (N_0 is the total number of superfluorescent atoms). Note the important role of δ in enhancing (suppressing) the ringing for small (large) values.

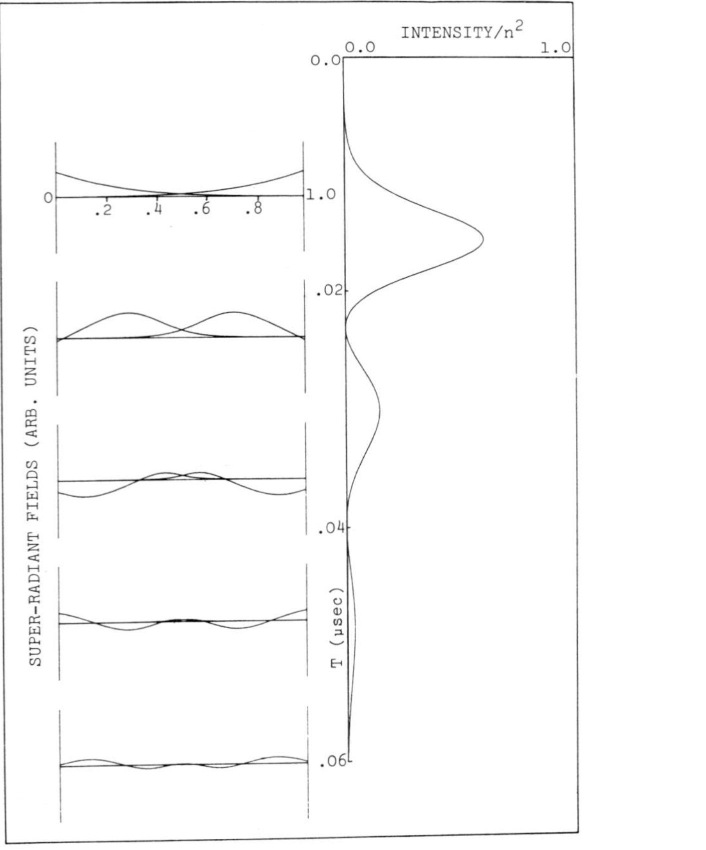

Figure 2a. $\tau_R = 0.73$ nsec, $\tau_E = 0.24$ nsec, $\tau_C = 0.30$ nsec, $\kappa L = 0.4$, $\delta = 10^{-5}$, $\tau_W = 16.19$ nsec, $t_D = 15.86$ nsec.

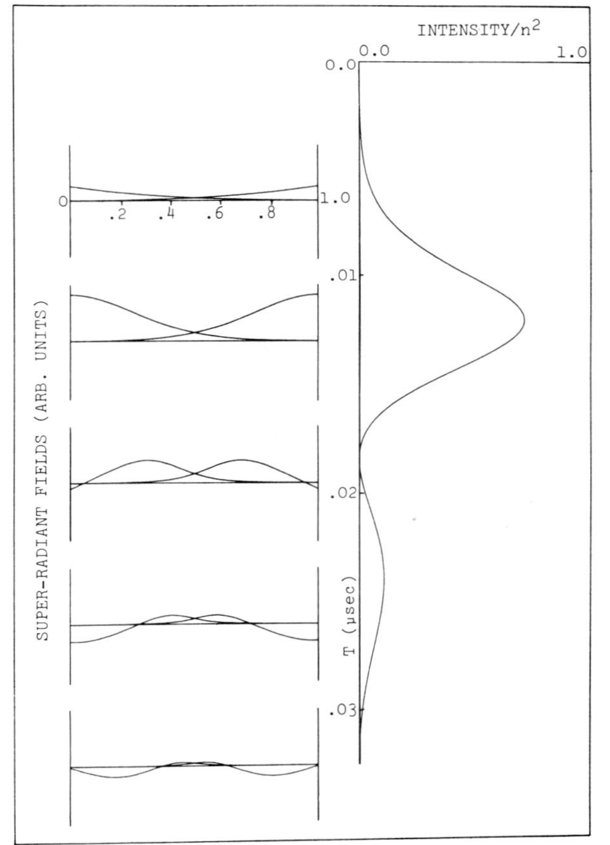

Figure 2b. $\tau_R = 0.70$ nsec, $\tau_E = 0.13$ nsec, $\tau_C = 0.21$ nsec, $\kappa L = 0.8$, $\delta = 10^{-4}$, $\tau_W = 5.33$ nsec, $t_D = 12.13$ nsec.

TABLE I

	τ_R	τ_E	δ	κL	t_D	τ_W
Expt.	0.73	0.24			15	7
Computation	0.73	0.24	10^{-5}	0.0	14.90	6.82
	0.73	0.24	10^{-5}	0.4	15.86	6.98
	0.73	0.24	10^{-5}	0.8	16.87	7.06

Finally we give new numerical data for Cs. Vrehen [6] observes single asymmetric peaks of superfluorescence by imposing a magnetic field of 2.8 k Gauss to isolate the $(M_J = -3/2, M_I = -5/2) \rightarrow (M_J = -1/2, M_I = -5/2)$ transition in the hyperfine $7^2P_{\frac{3}{2}} \rightarrow 7^2S_{\frac{1}{2}}$ transitions of Cs. Table 1 refers to the data for the atomic beam $(T_2^* = 18\,nsec)$ and L = 3.6 cms. The choice of $\delta = 10^{-5}$ needed for fitting the delay differs substantially from $N_0^{-1} \approx 10^{-8}$; the resultant widths (determined by δ) are then good. The choices $\kappa L = 0.4$ and 0.8 correspond to Fresnel numbers 0.9 and 0.4, respectively, compared with the value 1.0 used in the experiments: overall agreement is therefore good except for the ringing ratios (first ringing peak intensity compared with first peak intensity) which prove to be 29%, 24% and 20%, respectively, and are high.

Table 2 shows results for L = 2 cm$(T_2^* = 32$ n sec).

TABLE 2

	τ_R	τ_E	δ	κL	t_D	τ_W
Expt.	0.70	0.13			12.5	6
Computation	0.70	0.13	10^{-4}	0.0	10.53	5.03
	0.70	0.13	10^{-4}	0.4	11.31	5.16
	0.70	0.13	10^{-4}	0.8	12.13	5.33

The value $\delta = 10^{-5}$ leads to delays ~ 15 n sec and the value $\delta = 10^{-4}$ adopted seems the best choice. This large value is of order $N_0^{-\frac{1}{2}}$ rather than N_0^{-1}. The ringing ratios are lower: 21%, 18% and 15%, respectively. The data in Tables 1 and 2 corresponding most closely to the experimental results appear graphically in Figs. 2a and 2b. Notice that the delay determined by $t_D = 2\tau_R \ln N_0^{\frac{1}{2}}$ is in good agreement with the observations for both L = 2 cm and L = 3.6 cm (e.g. $2\tau_R = 1.4$, $N_0 = 4.5 \times 10^7$, t_D (calc) = 12.3 n sec),

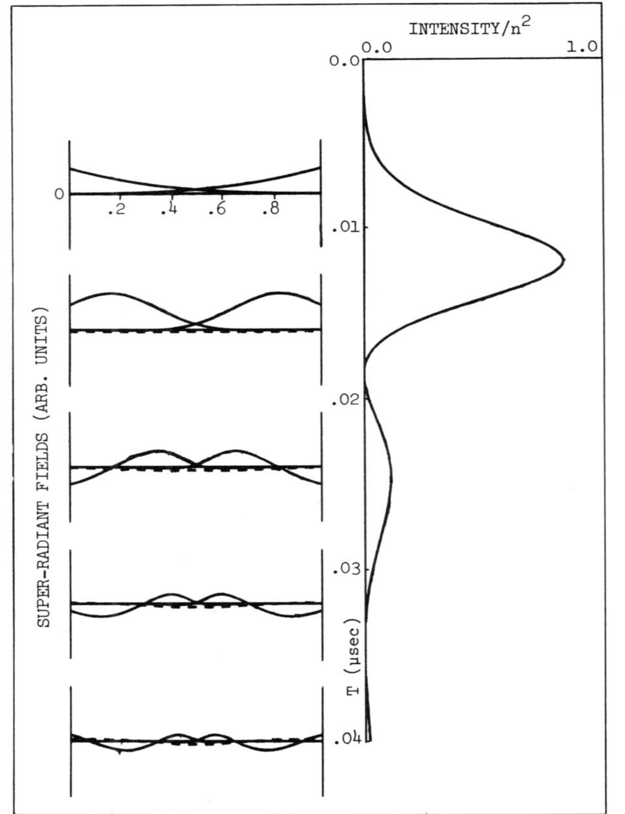

Figure 3a.　τ_R = 0.70 nsec, τ_E = 0.13 nsec, τ_C = 0.21 nsec, κL = 0.0, δ = 10^{-3}, τ_W = 5.04 nsec, t_D = 10.4 nsec.

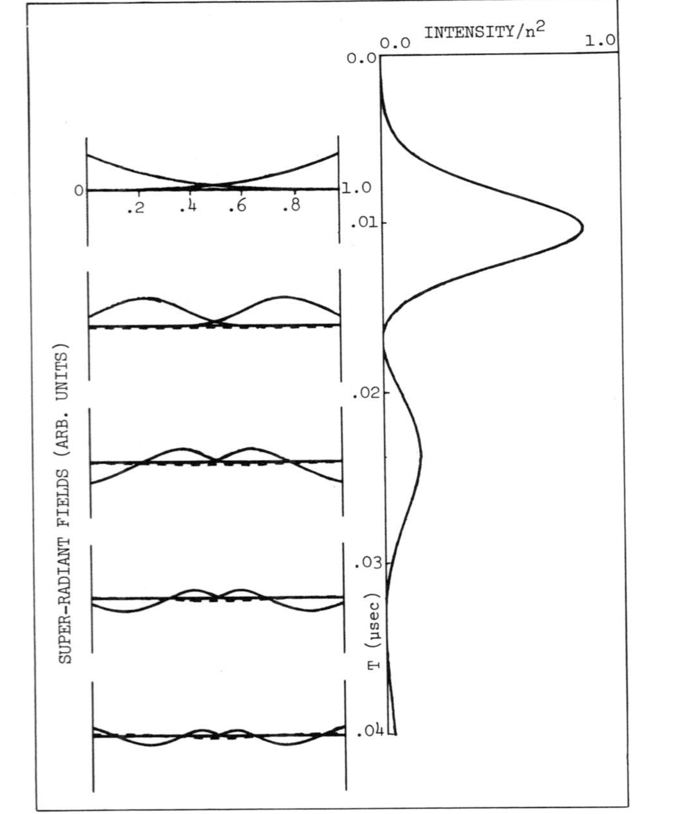

Figure 3b.　τ_R = 0.70 nsec, τ_E = 0.13 nsec, τ_C = 0.21 nsec, κL = 0.8, δ = 10^{-3}, τ_W = 5.41 nsec, t_D = 12.0 nsec.

TABLE 3   ($\kappa L = 0$ throughout)

	τ_R	τ_E	δ	t_D	τ_W	Ringing ratio (%)
Expt.*	0.53	0.13		11	5	<14**
Computation	0.53	0.13	10^{-8}	21.63	6.14	46.9
	0.53	0.13	10^{-6}	14.08	4.96	35.5
	0.53	0.13	10^{-4}	8.02	3.82	21.6
	0.53	0.13	10^{-2}	3.42	2.68	8.1

* Data taken from Figs. 11 and 12, Vrehen's paper, Ref. 6.
**Figure inferred from Fig. 6, Vrehen's paper, Ref. 6.

even though $\delta \sim N_0^{-\frac{1}{2}}$. The influence of δ on t_D and τ_W, and the ringing, is shown in Table 3 for $\tau_R = 0.53$ n sec. The choice $\delta = 10^{-5}$ $(10^{-4} - 10^{-6})$ again corresponds most closely to the observed t_D (11 n sec) and τ_W (5 n sec); $N_0 \sim 6 \times 10^7$ here and $2\tau_R \ln N_0^{\frac{1}{2}} = 9.5$ n sec.

The large value of $\delta \sim 10^{-4}$ means significant overlap between oppositely directed fields. Standing wave effects are important for large δ but not for very small δ. Figures 3a,b for $\tau_R = 0.7$ n sec shows results based on the equations from (3.2) which include standing wave effects. For the $\delta = 10^{-4}$ adopted to match t_D ringing is reduced from 21.4% (data of Table 2) to 17.1% ($\kappa L = 0$) and from 18.2% to 14.6% ($\kappa L = 0.4$). The best agreement with experiment is for $\kappa L = 0.8$: $t_D = 11.83$ n sec and the ringing is reduced from 14.9% to 12.0% by standing wave effects. The figure of 12% ringing is on the edge of that which could be concealed within the asymmetric pulse tails actually observed in almost all cases for $\tau_R \gtrsim 0.2$ n sec: these tails may be smeared-out rings arising through different chirps across the transverse wave profile. (All our intensity minima in the rings are zeros in disagreement with the observations). Notice (Table 3) that for $\delta = 10^{-2}$ ringing $\lesssim 8.1\%$ without standing waves: for this δ standing waves reduce this figure to a few per cent so very low ringing is possible in principle. We note that the densities n measured in Ref. 6 are uncertain by about 30% so to match the delays the chosen $\delta \sim 10^{-4}$ could be $\delta \sim 10^{-3}$ (it could also be about 10^{-6}). However, even the value $\delta \sim 10^{-4} \sim N_0^{-\frac{1}{2}}$ for Cs is large compared with the $\delta \sim 10^{-12} \sim N_0^{-1}$ for HF. Some indication why this should be so may lie in the following.

The A-coefficient for HF $\sim$ 1Hz whilst $\tau_R \sim$ n sec; the A-coefficient for Cs $\sim 2 \times 10^6$Hz. ($\tau_{SP} = 551$ n sec, $T_2' = 80$ n sec). The "classical delay" $t_D = 2\tau_R \ln N_0^{\frac{1}{2}} = \tau_R \ln N_0$ is derived by matching the radiation rate at $t = 0$ to $\tau_{SF}^{-1} = 2\tau_R^{-1}$: this is the spontaneous rate restricted to the solid angle $d\Omega = 3\lambda_s^2 A^{-1}$; $\lambda_s = 2\pi k_s^{-1} = 2\pi c\omega_s^{-1}$ and A is the cross-sectional area of the rod. The true initial rate is τ_{SP}^{-1} into the whole solid angle of 4π [5]. The effective N_0 therefore proves to be $3\lambda_s^2 N_0/4\pi A$; the factor on N_0 $\sim 10^{-5}$ for Cs so that reductions of N_0^{-1} from 10^{-8} to 10^{-3} become acceptable by this reasoning (Vrehen [6] refers to the related formula $t_D = \tau_R \ln \mu N_0$ (see his Ref. 16) and notes that factors ~ 2 on t_D are involved).

All these results suggest that in broad terms we have understood the relationship between the quantised theories [4] and semiclassical Bloch-Maxwell theory; but although the agreement with the HF data [2] and Cs data [6] is reasonably good there is still some way to go before we can simulate on the computer the exact details of superfluorescent emission as it is observed in the laboratory. Certainly the important role of δ in determining the delay, width and level of ringing is clear: δ also determines the measure of interaction between oppositely directed pulses and so determines the significance of standing wave effects. Why $\delta \sim N_0^{-1}$ for HF and $\sim N_0^{-\frac{1}{2}}$ for Cs is not at all clear, and all these points suggest that an exact study of the initial spontaneous emission process is urgently needed. Our use of damping to simulate the dynamical process of sideways loss is also rough — as rough as that approximation which replaces $\varepsilon_z = \sigma_{tz}$ in (3.6) by $\varepsilon/2L = \sigma_t/2L$. This, the problem of spontaneous emission, the effects of small degrees of degeneracy [12], and the problem of cascade from multi-level atoms raise intriguing theoretical problems for further work.

References

1. J.C. MacGillivray and M.S. Feld, Phys. Rev. A*14*, 1169 (1976).
2. N. Skribanowitz, I.P. Herman, J.C. MacGillivray and M.S. Feld, Phys. Rev. Lett. *30*, 309 (1973).
3. R. Saunders, R.K. Bullough and S.S. Hassan, J. Phys. A: Math. Gen. *9*, 1725 (1976).
4. G. Banfi and R. Bonifacio, Phys. Rev. Lett. *33*, 1259 (1974); G. Banfi and R. Bonifacio, Phys. Rev. A*12*, 2068 (1975); R. Bonifacio and L.A. Lugiato, Phys. Rev. A*11*, 1507, and *12*, 587 (1975); R.J. Glauber and F. Haake, in *Cooperative Phenomena,* H. Haken, ed. (North-Holland, Amsterdam, 1974) p. 71 (these authors discuss a single mode theory in both the semi-classical and quantised forms); R. Bonifacio, ibid, p. 97; N. Rehler and J.H. Eberly, Phys. Rev. A*3*, 1735 (1971).
5. R.K. Bullough and R. Saunders, in *Cooperative Effects in Matter and Radiation,* eds. C.M. Bowden, D.W. Howgate and H. Robl (Plenum Press, New York, 1977).
6. H.M. Gibbs, in *Cooperative Effects in Matter and Radiation,* and Q.H.F. Vrehen, in *Cooperative Effects in Matter and Radiation,*

eds. C.M. Bowden,. D.W. Howgate and H. Robl (Plenum Press, New York, 1977); Q.H.F. Vrehen, H.M.J. Hikspoors and H.M. Gibbs, Phys. Rev. Lett. *38*, 764 (1977).

7. M. Gross, C. Fabre, P. Pillet and S. Haroche, Phys. Rev. Lett. *36*, 1035 (1976).
8. S.R. Hartmann, in *Cooperative Effects in Matter and Radiation*, eds. C.M. Bowden, D.W. Howgate and H. Robl (Plenum Press, New York, 1977).
9. R. Bonifacio, F.A. Hopf, P. Meystre and M.O. Scully, Phys. Rev. A*12*, 2568 (1975).
10. F.T. Arecchi and E. Courtens, Phys. Rev. A*2*, 1730 (1970).
11. Co-operative Effects Meeting, Redstone Arsenal, Alabama, Dec. 1-2, 1976.
12. The effects of small hyperfine degeneracies (F = 2) on SIT pulses are briefly discussed in "Optical solitons and their spin wave analogues in ^{3}He," this volume, p. 767.

Note added in proof:

The single sentence below (3.12), that the pulse (2.3c) has width 3.52 τ_R is incorrect. As defined in (2.3c) this pulse has width 0.88 τ_R. However Bonifacio and Lugiato[4], using different definitions of the time constants from ours, reach (2.3c) with τ_{SF} replaced by τ_R. In (7.3) of Ref. 5 we reach the same result from a fundamental point of view via a decorrelation of a two-mode ansatz to the quantum theory. This result means a width of 1.76 τ_R. On this figure both the observed single pulses from Cs[6] and the calculations including diffraction loss damping have widths differing by factors up to ~5 from those predicted by mean field theory in the pure SF regime. Thus, although a good fit with a sech^2 is possible in both cases (if any ringing is neglected in the second case), we do not see this as evidence for mean field theory. As we show in this paper, mean field theory is a bad approximation compared with the approximation of the semiclassical theory. Accordingly an explanation of the details of the single pulse emission from Cs[6] needs an improvement on the semiclassical theory which *a fortiori* is an improvement on the mean field theory. Since going to press J.C. MacGillivray has sent us a fit of the values of t_D in Table 3 to a variant of his formula $t_D = \frac{1}{4} \tau_R |\ln(\sigma_0/2\pi)|^2$. We take this fit as evidence of good agreement between the numerical data of the M.I.T. Group and that reported in this paper. Further work is therefore needed for detailed agreement with the Cs data as we indicate in the paper.

DYNAMICAL APPROACH TO STEADY STATE AND FLUCTUATIONS IN OPTICALLY
BISTABLE SYSTEMS

G.S. Agarwal

University of Hyderabad, Hyderabad, India

L.M. Narducci[*] and Da Hsuan Feng

Drexel University, Philadelphia, Pa.

R. Gilmore

University of South Florida, Tampa, Fla.

1. INTRODUCTION

Optical bistability is a remarkable manifestation of cooperative atomic behavior.[1-3] A gas of resonant atoms under the action
of an incident driving field undergoes a first order phase transition, displaying discontinuity, hysteresis, and divergence. The
steady state behavior of the system is reminiscent of the equilibrium properties of a two-phase system: roughly speaking the bistable
transition may be interpreted as being due to the separation of the
gas into cooperative and uncorrelated phases in equilibrium.

Bonifacio and Lugiato have discussed an appealing quantum model
in the semi-classical approximation and derived explicit relations
among the atomic and the field parameters that control the bistable
behavior.[3] They have also outlined the qualitative features of
the fluorescence spectrum, and predicted a discontinuous transition
from a single line to the familiar three-peaked structure of conventional resonance fluorescence.[4]

In this paper we discuss a quantum dynamical model of optical
bistability,[5] derive the equations of motion of the atomic variables, and analyze the approach to equilibrium for different values
of the driving field amplitude. We also construct equations of motion for the atomic correlation functions, and propose a decorrela-

tion scheme that allows a closed-form solution for the steady state
spectrum of resonance fluorescence. As anticipated in Ref. 3, the
spectrum consists of a single peak centered at the resonant atomic
frequency, if the driving field amplitude is smaller than a certain
threshold value. Above threshold, the spectrum develops a pair of
side bands discontinuously, which are displaced from the central
peak by an amount proportional to the Rabi frequency of the driving
field. In addition, the bistable transition is heralded by a gra-
dual narrowing of the central line.[6]

We conclude our presentation with a discussion of the close
analogy between optical bistability and a second order phase transition for a certain range of atomic and field parameters.

2. DESCRIPTION OF THE MODEL AND ATOMIC MASTER EQUATION

The following physical processes are included in our analysis
of the atomic evolution:

  i)   interaction of the atoms with an external classical
       driving field in resonance with the atomic transition;

 ii)   irreversible atomic decay into vacuum;

iii)   cooperative interaction with a single quasi-mode of the
       interferometer containing the atoms.

Mathematically, the atomic evolution is governed by the reduced
density operator W solution of the master equation

$$\frac{dW}{dt} = -i\Omega_I \,[S^+ + S^-, W] + \Lambda_S W + \Lambda_A W \quad . \tag{2-1}$$

Equation (2-1) is derived under the Born and Markoff approximations
upon adiabatic elimination of the cavity field operators, following
a procedure similar to the one adopted in the early studies of superfluorescence.[7] The irreversible contributions

$$\Lambda_S W = \frac{2g^2}{\kappa} \,(S^- W S^+ - \tfrac{1}{2}WS^+ S^- - \tfrac{1}{2}S^+ S^- W) \quad , \tag{2-2}$$

and

$$\Lambda_A W = \sum_{i=1}^{N} \gamma_\perp \,\{[S_i^-, W S_i^+] + [S_i^- W, S_i^+]\} \quad , \tag{2-3}$$

describe the collective and single atom relaxation, respectively.
The operators $S^\pm$ are the collective polarization operators, Ω_I is
the Rabi frequency of the external field and g is the atom-quantum
field coupling constant. The parameters $\gamma_\perp$ and κ are the atomic
and field decay rates, respectively.

3. EVOLUTION OF THE ATOMIC EXPECTATION VALUES

A. Time Dependent Evolution

The exact evolution equations for the expectation values of the polarization and atomic energy operators can be easily derived from Eq. (2-1), with the result

$$d\langle S^+\rangle/dt = -\gamma_\perp\langle S^+\rangle - 2i\Omega_I\langle S_3\rangle + 2g^2/\kappa\ \langle S^+S_3\rangle \ ,$$

$$d\langle S^-\rangle/dt = -\gamma_\perp\langle S^-\rangle + 2i\Omega_I\langle S_3\rangle + 2g^2/\kappa\ \langle S_3S^-\rangle \ ,$$

$$d\langle S_3\rangle/dt = -2\gamma_\perp(\langle S_3\rangle+N/2) - i\Omega_I\langle S^+\rangle + i\Omega_I\langle S^-\rangle - 2g^2/\kappa\ \langle S^+S^-\rangle \ .$$

$$(3-1)$$

The above equations do not form a closed set: they can be closed by various decorrelation schemes, one of the most frequently used being the straight factorization ansatz (e.g. $\langle S^+S_3\rangle=\langle S^+\rangle\langle S_3\rangle$). This approximation is adopted here for the analysis of the mean value equations (3-1), but it will prove inadequate when dealing with the equations of motion for the atomic correlation functions (Sect. 4). For this reason a more suitable decorrelation ansatz will be suggested below. In view of the initial conditions $\langle S^+(0)\rangle=\langle S^-(0)\rangle = 0$, $\langle S_3(0)\rangle = -N/2$, it is clear that $\langle S^+(t)\rangle = \langle S^-(t)\rangle^*$ for all time. It is convenient to introduce the new dependent variables $S = -i\langle S^+\rangle = i\langle S^-\rangle$, $S_3 = \langle S_3\rangle$ and the dimensionless time variable $\tau = \gamma_\perp t$. With these replacements the equations of motion for the atomic expectation values become

$$\frac{dS}{d\tau} = -S - \sqrt{2}\ y\ S_3 + \frac{4c}{N}\ SS_3 \ ,$$

$$\frac{dS_3}{d\tau} = -2(S_3+N/2) + \sqrt{2}\ y\ S - \frac{4c}{N}\ S^2 \ ,\qquad\qquad (3-2)$$

where $y = \sqrt{2}\ \Omega_I/\gamma_\perp$ is proportional to the incident field amplitude, and the parameter $c = g^2N/2\kappa\ \gamma_\perp$ provides a measure of the density of atoms in the cavity.

Equations (3-2) have been solved numerically. An interesting feature of the time dependent solutions is their discontinuous dependence on the incident field amplitude y, a feature which has already been discussed in Ref. [3] An example is given in Fig. 1, where $S_3(\tau)$ is plotted as a function of the scaled time τ for different values of the incident field amplitude y. Even a small change in y around its threshold value can alter the dynamics

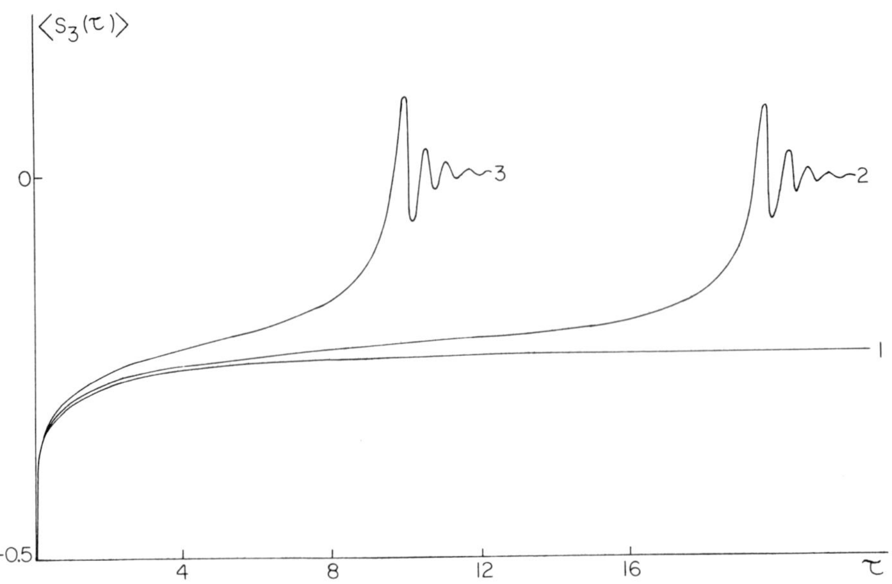

Fig. 1 Evolution of $<S_3(\tau)>$ as a function of the scaled time $\tau = \gamma_\perp t$ for c = 10 and different values of the driving field y. The three curves correspond to (1) y = 11.05; (2) y = 11.1; (3) y = 13.0. The threshold value for optical bistability is close to y = 11.055 for c = 10.

of the population inversion quite drastically, first of all by forcing non-monotonic behavior and, secondly, by altering the steady state value of the atomic variable.

B.   Steady State Behavior

The atomic equations (3-1) are supplemented by the adiabatic condition[7]

$$a = -g/\kappa \; S \; , \; a \equiv <a(t)> \; , \tag{3-3}$$

which is valid over the time scale $t \gg \kappa^{-1}$. If we define the transmitted field amplitude $\sqrt{2} \; (\Omega_I + ga)/\gamma_\perp$, it is simple to verify[3] that the steady state atomic expectation values take the form

$$S_3 = -N/2 \; \frac{1}{1+x^2} \; ,$$

$$S \; = N/\sqrt{2} \; \frac{x}{1+x^2} \; , \tag{3-4}$$

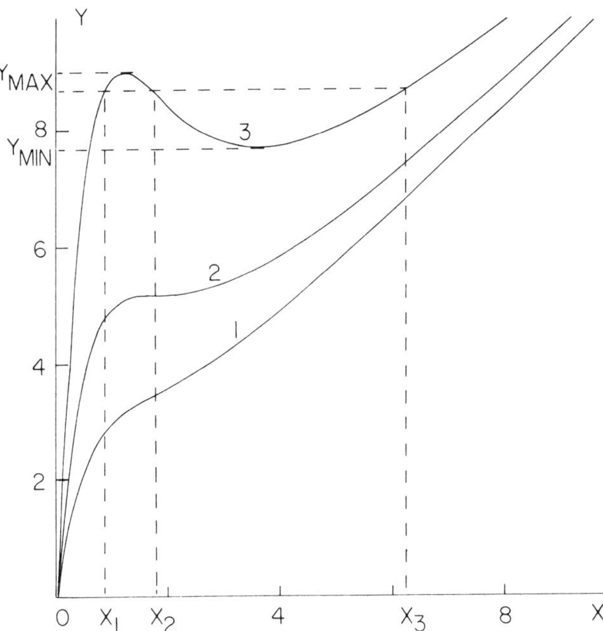

Fig. 2 Behavior of the function $y = x + 2cx/(1+x^2)$ for (1) c=2; (2) c=4; (3) c=8. For c=8 the output field amplitude is a multivalued function of the input field in the range $y_{min} < y < y_{max}$. The roots x_1 and x_3 are stable steady state values. The root x_2 is unstable.

while the transmitted field amplitude is related to the incident field by the cubic equation (Fig. 2),

$$y = x + \frac{2cx}{1+x^2} \quad .$$

(3-5)

For c < 4, Eq.(3-5) predicts a single-valued relation between x and y for all values of the incident amplitude y. For c > 4 and for $y_{min} < y < y_{max}$, the output field amplitude is a multivalued function of the input field. A simple stability analysis reveals that the roots x_1 and x_3 are stable steady-state values of the output field, while x_2 is unstable and cannot be physically realized. The discontinuous behavior of the transmitted field and its hysteresis properties as a function of the input y are related to the multi-valued nature of x[Ref. 3]. This behavior can also be understood in terms of simple catastrophe theory, where it is seen that optical bistability is just another example of cusp catastrophe,[8] and that the hysteresis of the transmitted field is a consequence of the so-called delay convention.

4. ATOMIC CORRELATION FUNCTIONS AND SPECTRUM OF RESONANCE FLUORESCENCE

The spectrum of resonance fluorescence is given by the Fourier transform of the field correlation function $\langle a^+(t+\tau)a(t)\rangle$. Since the source-field operators $a(t)$, $a^+(t)$ are directly proportional to the atomic polarization operators $S^-(t)$, and $S^+(t)$, we only need to consider the atomic correlation function $<S^+(t'+t)S^-(t')>$. In particular, we are interested in the steady-state value

$$\lim_{t' \to \infty} <S^+(t'+t)S^-(t')> \equiv <S^+(t)S^-> \quad .$$

To this end, we define the set of correlation functions

$$\psi_1 = <S^+(t)S^-> \quad ,$$

$$\psi_2 = <S^-(t)S^-> \quad ,$$

$$\psi_3 = <S_3(t)S^-> \quad . \tag{4-1}$$

With the help of the regression theorem,[10] Eqs.(3-1) lead to the following coupled differential equations

$$\frac{d\psi_1}{dt} = -\gamma_\perp \psi_1 - 2i\Omega_I \psi_3 + 2g^2/\kappa <S^+(t)S_3(t)S^-> \quad ,$$

$$\frac{d\psi_2}{dt} = -\gamma_\perp \psi_2 + 2i\Omega_I \psi_3 + 2g^2/\kappa <S_3(t)S^-(t)S^-> \quad ,$$

$$\frac{d\psi_3}{dt} = -2\gamma_\perp(\psi_3+N/2<S^->) - i\Omega_I \psi_1 + i\Omega_I \psi_2 - 2g^2/\kappa <S^+(t)S^-(t)S^-> \quad . \tag{4-2}$$

The expectation values of the products of three atomic operators are handled as follows. We define the fluctuation operators $X^\pm$, X_3, $(X^\pm(t)=S^\pm(t)-<S^\pm>$, $X_3(t) = S_3(t)-<S_3>)$ and assume that their odd-order moments vanish. This can be justified by an appropriate expansion of the equations of motion in powers of $1/N$[5]. The correlation functions of the fluctuation operators (i.e. $\chi_1(t) = <X^+(t)X^->$, etc.) satisfy the set of linear equations

$$d\chi_1/d\tau = -(1+2c/1+x^2)\chi_1 - i\sqrt{2} \, x \, \chi_3 \quad ,$$

$$d\chi_2/d\tau = -(1+2c/1+x^2)\chi_2 + i\sqrt{2} \, x \, \chi_3 \quad ,$$

$$d\chi_3/d\tau = -2\chi_3 + \frac{i}{\sqrt{2}} (y-2x)\chi_1 - \frac{i}{\sqrt{2}} (y-2x)\chi_2 \quad , \tag{4-3}$$

which is formally identical to the set of Eqs.(3-1) after factorization and linearization around equilibrium are carried out.

The initial conditions are given by

$$\chi_1(0) = \langle S^+ S^- \rangle - \langle S^+ \rangle \langle S^- \rangle = \frac{N}{2} \frac{x^4}{(1+x^2)^2} \quad ,$$

$$\chi_2(0) = N/2 \frac{x^2}{(1+x^2)^2} \quad ,$$

$$\chi_3(0) = \frac{i}{4} N \sqrt{2} \frac{x^3}{(1+x^2)^2} \quad . \tag{4-4}$$

Equations (4-3) can be solved conveniently by Laplace transformation. The solution of interest is given below.

$$\tilde{\chi}_1(z) = \frac{N}{2} \frac{x^4}{(1+x^2)^2} \frac{(z+A)(z+3)+x(2x-y)+2-A}{D(z)(z+A)} \quad ,$$

$$D(z) = (z+A)(z+2)+2x(2x-y) \quad ,$$

$$A \equiv 1 + 2c/1+x^2 \quad . \tag{4-5}$$

We observe that

$$\tilde{\psi}_1^{incoh}(z) \equiv \tilde{\psi}_1(z) - \frac{1}{z} \lim_{z \to 0} z \, \tilde{\psi}_1(z) = \chi_1(z) \quad .$$

Hence, the required spectrum is given by

$$S(\omega) = \mathrm{Re} \, \tilde{\psi}_1^{incoh}(z) \Big|_{z=i\omega} = \mathrm{Re} \, \tilde{\chi}_1(z) \Big|_{z=i\omega} \quad . \tag{4-6}$$

It is easy to see that, in the limit of vanishing atomic density
$(c \to 0)$, Eq.(4-6) reduces to the well known single atom resonance
fluorescence spectrum. For the interesting case $c > 4$, typical
spectra are shown in Figs. 3 and 4. As the incident field increases
from $y = 0$ to $y = y_{max}$, the spectrum consists of a single broadened
line with a monotonically decreasing half-width (Fig. 3). For
$y > y_{max}$ the spectrum develops sidebands discontinuously, and for
increasing driving field it approaches the usual single atom spectrum of resonance fluorescence. Upon decreasing the driving field
along the single atom branch, the sidebands merge continuously into
the central peak which becomes narrower and narrower as the lower
bistability threshold is approached. For values $y < y_{min}$ the atomic
system jumps back to the cooperative branch, thus exhibiting discontinuity and hysteresis.

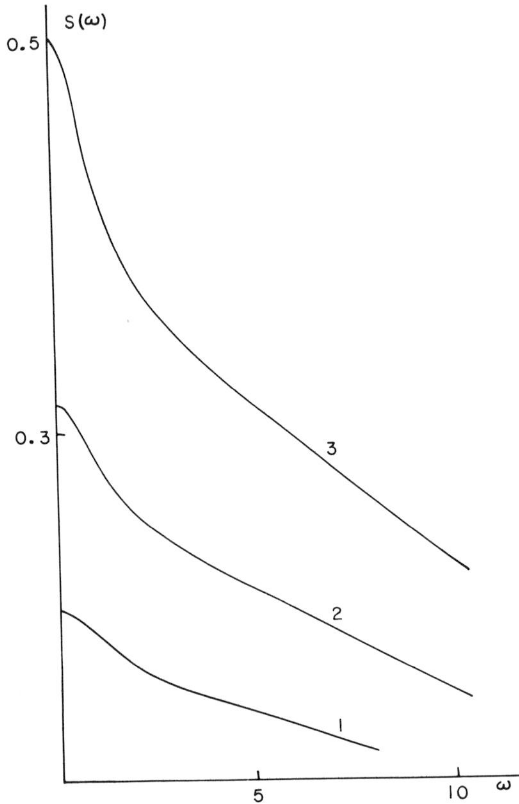

Fig. 3 Resonance fluores-
cence spectra for c = 10
and several values of the
driving field (1) y = 9;
(2) y = 9.5; (3) y = 10.
The stationary states of
the atomic system belong to
the cooperative branch.

5.   SECOND ORDER PHASE TRANSITION ANALOGY OF OPTICAL BISTABILITY

An interesting analogy can be drawn between optically bistable
systems and thermodynamic systems undergoing a second order phase
transition in the context of mean field theory. The analogy will
apply in a small neighborhood of a suitably defined critical point.

Consider Eq.(3-5) to be the thermodynamic equation of a state
of the bistable system with c playing the role of the temperature,
x being the external parameter and y the associated generalized
force. We define the critical point as the set of values x_0, y_0,
c_0 which satisfy the equation of state and the conditions

$$\left(\frac{\partial y}{\partial x}\right)_0 = 0 \quad , \qquad \left(\frac{\partial^2 y}{\partial x^2}\right)_0 = 0 \quad . \tag{5-1}$$

The derivatives are carried out along the "isotherms" (constant c).
It is easy to verify that the critical point is given by

$$y_o = 3\sqrt{3} \quad , \quad x_o = \sqrt{3} \quad , \quad c_o = 4 \quad . \tag{5-2}$$

In analogy with an ordinary fluid system, we cay say that the isotherms c < 4 identify all the single phase equilibrium points; for
c > 4 the isotherms consist of stable, metastable, and unstable
branches. The domain of instability for a single phase corresponds
to the domain of equilibrium of two coexisting phases: the cooperative and uncorrelated phases of the atomic system. Notice, however,
that unlike the case of an ordinary fluid, the coexistence region
lies above the critical "temperature" in this case. In terms of the
scaled variables

$$v = \frac{x-x_o}{x_o} \quad , \quad p = \frac{y-y_o}{y_o} \quad , \quad \varepsilon = \frac{c-c_o}{c_o} \quad , \tag{5-3}$$

the equation of state in the neighborhood of the critical point becomes

$$p(\varepsilon,v) \cong \frac{2}{3}\,\varepsilon - \frac{1}{3}\,\varepsilon\,v - \frac{1}{2}\,\varepsilon\,v^2 + \frac{1}{4}\,v^3 \tag{5-4}$$

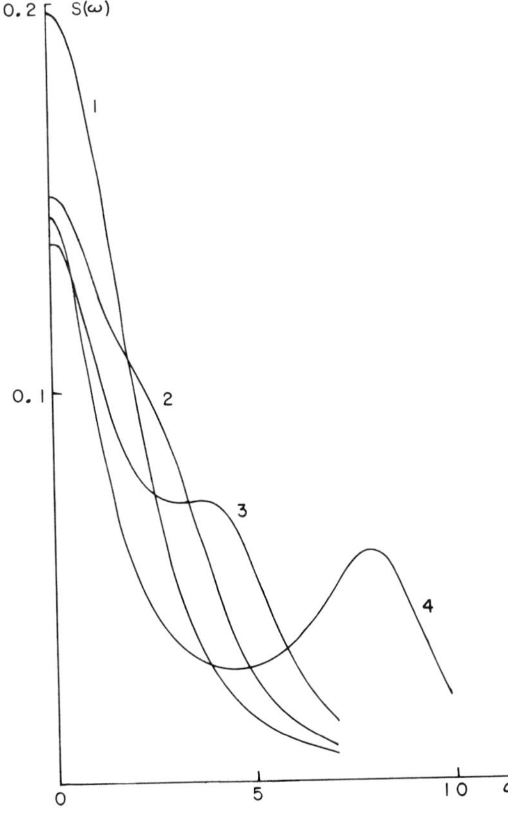

Fig. 4 Resonance fluorescence spectra for c = 10 and
several values of the driving
field: (1) y = 8.73; (2) y =
8.82; (3) y = 9; (4) y = 10;
the stationary states of the
system belong to the single
atom branch.

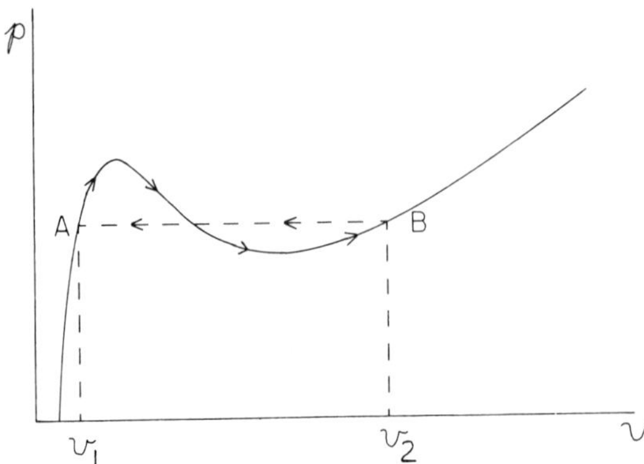

Fig. 5 Shape of a typical isotherm. The path of integration needed to calculate Eq.(5-6) consists of the segment of isotherm indicated by the arrows and by the straight line p = constant from B to A.

Thus the critical isotherm ($\varepsilon=0$) has the shape

$$p(0,v) = \frac{1}{4} v^3 \quad , \tag{5-5}$$

which is governed by the critical exponent $\delta = 3$, in analogy with the predictions of the ordinary Landau mean field theory.

The shape of the coexistence curve can be constructed on the basis of Maxwell's rule

$$\oint v \, dp = 0 \quad , \tag{5-6}$$

where the path of integration goes from A to B and back to A along the direction of the arrows in Fig. 5. Since the contribution along the straight line BA is identically zero, the required values v_1 and v_2 are given by

$$\oint v \, dp = \int_{v_1}^{v_2} v \left(\frac{\partial p}{\partial v}\right)_\varepsilon dv = 0 \quad , \tag{5-7}$$

where $(\partial p/\partial v)_\varepsilon$ is evaluated along the isotherm ε = constant, and p is given by Eq.(5-4). The result of the simple calculation is the coexistence curve

$$v_2 - v_1 = \frac{4}{\sqrt{3}} \varepsilon^{\frac{1}{2}} \quad . \tag{5-8}$$

We see that the asymptotic behavior of this curve is controlled by the exponent $\beta = 1/2$ in the neighborhood of the critical point.

A measure of the discontinuity suffered by the transmitted field amplitude as a function of the incident amplitude can be given in terms of the "isothermal susceptibility":

$$\chi_T = \frac{1}{3\sqrt{3}} \left(\frac{\partial p}{\partial v}\right)^{-1} . \tag{5-9}$$

Below the critical density ($\varepsilon < 0$) and, for example, along the critical isochore ($v=0$) we find

$$\chi_T = \frac{1}{\sqrt{3}} |\varepsilon|^{-1} \tag{5-10}$$

while, for $\varepsilon > 0$ and along the coexistence curve, we have

$$\chi_T = \frac{1}{2\sqrt{3}} \varepsilon^{-1} . \tag{5-11}$$

In either case the divergence of the susceptibility, as $\varepsilon \to 0$, is controlled by the critical exponent $\gamma = 1$. The numerical coefficients in Eqs. (5-10) and (5-11) are in the ratio 1:2 as predicted by the Landau mean field theory.

This analogy, of course, should not come as a surprise. The equilibrium points that control the steady state behavior of a bistable system are in one to one correspondence with the equilibrium manifold of a cusp catastrophe.[8] This, in turn, is the canonical manifold which is appropriate for a system undergoing first and second order phase transitions in the mean field theory approximation. The above considerations only make the analogy more transparent.

Acknowledgements

One of us (LMN) is indebted to Professors Arecchi, Bonifacio, Lugiato and Mollow for a very informative discussion.

*
Partially supported by the Office of Naval Research, Contract No. N0014-76-C-1082.

References

1. H.M. Gibbs, S.L. McCall, T.N.C. Venkatesan, Phys. Rev. Lett. *36*, 1135 (1976).
2. S.L. McCall, Phys. Rev. A *9*, 1515 (1974).

3.   R. Bonifacio, L.A. Lugiato, Opt. Comm. *19*, 172 (1976).
4.   B.R. Mollow, Phys. Rev. *188*, 1969 (1969); ibid. A *5*, 1522 (1972).
5.   G.S. Agarwal, L.M. Narducci, D.H. Feng, R. Gilmore (in preparation).
6.   In the jargon of phase transitions this feature of the spectrum is a manifestation of the onset of a soft mode.
7.   R. Bonifacio, P. Schwendimann, F. Haake, Phys. Rev. A *4*, 302 (1971).
8.   R. Gilmore, L.M. Narducci, Phys. Rev. A (to be published).
9.   G.S. Agarwal, *Springer Tracts in Modern Physics*, ed. G. Hohler et al. (Springer-Verlag, New York, 1974) Vol. 7, Sect. 18.
10. M. Lax, in *Statistical Physics, Phase Transitions and Superfluidity*, ed. M. Chretien, E.P. Gross and S. Deser (Gordon and Breach, New York, 1968) Vol. 2, p. 269.
11. G.S. Agarwal, A.C. Brown, L.M. Narducci and G. Vetri, Phys. Rev. A *15*, 1613 (1977).

ANALYTIC SOLUTIONS OF NON-MARKOVIAN MASTER EQUATIONS FOR SUPERFLUORESCENCE[*]

C. T. Lee[†]

Princeton University, Princeton, New Jersey

1. INTRODUCTION

In 1973 Skribanowitz et al. made the first observation of super-fluorescence in optically pumped HF gas [1]. One feature that was not predicted by the theoretical studies then available was ringing. Since its discovery, several authors [1-4] have tried to explain this phenomenon by taking into consideration the propagation effects such as field inhomogeneities and stimulated effects. In this article, we will focus our attention on the non-Markovian master equation derived from first principles by Bonifacio and Lugiato [2] in which the stimulated effects are incorporated to account for the ringing. We will use the Dicke state representation and obtain probability distributions for the case of no atom-field correlation and for the case when only first order atom-field correlation is considered. The expectation values of some basic operators will also be obtained explicitly.

2. BONIFACIO-LUGIATO MASTER EQUATION

In a recent paper Bonifacio and Lugiato [2] have obtained a non-Markovian master equation for the superfluorescence from a pencil-shaped active region as follows:

$$\frac{dW(t)}{dt} = \Lambda_F W(t) - ige^{-t/2T_2^*}[(aR^+ + a^\dagger R^-), e^{\Lambda_F t}W(0)]$$
$$-g^2 \int_0^t dt' e^{-(t+t')/2T_2^*}[(aR^+ + a^\dagger R^-), e^{\Lambda_F(t-t')}[(aR^+ + a^\dagger R^-), W(t')]],$$

$$(2.1)$$

where (a) $W(t)$ is the density matrix of the system composed of N
two-level atoms and a single axial mode of the internal field
(i.e., the field inside the active volume); (b) $a^\dagger$ and a are the
photon creation and annihilation operators; (c) R^+ and R^- are the
collective atomic excitation and de-excitation operators; (d)
$\Lambda_F W(t) = K\{[aW(t),a^\dagger] + \text{H.c.}\}$ where $K = c/2L$ (L being the length
of the active region) is the damping constant, so Λ_F describes the
irreversible escape of photons from the active region; (e) g is
the atom-field coupling constant and T_2^* is the smallest atomic
relaxation time which will be assumed to be infinite.

According to the theory of open systems developed by Lugiato
[5], Eq. (2.1) generates a hierarchy of equations for the quantities

$$N_{\alpha,\beta}(t) \equiv \text{Tr}_F\{(a^\dagger)^\alpha a^\beta W(t)\} , \qquad \alpha+\beta \text{ even}, \tag{2.2}$$

where Tr_F means the partial trace on the field Hilbert space. This
hierarchy of equations has been given explicitly by Gronchi and
Lugiato [6].

Using Dicke state representation, we define

$$P(n,t) \equiv <r,m|N_{0,0}(t)|r,m> , \tag{2.3}$$

$$Q(n,t) \equiv <r,m|N_{1,1}(t)|r,m> , \tag{2.4}$$

$$R(n,t) \equiv \frac{1}{2}\{<r,m-1|N_{0,2}(t)|r,m+1>+<r,m+1|N_{2,0}(t)|r,m-1>\} , \tag{2.5}$$

where r and m have the usual meaning and

$$n \equiv r - m . \tag{2.6}$$

We can see that $P(n,t)$ is the probability that the atomic system
exists in the Dicke state $|r,m>$ and $Q(n,t)$ and $R(n,t)$ are, respect-
ively, the diagonal and off-diagonal first order atom-field correlations. From Eq. (5) of Ref. [6], we can write down the following equations for $P(n,t)$, $Q(n,t)$ and $R(n,t)$:

$$\frac{d}{dt} P(n,t) = KI\int_0^t dt' e^{-K(t-t')}\{g(n-1)P(n-1,t') - g(n)P(n,t')$$

$$+ g(n-1)Q(n-1,t') - [g(n-1) + g(n)]Q(n,t')$$

$$+ g(n)Q(n+1,t) - \sqrt{g(n-2)g(n-1)}R(n-1,t')$$

$$+ 2\sqrt{g(n-1)g(n)}R(n,t') - \sqrt{g(n)g(n+1)}R(n+1,t')\}, \tag{2.7}$$

$$\frac{d}{dt} Q(n,t) = -2KQ(n,t)$$

$$+ KI \int_0^t dt' e^{-K(t-t')} \{ g(n-1)[P(n-1,t') + Q(n-1,t') - Q(n,t')]$$

$$- \sqrt{g(n-2)g(n-1)}R(n-1,t') + \sqrt{g(n-1)g(n)}R(n,t') \}$$

$$+ KI \int_0^t dt' e^{-3K(t-t')} \{ 2g(n-1)Q(n-1,t') - 2g(n)Q(n,t')$$

$$- \sqrt{g(n-2)g(n-1)}R(n-1,t')$$

$$+ \sqrt{g(n-1)g(n)}R(n,t') \} \tag{2.8}$$

$$\frac{d}{dt} R(n,t) = -2KR(n,t)$$

$$- KI \int_0^t dt' e^{-K(t-t')} \{ \sqrt{g(n-1)g(n)}[P(n-1,t')+Q(n-1,t')-Q(n,t')]$$

$$- \sqrt{g(n-2)g(n)}R(n-1,t') + g(n)R(n,t') \}$$

$$- KI \int_0^t dt' e^{-3K(t-t')} \{ 2\sqrt{g(n-1)g(n)}[Q(n-1,t') - Q(n,t')]$$

$$- 4\sqrt{g(n-2)g(n)}R(n-1,t')$$

$$+ \frac{1}{2}[g(n-1) + 3g(n+1)]R(n,t') \}, \tag{2.9}$$

where I is the emission rate of a single excited atom;

$$g(n) \equiv (2r-n)(n+1) , \tag{2.10}$$

and we have ignored higher order correlations, i.e. $N_{\alpha,\beta}(t)$ with $\alpha+\beta>2$.

We will assume the following initial conditions:

$$P(n,0) = \delta_{n,0}, \tag{2.11}$$

$$Q(n,0) = R(n,0) = 0, \tag{2.12}$$

and r = N/2, N being an even number. Let

$$S(n,t) \equiv g(n)Q(n,t) + \sqrt{g(n-1)g(n)}\,R(n,t); \tag{2.13}$$

then, from Eqs. (2.8) and (2.9), we have

$$\frac{d}{dt}S(n,t) = -2KS(n,t)$$

$$+ KI\int_0^t dt'\,e^{-3K(t-t')}\{g(n)S(n-1,t') - g(n+1)S(n,t')\}, \tag{2.14}$$

where we have used the approximation

$$g(n-1) + g(n+1) \simeq 2g(n). \tag{2.15}$$

Solving Eq. (2.14) with the initial condition Eq. (2.12), we obtain

$$S(n,t) = 0, \tag{2.16}$$

which means that, under the approximation of Eq. (2.15), we have

$$\sqrt{g(n-1)g(n)}\,R(n,t) = -g(n)Q(n,t), \tag{2.17}$$

which, in turn, means that the off-diagonal atom-field correlation
is almost equal to the negative of the diagonal one. This has
been pointed out by Gronchi and Lugiato [7]. Because of Eq. (2.17)
we can reduce the three equations, Eqs. (2.7-2.9), to two as follows:

$$\frac{d}{dt}P(n,t) = KI\int_0^t dt'\,e^{-K(t-t')}\{g(n-1)P(n-1,t') - g(n)P(n,t')$$

$$+ 2g(n-1)Q(n-1,t') - [g(n-1) + 3g(n)]Q(n,t')$$

$$+ [g(n) + g(n+1)]Q(n+1,t')\}, \tag{2.18}$$

$$\frac{d}{dt}Q(n,t) = -2KQ(n,t)$$

$$+ KI\int_0^t dt'\,e^{-K(t-t')}\{g(n-1)[P(n-1,t') + 2Q(n-1,t')]$$

$$- [g(n-1) + g(n)]Q(n,t')\}$$

$$+ 3KI\int_0^t dt'\,e^{-3K(t-t')}\{g(n-1)Q(n-1,t') - g(n)Q(n,t')\}. \tag{2.19}$$

3.   NO ATOM-FIELD CORRELATION

If we neglect the atom-field correlation completely, then the two equations, Eqs. (2.18) and (2.19), reduce to one only, namely,

$$\frac{d}{dt} P(n,t) = KI\int_0^t dt' e^{-K(t-t')}\{g(n-1)P(n-1,t') - g(n)P(n,t')\} \ . \quad (3.1)$$

Let the Laplace transform of $P(n,t)$ be

$$\tilde{P}(n,s) \equiv \int_0^\infty e^{-st}P(n,t)dt. \quad (3.2)$$

Then, from the Laplace transform of Eq. (3.1) together with the initial condition Eq. (2.11), we have [8]

$$\tilde{P}(n,s) = \frac{s+K}{s(s+K)+KIg(n)} \prod_{i=0}^{n-1} \frac{KIg(i)}{s(s+K)+KIg(i)} \ . \quad (3.3)$$

The inverse Laplace transform of Eq. (3.3) can be obtained by reducing the product to partial fractions. To save our effort, we can rewrite Eq. (3.3) as

$$\tilde{P}(n,s) = \frac{s+K}{KI} \tilde{F}(n,z) \ , \quad (3.4)$$

where

$$\tilde{F}(n,z) \equiv \frac{1}{z+g(n)} \prod_{i=0}^{n-1} \frac{g(i)}{z+g(i)} \ , \quad (3.5)$$

$$z \equiv s(s+K)/KI \ . \quad (3.6)$$

Now, the partial fractions of the expression $\tilde{F}(n,z)$ have been carried out in the solution of the superradiance master equation [9], so we can quote the result directly as follows:

$$\tilde{F}(n,z) = \begin{cases} \displaystyle\sum_{i=0}^{n} \frac{A_n^i}{z+g(i)} & \text{for } 0 \le n < N/2 \\[2em] \displaystyle\sum_{i=0}^{N-n-2} \frac{A_n^i}{z+g(i)} + \sum_{i=N-n-1}^{N/2-1} \left[\frac{B_n^i}{z+g(i)} + \frac{C_n^i}{[z+g(i)]^2} \right] & \text{for } N/2 \le n < N \\[2em] \displaystyle\frac{1}{z} + \sum_{i=0}^{N/2-1} \left[\frac{B_N^i}{z+g(i)} + \frac{C_N^i}{[z+g(i)]^2} \right] & \text{for } n=N, \end{cases} \tag{3.7}$$

where

$$A_n^i \equiv (-1)^i \, \frac{(N-2i-1)N!n!(N-n-i-2)!}{(N-n)!(n-i)!i!(N-i-1)!} \quad, \tag{3.8}$$

$$B_n^i \equiv (-1)^{N+n+1} \binom{N}{i+1}\binom{n}{i}\binom{i+1}{N-n} \left\{ 2 - (N-2i-1)\sum_{k=i+1}^{N-i-1}\left(\frac{1}{k} + \frac{1}{n-k+1} \right) \right\}, \tag{3.9}$$

$$C_n^i \equiv (-1)^{N+n+1} \binom{N}{i+1}\binom{n}{i}\binom{i+1}{N-n} (N-2i-1)^2 \quad. \tag{3.10}$$

Substitution of Eqs. (3.7) and (3.6) into Eq. (3.4) gives

$$\tilde{P}(n,s) = \begin{cases} \displaystyle\sum_{i=0}^{n} A_n^i \tilde{X}_0(i,s) & \text{for } 0 \le n < N/2 \\[2em] \displaystyle\sum_{i=0}^{N-n-2} A_n^i \tilde{X}_0(i,s) + \sum_{i=N-n-1}^{N/2-1}\left\{ B_n^i \tilde{X}_0(i,s) + C_n^i \tilde{Y}_0(i,s) \right\} & \text{for } N/2 \le n < N \\[2em] \displaystyle\frac{1}{s} + \sum_{i=0}^{N/2-1}\left\{ B_N^i \tilde{X}_0(i,s) + C_N^i \tilde{Y}_0(i,s) \right\} & \text{for } n=N, \end{cases} \tag{3.11}$$

where

$$\tilde{X}_0(i,s) \equiv \frac{s+K}{s(s+K) + KIg(i)} \quad, \tag{3.12}$$

$$\tilde{Y}_0(i,s) \equiv \frac{KI(s+K)}{[s(s+K) + KIg(i)]^2} \quad. \tag{3.13}$$

The inverse Laplace transforms of $\tilde{X}_0(i,s)$ and $\tilde{Y}_0(i,s)$ are

$$X_0(i,t) = e^{-Kt/2}[\cos\omega_i t + \frac{K}{2\omega_i} \sin\omega_i t] \quad , \tag{3.14}$$

$$Y_0(i,t) = \frac{K^2 I}{4\omega_i^2} e^{-Kt/2}[(1+2\omega_i t/K)\sin\omega_i t - \omega_i t\cos\omega_i t] \quad , \tag{3.15}$$

where

$$\omega_i \equiv [KIg(i) - K^2/4]^{\frac{1}{2}} \quad . \tag{3.16}$$

We are now ready to write the probability distribution as follows:

$$P(n,t)=\begin{cases} \displaystyle\sum_{i=0}^{n} A_n^i X_0(i,t) \quad \text{for } 0\leq n<N/2 \\[2em] \displaystyle\sum_{i=0}^{N-n-2} A_n^i X_0(i,t) + \sum_{i=N-n-1}^{N/2-1}\{B_n^i X_0(i,t) + C_n^i Y_0(i,t)\} \quad \text{for } N/2\leq n<N \\[2em] 1 + \displaystyle\sum_{i=0}^{N/2-1}\{B_n^i X_0(i,t) + C_N^i Y_0(i,t)\} \quad \text{for } n=N \; . \end{cases} \tag{3.17}$$

We can again use the result of Ref.[9] to write some expectation values as follows:

$$\langle n\rangle = N - \sum_{i=0}^{N/2-1} \frac{N!}{(N-i-1)!i!} \left\{[2-(N-2i-1)\sum_{k=i+1}^{N-i-1}\frac{1}{k}]X_0(i,t) \right.$$

$$\left. + (N-2i-1)^2 Y_0(i,t)\right\} \quad , \tag{3.18}$$

$$\langle R^+R^-\rangle \equiv \langle g(n)\rangle$$

$$= \sum_{i=0}^{N/2-1} \frac{N!}{(N-i-1)!i!}\{(N-i)(i+1)[2-(N-2i-1)\sum_{k=i+1}^{N-i-1}\frac{1}{k}]X_0(i,t)$$

$$- (N-2i-1)^2[X_0(i,t)-(N-i)(i+1)Y_0(i,t)]\} \quad . \tag{3.19}$$

4. FIRST ORDER ATOM-FIELD CORRELATION

In order to see the effects of atom-field correlation, we will try to solve Eqs. (2.18) and (2.19) simultaneously. In a brief discussion of the solution of such non-Markovian master equations, Bonifacio and Lugiato[2] neglected the off-diagonal correlation. So the pair of equations they considered are different from Eqs. (2.18)and (2.19). They also dropped the second integral in Eq. (2.19) because it contains rapidly decaying terms. We have some doubt about this justification. Perhaps, a better reason is that we expect $g(n)Q(n,t)$ to be a smooth function of n; hence, the difference of two consecutive values should be very small in comparison with $g(n-1)P(n-1,t)$. For the same reason, we could also drop the terms $2g(n-1)Q(n-1,t') - [g(n-1)+g(n)]Q(n,t')$ in the first integral of Eq. (2.19). But we chose not to do so because, strangely, we can obtain the solution easier with these terms included.

The Laplace transformation of Eqs. (2.18) and (2.19) are as follows:

$$s\tilde{P}(n,s) - \delta_{n,0} = \frac{KI}{s+K} \{g(n-1)[\tilde{P}(n-1,s) + 2\tilde{Q}(n-1,s)] - g(n)\tilde{P}(n,s)$$

$$- [g(n-1) + 3g(n)]\tilde{Q}(n,s) + [g(n) + g(n+1)]\tilde{Q}(n+1,s)\},$$

$$(4.1)$$

$$s\tilde{Q}(n,s) = -2K\tilde{Q}(n,s) + \frac{KI}{s+K} \{g(n-1)[\tilde{P}(n-1,s) + 2\tilde{Q}(n-1,s)]$$

$$- [g(n-1) + g(n)]\tilde{Q}(n,s)\} , \qquad (4.2)$$

where we have used the initial conditions Eqs. (2.11) and (2.12).

From Eq. (4.2), we have

$$\tilde{Q}(n,s) = \begin{cases} 0 \text{ for } n = 0 \\ \dfrac{KIg(n-1)}{(s+K)(s+2K)+KI[g(n-1)+g(n)]} [\tilde{P}(n-1,s)+2\tilde{Q}(n-1,s)] \text{ for } n>0 . \end{cases}$$

$$(4.3)$$

Using $\tilde{Q}(1,s)$ given by Eq. (4.3), we can solve Eq. (4.1) with n=0 to obtain

$$\tilde{P}(0,s) = \frac{(s+K)(s+2K) + KI[g(0)+g(1)]}{s(s+K)(s+2K) + 2KIg(0)(s+K) + KIg(1)s} . \qquad (4.4)$$

For n>0, we substitute Eq. (4.3) into Eq. (4.1) to obtain

$$\tilde{P}(n,s) = \frac{(s+2k)[(s+K)(s+2K) - KIg(n) + KIg(n+1)]}{s(s+K)(s+2K) + 2KIg(n)(s+K) + KIg(n+1)s} \tilde{Q}(n,s) \text{ for } n>0. \quad (4.5)$$

From Eqs. (4.3) and (4.5) we have

$$\tilde{P}(n,s) = \frac{KIg(n-1)(3s+2K)}{s(s+K)(s+2K) + 2KIg(n)(s+K) + KIg(n+1)s}$$

$$\times \frac{(s+K)(s+2K) - KIg(n) + KIg(n+1)}{(s+K)(s+2K) - KIg(n-1) + KIg(n)} \tilde{P}(n-1,s). \qquad (4.6)$$

Repeated applications of Eq. (4.6) together with Eq. (4.4) yield

$$\tilde{P}(n,s) = \frac{(s+2K)(3s+2K)^{n-1}[(s+K)(s+2K) - KIg(n) + KIg(n+1)]}{s(s+K)(s+2K) + 2KIg(n)(s+K) + KIg(n+1)s}$$

$$\times \prod_{i=0}^{n-1} \frac{KIg(i)}{s(s+K)(s+2K) + 2KIg(i)(s+K) + KIg(i+1)s} . \tag{4.7}$$

Substitution of Eq. (4.7) into Eq. (4.5) gives

$$\tilde{Q}(n,s) = (3s+2K)^{n-1} \prod_{i=0}^{n-1} \frac{KIg(i)}{s(s+K)(s+2K) + 2KIg(i)(s+K) + KIg(i+1)s}. \tag{4.8}$$

The inverse Laplace transforms of $\tilde{P}(n,s)$ and $\tilde{Q}(n,s)$ can now be obtained by reducing Eqs. (4.7) and (4.8) to partial fractions. To save the effort, we will introduce the approximation

$$g(i+1) \simeq g(i) . \tag{4.9}$$

Then Eqs. (4.7) and (4.8) can be rewritten as

$$\tilde{P}(n,s) = \frac{(s+K)(s+2K)^2}{KI(3s+2K)^2} \tilde{F}(n,z') , \tag{4.10}$$

$$\tilde{Q}(n,s) = \frac{s(s+K)(s+2K) + KIg(n)(3s+2K)}{KI(3s+2K)^2} \tilde{F}(n,z') , \tag{4.11}$$

where $\tilde{F}(n,z')$ is the same as defined by Eq. (3.5) and

$$z' \equiv \frac{s(s+K)(s+2K)}{KI(3s+2K)} . \tag{4.12}$$

Therefore, to obtain $\tilde{P}(n,s)$ for $n>0$, all we have to do is to replace $\tilde{X}_0(i,s)$ and $\tilde{Y}_0(i,s)$ in Eq. (3.11) by $\tilde{X}_1'(i,s)$ and $\tilde{Y}_1(i,s)$, respectively, as given in the following:

$$\tilde{X}_1'(i,s) \equiv \frac{(s+K)(s+2K)^2}{(3s+2K)[s(s+K)(s+2K) + KIg(i)(3s+2K)]}$$

$$= -\frac{2}{3s+2K} + \tilde{X}_1(i,s) , \tag{4.13}$$

where

$$\tilde{X}_1(i,s) \equiv \frac{(s+K)(s+2K) + 2KIg(i)}{s(s+K)(s+2K) + KIg(i)(3s+2K)} \,, \tag{4.14}$$

and

$$\tilde{Y}_1(i,s) \equiv \frac{KI(s+K)(s+2K)^2}{[s(s+K)(s+2K) + KIg(i)(3s+2K)]^2} \tag{4.15}$$

In the case of $\tilde{P}(N,s)$, we need also to replace the $1/s$ by $1/s - 2/(3s+2k)$. However, using the initial condition Eq. (2.11) in Eq. (3.17), we have the following identities:

$$\sum_{i=0}^{n} A_n^i = 0 \quad \text{for } 0<n<N/2 \,, \tag{4.16}$$

$$\sum_{i=0}^{N-n-2} A_n^i + \sum_{i=N-n-1}^{N/2-1} B_n^i = 0 \quad \text{for } N/2 \le n < N \,, \tag{4.17}$$

$$1 + \sum_{i=0}^{N/2-1} B_N^i = 0 \,. \tag{4.18}$$

Because of these identities, the sum of all terms involving $2/(3s+2K)$ in each of $\tilde{P}(n,s)$ for $n>0$ vanishes. Hence, we can replace $\tilde{X}_1'(i,s)$ by $\tilde{X}_1(i,s)$.

Similarly, $\tilde{Q}(n,s)$ can be obtained from Eq. (3.11) by replacing $\tilde{X}_0(i,s)$ and $\tilde{Y}_0(i,s)$ by $\tilde{V}'(n,i,s)$ and $\tilde{W}(n,i,s)$, respectively, as given in the following:

$$\tilde{V}'(n,i,s) \equiv \frac{s(s+K)(s+2K) + KIg(n)(3s+2K)}{(3s+2K)[s(s+K)(s+2K) + KIg(i)(3s+2K)]}$$

$$= \frac{1}{3s+2K} + \tilde{V}(n,i,s) \,, \tag{4.19}$$

where

$$\tilde{V}(n,i,s) \equiv \frac{KI[g(n) - g(i)]}{s(s+K)(s+2K) + KIg(i)(3s+2K)} \,, \tag{4.20}$$

and

$$\tilde{W}(n,i,s) \equiv \frac{KI[s(s+K)(s+2K) + KIg(n)(3s+2K)]}{[s(s+K)(s+2K) + KIg(i)(3s+2K)]^2} \quad . \qquad (4.21)$$

For $\tilde{Q}(N,s)$, the $1/s$ in Eq. (3.11) should be replaced by $1/(3s+2K)$. Again, because of identities (4.16 - 4.18), the sum involving $1/(3s+2K)$ in each $\tilde{Q}(n,s)$ vanishes; hence, $\tilde{V}'(n,i,s)$ can be replaced by $\tilde{V}(n,i,s)$.

We can now write down the final results for the probability distribution and diagonal first order correlation as follows:

$$P(n,t) = \begin{cases} \sum\limits_{i=0}^{n} A_n^i X_1(i,t) & \text{for } 0 \leq n < N/2 \\[2em] \sum\limits_{i=0}^{N-n-2} A_n^i X_1(i,t) + \sum\limits_{i=N-n-1}^{N/2-1} \{B_n^i X_1(i,t) + C_n^i Y_1(i,t)\} & \text{for } N/2 \leq n < N \\[2em] 1 + \sum\limits_{i=0}^{N/2-1} \{B_N^i X_1(i,t) + C_N^i X_1(i,t)\} & \text{for } n=N, \end{cases} \qquad (4.22)$$

$$Q(n,t) = \begin{cases} 0 & \text{for } n = 0 \\[1.5em] \sum\limits_{i=0}^{n} A_n^i V(n,i,t) & \text{for } 0 < n < N/2 \\[2em] \sum\limits_{i=0}^{N-n-2} A_n^i V(n,i,t) + \sum\limits_{i=N-n-1}^{N/2-1} \{B_n^i V(n,i,t) + C_n^i W(n,i,t)\} & \text{for } N/2 \leq n < N \\[2em] \sum\limits_{i=0}^{N-1} \{B_N^i(n,i,t) + C_N^i W(n,i,t)\} & \text{for } n=N, \end{cases} \qquad (4.23)$$

where $X_1(i,t)$, $Y_1(i,t)$, $V(n,i,t)$, and $W(n,i,t)$ are the inverse Laplace transforms of $\tilde{X}_1(i,s)$, $\tilde{Y}_1(i,s)$, $\tilde{V}(n,i,s)$ and $\tilde{W}(n,i,s)$, respectively.

Now we can easily obtain the expectation values of n as

$$\langle n \rangle = N - \sum_{i=0}^{N/2-1} \frac{N!}{(N-i-1)!i!} \left\{ [2-(N-2i-1) \sum_{k=i+1}^{N-i-1} \frac{1}{k}] X_1(i,t) \right.$$

$$\left. + (N-2i-1)^2 Y_1(i,t) \right\}, \qquad (4.24)$$

and of the emission intensity as

$$I \equiv \sum_{n=0}^{N} Q(n,t)$$

$$= \sum_{i=0}^{N/2-1} \frac{N!}{(N-i-1)!\,i!} \{(N-i)(i+1)[2 - (N-2i-1)\sum_{k=i+1}^{N-i-1}\frac{1}{k}]T(i,t)$$

$$- (N-2i-1)^2[T(i,t) - (N-i)(i+1)U(i,t)]\} , \qquad (4.25)$$

where $T(i,t)$ and $U(i,t)$ are the Laplace transforms of

$$\tilde{T}(i,s) \equiv \frac{KI}{s(s+K)(s+2K) + KIg(i)(3s+2K)} , \qquad (4.26)$$

and

$$\tilde{U}(i,s) \equiv (3s+2K)[\tilde{T}(i,s)]^2 , \qquad (4.27)$$

respectively.

We notice that $\tilde{X}_1(i,s)$, $\tilde{V}(n,i,s)$, and $\tilde{T}(i,s)$ have the same cubic form in their denominators; while $\tilde{Y}_1(i,s)$, $\tilde{W}(n,i,s)$ and $\tilde{U}(i,s)$ all have the same denominator which is the square of this cubic form. To obtain their inverse Laplace transforms by the method of partial fractions, we have to solve the cubic equation

$$s(s+K)(s+2K) + KIg(i)(3s+2K) = 0 . \qquad (4.28)$$

For small values of i, there are three negative real roots. The one with smallest absolute value can be approximated by

$$s_i \simeq - Ig(i). \qquad (4.29)$$

This implies that the short-time behavior is not much different from the solution of the Markovian superradiance master equation [9]. For large enough i, there are two complex roots and one negative real root. Let them be a+bi, a-bi, and c; then we have 2a+c=-3K. Since 0>c>-2K/3, we have -7K/6>a>-3K/2. This means that the oscillatory part will decay more rapidly than the monotonous part.

In comparison with the result of Sec. 3, we are tempted to draw the conclusion that the atom-field correlation has the effect of suppressing the ringing.

* Research supported by the Air Force Office of Scientific Research.

† Permanent Address: Department of Physics and Mathematics, Alabama A & M University, Normal, Alabama 35762.

References

1.   N. Skribanowitz, I. P. Herman, J. C. MacGillivray, and M. S. Feld, Phys. Rev. Lett *30*, 309 (1973); J. C. MacGillivray and M. S. Feld, Phys. Rev. A *14*, 1169 (1976).
2.   R. Bonifacio and L. A. Lugiato, Phys. Rev. A *12*, 587 (1975).
3.   R. Saunders, S. S. Hassan, and R. K. Bullough, J. Phys. A *10*, 1725 (1976).
4.   E. Ressayre and A. Tallet (preprint).
5.   L. A. Lugiato, Physica *81A*, 565 (1975).
6.   M. Gronchi and L. A. Lugiato, Phys. Rev. A *14*, 502 (1976).
7.   M. Gronchi and L. A. Lugiato, Phys. Rev. A *13*, 830 (1976).
8.   This expression has been obtained by R. Bonifacio, P. Schwendimann, and F. Haake, Phys. Rev. A *4*, 854 (1971).
9.   C. T. Lee, Phys. Rev. A *15*, 2019 (1977); Phys. Rev. A *16*, 301 (1977).

SUPERRADIANCE AND SUPERFLUORESCENCE OF A DICKE POINT SOURCE[*]

Martin E. Smithers

David Lipscomb College, Nashville, Tennessee

Spontaneous emission of electromagnetic radiation from an atomic (or molecular) system in a correlated state was first treated by Dicke [1]. Using a point source model, he defined collective energy eigenstates of two level atoms. The quantum numbers labeling these Dicke states are the energy quantum number M and the cooperation number J. If N is the total number of atoms in the sample, then $0 \leq J \leq N/2$ and $-J \leq M \leq J$. Dicke found that the initial radiation rate from a system in such a collective state is proportional to $(J+M)(J-M+1)$. A state with $J \simeq N/2$ and $M \simeq 0$ has the maximum emission rate, proportional to N^2. Such a state he called superradiant. A fully excited state has $M = J = N/2$, giving an initial spontaneous emission rate proportional to N. In the model J is a constant of the motion, but the average M decreases as the system loses energy through radiation, so that a fully excited state decays to a state with average $M \simeq 0$ and thus radiates at near superradiant levels [2].

For large cooperation number J, a coherently prepared super-position of superradiant Dicke states has macroscopically large dipole moment [3], so that the decay process is essentially classical [4]. Thus the radiation emitted from such a superradiant state is highly coherent (the photon statistics are Poisson [5]), at least initially. On the other hand, a highly excited initial state has macroscopically negligible dipole moment, and the decay process is inherently quantum-mechanical. The nature of the radiation pulse emitted in this case (referred to as superfluorescence [6]) is less apparent. Predictions previously obtained [7] for the dynamical be-havior of a general large J-system have been based on the master-equation approach [8,9]. The results given here are obtained by direct solution of the Schroedinger equation for the system. In agreement with the previous results [7], it is found that, while the pulse emitted by an initially superradiant Dicke state is highly coherent at or near peak intensity, the pulse emitted by an initially highly excited Dicke state is highly random, especially as the intensity reaches its maximum (near superradiant) level.

In the model approach, the dipole-dipole coupling between the atoms is neglected. As has been pointed out [10], this is a major weakness of the model since the dipole-dipole coupling in a dense sample results in a spatially varying frequency shift and consequent dephasing. The result is that J is no longer a constant of the motion. It has been shown [11], however, that the average intensity of the emitted radiation over a random distribution of atom positions remains unaffected. As it happens, the model holds for the special case of a ring of atoms, where the spatial variations are absent [10]. Recent treatments [12] have included dephasing effects.

The model system (Dicke point source [1]) consists of a collection of identical atoms or molecules. Each atom has two effective nondegenerate energy levels, a ground state and an excited state. Electric dipole transitions, with resonance frequency Ω, are allowed between the two levels. The atoms are contained in a volume of dimensions small compared with a radiation wavelength c/Ω, but the system is open, i.e., radiation is free to pass out of the region containing the atoms. The atoms are assumed to interact only via the electromagnetic field, and the dipole-dipole coupling between the atoms is neglected.

The Hamiltonian for the model system is $H = H^0 + H^1$, where H^0 is the non-interaction Hamiltonian,

$$H^0 = \hbar\Omega J_z + \hbar \sum_k \omega_k a_k^\dagger a_k , \tag{1}$$

and H^1 is the atom-field interaction Hamiltonian,

$$H^1 = \hbar \sum_k \{ g_k a_k J_+ + g_k^* a_k^\dagger J_- \} . \tag{2}$$

neglecting counter-rotating terms. Here g_k is the matrix element for the dipole transition. Treating each atom as a spin one-half system (spin-up corresponds to the atom being in its excited state, and spin-down, its ground state), the energy operator J_z is the z-component of the total angular momentum of the atoms, and J_+ (J_-) is the corresponding angular momentum raising (lowering) operator. Thus these atomic operators obey the angular momentum commutation relations. Here also $a_k^\dagger$ (a_k) is the photon creation (annihilation) operator for the field at frequency $\omega_k = ck$. These photon operators obey the canonical field commutation relation.

The atoms are taken to be initially in a collective Dicke state $|J,M\rangle$ (which is an eigenstate of J and J_z) [1], and the electromagnetic field, in the vacuum state $|\{0\}_k\rangle$ for all modes. The initial state of the atom-field system is then the product state,

$$|\Psi(o)\rangle = |J,M\rangle |\{o\}_k\rangle \ . \tag{3}$$

At later times, the state of the system has the following expansion,

$$|\Psi(t)\rangle = \sum_{n=0}^{J+M} (n!)^{-1} \sum_{k_1}\cdots\sum_{k_n} A_{k_1,\ldots,k_n}^{(J,M)}(n;t)$$

$$x \quad |J,M-n\rangle |1\rangle_{k_1}\cdots|1\rangle_{k_n} \ , \tag{4}$$

in terms of product states of atomic Dicke states $|J,M-n\rangle$, and photon number states $|1\rangle_{k_i}$. Here it is assumed that no more than one photon would be emitted in any one particular infinitesimal range of modes. From Eq. (3) it follows that $A^{(J,M)}(0;0) = 1$, while all other amplitudes vanish initially.

The expansion (4) is substituted into the Schroedinger equation, and the equations of motion for the amplitudes are obtained. Taking Laplace **transforms** of these equations, the resulting algebraic equations are solved for the Laplace transformed amplitudes. In obtaining such solutions, contributions which **result in slight frequency shifts** may be neglected (since we are not here concerned with the precise frequency of the emitted radiation). Also, since the sample is small and the radiation is allowed to escape, the small contributions arising from higher-order re-absorption effects may be neglected. However, the contribution of virtual emission processes to radiation damping are included. Finally, to find the time-dependent amplitudes, inverse Laplace transforms of these solutions must be taken.

The probability that n photons of modes $(k_1,\ldots,k_n)$ have been emitted at time t is given as

$$P_{k_1,\ldots,k_n}^{(J,M)}(n;t) = |A_{k_1,\ldots,k_n}^{(J,M)}(n;t)|^2 \ . \tag{5}$$

To find the probability $P^{(J,M)}(n;t)$ that n photons, whatever their modes, have been emitted at time t, this probability must be summed over all modes, that is,

$$P^{(J,M)}(n;t) = [n!]^{-1} \sum_{k_1}\cdots\sum_{k_n} P_{k_1,\ldots k_n}^{(J,M)}(n;t) \ . \tag{6}$$

The generalized form for these probabilities is found to be the following:

$$P^{(J,M)}(n;t) = (-1)^n \gamma_1^2 \gamma_2^2 \cdots \gamma_n^2$$

$$\times \sum_{m=1}^{n+1} \prod_{\substack{i=1 \\ i \neq m}}^{n+1} \frac{1}{(\gamma_m^2 - \gamma_i^2)} \; e^{-A\gamma_m^2 t} \;, \tag{7}$$

where $\gamma_m^2 = \gamma_m^2(J,M) \equiv (J+M-m+1)(J-M+m)$ and $(A)^{-1}$ is the lifetime of a single isolated atom in its excited state. The given form (7) for the probability holds only for $M \leq 0$, since when $M > 0$ certain of the terms become undefined. However, if the indices are extended to include non-integer values, the correct form for $M > 0$ could be obtained by taking the appropriate limit. The result would contain terms in which the coefficients of the exponentials are first (as well as zero'th) order in time. Instead of doing this, Eq. (7) is used to calculate the desired expectation values, which are then analytically continued to the $M > 0$ domain.

The probability function (7) is valid for all (positive) time. It is of interest, however, to find the short-time limit in order to make a comparison with results previously obtained [13] in the short-time limit for different cases of a closed resonant cavity. There are two cases of interest for which the short-time limit of Eq. (7) is readily calculated. First, for a highly inverted initial state, $J + M \gg n$ for short times, and the probability function has the limiting form,

$$P^{(J,M)}(n;t) \longrightarrow \frac{(J-M+n)!}{(n)!(J-M)!} \; e^{-A(J+M)(J-M+1)t} \; [1-e^{-A(J+M)t}]^n \;.$$

$$\tag{8}$$

The n-distribution is negative binomial (the photon statistics are super-Poisson) in this case. Second, for a low excited initial state, $J - M \gg n$ for short times, and the probability function has the limiting form,

$$P^{(J,M)}(n;t) \longrightarrow \frac{(J+M)!}{(n)!(J+M-n)!} \; e^{-A(J-M)(J+M-n)t}$$

$$\times [1-e^{-A(J-M)t}]^n \;. \tag{9}$$

The n-distribution is weighted binomial (the photon statistics are sub-Poisson) in this case. The statistics obtained for both these cases agree with what has been found [13] for the corresponding cases when the sample is enclosed in a resonant cavity. Such agreement is expected in the short-time limit only, since for longer times the emitted radiation itself begins to have a significant effect on the enclosed sample.

The general time behavior of the photon statistics for the open point source here considered is best obtained by direct calculation of the mean number $N(t) \equiv \langle n(t) \rangle$ of photons emitted and of the mean square variance $SV(t) \equiv \langle (\Delta n)^2 \rangle = \langle n^2(t) \rangle - \langle n(t) \rangle^2$ from the n-photon probability function (7). Thus for $M \leq 0$, $N(t)$ has the form,

$$\langle n(t) \rangle = (J+M) + \frac{(J+M)!}{(J-M)!} \sum_{m=1}^{J+M} (2m - 2M - 1)$$

$$\times \frac{(m-1-2M)!}{(m-1)!} (-1)^{J+M-m+1} e^{-A\gamma_m^2 t} . \tag{10}$$

For $M > 0$, $N(t)$ may be obtained from Eq. (10) by analytic continuation and is given by

$$\langle n(t) \rangle = (J+M) + \frac{(J+M)!}{(J-M)!} \left[\sum_{m=1}^{M} (-1)^{J+M} \frac{e^{-A\gamma_m^2 t}}{(m-1)!(2M-m)!} \right.$$

$$\times \left[(2M-2m+1)\Psi_m - 2 - At(2M-2m+1)^2 \right]$$

$$\left. + \sum_{m=2M+1}^{J+M} (2m-2M-1) \frac{(m-1-2M)!}{(m-1)!} (-1)^{J+M-m+1} e^{-A\gamma_m^2 t} \right] , \tag{11}$$

where

$$\Psi_m = \frac{1}{2M-m} + \frac{1}{2M-m-1} + \cdots + \frac{1}{m} . \tag{12}$$

The mean square number $\langle n^2(t) \rangle$ of photons emitted is found by direct calculation to be expressible in terms of $N(t)$ as

$$\langle n^2(t) \rangle = (J+M)(J-M+1) + (2M-1) \langle n(t) \rangle - \frac{1}{A}\frac{d}{dt} \langle n(t) \rangle . \tag{13}$$

Thus the explicit form of $SV(t)$ may be found by substituting Eq.(10) or Eq. (11) into Eq. (13) for the corresponding case.

One particular case is especially simple. If the atoms are in an initial Dicke state with $M = -J+1$ (from which only one photon can be emitted), then

$$\langle n(t) \rangle = 1 - e^{-A(2J)t}, \tag{14}$$

and

$$\langle (\Delta n(t))^2 \rangle = \langle n(t) \rangle [1 - \langle n(t) \rangle]. \tag{15}$$

Here the statistics are sub-Poisson for all finite time. The functional form of $N(t)$ in Eq. (14) (and of $SV(t)$ in Eq. 15 as well) is exactly the same as that obtained in the Wigner-Weisskopff treatment of emission from a single atom. Here, however, the decay time for the excited (collective Dicke) state is $[(2J)A]^{-1}$, which, for large

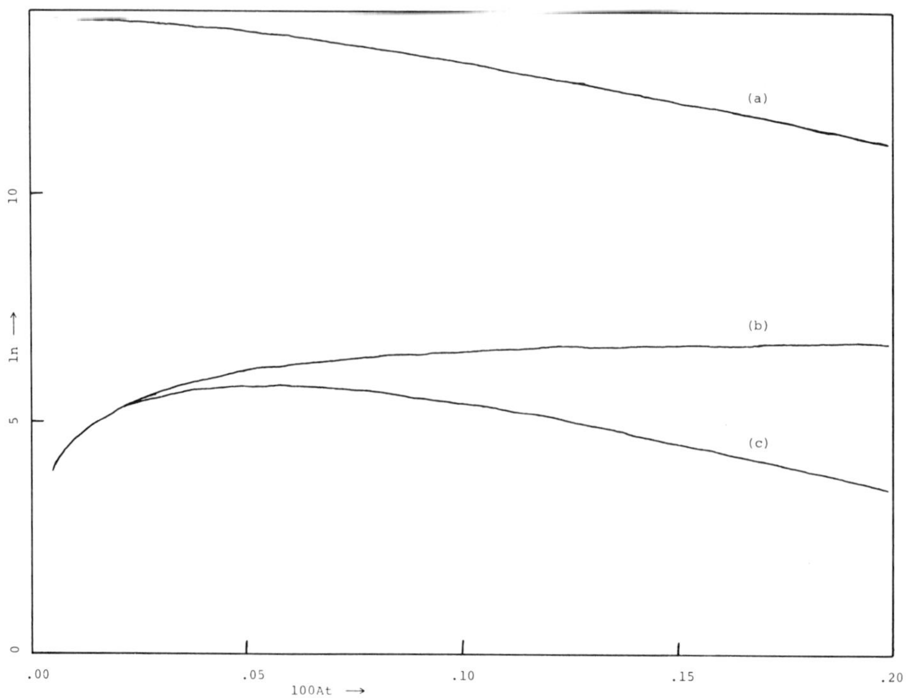

Fig. 1: Superradiance. The functions, (a) ln[I(t)], (b) ln[N(t)],
and (c) ln[SV(t)] are plotted vs. time for the case M=0 and J=1000.

J, is much shorter than for the single atom. Also SV(t) reaches a
maximum of N(t)/2 when $t = \ln(2)[(2J)A]^{-1}$, at which time N(t) = 1/2.

   In both the superradiant (M=0) and the superfluorescent (M=J)
cases, computer plots have been made of N(t), SV(t) and I(t) $\equiv$
$(A)^{-1}dN/dt$ using the above results for J=10, 50, 100, 500, 1000. Figs.
1 and 2 show the results (where the natural log of the functions
have been plotted against At) for J = 1000 with M = 0 and M = J,
respectively. For a superradiant (M = 0) initial state, the plots
reveal the following features. N(t) goes from zero steadily towards
its maximum value J. I(t) starts out at J(J+1) and decreases rapidly
to zero. SV(t) starts out at zero and increases to a maximum of
about (0.34)J before falling back to zero. At maximum SV(t), I(t)
has decreased to about 71% of its initial value. The ratio R(t) =
SV(t)/N(t) starts out at unity but decreases thereafter. R(t) remains, however, of order unity up until SV(t) reaches its maximum
value, at which time R(t) is about 0.64. Slightly over half the
photons have been emitted by this time. Thereafter, R(t) decreases
rapidly towards zero. Thus the photon statistics are near Poisson
over the first and most intense portion of the pulse but become
sub-Poisson over the latter portion.

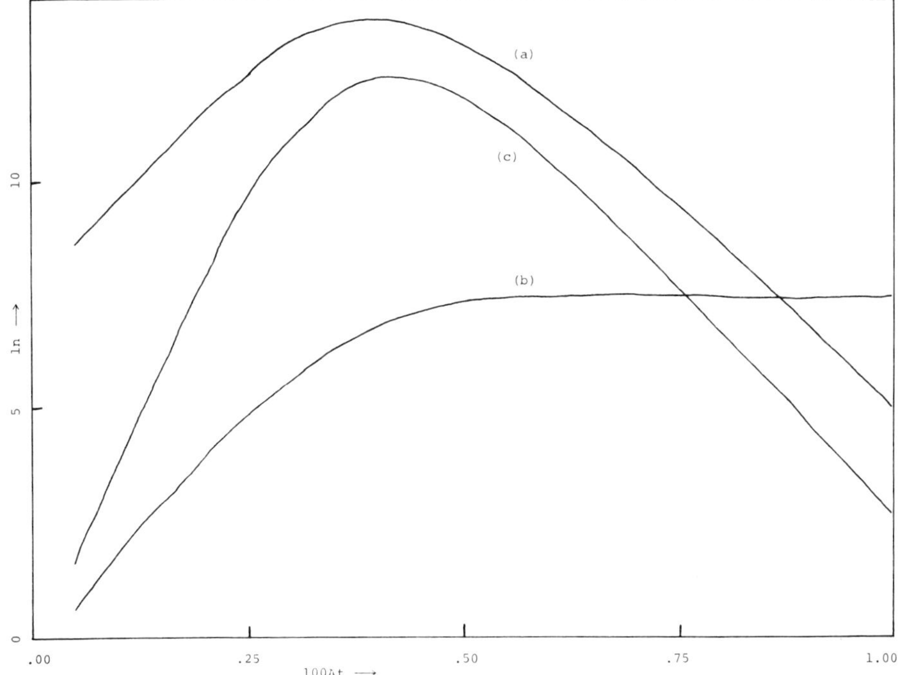

Fig. 2: Superfluorescence. The functions, (a) ln[I(t)], (b) ln[N(t)], and (c) ln[SV(t)], are plotted vs. time in the case M=J=1000.

For a fully excited (M=J) initial state, the plots reveal the following behavior. N(t) increases steadily from zero towards its maximum value 2J. I(t) starts out at 2J and *increases* to a maximum value of about (0.78)(J)(J+1), from which it decreases to zero. SV(t) goes from zero to a maximum of about (0.23)(J)(J+1) and then decreases back to zero. I(t) and V(t) peak at about the same time, with I(t) peaking slightly earlier, and in between, N(t) reaches half its maximum value. The ratio R(t) = SV(t)/N(t) remains greater than unity over most of the pulse and only becomes unity after almost all the photons have been emitted. When I(t) is a maximum, R(t) is about (0.23)(J+1). Thus the photon statistics remain super-Poisson over most of the pulse and near peak intensity are overwhelmingly so.

The character of superfluorescent emissions from a fully excited initial state is thus highly "nonclassical", that is, the photon statistics vary greatly from the Poisson statistics of coherent (superradiant) emissions. How this behavior depends on the excitation level of the system may be examined by computing values of the ratio R(t) = SV(t)/I(t) for different values of M at constant J. Table 1 gives the results for M=100, 99, 98, 97, 96, and 95 when

Table 1. This table indicates how the superfluorescent emission varies with the excitation level M for the case J = 100. The values of At correspond to the following situations: $t = t_1$ at maximum SV(t); $t = t_2$ at maximum I(t); $t = t_3$ when R(t) = 1.

M	At_1	$R(t_1)$	At_2	$R(t_2)$	At_3
100	.0300	21.5	.0280	23.7	.0520
99	.0255	11.7	.0240	13.0	.0390
98	.0225	8.6	.0210	9.5	.0345
97	.0210	6.6	.0195	7.4	.0315
96	.0195	5.6	.0180	6.2	.0285
95	.0180	5.0	.0180	5.0	.0270

J = 100. It is found that R(t) drops significantly with decreasing excitation, especially from the fully excited to the next lower state (where it drops by nearly half). The nature of the photon statistics does, however, remain super-Poisson over at least the range of excitations covered in Table 1. Of course, once the excitation of the initial state has been reduced to a superradiant ($M \simeq 0$) level, R(t) will have fallen to a value of order unity. From Table 1 it is also seen that the duration of the pulse decreases appreciably with decreasing M, which behavior is correlated with an increasing peak intensity.

The results obtained here agree substantially with the numerical results[7] obtained from the superradiance master equation.[8,9] In fact, the n-photon probability function (7) given here corresponds to a solution of the master equation in the Markovian approximation. Here we have obtained the explicit form of the expectation values N(t) $\left(\text{Eqs. (10) and (11)}\right)$ and SV(t) $\left(\text{Eq. (13)}\right)$. Also, our results were obtained by direct solution of the Schroedinger equation.

ACKNOWLEDGEMENT

I wish to thank Professor W. Ralph Butler for his assistance
in writing the computer program used to plot the curves presented
here.

* Work supported in part by a Faculty Fellowship Grant provided by
David Lipscomb College.

References

1.   R.H. Dicke, Phys. Rev. *93*, 99 (1954).
2.   V. Ernst and P. Stehle, Phys. Rev. *176*, 1456 (1968).
3.   F.T. Arecchi, et al., Phys. Rev. *A6*, 2211 (1972).
4.   C.R. Stroud, Jr., Phys. Rev. *A5*, 1094 (1972).
5.   R.J. Glauber, Phys. Rev. *131*, 2766 (1963).
6.   R. Bonifacio and L.A. Lugiato, Phys. Rev. *A11*, 1507 (1975);
     Phys. Rev. *A12*, 587 (1975).
7.   R. Bonifacio, et al., Phys. Rev. *A4*, 854 (1971); A.M. Ponte
     Concalves and A. Tallet, Phys. Rev. *A4*, 1319 (1971); Roy J.
     Glauber and Fritz Haake, Phys. Rev. *A13*, 357 (1976).
8.   G.S. Agarwal, Phys. Rev. *A2*, 2038 (1970); *Proc. Third
     Rochester Conference on Coherence and Quantum Optics*, eds.
     L. Mandel and E. Wolf (Plenum, New York, 1973), pp. 157-182.
9.   R. Bonifacio et al., Phys. Rev. *A4*, 302 (1971).
10. R. Friedberg and S.R. Hartmann, Phys. Rev. *A10*, 1728 (1974).
11. G.S. Agarwal, *Springer Tracts in Modern Physics*, vol. 10,
     (Springer-Verlag, New York, 1974), pp. 82-83.
12. P. Schwendimann, Z. Physik *265*, 267 (1973); R.H. Picard and
     Charles R. Willis, Phys. Rev. *A8*, 1536 (1973); R. Jodoin and
     L. Mandel, Phys. Rev. *A9*, 873 (1974); Phys. Rev. *A10*, 1898
     (1974); G. Banfi and R. Bonifacio, Phys. Rev. Lett. *33*, 1259
     (1974); Phys. Rev. *A12*, 2068 (1975), J.C. MacGillivray and
     M.S. Feld, Phys. Rev. *A14*, 1169 (1976).
13. Martin E. Smithers and Eugene Y.C. Lu, Phys. Rev. *A9*, 790
     (1974).

LIGHT BEATING STUDY OF THE TRANSITION TO TURBULENCE

H. L. Swinney, P. R. Fenstermacher

City College of CUNY, New York, New York

J. P. Gollub

Haverford College, Haverford, Pennsylvania

A light beating technique has been used to study the dynamical behavior of a fluid contained between concentric cylinders with the inner cylinder rotating (circular Couette flow). A laser was focused midway between the two cylinders, and backscattered Doppler-shifted light was mixed at the photocathode with unshifted laser light. The Doppler shift frequency, proportional to the radial component of the velocity for our geometry, is several orders of magnitude higher than the frequencies characteristic of the fluid flow; hence measurements of the frequency of the oscillating photocurrent for successive time intervals (long enough to contain many Doppler cycles) accurately reflect the time dependence of the velocity. The Doppler shift frequency was recorded for 8192 successive time intervals in a computer and then Fourier-transformed to obtain velocity power spectra and autocorrelation functions.

The velocity power spectra reveal several distinct dynamical regimes as the Reynolds number R is increased. [$R = \omega_{cyl} r_i (r_o - r_i)/\nu$, where ω_{cyl} is the angular frequency of the inner cylinder, r_i and r_o are the radii of the inner and outer cylinders, and ν is the kinetic viscosity. After each change of R, measurements were not made until the flow became statistically stationary.] The first time-dependent flow regime is characterized by a velocity spectrum with a single sharp frequency component and its harmonics. This regime has been studied photographically by Coles and others. At higher R a second (previously unobserved) fundamental frequency component appears. The linewidth of the two fundamental frequencies and their harmonics is quite narrow ($\Delta\omega/\omega < 0.001$), limited only by the instrumental resolution. Since the two frequencies appear to be incommensurate, this regime is quasiperiodic. At a higher R a broad component appears in the spectrum, indicating a chaotic element of the flow. At a yet higher R the sharp components disappear, leaving a broad spectrum (aperiodic flow). The three distinct dynamical regimes we have observed (periodic, quasiperiodic, and aperiodic) have been shown by Lorenz, Yorke, McLaughlin and Martin, and others to be exhibited by simple nonlinear models with only a few degrees of freedom, a result which suggests that excitation of many degrees of freedom is not necessary to produce aperiodic fluid flow.

SPATIAL COHERENCE OF LIGHT SCATTERED BY LIQUID CRYSTALS

M. Bertolotti and F. Scudieri

Università di Roma, Roma, Italy

1. INTRODUCTION

The study of spatial coherence of scattered light has received little attention until recently, because the short correlation length of fluctuations producing the scattering, made the scattering volume behave as a chaotic incoherent source. The coherence area of the scattered field was accordingly determined only by geometrical factors and gave no physical information upon the scattering medium.

Recently E. Wolf and W.A. Carter [1-4] and H.P. Baltes and collaborators [5] have studied the angular distribution of radiant intensity from sources of different degrees of spatial coherence. The analysis was made for radiation from large statistically quasi-homogeneous planar sources, a quasi-homogeneous source being one for which the intensity distribution across its plane can vary slowly with position, remaining sensibly constant over distances of the order of the correlation length of the light vibrations in the source plane. In particular, they found that the angular distribution of the radiant intensity depends in a simple manner on the two-dimensional spatial Fourier transform of the (spectral) degree of spatial coherence of the light emerging from the source, and that the degree of spatial coherence of the far field is proportional to the normalized two-dimensional spatial Fourier transform of the intensity distribution across the source. The second part of this reciprocity theorem is formally identical with the farzone form of the classic theorem of van Cittert and Zernike [6], that predicts the degree of coherence of light from a spatially incoherent source, the new fact now being that the far zone form of the van Cittert-Zernike theorem applies to light from any quasi-homogeneous source which may possess a high degree of coherence over arbitrarily large areas of the source plane provided the area of the source is much larger than its coherence area [3].

If this condition is not fulfilled it is to be expected that the van Cittert-Zernike theorem fails [7].

Light scattering from liquid crystals in which very large correlation lengths of the fluctuations producing the scattering can be obtained, is an example of such a situation and has recently been considered [8,9]. In these cases the correlation length of the fluctuations can be of the same order of magnitude as the dimensions of the scattering volume. The chaotic source model is therefore not valid, as in the scattering volume only one coherent source may be present if a coherent laser beam is used as the incident light.

In the time domain similar effects have been already considered and have importance, e.g. in the scattering from a turbulent fluid [10].

2. THEORY

Let us consider the scattering geometry shown in Fig. 1. On the screen along the x-direction, the correlation function between the fields scattered under two slightly different directions $\theta-\delta\theta$ and $\theta+\delta\theta$ is

$$J(P_1,P_2) = \langle E_s(P_1,t)\, E_s^*(P_2,t)\rangle \quad , \tag{1}$$

where $E_s(P_i,t)$ is the scattered field at point P_i and time t on the screen. By explicitly writing $E_s(P_i,t)$ as (see Ref. [10] Chaps. 1 and 5)

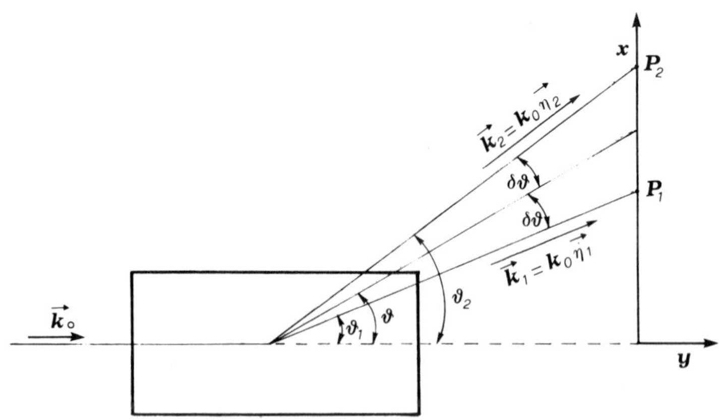

Fig. 1.   Geometry of the scattering experiment.

$$E_s(P_i,t) = -\frac{k_o^2}{4\pi R_i} \exp[-i\omega_o t]\; \tilde{\eta}_i \times (\tilde{\eta}_i \times \tilde{E}_o)$$

$$\times \int \Delta\varepsilon(r',t)\, \exp[-i(\underset{\sim}{k}_o-\underset{\sim}{k}_i)\cdot\underset{\sim}{r}']\; \underset{\sim}{dr'} \;, \tag{2}$$

where the incident field is assumed to be

$$\underset{\sim}{E}_i = \underset{\sim}{E}_o\, \exp[i(\underset{\sim}{k}\cdot\underset{\sim}{r} - \omega t)] \;\;,$$

$\underset{\sim}{\eta}_i = \underset{\sim}{R}_i/R$, $\underset{\sim}{k}_i = k_o\underset{\sim}{\eta}_i$ is the wave vector of the scattered field at $\underset{\sim}{R}_i$; $\varepsilon(\underset{\sim}{r},t) = \varepsilon_o+\Delta\varepsilon(\underset{\sim}{r},t)$ is the dielectric constant, one obtains

$$J(P_1,P_2) = \left(\frac{E_o\omega^2}{4\pi c^2}\right)^2 \frac{\sin\gamma_1\sin\gamma_2}{R_1 R_2} \exp[ik(R_1 - R_2)]$$

$$\times \underset{V}{\iint} <\Delta\varepsilon(\underset{\sim}{r}_1,t)\Delta\varepsilon(\underset{\sim}{r}_2,t)> \exp\{i[(\underset{\sim}{k}_0-\underset{\sim}{k}_1)\cdot\underset{\sim}{r}_1-(\underset{\sim}{k}_0-\underset{\sim}{k}_2)\cdot\underset{\sim}{r}_2]\}\underset{\sim}{dr}_1\underset{\sim}{dr}_2\;, \tag{3}$$

where γ_i is the angle between $\underset{\sim}{E}_o$ and $\underset{\sim}{\eta}_i$. The integrals are restricted to the scattering volume.

The integral for $J(P_1,P_2)$ has been computed in a special case and the value of the angle $\delta\theta_c$ at which it has its first zero has been considered. This value is the coherence angle under which the coherence dimension of the coherence area is seen. Computed values for $\delta\theta_c$ are shown as a function of the correlation length of refractive index fluctuations ℓ_n for different values of the scattering volume dimension d in Fig. 2, for the special case of a correlation function of dielectric permittivity fluctuations of the form

$$<\Delta\varepsilon(\underset{\sim}{r}_1,t)\Delta\varepsilon(\underset{\sim}{r}_2,t)>\alpha \exp\left\{-\frac{|x_1-x_2|+|y_1-y_2|+|z_1-z_2|}{\ell_n}\right\}\;. \tag{4}$$

Recently Baltes and collaborators [11,12] and Carter [13] have extended Wolf's analysis to the case examined here where the coherence dimension at the source is comparable with the linear dimension of the source itself, being able to solve analytically the problem in the case in which both the intensity emitted by the source and its coherence degree have a Gaussian distribution in space. Their results too indicate an increase in the coherence dimension of the spatial coherence area in the far field when the linear dimension of the source is of the same order of magnitude or smaller than the spatial coherence length of the field at the source plane.

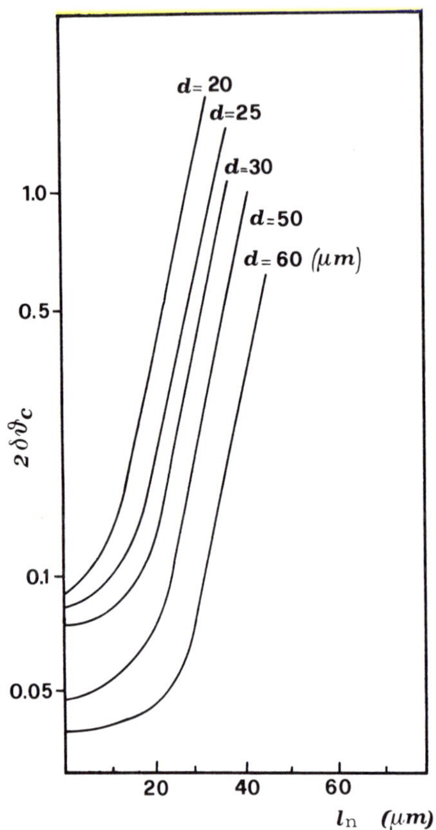

Fig. 2. Angle of coherence $\delta\theta_c$ as a function of the refractive index correlation length ℓ_n for different sizes d of the scattering volume.

3. EXPERIMENTAL PART

An excellent medium in which large values of the correlation length of the fluctuations producing the scattering can be produced, is offered by nematic liquid crystals. In these materials molecules can be ordered over dimensions of the order of the sample thickness (hundreds of microns). By application of some external field (electrical, elastic) a motion is produced, correlated over macroscopic dimensions (of the order of the cell thickness) that decreases down to very small lengths as the external field power increases.

A first example is offered by the application of a d.c. electric field [8]. Due to the applied voltage a molecular motion is produced into the cell [14]. The fluctuations of refractive index induced by the molecular motion are driven in some way by the motion, increasing with increasing voltage and their correlation length also changes. These fluctuations are responsible for the scattering of light which is produced by the application of a voltage to the cell larger than the threshold for William's domains.

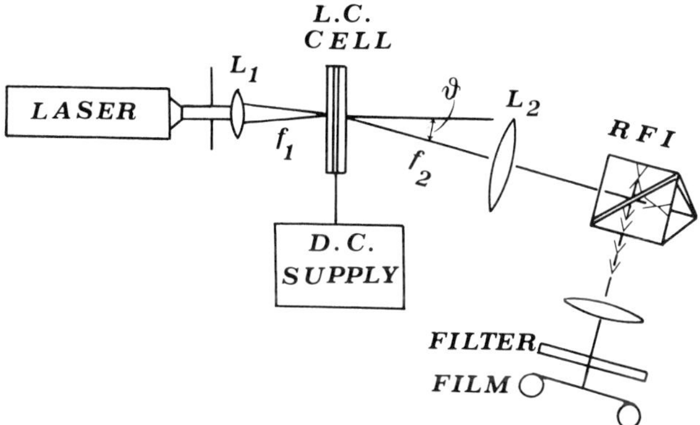

Fig. 3. Experimental disposition for the measurement of the coherence area of light scattered by a liquid crystal cell.

The experimental disposition is shown in Fig. 3 [8]. A reversing front interferometer is used for measuring the area of coherence of the scattered light, from a cell (25 μm thick) filled with MBBA, a nematic liquid crystal that has its nematic phase at room temperature, in which a d.c. electric field is applied parallel to the incident light direction. The cube-corner reversing-front interferometer splits the scattered beam into two parts, each one being the specular reflection of the other, and superimposes each beam on the other. The extension of the obtained fringes measures the coherence dimension ℓ_c which is shown as a function of the applied voltage in Fig. 4, starting from a voltage of 10 V corresponding to the threshold for dynamic scattering.

The coherence dimension calculated for a chaotic source with the same geometrical disposition would be constant, independent of the voltage and of a value

$$\ell_c^* \simeq 1.22 \frac{\lambda}{d} f_2 \cos\theta \quad , \tag{5}$$

where d is the diameter of the illuminated region, θ the scattering angle, f_2 the focal length of lens L_2 (see Fig. 3).

According to Eq. (5) with d = 40 μm, θ = 9°, f_2 = 20 cm, $\ell_c \simeq 4.5$ mm. The increasing value of ℓ_c with decreasing voltage shown in Fig. 4 is interpreted, according to our previous discussion, by assuming that for field strengths lower than about 30 V the correlation length of refractive index fluctuations inside the illuminated volume, is no longer negligible in comparison with the dimension of this volume, which in this case is of the order of 30 ÷ 40 μm. The continuous curve in Fig. 4 is the theoretical curve pertaining to the experimental situation, with d = 50 μm from the

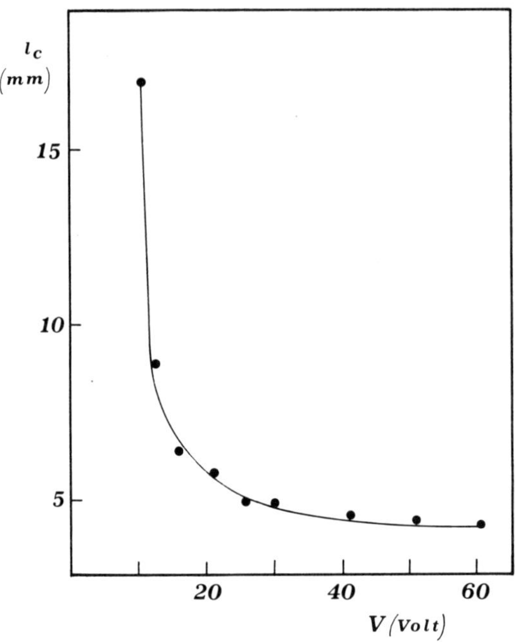

Fig. 4. Coherence dimension ℓ_c of light scattered by a nematic liquid crystal cell under a d.c. applied electric field as a function of the applied voltage V.

previously computed values. By considering the rather questionable form assumed for the dielectric permittivity correlation function, the agreement is surprisingly good.

A second example is offered by the study of disclination lines, generated in a nematic by the application of a suitable acoustic field [9]. A disclination line is a discontinuity in the molecular orientation located on a line [15,16]. By using ultrasounds [17] at a given threshold for the applied ultrasound power, disclinations are produced, a sudden increase in the intensity of scattered light is observed and fringes appear in the photographic plate plane. The measured coherence dimension ℓ_c is very large at the threshold and decreases by increasing the voltage applied to the piezoelectric transducer reaching a substantially constant value for high values of the applied voltage.

The dimension of the scattering volume in this case is very large and can be estimated d $\simeq$ 270 μm. The behavior of ℓ_c shows accordingly that the correlation length of fluctuations is very large, in this case of the order of the length of disclinations.

The last and clearest example of the possiblities of the method is given by a study of rolls produced by ultrasound in a nematic liquid crystal [18]. In orthoscopic vision between crossed polarizers, if an ultrasound field with shear components is applied to a nematic cell, above a given threshold for the voltage applied

to the piezoelectric transducer, a characteristic structure is
visible in the form of linear, bright zones on a dark background,
which displays some spatial periodicity (see Fig. 5a). By increasing the acoustic power (see Fig. 5b) these zones assume the aspect
of bright lines (rolls). Such a structure is characterized by a
cellular hydrodynamic motion, of a helicoidal type, obtained by
the superposition of a circular motion around the axis of the rolls
and a small drift motion along it.

The structure is unstable upon increasing the acoustic power
and for a voltage greater than a given threshold, rolls divide
along the direction of their length into smaller units (see Fig.
5c), and by further increasing the acoustic power a kind of "turbulent" motion sets up with consequent strong scattering of light
(see Fig. 5d).

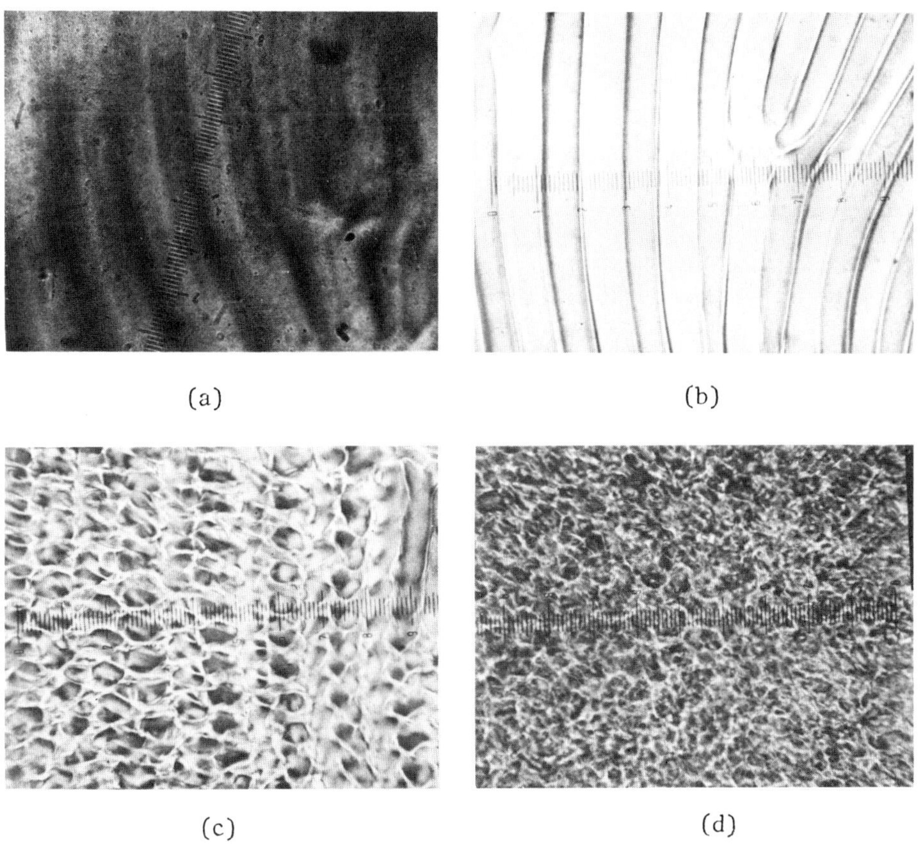

(a)   (b)

(c)   (d)

Fig. 5. Rolls in a liquid crystal cell at various voltages applied
to the piezoelectric transducer: a) $V_{rms} \simeq 10$ Volt; b) $V_{rms} \simeq 15$
Volt; c) $V_{rms} \simeq 25$ Volt; d) $V_{rms} \simeq 40$ Volt.

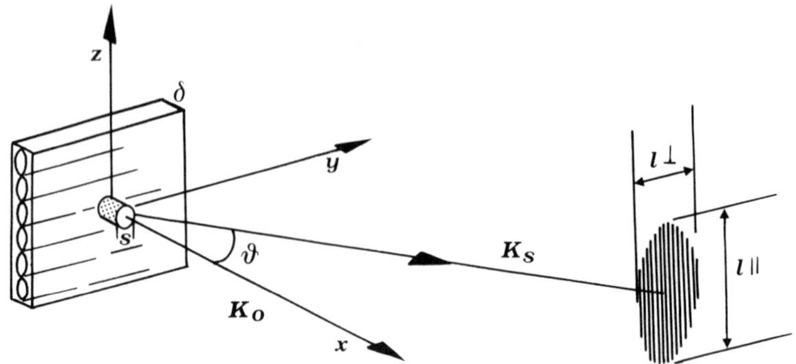

Fig. 6. Fringes produced by the reversing front interferometer
and roll's orientation.

The molecular director orientation is influenced by the motion
and this produces the index of refraction fluctuations which gives
rise to the scattering of light.

The study of the spatial coherence properties of the light
scattered by this roll structure, shows that the area of coherence
is asymmetric. The coherence dimension ℓ has accordingly been
measured in the parallel ($\ell_{\parallel}$) direction and in the perpendicular
($\ell_{\perp}$) direction to the fringe system. Due to the geometry of the
experiment $\ell_{\perp}$ was measured in a direction parallel to the roll's
axis, and $\ell_{\parallel}$ normal to it (see Fig. 6).

Figures 7 and 8 show the behavior of $\ell_{\parallel}$ and $\ell_{\perp}$ as a function
of the voltage V applied to the transducer, of cell thickness δ,
and of angle θ.

The behavior of ℓ is very characteristic; it first increases
as the voltage increases up to a voltage of about 25 V and then it
suddenly starts to decrease until at a voltage of about 70 V, it
approaches a constant.

Let us first consider the region of voltages lower than 25 V.
With the geometry and the material (MBBA) used, the threshold for
roll formation is about 9 V. By increasing the voltage from this
value on, the roll structure remains stationary and the cellular
velocity increases.

Table I shows for the first point at 9 V for different cell
thicknesses δ, and scattering angles θ, the values of the scattering
volume dimensions $s_{\parallel}$ and $s_{\perp} \simeq d \cdot \cos\theta + \delta\sin\theta$, the corresponding
coherence dimensions from the van Cittert-Zernike theorem $\ell_c^{(\parallel)}$,
$\ell_c^{(\perp)}$, the experimental values $\ell_{\parallel}(\exp)$ and $\ell_{\perp}(\exp)$, and the ratios
$\ell_{\parallel}(\exp)/\ell_c^{(\parallel)}$ and $\ell_{\perp}(\exp)/\ell_c^{(\perp)}$.

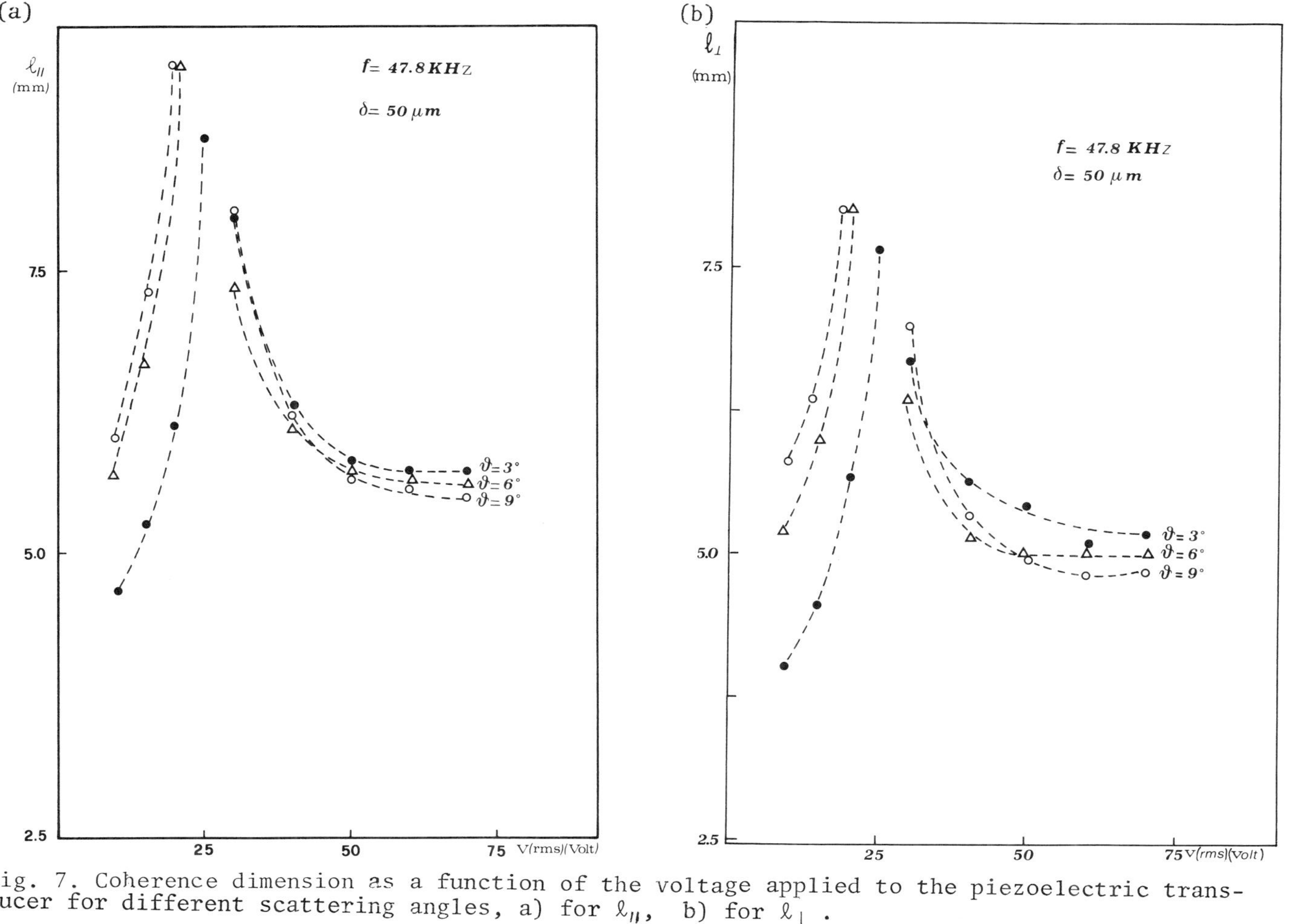

Fig. 7. Coherence dimension as a function of the voltage applied to the piezoelectric transducer for different scattering angles, a) for $\ell_{||}$, b) for $\ell_\perp$.

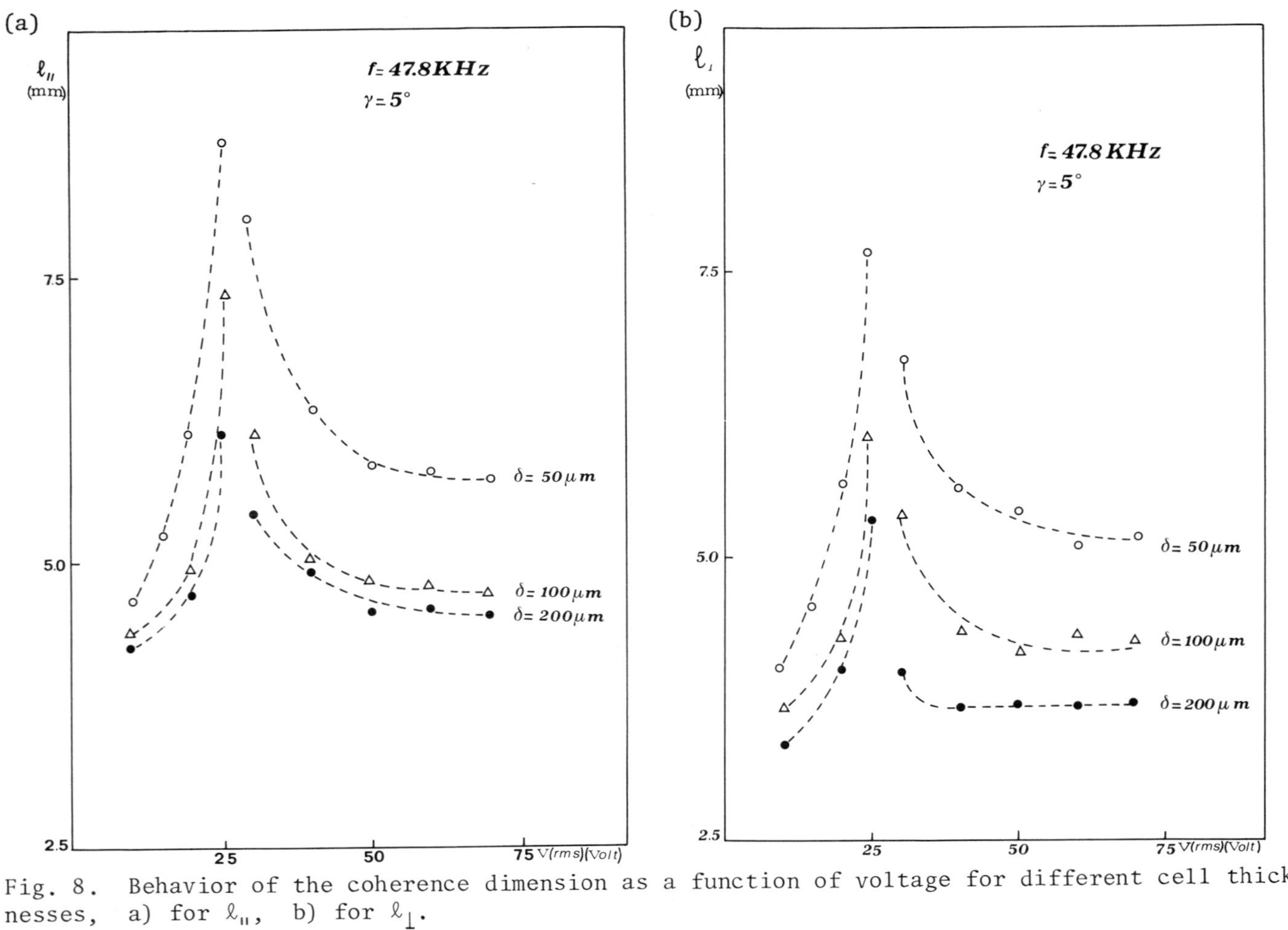

Fig. 8. Behavior of the coherence dimension as a function of voltage for different cell thicknesses, a) for $\ell_{\parallel}$, b) for $\ell_{\perp}$.

TABLE I

$\delta(\mu m)$	$\theta(°)$	$s_\parallel(\mu m)$	$\ell_c^{(\parallel)}(mm)$	$s_\perp(\mu m)$	$\ell_c^{(\perp)}(mm)$	V = 9 Volt				V = 70 Volt			
						$\ell_\parallel(exp)$	$\dfrac{\ell_\parallel(exp)}{\ell_c^{(\parallel)}}$	$\ell_\perp(exp)$	$\dfrac{\ell_\perp(exp)}{\ell_c^{(\perp)}}$	$\ell_\parallel(exp)$	$\ell_\perp(exp)$	$\dfrac{\ell_\parallel(exp)}{\ell_c^{(\parallel)}}$	$\dfrac{\ell_\perp(exp)}{\ell_c^{(\perp)}}$
50	3°	30	4.2	32	3.94	4.65	1.11	4.00	1.01	5.75	5.20	1.37	1.32
	6°	30	4.2	35	3.60	5.70	1.36	5.20	1.44	5.60	5.00	1.33	1.39
	9°	30	4.2	37	3.40	6.05	1.44	5.80	1.70	5.50	4.80	1.31	1.41
100	3°	30	4.2	35	3.60	4.40	1.05	4.20	1.17	4.40	4.25	1.05	1.18
	6°	30	4.2	40	3.15	4.95	1.18	4.70	1.49	4.60	4.00	1.09	1.27
	9°	30	4.2	42	3.00	5.65	1.34	5.00	1.69	5.05	3.60	1.20	1.20
200	3°	30	4.2	40	3.15	4.25	1.01	3.60	1.14	4.50	3.65	1.07	1.16
	6°	30	4.2	51	2.47	4.50	1.09	3.75	1.52	4.15	3.50	0.99	1.42
	9°	30	4.2	61	2.06	5.20	1.24	4.52	2.19	3.85	3.30	0.92	1.60

By simple inspection of the listed values for these last two quantities it is evident that, except for a few cases, these ratios are always significantly larger than unity. Therefore we must conclude that at the roll formation the material is already correlated over dimensions of the same order of the scattering volume dimension, which is to say, over lengths of the order of some tens of microns.

The increase of both $\ell_\parallel^{(exp)}$ and $\ell_\perp^{(exp)}$ with the applied voltage up to 25 V, means that the correlation length of the refractive index fluctuations increases with the voltage, or more physically with the velocity in the rolls. The conclusion is that the refractive index fluctuations are driven by the hydrodynamic motion inside the rolls with an increasing correlation as the velocity increases.

The behavior of ℓ as a function of δ and θ is simply explained as due to the asymmetry of ℓ_n inherent to the roll's structure.

The apparent correlation length $\ell_n^{(\perp)}$ in the direction parallel to the roll's axis can be written as $\ell_n^{(\perp)} \simeq b + L \sin\theta$, by considering that when rolls are present the motion is more correlated over the cell thickness for a length L which depends on δ. Therefore the ratio $\ell_n^{(\perp)}/s_\perp$ increases slowly with θ and so does $\ell_\perp^{(exp)}$ according to the theory.

The same consideration applies for $\ell_\parallel^{(exp)}$. In this case $\ell_n^{(\parallel)}$ increases with the angle due to the presence of the axial component of the motion which drives increasing correlation in this direction.

These considerations also explain the increase of both $\ell_\perp$ and $\ell_\parallel$ with decreasing thickness.

The fact that $\ell_\parallel > \ell_\perp$ means simply that

$$\ell_n^{(\parallel)}/s_\parallel > \ell_n^{(\perp)}/s_\perp \quad .$$

At a threshold of about 25 V a new fact is produced; the coherence dimension reaches a large maximum and then suddenly starts to decrease. At this moment the rolls become unstable and break themselves, producing an increase in the refractive index fluctuations and in the scattering of light (we enter a region of very strong light scattering analogous to the dynamic scattering produced by an electric field), and a decrease in ℓ_n. Therefore ℓ must also decrease, as is shown in Fig. 7. The last measured points at 70 V show some saturation. To understand this better we refer to the last columns of Table I. Within experimental errors $\ell_\parallel^{(exp)}$ is independent of θ and decreases with increasing δ, whilst $\ell_\perp^{(exp)}$ decreases with increasing θ and δ (remember that $s_\perp$ increases with increasing θ and δ). These facts are interpreted as indicating that in this voltage region ℓ_n decreases with increasing δ.

Quantitative information could be obtained on ℓ_n as a function of the different parameters. No theory however exists at the moment for its behavior, thus rendering any comparison impossible.

4. CONCLUSIONS

The exposed results show the potentialities of the method of the study of the area of coherence for obtaining information on the correlation length of fluctuations in a medium. It is worthwhile to point out that the increase in the spatial coherence obtained when $\ell_n/d \gtrsim 1$, is only one aspect of other important changes in the properties of the scattered radiation, as, for example, its statistical properties. If, in fact, the statistics of the fluctuations producing the scattering are not Gaussian by themselves, when in the scattering volume only one realization of the field is present, the statistics of the scattered field reflect the one of the medium. Cases of relevant interest where this happens are, for example, turbulence in fluids [19], scattering by objects moving with deterministic motion (e.g. a round-disk glass) [20], double, or more generally, multiple scattering [21].

References

1. E. Wolf and W.H. Carter, Optics Comm. *13*, 205 (1975).
2. W.H. Carter and E. Wolf, J. Opt. Soc. Am. *65*, 1067 (1975).
3. W.H. Carter and E. Wolf, J. Opt. Soc. Am. *67*, 785 (1977).
4. E. Wolf and W.H. Carter, Optics Comm. *16*, 247 (1976).
5. H.P. Baltes, B. Steinle and G. Antes, Optics Comm. *18*, 242 (1976).
6. M. Born and E. Wolf, *Principles of Optics* (Pergamon Press, London, 1965), Chap. X.
7. M. Bertolotti, F. Scudieri and S. Verginelli, Appl. Opt. *15*, 1842 (1976).
8. F. Scudieri, M. Bertolotti and R. Bartolino, Appl. Opt. *13*, 181 (1974).
9. M. Bertolotti, F. Scudieri, A. Ferrari and D. Apostol, Appl. Opt. *15*, 2468 (1976).
10. See, for example, B. Crosignani, P. Di Porto and M. Bertolotti, *Statistical Properties of Scattered Light* (Academic Press, New York, 1975).
11. H.P. Baltes, B. Steinle and G. Antes, this Volume, p. 431.

12. H.P. Baltes and B. Steinle, Lett, Nuovo Cimento *18*, 313 (1977).
13. W.A. Carter, private communication.
14. See for example S. Chandrasekhar, Reports on Progress in Physics *39*, 613 (1976).
15. F.C. Frank, Discuss. Faraday Soc. *25*, 1 (1958).
16. P.G. De Gennes, *The Physics of Liquid Crystals* (Clarendon Press, Oxford, 1974) Chap. 4.
17. F. Scudieri, Appl. Phys. Lett. *29*, 398 (1976).
18. M. Bertolotti, F. Scudieri and E. Sturla, J. Appl. Phys., to be published.
19. P. Di Porto, B. Crosignani and M. Bertolotti, J. Appl. Phys. *40*, 5083 (1969).
20. F. Scudieri and M. Bertolotti, J. Opt. Soc. Am. *64*, 776 (1974).
21. M. Bertolotti, in *Multiple Scattering in Photon Correlation Spectroscopy and Velocimetry*, ed. H.Z. Cummins and E.R. Pike (Plenum Press, New York, 1977) p. 22.

A SENSITIVE INFRARED IMAGING UPCONVERTER AND SPATIAL COHERENCE OF

ATMOSPHERIC PROPAGATION*

R.W. Boyd and C.H. Townes

University of California, Berkeley, California

Theoretical reasoning has indicated that atmospheric "seeing"
disturbances should allow sharper astronomical images at infrared
than at visible wavelengths [1], yet imaging devices for astronomy
have not generally been available at wavelengths beyond the near
infrared. Planets and their atmospheric structure provide interesting subjects for infrared imaging systems [2], in either broad
or narrow spectral regions. So do some more distant objects, such
as stars surrounded by thick dust shells, which emit strongly in the
infrared, and the nuclei of Seyfert galaxies. All of these objects
can be spatially resolved in the larger astronomical telescopes, and
hence are valuable subjects for infrared imaging. We describe here
the exploration of an infrared imaging technique based on the nonlinear interaction known as upconversion [3]. This technique was
used to obtain images of several astronomical objects in the 10 μm
spectral region, and to demonstrate quantitatively the sharper images
allowed for wavelengths beyond the visible region.

The upconversion process is shown schematically in Figure 1.
Infrared radiation is mixed with an intense laser beam in a nonlinear crystal. The nonlinearity in the optical susceptibility of
the crystal generates a signal at the sum frequency, which is in
the visible region. The upconversion process is, in prinicple,
noise free [4], and thus provides a sensitive method of detecting
infrared radiation by first converting it to the visible region
where sensitive, low noise detectors are readily available. Figure
1 also illustrated image conversion by upconversion. Photon momentum
conservation requires that for each infrared direction of propagation
the sum frequency radiation be emitted in a unique direction, and
thus a one-to-one correspondence exists between infrared angles of
propagation and sum frequency angles of propagation. In fact, the

UPCONVERSION PROCESS

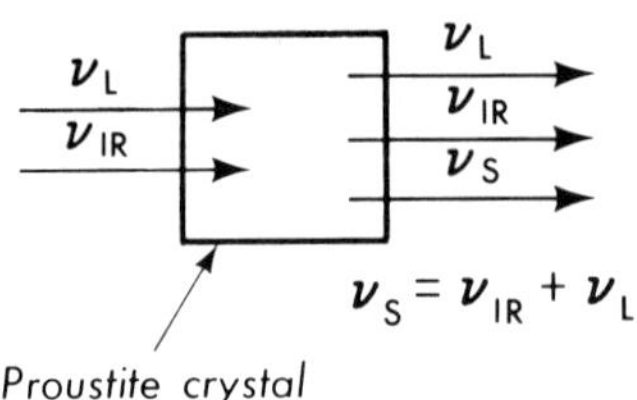

Figure 1. Schematic description of the upconversion process (upper). Imaging property of the upconversion process (lower).

IMAGING PROPERTIES

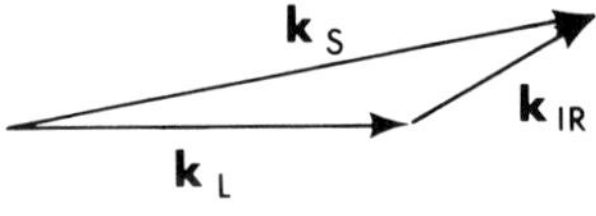

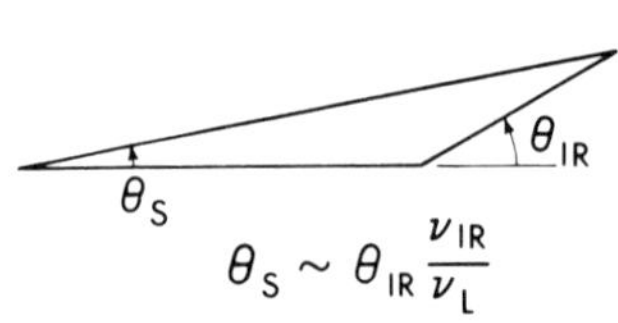

upconversion process simply demagnifies the sum frequency image by approximately the ratio of the sum frequency to the infrared frequency.

In our device, thermal infrared radiation is mixed in a 1-cm-long proustite crystal with an 0.25 watt, 0.7525 μm krypton ion laser beam to produce a sum frequency signal at about 0.7 μm (see Figure 2). The 2 cm^{-1} phase-matched band-pass of the upconverter can be angle-tuned by tilting the crystal about its y axis to allow operation at any wavelength from 9.0μ to 11.0μ. Type II phase-matching [5] is used. In order to provide a large usable field of view several hundred spatial modes of the infrared field are simultaneously upconverted. The quantum conversion efficiency obtained with our device was 2 x 10^{-7}, which was sufficient to allow useful astronomical measurements.

The sum frequency radiation is detected by a Varo 3-stage image intensifier tube, cooled to a temperature of -30 C. Cooling the tube decreases the dark count by a factor of 300 to a level of about 0.1 photoelectron per second within the 3 mm field utilized by the upconverter. The output of the image tube is lens-coupled at unit magnification to Kodak TRI-X film for short exposures and to Kodak

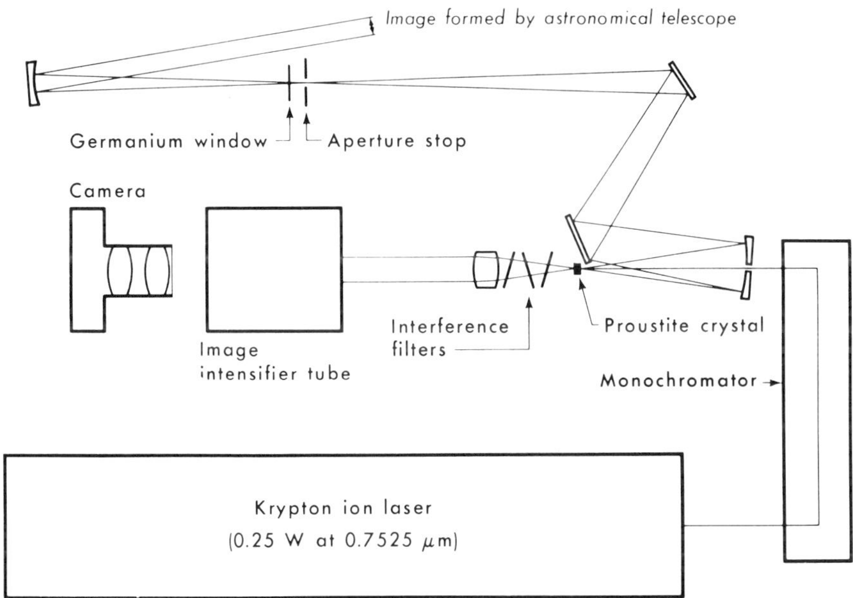

Figure 2. Optical layout of the 10 μm imaging upconverter. The monochromator is used to eliminate background light from the laser discharge tube. Collimated infrared radiation is mixed with the laser beam in the proustite crystal. The interference filters pass the sum frequency while rejecting the laser frequency, providing a factor of 10^{18} discrimination between the two frequencies. The sum frequency image is amplified by the image intensifier tube and recorded photographically.

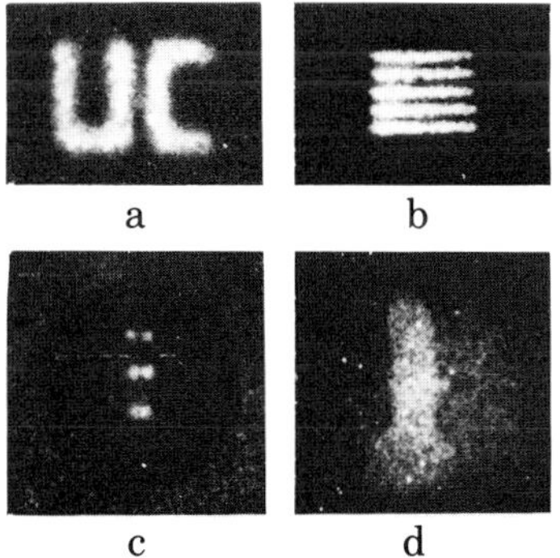

Figure 3. Upconverted images of several laboratoy test objects. Figures a, b and c show resolution test patterns. The smallest spacing in Figure c is 1.3 times the diffraction limit. Figure d shows a one minute integration on a soldering iron operating at 100 C.

IIa-D film for long exposures.   Some of the laboratory results are
shown in Figure 3.

This instrument has been used on the 150 cm McMath solar telescope of Kitt Peak National Observatory. The field of view on the
sky is then 40 seconds of arc, and the resolution is limited by
diffraction at the telescope aperture and by inhomogeneities in the
earth's atmosphere. Resulting images contain several hundred picture elements. Infrared pictures of the Sun, Moon, Mercury, and
the dust shell surrounding the star VY CMa have been obtained, and
are shown in Figure 4.   Integration times range from 4 seconds for
the Sun to 15 minutes for VY CMa.

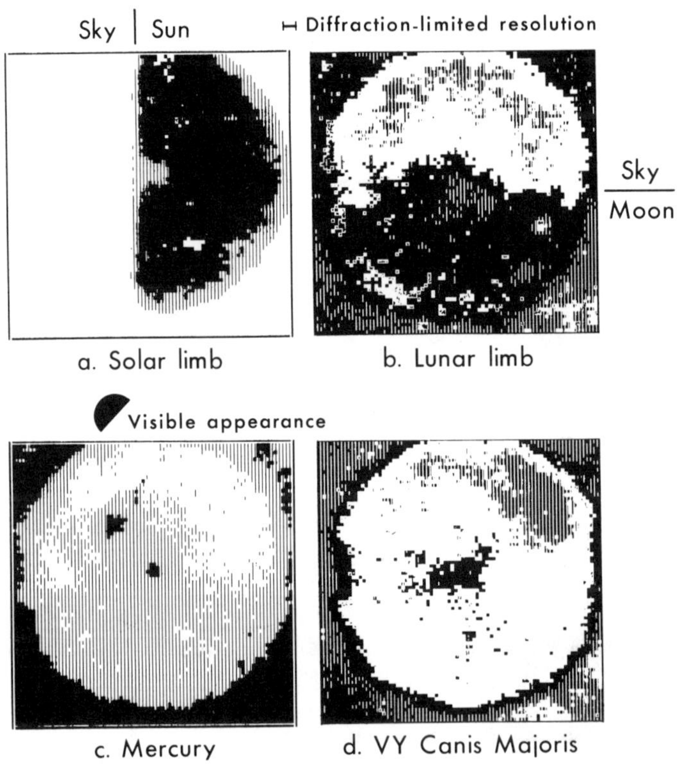

Figure 4.   Infrared images of a number of astronomical objects.
Computer generated plots of digitized photographic negatives are
shown. In each case the field of view is round, and a spurious
spot is introduced at the center of the field from the hole in the
collimating mirror shown in Figure 2. Note the enhancement in the
signal from the subsolar point on Mercury. The detection of VY
Canis Majoris is marginal.

The deleterious effects of atmospheric inhomogeneities on telescope resolution (or "seeing" effects) were studied in the infrared region using this apparatus. The theory of wave propagation in a turbulent medium [1] predicts a $\lambda^{-1/5}$ dependence to the angular resolution of a telescope limited by the effects of seeing when averaged over a sufficiently long time. Previous attempts at verification of this wavelength dependence have met with mixed results, due primarily to the lack of true imaging devices at long infrared wavelengths where seeing would be expected to be significantly better than in the visible. We have investigated the wavelength dependence of seeing by taking pictures of the sun's limb with the 1.5 m telescope simultaneously at 10 μm and at 0.55 μm, and comparing the sharpness of the solar limb in each case. A typical 10 μm limb profile is shown in Figure 5, and the corresponding visible limb profile is shown in Figure 6. This comparison was repeated a number of times under a variety of atmospheric conditions. After correcting the data for the differences in solar limb darkening [6] and diffraction at the two wavelengths, we found

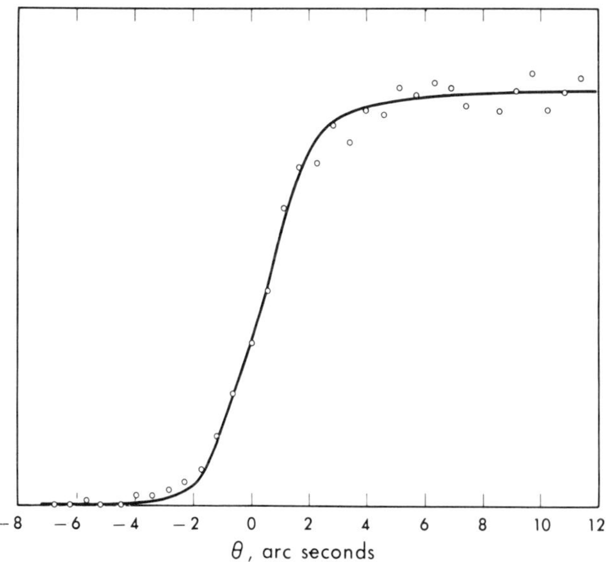

Figure 5. Solar Limb Profile at 10 μm. The solid curve is a theoretical limb profile formed by convolving the true limb profile with a Gaussian shaped point spread function representing the combined effects of "seeing", diffraction, and instrumental resolution. In this case the Gaussian has a full width to 1/e points of 4 arc seconds.

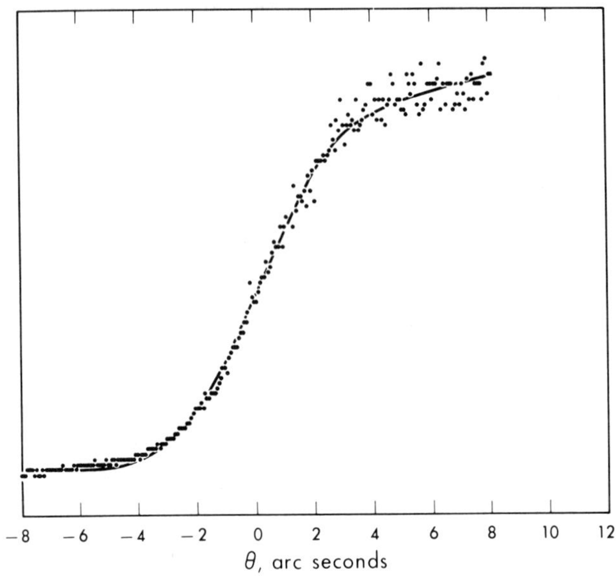

Figure 6.   Solar Limb Profile at visible wavelengths.   The solid
curve is a theoretical limb profile formed by convolving the true
limb profile with a Gaussian shaped point spread function representing the combined effects of "seeing", diffraction, and instrumental resolution.   In this case the Gaussian has a full width to
1/e points of 6 arc seconds.

a systematic improvement in seeing at the longer wavelength, as
shown in Figure 7.   The RMS width of the point spread function for
seeing is found to be 1.9 ± .2 times greater at .55 μ than at 10 μ
in good agreement with the theoretical value 1.80 predicted on the
basis of random turbulence.

These experiments represent the first successful application
of imaging upconversion to astronomy.   The low ($\sim$2 x 10^{-7}) quantum
efficiency of the device used severely limited its usefulness as
an astronomical detector; nevertheless, these measurements demonstrate the promise of infrared imaging and on upconversion techniques when materials which provide larger nonlinear effects become
available [7].

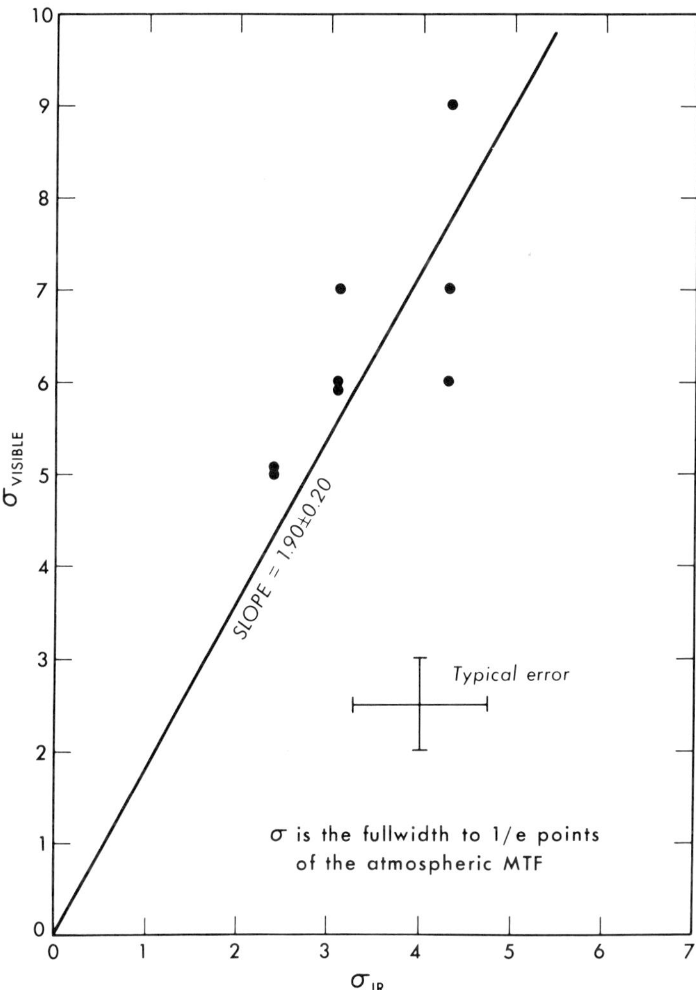

Figure 7. Visible Seeing vs. Infrared Seeing.

* This work is supported in part by NASA Grants NGL 05-003-272
and NGR 05-003-452.

References

1.   D.L. Fried, J. Opt. Soc. Am. *56*, 1372 (1966).
2.   J.A. Westphal, K. Matthews, and R.J. Terrile, Ap. J. Lett. *188*, L111 (1974).
3.   K.F. Hulme, Rep. Prog. Phys. *36*, 497 (1973) and F. Zernike and J.E. Midwinter, *Applied Nonlinear Optics*, (John Wiley and Sons, Inc., New York, 1973).
4.   C.L. Tang, Phys. Rev. *182*, 367 (1970).
5.   J.E. Midwinter and J. Warner, Br. J. Appl. Phys. *16*, 1135 (1965).
6.   P. Lena, Astron. Astrophys. *4*, 202 (1970) and J.E. Gaustad and J.R. Rogerson, Ap. J. *134*, 323 (1961).
7.   R.W. Boyd, to be published in Optical Engineering, 1977.

DENSITY DEPENDENT RAYLEIGH LINEWIDTH OF DEPOLARIZED DOUBLY

SCATTERED LIGHT[*]

J.G. Gallagher, Dae M. Kim and C.D. Armeniades

Rice University, Houston, Tex.

There is considerable theoretical and experimental interest in
the double scattering mechanism of fluids near their critical point
[1,2,3] and of macromolecules in solution [4,5]. Recently, Sorensen et al [4] reported the Rayleigh linewidth measurement of light,
doubly scattered from optically isotropic Rayleigh-Gans particles
in Brownian motion. Their results show a depolarized correlation
spectrum *independent* of particle number density and scattering
angle, and a correlation time, slightly greater than the correlation time for singly scattered light at 180°. Recently, Bertolotti
studied the double scattering process with Mie spheres, using two
separate cells [6]. In his work he considered the spatial coherence
properties of the scattered field and its effects on the second-
order correlation function. So far, these investigators have treated
the double scattering phenomena by means of the Van Hove space-time
density correlation function, applied successively to two statistically independent scattering events. This approach is based on
the assumption that the total phase fluctuation of doubly scattered
light is caused by both the first and second scatterers in Brownian
motion. In addition, this interpretation assumes that the complex
degree of coherence can be factorized into spatial and temporal
parts.

In this paper we report extensive and systematic measurements
and analysis of the intensity and Rayleigh linewidths of depolarized light, doubly scattered from a non-interacting system. In
particular, we present an explicit experimental test of the assumption that both the first and second scatterers in Brownian motion
contribute to the Rayleigh linewidth. On the basis of this test,
we report for the first time the strong dependence of the Rayleigh
linewidth (Γ_{HV}) of the depolarized, doubly scattered light on the

number density of the scatterers. We show that the density dependence of Γ_{HV} can be analyzed in terms of coupling between the temporal and spatial parts of the complex degree of coherence of the scattered light.

EXPERIMENTAL RESULTS AND ANALYSIS

In this work we used a model system of monodisperse polystyrene Rayleigh-Gans spheres (910 Å diameter), the concentration of which ranged from 2.7×10^{11} to 4.1×10^{12} particles per cm^3. Measurements were performed using a vertically polarized primary beam (diameter at the focal plane varying from 32 µm to 3 mm), and the depolarized component of the doubly scattered light was collected at 90°, using a Glan-Thompson polarizing prism (extinction ratio 5×10^{-6}). The secondary scattering volume as determined by the detector collection system was approximately the cylinder of revolution whose radius was $\sim$100 µm and 1 cm in length. The collection system consisted of two circular pinholes of radii 100 µm and 300 µm separated by distance 45 cm. The quartz scattering cell was 1 cm^2 by 4.5 cm.

Consider two statistically independent scattering events whose scattering geometry is shown in Fig. 1. For the spherical Rayleigh-Gans particle (radius a) the total integrated intensity of the depolarized component of doubly scattered light is given by [7]

$$I_{HV}^{d} = [\tfrac{4}{9} k_o^4 a^6 |m-1|^2/R]^2 \, N^2 I_o \int_{V_1} \int_{V_2} e^{-\alpha y} \left[\frac{\sin\phi\cos\phi\cos\gamma}{r_{12}} \right]^2$$

$$\times \; G^2(u) \, G^2(u') \, dV_1 \, dV_2 \quad . \tag{1}$$

Here, N, R, r_{12}, V_1, V_2, α, and m are, respectively, the particle number density, the distance to the detector, the distance between two scatterers, the primary and secondary scattering volumes, the linear attenuation coefficient, and the relative index of refraction. The form factor G(u) is given by

$$G(u) = (3/u^3)(\sin u - u \cos u) \quad , \tag{2a}$$

$$u = 2k_o a \, \sin(\theta/2) \quad . \tag{2b}$$

It can be seen from Eq.(1) that, because of the power law dependence, r_{12}^{-2}, and the polarization factor, the major contribution to I_{HV}^{d} comes from two scatterers in the vicinity of the overlapped

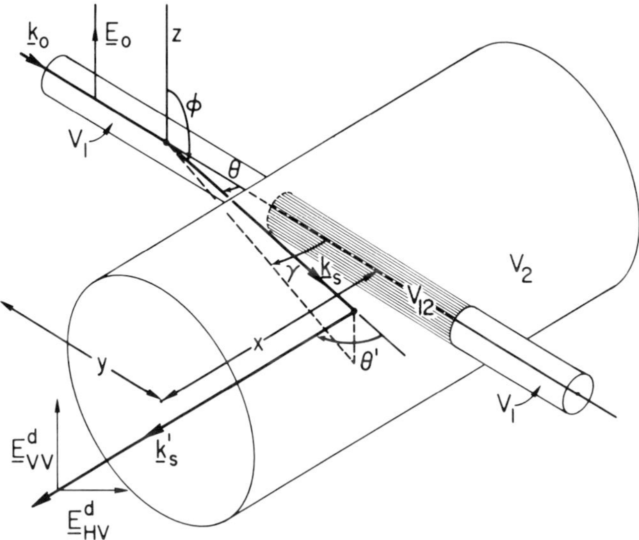

Figure 1. The double scattering geometry; the polarized $(\hat{z})$ primary laser beam is propagating toward y direction, and the scattered light is detected at $90°(\hat{x})$. λ_0 = 457.9 nm.

region between the primary and secondary scattering volumes. For the case where the cross-sectional dimension of the overlapped region, as viewed by the detector, is much less that the photon mean free path, α^{-1}, Eq. (1) reduces to

$$I_{HV}^{d} = K \, N^2 \, e^{-\alpha y} \quad . \tag{3}$$

K is a constant independent of N, y is the detector position. The measured depolarized intensity, corrected for the appreciable atten- uation of the primary beam is found to be proportional to the number density squared. This establishes the incoherent and non-interacting behavior of the double scattering event over the density range stud- ied. (See Fig. 2).

We next present our experimental test showing unambiguously for the first time that the random motion of the first scatterers does *not*, in general, contribute to the linewidth of Γ_{HV} in terms of phase fluctuations. The test is based on using the viscosity dependence of the translational diffusion coefficient and on sepa- rating the primary and secondary scattering volumes. The separation was done, using a two-compartment sample cell, one containing the

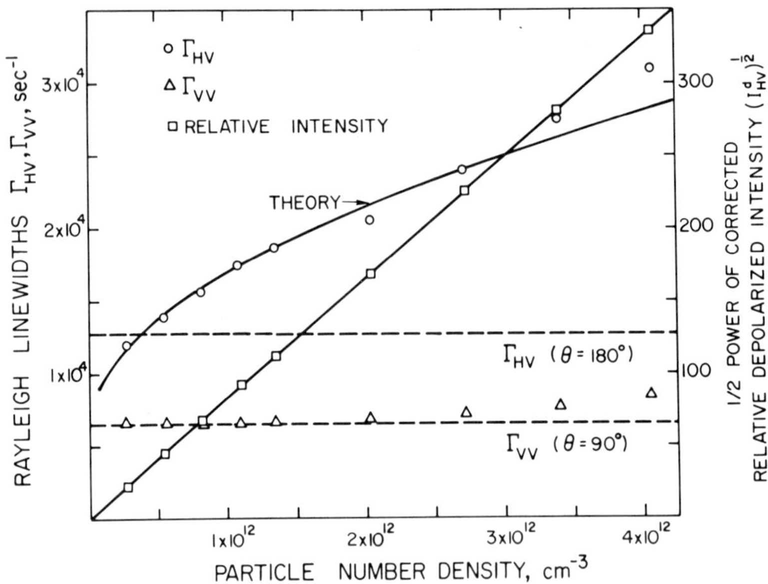

Figure 2. The depolarized doubly scattered intensity I_{HV}^d and the decay rates as a function of particle number density. (o) denotes the measured decay rate of the depolarized doubly scattered light, (Δ) the polarized singly scattered light at 90° scattering angle. The dashed lines represent the existing theories of the single and double scattering decay rates and the solid line present theory. λ_0 = 457.9 nm.

primary beam and the single scattering solution, the other the double scattering solution only. Γ_{HV} was measured for three different arrangements: (a) both the first and second scatterers in water, (b) the first scatterer in a glycerol-water mixture, while the second one in water, (c) the first scatterer in water, the second in the mixture. The value of Γ_{HV} was the *same* for the cases (a) and (b) (see Fig. 3), while the value *varied* for the cases (a) and (c) in accordance with the viscosity dependence of the diffusion coefficient, viz.

$$D = k_B T/6\pi\eta a \ ,\tag{4}$$

η being the viscosity. This test clearly establishes the fact that the decay rate (Γ_{HV}) of the depolarized doubly scattered light is *not* affected by the motion of the first scatterers in a typical double scattering experiment, in which the average distance between

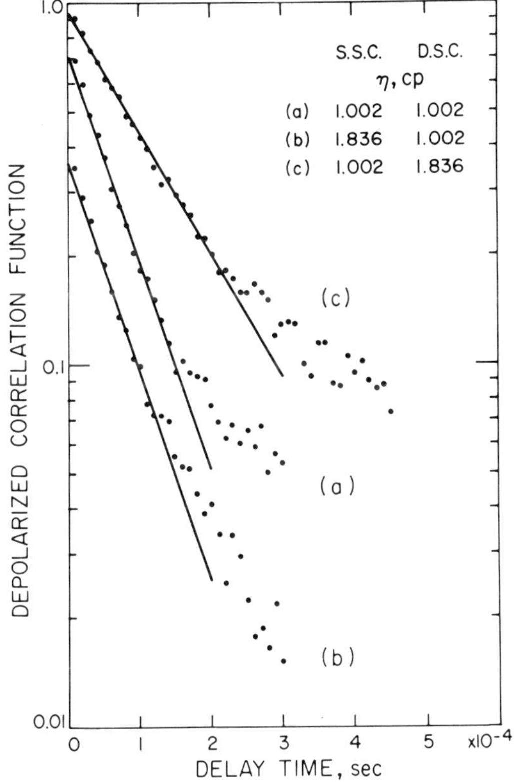

Figure 3. The measured depolarized autocorrelation functions for cases (a), (b) and (c). The water-glyceral mixture approximately 22% by weight. The single scattering compartment is 0.2 x 1 x 4.5 cm containing single scattering concentration ($N = 4.8 \times 10^{10}$/cm^3) and the double scattering compartment is 0.4 x 1 x 4.5 cm containing double scattering concentration ($N = 1.36 \times 10^{12}$/cm^3). The measured values of Γ_{HV} are 17930, 17777, 9875 sec^{-1}, for (a), (b) and (c) respectively. $\lambda_0 = 457.9$ nm.

the first and second scatterers is small. It is important to note that the interpretation of Γ_{HV}, by applying Van Hove space time correlation function to two successive scattering events [4,8], is not, in general, valid.

Over the same particle denisty range where I_{HV} was measured, the decay rate (Γ_{HV}) of the depolarized autocorrelation function was measured using a single clipped digital correlator with clipping level q = 0. For the particle number densities measured the average count per bin time varied from 0.05 to 0.25 with $g^{(2)}(0)/g^{(2)}(\infty)$ varying between 1.15 and 1.35. The depolarized autocorrelation functions were not found to be accurately described by a single exponential. The decay rate, Γ_{HV}, was obtained by using the method of cumulants with weighting factor favoring the head portion of the data [9].

In direct contrast with the previous result of Ref.[4], Γ_{HV} is found to be strongly dependent on the particle number denisty, as shown in Fig. 2. Clearly, our data for Γ_{HV} can not be explained by the density-independent, single scattering value at about 180°

scattering angle as predicted by the existing theories. Simultaneously presented in Fig. 2 are the decay rates of the polarized single scattering component (Γ_{VV} = 6540 sec^{-1}). The fact that Γ_{VV} is essentially independent of the particle density provides a further evidence for the non-interacting nature of Brownian particles.

The interpretation of our Γ_{HV} data can not be made by using the existing theories. In this paper, we propose a new approach to the problem. In our analysis the double scattering event is partitioned into statistically independent second scatterers acting as single scatterers as observed by the detector. Each second scatterer is illuminated by the singly scattered light from the entire first scatterers simultaneously, i.e. the entire single scatterers constitute an incoherent line source. This partition scheme incorporates the important fact that the random motion of the first scatterers does not affect Γ_{HV}.

Light emanating from an incoherent extended line source becomes spatially coherent in the process of propagation. For a typical double scattering concentration, the average distance ($<\ell>$) of a second scatterer from the source is small. (See our discussion on I_{HV}). A second scatterer located at a close distance will be illuminated by partially coherent light. Hence in this situation, the complex degree of coherence can not be factorized into spatial and temporal parts, as has been so far assumed. This means that a second scatterer in Brownian motion will experience an amplitude modulation associated with the spatial coherence function of partially coherent light [10,11].

The depolarized Rayleigh linewidth as measured by the detector will then be contributed by the usual phase fluctuation due to the motion of the second scatterer and by the above discussed amplitude modulation. Thus, the field-field autocorrelation function of the depolarized doubly scattered light is given by

$$<E_{HV}(0)E^*_{HV}(\tau)> = KNe^{i\omega_o\tau} \int_{V_2} dv \int d\Delta\underline{r} \; \frac{1}{(4\pi D\tau)^{3/2}} \; e^{-|\Delta\underline{r}|^2/4D\tau}$$

$$e^{-i[\kappa_x\Delta x + \kappa_z\Delta z + (\kappa_y + k_o(L/<\ell>))\;\Delta y]} \qquad . \quad (5a)$$

Here κ represents the scattering wave vector. L is the effective length of the line source along the y axis as viewed at close distances by a second scatterer at x_{12} and z_{12} (see Fig. 1). L is determined by the scattering indicatrix of the line source, which is a maximum in the neighborhood of $x_{12} \simeq y_{12}$ with the full width at half maximum point approximately equal to L. The average distance, $<\ell>$ of the second scatterer from the center of the line source is approximated in our analysis by

$$\langle \ell \rangle = (\pi a^2 / C_{sca}) \langle r \rangle \tag{5b}$$

where C_{sca}, $\langle r \rangle$ are, respectively, the scattering cross-section [7] and the mean interparticle distance at particle number density N, viz. $\langle r \rangle = N^{-1/3}$. From Eq.(5b) it can be **seen that, with increasing** particle density, a second scatterer on the average is moving closer to the line source. Upon inserting Eq.(5b) into Eq.(5a) it becomes clear that the coupling of spatial and temporal parts of the complex degree of coherence results in an additional density-dependent line-broadening term. In Eq.(5a) we used the fact that the particle displacement $|\Delta \underline{r}|$ is much less than $\langle \ell \rangle$ or L.

The integration w.r.t. $\Delta \underline{r}$ in Eq.(5a) can be readily performed. Summing over the second scatterers in V_2 corresponds to averaging over the scattering wave vector κ for the scattering geometry. We find

$$\langle E_{HV}(0)E_{HV}(\tau)^* \rangle = KN^2 e^{i\omega_0 \tau} \exp - \langle \kappa^2 \rangle D\tau \left[1 + \frac{2}{\sqrt{3}} \frac{k_0 L N^{1/3} C_{sca}}{\langle \kappa \rangle \pi a^2} \right.$$

$$\left. + \left(\frac{k_0 L N^{1/3} C_{sca}}{\langle \kappa \rangle \pi a^2} \right) \right] , \tag{6}$$

where $\langle \kappa \rangle$ is the average value of κ, which for our geometry corresponds to the scattering angle of about 82°. (See Fig. 4). It should be pointed out that the depolarized field-field autocorrelation function consists of many exponentials in general. Furthermore, the additional density-dependent line broadening term is contained explicitly in Eq.(6).

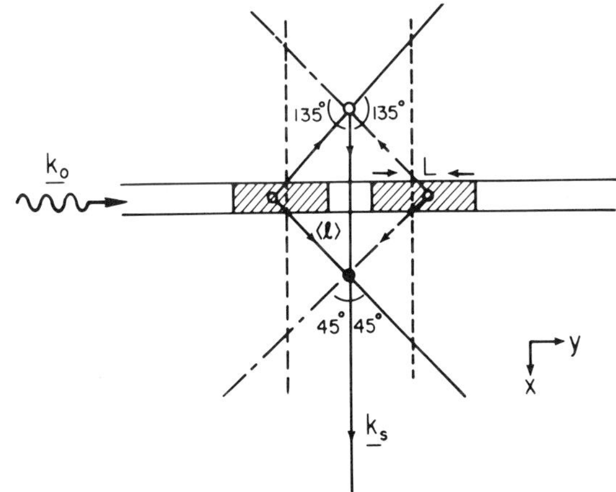

Figure 4. The topside view of the double scattering geometry for open scattering cell. The average scattering angles for particles symmetrically located w.r.t the primary beams are shown. The shaded region denotes the effective length of the line source as viewed by the second scatterer.

The intensity-intensity correlation function for depolarized doubly scattered light is given by

$$\langle I_{HV}(0) I_{HV}(\tau) \rangle = \Phi K^2 N^4 \{ \langle |E_{HV}(0)|^2 \rangle^2 + \gamma |\langle E_{HB}(0) E_{HB}^*(\tau) \rangle|^2 \} \quad (7)$$

Here Φ represents the quantum efficiency and the γ-factor describes the effect of spatial coherence on the detector from the extended secondary scattering volume, whose depth is 1 cm. Note that for the case where the distance between the first and the second scatterer is small, the scattered light is described by Gaussian statistics [6]. In Fig. 2 we present the theoretical curve of Γ_{HV}, Eq. (7), where the effective size of the source (L) is taken to be 98 μm. The agreement between theory and experimental data is excellent, as can be clearly seen from Fig. 2.

CONCLUSIONS

We have presented experimental data of Rayleigh linewidth for depolarized doubly scattered light, which is strongly dependent on particle number density. We have also presented an explicit experimental result, showing that the motion of the first scatterer does not contribute to Γ_{HV} in a typical double scattering experiment. We have analyzed Γ_{HV} in terms of the coupling between spatial and temporal parts of the complex degree of coherence. Our theory can be applied with proper modifications to Mie as well as Rayleigh scatterers, and may have an interesting application in fluids near the critical point, where the double scattering process is important.

* Supported in part by a grant from the Robert A. Welch Foundation.

References

1.   D.W. Oxtoby and W.M. Gelbart, J. Chem. Phys. *60*, 3359 (1974).
2.   L.A. Reith and H.L. Swinney, Phys. Rev. A*12*, 1094 (1975).
3.   A.J. Bray and R.F. Chang, Phys. Rev. A*12*, 2594 (1975).
4.   C.M. Sorensen, R.C. Mackler, and W.J. O'Sullivan, Phys. Rev. A*14*, 1520 (1976).
5.   R.C. Colby, L.M. Narducci, V. Bluemel, and J. Baer, Phys. Rev. A*12*, 1530 (1975).
6.   M. Bertolotti, *Photon Correlation and Velocimetry* (NATO Adv. Study Inst. Series) ed. by H. Cummins and R. Pike, (Plenum, New York, 1977).
7.   J.G. Gallagher, Dae M. Kim and C.D. Armeniades, *Colloid and Interface Science*, vol. V, ed. by M. Kerker, (Academic, New

York, 1976).

8. A.P. Ivanov, A. Ya Khairullina, and A.P. Chaikovokii, Opt. Spectroc. *35*, 668 (1973).

9. P.N. Pusey, *Photon Correlation and Light Beating Spectroscopy*, (NATO Adv. Study Inst. Series) ed. by H. Cummins and R. Pike (Plenum, New York, 1974).

10. M. Born and E. Wolf, *Principles of Optics*, 4th edition (Pergamon, New York, 1974).

11. S.G. Lipson and H. Lipson, *Optical Physics* (Cambridge University Press, 1969) (See Chap. 8).

THE STATISTICAL AND CORRELATION PROPERTIES OF LIGHT SCATTERED BY A

RANDOM PHASE SCREEN

G. Parry, P.N. Pusey, E. Jakeman and J.G. McWhirter

Royal Signals and Radar Establishment, Malvern, U.K.

1. INTRODUCTION

We report measurements of the statistical and correlation properties of laser light scattered by a random phase screen produced
in the laboratory by turbulent mixing of the convective hot air
flow above an electric heater and the surrounding cooler air. A
phase screen is a localized region of space containing refractive
index variations which cause random fluctuations in the phase of
electromagnetic radiation propagating through it. On further free
propagation of the emergent radiation, intensity fluctuations develop
through two distinct mechanisms. Firstly, the individual refractive
index inhomogeneities act as lenses which bend the light "rays" and
produce a random focussing of the radiation. Secondly, further from
the screen, the radiation scattered by different, independent inhomogeneities can overlap to form fringes and more complex interference
patterns (e.g. speckle). Phase screens have been of interest for
many years because of their role in various natural phenomena such
as the fading of radio signals reflected from the ionosphere [1],
the scintillation of distant radio sources due to the solar wind [2]
and the twinkling of starlight caused by atmospheric fluctuations
[3]. Nevertheless our measurements appear to be the first which
cover a wide range of screen-detector distance, thereby providing
focussing curves such as Figure 2.

The part of this work concerned with single-interval statistical
properties has been reported elsewhere [4] and is reviewed only
briefly here. Reference 4 can be consulted for further details.

2. EXPERIMENTAL

Figure 1 shows the layout of the experiment. The turbulent
flow was produced by two 1 kW electric bar heaters surrounded by
metal baffles having an opening about 25 cm x 1 cm at the top with
its long axis parallel to the bars. A beam of light of diameter
$W \sim 2.5$ cm from a 100 µW He - Ne laser passed about 15 cm above
this opening. The detectors were ITT FW130 photomultiplier tubes
operated in the photoncounting mode. The effects of dark count
and dead-time were small. The apertures at the detectors were
$\lesssim 25$ µm in diameter, much smaller than the spatial correlation
range of the scattered radiation. A pellicle beam splitter was
placed in front of the detectors, one of which could be translated
vertically and horizontally. The detectors could thus be optically
superimposed and then separated by a known distance d.

Photocount distributions P(n,T) (the probability of obtaining
n photocounts in sample time T) and correlations were measured by
a "Malvern" correlator and probability analyser. Normalized factorial moments $n^{[m]}$ of P(n,T) and normalized correlation functions
were computed by an on-line Hewlett-Packard 9830 calculator. The
factorial moments are equal to the normalized moments of the inten-

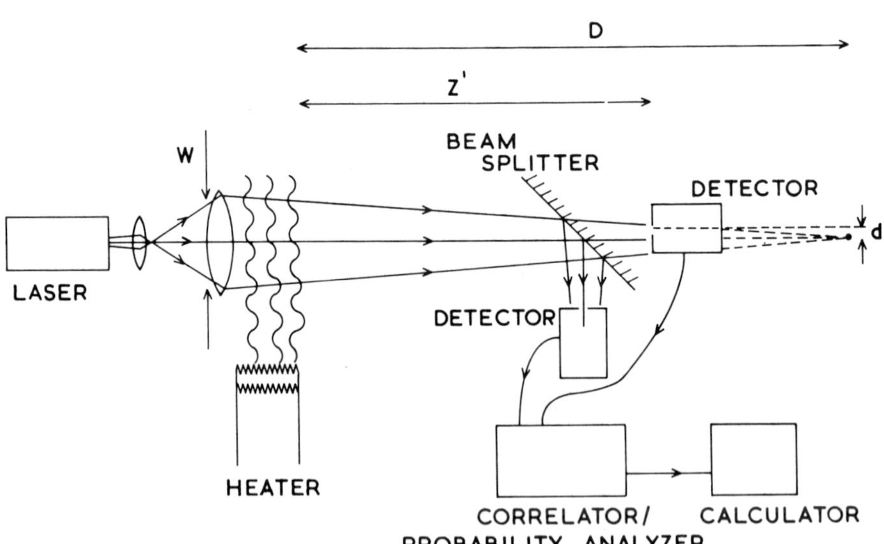

Figure 1. Layout of the experiment.

sity distribution $P(I)$:

$$n^{[m]} \equiv \frac{\langle n(n-1)\ldots(n-m+1)\rangle}{\langle n\rangle^m} \equiv \frac{\langle I^m\rangle}{\langle I\rangle^m} \quad . \tag{1}$$

To keep the actual experiments reasonably compact the laser beam was brought to a focus at a distance $D \sim 10$ m from the phase screen. The results obtained using this arrangement can be scaled to those for a plane wavefront using the relationships:

$$\frac{1}{Z} = \frac{1}{Z'} - \frac{1}{D}$$

and
$$\tag{2}$$
$$\frac{d}{Z} = \frac{d'}{Z'} \quad ,$$

where Z' is the actual screen-detector distance d', the actual separation of the detectors, and Z and d are the distances which would obtain with a plane wavefront. Thus Z ranges from 0 to ∞ as Z' goes from 0 to D.

3. THE SECOND MOMENT

Figure 2 shows a plot of the second moment $n^{[2]}$ against corrected screen-detector distance Z for two values of the mean square phase fluctuation $\overline{\phi^2}$, obtained by placing the heater baffle opening either parallel or perpendicular to the illuminating beam. The lines are the predictions of the theoretical treatment (see below). Theory and experiment were fitted at the peaks of the curves. The "best-fit" values of $\overline{\phi^2}$ and phase correlation length ξ were $\overline{\phi^2}$ = 10 rad^2, ξ = 3.4 mm and $\overline{\phi^2}$ = 2.5 rad^2, ξ = 3.7 mm. Errors in these values are probably $\sim 10\%$. The similar values of ξ in the two cases can be justified by simple arguments [4].

Figure 3 shows photographs of the intensity patterns at various distances Z for $\overline{\phi^2}$ = 10. These were taken by a lensless camera with a shutter speed of about 0.5 ms. Figures 2 and 3 can be discussed as follows: close to the screen (figure 3a) the intensity fluctuations have not developed much and $n^{[2]} \approx 1$. Farther from the screen the fluctuations increase to a maximum at $Z = Z_F$ due to the focussing effect of the individual phasefront inhomogeneities. The pattern (figure 3b) can be regarded as a random network of caustics. Due to the relatively small value of $\overline{\phi^2}$, diffraction broadening of the caustics is evident. For $Z > Z_F$ the contributions from the different inhomogeneities start to overlap leading to interference effects.

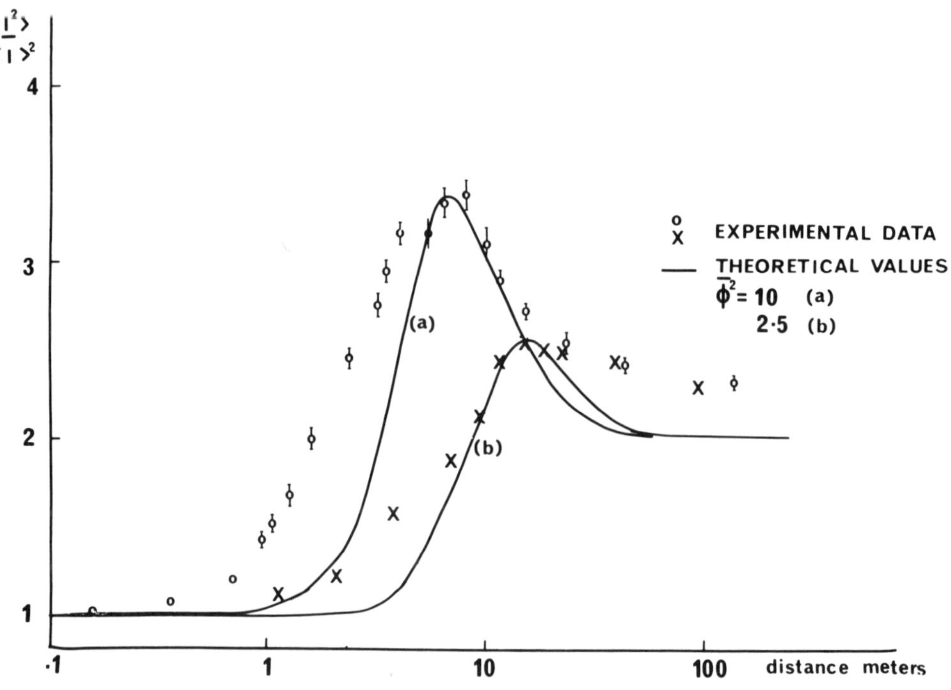

Figure 2. Normalized second moment $n^{[2]} \equiv \langle I^2\rangle/\langle I\rangle^2$ as a function of corrected screen-detector distance Z.

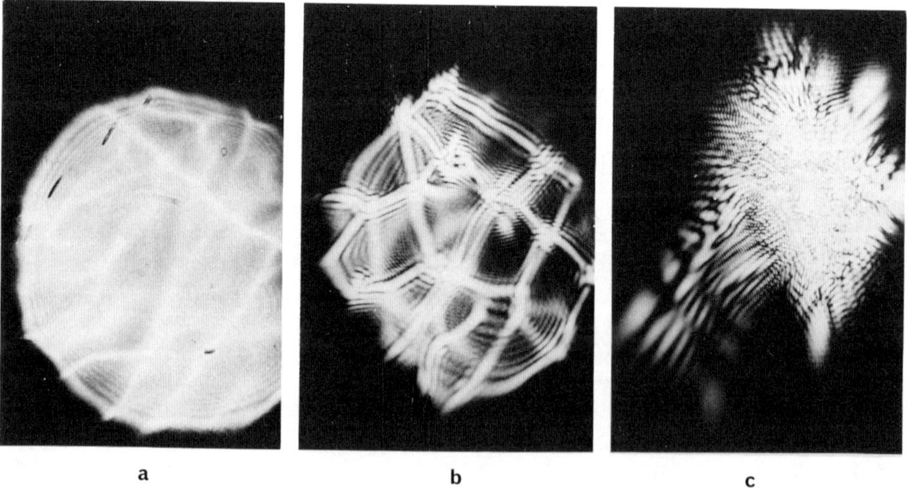

Figure 3. Short-exposure photgraphs of the intensity pattern at various distances Z from the phase screen: (a) Z = 0.7 m ($\ll Z_F$), (b) Z = 5.9 m ($\sim Z_F$), (c) Z = 37 m ($\gg Z_F$).

When many phase inhomogeneities contribute to a single detection point $Z \gg Z_F$, the illumination pattern tends towards the random interference or speckle pattern seen in the centre of Figure 3c. In this case, by the central limit theorem, negative exponential intensity statistics (Gaussian field distribution) are expected with $n^{[m]} = m!$, $n^{[2]} = 2$.

The theory involved averaging the fourth power of the scattered field (determined from the Kirchhoff integral) over the statistical properties of the phase screen [4,5]. It was assumed that the phase was joint-Gaussian distributed with a Gaussian spatial correlation function. The multiple integral was evaluated by a combination of analytical and numerical techniques. In these calculations it was possible to identify "many-scatterer" terms responsible for speckle and "single-scatterer" terms describing focussing effects. The calculations extend to much higher values of ϕ^2 than hitherto. The grosser features of theory and experiment are very similar and are not consistent with the theoretical predictions of Furuhama [6] and others based on the use of an unmodified power law phase autocorrelation function. However, significant differences in detail between the experimental results and our own predictions are clearly in evidence. These suggest that the wide range of eddy size present in the turbulence is not satisfactorily modelled by a Gaussian phase autocorrelation function.

4. PROBABILITY DISTRIBUTION $P(n,T)$

The experimental data for $P(n,T)$ can be compared with various two-parameter theoretical models for the intensity probability distribution $P(I)$. The normalized factorial moments of $P(n,T)$ are compared with the normalized moments of $P(I)$ with parameters chosen to give the same first and second moments. The log-normal intensity distribution, which figures prominently in theories of propagation through turbulence, only fitted the data over a limited range of $Z(Z \lesssim Z_F)$ and appears to have little fundamental significance in this problem.

For $Z > Z_F$ good fits of $P(n,T)$ were obtained with the W-distributions [7]:

$$P(n,T) = \left(\frac{Q}{<n>}\right)^{Q/2} \frac{\Gamma(Q+n)}{\Gamma(Q)} \exp\left(\frac{Q}{2<n>}\right) W_{-((Q/2)+n),\frac{1}{2}(Q-1)}\left[\frac{Q}{<n>}\right] , \qquad (3)$$

where $<n>$ is the mean photon count in time T, $W_{k,\ell}$ is the Whittaker function and Q is a variable parameter related to the second moment through:

$$n^{[2]} = 2 + \frac{2}{Q} \quad .$$

The W-distributions derive, by use of the Mandel relationship, from modified Bessel function (K_ν) distributions for the intensity [7,8]:

$$P(I) = \frac{2Q^{(Q+1)/2}}{<I>\Gamma(Q)} \left[\frac{I}{<I>}\right]^{(Q-1)/2} K_{Q-1} \left[2\sqrt{\frac{QI}{<I>}}\right] \quad .$$

These are exact solutions of a two-dimensional random-walk model in which the field received at the detector is the vector sum of a *finite* number of randomly-phased components of fluctuating amplitude.

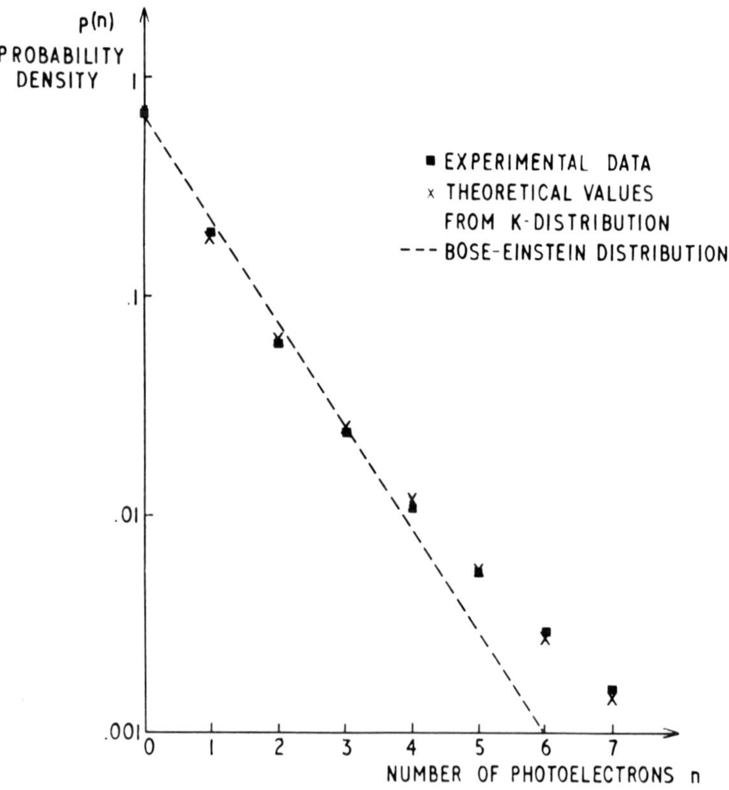

Figure 4. Photocount probability distribution $P(n,T)$ for $Z = 6.43$m. The theoretical values are equation (3) with $<n> = 0.511$ and $n^{[2]} = 3.29$. The dotted line is the Bose-Einstein photocount distribution (with the same value of $<n>$), which would obtain if the intensity were negative-exponentially distributed.

While the particular form assumed for the probability distribution of amplitude fluctuations cannot be fully justified at present, the general model is obviously reasonable for $\overline{\varphi^2} \gg 1$ and $Z > Z_F$, where the detector can receive contributions from an increasing number of phasefront inhomogeneities as $Z \to \infty$. Figure 4 shows a comparison between experimental data and a W-distribution; agreement is good.

5. CORRELATIONS

We discuss three types of correlation measurement: (A) Temporal correlations $\langle I(\underline{x}_1,t)\, I(\underline{x}_1,t+\tau)\rangle$ of the intensity at a single detection point $\underline{x}_1$; these measure the rate at which the intensity pattern evolves at or moves over the detector. (B) Correlations between two spatially separated detectors at zero delay time, $\langle I(\underline{x}_1,t)\, I(\underline{x}_2,t)\rangle$; these measure the typical size of a feature in the intensity pattern. (C) Spatio-temporal correlations $\langle I(\underline{x}_1,t)\, I(\underline{x}_2,t+\tau)\rangle$; if the scattered light field is cross-spectrally pure this correlation function can be separated into temporal and spatial components (A) and (B). However, in the absence of cross-spectral purity as in the present case where the whole intensity pattern moves bodily as well as evolving (see below), this separation cannot be made and a measurement of the spatio-temporal correlation function provides extra information not obtained from measurements of (A) and (B) alone. Of course, (A) and (B) are special cases of (C). All the data discussed below were taken with $\phi^2 \stackrel{\sim}{\scriptstyle\backsim} 10$.

The Near Field, $Z \ll Z_F$

In the near field, each point in the intensity pattern is associated with a region of the phasefront emerging from the screen of typical dimension $\ll \xi$, the phase correlation length. Thus it can be expected that the time-dependence of the intensity pattern will reflect simply that of the local emergent phase. This is, in turn, determined by motions within the turbulent region itself.

We consider first measurements of type (C) with the detector separation $\underline{d}$ in the direction of the flow (the vertical). Figure 5 shows the results of measurements for a series of detector separations $0 \leqslant d \leqslant 1$ mm. For $d \neq 0$ a peak is found in the correlation function at non-zero delay time. This indicates an overall motion of the intensity pattern and therefore of the instantaneous structure of the turbulence: the intensity feature which was at the first detector at time 0 arrives at the other detector at a later time. The reduction in correlation peak height with increasing d is a measure of the evolution of this intensity feature in transit between the two detectors and hence of the evolution of the turbulence within

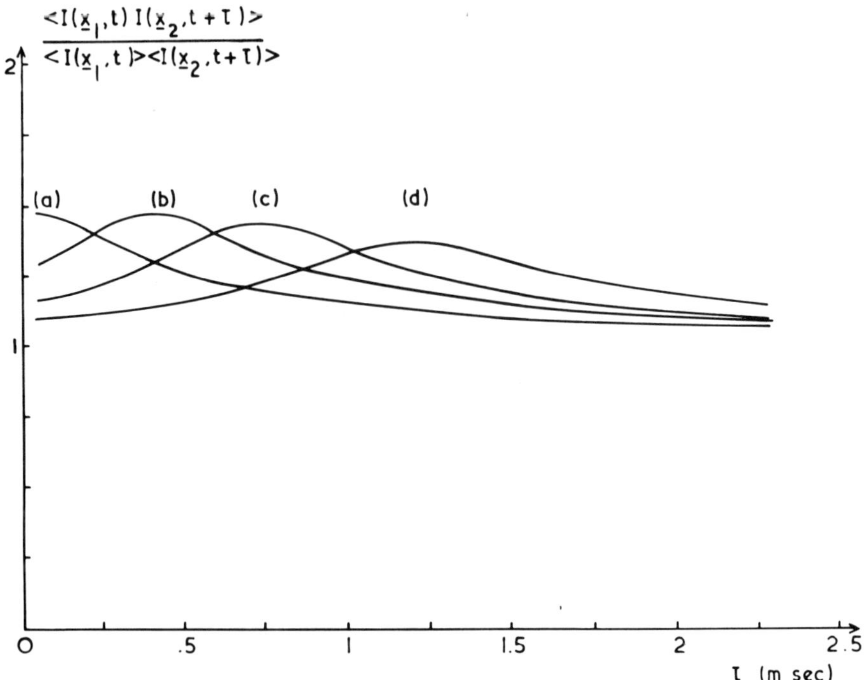

Figure 5. Spatio-temporal correlation functions for $Z = 1.08$ m and various (vertical) detector separations; (a) $d = 0$ mm, (b) $d = 0.34$mm, (c) $d = 0.64$ mm, (d) $d = 1.04$ mm.

the flow. Provided this evolution is not too rapid, the mean velocity of the flow is given by $\overline{V} \sim d/\tau_d$ where τ_d is the value of the correlation delay time at the peak. Thus by considering both peak positions and heights we conclude that a typical turbulent eddy in this flow travels with a mean velocity of about 0.85 m/s and decays after traversing one or two eddy dimensions ξ.

The $\tau = 0$ intercepts on Figure 5 constitute measurements of type (B). These are shown in curve (a) of Figure 6. Also shown are the results of measurements with detector separation in the horizontal direction. Here no displaced peak in the full spatio-temporal correlation function was found indicating the absence of a horizontal component of the mean velocity. The spatial correlation measurements for $\underline{d}$ horizontal nearly superimpose on those for d vertical. This implies that, despite the existence of a mean flow, the spatial structure of the turbulence is, on average, nearly isotropic. The half-width at half-height, ~ 0.4 mm, of the spatial coherence curve can be interpreted as the average "half-width" of

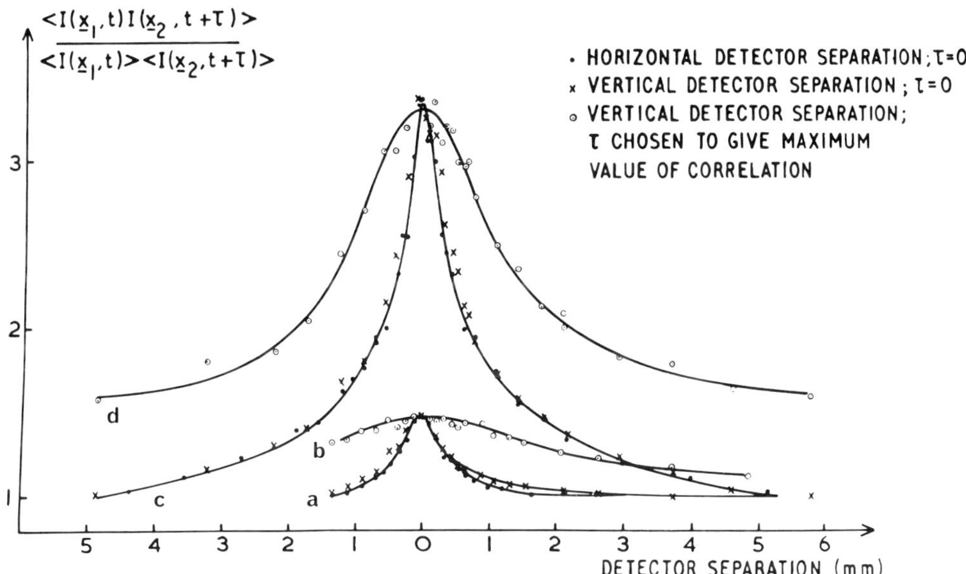

Figure 6. Curves (a) and (c): spatial correlation measurements for
(a) Z = 1.08 m and (c) Z = 6.42 m for vertical and horizontal detector separations d. (d is the "corrected" value obtained from equation (2)). Curves (b) and (d), not discussed in the text, show the
values of the maxima in the spatio-temporal correlation functions
for vertical d. The half-widths at half-height of these curves
provide a measure of the distance over which a feature in the moving
intensity pattern evolves significantly.

the bright lines in Figure 3(a). Curve (a) of Figure 5 is a measurement of type A. The correlation time of the light, about 0.4 ms,
is roughly the time taken by a line in Figure 3(a) to cross a "point"
detector.

The Focussing Region $Z \simeq Z_F$

In the focussing region each point in the intensity pattern
arises on average, from an area of emerging phasefront of dimension
$\sim\xi$. Thus the local motions of the intensity pattern can depend on
the relative motions of spatially separated regions of the flow and
the relationship between the time-dependences of the intensity pattern and the flow is likely to be more complicated than in the near
field. Nevertheless, displaced peaks were still observed in the
spatio-temporal correlation functions, indicating overall motion of

the intensity pattern. Figure 6 curve (c) shows the results of
spatial correlation measurements for $\underline{d}$, both parallel and perpendicular to the flow. Again isotropy of the intensity pattern is
indicated. The function has a narrow peak near d = 0 as well as
extensive wings. This, perhaps, reflects the average shape of a
diffraction-broadened caustic.

The Far Field, $Z \gg Z_F$

Far enough away from the phase screen the field or intensity
at a particular detection point can receive contributions from the
whole illuminated region of the scatterer. Then, as mentioned
previously, if $\overline{\phi^2} \gg 1$ and a large number of eddies is illuminated,
a "speckle" intensity pattern is expected with negative exponential
statistics P(I). We found little evidence of overall motion of the
intensity pattern. The one-detector temporal correlation function
of the intensity decayed in 1 or 2 ms, much faster than the transit
time of the flow through the beam, $W/\overline{V} \sim 25$ ms. This decay can
therefore be attributed to evolution of the turbulence in the flow.
(In the general case, "moving speckle" effects as discussed by
Jakeman [9] and Pusey [10] could occur even for $Z \gg Z_F$.)

The speckle size, determined by spatial correlation measurements, was found to be $\sim Z'\lambda/W$, dependent solely on the geometry of
the experiment, as expected in the far-field (see e.g. reference
[7]).

For detector separations comparable to the extent of the whole
intensity pattern it was found that the normalized spatial corre-
lation coefficient $\langle I(\underline{x}_1,t)\ I(\underline{x}_1+\underline{d},t)\rangle/\langle I(\underline{x}_1,t)\rangle\langle I(\underline{x}_1+\underline{d},t)\rangle$ could
be less than one, implying anticorrelation of the intensities at
the two detectors. This probably arises from "dancing" of the whole
light beam due to residual long-range correlations in the turbulence.
Thus if, at any instant, the beam is "on" one detector it is unlikely to be on the other, and anticorrelation results.

6. CONCLUDING REMARKS

We have reported what appear to be the first detailed measurements of radiation scattering by a phase screen which cover the
whole range of screen-detector distance $0 < Z < \infty$. We have identified two distinct mechanisms leading to intensity fluctuations in
the scattered radiation: diffraction-broadened, geometrical-optics
focussing effects which dominate near tc the screen and the familiar
random interference or speckle effects which dominate in the far
field. Two-detector spatio-temporal correlation measurements have the

potential to provide detailed information concerning the structure
of turbulent flows, e.g. the mean flow velocity, the rate of evolution of the turbulence and its degree of isotropy. Measurements
of this type may prove valuable in situations where more conventional
laser Doppler velocimetry techniques fail due, for example, to ex-
cessive wandering of the probing laser beam.

References

1. H.G. Booker, J.A. Ratcliffe and D.H. Shinn, Phil. Trans. Roy.
    Soc. *242*, 579 (1950).
2. L.T. Little and A. Hewish, Mon. Not. R. Ast. Soc. *134*, 221,
    (1966) and *138*, 393, (1968).
3. E. Jakeman, E.R. Pike and P.N. Pusey, Nature *263*, 215 (1976).
4. G. Parry, P.N. Pusey, E. Jakeman and J.G. McWhirter, Opt.
    Commun. *22*, 195 (1977).
5. E. Jakeman and J.G. McWhirter, J. Phys. A(GB) *10*, 1599 (1977).
6. Y. Furuhama, Radio Science *10*, 1037, (1975).
7. P.N. Pusey, in *Photon Correlation Spectroscopy and Velocimetry*,
    eds. H.Z. Cummins and E.R. Pike, (Plenum, New York, 1977).
8. E. Jakeman and P.N. Pusey, IEEE Trans. Antennas Propagat.
    AP-24, 806, (1976).
9. E. Jakeman, J. Phys. A(GB)*8*, L23, (1975).
10. P.N. Pusey, J. Phys. D(GB)*9*, 1399, (1976).

DOPPLER-FREE AND ULTRA-HIGH RESOLUTION SPECTROSCOPY

T. W. Hänsch and A. L. Schawlow

Stanford University, Stanford, California

We will report on recent progress in the development and improvement of laser methods which can overcome the Doppler broadening of
spectral lines of atoms and molecules.

Doppler-free two-photon excitation of the metastable hydrogen
2S state from the 1S ground state [1,2] promises ultimately a
(natural line-width limited) resolution of better than 1 part in
10^{15}. Intense light at the required wavelength of 2430Å can be
generated by frequency doubling of short dye laser pulses, but the
resolution is at best Fourier transform limited to about 1 part in
10^7. This difficulty may be overcome by coherently exciting the
atoms with two or more light fields, separated in space or time,
and leading to narrow spectral interference fringes (optical Ramsey
fringes). Very encouraging results have been achieved in preliminary
experiment at Stanford, [3] studying two-photon excitation of sodium
atoms inside an optical resonator with an injected recirculating
laser pulse which appears as a train of standing wave light pulses.
We have observed not only narrow fringes, much below the Fourier-
transform width of an individual pulse, but also a strong resonant
signal enhancement, proportional to the square of the number of
pulse roundtrips during the transverse atomic relaxation time.
Application of this technique to the hydrogen 1S-2S transition
promises accurate new measurements of fundamental constants and
stringent tests of quantum electrodynamics and special relativity.

The 1S-2S two-photon spectrum can be used to determine the
Lamb shift of the 1S ground state, by comparing it with a Doppler-
free spectrum of the Balmer beta line at 4860Å which is simultaneously recorded with the fundamental dye laser output. Initial

363

experiments recorded the Balmer line by saturated absorption spectroscopy, [2] but much improved sensitivity and resolution has more
recently been achieved with a new technique, laser polarization
spectroscopy, [4] which monitors the nonlinear interaction of two
counterpropagating monochromatic laser waves in an absorbing gas
via changes in light polarization rather than intensity. The effective suppression of fluctuating background probe laser light with
a blocking polarizer permits the observation of fewer atoms at
lower intensities. The advantages and limitations of this technique
will be discussed, and the status of a new 1S Lamb shift measurement,
using this method, will be reported.

Laser spectroscopy can also be used to unravel the complexities
of molecular absorption spectra. [5,6] In the polarization method,
a selected ground state level is labeled by orienting the rotation
axes of the molecules through optical pumping and all absorption
lines from the same lower state are identified through their resulting dichroism and birefringence. Doppler-free polarization spectroscopy of molecular sodium, using an accurate fringe-counting digital
wavemeter, is being used in our laboratory to determine improved
molecular structure parameters.

References

1.   T. W. Hänsch, S. A. Lee, R. Wallenstein, and C. Wieman, Phys.
      Rev. Lett. *34*, 307 (1975).
2.   S. A. Lee, R. Wallenstein, and T. W. Hänsch, Phys. Rev. Lett.
      35, 1262 (1975).
3.   R. Teets, J. Eckstein, and T. W. Hänsch, Phys. Rev. Lett.,
      to be published (1977).
4.   C. Wieman and T. W. Hänsch, Phys. Rev. Lett. *36*, 1170 (1976).
5.   M. E. Kaminsky, R. T. Hawkins, F. V. Kowalski, and A. L.
      Schawlow, Phys. Rev. Lett. *36*, 671 (1976).
6.   R. Teets, R. Feinberg, T. W. Hänsch, and A. L. Schawlow, Phys.
      Rev. Lett. *37*, 683 (1976).

TRANSIENT TWO-PHOTON COHERENT EFFECTS[*]

M. M. T. Loy

IBM-T.J. Watson Research Center, Yorktown Heights, N.Y.

Due to the availability of high power, single-mode laser sources,
it is now possible to study coherent interactions of light with
atoms and molecules via multi-photon processes. Here, we will describe the first experimental observation of transient two-photon
coherent precession - one of the many possible two-photon analogs
of well known one-photon coherent effects which have been predicted
by the two-photon vector model of Grischkowsky, Loy and Liao. In
terms of this vector model, the process can be visualized in a way
similar to its familiar one-photon analog: the two-photon polarization vector precesses about the effective field and, depending on
their relative phases, induces alternating absorption and emission
of light. There are, in addition, features unique to the two-photon
case, for example, the optical Stark effects and the Doppler width
reduction from counter-propagating beams are shown to be of crucial
importance in the experiment.

We used a pulsed, strong beam with a duration of 100 nsec and
peak intensity of 1 MW/cm^2 from a single-mode CO_2 TEA laser tuned
to the P18 line, together with a weaker (2 kw/cm^2) at the CO_2 P34
line. This weak beam, counter-propagating to the strong beam, has
a much longer pulse duration and can be considered CW in the experiment. The NH_3 two-photon transition used was (v_2J,K,M) =
$(0^-,5,4,\pm5) \rightarrow (2^-,5,4,\pm5)$ with the uniquely defined intermediate
state $(1^+,5,4,\pm5)$. This transition, at low laser intensities is
294 MHz away from the sum of the laser frequencies. However, by
the judicious choice of the TEA laser intensity and the M^2 dependence of the optical Stark effect, the M = ±5 states were selectively
shifted into resonance by the TEA laser intensity. With the weak
beam near resonant to the transition between the ground and intermediate states, and the strong beam near resonant to the transition

between the intermediate and the final states, parasitic one-photon
coherent effects were effectively eliminated. In the experiment,
the laser intensities were chosen so the two-photon transition was
shifted into resonance at the peak of the TEA laser pulse. The
precession signal was then observed at the trailing edge of the
pulse by monitoring the weak beam. The characteristic precession
signal - the alternating absorption and emission of light - was
unmistakable. A very interesting feature was that the precession
frequency increased with time, with the precession period decreasing from 20 nsec to 5 nsec; this was due to the fact that at the
trailing edge of the TEA laser pulse, the molecules were brought
away from resonance due to the decreasing Stark shift accompanying
the decreasing laser intensity. Note also that the observation
of the precession periods would have been impossible without the
Doppler width reduction since these periods were much longer than
the normal T_2^* of a few nsec.

An important advantage of the $CO_2 - NH_3$ system used in the
experiment is that all parameters relevant to the two-photon vector
model are known. A quantitative comparison between the prediction
of the model and the experimental result is therefore possible.
Without any free parameters in the calculation, we found excellent
agreement between theory and experiment. This is the first experimental verification of the two-photon vector model.

The decay of the coherent precession signal as a function of
gas pressure gives valuable information on the collision-induced
relaxation time of the two-photon polarization. For this measurement it is desirable to have reproducible frequency sweeping which
is not highly sensitive to the TEA laser intensity as in the above
experiment. Thus, instead of relying on the optical Stark effect
to shift the two-photon transition on resonance, we performed the
measurement in a Stark cell where the transition frequency can be
electronically controlled. Care was taken to keep the optical
Stark shifts small compared to the external field induced Stark
shifts. The two-photon coherent precession signal thus obtained
was similar to that observed in the above experiment. But unlike
the earlier experiment, stable coherent precession signal can now
be repeatably obtained in spite of laser intensity variations.
The collision-induced relaxation time was measured as a function
of NH_3 pressure, and also as a function of buffer gas pressure at
fixed NH_3 pressures. Results of our relaxation time measurements
will be discussed and compared with earlier pressure-broadened
linewidth measurements. Possible observation of other two-photon
coherent effects using this setup will also be discussed.

*
Work partially supported by the U.S. Office of Naval Research.

SEPARATED OPTICAL FIELDS IN DOUBLE QUANTUM TRANSITIONS*

M. M. Salour

Harvard University, Cambridge, Mass.

Interactions of atoms and molecules with intense monochromatic laser standing waves can give rise to very narrow Doppler-free two-photon resonances. This new high-resolution spectroscopic method has recently been extensively applied to the study of several atomic and molecular optical transitions[1]. The discovery and accurate measurement of these narrow, frequency-stable resonance lines in the emission and absorption spectra of substances, are significant because advances in this area will not only provide direct clues to the innermost processes of matter, but may also lead to novel applications in many areas of science.

The ultimate resolution achievable by the method of Doppler-free two-photon spectroscopy is limited in principle only by the natural linewidth of the excited state; in practice, however, the observed linewidth has thus far been limited by the laser. Thus, in order to take full advantage of the elimination of the Doppler width, the spectral linewidth of the laser light must be as small as possible, which explains the motivation for using cw lasers. But, on the other hand, two-photon resonances require an appreciable intensity, and it is well known that pulsed dye lasers deliver higher powers and furthermore, cover a broader spectral range. Recently, by using pulsed dye lasers pumped with an N_2 laser, it has been possible to observe two-photon resonances in several highly excited states of various atomic vapors which could not have been reached by cw excitation[2]. The resolution was limited by the spectral width $1/\tau$ of the pulse (τ: duration of each pulse), thus making it impossible to resolve many closely space two-photon resonance lines.

We have recently demonstrated[3] that, by exciting atoms with
two time-delayed coherent laser pulses, one can obtain interference
fringes in the profile of the Doppler-free two-photon resonances
with a splitting 1/2T (T: delay between the two pulses) much smaller
than the spectral width $1/\tau$ of the laser pulse. (τ: duration of each
pulse). This technique combined the advantages of pulsed dye lasers
(power, spectral range) with the high resolution usually associated
with a cw excitation, and could be considered as an extension, to
two-photon Doppler-free resonances in the optical range, of the well-
known Ramsey method of using two separated RF or microwave fields in
atomic beam experiments[4].

Such an idea was first suggested in a slightly different context
by Baklanov, Chebotayev, and Dubetskii[5], who proposed using two
spatially separated cw light standing waves. These authors considered two-photon Doppler-free resonances between two very sharp
levels, such as ground states or metastable states (an important
example being the $1S_{\frac{1}{2}}-2S_{\frac{1}{2}}$ transition of H). In such a case, the use
of a single laser beam leads to linewidths which are generally
limited by the inverse of the transit time of atoms through the laser
beam. Rather than increasing this time by expanding the laser beam
diameter (with a consequent loss of intensity), Baklanov et al.
proposed using two spatially separated beams to obtain structures,
in the profile of the resonance, having a width determined by the
time of flight between the two beams. The experiment described
below deals with short-lived atomic states (lifetime $\sim 5 \times 10^{-8}$ sec),
so that the transit time through the laser beam ($\sim 10^{-7}$ sec) plays
no role in the problem. Consequently, we use two time-delayed short
pulses instead of two spatially separated beams[6].

The interference structure appearing in the resonance profile
can be simply understood in the following way. Consider first a
single pulse of the form E(t) exp(-iωt) (τ: duration of the pulse).
The Fourier analysis of such a pulse leads to a frequency spectrum
spread over an interval $1/\tau$ around ω. This non-monochromatic
excitation acting upon the atom leads to a Doppler-free two-photon
resonance having a width $1/\tau$. Suppose now that E(t) consists of
two pulses of duration τ, separated by a delay T. Just as the
diffraction pattern through two spatially separated slits exhibits
interference fringes within the diffraction profile corresponding
to a single slit, the Fourier transform of this electric field
exhibits interference fringes within the spectrum profile of a
single pulse, with a frequency splitting of 1/T between two fringes.

Two important requirements must be fulfilled in order to obtain
such interference fringes in the profile of the two-photon resonance.
First, each pulse must be reflected against a mirror placed near the
atomic cell in order to expose the atoms to a pulsed standing wave
and in this way to suppress any dephasing factor due to the motion
of the atoms. The probability amplitude for absorbing two

counterpropagating photons is proportional to $\exp[i(\omega t - kz)]$
$\exp[i(\omega t + kz)] = \exp(2i\omega t)$ and does not depend on the spatial position
z of the atoms. Second, the phase difference between the two pulses
must remain constant during the entire experiment. Significant phase
fluctuations between the two pulses will wash out the interference
fringes, since any phase variation produces a shift of the whole
interference structure within the diffraction background. To avoid
such fluctuations, the experiment must be done not with two independent pulses, but with two time-delayed pulses having a constant
phase difference during the entire experiment. Furthermore, these
two pulses have to be Fourier-limited, that is, their coherence
times must be not shorter than their duration.

From a more physical point of view, one can think of a two-
level system, $|g\rangle$ and $|e\rangle$. After the first pulse the atomic state
is a coherent superposition of the two levels. Atoms in this superposition freely precess and, depending on the point in time at
which the delayed pulse arrives (while the excited atom is still
freely precessing), one can see either constructive or destructive
interference with the atomic precession.

Note again that it is important that the initial and the delayed
pulses be phase-locked. Otherwise, when the delayed pulse arrives,
the random phase difference between initial and delayed pulse will
give rise randomly to constructive or destructive interference, and
when averaged the interference signal will be washed out.

One method for obtaining such a sequence of two phase-locked
and Fourier-limited pulses is to start with a cw dye laser wave
having a very long coherence time and to amplify it with two time-
delayed pulsed amplifiers. Even if the cw laser delivers a very
weak intensity in the spectral range under study, with a sufficient
number of synchronized pulsed amplifiers one can obtain enough power
to observe the two-photon resonance. The time dependence of the
amplified wave has the shape represented by the solid lines of
Fig. 1-I. The pulses are two portions of the same sinusoid (represented in dotted lines), and have the same phase, that of the cw
carrier wave. In such a case, a very simple calculation [7] shows
that the Doppler-free probability $|b_2|^2$ for the atom to be excited
in the upper state after the second pulse is related to the
probability $|b_1|^2$ of excitation after the first pulse by:

$$|b_2|^2 = 4|b_1|^2 [\cos(\omega_0 - 2\omega)T/2]^2 \ ,$$

where ω_0 is the Bohr frequency of the atomic transition, ω the laser
frequency and T the time delay between the two pulses. If ω is
varied, with ω_0 and T held constant, the diffraction profile
associated with $|b_1|^2$ includes interference fringes described by
the last oscillatory term, with a splitting in ω determined by

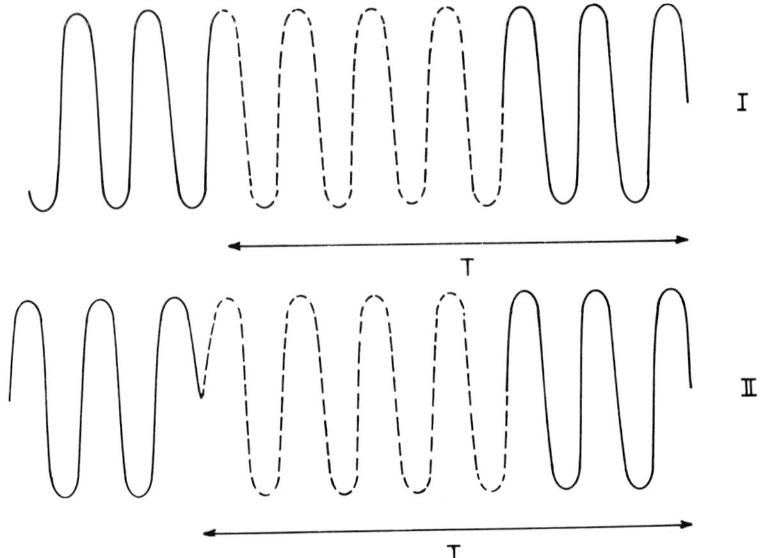

Fig. 1. (I) Two Fourier-limited time-delayed pulses generated by
amplifying the same cw wave by two time-delayed independent pulsed
amplifiers. The two pulses clearly have the same phase, that of
the cw carrier. (II) Two pulses originated from the same Fourier-
limited pulse by using a delay line. The second pulse is the time
translation of the first one by an amount T. The phase difference
between the two pulses clearly depends on ωT

$1/T_{eff}$, where $T_{eff} = 2T$. The central fringe is exactly centered at
half the Bohr frequency $(\omega=\omega_0/2)$. As long as the two pulses are
phase-locked, any small variation of T is not important, since it
does not change the position of the central fringe.

A second method, which is experimentally simpler and which we
have used[3], is to start with a Fourier-limited pulse obtained by
amplifying a cw wave and to generate two pulses from it in an optical
delay line. The time dependence of the resulting wave has the shape
represented by the solid lines of Fig. 1-II. The second pulse is
the time translation of the first one by an amount T, $(T>\tau)$.
However, the sinusoid extrapolated from the first pulse (dotted
lines) does not generally match the second pulse, which means that
the two pulses do not generally have the same phase, unless T is
an integral number of optical periods $2\pi/\omega$ [i.e., $\exp(i\omega T) = 1$].
The phase difference between the two pulses is ω dependent, in such
a way that the interference fringes would disappear if ω is varied
with T fixed[8]. One possible scheme (which will be described later)
for locking the phases of the two pulses, and consequently causing

the interference fringes to reappear, is to simultaneously vary T
and ω in such a way that $\exp(i\omega T)$ remains equal to one.

The interference fringes discussed above are useful when the
duration τ of the pulse is shorter than the radiative lifetime τ_R
of the excited state. By choosing T of the order of τ_R, one can
obtain narrow fringes in the profile of the two-photon resonance
with a good contrast. Note that, in principle, it would be possible
to choose T longer than τ_R, in which case the resolution would be
better than the natural linewidth. In such a case, however, only
atoms with a lifetime at least equal to T would contribute to the
interference effect, so the fringes would have a very poor contrast.

Figure 2 shows our experimental setup. We utilize a single mode,
pressure-tuned and servo-looped, cw dye laser (Spectra Physics 580A,
pumped by a Spectra Physics 164 Ar^+ laser) which is tuned to the
3^2S-4^2D two-photon transition in sodium (5784.3 Å). (Note: lifetime
$4^2D \sim 5 \times 10^{-8}$ sec.) The output of this cw dye laser (40 mW, 2 MHz)
is then amplified in three synchromized stages of dye amplifiers
pumped by a one-megawatt nitrogen laser. (Three stages of amplification are necessary to boost the peak output power to an
acceptable level for this experiment.) To maximize the length of
the output pulse, the pump light for the amplifiers is geometrically
divided between the amplifiers and optically delayed in order to
arrive slightly later than the input signal for the cw dye laser.
To avoid undesirable amplified spontaneous emission from the amplifiers, suitable spectral and spatial filters are inserted between
stages. (A more detailed description of this oscillator-amplifier
dye laser system has appeared elsewhere.[9]) The 75 kW output of
the third stage amplifier is split into two parts, one of which goes
into a reference sodium cell, and the other is further split into
two parts. The first of these goes directly into a sample sodium
cell, and the second is optically delayed in a delay line by a time
T(2T = 17, 25, 33 nanoseconds) and subsequently recombined with the
first and focused into the sample sodium cell.

We have stabilized the delay line by locking the center fringe
of the two cw carrier beams to the motion of the mirror M which is
mounted on a piezoelectric ceramic. The center fringe is at
maximum when the direct and the delayed cw carrier beams are in
phase, i.e., when $\exp(i\omega T) = 1$. In this way, the two time-delayed
and Fourier-limited pulses are phase-locked during the entire
experiment. The delayed beam, with its longer optical path, has
wavefronts with a larger radius of curvature than those of the direct
beam, so a high quality negative lens (L) is used to match the
curvature of the two wavefronts. Both sodium cells are kept at
$130°C (2 \times 10^{-6}$ Torr of vapor pressure); they were made of Corning 1720
calcium aluminosilicate glass with optically flat windows and were
baked at high temperature and later filled with sodium under very

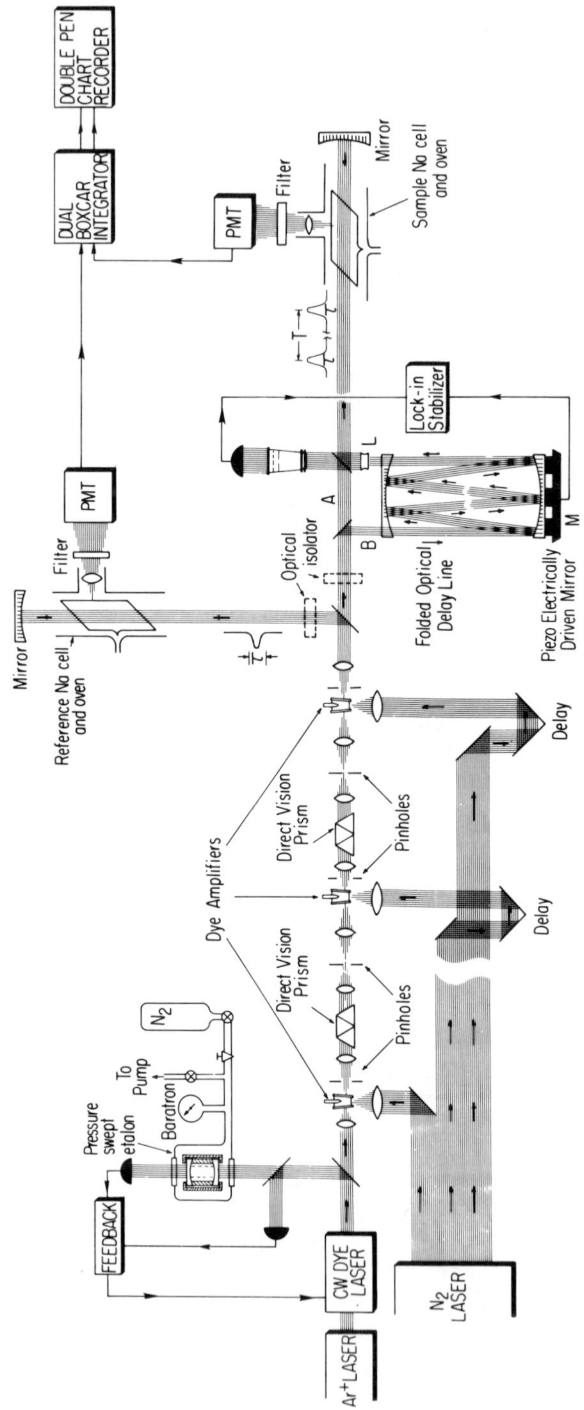

Fig. 2 : Schematic of the experimental setup.

high vacuum to avoid any foreign gas contamination. To generate
the laser standing waves with short pulses, the back reflecting
mirror is adjusted so that the time delay between the original pulse
and the back reflected pulse is significantly shorter than the pulse
duration τ. The outputs of the two photomultipliers (EMI 9635QB)
monitoring the fluorescence from 4P to 3S at 3303 Å are simultaneously
processed by a dual channel boxcar integrator, and the results are
recorded simultaneously by a dual pen chart recorder.

Figure 3-I shows the four well-known two-photon resonances of
the 3^2S-4^2D transition of Na, observed in the reference cell with
a single pulse. Figure 3-II shows the same four resonances observed
in the sample cell which is excited by the two time-delayed coherent
pulses with $2T \simeq 17$ nsec. An interference structure clearly appears
on each resonance. We have verified that the splitting between the
fringes is inversely proportional to the effective delay time T_{eff}
($T_{eff} = 2T$) between the two coherent pulses. Figures 3-III and
3-IV show the same experimental traces as those of Figure 3-II,
except that the effective delays are $T_{eff} = 25$ and 33 nsec,
respectively. Note that in these cases (Figs. 3-III and 3-IV) the
contrast of the fringes is progressively less than that of
Fig. 3-II, where a shorter delay was used. We have also verified
that with the delay line unlocked, the interference fringes
disappear (when ω is varied).

In order to take full advantage of the high resolution
associated with this technique, the delay time T between the two
pulses must be made as long as possible. But because of the
radiative lifetime ($\tau_R = 1/\Gamma$) of the excited state, a simple
calculation[7] shows that the contrast C of the fringes is given by:

$$C = \frac{2e^{-\Gamma \frac{T}{2}}}{1 + e^{-\Gamma T}} \ .$$

Clearly, increasing the delay time T between the two pulses
causes both the fringe spacing and the fringe contrast to decrease;
consequently, one obtains better resolution, but a smaller signal-
to-noise ratio. This can be understood by noting that when
$T > 1/\Gamma$, a smaller number of atoms excited by the first pulse can
live for a sufficiently long time to experience the second pulse;
since it is the interference due to the second pulse that provides
the high resolution, only the still-excited atoms contribute to
the signal (i.e., only those atoms that remain in the excited state
for a time at least equal to T contribute to the interference effect,
and consequently if T is large, the fringes will have a very poor
contrast).

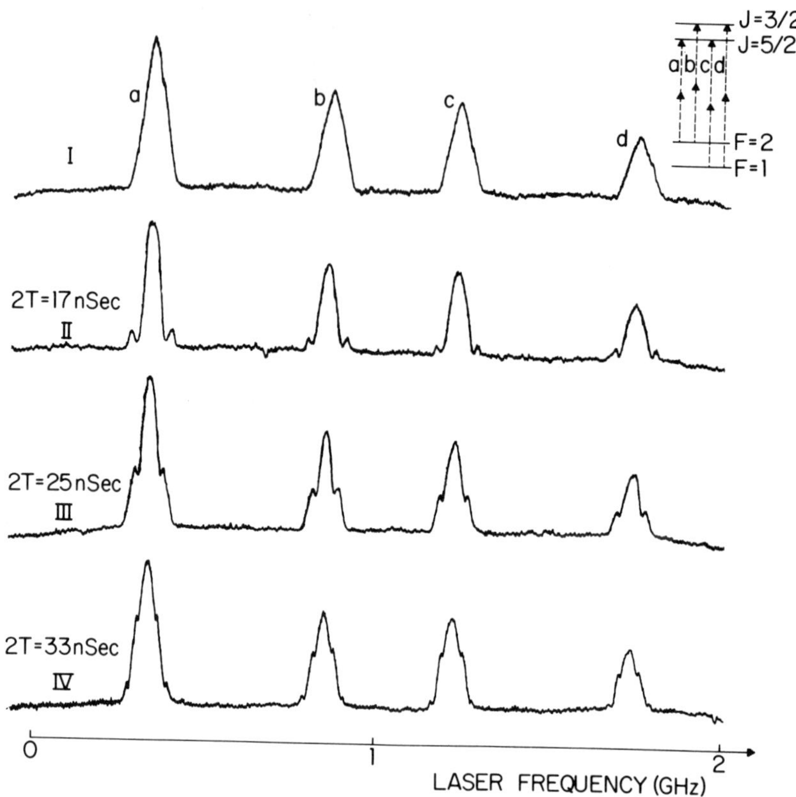

Fig. 3.   (I) Recording of the four Doppler-free two-photon resonances corresponding to the 3^2S-4^2D transitions of Na (transitions F = 2, F = 1, $3^2S_{1/2} \to 4^2D_{3/2}$, $4^2D_{5/2}$) observed in the reference cell excited with a single pulse. (II, III, and IV) Same resonances observed in the sample cell excited with two time-delayed coherent pulses. Width of each pulse, τ = 8 nsec; effective delay between the two pulses, T_{eff} = 2T = 17, 25, and 33 nsec, respectively. The spacings between the fringes, $\Delta\nu \approx$ 60, 40, and 30 MHz, respectively, are in good agreement with the theoretical prediction 1/2T.

The sum of the quantum mechanical amplitudes A_1 and A_2 for excitation by the first and second pulses, respectively, represents the total amplitude $A = A_1 + A_2$ for reaching the excited state $|e\rangle$ from the ground state $|g\rangle$. When one introduces a phase shift $\exp(i\delta)$ on the second pulse only, then the total amplitude is $A_1 + \exp(i\delta)A_2$, and the probability of reaching the $|e\rangle$ state is

$$|A|^2 = |A_1 + e^{i\delta}A_2|^2 = \underbrace{|A_1|^2 + |A_2|^2}_{\substack{\text{diffraction}\\\text{terms}}} + \underbrace{2A_1A_2 \cos\delta}_{\substack{\text{interference}\\\text{term}}} \quad .$$

In order to eliminate the diffraction background appearing in each of the four resonances of Figs. 3-II, 3-III, and 3-IV, one could induce a 90° phase shift (so that $\delta = 180°$ for the two-photon excitation) between every other pair of coherent pulses into the sample cell. By subtracting the resulting fluorescence from pairs of coherent pulses with and without the phase shift

$$|A|^2_{\delta=0} - |A|^2_{\delta=\pi} = 4A_1A_2 \quad ,$$

one can eliminate the diffraction background and thus isolate the interference fringes. Note that in this way not only does one get rid of the diffraction background, but also one doubles the interference signal automatically. Such a technique seems promising for situations where the contrast of the interference fringes is poor, for example, when T is chosen much longer than the radiative lifetime in order to get a fringe splitting smaller than the natural linewidth.

One can also demonstrate this analytically. The Doppler-free probability $|b_2|^2$ that an atom is excited in the upper state after the second pulse, when one has introduced a phase shift $e^{i\delta}$ on the second pulse only, is related to the probability $|b_1|^2$ of excitation after the first pulse by[7]:

$$|b_2|^2 = |b_1|^2 (1 + e^{-\Gamma T}) + 2|b_1|^2 e^{-\Gamma\frac{T}{2}} \cos[(\omega_0 - 2\omega)T + 2\delta] \quad .$$

$$(1)$$

The phase shift δ produces a frequency shift by $(\frac{1}{2}\pi)(\delta T)$ of the whole interference structure within the diffraction profile; if δ is chosen equal to $\pi/2$, this shift corresponds to half-a-fringe spacing. By subtracting the signals corresponding to $\delta = 0$ and $\delta = \pi/2$, one obtains

$$|b_2(\delta = 0)|^2 - |b_2(\delta = \frac{\pi}{2})|^2 = 4|b_1|^2 e^{-\Gamma\frac{T}{2}}[\cos(\omega_0 - 2\omega)T] \quad .$$

$$(2)$$

Thus not only is the diffraction background [the first term of (1)] suppressed, but the interference signal is automatically doubled. Note that the central fringe remains centered at $\omega = \omega_0/2$.

Figure 4 shows our experimental setup[10]. It is identical to the experimental arrangement of Fig. 2, except that in the arrangement discussed here the N_2 laser was externally triggered by a home-made trigger generator. The trigger generator produced a train of pulses. For simplicity, let us distinguish between the first and the second pulses by calling the first pulse A and the second pulse B. Thus, the train of pulses from the trigger generator consisted of a train of A, B pulses. (A and B pulses were each 8 nsec long, 50 msec apart.)

While a train of A, B pulses triggered the N_2 laser, a flip-flop counter discriminated A pulses from B pulses, and after a proper delay the A pulses triggered boxcar I, and the B pulses triggered boxcar II. Since each A pulse from the dye laser gave rise to a pair of A pulses at the sodium sample cell (and each B pulse similarly gave rise to a pair of B pulses), the electronic delay was adjusted so that the 100 nsec wide gates of both boxcars were opened immediately after the delayed pulses, as illustrated in Fig. 5. Meanwhile, the B pulses also triggered an EG&G KN-22 Krytron switch which turned on a Pockels cell (Interactive Radiation Model 212-080, or Isomet Model EON4203). The voltage on the Pockels cell was adjusted so that it would produce a 90° phase shift on the delayed B pulses. The Krytron switch turned slightly ahead of the delayed B pulses and turned off slightly after the delayed B pulses, as illustrated in Fig. 5; for this reason, proper care had to be taken in the design of the trigger generator so that the Pockels cell remained on before and after the delayed B pulses to allow for any possible jitter associated with the rise time of the KN-22 Krytron and the Pockels cell. The outputs of the two boxcar integrators were subtracted in a linear differential amplifier and the result was recorded on a chart recorder.

Figure 6-II shows the experimental results for a delay of 25 nsec. The upper trace (i.e., Fig. 6-I) corresponds to the signal obtained when the atoms are excited by a single pulse in the reference sodium cell. This signal not only made it possible to calibrate the frequency accurately (since the cw oscillator was frequency-stabilized to a pressure-tuned reference etalon, thus allowing a highly linear frequency scanning mechanism), but also allowed us to optimize the signal from the sample sodium cell, so that by properly adjusting the electronic gain in each channel of boxcars, any spurious signal building up between the two-photon resonances could be eliminated. The output of the reference PMT(EMI-9635 QB) monitoring the resulting $4P \rightarrow 3S$ fluorescence at 3302 Å due to single pulse excitation (train of A, B pulses) was processed by the other channel of the PAR 162 boxcar integrator whose gate was triggered just after the A and B pulses. The resulting fluorescence was recorded by the same dual pen chart recorder.

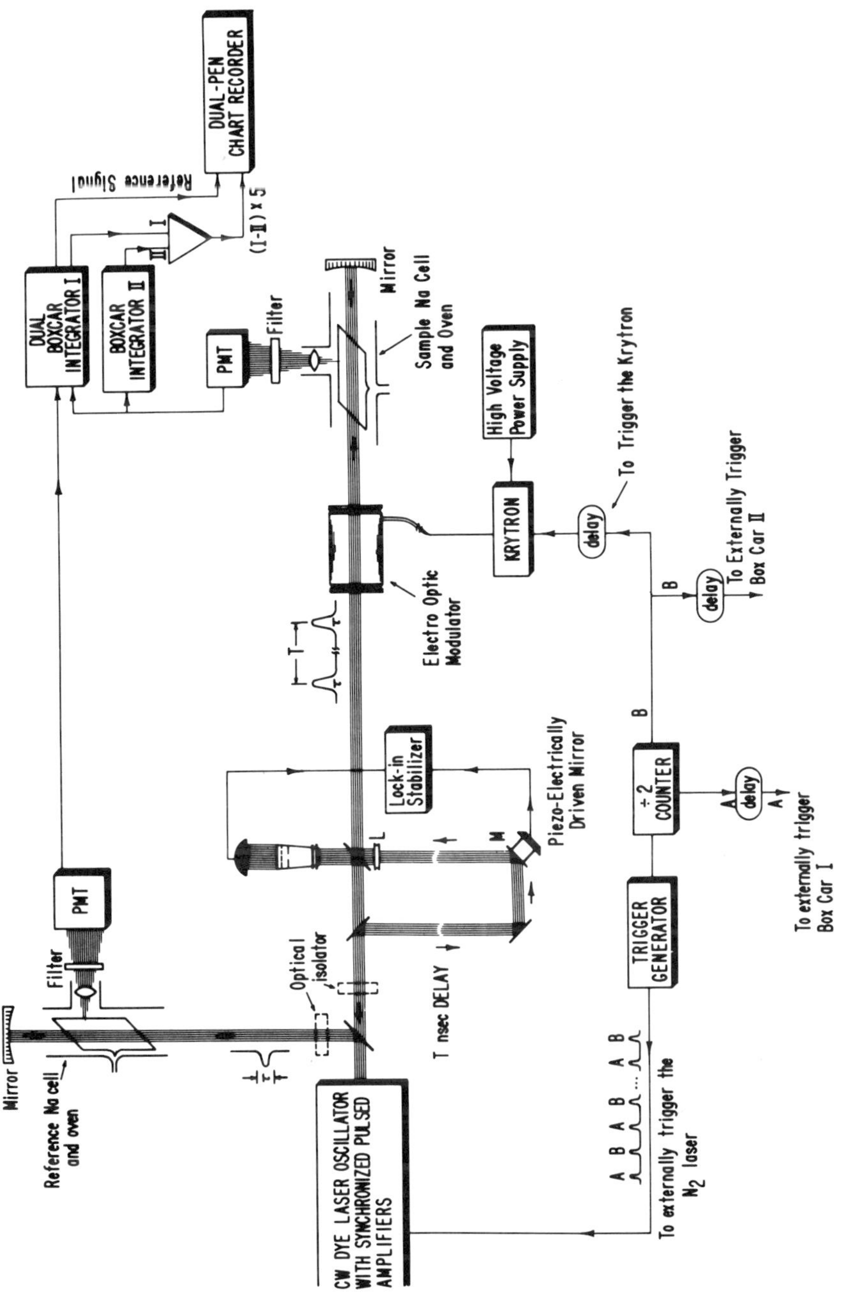

Fig. 4. The experimental setup.

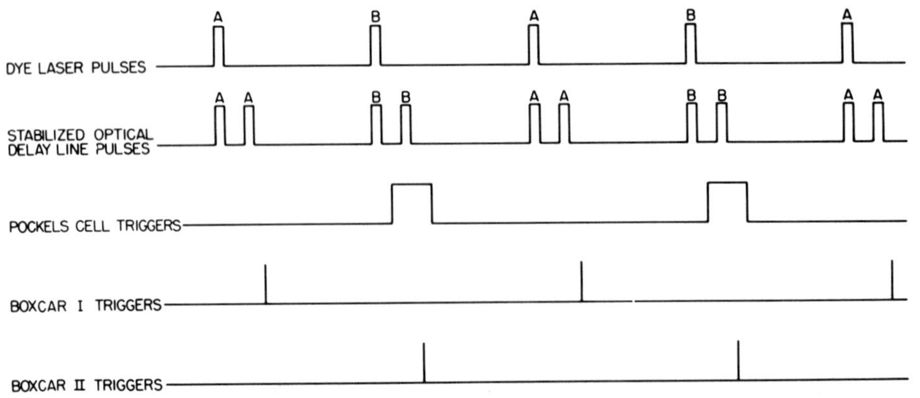

Fig. 5. The timing sequence of the electronic triggering system for triggering the Pockels cell, boxcar I, and boxcar II.

A number of important tests were performed to assure that the resulting data corresponded to real isolated Ramsey fringes. First, it was verified that no signal was observed when zero phase shift was introduced between B pulses (by turning off the high voltage supply to the Kryton). Figure 6-III shows this result. (An equivalent test would have been to induce a 180° phase shift between B pulses, in which case one would again expect a null signal. However, the electronic circuitry for triggering the Krytron did not allow us to increase the voltage sufficiently to produce a 180° phase shift. A slight modification in the electronic design should make this test possible also.) Second, by slightly varying the voltage supplied to the electro-optic modulator, and thus in effect slightly varying the phase of the phase-shifted delayed B pulses, the signal was optimized so that at the voltage corresponding to the exact 90° phase shift, the diffraction background was completely eliminated. There was still some small background in some of the experimental traces, which could be eliminated only by making a slight adjustment in the Krytron power supply voltage. We attribute this to the induced phase jitter associated with the electro-optic modulator, that is, the pulse-to-pulse jitter in the phase shift induced by the Pockels cell.

Figure 6-IV shows similar experimental results for a delay of 33 nsec. Note that in this case the fringe spacings are much smaller than in the case of T_{eff} = 25 nsec delay. It was verified that the fringe spacing was proportional to $1/T_{eff}$. In comparison with the results of Fig. 3, note the enormous gain in constrast associated with the traces of Figs. 6-II and 6-IV.

Clearly, using this technique of phase modulation between alternate coherent pulse pairs, it should be possible to achieve a resolution much smaller than the natural linewidth of the excited

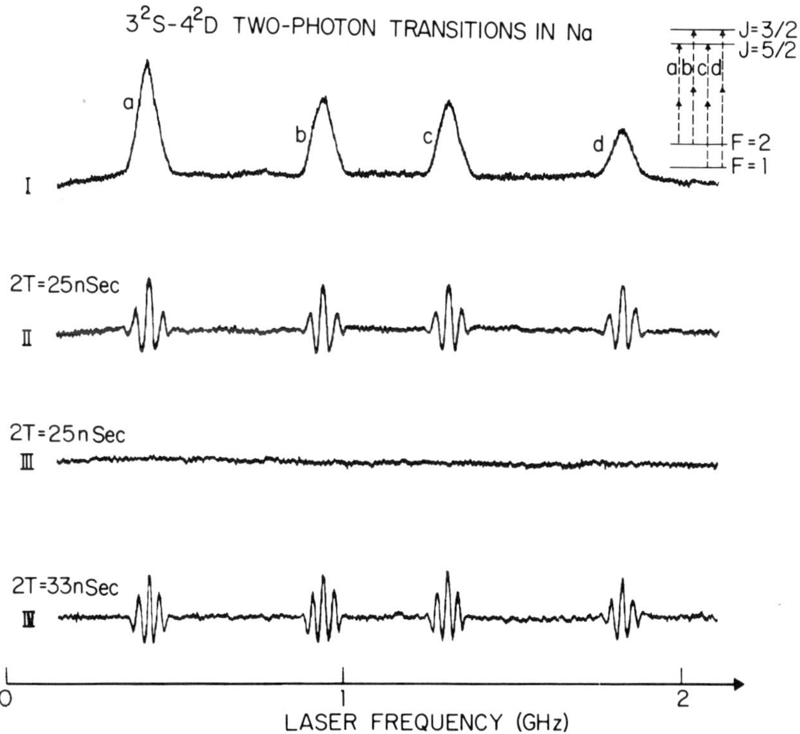

Fig. 6.   (I) The four Doppler-free $3^2S\text{-}4^2D$ two-photon resonances of
Na observed in the reference cell excited with a single pulse of
τ = 8 nsec duration.   (II) and (IV) The same resonances observed
in the sample cell excited with two time-delayed coherent pulses
(with 2T = 25 nsec and 2T = 33 nsec, respectively), where a 90° phase
shift is introduced between alternate pairs of pulses, and with
amplification of the difference between the fluorescence resulting
from pairs of coherent pulses with and without the phase shift.
(III) The same resonances and procedure as for Fig. 6-II except that
zero phase shift is introduced between alternate pairs of coherent
pulses.

state.   From Equation (2), one can note that while increasing T does
not change the contrast, it does, however, exponentially decrease
the amplitude of the signal.   The condition for resolution smaller
than the natural width is:

$$(\text{Fringe spacing}) = \frac{1}{T_{eff}} = \frac{1}{2T} < \frac{1}{2\pi\tau_R} = (\text{Natural linewidth of the excited state})$$

or

$$T_{eff} = 2T > 2\pi\tau_R \quad .$$

For the case of the 4^2D excited state, $\tau_R \simeq 50$ nsec.   Thus:

$$T > 160 \text{ nsec.}$$

In order to obtain such a delay, one would need an ultra-stable
delay line with mirrors of ultra-high optical quality, so that
locking the center fringe of the two cw beams (direct and delayed)
to the motion of the piezoelectrically driven mirror would become
less troublesome. Furthermore, for such a long delay the wavefront curvature might be a serious problem; we have observed that
when the wavefronts are not mode-matched, the contrast of the fringes
is reduced, and some asymmetry of the side fringes relative to the
position of the central fringe develops.   However, these difficulties
may be overcome with a reasonable amount of effort. A more
sophisticated detection system is needed, however, since otherwise
the signal from the small number of longer lived atoms is likely to
be hopelessly buried in laser fluctuations and detector noise.

In conclusion, we have introduced a technique which could mark
a logical beginning of a stimulating and exciting new generation of
optical experiments, in which through the use of delayed pulses,
pulse-length broadening can be overcome. Thus, pulsed dye lasers,
the only spectroscopic tool presently available in many spectral
regions, can be used for ultra-high resolution spectroscopy of
atomic and molecular systems. Our technique can be generalized.
Instead of two pulses, one could use a sequence of N equally spaced
pulses (obtained, for example, by sending an initial pulse into a
confocal resonator[11]), and thus submit atoms to a series of N
equally spaced time-delayed and phase-locked pulsed standing waves.
An argument parallel to the discussion of two-pulse interference
would show that the optical analogue of such a system is a grating
with N lines; the total length of the grating corresponds to the
total duration NT of the pulse train. In such a case, the frequency spectrum experienced by the atom would be a series of
narrow peaks, N times narrower in width and N^2 times stronger in
intensity separated by a frequency interval T_{eff}^{-1} .

ACKNOWLEDGEMENTS

The work reported here represents contributions by Professor
C. Cohen-Tannoudji and expert technical assistance of P.H. Cox,
L. Donaldson, S.S. Mertz, R.W. Stanley, and D.A. Van Baak. The work
was supported in part by the Joint Services Electronics Program.

*Supported in part by the Joint Services Electronics Program.

References

1. L.S. Vasilenko, V.P. Chebotayev, and A.V. Shishaev, Pis'ma Zh. Eksp. Teor. Fiz. *12*, 161 (1970)[JEPT Lett. *12*, 113 (1970)]; B. Cagnac, G. Grynberg, and F. Biraben, J. Phys. (Paris)*34*, 56 (1973) and Phys. Rev. Lett. *32*, 645 (1974); M.D. Levenson and N. Bloembergen, Phys. Rev. Lett. *32*, 645 (1974); T. W. Hansch, K. Harvey, G. Meisel and A.L. Schawlow, Opt. Comm. *11*, 50 (1974); M.M. Salour, Opt. Comm. *18*, 377 (1976).

2. M.M. Salour, Opt. Comm. *18*, 377 (1976); M.M. Salour, Phys. Rev. A *17*, 614 (1978).

3. M.M. Salour, Bull. Am. Phs. Soc. *21* 1245 (1976); M.M. Salour and C. Cohen-Tannoudji, Phys. Rev. Lett. *38*, 757 (1977).

4. N.F. Ramsey, Phys. Rev. *76*, 996 (1949) and *Molecular Beams* (Oxford University Press, New York, 1956); P.B. Kramer, S. R. Lundeen, B.O. Clark and F.M. Pipkin, Phys. Rev. Lett. *32*, 635 (1974).

5. Ye. V. Baklanov, B. Ya. Dubetskii, and V.P. Chebotayev, Appl. Phys. *9*, 171 (1976); Ye. V. Baklanov, B. Ya. Dubetskii, and V.P. Chebotayev, Appl. Phys. *11*, 201 (1976); B. Ya. Dubetskii, Kvantovayev Elektron (Moscow) *3*, 1258 (1976) [Sov. J. Quantum Electron. *6*, 682 (1976)]. Note that the method of using spatially separated laser fields has recently been applied to saturated absorption spectroscopy by J.C. Bergquist, S.A. Lee, and J.L. Hall, Phys. Rev. Lett. *38*, 159 (1977).

6. Note that the method of using two coherent time-delayed electromagnetic pulses has been already used in the microwave domain. See for example, C.O. Alley, in *Quantum Electronics, A Symposium* (Columbia University Press, New York, 1960) p. 146; M. Arditi and T.R. Carver, IEEE Trans. Instrumentation and Measurement, IM-13(2), (3), (1964), p. 146.

7. M.M. Salour, Ann. Physics (to be published, 1978).

8. For the two pulses of Fig. 1-II, the result of the calculation is:

$$|b_2|^2 = 4|b_1|^2 [\cos \omega_0 T/2]^2$$

so that $|b_2|^2$ does not depend on ω, with ω_0 and T being fixed. Note, however, that the interference fringes reappear when T alone (or ω_0 alone) is varied.

9. M.M. Salour, Opt. Comm. *22*, 202 (1977).

10. M.M. Salour, Appl. Phys. Lett. *31*, 394 (1977).

11. R. Teets, J. Eckstein, and T. W. Hänsch. Phys. Rev. Lett. *38*, 760 (1977).

ADIABATIC FOLLOWING IN TWO-PHOTON TRANSITION

Munir H. Nayfeh

*University of Illinois at Urbana-Champaign,
Urbana, Illinois*

Ali H. Nayfeh

*Virginia Polytechnic Institute and State University
Blacksburg, Virginia*

Renewed interest has been generated recently in coherent
multiphoton transitions in many-level systems [1-5]. Coherent
two-photon propagations, exhibiting self-induced transparency when
twice the propagating frequency is resonant with a two-level system, have been analyzed [1,2], and the one- and two-photon resonant
fields were shown to obey together an area theorem [2]. Tan-no
et al. [3] reported the observation of coherent propagations based
on a two-photon transition which involved the interaction of two
optical pulses of different frequencies with a gaseous three-level
system. Further, self-transparency in a resonant two-photon
absorption with a two-level system was observed [4]. In this work,
we consider the effect of relaxation on the response of a three-
level system to a smoothly varying, near resonant, two-photon field.

Consider a three-level system (Fig. 1) with only level 2
optically connected to the ground state and the resonance frequencies ω_{12} and ω_{23} are far away from each other. The interaction of
two smooth pulses with the above system can be described by the
equation of motion of the density matrix

$$\frac{\partial \rho}{\partial t} = - \frac{i}{\hbar} [H, \rho] , \qquad (1)$$

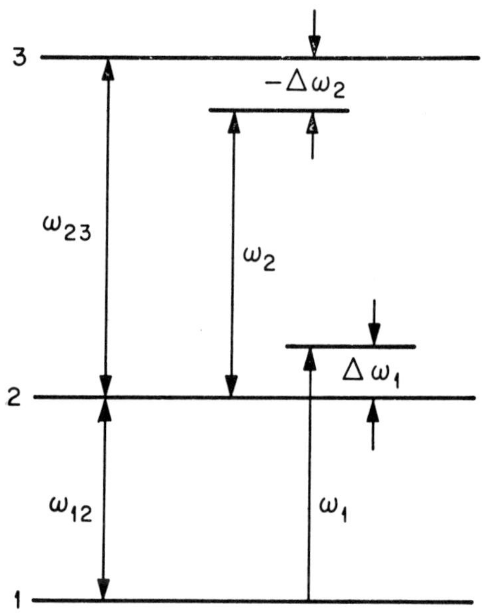

Figure 1.   Energy level diagram of the three-level system.

where $H = H_0 + H_1$, H_0 is the unperturbed Hamiltonian, and H_1 is the interaction Hamiltonian which can be written as

$$H_1 = \begin{bmatrix} 0 & v_{12} & 0 \\ v_{21} & 0 & v_{23} \\ 0 & v_{32} & 0 \end{bmatrix} , \tag{2}$$

where $v_{12} = \underline{\mu}_{12} \cdot \underline{E}_1$, $v_{23} = \underline{\mu}_{23} \cdot \underline{E}_2$, and the electric field is given by

$$\underline{E} = \underline{E}_1 + \underline{E}_2 = \underline{\varepsilon}_1 \cos\omega_1 t + \underline{\varepsilon}_2 \cos\omega_2 t . \tag{3}$$

Note that E_1 and E_2 are taken to induce transitions between the levels 1-2 and 2-3 respectively, due to the assumption that ω_{12} is much different from ω_{23}. The equation of motion of the diagonal and off-diagonal elements of the density matrix are derived by substituting Eq. (2) in Eq. (1). The result is

$$\frac{d}{dt} \rho_{33} = -(i/\hbar) \, (v_{32}\rho_{23} - \rho_{23}^* v_{23}) , \tag{4}$$

$$\frac{d}{dt} \rho_{11} = -(i/h) \, (v_{12}\rho_{12}^* - \rho_{12}v_{21}) , \tag{5}$$

$$\frac{d}{dt}\,\rho_{13} = -i(\omega_{12}+\omega_{23})\rho_{13} - (i/\hbar)(v_{12}\rho_{23} - \rho_{12}v_{23})\ ,\qquad (6)$$

$$\frac{d}{dt}\,\rho_{23} = -i\omega_{23}\rho_{23} - (i/\hbar)(\rho_{33} - \rho_{22})v_{23} - (i/\hbar)v_{21}\rho_{13}\ ,\qquad (7)$$

$$\frac{d}{dt}\,\rho_{12} = -i\omega_{12}\rho_{12} - (i/\hbar)(\rho_{22} - \rho_{11})v_{12} - (i/\hbar)v_{32}\rho_{13}\ .\qquad (8)$$

The conservation law for the number of atoms implies that

$$\rho_{11} + \rho_{22} + \rho_{33} = 1\ .\qquad (9)$$

Neither of the fields is taken to be resonant with the corresponding transitions so that level 2 is sufficiently far away from resonance, and thus its amplitude can be taken to be slowly varying. We write

$$\rho_{23} = \sigma_{23}\,e^{-i\omega_{23}t}\ ,\qquad \rho_{12} = \sigma_{12}\,e^{-i\omega_{12}t}\ ,$$

$$\rho_{13} = \sigma_{13}\,e^{-i(\omega_{12}+\omega_{23})t}\ ,\qquad (10)$$

integrate Eqs. (6) and (7) by parts, and in the rotating wave approximation obtain

$$\sigma_{23} = \frac{(\rho_{33} - \rho_{22})\varepsilon_2\mu_{23}}{2\hbar\Delta\omega_2}\,e^{i\Delta\omega_2 t} + \frac{\mu_{21}\varepsilon_1\sigma_{13}}{2\hbar\Delta\omega_1}\,e^{i\Delta\omega_1 t}\qquad (11)$$

$$\sigma_{12} = \frac{(\rho_{22} - \rho_{11})\varepsilon_1\mu_{12}}{2\hbar\Delta\omega_1}\,e^{i\Delta\omega_1 t} + \frac{\mu_{32}\varepsilon_2\sigma_{13}}{2\hbar\Delta\omega_2}\,e^{i\Delta\omega_2 t}\ ,\qquad (12)$$

where $\Delta\omega_1 = \omega_1 - \omega_{12}$ and $\Delta\omega_2 = \omega_2 - \omega_{23}$. The transformation of σ_{23} and σ_{12} in terms of σ_{13} results in an induced two-photon polarization which, when substituted in Eqs. (4)-(6), results in an explicit form for the two-photon coupling. In evaluating the two-photon coupling and the induced level shifts, we evaluate the amplitudes of σ_{23} and σ_{12} when $|\Delta\omega_2| = |\Delta\omega_1| = \Delta\omega_0$, which implies a slowly varying condition of the off-resonant frequency $\Delta\omega_0$ with respect to the virtual state 2. Substituting Eqs. (11) and (12) in (4) and (5) and using the rotating wave approximation, we obtain

$$\frac{d}{dt}(\rho_{33} - \rho_{11}) = k\varepsilon^2[i(\sigma_{13} - \sigma_{13}^*)\cos\Delta_0 t - (\sigma_{13} + \sigma_{13}^*)\sin\Delta_0 t]\ ,$$

$$\qquad (13)$$

$$\frac{d}{dt}(\sigma_{13} + \sigma_{13}^*) = k\varepsilon^2(\rho_{33} - \rho_{11})\,\sin\Delta_0 t$$

$$\qquad + \frac{i(\mu_{12}^2\varepsilon_1^2 + \mu_{23}^2\varepsilon_2^2)}{4\hbar^2\Delta\omega_0}(\sigma_{13} - \sigma_{13}^*)\ ,\qquad (14)$$

$$\frac{d}{dt} (\sigma_{13} - \sigma_{13}^*) = ik\varepsilon^2 (\rho_{33} - \rho_{11}) \cos\Delta_0 t$$

$$+ \frac{i(\mu_{12}^2 \varepsilon_1^2 + \mu_{23}^2 \varepsilon_2^2)}{4\hbar^2 \Delta\omega_0} (\sigma_{13} + \sigma_{13}^*) \quad , \tag{15}$$

where $\Delta_0 = \Delta\omega_1 + \Delta\omega_2$ is the two-photon resonance detuning, $\varepsilon^2 = \varepsilon_1 \varepsilon_2$, and $k = \mu_{23}\mu_{12}/2\hbar^2 \Delta\omega_0$.

Equations (13)-(15) can now be written in terms of the Bloch vector components by defining

$$w = \rho_{33} - \rho_{11} \ , \qquad \sigma_{13} + \sigma_{13}^* = u \cos(\Delta_0 t) + v \sin(\Delta_0 t)$$

$$\sigma_{13} - \sigma_{13}^* = -i[(u \sin(\Delta_0 t) - v \cos(\Delta_0 t)] \quad , \tag{16}$$

where w is the two-photon inversion and u and v are the dipole dispersive and absorptive components, respectively. Substituting Eq. (16) in Eqs. (13)-(15) and equating the coefficients of $\cos\Delta\omega$ and $\sin\Delta\omega$, we obtain

$$\frac{du}{dt} + \gamma_2 u = -\Delta(t)v \ , \tag{17}$$

$$\frac{dv}{dt} + \gamma_2 v = \Delta(t)u + k\varepsilon^2 w \ , \tag{18}$$

$$\frac{dw}{dt} + \gamma_1 w = -k\varepsilon^2 v - \gamma_1 \ , \tag{19}$$

where $\Delta(t) = \Delta_0 + s_1 \varepsilon_1^2 + s_2 \varepsilon_2^2$; $s_1 = \mu_{12}^2/4\hbar^2 \Delta\omega_0$; $s_2 = \mu_{23}^2/4\hbar^2 \Delta\omega_0$; and $\gamma_1 = 1/T_1$ and $\gamma_2 = 1/T_2$ are the atomic energy and phase relaxation widths, respectively.

Damping effects have been incorporated in the above equations phenomenologically by introducing a longitudinal relaxation time T_1 of the upper level and a homogeneous transverse relaxation time T_2. The relaxation effects of the intermediate level may be neglected if it is sufficiently far away from resonance.

Note that the interaction of two pulses with a three-level system, when the intermediate state is sufficiently far away from resonance, can be described by the above "two-photon Bloch equations". These are analogous to the equations governing a one-photon transition in a two-level system, except for the presence of a two-photon coupling and a frequency shift [5].

An adiabatic following model based on the above approximation was used to derive and explain parametric generation processes [6]. Power-dependent nonlinear susceptibilities were obtained, and pulse propagation was studied. However, in the above study, relaxation of the atomic levels was neglected. In this paper we determine the effect of relaxation on the model and calculate the relaxation-dependent contributions to the nonlinear refractive index (n). The leading contribution to $(n-1)\varepsilon_i^{-2}$ is intensity dependent caused by the level shifts inherent in multiphoton processes; it includes a relaxation-dependent part which is important at times shorter than γ_1^{-1}. The second-order contributions depend on the square of the intensity and the time-integrated square of the intensity. The latter contribution, which is relaxation dependent, causes line asymmetry at the long wavelength wing; it consists of a term that is proportional to $(\gamma_2-\gamma_1)$ and only important at early times and a term that is proportional to $(2\gamma_2-\gamma_1)$.

In the intervals in which the amplitudes of the fields vary slowly with time, approximate solutions of Eqs. (17)-(19) can be obtained by using either the method of multiple scales [7] or the factorization technique [8-10] in which one first reduces these equations to three uncoupled third-order equations with time-dependent coefficients. To use the latter technique, we first eliminate the variable v from Eq. (19) by multiplying it by $k\varepsilon^2[(d/dt)+\gamma_2](1/k\varepsilon^2)$ and then substituting for $[(d/dt)+\gamma_2]v$ from Eq. (18). The result is

$$\left[\left(\frac{d}{dt} + \gamma_2 - \frac{d}{dt}\,\ell n k\varepsilon^2\right)\left(\frac{d}{dt} + \gamma_1\right) + k^2\varepsilon^4\right] w$$

$$= -k\varepsilon^2\Delta u - k\varepsilon^2\left(\frac{d}{dt} + \gamma_2\right)\frac{\gamma_1}{k\varepsilon^2} \quad . \tag{20}$$

The variable u can be eliminated from Eq. (20) by multiplying it by $k\varepsilon^2\Delta(d/dt + \gamma_2)(1/k\varepsilon^2\Delta)$ and substituting for $(d/dt + \gamma_2)u$ from Eq. (17) and for v from Eq. (19). We neglect terms the order of $(d^2/dt^2)\ell n\varepsilon$, $[(d/dt)\ell n\varepsilon] \times [(d/dt)\ell n\varepsilon^2\Delta]$, and $[(d/dt)\ell n\varepsilon]^2$ in the homogeneous part of the equations and terms the order of $(d/dt)\ell n\varepsilon$ in the sources. Terms the order of $(d/dt)\ell n\varepsilon$ are kept in the homogeneous parts because their integrals over time determine the solutions. Consequently, these terms are not negligible for non-constant field amplitudes even as the rate of change of the amplitudes becomes arbitrarily small. Using a similar procedure for Eq. (17), one can show that the results for w and u are

$$\left\{ \frac{d^3}{dt^3} + [2\gamma_2 + \gamma_1 - \frac{d}{dt}\ln(k^2\varepsilon^4\Delta)]\frac{d^2}{dt^2} + [(\gamma_1 + \gamma_2)^2 - \gamma_1^2 + \Omega^2 \right.$$

$$- (\gamma_2 + \gamma_1)\frac{d}{dt}\ln(k^2\varepsilon^4\Delta)]\frac{d}{dt} + \gamma_1(\Delta^2 + \gamma_2^2)$$

$$\left. - (k^2\varepsilon^4 + \gamma_1\gamma_2)\frac{d}{dt}\ln(k\Delta\varepsilon^2) + (\gamma_2 + \frac{d}{dt}\ln k^2\varepsilon^4)k^2\varepsilon^4 \right\} w$$

$$= -\gamma_1(\Delta^2 + \gamma_2^2) \tag{21}$$

and

$$\left\{ \frac{d^3}{dt^3} + (2\gamma_2 + \gamma_1 - \frac{d}{dt}\ln k\varepsilon^2\Delta^2)\frac{d^2}{dt^2} + [(\gamma_1 + \gamma_2)^2 - \gamma_1^2 + \Omega^2 \right.$$

$$- (\gamma_1 + \gamma_2)\frac{d}{dt}\ln\Delta - 2\gamma_2\frac{d}{dt}\ln k\varepsilon^2\Delta]\frac{d}{dt} + \gamma_1(\gamma_2^2 + \Delta^2)$$

$$\left. + \gamma_2 k^2\varepsilon^4 + (\gamma_2^2 + \Delta^2)\frac{d}{dt}\ln k\varepsilon^2\Delta + (2\Delta^2 - \gamma_1\gamma_2)\frac{d}{dt}\ln\Delta \right\} u$$

$$= \gamma_1\Delta k\varepsilon^2 \ , \tag{22}$$

where

$$\Omega^2(t) = \Delta^2(t) + k^2\varepsilon^4 \ . \tag{23}$$

The general solution of Eq. (21) is the sum of a particular solution and a linear combination of three linearly independent homogeneous solutions, two of which are oscillatory. For adiabatically switched fields, the oscillatory solutions vanish and for small γ_1 and $\gamma_2 k^2\varepsilon^4\Omega^{-2}$ the nonoscillatory solution is slowly varying. Thus, to determine an approximate solution for the case of adiabatically switched fields when γ_1 and $\gamma_2 k^2\varepsilon^4\Omega^{-2}$ are small, we need to determine only the particular solution of Eq. (19) and its slowly varying homogeneous solution. This can be accomplished by dropping the second and third derivatives of w, the products of γ_1 and the derivatives of w or ε, and the products of the derivatives of w and the derivatives of ε and Δ. Under these approximations, Eq. (19) becomes

$$(\gamma_2^2 + \Omega^2)\frac{dw}{dt} + [\gamma_1(\Delta^2 + \gamma_2^2) - k^2\varepsilon^4\frac{d}{dt}\ln(k\Delta\varepsilon^2) + (\gamma_2$$

$$+ \frac{d}{dt}\ln k^2\varepsilon^4)k^2\varepsilon^4]w = -\gamma_1(\Delta^2 + \gamma_2^2) \ . \tag{24}$$

Similarly, under similar approximations, Eq. (22) becomes

$$(\gamma_2^2 + \Omega^2)\, \frac{du}{dt} + [\gamma_1(\Delta^2 + \gamma_2^2) + \gamma_2 k^2\varepsilon^4 - (\Delta^2 + \gamma_2^2)\frac{d}{dt}\,\ell n(k\varepsilon\Delta^2)$$

$$+\ 2\Delta\,\frac{d\Delta}{dt}\,]\ u\ =\ \gamma_1\Delta k\varepsilon^2 \tag{25}$$

We note that Eq. (24) can be derived alternatively by using the above approximations while eliminating u and v from Eqs. (17)-(19). Thus, this technique can be readily used to determine the influence of adiabatically switched fields on many-level systems when the relaxations are small.

For an atom that is initially in the ground state, the solution of Eq. (24) is

$$w\ =\ -\ \frac{\Delta(\Delta_0^2 + \gamma_2^2)^{\frac{1}{2}}}{\Delta_0(\Omega^2 + \gamma_2^2)^{\frac{1}{2}}}\ e^{-\gamma_1 t - \Gamma(t)}$$

$$-\ \frac{\gamma_1\Delta}{(\Omega^2+\gamma_2^2)^{\frac{1}{2}}}\ e^{-\gamma_1 t - \Gamma(t)}\int_{-\infty}^{t}\frac{\Delta^2+\gamma_2^2}{\Delta(\Omega^2+\gamma_2^2)^{\frac{1}{2}}}\ e^{\gamma_1 t + \Gamma(t)}\,dt \tag{26}$$

where

$$\Gamma\ =\ (\gamma_2 - \gamma_1)\int_{-\infty}^{t}\frac{k^2\varepsilon^4}{\Omega^2+\gamma_2^2}\,dt\ +\ \gamma_2^2\int_{-\infty}^{t}\frac{1}{\Delta(\Omega^2+\gamma_2^2)}\ \frac{d\Delta}{dt}\,dt\ . \tag{27}$$

When γ_2 is small, Eqs. (26) and (27) become

$$w\ =\ -\ \frac{\Delta(t)}{\Omega(t)}\ \exp[-\gamma_1 t - q(t)]\ -\gamma_1\exp[-q(t)]\,\frac{\Delta(t)}{\Omega(t)}\int_{-\infty}^{t}\frac{\Delta(\tau)}{\Omega(\tau)}$$

$$\times\ \exp[\gamma_1(\tau - t) + q(\tau)]\ d\tau\ , \tag{28}$$

where

$$q(t)\ =\ (\gamma_2 - \gamma_1)\int_{-\infty}^{t}\frac{k^2\varepsilon^4}{\Omega^2}\,dt\ . \tag{29}$$

When $\gamma_2 = 0(\Omega)$ and $k^2\varepsilon^4 \ll \Omega^2$, Eqs. (26) and (27) reduce to the rate equation limit

$$w\ =\ -e^{-\gamma_1 t - \tilde{\Gamma}(t)}\left[\,1\ +\ \gamma_1\int_{-\infty}^{t}e^{\gamma_1 t + \tilde{\Gamma}(t)}\,dt\,\right]\ , \tag{30}$$

where

$$\tilde{\Gamma}(t) = \gamma_2 \int_{-\infty}^{t} \frac{k^2\varepsilon^4 \, dt}{\Delta_0^2 + \gamma_2^2} \quad . \tag{31}$$

In this limit, the functions u and v are given in terms of w as

$$u = -\Delta_0 k\varepsilon^2 w/(\Delta_0^2 + \gamma_2^2) \quad , \tag{32}$$

$$v = \gamma_2 k\varepsilon^2 w/(\Delta_0^2 + \gamma_2^2) \quad . \tag{33}$$

When the relaxations are not small, we cannot determine an approximate solution to Eq. (21) by using the above described simplified scheme. Instead, we use the factorization technique and express Eq. (21) as

$$\left(\frac{d}{dt} + g_3\right)\left(\frac{d}{dt} + g_2\right)\left(\frac{d}{dt} + g_1\right) w = -\gamma_1(\Delta^2 + \gamma_2^2) \quad . \tag{34}$$

Equation (28) can be solved by letting $g_i = G_i + \delta_i$ where $\delta_i \ll G_i$. Expanding Eq. (34) and comparing the zeroth- and first-order terms, results in equations for the G_i and the amplitude functions δ_1, δ_2, and δ_3. The equation for G_i is cubic. In some cases, this cubic equation has simple solutions. These include the case of equal decay widths, the case where the decay widths are smaller than, or equal to, $O[(d/dt)\ln\varepsilon]$, and the case of small fields and small γ_1. Note that, in the latter case, there is no restriction on the magnitude of γ_2.

When $\gamma_1 = \gamma_2$, one finds [10] that the general solution consists of two homogeneous solutions which oscillate with the Rabi frequency, a homogeneous, non-oscillatory solution, and a particular solution. For adiabatic fields and an atom that is initially in the ground state, the oscillatory solutions vanish and w takes the simple form

$$w = -\frac{\Delta}{\Omega} e^{-\gamma t} - \gamma \frac{\Delta}{\Omega} e^{-\gamma t} \left[\int_{-\infty}^{t} \frac{\Omega e^{\gamma\tau}(\gamma^2+\Delta^2)}{\Delta(\gamma^2 + \Omega^2)} \, d\tau\right] \quad . \tag{35}$$

When $\gamma_1 \neq \gamma_2$ it follows from Eq. (28) that the inversion has an induced damping q(t). For weak fields, the lowest-order term in q(t) depends on the integrated square of the intensity, an indication of the two-photon process. For sufficiently strong fields, q(t) has no field dependence, and the inversion is damped only by the phase relaxation γ_2, as in the one-photon transition. However, the population inversion given by Eq. (35) does not have any power-dependent decay. As $\gamma \to 0$, Eq. (35) reduces to

$$w = - \Delta/\Omega \quad , \tag{36}$$

in agreement with the calculation of Ref. 6.

It is interesting to compare the above results with the results for two-level systems [9,11,12]. The first contribution in Eq. (35) is identical in form to the corresponding contribution in two-level systems, except that the detuning Δ is a function of time due to the frequency shift and the Rabi frequency Ω is modified by the two-photon coupling and the frequency shift. The presence of the frequency shift causes a shift in the crossing point of the inversion $[\Delta(t) = 0]$. However, the integral contribution form is different from that for two-level systems because Δ is a function of time in three-level systems.

Since Eq. (26) was derived with the assumption that $\gamma_2 k^2 \varepsilon^4 \ll \Omega^3$, which involves the product of γ_2 and $k^2 \varepsilon^4$, it is less restrictive on each of them individually. This implies that solution (26) is valid in the intermediate regime in which neither γ_2 nor $k^2 \varepsilon^4$ takes extreme values. Therefore, it bridges the gap between the rate equation limit (30) and the coherent limit.

In addition to the line shift, which is inherent in two-photon processes due to the involvement of virtual intermediate levels, there is a relaxation-dependent line asymmetry due to the buildup of the pulsed field. To see this, we derive an expression for the refractive index of three-level systems in the weak field limit. For an atom that is initially in the ground state, one can easily show that the solution of Eq. (25) is

$$u = \frac{k\varepsilon^2}{\Omega} e^{-\gamma_1 t - q(t)} \left[1 + \gamma_1 \int_{-\infty}^{t} \frac{\Delta(\tau)}{\Omega(\tau)} e^{\gamma_1 \tau + q(\tau)} \, d\tau \right] \tag{37}$$

for small γ_2. When $\gamma_1 = \gamma_2 = 0$, u reduces to $k\varepsilon^2/\Omega$, in agreement with the result of Ref. 6.

Since the refractive index n of the medium at either excitation frequency due to the nonlinear mixing of the field components is related to u by

$$n = 1 + 2\pi N k \hbar u \, \varepsilon_i/\varepsilon_j \quad , \quad i \neq j \tag{38}$$

it is given for weak fields by

$$\frac{n-1}{\varepsilon_i^2} = - \frac{2\pi N k^2 \hbar}{\Delta_0} \left\{ \left[1 - \frac{s_1 \varepsilon_1^2 + s_2 \varepsilon_2^2}{\Delta_0} + \frac{(s_1 \varepsilon_1^2 + s_2 \varepsilon_2^2)^2}{\Delta_0^2} - \frac{1}{2} \frac{k^2 \varepsilon^4}{\Delta_0^2} \right] \right.$$

$$\times (1 + e^{-\gamma_1 t}) - \frac{1}{2}(2\gamma_2 - \gamma_1) \int_{-\infty}^{t} \frac{k^2 \varepsilon^4(\tau)}{\Delta_0^2} e^{\gamma_1(t-\tau)} \, d\tau$$

$$\left. - (\gamma_2 - \gamma_1) e^{-\gamma_1 t} \int_{-\infty}^{t} \frac{k^2 \varepsilon^4(\tau)}{\Delta^2} \, d\tau \right\} + 0(\varepsilon^6) \, . \qquad (39)$$

We note that the lowest-order nonlinear contribution to
$(n-1)\varepsilon_i^{-2}$ is intensity dependent. It is equal to $2\pi N k^2 \hbar \Delta_0^{-2}$
$(s_1 \varepsilon_1^2 + s_2 \varepsilon_2^2)[1 + \exp(-\gamma_1 t)]$, which is inversely proportional to
the square of the frequency shift of the transition. The next-
order contribution is $\pi N k^2 \hbar \Delta_0^{-3}[k^2 \varepsilon^4 - 2(s_1 \varepsilon_1^2 + s_2 \varepsilon_2^2)][1+\exp(-\gamma_1 t)]$;
part of this contribution is due to the two-photon level shift.
The last two contributions depend on the integrated square of the
intensity. These contributions cause asymmetry in the refractive
index at the long wavelength wing of the transition due to the
accumulation of the square of the intensity at later times. We
note that for pulses having the same total energy, the asymmetry
depends on the pulse length. For long pulses, the term propor-
tional to $(\gamma_2-\gamma_1)$ vanishes, while the term proportional to $(2\gamma_2-\gamma_1)$
increases with increasing times. Thus, it is expected that there
exists a pulse length for which the asymmetry is maximum.

All of these contributions are dependent on the relaxation.
In the absence of relaxation (i.e., $\gamma_1 = \gamma_2 = 0$), it follows from
Eq. (39) that

$$\frac{n-1}{\varepsilon_i^2} = - \frac{4\pi N k^2 \hbar}{\Delta_0} \left[1 - \frac{s_1 \varepsilon_1^2 + s_2 \varepsilon_2^2}{\Delta_0} + \frac{2(s_1 \varepsilon_1^2 + s_2 \varepsilon_2^2)^2 - k^2 \varepsilon^4}{2\Delta_0^2} \right] .$$

$$(40)$$

When the relaxation is due entirely to spontaneous emission (i.e.,
$2\gamma_2 = \gamma_1$), the second contribution in Eq. (39) vanishes. In this
case the two-photon level shifts contribute to the asymmetry of
the line to $0(\varepsilon^4)$ at early times only. At times larger than the
relaxation time T_1, it follows from Eq. (39) that

$$\frac{n-1}{\varepsilon_i^2} = - \frac{2\pi N k^2 \hbar}{\Delta_0} \left[1 - \frac{s_1 \varepsilon_1^2 + s_2 \varepsilon_2^2}{\Delta_0} + \frac{2(s_1 \varepsilon_1^2 + s_2 \varepsilon_2^2) - k^2 \varepsilon^4}{2\Delta_0^2} \right.$$

$$\left. - \frac{1}{2}(2\gamma_2 - \gamma_1) \int_{-\infty}^{t} \frac{k^2 \varepsilon^4(\tau)}{\Delta_0^2} e^{\gamma_1(\tau-t)+q(\tau)} \, d\tau \right] . \qquad (41)$$

In conclusion, we have shown that the coherent interaction of two smoothly-varying, near-resonant, two-photon pulses with a three-level system can be described by "two-photon damped Bloch equations" which are analogous to those for a one-photon transition in a two-level system except for the presence of a two-photon coupling and a frequency shift. These equations are solved for the cases γ_1, $\gamma_2 \ll \Omega$, $\gamma_1 = \gamma_2$, $\gamma_2 k^2 \varepsilon^4/\Omega^2$, and $\gamma_1 \ll \Omega$, where γ_1 and γ_2 are the atomic energy and phase relaxation widths and Ω is the Rabi frequency. The leading contribution to $(n-1)\varepsilon_1^{-2}$ is intensity dependent caused by the level shifts inherent in multi-photon processes; it includes a relaxation-dependent part which is important at times shorter than γ_1^{-1}. The second-order contributions depend on the square of the intensity and the time-integrated square of the intenstiy. The latter contribution, which is relaxation dependent, causes line asymmetry at the long wavelength wing; it consists of a term that is proportional to $(\gamma_2-\gamma_1)$ and only important at early times and a term that is proportional to $(2\gamma_2-\gamma_1)$. The case $\gamma_2 k^2 \varepsilon^4/\Omega^2$, $\gamma_1 \ll \Omega$ bridges the gap between the coherent limit and the rate equation limit.

References

1. E.M. Belenov and I.A. Polucktov, Sov. Phys. - JETP *29*, 754 (1969).
2. M. Takatsuji, Phys. Rev. A *4*, 808 (1971).
3. N. Tan-no, K. Yokoto and H. Inaba, Phys. Rev. Lett. *29*, 1211 (1972).
4. T.L. Gvardzhaladze, A.Z. Grasyuk and V.A. Kovalenko, Sov. Phys. - JETP *37*, 227 (1973).
5. N. Tan-no, K. Yokoto and H. Inaba, J. Phys. B *8*, 339 (1975).
6. D. Grischkowsky, M.M.T. Loy and P.F. Liao, Phys. Rev. A *12*, 2514 (1975).
7. A.H. Nayfeh, *Perturbation Methods* (Wiley-Interscience, New York, 1973) Chap. 6.
8. M.G. Payne and M.H. Nayfeh, Phys. Rev. A *13*, 595 (1976).
9. M.H. Nayfeh, Phys. Rev. A *14*, 1304 (1976).
10. M.H. Nayfeh and A.H. Nayfeh, Phys. Rev. A *15*, 1169 (1977).
11. R.H. Lehmberg and J. Reintjes, Phys. Rev. A *12*, 2574 (1975).
12. M.H. Nayfeh and A.H. Nayfeh, J. Appl. Phys. *47*, 2528 (1976).

SUBPICOSECOND PULSE GENERATION IN RESONANT TWO-PHOTON PROPAGATION*

H. P. Yuen and F. Y. F. Chu

Massachusetts Institute of Technology

Cambridge, Massachusetts

A novel pulse-shortening behavior in traveling wave degenerate two-photon amplification has been theoretically demonstrated for long pulses. [1] Here we describe a corresponding pulse-shortening effect [2,3] in ultrashort pulse propagation in an ideal two-photon medium. The previous result [2,3] will be extended and applied to the 1S-2S two-photon transition in atomic hydrogen. It appears that subpicosecond pulses may be generated through this medium with slight extension of current technology.

Consider a homogeneously broadened two-photon medium of identical two-level atoms with energy separation $2\hbar\omega$ so that each atomic transition gives rise to the absorption or emission of two photons at the same frequency ω. The equations governing the propagation of plane-wave ultrashort pulses in the slowly-varying envelope approximation are [1-5].

$$\frac{\partial E}{\partial z} = -i \ \frac{2\xi N_0}{c} \ E^*m + \frac{i\eta N_0}{c} \ E(D - 1) - \frac{\gamma}{2c} \ E \tag{1}$$

$$\frac{\partial m}{\partial \tau} = i\xi E^2 D + i2\eta |E|^2 m \tag{2}$$

$$\frac{\partial D}{\partial \tau} = i2\xi (mE^{*2} - E^2 m^*) \tag{3}$$

where $\tau = t - z/c$, N_0 the total number of atoms per unit volume in the system, D the difference between the upper and lower state population per unit atom, M the complex envelope of the atomic polarization per unit atom, E the complex field envelope, and $1/\gamma$ the photon lifetime in the medium. The resonant frequencies of E and M have been removed and the coupling parameters are given by

$$\xi = \frac{2\pi\omega e^2}{\hbar} \sum_{k\neq 1,2} \frac{r_{1k}r_{2k}}{\omega_{k2}+\omega} \tag{4}$$

$$\eta = \frac{2\pi\omega e^2}{\hbar} \sum_{k\neq 1,2} \left(\frac{\omega_{k2}|r_{2k}|^2}{\omega_{k2}^2-\omega^2} - \frac{\omega_{k1}|r_{1k}|^2}{\omega_{k1}^2-\omega^2} \right) \tag{5}$$

with $\omega_{ki} = (E_k - E_i)/\hbar, r_{ik}$ the dipole matrix elements. Certain conditions of validity for (1)-(3) as a model of two-photon pulse propagation will be discussed later in connection with hydrogen.

If the incident pulse is not phase-modulated, phase modulation effects can be neglected and we may assume E to be real. Defining the photon density $P \equiv E^2$ and introducing a new variable $M \equiv 2\mathfrak{m}$, (1)-(3) become

$$\frac{\partial P}{\partial z} = -i \frac{2\xi N_0}{c} PM + \frac{i2\eta N_0}{c} P(D - 1) - \frac{\gamma}{c} P \tag{6}$$

$$\frac{\partial M}{\partial \tau} = i2\xi PD + i2\eta PM \tag{7}$$

$$\frac{\partial D}{\partial \tau} = i\xi P(M - M^*). \tag{8}$$

Equations (6)-(8) are to be solved for $z \geq 0$ under the condition that the medium is excited at $z = 0$ by a pulse $P_0(t)$ starting from $t = t_0$:

$$D(z,t_0) = \pm 1, \quad M(z,t_0) = 0, \quad P(0,\tau) = P_0(\tau). \tag{9}$$

The initial population inversion $D(z,t_0)$ in (9) takes a positive sign for an inverted (amplifying) medium and a negative sign for a ground state (absorbing) medium.

Introducing the transformation

$$\psi(z,\tau) = 2\xi(1 + \eta^2/\xi^2)^{1/2} \int_{t_0}^{\tau} P(z,t)dt \tag{10}$$

$$P(z,\tau) = \frac{1}{2\xi(1 + \eta^2/\xi^2)^{1/2}} \frac{\partial\psi}{\partial\tau} \tag{11}$$

one can readily show that

$$D(z,\tau) = \pm\left[\frac{1}{1 + \eta^2/\xi^2}\cos\psi + \frac{\eta^2/\xi^2}{1 + \eta^2/\xi^2}\right] \tag{12}$$

$$M(z,\tau) = \pm\left[\frac{\eta/\xi}{1 + \eta^2/\xi^2}(\cos\psi-1) + i\frac{1}{(1 + \eta^2/\xi^2)^{1/2}}\sin\psi\right] \tag{13}$$

$$\frac{\partial^2\psi}{\partial z\partial\tau} = \left(-\frac{\gamma}{c} + g\,\sin\psi\right)\frac{\partial\psi}{\partial\tau} \tag{14}$$

$$g \equiv \pm\frac{2\xi N_0}{c(1 + \eta^2/\xi^2)^{1/2}}\,. \tag{15}$$

The sign of g in (15) follows that of $D(z,t_0)$ in (9). Equation (14) yields

$$\frac{\partial\psi}{\partial z} = -\frac{\gamma}{c}\psi + g(1 - \cos\psi) \tag{16}$$

from which an "area theorem" follows immediately:

$$\frac{d\theta}{dz} = -\frac{\gamma}{c}\theta + g(1 - \cos\theta) \tag{17}$$

$$\theta(z) \equiv \psi(z,\infty) \tag{18}$$

The solution (11)-(13) is obtained once (16) is solved under the conditions

$$\psi(z,t_0) = 0\,,\quad \psi(0,\tau) = 2\xi(1 + \eta^2/\xi^2)^{1/2}\int_{t_0}^{\tau} P_0(t)dt \equiv \psi_0(\tau) \tag{19}$$

The exact solution for (16) is easily obtained when $\gamma = 0$, [2-4] from which

$$P(z,\tau) = P_0(\tau)A(z,\tau) \tag{20}$$

$$A(z,\tau) = A(z,\psi_0) \equiv [1 - 2gz \sin\frac{\psi_0}{2}\cos\frac{\psi_0}{2} + g^2 z^2 \sin^2\frac{\psi_0}{2}]^{-1}. \quad (21)$$

The pulse propagation behavior can be evaluated numerically from (19)-(21) for any input $P_0(t)$. Some general features can be concluded analytically as summarized in the following.

For the absorbing case, $D(z,t_0) = -1$, an input pulse with a bounded support (i.e., a pulse that is only nonvanishing over a finite interval) and an input area $\theta_1(0)$ will be reduced to the nearest $\theta_1(\infty) = 2p\pi$, for an integer p, when z grows large. In particular, the pulse energy $\theta_1(z)$ is broken up into a sequence of continuously sharpened pulses around the position τ_n defined by

$$\psi_0(\tau_n) = 2n\pi; \; n = 1,2, \ldots \quad (22)$$

In the limit $z \to \infty$, a δ-function pulse train is formed

$$P(\infty,\tau) = \frac{\pi}{\xi(1 + \eta^2/\xi^2)^{1/2}} \sum_{n=1}^{p} \delta(\tau - \tau_n) \quad (23)$$

The equilibrium values $2p\pi$ are unstable against a loss of 2π.

The behavior for the amplifying case, $D(z,t_0) = 1$, is similar to the absorbing case except that an additional sharp pulse is formed at $t = t_0$. The equilibrium values of θ are also unstable against a gain of 2π. Note the contrast of the present two-photon behavior from the usual one-photon self-induced transparency.

For both square input pulse and hyperbolic secant input pulse cutoff at a finite interval, an order of magnitude shortening of the input pulse width in an absorbing medium occurs at 1 gain length $z_1 \sim 1/|g|$. Furthermore, two order of magnitude shortening of the input width occurs at about 10 gain lengths. If the parameters of the system and the input pulse are favorable, this indicates the possibility of generating femtosecond pulse from a two-photon medium.

When the effect of loss is taken into account, no analytic solution is found. It is still possible, however, to demonstrate analytically that a similar pulse shortening effect occurs when the loss is not too big by examining (16), which can be reduced to quadrature

$$\int_{\psi_0(\tau)}^{\psi(z,\tau)} \frac{d\psi}{\psi_0(\tau) - \frac{\gamma}{c}\psi + g(1 - \cos\psi)} = z \quad (24)$$

The equilibrium values $\theta(z)$ for the amplifier case are determined by the zeros of the following equation for θ_n:

$$(\gamma/c)\theta_n = g(1 - \cos\theta_n) \tag{25}$$

For $\gamma/c > 0.725\ g$, no nonzero equilibrium value exists. For $\gamma/c \ll g$, the zeros of (25) occur at approximately $\theta_n = 0$ and $\theta_n \simeq 2p\pi \pm \gamma/cg$ for integer p. The number of equilibrium values in the general region $\gamma/cg < 0.725$ is always finite. In contrast to the lossless case, half of the equilibrium values, $\theta_n = 0$ and those satisfying

$$\sin\theta_n < \gamma/cg \tag{26}$$

are stable in the present situation. By a careful examination of the regions where $\partial\psi/\partial z > 0$, $\partial\psi/\partial z < 0$, we conclude that for $z > 0$, ψ as a function τ wiggles around the τ_n determined by

$$\frac{\gamma}{c}\psi_0(\tau_n) = g[1 - \cos\psi_0(\tau_n)], \ \sin\psi_0(\tau_n) < \tau/cg \tag{27}$$

and approaches a sequence of steps as $z \to \infty$. Thus, a δ-function pulse train for $P(z,\tau)$ obtains asymptotically at these τ_n.

In the absorption case, all pulses are attenuated as $z \to \infty$ no matter how small γ is. However a similar, though not as sharp, series of steps for ψ also occurs around $\psi_0(\tau_n) = 2p\pi$ for sufficiently small $\gamma/c|g|$, as one can again show from (24). This leads to pulse sharpening in the same way as the above amplifying case. For square and hyperbolic secant input pulse, an order of magnitude shortening of the pulse width occurs at $z_1 \sim 1/|g|$ if $\gamma/c|g| \leq 10^{-2}$.

In Fig. 1, the formation of two sharp pulses at $\tau = 0$ and $\tau = \tau_0$ for a 4π input

$$2\mu P_0(\tau) \equiv 2\xi(1 + \eta^2/\xi^2)^{1/2}P_0(\tau) = \frac{2\pi}{\tau_0}\frac{\mathrm{sech}^2\tau/\tau_0}{\tanh 1}, \ \tau_0 \geq \tau \geq -\tau_0. \tag{28}$$

to an absorbing medium is plotted from (20) - (21). The presence of an γ, $\gamma/c|g| \leq 10^{-2}$ does not alter these pulse shapes in any significant way.

The above results can be applied to the 1S-2S transition in atomic hydrogen, for which Hänsch et al. [6] have performed many degenerate two-photon absorption experiments with a dye laser. Equations (1) - (3) appear to provide a realistic description of

subpicosecond two-photon pulse propagation in this system. First
of all, lasing to intermediate states is insignificant because only
the Lamb-shifted 2P state lies between the 1S and 2S states. Secondly, third harmonic generation and any self-defocusing due to
two-photon resonance can be eliminated by using a circularly polarized input pulse. Since the Doppler broadening is only of the order
GHz at nitrogen temperature, the above unbroadened or homogeneously
broadened model should apply for a picosecond input pulse.

The coupling parameters ξ and η can be computed exactly in
hydrogen, with $\xi \sim 3 \times 10^{-7}$ cm^3/sec which also agrees well with experiment. [6] The parameter η does not significantly alter the result,
being of the value $\eta \sim \xi$. At the required input wavelength $\lambda = 0.243\mu$,
the 2π energy threshold is determined by assuming optimum focusing
of the fundamental Gaussian mode over a distance $z_1 \sim 1/|g|$. Of
course, this threshold is not required for an amplifying medium,

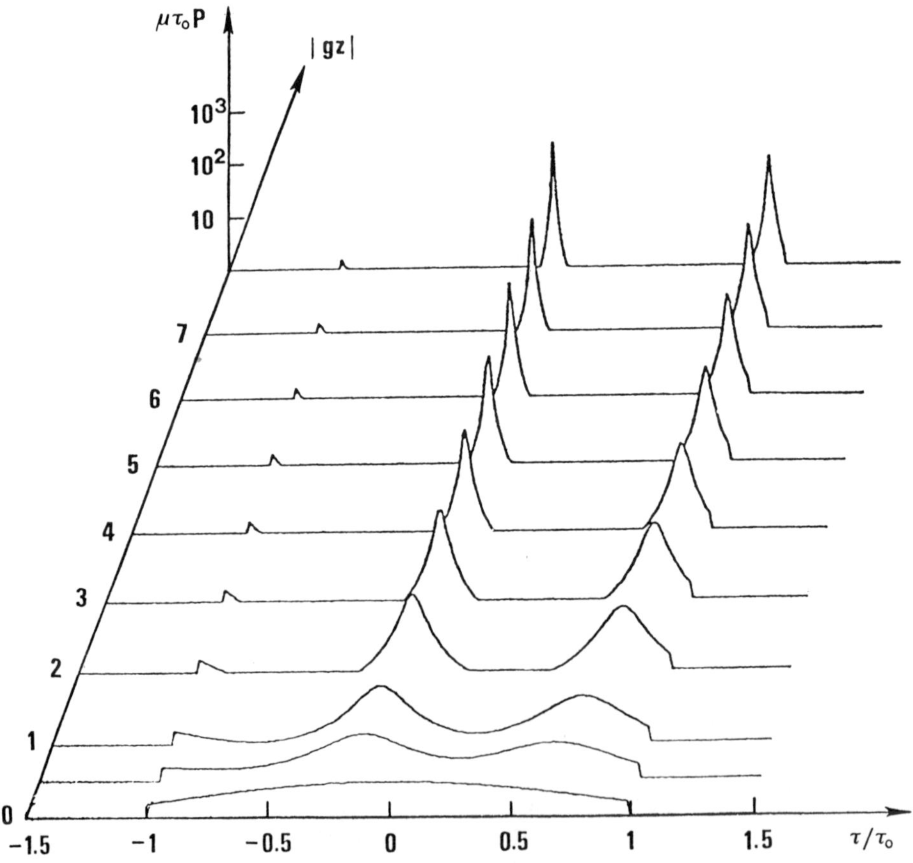

Fig. 1   Propagation of 4π Hyperbolic Secant Input Pulse

but it does not seem possible at present to produce good samples of
metastable hydrogen at high enough density and length for significant short pulse generation. On the other hand, a density
$\sim 10^{15}/cm^3$ of ground state hydrogen can be obtained in gaseous discharge. Since the photon lifetime in the medium would surely satisfy
$\gamma/c|g| \leq 10^{-2}$ in this case, only one meter length of the medium is
required for an order of magnitude shortening of the input. The 2π
requirement becomes $\sim$ 1mJ/pulse without taking into account any
possible gain from self-focusing. The photoionization transition
is several orders of magnitude weaker than the two-photon transition
at this power level.

To guarantee degenerate two-photon transition, one can use
transform-limited stable psec pulse from a dye laser. The required
characteristics of such a laser, including the power level, is not
far from some existing laser systems. If this suggested experiment
is successful, we will also obtain two-photon radiation [7] with
novel quantum statistical behavior because stimulated two-photon
re-emission has taken place in the pulse propagation.

Acknowledgment

The authors thank H. A. Haus, C. F. F. Karney, R. S. Kennedy,
D. Kleppner, and M. G. Littman for stimulating discussions.

*Work supported by the Joint Service Electronics Program under
Contract DAAB07-76-G1400 and by the National Aeronautics and Space
Administration under Grant NGL 22-009-013.

References

1. H. P. Yuen, Appl. Phys. Letters *26*, 505 (1975).
2. H. P. Yuen and F. Y. F. Chu, Research Laboratory of Electronics,
    Massachusetts Institute of Technology, Progress Report *118*, 190
    (1976).
3. I. A. Poluektov, Yu. M. Popov, and V. S. Roitberg, Sov. J. Quant.
    Electron. *5*, 620 (1975).
4. L. E. Estes, L. M. Narducci, and B. Shammas, Nuovo Cimento
    Letters *2*, 775 (1971).
5. D. Grischkowsky, M. M. T. Loy, and P. F. Liao, Phys. Rev. A*12*,
    2514 (1975).
6. T. W. Hänsch, S. A. Lee, R. Wallenstein, and C. Weiman, Phys.
    Rev. Letters *6*, 307 (1975).
7. H. P. Yuen, Phys. Rev. A*13*, 2226 (1976).

EXACT SOLUTION AND ANALYTIC PROPERTIES OF COHERENT TWO-PHOTON NEAR-

RESONANT PULSE PROPAGATION UNDER ADIABATIC FOLLOWING CONDITIONS

F. H. M. Faisal

University of Bielefeld

Bielefeld, Federal Republic of Germany

Coherent propagation of optical pulses in non-linear media is one of the most interesting and difficult problems in quantum optics. Considerable progress has been made in the last decade or so, in understanding the nature of optical pulse propagation under coherent conditions. A number of very interesting and by now well-known phenomena have been predicted and discovered, most of which exhibit self-regulating characteristics. They include such phenomena as self-induced transparency, self-focusing and de-focusing, self-trapping, self-phase modulation, self-broadening; self-steepening and optical-shock formation. They also include such interesting local coherent phenomena as adiabatic inversion, optical nutation and free induction decay and photon-echo. Most of these phenomena have been investigated when the pulse carrier frequency was taken to be resonant or near-resonant with a single-photon transition in the medium.

With the availability of high intensity optical radiation from lasers it has become possible for some time to cause resonant or near-resonant two-photon transitions with considerable probability. Already several local two-photon non-linear coherent phenomena (having their analogy in the single-photon case) have been observed experimentally. Furthermore, it is expected that, because of its intrinsic non-linearity which persists even at lower intensities, and because of its involvement with more than two-levels of the atomic system, two-photon coherent processes should be capable of showing interesting differences from the corresponding single-photon case and may even bring out completely new phenomena, without analogy in the single-photon case. One of the places in which such differences may be looked for is certainly the two-photon resonant or near-resonant pulse propagation phenomena. Much interest and efforts are currently being directed in the study of such propagation problems in the two-photon case. The subject merits special studies if only because it is the simplest of a whole class of multiphoton coherent phenomena that are already making their appearance in one or other guise in the field of coherent optics.

We report here some results of theoretical studies of two-photon near-resonant pulse propagation under adiabatic following conditions. Such conditions may be easily achieved by taking a light pulse with large enough rise time so that it is made to rise adiabatically slowly with respect to the detuning-period of the intermediate transition (single-photon, virtual) in the two-photon process.

As is well-known, one of the more powerful theoretical tools of coherent optics in the single-photon case has been the vector-model of Feynman, Vernon and Hellwarth [1], which succinctly describes the atomic Bloch-equations and supplies the atomic polarisations as source for the Maxwell-equations. A fruitful development in the present connection has been the derivation of an equivalent vector-model for the two-photon transitions by Takatsuji [2]. Solution of this model has been given for slowly varying pulses under adiabatic following conditions by Grischkowsky, Loy and Liao [3], which allowed them to derive the corresponding pulse propagation equations for the interesting case of a circularly polarized light pulse with its carrier frequency in near-resonance with a two-photon atomic transition. The term 'near-resonance' (used synonimously with the term 'quasi-resonance' elsewhere) is defined to mean a two-photon detuning much smaller than the two-photon transition frequency but is much greater than the line widths involved.

Here we describe briefly first the method of solving the relevant pair of coupled quasi-linear, first order, partial differential equations. Second, we present the exact results corresponding to the boundary conditions appropriate for pulse propagation and then, finally, we discuss analytically the evolution of envelope and phase of two-photon quasi-resonant pulses (of more or less general input shapes) at arbitrary intensities and distances of propagation.

The two-photon pulse propagation equations of interest are [3]

$$\frac{\partial \varepsilon}{\partial z} + \frac{1}{c}\frac{\partial \varepsilon}{\partial t} = -\lambda(z,t)\left[(\Omega_{12}-2\omega')\varepsilon^2\frac{\partial \varepsilon}{\partial t} + \varepsilon^3\frac{\partial \omega'}{\partial t}\right] \tag{1}$$

$$\frac{\partial \omega'}{\partial z} + \frac{1}{c}\frac{\partial \omega'}{\partial t} = -\lambda(z,t)\left[(\Omega_{12}-2\omega')\varepsilon^2\frac{\partial \varepsilon}{\partial t} + \varepsilon^3\frac{\partial \omega'}{\partial t}\right]\frac{(\Omega_{12}-2\omega')}{\varepsilon} \tag{2}$$

where $\varepsilon = \varepsilon(z,t)$ and $\omega' = \omega'(z,t) = \omega + \partial\phi(z,t)/\partial t$ are the slowly varying electric field amplitude and the instantaneous frequency of the laser pulse, respectively. ω is the nominal carrier frequency and $\phi(z,t)$ is the slowly varying phase.

$$\lambda(z,t) = \frac{2\pi N \chi^{(3)} \omega |\delta|}{c\gamma^3(z,t)} \tag{3a}$$

$$\gamma(z,t) = [\{s\varepsilon^2(z,t) + (\Omega_{12} - 2\omega'(z,t))\}^2 + k^2\varepsilon^4(z,t)]^{\frac{1}{2}} \tag{3b}$$

and $\delta = \Omega_{12} - 2\omega$ is the two-photon resonance detuning. $s = (1/2\hbar)(\chi_2(\omega) - \chi_1(\omega))$ is the dynamic stark-shift constant and $k = (1/\hbar^2)|\sum_n (p_{1n}p_{n2}/\Omega_{n2} - \omega)|$ is the two-photon gyro-electric ratio;

$$\chi_1(\omega) = \sum_n \frac{p_{1n}p_{n1}}{\hbar(\Omega_{n1} + \omega)}, \qquad \chi_2(\omega) = \sum_n \frac{p_{2n}p_{n2}}{\hbar(\Omega_{n2} - \omega)}$$

are the linear electric susceptibilities where $\hbar\Omega_{nj}$ and p_{nj} are the energy difference and the dipole matrix element, between the levels n and j, respectively.

$$\chi^{(3)} = \frac{2}{\hbar^3} \sum_{nm} \frac{p_{2n}p_{n1}p_{1m}p_{m2}}{\delta(\Omega_{m2} - \omega)(\Omega_{n2} - \omega)}$$

is the non-linear susceptibility constant. N is the atomic number density.

We note that in the coupled partial differential equations (1) and (2), the dependent variables $\varepsilon(z,t)$ and $\omega'(z,t)$ appear in highly non-linear fashion, Nevertheless, these equations (and for that matter most equations of propagations of coherent optics) are quasi-linear in character; in other words, the partial derivatives still appear in the linear form. This is a relative simplicity that should not be underestimated, for the general properties of the quasi-linear equations are better known than that of fully non-linear equations. The first hindrance to the solution of the above equations is the fact that they are not independent but are coupled to each other. The coupling may be eliminated by going to a higher order than the first but this leads to complicated forms. The strategy is therefore to find suitable transformations that may decouple them in *the first order* itself. To this purpose we first rewrite the equations in terms of slightly modified dependent and independent variables. Thus we define

$$\tau = t-z/c, \quad \xi = z, \quad \varepsilon(z,t) = \varepsilon(\xi,\tau), \quad D(\xi,\tau) = -\frac{1}{2}(\Omega_{12}-2\omega') \; .$$

Then we rearrange the equations by crosswise addition and subtraction in order to exhibit this maximum symmetry, and obtain

$$\frac{1}{\varepsilon}\frac{\partial \varepsilon}{\partial \xi} = -\frac{1}{2D}\frac{\partial}{\partial \xi} D \tag{4}$$

$$\frac{1}{\varepsilon}(\frac{\partial}{\partial \xi} - 4\lambda D\varepsilon^2 \frac{\partial}{\partial \tau})\varepsilon = \frac{1}{2D}(\frac{\partial}{\partial \xi} - 4\lambda D\varepsilon^2 \frac{\partial}{\partial \tau})D \; . \tag{5}$$

The form of eqs. (4) and (5) suggests that we introduce two transformations

$$F_1(\xi,\tau) = D(\xi,\tau) \; \varepsilon^2(\xi,\tau) \tag{6}$$

and

$$F_2(\xi,\tau) = D(\xi,\tau)/\varepsilon^2(\xi,\tau) \; . \tag{7}$$

In terms of (6) and (7) eqs. (4) and (5) becomes

$$\frac{1}{F_1}\frac{\partial}{\partial \xi} F_1 = 0 \tag{8}$$

and

$$\frac{1}{F_2}(\frac{\partial}{\partial \xi} - 4\lambda D\varepsilon^2 \frac{\partial}{\partial \tau})F_2 = 0 \; . \tag{9}$$

With eqs. (8) and (9) the system is now decoupled at the level of first order equations. Eq. (8) immediately integrates to

$$F_1(\xi,\tau) = F_1(0,\tau) \; , \tag{10}$$

which is the same as

$$(\Omega_{12}-2\omega'(z,t))\varepsilon^2(z,t) = (\Omega_{12}-2\omega'(0,t))\varepsilon^2(0,t) \; . \tag{11}$$

This is an interesting relation showing intrinsic coupling of the
self-phase modulation and the pulse envelope, and has already been
found by Grischkowsky et al. [3]. The solution of (9) satisfying
the pulse propagation boundary conditions is discussed in greater
detail elsewhere [6]; here we only write down the solution in the
following form

$$F_2(\xi,\tau) = F_2(0,\tau+a) \tag{12}$$

where

$$a \equiv a(\xi,\tau) = 4\lambda D\epsilon^2\xi \quad . \tag{13}$$

The desired pulse shape and the phase modulation are obtained
directly by remembering the definitions of F_1 and F_2 and combining
(10) and (12). This yields for the pulse envelope

$$\epsilon^2(\xi,\tau) = [D(0,\tau)D(0,\tau+a)]^{\frac{1}{2}}\epsilon(0,\tau)\epsilon(0,\tau+a) \tag{14}$$

and for the phase modulation

$$D(\xi,\tau) = [D(0,\tau)D(0,\tau+a)]^{\frac{1}{2}}\epsilon(0,\tau)/\epsilon(0,\tau+a) \quad . \tag{15}$$

It is clear from the form of these results that the evolution
of pulse envelope and phase is critically controlled by the 'deformation' parameter $a \equiv a(\xi,\tau)$, for if a = 0, the pulse reduces to its
initial input form automatically. The analytical solutions of
course permit one to calculate explicitly the development of any
particular pulse of interest with much less effort than that required
to do the same by numerical integration of the coupled partial differential equations (1) and (2). Perhaps more importantly they allow us to gain valued insights into the general behaviour of quite
arbitrary pulse shape and phase propagating under two-photon near-
resonant conditions. It is to some of these considerations that
we devote the rest of this brief report.

(I)   Small Deformation Limit

As noted above the parameter $a(\xi,\tau)$ controls the possible deformation of the propagating pulse. From (13) it would be seen that
this limit may be reached under several interesting and controllable
situations. First assume that $a(\xi,\tau)$ is indeed small compared to
the characteristic duration of the pulse. Developing (14) and (15)
in Taylor series around τ and retaining only the leading term one
finds

$$\varepsilon(\xi,\tau) = \varepsilon(0,\tau + \tfrac{1}{2}a) + 0(a^2)$$

and

$$D(\xi,\tau) = D(0,\tau)(1 - a\frac{\partial\varepsilon(0,\tau)}{\partial\tau}/\varepsilon(0,\tau)) + 0(a^2) \quad . \qquad (17)$$

Equation (16) yields the instantaneous pulse velocity v_p under small deformation condition to be

$$\frac{1}{v_p} = \frac{1}{c} - \frac{1}{2}\frac{a}{\xi} = \frac{1}{c}(1 + \varepsilon_2\omega|\delta|(\Omega_{12}-2\omega')\varepsilon^2/\gamma^3) \quad , \qquad (18)$$

where ε_2 is the non-linear susceptibility. The deformation, a, and the velocity, v_p, are thus directly related to each other, as may be expected on physical grounds. The velocity-change and, a, are small, as may be expected when the pulse intensity is small or the propagation distance z is small. It is to be shown below that in the opposite limit of very high intensity too the change in velocity as well as the deformation, a, are small.

(Ia)   Low Intensity Limit

It would be seen that for a pulse with constant phase at the cell entrance, eq. (17) reduces to

$$\frac{\partial\phi(\xi,\tau)}{\partial\tau} = -\frac{1}{2}\left(\frac{\omega}{c}\right)\varepsilon_2\xi\left|\frac{\delta}{\gamma}\right|^3\frac{\partial}{\partial\tau}\varepsilon^2(0,\tau) \quad , \qquad (19)$$

which is the approximate result for self-phase modulation given in Ref. [3]. In the limit of low intensity such that $s\varepsilon^2/(\Omega_{12}-2\omega') \ll 1$ this result goes over to the usual perturbation theoretical predictions [3]. From (18) it is seen that in the same limit the pulse envelope tends towards a steady-state-like simple-wave. It is also interesting to note [4,5,9] that the single-photon quasi-resonant pulses also tend toward simple-wave forms but the limiting velocity in the case is $v_g = c/(1+s)$;   in the present two-photon case the low intensity limiting velocity is c. Furthermore the departure from the low intensity pulse velocity occurs in the single-photon case as a linear function of ε, while in the present case it is proportional to ε^2, and reflects the intrinsic non-linearity of the two-photon processes.

(Ib) High Intensity Limit

It may appear at first that with increasing intensity the deformation of the propagating pulse would increase. A closer examination shows, perhaps unexpectedly, that the situation is quite the opposite and in fact the pulse once again approaches a steady-state-like simple-wave solution with velocity c, in the limit of extremely large intensity. Thus, so long

$$\frac{\Omega_{12}-2\omega'(z,t)}{s\varepsilon^2(z,t)} \ll 1 \; ,$$

$$\tau + a(\xi,\tau) = t - \frac{1}{c}\left[1 + \frac{2\varepsilon_2\omega}{s}\left(\frac{\Omega_{12}-2\omega'}{s\varepsilon^2}\right)\left(\frac{\delta}{s\varepsilon^2}\right)\right]z$$

$$\Rightarrow t - \frac{1}{c}z \; .$$

This result combined with Eq. (14) (for input phase $\phi(0,\tau)$ = Const.) shows that the two-photon near-resonant pulse tends to propagate freely because of its own very high intensity. The limiting velocity c in the high intensity limit above, is reached also faster than in the low intensity case of (Ia). We may add that a similar tendency towards free-propagation at very high intensities is expected to occur in the case of single-photon near-resonant pulse propagation as well [7]. That this velocity of coherent pulse propagation should tend to c, is perhaps to be expected. The well-known relation due to Courtens [8] show that if the energy of the electromagnetic field U_{em} is so large that it exceeds the coherent energy U_c of the atomic system then

$$\nu_p = \frac{c}{n_0(1+u_c/u_{em})} \Rightarrow c/n_0 \; .$$

(Note that, $n_0= 1$, has been assumed formally through the present discussion).

(II) Upper Bound of the Pulse Maximum

We now look at the general question of the maximum height of the propagating pulse. It is well-known that under single-photon near-resonant condition the maximum of the deformed pulse can exceed the input maximum [3]. In the present two-photon near-resonant condition the situation is opposite in character. The pulse peak-height is found to be bounded by the maximum of the input pulse. Once again, let the input phase be a constant. Equation (14) gives

$$\varepsilon^2(\xi,\tau) = \varepsilon(0,\tau)\varepsilon(0,\tau+a) \quad . \tag{20}$$

Let also the input pulse maximum   be defined to occur at $\tau = 0$. Examination of the arguments of both the factors on the right hand side of (20) shows that once the pulse has propagated away from the cell entrance ($z = 0$), they cannot be both simultaneously made to vanish. Consequently, the propagating pulse cannot again attain the input maximum in the course of its evolution, but remains bounded by it from above. (Note that the limit itself is only reached asympotically in the vanishingly small and infinitely large values of intensity as discussed in section (I)).

(III)   'Intersection-Time' of the Output Pulse with the Input Pulse

Although the output pulse maximum fails to exceed that of the input (as noted above) this does not prevent the output pulse from deforming greatly from its initial form. The deformation can be quite conveniently described by means of a parameter which may be called the 'intersection-time'. This we define as the time in the rest frame of the pulse at which it attains the same value as that attained by the input pulse at the same time. For simplicity let us consider only symmetric pulses here, which may be otherwise quite arbitrary. From (20) it is easily seen that at

$$\tau = \tau_+ = -\frac{1}{2}a = -2\lambda D\varepsilon^2\xi = \delta\frac{\varepsilon_2\omega|\delta|}{c\gamma^3}\varepsilon^2\xi \quad , \tag{21}$$

$$\varepsilon^2(\xi,\tau_+) = \varepsilon(0,-\tau_+)\varepsilon(0,\tau_+) = \varepsilon^2(0,\tau_+) \quad . \tag{22}$$

Equation (22) allows the identification of τ_+ as the 'intersection-time'. Hence from (21) it follows that:

(a)   if the initial two-photon detuning $\delta > 0$, then $\tau_+ > 0$. In other words the intersection occurs on the trailing edge of the pulse, past the maximum.

(b)   if, however, $\delta < 0$ then $\tau_+ < 0$. In this case the deformed pulse intersects the input pulse on the leading edge of the pulse, before the arrival of the input maximum.

The reader will notice that this simple consideration coupled with the result of section (II) which shows that the pulse maxima remain under the input maximum, allows one to visualize quite conveniently the general trend of deformation of initially symmetric pulses.

(IV)   The Zero of the Self-Phase Modulation

An invariant relation between the pulse envelope and the self-phase modulation may now be shown to exist in the two-photon quasi-resonant conditions. Assume as before that the phase at the cell entrance is a constant. It follows from (15) that $D(\xi,\tau) = \varepsilon(0,\tau)/\varepsilon(0,\tau+a)$. If the pulse is assumed further to be symmetric (but otherwise quite arbitrary) then at

$$\tau = \tau_0 = -\frac{1}{2}a = -2\lambda D\varepsilon^2\xi \; ; \tag{23}$$

$$D(\xi,-\frac{1}{2}a) = D(0,-\frac{1}{2}a) \; .$$

Remembering the definition of D this is rewritten to give

$$\left.\frac{\partial\phi(z,\tau)}{\partial\tau}\right|_{\tau = \tau_0} = 0 \; . \tag{24}$$

Equation (24) shows that at any distance z along the direction of propagation the self-phase modulation vanishes whenever τ reaches the value given by (23). But from eq. (21) one already knows that $\tau_+ = -1/2a$ which yields

$$\tau_0 = \tau_+ = -\frac{1}{2}a \; . \tag{25}$$

Relation (25) shows that the time of intersection of the deformed pulse in its rest frame with the input pulse is the same as the time when the self-phase modulation passes through its zero, and that this relation is *invariant* with respect to the propagation distance z. (Note that this exact relation replaces the small intensity perturbative result which predicts the zero of the self-pahse modulation to be coincident with the peak of the pulse; see eq. (19)).

(V) Two-Photon Optical Shocks

One can go further in this discussion of the general properties of the (two-photon quasi-resonant) pulse deformation. Extensive theoretical investigations and not so extensive experimental results for single-photon resonant or near-resonant pulses show the existence of shock-like discontinuities [9,10] in the output pulses that propagate a sufficiently long distance along the cell. The early

theorectical work of DeMartini, Townes, Gustavson and Kelly [9] and
the experimental observation of Grischkowsky, Courtens and Armstrong
[5] are particularly relevant for present comparison. The present
two-photon solution predicts (as does a numerical solution of the
coupled partial differential equations (1) and (2), - for a Gaussian
pulse - by Grischkowsky, Loy and Liao [3]) the existence of shock-
like discontinuities in the two-photon case as well. What is more
is that the presence of strong coupling between the envelope and
phase of the propagating pulse causes the shock-like discontinuity
to occur also in the instantaneous self-phase modulation and the
instantaneous frequency detuning, all at the same time. The shock-
like discontinuities appear at a distance $z_s = \xi_s$ that can be
obtained from the characteristic equation of the solution for $F_2(\xi,\tau)$
given by

$$F_2(\xi,\tau) = F_2(0,t_0) \tag{26a}$$

$$\tau - t_c = -4\lambda D\varepsilon^2\xi \quad . \tag{26b}$$

This result is equivalent to that given in eq. (12). In this form
of the result it is easier to see the mathematical nature of the
discontinuity more clearly. Eq. (26a) shows that $F_2(\xi,\tau)$ remains
constant along a characteristic t_c. A discontinous change in magnitude is therefore involved in going from one constant magnitude
to another at the point where two adjacent characteristics may
happen to cross. One may identify such discontinuities with the
phenomena of optical shocks [9].   From (26b) the distance ξ_s may
be obtained by taking the difference of two characteristics t_s and
t_s' and demanding it to vanish in the limit $t_s \rightarrow t_s'$. Thus one finds

$$\xi_s^{-1} = \lim_{t_s \rightarrow t_s'} [a(0,t_s) - a(0,t_s')]/(t_s - t_s')$$

$$= + \frac{\partial}{\partial t_s} [a(0,t_s)] \quad .$$

In other words,

$$z_s^{-1} = 4\frac{\partial}{\partial t_s} [\lambda(0,t_s)D(0,t_s)\varepsilon^2(0,t_s)]. \tag{27}$$

The physical content of this result may be appreciated by noticing
that the deformation parameter a, as noted before, is associated
with the velocity of the pulse. Since $a(0,t_s)$ depends both on
intensity as well as on the instantaneous detuning, two portions
of the pulse that are initially separated in time by $\delta t = t_s - t_s'$
travel at different velocities and eventually pile-up onto one

another at a distance $z = z_S$.

It is instructive to consider the shock-distance z_S in the following two limiting cases

(Va)   Low Intensity Limit:

In this limit $s\varepsilon^2/(\Omega_{12}-2\omega') \ll 1$ and

$$\frac{1}{z_s} \Rightarrow -2\,\frac{\varepsilon_2}{s}\left(\frac{\omega}{c}\right)\delta\,\frac{\frac{\partial}{\partial t_s}(s\varepsilon^2(0,t_s))}{|\delta|^2} \quad . \tag{28}$$

(Vb)   High Intensity Limit:

In this limit $(\Omega_{12}-2\omega')/s\varepsilon^2 \ll 1$ and

$$\frac{1}{z_s} \Rightarrow -\delta\,\frac{\varepsilon_2}{s}\left(\frac{\omega}{c}\right)\frac{|\delta|}{s\varepsilon^2(0,t_s)}\,\frac{\frac{\partial}{\partial t_s}(s\varepsilon^2(0,t_s))}{s\varepsilon^2(0,t_s)} \quad . \tag{29}$$

Note that for $\delta>0$ (and since z is by definition positive), shock-like discontinuities form on the leading edge of the pulse, in both the cases, (Va) and (Vb). Since the shock-distance in both (28) and (29) is related to the slope of the intensity, for $\delta<0$, the discontinuity appear on the trailing edge of the pulse (after the input pulse maximum). It is also interesting to consider the very low and very high values of the intensity and its relation to shock-formation. From (28) and (29) one sees that as $\varepsilon \rightarrow 0$ and as $\varepsilon \rightarrow \infty$, respectively, the right hand sides of both the expressions tend towards zero and the shock-distance therefore receeds to infinity (that is to say, beyond the range of practical propagation) and in so doing it allows the respective low and high intensity steady-state-like propagation (as discussed in section (I)) to occur. Note also that the smallest value of z_S is associated with the point of maximum slope of the input pulse. It is therefore expected that the minimum shock-distance would be associated with the point of inflection of the pulse intensity and that for steeper input pulses, the minimum shock-distance is smaller than that of relatively flatter pulses.

ACKNOWLEDGEMENT

It is a pleasure to thank Dr. D. Grischkowsky for drawing my attention, in a stimulating private communication, to the argument given at the end of section (Ib).

References

1. R.P. Feynman, F.L. Vernon, Jr., and R.W. Hellwarth, J. Appl. Phys. *28*, 49 (1957).
2. M. Takatsuji, Phys. Rev. A *4*, 808 (1971), Phys. Rev. A *11*, 919 (1975); E.M. Belenov and I.A. Poluektov, Zh. E.T. Fiz. *56*, 1407 (1969), Sov. Phys. JETP *29*, 754 (1969); R.G. Brewer and E.L. Hahn, Phys. Rev. A *11*, 1641 (1976); D. Grischkowsky and R.G. Brewer, Phys. Rev. A *15*, 1789 (1977); J.C. Garrison, T.H. Einwohner and J. Wong, Phys. Rev. A *14*, 731 (1976).
3. D. Grischkowsky, M.M.T. Loy and P.F. Liao, Phys. Rev. A *12*, 2514 (1975).
4. D. Grischkowsky, Phys. Rev. Lett. *24*, 866 (1970); D. Grischkowsky and J.A. Armstrong, Phys. Rev. A *6*, 1566 (1972); D. Grischkowsky, Phys. Rev. A *7*, 2096 (1973).
5. D. Grischkowsky, E. Courtens, and J.A. Armstrong, Phys. Rev. Lett. *31*, 422 (1973).
6. F.H.M. Faisal, J. Phys. B: Atom. Molec. Phys. *10*, 2003 (1977).
7. F.H.M. Faisal, Phys. Lett. *60A*, 295 (1977).
8. E. Courtens, Phys. Rev. *21*, 3 (1968); S.L. McCall and E.L. Hahn, Phys. Rev. *183*, 457 (1969); D. Grischkowsky, Phys. Rev. A *7*, 2096 (1973).
9. F. DeMartini, C.H. Townes, T.K. Gustavson, and P.L. Kelley, Phys. Rev. *164*, 312 (1967).
10. L.A. Ostrovski, Zh. E. Th. Fiz. *51*, 1189 (1966), Sov. Phys. JEPT *24*, 797 (1967); R.J. Joenk and R. Landauer, Phys. Lett. *24A*, 228 (1967); T.K. Gustavson, J.P. Taran, H.A. Haus, R.J. Lifsitz, and P.L. Kelley, Phys. Rev. *177*, 306 (1969); V.M. Arutynunyan, N.N. Badalyan, V.A. Iradyan, and M.E. Movsesyan, Zh. E. Th. Fiz. *58*, 37 (1970), Sov. Phys. JETP *31*, 22 (1970); B.Y. Zeldovich and I.I. Sobelman, Pis'ma Zh. E. Th. Fiz. *13*, 182 (1971), JETP Lett. *13*, 129 (1971).

ON THE RADIATION EFFICIENCY OF QUASI-HOMOGENEOUS SOURCES OF

DIFFERENT DEGREES OF SPATIAL COHERENCE

E. Wolf

University of Rochester, Rochester, N.Y.

W.H. Carter[†]

Naval Research Laboratory, Washington, D.C.

1. INTRODUCTION

Recent researches on the foundation of radiometry[1,2] have made it possible to provide answers to a number of puzzling questions relating to radiation from sources of different states of coherence. In particular, a relationship between the coherence properties of a source and the directionality of the light that the source generates has been established[3,4] at least for a wide class of sources of practical interest, and the coherence properties of Lambertian sources have been clarified[5] to a large extent.

In the present paper we consider another problem in this general area, which may be stated as follows. Consider a planar source that emits radiation at a (macroscopically) steady rate and let $I_\omega^{(0)}(\underline{r})$ be the distribution, at frequency ω, of the intensity across the source. The total radiant flux F_ω, generated by the source at frequency ω, may be expected to depend not only on $I_\omega^{(0)}(\underline{r})$ but also on the degree of coherence $\mu_\omega^{(0)}(\underline{r}_1,\underline{r}_2)$ of the light in the source plane. The problem is to determine the total radiant flux generated by sources of identical intensity distributions as a functional of the degree of coherence.

In a recent paper[6] we derived an expression for the total radiant flux F_ω generated by so-called quasi-homogeneous sources. Sources of this type are characterized by the property that the degree of coherence $\mu_\omega^{(0)}(\underline{r}_1,\underline{r}_2)$ of the light in the source plane depends on the position of the two points $\underline{r}_1$ $\underline{r}_2$ only through the difference $\underline{r}_1 - \underline{r}_2$, and that the intensity $I_\omega^{(0)}(\underline{r})$, considered as a function of position $(\underline{r})$, varies so slowly across the source that it remains sensibly constant over intervals of the order of the correlation length of the light (the effective width of $\mu_\omega^{(0)}$. In addition, the linear dimensions of such a source are assumed to be large compared to the correlation length. In this paper we present a more direct derivation of the expression for the total flux generated by sources of this type and we discuss some of its consequences. In particular, we show that among quasi-homogeneous sources of the same intensity distribution a spatially fully coherent and cophasal source generates maximum possible total flux, whereas a source which is spatially completely incoherent (in a sense to be defined precisely later) does not radiate at all. These conclusions are in agreement with the limiting forms of results derived previously for the special class of Gaussian correlated quasi-homogeneous sources. We also consider quasi-homogeneous Lambertian sources and find that the amount of total flux which they generate is precisely one-half of that from a fully coherent and cophasal quasi-homogeneous source of the same intensity distribution.

2. A GENERAL EXPRESSION FOR THE TOTAL FLUX RADIATED BY A PARTIALLY COHERENT PLANAR SOURCE

We consider a (primary or secondary) planar source σ, occupying a finite portion of the plane $z = 0$ and radiating into the half-space $z > 0$. Let $V(\underline{r},t)$ be a member of the statistical ensemble that represents the light vibrations at a point P, specified by the position vector $\underline{r}$, at time t. We assume a steady source, so that the ensemble of V will be a stationary one and we assume it also to be ergodic. We will ignore polarization effects and hence take $V(\underline{r},t)$ to be scalar.

Let $v(\underline{r},\omega)$ be the Fourier frequency transform[7] of $V(\underline{r},t)$

$$v(\underline{r},\omega) = \int_{-\infty}^{\infty} V(\underline{r},t) e^{i\omega t} dt \quad . \tag{2.1}$$

The cross-spectral density function $W(\underline{r}_1,\underline{r}_2;\omega)$ of the light at two points $P_1(\underline{r}_1)$ and $P_2(\underline{r}_2)$ in the half-space $z \geq 0$ may then be defined by the formula

$$\langle v(\underline{r}_1,\omega) \, v^*(\underline{r}_2,\omega') \rangle = W(\underline{r}_1,\underline{r}_2;\omega) \, \delta(\omega-\omega') \quad . \tag{2.2}$$

where the asterisk denotes the complex conjugate, the angular brackets
denote the ensemble average and δ is the Dirac delta function. It
follows from the generalized Wiener-Khintchine theorem that the
cross-spectral density function is equal to the Fourier transform
of the mutual coherence function,

$$\Gamma(\underline{r}_1,\underline{r}_2,\tau) = \langle V(\underline{r}_1,t+\tau)V^*(\underline{r}_2,t)\rangle , \qquad (2.3)$$

i.e. that

$$W(\underline{r}_1,\underline{r}_2,\omega) = \int_{-\infty}^{\infty} \Gamma(\underline{r}_1,\underline{r}_2,\tau)e^{i\omega\tau}d\tau. \qquad (2.4)$$

Since according to Eq. (2.2) the different frequency components
are uncorrelated, it is sufficient to analyze the problem for a
single frequency component only and we will do so from now on. To
simplify the notation we will omit the argument ω, it being understood that all our subsequent formulae refer to one and the same
frequency component.

To obtain an expression for the total flux radiated by the
source we recall first a relation that connects the cross-spectral
density $W^{(\infty)}(\underline{R}_1,\underline{R}_2)$ in the far zone and the spectral density
$W^{(0)}(\underline{r}_1,\underline{r}_2)$ in the source plane[8]. If $\underline{R}_1 = R_1\underline{s}_1$, $\underline{R}_2 = R_2\underline{s}_2$, where
$\underline{s}_1$ and $\underline{s}_2$ are unit vectors [see Fig. 1], this relation may be written
in the form [9]

$$W^{(\infty)}(R_1\underline{s}_1,R_2\underline{s}_2) = (2\pi k)^2\cos\theta_1\cos\theta_2\, \tilde{W}^{(0)}(k\underline{s}_{1\perp},-k\underline{s}_{2\perp})\frac{e^{ik(R_1-R_2)}}{R_1R_2}.$$

$$(2.5)$$

In this formula, $\tilde{W}^{(0)}$ is the four-dimensional spatial Fourier transform of the cross-spectral density function $W^{(0)}$ of light in the
source plane, viz.

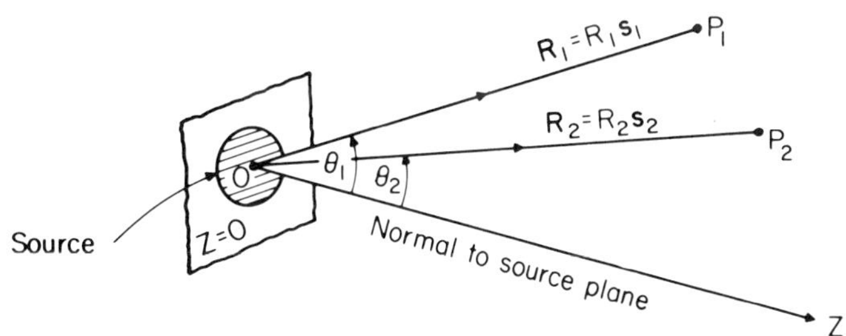

Fig. 1: Illustration of the notation relating to formula (2.5)

$$\tilde{W}^{(o)}(\underline{f}_1,\underline{f}_2) = \frac{1}{(2\pi)^4}\iint W^{(o)}(\underline{r}_1,\underline{r}_2)\, e^{-i(\underline{f}_1\cdot\underline{r}_1+\underline{f}_2\cdot\underline{r}_2)}\, d^2r_1\, d^2r_2,$$

$$(2.6)$$

where $\underline{f}_1,\underline{f}_2$ are two-dimensional (spatial-frequency) vectors and the integration extends twice independently over the source plane $z = 0$. Further, $\underline{s}_{1\perp}$ and $\underline{s}_{2\perp}$ are the projections (considered as vectors) of the unit vectors $\underline{s}_1$ and $\underline{s}_2$, respectively, onto the plane of the source and θ_1 and θ_2 are the angles that $\underline{s}_1$ and $\underline{s}_2$ make with the z-axis (the normal to the source). The formula (2.5) is to be understood as an asymptotic approximation valid as $kR_1 \to \infty$ and $kR_2 \to \infty$, with $\underline{s}_1$ and $\underline{s}_2$ being kept fixed and with $k = \omega/c$ (c = speed of light in vacuo) being, of course, the wave number associated with the frequency ω.

The radiant intensity $J(\underline{s})$, which represents the energy flux in the far zone per unit solid angle around the s-direction is equal to $R^2 I^{(\infty)}(R\underline{s})$, where $I^{(\infty)}(R\underline{s}) \equiv W^{(\infty)}(R\underline{s},R\underline{s})$ is just the optical intensity [10]; hence, using (2.5),

$$J(\underline{s}) = (2\pi k)^2\cos^2\theta\, \tilde{W}^{(o)}(k\underline{s}_\perp, - k\underline{s}_\perp),\qquad (2.7)$$

where θ is the angle that the s-direction makes with the source plane and $\underline{s}_\perp$ is the projection of the unit vector $\underline{s}$, (considered as a vector), onto the source plane $z = 0$ (see Fig. 2). Hence, on using (2.7), we see that the total flux

$$F = \int_{(2\pi)} J(\underline{s})\, d\Omega \qquad (2.8)$$

radiated by the source is given by

$$F = (2\pi k)^2 \int_{(2\pi)} \tilde{W}^{(o)}(k\underline{s}_\perp, - k\underline{s}_\perp)\, \cos^2\theta\, d\Omega. \qquad (2.9)$$

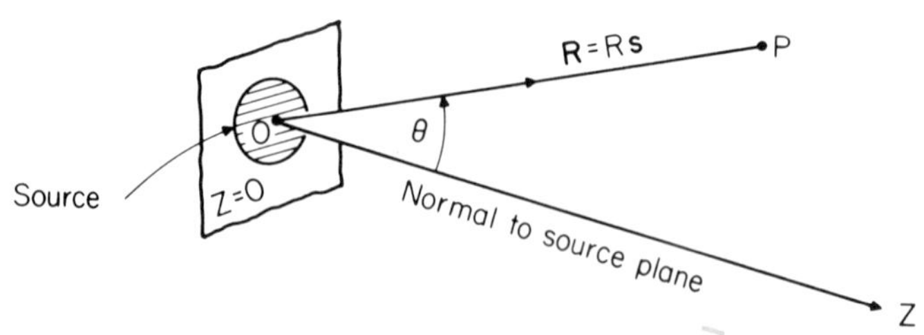

Fig. 2: Notation relative to formula (2.7).

The factor (2π) under the integral signs in Eqs. (2.8) and (2.9)
indicates that the integrals extend over the solid angle subtended
by the hemisphere in the half-space $z > 0$, i.e., with $\underline{s}_\perp \equiv (\cos\theta\cos\phi,$
$\cos\theta\sin\phi)$, $d\Omega = \sin\theta\ d\theta d\phi$, the ranges of the integrations are
$0 \leq \theta < \pi/2,\ 0 \leq \phi < 2\pi$.

3.   THE TOTAL FLUX RADIATED BY A QUASI-HOMOGENEOUS SOURCE

The formula (2.9) gives the flux radiated by any finite planar
source, whatever its state of coherence. We will now specialize it
to a class of sources, known as *quasi-homogeneous sources*, that
have recently been investigated [6]. We begin by briefly explaining
what is meant by a source of this kind.

Let $\mu^{(o)}(\underline{r}_1,\underline{r}_2)$ be the complex degree of spectral coherence [11]
of the light at points $P_1(\underline{r}_1)$ and $P_2(\underline{r}_2)$ in the source plane $z = 0$,
i.e.

$$\mu^{(o)}(\underline{r}_1,\underline{r}_2) = \frac{W^{(o)}(\underline{r}_1,\underline{r}_2)}{[I^{(o)}(\underline{r}_1)]^{\frac{1}{2}}\,[I^{(o)}(\underline{r}_2)]^{\frac{1}{2}}} \quad , \qquad (3.1)$$

where

$$I^{(o)}(\underline{r}_j) = W^{(o)}(\underline{r}_j,\underline{r}_j) \quad , \qquad (j = 1,2) \quad , \qquad (3.2)$$

is the optical intensity at the point $\underline{r}_j$.

For many sources of practical interest the complex degree of
spectral coherence will, to a good approximation, depend on $\underline{r}_1$ and
$\underline{r}_2$ only through the difference vector $\underline{r}_1 - \underline{r}_2$, i.e. it will be of
the form

$$\mu^{(o)}(\underline{r}_1,\underline{r}_2) = g^{(o)}(\underline{r}_2-\underline{r}_1) \quad , \qquad (3.3)$$

where $g^{(o)}(\underline{r}')$ is some function of a two-dimensional vector
variable $\underline{r}'$. It follows from (3.1) and (3.3) that the cross-
spectral density function of the light in the source plane is then
of the form

$$W^{(o)}(\underline{r}_1,\underline{r}_2) = [I^{(o)}(\underline{r}_1)]^{\frac{1}{2}}\,[I^{(o)}(\underline{r}_2)]^{\frac{1}{2}}\,g^{(o)}(\underline{r}_1-\underline{r}_2) \quad . \qquad (3.4)$$

In many cases of practical interest the linear dimensions of the
source are large compared to the correlation length across the source

(the effective width of $|g^{(o)}(\underline{r}')|$) and, moreover, the correlation
coefficient $g^{(o)}(\underline{r}')$ will, as a rule, vary much more rapidly with
$\underline{r}'$ than the optical intensity $I^{(o)}(\underline{r})$ will vary with $\underline{r}$. An example
of this situation is a primary thermal source, e.g. a blackbody
source. For such a source $|g(\underline{r}')|$ will only have appreciable values
when the separation $r' = |\underline{r}_1 - \underline{r}_2|$ of the two points is of the order
of, or less than, the mean wavelength of the radiation emitted by
the source. On the other hand the optical intensity $I^{(o)}(\underline{r})$ will
remain sensibly constant over a large portion of the source area,
often of linear dimensions many thousands or millions of wavelengths.
In analogy with the terminology employed in the theory of locally
homogeneous turbulence, we will say that in such a case $g(\underline{r}')$ is a
fast function of $\underline{r}'$ and that $I(\underline{r})$ is a *slow* function of $\underline{r}$. This
situation is illustrated in Fig. 3, which for simplicity refers to
a one-dimensional source.

It is clear that under the circumstances that we have just
discussed, we may replace the factor $[I^{(o)}(\underline{r}_1)]^{\frac{1}{2}} [I^{(o)}(\underline{r}_2)]^{\frac{1}{2}}$ in
(3.4) by $I^{(o)}((\underline{r}_1+\underline{r}_2)/2)$ and the expression for the cross-spectral
density then becomes

$$W^{(o)}(\underline{r}_1,\underline{r}_2) \approx I^{(o)}\left(\frac{\underline{r}_1+\underline{r}_2}{2}\right) g^{(o)}(\underline{r}_1-\underline{r}_2) . \tag{3.5}$$

When the cross-spectral density function of the light in the source
plane may be approximated in the form (3.5), where $I^{(o)}(\underline{r})$ is a slow
function of $\underline{r}$ and $g^{(o)}(\underline{r}')$ is a fast function of $\underline{r}'$ and when, moreover,
the linear dimensions of the source are large compared to the

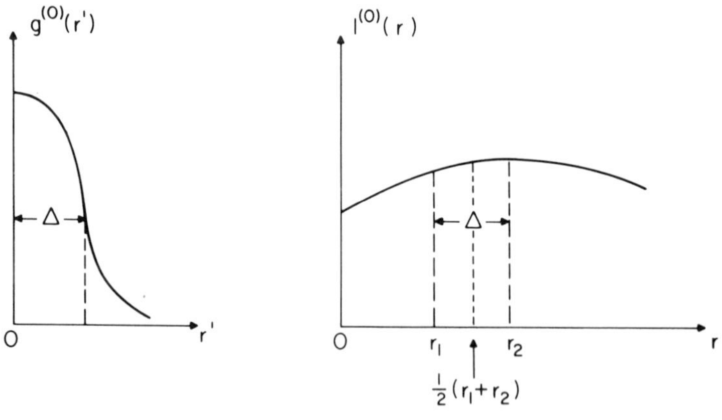

Fig. 3: Concept of a quasi-homogeneous source.

correlation length across the source, the source is said to be
quasi-homogeneous.

Let us now determine the total radiant flux generated by such
a source. According to Eq. (2.9) we must first obtain an expression
for the four-dimensional spatial Fourier transform $\tilde{W}^{(o)}(\underline{f}_1,\underline{f}_2)$ of
the cross-spectral density. On substituting from (3.5) into (2.6)
we readily obtain after a straightforward calculation, (carried out
in full in Sec. 3 of ref. 6), the following simple expression for
$\tilde{W}^{(o)}(\underline{f}_1,\underline{f}_2)$:

$$\tilde{W}^{(o)}(\underline{f}_1,\underline{f}_2) \approx \tilde{I}^{(o)}(\underline{f}_1+\underline{f}_2)\ \tilde{g}^{(o)}\left(\frac{\underline{f}_1-\underline{f}_2}{2}\right) \quad , \tag{3.6}$$

where $\tilde{I}^{(o)}$ and $\tilde{g}^{(o)}$ are two-dimensional spatial Fourier transforms
of the optical intensity $I^{(o)}$ and of the degree of spectral coherence
$g^{(o)}$, respectively, viz.

$$\tilde{I}^{(o)}(\underline{f}) = \frac{1}{(2\pi)^2} \int I^{(o)}(\underline{r})e^{-i\underline{f}\cdot\underline{r}}\ d^2r \quad , \tag{3.7}$$

$$\tilde{g}^{(o)}(\underline{f}) = \frac{1}{(2\pi)^2} \int g^{(o)}(\underline{r}')e^{-i\underline{f}\cdot\underline{r}'}\ d^2r' \quad . \tag{3.8}$$

It follows from (3.6) that

$$\tilde{W}^{(o)}(k\underline{s}_\perp, - k\underline{s}_\perp) = \tilde{I}^{(o)}(0)\tilde{g}^{(o)}(k\underline{s}_\perp) \quad . \tag{3.9}$$

On substituting from (3.9) into (2.9) we finally obtain the following
expression for the total radiant flux generated by a quasi-homogeneous
planar source:

$$F = (2\pi)^2\ \tilde{I}^{(o)}(0)C_g \quad , \tag{3.10}$$

where

$$C_g = k^2 \int_{(2\pi)} \tilde{g}^{(o)}(k\underline{s}_\perp)\ \cos^2\theta d\Omega \quad . \tag{3.11}$$

In view of the relationship (3.7), the formula (3.10) may be
expressed in the alternative form

$$F = C_g \int_\sigma I^{(o)}(\underline{r})d^2r \quad . \tag{3.10a}$$

The formula (3.10a) shows that the total flux radiated by a quasi-homogeneous source is a product of two terms. One of them, C_g, is, according to (3.11), entirely determined by the complex degree of spectral coherence of the source. The other factor is seen to be just equal to the integral of the optical intensity across the source. In particular, we see that two sources, with identical distribution of optical intensity but with different degrees of spatial coherence $g_1(\underline{r}')$ and $g_2(\underline{r}')$ generate total radiant fluxes F_1 and F_2 whose ratio is

$$\frac{F_2}{F_1} = \frac{C_{g_2}}{C_{g_1}} \; . \tag{3.12}$$

Since the factor C_g, defined by Eq. (3.11), is a measure of how "efficient" the source is regarding the generation of total flux, we will call C_g "the efficiency factor" of the quasi-homogeneous source [12].

The formula (3.11) expresses the efficiency factor C_g in terms of the spatial Fourier transform $\tilde{g}^{(o)}$ of the complex degree spectral coherence $g^{(o)}$ of the source. It is possible to express C_g directly in terms of $g^{(o)}$. The derivation is given in Section 6 of ref. 6; the result is

$$C_g = \frac{k^2}{2(2\pi)^{\frac{1}{2}}} \int g^{(o)}(\underline{r}') \frac{J_{3/2}(kr')}{(kr')^{3/2}} \, d^2r' \; , \tag{3.13}$$

where [13]

$$J_{3/2}(x) = \left(\frac{2}{\pi x}\right)^{\frac{1}{2}} \left(\frac{\sin x}{x} - \cos x\right) \tag{3.14}$$

is the Bessel function of the first kind and order 3/2 and the integration in (3.13) is carried over the plane $z = 0$.

It is also shown in reference 6 that the efficiency factor is always real and is bounded by the values zero and unity:

$$0 \le C_g \le 1 \; . \tag{3.15}$$

We will now show that for a source that is spatially completely coherent and cophasal the efficiency factor has the maximum possible value, unity; and that for a spatially completely incoherent source (in a sense that is defined precisely below) the efficiency factor has the other extreme value, zero. We will also show that for a Lambertian source the efficiency factor takes on the value $\frac{1}{2}$.

4. THE EFFICIENCY FACTOR OF COHERENT, INCOHERENT
AND LAMBERTIAN SOURCES

4.1 Spatially Coherent Source

For a source that is spatially coherent at some particular
frequency, the absolute value of its degree of spectral coherence
at that frequency has the extreme value, unity, i.e. $\left|g^{(o)}(\underline{r}')\right| \equiv 1$.
If, in addition, the source is also cophasal, the phase of $g^{(o)}(\underline{r}')$
is a constant and, since necessarily $g^{(o)}(0) = 1$, this constant
must have the value zero, i.e. we then have

$$g^{(o)}(\underline{r}') \equiv 1 \ . \tag{4.1}$$

It follows at once on substituting from (4.1) into (3.8) that
the two-dimensional spatial Fourier transform $\tilde{g}^{(o)}(\underline{f})$ of $g^{(o)}(\underline{r}')$
is now a two-dimensional Dirac delta function:

$$\tilde{g}^{(o)}(\underline{f}) = \delta^{(2)}(\underline{f}) \ . \tag{4.2}$$

With $g^{(o)}(\underline{f})$ given by (4.2) the expression (3.11) for the efficiency
factor becomes

$$C_g = k^2 \int_{(2\pi)} \delta(ks_x)\delta(ks_y)\cos^2\theta d\Omega \ , \tag{4.3}$$

where δ is the one-dimensional Dirac delta function and s_x and s_y
are the components of the unit vector $\underline{s}_\perp$ in two mutually orthogonal
directions in the source plane $z = 0$. Now

$$\cos^2\theta = 1 - s_x^2 - s_y^2 \tag{4.4}$$

and the element $d\Omega$ of the solid angle may readily be shown to be
expressible in the form

$$d\Omega = \frac{ds_x \, ds_y}{[1 - s_x^2 - s_y^2]^{\frac{1}{2}}} \ . \tag{4.5}$$

Using (4.4) and (4.5) in (4.3) and also making use of the relations
$\delta(ks_x) = (1/k)\delta(s_x)$, $\delta(ks_y) = (1/k)\delta(s_y)$ we obtain for C_g the
expression

$$C_g = \iint_{s_x^2 + s_y^2 \leq 1} \delta(s_x)\delta(s_y) \, [1-s_x^2-s_y^2]^{\frac{1}{2}} \, ds_x ds_y \ ,$$

which gives at once

$$C_g = 1 \ . \tag{4.6}$$

Thus we see that for a spatially coherent and cophasal source, the efficiency factor takes on the maximum possible value, unity [cf (3.15)], and the total radiant flux generated by such a source may according to (4.6) and (3.10a) be expressed in the form

$$F = \int_\sigma I^{(o)}(\underline{r})d^2r \ . \tag{4.7}$$

4.2 Spatially Incoherent Source

Several different definitions of a spatially incoherent source can be found in the literature. The most commonly used definition is the following expression for the cross-spectral density in the source plane:

$$W^{(o)}(\underline{r}_1,\underline{r}_2) = i(\underline{r}_1) \ \delta^{(2)}(\underline{r}_1-\underline{r}_2) \ . \tag{4.8}$$

Here $i(\underline{r})$ is some non-negative function of position, which is identically zero when $\underline{r}$ represents a point outside the area occupied by the source. The definition (4.8) implies that the optical intensity $I^{(o)}(\underline{r}) = W^{(o)}(\underline{r},\underline{r})$ is infinite at each source point, and clearly the associated degree of spectral coherence is now undefined, since the cross-spectral density function (4.8) cannot be normalized in the manner indicated by Eqs. (3.1) and (3.2).

We will use an alternative definition of a spatially incoherent source which ensures that the distribution of optical intensity $I^{(o)}(\underline{r})$ across it is finite, and which, nevertheless, contains the essential aspect of spatial incoherence, namely the complete absence of correlation between the values of the field variable at any two points of the source. These conditions are exhibited by the cross-spectral density function

$$\begin{aligned} W^{(o)}(\underline{r}_1,\underline{r}_2) &= I(\underline{r}_1) \qquad &\text{if } \underline{r}_2 = \underline{r}_1 \\ &= 0 \qquad &\text{if } \underline{r}_2 \neq \underline{r}_1 \ , \end{aligned} \tag{4.9}$$

where $I(\underline{r})$ is some non-negative function of position. We see that (4.9) is a degenerate form of Eq. (3.5), with

$$g^{(o)}(\underline{r}') = 1 \qquad \text{if} \qquad \underline{r}' = 0$$
$$\phantom{g^{(o)}(\underline{r}')} = 0 \qquad \text{if} \qquad \underline{r}' \neq 0. \tag{4.10}$$

Since $g^{(o)}(\underline{r}')$ differs from zero only at one point ($\underline{r}' = 0$) in the $\underline{r}'$-plane, it follows at once from (3.13) that in this case

$$C_g \equiv 0 . \tag{4.11}$$

Thus, for a source that is spatially incoherent in the sense of the definition (4.9) the efficiency factor has zero value and hence according to (3.10a)

$$F = 0 , \tag{4.12}$$

showing that no flux of radiation is generated by such a source.

Obviously our model of an incoherent source is an idealization that is never satisfied in practice. Nevertheless, it may be regarded as a limiting case of realizable sources. Consider, for example, radiation from a planar quasi-homogeneous source whose degree of spectral coherence is described by a Gaussian distribution:

$$g^{(o)}(\underline{r}') = e^{-r'^2/2\sigma^2} . \tag{4.13}$$

This case has been studied in detail in ref. 6, where a curve was given (reproduced here as Fig. 4), that shows the behavior of the efficiency factor C_g for this class of sources as a function of the root-mean-square width σ. Evidently σ is a rough measure of the correlation distance of the light in the source plane. As $\sigma \to \infty$,

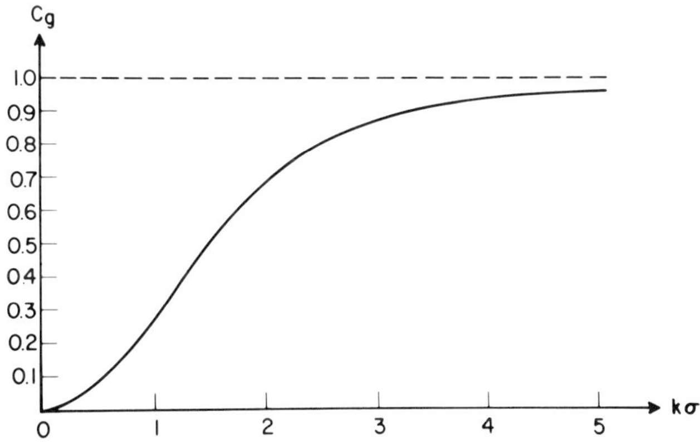

Fig. 4: The efficiency factor C_g of a Gaussian correlated quasi-homogeneous source [$g^{(0)}(\underline{r}') = e^{-r'^2/2\sigma^2}$] as a function of the normalized r.m.s. width $k\sigma$. [After W.H. Carter and E. Wolf, J. Opt. Soc. Amer. *67*, 785 (1977)].

the source becomes spatially fully coherent and cophasal in the sense of our definition (4.1), and as $\sigma \to 0$, it becomes spatially incoherent in the sense of the definition (4.10). Figure 4 shows

$$C_g \to 1 \quad \text{as} \quad \sigma \to \infty$$

and

$$C_g \to 0 \quad \text{as} \quad \sigma \to 0$$

in agreement with our general results expressed by Eqs. (4.6) and (4.11), respectively.

4.3 Lambertian Source

Another class of sources of special interest are Lambertian sources, i.e. sources for which the radiant intensity $J(\underline{s}) \equiv R^2 I^{(\infty)}(R\underline{s})$, considered as a function of the angle θ which the $\underline{s}$-vector makes with the normal to the source plane, falls off in proportion to $\cos \theta$ (see Fig. 2), i.e.

$$J(\underline{s}) \equiv R^2 I^{(\infty)}(R\underline{s}) = C \cos\theta , \qquad (kR \to \infty) , \qquad (4.14)$$

where C is a constant. We will now determine the efficiency factor for sources of this type.

It follows from the analysis presented in Sec. II of ref. 5 that for a quasi-homogeneous Lambertian source, the cross-spectral density function of the light in the area occupied by the source is necessarily of the form

$$W^{(o)}(\underline{r}_1,\underline{r}_2) = C' \left[\frac{\sin k|\underline{r}_1-\underline{r}_2|}{k|\underline{r}_1-\underline{r}_2|} \right] + W^{HF}(\underline{r}_1-\underline{r}_2) , \qquad (4.15)$$

where C' is a constant and $W^{HF}(\underline{r}_1 - \underline{r}_2)$ is the so-called high-frequency part of the cross-spectral density function, that is, a function with the property that its two-dimensional spatial Fourier transform $\tilde{W}^{HF}(\underline{f})$ is identically zero when $\underline{f}^2 < k^2$. The term $W^{HF}(\underline{r}_1 - \underline{r}_2)$ in the expression (4.15) has been shown not to contribute at all to the far field, being associated with effects of evanescent waves. Hence in calculating the efficiency factor of Lambertian sources, this term may be omitted in (4.15), With this being understood the complex degree of spectral coherence of a quasi-homogeneous source that radiates as a Lambertian source may be taken according to (4.15), (3.1) and (3.3), to be of the form

$$g^{(o)}(\underline{r}') = \frac{\sin kr'}{kr'} \quad . \tag{4.16}$$

The two-dimensional spatial Fourier transform of (4.6) is
evaluated in the Appendix. The result is

$$\tilde{g}^{(o)}(\underline{f}) = \frac{1}{2\pi k} \frac{1}{(k^2 - f^2)^{\frac{1}{2}}} \quad \text{when} \quad \underline{f}^2 < k^2$$

$$= 0 \qquad \text{when} \quad \underline{f}^2 > k^2 \quad . \tag{4.17}$$

It follows on substituting from (4.17) into (3.11) that the efficiency
factor is now given by the integral

$$C_g = \frac{1}{2\pi} \int \frac{1}{(1-\underline{s}_\perp^2)^{\frac{1}{2}}} \cos^2\theta d\Omega \quad . \tag{4.18}$$

On using the fact that $\underline{s}^2 = s_x^2 + s_y^2$ and on using Eqs. (4.4) and
(4.5), the expression (4.18) is readily seen to reduce to

$$C_g = \frac{1}{2\pi} \iint_{s_x^2 + s_y^2 \leq 1} ds_x ds_y \quad ,$$

which implies that for a quasi-homogeneous Lambertian source the
efficiency factor C_g has the value $\frac{1}{2}$,

$$C_g = \tfrac{1}{2} \quad . \tag{4.19}$$

Hence, using (3.10a) it follows that the total flux radiated by such
a source is given by

$$F = \tfrac{1}{2} \int_\sigma I^{(o)}(\underline{r}) d^2 r \quad . \tag{4.20}$$

Recalling our earlier results about coherent sources [Eqs. (4.6)
and (4.7)] it follows that a Lambertian source is half as efficient
as a fully coherent and cophasal source of the same intensity
distribution.

Let us briefly summarize our results. We derived an expression
for the total radiant flux generated by a quasi-homogeneous source.
We found that the total flux is expressible as the product of two
factors: the first factor is the integrated optical intensity
across the source and the second factor (which we call the efficiency

factor), depends entirely on the degree of coherence of the source. We showed that among quasi-homogeneous sources with the same intensity distribution, a completely coherent and cophasal source is the most efficient one, a completely incoherent source is the least efficient one, and a Lambertian source is exactly half as efficient as the fully coherent source.

Appendix: EVALUATION OF THE TWO-DIMENSIONAL FOURIER TRANSFORM OF THE EXPRESSION (4.16).

We have

$$g^{(o)}(r') = \frac{\sin kr'}{kr'} \quad .   \tag{A1}$$

The two-dimensional spatial Fourier transform of this expression is given by

$$\tilde{g}^{(o)}(\underline{f}) = \frac{1}{(2\pi)^2} \int \left(\frac{\sin kr'}{kr'} \right) e^{-i\underline{f}\cdot\underline{r}'} d^2 r' \quad .   \tag{A2}$$

Let θ and ϕ' be the angles that the two-dimensional vectors $\underline{f}$ and $\underline{r}'$ make with the x-axis in the source plane. Then

$$\underline{f}\cdot\underline{r}' = fr'\cos(\theta-\phi'), \quad d^2 r' = r'dr'd\phi' \ ,   \tag{A3}$$

and (A2) becomes

$$\tilde{g}^{(o)}(\underline{f}) = \frac{1}{k(2\pi)^2} \int_0^\infty \int_0^{2\pi} \sin kr' \ e^{-ifr'\cos(\theta-\phi')} dr'd\phi' \quad .   \tag{A4}$$

The integration with respect to ϕ' can be carried out at once and gives [14]

$$\int_0^{2\pi} e^{-ifr'\cos(\theta-\phi')} d\phi' = 2\pi J_0(fr') \ ,   \tag{A5}$$

where J_0 is the Bessel function of the first kind and zero order. Hence (A4) reduces to

$$\tilde{g}^{(o)}(\underline{f}) = \frac{1}{2\pi k} \int_0^\infty \sin kr' \ J_0(fr')dr' \quad .   \tag{A6}$$

The value of the integral that occurs in (A6) is known to be [15]

$$\int_0^\infty \sin kr' \; J_o(fr')dr' = \frac{1}{(k^2-f^2)^{\frac{1}{2}}} \quad \text{when } f^2 < k^2$$
$$= 0 \quad \text{when } f^2 > k^2 \, , \qquad\qquad (A7)$$

and it finally follows from (A6) and (A7) that

$$\tilde{g}^{(o)}(\underline{f}) = \frac{1}{2\pi k} \frac{1}{(k^2-f^2)^{\frac{1}{2}}} \quad \text{when } f^2 < k^2$$
$$= 0 \quad \text{when } f^2 > k^2 \, .$$

† Visiting Research Fellow, Physics Department, University of
Reading, Reading, England during the academic year 1976-1977.

References

1. A. Walther, J. Opt. Soc. Am. *58*, 1256 (1968).
2. E.W. Marchand and E. Wolf, J. Opt. Soc. Am. *64*, 1219 (1974).
3. E. Wolf and W.H. Carter, Opt. Commun. *13*, 205 (1975).
4. B. Steinle and H.P. Baltes, J. Opt. Soc. Am. *67*, 241 (1977).
5. W.H. Carter and E. Wolf, J. Opt. Soc. Am. *65*, 1067 (1975).
6. W.H. Carter and E. Wolf, J. Opt. Soc. Am. *67*, 785 (1977).
7. For reasons well known in the theory of stationary random
   processes, the Fourier transform $v(\underline{r},\omega)$ does not exist in the
   sense of the ordinary function theory, and must be understood
   to be a generalized function.
8. Superscripts (∞) and (o) label quantities pertaining to the
   far zone and to the source plane, respectively.
9. E.W. Marchand and E. Wolf, J. Opt. Soc. Am. *62*, 379 (1972),
   Eq. (34).
10. The quantity $I(\underline{r}) = W(\underline{r},\underline{r})$ represents what is traditionally
    known in physical optics as simply the intensity. [The rather
    inappropriate term 'irradiance', which indicates a confusion
    between radiometry and physical optics, has also been frequently
    employed in recent literature]. Throughout this paper we refer
    to $I(\underline{r})$, as the *optical intensity* to distinguish it clearly from
    the radiometric concept of *radiant intensity* that we denote by
    $J(\underline{s})$.
11. L. Mandel and E. Wolf, J. Opt. Soc. Am. *66*, 529 (1976).
12. The efficiency factor corresponds to the so-called "transfer
    factor" that was recently discussed in the context of another
    class of sources by H.P. Baltes, B. Steinle and G. Antes,

this Volume, p. 431.

13. There is a misprint in the corresponding formula (6.7) of Ref.
     6. On the right-hand side of that formula the factor $(2/\pi x)^{3/2}$
     should be replaced by $(2/\pi x)^{1/2}$.

14. See, for example, G.N. Watson, *A Treatise on the Theory of
     Bessel Functions* (Cambridge University Press, 1922) p. 20,
     Eq. (5) (with an obvious substitution).

15. I.S. Gradsteyn and I.M. Ryshik, *Tables of Integrals, Series
     and Products* (Academic Press, New York, 1965) p. 730, formula
     1 of 6.671.

RADIOMETRIC AND CORRELATION PROPERTIES OF BOUNDED PLANAR SOURCES

H. P. Baltes, B. Steinle, and G. Antes

Landis and Gyr Zug AG, Zug, Switzerland

1. INTRODUCTION

Since the important seminal paper by Walther [1] there has
been a growing interest in radiometry [2-10] and interferometry
[10-16] of partially coherent sources of time harmonic electromagnetic radiation, as well as in the inverse problem [4,6,9,14-16] of
specifying the spatial correlation or intensity profile in the
source plane from optical far-field data. These optical studies
are complemented by current investigations of partial coherence in
electron microscopy [16-18]. Statistically homogeneous [3-5,9] and
quasi-homogeneous [1,7,10,14-16] model sources have been considered
hitherto. The calculation [10] of the transfer factor that controls
the radiant flux is of current interest.

The quasi-homogeneous model can, however, not account for situations where the intensity profile in the source plane varies appreciably over the distance for which the correlation differs sensibly
from zero, as described e.g. in [19]. In Sec. 2 of the present
contribution, we calculate the far-zone degree of angular coherence,
the radiant intensity, and the related radiometric properties of
primary or secondary sources of arbitrary size and correlation length
in terms of the Schell-type model source [8,12,15]. We present analytical results in the case of Gaussian-correlated sources with a
Gaussian intensity profile and establish simultaneous scaling properties of the coherence angle and the intensity spread in the far
zone. Our theory comprizes previous results [1,10,14-16] in the
limit of small correlation length and large beam width. Furthermore,
we calculate the radiant emittance of Bessel correlated sources
with slowly varying intensity profile.

431

The current studies of the radiometry of sources of any state
of coherence [1-18] are concerned with the radiant intensity and
the degree of first-order spatial coherence, but do not take into
consideration the fluctuations of the radiant intensity and the related second-order correlations. In Sec. 3 of the present note
we introduce the basic concepts and relations of second-order radiometry concerned with the radiant noise and the radiant intensity
autocorrelation. We propose the new radiometric concepts of second-
order radiance and second-order radiant emittance and show how these
quantities may be expressed in terms of the second-order spatial
correlation function of the source. We use the standard quasi-
monochromatic approach and ignore polarization effects.

2.   FIRST-ORDER PROPERTIES

2.1 Radiant Intensity and Angular Coherence

We denote the stochastic scalar field amplitude and its conjugate by u and $u^{\dagger}$. We describe the source in terms of the spatial
coherence function

$$W_{\omega}(\underline{r}_1,\underline{r}_2) \equiv \langle u^{\dagger}(\underline{r}_1)\, u(\underline{r}_2)\rangle_{\omega} \tag{1}$$

with $\underline{r}_1$ and $\underline{r}_2$ denoting positions in the source plane $z = 0$. We
characterize far-zone positions in the half-space $z > 0$ by their
distance R from the origin (i.e. the center of the source) and by
their direction $\underline{s} = \sin\theta\,(\cos\psi, \sin\psi)$. We are interested in
the far-zone angular coherence function

$$W_{\omega}(\underline{s}_1,\ \underline{s}_2) \equiv R^2 \langle u^{\dagger}(R,\underline{s}_1)u(R,\underline{s}_2)\rangle_{\omega} \tag{2}$$

and in the radiant intensity

$$J_{\omega}(\underline{s}) \equiv W_{\omega}(\underline{s},\underline{s}). \tag{3}$$

In agreement with Schell's theorem [20,21] we account for the effective size of the source in terms of [8,12,15]

$$W_{\omega}(\underline{r}_1,\underline{r}_2) = [I(\underline{r}_1)I(\underline{r}_2)]^{1/2}\, \mu(\underline{r}), \qquad \underline{r} = \underline{r}_1 - \underline{r}_2, \tag{4}$$

where $I(\underline{r}_i)$ denotes the intensity profile and $\mu(\underline{r})$ denotes the
degree of spectral coherence in the source plane with $\mu(0) = 1$. The
corresponding far-zone angular coherence function is given by

$$W_{\omega}(\underline{s}_1,\underline{s}_2) \propto \cos\theta_1\ \cos\theta_2 \int_{z=0} d^2\underline{r}\, e^{-ik\underline{s}\underline{r}}\, \mu(\underline{r})$$

$$\times \int_{z=0} d^2\rho\, e^{-ik\underline{\sigma}\underline{\rho}}\, [I(\underline{\rho} + \underline{r}/2)\, I(\underline{\rho} - \underline{r}/2)]^{1/2} \tag{5}$$

with $k = \omega/c$,

$$\underline{s} = (\underline{s}_1 + \underline{s}_2)/2, \quad \underline{\sigma} = \underline{s}_1 - \underline{s}_2, \tag{6}$$

and

$$\underline{r} = \underline{r}_1 - \underline{r}_2, \quad \underline{\rho} = (\underline{r}_1 + \underline{r}_2)/2. \tag{7}$$

The corresponding radiant intensity reads

$$J_\omega(\underline{s}) \propto \cos^2\theta \; FT\{\mu\} \otimes [FT\{I^{1/2}\}]^2, \tag{8}$$

where FT denotes the Fourier transform and $\otimes$ the convolution. In the case of a source with isotropic Gaussian intensity profile of width a, viz.

$$I(\underline{r}_i) = A \exp (-\underline{r}_i^2/2a^2), \quad A = const, \quad i = 1,2, \tag{9}$$

and with Gaussian correlation of correlation length b, viz.

$$\mu(\underline{r}) = \exp (-\underline{r}^2/2b^2), \tag{10}$$

the convolutions (5) and (9) can be evaluated analytically and lead to the radiant intensity

$$J_\omega(\theta) = J_\omega(0) \cos^2\theta \; \exp [-\sin^2\theta /2(\Delta I)^2] \tag{11}$$

with the "beam-spread"

$$\Delta I = (bk)^{-1}[1 + (b/2a)^2]^{1/2} \tag{12}$$

and the degree of angular coherence

$$\mu(\sigma) \equiv W_\omega(\underline{s}_1,\underline{s}_2) [J_\omega(\underline{s}_1)J_\omega(\underline{s}_2)]^{-1/2} = \exp [-\sigma^2/2(\Delta\mu)^2] \tag{13}$$

with the "coherence angle"

$$\Delta\mu = (ak)^{-1}[1 + (b/2a)^2]^{1/2} = (b/a)\Delta I, \tag{14}$$

where $\sigma = |\underline{\sigma}| = |\underline{s}_1 - \underline{s}_2|$. We observe that the variances (12) and (14) show the same scaling factor $[1+(b/2a)^2]^{1/2}$ with respect to the quasi-homogeneous limit $b \ll a$.

2.2 Radiance and Radiant Emittance

The radiant intensity (3) can be written as

$$J_\omega(\underline{s}) = \cos\theta \int_{z=0} d^2\underline{\rho} \; B_\omega(\underline{\rho},\underline{s}), \tag{15}$$

where $B_\omega(\underline{\rho},\underline{s})$ denotes Walther's generalized radiance [1,2,10] that can be expressed in terms of the first-order spatial coherence function (1), viz.

$$B_\omega(\underline{\rho},\underline{s}) = \left(\frac{k}{2\pi}\right)^2 \cos\theta \int_{z=0} d^2\underline{r}\, e^{-ik\underline{r}\underline{s}} W_\omega(\underline{\rho} + \frac{\underline{r}}{2}, \underline{\rho} - \frac{\underline{r}}{2}) \quad . \quad (16)$$

This relation leads to the generalized radiant emittance [2,10]

$$E_\omega(\underline{\rho}) \equiv \int_{2\pi} d\Omega\, \cos\theta\, B_\omega(\underline{\rho},\underline{s}) \qquad (17)$$

$$= (8\pi)^{-1} k^{-2} \int d^2\underline{r}\, (kr)^{-3/2} J_{3/2}(kr) W_\omega(\underline{\rho} + \frac{\underline{r}}{2}, \underline{\rho} - \frac{\underline{r}}{2}) \quad ,$$

where the first integration extends over the solid angle 2π formed by all the directions $\underline{s}$ that point into the half space $z > 0$, $J_{3/2}$ denotes the Bessel function of the first kind of order 3/2, and $r = |\underline{r}|$. We evaluate (16) and (17) for the Schell-type source (4) with Gaussian intensity profile (9) and Gaussian degree of spatial coherence (10). We obtain the radiance

$$B_\omega(\underline{\rho},\underline{s}) = (2\pi)^{-1} I(\underline{\rho}) \cos\theta\, (\Delta I)^{-2} \exp[-\sin^2\theta /2(\Delta I)^2] \qquad (18)$$

in terms of the beam spread (12). We find the emittance

$$E_\omega(\underline{\rho}) = C'_{ab} I(\underline{\rho}) \qquad (19)$$

with

$$C'_{ab} = C(1/\Delta I) = C\{bk[1 + (b/2a)^2]^{-1/2}\} \quad . \qquad (20)$$

The function $C(x)$ is defined as

$$C(x) = 1 - (\pi/2)^{1/2} x^{-1} \exp(-x^2/2) \operatorname{erfi}(2^{-1/2}x) \qquad (21)$$

and plotted in the upper part of Fig. 1. The transfer factor C'_{ab} depends on both the width a and the correlation length b of the source. We find $C'_{ab} \to 1$ provided that both $ka \gg 1$ and $kb \gg 1$, whereas $C'_{ab} \to 0$ if $a \to b$ or $b \to 0$. The relations (18) - (20) comprise previous results [10] in the limit $a/b \to \infty$ or $\Delta I \to (kb)^{-1}$. In Fig. 2 we show C'_{ab} as a function of kb for several values of the parameter $[1 + (b/2a)^2]^{1/2}$ (full curves). We observe that C'_{ab} decreases with both decreasing correlation length b and decreasing effective diameter a. The total flux is reduced accordingly.

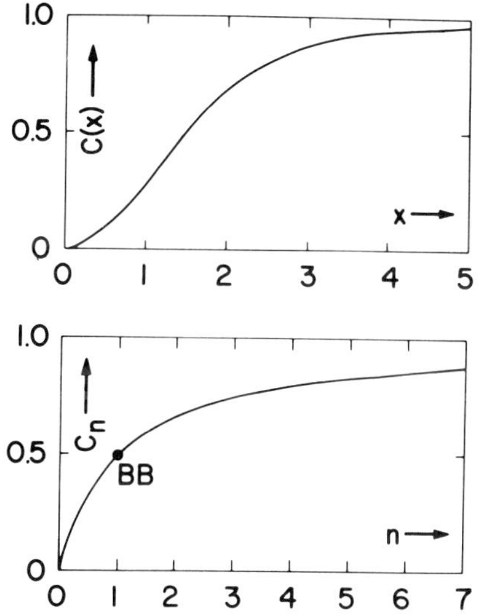

Fig. 1. Illustration of the transfer factors. Upper curve: the function $C(x)$ defined by (21). Lower curve: the transfer factor of Bessel-correlated sources with BB indicating the case of the blackbody.

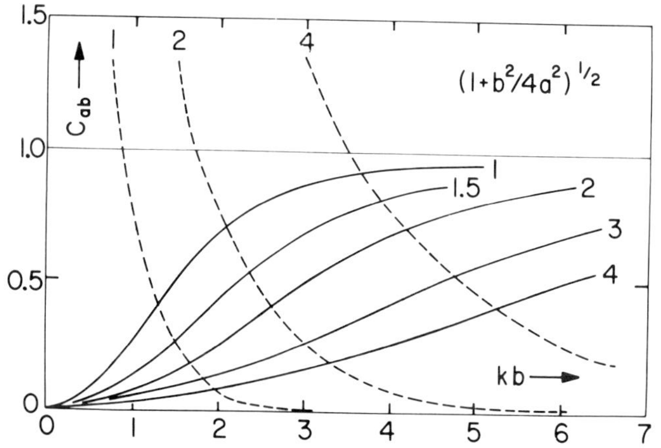

Fig. 2. The real part (20) (full curves) and the imaginary part (30) (dashed curves) of the transfer factor occurring in (29) for various values of the parameter $[1 + (b/2a)^2]^{\frac{1}{2}}$.

As an illustration of the above theory we consider the radiometric properties of a deep random phase screen which is illuminated by a coherent beam with Gaussian intensity profile. The phase $\Phi(\underline{r}_i)$ is supposed to be Gaussian distributed with large mean square phase deviation $<\Phi^2> >>1$. The resulting indirect source can be described by (4) with $I(\underline{r}_i)$ as given by (9), but with [22]

$$\mu(\underline{r}) = \exp\left(-<\Phi^2>\right) + \left[1 - \exp\left(-<\Phi^2>\right)\right] \exp\left(-\underline{r}^2/2b^2\right) \qquad (22)$$

instead of (10). The description (22) is valid if $<\Phi^2> >>1$ and if the phase correlation can be expanded around the origin in terms of the correlation length ξ, viz.

$$<\Phi(\underline{r}_1)\ \Phi(\underline{r}_2)><\Phi^2>^{-1} = 1 - (r/\xi)^2 + \ldots \quad . \qquad (23)$$

The effective correlation length b is now given by

$$b = \xi\ (2<\Phi^2>)^{-1/2}. \qquad (24)$$

From (22) we obtain the radiance

$$B_\omega(\underline{\rho},\underline{s}) = (2\pi)^{-1}\ I(\underline{\rho})\ \cos\ \theta$$

$$\times\ \{\exp(-<\Phi^2>)(2ak)^2\ \exp[-\tfrac{1}{2}(2ak)^2\sin^2\theta\ ]$$

$$+\ [1-\exp(-<\Phi^2>)](\Delta I)^{-2}\ \exp[-\tfrac{1}{2}(\Delta I)^{-2}\sin^2\theta\ ]\} \ , \qquad (25)$$

where (12) and (24) implies

$$\Delta I = [2<\Phi^2>(\xi k)^{-2} + (2ak)^{-2}]^{1/2}. \qquad (26)$$

The corresponding emittance is described by (19), but with the modified transfer factor

$$C'_{ab} = \exp(-<\Phi^2>)\ C(2ak)$$

$$+\ [1-\exp(-<\Phi^2>)]C(1/\Delta I) \qquad (27)$$

with $C(x)$ as defined by (21). The first term in (25) and (27) corresponds to the classical coherent diffraction limit and can be interpreted as the "specular component". The second term in (25) and (27) is equivalent to the Schell-model results (18) and (20), respectively. The specular component is suppressed, and (18) and (20) are reproduced from (25) and (27), only in the hypothetical limit $<\Phi^2> \to \infty$. In the case of very narrow illuminant beam width, i.e. $a << b$ or $a<\Phi^2>^{\frac{1}{2}} << \xi$, (25) and (27) reproduce the classical diffraction result, i.e. $C''_{ab} \to C\ (2ak)$. In the quasi-homogeneous limit $a >> b$ we obtain (25) and (27) with $1/\Delta I$ replaced by $bk = (2<\Phi^2>)^{-\frac{1}{2}}\ \xi k$. We notice that (27) does not vanish in the "incoherent limit" $bk \ll 1$ with $ak \gg 1$, but one still has

$$C'_{ab} \to \exp \; (-<\Phi^2>) \tag{28}$$

from the specular component.

We observe that the real-valued transfer factor (20) corresponds to the low spatial frequencies below k (real angles θ in (18)). The formal extension to the high spatial frequencies (evanescent waves, complex angles θ) leads to

$$E_\omega(\underline{\rho}) = C_{ab} I(\underline{\rho}) = (C'_{ab} + i \; C''_{ab}) I(\underline{\rho}) \tag{29}$$

with the complex transfer factor $C_{ab} = C'_{ab} + i \; C''_{ab}$. The real part C'_{ab} is given by (20). The imaginary part

$$C''_{ab} = (\pi/2)^{1/2} \; \Delta I \; \exp \; [-\tfrac{1}{2} \; (\Delta I)^{-2}] \tag{30}$$

corresponds to the spatial frequencies above k and becomes rapidly negligible with increasing beam width a and correlation length b (see Fig. 2, dashed curves). The imaginary term in the emittance (29) may tentatively be interpreted as a wattless power flux. The relevance of the high spatial frequencies for the far field of a coherent Gaussian beam was demonstrated recently [23].

For comparison we study the emittance of quasi-homogeneous sources with Bessel correlation [5,6,9] described by the spatial coherence function

$$W_\omega(\underline{r}_1,\underline{r}_2) = I(\underline{\rho}) \; \mu_n(\underline{r}),$$

$$\mu_n(\underline{r}) = (n/2)! (kr/2)^{-n/2} J_{n/2}(kr) \tag{31}$$

with $\underline{\rho}$ and $\underline{r}$ as defined by (7), $J_{n/2}$ denoting the Bessel function of the first kind of index $n/2$, and $n > -2$. The intensity profile is assumed to vary slowly in comparison with $\mu_n(\underline{r})$. Here we calculate the emittance

$$E_\omega(\underline{\rho}) = C_n I(\underline{\rho}) = n \; (n + 1)^{-1} I(\underline{\rho}). \tag{32}$$

In the special case $n = 1$ we obtain the transfer factor

$$C_{BB} \equiv C_1 = \tfrac{1}{2} \tag{33}$$

of the blackbody with sufficiently large aperture. We observe that $C_0 = 0$ although the related correlation length in the source plane does not vanish. The transfer factor C_n is real, since the high-frequency part of the spatial-frequency spectrum of $\mu_n(\underline{r})$ vanishes [5,6], in contrast to the Gaussian correlated source (10). The Bessel-type transfer C_n is shown in Fig. 1 and compared with the function (21) which determines the Gauss-type transfer factor.

3. SECOND-ORDER RADIOMETRY

The relations (15) - (17) are basic to the first-order radiometry of partially coherent planar sources. In second-order radiometry we are interested in the mean-square radiant intensity

$$<J^2(\underline{s})>_\omega \equiv R^4 <|u(R,\underline{s})|^4>_\omega . \tag{34}$$

We aim at introducing a second-order radiance $B_\omega^{(2)}(\underline{\rho},\underline{s})$ such that (34) can be written as

$$<J^2(\underline{s})>_\omega = \cos^2\theta \int_{z=0} d^2\underline{\rho}\ B_\omega^{(2)}(\underline{\rho},\underline{s}) . \tag{35}$$

To this end we consider the second-order spatial correlation function

$$W_\omega^{(2)}(\underline{r}_1,\underline{r}_2,\underline{r}_3,\underline{r}_4) = <u^\dagger(\underline{r}_1)u^\dagger(\underline{r}_2)u(\underline{r}_3)u(\underline{r}_4)>_\omega \tag{36}$$

of positions $\underline{r}_i$, $i = 1,2,3,4$, in the source plane $z = 0$. The corresponding radiant intensity autocorrelation can be shown to read

$$<J(\underline{s}_1)J(\underline{s}_2)>_\omega \equiv R^4 <|u(R,\underline{s}_1)|^2|u(R,\underline{s}_2)|^2>_\omega$$
$$= (k/2\pi)^4 \cos^2\theta_1 \cos^2\theta_2 \int d^2\underline{r}_1 \int d^2\underline{r}_2 \int d^2\underline{r}_3 \int d^2\underline{r}_4 \tag{37}$$
$$\times \exp\{-ik[\underline{s}_1(\underline{r}_1-\underline{r}_4) + \underline{s}_2(\underline{r}_2-\underline{r}_3)]\}W_\omega^{(2)}(\underline{r}_1,\underline{r}_2,\underline{r}_3,\underline{r}_4)$$

and leads to the mean-square radiant intensity

$$<J^2(\underline{s})>_\omega = (k/2\pi)^4 \cos^4\theta \int d^2\underline{r}_1 \int d^2\underline{r}_2 \int d^2\underline{r}_3 \int d^2\underline{r}_4$$
$$\times \exp[-ik\underline{s}(\underline{r}_1+\underline{r}_2-\underline{r}_3-\underline{r}_4)]W_\omega^{(2)}(\underline{r}_1,\underline{r}_2,\underline{r}_3,\underline{r}_4) , \tag{38}$$

where all the integrals extend over the source plane. The results (37) and (38) suggest the transformations
$$\underline{s} = (\underline{s}_1 + \underline{s}_2)/2, \quad \underline{\sigma} = \underline{s}_1 - \underline{s}_2, \quad \text{and}$$

$$\underline{r} = (\underline{r}_1+\underline{r}_2-\underline{r}_3-\underline{r}_4)/2, \quad \underline{r}' = (\underline{r}_1-\underline{r}_2+\underline{r}_3-\underline{r}_4)/2$$

$$\underline{r}'' = (\underline{r}_1-\underline{r}_2-\underline{r}_3+\underline{r}_4)/2, \quad \underline{\rho} = (\underline{r}_1+\underline{r}_2+\underline{r}_3+\underline{r}_4)/2 \tag{39}$$

that lead to the exponentials $- ik(2\ \underline{s}\underline{r} + \underline{\sigma}\underline{r}')$ and $-2ik\underline{s}\underline{r}$ in (37) and (38), respectively. In terms of the new coordinates, the relation (37) represents a second-order analogue of the generalized Van Cittert-Zernike theorem [1,10,13,15] used in first-order radiometry. From (35), (38) and (39) we derive the expression

$$B_\omega^{(2)}(\underline{\rho},\underline{s}) = (k/2\pi)^4 \cos^2\theta \int d^2\underline{r}\ e^{-ik\underline{s}\,\underline{r}}$$

$$\times \int d^2\underline{r}'\ \int d^2\underline{r}''\ W_\omega^{(2)}[(\underline{\rho}+\underline{r}+\underline{r}'+\underline{r}'')/2,\ (\underline{\rho}+\underline{r}-\underline{r}'-\underline{r}'')/2,$$

$$\times\ (\underline{\rho}-\underline{r}+\underline{r}'-\underline{r}'')/2,\ (\underline{\rho}-\underline{r}-\underline{r}'+\underline{r}'')/2] \tag{40}$$

for the second-order radiance. The corresponding second-order
radiant emittance is readily obtained as

$$E_\omega^{(2)}(\underline{\rho}) \equiv \int_{2\pi} d\Omega\ \cos^2\theta\ B_\omega^{(2)}(\underline{\rho},\underline{s}) = [3k^4/2(2\pi)^{5/2}]$$

$$\times \int d^2\underline{r}(2k\underline{r})^{-5/2}\ J_{5/2}\ (2k\underline{r})\ \int d^2\underline{r}'\ d^2\underline{r}''\ W_\omega^{(2)}[(\underline{\rho}+\underline{r}+\underline{r}'+\underline{r}'')/2,$$

$$\times\ (\underline{\rho}+\underline{r}-\underline{r}'-\underline{r}'')/2,(\underline{\rho}-\underline{r}+\underline{r}'-\underline{r}'')/2,(\underline{\rho}-\underline{r}-\underline{r}'+\underline{r}'')/2] \tag{41}$$

The expressions (35),(40), and (41) are the second-order analogues
of the first-order relations (15) - (17).

Let us apply the above concepts to chaotic sources where

$$W_\omega^{(2)}(\underline{r}_1,\underline{r}_2,\underline{r}_3,\underline{r}_4) =$$

$$W_\omega(\underline{r}_1,\underline{r}_4)W_\omega(\underline{r}_2,\underline{r}_3) + W_\omega(\underline{r}_1,\underline{r}_3)W_\omega(\underline{r}_2,\underline{r}_4). \tag{42}$$

We find that the corresponding radiances (16) and (40) obey

$$\int d^2\underline{\rho}\ B_\omega^{(2)}(\underline{\rho},\underline{s}) = 2[\int d^2\underline{\rho}\ B_\omega(\underline{\rho},\underline{s})]^2 \tag{43}$$

in agreement with the well-known result $\langle J^2\rangle_\omega = 2(J_\omega)^2$.
Statistically homogeneous chaotic sources are characterized by (42)
with $W_\omega(\underline{r}_i,\underline{r}_j) = W_\omega(\underline{r}_i-\underline{r}_j)$ and lead to position-independent
radiances with

$$B_\omega^{(2)}(\underline{\rho},\underline{s}) = 2S[B_\omega(0,\underline{s})]^2 \tag{44}$$

where $S=\int d^2\rho$ denotes the (sufficiently large) source area. A
particular example is the blackbody source where

$$B_\omega^{(2)}(\underline{\rho},\underline{s}) = 2S\{(\hbar\omega^3/\pi c^3)\,[\exp(\hbar\omega/KT)\,-1]^{-1}\}^2. \tag{45}$$

The Schell-type chaotic source is described by (42) with
$W_\omega(\underline{r}_1,\underline{r}_2)$ obeying (4). In the special case of isotropic Gaussian
intensity profile (9) and Gaussian degree of first-order spatial
coherence (10) we find the second-order radiance

$$B_\omega^{(2)}(\underline{\rho},\theta) = (2\pi)^{-1} k^4 A^2 a^2 b^4 [1+(b/2a)^2]^{-2}$$

$$\times \exp(-\underline{\rho}^2/4a^2) \cos^2\theta \, \exp\{-k^2 b^2 [1+(b/2a)^2]^{-1} \sin^2\theta\} \qquad (46)$$

in agreement with (43) and the first-order result (18). The second-order radiance of the corresponding quasi-homogeneous chaotic source is obtained from (46) in the limit $b/a \to 0$. The second-order radiometry of non-chaotic sources can be developed in terms of correlation models similar to those used in Ref. 22 for studying the fluctuations in the radiation scattered by random phase screens.

Let us finally touch upon the physical interpretation and possible application of the second-order radiance. Intuitively, $B_\omega^{(2)}(\underline{\rho},\underline{s})$ measures the effective contribution to the mean-square radiant intensity under the direction $\underline{s}$ from the position $\underline{\rho}/2$ in the source plane, whereas $E_\omega^{(2)}(\underline{\rho})$ indicates the corresponding effective contribution to the total mean-square flux $\int d\Omega \, <J^2(\underline{s})>_\omega$. The second-order radiation efficiency, i.e. the transfer of the noise properties, can be studied by evaluating (41). The concepts of optical second-order radiometry are perhaps useful for the study of second-order coherence in electron microscopy. The thermionic emission of electrons by a planar cathode seems to be an example for a quasi-homogeneous chaotic source obeying (46) with $a \gg b$. In view of the success of optical second-order interferometry, we wonder whether the analogous technique is feasible in electron microscopy.

References

1. A. Walther, J. Opt. Soc. Am. *58*, 1256 (1968).
2. E. W. Marchand, E. Wolf, J. Opt. Soc. Am. *64*, 1219 (1974).
3. E. Wolf, W. H. Carter, Optics Comms. *13*, 205 (1975).
4. W. H. Carter, E. Wolf, J. Opt. Soc. Am. *65*, 1067 (1975).
5. H. P. Baltes, B. Steinle, G. Antes, Opt. Comms. *18*, 242 (1976).
6. H. P. Baltes, Appl. Phys. *12*, 221 (1977), Sec. 3 (review).
7. A. S. Marathay, Optica Acta *23*, 785 (1976).
8. B. Steinle, H. P. Baltes, J. Opt. Soc. Am. *67*, 241 (1977).
9. C. Pask, Optica Acta *24*, 235 (1977).
10. W. H. Carter, E. Wolf, J. Opt. Soc. Am. *67*, 785 (1977).
11. E. W. Marchand, E. Wolf, J. Opt. Soc. Am. *62*, 379 (1972).
12. A. K. Jaiswal, G. P. Agrawal, and C. L. Mehta, Nuovo Cim. *15B*, 295 (1973).
13. E. Wolf, W. H. Carter, Opt. Comms. *16*, 297 (1976).
14. S. Wadaka, T. Sato, J. Opt. Soc. Am. *66*, 145 (1976).
15. H. P. Baltes, B. Steinle, Lett. Nuovo Cim. *18*, 313 (1977).
16. H. A. Ferwerda, M. G. van Heel, Optik *47*, 357 (1977).
17. H. A. Ferwerda, Optik *45*, 411 (1976).
18. P. W. Hawkes, Optik *47*, 453 (1977).

19. M. Bertolotti, F. Scudieri, S. Verginelli, Appl. Opt. *15*, 1842
     (1976).
20. A. C. Schell, Doctoral Dissertation, Massachusetts Institute
     of Technology (1961).
21. J. T. Winthrop, J. Opt. Soc. Am. *62*, 1234 (1972).
22. E. Jakeman, P. N. Pusey, J. Phys. A *8*, 369 (1975).
23. H. P. Baltes, H. G. Schmidt-Weinmar, Phys. Letts. *60A*, 275
     (1977).

DETERMINATION OF COHERENCE LENGTH FROM DIRECTIONALITY

H.A. Ferwerda and M.G. van Heel

State University at Groningen, Groningen, Netherlands

1. INTRODUCTION

In this contribution we shall present a method for obtaining
the coherence conditions on a planar, quasi-monochromatic, possibly
virtual, source from the angular distribution of its emission. The
result is obtained by assuming a special functional dependence of
the cross-spectral density function [cf. Eq. (2.13)]. As an
example of current interest we will discuss the coherence properties of thermionic emission sources which are widely used in
electron microscopy and whose coherence properties become of increasing importance for the image evaluation problem in electron
microscopy.

As long as we stay within the framework of classical coherence
theory [1], by which we mean that we may neglect such typical
quantum features as Bose-Einstein and Fermi-Dirac statistics, we
can treat a beam of electrons in the same way as a light beam.
Criteria for the validity of this assumption may be found in [2].

In order to keep the presentation self-contained we shall
develop the necessary formalism in section 2 in spite of the fact
that the required formulas are available in the literature [3].
In section 3 the theory developed in section 2 will be applied
to the coherence properties of thermionic sources.

2. THEORY

Both in light and electron optics the (scalar) complex
amplitude or wave function satisfies the Helmholtz equation when

443

we restrict ourselves to quasi-monochromatic radiation:

$$(\nabla^2 + \frac{\omega^2}{c^2}) \, U(\vec{r}) = 0 \quad . \tag{2.1}$$

ω is the angular time frequency, c is the velocity of light. For simplicity we confine ourselves to radiation propagating in free space. $U(\vec{r})$ is the scalar complex amplitude in light optics and the wave function in electron optics. When we want to study the angular coherence of the radiation it is convenient to use the Fourier transform $\tilde{U}(\vec{k})$ of $U(\vec{r})$:

$$U(\vec{r}) = \int \tilde{U}(\vec{k})\exp(i\vec{k}.\vec{r})d\vec{k} \quad . \tag{2.2}$$

Substituting (2.2) into (2.1) gives

$$(k^2 - \frac{\omega^2}{c^2}) \, \tilde{U}(\vec{k}) = 0 \quad ,$$

where $k^2 = k_1^2 + k_2^2 + k_3^2$, so that $\tilde{U}(\vec{k})$ is only different from zero on the sphere $k_1^2 + k_2^2 + k_3^2 = \omega^2/c^2$ in k-space. Let us suppose the wave function $U(\vec{r})$ is known on the plane z=o. $\vec{r}$ is a vector which denotes the position of a point in this plane. We may Fourier transform $U(\vec{r})$ as:

$$U(\vec{r}) = \int v(k_1,k_2)\exp[i(k_1 x_1 + k_2 x_2)] \, dk_1 \, dk_2 \quad . \tag{2.3}$$

On another plane $z = z_o > o$ $U(\vec{r})$ can be found from [4]

$$U(\vec{r}) = \int v(k_1,k_2)\exp[i(k_1 x_1 + k_2 x_2 + x_3 (\frac{\omega^2}{c^2} - k_1^2 - k_2^2)^{1/2}] \, dk_1 dk_2 \quad . \tag{2.4}$$

Let us consider a planar source, whose normal is taken along the z - axis while the plane of the source is assigned the z - co-ordinate zero. The complex amplitude on the source is then given by (2.3) and will now be written as:

$$U(\vec{r}) = \int v(\vec{k}')\exp(i\vec{k}'.\vec{r})d\vec{k}' \quad . \tag{2.5}$$

$\vec{k}'$ denotes a two-dimensional vector with components k_1,k_2 (the prime reminds of the restriction to two dimensions), $\vec{r}$ is a vector which specifies a point on the source (strictly speaking, $\vec{r}$ should also be considered as a two-dimensional vector in order that the inner product $\vec{k}'.\vec{r}$ makes sense). The cross-spectral coherence function $W(\vec{r}_1,\vec{r}_2)$ on the source is defined as the ensemble average

$$W(\vec{r}_1,\vec{r}_2) = \langle U(\vec{r}_1)U^*(\vec{r}_2)\rangle \quad . \tag{2.6}$$

Substituting (2.5) into (2.6) we find

$$W(\vec{r}_1,\vec{r}_2) = \iint \langle v(\vec{k}_1')v^*(\vec{k}_2')\rangle \exp[i(\vec{k}_1'.\vec{r}_1 - \vec{k}_2'.\vec{r}_2)]d\vec{k}_1'\, d\vec{k}_2' \quad . \tag{2.7}$$

For the study of angular coherence we introduce the vectors $\vec{s}_1$ and $\vec{s}_2$ by

$$\vec{k}_1' = k\,\vec{s}_1 \;\; ; \;\; \vec{k}_2' = k\,\vec{s}_2 \tag{2.8}$$

where
$$k = \frac{\omega}{c} = \frac{2\pi}{\lambda} \quad . \tag{2.9}$$

$\vec{s}_1$ consists of the first two components of the unit vector $k^{-1}(k_1,\ k_2,\ (k^2-k_1{}^2-k_2{}^2)^{1/2})$. So $\vec{s}_1$ defines a direction[†].

Denoting $\langle v(k\vec{s}_1)v^*(k\vec{s}_2)\rangle$ by $k^{-4}A(\vec{s}_1,\vec{s}_2)$ we find from (2.7):

$$W(\vec{r}_1,\vec{r}_2) = \iint A(\vec{s}_1,\vec{s}_2)\exp[ik(\vec{s}_1.\vec{r}_1 - \vec{s}_2.\vec{r}_2)]d\vec{s}_1\, d\vec{s}_2 \quad . \tag{2.10}$$

$A(\vec{s}_1,\vec{s}_2)$ describes the coherence between the directions $\vec{s}_1$ and $\vec{s}_2$ and is obtained by inverting (2.10):

$$A(\vec{s}_1,\vec{s}_2) = \lambda^{-4}\iint W(\vec{r}_1,\vec{r}_2)\exp[-ik(\vec{s}_1.\vec{r}_1 - \vec{s}_2.\vec{r}_2)]d\vec{r}_1\, d\vec{r}_2 \quad . \tag{2.11}$$

(2.10) and (2.11) are the basic formulas from which our results are derived.

Let us *assume* the following expression for the cross-spectral coherence function on the source:

$$W(\vec{r}_1,\vec{r}_2) = [\, I(\vec{r}_1)I(\vec{r}_2)]^{1/2}K(\vec{r}_1 - \vec{r}_2) \quad . \tag{2.12}$$

$K(\vec{r})$ has to satisfy $K^*(\vec{r}) = K(-\vec{r})$ and $K(\vec{o}) = 1$, because $W(\vec{r}_1,\vec{r}_2) = W^*(\vec{r}_2,\vec{r}_1)$, $W(\vec{r}_1,\vec{r}_1) = I(r_1)$. $I(\vec{r}_1)$ and $I(\vec{r}_2)$ are the intensities at the points $\vec{r}_1$ and $\vec{r}_2$ on the source: $I(\vec{r})d\vec{r}$ is the energy which leaves the area $d\vec{r}$ around the point $\vec{r}$ on the source. $K(\vec{r}_1 - \vec{r}_2)$ expresses the correlation between the values of the wave function at $\vec{r}_1$ and $\vec{r}_2$. This correlation has been assumed to be spatially stationary. The assumption (2.12) is more general than the

---

[†] There is no danger of confusion when we assign three components to $\vec{s}_1$ as $\vec{s}_1$ only occurs in scalar products with two-dimensional vectors (perpendicular to the z-axis), which can be considered as three-dimensional vectors with zero 3rd component.

customary assumption of statistical homogeneity [5]. We do not
know whether the assumption (2.12) can be deduced from first
principles. An expression reminiscent of (2.12) has already
briefly been mentioned by Walther [6]. We now make two additional
assumptions:

    i)   the intensity on the source does not vary appreciably
over distances $\vec{r}$ for which $K(\vec{r}) \neq 0$. This distance,
to be defined more precisely in section 3, will be
denoted as the "correlation length".

    ii)   the size of the source is large compared with the
correlation length.

Under these restrictions (2.12) reduces to

$$W(\vec{r}_1, \vec{r}_2) = I(\vec{r}_1) K(\vec{r}_1 - \vec{r}_2) \quad . \tag{2.13}$$

Substituting (2.13) into (2.11) we obtain for the angular
coherence function:

$$A(\vec{s}_1, \vec{s}_2) = \lambda^{-4} (2\pi)^4 \, \hat{I}(\vec{s}_2 - \vec{s}_1) \, \hat{K}(\vec{s}_2) \quad , \tag{2.14}$$

where $\hat{I}$ and $\hat{K}$ are the two-dimensional Fourier transforms of I
and K, respectively.

$$K(\vec{s}) = (2\pi)^{-2} \int K(\vec{r}) e^{-i k \vec{s} \cdot \vec{r}} \, d\vec{r} \quad . \tag{2.14a}$$

We note the remarkable result that the angular correlation is
described by the Fourier transform of the intensity on the source
and the angular distribution is expressed by the Fourier transform
of the correlation function on the source. In this way we obtained
a simple extension of the van Cittert-Zernike theorem [1].

For future use we mention the following relation between
$A(\vec{s}_1, \vec{s}_2)$ and the energy flux $I(\vec{s}) d\Omega$ in the solid angle $d\Omega$
around the direction $\vec{s}$ [3]:

$$I(\vec{s}) d\Omega = \lambda^2 \cos^2 \Theta \, A(\vec{s}, \vec{s}) d\Omega \quad , \tag{2.15}$$

where Θ is the angle between $\vec{s}$ and the normal on the source
(z-axis).

3.   APPLICATION TO THERMIONIC EMISSION SOURCES

We assume a planar electron source which emits electrons
which are subsequently accelerated by an electrostatic potential V
often of the order of 100kV. After acceleration the electrons

appear to emerge from a virtual source and obey the law of
propagation in free space. The virtual source can also be shown
to be planar [7]. The electrons leaving the cathode have approximately a Maxwell-Boltzmann distribution which is altered by
the subsequent acceleration. If the additional velocity which
the electrons acquire during the acceleration by the electrostatic
potential is large compared with their thermal velocities, we can
easily write down the expression for the angular distribution of
the electrons:

$$I(\vec{s}) = I_o \exp\left(-\frac{eV}{KT}\theta^2\right) \tag{3.1}$$

where -e is the charge of the electron, K is Boltzmann's constant
and T is the absolute temperature of the source. θ is the angle
between the direction $\vec{s}$ and the z-axis. In this way we find from
(2.14), (2.15) and (3.1) for the Fourier transform of the correlation function on the source:

$$\hat{K}(\vec{s}) = (2\pi)^{-4}[\hat{I}(\vec{o})]^{-1} I_o \lambda^2 \exp\left[-\frac{eV}{KT}(s_1^2 + s_2^2)\right] \tag{3.2}$$

where $s_3 = \cos\theta$ has been put equal to unity which is a reasonable
approximation for a customary cathode temperature of T = 3000 ^{o}K.
In this way we find for the correlation function $K(\vec{r})$ on the source:

$$K(\vec{r}) = (4\pi)^{-1} I_o[\hat{I}(\vec{o})]^{-1} \frac{KT}{eV} \exp\left[-\frac{KT}{4eV}k^2r^2\right] ; k = \frac{2\pi}{\lambda} . \tag{3.3}$$

We define a correlation length ℓ_c on the source by

$$\ell_c \triangleq \left[\frac{4eV}{k^2KT}\right]^{1/2} = 2^{1/2}\lambda_{th} \tag{3.4}$$

where λ_{th} is the thermal wavelength: $\lambda_{th} = h(mKT)^{-1/2}$. For
T = 3000 ^{o}K we find $\ell_c \simeq 5$ Å. If the virtual source were fully
incoherent the restriction to homogeneous waves would introduce
correlation lengths of the order of the wavelength which is 0.037 Å
for 100 keV electrons. So the virtual source is certainly not
incoherent in this case. The expression (3.4) for the coherence
length can be understood in physical terms: the extent Δr of an
electron wave packet leaving the source is given approximately
by (Heisenberg's uncertainty relation) $\Delta r \simeq \hbar(mKT)^{-1/2} = \lambda_{th}$.
As the electrons can only interfere when their wave packets overlap
it is now clear why the correlation length is of the order of
the thermal wavelength.

REFERENCES

1.   M. Born and E. Wolf, *Principles of Optics*, (Pergamon,
     Oxford, 1970), Ch. 10.
2.   H.A. Ferwerda, Optik 45, 411 (1976).
3.   See e.g. E.W. Marchand and E. Wolf, J. Opt. Soc. Am. 64,
     1219 (1974).
4.   J.W. Goodman, *Introduction to Fourier Optics*, (McGraw Hill,
     New York, 1968), p. 49
5.   E. Wolf, Phys. Rev. D 13, 869 (1976).
6.   A. Walther, J. Opt. Soc. Am. 58, 1256 (1968), section I.
7.   M.G. van Heel, Master's Thesis, State University of
     Groningen, July 1976, unpublished.

ON THE EXISTENCE OF A RADIANCE FUNCTION FOR A PARTIALLY COHERENT
PLANAR SOURCE*

Ari T. Friberg

University of Rochester, Rochester, New York

1. INTRODUCTION

Since the publication of an important paper by A. Walther [1]
in 1968, several authors [2] have studied radiometry taking into
account the random nature of the optical wave field. The traditional radiometric quantities, commonly believed to apply to incoherent sources, have been generalized to pertain to planar sources
of an arbitrary state of coherence [1,3]. The generalized formalism
of radiometry is based on the second-order coherence theory and the
generalized radiometric quantities are linear in the correlation
function of the wave field at two points in the source plane.

Traditionally the basic radiometric quantity associated with
a source is the radiance function. At least two different definitions [1,4] have been given for radiance in the context of generalized radiometry with partially coherent planar sources (and many
more could be given). It has been pointed out, however, by Marchand
and Wolf [3,4b] that neither of the two above mentioned functions
can be considered as a true measure of energy distribution in the
traditional sense, since they may occasionally take on negative
values.

In this paper we take up the general question of the existence
of a radiance function for a partially coherent planar source of
finite size, which would have all the properties that one normally
attributes to such a function in traditional radiometry. As the
main result of this investigation we find that no such radiance
function (assumed to be linear in the correlation function of the
wave field in the source plane) can exist for sources of any state

of coherence. Only a brief sketch of the proof of this result is
given in this note. A detailed proof will be presented in another
publication.

2. PROPERTIES OF RADIANCE

We consider a planar source σ occupying a finite area in the
plane $z = 0$ and radiating into the half-space $z > 0$, (see Fig. 1).
We assume that the field generated by the source may be characterized
by a stationary ensemble. Since the different temporal frequencies
of a stationary field are uncorrelated [5], we may restrict our
discussion to a single frequency component, ω say, of the light and
we may represent the optical wave field generated by this source
by an ensemble of monochromatic fields of this frequency, viz.

$$V(\underline{R},t) = v(\underline{R})\,e^{-i\omega t} , \tag{1}$$

where $\underline{R}$ is a vector specifying a typical field point. The source
σ, which may be either a primary or a secondary source, is identified with that region in the plane $z = 0$ where the optical intensity
does not vanish identically. The second-order coherence properties
of the source may be characterized by the correlation function

$$W(\underline{r}_1,\underline{r}_2) = \langle v(\underline{r}_1)\, v^*(\underline{r}_2) \rangle , \tag{2}$$

where the asterisk denotes the complex conjugate and the brackets

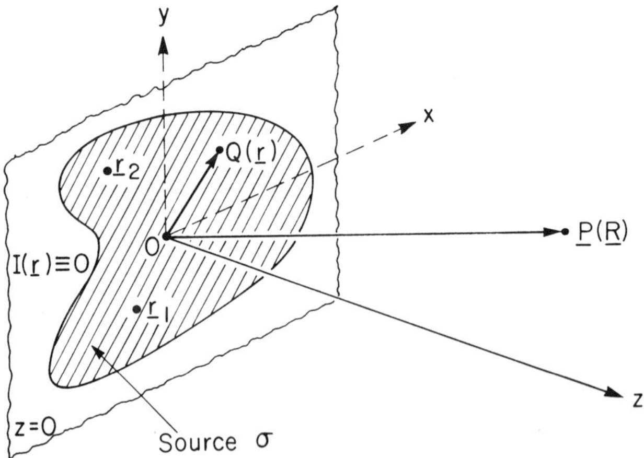

Figure 1. Illustration of the notation relating to the geometry of
the problem. The optical intensity $I(\underline{r})$ is identically zero in the
plane $z = 0$ outside the source σ.

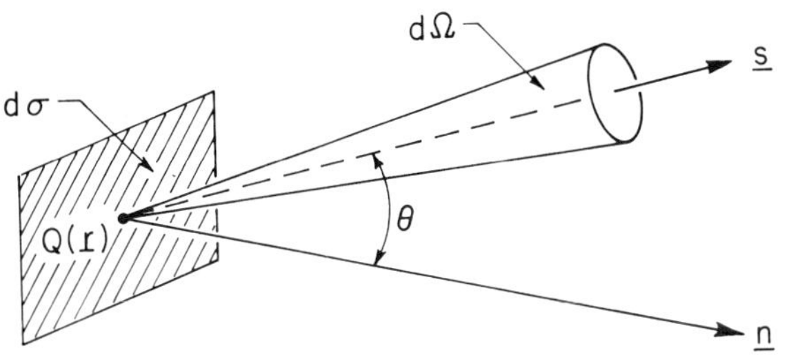

Figure 2. Illustration of the notation relating to the customary definition of radiance.

denote ensemble averaging. The lower case symbols $\underline{r}_1$ and $\underline{r}_2$ denote position vectors of two typical points in the source plane $z = 0$.

To discuss the radiance associated with the source we first recall the customary definition of this quantity [6]. Let $d\sigma$ be an element of the source area around a point Q specified by a position vector $\underline{r}$, and let $d\Omega$ be an element of solid angle around a direction specified by a unit vector $\underline{s}$ (see Fig. 2). Moreover, let θ be the angle between the direction $\underline{s}$ and the unit normal $\underline{n}$ to the source element. If dP denotes the average power radiated by the source element $d\sigma$ into the solid angle $d\Omega$, then the formula

$$dP = B(\underline{r},\underline{s}) \, \cos\theta \, d\sigma \, d\Omega \tag{3}$$

defines the radiance $B(\underline{r},\underline{s})$, at frequency ω, at a point $Q(\underline{r})$ and in the direction $\underline{s}$. Implicit in this definition are two assumptions: first, the radiance is a non-negative quantity, i.e.

$$B(\underline{r},\underline{s}) \geq 0 \tag{4}$$

for all possible values of $\underline{r}$ and $\underline{s}$, and second, the radiance vanishes in the source plane outside the source area, i.e.

$$B(\underline{r},\underline{s}) = 0 \, , \quad \text{when } \underline{r} \notin \sigma \tag{5}$$

for all possible values of $\underline{s}$.

To connect these properties of radiance with the coherence existing in the source plane we can use a result due to Marchand and Wolf [3] relating to the angular distribution of radiant intensity. They showed that, no matter what the state of coherence of the source is, the radiant intensity in the direction $\underline{s}$ is always proportional to a certain "anti-diagonal element" of the four-folded spatial Fourier transform of the source correlation function $W(\underline{r}_1,\underline{r}_2)$, [Eq. (41) of Ref. 3]. For brevity we will denote the Fourier transform simply by $\mathcal{F}_{\underline{f}_1,\underline{f}_2}\{W(\underline{r}_1,\underline{r}_2)\}$, indicating that the variables conjugate to $\underline{r}_1$ and $\underline{r}_2$ are $\underline{f}_1$ and $\underline{f}_2$, respectively. The relevant "anti-diagonal element" is obtained setting $\underline{f}_1 = k\underline{s}_\perp$ and $\underline{f}_2 = -k\underline{s}_\perp$, i.e. it is given by $\mathcal{F}_{k\underline{s}_\perp,-k\underline{s}_\perp}\{W(\underline{r}_1,\underline{r}_2)\}$. Here $\underline{s}_\perp$ is the projection (considered as a two-dimensional vector) of the three-dimensional unit vector $\underline{s}$ onto the source plane and $k = \omega/c$, c being the speed of light in free space. If we recall, moreover, that according to traditional radiometry the radiant intensity is proportional [7] to the integral of radiance over the source area σ, we are led to the following constraint on the possible radiance functions:

$$\int_\sigma B(\underline{r},\underline{s})\,d\sigma \;\alpha\; \mathcal{F}_{k\underline{s}_\perp,\,-k\underline{s}_\perp}\{W(\underline{r}_1,\underline{r}_2)\} \quad , \tag{6}$$

where the symbol α stands for "is proportional to".

Finally, since the radiance describes the distribution of energy, it seems reasonable to assume that it involves only terms that are quadratic in the field variable. Therefore we only consider possible radiance functions that are such that for all values of $\underline{r}$ and $\underline{s}$

$$B(\underline{r},\underline{s}) \;=\; \mathcal{L}\{W(\underline{r}_1,\underline{r}_2)\} \quad , \tag{7}$$

where $\mathcal{L}$ is a linear transformation.

3. EXAMPLES OF RADIANCE FUNCTIONS

From Eqs. (4)-(7) it is clear that the radiance function $B(\underline{r},\underline{s})$ for a partially coherent planar source σ should fulfill the following requirements, relabeled here as (A)-(D), for all possible values of the arguments $\underline{r}$ and $\underline{s}$ and for all possible states of coherence of the source:

(A) $B(\underline{r},\underline{s}) = \mathcal{L}\{W(\underline{r}_1,\underline{r}_2)\}$,

(B) $B(\underline{r},\underline{s}) \geq 0$,

(C) $B(\underline{r},\underline{s}) = 0$,   when $r \notin \sigma$,

(D) $\int_\sigma B(\underline{r},\underline{s})\,d\sigma \;\alpha\; \mathcal{F}_{k\underline{s}_\perp,\,-k\underline{s}_\perp}\{W(\underline{r}_1,\underline{r}_2)\}$.

As mentioned in the introduction, at least two "candidates" for a radiance function associated with a partially coherent planar source have been proposed. The first one [Eq. (21) of Ref. 1] reads, in the present notation,

$$B(\underline{r},\underline{s}) = \left(\frac{k}{2\pi}\right)^2 \cos\theta \int_{-\infty}^{\infty} W(\underline{r} + \tfrac{1}{2}\,\underline{r}',\underline{r} - \tfrac{1}{2}\,\underline{r}')\; e^{-ik\underline{s}_\perp \cdot \underline{r}'}\,d^2r' \; . \tag{8}$$

This definition of the radiance has been employed almost exclusively in recent researches on radiometry with partially coherent

sources [2,8]. The second definition of the radiance function
[Eq. (11) of Ref. 4] can be written as

$$B(\underline{r},\underline{s}) = \left(\frac{k}{2\pi}\right)^2 \cos\theta \ \mathrm{Re} \int_{-\infty}^{\infty} W(\underline{r}',\underline{r}) \ e^{-ik\underline{s}_{\perp}\cdot(\underline{r}'-\underline{r})} \ d^2 r' \ , \qquad (9)$$

where Re stands for the real part. Neither of these two proposed
radiance functions satisfies all the four requirements (A)-(D).
In the table below we indicate which of the four requirements each
of these two functions does or does not satisfy.

TABLE. Summary of the properties of the two radiance functions
defined by Eqs. (8) and (9) in the text. The entries YES and NO
indicate whether or not the corresponding requirement (A) through
(D) is satisfied.

Requirement	Eq.(8)	Eq.(9)
(A)	YES	YES
(B)	NO	NO
(C)	NO	YES
(D)	YES*	YES

(*) It is understood here that the integration on the left-
hand side of the requirement (D) is extended over the entire source
plane, rather than just over the source area σ.

4. INCONSISTENCY OF REQUIREMENTS (A)-(D)

We will now return to the four requirements (A)-(D) on the
radiance function. The main result to which we are led is that
these four requirements are incompatible. This means that no
radiance function that is linear in the correlation function
$W(\underline{r}_1,\underline{r}_2)$, [requirement (A)], can simultaneously fulfill the requirements (B)-(D), for all possible states of coherence of the source.
We will only briefly indicate how this result can be established.
A fuller analysis will be published elsewhere.

Let us consider an ensemble of source distributions of the form

$$v(\underline{r}) = a_1 v_1(\underline{r}) + a_2 v_2(\underline{r}) \ . \qquad (10)$$

Here $v_1(\underline{r})$ and $v_2(\underline{r})$ are the field distributions across two finite,
non-overlapping areas σ_1 and σ_2, respectively, in the plane $z = 0$
and a_1 and a_2 are arbitrary complex numbers. One can show that if
the four requirements (A)-(D) on the radiance are to be satisfied
for this type of a source, for all pairs of values of a_1 and a_2
and for all possible states of coherence of the field in the domains
σ_1 and σ_2, then a contradiction follows. The contradiction arises
because the cross terms in the source correlation function $W(\underline{r}_1,\underline{r}_2)$,
to which the distribution (10) gives rise, cannot give any contri-
bution to the radiance function, if the conditions (B) and (C) are
to be satisfied. Consequently, they cannot contribute anything to
the radiant intensity either. In light of the requirement (D) this
means that the Fourier transforms of the cross terms in $W(\underline{r}_1,\underline{r}_2)$
have to vanish, for all values of $\underline{s}_\perp$ such that $|\underline{s}_\perp| \leq 1$ and for all
states of coherence of the field in the domains σ_1 and σ_2. This
is, of course, impossible, proving the inconsistency of the requirements (A)-(D).

5. ANALOGY WITH QUANTUM MECHANICS IN PHASE-SPACE

It seems worthwhile to mention that the present situation in
radiometry is somewhat analogous to that encountered a long time
ago in the so-called phase-space formulation of quantum mechanics
[9]. In that representation of quantum mechanics the expectation
values of quantum mechanical operators are calculated in phase-
space using the methods of classical statistical mechanics. The
two radiance functions (8) and (9) of radiometry with partially
coherent planar sources are in many respects analogous to two
commonly used quantum mechanical phase-space distribution functions:
the radiance function defined by Eq.(8) should be compared to the
well-known Wigner distribution function [10],

$$f(x,p) = \int_{-\infty}^{\infty} \rho(x + \tfrac{1}{2} x', x - \tfrac{1}{2} x')\ e^{-ipx'/\hbar}\ dx' \quad , \tag{11}$$

and that defined by Eq. (9) should be compared to another widely
used phase-space distribution function, introduced by Margenau and
Hill [11],

$$f(x,p) = \mathrm{Re} \int_{-\infty}^{\infty} \rho(x',x)\ e^{-ip(x'-x)/\hbar}\ dx' \quad , \tag{12}$$

where Re denotes the real part. In Eqs. (11) and (12) x and p are,
of course, the position and momentum variables of a particle moving
in one dimension, and $\rho(x,x')$ is the coordinate representation of
the density operator.

It is also well-known that, even though quantum mechanics can, in principle, be formulated in phase-space, the phase-space distribution functions (such as those defined by Eqs. (11) and (12)) are not true probability densities, because they may in general become negative (Wigner's theorem [9]); for this reason they are sometimes referred to as quasi-probabilities. This feature of the phase-space representation of quantum mechanics is to some extent analogous to the main result of this paper; namely, that no radiance function associated with a finite partially coherent source can have, irrespective of the state of coherence of the source, all the properties of radiance of traditional radiometry.

Acknowledgement

I wish to express my appreciation to Professor E. Wolf for having suggested the problem treated in this paper and for helpful discussions.

* Research supported in part by a grant from the Alfred Kordelin Foundation (Finland) and by the U.S. Army Research Office.

References

1.   A. Walther, J. Opt. Soc. Am. *58*, 1256 (1968).
2.   See, for example, the review article by H.P. Baltes, Appl. Phys. *12*, 221 (1977), Sec. 3, and the references cited therein.
3.   E.W. Marchand and E. Wolf, J. Opt. Soc. Am. *64*, 1219 (1974).
4.   a) A. Walther, J. Opt. Soc. Am. *63*, 1622 (1973). See also
     b) E.W. Marchand and E. Wolf, J. Opt. Soc. Am. *64*, 1273 (1974),
     and c) A. Walther, J. Opt. Soc. Am. *64*, 1275 (1974).
5.   See, for example, R.L. Stratonovich, *Topics in the Theory of Random Functions* (Gordon and Breach, New York, 1963) Vol.1, p.28.
6.   The radiance is also known as the specific intensity or the brightness. For the customary definition of the radiance see, for example, S. Chandrasekhar, *Radiative Transfer* (Dover, New York, 1960) Section 1.
7.   Actually the proportionality factor is trivially dependent on s itself, but this dependence is of no consequence here.
8.   The radiance function defined by Eq. (8) is also used, for example, by a) P.W. Hawkes, Optik *47*, 453 (1977); and by b) W.H. Carter and E. Wolf, J. Opt. Soc. Am. *67*, 785 (1977).
9.   See, for example, a) E.P. Wigner, contribution in *Perspectives in Quantum Theory*, ed. by W. Yourgrau and A. van der Merwe, (M.I.T. Press, Cambridge, Mass., 1971) pp. 25-36; b) M.D. Srinivas and E. Wolf, Phys. Rev. D *11*, 1477 (1975).
10. E.P. Wigner, Phys. Rev. *40*, 749 (1932).
11. H. Margenau and R.N. Hill, Progr. Theoret. Phys. (Kyoto) *26*, 722 (1961).

RADIATIVE ENERGY TRANSFER IN SCALAR WAVE FIELDS*

E. Wolf and M. S. Zubairy

University of Rochester, Rochester, New York

In recent years several attempts have been made [1] to provide
a satisfactory basis for the theory of radiative energy transfer in
optical fields. The conventional theory [2], on which many investigations in astrophysics and in the field of atmospheric propagation
are based, was developed quite independently of modern theories of
radiation; consequently its range of validity is not understood
at the present time.

We recently put forward a new theory of radiative energy transfer in free electromagnetic fields, based on second-order coherence
theory. [3,4] This theory, which takes into account both correlation and polarization properties of the field differs in some
respects from the conventional theory, but it reduces to it in certain limiting cases. Moreover, it provides a precise interpretation of some of the basic concepts (such as the specific intensity
of radiation), employed in the traditional heuristic model for
energy transport. In attempting to extend this theory to fields
that interact with matter, it seems advisable to consider first a
scalar model, i.e., to ignore specific effects arising from the
state of polarization of the field.

In the first part of this paper we outline a simplified scalar
analogue of the theory that we developed in references 3 and 4. We
then show how the concepts of angular components of the average
energy density and of the average flux vector, that are basic in
our earlier work, may be introduced for any stationary ensemble
of scalar wave fields. These quantities will be shown to be expressible in terms of certain second order correlation functions of

the field. Some preliminary work relating to the derivation of the
appropriate transport equations for these quantities will also be
discussed.

*Research supported by the Army Research Office.

References

1.   References to some of the relevant papers are given in E. Collett,
     J. T. Foley and E. Wolf, J. Opt. Soc. Amer. *67*, 465 (1977).
2.   See, for example, S. Chandrasekhar, *Radiative Transfer* (Dover,
     New York, 1960).
3.   E. Wolf, Phys. Rev. D *13*, 869 (1976).
4.   M. S. Zubairy and E. Wolf, Opt. Commun. *20*, 321 (1977).

RADIATIVE ENERGY TRANSFER IN THE PRESENCE OF RANDOM SOURCE
DISTRIBUTIONS[*]

M.S. Zubairy

University of Rochester, Rochester, New York

1. INTRODUCTION

Radiative energy transfer is generally treated on the basis of
a phenomenological theory[1] that bears no apparent relationship to
the theory of the electromagnetic field. In recent years, some progress has been made towards elucidating the foundations of this
theory.[2]

The central quantity of the phenomenological theory is the so-
called specific intensity of radiation $I(\underline{r},\underline{s})$ which satisfies the
radiative transfer equation

$$\underline{s}\cdot\nabla_r I(\underline{r},\underline{s}) = -\sigma_a(\underline{r},\underline{s})I(\underline{r},\underline{s}) + \int_{(4\pi)} \sigma_s(\underline{r},\underline{s},\underline{s}')I(\underline{r},\underline{s}')d\Omega_{s'} + D(\underline{r},\underline{s}).$$

$$(1.1)$$

The left-hand side represents the rate of change in the specific
intensity $I(\underline{r},\underline{s})$ in the direction $\underline{s}$, the first term on the right describes the extinction, i.e. the decrease of energy as a result of
scattering and absorption, the second term represents the increase
in energy in the direction $\underline{s}$ as a result of scattering from all
other directions and $D(\underline{r},\underline{s})$, (known as the source function), is the
rate of energy emission due to spontaneous processes. The functions
$\sigma_a(\underline{r},\underline{s})$ and $\sigma_s(\underline{r},\underline{s},\underline{s}')$ are the so-called extinction coefficient and
differential scattering coefficient, respectively, which are intro-
duced rather formally from considerations based on the conservation
of energy.

Recent researches have sought to elucidate the meaning of the
specific intensity of radiation within the framework of stochastic
wave theory and of the true significance of the extinction coef-

ficient, the differential scattering coefficient and the source
function. Attempts in this direction and, in particular, efforts
to derive the equation of radiative transfer or to deliminate its
range of validity have met so far with limited success.

A theory of radiative transfer, based on the correlation
theory of full electromagnetic field, but restricted to free
field, was formulated recently by Wolf[3] and Zubairy and Wolf.[4]
Earlier, Ovchinnikov and Tatarskii[5] proposed a definition of the
specific intensity of radiation in terms of the second order
correlation function of a stochastic scalar wavefield. In the
present paper, we employ their definition of the specific intensity
to derive the equation of radiative transfer in the presence of
random sources in free space. We also derive a related equation
associated with the transport of energy flux. Our analysis reveals
the true significance of the source function $D(r,s)$ that is usually
introduced in a purely formal manner in the phenomenological
theory.

2.   THE ANGULAR COMPONENTS OF THE AVERAGE ENERGY DENSITY AND THE AVERAGE ENERGY FLUX

We consider a scalar field $v(\underline{r})$ whose statistical properties
are characterized by a stationary ensemble. Let

$$\bar{\Gamma}(\underline{r}_1,\underline{r}_2) = \langle v(\underline{r}_1)\, v^*(\underline{r}_2) \rangle \qquad (2.1)$$

be the second order coherence function of the wavefield. In
Eq. (2.1), the sharp brackets denote the ensemble average and the
asterisk denotes the complex conjugate. Introducing the new
variables

$$\underline{r} = \frac{1}{2}(\underline{r}_1 + \underline{r}_2),$$

$$\underline{r}' = \underline{r}_1 - \underline{r}_2, \qquad (2.2)$$

we may write

$$\bar{\Gamma}(\underline{r}_1,\underline{r}_2) \equiv \Gamma(\underline{r},\underline{r}'). \qquad (2.3)$$

The spectral density $f(\underline{r},\underline{K})$ of the function $\Gamma(\underline{r},\underline{r}')$ with
respect to the variable $\underline{r}'$ is defined by the formula

$$\Gamma(\underline{r},\underline{r}') = \int f(\underline{r},\underline{K})\, e^{i\underline{K}\cdot\underline{r}'}\, d^3K. \qquad (2.4)$$

The ensemble average of the energy density $U(\underline{r})$ and the
ensemble average of the energy flux vector $\underline{F}(\underline{r})$ are given by

$$U(\underline{r}) = \frac{1}{c} <v(\underline{r})v^*(\underline{r})> , \qquad (2.5)$$

$$\underline{F}(\underline{r}) = \frac{1}{2ik_o} [<v^*(\underline{r}) \nabla_{\underline{r}} v(\underline{r})> - <v(\underline{r})\nabla_{\underline{r}} v^*(\underline{r})>], \qquad (2.6)$$

where c denotes the speed of light in vacuo, k_o is the wave
number in free space and $\nabla_{\underline{r}}$ denotes the gradient operator acting
with respect to the variable r. If we use Eqs. (2.1) and (2.3),
$U(\underline{r})$ and $\underline{F}(\underline{r})$ can be expressed in terms of the coherence function
Γ as

$$U(\underline{r}) = \frac{1}{c} \Gamma(\underline{r},o), \qquad (2.7)$$

$$\underline{F}(\underline{r}) = \frac{1}{ik_o} [\nabla_{\underline{r}'} \Gamma(\underline{r},\underline{r}')]_{\underline{r}'=o}. \qquad (2.8)$$

Let us now substitute the spectral representation (2.4) of
the coherence function $\Gamma(\underline{r},\underline{r}')$ in Eqs. (2.7) and (2.8) and change
to spherical polar coordinates in K-space. Then $d^3K = K^2 dKd\Omega$,
$d\Omega = \sin\theta\, d\theta\, d\phi$. If we next interchange the order of integration
and ensemble averaging, we obtain the following expressions for
the average energy density $U(\underline{r})$ and the average energy flux $\underline{F}(\underline{r})$:

$$U(\underline{r}) = \int Q(\underline{r},\underline{s})\, d\Omega, \qquad (2.9)$$

$$\underline{F}(\underline{r}) = \int \underline{R}(\underline{r},\underline{s})\, d\Omega, \qquad (2.10)$$

where

$$\underline{s} = \frac{\underline{K}}{|\underline{K}|} \equiv (\sin\theta\cos\phi,\sin\theta\sin\phi,\cos\theta), \qquad (2.11)$$

and

$$Q(\underline{r},\underline{s}) = \frac{1}{c} \int_o^\infty f(\underline{r},\underline{K})\, K^2\, dK, \qquad (2.12)$$

$$\underline{R}(\underline{r},\underline{s}) = \frac{1}{k_o} \underline{s} \int_o^\infty f(\underline{r},\underline{K})\, K^3\, dK . \qquad (2.13)$$

The integrations in Eqs. (2.9) and (2.10) extend over the whole
4π-solid angle generated by the unit vector s. The quantities
$Q(\underline{r},\underline{s})$ and $\underline{R}(\underline{r},\underline{s})$, defined by the relations (2.9) and (2.10),
can be called the "angular components" of the average energy
density and the average energy flux vector, respectively, in analogy
with the corresponding quantities, first introduced by Wolf.[3]

Equations (2.9) and (2.10) are analogous to the expressions
of the phenomenological theory of radiative energy transfer for

the "space density" of radiation $U(\underline{r})$ and for the net flux $F(\underline{r})$,
at $\underline{r}$, in terms of the specific intensity of radiation $I(\underline{r},\underline{s})$, viz.

$$U(\underline{r}) = \frac{1}{c} \int I(\underline{r},\underline{s}) \, d\Omega, \tag{2.14}$$

$$\underline{F}(\underline{r}) = \int \underline{s} \, I(\underline{r},\underline{s}) \, d\Omega. \tag{2.15}$$

A comparison of (2.14) with (2.9) suggests that the specific
intensity of radiation $I(\underline{r},\underline{s})$ can be related to $Q(\underline{r},\underline{s})$ by the
simple equation

$$I(\underline{r},\underline{s}) = c \, Q(\underline{r},\underline{s}). \tag{2.16}$$

Using Eq. (2.12), we can thus write $I(\underline{r},\underline{s})$ in terms of the spectral
density $f(\underline{r},K)$ as follows:

$$I(\underline{r},\underline{s}) = \int_0^\infty f(\underline{r},K) \, K^2 \, dK. \tag{2.17}$$

This definition of the specific intensity was first introduced by
Ovchinnikov and Tatarskii[5]. It may be shown, using the Fourier
inverse of Eq. (2.4), that the relationship between $I(\underline{r},\underline{s})$ and
$\Gamma(\underline{r},\underline{r}')$ is given by

$$I(\underline{r},\underline{s}) = \frac{-1}{(2\pi)^2} \int \Gamma(\underline{r},\underline{r}') \, \delta_+'' (\underline{s}\cdot\underline{r}') \, d^3r', \tag{2.18}$$

where δ_+ is the positive frequency part of the Dirac delta func-
tion [6].   The primes on $\delta_+(\underline{s}\cdot\underline{r}')$ denote differentiation with
respect to the argument.

   It is clear from Eqs. (2.10) and (2.13) that, with this definition of the specific intensity, the usual flux relation (2.15) is
not obeyed.

3.   RADIATIVE TRANSFER EQUATIONS FOR THE PROPAGATION OF THE ANGULAR COMPONENTS $Q(\underline{r},\underline{s})$ and $R(\underline{r},\underline{s})$

   The wave propagation in a homogeneous medium with random
source distribution $\rho(\underline{r})$ is governed by the stochastic scalar
wave equation (with the time dependent factor $e^{-i\omega t}$ being omitted)

$$(\nabla^2 + k_0{}^2) \, v(\underline{r}) = -4\pi \, \rho(\underline{r}), \tag{3.1}$$

where, as before, $k_0 = \frac{\omega}{c}$ is the wave number of the field in free
space. The field $v(\underline{r})$ generated by $\rho(\underline{r})$ is identified with that
particular solution of Eq. (3.1) which behaves at infinity as an
outgoing spherical wave i.e.

$$v(\underline{r}) = \int \rho(\underline{p}) \, G_0(\underline{r} - \underline{p}) \, d^3p. \tag{3.2}$$

Here

$$G_o(\underline{r} - \underline{p}) = \frac{e^{ik_o|\underline{r}-\underline{p}|}}{|\underline{r} - \underline{p}|} \tag{3.3}$$

is the (outgoing) free-space Green's function of the Helmholtz equation with wave number k_o.

If we recall the definition (2.1) of the second-order coherence function of the wavefield $\bar{\Gamma}(\underline{r}_1,\underline{r}_2)$, it follows from Eq. (3.1) that $\bar{\Gamma}(\underline{r}_1,\underline{r}_2)$ satisfies the equations

$$(\nabla_1^2 + k_o^2)\,\bar{\Gamma}(\underline{r}_1,\underline{r}_2) = -4\pi\,\,<\rho(\underline{r}_1)\,v^*(\underline{r}_2)>, \tag{3.4}$$

$$(\nabla_2^2 + k_o^2)\,\bar{\Gamma}(\underline{r}_1,\underline{r}_2) = -4\pi\,\,<v(\underline{r}_1)\,\rho^*(\underline{r}_2)>, \tag{3.5}$$

where ∇_i^2, $(i = 1,2)$, denotes the Laplacian operator with respect to the point $\underline{r}_i$. On subtracting Eq. (3.5) from Eq. (3.4), we obtain

$$(\nabla_1^2 - \nabla_2^2)\,\bar{\Gamma}(\underline{r}_1,\underline{r}_2) = -4\pi[<\rho(\underline{r}_1)v^*(\underline{r}_2)>-<(v(\underline{r}_1)\rho^*(\underline{r}_2)>]. \tag{3.6}$$

Let

$$\bar{B}(\underline{r}_1,\underline{r}_2) = <\rho(\underline{r}_1)\rho^*(\underline{r}_2)> \tag{3.7}$$

be the second-order correlation function of the source distribution. In view of the Eq. (3.2), Eq. (3.6) can be rewritten in the form

$$(\nabla_1^2 - \nabla_2^2)\bar{\Gamma}(\underline{r}_1,\underline{r}_2) = -4\pi\!\int[\bar{B}(\underline{r}_1,\underline{p})G_o^*(\underline{r}_2-\underline{p})-\bar{B}(\underline{p},\underline{r}_2)G_o(\underline{r}_1-\underline{p})]d^3p. \tag{3.8}$$

Next we apply to Eq. (3.8) the transformation (2.2) and recall that $\bar{\Gamma}(\underline{r}_1,\underline{r}_2) \equiv \Gamma(\underline{r},\underline{r}')$. The correlation function $\bar{B}(\underline{r}_1,\underline{r}_2)$, defined by Eq. (3.7), when expressed in terms of the new variables $\underline{r}$ and $\underline{r}'$, will be denoted by $B(\underline{r},\underline{r}')$ i.e.

$$B(\underline{r},\underline{r}') \equiv \bar{B}(\underline{r}_1,\underline{r}_2) = \bar{B}(\underline{r} + \tfrac{1}{2}\underline{r}',\ \underline{r} - \tfrac{1}{2}\underline{r}'). \tag{3.9}$$

One may readily show that $\nabla_1^2 - \nabla_2^2 = 2\,\nabla_{\underline{r}}\cdot\nabla_{\underline{r}'}$. In terms of the variables $\underline{r}$ and $\underline{r}'$, Eq. (3.8) may then be shown to take the form

$$2\nabla_{\underline{r}}\cdot\nabla_{\underline{r}'}\Gamma(\underline{r},\underline{r}') = -4\pi\!\int[B(\tfrac{1}{2}\underline{r}+\tfrac{1}{4}\underline{r}'+\tfrac{1}{2}\underline{p},\ \underline{r}+\tfrac{1}{2}\underline{r}'-\underline{p})$$

$$\times\,G_o^*(\underline{r}-\tfrac{1}{2}\underline{r}'-\underline{p}) - B(\tfrac{1}{2}\underline{p}+\tfrac{1}{2}\underline{r}-\tfrac{1}{4}\underline{r}',\ \underline{p}-\underline{r}+\tfrac{1}{2}\underline{r}')\,G_o(\underline{r}+\tfrac{1}{2}\underline{r}'-\underline{p})]\,d^3p.$$

$$\tag{3.10}$$

After changes of the variable of integration ($\underline{r}-\tfrac{1}{2}\underline{r}'-\underline{p} \to \underline{p}$ in first integral and $\underline{r}+\tfrac{1}{2}\underline{r}'-\underline{p} \to \underline{p}$ in second integral), Eq. (3.10) can be written in the form

$$2\nabla_r \cdot \nabla_{r'} \Gamma(\underline{r},\underline{r}') = -4\pi \int [B(\underline{r}-\tfrac{1}{2}\underline{p},\underline{r}'+\underline{p})\, G_o^*(\underline{p}) - B(\underline{r}-\tfrac{1}{2}\underline{p},\underline{r}'-\underline{p})$$

$$\times\, G_o(\underline{p})]\, d^3p \quad . \tag{3.11}$$

Next we derive an equation of transfer for the spectral density $f(\underline{r},\underline{k})$ of the function $\Gamma(\underline{r},\underline{r}')$. For this purpose we also define the spectral density $h(\underline{r},\underline{k})$ of the correlation function $B(\underline{r},\underline{r}')$ with respect to the variable $\underline{r}'$ by the formula

$$B(\underline{r},\underline{r}') = \int h(\underline{r},\underline{K})\, e^{i\underline{K}\cdot\underline{r}'}\, d^3K \quad . \tag{3.12}$$

If we now substitute for Γ from Eq. (2.4) into Eq. (3.11) and use Eq. (3.12), we obtain

$$\int 2i\underline{K} \cdot \nabla_r\, f(\underline{r},\underline{K})\, e^{i\underline{K}\cdot\underline{r}'} d^3K = -4\pi\!\int\!\!\int h(\underline{r}-\tfrac{1}{2}\underline{p},\underline{K})\,[e^{i\underline{K}\cdot\underline{p}}\, G_o^*(\underline{p})$$

$$-e^{-i\underline{K}\cdot\underline{p}}\, G_o(\underline{p})]\, e^{i\underline{K}\cdot\underline{r}'}\, d^3p\, d^3K \quad . \tag{3.13}$$

Since this equation holds for all values of $\underline{r}'$, it follows that

$$\underline{K}\cdot\nabla_r\, f(\underline{r},\underline{K}) = 2\pi i \int h(\underline{r}-\tfrac{1}{2}\underline{p},\underline{K})\,[e^{i\underline{K}\cdot\underline{p}}\, G_o^*(\underline{p}) - e^{-i\underline{K}\cdot\underline{p}}\, G_o(\underline{p})]\, d^3p \quad . \tag{3.14}$$

Expressing h in terms of B via the Fourier inverse of the relation (3.12) in this equation, we obtain the following transport equation for $f(\underline{r},\underline{K})$:

$$\underline{K}\cdot\nabla_r\, f(\underline{r},\underline{K}) = \frac{i}{(2\pi)^2}\int\!\!\int B(\underline{r}-\tfrac{1}{2}\underline{p},\underline{q})\,[e^{i\underline{K}\cdot(\underline{p}-\underline{q})}\, G_o^*(\underline{p}) - e^{-i\underline{K}\cdot(\underline{p}+\underline{q})}$$

$$\times\, G_o(\underline{p})]\, d^3p\, d^3q \quad . \tag{3.15}$$

If we now recall the definitions (2.12) and (2.13) of the angular components $Q(\underline{r},\underline{s})$ and $R(\underline{r},\underline{s})$ of the average energy density and of the average energy flux, respectively, we find that, as a consequence of Eq. (3.15), they satisfy the following equations of transfer:

$$\underline{S}\cdot\nabla_r\, Q(\underline{r},\underline{s}) = \int\!\!\int B(\underline{r}-\tfrac{1}{2}\underline{p},\underline{q})\, L_Q(\underline{p},\underline{q},\underline{s})\, d^3p\, d^3q \quad , \tag{3.16}$$

$$(\underline{S}\cdot\nabla_r)R(\underline{r},\underline{s}) = \int\!\!\int B(\underline{r}-\tfrac{1}{2}\underline{p},\underline{q})\, L_R(\underline{p},\underline{q},\underline{s})\, d^3p\, d^3q \quad , \tag{3.17}$$

where the kernels L_Q and L_R are given by

$$L_Q(\underline{p},\underline{q},\underline{s}) = \frac{1}{2\pi c} [\delta'_-(\underline{s}\cdot(\underline{p}-\underline{q}))\ G_o^*(\underline{p}) + \delta'_+(\underline{s}\cdot(\underline{p}+\underline{q}))\ G_o(\underline{p})]\ , \quad (3.18)$$

$$\underline{L}_R(\underline{p},\underline{q},\underline{s}) = \frac{-i}{2\pi k_o}\ \underline{s}[\delta''_-(\underline{s}\cdot(\underline{p}-\underline{q}))\ G_o^*(\underline{p}) - \delta''_+(\underline{s}\cdot(\underline{p}+\underline{q}))\ G_o(\underline{p})]\ , \quad (3.19)$$

and primes denote differentiation with respect to the argument.

In view of the relations (2.9) and (2.10), the transport equations (3.16) and (3.17) can be interpreted as rigorous equations for radiative transfer of energy density and of energy flux in the presence of random source distribution. Clearly the source function $D(\underline{r},\underline{s})$ that is introduced formally in the radiative energy transfer equation (1.1), can now be identified as the following transform of the correlation function $B(\underline{r},\underline{r}')$ of the source distribution:

$$D(\underline{r},\underline{s}) = c \iint B(\underline{r}-\tfrac{1}{2}\underline{p},\underline{q})\ L_Q(\underline{p},\underline{q},\underline{s})\ d^3p\ d^3q\ . \quad (3.20)$$

The preceding equations are exact. We will now make certain simplifying assumptions concerning the source correlation function that will lead to approximate (but much simpler) forms of the transport equations.

Using the explicit expression of the Green's function (3.3), Eq. (3.14) can be rewritten as

$$\underline{K}\cdot\nabla_r f(\underline{r},\underline{K}) = 4\pi\ \int h(\underline{r}-\tfrac{1}{2}\underline{p},\underline{K})\ \frac{\sin(k_o|\underline{p}|-\underline{K}\cdot\underline{p})}{|p|} \quad (3.21)$$

Let ℓ be the distance over which $|B(\underline{r},\underline{r}')|$ and $h(\underline{r},\underline{k})$ remain sensibly constant with respect to their first argument. If we assume that ℓ is large compared to the wavelength i.e. if

$$\ell \gg \frac{2\pi}{k_o}\ , \quad (3.22)$$

then Eq. (3.21) can be simplified to

$$\underline{K}\cdot\nabla_r f(\underline{r},\underline{K}) = 4\pi h(\underline{r},\underline{K})\ \int \frac{\sin(k_o|\underline{p}|-\underline{K}\cdot\underline{p})}{|\underline{p}|}\ d^3p\ . \quad (3.23)$$

It can be shown that (Ref. 2, p. 370)

$$\int \frac{\sin(k_o|\underline{p}|-\underline{K}\cdot\underline{p})}{|\underline{p}|}\ d^3p = \frac{2\pi^2}{k_o}\ \delta(K-k_o)\ . \quad (3.24)$$

On substituting from Eq. (3.24) on the r.h.s. of Eq. (3.23) and
recalling the definitions (2.12) and (2.13) of the angular com-
ponents $Q(\underline{r},\underline{s})$ and $\underline{R}(\underline{r},\underline{s})$, we obtain the following simplified
transfer equations:

$$\underline{s} \cdot \nabla_r \, Q(\underline{r},\underline{s}) = \frac{(2\pi)^3}{c} \, h(\underline{r},k_o\underline{s}), \tag{3.25}$$

$$(\underline{s} \cdot \nabla_r) \, \underline{R}(\underline{r},\underline{s}) = (2\pi)^3 \, \underline{s} \, h(\underline{r},k_o\underline{s}). \tag{3.26}$$

Moreover, the source function $D(\underline{r},\underline{s})$ is seen to be given by

$$D(\underline{r},\underline{s}) = (2\pi)^3 \, h(\underline{r},k_o\underline{s}). \tag{3.27}$$

An interesting case is that of uncorrelated random sources,
previously investigated by Keller.[7] The correlation function of
the source distribution is of the form

$$B(\underline{r},\underline{r}') = M(\underline{r}) \, \delta(\underline{r}') \quad . \tag{3.28}$$

Then, in view of Eq. (3.12), we obtain

$$h(\underline{r},k_o\underline{s}) = \frac{1}{(2\pi)^3} \, M(\underline{r}). \tag{3.29}$$

It thus follows from Eqs. (3.25) and (3.26) that the angular
components satisfy the equations

$$\underline{s} \cdot \nabla_r \, Q(\underline{r},\underline{s}) = \frac{1}{c} \, M(\underline{r}), \tag{3.30}$$

$$(\underline{s} \cdot \nabla_r) \, \underline{R}(\underline{r},\underline{s}) = \underline{s} \, M(\underline{r}), \tag{3.31}$$

and the source function $D(\underline{r},\underline{s})$ is given by

$$D(\underline{r},\underline{s}) = M(\underline{r}). \tag{3.32}$$

4. CONCLUSION

In this paper, we derived rigorous equations for radiative
transfer of energy and flux [Eqs. (3.16) and (3.17)] in the
presence of random source distributions on the basis of the
stochastic scalar wave equation (3.1).   Our results relate the
source function, which is introduced rather heuristically in the
conventional theory of radiative energy transfer, to the stochastic

characteristics of the source distribution [Eq. (3.20)]. We have
shown, that under certain assumptions [Eqs. (3.22) and (3.27)] the
transfer equations (3.16) and (3.17) reduce to much simpler forms.

Finally, it is clear that the present theory can be generalized
to the case of stochastic electromagnetic fields.

ACKNOWLEDGMENT

It is a pleasure to thank Professor Emil Wolf for help
and guidance.

*Research supported by U.S. Army Research Office.

References

1.   See, for example, G. C. Pomraning, *The Equations of Radiation
      Hydrodynamics*, (Pergamon Press,1973).
2.   References to some of the relevant papers are given by
      V.I. Tatarskii, *The Effects of Turbulent Atmosphere on Wave
      Propagation*, (U.S. Department of Commerce, National Technical
      Information Service, Springfield, Va., 1971), Sec. 63.
3.   E. Wolf, Phys. Rev. *D 13*, 869 (1976).
4.   M.S. Zubairy and E. Wolf, Opt. Commun. *20*, 321 (1977).
5.   G.I. Ovchinnikov and V.I. Tatarskii, Radiophysics and Quantum
      Electronics *15*, 1087 (1972)(Translated from Russian).
6.   For a description of the functions $\delta_+(x)$, see, for example,
      W. Heitler, *The Quantum Theory of Radiation*, (Oxford Press,
      1954) pp. 69-70.
7.   J.B. Keller, "Stochastic Equations and Wave Propagation in
      Random Media", *Proc. Sympos. Appl. Math.*,ed. R. Bellman,
      Vol. 16, (Amer. Math. Soc., Providence, R.I., 1964) pp. 145-170.

THE QUANTUM MECHANICAL FOUNDATIONS OF SEMICLASSICAL RADIATION
THEORIES*

I. R. Senitzky

Technion, Haifa, Israel

I.   INTRODUCTION

The wide interest, in recent years, in the validity of semi-
classical radiation theory (there appear to be several such theories, if one examines their - usually implicit - definitions
carefully) may be attributed to two causes. The first is the successful use of such theories in a number of important phenomena,
such as maser and laser operation, self-induced transparency,
various atomic and nuclear resonance effects, and photoelectric
detection. The second is the attempt, on the one hand, to make
semiclassical theory a fundamental theory in place of quantum
electrodynamics,[1] and, on the other hand, to question its validity under conditions for which its use had been considered acceptable.[2] It is the purpose of the present discussion to examine,
from an orthodox quantum mechanical viewpoint, the conditions for
which the use of semiclassical radiation theory is valid.

II.   DEFINITIONS

It is essential, for clarity, to define the several semiclassical theories that have received wide usage or have been proposed
as improved versions. They will be designated as SCTI, SCTII, etc.

SCTI, the oldest theory, is, in the words of Schiff,[3] a
theory in which "we treat the electromagnetic field classically
and the particles with which the field interacts by quantum
mechanics."

SCTII, which has been used widely in the analysis of almost

all coherent phenomena when the mutual interaction of atoms and
field is involved,[4] couples Schrodinger's equation and Maxwell's
equations; the field is described classically in both sets of
equations, the atoms are described quantum mechanically by
Schrodinger's equation, while the atomic - or matter - variables
in Maxwell's equation (for most applications, only the polarization,
or atomic dipole-moment, is of interest) are replaced by their
expectation values.

SCTIII, which is the latest version of Jaynes' "neoclassical
theory,"[5] introduces a classical model for the atom based on its
natural frequencies (or energy spectrum) and the associated oscillating dipole moments, with the latter providing the atomic polarization in the classical Maxwell's equations. An essentially similar version has been proposed by Eberly.[6]

SCTIV is the classical-limit form of a fully quantum mechanical theory in which the atoms are described by a boson-second-
quantization formalism.[7]

III. DISCUSSION

A. Semiclassical Theory I

It has been shown some time ago that the mutual interaction
of two systems, one strictly quantum mechanical and the other
strictly classical, cannot be described by a self consistent dynamical formalism.[8] In the present instance, one reason is the fact
that quantum-mechanical atoms generate a quantum mechanical field,
so that the field, even if it is classical initially, cannot remain classical. Another reason is the fact that, to the classical
system, the zero-point motion of the quantum mechanical system
looks like motion that can do work, which obviously leads to absurd
results.[8] In order to place SCTI within a quantum mechanical
framework, we consider a field (in the Heisenberg Picture) which is
the sum of two parts, one purely quantum-mechanical and the other
classical. The total field is, of course, fully quantum-mechanical.
The purely quantum-mechanical part is due to the atoms under consideration and to the loss mechanism with which the field that is
interacting with the atoms is coupled. The classical part is due
to external sources which are unaffected by the atomic behavior or
by the loss mechanism. (It has been shown that such an external
field, no matter what its sources are, may be described classically
as far as the atoms are concerned.[9]) SCTI is a valid approximation in the case where the quantum-mechanical part of the field is
negligible compared to the classical part, and also in the case
where the questions asked refer only to atomic effects produced by
the classical part. An example of the first case is the behavior
of atoms in a strong laser field - assuming, of course, that the

laser operation is unaffected by the atoms under consideration.
An example of the second case is (lowest order) induced emission
or absorption; [10] here, only the external field determines the
result - assuming that the loss mechanism does not produce a significant thermal field - since the quantum-mechanical part of the
field contributes, in lowest order, only to spontaneous emission.
An important application - entirely valid - of SCTI is the theory of
photoelectric detection. [10] In the usual approach to photoelectric phenomena, the reaction of the photoelectrons on field being
detected is ignored. The induced absorption by the photosensitive
atoms is investigated to lowest order in perturbation theory, in
order to obtain information about the field produced by external
sources. This field is the classical part of the total field, the
only part needed to describe the production of photoelectrons.

B. Semiclassical Theory II

In contrast to SCTI, SCTII provides a prescription for analyzing the mutual interaction between the atoms and the field. [11]
It allows the field to remain classical even though the atoms are
treated quantum-mechanically; however, this is accomplished by an
arbitrary requirement, namely, that the matter variables in Maxwell's
equations be replaced by their expectation values. The conditions
for the validity of such a procedure from a quantum-mechanical viewpoint will be discussed later. Presently, it is instructive to illustrate a case for which such a procedure is invalid. Let the
matter under consideration consist of a single, highly excited harmonic oscillator with an electric dipole moment proportional to its
displacement. One can, in principle, specify its quantum-mechanical
state to be an energy state (corresponding to a high quantum number).
Since, in this state, the expectation value of the dipole moment is
zero, this highly excited oscillator will not radiate at all according to SCTII! The reason for such an incorrect result is not difficult to find. The description of the oscillator by means of an
energy state is a statistical description. In the limit of high
quantum numbers (the classical limit), the energy state becomes
equivalent to a classical description in which the amplitude of
oscillation is well defined but the probability distribution of
the phase of oscillation is constant. [12] We see that SCTII ignores
the purely statistical aspect of a quantum-mechanical description,
even if the description refers to a macroscopic system. (A harmonic oscillator in a high energy state is essentially such a system.) SCTII is, therefore, invalid when the statistical aspects
of the description of the matter - in the sense of being nondeter-
ministic - are significant.

C. Semiclassical Theory III

SCTIII achieves the same results as those of SCTII, for certain types of problems, without an arbitrary prescription for joining quantum-mechanical and classical theories. In SCTIII the atom is described schematically by its natural frequencies and associated dipole moments, as follows:[5] let the natural frequencies of the atom be those determined by the set of energy levels $\hbar\omega_n$, n = 1,2,... The (classical) Hamiltonian describing the free atom is then given by

$$H_o = \sum_n \hbar\omega_n a_n{}^* a_n \, , \tag{1}$$

where a_n and $a_n{}^*$ are independent (complex) dynamical variables. Corresponding to each atomic frequency $|\omega_i - \omega_j|$, there exists a dipole moment, the components of which are linear superpositions of the quantities

$$d_{ij}{}^{(1)} = \frac{1}{2} (a_j{}^* a_i + a_i{}^* a_j),$$

$$d_{ij}{}^{(2)} = -\frac{1}{2} i(a_j{}^* a_i - a_i{}^* a_j),$$

$$d_{ij}{}^{(3)} = \frac{1}{2} (a_j{}^* a_j - a_i{}^* a_i) \, . \tag{2}$$

The coupling between the atom and the field is assumed to be of the dipole-moment type described most generally by the interaction Hamiltonian

$$H' = \hbar \sum_{m=1}^{3} \sum_{i<j} d_{ij}{}^{(m)} F_{ij}{}^{(m)} \, , \tag{3}$$

where $F_{ij}{}^{(m)}$ is a linear superposition of the electromagnetic field components that contain appropriate coupling constants. The total Hamiltonian for the atom and field is given by

$$H = H_o + H' + H_f \, , \tag{4}$$

H_f being the classical Hamiltonian for the field only. The "canonical" equations for the atomic variables are [5]

$$i\hbar\dot{a}_n = \frac{\partial H}{\partial a_n{}^*} \, , \qquad i\hbar\dot{a}_n{}^* = -\frac{\partial H}{\partial a_n} \, , \tag{5}$$

and the equations of motion for the field variables are those of the conventional Hamiltonian formalism for the classical electromagnetic field. It can be shown that the equations of motion for the a_n's, together with the definition of dipole moment by means of the $d_{ij}{}^{(m)}$'s, yield the same equations of motion for the dipole moment in SCTIII as those for the expectation value of the dipole moment in SCTII. The initial values of the a_n's in SCTIII are chosen so that the initial dipole moment obtained from Eqs. (2) is equal to the initial expectation value of the dipole moment in SCTII. Thus, SCTIII becomes a formal Hamiltonian theory that gives the same results as SCTII, as far as the field is concerned, without an *ad hoc* prescription to connect the quantum-mechanical atomic equations with the classical field equations. SCTIII exhibits the same lack of validity as SCTII when applied to states that are not sufficiently deterministic.

D. Semiclassical Theory IV

In order to *derive* a semiclassical theory from quantum mechanics, we begin with a boson-second-quantization formalism for the description of a number of identical atoms that couple similarly through their dipole moment to the electromagnetic field. Let the relevant spectrum of each atom consist of n levels with energies $\hbar\omega_i$, i = 1,2,...n. The atomic state vectors that describe the entire collection lie in a space spanned by the vectors $|r_1...r_i..r_n\rangle$ where the r_i's are non-negative integers. The fundamental operators, from which the pertinent dynamical variables may be constructed, are - using the Heisenberg Picture - $a_1(t),...a_n(t)$ and $a_1{}^\dagger(t)$, ... $a_n{}^\dagger(t)$, such that

$$a_i(0)|r_1...r_i...r_n\rangle = r_i{}^{1/2}|r_1...r_i-1...r_n\rangle \tag{6}$$

and

$$a_i{}^\dagger(0)|r_1...r_i...r_n\rangle = (r_i+1)^{1/2}|r_1...r_i+1...r_n\rangle. \tag{7}$$

The commutation relationships are $[a_i(t), a_i{}^\dagger(t)] = 1$, with all other equal-time commutators vanishing. The Hamiltonian for the collection of atoms is given by the expression

$$H_o = \sum_{i=1}^{n} \hbar\omega_i a_i{}^\dagger a_i . \tag{8}$$

The collective atomic dipole moment is a linear superposition of the operators

$$d_{ij}^{(1)} = \frac{1}{2} (a_j^\dagger a_i + a_i^\dagger a_j) ,$$

$$d_{ij}^{(2)} = -\frac{1}{2} i (a_j^\dagger a_i - a_i^\dagger a_j),$$

$$d_{ij}^{(3)} = \frac{1}{2} (a_j^\dagger a_j - a_i^\dagger a_i) ,$$

(9)

$i \neq j$, to which we will refer as the dipole moment components. The coupling to the electromagnetic field is described by the inter- action Hamiltonian

$$H' = \hbar \sum_{m=1}^{3} \sum_{i<j} f_{ij}^{(m)} d_{ij}^{(m)} ,$$

(10)

where the $f_{ij}^{(m)}$'s are linear superpositions of the components of the electromagnetic field (considered to be Heisenberg-Picture operators) that contain appropriate coupling constants. The total Hamiltonian for atom and field is given by

$$H = H_o + H' + H_f ,$$

(11)

H_f being the field Hamiltonian, which we do not need in explicit form.

Using the notation $n_i \equiv a_i^\dagger a_i$, we note that the basis vectors are eigenvectors of $n_i(0)$ with eigenvalue r_i, and that $\Sigma_i n_i(t)$ is a constant of motion, such that [7]

$$[\Sigma_i n_i(t)] | r_1 \ldots r_n > = N | r_1 \ldots r_n > ,$$

(12)

where $N = \Sigma_i r_i$. We take the state of the atomic collection (the initial state, in the Schrodinger Picture) to be that for which N is the number of atoms under consideration. The bosons in this descrip- tion are, therefore, the atoms themselves, and any state in this description corresponds to a fully symmetrized many-atom state in a first quantization formalism.

The basis vectors are, clearly, energy states, that is, eigen- vectors of $H_o(0)$ with eigenvalues $\Sigma_i r_i \hbar \omega_i$. It is useful to define another set of states, coherent states, [7,13-15] described by the complex set of numbers $C_1 \ldots C_n$ (designated as $\{C\}$, for brevity), with $\Sigma_i |C_i|^2 = 1$, as follows:

$$| \{C\}_N > = \sum_{r_1 \ldots r_n}^{(N)} \left(\frac{N}{r_1! \ldots r_n!} \right)^{1/2} C_1^{r_1} \ldots C_n^{r_n} | r_1 \ldots r_n > ,$$

(13)

the superscript (N) indicating that the summation is taken over all values of $r_1, \ldots r_n$ for which $\Sigma_i r_i = N$. These states obey the simple relationship

$$a_k(0) | \{C\}_N \rangle = N^{1/2} \, C_k | \{C\}_{N-1} \rangle \; . \tag{14}$$

In the case of two levels, and translated into a first quantization formalism, the energy states have been referred to in the literature as correlated incoherent states,[16] or as fully symmetrized Dicke states,[14] and the coherent states have been referred to as uncorrelated coherent states,[16] or as fully symmetrical Bloch states.[14] It can be shown that, for the free atoms (uncoupled from the field),

$$\langle r_1 \ldots r_n | d_{ij}^{(m)} | r_1 \ldots r_n \rangle = 0, \; m = 1, 2,$$

$$\langle r_1 \ldots r_n | d_{ij}^{(3)} | r_1 \ldots r_n \rangle = \frac{1}{2}(r_j - r_i), \tag{15}$$

and

$$\langle \{C\}_N | d_{ij}^{(1)} | \{C\}_N \rangle = N |C_j C_i| \cos[(\omega_j - \omega_i)t + \theta_{ji}]$$

$$\langle \{C\}_N | d_{ij}^{(2)} | \{C\}_N \rangle = N |C_j C_i| \sin[(\omega_j - \omega_i)t + \theta_{ji}]$$

$$\langle \{C\}_N | d_{ij}^{(3)} | \{C\}_N \rangle = \frac{1}{2} N(|C_j|^2 - |C_i|^2) \; , \tag{16}$$

where θ_{ji} is defined by $C_j{}^* C_i = |C_j C_i| \exp(i\theta_{ji})$. Thus, $\langle d_{ij}^{(1)} \rangle$, for instance, vanishes for an energy state but oscillates with frequency $|\omega_j - \omega_i|$ and well defined phase for a coherent state (provided C_i, $C_j \neq 0$). On the other hand, if we look at $\langle d_{ij}^{(1)2} \rangle$, we obtain [7]

$$\langle r_1 \ldots r_n | d_{ij}^{(1)2} | r_1 \ldots r_n \rangle = \frac{1}{4}[2 r_i r_j + r_i + r_j], \tag{17}$$

and

$$\overline{\langle \{C\}_N | d_{ij}^{(1)2} | \{C_N\} \rangle} = \frac{1}{4} N[2(N-1)|C_i|^2 |C_j|^2 + |C_i|^2 + |C_j|^2], \tag{18}$$

where the bar indicates time average. We see that for $r_k = N|C_k|^2$, the terms of order N^2 are equal.

It may be remarked that the physical significance of treating atoms as bosons in the present manner lies in their cooperative behavior. Complete symmetrization in a first quantization

formalism - as is required by such a treatment - may be regarded,
intuitively, as maximum cooperation among the atoms under consideration. The present method may be generalized to the case of
less than maximum atomic cooperation by dividing a number of
atoms into two or more collections, each collection being bosons
of a different kind.

The collection of atoms is described completely, in the
present formalism, by the coordinates of n harmonic oscillators.
In order to investigate the classical-limit conditions, we merely
need to consider the classical-limit conditions pertaining to the
harmonic oscillators. These are well known to be the limit of
high (energy) quantum numbers. Since a description need not be
associated with a definite energy state, a more general formulation can be expressed by the inequalities,

$$\langle n_j \rangle \gg 1, \quad j = 1,2,\ldots n \;.$$
(19)

When these inequalities are fulfilled, the oscillators may be
treated classically, the commutators $[a_j, a_j\dagger]$ are relatively negligible (the last statement can be considered an alternate formulation
of the classical-limit condition), and the operators a_j and $a_j\dagger$
may be treated as c-number variables. For clarity, where necessary,
we will use a tilde henceforth to indicate such variables.

One also needs to look at the field under these limiting conditions. If the atoms under consideration are classical, the field
they generate is classical. External fields are also classical,
since they must be prescribed.[9] The only part of the field that
remains quantum-mechanical is the zero-point field. However,
classical systems must not "see" the zero point field, since,
formally, it appears to them as a field which can do work.[8]
Thus, as far as the field that interacts with the collection of
atoms under consideration is concerned, it may be described classically if the atoms are described classically.

Thus, the Hamiltonian of Eq. (11) becomes a classical Hamiltonian. Equations of motion for the dynamical variables are
obtained according to classical dynamics by means of Poisson
brackets. Noting that $a_j = 2^{-1/2}(q_j + ip_j)$, where q_j and p_j are
the dimensionless coordinate and momentum of the j'th oscillator,
such that $H_0 = 1/2 \sum_j \hbar\omega_j (q_j^2 + p_j^2)$, the canonical equations of
motion for $\tilde{a}_j$ and $\tilde{a}_j^*$ become

$$\frac{d\tilde{a}_j}{dt} = \frac{1}{i\hbar}\frac{\partial\tilde{H}}{\partial\tilde{a}_j^*}, \quad \frac{d\tilde{a}_j^*}{dt} = -\frac{1}{i\hbar}\frac{\partial\tilde{H}}{\partial\tilde{a}_j},$$
(20)

while the canonical equations of motion for the field variables
become essentially equivalent to the classical Maxwell's equation.
In order to complete this theory, which was obtained as the classical limit of a fully quantum-mechanical theory, we need, in
addition to the equations of motion, a method of prescribing
initial conditions that is consistent with the classical-limit
procedure.

In the quantum-mechanical theory, initial conditions are
described by means of a quantum state. What is the classical limit
of such a description? As is well known, a quantum-state description yields information in a statistical form, in general. In the
classical limit, therefore, a quantum-state description must become
a statistical description, in which $\tilde{a}_j(0)$ and $\tilde{a}_j{}^*(0)$ are considered
to be random variables. One method of providing such a description
is that of specifying moments. Consider the moments
$\langle O\{\pi_i a_i{}^{\dagger v_i}(0) a_i{}^{w_i}(0)\}\rangle$, where O is an order operator that provides
for some specific ordering arrangement of the operators inside the
curly bracket. Moments corresponding to different arrangements
may have different values because of non-vanishing commutators.
In the classical limit, these commutators are relatively negligible,
provided

$$v_j + w_j \ll \langle n_j \rangle , \tag{21}$$

for all j. We restrict the number and kind of specified moments
by these inequalities (thus making the statistical description
somewhat non-unique - or somewhat approximate) and specify the
various moments of the random variables $\tilde{a}_j$, $\tilde{a}_j{}^*$ by

$$\langle \pi_i \tilde{a}_i{}^{*v_i}(0) \tilde{a}_i{}^{w_i}(0) \rangle = \langle \psi | O\{\pi_i a_i{}^{+v_i}(0) a_i{}^{w_i}(0)\} | \psi \rangle \tag{22}$$

where $|\psi\rangle$ is the (initial) state of the collection of atoms, and
O indicates an arbitrary ordering arrangement that may be chosen
for convenience of evaluation. [Note that all ordering arrangements of moments subject to inequality (21) yield results that differ by negligible amounts in the classical limit.] Probability
distributions consistent with this definition of moments provide
an alternate statistical description.

Comparison of the equations of motion of SCTIV with those of
SCTIII shows that they are identical; however, the method of
determining initial conditions in the two theories is different.
While in SCTIII the initial conditions for the dipole moments are
obtained by equating their values to the corresponding quantum-
mechanical expectation values, in SCTIV the initial conditions
consist of a statistical description obtained from the quantum-
mechanical state. SCTIII and SCTII are, thus, a special case of

SCTIV; they may be applied only in those situations where not only
the classical-limit conditions for the application of SCTIV are
met, but also where the (initial) state of the collection of atoms
describes the atomic dipole moment in a sufficiently deterministic
manner.

A simple argument shows that the above conditions for the
application of SCTII - IV are both necessary and sufficient. If
we begin with SCTIII as a classical Hamiltonian theory, and convert
it, by usual methods, into a quantum theory, we obtain precisely
the boson-second-quantization formalism from which we derived
SCTIV. Thus, only this formalism can give, in the classical limit,
a theory in which SCTIII may be included, and if SCTIII (and SCTII,
for which SCTIII is the Hamiltonian formalism) is to be derived
from quantum mechanics, it must meet the aforementioned conditions.
Needless to say, these conditions do not allow the application of
SCTII-IV to a single atom.

It is instructive to illustrate the application of SCTIV to
an energy-state description and to a coherent-state description.
For simplicity, we ignore the reaction of the field as well as the
effect of external fields. For an energy state that meets the
classical-limit conditions ($r_i \gg 1$ for all r_i), Eq. (22) yields,
using an ordering arrangement in terms of n_i's for easy evaluation,

$$\langle \pi_i \tilde{a}_i^{*\, v_i}(0) \tilde{a}_i^{\, w_i}(0) \rangle = \pi_i r_i^{\, v_i} \delta_{w_i v_i} \tag{23}$$

A simple probability distribution consistent with these moments
is given by

$$\tilde{a}_j(0) = r_j^{1/2} e^{-i\theta_j} , \tag{24}$$

where θ_j has a uniform probability distribution. Note that the θ_j's
and, therefore, the $\tilde{a}_j(0)$'s, are *independent* random variables.
For a coherent state which meets the classical limit conditions,
Eq. (22) yields, using a normal ordering arrangement for easy evaluation,

$$\langle \pi_i \tilde{a}_i^{*\, v_i}(0) \tilde{a}_i^{\, w_i}(0) \rangle$$

$$= [N(N-1)\ldots(N-V+1)] \pi_i C_i^{*\, v_i} C_i^{\, w_i} \delta_{VW}$$

$$\approx N^V \pi_i C_i^{*\, v_i} C_i^{\, w_i} \delta_{VW} ,$$

where $V = \Sigma_i v_i$, $W = \Sigma_i w_i$, and the inequality (21) was utilized in
the approximation. A simple probability distribution consistent

with these moments is

$$\tilde{a}_j(0) = N^{1/2} C_j e^{i\theta},$$

where θ is a random variable with a uniform probability distribution, but it has the *same* value for all $\tilde{a}_j(0)$'s. The $\tilde{a}_j(0)$'s are, therefore, *dependent*, random variables. Consider, now, the dipole moment of the collection of atoms. If we take $r_j = N|C_j|^2$, the amplitude of oscillation of $d_{ij}^{(1)}$ and $d_{ij}^{(2)}$ will have the same value for both the energy state and the coherent state. However, the phase of oscillation for the energy state will be a random variable with all values equally probable, while that for the coherent state will be well defined. It is clear that, in the case of these two states, SCTII-III can be applied only to the coherent state (in the classical limit) while SCTIV can be applied to both the coherent state and the energy state. The former is an example of a deterministic description (of the pertinent matter variables in Maxwell's equation) while the latter is an example of a statistical - in the sense of non-deterministic - description.

The conclusion of the present section may be expressed, loosely speaking, by the statement that SCTII-IV are applicable only to those situations where the atoms belong to one or more groups in each of which they behave cooperatively and their number is large; SCTII-III are further restricted to cases where the description of the dipole moment is deterministic.

*Sponsored in part by the U.S. Army through its European Research Office.

References

1. For review articles on this subject, see L. Mandel, in *Progress in Optics*, Vol. XIII, ed. by E. Wolf (North-Holland Publishing Co., Amsterdam, 1976) Ch. II; P. W. Milonni, Phys. Reports *25*, 1 (1976).
2. R. J. Glauber, Phys. Rev. Letters *10*, 84 (1963).
3. L. J. Schiff, *Quantum Mechanics* (McGraw-Hill Book Co., New York, 1949) p. 240.
4. See, for instance, H. Haken, *Handbuch der Physik*, Vol. XXV/2c (Springer Verlag, Berlin, 1970).
5. E. T. Jaynes, in *Coherence and Quantum Optics*, ed. by L. Mandel and E. Wolf (Plenum Press, New York, 1973) p. 35.
6. J. H. Eberly, in *Physics of Quantum Electronics*, Vol. IV, ed. by S. Jacobs, M. Sargent III, and M. O. Scully (Addison-Wesley, Reading, Mass., 1976) p. 421.
7. I. R. Senitzky, Phys. Rev. A*10*, 1868 (1974); *15*, 284 (1977).
8. I. R. Senitzky, Phys. Rev. Letters *20*, 1062 (1968).

9.   I. R. Senitzky, Phys. Rev. Letters *15*, 233 (1965); *16*, 619
     (1966).
10. I. R. Senitzky, Phys. Rev. *174*, 1588 (1968).
11. For an illustration of the application of SCTII to a gaseous
     laser, see W. E. Lamb, Jr., Phys. Rev. *134*, A1429 (1964).
12. A. Messiah, *Quantum Mechanics* (North-Holland Publishing Co.,
     Amsterdam, 1961) p. 460.
13. R. Gilmore, Ann. Phys. (N.Y.) *74*, 391 (1972).
14. F. T. Arecchi, E. Courtens, R. Gilmore and H. Thomas, Phys.
     Rev. A*6*, 221 (1972).
15. R. Gilmore, C. M. Bowden and L. M. Narducci, Phys. Rev. A*12*,
     1019 (1975).
16. I. R. Senitzky, Phys. Rev. *111*, 3 (1958).

THE $\hbar \to 0$ LIMIT OF QED

G.C. Dente

St. Louis University, St. Louis, Missouri

Covariant perturbation theory solutions of quantum electrodynamics indicate that charged particles interact through the Feynman
photon propagator, D_F. The difficulty of establishing the classical-
correspondence limit for QED interactions exists because these D_F
propagators propagate interactions into the past and future; these
interactions have no obvious connection with the classical physics
of particles interacting via retarded Lienard-Wiechert potentials.

Two routes were sometimes followed when faced with this apparent
lack of correspondence: 1) alternatives to QED were sought; 2) alternatives to conventional classical theory were sought. Redeveloping
the classical theory as a theory of particle currents interacting
through the D_F propagator is one example of the second of these
alternatives. Wheeler and Feynman approximated this approach when
they developed their "absorber theory" of particles interacting
through the real part of D_F, $Re(D_F) = 1/2(D_{Ret} + D_{adv})$, which was
then supplemented by the complete absorber condition [1]. Although
absorber theory was proposed prior to the modern development of QED
between 1948 and 1950, realization of the important part played by
the D_F propagator in QED has subsequently been used as a motivation
for the absorber theory [2].

I feel that the unusual assumptions of absorber theory and of
most QED alternatives are a high price to pay in order to establish
a correspondence limit. There must be a more natural way to solve
the problem.

We will obtain the $\hbar \to 0$ of QED in two steps. 1) We will calculate transition amplitudes from an initial state of particles with
a transverse photon vacuum to a final state of particles accompanied

by radiation. (The essentials of this step are contained in an
earlier paper [3]). 2) We will take the $\hbar \to 0$ limit, being careful
to vary only those paths which have not been specified. The meaning
of this statement will be clarified in what follows.

2. TRANSITION AMPLITUDE CALCULATION

As a first step towards extracting the classical limit, we
must calculate a QED transition amplitude between relevant initial
and final states. The relevant initial state is a particle configuration and the transverse photon vacuum, while the relevant final
state is a particle configuration and the coherent radiation state
generated by the particles moving on definite trajectories, $\{\varepsilon_n(t)\}$.
This particular choice of final state is motivated by our anticipa-
tion of the $\hbar \to 0$ limit.

Using the path-integral formulation of QED, we can perform a
functional integration to eliminate the photon degrees of freedom
from this QED transition amplitude. This functional integration
must be done subject to the end-point conditions of an initial photon
vacuum and the final coherent radiation mentioned above. The details
for eliminating the photon coordinates in this fashion are contained
in Ref. 3. Having done this, the relevant amplitude can be written
as a sum over particle paths as:

$$\text{Amplitude} = \sum_{\text{Paths}\{\underline{x}_n(t)\}} \exp\left[\frac{i}{\hbar} S_{\text{eff}}\right]$$

where S_{eff} = (action contributed by particle kinetic energies) +
(action due to instantaneous Coulomb interaction) + (I = action due
to interaction with the transverse quantized modes of the electromagnetic field).

Allowing for the classically anticipated radiation in the final
photon state will affect only the last of these three pieces in S_{eff}.
The explicit expression for the last piece is:

$$I = i \sum_{n,m} \int_{-\infty}^{\infty} dt \int_{-\infty}^{\infty} ds \int \frac{d^3k}{(8\pi^2)|k|} e_n e_m$$

$$\times \{\exp(-ik|t-s|)[\dot{\underline{x}}_n(t) \cdot \dot{\underline{x}}_m(s) - k^{-2}\underline{k} \cdot \dot{\underline{x}}_n(t)\,\underline{k} \cdot \dot{\underline{x}}_m(s)]$$

$$\times \cos[\underline{k} \cdot (\underline{x}_n(t) - \underline{x}_m(s))]$$

$$+ \exp(-ik(t-s))[\dot{\underline{\varepsilon}}_n(t) \cdot \dot{\underline{\varepsilon}}_m(s) - k^{-2}\underline{k} \cdot \dot{\underline{\varepsilon}}_n(t)\, \underline{k} \cdot \dot{\underline{\varepsilon}}_m(s)]$$

$$\times \cos[\underline{k} \cdot (\underline{\varepsilon}_n(t) - \underline{\varepsilon}_m(s))]$$

$$- 2\exp(-ik(t-s))[\dot{\underline{\varepsilon}}_n(t) \cdot \dot{\underline{x}}_m(s) - k^{-2}\underline{k} \cdot \dot{\underline{\varepsilon}}_n(t)\, \underline{k} \cdot \dot{\underline{x}}_m(s)]$$

$$\times \cos[\underline{k} \cdot (\underline{\varepsilon}_n(t) - \underline{x}_m(s))] \} \quad ,$$

in which, we once again emphasize, $\{\underline{\varepsilon}_n(t)\}$ is the anticipated classical trajectory for particle n (of charge e_n), while only the paths $\{\underline{x}_m(t)\}$ are unspecified and free to vary in the sum over paths. Notice that if we drop the distinction between $\{\underline{\varepsilon}_m(t)\}$ and $\{\underline{x}_m(t)\}$, this expression collapses to Eq.(13) of Ref. 3.

It is important to realize that in constructing this relevant amplitude, we have assumed that the $\hbar \to 0$ limit involves particles moving on definite trajectories. It does not require that these trajectories be explicitly known, but rather, only that they exist. Therefore, the program which we are presenting is more than a consistency check; we are constructing the classical limit, and in this first step we have assumed only that the limit exists.

3. $\hbar \to 0$ LIMIT

For $\hbar << |S_{eff}|$, the paths near the path giving stationary action, $\delta S_{eff}/\delta x = 0$, dominate in the sum over paths; in fact, as $\hbar \to 0$, we need only consider the path of stationary action. We must perform all variations within the set $\{\underline{x}_n(t)\}$, being careful to notice that we are not free to vary the $\{\underline{\varepsilon}_n(t)\}$. These $\{\underline{\varepsilon}_n(t)\}$ comprise the trajectories anticipated in the $\hbar \to 0$ limit, and they are, in principle, specified by the radiation field to which they give rise. Specifically, a variation in I would yield

$$\delta I = ie_n \sum_m \int_{-\infty}^{\infty} dt \int_{-\infty}^{\infty} ds \int \frac{d^3k}{(8\pi^2)k}\, e_m$$

$$\times \{ 2\exp(-ik|t-s|)[\dot{\underline{x}}_m(s) - k^{-2}\, k_x\, \underline{k} \cdot \dot{\underline{x}}_m(s)]$$

$$\times \cos[\underline{k} \cdot \{\underline{x}_n(t) - \underline{x}_m(s)\}] \, \delta\dot{\underline{x}}_n(t)$$

$$- 2k_x \exp(-ik|t-s|)[\dot{\underline{x}}_n(t) \cdot \dot{\underline{x}}_m(s) - k^{-2}\underline{k} \cdot \dot{\underline{x}}_n(t)\ \underline{k} \cdot \dot{\underline{x}}_m(s)]$$

$$\times \sin[\underline{k} \cdot \{\underline{x}_n(t) - \underline{x}_m(s)\}]\delta x_n(t)$$

$$- 2 \exp[-ik(t-s)][\dot{\varepsilon}_{xm}(t) - k^{-2} k_x \underline{k} \cdot \dot{\underline{\varepsilon}}_m(t)]$$

$$\times \cos[\underline{k} \cdot \{\underline{x}_n(s) - \underline{\varepsilon}_m(t)\}]\delta\dot{x}_n(s)$$

$$+ 2k_x \exp[-ik(t-s)][\dot{\underline{\varepsilon}}_m(t) \cdot \dot{\underline{x}}_n(s) - k^{-2}\underline{k} \cdot \dot{\underline{\varepsilon}}_m(t)\ \underline{k} \cdot \dot{\underline{x}}_n(s)]$$

$$\times \sin[\underline{k} \cdot \{\underline{x}_n(s) - \underline{\varepsilon}_m(t)\}]\delta x_n(s)\ \}\qquad .$$

We have now performed the variations, and it is safe to drop the
distinction between $\{x_n(t)\}$ and $\{\varepsilon_n(t)\}$. If we make the redefinition
$S \rightleftarrows t$ in the last two terms, then we can rewrite the variation as

$$\delta I = e_n \sum_m \int dt \int^t ds \int \frac{d^3k}{2\pi^2 k} e_m \sin[k(t-s)]$$

$$\times \{[\dot{\varepsilon}_{xm}(s) - k^{-2} k_x \underline{k} \cdot \dot{\underline{\varepsilon}}_m(s)]\cos[\underline{k} \cdot \{\underline{\varepsilon}_n(t) - \underline{\varepsilon}_m(s)\}]\delta\dot{\varepsilon}_{xn}(t)$$

$$- k_x[\dot{\underline{\varepsilon}}_n(t) \cdot \dot{\underline{\varepsilon}}_m(s) - k^{-2}\underline{k} \cdot \dot{\underline{\varepsilon}}_n(t)\ \underline{k} \cdot \dot{\underline{\varepsilon}}_m(s)]$$

$$\times \sin[\underline{k} \cdot \{\underline{\varepsilon}_n(t) - \underline{\varepsilon}_m(s)\}]\delta\varepsilon_{xn}(t)\ \}$$

$$= e_n \int dt\ \delta\varepsilon_{xn}(t)\{[-\frac{\partial}{\partial t} + \dot{\underline{\varepsilon}}_n(t) \cdot \nabla_{\underline{\varepsilon}_n}]\ A_x(\varepsilon_n,t)$$

$$+ \underline{\varepsilon}_n(t) \cdot \frac{\partial}{\partial\varepsilon_{xn}}\ \underline{A}(\varepsilon_n,t)\ \}\ ,$$

where

$$\underline{A}(\varepsilon_n,t) = + \sum_m e_m \int \frac{d^3k}{2\pi^2 k} \int^t ds\ \sin[k(t-s)]$$

$$\times [\dot{\underline{\varepsilon}}_m(s) - k^{-2}\underline{k}\ \underline{k} \cdot \dot{\underline{\varepsilon}}_m(s)]\cos[\underline{k} \cdot \{\underline{\varepsilon}_n(t) - \underline{\varepsilon}_m(s)\}]$$

is the retarded transverse vector potential generated by the charged
particles on trajectories $\{\varepsilon_m(s)\}$. A final manipulation leads to

$$\delta I = e_n \int [- \frac{\partial}{\partial t} \underline{A}(\varepsilon_n,t) + \dot{\underline{\varepsilon}}_n \times \underline{B}(\varepsilon_n(t),t)]_x \, \delta\varepsilon_{xn} dt \quad .$$

When this variation is added to the other variations in $\delta S_{eff}/\delta x_n$,
it is obvious that we have derived the classical theory of retarded
interaction as the $\hbar \to 0$ limit of QED interaction.

The classical-correspondence limit for particle interaction in
QED has thus been established: the key to resolution of the problem
is to first incorporate the anticipated classical radiation, and then
to take the $\hbar \to 0$ limit.

There is at least one important consequence to having established
a realistic, causal correspondence limit for QED. For years, it has
been known that particle interaction via the Feynman photon propagator is a direct result of the QED S-matrix expansion. The apparent
incompatibility of this interaction via D_F with classical pictures
of particle interaction left physicists in a quandary, at least one
manifestation of which may have been the recurring interest in alternative theories to QED and classical theory. The results given here
show us that we can avoid the interpretational problems associated
with the Feynman photon propagator. If we allow for classically
reasonable final state radiation, then the $\hbar \to 0$ limit of QED is the
familiar classical theory of charged particles interacting through
retarded Lienard-Wiechert potentials.

References

1.   J.A. Wheeler and R.P. Feynman, Rev. Mod. Phys. *17*, 157 (1945);
     also *21*, 425 (1949).
2.   R.P. Feynman and A.R. Hibbs, *Quantum Mechanics and Path Integrals*
     (McGraw-Hill Publishing Co., 1965).
3.   G.C. Dente, Phys. Rev. *D12*, 1733 (1975).

THE LAMB SHIFT IN ATOMIC HELIUM AND THE ANOMALOUS MAGNETIC MOMENT

OF THE ELECTRON FROM A NEW THEORY OF ELEMENTARY MATTER

Mendel Sachs

SUNY at Buffalo, Amherst, N.Y.

A new theory of elementary matter and light, and several of
its physical implications, were discussed at the Third Rochester
Conference on Coherence and Quantum Optics.[1] This theory attempts
to fully exploit the basis of the theory of relativity by asserting
the following three underlying axioms: a) the principle of general
relativity, b) the generalized Mach principle, and c) the principle
of correspondence. General consequences of this theory are 1) the
abandonment, *in principle*, of the atomistic view of matter *and
quantization,* and 2) the abandonment of the photon model of light.
The first of these consequences implies that the formalism of nonrelativistic quantum mechanics serves here only as a mathematical
approximation for an exact formalism that is a nonlinear, deterministic, self-consistent field theory for a closed system. The
second consequence implies that the "radiative corrections" to
the energies and other physical properties of elementary matter do
not appear here. The effects that are conventionally attributed
to radiative processes must then be re-derived from the generalization achieved in the nonlinear and spinor structure of this theory.

In the previous Conference, the following rigorous predictions
of the field equations of this theory were discussed: the Pauli
exclusion principle, the Planck distribution law for blackbody
radiation, the prediction of a bound state for the particle-anti-
particle pair (from an exact solution of the nonlinear formalism that
gives all of the physical effects that are conventionally interpreted
as pair annihilation and creation), the fine structure of hydrogen,
including the Lamb shift, the lifetimes of the excited states of
hydrogenic atoms, and general relativity effects relating to the
inertial masses of elementary particles.

Since that time, studies have continued in electrodynamics, relating to 1) the Lamb shifts in He$^+$ and deuterium, 2) high energy p-p scattering, 3) corrections to the energies for the low lying levels of the helium atom, and 4) the anomalous magnetic moment of the electron. The initial stages of the latter two investigations have been reported in the literature and will be discussed at this meeting.[2],[3]

I. ON THE LAMB SHIFT

The origin of the Lamb shift, according to this theory, is *not* radiative effects, as is the case in quantum electrodynamics, since there is no free radiation according to this approach. Atomic structure is represented entirely by matter fields - coupled nonlinear spinor field equations. The Lamb shift results here from a generalization of the electromagnetic interaction, appearing in a natural fashion in the theory, rather than as an ad hoc insertion of an extra interaction into the electron Hamiltonian.

The electromagnetic generalization comes about in this theory for the following reasons: the basic conceptual change from quantum electrodynamics to this theory is the replacement of the elementarity of the particle with the elementarity of the interaction - that is, it is the closed system, without actual separable parts, that one starts with here, rather than the variables of free particles, later to be perturbed by each other in terms of their interactions. Thus, according to this view, there are no free electrons or other free elementary bit of matter, nor are there free photons of the electromagnetic radiation field. Because of the latter rejection, it would be redundant to maintain the vector form of Maxwell's equations, if they could be factorized into a lower dimensional form. Indeed, according to the definition of the symmetry group of special relativity theory - the Poincaré group - as a set of only continuous transformations in space-time, there is an implication that the irreducible basis functions of any physical theory that is consistent with special relativity theory are two-component spinor variables. [A similar result also follows from the symmetry group of general relativity theory]. The implication then follows that the vector form of Maxwell's equations should factorize into a pair of two-component spinor field equations. This was shown to happen, in the early stages of my research program [4].

It was found that the pair of equations

$$\sigma^\mu \partial_\mu \phi_\alpha = T_\alpha \, , \tag{1}$$

where $\alpha = 1,2,$ are two separate (uncoupled) two-component spinor field equations that incorporate the Maxwell field equations.

Equations(1)were then found to duplicate all of the physics that
is predicted by Maxwell's equations.[5] They yield more invariants
and conservation equations than Maxwell's equations do; *some* of
these are in one-to-one correspondence with *all* of those of the
Maxwell formalism; thus there are *extra* physical predictions here
that are not predicted by the standard form of the Maxwell theory.
The spinor form (1) of the electromagnetic equations is then a
true generalization of the Maxwell form of electromagnetism, that
can be tested in physical experimentation. One of these tests,
according to this theory, is in regard to the Lamb shift.

Because of the relativistic covariance of the spinor field
equations (1), it follows that when the Poincaré transformations
of special relativity are applied to the space-time coordinates,

$$\phi_\alpha(x) \rightarrow \phi_\alpha'(x') = S\phi_\alpha(x), \quad T_\alpha(x) \rightarrow T_\alpha'(x') = S^{\dagger-1}T_\alpha(x) , \qquad (2)$$

where S is a two-dimensional representation of the Poincaré group.
It then follows that the four terms, that are the hermitian products:

$$\phi_\alpha^{\dagger}T_\beta \qquad (\alpha,\beta = 1,2)$$

are complex invariants. Thus, this theory entails eight real invariants in terms of the coupling of the spinor field intensities,
ϕ_α, to the spinor current terms, T_α.

The spinor source current terms, T_α, of eq. (1) depend bilin-
early on the Dirac functions. Using the usual four-component Dirac
representation, the generalization from the standard vector source
current to the spinor source current, in Maxwell's electromagnetic
field generalization, is:

$$j_\mu = e\bar{\psi}\gamma_\mu\psi \rightarrow T_\alpha = e\bar{\psi}\Gamma_\alpha\psi , \qquad (3)$$

where, by definition,

$$\bar{\psi}\Gamma_\alpha\psi \equiv \begin{pmatrix} \bar{\psi}\Gamma_\alpha(1)\psi \\ \\ \bar{\psi}\Gamma_\alpha(2)\psi \end{pmatrix}$$

and

$$\Gamma_1(1) = 4\pi i(-\gamma_0 + i\gamma_3) \quad \Gamma_1(2) = 4\pi i(i\gamma_1 - \gamma_2)$$

$$\Gamma_2(1) = 4\pi i(-i\gamma_1 - \gamma_2) \quad \Gamma_2(2) = 4\pi i(\gamma_0 + i\gamma_3)$$

is the corresponding matrix structure, in any particular Lorentz frame.

Recall that $\bar{\psi}\Gamma_\alpha\psi$ is a two-component spinor variable. Thus, it is not form-invariant with respect to the vector components of the Dirac matrices γ_μ indicated above. According to eq. (2), when $x \to x'$, then

$$\bar{\psi}(x)\Gamma_\alpha\psi(x) \to \bar{\psi}(x')\Gamma_\alpha{}'\psi(x') = S^{\dagger-1}(\psi(x)\Gamma_\alpha\psi(x)) \ .$$

The Lagrangian density, whose variation with respect to the electromagnetic field variables, ϕ_α, yields the spinor form of the electromagnetic equations (1), is

$$\mathcal{L}_M = ig_M\Sigma(-1)^\alpha\phi_\alpha{}^\dagger(\sigma^\mu\partial_\mu\phi_\alpha - 2T_\alpha) + \text{h.c.} \ ,$$

where g_M is the new fundamental constant of this theory, resulting from the factorization of the Maxwell formalism into a pair of uncoupled spinor equations (1). Since the Lagrangian density is an energy density, and since the terms $\phi_\alpha{}^\dagger\partial_\mu\phi_\alpha$ have the dimension of energy density per length, it follows that the constant g_M must have the dimension of length.

With the replacement (3) of the source spinor current T_α in terms of the Dirac matter field spinors, the variation of the Lagrangian density $\mathcal{L}_M$ with respect to Dirac bispinor variables yields an extra interaction term in the Hamiltonian for the hydrogenic electron. This extra interaction Hamiltonian has no counterpart in the standard vector form of electromagnetic theory, when it is combined with the Dirac theory in the usual way.

When the coupled nonlinear electron-proton spinor equations of this theory are then approximated by letting the ratio of the electron mass to the proton mass be zero, the resulting linear approximation for the Dirac Hamiltonian is not the same as the usual equation for hydrogen; it contains the following generalization of the Coulomb term:

$$e^2/r \to e^2[1/r + 16\pi g_M(\underline{r}\times\underline{\alpha})_3/r^3] \ ,$$

where $\underline{\alpha}$ is the Dirac matrix, equal to $i\gamma^0\gamma$.

Note that it is the linear approximation in the ordinary Dirac theory of hydrogen, which is equivalent to assuming that the proton is at a fixed spatial location, that yields the change from the covariant interaction Hamiltonian to the noncovariant form:

$$\gamma^\mu eA_\mu \to \gamma^0 e^2/r \ .$$

The corresponding noncovariant approximation for the covariant generalization of the electromagnetic interaction Hamiltonian in this theory is:

$$e[\gamma^\mu A_\mu + ig_M(\Sigma(-1)^\alpha \phi_\alpha^\dagger \Gamma_\alpha + h.c.] \to e^2[\gamma^0/r + 16\pi g_M(\underline{r}\times\underline{\gamma})_3/r^3] . \quad (4)$$

The last term in eq. (4) predicts a lifting of the accidental degeneracy of the Dirac states for hydrogen. It was treated as a perturbation on the states of the remainder of the Hamiltonian - which are the conventional Dirac states for Hydrogen. The calculation for the ratio of splittings, $(3S_{1/2} - 3P_{1/2})/(2S_{1/2} - 2P_{1/2})$, which is independent of the constant g_M (to the accuracy required to compare the theory with the experimental ratio) was then found to compare favorably with the experimental ratio.[6] The constant, g_M, was then determined from a comparison between the experimental splitting, $(2S_{1/2} - 2P_{1/2})$ and the theoretical result. This gave

$$g_M = (2.087 \pm .001) \cdot 10^{-14} \text{ cm} .$$

The same interaction also predicts a shift in the states of atomic helium - though such an effect is not as spectacular as the appearance on new lines in a spectrum! The attempt has been made by H. Yu and myself,[2] to see if the physical source of the Lamb shift in hydrogen, according to this theory, will also predict the contribution to the energies of the states of helium that are conventionally attributed to radiative corrections. Starting with the relativistic two-electron Breit Hamiltonian for Helium, and perturbing its energy states with the new interaction of this theory, the calculations were carried out with the use of the Foldy-Wouthuysen transformation, with the generalization of Charplvy.[7] The level shifts were then determined in a first approximation, for the 1^1S state of parahelium, and the 2^3S and 2^1S states of orthohelium. The results are shown below (in units of cm^{-1}):

	Experiment	QED	Yu-Sachs
Parahelium			
1^1S	-1.21 ± 0.15	$-1.34 \pm .05$	-0.9
Orthohelim			
2^3S	$-0.10 \pm .05$	$-0.109 \pm .009$	-0.10
2^1S	$-0.06 \pm .03$	$-0.104 \pm .014$	-0.10

While our results do not come as close to the data as do the calculations of quantum electrodynamics, they are based on a

mathematically consistent formalism, that is finite at all stages.
In view of the assumptions that we have made to arrive at these,
having to do with the neglect of nonlinear terms in our Hamiltonian,
we did not anticipate at this stage that we would have any closer
agreement. In the next stage, the attempt will be made to take the
nonlinear terms into account. It is interesting to see at this
stage, however, that our results are within the order of magnitude
of the experimental results.

II.   ON THE ANOMALOUS MAGNETIC MOMENT OF THE ELECTRON

The source of the anomalous part of the magnetic moment of the
electron is again not due to radiative effects (which do not exist
in this theory). The effect in this theory comes from the coupling
of the magnetic moment of the Dirac electron to the combination of
the applied magnetic field *and* the diamagnetic field that is induced
in the background of electron-positron pairs, implied by this
theory. The "physical vacuum," as a real, countable set of pairs,
was already shown in this theory to be responsible for the black-
body radiation spectral curve, the lifetimes of excited atomic
states, and the inertial properties of elementary particles.[1]
The zero energy and momentum of the pairs was derived from an exact
solution.

Because these pairs form a background for any "observed" matter,
the application of an external magnetic field to measure the magnetic moment of the electron would automatically induce a diamagnetic contribution to add to the net field that acts on this electron's moment. Thus, the total magnetic field is $H + H_D$, where H
is the applied field and

$$H_D = (Hr_0/4r)(1 - 3\cos^2\theta)$$

is the diamagnetic contribution, H is assumed to be oriented in the
z-direction, at an angle θ from the radial direction of r, r_0 is
the constant e^2/mc^2, and r is the radial separation between the
observed electron and an interacting pair of the background.

The magnetic energy Hamiltonian may then be expressed in the
form $\mathcal{H}_{mag} = -\hat{\mu}_{eff} \cdot H$, where

$$(\hat{\mu}_{eff})_z = \hat{\mu} + \hat{\mu}(r_0/r)(1 - 3\cos^2\theta) = \hat{\mu} + \delta\hat{\mu} .$$

Using the feature of this theory, that the ground state of the
electron-positron bound pair corresponds to zero energy, the hydro-
gen-like Dirac function of the electron that is observed, bound to
a positron of the background, is proportional to $\exp(-r/\lambdabar)$, where

λ is the (reduced) Compton wavelength of the electron. It then follows that the ratio of the anomalous part of the magnetic moment to the Dirac value is:

$$<\delta\mu>/<\mu> \simeq (\alpha/4)<1 - 3\cos^2\theta> ,$$

where α is the fine structure constant, and $<>$ denotes electron expectation values.

For the symmetry of an S-state, $<\cos^2\theta> = 1/3$, giving $<\delta\mu>/<\mu> = 0$. On the other hand, if the bound electron, coupled to the positron of a background pair, would be in a state such that $<1 - 3\cos^2\theta> = 2/\pi$, then the result for the ratio, $\alpha/2\pi$, would agree precisely with the data, and with the prediction of quantum electrodynamics.

At the present stage of this theory, it is interesting to note that the prediction for the anomalous magnetic moment of the electron (that doesn't entail any radiative effects) is the order of the magnitude of $\alpha<\mu>$, since the electron state in the background of pairs is far from being an S-state. A more precise determination of the anomalous magnetic moment of the electron will depend on taking better account of the nonlinear features of this theory with respect to their influence on the actual electron wave function, in the presence of the externally applied magnetic field.

III. ON ELECTRODYNAMIC SCATTERING

The other implications of this theory, mentioned above, on the Lamb shifts in He$^+$ and D, and high energy p-p scattering, are more sensitive to the nonlinear aspects of the theory than were the effects reported. The studies of p-p scattering are an extension of earlier work, also carried out in a linear approximation, on e-p and e-α scattering.[8],[9] An interesting result in the latter studies was that when the effective "impact parameter" in the scattering process approaches the order of $g_M \sim 10^{-14}$ cm, the effective mutual interaction of the projectile and target particles weakens, according to the factor $\exp(-g_M/r)$, approaching zero as the mutual separation approaches zero. Thus, when r *increases from zero* to separations that are the order of g_M, the mutual interaction *strengthens*, and when r becomes greater than g_M the interaction approaches the normally expected 1/r dependence. That is, as r becomes large compared with g_M, $(1/r)\exp(-g_M/r) \rightarrow (1/r)$. Such an effect, whereby an interaction between particles of matter increases (rather than decreasing) as their mutual separation increases from zero, has been used in models to explain the dynamics of quark interactions within nucleons, in recent particle physics speculations.

IV. CONCLUSIONS

It may be concluded that, thus far, the (*in principle* unquan-
tized) nonlinear field theory that I have been investigating yields
many of the numerical results that quantum electrodynamics gives.
The Lamb shifts in hydrogen and the lifetimes of the excited hydrogenic states are in equally good agreement with the data as are
the results of QED. The corrections predicted for the low lying
energy levels of atomic helium and the anomalous magnetic moment
of the electron are the same order of magnitude as the predictions
of QED, though not as close to the data - yet! But in addition to
these results, this theory makes other predictions, not made by
QED. Thus, there is enough encouragement here to continue further
along this course of investigation. For this is indeed a test of
fully exploiting Einstein's fundamental approach to matter, in
contrast with the approach of the Copenhagen school. These researches make clear to me the fact that the Copenhagen interpretation of atomic physics in terms of atomism and wave-particle
dualism, and its fundamental view of the probability calculus and
nondeterminism, have by no means yet succeeded in ruling out
Einstein's view of matter in terms of the continuous field concept
and determinism.

References

1. M. Sachs, *Coherence and Quantum Optics:* Third Rochester
    Conference, ed. L. Mandel and E. Wolf (Plenum, 1973) p. 605.
2. H. Yu and M. Sachs, Int. Jour. Theoret. Phys. *13*, 73 (1975);
    H. Yu, Ph.D. thesis, Department of Physics and Astronomy,
    State University of New York at Buffalo, 1974.
3. M. Sachs, Lettr. Nuovo Cimento *13*, 169 (1975).
4. M. Sachs and S. L. Schwebel, Jour. Math. Phys. *3*, 843 (1962).
5. M. Sachs, Int. Jour. Theoret. Phys. *4*, 145, 453 (1971).
6. M. Sachs, Int. Jour. Theoret. Phys. *5*, 161 (1972).
7. Z. V. Charplvy, Phys. Rev. *91*, 388; *92*, 1310 (1953).
8. M. Sachs and S. L. Schwebel, Nuclear Physics *43*, 204 (1963).
9. M. Sachs, Nuovo Cimento *53A*, 561 (1968).

ELECTRODYNAMICS TODAY

E. T. Jaynes

Washington University, St. Louis, Missouri

1. BACKGROUND

"Für den Rest meines Lebens will Ich darüber nachdenken,
was das Licht ist." -- Albert Einstein [1]

In 1966, QED was in one of its recurrent states of pessimism,
just as it had been twenty years earlier. In 1946, the teacher
from whom I first learned about it - J. R. Oppenheimer - described
it publicly as "a monumental flop." But starting just at that
time, the work of Tomonaga, Schwinger, Feynman, Dyson and others
quickly restored the patient to vigorous health. These developments were pretty well consolidated by 1953 with the successful
treatment of the Lamb shift and the anomalous moment, and optimism
again ran high, with the belief that QED had been vindicated and
could now advance to many new applications and elegant formulations.

However, the next decade saw very little of that advance.
Wigner,[2] Schwinger,[3] Feynman,[4] and Weisskopf [5] expressed
dissatisfaction with the theory on grounds of logical consistency
and lack of conceptual clarity. On the mathematical side, far
from advancing to new applications, we became aware of more and
more difficulties, whose discussion left room for fewer and fewer
real applications in textbooks.

Perhaps the low point came just before 1966. Arthur
Wightman,[6] in his 1964 Cargèse lectures, said that the problem,
not of solving, but only of proving the *existence* of solutions
for the standard models had "conspicuously and completely defeated
two generations of theoretical physicists." To appreciate the
kind of difficulty that made the future seem bleak, see Wightman's

discussion of strange representations and Haag's theorem. Some
had suggested that the whole apparatus of fields and Hamiltonians
ought to be abandoned in favor of the S-matrix. Dirac,[7] in his
Belfer lectures of 1963-64, described the usual treatment of
quantum field theory as "a stopgap, without any lasting future."
So we were back just about to Oppenheimer's remark.

Now as I like to put it, any modern physical theory is a rather
complicated blend, containing important elements of truth, but all
scrambled up, inevitably, with some elements of nonsense. Each
major advance in understanding comes when we accomplish one more
step in disentangling them. In 1966, then, we faced a seemingly
desperate problem of separating the truth from the nonsense. What
parts of QED are really required by experimental facts, what parts
might be modified?

Undoubtedly, the most sacred part of QED was field quantization itself; indeed, since the 1927 work of Dirac[8] that first used
field quantization and derived the Einstein A-coefficients, QED was
in the minds of most physicists *defined* as the theory which starts
out by quantizing the EM field. And there was a weight of author-
ity supporting the belief that field quantization and the resulting
vacuum fluctuations were the essential physical cause of the Lamb
shift and the anomalous moment. Schwinger [9] and Weisskopf [5]
had stated this very explicitly. Dyson,[10] in concurring,
pictured the quantized field as something akin to hydrodynamic flow
with superposed random turbulence, and said, "The Lamb-Retherford
experiment is the strongest evidence we have for believing that our
picture of the quantum field is correct in detail."

Furthermore, we had Welton's [11] elementary derivation of the
Lamb shift directly from field fluctuations, and as if to emphasize
the point, on the day the 1966 Conference opened there arrived in
the mail the paper of E. A. Power [12] which derived the Lamb shift
directly from the change of zero-point energy in the fields surrounding a hydrogen atom in its 2S state. At the 1966 Conference,
Roy Glauber told us that vacuum fluctuations are "very real things."

Yet my own thinking had led me to doubt whether vacuum fluctuations are the real cause of the Lamb shift, or indeed whether
they could be said to be "real" at all, compared to the unquestioned
reality of the thermal fluctuations that we observe as Nyquist
noise in electrical circuits or Planck black-body radiation. It
seems to me that, if you say radiation is "real," you ought to mean
by that, that it can be detected by a real detector. But an optical pyrometer sees only the Planck term, and not the zero-point
term, in black-body radiation. And, if the Einstein A-coefficients
arise physically from zero-point fluctuations, then why is it that
the derivation of the black-body radiation density from the A-
coefficients gives only the Planck term, and not the zero-point

term? Some further facts about the fantastic numerical values of
the zero-point energy density and the resulting turbulent power
flow in space ($\sim 10^{20}$ megawatts/cm^2) required by the cutoff at the
Compton wavelength used in Welton's and Bethe's [13] calculations,
were noted in my paper at the last conference.[14]

Of course, a staunch defender of present theory will say immediately that such objections reflect only naive metaphysical
preconceptions of "reality," not unlike pre-relativistic notions
of absolute simultaneity, of just the kind that the Copenhagen
interpretation of quantum theory has recognized, and rightly
removed from science. To this I can only reply with what Heinrich
Hertz [15] said on a similar occasion: "A doubt which makes an
impression on our mind cannot be removed by calling it metaphysical." It is a supple ontology which supposes that vacuum
fluctuations are just real enough to shift the hydrogen 2s level
by 4 microvolts, but not real enough to be seen by our eyes,
although in the optical band they correspond to a flux of over
100 kilowatts/cm^2. Nevertheless, the dark-adapted eye, looking
for example at a faint star, can see *real* radiation of the order
of 10^{-15} watts/cm^2.

Another piece of evidence strengthened these doubts. The
first volume of Bjorken and Drell was just out, with its development of the Feynman rules of calculation and the usual first applications (Compton effect, Bethe-Heitler formula, Lamb shift, anomalous moment, vacuum polarization). But how many readers were
surprised to note that this volume contains no mention at all of
EM field quantization! Mathematically, the propagator $D_F(x-y)$ is
equally well a Green's function for the classical Maxwell equations,
and its role as the elementary response function is the same whether
the EM field is or is not quantized.

But while the theoretical picture was bleak, the experimental
picture had never been brighter, with the creation of new optical
technology beyond the dreams of a few years before. Any physicist
worthy of the name must have an interest in fundamental questions;
but to raise and pursue them actively when we see no way to settle
them is not a profitable occupation and may be left to philosophers
(who, as a colleague of mine remarked, "are free to do whatever
they please, because they don't have to do anything right"). But
when we do see the means by which deep fundamental issues can be
removed from the realm of philosophical debate and settled on the
level of demonstrable fact, then we ought to retrieve them from
the philosophers and see what we can learn about them.

Now from a pragmatic standpoint, our present quantum theory
of electrons has been an unqualified success, yielding thousands
of quantitatively correct predictions from straightforward, relatively easy calculations. It was in the extension of that theory

to include radiation phenomena that we got into a seemingly endless
series of difficulties. Yet probably 95% of the clues that led
to present quantum theory were provided by optical experiments
performed in the period 1880-1925. It would be astonishing if all
this new optical capability could not provide any new fundamental
tests of the theory which grew out of the original crude optical
experiments. But I could see no sign of recognition of this; the
prevailing opinion was that QED had in it the complete and final
answer to all questions in the optical region, and the remaining
difficulties could be resolved only by more evidence from high-
energy experiments. This made it surprising that so many physicists took up Quantum Optics; for I personally would not choose
to work in this field if I believed there was no new fundamental
knowledge to be had from it.

Since provocation is much more effective than exhortation,
I gave a talk at the 1966 Rochester Coherence Conference entitled,
"Is QED Necessary?" which marshalled as many arguments against QED
as possible, noted that semiclassical theory has far more truth
in it than was generally recognized, and suggested that optical
experiments just then feasible might provide important new evidence
about the range of validity of QED, as well as that of semiclassical
theory.

However, it seems that those remarks succeeded only in pro-
voking Peter Franken, although he could hardly be classed as a
defender of QED (see the Conference report [16] of a year earlier,
which has him "postulating that quantum mechanics applies only to
the matter and not to the light"). The result was a bet, recorded
by D. L. MacAdam.[17] The key issue, on which my statements contrasted most strongly with the prevailing view (the above quotation
from Dyson) was the Lamb shift, as Franken correctly saw. The
utter certainty with which the defenders of QED believed that this
was a direct proof of the reality of vacuum fluctuations arising
from field quantization, made this the favored ground on which to
challenge my own speculation that the effect would be found already
in semiclassical theory, if we complete it by adding terms giving
the effect of the atom on the field, which had been left out in
the first semiclassical theories.

The issue was then whether the Lamb shift could be calculated
using any commonly accepted formalism (i.e., nonrelativistic
Schrödinger equation, Pauli equation, Dirac equation) for the
electron, but without quantizing the electromagnetic field.

2. RADIATION REACTION AND SOURCE-FIELD THEORY

Soon after the 1966 meeting, my students and I had realized that the usual modal expansions of the EM field, although correct in principle, were complicated and tended to obscure the physics if one wants only the field in the immediate vicinity of an oscillating charge distribution. For the sum of all mode contributions is there just the radiation reaction field. Usually one sees only the term $(2e^2/3c^3)\dddot{x}$, found by Lorentz, which is independent of the exact charge distribution and gives rise to radiation damping. But there is another term proportional to $\ddot{x}$ which was sometimes held to be physically meaningless on the grounds that it depends on the charge distribution and diverges in the limit of a point charge. For an extended charge, however, it is finite and calculable. Being 90° out of phase with the damping component, it gives rise to reactive, frequency shift effects. M. Crisp [18] studied the effect of this term in NCT, using the nonrelativistic spinless Schrödinger equation and the two-level approximation. For the Lyman alpha line $[\psi(t) = a(t)\psi_{1s} + b(t)\psi_{2p}]$, where one would expect the two-level approximation to be best, he found a result slightly different from the QED prediction, but within the experimental error. But for the Balmer alpha ending on 2s the result was only about two-thirds of the hoped-for $1058 + 27 = 1085$ MHz. The exact numerical value is unimportant, because in any event the two-level approximation is basically inconsistent; i.e., if we demand that $\psi(t) = a_1(t)\psi_1 + a_2(t)\psi_2$ with no other terms present, then the charge density $\rho(x,t) = e|\psi|^2$ cannot oscillate at any other frequency than $(E_1 - E_2)/\hbar$ without violating charge conservation, $\nabla\cdot J+\dot{\rho} = 0$.

A more complete calculation is then needed; this was done by J. Mahanty [19] by an elegant contour integral method. In first order, the result agreed with the original "Bethe logarithm" term; the last several equations of Mahanty are identical with those of E. A. Power.[12]

Nevertheless, we are not entitled to claim success; for the NCT equations also predict the "dynamic Lamb shift" chirp discussed before,[14] the emitted frequency varying from $\omega_0 + \Delta\omega_{Lamb}$ to $\omega_0 - \Delta\omega_{Lamb}$ as the atom moves down from the excited state to the ground state. There is now very convincing experimental evidence of Citron, Gray, Gabel, and Stroud [20] indicating that this chirp does not, after all, exist. Their experiment is in principle identical with one I performed in 1951, observing unsymmetrical resonance curves of piano strings due to nonlinearities that cause the pitch to rise with amplitude. But in the optical case no asymmetry could be detected. The mere fact of getting the right numerical magnitude of $\Delta\omega_{Lamb}$ cannot be claimed as a valid "derivation" of the Lamb shift if we do not get also the correct qualitative behavior.[17]

However, this emphasis on the radiation reaction field did point to what now appears as the correct answer. Let us use the radiation reaction field, but interpret it as an operator. However, it is an operator not on the "Maxwell Hilbert space" of a quantized field, but on the "Dirac Hilbert space" of the electrons. This is the "source field" approach, which need not be discussed at length here, since it has already moved out of the research journals and become textbook material. Allen and Eberly [21] give a unified discussion of the work of Series,[22] Senitzky,[23] Milonni, Ackerhalt and Smith,[24] and Fain and Khanin.[25] When extended from two-level systems to real systems, it now appears that source fields will give a proper account of the Lamb shift, the anomalous moment, and presumably all of the usual "electrodynamic" effects. This approach is being carried much further in a series of articles by R. K. Bullough and collaborators.[26]

Of course, source-field theory is not in conflict with QED; it is a truncated form of QED in which one notices that, with proper ordering of the operators at t = 0, the vacuum fluctuations of the quantized EM field play no role in the phenomenon; and so the quantized free field and its whole Maxwell-Hilbert space need never be introduced at all. This should be adequate for any problem of electrodynamics, since in a very fundamental sense *every* EM field is a source field from somewhere.

On the other hand, if we define QED as the theory based on quantizing the radiation field, representing it by operators on a new Maxwell-Hilbert space, then source-field theory could hardly be called "Quantum Electrodynamics." If the above speculations should prove correct, the electrodynamics of the future will be far simpler than QED, having no use for the quantized free field, its Maxwell-Hilbert space, and its vacuum fluctuations.

3.   WHERE ARE THE VACUUM FLUCTUATIONS?

Why then, was there so much early confidence that vacuum fluctuations are "very real things," essential to account for experimental facts? Why did calculations like those of Welton [11] and Power [12] succeed? Part of the answer is well-known, and is discussed at length in references 21-25. At time t = 0 the field and current operators commute. Whatever order we use, the results of the calculation must be the same, but the physical interpretation is different. With one ordering, it appears that the effects are due only to the source field. With any other ordering, vacuum fluctuations play a role, but there is no ordering for which vacuum fluctuations are the sole mechanism at work.

This independence of the initial order is, then, just a very

simple, general, and elegant fluctuation-dissipation theorem; but
let me suggest a different physical interpretation from the usual
one. This complete interchangeability of source-field effects and
vacuum-fluctuation effects does not show that vacuum fluctuations
are "real." It shows that source field effects are the same *as if*
vacuum fluctuations were present. For many years, starting with
Einstein's relation between diffusion coefficient and mobility,
theoreticians have been discovering a steady stream of close
mathematical connections between stochastic problems and dynamical
problems. It has taken us a long time to recognize that QED was
just another example of this.

To show this equivalence directly and see the sense in which
vacuum fluctuations *are* "very real things", consider an atom
emitting light. The energy density of these hypothetical zero-
point fluctuations, in a small frequency band $\Delta\omega$, is

$$W_{ZP} = \rho(\omega)\Delta\omega = \frac{1}{2}\hbar\omega \cdot \frac{\omega^2}{\pi^2 c^3}\Delta\omega \ \text{ergs/cm}^3 \ . \tag{1}$$

Over what bandwidth $\Delta\omega$ should this be effective in causing the
atom to radiate? Presumably, over the width of the natural emission line, as determined by the Einstein A-coefficient

$$A = \frac{4\mu^2\omega_o^3}{\hbar c^3} \ ; \tag{2}$$

where μ is the electric dipole moment matrix element for the transition, ω_o the natural line frequency. Exponential decay at this
rate, energy $\sim\exp(-At)$, leads to the usual Lorentzian spectral
density of the radiation: $I(\omega)\sim[(\omega-\omega_0)^2 + (A/2)^2]^{-1}$, which has no
sharply defined width; but the effective width $\Delta\omega$ as far as energy
is concerned, is determined by the condition that $I(\omega_0)\Delta\omega$ shall
equal $\int I(\omega)d\omega$, the total energy radiated. This yields the result
$\Delta\omega = \pi A/2$.

W_{ZP} is the sum of six equal contributions from the averages of
$\{E_x^2, E_y^2, E_z^2, H_x^2, H_y^2, H_z^2\}$, only one of which (say E_z, the component
parallel to the atom's dipole moment) interacts with the atom. The
energy density in the effective field E_z is then $(W_{ZP})_{eff} =$
$(1/6)\rho(\omega)(\pi A/2)$, or

$$(W_{ZP})_{eff} = \frac{1}{18\pi}\mu^2(\frac{\omega}{c})^6 \ \text{ergs/cm}^3, \tag{3}$$

and we note with interest that Planck's constant has cancelled out.

Now in classical electromagnetic theory, radiation from an

oscillating dipole $\mu(t)$ is not attributed to "zero-point fluctuations" but to the radiation reaction field against which the dipole must do work:

$$E_{RR} = \frac{2}{3c^3} \frac{d^3\mu}{dt^3} = \frac{2\omega^3}{3c^3} \mu \; . \tag{4}$$

This provides an energy density at the position of the atom, of

$$W_{RR} = \frac{E_{RR}^2}{8\pi} = \frac{1}{18\pi} \mu^2 \left(\frac{\omega}{c}\right)^6 \; . \tag{5}$$

But this is identical with (3)! The radiating atom is indeed interacting with an EM field of the intensity predicted by the zero-point energy, but this is just the atom's own radiation reaction field.

The fantastic numbers noted before [14] disappear as soon as we realize that, in order to account for spontaneous emission, there is no need for this energy density to be present in all space, at all times, in all frequency bands. It is produced automatically by the radiating atom, but in a more economical way; only the field component that is needed, where it is needed, when it is needed, and in the frequency band needed.

But are there other phenomena which require zero-point energy throughout space? What are we to make of the calculation of E. A. Power [12] obtaining the Lamb shift from the change in total zero-point energy of a region due to coupling the field to a hydrogen atom in its 2s state?

4.   LAMB SHIFT IN CLASSICAL MECHANICS

Space not permitting an explicit, detailed answer to this question, let us note a quite general relation between two methods of calculating line shifts, which holds even in classical mechanics. We have a set of classical harmonic oscillators (p_i, q_i), the "field oscillators," coupled to one "extra oscillator" (P, Q) via coupling constants α_i, the total Hamiltonian being

$$H = \sum_{i=1}^{n} \frac{1}{2}(p_i^2 + \omega_i^2 \, q_i^2) + \frac{1}{2}(P^2 + \Omega^2 \, Q^2) + \sum_{i=1}^{n} \alpha_i \, q_i \, Q \; , \tag{6}$$

and define the dispersion function

$$K(\nu) \equiv \sum_i \frac{\alpha_i^2}{\omega_i^2 - \nu^2} = \int_0^\infty K(t) e^{-st}\, dt, \quad s = i\nu \quad . \tag{7}$$

If $q_i(0) = \dot{q}_i(0) = 0$, the extra oscillator decays according to a Volterra equation:

$$\ddot{Q} + \Omega^2 Q = \int_0^t K(t-t')Q(t')dt' \quad , \tag{8}$$

which has the exact solution

$$Q(t) = Q(0)\dot{G}(t) + \dot{Q}(0)G(t) \quad , \quad t>0 \tag{9}$$

with the Green's function

$$G(t) = \frac{1}{2\pi} \int_{-\infty}^\infty \frac{e^{i\nu t}\, d\nu}{\Omega^2 - \nu^2 - K(\nu)} \quad , \tag{10}$$

the integration contour passing below all the poles ν_k, which lie only on the real axis [since $\nu^2 + K(\nu)$ cannot be real unless ν is real] and represent the normal mode frequencies.

Method 1. In the limit of large mode density, $\sum_i (\ )_i \to \int (\ )\rho_0(\omega)d\omega$, then on the path of integration $\mathrm{Im}(\nu) < 0$, $K(\nu)$ goes into

$$K(\nu) \to \int_0^\infty \frac{\alpha^2(\omega)\rho_0(\omega)}{\omega^2 - \nu^2}\, d\omega \quad . \tag{11}$$

Supposing $\alpha(\omega)$ a smooth function, and assuming a sharp resonance, certain small terms may be neglected, and $G(t)$ goes into

$$G(t) \to \exp(-\Gamma t) \frac{\sin(\Omega+\Delta)t}{(\Omega+\Delta)} \quad , \quad t>0 \quad , \tag{12}$$

where we have defined the "spontaneous emission rate"

$$\Gamma \equiv \frac{\alpha^2(\Omega)\rho_0(\Omega)}{4\Omega^2} \tag{13}$$

and the "radiative frequency shift"

$$\Delta \equiv \frac{1}{2\Omega} P \int_0^\infty \frac{\alpha^2(\omega)\rho_0(\omega)}{\Omega^2 - \omega^2}\, d\omega \quad . \tag{14}$$

Thus, for example, the vibrations of a plucked guitar string are damped and shifted by its coupling to the acoustical radiation field, the expressions for these effects having a rather familiar appearance. Obviously, we have invoked no field "vacuum fluctuations," having started with the explicit initial conditions of a quiescent field, $q_i = \dot{q}_i = 0$. The damping and shifting are due entirely to the source field reacting back on the extra oscillator.

Method 2: The mode density $\rho_0(\omega)$ of the free field is changed by the added oscillator by a small increment: $\rho_0(\omega) \rightarrow \rho(\omega) = \rho_0(\omega) + \rho_1(\omega)$. By a somewhat delicate analysis of the limiting behavior of $K(\nu)$ on the real axis, to be given elsewhere, we can deduce

$$\rho_1(\nu)d\nu = \frac{1}{\pi}\frac{\Gamma\, d\nu}{(\nu-\Omega-\Delta)^2 + \Gamma^2}\;, \tag{15}$$

which is just the spectrum of the damped oscillation (12) of the dynamical solution:

$$\int \rho_1(\nu)e^{i\nu t}\, d\nu = \exp(i\Omega + i\Delta -\Gamma)t),\quad t \geq 0, \tag{16}$$

a connection which holds generally, even when [due to variations in $\alpha(\omega)$] $\rho_1(\omega)$ is not Lorentzian and the damping is not simple exponential. Every detail of the transient decay of the dynamical problem (9) is, so to speak, "frozen into" the static mode density increment function $\rho_1(\omega)$. It is normalized [$\int\rho_1(\omega)d\omega = 1$], since the "global" effect of the coupling is to add one mode to the system.

Recognizing this, we could as well calculate the line shift Δ by the time-honored methods of "subtraction physics." Before the coupling is turned on, the total frequency of all modes is a badly divergent expression:

$$\Omega + \int_0^\infty \omega\rho_0(\omega)d\omega \equiv (\infty)_1\;. \tag{17}$$

Afterward, it is

$$\int_0^\infty \omega[\rho_0(\omega) + \rho_1(\omega)]d\omega \equiv (\infty)_2\,, \tag{18}$$

which is no better. But then the change in total mode frequency due to the coupling is

$$(\infty)_2 - (\infty)_1 = \int_0^\infty \omega\rho_1(\omega)d\omega - \Omega = \Delta\;. \tag{19}$$

It is an awkward way of asking the question, but it leads to the
same answer.

Perhaps from this one can understand why Power [12] and
Mahanty [19] were able to calculate the Lamb shift from the total
change in the (infinite) zero-point energy, without any need for
the zero-point energy to be physically real. In fact, they calculated the total change in all mode *frequencies*, a quantity that is
equal to the shift in the dynamical problem even if all modes are
perfectly quiescent. Perhaps also, one can now look at calculations of the Casimir attraction effect through new eyes.

5. CONCLUSION

It is clear that, over the past ten years, theoretical and
experimental work in Quantum Optics has yielded - just as I had
hoped it would - important and, to all of us, surprising fundamental new information about electrodynamics. In both QED and classical theory, our judgment as to which parts are elements of truth,
which are elements of nonsense, which parts are necessary to account
for experimental facts, which parts were unnecessary complications,
are very different today. In QED, both the pessimism at the high-
brow level and the childlike faith at lowbrow level have been
greatly reduced, and as a result, I think we can now continue the
pursuit of truth in a more rational way, with more emphasis on
demonstrable fact, less on ideology.

But it is equally clear that we are still very far from having
separated all the truth from all the nonsense. Having shown that
the Maxwell Hilbert space and vacuum fluctuations are not necessary
for the Lamb shift does not prove that they are not needed at all.
But it does lead us to raise again the question of 1966: Is there
now any experimental fact in electrodynamics which still requires
the Maxwell Hilbert space and/or vacuum fluctuations for its
explanation? If not, then the Electrodynamics of the future can
be considerably simpler than QED, for source field theory can be
re-cast in a mathematical form appropriate to its own nature, and
not dictated by Ancient History.

References

1. A. Einstein, as quoted by W. Pauli, *Aufsätze und Vorträge über
    Physik und Erkenntnistheorie*, Fr. Vieweg u. Sohn, Braunschweig
    (1961), p. 88.
2. E. P. Wigner, *Symmetries and Reflections*, Indiana Univ. Press,
    Bloomington (1967), p. 46.
3. J. Schwinger, *Stanford Lectures on Quantum Field Theory*,
    Stanford Research Institute (1957); also *Quantum Electrodynamics*,
    Dover Publications, Inc. N. Y. (1958), p. xv.
4. R. P. Feynman, in *The Quantum Theory of Fields*, Interscience
    Pub., Inc. N. Y.(1961).
5. V. F. Weisskopf, Rev. Mod. Phys. *21*, 305 (1949); see also
    Physics in the Twentieth Century, M.I.T. Press, Cambridge,
    Mass. (1972), pp. 97-111 and 120-129.
6. A. S. Wightman, "Introduction to Some Aspects of the Relativ-
    istic Dynamics of Quantized Fields," French Summer School of
    Theoretical Physics, Cargèse, Corsica (July, 1964).
7. P. A. M. Dirac, *Lectures on Quantum Field Theory*, Belfer Graduate
    School of Science, Yeshiva Univ., New York (1966).
8. P. A. M. Dirac, Proc. Roy. Soc. London A*114*, 243 (1927).
9. J. Schwinger, Phys. Rev. *74*, 1439 (1948); *75*, 651 (1949).
10. F. J. Dyson, Scientific American (April, 1953).
11. T. A. Welton, Phys. Rev. *74*, 1157 (1948).
12. E. A. Power, Am. Jour. Phys. *34*, 516 (1966).
13. H. A. Bethe, Phys. Rev. *72*, 339 (1947).
14. E. T. Jaynes, in *Coherence and Quantum Optics*, L. Mandel and
    E. Wolf, Editors, Plenum Press, New York (1973) p. 35-81.
15. H. Hertz, *Die Prinzipien der Mechanik*, Leipzig (1894).
16. Feature article in Physics Today (Aug., 1965), p. 41.
17. D. L. MacAdam, J.O.S.A. *56*, 1148 (1966). Following the present
    paper, Professor Lamb awarded the prize to Franken. Had we
    realized that the numerical magnitude in NCT would be considered
    all-important, and the qualitative nature of the phenomenon
    unimportant, a different outcome could easily have been arranged.
18. M. D. Crisp, Phys. Rev. *179*, 1253 (1969).
19. J. Mahanty, Il Nuovo Cimento *22B*, No. 1, 110 (1974).
20. M. Citron, H. Gray, C. Gabel, and C. Stroud, "Experimental Study
    of Power Broadening in a Two-Level Atom," Phys. Rev. (to be
    published).
21. L. Allen and J. H. Eberly, *Optical Resonance and Two-Level Atoms*,
    J. Wiley & Sons, New York (1975), Chap. 7.
22. G. W. Series, in *Optical Pumping and Atomic Line Shape*, T.
    Skalinski, Ed., Panst. Wdaw. Nauk Warsawa (1969); and "Radiative
    Damping and Level Shifts without Field Quantization," (unpub-
    lished, 1975).
23. I. R. Senitzky, Phys. Rev. Lett. *31*, 954 (1973).
24. P.W. Milonni, J.R. Ackerhalt and W.A. Smith, Phys. Rev. Lett.
    31, 958 (1973).

25. V. M. Fain and Ya. I. Khanin, in *Quantum Electronics*, MIT
     Press (1969).
26. R. K. Bullough, et al., J. Phys. A7, 1647 (1974); *8*, 759
     (1975): J. Phys. B*8*, L147 (1975).

Addendum

POST MORTEM ON THE BET

This note is added because the final moments of Peter Franken's
little circus act were complicated, and several eyewitnesses came
away with quite different impressions of what had actually happened.
The decision imposed a considerable - and unforeseen - burden on
Professor Lamb, for which I alone am responsible, and for which I
apologize to him. I know that Peter would have been just as happy
as I to hear a different verdict, provided this could have been
clear and unequivocal. What is important now is to understand the
technical situation as it emerges from this. Those who came away
confused about the facts had plenty of company. Briefly, the
present situation is this:

(1) NCT *did* fail to predict the facts for the Lamb shift, as
we now believe them to be. Professor Lamb's verdict was therefore
entirely proper and just according to the conditions of the bet.

(2) However, the failure of NCT lay not in the numerical value
of the shift, but in the qualitative matter of the "dynamical Lamb
shift" chirp. At the 1972 meeting [14] I stressed the importance
of obtaining experimental evidence about this, because a dozen
theoretical decisions, and the interpretation of several other
experiments, all hung on the issue whether this chirp does or does
not exist. The Citron [20] experiment now seems to show that it
does not.

(3) The implications of this for the Lamb shift are the following. NCT will still agree with existing experiments, which
measure only the stimulating frequency needed to initiate a transition starting from a metastable S state.[18] But if the Lamb
shift could be measured starting from a P state, it now appears
that NCT would predict not the wrong magnitude, but the wrong sign,
of the shift. It is for this reason that I stated in my presentation, "we are not entitled to claim success."

(4) But this leads us into another mystery; for the sign reversal that leads to this discrepancy has nothing to do with the
anti-quantum heresies of NCT. It has been a hitherto unquestioned
part of quantum theory, appearing already in the Kramers-Heisenberg

dispersion formula. Here an atom in state n irradiated with frequency ν has an electric polarizability

$$\alpha_n(\nu) = \frac{2}{\hbar} \sum_m \frac{|\mu_{mn}|^2 \, \omega_{mn}}{\omega_{mn}^2 - \nu^2} \, ,$$

The factor ω_{mn} has opposite signs for upward and downward transitions, as analyzed by Ladenburg [Zeit. f. Phys. *48*, 15 (1928)].

This same sign reversal appears in a more modern context. An atom, in absorbing energy $E = \hbar ck$ from an incident plane wave k, picks up momentum (on either classical or quantum theory) $E/c = \hbar k$. An atom stimulated to emit by the same plane wave responds in opposite phase, feels the opposite (JxB) Lorentz force, and ought to experience the opposite recoil ($-\hbar k$) if the stimulated emission is fast compared to the spontaneous. In photon language, the photon is emitted in the forward direction. The same is true in NCT; an atom absorbs radiation by emitting a spherical wavelet with such phase that it partially cancels the incident wave in the forward direction (incipient shadow formation). In stimulated emission the wavelet has the opposite phase, and makes instead a bright spot in the forward direction. Indeed, it was Professor Lamb who pointed out this directional property of stimulated emission, at the 1961 Quantum Electronics Conference.

(5) We see, then, how acute the mystery is: In NCT, this same sign reversal carries through to reactive as well as dissipative effects; i.e., reversing the sign of the atomic currents reverses the sign of the frequency shift, as well as the direction of energy flow. Does this seem right on physical grounds? It seems to me that it does; for at low frequencies this becomes: if you reverse the sign of the current in an inductor, it becomes a capacitor and has the opposite tuning effects. Evidently, we are still far from understanding how a real atom emits or absorbs light, in the sense of having any self-consistent picture.

(6) Could we, then, get more experimental evidence trying to pin down the exact role of this sign reversal? A direct experimental test of the directional properties of stimulated emission would establish whether it is still operative there. Also, if some experimentalist could figure out how to measure the Lamb shift via a P $\rightarrow$ S transition instead of S $\rightarrow$ P, this would check what the Citron experiment seems to imply. The difficulty thus far has been that metastable P states are hard to come by.

It would be important to get such evidence, because as we noted, a two-level wave function cannot oscillate at other than its natural frequency without violating charge conservation. Perhaps

in the Citron experiment the pumping, so carefully adjusted to
give a pure two-level state, also inadvertently wiped out the
very effect that one was trying to observe!

(7) Finally, it should be noted that there was considerable
uncertainty in all our minds as to just what the real issue of the
bet had been, the only written record being that of MacAdam.[17]
Two quite different issues are: (1) whether QED and vacuum fluctuations are necessary; and (2) whether NCT is adequate, to account
for the Lamb shift. In fact, it was issue (1) that I stressed in
my 1966 talk, just as MacAdam's report suggests. At the time, of
course, issue (2) was nothing but a barely formulated conjecture.
Today it is clear that, if I was wrong on issue (2), then the
defenders of QED were equally - and perhaps more importantly -
wrong on issue (1), as shown by the source field theory that none
of us foresaw. However, all had agreed to accept Professor Lamb's
decision, and so he had the burden not only of choosing the winner,
but also of choosing the issue.

ANTIBUNCHING IN LIGHT HARMONICS GENERATION FROM FIELD QUANTISATION

S. Kielich, M. Kozierowski and R. Tanaś

A. Mickiewicz University, Poznań, Poland

1. INTRODUCTION

In quantum optics, one usually deals with light which can be
described in terms of a Sudarshan-Glauber phase-space distribution
function. Light of this kind exhibits a positive, or at the most,
a zero photon correlation (Hanbury Brown-Twiss or bunching) effect.
Recently, however, it has been shown that nonlinear processes such
as degenerate parametric amplification[1] as well as two-[2] and
many-photon absorption,[3] can give rise to light with a negative
Hanbury Brown-Twiss (antibunching) effect. Such light has no
classical counterpart, and has to be treated by quantum mechanics.

The problem of light statistics in second harmonic generation
has been considered in various papers;[4-8] nonetheless, even if
based on a quantal approach (dealing with the field as an operator
rather than a c-number), their results do not go beyond formal
quantum-classical equivalence in their description of the field.
Walls[9] drew attention to the fact that, if spontaneous decay of
the second harmonic into two photons of the fundamental beam is
taken into account, an oscillating solution given by an elliptical
Jacobi function is obtained for the number of photons of the second
harmonic. Classically, omitting fluctuations one obtains a monotonically increasing solution in the form of a hyperbolic tangent.
Dewael[10] first attempted to take into account effectively the
influence of the quantal properties of light on the degree of
second-order coherence of the generated beam. Albeit his result,
a positive value of the Hanbury Brown-Twiss effect, is hardly
correct.

In this paper we show that if the field is dealt with quantally, a negative Hanbury Brown-Twiss effect results for processes of harmonics generation also. This amounts to the statement that harmonics generation can be the source of nonclassical fields to the same degree as the above mentioned nonlinear processes.

2.   THEORY

Quantum mechanically, k'th harmonic generation can be described starting from the Hamiltonian of interaction between the fundamental and generated beams, in the form:[5]

$$H_I = \hbar\, c\, L_{k\omega}\, a_h^\dagger\, a_f^k + h.c. \tag{1}$$

with $L_{k\omega}$ - the coupling constant - dependent on the nonlinear properties of the medium and the state of polarisation of the incident beam, and $a_f(a_f^\dagger), a_h(a_h^\dagger)$ - annihilation (creation) operators for the fundamental (f) and harmonic (h) beams. Using the interaction Hamiltonian (1), one readily derives the Heisenberg equations of time-evolution for the four operators. However, in processes of harmonics generation, we deal with travelling waves rather than fields in a cavity. By substitution of $t = - z/c$, where z is the path traversed by the wave in the medium, the cavity problem reduces formally to a travelling waves problem.[11,5] After the above substitution, the equations of motion of the slowly-variable part (free evolution is eliminated) of the annihilation operators for both beams, for perfect phase matching, become:

$$\frac{da_h(z)}{dz} = i\, L_{k\omega}\, a_f^k(z) \quad ,$$

$$\frac{da_f(z)}{dz} = i\, k\, L_{k\omega}^*\, [a_f^\dagger(z)]^{k-1}\, a_h(z) \quad . \tag{2}$$

Equations (2), jointly with the Hermitian-conjugate equations of the creation operators, form a set of differential operator equations, inaccessible to strict solution. Applying Eqs.(2) and their operator properties, it is possible to calculate approximately the variations in field correlation function for the generated and fundamental beams on traversal of the path z in the medium.[12]

On expanding the correlation functions in z, one has:

$$G^{(n)}(z) = G_0^{(n)} + \sum_{k=1}^{} \frac{z^k}{k!}\, \frac{d^k}{dz^k}\, G^{(n)}(z)\, \Big|_{z=0} \quad , \tag{3}$$

where the correlation functions $G^{(n)}(z)$ are defined as:

$$G^{(n)}(z) = <[a^{\dagger}(z)]^n [a(z)]^n> \ . \tag{4}$$

The symbol $< \ldots >$ in (4) stands for the quantum mechanical mean. Provided the correlation functions (4) vary but little along the path z in the medium, it is sufficient to take but the first few terms of (3) in order to approximate the values of these functions satisfactorily. The procedure, in fact, corresponds to that of short-time solutions in a cavity problem. On differentiating Eqs. (2) and putting z=0, one obtains the values of the successive derivatives of the creation and annihilation operators. By insertion of the thus calculated values of the field operator derivatives into the right-hand term of (3) and on reduction of all terms to normal ordering by the use of boson commutation rules, one obtains the successive terms of the expansion of the functions $G^{(n)}(z)$. The expressions thus obtained for $G^{(n)}(z)$ depend, in successive approximations, on higher powers of z and on correlation functions of the incident beam $G_{f0}^{(n)}$ of higher and higher orders.

Applying the above procedure for the magnitude of the Hanbury Brown-Twiss effect, proportional to $G^{(2)}(z) - [G^{(1)}(z)]^2$, in the experimentally most highly relevant case of second-harmonic generation, we have obtained[13] the expressions:

$$G_{2\omega}^{(2)}(z) - \left[G_{2\omega}^{(1)}(z)\right]^2 = \left|L_{2\omega}\right|^4 \{G_{f0}^{(4)} - \left[G_{f0}^{(2)}\right]^2\} \ z^4$$

$$- \frac{4}{3} \left|L_{2\omega}\right|^6 \{2\left[G_{f0}^{(5)} - G_{f0}^{(3)} G_{f0}^{(2)}\right] + 3 G_{f0}^{(4)} - \left[G_{f0}^{(2)}\right]^2\} \ z^6 + \ldots \ ,$$

$$G_f^{(2)}(z) - \left[G_f^{(1)}(z)\right]^2 = G_{f0}^{(2)} - \left[G_{f0}^{(1)}\right]^2$$

$$- 2\left|L_{2\omega}\right|^2 \{2\left[G_{f0}^{(3)} - G_{f0}^{(2)} G_{f0}^{(1)}\right] + G_{f0}^{(2)}\} \ z^2 + \ldots \ . \tag{5}$$

In (5), $G_{f0}^{(n)} = <a_{f0}^{\dagger n} a_{f0}^{n}>$ are correlation functions of the incident beam. The absence of second-harmonic photons at the input (z=0) is assumed, $G_{2\omega 0}^{(n)} = 0$.

Similarly, expressions for the second-order coherence variations in arbitrary k'th harmonic generation processes can be derived. The formulae, however, are rather bulky. We restrict ourselves to adducing the formula of the generated beam:

$$G_{k\omega}^{(2)}(z) - \left[G_{k\omega}^{(1)}(z)\right]^2 = |L_{k\omega}|^4 \, z^4 \, \left\{G_{f0}^{(2k)} - \left[G_{f0}^{(k)}\right]^2\right\}$$

$$- \frac{k}{3} \, |L_{k\omega}|^6 \, z^6 \, \left\{ \sum_{p=0}^{k-1} \sum_{s=0}^{p} s! \, \binom{p}{s} \binom{k-1}{s} \left[G_{f0}^{(3k-s-1)} - 2\,G_{f0}^{(2k-s-1)} \, G_{f0}^{(k)}\right]\right.$$

$$\left. + \sum_{p=0}^{k-1} \sum_{s=0}^{k-1} s! \, \binom{p+k}{s} \binom{k-1}{s} G_{f0}^{(3k-s-1)}\right\} \quad ,$$

$$(6)$$

where the $\binom{p}{s}$ are Newton binomial coefficients.

3. DISCUSSION

The expressions (5) and (6), obtained by us for the variations in second-order coherence functions, take a particularly simple and interesting form if the incident beam is coherent, i.e., if its field is given by the coherent state: $a|\alpha\rangle = \alpha|\alpha\rangle$. One then has $G_{f0}^{(n)} = \langle n_{f0}\rangle^n$, where $\langle n_{f0}\rangle$ is the mean number of photons of the incident beam. Formulae (5), in the lowest non-vanishing approximation, now reduce to:

$$G_{2\omega}^{(2)}(z) - \left[G_{2\omega}^{(1)}(z)\right]^2 = -\frac{8}{3} \, |L_{2\omega}|^6 \, \langle n_{f0}\rangle^4 \, z^6 + \dots \, ,$$

$$G_{f}^{(2)}(z) - \left[G_{f}^{(1)}(z)\right]^2 = -2 \, |L_{2\omega}|^2 \, \langle n_{f0}\rangle^2 \, z^2 + \dots \, ,$$

$$(5a)$$

yielding a negative Hanbury Brown-Twiss effect, both for the beam generated at the frequency 2ω and for the fundamental beam of frequency ω , on traversal by them of the path z in the medium.

If the incident beam is coherent, formula (6) takes the form:

$$G_{k\omega}^{(2)}(z) - \left[G_{k\omega}^{(1)}(z)\right]^2 = -\frac{k}{3} \, |L_{k\omega}|^6 \, z^6 \, \sum_{p=0}^{k-1} \left\{ \sum_{s=0}^{k-1} s! \, \binom{p+k}{s} \binom{k-1}{s}\right.$$

$$\left. - \sum_{s=0}^{p} s! \, \binom{p}{s} \binom{k-1}{s}\right\} \, \langle n_{f0}\rangle^{3k-s-1} \quad .$$

$$(6a)$$

Since the expression in parentheses $\{\dots\} \geqslant 0$, the right-hand term of (6a) is negative. Thus, for arbitrary $k \geqslant 2$ we obtain a negative correlation of photons, if the incident beam is coherent. It is worth stressing that for classical fields, i.e., if a_{f0} and $a_{f0}^{\dagger}$ are dealt with as c-numbers, s takes but the one value s=0 and the right-hand term of Eq.(6a) vanishes. All the terms with $s \geqslant 1$ emerged owing to our application of boson commutation rules to the field operators. Consequently, the negative photon correlation reflects the quantal properties of the fields. The effect is negative, as apparent from (5a), not only for the generated beam but, as well, for the fundamental beam on traversal of the medium.

One easily finds from (5) and (6) that in the case of a chaotic incident beam, $G_{f0}^{(n)}=n!<n_{f0}>^{n}$, the photon correlation is positive. Quantisation of the fields, however, reduces the effect in magnitude.

It is to be expected that all nonlinear interactions of quantized electromagnetic fields and matter have the ability to produce fields having no classical counterpart.

Quite recently, Mišta and Peřina[14] have discussed the problem of photon statistics in nondegenerate, parametric amplification process. Their results point to a negative correlation effect as well, and, in particular, go over into ours, (5a), for the fundamental beam.

References

1.   D. Stoler, Phys. Rev. Lett. *33*, 1397 (1974).
2.   H.D. Simaan and R. Loudon, J. Phys. *A8*, 539 and 1140 (1975).
3.   W. Brunner, U. Mohr and H. Paul, Opt. Comm. *17*, 145 (1976).
4.   J.A. Armstrong, N. Bloembergen, J. Ducuing and P.S. Pershan, Phys. Rev. *127*, 1918, (1962).
5.   Y.R. Shen, Phys. Rev. *155*, 921 (1967).
6.   G.S. Agarwal, Opt. Comm. *1*, 132 (1969).
7.   B. Echtermeyer, Z. Physik *246*, 225 (1971).
8.   J. Peřina, Chech. J. Phys. *B26*, 140 (1976).
9.   D.F. Walls, Phys. Lett. *A32*, 476 (1970).
10. P. Dewael, J. Phys. *A8*, 1614 (1975).
11. L. Mandel, Phys. Rev. *144*, 1071 (1966).
12. R. Tanaś, Optik *40*, 109 (1974).
13. M. Kozierowski and R. Tanaś, Opt. Comm. *21*, 229 (1977).
14. L. Mišta and J. Peřina, Internationale Tagung Laser und Ihre Anwendungen, Dresden, 1977, Acta Phys. Polonica (in press).

STUDY OF PHOTON CORRELATION IN THE 100-NANOSECOND RANGE WITH

PHOTOELECTRON TIME-OF-ARRIVAL TECHNIQUE*

Shen Jen[†] and Benjamin Chu

State University of New York, Stony Brook, NY

Among the many advantages of studying photon correlation with
the photoelectron time-of-arrival technique, one of the apparent
unsurpassing capabilities is the measurement of the correlation
time in the range around 10^{-6} to 10^{-7} seconds. This is the region
for which both the conventional light beating and the interferometric
(Fabry-Perot) techniques are likely to become unreliable. We
present in this paper such a study encountered in Rayleigh scattering
experiments. We first review briefly the historical development,
ranging from single-stop to multi-stop measurements, of this time-
of-arrival technique. This is followed by an examination on the
dead-time effect of the photomultiplier in general counting
experiments. The analysis is based upon the single time-interval
counting statistics developed by Glauber. Some close-form analytical
expressions are derived for the case of intensity-stabilized laser
light. This consideration of the dead-time effect is of crucial
importance for correlation time measurements in the time scale
mentioned above. It indicates that in this particular case a single-
stop measurement is the only appropriate approach. We also discuss
for this technique the effect due to the imperfection of spatial
coherence, which is of practical interest but has not been mentioned
before. Finally experimental evidence is presented to substantiate
these discussions. With the combination of fast gating circuitry,
multiple-input time-to-amplitude converter and a pulse height
analyser, we made measurements on a full scale of 510 nanoseconds.
The speed of data transfer was made much faster than those of the
earlier reports, which greatly improves the probable deficiency of
long accumulation time in this technique. Both single-stop and
multi-stop results from an intensity-stabilized laser show excellent
agreement with the calculation. Correlation measurements were made
for two physical systems. The relaxation of the orientational

fluctuations due to the short-range ordering in a nematic liquid
crystal above the nematic-isotropic transition temperature was
measured with a common photomultiplier tube to up to 3 MHz. The
use of a faster photomultiplier tube demonstrated that the measurement could be carried further up to 6 MHz. Also the Rayleigh
scattering due to the mutual diffusion in a binary mixture was
experimented at a scattering angle of 90 degrees, which is otherwise impossible to observe with the traditional light beating
technique.

*Work supported by the National Science Foundation, the National
 Institute of Health and the U.S. Army Research Office.

†Present address: Xerox Corporation, 800 Philips Road, Webster,
 New York 14580.

MEASUREMENT OF THE SECOND ORDER CORRELATION TENSOR

G. Bose and C. L. Mehta

Indian Institute of Technology, New Delhi, India

1. INTRODUCTION

It is well known that the second order coherence functions are
closely related to the visibility of the interference fringes produced in an interference experiment [1]. There have been several
attempts [2,3] to determine the diagonal components of the correlation tensor from such visibility measurements. However, no attempts
have been made to determine all the components of the second order
correlation tensor. In this paper we report the results of the
visibility measurements which may be used to obtain the various components of the second order correlation tensor for a stationary,
homogeneous and isotropic optical field.

The second order correlation tensor for an optical field is
defined as [1]

$$\Gamma_{ij}(\underline{r}_1,t_1;\underline{r}_2,t_2) = \langle V_i^*(\underline{r}_1,t_1)V_j(\underline{r}_2,t_2)\rangle , \tag{1}$$

where $V(\underline{r},t)$ is the instantaneous vector field at the point $\underline{r}$ at time
t, and the sharp brackets denote the ensemble averaging. We con-
sider the normalized correlation tensor

$$\gamma_{ij}(\underline{r}_1,t_1;\underline{r}_2,t_2) = \frac{\Gamma_{ij}(\underline{r}_1,t_1;\underline{r}_2,t_2)}{\{\Gamma_{ii}(\underline{r}_1,t_1;\underline{r}_1,t_1)\Gamma_{jj}(\underline{r}_2,t_2;\underline{r}_2,t_2)\}^{1/2}} .\tag{2}$$

For a polarized optical field, one needs only a scalar field to
describe it and one refers $|\gamma(\underline{r}_1,t_1;\underline{r}_2,t_2)|$ as the degree of coher-
ence.

519

As mentioned earlier, we restrict our discussion to stationary, homogeneous and isotropic optical fields. It is well known that the correlation tensor for such a field may, in general, be expressed in the form [4]

$$\gamma_{ij}(\underline{r}_1,t_1;\underline{r}_2,t_2) = A(r,\tau)\delta_{ij} + B(r,\tau)\,\frac{r_i r_j}{r^2}\ , \tag{3}$$

where

$$\underline{r} = \underline{r}_2 - \underline{r}_1,\quad \tau = t_2 - t_1\ . \tag{4}$$

It is readily seen that the scalar functions $A(r,\tau)$ and $B(r,\tau)$ are related to the lateral and longitudinal coherence functions [5]. Thus if we consider $i=j=x$ and take $\underline{r}$ along the z-axis, we obtain the lateral coherence function

$$\gamma_{xx}(r\hat{z},\tau) \equiv \gamma_{lat}(r,\tau) = A(r,\tau). \tag{5}$$

On the other hand the longitudinal coherence function is obtained by setting $i=j=x$ and taking $\underline{r}$ along the x-axis. Thus

$$\gamma_{xx}(r\hat{x},\tau) \equiv \gamma_{long}(r,\tau) = A(r,\tau) + B(r,\tau)\ . \tag{6}$$

Note that isotropy implies $\gamma_{xx}(r\hat{z},\tau) = \gamma_{zz}(r\hat{x},\tau) = \gamma_{zz}(r\hat{y},\tau)$, etc. In general if we set $i=j=x$ and take $\underline{r}$ in the direction making an angle θ with the x-axis, we obtain from Eq. (3) the relation

$$\gamma_\theta(r,\tau) = A(r,\tau) + B(r,\tau)\,\cos^2\theta\ , \tag{7}$$

$\theta = 0$ corresponds to the longitudinal coherence and $\theta = \pi/2$ corresponds to the lateral coherence. We thus find that both $A(r,\tau)$ and $B(r,\tau)$ and hence the complete second order correlation tensor $\gamma_{ij}(r,\tau)$ may be evaluated from the knowledge of γ_θ for two different values of θ.

It is of course not necessary that γ_{lat} and γ_{long} are identical or equivalently that γ_θ does not vary with θ. In particular, if in Eq. (3) we set $i=x$, $j=y$ and take $\underline{r}$ along the direction x=y, z=0, we obtain

$$\gamma_{xy}(r,0) = \frac{1}{2}\,\{\gamma_{long}(r,0) - \gamma_{lat}(r,0)\}\ . \tag{8}$$

This implies that even for an unpolarized radiation, the two orthogonal components of the electric field may be correlated provided one chooses the orientation and the spatial separation properly.

As an illustration we show in Fig. 1, the variation of $\gamma_{xy}(r,0)$ for blackbody radiation [6].

2. EXPERIMENTAL SETUP AND THE RESULTS

In this paper we report the results of the measurement of $\gamma_{\theta}(r,0)$ and observe a definite variation with θ. The method employed is an extension of that suggested by Fracon and Mallick [2] for the measurement of the second order degree of coherence using a polarization interferometer. The essential experimental setup is illustrated in Fig. 2. An aperture S is illuminated by a sodium lamp, and one is interested in the measurement of the correlation functions in the plane T. A ray from $\underline{r}_1$ is split into two rays O_1 and E_1 by the Wollaston prism W. Another ray from $\underline{r}_2$ is split in a similar way into O_2 and E_2 . The rays E_1 and O_2 superimpose and interfere at Q', the image through the lense L of the point Q where the rays E_1, O_2 appear to be coming from. Of course E_1 and O_2 are orthogonally polarized and therefore a polarizer P_1 is placed before the Wollaston prism and another polarizer P_2 is placed after it. These polaroids enable the "Ordinary" and the "Extraordinary" rays to interfere and produce fringes in the neighborhood of Q'.

Let $\pi/2 - \theta$ be the angle which the axis of polaroid P_1 makes with the axis of the Wollaston prism. Having fixed the orientation of P_1, we now adjust the orientation of P_2, such that we get the

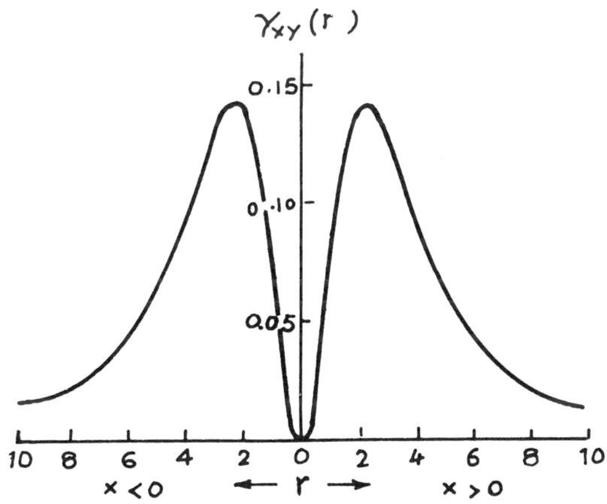

Fig. 1. Variation of $\gamma_{xy}(r,0)$ with r for a blackbody radiation, when $\underline{r}$ is in the direction x=y, z=0. The abscissa is marked in units of $\hbar c/(\pi kT)$. {After C.L. Mehta and E. Wolf, Phys. Rev. *134*, A1143 (1964)}.

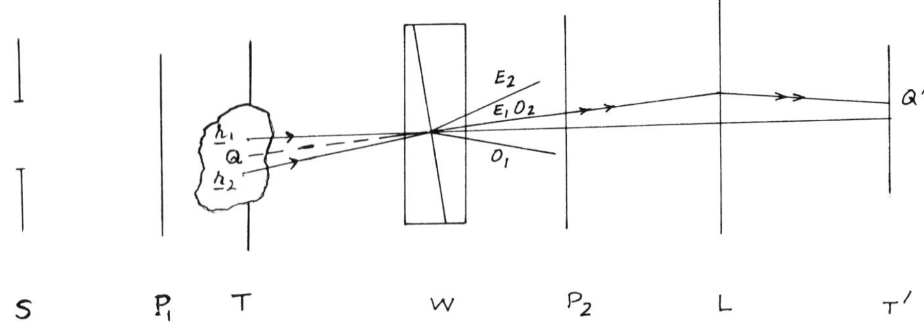

Fig. 2. Experimental setup for the measurement of $\gamma_\theta(r,0)$.

best fringe contrast in the plane T'. This will be so when the
axis of P_2 makes an angle θ with the axis of the Wollaston prism.
Following the arguments of Fracon and Mallick, one may show in this
case that the visibility of the fringes in the neighborhood of Q'
is given by

$$v = |\gamma_\theta(r,0)| = \frac{\Gamma_\theta(r,0)}{\Gamma_\theta(0,0)} \; ,\qquad (9)$$

where

$$\Gamma_\theta(r,0) = \langle E_\theta^*(\underline{r}_1,t)E_\theta(\underline{r}_2,t)\rangle \qquad (10)$$

and E_θ denotes the component of the electric field along the axis
of the polaroid P_1. The direction of $\underline{r} = \underline{r}_2 - \underline{r}_1$ is perpendicular
to both the axis of the Wollaston prism as well as the axis of the
experimental setup. The distance r is such that it subtends an
angle α at the Wollaston prism

$$r = \alpha d \; ,\qquad (11)$$

where d is the distance between the planes T and W and α is the angle
of separation between the ordinary and extraordinary rays caused by
the Wollaston prism. The separation r may be changed by simply translating the Wollaston prism along the axis of the experimental setup.
The visibility of the finges is evaluated by scanning the fringe
pattern with the help of a photocell coupled to a microvoltmeter.

Figure 3 shows the observed visibility of the fringes in the
case when the axes of P_1 and P_2 are held at $\pi/4$ to the axis of the
Wollaston prism. The aperture S is a circle of diameter a. Figure 4

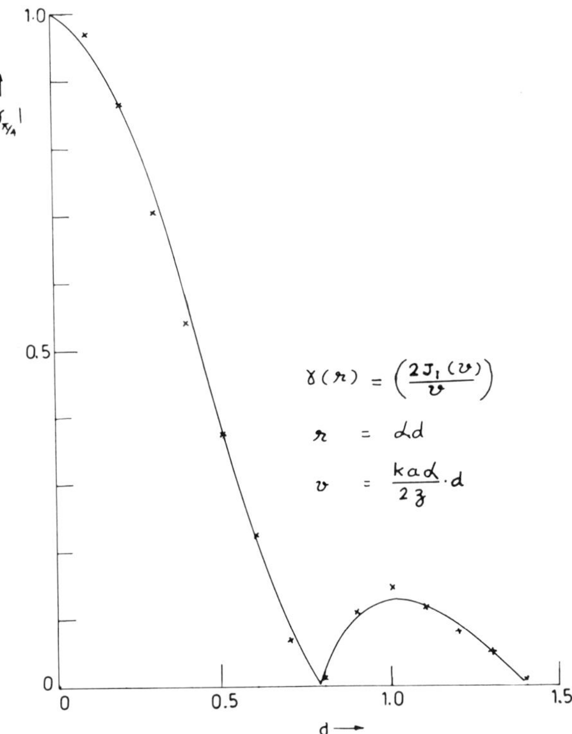

Fig. 3. Visibility of the fringes $\{|\gamma_{\pi/4}(r,0)|\}$ plotted against the separation r. The aperture S is a circle of diameter a.

shows the results of a similar measurement when the source aperture is rectangular. It is readily seen that the visibility of the fringes in this case is given by $|\gamma_{\pi/4}(r,0)|$, and is essentially the average of the lateral and the longitudinal coherence functions.

Making use of the van Cittert-Zernike theorem as applied to vector fields, one may show that in the case when the source aperture is symmetric, incoherent and uniformly illuminated, γ_θ does not depend on θ. In order to observe a marked dependence of γ_θ on θ, we must therefore use an asymmetric source. For this purpose we take the circular aperture with half its portion covered with another polaroid. We have plotted the visibility curves $\{|\gamma_\theta(r,0)|\}$ for various values of θ and observe a definite dependence of $\gamma_\theta(r,0)$ on θ, as shown in Fig. 5. The results of these measurements may then be used in obtaining all the components of $\gamma_{ij}(r,0)$ as discussed earlier.

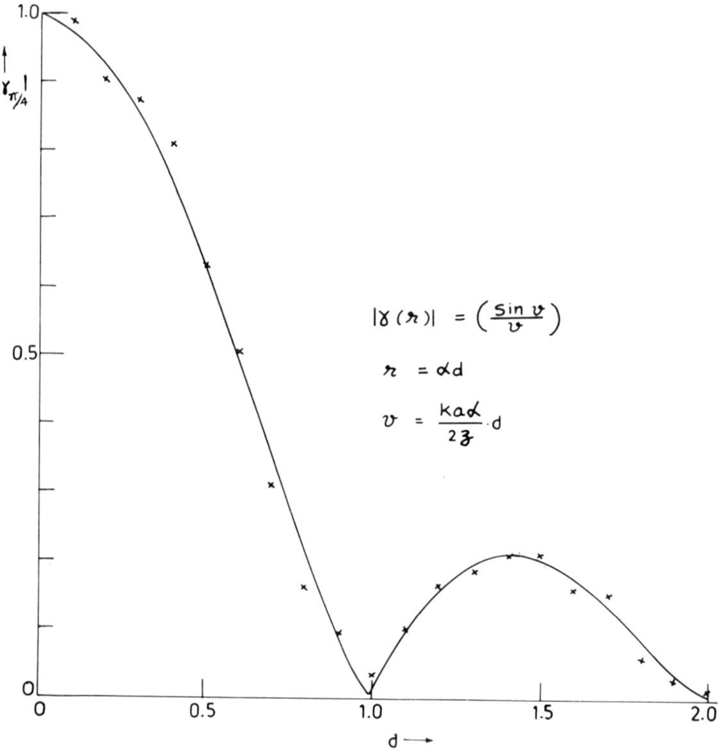

Fig. 4. Same as in Fig. 3 when the aperture S is a rectangle of width a.

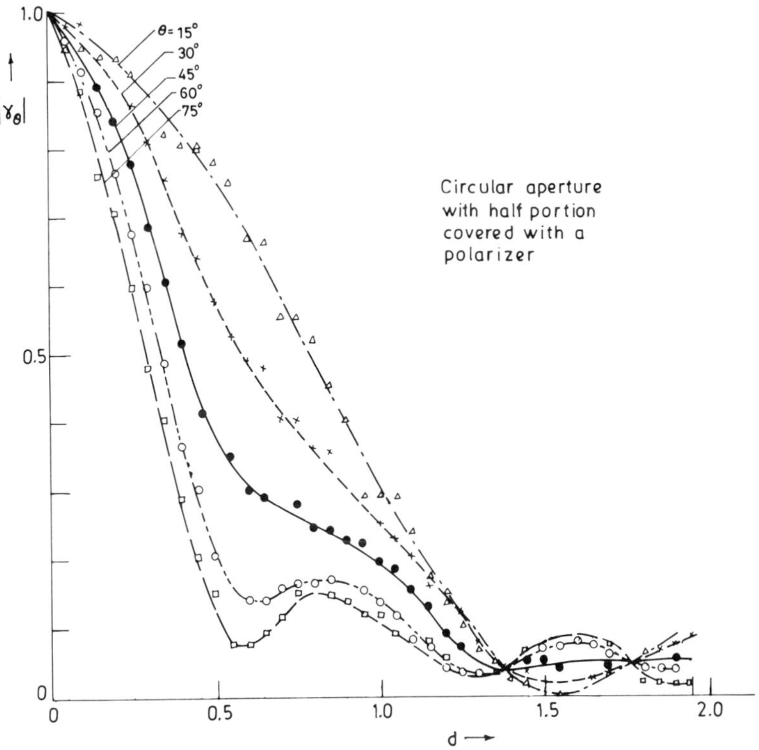

Fig. 5. Visibility curves $\{|\gamma_\theta(r,0)|\}$ for various values of θ. The aperture S is a circle with half its portion covered with a polaroid.

References

1. M. Born and E. Wolf, *Principles of Optics*, 5th edition, (Pergamon Press, New York, 1975), Chapter X.
2. M. Francon and S. Mallick, *Polarization Interferometers*, (Wiley Interscience, London, 1971) Chapter 9. See also *Progress in Optics*, Vol. VI, editor E. Wolf (North-Holland Pub., Amsterdam, 1967) p. 73.
3. D. Kohler and L. Mandel, J. Opt. Soc. Am. *63*, 126 (1973).
4. D. Dialetis and C. L. Mehta, Nuovo Cimento *56B*, 89 (1968).
5. For a similar concept, see for example, G. K. Batchelor, *The Theory of Homogeneous Turbulence*, (Cambridge Union Press, 1953).
6. C. L. Mehta and E. Wolf, Phys. Rev. *134*, A1143 (1964).

TWO-PHOTON TIME CORRELATIONS IN DIVERGING SPATIALLY COHERENT LIGHT

Donald Scarl and Sigrid McAfee*

Polytechnic Institute of New York, Farmingdale, N.Y.

We have measured the time correlation of pairs of photons from a neon discharge source limited by a 2 micrometer diameter aperture.[1] Light from this small source aperture is coherent over a wide range of angles so that the second order correlation function shows an excess counting rate (the Hanbury Brown-Twiss effect) over a wide range of propagation directions. This geometry allows the measured two-photon counting rate to be compared with the predictions of a maximum entropy density matrix over a wider range of angles than was previously possible.

The field from a chaotic light source can be described by a density matrix which factors into a product of the matrices for each mode of the field[2]

$$\rho = \prod_{\ell} \rho_{\ell} \tag{1}$$

This factorization implies that field modes of different index ℓ are independent. In addition, since the single mode density matrices can be written

$$\rho_{\ell} = \int P_{\ell}(\alpha_{\ell}) |\alpha_{\ell}\rangle\langle\alpha_{\ell}| \; d^2\alpha_{\ell} \tag{2}$$

and, for chaotic light $P(\alpha)$ is a Gaussian functions of α,

$$P_{\ell}(\alpha_{\ell}) = \frac{1}{\pi\langle n\rangle_{\ell}} e^{-|\alpha_{\ell}|^2/\langle n_{\ell}\rangle} \quad , \tag{3}$$

the higher order moments of the chaotic field are completely
determined by the first order moments. In particular, the second
order correlation function that describes two-photon experiments
reduces to

$$G^{(2)}(\vec{r}_1,t_1;\vec{r}_2,t_2) = G^{(1)}(\vec{r}_1,t_1)G^{(1)}(\vec{r}_2,t_2) + |G^{(1)}(\vec{\Delta r},\Delta t)|^2$$

$$(4)$$

and can be predicted from a knowledge of the first order correlation
function. The term containing the first order correlation function
gives rise to the Hanbury Brown-Twiss excess counting rate near
$\Delta r = \Delta t = 0$.

For cross spectrally pure light, $G^{(1)}(\Delta r,\Delta t)$ factors into the
product of a spatial coherence function and a temporal coherence
function. The spatial coherence function is usually written in
terms of the distance between field points on a plane downstream of
the source aperture. For a small source, the spatial coherence
function can also be written in terms of the difference between
propagation vectors Δk from the source to the field points, or in
terms of the angle θ, subtended at the source by each pair of field
points. When written in terms of propagation vector or angles, the
correlation function is also applicable to scattering measurements
and to correlated emission processes such as atomic cascades and
positron annihilation.

Although a source with no preferred spatial direction must produce a spherically symmetric intensity distribution, such a source
is not precluded from producing a two-photon intensity correlation
distribution that is a function of the difference in propagation
directions of the two photons. In fact, completely chaotic sources
lead to one such distribution because of the dependence of the
second order correlation function $G^{(2)}$ on the square of the first
order correlation function $G^{(1)}$. A departure of the two-photon
counting rate from the rate predicted by the above chaotic density
matrix for the field may give information about two-photon processes
in the source.

For chaotic sources, the first order correlation function has
been measured many times, both directly and by use of the
van Cittert-Zernike theorem. Beginning with the original experiment
of Hanbury Brown and Twiss, Table I shows some of the measurements
of the second order correlation function from chaotic sources.
These measurements of the second order correlation function in
chaotic light have explored this function for small values of θ
and therefore for small values of $\Delta k/k$. The dependence of $G^{(2)}$ on
$G^{(1)}$ can best be measured for regions in which $G^{(1)}$ is non-zero,
so that measurement of $G^{(2)}$ at large Δk can be done best if a small

Table 1 Two-photon correlation measurements for fields from chaotic sources.

AUTHORS	YEAR	WAVELENGTH (μm)	d_1 (μm)	$\theta_c = \lambda/d_1$ (mr)
Hanbury Brown and Twiss[3]	1956	.4358	130x150	3.36
Rebka and Pound[4]	1957	.4358	200	2.18
Hanbury Brown and Twiss[5]	1958	.4358	190	2.29
Twiss and Little[6]	1959	.5461	360	1.52
Martienssen and Spiller[7]	1964	.6328	100	6.33
Morgan and Mandel[8]	1966	.5461	540	1.01
Scarl[9]	1966	.4358	140	3.11
Phillips, Kleiman and Davis[10]	1967	.4358	300	1.45
Scarl[11]	1968	.4358	180	2.42
This Experiment	1977	.5852	2	292

source aperture is used, leading to large values of $G^{(1)}$ at large Δk. If the coherence angle θ_c is defined as the reciprocal of the source aperture diameter measured in wavelengths,

$$\theta_c = \frac{\lambda}{d} \ . \tag{5}$$

The maximum coherence angles used in these measurements was 6.3 milliradians.

The present experiment, with its coherence angle of 292mr, allowed the detectors to sample a large fraction of the light leaving the source aperture with differing propagation directions. A detector aperture subtending a full-angle of 254 milliradians allowed second order correlation measurements to be made in a spatially coherent beam filling a solid angle of 0.051 steradians and including a wide range of photon propagation directions. This geometry helps to test the assumption that the simple constant-energy maximum-entropy chaotic density matrix completely describes the light leaving a gas discharge.

Because of the small number of photons in a single cell of
phase space in the radiation from natural chaotic sources, it is
convenient to average $G^{(2)}$ over a large detector aperture rather
than to attempt to measure $G^{(2)}$ using two small separate detector
apertures. This type of averaging was done in the preliminary
measurements reported here.

EXPERIMENT

A 2mm inside diameter DC discharge tube filled with 1.6 torr of
Ne[20] and viewed side-on was placed behind a 2μm aperture in a 1 μm
thick nickel foil. The light leaving the 2 μm aperture passed
through a 12.7mm aperture, 50mm downstream. A lens collected the
light from the 12.7mm aperture and directed it through a 585nm
interference filter, a polaroid linear polarizer, and a beam
splitter to the photocathodes of two RCA 8852 photomultipliers.
The time difference between photons arriving at the two phototubes
was measured and recorded on a pulse height analyzer. The singles
counting rate in each photomultiplier was about 5500 counts per
second corresponding to about 7×10^5 photons per second in the
beam passing through the detector aperture.

The two-photon counting rate as a function of Δt, the time
between detected photons, is gotten by integrating equation (4)
over the detector aperture and convolving with the detector time
response function, $D(\Delta t)$,

$$R(\Delta t) = \int \int \varepsilon_1 \varepsilon_2 I_1 I_2 (1 + N \frac{|G^{(1)}(r_1,r_2)|^2 |G^{(1)}(\Delta t)|^2}{I_1 I_2}) dr_1 dr_2 * D(\Delta t)$$

$$= R_o [1 + SNW \exp(-\Delta t^2/2\sigma^2)] \tag{6}$$

The spatial coherence factor S as a function of coherence angle
and detector aperture diameter, is evaluated in Reference 11. N
is the product of the signal-to-noise ratios in each of the detectors.
W and σ are functions of the square of the temporal correlation
function and of $D(\Delta t)$. These were measured in separate experiments
to be Gaussian with standard deviations of .13 ns and .55 ns,
respectively.

Figure 1 shows a plot of the theoretical $R(\Delta t)$ derived fron
the usual maximum entropy density matrix, along with preliminary
experimental results. Agreement between the measured and calculated
points can be classified as almost acceptable. Plans are being

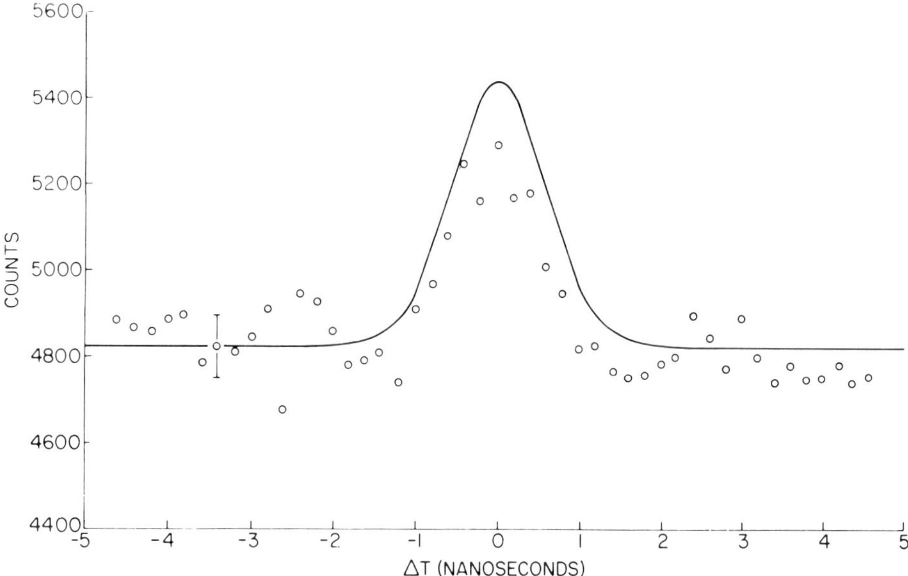

Fig. 1 The two-photon counting rate, $R(\Delta t)$, in a field with a coherence angle of 292 mr. The solid line is the counting rate predicted from a maximum entropy density matrix. The error bar is common to all of the experimental points.

made to improve the statistical accuracy of the measurements and to refine the procedure by which the second order correlation function is used to predict the experimental counting rate.

*Present address: Bell Telephone Laboratories, Murray Hill, New Jersey 07974.

References

1. Sigrid R. McAfee, "Two-photon Correlations in a Widely Divergent Light Beam", Ph.D. Dissertation (Physics), Polytechnic Institute of New York, 1976.
2. R.J. Glauber, in *Quantum Optics and Electronics*, C. DeWitt, A. Blandin, and C. Cohen-Tannoudji, Eds. (Gordon and Breach, New York, 1964).
3. R. Hanbury Brown and R.W. Twiss, Nature *177*, 27 (1956).
4. G.A. Rebka and R.V. Pound, Nature *180*, 1035 (1957).
5. R. Hanbury Brown and R.W. Twiss, Proc. Roy. Soc. *A243*, 291 (1958).
6. R.Q. Twiss and A.G. Little, Australian J. of Physics *12*, 77 (1959).

7. W. Martienssen and E. Spiller, Am. J. Phys. *32*, 919 (1964).
8. B.L. Morgan and L. Mandel, Phys. Rev. Lett. *16*, 1012 (1966).
9. D.B. Scarl, Phys. Rev. Lett. *17*, 663 (1966).
10. David T. Phillips, Herbert Kleiman, and Sumner P. Davis, Phys. Rev. *153*, 113 (1967).
11. D.B. Scarl, Phys. Rev. *175*, 1661 (1968).

ABOUT THE ATOMIC CORRELATION FUNCTIONS IN HIGHER ORDER COHERENCE

Andrei N. Weiszmann

College of Staten Island, CUNY, New York

The system under consideration in two recent papers [1,2] is
an ensemble of N identical 3-level atoms: (ground state) $|g>$;
(intermediate state) $|h>$; (final state) $|f>$, interacting inelas-
tically with an incident coherent radiation. The reduced field
density operator:

$$\sigma(t) = Tr_A \chi(t) \tag{1}$$

where Tr_A stands for trace over the atomic system, was obtained by
the so-called projection technique [2] and was used in defining the
j-th order field correlation functions:

$$G^{(j)}(x_1;\ldots;x_{2j}) = tr\{\sigma E^{(-)}(x_1)\ldots E^{(-)}(x_j)E^{(+)}(x_{j+1})\ldots E^{(+)}(x_{2j})\} \tag{2}$$

where $x_j = (\bar{r}_j;t_j)$ and $E^{(-)}(x_j);E^{(+)}(x_j)$ are the negative and positive frequency parts, respectively, of the $\bar{E}(\bar{r}_j;t_j)$ field.

These correlation functions play a fundamental role in the concept of higher order coherence in the Glauber [3] sense, namely,
the necessary and sufficient conditions for coherence are connected
to the normalization and factorization properties of $G^{(j)}(x_1;\ldots;x_{2j})$.

The question imposed was whether an initially coherent laser
beam preserves its coherence (and if so, to what order), after being
inelastically scattered by an atomic ensemble.

The interaction hamiltonian in the dipole-dipole approximation was written in the form:

$$V = \sum_k \sum_m i\hbar\mu_{hf}[a_k^+|h\rangle\langle f|_m - a_k|f\rangle\langle h|_m] \tag{3}$$

with μ_{hf} the component of the dipole moment in the direction of the field. The normalized field correlation functions were found:

$$g^{(j)}(x_1;\ldots;x_{2j}) = \sum_{k^2}\sum_n \exp\{-\frac{k^2}{8}F[t_{j+1}+\ldots+t_{2j})-3(t_1+\ldots+t_j)]\}$$

$$\times\ \exp\,-i(\phi_{j+1}+\ldots+\phi_{2j}-\phi_1-\ldots-\phi_j)\cdot\frac{n(n+1)\ldots(n+j-1)}{n^j}$$

$$\times\ \frac{\exp\{(j-1)\frac{B-F}{F}(n+1)\}}{J_o^{(j-1)}}\ . \tag{4}$$

where: $\phi_j=(\bar{k}\cdot\bar{r}_j-\omega_k t_j)$; $k^2=2(n_x+n_y)$; $n_x,n_y=0,1,2\ldots$; $n=1,2\ldots3^N$, and $J_o(k\sqrt{n+1})$ = the zeroth order Bessel functions.

The conclusion of these papers was that an initially coherent laser beam after being inelastically scattered by a three-level atomic ensemble is — as it could be expected — in general, incoherent.

It was found, however, that in some specific phase and time conditions the scattered field is coherent; the order of coherence is entirely determined by the atomic correlations due — in the first approximation — to the dipole-dipole interactions among the particles in the scattering system.

Thus to analyse the behaviour of the $g^{(j)}$-s we have to concentrate our attention on the atomic correlation functions B and F. These functions were found [1] in the form:

$$B = \phi(\omega_k)(g,h)\quad;\qquad F = \phi(\omega_k)(g,f) \tag{5}$$

with
$$\phi(\omega_k) = \hbar^2\mu_{hf}\mu'_{fh}\,2\pi g(\omega_k)$$

where $g(\omega_k)$ is a weight function [4].

The $(g,h); (g,f)$ are:

$$(g,h) = \text{Tr}_A \sum_{\ell,m=1}^{N} \left[|h><f|_m |f><h|_\ell \frac{\exp(-\beta H_A)}{\text{Tr}_A \exp(-\beta H_A)} \right]$$

$$(g,f) = \text{Tr}_A \sum_{\ell,m=1}^{N} \left[|h><f|_m \frac{\exp(-\beta H_A)}{\text{Tr}_A \exp(-\beta H_A)} |f><h|_\ell \right]$$

(6)

with

$$H_A = \sum_{\ell=g}^{f} \varepsilon_\ell n_\ell \; ; \quad n_\ell = \sum_{m=1}^{N} |\ell><\ell|_m \; .$$

These functions are the key elements in the atomic correlation functions; thus they play a basic role in the higher order coherence. It is well known [3] that the necessary conditions for coherence in the Glauber sense are that the normalized field correlations functions:

$$|g^{(j)}(x_1; \ldots; x_{2j})| = 1 \tag{7}$$

have unit absolute magnitude.

To find out whether this is the case in our system, let us split (4) in two parts, the product of two exponentials, and the factor involving the summation over n and k^2.

$$\text{The first part} = \sum_{k^2} \exp\{- \frac{k^2}{8} F(t_{j+1} + \ldots + t_{2j})$$

$$- 3(t_1 + \ldots + t_j)\} \exp\{-i(\phi_{j+1} + \ldots + \phi_{2j} - \phi_1 - \ldots - \phi_j)\} \tag{4a}$$

and has a unit absolute magnitude whenever the conditions:

$$(t_{j+1} + \ldots + t_{2j}) = 3(t_1 + \ldots + t_j) \; ; \quad (\phi_{j+1} + \ldots + \phi_{2j}) = (\phi_1 + \ldots + \phi_j)$$

(8)

are met.

$$\text{The second part} = \sum_{k^2} \sum_{n} \frac{n(n+1)\ldots(n+j-1)}{n^j} \cdot \frac{\exp\{(j-1)\frac{B-F}{F}(n+1)\}}{J_o^{(j-1)}}$$

(4b)

and for j=1 is always equal to unity independent of the values taken
by n and k^2. For higher order correlation functions the limiting
value for this term depends entirely on the value of B and F. By
definitions (5) and (6), B and F are positive and F > B. If BαF
then B-F/F is always negative. Furthermore we can see that
$\exp\{(j-1)B-F/F(n+1)\}$ with increasing values of n tends to zero
together with $J_0^{(j-1)}$. Thus if BαF the whole second part (4b) tends
to unity.

Performing the indicated traces in (6) we have:

$$(g,h) = \frac{\displaystyle\sum_{A=1}^{N}\sum_{P=0}^{N-1}\frac{N!}{P!Q!(A-1)!}\exp[-\beta(Q\varepsilon_g+A\varepsilon_h+P\varepsilon_f)]}{\displaystyle\sum_{A=1}^{\mathrm{INT}\,N/2}\sum_{P=A}^{N}\frac{N!}{P!Q!A!}\begin{bmatrix}\exp[-\beta(Q\varepsilon_g+A\varepsilon_h+P\varepsilon_f)]+\exp[-\beta(Q\varepsilon_g+A\varepsilon_f+P\varepsilon_h)]\\ +\exp[-\beta(Q\varepsilon_h+A\varepsilon_f+P\varepsilon_g)]+\exp[-\beta(Q\varepsilon_h+A\varepsilon_g+P\varepsilon_f)]\\ +\exp[-\beta(Q\varepsilon_f+A\varepsilon_g+P\varepsilon_h)]+\exp[-\beta(Q\varepsilon_f+A\varepsilon_h+P\varepsilon_g)]\end{bmatrix}} \tag{9}$$

where

$$Q = N - (P+A)$$

and (g,f) becomes the same as (9) having ε_h interchanged with ε_f.
The numerical values of these functions can be easily obtained by a
relatively simple computer program for any desired value of N.

However, from the form of (g,h) and (g,f) it immediately follows that B and F are always proportional. As a matter of fact,
expanding the numerator of (9) in a multinomial series we have:

$$(g,h) = \frac{Ne^{-\beta\varepsilon_h}[e^{-\beta\varepsilon_g}+e^{-\beta\varepsilon_h}+e^{-\beta\varepsilon_f}]^{N-1}}{\mathrm{Tr}_A[\exp(-\beta H_A)]} \tag{10}$$

where by interchanging ε_h with ε_f it follows that BαF. This proves
the existence of coherence for the considered inelastically scat-
tered radiation, subject to the conditions (8).

Further considerations will be needed for multipole interactions and exciton generations which may alter the conditions (8),
leaving, however, the basic conclusions about the coherent properties of the scattered quantum radiation unchanged.

Acknowledgment

The author wishes to thank Professor Kaya Imre for many helpful
discussions.

References

1. A.N. Weiszmann, submitted for publication to Journal of Statistical Physics, New York (1977).
2. A.N. Weiszmann, submitted for publication to Journal of Mathematical Physics, New York (1977).
3. R.J. Glauber, Phys. Rev. *131*, 2766 (1963).
4. W.H. Louisell, in *Quantum Optics*, ed. R.J. Glauber (Academic Press, New York, 1969) p. 680.

SUPERRADIANCE: THEORIES VERSUS EXPERIMENTS

S. Haroche

l'Ecole Normale Supérieure, Paris, France

Superradiance has been in the past years subject to renewed
interest due to the development of experiments demonstrating the
effect in the far-[1] and near-infrared[2,3,4] ranges of the
spectrum. Theoretical interpretation of these experiments raises
serious questions because two kind of superradiance theories with
different predictions do exist:

(i) Semi-classical theories[5], describing classically the
    electromagnetic field via Maxwell equation and simulating
    spontaneous emission in a phenomenological way by a
    "noise" term in the equations.

(ii) Quantum-mechanical theories[6] describing in a fully
     quantum mechanical picture the "atoms + field system"
     and thus dealing more satisfactorily with spontaneous
     emission.

These two kinds of theories appear to yield quite different
results when applied to the experimental situations, with a better
fitting for the semi-classical approach[7]. This result seems
paradoxical since the theory which describes best the
spontaneous emission process is the one which seems to fail to
describe accurately a phenomenon in which spontaneous emission has
obviously a great importance.

To solve this difficulty, we present in this communication a
careful analysis of the quantum-mechanical description of superradiance, in order to discuss the approximations made to derive the
equations and to study if these approximations are consistent with
the parameters of the performed experiments.

The quantum mechanical description of superradiance in an N atom system is most conveniently expressed in the basis of the collective Dicke states[8] representing symmetric superposition of atomic levels in which the excitation is equally shared between the N atoms of the medium. The evolution of the system is generally described by the so-called "Dicke equations" connecting the populations of the Dicke states and describing a "cascade" emission by the atomic system going down the "ladder" of the N Dicke states from the totally inverted level to the collective ground state. The characteristic emission time of superradiance is shown to be:

$$T_{SR} = (\Gamma N \mu)^{-1} , \tag{1}$$

where Γ is the natural linewidth of the transition and μ is a "form factor"[9] accounting for the geometry of the excitation.

A basic approximation in this theory is the so-called "Born-Markov" one, which amounts to assuming that the atomic sample radiates via *pure spontaneous* emission in the vacuum field, without any reabsorption or stimulated emission effect (i.e. neglecting "propagation" effects). It was generally assumed so far[10] that this approximation holds so long as the "escape time" L/c of the photons (L: characteristic length of the sample) is shorter than the characteristic superradiance time T_{SR}, that is if the condition:

$$T_{SR} > \frac{L}{c} \tag{2}$$

is fulfilled. This condition may also be expressed as a limit on the number of emitting atoms:

$$N \lesssim \frac{c}{\lambda \Gamma} \mathcal{F} \tag{3}$$

(where $\mathcal{F}$ is the "Fresnel number" of the emitting volume). Condition (3) corresponds to a limit of the order of 10^9 - 10^{13} atoms for most situations of experimental interest. This condition is fulfilled - or nearly fulfilled - in the superradiance experiments performed so far, which should thus be understood - at least qualitatively - in terms of "Markovian" Dicke equations. It is however not the case and observation shows that propagation effects are in the experiments much more important than would be predicted according to the above reasoning. These propagation effects manifest themselves by an important lengthening of the superradiance delays and pulse widths, by existence of slow coherent ringings in the superradiance pulse ...

We show that the explanation for the underestimation of propagation effects in the above mentioned quantum mechanical theories

comes from the fact that the limits (2) and (3) are not the correct
ones and that *"Non-Markovian"* behaviour does appear for much
smaller number of initially excited atoms. Collective emission
in a sample of dimensions $L \gg \lambda$ is indeed a process having a
correlation (memory) time of the order L/c. This implies that
each of the N steps down the "ladder" of the Dicke states adds to
the emission time an intrinsic "Non-Markovian" delay of the order
of L/c. The condition for "Markovian-behaviour" of the system is
expressed by the fact that the *cumulated* delay NL/c should be
shorter than the expected superradiance time T_{SR}. One thus gets:

$$T_{SR} > \frac{NL}{c} \tag{4}$$

instead of (2) or:

$$N < \sqrt{\frac{c}{\Gamma \lambda} \mathcal{F}} \tag{5}$$

instead of (3).

Conditions (4) and (5) are much more drastic than the previously
assumed ones and practically limits to about $10^4 - 10^6$ atoms the
size of a sample superradiating according to *pure collective
spontaneous emission* (without propagation effects). All experiments
so far have been performed with much larger atomic samples
($N \sim 10^{10}$ to 10^{13}) obeying the condition:

$$N \gg \sqrt{\frac{c}{\Gamma \lambda} \mathcal{F}} \tag{6}$$

This explains the strong "Non-Markovian" behaviour of the
observed signals.

To completely resolve the paradox mentioned at the beginning
of this paper, we have also carefully analyzed the connection
between the quantum mechanical and semi-classical theories of
superradiance and checked that - when the *Non-Markovian condition* [6]
is fulfilled - most of the evolution of the system is well described
by the semi-classical Maxwell-Schrödinger equations which are very
well suited for the description of propagation effects. In this
case, superradiance may be described as a two-stage process:

(i) *in an initial "quantum mechanical" stage* (lasting a time
     of the order of a few T_{SR}), the first photons are emitted
     according to a Markovian process during which strong
     interatomic-correlations and atom-field correlations
     build up in the system;

(ii) *in a subsequent "classical" stage* (lasting a time of the order of several tens of T_{SR}), the atomic polarization and field built up in the first stage evolve in a "Non-Markovian" way according to classical Maxwell equations throughout the sample, with strong self-induced nutation and propagation effects.

Matching between the initial quantum mechanical stage and the subsequent classical evolution does not depend drastically on the details of the system evolution during the first stage[5], provided condition (6) is fulfilled. This explains why the semiclassical theories are well adapted to the description of Non-Markovian superradiance[11,12].

In conclusion, we may say that superradiance has been in the past literature described in two different ways: the quantum mechanical description sees the phenomenon mainly as "collective spontaneous emission" in the vacuum field, whereas the semi-classical picture views it as the transient form of stimulated emission from a high gain medium. These two aspects of the phenomenon correspond to two limiting regimes depending on the number of radiating atoms. The original contribution of this paper is to reconsider the limit between these two regimes and to show that it corresponds to much smaller atom numbers than previously assumed. This explains the agreement between the experiments and the semi-classical approach of the phenomenon.

References

1.   N. Skribanowitz, I.P. Herman, J.C. McGillivray and M.S. Feld, Phys. Rev. Lett. *30*, 309 (1973).
2.   M. Gross, C. Fabre, P. Pillet and S. Haroche, Phys. Rev. Lett. *36*, 1035 (1976).
3.   A. Flusberg, T. Mossberg and S.R. Hartmann, Phys. Lett. *58A*, 6, 373 (1976).
4.   H.M. Gibbs, Q.H.F. Vrehen and H. Hikspoors, (to be published).
5.   J.C. McGillivray and M.S. Feld, Phys. Rev. A *14*, 1169 (1976).
6.   R. Bonifacio and L.A. Lugiato, Phys. Rev. A *11*, 1507 (1975); A *12*, 587 (1975).
7.   R. Friedberg and B. Coffey, Phys. Rev. A *13*, 1645 (1976); R. Bonifacio, L.A. Lugiato and A. Crescentini, Phys. Rev. A *13*, 1648 (1976).
8.   R.H. Dicke, Phys. Rev. *93*, 99, (1954).
9.   N.E. Rehler and J.H. Eberly, Phys. Rev. A *3*, 1735 (1971).
10.   F.T. Arecchi and E. Courtens, Phys. Rev. A *2*, 1730 (1970).
11.   I.P. Herman, J.C. McGillivray, N. Skribanowitz and M.S. Feld, in *Laser Spectroscopy*, eds. R.G. Brewer and A. Mooradian (Plenum Press, New York, 1975).
12.   P. Pillet, Thèse de 3e cycle, Paris (1977) unpublished.

EXPERIMENTS ON SUPERFLUORESCENCE IN CESIUM

Q.H.F. Vrehen and H.M.J. Hikspoors

Philips Research Laboratories, Eindhoven, The Netherlands

H.M. Gibbs

Bell Laboratories, Murray Hill, New Jersey

1. INTRODUCTION

Superfluorescence (SF), i.e., the cooperative or superradiant[1] emission of an initially inverted system, has been the subject of extensive theoretical investigation[2-9]. The first experiment[10] and its interpretation[11] have aroused further interest in the phenomenon. The SF decay of a pencil-shaped volume of density n is characterized by the emission of a delayed pulse (or pulses) in a narrow cone along the sample axis; the delay time τ_D is nearly proportional to n^{-1}, the peak power is nearly proportional to n^2. The quantitative description of the initial generation of atomic coherence and of the subsequent coherent pulse evolution is still under discussion and so are the precise conditions for SF to occur. Stringent conditions have been specified by Bonifacio and Lugiato[6] for what these authors call "pure SF": (a) a pure two-level system, (b) a pencil-shaped volume of Fresnel number $\mathcal{N} \equiv A/\lambda L = 1$, (c) transverse excitation with a pump pulse of duration $\tau_p \ll \tau_D$, and (d) $\tau_E < \tau_c < \tau_R < \tau_D < T_1, T_2', T_2^*$. Here A is the cross-sectional area and L is the length of the sample; $\tau_E \equiv L/c$, with c the velocity of light; $\tau_R \equiv \tau_0/\mu N$, with τ_0 the spontaneous decay time, N the number of atoms and μ the shape factor introduced by Rehler and Eberly[4,12]; τ_c is the Arecchi-Courtens[5] cooperation time $\tau_c \equiv \sqrt{\tau_E \tau_R}$; τ_D is the delay time; T_1 and T_2' are the longitudinal and transverse homogeneous relaxation times and T_2^* is the inhomogeneous dephasing time.

The Cs experiment has been designed to meet the conditions

above, with the exception of the transverse excitation; instead
longitudinal pumping is used with $\tau_P \gg \tau_E$ to avoid swept excitation.
A typical set of parameters is τ_E = 0.07 ns, τ_c = 0.18 ns,
τ_R = 0.5 ns, τ_D = 10 ns, T_1 = 70 ns, T_2' = 80 ns, T_2^* = 32 ns and
τ_P = 2 ns. Under such conditions single pulse outputs are observed.

Experimental parameters have also been changed to violate
various conditions. Excitation of several, nearly degenerate, upper
levels leads to the observation of quantum beats. At high densities
such that $\tau_R < 2\tau_E$, the output consists of multiple pulses
(for $\mathcal{N} \approx 1$). For a large Fresnel number the output has "multi-mode"
character, and the multiple pulses nearly disappear. For $\tau_D \gg T_2^*$
the SF signal is quenched by dephasing. Some information about
fluctuations is obtained from the correlations between pulses
emitted in the forward and backward direction. The possible effect
of feedback has been studied by placing the sample in a plane mirror
cavity.

Many details of the experiment, of the quantum beat results,
and of the results for $\mathcal{N}$ = 1 have been reported previously by
Gibbs[13a] and Vrehen[13b]; in this paper reference to these earlier
papers will be made as [[1]] and [[2]], respectively. The earlier
results are summarized here. The remainder of the paper is devoted
to more recent experiments.

2. THE CESIUM SYSTEM AND EXPERIMENTAL APPARATUS

A simplified level scheme of Cs is shown in Fig. 1, together
with a diagram of the experiment. The 7P level is saturated from
the ground state by a short dye-laser pulse of 2 ns duration
and a bandwidth of 400-500 MHz. It then decays by SF emission to
7S. The decay on competing transitions (7P to 6S and 7P to 5D),
and on the cascade transition (7S to 6P) is negligible [[1]] and
[[2]]. From atomic parameters[14] τ_0 = 551 ns (for linear σ
emission), T_1 = 70 ns and T_2' = 80 ns. The Cs atoms are contained in
a cell (T_2^* = 5 ns), or in an atomic beam (T_2^* = 18 ns for L = 3.6 cm,
and T_2^* = 32 ns for L = 2.0 cm). The SF emission is focused onto an
InAs-detector of 1 ns response time; the resulting signals are
displayed on a Tektronix Transient Digitizer and recorded on video-
tape for later inspection. In zero magnetic field several hyperfine
levels are excited [[1]]. In a transverse magnetic field of about
2.8 kOe an inverted two-level system is prepared [[2]]. For a
quantitative analysis of the data the density of initially inverted
atoms must be known accurately, in particular because the delay time
depends linearly on the density and only logarithmically on the
initial tipping angle θ_0. Much attention has been given to the
determination of the density[[2]]. Recently a hot-wire detector has

has been installed in the atomic beam apparatus; with it the
earlier measurements have been confirmed. Densities quoted are
accurate to plus 60% or minus 40%.

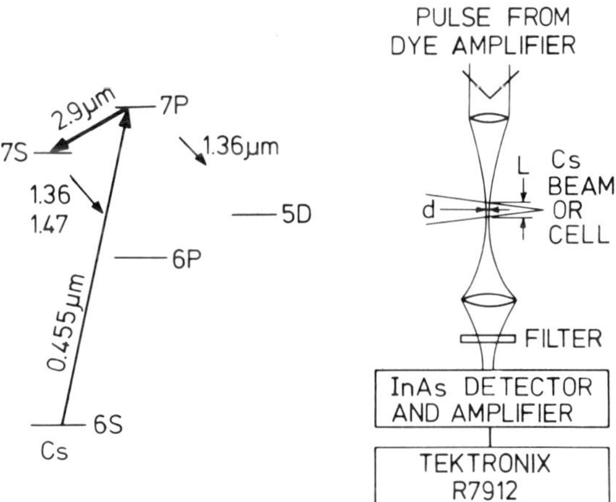

Fig. 1 Simplified level scheme of Cs and diagram of the experimental
apparatus.

3.   QUANTUM BEATS

When two or more upper levels are populated, beats are
observed in the output ([1] and ref.[15]). Two different cases
can be distinguished, depending on whether the upper levels are
excited as a coherent superposition or as an incoherent mixture.
The first case applies for H = 0. With polarized pump light a
coherent superposition of hyperfine levels is excited, either
on $7P_3/2$ or on $7P_{1/2}$. These levels decay to (a) common final
level(s). The resulting beats are in principle similar to those
in single-atom emission[16]; the highly nonlinear process of SF
emission, however, can modify the observed time-dependence of the
intensity considerably. The second case has been realized in a
magnetic field of about 2.8 kOe. At high excitation intensities
two different upper levels are excited incoherently and they decay
to different final levels. No beats should be observed in single-
atom emission. In SF, however, coherent optical fields develop at
both transition frequencies and these fields beat in the detector.

Quantum beats in SF may be useful both for spectroscopic
applications and for the study of SF itself. First, the intensities
in SF emission are many orders of magnitude larger than in

spontaneous emission. Second, in SF beats lower level splittings can be observed in addition to upper level splittings. Finally, a careful study of the beat frequencies may reveal chirps in the SF emission. These applications will require a further development of the theory of quantum beats in SF[17].

4. RESULTS WITH FRESNEL NUMBER ONE

In this and the next section results are presented for a two-level system. In a transverse field of about 2.8 kOe the level $7P_{3/2}$ ($m_j = -3/2$, $m_I = -5/2$) is excited, which then decays to $7S_{1/2}$ ($m_j = -\frac{1}{2}$, $m_I = -5/2$). The experimental results obtained with $\mathcal{N} = 1$ have been described in detail in [[2]]. In this section the earlier results are summarized, recent computer simulations are briefly described and some new data are presented.

A. Single-Pulse SF

The most significant result of the experiment is the observation of single-pulse outputs whenever the Bonifacio-Lugiato conditions are well satisfied. An example is shown in Fig. 2 for an atomic beam of 2.0 cm length. Similar results are obtained for a 3.6 cm beam and for a 5.0 cm cell [[2]]. The data indicate that the best single-pulse results are obtained in an atomic beam with $\tau_D \gtrsim 10$ ns. The pulse width τ_W is proportional to the delay time, $\tau_W \approx 0.5\ \tau_D$. The delay time itself is close to $\tau_D = \tau_R\ \ln N$ (Fig. 3) and the peak powers are proportional to τ_R^{-2} (Fig. 4). The absence of ringing cannot be explained by homogeneous relaxation or inhomoheneous dephasing ([[2]] and section 5d).

B. Computer Simulations

On the basis of a semiclassical model the SF output can be simulated as follows. The sample is treated as a coherent amplifier described by the coupled Maxwell-Bloch equations. An input pulse of area Θ_0 is fed into the amplifier and the output is calculated numerically. Such simulations have been made extensively by MacGillivray and Feld.[11] These authors present the following expression for the delay time $\tau_D = (\tau_R/4)\ (\ln(\Theta_0/2\pi)^2$; this relation describes the numerical results accurately[18].   Taking into account the uncertainty in the excited state density the experiment yields $14\tau_R < \tau_D < 37\tau_R$, which corresponds to $3 \times 10^{-3} < \Theta_0 < 3 \times 10^{-5}$. This range of Θ_0 includes the value $(\sqrt{N})^{-1} \simeq 10^{-4}$; it is somewhat above the value $(\sqrt{\mu N})^{-1} \approx 3 \times 10^{-2}$ and somewhat below the value calculated by MacGillivray and Feld [11] $\Theta_0^{MF} = N^{-\frac{1}{2}}(2\pi)^{-\frac{1}{4}}(\alpha L)^{-3/4}$, where αL is the amplitude gain. For the present experiment $\Theta_0^{MF} \approx 3 \times 10^{-6}$.

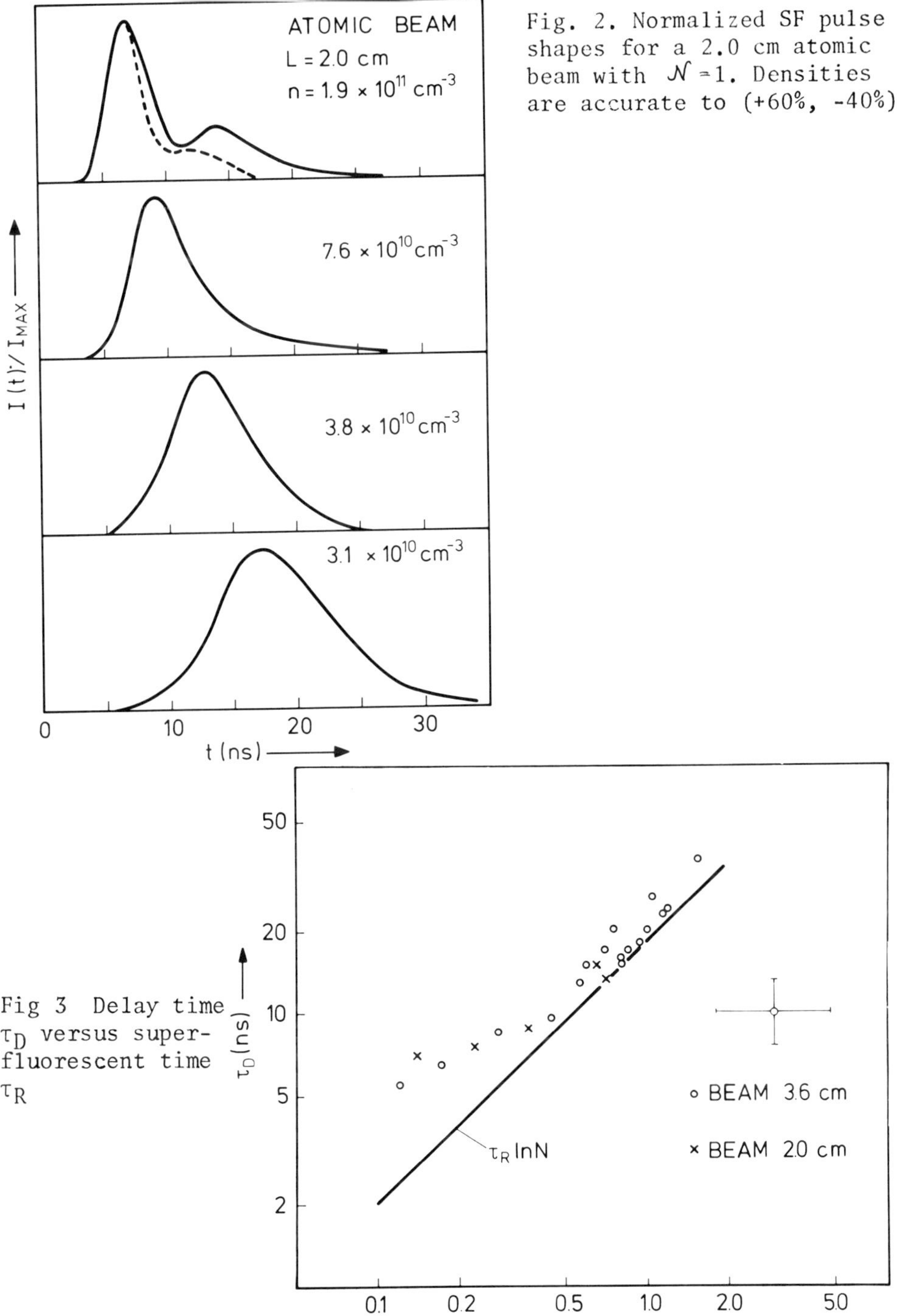

Fig. 2. Normalized SF pulse shapes for a 2.0 cm atomic beam with $\mathcal{N}$ =1. Densities are accurate to (+60%, -40%).

Fig 3 Delay time τ_D versus super-fluorescent time τ_R

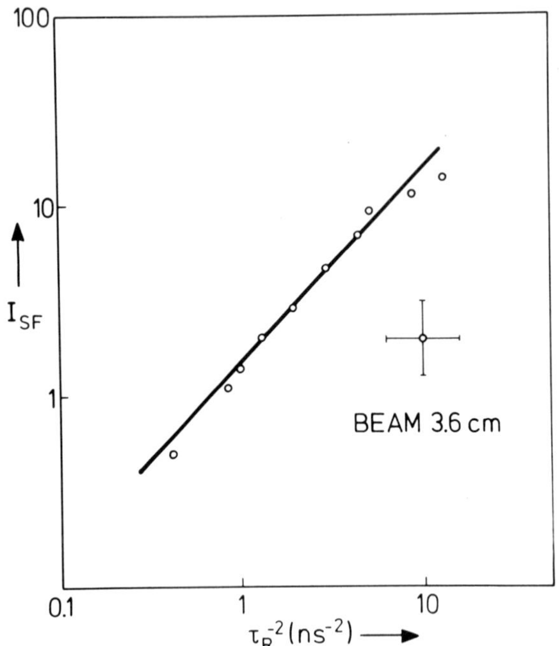

Fig. 4   SF signal strength versus τ_R^{-2}

C.   Multiple-Pulse SF

For the highest densities multiple-pulse output is observed
as seen in Fig. 2. The tendency toward multiple-pulse generation
increases with length L for constant $\mathcal{N}$ = 1 (see [[2]]), but it
descreases with increasing $\mathcal{N}$ at constant L (see section 5a). Recent
experiments show that the observed pulse shapes depend on the position
of the detector in the image plane (the converging SF beam formed by
the collecting lens reaches its minimum diameter near the image of
the center of the emitting pencil). The effect is clearly demonstrated
as follows. With the help of a beam splitter two images are formed
(Fig. 5); a detector (diameter 150 μm) is placed in each image and
the signals of both detectors are displayed on the same trace of the
Transient Digitizer (one signal is suitably delayed). When the
detectors occupy equivalent positions in the image planes, their
signals are very similar. When, however, the detectors occupy non-
equivalent positions they show quite different signals. Examples are
given in Fig. 5. The experiment suggests that the multiple-pulses
cannot be described adequately in a plane-wave approximation. The
shapes of the single pulses observed at longer delays do not depend
on detector position significantly

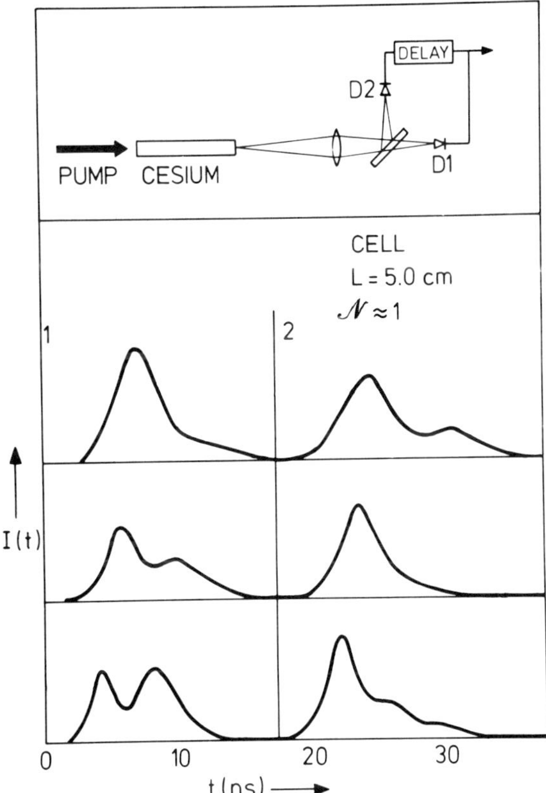

Fig. 5 Pulse shapes observed simultaneously with two detectors at different positions in the image plane, as indicated at the top. Cell, L = 5.0 cm, $\mathcal{N} \approx 1$.

5.   RESULTS WITH LARGE FRESNEL NUMBER

A number of experiments has been performed with a cell ($T_2^* = 5$ ns) of 5.0 cm length and with a large Fresnel number $\mathcal{N} = 15 \pm 5$.

A.   Pulse Shapes and Delay Times

Pulse shapes are shown in Fig. 6, for various vapour temperatures. For low densities (low temperatures, large delays) the pulse shapes are similar to those for $\mathcal{N} = 1$. At higher densities the output retains largely its single-pulse character; only weak second pulses are observed at the highest densities. Delay times are about 1.5 to 2 times shorter for $\mathcal{N} = 15$ than $\mathcal{N} = 1$.

                                          VREHEN et al.

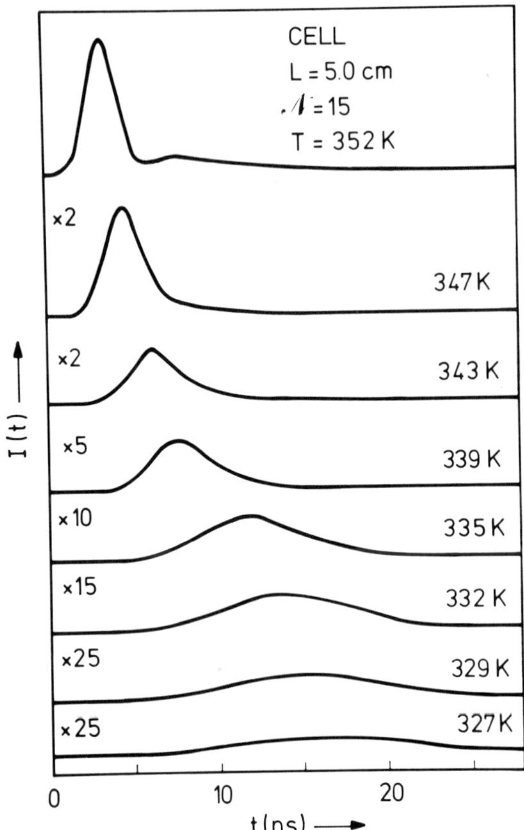

Fig. 6   SF pulse shapes for a 5.0 cm cell, $\mathcal{N} \approx 15$.

B.   Beam Characteristics

For $\mathcal{N} \gg 1$ the output has multimode character as demonstrated by the following experiments. The (half-intensity) diameter of the pump beam is d = 1.7 mm, the sample length L = 50 mm. The full divergence angle of the SF output is measured to be θ = 0.06, i.e., close to the geometrical angle 2d/L = 0.07. The full 60 mrad angle is essentially filled in each individual shot, as can be concluded from the high correlation between the signals from two different detectors, receiving radiation from different halves of the solid angle. The 1:1 image of the SF pencil has a waist diameter of 1.2 mm, comparable to the diameter of the pumped volume. The number of modes is of order $\mathcal{N}^2$ (solid angle times area divided by λ^2).

C.   Fluctuations

Shot to shot fluctuations in pulse amplitude and delay are mainly determined by fluctuations in the pump beam. An indication

of intrinsic fluctuations is obtained by observing simultaneously
the pulses emitted in the forward and in the backward direction.
The delays of the two pulses are always equal to within the
experimental uncertainty, which is roughly equal to one-tenth of
the pulse width, or, equivalently, to $\tau_R (\tau_W \approx 10 \tau_R)$. The pulse
amplitudes are highly correlated (0.8). The experiment also confirms
that "swept gain" can be neglected.

D. Dephasing

When all relaxation can be neglected one would expect the peak
power of the SF pulse, I_{SF}, to be proportional to n^2, or equivalently,
to τ_R^{-2}, at least in the "pure SF" regime. As shown in Fig. 7
$I_{SF}\tau_R^2$ is essentially constant for a 3.6 cm beam ($T_2^* = 18$ ns) for
$\tau_D > 9$ ns, but it drops off in a cell ($T_2^* = 5$ ns) for delay times
beyond 8 ns. The drop-off is attributed to inhomogeneous dephasing.

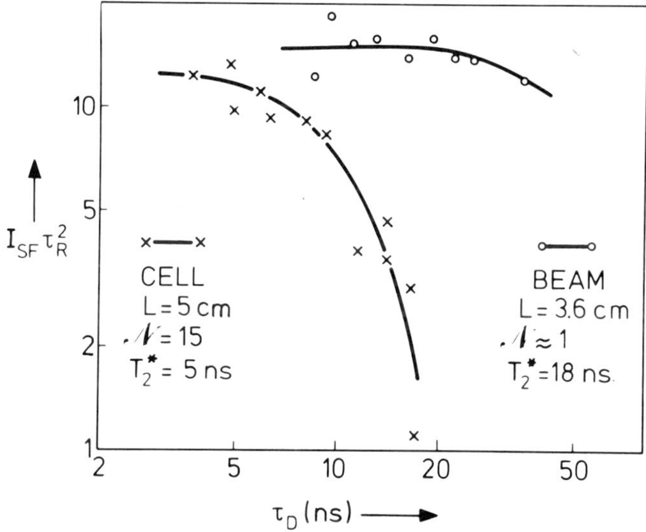

Fig. 7 Quenching of the SF signal by dephasing in a cell (L = 5.0 cm,
$\mathcal{N} \approx 15$); dephasing is negligible in an atomic beam (L = 3.6 cm,
$\mathcal{N} \approx 1$).

E.   Feedback

The cell has been placed in a plane-mirror cavity, mirror
reflectivities 10%, length of cavity 30 cm, round-trip time 2 ns.

For short delay times ($\tau_D \approx 5$ ns) the presence of the mirrors does
not affect the SF signal. For long delay times, such that dephasing
quenches the SF signal ($\tau_D \sim 15$ ns) the signals are recovered in the
presence of the mirrors, the delay time is somewhat increased (10-20%)
and the pulse width is reduced (a factor 2). For intermediate delay
time (8-10 ns) the feedback somewhat increases the peak power; it
never decreases the delay time. To observe the effects of feedback
the mirrors have to be adjusted to better than 5 mrad. In the SF
experiments care is always taken that reflections from windows, etc.
are off-axis by at least 200 mrad. It may be concluded that
unintentional feedback plays no part in the SF measurements.

6. SUMMARY

The 7P $\rightarrow$ 7S transition of Cs is ideally suited for experimental
studies of SF. The 7P level is easily excited. The 3 μm wavelength
of the transition is short enough for convenient detection and long
enough for strong superradiant effects and large values of T_2^x;
dephasing can be reduced further in an atomic beam. Application
of a magnetic field allows the preparation of a pure two-level
system. The stringent conditions of Bonifacio and Lugiato for
"pure SF" can be readily fulfilled. Relevant experimental parameters
such as sample length, Fresnel number, density, and inhomogeneous
dephasing time can be varied over wide ranges.

Under ideal conditions singles pulses are observed. Computer
simulations based on a pulse propagation model require an input
pulse of area of order $(\sqrt{N})^{-1}$ to give a good fit of the delay.
The delay time is essentially given by $\tau_D = \tau_R \ln N$. At higher
densities multiple-pulses are emitted. The different pulses are
probably generated in different parts of the sample and/or emitted
in different directions.

For $\mathcal{N} \gg 1$ the output beam has multi-mode character. Only
weak second pulses are observed at the shortest delay times. Pulses
emitted in forward and backward directions are strongly correlated,
both in amplitude and delay time. Inhomogeneous dephasing reduces
the signal intensity in the cell, but not in the atomic beam. For
the cell feedback influences the signal at long delays, where it
helps overcome the dephasing; for short delays the effect of feedback is negligible.

Acknowledgements

The authors would like to thank R. Bonifacio, R.K. Bullough,
J.H. Eberly, M.S. Feld, L.A. Lugiato, J.C. MacGillivray, D. Polder
and M.F.H. Schuurmans for illuminating discussions.

References

1. R.H. Dicke, Phys. Rev. *93*, 99 (1954).
2. V. Ernst and P. Stehle, Phys. Rev. *176*, 1456 (1968).
3. R. Bonifacio, P. Schwendimann, and F. Haake, Phys. Rev. *A4*, 302 and 854 (1971).
4. N.E. Rehler and J.H. Eberly, Phys. Rev. *A3*, 1735, (1971).
5. F.T. Arecchi and E. Courtens, Phys. Rev. *A2*, 1730 (1970).
6. R. Bonifacio and L.A. Lugiato, Phys. Rev. *A11*, 1507 (1975) and *12*, 587 (1975).
7. E. Ressayre and A. Tallet, Phys. Rev. Lett. *37*, 424 (1976); C.T. Lee, Phys. Rev. *A14*, 1926 (1976).
8. R. Friedberg and S.R. Hartmann, Phys. Lett. *37A*, 285 (1971), and Phys. Rev. *A13*, 495 (1976).
9. R. Saunders, R.K. Bullough and F. Ahmad, J. Phys. A: Math. Gen. *8*, 579 (1975).
10. N. Skribanowitz, I.P. Herman, J.C. MacGillivray, and M.S. Feld, Phys. Rev. Lett. *30*, 309 (1973); I.P. Herman, J.C. MacGillivray, N. Skribanowitz and M.S. Feld, in *Laser Spectroscopy*, R.G. Brewer and A. Mooradian, editors (Plenum, NY, 1974).
11. J.C. MacGillivray and M.S. Feld, Phys. Rev. *A14*, 1169 (1976).
12. The shape-factor can be written as $\mu = \xi(3\lambda^2/8\pi A)$. For $\mathcal{N} \gg 1$ the parameter ξ is constant, $\xi = 1$, so that $\tau_R = (8\pi\tau_0/3\lambda^2 Ln)$. For $\mathcal{N} = 1$ ξ is calculated from ref. 4, eq. (5.16). The result is $\xi = 0.48$, 0.67, 0.79, 0.85, 0.86, and 0.90 for L = 0.5, 1.0, 2.0, 3.0, 5.0 and 10.0 cm. In view of the unavoidable experimental uncertainty in $\mathcal{N}$ the value $\xi = 1$ is used throughout in this paper.
13. (a) H.M. Gibbs in *Proceedings of the Meeting on Cooperative Effects*, Redstone Arsenal, December 1976, C.M. Bowden, D. Howgate, and H. Robl, editors, (Plenum, NY, 1977); (b) Q.H.F. Vrehen, ibidem.
14. S. Svanberg and S. Rydberg, Z. Physik *227*, 216 (1969); P.W. Pace and J.B. Atkinson, Can. J. Phys. *53*, 937 (1975); O.S. Heavens, J. Opt. Soc. Am. *51*, 1058 (1961).
15. Q.H.F. Vrehen, H.M.J. Hikspoors and H.M. Gibbs, Phys. Rev. Lett. *38*, 764 (1977).
16. S. Haroche, J.A. Paisner, and A.L. Schawlow, Phys. Rev. Lett. *30*, 948 (1973).
17. J.H. Eberly, Lett. Nuovo Cimento *1*, 182 (1971); I.R. Senitzky, Phys. Rev. Lett. *35*, 1755 (1975).
18. J.C. MacGillivray, private communication.

FAR-INFRARED SUPERRADIANCE IN METHYL FLUORIDE *

A.T. Rosenberger, S.J. Petuchowski and T.A. DeTemple

University of Illinois, Urbana, Illinois

1. INTRODUCTION

The cooperative spontaneous emission of radiation, or super-radiance [1], should be observable from initially fully inverted molecular systems which have high gain and are either 1) shorter than a cooperation length [2], or 2) inverted by means of swept excitation [3]. The emission from an extended cylindrical sample should occur as delayed pulses, the intensity, width and delay of which depend on the density ρ of excited molecules and on the length L and cross-sectional area A of the sample. The dependence of the delay t_0, width Δt, and peak intensity I of the superradiant pulses on the characteristic time T_S is as follows:

$$t_0 \propto T_S , \tag{1}$$

$$\Delta t \propto T_S , \tag{2}$$

$$I \propto N/T_S , \tag{3}$$

where $T_S \equiv 3\tau_N$ and [4]

$$\tau_N = T_{sp}/\mu N . \tag{4}$$

Here T_{sp} is the spontaneous lifetime, $N = \rho AL$ is the number of coop-erating molecules, and μ is a wavelength- and geometry-dependent shape factor [4]:

$$\mu = 3\lambda^2/8\pi A , \quad (F \gg 1; \text{ disk}) \tag{5}$$

$$\mu = 3\lambda/8L \, , \qquad (F \ll 1; \text{ needle}) \tag{6}$$

where the two expressions for μ apply in the limits of large and
small Fresnel number $F \equiv 2A/\lambda L$. This implies qualitatively different
geometrical dependences of the pulse characteristics at constant
density in the large-F and small-F limits:

	disk	needle
$t_o \propto$	L^{-1}	A^{-1}
$\Delta t \propto$	L^{-1}	A^{-1}
$I \propto$	AL^2	$A^2 L$

For a system of Fresnel number of order one, an intermediate geometrical behavior might be expected.

The far infrared should be a regime well suited to the observation of superradiance, because the radiative lifetimes of rotational transitions are long, the gains on these transitions can be
high, and the possibility of swept excitation by optical pumping
exists. The first observation of superradiance, in fact, took place
on FIR rotational transitions in HF [5], and the present work on
CH_3F is, in a sense, its complement. In the pressure range of
interest, the FIR emission from CH_3F is homogeneously broadened.
In spite of this, the superradiant pulses extract a large fraction
of the energy available at the time of emission, and so we claim
that the process involved is more correctly labeled "strong superradiance" than "limited superradiance" [7]. As can be seen from
Table 1, the conditions [7] for "strong superradiance" are satisfied:

$$T_s \ll T_2 \, , \tag{7}$$

$$\alpha_o L \gg |\ln\phi_o| \, . \tag{8}$$

In addition, $T_1 = T_2$ for the FIR transition, so the analysis of
Ref. 3 for swept-gain superradiance applies, provided that the pump
width is small enough and the sample long enough.

In Table I are defined and listed the various times and other
parameters associated with a particular set of experimental conditions. The experiments and results are described in Section 2, the
results are discussed and compared with theoretical predictions in
Section 3, and in Section 4 the assumptions of the theory and the
implications of the experimental observations are summarized.

Table 1. Experimental parameters for the case $L = 3.5$ m, $A \simeq 11$ cm^2, and $p = .250$ torr.

λ	496 μm
ρ	2.61×10^{13} cm^{-3}
N	1.01×10^{17}
$\alpha_o L$	125
ϕ_o	2.6×10^{-8}
$|\ell n \phi_o|$	17.5
R_e	-4.3×10^{-3}
L_c	53 cm
T_{sp}	229 sec
T_s	0.256 nsec
T_2	32 nsec
T_2^*	366 nsec

Above, ρ is the excited density and $N = \rho AL$ the excited number at pump cutoff, $\alpha_o L$ is the field gain per pass, $R_e = (N_2 - N_1)_e/N$ is the equilibrium population fraction, $L_c = (8\pi^2 c T_{sp}/3\rho\lambda^2)^{\frac{1}{2}}$ is the cooperation length, and $T_s = 8\pi T_{sp}/\rho\lambda^2 L$ is the superradiant lifetime.

2. EXPERIMENTS

The energy level diagram in Fig. 1 and the experimental apparatus of Fig. 2 have been described in detail in Ref. 6, and only the main points of interest will be repeated here. The pump in Fig. 1 can be preferentially absorbed in either K (1 or 2) manifold, although K = 2 is more probably dominant because of the larger matrix element. The resultant FIR emission will be at 496 μm with a 40 MHz frequency difference between the two possible FIR lines. Rotational (ΔJ) relaxation yields a 40 MHz/torr homogeneous linewidth and a nonzero equilibrium population difference. (Table I).

The CO_2 TEA Laser in Fig. 2 has been mounted in an Invar frame for improved stability and emits a single-mode, 120 nsec, 1.2 MW pulse which is truncated, 35 nsec before its peak, to a width of about 30 nsec [8]. This early cutoff extends the pressure range in which the pump and FIR pulses do not temporally overlap. The positions of the absorber and the detector system may be interchanged to allow observation of FIR emission in the direction opposite that of the propagation of the pump.

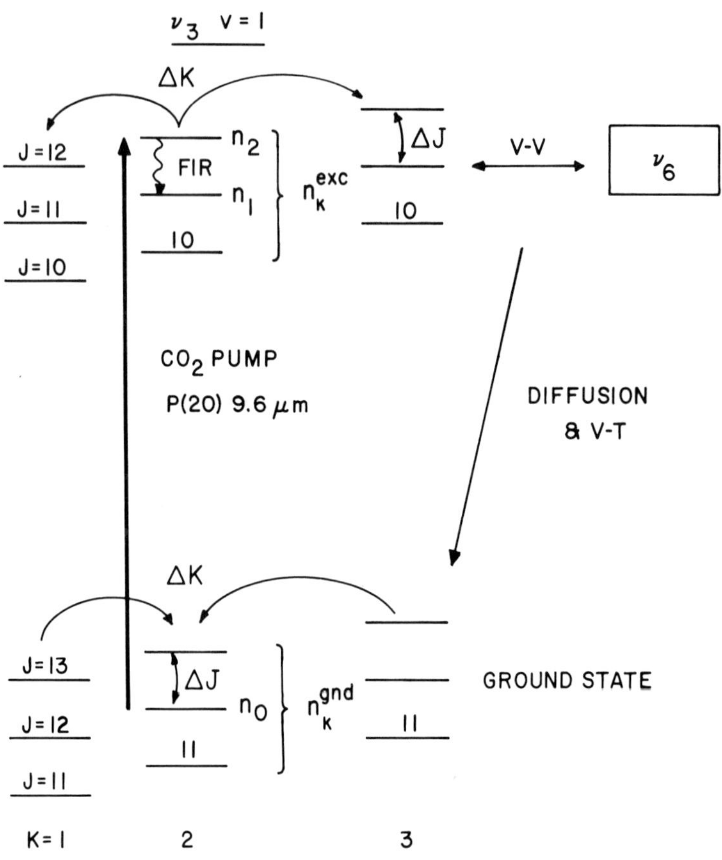

Figure 1. Partial energy level diagram for CH_3F pumped by a CO_2 laser. The two fastest relaxation rates are $\Gamma_J = 1.3 \times 10^8 sec^{-1}torr^{-1}$ and $\Gamma_K = 1.2 \times 10^7 sec^{-1}torr^{-1}$. The fractional populations in the initial ground states, $f(J,K)$, are $f(12,1) = 6.89 \times 10^{-3}$, $f(12,2) = 6.49 \times 10^{-3}$.

The cutoff pump pulse and two typical FIR superradiant pulses are shown in Fig. 3 in synchronous oscilloscope traces. The low tail on the pump pulse is the result of a slignt impedance mismatch between a pulse amplifier and the oscilloscope. The FIR pulses are multiple traces to show shot-to-shot fluctuations; notice that there is no ringing (ringing has not been observed on any FIR pulses occuring after cutoff of the pump) and that the pulses are asymmetric. The qualitative pressure (hence density, since the pump is saturating the IR absorption) dependence is also evident.

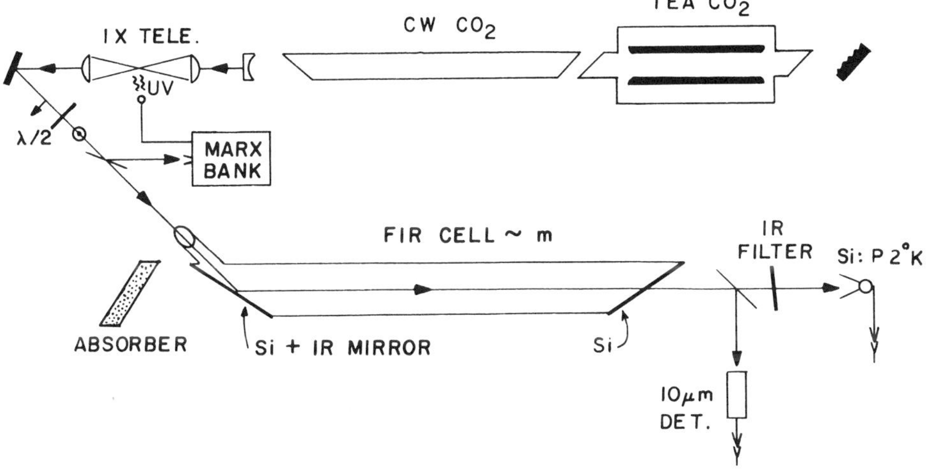

Figure 2. Experimental apparatus.

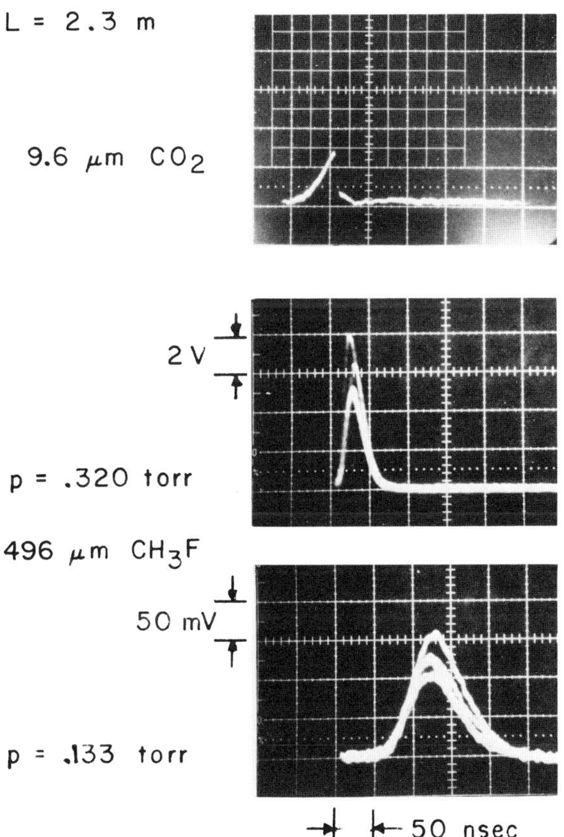

Figure 3. Typical pump (upper photo) and FIR pulses. All scales are 50 nsec/division, and the displays are synchronized.

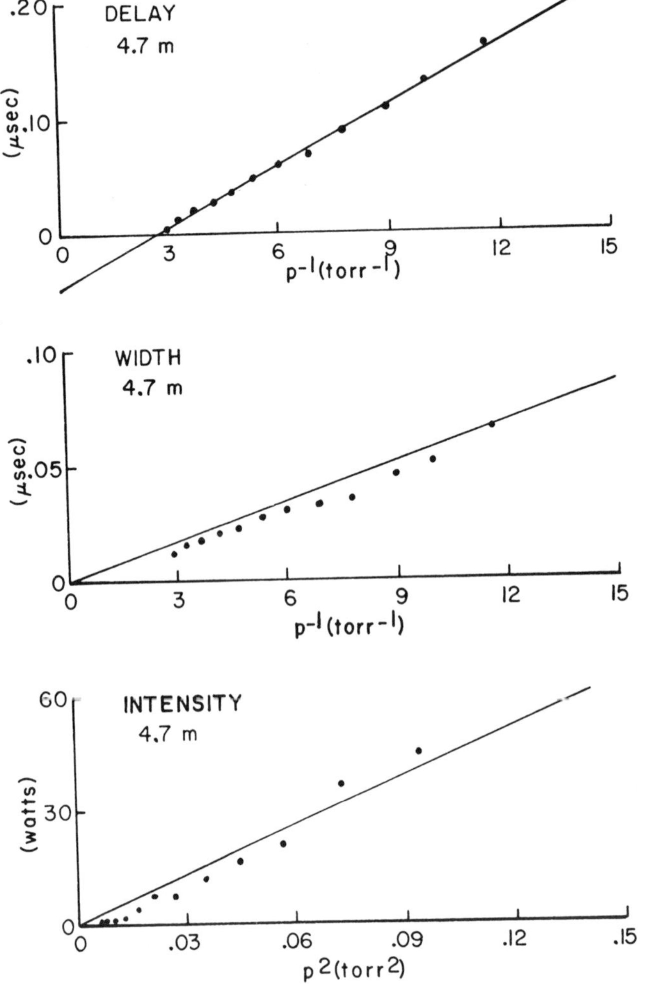

Figure 4. Results for a sample with $L = 4.7$m, $A \simeq 7$ cm^2. The straight lines are least-square fits to the data; their slopes are $pt_0 = 17.7$ nsec-torr, $p\Delta t = 5.82$ nsec-torr, $I/p^2 = 421$ W/torr^2. The homogeneously broadened regime lies below $p^{-1} = 12$ torr^{-1}, or above $p^2 = 0.007$ torr^2.

The pressure dependence is better shown in Fig. 4, where the data plotted are for a cell 4.7 m long and the lines are least-square fits to the data. The linear variation of delay and width with inverse pressure, and of the average peak intensity with pressure squared, is evident as expected, with slight discrepancies which are qualitatively understood: negative delay at high pressure, and deviation from linear behavior at low pressure. The delay is measured from the cutoff of the pump pulse, so that the finite width of the pump and the onset of swept gain at higher pressure are responsible for the negative delay and deviations in width and intensity [9]. The onset of Doppler broadening below 0.08 torr is thought responsible for the deviation at low pressures. All the pulses are less than T_2 in width.

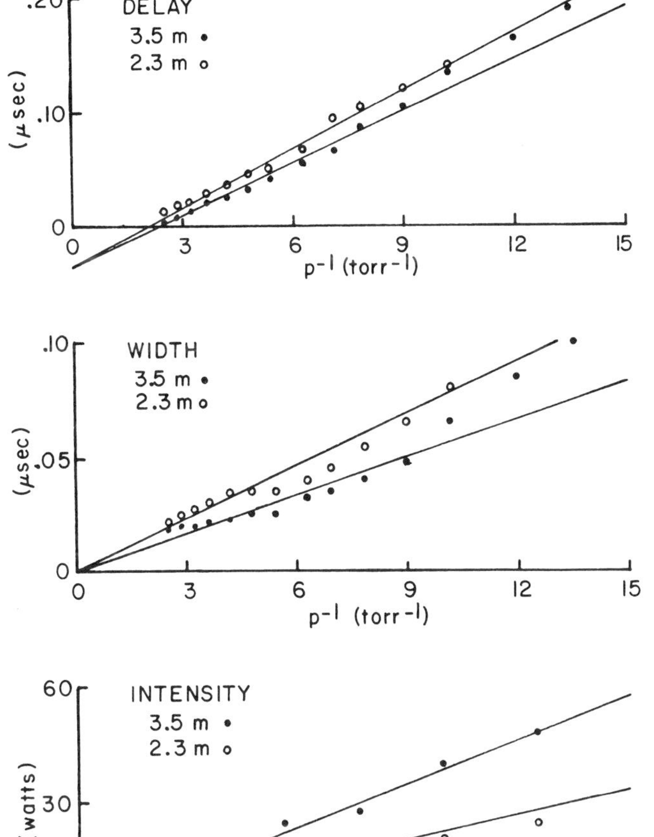

Figure 5. Results for samples with $A = 11$ cm^2, $L = 3.5$ and 2.3 m. The least-square slopes are $pt_0(3.5m) = 15.4$ nsec-torr, $pt_0(2.3m) = 16.9$ nsec-torr; $p\Delta t(3.5m) = 5.08$ nsec-torr, $p\Delta t(2.3m) = 6.99$ nsec-torr; $I/p^2(3.5m) = 384$ W/torr, $I/p^2(2.3m) = 212$ W/torr^2.

Comparison of the results for two samples identical except for length is shown in Fig. 5. The qualitative behavior is again the same, but a comparison of the slopes of the least-square linear fits for the two sample lengths shows a behavior intermediate between that of the disk and that of the needle, as given following Eq. (6). The ratio of the intensity of the forward pulse to the intensity of the backward pulse was investigated as a function of pressure, but the results were inconclusive. However, a similar measurement on a 5.9 m sample confirmed the findings of Ref. 9: that the ratio is nearly equal to one at low pressures and becomes large at higher pressures, when the excitation develops a swept character. One would expect a ratio of one at even higher pressures in a shorter cell.

3.   COMPARISON WITH THEORY

In comparing our results with theory, pump effects and level degeneracy are neglected and a Maxwell-Bloch approach with an input field fluctuating randomly in phase to simulate spontaneous and black-body emission, much as in Ref.[7], is used. The coupled equations are written in complex form as follows:

$$\frac{\partial}{\partial t} P = -P/T_2 + ER \ , \tag{9}$$

$$\frac{\partial}{\partial t} R = -(R-R_e)/T_1 - Re(EP^*) \ , \tag{10}$$

$$\frac{\partial}{\partial x} E = -\kappa E + \alpha P \ , \tag{11}$$

where $P = -\mathcal{P}/\rho\mu_{12}$, $E = \mu_{12}\varepsilon/\hbar$, with μ_{12} being the dipole matrix element between the two FIR levels and $\mathcal{P}$ and ε the complex polarization and electric field amplitudes; $R = (N_2-N_1)/N$, the normalized population difference, having an equilibrium value $R_e < 0$. The times T_1 and T_2 are equal for the transition under consideration, $\alpha = (T_SL)^{-1}$ (T_S is here, and in the following, evaluated in the disk limit), and κ is a linear loss to approximate diffraction. It was found that a reasonable fit could be obtained by choosing a credible value for either κ or ϕ_0 and then treating the other as a free parameter. Two representative examples are given in Table 2. The observed delay listed in Table 2 is that delay measured not from pump cutoff but from the $p^{-1} = 0$ intercept and thus is an overestimation.

The Maxwell-Bloch equations (9)-(11) were modified to treat propagation in both directions, qualitatively confirming the pressure dependence of the forward/backward intensity ratio observed in longer cells and discussed in more detail in Ref.[9]. The ratio was calculated to increase with increasing pressure. In the numerical calculations, the asymmetry of the resulting pulses matches that observed, and ringing is predicted only for unphysically large densities or initial tipping angles [10].

If one assumes no spatial variation in the electric field, and $R_e = 0$, the intensity predicted by Eqs. (9)-(11) has the form

$$I(t) = (\hbar\omega_0\mu N^2/4T_{sp}) \ z^2 \mathrm{sech}^2[\beta(1-z) - \alpha], \tag{12}$$

where $z = \exp(-t/T_2)$, $\beta = T_2/2T_S$, and $\alpha = \ell n(2/\phi_0)$. This functional form applies both to the pure Dicke case $L \ll L_c$ and to the steady-state "swept superradiant" case [3]. To see the connection with the

Table 2. Sample comparison of experiment with theory for the case $L = 3.5$ m, $A \simeq 11$ cm^2.

p(torr)		t_o(nsec)	Δt(nsec)	I(W)
	Observed	125±6	41±5	6±2
.123	Calculated (1)	109	49	10
	Calculated (2)	99	42	13
	Observed	62±5	22±3	24±8
.250	Calculated (1)	56	25	32
	Calculated (2)	53	23	38

(In this table, the delay t_o is measured from the vertical intercept of t_o vs. p^{-1} in Fig. 5 and is therefore an overestimate, as noted in the text).

Calculation (1)

$$\kappa = 2.15 \times 10^{-3} \text{cm}^{-1}$$
$$\phi_o(.123) = 2.3 \times 10^{-8}$$
$$\phi_o(.250) = 2.6 \times 10^{-8}$$

Calculation (2)

$$\kappa = 3.00 \times 10^{-3} \text{cm}^{-1}$$
$$\phi_o(.123) = 6.2 \times 10^{-8}$$
$$\phi_o(.250) = 4.4 \times 10^{-8}$$

latter, note that by using the disk expression for μ, Eq. (5), and replacing L by κ^{-1}, an effective loss length, one may recover the intensity predicted for the steady-state case with, of course, t being the retarded time [3]. This applies to $L \gg L_c$ and $L \gg \kappa^{-1}$. It is reasonable to seek, then, a fit to this form for the case $L \gtrsim L_c$ if an effective length L_{eff} is used in Eq. (14); this is, in fact, possible, and with the experimental data a fit is found for $L_{eff} \simeq$ .225 L. This may be compared with the Maxwell-Bloch prediction that shows spatial variation in the fields and polarization: only the forward 1/5 to 1/4 of the sample contributes appreciably to the emission.

4. DISCUSSION

It has been observed that the FIR emission is sensitive to the exact nature of the shape and frequency of the pump pulse, and this is reflected in a sensitivity of the Maxwell-Bloch model to the specific choice of initial conditions. A prescription for ϕ_0 is not yet evident, but it appears to depend more on $N^{-\frac{1}{2}}$ than $(\mu N)^{-\frac{1}{2}}$ in these experiments [2], [3], [4]. The field losses arrived at in Table 2 are not consistent with the FIR evolving as a single mode (EH_{11} guided wave loss $< 5 \times 10^{-5} cm^{-1}$) but are comparable to multimode diffraction losses ($\kappa \sim 5 \times 10^{-3} cm^{-1}$), and to absorption of the thermally populated levels in unexcited regions ($\kappa \sim 10^{-3} cm^{-1}$) [11]. From these studies, the experimental uncertainties are such that A, ϕ_0, κ and perhaps ρ can be chosen consistently to yield calculated results in good agreement with experiments. Infrared nutation due to the large-area coherent pump also may play a role and should be included in the semiclassical model, as should the finite width of the pump pulse. A first step to do this has been taken [9].

To summarize, FIR superradiance in homogeneously broadened CH_3F is "strong" superradiant emission of single asymmetric pulses of width less than T_2, showing the expected dependence on density and exhibiting shot-to-shot fluctuations which are probably due to variation in the pump. The geometrical character of the samples is calculated and observed to be intermediate between the two limits of large and small Fresnel number. The ratio of forward to backward intensities supports the hypothesis of the onset of swept gain in long cells at high pressures. A Maxwell-Bloch approach, with either ϕ_0 or linear loss κ a free parameter, allows a quantitative fit to the observations, and, by including bidirectional propagation, the pressure dependence of the forward/backward ratio is qualitatively reproduced for very long cells. The relation to swept excitation is also shown in the effective-length fit of what was called in Ref. 6 the "semiclassical mean-field theory", shown here in Eq. (12). The possibility thus exists in CH_3F of observing superradiance in regimes ranging from $L < L_c$ to the steady-state swept case.

Acknowledgements

The authors wish to thank Drs. R. Bonifacio, C.M. Bowden, R.K. Bullough, F.A. Hopf, and A. Zardecki for stimulating conversation and correspondence, and P. Norton for generous loan of the Si detector.

*Research supported by U.S. Army Research Office, Durham, N.C.

References

1.   R.H. Dicke, Phys. Rev. *93*, 99 (1954).
2.   R. Bonifacio and L.A. Lugiato, Phys. Rev. A *11*, 1507 (1975).
3.   R. Bonifacio, F.A. Hopf, P. Meystere, and M.O. Scully, Phys. Rev. A *12*, 2568 (1975).
4.   N.E. Rehler and J.H. Eberly, Phys. Rev. A *3*, 1735 (1971).

5. N. Skribanowitz, I.P. Herman, J.C. MacGillivray, and M.S. Feld, Phys. Rev. Lett. *30*, 309 (1973).
6. A.T. Rosenberger, S.J. Petuchowski, and T.A. DeTemple, in *Cooperative Effects in Matter and Radiation*, C.M. Bowden, D.W. Howgate, and H.R. Robl, editors (Plenum, N.Y., 1977)p. 15.
7. J.C. MacGillivray and M.S. Feld, Phys. Rev. A *14*, 1169 (1976).
8. E. Yablonovitch, Phys. Rev. A *10*, 1888 (1974); H.S. Kwok and E. Yablonovitch, Appl. Phys. Lett. *27*, 583 (1975).
9. J.J. Ehrlich, C.M. Bowden, D.W. Howgate, S.H. Lehnigk, A.T. Rosenberger, and T.A. DeTemple, "Swept-gain Superradiance in CO_2-Pumped CH_3F", this Volume, p. 923.
10. This is a consequence of the relatively small value of T_2; the ringing is more effectively suppressed for $t_o = 2T_2$, as in this experiment, than it is for $t_o = 2T_2^*$, as shown in the calculations of Ref. 7 and those done by R. Saunders and R.K. Bullough in *Cooperative Effects in Matter and Radiation*, C.M. Bowden, D.W. Howgate, and H.R. Robl, editors (Plenum, N.Y., 1977)p. 209.
11. E.A.J. Marcatili and R.A. Schmeltzer, Bell Syst. Tech. J. *43*, 1783 (1964).

LEVEL-DEGENERACY EFFECTS IN SUPERRADIANCE THEORY

Anne Crubellier and Marie-Gabrielle Schweighofer

C.N.R.S. II, Orsay, France

1. INTRODUCTION

Until now, all superradiance studies have been concerned with
non-degenerate levels. So there is an interesting problem remaining
to be solved: the influence of level-degeneracy on superradiance.
Because of the coherent nature of the phenomenon, important interference effects can be expected: experimental evidence for the
existence of such effects has been given in a recent work [1], in
which quantum beats are observed on superradiant signals even where
no coherent superposition of states has been initially prepared.

In this paper, a first approach to this problem of the influence
of level-degeneracy on superradiance is given in the frame of the
"small-system" model [2], which is the simplest quantum markovian
model. The imperfections of this model are numerous and well-known.
Let us recall only the fact that when the atoms are confined in a
very small volume, the dipole-dipole interactions would, in practice,
destroy superradiance, [3] and, as has been proved [4], the physical
conditions for observing markovian superradiance would be almost
impossible to obtain. However, it is known that for two-level (non-
degenerate) systems, this model describes qualitatively the phenomenon, so long as the non-markovian effects are not too important (in
particular, when no "ringing" appears). Moreover, the relative simplicity of the formalism is convenient if one wants to understand
the origin of the different effects which are due to level-degeneracy
and to describe them qualitatively. Finally, the most important
point in favor of such a study is probably that a quantum equation
is needed if one wants to account for the beginning of the superradiance phenomenon.

In this paper we first describe the generalization of the
formalism of the small-system model to the case of a collection of
N atoms with two degenerate levels of angular momenta j and j' (j
for the upper level), connected by an electric or magnetic dipole
transition. We give the form of the master equation and expressions
for the expectation values of the radiated field (section 2), then
we evoke some problems related to the choice of a basis of collective
states (section 3). A comparison between this master equation and
similar ones corresponding to the case of two, or several, non-
degenerate levels allows us to show that the difference between the
degenerate and non-degenerate two-level cases is due to two types
of effects. The first effect is present in the case of more than
two non-degenerate levels and consists in a competition between
different transitions sharing a common upper or lower level; the
second effect is quite specific to level degeneracy and consists in
interferences between the transitions of the collective system
having the same frequency and the same polarization. These two
types of effects are qualitatively studied in sections 4 and 5.
As an application of this formalism, we have studied in more detail
the case of a $\frac{1}{2} \rightarrow \frac{1}{2}$ transition. This study has been done with the
help of group theory which allows some important formal simplifi-
cations, and it will be published elsewhere in more detail. In this
paper (section 6) we describe rapidly the main results: in particular, we solve numerically the master equation for various excitation conditions and discuss some of the results.

2. BASIC EQUATIONS FOR COLLECTIVE SPONTANEOUS EMISSION

The basic assumptions of the small-system model are well-known
[1]. One considers a collection of N atoms confined in a volume
smaller than the wavelength of the transition, and studies the
spontaneous evolution of this system in the markovian approximation,
that is with the assumption that the emitted photons escape rapidly
from the active volume and do not interact with the atoms again.

For two-level (non-degenerate) atoms this evolution is given
by the well-known master equation:

$$\dot{\rho} = \Gamma \; \{R\rho R^{+} - \frac{1}{2} \; [R^{+}R\rho + \rho R^{+}R]\} \quad , \tag{1}$$

where ρ is the reduced atomic density operator, Γ the transition
probability. The energy-decreasing and -increasing operators R and
R^{+} are sums of monatomic operators:

$$R = \sum_{\alpha=1,N} r_{\alpha} = \sum_{\alpha=1,N} |->_{\alpha} \, {}_{\alpha}<+| \tag{2}$$

$|+>$ and $|->$ being, respectively, the monatomic upper and lower states.

The collective character of the emission is interpreted in this model as being due to the fact that the atoms are so close to one another that it is impossible to tell which atom has emitted a photon. The mathematical consequence is that R or R^+ is written as a sum of monatomic operators, so Eq. (1) contains products of terms r_α and r_β^+ with $\alpha \neq \beta$ which appear as interatomic interference terms [4], these interference effects being directly responsible for the build-up of the superradiant signal.

It is straightforward to generalize this master equation to the case of atoms with several non-degenerate levels:

$$\dot{\rho} = \sum_{ij} \Gamma_{ij} \{ R_{ij} \, \rho R_{ij}^+ - \frac{1}{2} [R_{ij}^+ \, R_{ij} \, \rho + \rho R_{ij}^+ \, R_{ij}] \} \quad . \qquad (3)$$

The sum runs over the transitions $i \to j$ (i being the upper level), the Γ_{ij}'s are the corresponding transition probabilities and the operators R_{ij} and R_{ij}^+ are energy-decreasing and -increasing operators which are sums of monatomic operators. One then has:

$$R_{ij} = \sum_{\alpha=1,N} |j\rangle_\alpha \, {}_\alpha\langle i| \, , \qquad (4)$$

$|i\rangle$ and $|j\rangle$ being monatomic states.

For generalizing Eq. (1) to the case of atoms with degenerate levels one must take into account the presence of "degenerate transitions", that is, transitions between pairs of quantum collective states which are resonant with the same electromagnetic field mode. The manner of obtaining the master equation for the case of atoms with two degenerate levels of angular momenta j and j' (j for the upper level) connected by an electric or magnetic dipole transition, is detailed in a preceding paper [5]. We recall here the form of this master equation:

$$\dot{\rho} = \Gamma \sum_q \{ R_q \rho R_q^+ - \frac{1}{2} [R_q^+ \, R_q \, \rho + \rho R_q^+ \, R_q] \} \quad , \qquad (5)$$

where Γ is the decay constant of the upper level. The sum runs over the values 1, 0 and -1 of q which characterize the three polarizations σ_+, π and σ_- of the emitted photons. The operators R_q and R_q^+ are linear combinations of the different collective energy-decreasing and -increasing operators corresponding to all different transitions of polarization q (see Ref. [5]).

A remark must be made concerning Eq. (5). Because of the geometrical symmetry of the model and of the isotropic nature of spontaneous emission, this equation is invariant under rotations; the system will radiate in all directions and with all polarizations.

Consequently this equation does not apply to the experimental situation in which the observation of superradiance was made [6,7]. However, the aim of this work is to understand and to describe qualitatively the effects due to level degeneracy and from that point of view, this simple model has been found to be sufficient. (Let us add that the rotational invariance allows formal simplifications.)

The evaluation of expectation values for the radiated field is well-known for the two-level (non-degenerate) case; the generalization to the considered case is not difficult (see Ref. [5]), particularly the intensity radiated in all directions with a polarization $\underline{\varepsilon}$ which is given by:

$$I_{\underline{\varepsilon}} = I_0 \; <(\underline{R} \cdot \underline{\varepsilon})^+ \; (\underline{R} \cdot \underline{\varepsilon})> \; . \qquad (6)$$

Here $<X> = \text{tr}(\rho X)$ is the mean value of operator X, I_0 is the intensity radiated in all directions and with all polarizations by an isolated atom, and $\underline{R}$ is the vector operator whose spherical components are R_q.

It is interesting to compare the three master equations (1), (3) and (5) corresponding respectively to two non-degenerate levels, several non-degenerate levels and two degenerate levels. They have essentially the same structure and they all contain some energy-decreasing and energy-increasing operators R and R^+, which indicate the existence of interatomic interferences and show the collective nature of the phenomenon. However in Eqs. (3) and (5) some new features appear.

In Eq. (3), one has to consider one pair of R and R^+ operators for each allowed transition and the right-hand side contains a sum over these transitions. It is easy to show that, if these transitions do not share any common (upper or lower) state, the operator ρ can be written as a sum of terms corresponding to each transition, each having independent evolution. If, however, some transitions share a common level, this is impossible and superradiance on different transitions will influence one another. We call this phenomenon "competition effects" and we shall discuss it qualitatively in section 4.

These effects also will be present in Eq. (5) since it consists of a sum over the Zeeman transitions, but there is a difference from the preceding case in that the operators R_q and R_q^+ already contain from their definition the sum over the different Zeeman transitions of given polarization q: this sum therefore plays exactly the same role as the sum over the atoms and it expresses the same physical phenomenon. The emitted photons of a given polarization are indiscernible not only if they come from different atoms but also if they correspond to different Zeeman transitions. This effect is the most

specific consequence of level degeneracy and leads to supplementary
interference effects which will be discussed in section 5.

3. COLLECTIVE STATES

In order to use the master equation (5) one has to define a
basis of collective states. This problem is more complicated in the
degenerate case than for two non-degenerate levels. It is shown in
Ref. [5] that the Lie group - the cooperation group - which plays
the same role as the Dicke's basis $SU(2)$ group [1] is a $SU(2j+2j'+1)$
group. The operators R_q and R_q^+ are infinitesimal operators of this
group and consequently do not mix states of different irreducible
representations. Therefore these representations play the same role
as the cooperation numbers r of the Dicke model. Their conservation
has the same interpretation as the conservation of r. The system
evolves into coherent superpositions of products of monatomic states
so that interatomic interference may appear. When the interferences
are constructive, such a coherent superposition is a superradiant
state in Dicke's sense; for example, for two atoms with two non-
degenerate levels the interferences are constructive for the super-
radiant symmetric state $(1/\sqrt{2})\{|+-> + |-+>\}$, which radiates twice more
than each constituting state, and the interferences are destructive
for the subradiant antisymmetric state $(1/\sqrt{2})\{|+-> - |-+>\}$ which does
not radiate at all. However, whereas the states of the representations of $SU(2)$ are easy to label, the same is not true for the representations of $SU(2j + 2j' + 2)$, especially for large values of j and
j', except for the symmetric representation $\{N\}$.

Therefore, for the sake of simplicity, we shall restrict ourselves in the following to the symmetric representation of the
cooperation group, that is, to collective states that are symmetric
under the permutations of atoms. In the two non-degenerate level
case, this corresponds to the value $N/2$ of r, which is always obtained if the initial population inversion is complete; however,
in the degenerate case, collective states corresponding to complete
population inversion are not necessarily symmetric. The above
assumption however is rather natural, because of the indiscernibility
of the atoms and, in the applications (see section 6), we shall
always assume that the initial density matrix is defined by symmetric
states only.

Symmetric states may be written in the occupation number representation, which reads in the case of two levels of angular momenta
j and j':

$$|\ldots N_{jm} \ldots N_{j'm'} \ldots> \quad .$$

It must be noted that two physically significant quantum numbers are simply related to the occupation numbers. The first one, μ, is the equivalent of Dicke's quantum number m and characterizes the energy of the collection of atoms; in effect the eigenvalues of the atomic hamiltonian H_0 (see Ref. [5]) can be written $E = E_0\mu$ where E_0 is the energy difference between the two levels and μ is given by:

$$\mu = \frac{1}{2} \left[\sum_m N_{jm} - \sum_{m'} N_{j'm'} \right] \quad . \tag{7}$$

The second one, M, is the projection on the quantization axis of the total angular momentum $\underline{J}$ of the collection of atoms and one has:

$$M = \sum_m m\, N_{jm} + \sum_{m'} m'\, N_{j'm'} \quad . \tag{8}$$

4. COMPETITION EFFECTS

As shown before, these effects are due to the presence of several transitions sharing a common (upper or lower) state. In order to discuss them qualitatively we consider two cases of three-level atoms, the levels being connected by two transitions having either the same upper level (case (a)), or the same lower level (case (b)). They are schematically represented in Figs. 1a and 1b. The master equation for each case is given by Eq. (3). The cooperation group is a SU(3) group and the evolution of the system starting from a symmetric state, consists in cascading emission between the different collective symmetric states which are represented, in the occupation number representation, in Figs. 2a and 2b. The transition probability from a state ϕ to a lower state Ψ and corresponding to the emission of a photon on the transition $i \rightarrow j$, is given by:

$$\left| (\Psi | R_{ij} | \phi) \right|^2 \quad .$$

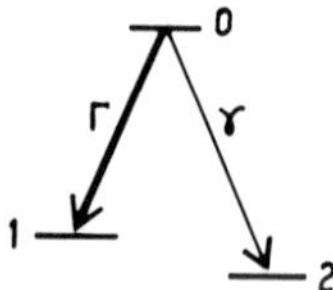

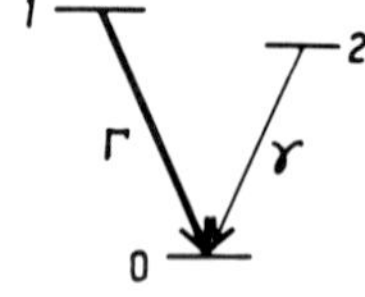

Figure 1a. Energy-level diagram of a single atom (case (a)).   Figure 1b. Energy-level diagram of a single atom (case (b)).

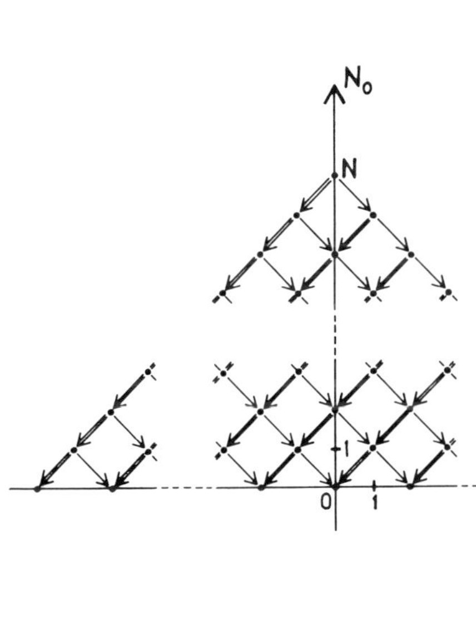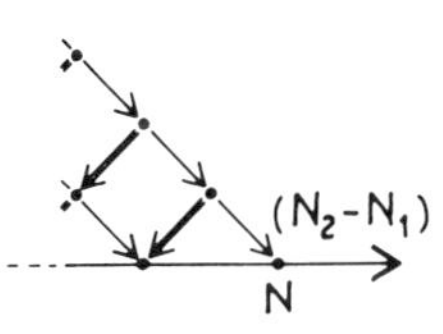

Figure 2a. Symmetric collective state diagram for N three-level atoms (case (a)); each point represents a state $|N_0 N_1 N_2\rangle$ with $N_0 + N_1 + N_2 = N$, the coordinates being N_0 and $(N_2 - N_1)$. The single and double arrows correspond respectively to the emission of a photon on the transitions $0 \to 2$ and $0 \to 1$.

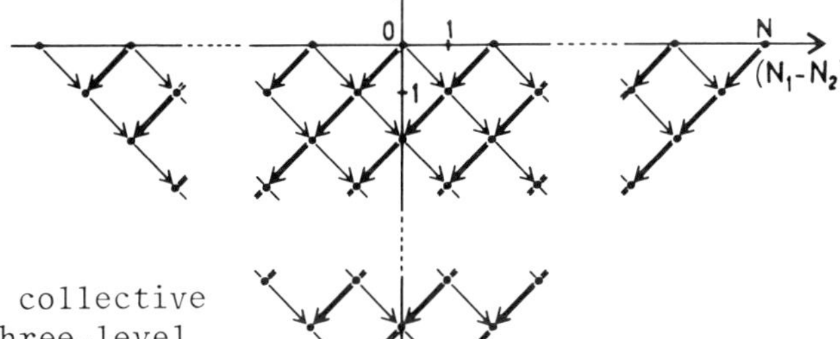

Figure 2b. Symmetric collective state diagram for N three-level atoms (case (b)); each point represents a state $|N_1 N_2 N_0\rangle$, with $N_0 + N_1 + N_2 = N$; the coordinates are N_0 and $(N_1 - N_2)$. The single and double arrows correspond respectively to the emission of a photon on the transitions $2 \to 0$ and $1 \to 0$.

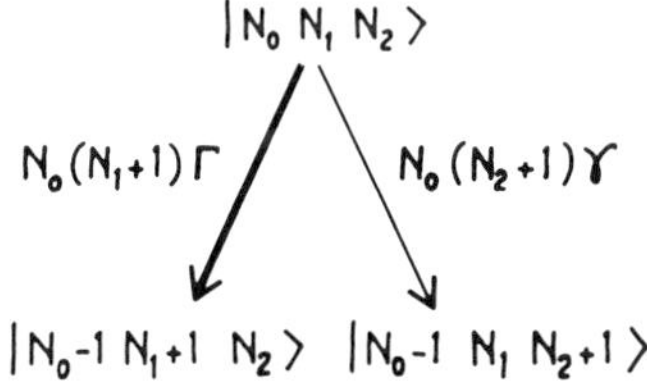

Figure 3a. Collective transition probabilities (case (a)).

Figure 3b. Collective transition probabilities (case (b)).

By expressing R_{ij} in terms of annihilation and creation operators
it is easy to obtain the general expressions for the different
transition probabilities of Figs. 2a and 2b, which are given in
Figs. 3a and 3b. The evolution of the system is very different for
for the two considered cases.

(a) Common Upper Level; Inhibition

The branching ratio between the two transitions corresponding
to the emission of the two sorts of photons at a given point of the
Fig. 2a is given by $(N_1+1)\Gamma/(N_2+1)\gamma$. If the transition $0 \to 1$ is
the most probable, that is, if $\Gamma > \gamma$, and if one starts from a
complete population inversion, this ratio is always favourable to
the emission of a photon on the transition $0 \to 1$, especially since
each cascade on this transition increases N_1. The process is then
cumulative and few photons will be emitted finally on the other
transition. In other words the appearance of superradiance on the
transition $0 \to 1$ empties the upper level before the appearance of
superradiance on the other transition, the latter being therefore
"inhibited".

(b) Common Lower Level; Initiation

In this case, the branching ratio of the emission probabilities
of the two sorts of photons at a given point of the Fig. 2b is given
by $(N_1\Gamma)/(N_2\gamma)$. If $\Gamma > \gamma$, this ratio is favourable to the transition
$1 \to 0$ only if N_2/N_1 is not too large; however each cascade on $1 \to 0$
decreases N_1, so that the emission on the other transition may become
more probable. It is then expected that superradiance on the transition $2 \to 0$ will be "initiated": it will appear sooner and larger
than if alone. The emission on $1 \to 0$ prepares coherent superpositions of states which are superradiant states as well for the transition $2 \to 0$ as for $1 \to 0$. Let us remark that this phenomenon does
not imply that a coherent superposition of the two upper levels has
been prepared; it also appears if the mean value of the initial
coherence between the two states is zero at $t = 0$.

5. INTERFERENCE EFFECTS

As shown in section 2, by interference effects we mean only
those effects which are specific to the level degeneracy; these are
due to the fact that photons emitted on different Zeeman transitions
corresponding to the same polarization are indiscernible. It is
interesting to discuss their consequences in detail, and to do so
we analyse Eq. (5) again.

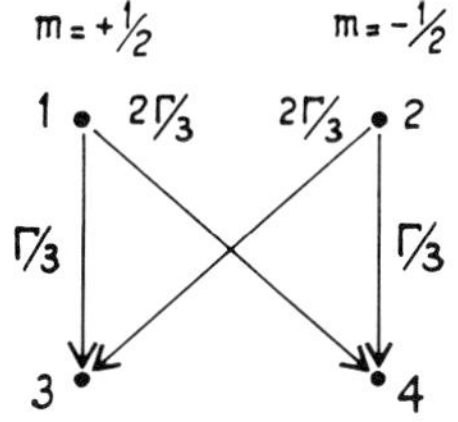

Figure 4. State diagram for an atom with two degenerate levels of angular momentum ½.

The matrix form of this equation can be written:

$$(\phi|\dot{\rho}|\Psi) = \Gamma \left\{ \sum_{q\bar{\phi}\bar{\Psi}} (\phi|R_q|\bar{\phi})(\Psi|R_q|\bar{\Psi})(\bar{\phi}|\rho|\bar{\Psi}) \right.$$

$$\left. - \frac{1}{2} \sum_{q\phi'} (\phi|R_q^+ R_q|\phi')(\phi'|\rho|\Psi) - \frac{1}{2} \sum_{q\Psi'} (\phi|\rho|\Psi')(\Psi'|R_q^+ R_q|\Psi) \right\} .$$

$$(9)$$

The states $\bar{\phi}$ and $\bar{\Psi}$ belong respectively to energy levels just above ϕ and Ψ to which they respectively lead by emission of a real photon of polarization q; the states ϕ' and Ψ' belong to the same energy levels as respectively ϕ and Ψ to which they can lead by emitting and reabsorbing a virtual photon of polarization q.

First, since R_q is a sum over different transitions, the emission of a photon of given polarization will in general lead to a linear superposition of several states, at least in the occupation number representation. This can be made more precise in the case of atoms with two levels of angular momentum ½ (see Fig. 4), for which the phenomenon appears using π photons only. One has:

$$R_0|N_1 N_2 N_3 N_4> \to \sqrt{N_1(N_3+1)}\ |N_1-1\ N_2\ N_3+1\ N_4>$$

$$+ \sqrt{N_2(N_4+1)}\ |N_1\ N_2-1\ N_3\ N_4+1> . \qquad (10)$$

The particular case of two such atoms is illustrated in Fig. 5. The consequence of this fact is that Eq. (5) will mix the evolution of the populations with that of some coherences, at least in the considered basis. This problem is studied in more detail in Ref. [5] and we show that it is in fact true whatever the choice of the basis.

Secondly, the absorption and re-emission of a photon of a given polarization will also lead, in general, to superposition of states. In the case of two levels of angular momentum ½ one has:

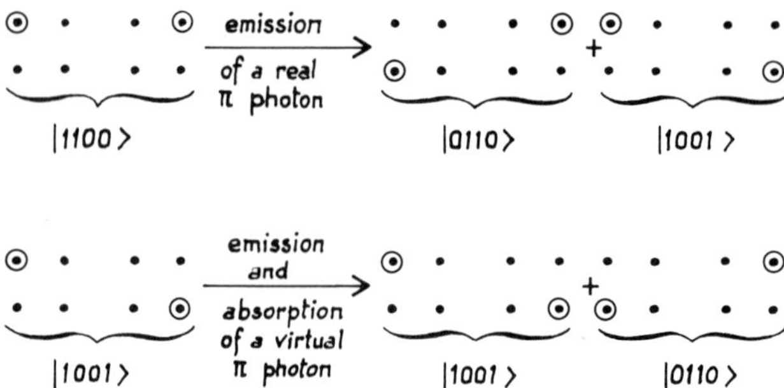

Figure 5. Elementary level-degeneracy interference processes for atoms with two levels of angular momentum ½ (the represented states are symmetric).

$$R_0^+ R_0 \left| N_1 N_2 N_3 N_4 \right\rangle \rightarrow \left[N_1(N_3+1) + N_2(N_4+1) \right] \left| N_1 N_2 N_3 N_4 \right\rangle$$

$$+ \sqrt{N_1(N_2+1)(N_3+1)N_4} \left| N_1-1 \ N_2+1 \ N_3+1 \ N_4-1 \right\rangle$$

$$+ \sqrt{(N_1+1)N_2 N_3(N_4+1)} \left| N_1+1 \ N_2-1 \ N_3-1 \ N_4+1 \right\rangle \ . \tag{11}$$

(See also Fig. 5 for the case of two atoms.) So there appears an interesting phenomenon: due to interference effects the population of an upper state may increase, although only markovian processes are considered and no real absorption is taken into account. Another consequence, mainly formal, is that the occupation number representation is not the one best adapted to the solution of the master equation, since the derivative of a given density matrix element is mixed with other density matrix elements of the same energy. We show in Ref. [5] that a better basis can be obtained, in which the operator $\sum_q R_q^+ R_q$ is diagonal, and for which such a mixing does not appear.

6. APPLICATION TO ½ → ½ TRANSITION

As an illustration of the preceding formalism we have studied the case of atoms with two degenerate levels of angular momentum ½, which is schematically represented in Fig. 4. In this case it has been possible to solve numerically the master equation for small values of N. The resolution has been greatly facilitated by the use of group theoretical methods and of tensorial formalism.

(a) Initial Conditions

The density operator ρ for $t = 0$ results from the excitation
conditions, in particular from the polarization properties of the
exciting light. We have considered only cases of complete population inversion and we have also assumed that the initial density
matrix is defined on symmetric states only (see section 3). More
precisely we have considered four cases for the initial conditions.
Three of them are obtained by assuming that the distribution of the
atoms in the two states of the upper level is given by a boson
statistics, that is:

$$\rho\big|_{t=0} = \sum_{\substack{N_1,N_2 \\ N_1+N_2=N}} \alpha^{N_1} \beta^{N_2} \frac{N!}{N_1!N_2!} |N_1\ N_2\ 00><N_1\ N_2\ 00| , \qquad (12)$$

the states being written in the occupation number representation
and $\alpha + \beta$ being equal to 1. In these conditions, it is easy to show
that the mean values of the populations of the upper states and of
the coherence between them are, for $t = 0$:

$$\begin{cases} <N_1>\big|_{t=0} = \alpha\ N \\[2ex] <N_2>\big|_{t=0} = \beta\ N \\[2ex] <N_{12}>\big|_{t=0} = 0\ . \end{cases} \qquad (13)$$

Therefore three particular pairs of values of α and β can be interpreted as resulting from simple excitation conditions:

(i) $\alpha = 1$, $\beta = 0$; this situation could be obtained by pumping in σ_+
circular polarization from a third level of angular momentum 1/2;

(ii) $\alpha = 3/4$, $\beta = 1/4$; it could be obtained by pumping also in σ_+
polarization but from a level of angular momentum 3/2;

(iii) $\alpha = \beta = 1/2$; this case results from pumping in linear polarization from a third level of angular momentum 1/2 or 3/2.

In addition to these three cases we have considered a fourth one:

$$(iv)\ \rho\big|_{t=0} = \sum_{\substack{N_1,N_2 \\ N_1+N_2=N}} |N_1\ N_2\ 00><N_1\ N_2\ 00| , \qquad (14)$$

in which all collective symmetric states corresponding to complete
population inversion are equally populated (corresponding to thermal

equilibrium); in this case, the only non-zero mean values are those
of scalar operators and the emitted light is isotropic.

(b) Discussion of Numerical Results

Numerical results have been obtained for the four cases of
initial conditions described above and for values of N up to 50,
for the cases (i), (ii) and (iii), and up to 110, for the case (iv).
Moreover, in order to better discuss these results, we have made
the corresponding calculations for the case where the level-degeneracy is removed, that is, by solving a master equation analogous to
Eq. (3); however, in this case the group theoretical simplifications
do not appear and the values of N have been limited to N = 20.

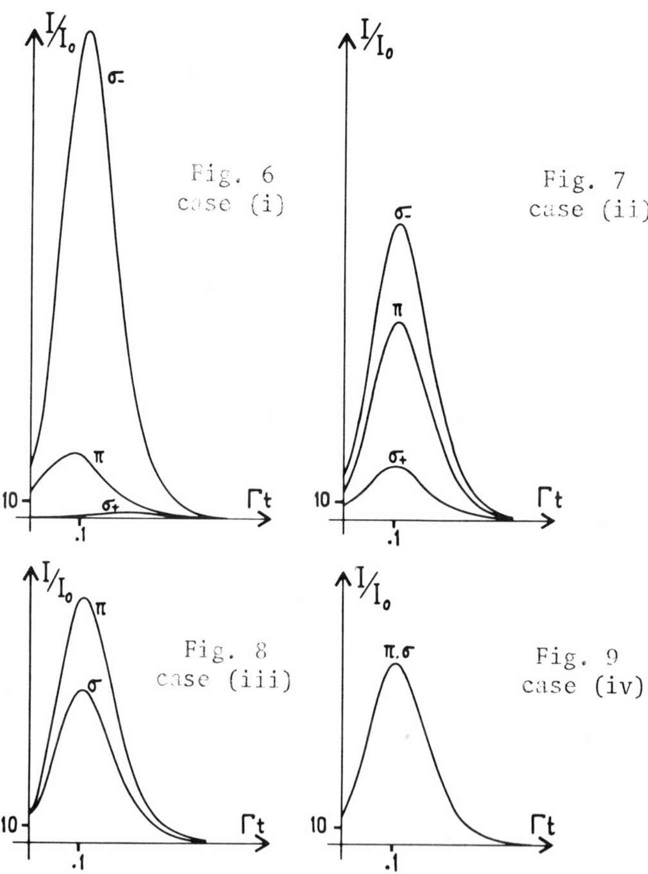

Figures 6-9. Intensity radiated with π, σ_+ and σ_- polarizations by
a collection of N = 50 atoms with two degenerate levels of angular
momentum ½; the different figures correspond to different initial
conditions.

Figures 6, 7, 8 and 9 show typical examples of the calculated pulses corresponding to the different polarizations and for the four considered initial conditions. It appears in particular that the polarization properties of the emitted light are very different in the four cases and that the relative values of the intensities corresponding to the different polarizations may be very different from those one would obtain in non-collective spontaneous emission (which are given by the values for t = 0).

Two comments can be made on case (i). First, since the upper state $m = -\frac{1}{2}$ is empty for t = 0, it is essentially a case of competition between two transitions, $1 \rightarrow 3$ and $1 \rightarrow 4$, sharing a common upper level, the σ_- transition being twice more probable than the π transition. As expected the π transition is inhibited: the ratio of the maxima of the σ_- and π pulses is much greater than 2, as would be expected if the two transitions were independent (see Fig. 6). Moreover it appears that this effect increases with N: more precisely, the ratio of the σ_- and π maxima increases as $\sqrt{N}$ (see Fig. 10). Secondly this case illustrates clearly an effect described in section 5, which is due to level-degeneracy interferences: the calculated mean value of the population of the upper state 2 does not remain zero but has a small maximum; this effect is small because of the inhibition of the superradiance on the π transition, but it could be much more important for other values of j and j'.

The relative influence of the two types of competition effects, inhibition and initiation, can be estimated by the value, for the non-degenerate case, of the ratio:

$$x = \frac{I_\pi}{I_{\sigma_+} + I_{\sigma_-}} \quad ,$$

where we note I_ε as the maximal intensity of polarization ε. The importance of the initiation effect depends on the ratio of upper

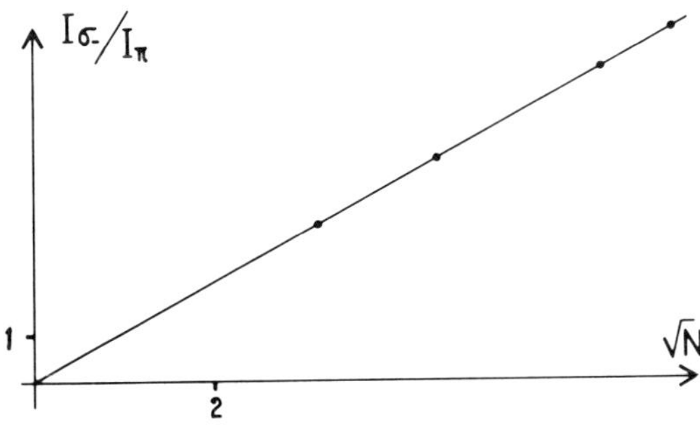

Figure 10. Ratio of the σ_- and π maxima versus $\sqrt{N}$ ($\frac{1}{2} \rightarrow \frac{1}{2}$ transition, degenerate case, initial conditions (i)).

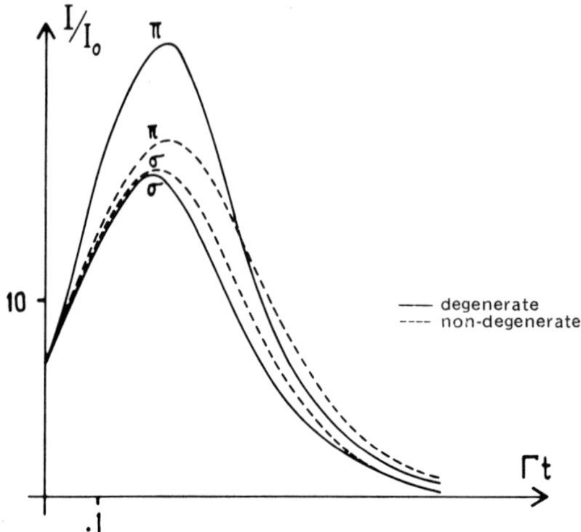

Figure 11. Intensity radiated with the different polarizations by a collection of N = 20 atoms with two levels of angular momentum ½, degenerate or non-degenerate, with initial conditions (iii).

Figure 12. π , σ_+ or σ_- maximum divided by N^2 versus N ($\frac{1}{2} \to \frac{1}{2}$ transition, degenerate case and initial conditions (iv)); the asymptotic behavior is $I \propto N^2$.

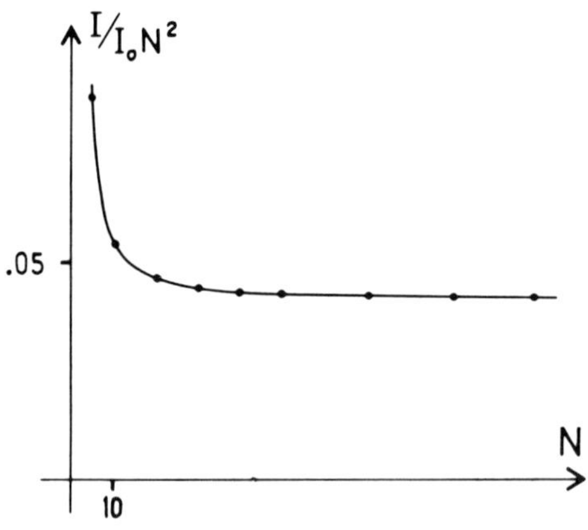

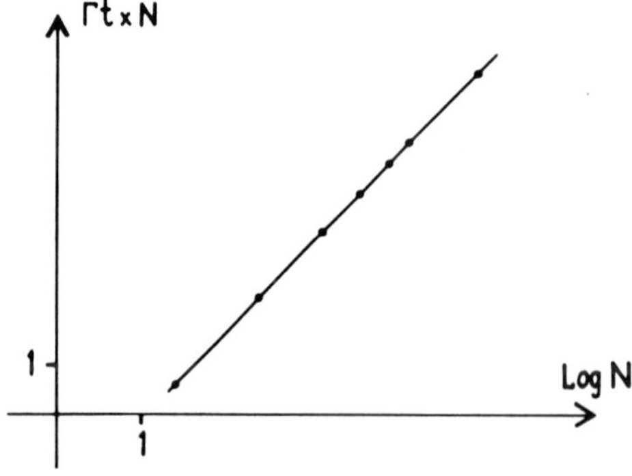

Figure 13. π , σ_+ or σ_- delay time multiplied by N versus Log N (for the same case as above); the asymptotic behavior is $t = \alpha/N \, \text{Log} \, \beta N$.

state initial populations; it seems to be maximum and to prevail
upon the inhibition effect in case (iii), where these two populations are equal. In this latter case x is greater than ½ and increases with N; on the contrary in the other cases x is smaller than
½ and decreases when N increases.

The interference effects appear to be always mainly constructive and the pulses of π polarization are always greater in the
degenerate case than in the non-degenerate case (see for example
Fig. 11). Moreover, these effects seem to increase with N: for
example, in case (iii) the ratio x increases with N much more
quickly in the degenerate case than in the non-degenerate case.

Concerning the delay times, some qualitative remarks can be
made. First, in all cases the delay times of the pulses corresponding to the different polarizations are very close together.
The different values of these delay times are always greater (by a
factor which is typically of the order of 1.5) than the values one
would obtain by considering the two levels as non-degenerate (see
for example the numerical values given in Ref. [8]). Finally one
may remark that the interference effects on the π light, being
constructive, decrease the corresponding delay time.

More quantitative conclusions can be made on case (iv) where
results are obtained for N up to 110: asymptotic formulae can be
obtained for the delay time and for the maximal intensity (see
Figs. 12 and 13).

7. CONCLUSION

In conclusion, we recall that the present work is an approach
to the problem of the influence of level-degeneracy on superradiance.
The theory of superradiance between degenerate levels, without being
in essence more difficult than for the two-level (non-degenerate)
case, is much more complex. Therefore we have been unable to introduce the numerous recent improvements of superradiance theory, and
we have thus chosen a simple quantum markovian model. It has permitted us to make clearer different types of effects due to level-
degeneracy, with complete understanding of their physical origin;
the corresponding influence on the emitted light has been qualitatively studied and some numerical results have been obtained for
the particular case of a j = j' = ½ transition. Finally, since a
quantum model only can account rigorously for the beginning of
superradiance, the master equation which is given here should provide a good starting point for more sophisticated studies of superradiance between degenerate levels.

References

1. Q.H.F. Vrehen, H.M.J. Hikspoors and H.M. Gibbs, Phys. Rev. Lett. *38*, 764 (1977).
2. R.H. Dicke, Phys. Rev. *93*, 99 (1954);
    D. Dialetis, Phys. Rev. A *2*, 599 (1970);
    G.S. Agarwal, Phys. Rev. A *2*, 2038 (1970);
    R. Bonifacio, P. Schwendimann and F. Haake, Phys. Rev. A *4*, 302 (1971); *4*, 854 (1971).
3. R. Friedberg, S.R. Hartmann and J.T. Manassah, Phys. Lett. *40A*, (1972); R. Friedberg and S.R. Hartmann, Phys. Rev. A *10*, 1728 (1974).
4. P. Pillet, Thèse de 3ème cycle, Paris, 1977.
5. A. Crubellier, Phys. Rev. A, to be published.
6. N. Skribanowitz, I.P. Herman, J.C. McGillivray and M.S. Feld, Phys. Rev. Lett. *30*, 309 (1973).
7. M. Gross, C. Fabre, P. Pillet and S. Haroche, Phys. Rev. Lett. *36*, 1035 (1976).
8. R. Bonifacio, P. Schwendimann and F. Haake, Phys. Rev. A *4*, 854 (1971).

CHARACTERISTICS OF SPONTANEOUS EMISSIONS AMPLIFIED BY A HIGH-GAIN

LASER AMPLIFIER

H. Gamo and H. Osada

University of California, Irvine, California

INTRODUCTION

The line-narrowing and line-rebroadening of the Amplified
Spontaneous Emission (ASE) in the high-gain He-Ne 3.39 μm laser
amplifier and statistics of intensity fluctuations of the ASE were
studied with emphasis on the transition region from the linear amplification to saturation.

First, we shall describe the formulas developed for describing
the population inversion density, the total gain for broad-band ASE,
and line narrowing for the case in which the line-narrowed radiation
with various intensity levels is incident on the laser amplifier.
These formulas will be applied to the analysis of the experimental
results, which are obtained by a scanning Fabry-Perot interferometer and gain measurements.

Second, we have measured the histogram of intensity fluctuations
of the ASE and determined the moments, central moments and cumulants
up to the eighth order. Coefficients of variation, skewness, excess
and their error estimates were determined under the well-defined
amplifier conditions.

Compared to our experimental results presented at the Third
Rochester Conference, [1] the new experimental results were obtained
under significantly improved conditions. We confirmed that the
frequency bandwidth of the photodetector-amplifier is broader than
that of the ASE by detecting the mode-locked laser pulse, and the
histograms of fluctuating photocurrents were obtained by using a
new sampling scope as a sample-hold amplifier. The time resolution

of the new system is far better than that which is required to process the line-narrowed ASE signals.

We utilized the unilaterally amplified spontaneous emissions
obtained by using infrared isolators, [2] because such a radiation
source is much more stable and more convenient for comparing the
experimental results with theoretical analysis.

APPROXIMATE FORMULAS FOR POPULATION INVERSION, REFRACTIVE
INDEX AND INCREMENTAL GAIN

We shall derive the formulas for the population inversion, gain
and refractive index for a partially homogeneously broadened laser
medium in the presence of a wide-band incident radiation by using
a semi-classical approach; that is, the incident radiation is
treated as classical electromagnetic waves and the atomic system
is treated quantummechanically. [3] A macroscopic polarization
p(t) produced by the incident radiation in the laser medium can be
described by the product of the atomic density N and the expectation
value of the electric dipole moment operator $\hat{\mu}$ of the atomic systems:

$$P(t) = N \, Tr(\rho \, \hat{\mu}) = N \, \mu \, (\rho_{21} + \rho_{12}) \, , \tag{1}$$

where ρ is the density matrix which describes a laser medium consisting of two level atomic systems with resonant frequency ν_0. The
equations of motion for each component of the density matrix ρ
derived by using the Heisenberg picture are given by:

$$\frac{d\rho_{22}}{dt} = r_2 - \frac{\rho_{22}}{\tau_2} - \frac{i\mu E(t)}{\hbar} (\rho_{21} - \rho_{12}) \, , \tag{2}$$

$$\frac{d\rho_{11}}{dt} = r_1 - \frac{\rho_{11}}{\tau_1} + \frac{\rho_{22}}{\tau_s} + \frac{i\mu E(t)}{\hbar} (\rho_{21} - \rho_{12}) \, , \tag{3}$$

$$\frac{d\rho_{21}}{dt} = -(i\omega_0 + 1/\tau) \, \rho_{21} - \frac{i}{\hbar} \mu E(t) (\rho_{22} - \rho_{11}) \, , \tag{4}$$

and

$$\rho_{12} = \rho_{21}^{*} \, , \tag{5}$$

where r_1, r_2 are excitation rates and τ_1, τ_2 are the lifetimes of
the upper and lower levels, respectively, and τ_s is the spontaneous
emission lifetime associated with the transition and is equal to
the inverse of Einstein A coefficient, and

$$1/\tau = 1/2(1/\tau_1 + 1/\tau_2),\tag{6}$$

and $E(t)$ represents the incident electromagnetic radiation due to the amplified spontaneous emission. First, we solve Eq. (5) with respect to ρ_{21} by assuming that $\rho_{22} - \rho_{11}$ is given by its ensemble average:

$$\rho_{21}(t,\ t_o) =$$

$$- \frac{i\mu}{h} <\rho_{22} - \rho_{11}> \int_{t_o}^{t} dt'\ E(t')\exp[-(i\omega_o+1/\tau)(t-t')]\ .\tag{7}$$

We insert Eq. (7) into Eqs. (3) and (4) and solve them with respect to ρ_{22} and ρ_{11}. Evaluating the ensemble average of $\rho_{22} - \rho_{11}$ by assuming that the incident spontaneous emission is a stationary random process, similar to the shot noise, we obtain:

$$<\rho_{22} - \rho_{11}> = \Delta\rho/[1 + I(\bar{\omega} - \omega_o)/I_s]\ ,\tag{8}$$

where

$$\Delta\rho = r_2\tau_2(1 - \tau_1/\tau_s) - r_1\tau_1,\tag{9}$$

$$I(\bar{\omega} - \omega_o) = \frac{1}{2\pi\eta_o}\int_{o}^{\infty} d\omega\ S(\omega)\ 1/\tau^2\ /\ [(\bar{\omega} - \omega_o + \omega)^2 + 1/\tau^2],\tag{10}$$

and the saturation intensity I_s for $\tau_2 \ll \tau_s$ is given by

$$I_s = 2\hbar^2/\mu^2\tau_1\tau_2\eta_o\ .\tag{11}$$

$\eta_o = \sqrt{\mu_o/\varepsilon_o}$ is the intrinsic wave impedance of the vacuum.

The power spectrum $S(\omega)$ of the incident electromagnetic radiation $E(\tau)$ with the center frequency $\bar{\omega}$ and auto-covariance function $<E(t+\tau)\ E(t)>$ form a Fourier transform pair [4]:

$$<E(t+\tau)E(t)> = \frac{1}{2\pi}\int_{-\infty}^{\infty} S(\omega)\ \exp(i\omega\tau)d\omega\ .\tag{12}$$

By introducing the first order Doppler shift due to the thermal motion of an atom/molecule, $\nu = \nu_o(1 + v/c)$, and using the Maxwell-Boltzmann velocity distribution,

$$P(v)dv = (\frac{m}{2kT})^{1/2}\ \exp\left\{-(\frac{mv^2}{2kT})\right\} dv,\tag{13}$$

we obtained the population inversion density

$$N(v)dv = NP(v)[\rho_{22}(v) - \rho_{11}(v)]dv$$

$$= \frac{\Delta N(m/2kT)^{1/2} \exp[-mv^2/2kT]}{1 + I(v)/I_s} \, dv, \tag{14}$$

where

$$\Delta N = N\Delta\rho, \tag{15}$$

and N is the density of atoms, $\Delta\rho$, is given by Eq. (9), and

$$I(v) = \frac{2}{\eta_o} \int_o^\infty \frac{(1/2\pi\tau)^2 S(2\pi v) dv}{[v-v_o(1+v/c)]^2 + (1/2\pi\tau)^2} \tag{16}$$

and I_s is defined by Eq. (11).

We shall introduce the normalized variable ξ,

$$\xi = 2(\ell n2)^{1/2} v_o v/(c\Delta v_D)^{1/2} , \tag{17}$$

and define the Doppler linewidth as the full width at half intensity as

$$\Delta v_D = 2(\ell n2)^{1/2} v_o(2kT/mc^2)^{1/2}. \tag{18}$$

The population inversion density $N(\xi)$ can be expressed as

$$N(\xi) = \frac{N}{(\pi)^{1/2}} \exp(-\xi^2)/[1 + I(\xi)/I_s] , \tag{19}$$

and $I(\xi)$ is equal to $I(v)$ in Eq. (16) by changing the variable from v to ξ defined above:

$$I(\xi) = \frac{\Delta v_D}{2(\ell n2)^{1/2}} \frac{1}{\eta_o} \int_{-\infty}^\infty \frac{y^2 S(x) \, dx}{(x-\xi)^2 + y^2} . \tag{20}$$

The polarization of laser medium is obtained by inserting Eqs. (7) and (8) into Eq. (1). It is more convenient for later formulation to utilize the Fourier transform of polarization:

$$\tilde{P}(\omega) = \varepsilon_o \, \tilde{\chi}(\omega) \tilde{E}(\omega). \tag{21}$$

Since the index of refraction $n(x)$ and the incremental gain γ_A are respectively given by real and imaginary parts of the complex susceptibility $\tilde{\chi}(\omega) = \chi'(\omega) + i\chi''(\omega)$ as $n \simeq 1 + 1/2 \, \chi'(\omega)$ and $\gamma_A \simeq 1/2 \, \chi''(\omega)$, we obtain the refractive index and incremental amplitude gain in terms of the normalized frequency x:

$$n(x) \simeq n_\infty + \frac{c}{2\pi\nu_o} \left[\frac{\Delta\nu_D}{2(\ell n2)^{1/2}} \, x + \frac{G}{\pi} \int_{-\infty}^{\infty} d\xi \right.$$

$$\left. \times \frac{(x-\xi) \, \exp(-\xi^2)}{[(x-\xi)^2 + y^2][1 + I(\xi)/I_s]} \right] , \tag{22}$$

and

$$\gamma_A(x) \simeq \frac{G}{\pi} \int_{-\infty}^{\infty} d\xi \, \frac{y \, \exp(-\xi^2)}{[(x-\xi)^2 + y^2][1 + I(\xi)/I_s]} , \tag{23}$$

where

$$G = \frac{\Delta N \lambda^2}{8\pi\tau_s} \, \frac{2(\ell n2)^{1/2}}{\Delta\nu_D} \, \frac{1}{(\pi)^{1/2}} \tag{24}$$

and ξ is given by Eq. (17) and

$$x = 2(\ell n2)^{1/2} \, (\nu-\nu_o)/\Delta\nu_D, \tag{25}$$

$$y = (\ell n2)^{1/2} \, \Delta\nu_L/\Delta\nu_D , \tag{26}$$

and the Lorentz linewidth $\Delta\nu_L$ is given by

$$\Delta\nu_L = 1/(\pi\tau), \tag{27}$$

where τ is defined by Eq. (6).

The approximate formulas for population inversion density, incremental amplitude gain and the refractive index can be derived as follows. First, the power spectrum of an incidental ASE is assumed as a Gaussian function for the convenience of analytical formulation

$$S(\nu)d\nu = \eta_o \, I_i \, \exp(-\frac{x^2}{s^2}) \, \frac{dx}{\pi^{1/2} s} \, , \tag{28}$$

where

$$s = \Delta\nu_i / \Delta\nu_D \tag{29}$$

and I_i is the total intensity of incident radiation. Then, $I(\xi,y,s)$, defined in Eq. (20), can be expressed in terms of the Voigt function $\psi(x,y)$:

$$I(\xi,y,s) = I_i \, \pi^{1/2} (\frac{y}{s}) \, \psi(\xi/s, \, y/s) \, , \tag{30}$$

where

$$I_i = \frac{S(\nu_o)}{\eta_o} \, \frac{\Delta\nu_i}{8\hbar(\ell n 2)^{1/2}} \, . \tag{31}$$

The Voigt function $\psi(x,y)$ and its Hilbert conjugate $\phi(x,y)$ are tabulated in reference [5] and also computer programs of these functions are available. [6] Second, we shall expand $I(\xi,y,s)$ in a power series up to the second order of variable ξ, because this approximation enables us to utilize the Voigt function for further development of approximate formula:

$$I(\xi,y,s) \simeq I_i \pi^{1/2} (\frac{y}{s}) [\psi(0,y/s) \, - \, |\psi''(0,y,s)| \xi^2/(2s^2)]. \tag{32}$$

Inserting this into Eq. (19) and using $1/[1+I(\xi)/I_s] \simeq \exp[-I(\xi)/I_s]$, we obtain the population inversion density

$$N(\xi) \simeq \Delta N \, \exp\left\{-\pi^{1/2}(\frac{I_i}{I_s}) \, (\frac{y}{s}) \, \psi(0,y/s) \, - \, \frac{\xi^2}{\sigma^2}\right\} \tag{33}$$

where

$$\frac{1}{\sigma^2} = 1 \, - \, \pi^{1/2}(\frac{I_i}{I_s}) \, (\frac{y}{s}) \, |\psi''(0,y/s)| \frac{1}{2s^2} \, . \tag{34}$$

Owing to the saturation effect, the population inversion density decreases uniformly over the entire velocity region and the standard deviation of the velocity distribution also increases.

Inserting Eq. (30) into Eqs. (22) and (23), respectively, and adopting the same approximation procedure, we obtain the refractive index of gain medium under weak saturation

$$n(x) \simeq n_\infty + \frac{c}{2\pi\nu_0} \left\{ \frac{\Delta\nu_D}{2(\ln 2)^{1/2}} x + \frac{G}{\sigma} \phi(0,y/\sigma) \right.$$

$$\left. \times \exp\left[-\pi^{1/2} \left(\frac{I_i}{I_s}\right)\left(\frac{y}{s}\right) \psi(0,y/s)\right]\right\} \tag{35}$$

and the incremental gain for wave amplitude in weakly saturated medium

$$\gamma_A(x) \simeq G[\psi(0,y/\sigma) - |\psi''(0,y/\sigma)| x^2/2\sigma^2)]$$

$$\times \exp\left[-\pi^{1/2}\left(\frac{I_i}{I_s}\right)\left(\frac{y}{s}\right)\psi(0,y/s)\right] \quad . \tag{36}$$

It should be noted that the refractive index in weak saturation given by Eq. (35) will be used in our later paper [7] on the pulse propagation in a high-gain laser amplifier, in conjunction with the group velocity. The functions $\psi(0,y)$, $|\psi''(0,y)|$ and $\phi'(0,y)$ are tabulated in reference [8].

TOTAL POPULATION INVERSION, TOTAL GAIN AND LINE-NARROWING

Since the total gain and line-narrowing for the incident ASE in a laser amplifier are measured conveniently, we have derived the following approximate formulas from those described in the preceding section.

First, by integrating Eq. (33) for population inversion density $N(\xi)$ over the entire normalized velocity region, we obtain the total population inversion $\tilde{N}$,

$$\tilde{N} \equiv \int_{-\infty}^{\infty} N \, d(\xi)d\xi \simeq \Delta N\sigma \, \exp\left\{-\pi^{1/2}\left(\frac{I_i}{I_s}\right)\left(\frac{y}{s}\right)\psi(0,y/s)\right\} \quad . \tag{37}$$

In the limit of $I_i/I_s \to 0$, we find that $\tilde{N} = N$.

We shall define the total power gain Γ for the ASE by the ratio of the total intensity of the electromagnetic waves amplified by a laser amplifier of plasma length z to the total intensity of the incident wave with the Gaussian power spectrum:

$$\Gamma(z) \equiv I_{out}/I_i \quad , \tag{38}$$

where

$$I_{out} = \frac{I}{\pi^{\frac{1}{2}}s} \int_{-\infty}^{\infty} \exp(2\gamma_A z - x^2/s^2) \, dx \quad . \tag{39}$$

By inserting the quadratic approximation of the incremental gain in Eq. (36), and by using the known Gaussian integral, we obtain the total intensity gain:

$$\Gamma(z) \simeq \exp(\gamma_0 z) \Big/ \left[1 + \frac{s^2}{2\sigma^2} \frac{|\psi''(0,y/\sigma)|}{\psi(0,y/\sigma)} \gamma_0 z \right]^{\frac{1}{2}}$$

$$\simeq \exp\left\{ \gamma_0 z \left[1 - \frac{1}{4} \frac{s^2}{\sigma^2} \frac{|\psi''(0,y/\sigma)|}{\psi(0,y/\sigma)} \right] \right\} , \tag{40}$$

where γ_0 is incremental intensity gain at line center given by

$$\gamma_0 = 2 \, G\psi(0,y/\sigma) \, \exp\left[-\pi^{\frac{1}{2}} \left(\frac{y}{s} \right) \left(\frac{I_i}{I_s} \right) \psi(0,y/s) \right] \quad . \tag{41}$$

The line-narrowing defined by the ratio of the full-width at half intensity of outgoing wave $\Delta\nu_s$ to the full-width at half intensity of the incoming wave $\Delta\nu_i$ can be approximated as follows:

$$\frac{\Delta\nu_s}{\Delta\nu_i} = \frac{1}{\sqrt{1 + \frac{s^2}{2\sigma^2} \frac{|\psi''(0,y/\sigma)|}{\psi(0,y/\sigma)} \gamma_0 z}} \quad . \tag{42}$$

This new formula is more general and should be more convenient from a practical standpoint, compared to the formula derived in reference [9], where the spontaneous emission from the atom characterized by the Doppler broadened Lorentz line is assumed as an input signal and very often the linewidth is an unknown parameter. The determination of linewidth of spontaneous emission requires an independent measurement which may not be convenient due to the line-narrowing effect. On the other hand, in the above formulation, we can specify both linewidth and intensity of incident radiation which can be directly measured.

The quantity σ defined by Eq. (34) can be treated as a measure of the weak saturation effect, because it depends on the intensity of incident radiation and saturation intensity, Doppler and Lorentz linewidths of the laser amplifier. According to Eq. (42), the line-narrowing will cease when σ is very large, even if the

plasma length z is increased. From the measurement of line-
narrowing under the influence of saturation, we can define a
saturation intensity by using a specific intensity of incident
radiation, I_{im}, which minimizes the linewidth at which σ is assumed
to be infinity:

$$I_s = I_{im} \frac{\pi^{1/2}}{2s^2} \left(\frac{y}{s}\right) \left| \psi''(0, y/s) \right| \quad . \tag{43}$$

According to Eq. (38), the line-narrowing depends on both s and σ.
The larger s is, the faster the line narrows. We can observe
appreciable line-narrowing by increasing the plasma length of a
high-gain laser amplifier.

EXPERIMENTAL PROCEDURES

     To measure the line-narrowing and line-rebroadening effects,
the following experimental set-up has been used. (See Fig. 1(a)).
The original incident radiation was generated by the series of four
laser tubes which are filled with He-Ne gases of 2 Torr (He:Ne=5:1).
The total length of plasma is 240 cm. With these, we obtained ASE
of 1.12 $\mu W/mm^2$ power with linewidth of 89.3 MHz. This radiation
is incident upon the high-gain laser amplifier which is filled with
the same He-Ne gases of 2 Torr. This amplifier is equipped with
12 electrodes with spacing of 5 or 10 cm up to 105 cm. Using a
scanning Fabry-Perot interferometer with mirror separation 50 cm,
the linewidth of the incident radiation is measured. The intensity
was measured with the Eppley thermopile (Eppley Laboratory, KRS-5
window, circular type), and InSb photo-voltaic detector cooled by
liquid nitrogen (Philco-Ford L 4540) with termination of 1 KΩ. In
the figure YIG stands for an optical isolator using the Faraday
Rotation in Yttrium Iron Garnet (YIG) crystal [2]. Isolations for
YIG 1 and YIG 2 are 33.0 dB and 33.2 dB, respectively, and insertion
losses for YIG 1 and YIG 2 are 3.8 dB and 1.8 dB, respectively.

     The histograms and cumulants of intensity fluctuations were
measured by using the experimental set-up illustrated in Fig. 1(b).
The essential features of the instrumentation are given in the
following. First, the detector output was introduced into a unity
gain preamplifier (Keithly 110, 180 MHz bandwidth) which operated
as an impedance transformer. Second, the output from the preamplifier was sampled and held by the high speed Sampling Unit
(Tektronix 7S12 with S-6 Sampling Head, S-5 Trigger Recognizer,
0.1 n sec sampling interval), mounted in Tektronix 7704A oscilloscope. The voltage from the Sampling Unit was amplified by
amplifiers (Tektronix AM 502 and DANA 280 IR) and then introduced
into the Analog-to-Digital converter interfaced with the data

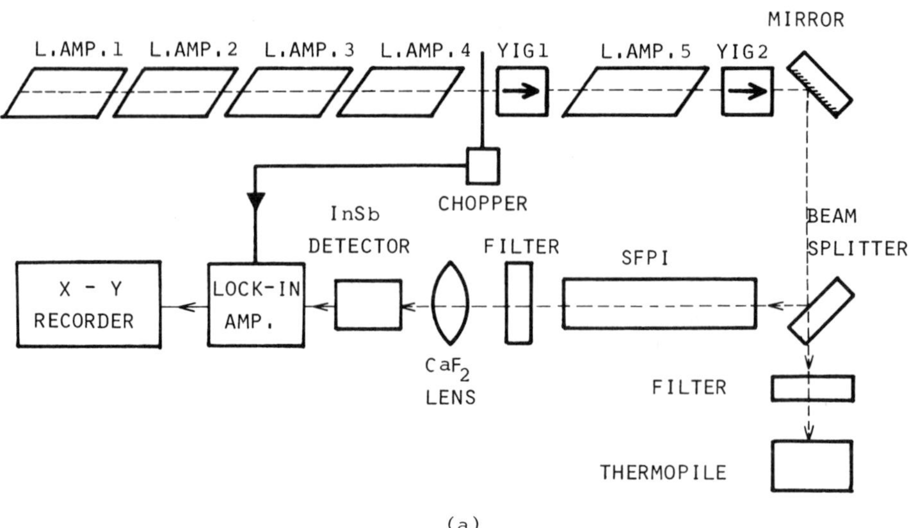

(a)

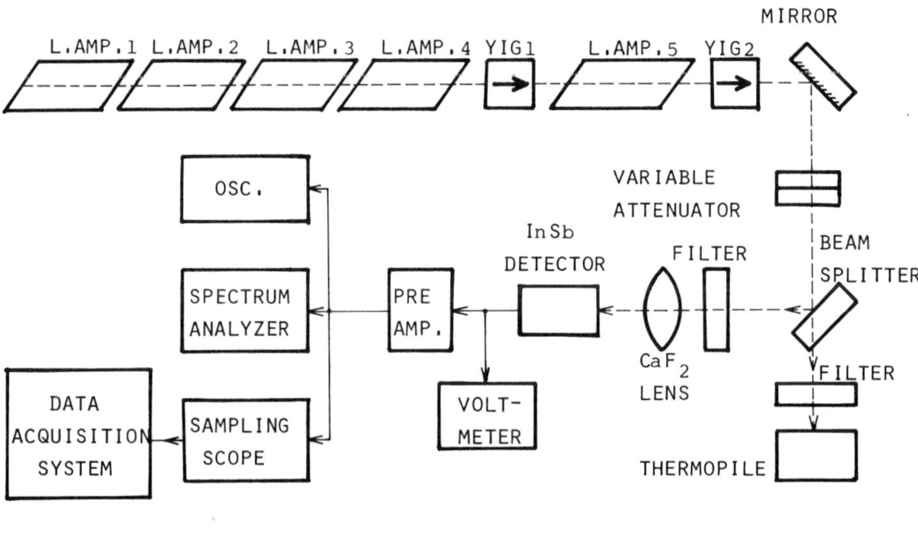

(b)

Figure 1.
(a) Experimental set-up for the spectral linewidth measurements.
(b) Experimental set-up for histograms, cumulants and power spectrum measurements.

acquisition system, PDP 15/20 with 24 K core memory. Third, by processing the sampled signals, we obtained histograms, moments, central moments and cumulants up to the 8th order. From these quantities, we calculated the coefficients of variation, skewness and excess with error estimates.

EXPERIMENTAL RESULTS

Figure 2 shows the change of linewidth and intensity with increase of the plasma length. The incident radiation has a linewidth of 89.3 MHz and intensity of 1.12 $\mu W/mm^2$. In the linear amplification and weakly saturated regions, the line narrows. The linewidth decreases down to 79.3 MHz of the intensity of 35 $\mu W/mm^2$. Applying Eq. (43), we obtained the saturation intensity of 88 $\mu W/mm^2$ by using the homogeneous linewidth of 146 MHz [10] and our estimated Doppler linewidth of 277 MHz of amplifier. By analyzing the total gain Γ and the change of linewidth at various intensities, we obtained the incremental gain γ_0 at the line center frequency. (See Fig. 3.) Then, the incremental gain is equal to 0.058 cm^{-1} at the linear amplification region. Using these numbers, we can draw a solid line for the theoretical curve of line-narrowing in the weakly saturated region in Fig. 4. This is fitted reasonably well within the experimental error.

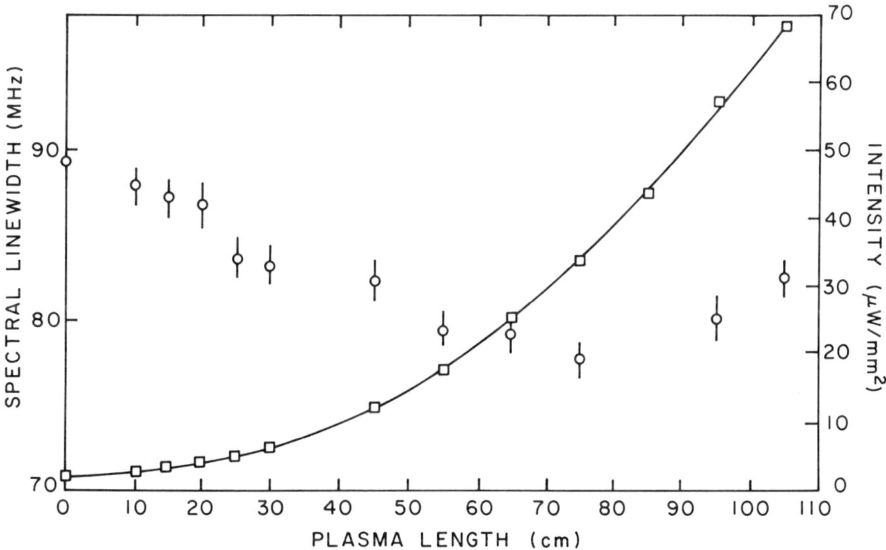

Figure 2. Linewidth and intensity versus plasma length, where cir- and square indicate linewidth and intensity, respectively.

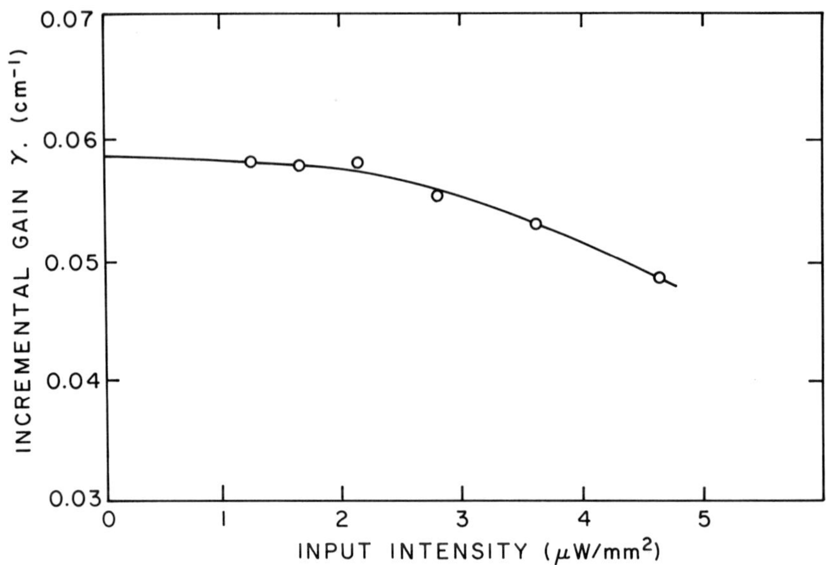

Figure 3.   Incremental gain at line center frequency versus input intensity.

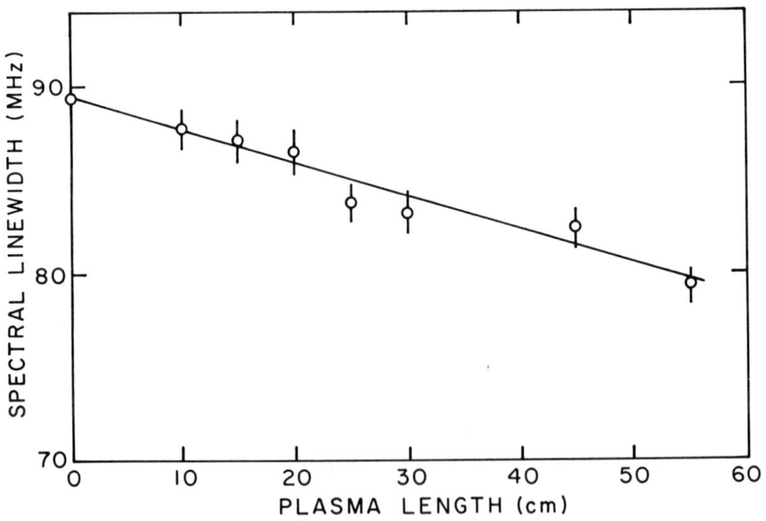

Figure 4.   Comparison of the theoretical calculation of line-narrowing with the experimental results.

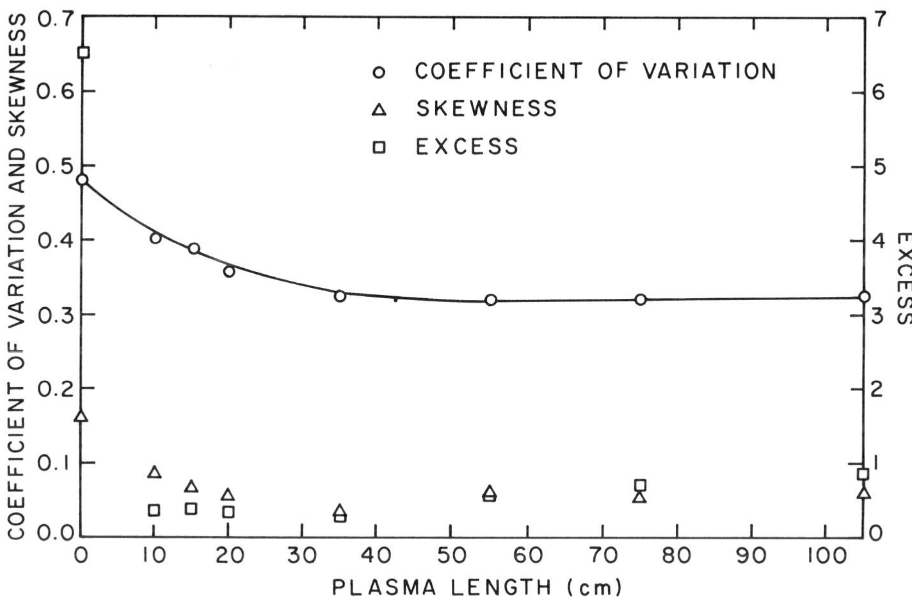

Figure 5. Coefficients of variation, skewness and excess versus
plasma length.

To study the statistics of the intensity fluctuations of ASE,
we measured the histograms of photocurrent fluctuation for different
plasma length, that is, different intensity. From these, we calculated moments, central moments and cumulants up to the 8th order.
By using the cumulants, we also calculated coefficients of variation, skewness and excess (Fig. 5). The coefficients of variation
decreased in the unsaturated region and after saturation stayed
constant. The excess coefficient also first decreased from 0.26
to 0.03, then increased very slowly to 0.06. The skewness coefficients were almost constant in the linear amplification region and
increased slowly.

CONCLUSION

The approximate formulas for both total gain and line-narrowing
of the ASE in the weakly saturated region, (Eqs. 40, 42) are shown
to agree with the experimental results, and these new formulas will
conveniently be used for analyzing the ASE and determining the
saturation intensity, the gain parameter and the Doppler linewidth
of a high-gain transition without using the radiation from a single-
mode laser [11]. The estimated saturation intensity I_S of the He-Ne

3.39 µm line is 88 µW/mm^2 and the measured incremental gain γ_0 is
0.058 cm^{-1}. Although we have not performed the dispersion experiment for analyzing the refractive index of the gain medium as a
function of frequency, when it is combined with the gain and line-
narrowing experiments, we should be able to determine the Lorentz,
Doppler linewidths, gain and saturation parameters of the high gain
transition.

As is shown in Fig. 2, the line-rebroadening effect was observed in the strongly saturated He-Ne 3.39 µm laser amplifier in
a manner similar to the case of He-Xe 3.5 µm high-gain laser amplifier reported in the previous conference[1].

By measuring the histogram of photocurrent fluctuations produced by the ASE using the wide band detector-amplifier and high-
speed sampling unit, we observed the change of statistics of intensity fluctuations in the transition region from linear amplification
to saturation. The coefficients of variation, excess and skewness
will tend to decrease in the weakly saturated amplifier region. In
the strongly saturated region, both skewness and excess coefficients
increase gradually but the coefficient of variation stays constant.

The results of these statistical experiments performed by using
a unilaterally propagating ASE with infrared isolators, should be
analyzed much more conveniently than the case of bilateral transmission presented at the last conference[1].

Details of derivation of approximate formulas and experiments
of Xenon 3.5 µm gas laser amplifier are described in Osada's Ph.D.
Thesis[12].

*This work was supported by the National Science Foundation.

References

1.   H. Gamo and S.S. Chuang, *Coherence and Quantum Optics* (Proc.
      Third Rochester Conference on Coherence and Quantum Optics)
      eds. L. Mandel and E. Wolf (Plenum Press, New York, 1973)
      pp. 491-507.
2.   H. Gamo and S.S. Chuang, AFOSR 70-1956, TR-1970, Available as
      AD-716468, National Information Service, Springfield, VA.
3.   M. Sargent, M.O. Scully and W.E. Lamb, Jr., *Laser Physics*
      (Addison-Wesley, Reading, Mass. 1974) Chapter 8.
4.   M. Born and E. Wolf, *Principles of Optics* (Pergamon Press,
      New York, 1970) Chapter 10.
5.   V.N. Faddeyeva and N.M. Terent'ev, *Tables of Probability
      Integral for Complex Argument* (Pergamon Press, New York, 1961).

6.   B.H. Armstrong and R.W. Nicholls, *Emission, Absorption and Transfer of Radiation in Heated Atmospheres* (Pergamon Press, New York, 1972).
7.   H. Gamo, N. Takahashi, H. Osada and H. Sato, this Proceedings, p. 613.
8.   H. Gamo and H. Sato, J. Appl. Phys. *46*, 3585 (1975).
9.   H. Gamo, J.S. Ostrem and S.S. Chuang, J. Appl. Phys. *44*, 2750 (1973).
10.   Misao Ohi, Ph.D. Dissertation (Tokyo Institute of Technology, Tokyo, Japan, 1974). Unpublished.
11.   H. Gamo and J.S. Ostrem, J. Opt. Soc. Am. *65*, 29 (1975)
12.   H. Osada, Ph.D. Dissertation (University of California, Irvine, Calif., 1978).

WHAT DOES SPONTANEOUS EMISSION LOOK LIKE?

L. Allen, S.P. Kravis and J.S. Plaskett

University of Sussex, Brighton, England

It appears that Amplified Spontaneous Emission [1], A.S.E.,
the emission process which occurs when a population inversion
exists in an extended distribution of atoms in the absence of a
laser cavity, is singularly well named. Sources of A.S.E. can be
c.w. or pulsed. It is found in some pulse systems that the output
has a granular structure (Fig. 1), or what another literature [2]
describes as speckles. Such a structure does not occur in c.w.
sources of A.S.E., in certain pulsed A.S.E. systems or in lasers;
it is, however, well known that laser light scattered from rough
surfaces gives rise to the speckles mentioned above. We are able
to explain why speckles are, or are not, seen. The explanation
forces a re-examination of the nature of the radiation from a
source of spontaneous emission, derives the van Cittert-Zernike
theorem [3] in an alternative way and relates the predicted coherence area to the size of the speckle structure. The explanation
may be extended to demonstrate why different frequencies appear in
different parts of the A.S.E. output [4].

In the régime where saturation effects are absent, we demonstrate that a pulsed A.S.E. source may be thought of simply as a
planar source of spontaneous emission feeding a linear amplifier.
Thus for the 540 nm transition in neon, for example, where the saturation parameter is small and the pulse duration is <1 nsec, a photograph of the output of a single pulse is a record of the spontaneous emission pattern, suitably amplified, integrated over a time
short or comparable with the lifetime of the transition (Fig. 1).
That is, the granularity recorded in a photograph of a single pulse
is simply the diffraction pattern arising from a random array of
spontaneously emitting atoms observed for a time determined by the

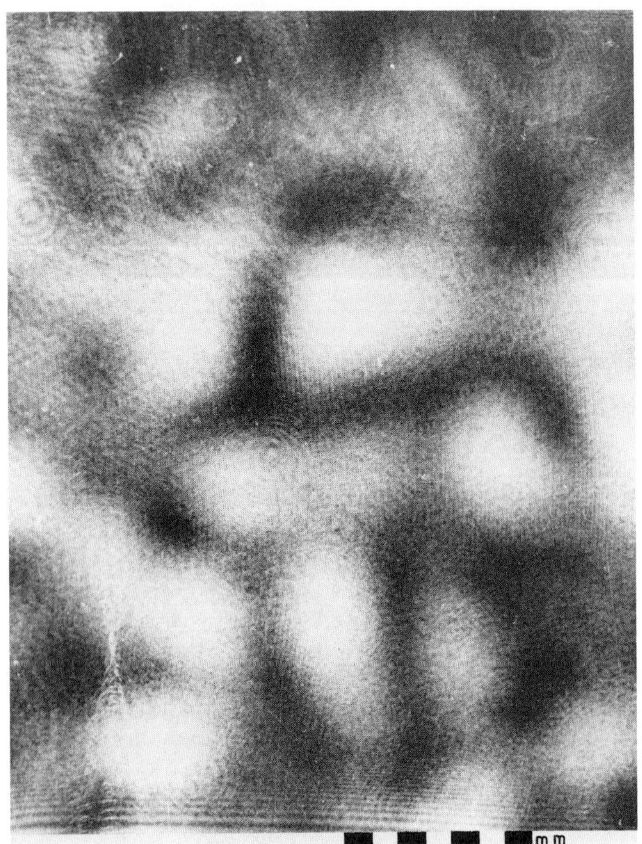

Figure 1. Spatial speckles from a 540 nm Ne A.S.E. source photographed at 700 cms. The speckle size agrees with that expected
from $|r-r'| = 0.61\ \lambda\ell/a$ when $\mu = 0$.

gating effect of the transient amplification; the grains correspond
to regions of constructive interference. Atoms emitting spontaneously before or after the amplification process contribute nothing
because their radiation is so weak.

A source consisting of independently emitting atoms at the
points $(x_j,\ y_j,\ z_j)$ gives rise to a field at the point $(x,\ y,\ \ell)$
in some distant viewing plane, one polarization component of which
is given by

$$E(\underline{r},t) = \sum_j A_j\ e^{-i\omega_j t}\ e^{2\pi i(\ell-z_j)/\lambda}\ e^{i\pi(\underline{r}-\underline{r}_j)^2/\lambda\ell} \qquad (1)$$

provided ℓ is large. Here λ is the wavelength, ω_j is the Doppler-shifted frequency and $\underline{r} = (x,y)$, $\underline{r}_j = (x_j,y_j)$. In the case of a real source of A.S.E. the source of atoms is distributed, in reality $A_j(z_j) = A_j(0) \exp -\alpha z_j$, but provided $z_j < \ell$ this does not present any limitation to the theory. If a is the transverse radius of the source, then all atoms within a transverse distance δ of one another, $|\underline{r}_j - \underline{r}_{j'}| < \delta$, will produce the same variation of the field over a circle of radius d, that is $|\underline{r}| \leq d$ provided

$$\delta << \lambda \ell / (d + a) \; . \tag{2}$$

It follows that the actual set of emitting atoms may be replaced by a plane array of regularly arranged sources at the points $(x_k, y_k, 0)$ provided these sources are not further apart than δ, and the field can then be written

$$E(\underline{r},t) = e^{-i\omega t} \sum_k C_k \; e^{i\pi(\underline{r}-\underline{r}_k)^2/\lambda \ell} \; .$$

In this expression C_k is the sum of

$$A_j \; e^{-i(\omega_j - \omega)t} \; e^{2\pi i(\ell - z_j)/\lambda}$$

for the atoms within a radius δ. This sum is a random walk in the complex plane in which the steps have random length due to the random polarization of the atoms, and random direction due to the random phase and longitudinal position of the atoms. If there are enough steps in this walk the probability that $C_k = u + iv$ within the rectangle of sides du, dv is

$$\frac{1}{\pi I_k} \; e^{-(u^2+v^2)/I_k} \; du \; dv$$

where I_k is proportional to the number of steps, which is the same as the number of atoms in a cylinder of radius δ centered on (x_k,y_k) and now replaced by one source. At this stage we neglect the slow variation of C_k with time over the duration of a pulse. This is equivalent to making all atoms emit with the same frequency.

The single source at $(x_k, y_k, 0)$ produces a field in the image plane (x,y,ℓ) of uniform intensity $|C_k|^2$. The probability that $|C_k|^2 = I$ within dI is found from the normal distribution of C_k to be

$$\frac{1}{I_k} \; e^{-I/I_k} \; dI \; ,$$

the mean of which is I_k. The intensity $|E(\underline{r},t)|^2$ due to all the sources k, is not uniform and the probability, at any given point $\underline{r}$, that $|E(\underline{r},t)|^2 = I$ within dI is

$$\frac{1}{I_0} e^{-I/I_0} dI \tag{3}$$

where $I_0 = \Sigma I_k$. This leads to a uniform mean field of ΣI_k if a large number of such field distributions are superimposed; this is what is found in photographs of pulsed A.S.E. integrated over many pulses when the speckles disappear. A typical variation of $|E(\underline{r},t)|^2$ with $\underline{r}$ may be obtained by means of a computer simulation for a regular array of points (x_k,y_k) with the C_k's chosen from a sequence of random numbers each with the normal distribution given above and the same I_k. If the parameters chosen are the same as for the neon source output displayed in Fig. 1, the computer-generated pattern is similar and comparable in size.

Given the normal probability distribution for each C_k it is easy to work out the joint probability distribution for $E(\underline{r},t)$ at two different points. For example, the probability that $E(\underline{r},t) = u + iv = w$ within du, dv, and $E(\underline{r}',t) = u' + iv' = w'$ within du', dv' is

$$\frac{1}{\pi^2} \frac{1}{I_0^2(1-|\mu|^2)} \exp\left(- \frac{1}{I_0(1-|\mu|^2)} \left[|w|^2+|w'|^2-\mu w^*w'-\mu^* ww'^*\right]\right) du\, dv\, du'dv'$$

where

$$\mu(\underline{r},\underline{r}') = \frac{1}{I_0} \sum_k I_k \exp\left\{i\pi[(\underline{r}-\underline{r}_k)^2 - (\underline{r}'-\underline{r}_k)^2]/\lambda\ell\right\} .$$

Dividing this joint probability distribution by the probability that $E(\underline{r}',t) = w'$ within du', dv', namely

$$\frac{1}{\pi I_0} \exp\left(-|w'|^2/I_0\right) \ du'\ dv' \quad ,$$

gives the conditional probability that $E(\underline{r},t) = w$ within du, dv is

$$\frac{1}{\pi I_0(1-|\mu|^2)} \exp\left(- \frac{1}{I_0(1-|\mu|^2)} |w-\mu w'|^2\right) \ du\, dv \quad .$$

It follows that the most probable value, and also the mean value for many pulses, for $E(\underline{r},t)$, given $E(\underline{r}',t)$, is simply $\mu E(\underline{r}',t)$. When $|\mu|$ is nearly unity, the most probable value is extremely likely.

From the conditional probability distribution it is easy to
obtain the mean value, over many pulses, of the products
$E(\underline{r},t)E^*(\underline{r}',t)$ and $|E(\underline{r},t)|^2|E(\underline{r}',t)|^2$. These are

$$<E(\underline{r},t)E^*(\underline{r}',t)> = I_o\mu$$

and

$$<|E(\underline{r},t)|^2|E(\underline{r}',t)|^2> = I_o^2(1 + |\mu|^2) \ . \qquad (4)$$

We may note that for a uniform circular source of radius 'a', when
all the I_k's are equal,

$$\mu = e^{\pi i(r^2-r'^2)/\lambda\ell} \ \frac{2J_1(\theta)}{\theta} \ ,$$

where

$$\theta = 2\pi a|\underline{r}-\underline{r}'|/\lambda\ell \ \ .$$

It is of course true that the equivalent of the above formulation
is to be found in the laser speckle pattern literature [2], where
the assumption is made that the scattered light follows a Gaussian
distribution of complex amplitude, but here the Gaussian distribution is established.

Both correlations may be evaluated analytically while only the
latter, the intensity correlation, may be measured from a speckle
pattern photograph of a real A.S.E. source. It is reasonable to
suppose that these mean values could be computed by integration
over a reasonably homogeneous area A of one pulse, for example

$$<|E(\underline{r},t)|^2|E(\underline{r}+\underline{a},t)|^2> = \frac{1}{A}\int_A |E(\underline{r},t)|^2|E(\underline{r}+\underline{a},t)|^2 d\underline{r} \ .$$

The amplitude correlation function is exactly the same as that
given by the van Cittert-Zernike theorem [3], which predicts the
coherence area at a distance due to a quasi-monochromatic source
of particular area. It may be noted that the coherence length
predicted by the van Cittert-Zernike theorem corresponds to the
central core of a speckle and not to a whole speckle, in those
areas where the field amplitude is high. The concept of 'coherence
length' is, of course, equally meaningful in the dark regions of
the resulting intensity pattern. It is not surprising that the
van Cittert-Zernike theorem applies here because the introduction
of exponential growth into each of the amplitudes in the Born and
Wolf [3] derivation leads to a value for the coherence area identical with that of the unamplified case.

The theory discussed above is for a single frequency source;
but the spontaneous emission driving the neon A.S.E. source is
Doppler and collision-broadened to $\sim$3 GHz spectral width. This may

be accounted for by allowing each atom to have its own frequency ω_j taken from an overall distribution $f(\omega)$. However, the radiation from those atoms which are highly detuned from the centre will be only slightly amplified; so the width of $f(\omega)$ will be narrower [1, 5] than the full 3 GHz. In addition, the field is not now given by Eq. (1) but has an envelope $g(t)$ of width τ and can therefore be written

$$F(t) = E(\underline{r},t)\,g(t)$$

$$= \sum_j A'_j \; e^{-i\omega_j t} \; g(t) \; .$$

Expression (2) indicated the degree of closeness the atoms must have to allow us to replace them with a regular array. Similarly, if two frequencies satisfy $|\omega_j - \omega_{j'}| \ll 1/\tau$ they make the same contribution to $F(t)$. We may thus write

$$F(t) = \sum_k C'_k \; e^{-i\omega_k t} \; g(t) \quad ,$$

where ω_k is a sequence of regularly spaced frequencies and C'_k is the sum of A'_j's corresponding to all atoms within the neighbourhood of ω_k. As before the C'_k are normally distributed with mean I'_k, where I'_k is proportional to the number of atoms emitting with frequency near ω_k, and is thus proportional to $f(\omega_k)$.

The frequency distribution at point $\underline{r}$ may be found from

$$S(\omega) = \int F(t) \; e^{i\omega t} \; dt$$

$$= \sum_k C'_k \; \tilde{g}(\omega - \omega_k) \; ,$$

where

$$\tilde{g}(\omega) = \int g(t) \; e^{i\omega t} \; dt \; .$$

Korolev et al [6] examined this case originally for a square pulse of length τ when $\tilde{g}(\omega) = 2/\omega \; \sin\omega\tau/2$. We should note, perhaps, that we are no longer discussing a stationary process, even in the wide sense, that pertained in the single frequency discussion.

An analogous argument to that presented for the field variation with position in the single frequency case shows that the probability that $S(\omega) = u + iv$ within du, dv is

$$\frac{1}{\pi I(\omega)} \; e^{-(u^2+v^2)/I(\omega)} \; du \; dv \; ,$$

where

$$I(\omega) = \sum_k I'_k \, |\tilde{g}(\omega-\omega_k)|^2 = K \sum_k f(\omega_k) \, |\tilde{g}(\omega-\omega_k)|^2$$

if

$$I'_k = K \, f(\omega_k) \ .$$

We may show in a way analogous to the discussion of $|E(\underline{r},t)|^2$ given earlier that,

$$<|S(\omega)|^2> = I(\omega)$$

$$<|S(\omega)|^2 |S(\omega')|^2> = I(\omega) \, I(\omega') + |J(\omega,\omega')|^2 \qquad (5)$$

where

$$J(\omega,\omega') = K \sum_k f(\omega_k) \, \tilde{g}(\omega-\omega_k) \, \tilde{g}(\omega'-\omega_k)^* \ .$$

Equation (5) is the frequency analogue of Eq. (4). For a single pulse, just as the intensity $|E(\underline{r},t)|^2$ is not I_0 but is spatially speckled, so the spectrum $|S(\omega)|^2$ is not $I(\omega)$ but is frequency speckled. The width of these frequency speckles is the width of $\tilde{g}(\omega)$, which is the reciprocal of τ. They will only be prominent if their width is less than that of $f(\omega)$; this correponds to the case for which no spatial speckles are anticipated. Conversely, prominent spatial speckles imply an absence of frequency speckles.

The conditions for frequency speckles may be understood in the following way. Consider Fig. 2(a). Here the pulse duration is long and so $1/\tau$ is narrow. Inspection shows that the product of $f(\omega)\tilde{g}(\omega-\omega_k)\tilde{g}(\omega'-\omega_k)^*$, that is, the correlation term, is zero. This means that at one space point during a single pulse the intensities of the spectrum at ω and ω' are independent, each with an exponential distribution. Indeed, one may be large and the other small and so we have frequency speckles. In exactly the same way in the single frequency treatment, if μ is zero at two positions $\underline{r}$ and $\underline{r}'$, the intensities at these points are independent and this gives spatial speckles. In the case shown in Fig. 2(b) the correlation term is large, and when one frequency appears at a space point so do all other frequencies. Thus, there are no frequency speckles and all spatial points with any light present have the full range of frequencies. Different spectra have been experimentally observed at two space points; also the spectrum exhibits fluctuations from pulse to pulse [4,7]. The spectrum integrated over many pulses is $I(\omega)$, which is the convolution of $f(\omega)$ with $|\tilde{g}(\omega)|^2$.

Laser light does not show a speckle pattern when photographed at a distance. This is because the relative phase between the

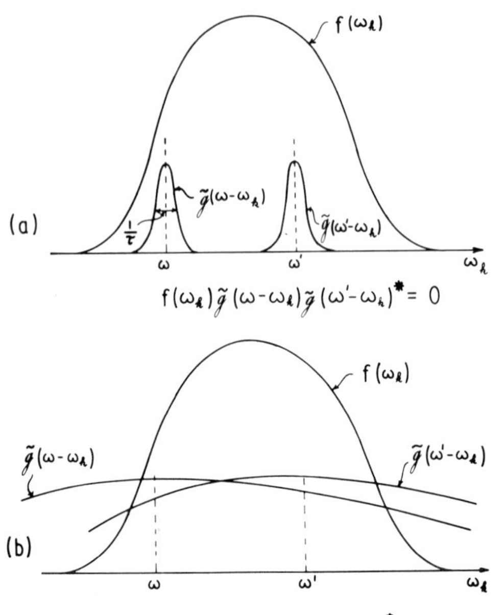

Figure 2.

(a) A long pulse with $\tilde{g}(\omega-\omega_k)$ and $\tilde{g}(\omega'-\omega_k)$ well separated when frequency speckles occur because the correlation term is zero.

(b) The pulse is short and there are no frequency speckles.

emitting atoms is not random but is well organised, and small intensity fluctuations will occur but will be superimposed on a high mean intensity. Speckle patterns are not seen from c.w. sources of A.S.E. because the observation times are long compared with the coherence time of the radiation involved and multiple independent patterns are effectively superimposed, thus averaging out the fluctuations.

For the case when a single pulse from the device embraces two partially correlated amplitude distributions, the distribution of intensity is no longer exponential [2], as in Eq. (3), but is

$$\frac{1}{\phi I_o} \left[\exp \frac{-2I}{(1+\phi)I_o} - \exp \frac{-2I}{(1-\phi)I_o} \right] dI \quad ,$$

where $0 \leqslant \phi \leqslant 1$; $\phi = 1$ represents completely correlated distributions and $\phi = 0$ represents two totally independent distributions.

A two component distribution could arise if, for example, during the course of a single A.S.E. pulse an appreciable number of atoms changed phase because of atomic collisions. The finite spread of frequencies will also lead to a two, or many, component distribution when the width of $f(\omega)$ is greater than $1/\tau$. This is, of course, the case when frequency speckles would be anticipated.

In neon where $f(\omega)$ is $\sim$ 700 MHz wide and $\tau < 1$ nsec, we should
expect a single component distribution to work well. Indeed,
Fig. 3 shows that the one component theory satisfying Eq. (3) is
sufficient to explain the behaviour of the neon source.

We find, too, that the joint distribution for the intensity
I and I' at $\underline{r}$ and $\underline{r}'$ for the two component case, differs from that
for one component. For example, instead of Eq. (4) we have,

$$< I\ I'> = \left(1 + \tfrac{1}{2}\ (1+\phi^2)\,|\mu|^2\right)\ I_o^2\ ,$$

while

$$< I^2 > = (3 + \phi^2)\ I_o^2/2\ ,$$

$$<I^2 I'> = \left[\,(3 + \phi^2)+(3 + 5\phi^2)\,|\mu|^2\,\right]\ I_o^3/2\ ,$$

$$<I^3 I'> = 3\left[\,2(1+\phi^2)+(3+8\phi^2+\phi^4)\,|\mu|^2\,\right]\ I_o^4/2\ .$$

Within the framework of this theory some sense can be made of
experiments which measure the degree of coherence from pulsed A.S.E.
sources [8] by examining the interference patterns produced by
pairs of pin-holes placed at various separations in the output beam.
Because we have assumed that during the course of a single pulse
the phase of the sources does not change, the relative phase between
any two points in the image plane is a constant. Consequently,

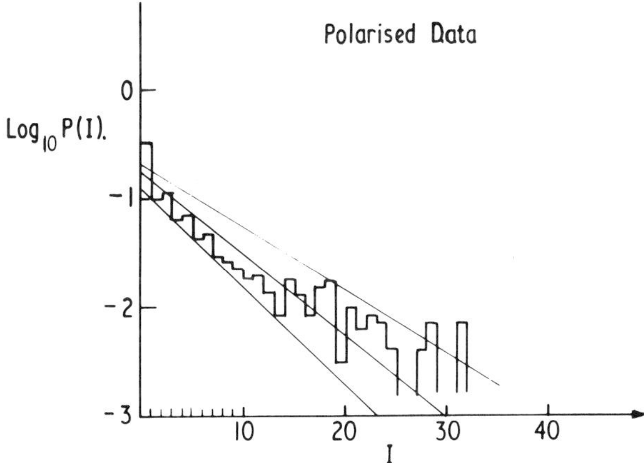

Figure 3. $Log_{10}P(I)$ plotted against I yields a straight line for
light from a Ne A.S.E. source observed through a linear polarizer.
This demonstrates the single component exponential distribution
of the light.

whether or not visible fringes occur depends entirely on the
intensity distribution and the probability that the pin-holes see
intensities that are not too dissimilar. As the distribution of
intensity is exponential for the single frequency case, we may
show the probability of $\alpha I_2 \leqslant I_1 \leqslant (\alpha+d\alpha)I_2$ and $I_1+I_2 > \beta I_0$ to be

$$P = (1 + \beta) \, e^{-\beta} \, \frac{1}{(1+\alpha)^2} \, d\alpha \, ,$$

where I_1 and I_2 are the intensities at each of the pin-holes
supposed far apart so that I_1 and I_2 are independent. The fringe
visibility for a single pulse is given by $V = 2\sqrt{I_1} \, \sqrt{I_2}/(I_1+I_2)$
which reduces to $V = 2\sqrt{\alpha}/(\alpha+1)$. When $I_1+I_2 > I$ the probability
for the visibility to be at least V is

$$(1 - V^2)^{\frac{1}{2}} \, (1 + I/I_0) \, \exp -I/I_0 \, .$$

We may note that high visibility fringes are to be anticipated.
For example the probability that $V > 0.6$ for any intensity is 0.8.
If $I > I_0$, the value is still as high as 0.58. Unfortunately it
is difficult to obtain fringes from a single pulse of an unsaturated
A.S.E. source and a more physical problem would be to calculate
the visibility of the fringes when many independent pulses are
superposed. When this occurs it is as if we were examining the
field of a uniformly intense quasi-monochromatic source. Conse-
quently the usual van Cittert-Zernike result will hold and $V = |\mu|$.
It may be noted that Abrosimov [8] was aware that coherence existed
strongly only within a grain. If the density of excited atoms in
a real A.S.E. source is not uniform across the diameter of the
tube, a, but follows a distribution

$$g(s) = \begin{cases} (a^2 - s^2)^{\nu} & , \quad s < a \\ 0 & , \quad s > a \end{cases}$$

then

$$V = 2^{\nu+1} \, \Gamma(\nu+2)\theta^{-\nu-1} \, J_{\nu+1}(\theta) \, ,$$

where

$$\theta = 2\pi a |\underline{r} - \underline{r}'|/\lambda \ell \ .$$

A non-uniform distribution together with the effect of
reflections from the walls of the tube containing the active medium
may be shown to have a powerful effect on the output of the A.S.E.
Although we have no data to present for neon it is interesting to
see in Fig. 4 the output from a nitrogen A.S.E. source where such
effects are almost certainly present. Although Fig. 4 is the
result of emission from a partially saturated medium, and as such

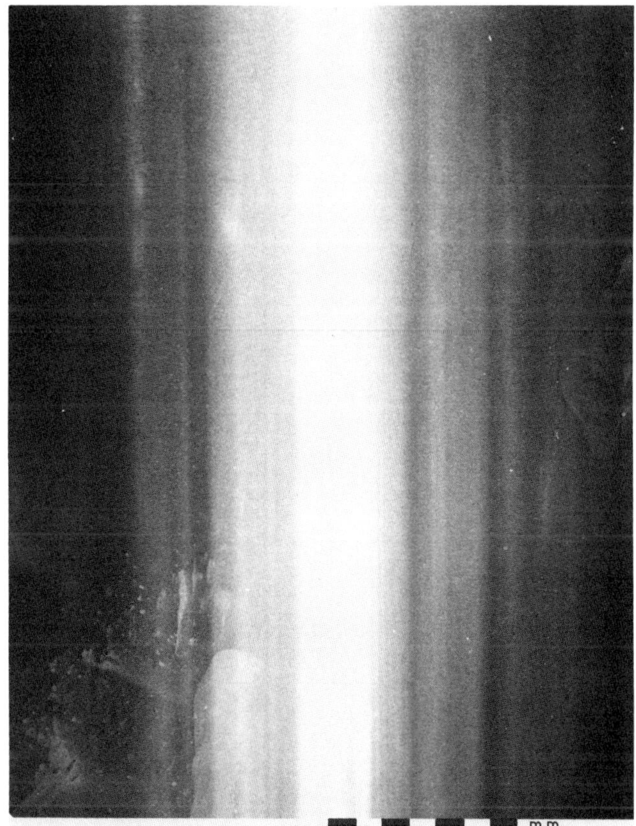

Figure 4.   The output from a 337 nm N_2 A.S.E. source with b = 40 cms, a = 1 mm, a' = 2 cms, photographed at 38 cms so that ℓ = 2b.

is not strictly susceptible to an analysis of the kind we have
carried out earlier in this work, it contains certain interesting
features.   There is a well defined geometric distribution of intensity, and it possesses some rather finer "fringe-like" features.
Past experience of A.S.E. devices suggests that intensity distributions like this occur even in the linear, or non-saturated,
regime.   The most convenient case to consider in the linear regime
is a source of A.S.E. of rectangular cross-section, and this has
been investigated.   The calculation leads to an expression very
similar to Hopkin's formula [3] for the complex degree of coherence
in a heterogeneous medium and an intensity correlation which is a
generalization of our equation (4).   The expression for $I(x,y)$ is
itself simply the formula for the diffraction from an extended source.

Figure 5. Elongated speckles from the same N_2 A.S.E. source photographed at 700 cms.

Indeed, we predict that the rectangular geometry will lead to elongated speckles and these may be seen in Fig. 5. At small distances these speckles are unimportant and one just observes the mean intensity which is, of course, what is observed at all ℓ when a photograph records many pulses. We find that we can, in broad terms, account for the structure found in Fig. 4; the calculation and results will be presented elsewhere.

It should be realized that any measurement of intensity, intensity correlation or visibility, should be corrected for the effect of finite aperture [9] at the detector. It is possible, too, that these results may indirectly be another contribution to the current debate [10] concerning the roles of spontaneous and

stimulated emission in the Hanbury Brown-Twiss effect. Finally,
we may summarize: because of the short pulse duration and high
gain of many-pulsed sources of A.S.E., spontaneous emission may in
effect be viewed with its fluctuations, which normally completely
dominate it, "frozen out". This remains the case even when wall
reflections are present. The answer to the question posed in the
title of this paper is thus "speckled", with either spatial or
frequency speckles.

ACKNOWLEDGEMENT

One of us (S.P.K.) acknowledges the Science Research Council
for a Research Studentship.

References

1.   L. Allen and G.I. Peters, Phys. Rev. A *8*, 2031 (1973).
2.   J.C. Dainty, Ed: *Laser Speckle and Related Phenomena* (Springer-
     Verlag, Heidelberg, 1975); J.C. Dainty, in *Progress in Optics*
     Vol. XIV, ed. E. Wolf (North-Holland, Amsterdam, 1976).
3.   M. Born and E. Wolf, *Principles of Optics* (Pergamon, Oxford,
     1964), see Section 10.4.2.
4.   F.A. Korolev, G.V. Abrosimov, A.I. Odintsov and V.P. Yakunin,
     Opt. and Spect. *28*, 290 (1970).
5.   H. Maeda and A. Yariv, Phys. Letts. *43A*, 383 (1973); H. Gamo,
     J.S. Ostrem and S. Chuang, J. Appl. Phys. *44*, 2750 (1973).
6.   F.A. Korolev, A.I. Odintsov, N.G. Turkin and V.P. Yakunin,
     Sov. J. Qu. Elec. *5*, 237 (1975).
7.   V.I. Ischenko, V.N. Lisitsyn, A.M. Razhev, S.E. Rautian and
     A.M. Shalagin, J.E.T.P. Letts. *19*, 346 (1974).
8.   D.A. Leonard and W.R. Zinky, Appl. Phys. Letts. *12*, 113 (1968);
     G.I. Peters and L. Allen, J. Phys. A *5*, 546 (1972); G.V.
     Abrosimov, Opt. and Spect. *31*, 54 (1971).
9.   A. Bark and S.R. Smith, Phys. Rev. A *15*, 269 (1977).
10. N.B. Abraham and S.R. Smith, Phys. Rev. A *15*, 421 (1977);
     L. Mandel, Phys. Rev. A *14*, 2351 (1976).

PULSE PROPAGATION IN A HIGH-GAIN LASER AMPLIFIER: PULSE WAVEFORM
AND SATURATION

H. Gamo, N. Takahashi and H. Osada

University of California, Irvine, California

and

H. Sato

National Defense Academy, Yokosuka, Japan

INTRODUCTION

The resonant pulse propagation is applied to a high-gain helium
xenon 3.5 µm laser amplifier with a partially homogeneously broadened
lineshape, with emphasis on the saturation effect and pulse waveform. This problem has been studied in a linear gas laser amplifier for determining the line-shape parameters in the limited case
when the spectrum of the incident pulse is narrow enough near and
around the line-center frequency of the gain medium.[1] Here, we
shall find the formulas of resonant pulse propagation which are applicable to the general case, even if the width of the spectrum is
comparable with or greater than the spectral width of the gain
medium.[2] Including the saturation, our particular interests are
in the formulas of the incremental intensity gain at the pulse
peak γ_I, the increments of the pulse delay time $\partial T/\partial z$, and of the
pulse duration $\partial \tau'/\partial z$ per unit plasma length. Using the experimental
results of the value of $\partial T/\partial z$, we shall determine the saturation
intensity I_s of the gain medium. The value of I_s, obtained by applying a quasi-CW treatment to pulse propagation, is compared with the
above results.

We shall also discuss the characteristics of pulse propagation,
comparing them with those of line-narrowing and rebroadening of the
amplified spontaneous emission in the high-gain laser amplifier.
Using a self-mode-locked pulse as an incident pulse, the amplified

pulse waveforms as a function of plasma length are also evaluated
in terms of the coefficients of the skewness κ_1 and excess κ_2 by
using the curve fitting to the experimental waveform and the
property of a quasi-Gaussian pulse [3].

THEORETICAL PREPARATION ON SATURATION AND WAVEFORM IN RESONANT PULSE PROPAGATION

For the waveform analysis of the resonant pulse amplified
through the gain medium, important parameters which express the
characteristics of the amplified pulse waveform are the pulse peak
intensity, the pulse delay time at the pulse peak and the pulse
duration defined by the half intensity full-width as functions of
the pulse intensity and plasma length. The general formulas of
pulse progagation in a partially homogeneously broadened, linear
gain medium have been conveniently derived by H. Gamo and H. Sato[1]
for determining the line-shape parameters. However, its validity
of usefulness is limited to the case in which the pulse spectrum is
narrow enough around the line center frequency ν_o. Considering
that the incident pulse has a certain spectrum range comparable to
or greater than that of the gain medium, the formulas of the intensity gain, the group velocity and the line-narrowing have been derived for a weakly saturated, partially homogeneously broadened
gain medium.[2] According to the definition in Ref.[1] and the
above derivation for the intensity gain, the incremental pulse-
peak intensity gain γ_I at the line center can be expressed

$$\gamma_I \simeq 2G\ \psi(0,\ y/\sigma)\ \exp\left[-\pi^{\frac{1}{2}}\ \left(\frac{y}{s}\right)\ \left(\frac{I_i}{I_s}\right)\psi(0,\ y/s)\right], \tag{1}$$

where G is the gain parameter of the medium, $\psi(0,\ y/s)$ is the Voigt
function at the line center with the parameters y (the ratio of
the Lorentzian lenewidth $\Delta\nu_L$ to the Doppler linewidth $\Delta\nu_D$) and σ:

$$y = (\ell n2)^{\frac{1}{2}}\ \Delta\nu_L/\Delta\nu_D, \tag{2}$$

and

$$\frac{1}{\sigma^2} = 1 - \pi^{\frac{1}{2}}\ \left(\frac{y}{s}\right)|\psi''(0,\ y/s)|\ \frac{1}{2s^2}\left(\frac{I_i}{I_s}\right). \tag{3}$$

The parameter s is the ratio of the pulse spectrum width defined by
the half intensity full-width $\Delta\nu_i$ to $\Delta\nu_D$, ψ'' is the second derivative of ψ with respect to $x = 2(\ell n2)^{\frac{1}{2}}\ (\nu - \nu_o)/\Delta\nu_D$, I_i is the incident pulse-peak intensity and I_s is the saturation intensity of the
medium.[4]

In the weakly saturated region the pulse duration will become
longer and the spectral linewidth will become narrow in accordance

with the uncertainty relation between the pulse duration and the
spectral linewidth. Hence, the line-narrowing formula described in
our previous paper,[2] can be reformulated in terms of pulse duration,

$$\tau'^2 - \tau_i'^2 = \frac{Gs^2 |\psi'' (0,y/\sigma)| (2\ell n2)^2}{\sigma^2 (\pi\Delta\nu_D)^2} z \, , \tag{4}$$

where τ' is the pulse duration at plasma length z and τ_i' is the
incident pulse duration. The increment of the pulse duration
$\partial\tau'/\partial z$ per unit plasma length can be approximated from the above
formula in the form:

$$\frac{\partial\tau'}{\partial z} \simeq \frac{Gs^2 |\psi''(0, y/\sigma)| (2\ell n2)^2}{2\tau_i' \sigma^2 (\pi\Delta\nu_D)^2} \, . \tag{5}$$

This formula, which can be applied to the weakly saturated region,
is reduced to the formula developed for a linear gain medium,[1]
in the limit of $\sigma = 1$ for a very weak incident pulse-peak intensity.

From the refractive index for a weakly saturated laser amplifier, [Eq. (35) in our preceding paper, Ref.[2]], we can calculate
the group velocity v_g, from which the pulse delay time T can be
derived as z/v_g. The increment of the pulse delay time, $\partial T/\partial z$, per
unit plasma length can be given as follows:

$$\frac{\partial T}{\partial z} = \frac{G(\ell n2)^{1/2}}{\pi\Delta\nu_D} \phi' (0,y/\sigma)\left\{ 1 - \pi^{1/2}\left(\frac{y}{s}\right)\right.$$

$$\left. \times \left[\psi(0,y/s) + \frac{|\psi''(0,y/s)|}{4s^2}\right] \left(\frac{I_i}{I_s}\right)\right\} \, , \tag{6}$$

where $\phi' (0,y/\sigma)$ is the first derivative of the Hilbert conjugate
of the Voigt function.

EXPERIMENTS

1. Experimental Set-Up

The schematic diagram of the experimental set-up is illustrated
in Fig. 1. A self-mode locked helium-xenon 3.5 μm laser with 9 m
mirror separation is used as an optical pulse source, providing
10 nsec pulse duration and 16 MHz repetition frequency. By a beam
splitter inside the resonator two output beams are coupled out, one
for the incident pulse on the laser amplifier and the other for
generating a trigger pulse for the sampling scope. Two infrared
isolators using the Faraday rotation in Yttrium Iron Garnet (YIG)
with the calcite dichroic polarizers are inserted before and after
the laser amplifier to prevent any spurious coupling between the

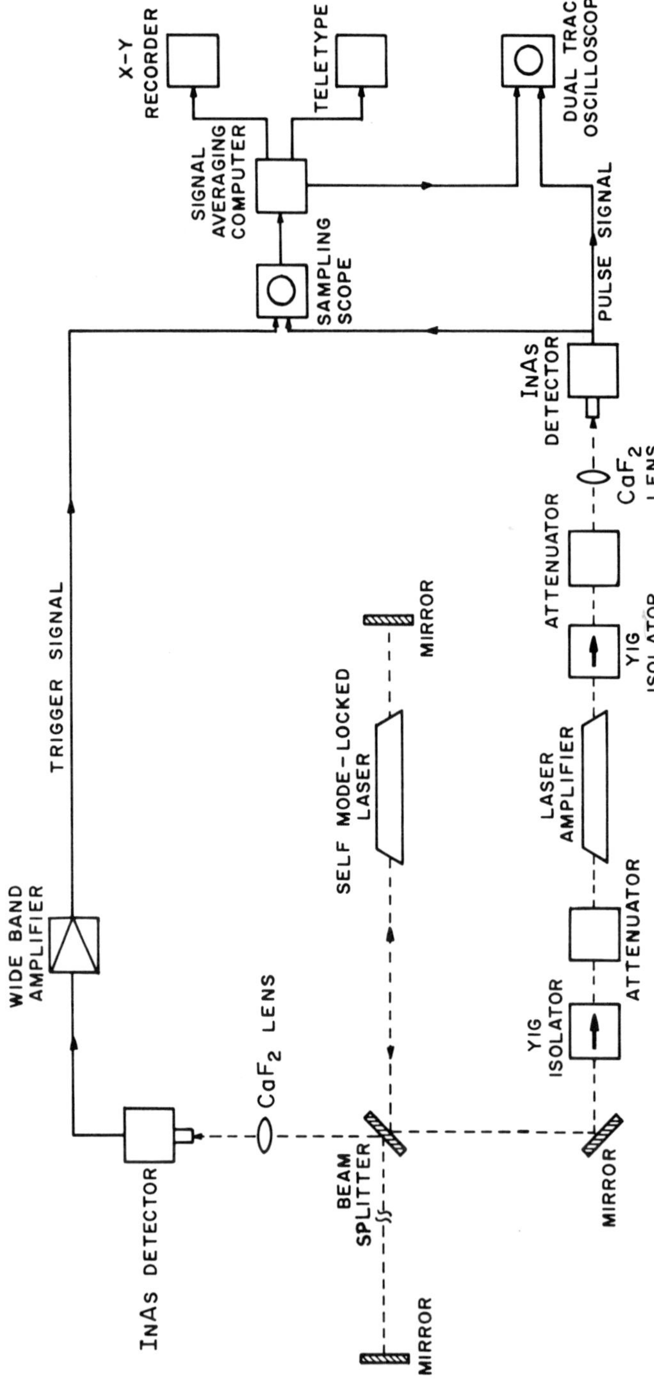

Fig. 1. Schematic diagram of experimental set-up for resonant pulse propagation in a laser amplifier.

amplifier and the mode-locked laser, and the detector. In order to
change the input pulse intensity, a variable attenuator using the
polarizers is set right before the laser amplifier. The plasma
tube of a laser amplifier has 14 tungsten anodes and 2 aluminum
cathodes spaced along the discharge tube, enabling us to choose
various plasma lengths from 5 cm to 100 cm. The laser source for
mode-locking consists of two laser tubes filled with the same pressure, 120 m Torr of xenon. The laser pulses are detected by InAs
photovoltaic detectors (Philco L4530) through the calcium fluoride
(CaF_2) lenses. It should be noticed that the YIG isolator plays an
essential role in eliminating the undesirable coupling between the
laser amplifier and the mode-locked laser, especially in the case
of a high-gain medium. In applying the formulas to the experimental
results it is also necessary that the line center of the laser amplifier be tuned to the mean frequency of the self-mode locked pulse,
i.e. the so-called resonant condition. This can be done by adjusting
the discharge current within the experimental error, when all plasma
tubes are filled with the same gas pressure mentioned above. Another
important aspect of the experimental technique is to adjust the intensity of the incident pulse in the region of the linearity of the
photovoltaic detector. The characteristic of the detectors has been
confirmed as responding sufficiently for the optical pulse used in
the experiments.[5,6]

For measuring the amplified pulse through the amplifier, the
signal averaging technique was applied to the computer (PDP 15/20).
Using this technique the sampling scope (Tektronix 1S1) is triggered by the reference pulse mentioned above, and the signal sampled
out at the sampling unit is transferred to the computer buffer (512
words) through a DC amplifier and an analog-to-digital converter.
Each scan of the waveform is summed up and averaged at the accumula-
tor under the control of the index and scan generator, which produces
the horizontal timing signal. The averaged signal is displayed on
the oscilloscope through a digital-to-analog converter with the horizontal timing signal. The scanning time on the sampling unit is also
controlled by the scan generator, based on the preset scan number.

Since the summed signal increases linearly as the scan number
but the noise increases as the square root of the scan number, the
signal-to-noise ratio of the averaged signal can be improved in proportion to the square root of the scan number. In order to take full
advantage of signal averaging of the sampled pulse, it is extremely
important to secure stable operation of a mode-locked laser. In
the case of the He-Xe 3.5 μm laser, up to a mirror separation of
nine meters, we can easily obtain stable self-mode locked operation.
For measuring the linewidth of the incident pulse $\Delta\nu_i$ in the stable
mode-locking condition, the scanning Fabry-Perot interferometer
(Lansing No. 21-334) was used. By taking the deconvolution of the
power spectrum obtained, the real linewdith of the incident pulse
was measured.

2. Experimental Results

For the pulse waveform analysis, the incident mode-locked
pulse amplified through the laser amplifier was measured for various plasma lengths in increments of 10 cm by keeping the incident
pulse intensity at a certain value. Then, changing the incident pulse
intensity on the amplifier by adjusting the angle of polarizers in
the attenuator, we obtained several groups of amplified pulse
waveforms as functions of plasma length z, depending on the numbers
of the changed input pulse intensity. A series of typical pulse
waveforms obtained by the signal averaging process is shown in Fig.
2, as an example of the input pulse peak intensity I_i = 12.5
μW/mm^2. Particularly, focusing on the pulse-peak intensity, the
pulse delay at the pulse peak, and the pulse duration defined by the
half intensity fullwidth, Fig. 3 shows the incremental intensity
gain γ_I, the increments of the pulse delay time, $\partial T/\partial z$, and of the
pulse duration, $\partial \tau'/\partial z$, per unit plasma length.[7] Since these
are monotonically decreasing functions with increase of the input
pulse-peak intensity I_i, γ_I, $\partial T/\partial z$ and $\partial \tau'/\partial z$ can be expressed as

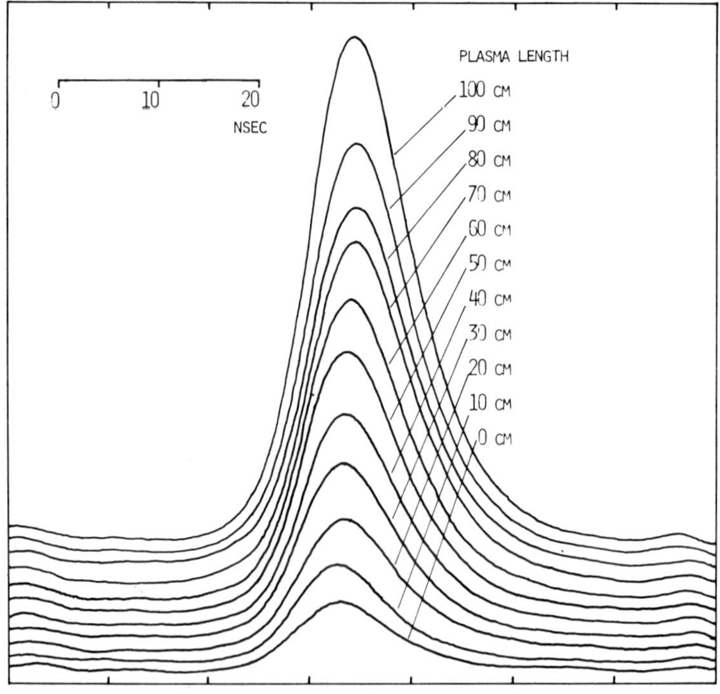

Fig. 2 A series of pulse intensity waveforms for various plasma
lengths of a He-Xe 3.5 μm laser amplifier at the input pulse-peak
intensity I_i = 12.5 μW.mm^2. The vertical scale is arbitrary.

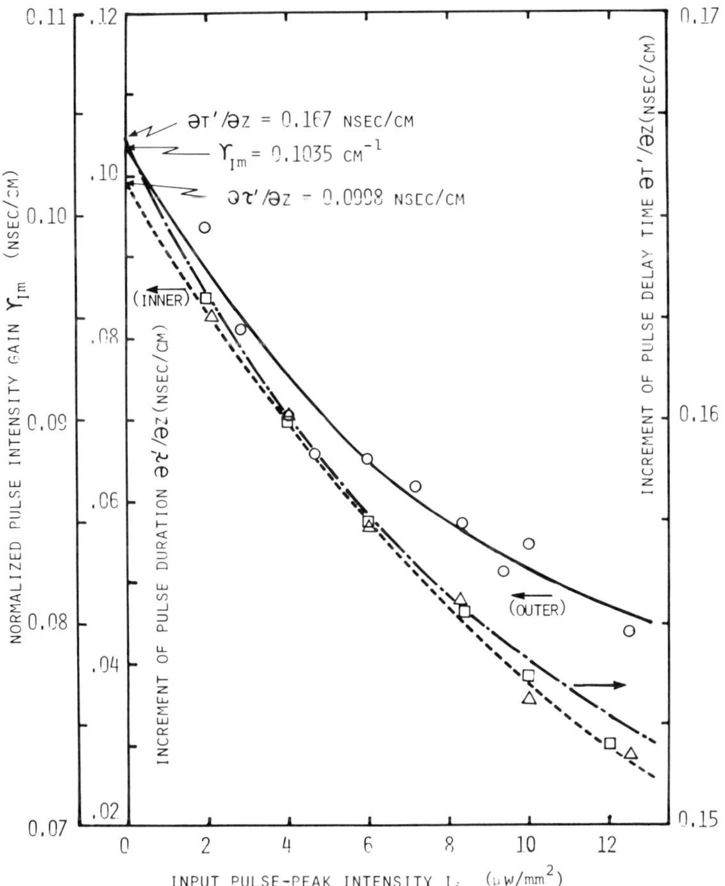

Fig. 3. Experimentally obtained incremental intensity gain γ_I, increments of pulse delay time $\partial T/\partial z$, and of pulse duration $\partial \tau'/\partial z$, per unit of plasma length versus input pulse-peak intensity I_i.

functions of the input pulse-peak intensity I_i by fitting the experimental curve by the method of least squares. Their coefficients up to the second order in the power series expansion, are summarized as follows:

$$\gamma_I \simeq 0.1035 - 0.0033\ I_i + 0.00011\ I_i^2\ (\mathrm{cm}^{-1}), \tag{7.a}$$

$$\partial T/\partial z \simeq 0.1673 - 0.0163\ I_i + 0.00069\ I_i^2\ (\mathrm{nsec\ cm}^{-1}), \tag{7.b}$$

$$\partial \tau'/\partial z \simeq 0.0998 - 0.0084\ I_i + 0.00021\ I_i^2\ (\mathrm{nsec\ cm}^{-1}). \tag{7.c}$$

Based on the power spectrum measured by the scanning Fabry-Perot interferometer we also obtained the spectral linewidth of the incident pulse $\Delta\nu_i \simeq 58.9$ MHz in the well-defined mode-locked condition.

SATURATION INTENSITY AND PULSE WAVEFORM ANALYSIS

The experimental results for γ_I, $\partial T/\partial z$ and $\partial\tau'/\partial z$ described in the preceding section include some information on the saturation intensity I_s of the gain medium. Thus, we can determine the saturation intensity I_s by using either one of these three experimental results. However, it must be noticed that the measured increment of the pulse delay time $\partial T/\partial z$ is more reliable for obtaining the saturation intensity I_s, so long as the trigger signal to the sampling scope works well. This is one of the advantages of the pulse propagation method for determining I_s. For this reason, we intentionally use Eq. (6). Normalizing Eq. (6), we obtain the normalized increment of pulse delay time $\partial\hat{T}/\partial z$ as

$$\frac{\partial\hat{T}}{\partial z} = \frac{\partial T/\partial z}{G(\ell n2)^{\frac{1}{2}}\ \phi'(0,\ y/\sigma)/\pi\Delta\nu_D}$$

$$\simeq 1 - \pi^{\frac{1}{2}}(\frac{y}{s})\ [\psi(0,\ y/s) + |\psi''(0,\ y/s)|/4s^2]\ (\frac{I_i}{I_s}). \qquad (8)$$

Using the obtained values of $\Delta\nu_L = 11.8$ MHz and $\Delta\nu_D = 104$ MHz, the corresponding values of ψ and ψ'' to $y = 0.094$ and $s = 0.564$ with measured $\Delta\nu_i = 58.9$ MHz, we obtain

$$\frac{\partial\hat{T}}{\partial z} \simeq 1 - 0.566\ (\frac{I_i}{I_s}). \qquad (9)$$

From Eq. (7b), we have also the normalized power series expansion up to the second order for the experimentally obtained increment of pulse delay time $\partial T_{exp}/\partial z$, as

$$\frac{\partial\hat{T}_{exp}}{\partial z} \simeq 1 - 0.09742\ I_i + 0.004127\ I_i^2. \qquad (10)$$

Then, equating both first order coefficients of Eqs. (8) and (9), we find the saturation intensity I_s:

$$I_s \simeq 5.79\ \mu W/mm^2. \qquad (11)$$

This is in good agreement with the value obtained by using the CW method [8] within the experimental error.

On the other hand, when the spectrum of the incident pulse

is narrow enough around the line center of the medium compared with
that of the gain medium, we can treat the problem as a quasi CW-
propagation. In this case, we can express γ_I, $\partial T/\partial z$ and $\partial \tau'/\partial z$
by the Voigt function and its Hilbert conjugate, including the
saturation effect. Thus, these three formulas can be expanded into
a power series as a function of the normalized intensity divided by
the saturation intensity, I_i/I_s. Similarly, by using curve fitting
and the numerically calculated $\psi(0,y)$, $\psi''(0,y)$ and $\phi'(0,y)$ and the
y value obtained, we can also obtain the saturation intensity I_s.
In a trial, we applied this to evaluate the saturation intensity
and obtained $I_s = 6.3\ \mu W/mm^2$, which was slightly off from that obtained above.

Based on one group of experimentally obtained pulse waveforms
as a function of plasma length for the incident pulse peak intensity
$I_i = 12.5\ \mu W.mm^2$, the coefficients of the intensity skewness and excess, κ_1 and κ_2 were evaluated by curve fitting as shown in Fig. 4.

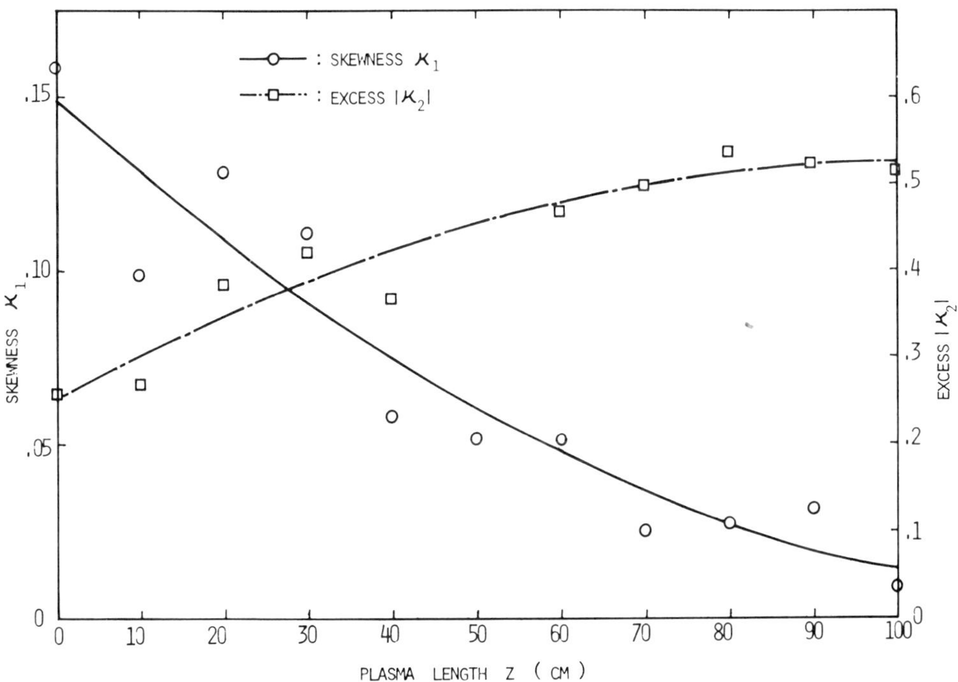

Fig. 4 Evaluated skewness κ_1 and excess κ_2 of the amplified pulse
waveform as a function of the plasma length of a laser amplifier.

This was done by equating each coefficient obtained by the curve
fitting method to the coefficients including the skewness κ_1 and
excess κ_2 and the Hermite polynomials, since a quasi-Gaussian pulse
can be conveniently expressed in terms of κ_1 and κ_2 with third and
fourth order Hermite polynomials.[3] From the values illustrated
in Fig. 4 it is seen that κ_1 decreases and the absolute value of
κ_2 increases with increasing of plasma length. This indicates that
the pulse waveform tends to become symmetric but off-Gaussian with
increasing flatness.

CONCLUSION

From the formulas derived for the group velocity and line-
narrowing of amplified spontaneous emission (ASE) in the weakly
saturated laser amplifier,[2] we derived those formulas of the incremental intensity gain γ_I, the increments of the pulse delay time
$\partial T/\partial z$ and of the pulse duration $\partial \tau'/\partial z$ to the pulse delay in a He-Xe
3.5 μm laser amplifier, we determined the saturation intensity $I_S =$
5.79 μW/mm^2. This value of I_S is consistent with that measured by
the CW operation method within a small difference comparable to the
possible experimental error.[8]

A series of pulse intensity waveforms experimentally obtained
as a function of plasma length were analyzed in terms of the coefficients of skewness κ_1 and excess κ_2.

It is desirable to extend the theoretical analysis to the heavily saturated region and to compare the experimental results with
the numerically calculated waveforms [9] and also the stationary
waveforms known as solitons.[10]

References

1.   H. Gamo and H. Sato, J. Appl. Phys. *46*, 3585 (1975).
2.   H. Gamo and H. Osada, see this Volume, p. 583.
3.   H. Cramer, *Mathematical Methods of Statistics* (Princeton Uni-
     versity Press, Princeton, N.J., 1951) p. 223.
4.   D.C. Sinclair and W.E. Bell, *Gas Laser Technology* (Holt, Rine-
     hart and Winston, New York, 1969).
5.   H. Sato and H. Gamo, Jpn. J. Appl. Phys. *16*, No. 7, 1213 (1977).
6.   H. Sato, Ph.D. Dissertation, University of California, Irvine,
     Calif., 1975. Unpublished; Appendix C., p. 81.
7.   These experimental results are exactly the same as shown in
     Fig. 8 (a), (b), (c) of Ref. 1.
8.   H. Gamo and J.S. Ostrem, J. Opt. Soc. Am. *65*, 29 (1975).
9.   A. Icsevgi and W.E. Lamb, Jr., Phys. Rev. *185*, 517 (1969).
10. G.L. Lamb, Jr., see this Volume, p. 747.

FREQUENCY SELECTIVE INTENSITY STATISTICS OF AMPLIFIED SPONTANEOUS EMISSION

N.B. Abraham[*] and S.R. Smith

Bryn Mawr College, Bryn Mawr, Pa.

The intensity fluctuations of lasers and thermal sources have been extensively studied and are well understood. A saturating laser amplifier is an intermediate case which merits further study. In a laser amplifier distributed spontaneous emission is amplified by the medium. With linear amplification the resulting signal should display the Gaussian amplitude fluctuations characteristic of spontaneous emission. For sufficiently long high-gain amplifiers the intensity may reach a level that significantly saturates the gain. The resulting non-linear amplification tends to reduce the fluctuations of the signal.

Amplified spontaneous emission (ASE) is a broadband signal. As a result, a calculation of the effects of saturation is generally more complicated than in the single mode case[1,2]. For inhomogeneously broadened media, however, the level of saturation of a given frequency component of the gain depends mainly on the spectral intensity within one homogeneous width of that frequency. This causes the well-understood rebroadening of the initially gain-narrowed spectral profile. The frequency-selective saturation also means that the degree of reduction of intensity fluctuations would be expected to vary across the spectral profile. The most pronounced effects should occur at the higher intensity levels near line center.

Homogeneously broadened transitions are expected to display less frequency variation in intensity fluctuations, since the various frequency components of the gain are equally saturated by any signal. This simple view is, however, likely to be complicated by coherent interactions among the frequency components which may lead to cross-spectral modulation or even to strong pulsing in the output[3,4].

Theoretical work on the intensity statistics of broadband amplified spontaneous emission has been limited, although saturation reduction of fluctuations, cross-spectral modulation (which increases fluctuations), and other coherent interactions leading to pulsing have been predicted[3-7].

Gamo and coworkers have reported experimental evidence of non-Gaussian fluctuations in the output of saturating amplifiers[8-10]. These measurements, however, averaged the statistical fluctuations over the spectral profile.

We report here a series of measurements of the fluctuations of the different frequency components of saturated signals. The high gain transition at 3.51 µm in xenon was chosen because it is predominantly inhomogeneously broadened, having a Doppler width of order 110 MHz and a natural linewidth of about 4 MHz[11]. The gain of this transition can be increased by adding helium to the discharge[12], The resulting collision broadening of the transition causes significant homogeneous broadening. The intensity statistics which result may then be contrasted with those of the inhomogeneously broadened case.

Frequency selectivity is achieved by heterodyning the ASE with the output of a tunable single-mode laser. Fluctuations in the low frequency heterodyne signal are then characteristic of the statistics of that portion of the broadband ASE signal having optical frequencies near the laser frequency.

A diagram of the optical bench layout is shown in Fig. 1. Part of the signal from the tunable laser (Laser 1) was heterodyned with the output of a frequency-stabilized laser (Laser 2). The resulting beat note, which was stable to about $\pm \frac{1}{2}$ MHz, served as a monitor of the relative frequency of the tunable laser. Laser 2 was feedback-stabilized to the gain-narrowed peak of a Xe-He amplifier. The entire optical system was mounted on a vibration-isolated table.

The ASE sources were made of heavy-wall pyrex tubing with a 4 mm. inside diameter. The discharge length could be varied by choosing among anodes spaced 10 or 20 cm. apart. Two sources were used. One was variable in length from 10 to 120 cm., and was used for measurements on Xe-He. A longer source tube, variable from 40 to 200 cm., was used for the lower gain xenon discharge. The longer source tube had a bypass channel for pressure equalization. Both tubes used a cold cathode and were excited by a D.C. supply with a 300 KΩ series ballast resistor to damp plasma oscillations. To overcome gas cleanup problems, the discharge tubes were kept attached to a vacuum system and gas filling station. Single-isotope enriched xenon (92% Xe^{136}) was used to avoid the effects of isotope broadening.

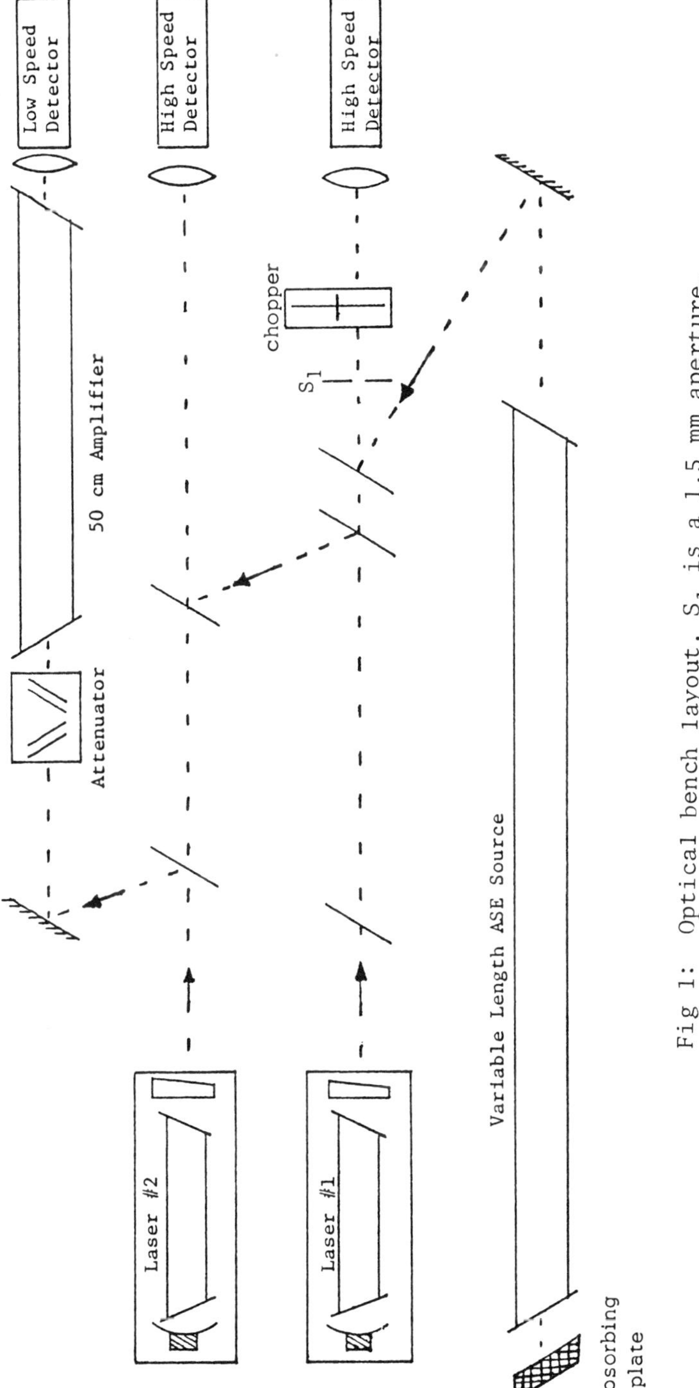

Fig 1: Optical bench layout. S_1 is a 1.5 mm aperture.

A dielectric filter was used to select the 3.51 μm line. The heterodyne signal was detected with an InAs solid state diode which was reverse-biased for high frequency response. The signal passed through a high-pass filter (300 kHz cut-off frequency), two Hewlett-Packard 161A amplifiers (80 db total gain), and then through a low-pass filter (1.0 MHz cut-off frequency). The filtered signal was sampled by a Tektronix Model 3S76 sampling oscilloscope used as a fast sample and hold amplifier. Histograms of the sampled intensity were compiled by a LeCroy Model 3001 Multichannel Analyzer. The sampling system was triggered by a pulse generator at a rate of about 10 kHz. Approximately 200,000 samples were taken for each histogram. The histograms were printed out and then analyzed by computer for their central moments and cumulants. The gain of the system was set so that the MCA had an effective resolution at the detector of about $\frac{1}{4}$μV/channel. The equivalent noise of the detector and electronics was of order 1.8μV RMS. The overall signal histograms typically involved about 150 channels. Signal-to-noise intensity ratios ranged from 1:4 to 4:1 but the sensitivity for higher moment determinations was enhanced by the fact that the heterodyne signals were typically asymmetric while the noise was Gaussian.

For each frequency setting of laser, histograms were compiled for the combined laser and ASE sources, the ASE source alone, the laser source alone, and the detector and electronic noise. Subtracting the cumulants of the laser, ASE, and the noise signals from the corresponding cumulants of the combined signal leaves terms that depend on the statistical properties of the laser and the spectral components of the ASE signal in two 700 kHz bands symmetrically placed 650 kHz on each side of the laser frequency. The resulting second difference cumulant has the form

$$DK_2(\nu) = 2\gamma^2 \, I_L(\nu) \, I_S(\nu) \, ,$$

where $I_L(\nu)$ is the laser intensity, $I_S(\nu)$ is that portion of the intensity of the ASE in the two bands on either side of the laser frequency, and γ is the mutual coherence factor in the superposition of the laser and ASE. This is just the usual heterodyne term in the superposition of two signals which have been electronically filtered after detection.

The difference in the third cumulants is given by

$$DK_3(\nu) = 6\gamma^2 \left[K_{2L}(\nu) \, I_S(\nu) + K_{2S}(\nu) \, I_L(\nu) \right] ,$$

where K_{2L} and K_{2S} represent the second cumulants of the laser intensity and the selected part of the ASE, respectively. Thus calculation of the cumulants of the histograms together with measurements of the laser intensity allows the determination of relative values of I_S and K_{2S}. It should be noted that the second and third order cumulants are equal to the second and third central moments of any distribution.

The second cumulant K_2 is more commonly known as the variance of the signal,

$$K_2 = <I^2> - <I>^2 \ .$$

To compare the statistical fluctuations of signals of different intensities, it is common to use the normalized variance:

$$Q \equiv \frac{K_2}{M_1^{\ 2}} = \frac{<I^2> - <I>^2}{<I>^2} \ .$$

The normalized variance of the ASE signal is given by

$$Q_S(\nu) = \frac{2\gamma^2 \ I_L(\nu) \ \left[\frac{1}{3} \frac{DK_3(\nu)}{DK_2(\nu)} - \frac{K_{2L}(\nu)}{I_L(\nu)} \right]}{DK_2(\nu)} \ ,$$

and the spectral intensity of the ASE signal is given by

$$I_S(\nu) = \frac{DK_2(\nu)}{2I_L(\nu) \ \gamma^2} \ .$$

Only relative values of these quantities could be calculated because of the undetermined coherence factor. The laser and ASE intensities were measured by use of a mechanical chopper and a lock-in amplifier.

RESULTS

Results for an inhomogeneously broadened amplifier are shown in Figs. 2-5. The source was filled with 108 mTorr of xenon and run at a discharge current of 5 mA. Figure 2 shows the variation of total ASE intensity and spectral width with discharge length. The variation in the slope of the intensity curve and the rebroadening of the spectral width are clear indications of saturation[11].

Figures 3 and 4 show spectral density and normalized variance as functions of frequency for several amplifier lengths. It is clear that the variance decreases with increasing spectral density. This relation is emphasized in Fig. 5 where a log-log plot of

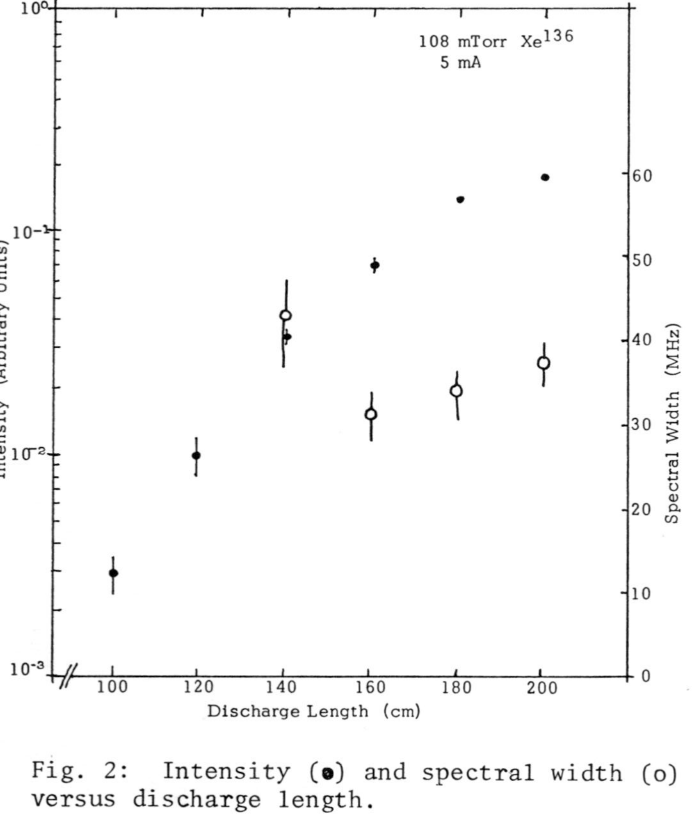

Fig. 2: Intensity (●) and spectral width (o) versus discharge length.

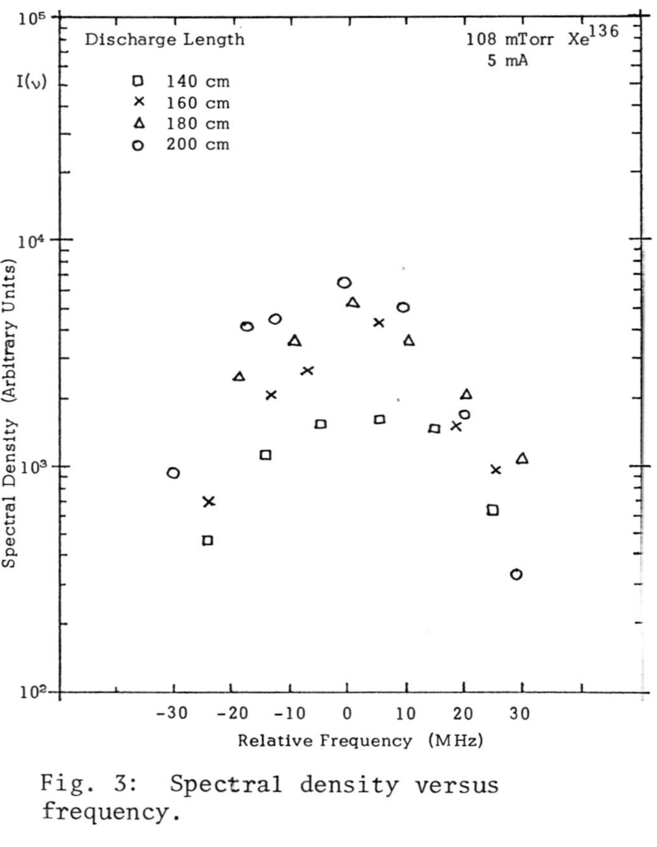

Fig. 3: Spectral density versus frequency.

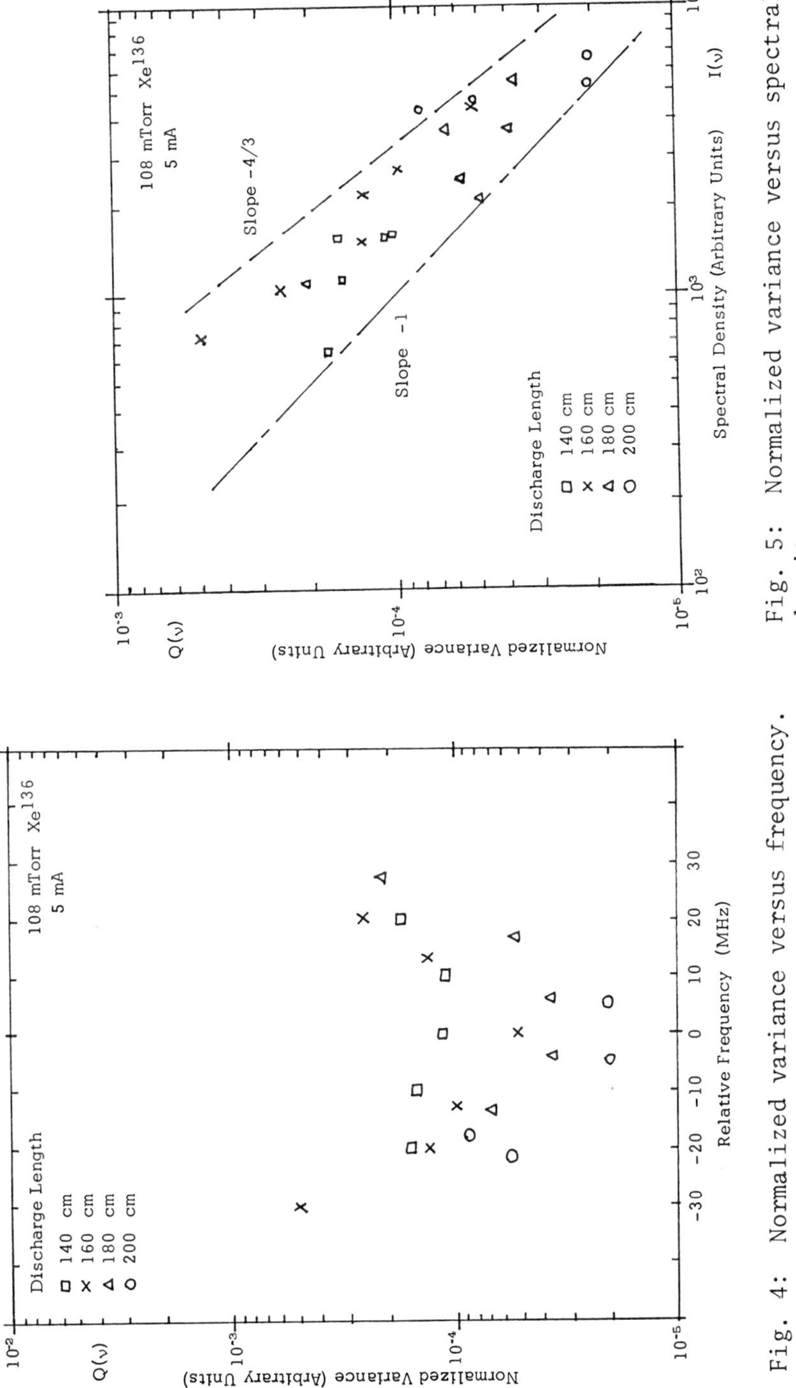

Fig. 5: Normalized variance versus spectral density.

Fig. 4: Normalized variance versus frequency.

normalized variance versus spectral density shows a slope of approximately -4/3. Note that the values of Q cover a range of almost
1½ orders of magnitude. The agreement of values for different discharge lengths indicates that the dominant factor in determining
the normalized variance is simply the spectral density. These results also show the existence of different fluctuation characteristics for different parts of each broadband signal.

To achieve a relatively homogeneously broadened medium 4 Torr
of helium was added to 178 mTorr of xenon. The discharge was operated at 3 mA. The pressure broadening due to helium has been measured to be about 20 MHz/Torr[13]. Thus, the homogeneous width
(about 80 MHz) is larger than the width of the measured spectral
profiles but not quite as large as the Doppler width.

Figures 6-9 display our results for the homogeneous case.
The intensity measurements in Fig. 6 indicate saturation, while
the continued narrowing of the spectral width shown in the same
figure is expected for a homogeneously broadened amplifier.

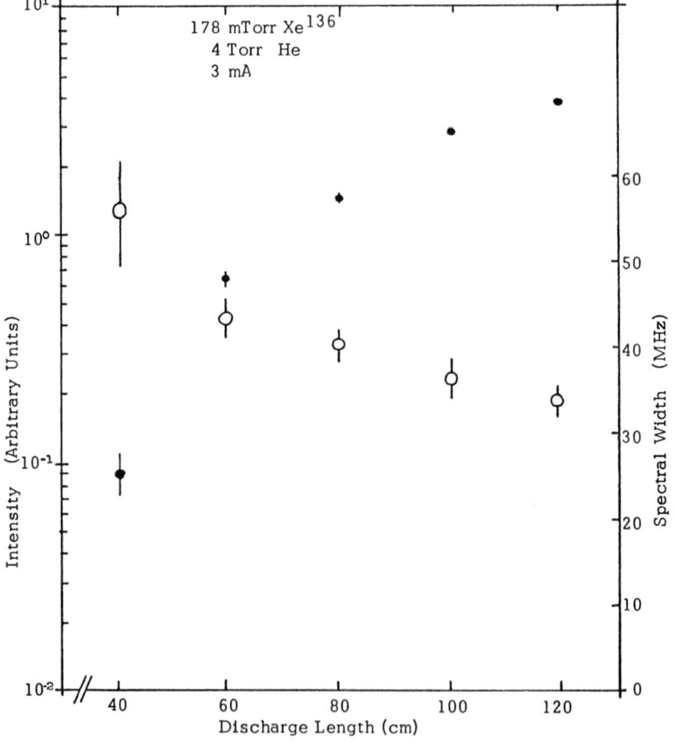

Fig. 6: Intensity (●) and spectral width (o) versus discharge
length.

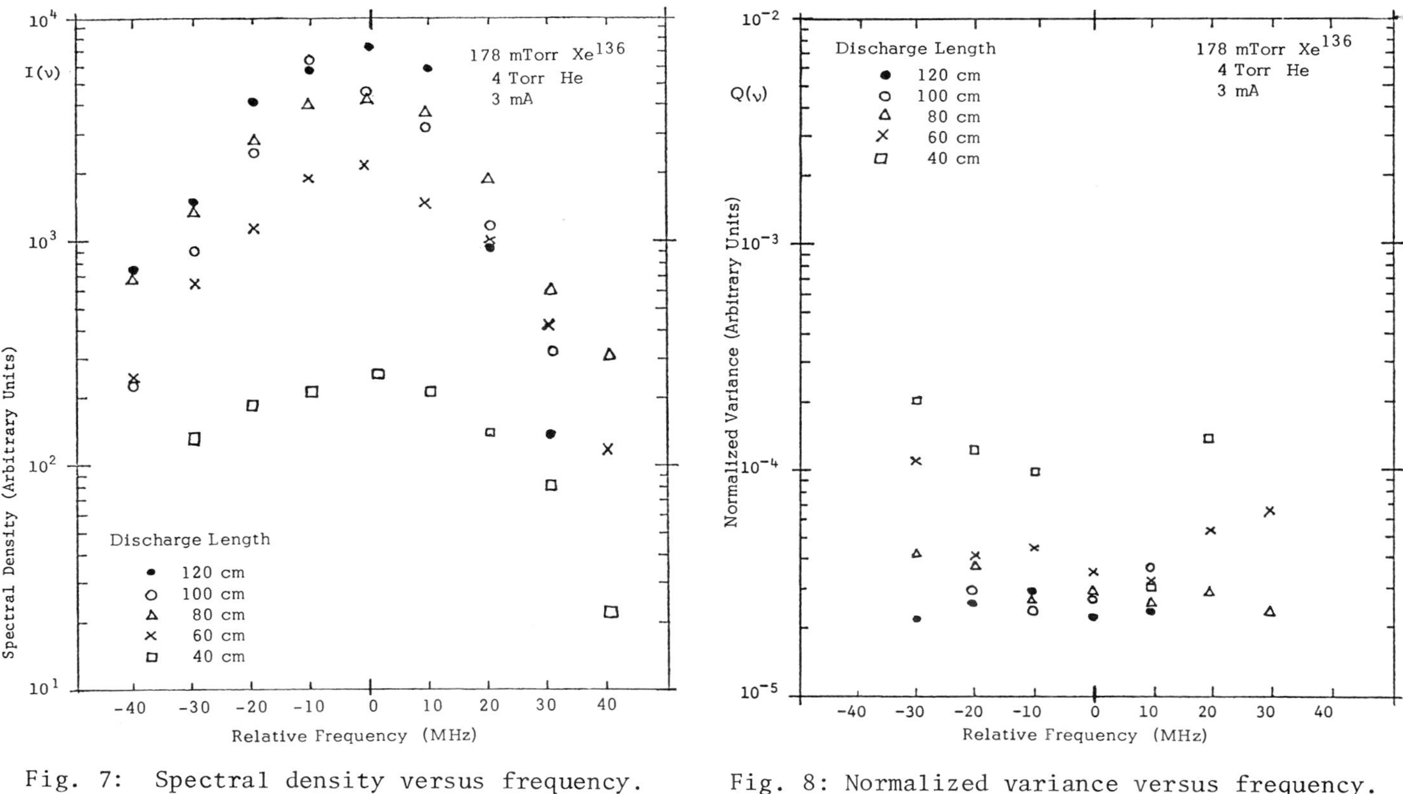

Fig. 7: Spectral density versus frequency.

Fig. 8: Normalized variance versus frequency.

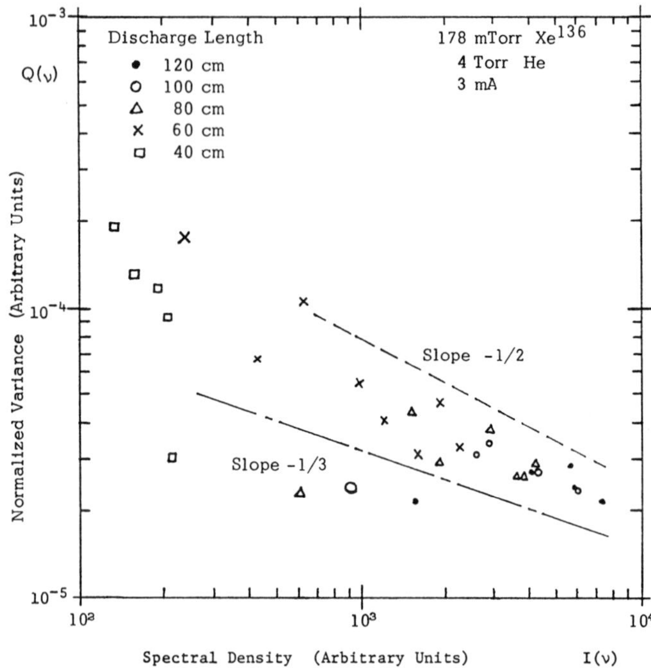

Fig. 9: Normalized variance versus spectral density. For spectral
densities less than 10^3 the signal-to-noise was less than 20%. The
validity of these points is thus less certain.

The variations of spectral density shown in Fig. 7 should be
compared to the variations shown in Fig. 8, which show the corresponding behavior of the normalized variance. The variance depends
on frequency for the weak signals but becomes almost independent of
frequency for higher intensities even over a range larger than the
spectral width. For the shorter discharge lengths the variance
changes by a factor of 5 over the spectral width. At lengths of
80 cm. or more the variance changes by no more than a factor of 2
over a frequency range of at least $1\frac{1}{2}$ times the spectral width.

Figure 9 displays the dependence of the variance on spectral
density. The general trend indicates a considerably weaker dependence than was observed in the inhomogeneous case. (Compare the
slope, approximately -0.4, to the inhomogeneous value -4/3.)

In summary, the data presented show that, for saturating inhomogeneously broadened media, ASE intensity statistics vary significantly across the spectral profile. For the homogeneously broadened case there is less variation. In both cases it appears there exists a close correlation between the spectral density and the variance of the fluctuations.

*New address: Physics Dept., Swarthmore College, Swarthmore, Pa.

References

1. M. Sargent, M.O. Scully, and W.E. Lamb, Jr., Appl. Opt. *9*, 2423 (1970).
2. F.A. Hopf, IEEE J. Quant. Electron. *QE-10*, 717 (1974) and J. Opt. Soc. Am. *61*, 659 (1971).
3. R. Graham and H. Haken, Z. Physik *213*, 420 (1968).
4. C.H. Willis, Phys. Lett. *36A*, 187 (1971); see also D.B. Ostrowsky and C.R. Willis, Phys. Lett. *29A*, 707 (1969).
5. H. Gamo, IEEE J. Quant. Electron. *QE-2*, xii (1966); also H. Gamo, private communication.
6. J.H. Parks, J. Opt. Soc. Am. *61*, 658 (1971).
7. M.M. Miller and A. Szöke, *Coherence and Quantum Optics*, ed. L. Mandel and E. Wolf (Proc. of Third Rochester Conference on Coherence and Quantum Optics) (Plenum Press, New York, 1973) p. 885.
8. H. Gamo and S.S. Chuang, *Coherence and Quantum Optics*, ed. L. Mandel and E. Wolf (Proc. of Third Rochester Conference on Coherence and Quantum Optics) (Plenum Press, New York, 1973) p. 491.
9. H. Gamo and S.S. Chuang, Final Report AFCRL-71-0612 (1971).
10. H. Gamo, H. Osada and A.S. Liao, J. Opt. Soc. Am. *66*, 1076 (1976); also H. Gamo, private communication.
11. L.W. Casperson, J. Appl. Phys. *48*, 256 (1977).
12. R.A. Paananen and D.L. Bobroff, Appl. Phys. Lett. *2*, 99 (1963).
13. S.C. Wang, Ph.D. Thesis, Stanford University, available from University Microfilms, 1971.

NON-MARKOVIAN EFFECTS IN SPONTANEOUS EMISSION: DEVIATIONS FROM
THE EXPONENTIAL DECAY 'LAW'

Peter L. Knight

Royal Holloway College, University of London, England

1.   INTRODUCTION

In this paper we re-examine the validity of the exponential
decay law in atomic spontaneous emission theory. The observation
of exponential decay is considered by many to test quantum electrodynamics (QED) in a fundamental dynamical sense. We have to date
no evidence which contradicts QED, but the increasing utilization
of tunable coherent light sources has made possible optical tests
in totally new regimes until very recently inaccessible to the
experimentalist. It is now possible to investigate the detailed
dynamical evolution of strongly excited atoms and to examine with
some precision the associated spectrum of the radiated field. In
such an investigation it is important to discriminate between a
test of the basic principles of QED and a test of commonly used
but subsidiary approximations made within QED.

Standard quantum electrodynamic descriptions of atomic spontaneous emission rely on the Markov approximation. The Markov approximation in decay theory is entirely equivalent to a Weisskopf-Wigner
"pole" or on-shell approximation [1] and we will use it in this
sense. The use of this approximation is widespread in the study of
the interaction of atoms with a quantized electromagnetic field and
for this reason it is worth examining those effects omitted in a
Markovian treatment. Once the Markov approximation has been made,
we are led immediately to an exponential decay of the excited state
probability in spontaneous emission problems. The exponential
decay law seems to be verified experimentally, and particularly
well verified for weakly excited atoms. Nevertheless it has been
known for some time [2] that any unstable quantum system whose

635

energy spectrum is bounded from below cannot spontaneously decay
with a purely exponential law. This is only an apparent contradiction, resolved by a retention of the leading effects eliminated
by the Markov approximation. The QED equations of motion for
spontaneously emitting atom coupled to the vacuum radiation field
do possess some short-time memory features. An analysis of the
QED equations of motion which recognises the existence of non-
Markovian features is outlined in this note.

We find, as expected, corrections to the exponential law at
short and at long times. Short-time corrections are related to
the existence of a stationarity time governed by the mean square
dispersion of the energy of the decaying quantum system $(\Delta H)^2$. At
long times corrections proportional to inverse powers of the time
dominate and exceed the exponential decay contribution.[3] Exponential decay observations are thus better described as a test of
the validity of the Markov approximation than of QED. The power
law corrections to the decay law turn out to be very small for an
optical transition [4] and *in practice* unobservable. The Markov
approximation is quite reliable for purely spontaneous decay
problems.

There remains the possibility that deviations from an exponential decay law could be observed *in principle* at long times.
We demonstrate that in some situations designed to test atomic decay
laws, a purely exponential behaviour is forced upon the system by
the nature of the observation. Deviations from an exponential law
in this case seem to be unobservable not only in practice but in
principle. The measurement process eliminates the non-Markovian
power-law decay terms and re-instates exponential decay as a general
characteristic of quantum decay theory.[5]

2.   GENERAL FEATURES OF SPONTANEOUS DECAY

It is instructive to examine the general features underlying
the decay of an unstable state without resorting to specific
models.[5] We imagine the *total* system to be described by a
Hamiltonian H so that

$$H\phi_E(x) = E\phi_E(x), \tag{1}$$

where $\phi_E(x)$ are continuous energy eigenfunctions of H. Then the
initial unstable state, which we imagine to be described by a
localized or "peaked" wavefunction $\psi(x,t)$ may be expanded into the
complete set $\phi_E(x)$

$$\psi(x,t) = \int_0^\infty dE \; c(E)\phi_E(x)\exp(-iEt) \, , \tag{2}$$

where $E \geq 0$. We use units such that $\hbar = c = 1$. A sensible definition of the "non-decay" probability is

$$P(t) = |A(t)|^2 , \tag{3}$$

where the non-decay amplitude $A(t)$ is

$$A(t) = \int dx \; \psi^*(x,0)\psi(x,t) = \langle\psi(0)|e^{-iHt}|\psi(0)\rangle . \tag{4}$$

Using eq. (2) in eq. (4) we obtain the *exact* result [2]

$$A(t) = \int_0^\infty dE \; |c(E)|^2 \exp(-iEt) \quad . \tag{5}$$

If $c(E)$ is a real function, then since $A(t)$ is the Fourier transform
of a square integrable function the Riemann-Lebesgue lemma tells
us that $A(t)$ vanishes at large times. In other words, the system
"decays."

The unstable system is described by an energy spectrum $p(E) = |c(E)|^2$ which is bounded from below: $E \geq 0$. We define the modified
spectrum $p'(E) = p(E)$ for $E \geq 0$ and $p'(E) = 0$ for $E < 0$. Then we
may obtain $p'(E)$ from the Fourier inversion

$$p'(E) = \frac{1}{2\pi} \int_{-\infty}^\infty A(t)\exp(iEt)dt \quad . \tag{6}$$

There are strong restrictions on the form of $A(t)$ if $p'(E) = 0$ for
$E < 0$. If this condition is to be met, the Paley-Wiener theorem [2]
demands that $A(t)$ must satisfy

$$\int_{-\infty}^\infty \frac{|\ell n|A(t)||}{1 + t^2} \; dt < \infty \quad . \tag{7}$$

This in turn forbids *purely* exponential decay for all times t.

If we assume a Lorentzian spectral lineshape

$$|c(E)|^2 = (\gamma/2\pi)/[(E - E_i)^2 + \gamma^2/4] \; , \tag{8}$$

where E_0 is the resonance energy and γ its width, and then take the
unjustified step of taking the lower limit in eq. (5) to $(-\infty)$, the
conventional result $A(t) = \exp(-iE_0 t - \gamma t/2)$ is immediately obtained. However there is certainly no justification for allowing
$E < 0$, and no *a priori* reason for choosing a Lorentzian for $|c(E)|^2$.
We allow that a Lorentzian spectrum is physically reasonable, but
incorporate a cut-off function $g(E)$ to ensure $p(E) = 0$ for $E < 0$,

and admit a degree of skewness in allowing $p(E)$ an overall E-dependence

$$p(E) = (\gamma/2\pi) \; \frac{(E/E_o)^m g(E)}{(E-E_o)^2 + \gamma^2/4} \quad , \tag{9}$$

where m is a number to be determined by the decay model. Then from eq. (5),

$$A(t) = \frac{\gamma}{2\pi} \int_o^\infty dE \; e^{-iEt} \; \frac{(E/E_o)^m g(E)}{(E-E_o)^2 + \gamma^2/4} \quad . \tag{10}$$

This is equivalent to the result of Messiah [6] obtained by resolvent methods. The integral in eq. (10) may be performed by adding and subtracting the integration from $(-\infty)$ to zero, performing the resultant contour integral by picking up the pole at $E = E_o - i\gamma/2$, to give

$$A(t) = C\exp[-iE_o t - \gamma t/2]$$

$$+ \frac{\gamma}{2\pi} \int_o^\infty dx \; (-1)^{m+1} \; \frac{e^{-xt}(x/E_o)^m g(-ix)}{(ix+E_o)^2 + \gamma^2/4} \quad , \tag{11}$$

where C is a constant. This result is similar to that of Mostowski and Wódkiewicz [3] and that obtained using Källén's approach [7] to spontaneous emission.[4] The second term will lead to inverse power-law corrections to exponential decay.[8]

The corrections to the exponential "law" are expected to be important only at very large times, if at all. At large t the integrand is dominated by the fast oscillating exponential and will be significant only for small x. For such small x the variations of the resonance denominator and the cut-off are unimportant. Using

$$\int_o^\infty dx \; e^{-xt} \; x^m = m! \, t^{-(m+1)} \quad , \tag{12}$$

we obtain for large times

$$A(t) \simeq C \exp[-iE_o t - \gamma t/2] + Dt^{-(m+1)} \quad . \tag{13}$$

The two constants are

$$c \simeq g(E_o - i\gamma/2),$$

$$D = (\gamma/2\pi E_o^m) g(0) m! (-1)^{m+1} / [E_o^2 + \gamma^2/4].$$

If m = 1, we recover the t^{-2} dependence predicted in earlier
work.[4] But why should p(E) be given by eq. (9) with m = 1?
Certainly the spectral lineshape for spontaneous emission *is* a skew
Lorentzian [9] with m = 1 if the minimal substitution Hamiltonian
$V = -(e/m)\underline{p}\cdot\underline{A}$ is used to describe the decay (as in Refs. 3 and 4).
But we could have used the dipole form $V' = -\underline{d}.\underline{E}$, which leads to a
skewed spectrum with m = 3. This leave the pole term unchanged
but at very large times $\Lambda(t)$ is proportional to t^{-4}. There are
reasons for preferring V' and for believing m = 3 in lineshape
calculations.[9] But no lineshape theory is reliable far from
resonance so this point needs careful investigation.

There are problems with the exponential "law" at short times
as well as at these very long times.[5] If we return to the
exact result eq. (5) and expand the exponential assuming t is
small, we obtain

$$A(t) \simeq \int_0^\infty dE \; |c(E)|^2 (1 - iEt - E^2 t^2/2 \ldots). \tag{14}$$

Then with *no* assumptions concerning c(E), the non-decay probability
at short times is

$$P(t) = |A(t)|^2 \simeq 1 - (\Delta H)^2 t^2, \tag{15}$$

where $(\Delta H)^2 = <E^2>-<E>^2$, which we might expect would be of order
γ^2. An expansion of the exponential "law" for comparison is
$\exp(-\gamma t) \simeq 1 - \gamma t + (\gamma t)^2 \ldots$ We note the general result has *no*
term linear in t. In fact the *rate* of decay vanishes near t = 0:

$$\frac{dp}{dt}\Big|_{t\sim 0} = -2(\Delta H)^2 t \to 0. \tag{16}$$

In what follows we explore the decay behaviour predicted by both
Schrödinger and Heisenberg operator picture methods applied to
the two-level atom (TLA), and compare the results where appropriate
with those from the model-independent approach sketched in this
section.

3. THE EXPONENTIAL DECAY 'APPROXIMATION' IN THE SCHRÖDINGER PICTURE

Consider a discrete initial state $|i>$ which we consider decays
into a set of continuum states $|f>$ under the influence of a perturba-
tion V. Our language already implies some sort of perturbation
theory is to be used: we take $|i>$ and $|f>$ to be basis states of
the unperturbed Hamiltonian (H - V). The wavefunction of the total
system at time t we write as

$$|\psi(t)> = a(t)e^{-iE_i t}|i> + \int df \; b(f,t)e^{-iE_f t}|f>, \tag{17}$$

with initial conditions $a(0) = 1$, $b(f,0) = 0$. Equation (17) is the
"essential states" approximation, where we neglect multiple excitations of the continuum and all other levels. The time-dependent
Schrödinger equation and orthonormality then leads to the equations
of motion for the probability amplitudes

$$i\dot{a} = \int df\, V_{if}\, \exp(-iE_{fi}t)b \, , \tag{18}$$

$$i\dot{b} = V_{fi}\, a\, \exp(iE_{fi}t), \tag{19}$$

where $V_{fi} \equiv \langle i|V|f\rangle$, $E_{fi} = E_f - E_i = \omega-\omega_0$ where ω_0 is the TLA
resonance frequency and ω the radiated field frequency.

We know that over a wide time range, spontaneous decay *is*
exponential. So if we use

$$a(t) = \exp(-\gamma t/2) \tag{20}$$

as a trial solution in eqs. (18) and (19), with γ some number to be
determined, what happens? We obtain from eq. (19)

$$b(f,t) = \frac{V_{fi}[e^{iE_{fi}t-\gamma t/2}-1]}{E_{fi} + i\gamma/2} \, . \tag{21}$$

Demanding that probability be conserved as $t \to \infty$ and that γ is small,
fixes γ to be

$$\gamma = 2\pi \int df. \left|V_{fi}\right|^2 \delta(E_{fi}) \, . \tag{22}$$

These results can be obtained using more formal methods capable
of going beyond eqs. (20) and (21). Using the Laplace transform
method of Källen [7] we obtain, [4]

$$a(t) = \frac{1}{2\pi i} \int_\infty^\infty \frac{dy\, \exp(iyt)}{[y-iA(iy+\epsilon)]} \, , \tag{23}$$

where the self-energy function $iA(iy +\epsilon)$ is

$$iA(iy + \epsilon) = \lim_{\epsilon\to0} \frac{\gamma}{2\pi\omega_0} \int_0^\Lambda \frac{\omega d\omega}{\omega-\omega_0+y-i\epsilon} \tag{24}$$

To obtain eq. (24) we have used the minimal substitution Hamiltonian
$V = -(e/m)\underline{p}\cdot\underline{A}$ in dipole approximation and Λ is some high frequency

cut-off. We assume that spontaneous emission is a weak perturbation of the atom. The lifetime of an excited atomic state is of
the order of 10^{-8} seconds and ω_0 is of the order of 10^{14} Hz, so
that an atom executes about 10^6 Bohr orbits almost unaffected by
radiation forces. We therefore expect the principle oscillation
frequency to be very little shifted from ω_0 and we anticipate a
pole in the inversion integral eq. (23) very close to the pole at
$y = 0$ that occurs if $V = 0$. The Weisskopf-Wigner "pole approximation" consists of assuming $y \ll \omega_0$ is the most important region of
the integrand and dropping the y-dependence in the self-energy term
$iA(iy + \varepsilon)$. Then we obtain the usual result

$$a(t) = \exp(-\gamma t/2)\,\exp(-i\Delta t), \tag{25}$$

where $(\gamma/2) = \mathrm{IM}[iA(\varepsilon)]$ as before, and $\Delta = \mathrm{Re}[iA(\varepsilon)]$ is the TLA
"Lamb shift" of the initial state, a shift omitted in the simplest
WW approach.

If we retain the y-dependence of $iA(iy + \varepsilon)$ but assume $y < \omega_0 \ll \Lambda$
is the dominant range of y's, then we may still perform the Laplace
inversion. For times $t > \gamma^{-1}$ we find [4]

$$a(t) \simeq e^{-\gamma t/2} - (\gamma/2\pi\omega_0)(1/\omega_0 t)^2\, e^{i\omega_0 t}. \tag{26}$$

The exponential decay "law" is replaced by a slower t^{-2} decay at
long times in agreement with the combined requirements of energy
positivity and the Paley-Wiener theorem. To see that the traditional
Weisskopf-Wigner result eq. (20) is inadequate at short times is
straightforward. If we substitute eq. (20) for $a(t)$ and eq. (21)
for $b(f,t)$ into eq. (18) we should have an equality. In fact we
find

$$\frac{\gamma}{2}\, e^{-\gamma t/2} \stackrel{?}{=} (-1) \int df\; \frac{|V_{if}|^2 [e^{-\gamma t/2} - e^{-iE_{fi}t}]}{E_{fi} + i\gamma/2}, \tag{27}$$

and near $t \sim 0$ this is obvious nonsense.

Finally, in this section, it is useful to examine the time-
energy uncertainty relation and stationarity times in the context
of unstable state decay.[10] We have already noted the importance
of the mean square dispersion $(\Delta H)^2$ in decay theory. Defining an
energy dispersion ΔE by

$$\Delta E = [\langle H^2 \rangle - \langle H \rangle^2]^{1/2} \tag{28}$$

then ΔE satisfies the time-energy uncertainty inequality [11]

$$T_A \Delta E \geq \frac{1}{2} \ . \tag{29}$$

Here T_A is the Mandelstam-Tamm stationarity time

$$T_A = \Delta A / \left| d<A>/dt \right| \ , \tag{30}$$

the characteristic time taken for some observable represented by A to significantly evolve. We take $A = \sigma_{ii} = \left| i><i \right|$, so that $<A> = \left| a(t) \right|^2 = <A^2>$ and

$$\Delta A = e^{-\gamma t/2} \left[1 - e^{-\gamma t} \right]^{1/2}, \tag{31}$$

$$\left| d<A>/dt \right| = \gamma e^{-\gamma t} \ . \tag{32}$$

Then

$$T_A = \gamma^{-1} \left[\exp(\gamma t) - 1 \right]^{1/2} \ . \tag{33}$$

This is a surprising result not only because $T_A \to \infty$ as $t \to \infty$ (this may be expected since $T_A \to \infty$ for a pure state) but also because $T_A \to 0$ as $t \to 0$. The energy dispersion ΔE within the Weisskopf-Wigner approximation strictly diverges: the energy distribution function is Lorentzian and the second moment of a Lorentzian diverges. This enables the basic inequality

$$\Delta E . \Delta A \geq \frac{1}{2} \left| d<A>/dt \right| \tag{34}$$

to be satisfied, but is, in itself, of no value. To exploit eq. (34) it is necessary to go beyond the Weisskopf-Wigner approximation. This failure of the exponential law is most interesting: the earlier breakdowns happened at awkward times or represented small corrections, whereas here it is a total failure everywhere.

4. THE HEISENBERG OPERATOR METHOD

An alternative approach to atomic spontaneous emission is to work directly with the relevant Heisenberg operator of motion.[1,11] We write the Hamiltonian in terms of TLA pseudospin operators

$$H = \frac{1}{2} \omega_o \sigma_3 + (\omega_o d) A \sigma_2 + \sum_\lambda \omega_\lambda a_\lambda^\dagger a_\lambda \tag{35}$$

where d is the atomic transition dipole matrix element, and A is the component of the vector potential operator in the direction of the dipole moment evaluated at the centre of the atom:

$$A(t) = \sum_\lambda g_\lambda [a_\lambda(t) + a_\lambda^\dagger(t)] \; . \tag{36}$$

Here $g_\lambda \equiv (2\pi/\omega_\lambda V)^{1/2} \, \underline{\varepsilon}_\lambda \cdot \underline{d}$, V is the quantization volume, a_λ and $a_\lambda^\dagger$ are the usual annihilation and creation operators, and λ indexes both wavevector and polarization.

To allow a fair comparison with the Weisskopf-Wigner results we will make the rotating wave approximation (RWA). Then the equations of motion for the field mode operators can be formally integrated to give, for example,

$$a_\lambda^+(t) = a_\lambda^+(0)e^{i\omega_\lambda t} + g_\lambda \int_0^t dt' \sigma_+(t') e^{i(\omega_\lambda + i\varepsilon)(t-t')} \; . \tag{37}$$

Substituting this into the equation of motion for the atomic raising operator we obtain the non-Markovian form

$$\dot{\sigma}_+ = i\omega_o\sigma_+ + \sum_\lambda \omega_o d\{g_\lambda [a_\lambda^+(0) e^{i\omega_\lambda t}$$

$$+ \; \omega_o d g_\lambda \int_0^t dt' \sigma_+(t') e^{i(\omega_\lambda + i\varepsilon)(t-t')}]\}\sigma_3(t) . \tag{38}$$

If we now exploit the slow time dependence of $\sigma_3(t)$ to bring it inside the integral as $\sigma_3(t')$, use the equal-time result $\sigma_+(t)\sigma_3(t) = -\sigma_+(t)$ and take expectation values in the state $|\Psi(0)\rangle = |\phi_A(0)\rangle|\{0_\lambda\}\rangle$, we obtain the linear Volterra equation

$$\langle\dot{\sigma}_+\rangle = i\omega_o\langle\sigma_+(t)\rangle - \sum_\lambda g_\lambda^2(\omega_o d)^2 \int_0^t dt' \langle\sigma_+(t')\rangle e^{i(\omega_\lambda + i\varepsilon)(t-t')} \; . \tag{39}$$

This may be solved by Laplace transforms to give

$$\langle\sigma_+(t)\rangle = \frac{\langle\sigma_+(0)\rangle}{2\pi i} \int_{-i\infty+\varepsilon}^{i\infty+\varepsilon} \frac{ds.e^{st}}{s+i\omega_o+\Delta(s)} \; , \tag{40}$$

where the complex off-shell self-energy $\Delta(s)$ is

$$\Delta(s) = \sum_\lambda g_\lambda^2(\omega_o d)^2 / (s+i\omega_\lambda) . \tag{41}$$

If we put $s = iy + \varepsilon$ and observe that the unperturbed "pole" value of $\Delta(s)$ in eq. (41) occurs at $y = -\omega_o$ instead of at $y = 0$, then the integral to be evaluated in eq. (40) is identical to eq. (23)

using the extended Weisskopf-Wigner result. So for $t>\gamma^{-1}$, we find

$$<\sigma_+(t)> = <\sigma_+(0)>[e^{-\gamma t/2 - i\omega_o t} - \frac{\gamma}{2\pi\omega_o}\frac{1}{(\omega_o t)^2}] \, , \qquad (42)$$

as expected. Wódkiewicz and Eberly [13] have presented a general Liouville approach to non-Markovian effects in spontaneous emission and have arrived at similar results.

5. THE OBSERVABILITY OF NON-EXPONENTIAL DECAY

The power law corrections to the decay law are very small for an optical transition and are unobservable. As Mostowski and Wódkiewicz [3] have demonstrated, the power law term will be dominant only after perhaps 100 lifetimes. But it is not inconceivable that the power law could be of measurable size for some system. Would the experimenter *in principle* see non-exponential decay? There is an argument suggesting he would not.[5] To measure the non-decay probability $P(t)$ we must prepare $\psi(x,0)$, leave the system for a time t to freely evolve and then at time t make a measurement of the survival of the unstable state. If that is what the experimenter does, he could, in principle, detect deviations from an exponential decay law.

Most experimentalists leave their measuring apparatus on all the time. If we divide up the observation time T into short times (T/N) and make a sequence of measurements at t = 0, T/N, 2T/N...t=T, allowing the system to freely evolve between measurements, the probability of the system surviving in the unstable initial state just up to (T/N) is

$$P_1(T/N) = 1 - \gamma(T/N), \qquad (43)$$

where we have assumed $(T/N) \ll \gamma^{-1}$, expanded the exponential decay term and since t is small, ignored the non-exponential correction. We consider the measurement to have totally disrupted the system. The probability that the system has not decayed up to the Nth measurement at T is

$$P_N(T) = [1 - \gamma(T/N)]^N, \qquad (44)$$

and as $N \to \infty$

$$P(T) = \exp(-\gamma T) \, . \qquad (45)$$

The interval between measurements is never large enough for a

power-law correction to develop. For this type of experiment a
purely exponential decay law is forced upon the system by the nature
of the observation.

Acknowledgements

     I am grateful to G. Barton and P. W. Milonni for many conversa-
tions and a great deal of instruction concerning the decay "law"
of unstable systems.

References

1.   P. W. Milonni, Phys. Reports *25C*, 1 (1976); T. Von Foerster,
     Amer. J. Phys. *40*, 854 (1972).
2.   L. A. Khalfin, Sov. Phys. JETP *6*, 1053 (1958).
3.   J. Mostowski and K. Wódiewicz, Bull. de l'Acad. Polonaise des
     Sciences *21*, 1027 (1973).
4.   P. L. Knight and P. W. Milonni, Phys. Lett. *56A*, 275 (1976)
     and references therein.
5.   L. Fonda, Trieste preprint IC/76/20, and Proc. XIII Winter
     School of Theoretical Physics, February 1976, Karpacz, Poland.
6.   A. Messiah, *Quantum Mechanics*, Vol. II (North-Holland, Amster-
     dam, 1961) pp. 994-1001.
7.   G. Källen, in *Encyclopaedia of Physics*, Vol I, ed. S. Flügge
     (Springer, Berlin, 1958) pp. 274-279. W. H. Louisell, *Quantum
     Statistical Properties of Radiation* (Wiley, New York, 1973)
     p. 285-296.
8.   P. L. Knight, Phys. Lett. *61A*, 25 (1977).
9.   E. A. Power, *Introductory Quantum Electrodynamics* (Longmans,
     London, 1964) pp. 113-116.
10. J. H. Eberly and L. P. S. Singh, Phys. Rev. *D7*, 359 (1973).
11. A. Messiah, *Quantum Mechanics*, Vol. I (North-Holland, Amsterdam,
     1961) p. 320.
12. J. R. Ackerhalt, P. L. Knight and J. H. Eberly, Phys. Rev.
     Lett. *30*, 456 (1973). J. R. Ackerhalt and J. H. Eberly, Phys.
     Rev. *D10*, 3350 (1974).
13. K. Wódkiewicz and J. H. Eberly, Ann. Phys. (N.Y.) *101*, 574
     (1976).

THEORY OF A FREE ELECTRON LASER

M.O. Scully, F.A. Hopf, and P. Meystre

University of Arizona, Tucson, Arizona

In a free electron laser a relativistic electron beam is
passed through a periodic static field, and amplification is
achieved via stimulated emission in the direction of the electron
beam. The wavelength of the emitted radiation is related to the
periodicity of the static field and can be tuned continuously over
a broad region by changing the energy of the incident electron
beam. Laser oscillations at 3.4 microns in such a device has
recently been reported by the Stanford group[1].

The small-signal theory of a free electron laser is a "textbook
example" of classical physics which can be carried out from beginning
to end via classical mechanics and electrodynamics, together with
some elementary knowledge of laser physics. The coupled Maxwell
and relativistic Boltzmann equations describing the system have
the same formal structure as those describing usual stimulated
scattering. The small-signal gain is obtained by expanding the
Boltzmann distribution function in powers of the product of the
incident and scattered field and keeping contributions only up to
first order.

The strong signal regime, on the other hand, is best analyzed
in terms of a set of "generalized Bloch equations", obtained as a
result of a harmonic expansion which will be described in detail.
This formalism allows one to show that the effects that lead to
gain (i.e., electron recoil) play a relatively minor role compared
to other nonlinear effects, e.g., a considerable spread of the
electron distribution function. This poses limitations on the
efficiency of the device. In particular, since the free electron
laser does not appear to be easily saturable in the conventional

sense, it has been proposed to recycle the electrons (i.e., put them back in the accelerator and replace the energy lost to recoil). The spread limits the number of times that this can be done before the electrons can no longer contribute to the gain.

References

1. D.A.G. Deacon, L.R. Elias, W. Fairbank, J.M.J. Madey, H.A. Schwettmann and T.J. Smith, VII Winter Symposium on high power lasers, Park City (Utah) February 1977.

A DESCRIPTION OF LASER STABILITY NEAR THRESHOLD BY LIAPOUNOV'S SECOND METHOD

Adi R. Bulsara and William C. Schieve

University of Texas, Austin, Texas

1. INTRODUCTION

The theory of the generalized entropy developed by Prigogine, George and Henin [1] in Brussels and Austin suggests a form for an N-body Liapounov function which has already been discussed for closed systems [2-4]. Starting from the Liouville-von Neumann equation for the density operator ρ, Prigogine introduces a causal or physical particle representation wherein he defines a "physical" density matrix $\rho^{(p)}$ which is generated from ρ by a non-unitary transformation in strongly coupled systems. In this representation, the retarded and advanced components of $\rho^{(p)}$ obey separate evolution equations. For weakly coupled systems, $\rho^{(p)}$ reduces to the untransformed ρ which is a solution to the Pauli equation. This theory has been applied by Hubert [5] to a dense hard sphere gas obeying the Enskog equation. Further, Bulsara and Schieve have demonstrated [2] that infinitesimally close to thermal equilibrium, the generalized form of the entropy proposed by Prigogine et al. has the same form as the Liapounov function referred to above, for weakly coupled systems, a fact which we shall exploit in section 5 of this paper.

The above properties are exploited here in a model of a c.w. solid state laser near threshold, using Lamb's semi-classical theory [6]. In section 2, we briefly review Lamb's semi-classical theory and outline the results of a linear stability analysis (Liapounov's first method) which have already been derived elsewhere [7,8]. This is followed by a development of the rate equation approximation in section 3 which we then apply to the construction of a Liapounov function for the nonequilibrium branch in section 4. Finally, the entropy production for our model, close to equilibrium, is computed

in section 5 by linearization of the semi-classical equations, whence
we arrive at the results of linear irreversible thermodynamics. A
possible extension of Einstein's near-equilibrium fluctuation formula
to nonequilibrium steady states is also discussed. It should be
noted that Hofelich-Abate and Hofelich [9] have proved that the equilibrium branch is asymptotically stable in the entire physical region
of the phase space (corresponding to positive photon numbers). They
construct a functional $\cup(x,y)$ which vanishes only at the origin and
is positive definite, and show that all trajectories of the rate
equations, originating in this domain of positivity of $\cup$ tend asymptotically to the origin. Hence, $\cup$ is a Liapounov function, fulfilling
Liapounov's theorem on asymptotic stability. Further, Walgraef [10]
has used a fully quantum kinetic equation to derive a Gibbs entropy
for the near-equilibrium branch and discusses its stability within
the framework of the Glansdorff-Prigogine stability theory [11].
In contrast, we show that it is possible to obtain Walgraef's results
as well as the stability properties of the non-equilibrium branches
without using a quantum (Fokker-Planck) description of the system.

2A. SEMI-CLASSICAL LASER THEORY

The essence of the semi-classical approach is to treat the
electric field $\underline{E}$ in the laser cavity as being composed of a classical
part (describable by Maxwell's equations) and a smaller additional
part describing quantum fluctuations (this latter part may be removed
by averaging the fully quantum-mechanical equations over fluctuations). The classical field $\underline{E}$ is assumed to induce microscopic
dipole moments $\underline{p}_\alpha$ in the medium according to quantum-mechanical laws.
These moments are then summed to yield the macroscopic polarization
$\underline{P}$ which serves as the source term in Maxwell's equations.

We will consider a model of a c.w. solid state laser. Our model
consists of N homogeneously broadened two-level atoms interacting
with a single transverse mode of the radiation field in the cavity.
It should be noted in passing that the equations (2a.3-6) adequately
describe an organic laser [12] if we consider multimode effects and
neglect spectral cross-relaxation.

Let us write for the electric field,

$$\underline{E}(\underline{x},t) = \sum_\lambda E_\lambda(t)\underline{u}_\lambda(\underline{x}) \qquad (2a.1)$$

where

$$E_\lambda^{(+)} = i\sqrt{2\pi\hbar\omega_\lambda}\ b_\lambda$$

$$E_\lambda^{(-)} = -i\sqrt{2\pi\hbar\omega_\lambda}\ b_\lambda^* \qquad (2a.2)$$

are the positive and negative frequency components of the field. Here, b_λ, b_λ^* are (complex) amplitudes, ω_λ being the photon frequency in the λ-th mode. E_1 and E_2 are the energies of the atom in the ground and excited states respectively, $\omega_{21} = (E_2-E_1)/\hbar$ being the frequency separation of the levels. Without further ado, we now write down the complete set of semi-classical equations [6,14] in weak coupling (for the case of a single photon mode) :

$$\dot{\rho}_{12,\mu}(t) = (i\omega_{21,\mu} - \Gamma_{21,\mu})\rho_{12,\mu} - ig_{\lambda,\mu}b_\lambda^* d_\mu \qquad (2a.3)$$

$$\dot{\rho}_{21,\mu}(t) = -(i\omega_{21,\mu} + \Gamma_{21,\mu})\rho_{21,\mu} + ig_{\lambda,\mu}^* b_\lambda d_\mu \qquad (2a.4)$$

$$\dot{b}_\lambda^*(t) = (i\omega_\lambda - x_\lambda)b_\lambda^* + ig_{\lambda,\mu}^* \rho_{12,\mu} \qquad (2a.5)$$

$$\dot{b}_\lambda(t) = - (i\omega_\lambda + x_\lambda)b_\lambda - ig_{\lambda,\mu}\rho_{21,\mu} \qquad (2a.6)$$

$$\dot{d}_\mu = \frac{d_{0,\mu} - d_\mu}{T} + 2i(g_{\lambda,\mu}b_\lambda^* \rho_{21,\mu} - g_{\lambda,\mu}^* b_\lambda \rho_{12,\mu}) \qquad (2a.7)$$

where

$d_\mu \equiv N_{2,\mu} - N_{1,\mu} \equiv \rho_{22,\mu} - \rho_{11,\mu}$ is the population difference between the levels,

$d_{0,\mu} \equiv d_\mu$ at equilibrium,

$\Gamma_{12,\mu} \equiv$ atomic half-width,

$x_\lambda \equiv$ cavity half-width,

$T \equiv$ atomic relaxation time,

$$g_{\lambda,\mu} = i|\vec{\theta}_{21}|\sqrt{\frac{2\pi\omega_\lambda}{\hbar}}\, u_\lambda(\underline{x}) \;\; ; \;\; \vec{\theta}_{21,\mu} \equiv \text{atomic dipole moment.}$$

ρ_{12} and ρ_{21} are density matrices representing correlations between the levels, ρ_{11} and ρ_{22} being probabilities. The subscript μ representing the μ-th atom will be dropped henceforth (a sum over μ is implied throughout this work). We also have

$$\rho_{12} = \rho_{21}^* \qquad (2a.8)$$

which follows from hermiticity of the hamiltonian and,

$$\Gamma_{12} = \Gamma_{21} \, , \qquad (2a.9)$$

to this order (weak coupling). Further,

$$\rho_{11} + \rho_{22} = 1 ,$$ (2a.10)

which implies probability conservation. It is easily seen that,

$$\rho_{22} = \frac{1}{2} (1 + d) ,$$

$$\rho_{11} = \frac{1}{2} (1 - d) .$$ (2a.11)

It should be noted that the density matrix ρ introduced above is a
solution of the weak coupling Master equation (Pauli equation) and
constitutes a special case of the more general N-body density operator. Thus it is not necessary in the following to use the Λ-trans-
formation to construct the generalized entropy and Liapounov function
[1,3].

2B. LINEAR STABILITY ANALYSIS

A study of the semi-classical equations (2a.3-7) reveals that
we may in general have at least two steady states:

(a) $b_\lambda^* = 0 = b_\lambda$, $\rho_{12} = 0 = \rho_{21}$, $d_0 = d$.

This is the equilibrium branch.

(b) $\dot{\rho}_{12} = 0 = \dot{\rho}_{21}$, $\dot{b}_\lambda^* = 0 = \dot{b}_\lambda = \dot{d}$.

This is a non-equilibrium steady state which may be multiply-branched.
The linear analysis of these states has been carried out in detail
[7,8] and the stability criteria obtained have importance in the
succeeding sections. For this reason, we briefly outline the procedure and results here. Let us first consider the steady state (a).
We linearize the semi-classical equations about equilibrium by setting,

$$\rho_{12} = \rho_{12}^e + \varepsilon\phi_{12} ,$$ (2b.1)

$$b_\lambda = b_\lambda^e + \varepsilon b_{\lambda 1} ,$$ (2b.2)

$$d = d_0 + \varepsilon d_1 ,$$ (2b.3)

where $\rho_{12}^e = 0 = b_\lambda^e$. We then obtain the linearized equations of
motion,

$$\dot{\phi}_{12} = (i\omega_{21} - \Gamma_{21})\phi_{12} - ig_\lambda d_o b^*_{\lambda 1} \qquad (2b.4)$$

$$\dot{b}^*_{\lambda 1} = (i\omega_\lambda - x_\lambda)b^*_{\lambda 1} + ig^*_\lambda \phi_{12} \qquad (2b.5)$$

$$\dot{d}_1 = -d_1/T \quad . \qquad (2b.6)$$

If λ_i are the eigenvalues of these equations, one obtains directly,

$$\lambda_3 = -1/T \qquad (2b.7)$$

and,

$$\lambda^2 - \lambda(i\omega_\lambda + i\omega_{21} - \Gamma_{21} - x_\lambda) + (i\omega_{21} - \Gamma_{21})(i\omega_\lambda - x_\lambda) - g^2_\lambda d_o = 0 , \qquad (2b.8)$$

where we invoke the rotating wave approximation [6] to set $\omega_\lambda + \omega_{21} \approx 0$. Then, the system (2b.4-6) will be stable if the eigenvalues λ have negative real parts. It is easy to see that this is the case with (2b.8) if the inequality

$$d_o < \frac{\Gamma_{21}x_\lambda}{g^2_\lambda} , \qquad (2b.9)$$

is fulfilled. This is the condition for the pumping state to be stable near equilibrium. It has been derived by us without using an on-resonance assumption in contrast to references 7 and 8. The reverse inequality holds for lasing.

We now consider the lasing steady state(s) (b). Once again we linearize about the steady state by setting,

$$\rho_{12} = \rho^S_{12} + \varepsilon\phi'_{12} , \qquad (2b.10)$$

$$b_\lambda = b^S_\lambda + \varepsilon b'_{\lambda 1} , \qquad (2b.11)$$

$$d = d^S + \varepsilon d'_1 , \qquad (2b.12)$$

where ρ^S_{12}, b^S_λ and d^S are obtained from the steady state solutions of the semi-classical equations by setting their left-hand sides equal to zero. The results are:

$$\rho^S_{12} = \frac{ig_\lambda b^{*S}_\lambda d^S}{i\omega_{21} - \Gamma_{21}} , \qquad (2b.13)$$

$$b_\lambda^{*S} = \frac{-ig_\lambda^* \rho_{12}^S}{i\omega_\lambda - x_\lambda} \quad , \tag{2b.14}$$

and

$$\left(\frac{d_0 - d}{T}\right)^S = 4x_\lambda |b_\lambda^S|^2 \quad . \tag{2b.15}$$

The linearized equations of motion are,

$$\dot{\phi}'_{12} = (i\omega_{21} - \Gamma_{21})\phi'_{12} - ig_\lambda b_\lambda^{*S} d_1 \quad , \tag{2b.16}$$

$$\dot{b}'^*_{\lambda 1} = (i\omega_\lambda - x_\lambda)b'^*_{\lambda 1} + ig_\lambda^* \phi'_{12} \quad , \tag{2b.17}$$

$$\dot{d}'_1 = -\frac{d'_1}{T} + 2i(g_\lambda b_\lambda^{*S} \phi'_{21} - g_\lambda^* b_\lambda^S \phi'_{12}) \quad . \tag{2b.18}$$

Here we have made the important assumption that b_λ varies slowly compared to ρ_{12} and d. Hence, only the lowest order terms in the expansion of b_λ are retained even though other terms in $b_{\lambda 1}'$ may be present to $O(\varepsilon)$. As an example, we note that the term $-ig_\lambda d^S b'^*_{\lambda 1}$ is not included in (2b.16) even though it is of the same order as ϕ'_{12}. The analysis of the system (2b.16-18) has been carried out in detail by Risken and Nummedal [7]. The stability criterion is

$$T^{-1} < x_\lambda \text{ and } \Gamma_{21} \tag{2b.19}$$

The criteria (2b.9) and (2b.19) constitute sufficient conditions for the stability of the steady states (a) and (b) within the framework of Liapounov's first method.

3. RATE EQUATION APPROXIMATION

Following Tang [13] we write (2a.3) in the form,

$$\dot{\rho}_{12} = i\omega_{12}\rho_{12} - \frac{1}{T'}\,\rho_{12} - ig_\lambda b_\lambda^* d \tag{3.1}$$

where

$$T'^{-1} = \Gamma_{12} \quad ,$$

and consider the case,

$$T'^{-1} \gg T^{-1} \quad . \tag{3.2}$$

The system is then in a quasi-steady state near (non-equilibrium) threshold, wherein the correlations decay very rapidly compared with the probabilities. Hence, $\dot{\rho}_{12} \approx 0$ and we may set $\rho_{12} \equiv \rho_{12}^{S}$ in equations (2a.5-7). Alternatively, by assuming that the lasing process causes the cavity field to be built up according to [14]

$$b_\lambda^* = B_\lambda^* \, e^{i\Omega_\lambda t} \quad , \tag{3.3}$$

and assuming further that b_λ varies slowly compared to ρ_{12} and d, one may integrate equation (3.1) using Laplace's method [15]. We finally arrive at the rate equations:

$$\dot{d} = \frac{d_o - d}{T} - \frac{4\Gamma g_\lambda^2 n_\lambda d}{(\Omega-\omega)^2+\Gamma^2} \quad , \tag{3.4}$$

$$\dot{n}_\lambda = -2x_\lambda n_\lambda + \frac{2\Gamma g_\lambda^2 n_\lambda d}{(\Omega-\omega)^2+\Gamma^2} \quad , \tag{3.5}$$

where we have set $n_\lambda \equiv b_\lambda^* b_\lambda$.

4. LIAPOUNOV FUNCTION FOR THE NON-EQUILIBRIUM STEADY STATES

Let us define a functional [1,2] suggested by the general theory,

$$\Omega = \frac{1}{2} \, \mathrm{Tr}\rho^\dagger\rho \tag{4.1}$$

and

$$\Omega_T = \frac{1}{2} \, (\mathrm{Tr}\rho_S^\dagger\rho_S + \mathrm{Tr}\rho_F^\dagger\rho_F)$$

$$\quad = \Omega_S + \Omega_F \quad \text{say} \quad , \tag{4.2}$$

where the subscripts S and F refer to the system and cavity field, respectively. It is apparent that,

$$\Omega_T > 0 \quad . \tag{4.3}$$

Differentiating with respect to time yields,

$$\dot{\Omega}_T = \dot{\Omega}_S + \dot{\Omega}_F$$

$$= \text{Tr}(\dot{\rho}_S^\dagger \rho_S + \rho_S^\dagger \dot{\rho}_S) + \text{Tr}(\dot{\rho}_F^\dagger \rho_F + \rho_F^\dagger \dot{\rho}_F) \ . \tag{4.4}$$

Using (2a.8) and (2a.11) we obtain,

$$\dot{\Omega}_S = \frac{d\dot{d}}{2} + \dot{\rho}_{12}^\dagger \rho_{12} + \rho_{12}^\dagger \dot{\rho}_{12} \ ,$$

$$= \frac{d}{2T}(d_o - d) - \frac{2\Gamma g_\lambda^2 n_\lambda d^2}{(\Omega - \omega)^2 + \Gamma^2} \ , \tag{4.5}$$

where we have used (3.4), (2a.3-4) and (2b.13). Similarly we obtain
for the cavity field,

$$\dot{\Omega}_F = \dot{n}_\lambda n_\lambda = -2x_\lambda n_\lambda^2 + \frac{2\Gamma g_\lambda^2 n_\lambda^2 d}{(\Omega - \omega)^2 + \Gamma^2} \ , \tag{4.6}$$

where we have used (3.5). Thus,

$$\dot{\Omega}_T = -\frac{d}{2}\left\{ \frac{4\Gamma g_\lambda^2 n_\lambda d}{(\Omega - \omega)^2 + \Gamma^2} - \frac{d_o - d}{T} \right\} - 2n_\lambda^2\left\{ x_\lambda - \frac{\Gamma g_\lambda^2 d}{(\Omega - \omega)^2 + \Gamma^2} \right\}. \tag{4.7}$$

Let us first consider the pumping branch near threshold. In this
case we have [14],

$$d_o - d < 0$$

and

$$x_\lambda > \frac{\Gamma g_\lambda^2 d}{(\Omega - \omega)^2 + \Gamma^2} \ . \tag{4.8}$$

Hence,

$$\dot{\Omega}_T < 0 \ , \tag{4.9}$$

in this case. So, the pumping branch is stable in the region of
validity of the rate equations, which agrees with the results of the
preceding sections, Ω_T being a good Liapounov function in this case.
It may be noted that for the near equilibrium branch, $d_o \sim d$ and the
conditions (4.8) reduce to the stability criterion (2b.9), so that
our results are consistent with the linear analysis.

To consider the lasing branches we recall that our model is of a c.w. laser. Then, above threshold, the lasing state may be regarded as a set of multiple nonequilibrium steady states (the multiplicity of which we shall not discuss here), which persist as long as the cavity field $E \neq 0$. Hence, using the steady state solutions (2b.13-15) we obtain directly,

$$\dot{\Omega}_T = 0 \quad , \tag{4.10}$$

so that we finally obtain,

$$\Omega_T > 0$$

and,

$$\dot{\Omega}_T \leqslant 0 \quad , \tag{4.11}$$

where the equality holds for lasing. Ω_T is then a Liapounov function for both the (nonequilibrium) pumping and lasing branches. These states are stable near threshold, the pumping state being characterized by the inequalities (4.8) and the reverse inequalities holding for lasing. These inequalities are also seen to be consistent with the criterion (2b.9) for the equilibrium branch. The linear analysis of section 2 has also shown that a sufficient condition for nonequilibrium stability is the validity of the rate equation treatment. As noted by Tang [13], the approximation is indeed a good one for most solid state lasers and the fact that a Liapounov function may be constructed within its framework lends credence to his arguments. It should be noted that the inequalities (4.10-11) may also be obtained by linearizing Ω around the steady state through the expansions (2b.10-12) and using the linearized semi-classical equations (2b.16-18).

Finally, one may construct an N-body generalized entropy [1] for the nonequilibrium branch by setting,

$$S = - \frac{1}{2} \ln\Omega_S - \frac{1}{2} \ln\Omega_F \quad . \tag{4.12}$$

Differentiating with respect to time yields

$$\dot{S} = - \left(\frac{\dot{\Omega}_S}{\Omega_S} + \frac{\dot{\Omega}_F}{\Omega_F} \right) \quad . \tag{4.13}$$

Noting that $\dot{\Omega}_S \leqslant 0$ and $\dot{\Omega}_F \leqslant 0$ one immediately obtains for the entropy,

$$\dot{S} \geqslant 0 \quad . \tag{4.14}$$

Thus, one may construct a functional S which has the same properties as the entropy in a steady state, far from thermal equilibrium.

5. THE EQUILIBRIUM BRANCH: ENTROPY AND THERMODYNAMICS

The results of the preceding section hold immediately for the equilibrium branch. However, in order to consider the thermodynamics of this branch, it is necessary to depart from the rate equation treatment and consider the linearized equations of motion. Before doing so, however, let us introduce the quantities [16],

$$\rho_{12}(t) = \tilde{\rho}_{12}(t)e^{i\omega_{12}t} , \tag{5.1}$$

$$b_\lambda(t) = \tilde{b}_\lambda(t)e^{i\omega_\lambda t} . \tag{5.2}$$

Substituting in the semi-classical equations (2a.3-7) we obtain,

$$\dot{\tilde{\rho}}_{12} = -\Gamma_{12}\tilde{\rho}_{12} - ig_\lambda b_\lambda^* d , \tag{5.3}$$

$$\dot{\tilde{b}}_\lambda = -x_\lambda b_\lambda - ig_\lambda \tilde{\rho}_{21} , \tag{5.4}$$

$$\dot{d} = \frac{d_o - d}{T} + 2i(g_\lambda \tilde{b}_\lambda^* \tilde{\rho}_{21} - c.c.) , \tag{5.5}$$

with corresponding equations for $\tilde{b}_\lambda^*$ and $\tilde{\rho}_{21}$. Here we have set $\omega_\lambda = \omega_{12}$ so that we are now assuming the interaction of only one mode of the radiation field at resonance with the atom. The linearized equations (2b.4-6) now become (we drop the tilde on ρ_{12} and b_λ),

$$\dot{\phi}_{12} = -\Gamma_{12}\phi_{12} - ig_\lambda d_o b_{\lambda 1}^* , \tag{5.6}$$

$$\dot{b}_{\lambda 1}^* = -x_\lambda b_{\lambda 1}^* + ig_\lambda^* \phi_{12} , \tag{5.7}$$

$$\dot{d}_1 = -d_1/T . \tag{5.8}$$

Further, it should be noted that the quantities ϕ_{12}, $b_{\lambda 1}$, and d_1 are now macroscopic, a fact which allows us to describe this state using the laws of irreversible thermodynamics. As in the last section, let us introduce the generalized entropy,

$$S_S = -\frac{1}{2} \ln \Omega_S \tag{5.9}$$

for the system ($K_B = 1$). It has been shown [2] that close to equilibrium we may write (apart from positive multiplicative constants),

$$\dot{S}_S = -\dot{\Omega}_S \quad . \tag{5.10}$$

Using the expansions (2b.1-3) we have to $O(\varepsilon^2)$,

$$\dot{S}_S = -\left(\phi_{12}\dot{\phi}_{21} + \dot{\phi}_{12}\phi_{21} + \frac{d_1\dot{d}_1}{2}\right) \quad , \tag{5.11}$$

where the $O(\varepsilon)$ contribution vanishes due to probability conservation (2a.10). Using (5.6) and (5.8) we obtain,

$$\dot{S}_S = 2\Gamma_{21}|\phi_{21}|^2 + id_0(g_\lambda b^*_{\lambda 1}\phi_{21} - g^*_\lambda b_{\lambda 1}\phi_{12}) + \frac{d_1^2}{2T} \quad . \tag{5.12}$$

For the radiation field we have

$$\dot{S}_F = -n_\lambda \dot{n}_\lambda \quad . \tag{5.13}$$

However, this quantity has a leading term of $O(\varepsilon^4)$ and does not contribute close to equilibrium. Hence, to $O(\varepsilon^2)$ we may take

$$\dot{S}_S \equiv \sigma_S \quad , \tag{5.14}$$

where σ_S is the entropy production of the system. We may put (5.12) in the form,

$$\sigma_S = 2\Gamma_{21}\left|\phi_{21} - \frac{ig^*_\lambda d_0}{2\Gamma_{21}}b_{\lambda 1}\right|^2 + \frac{d_1^2}{2T} - \frac{g_\lambda^2 d_0^2}{2\Gamma_{21}}|b_{\lambda 1}|^2 \quad . \tag{5.15}$$

Let

$$A = \frac{d_1^2}{2T} - \frac{g_\lambda^2 d_0^2}{2\Gamma_{21}}|b_{\lambda 1}|^2 \quad . \tag{5.16}$$

For low photon numbers we may write [14],

$$ng_\lambda^2 < \Gamma_{12} \quad . \tag{5.17}$$

Typically one has $g_\lambda^2 = 10^6 \text{sec}^{-2}$; $\Gamma_{12} = 10^{11}\text{sec}^{-1}$ so that $n < 10^5$ photons, which is certainly the case near threshold. It is easy to show from (5.17) that [17]

$$T^{-1} > \frac{g_\lambda^2 |b_{\lambda 1}|^2}{\Gamma_{21}} \quad , \tag{5.18}$$

so that,

$$A > g_\lambda^2 |b_{\lambda 1}|^2 \Gamma_{21}^{-1}(d_1^2 - d_o^2) \quad . \tag{5.19}$$

Recalling that $d_o < 0$ and $|d_1| > |d_o|$ for pumping (the system is very close to saturation which corresponds to an over-population of the excited state) we see that,

$$A > 0 \quad . \tag{5.20}$$

Hence we arrive at

$$\sigma_S \geqslant 0 \quad , \tag{5.21}$$

consistent with the results of near equilibrium linear thermodynamics. The equality above holds trivially at equilibrium. A glance at (5.9) and (5.21) reveals that the entropy production is also a Liapounov function for this case.

We define "generalized forces," [18],

$$X_1 = \phi_{12} \; , \quad X_2 = \phi_{21} \; , \quad X_3 = b_{\lambda 1} \; , \quad X_4 = b_{\lambda 1}^* \; , \quad X_5 = d_1 \quad . \tag{5.22}$$

Then, we may write

$$\sigma_S = \sum_i J_i X_i \quad , \tag{5.23}$$

where J_i is the "flux" corresponding to the force X_i. From (5.12) we readily obtain

$$J_1 = \Gamma_{21}\phi_{21} - \frac{i d_o g_\lambda^*}{2} b_{\lambda 1} \quad , \tag{5.24}$$

$$J_2 = J_1^* \quad , \tag{5.25}$$

$$J_3 = \frac{-i d_o g_\lambda^*}{2} \phi_{12} \quad , \tag{5.26}$$

$$J_4 = J_3^* \quad , \tag{5.27}$$

$$J_5 = d_1/T \quad . \tag{5.28}$$

Further we have

$$J_i = \sum_k L_{ik} X_k \quad , \tag{5.29}$$

where L_{ik} are the Onsager coefficients. The Onsager matrix then reads,

$$
L = \begin{pmatrix}
0 & \Gamma_{21} & \dfrac{-id_0 g_\lambda^*}{2} & 0 & 0 \\[2em]
\Gamma_{21} & 0 & 0 & \dfrac{id_0 g_\lambda}{2} & 0 \\[2em]
\dfrac{-id_0 g_\lambda^*}{2} & 0 & 0 & 0 & 0 \\[2em]
0 & \dfrac{id_0 g_\lambda}{2} & 0 & 0 & 0 \\[2em]
0 & 0 & 0 & 0 & \dfrac{1}{2T}
\end{pmatrix} \tag{5.30}
$$

It is readily apparent that,

$$L_{ik} = L_{ki} \quad (i \neq k)$$

and

$$L_{ii} \geq 0 \quad . \tag{5.31}$$

These are sufficient conditions [11] for our near-equilibrium state to be stable and have a minimum entropy production associated with it.

A final remark about equations (5.22)-(5.28) is in order. As originally defined by Onsager, the forces (5.22) and their associated flows are macroscopic quantities, the flows being linearly dependent

on the forces according to (5.29). The forces appearing in (5.22)
are indeed macroscopic, but we have as many flows as there are
levels of description in the system. This was first discussed by
Klein and Meijer [19] who showed that the minimum entropy theorem
held in a microscopic description wherein the entropy was considered
a function of the microscopic probabilities (and correlations) and
minimized with respect to these parameters. A general proof has
been given by Callen [20] and our results bear it out.

To summarize briefly, we have demonstrated that the nonequilibrium lasing states are stable near threshold by constructing a generalized Liapounov function and using the rate equation approximation. The stability of the equilibrium branch is proved as a consequence of a minimum entropy production theorem which holds in this
case, the entropy production σ_S defined in (5.14) being a Liapounov
function for this branch. It should be re-emphasized that our
results of section 5 are valid only *close to equilibrium* $(O(\varepsilon^2))$
where (5.14) is true. To higher orders in ε, the contributions to
σ from the cavity field (5.13) must be considered. Little is known
about the entropy production of this system or for that matter any
other open system, far from equilibrium. This will be the subject
of continuing investigation.

6. NON-EQUILIBRIUM FLUCTUATION THEORY BY THE EINSTEIN METHOD

The theory of near-equilibrium fluctuations is based on the
celebrated Einstein formula [11],

$$P \sim \exp \Delta S/K_B \quad , \tag{6.1}$$

where P is the fluctuation probability, ΔS being the entropy change
in the fluctuation. Let us now expand the entropy in a Taylor
series,

$$S = S_e + (\delta S)_e + \frac{1}{2} (\delta^2 S)_e + \ldots \ldots \quad , \tag{6.2}$$

with $(\delta S)_e = 0$ for an isolated system, S_e being the equilibrium
entropy. Then, in terms of the "excess entropy" we may write (6.1)
as [11],

$$P \sim \exp \frac{1}{2K_B} (\delta^2 S)_e \tag{6.3}$$

where,

$$(\delta^2 S)_e < 0 \tag{6.4a}$$

and,

$$\frac{\partial}{\partial t} (\delta^2 S)_e > 0 \quad , \tag{6.4b}$$

i.e., $\delta^2 S$ must be a Liapounov function for this state. The extension of (6.3) to steady states far from thermal equilibrium has been conjectured by Glansdorff and Prigogine [11] and specific systems have been treated using this technique with a local equilibrium assumption (see ref. 21 for a recent survey). It must be noted that far away from thermal equilibrium, one has in general, $\delta S \neq 0$. Let us now expand Ω about the steady state:

$$\Omega = \Omega^S (1 + \varepsilon \frac{(\delta\Omega)^S}{\Omega^S} + \varepsilon^2 \frac{(\delta^2\Omega)^S}{\Omega^S} + \dots) \quad . \tag{6.5}$$

Using (4.12) we find for the entropy,

$$S = -\ln\Omega^S - \varepsilon \frac{(\delta\Omega)^S}{\Omega^S} - \varepsilon^2 \left\{ \frac{(\delta^2\Omega)^S}{\Omega^S} - \frac{1}{2}\left|\frac{(\delta\Omega)^S}{\Omega^S}\right|^2 \right\} \quad , \tag{6.6}$$

where we retain only $O(\varepsilon^2)$ terms. Noting that,

$$\Omega = \frac{1}{2}|n_\lambda|^2 + \frac{1}{2}|\rho_{12}|^2 + \frac{1}{4}d^2 \quad , \tag{6.7}$$

and using the expansions (2b.10-12) we readily find,

$$-(\delta^2 S)_S = \frac{2}{n_\lambda^S} |b_{\lambda 1}|^2 + \frac{d^{S2}}{4\Omega_S^A} |\phi_{12}|^2 + \frac{1}{4\Omega_S^A} d_1^2$$

$$+ \frac{1}{2|n_\lambda^S|^2} (b_\lambda^S b_\lambda^S b_{\lambda 1}^* b_{\lambda 1}^* + c.c.) = A + B \quad , \tag{6.8}$$

where we define for the atomic system,

$$\Omega_S^A = \frac{1}{2}|\rho_{12}^S|^2 + \frac{1}{4}d^{S2} \quad . \tag{6.9}$$

Here,

$$A \equiv \frac{2}{n_\lambda^S} |b_{\lambda 1}|^2 + \frac{d^{S2}}{4\Omega_S^A} |\phi_{12}|^2 + \frac{d_1^2}{4\Omega_S^A} \tag{6.10a}$$

and

$$B \equiv \frac{1}{2|n_\lambda^S|^2} \left(b_\lambda^S b_\lambda^S b_{\lambda 1}^* b_{\lambda 1}^* + \text{c.c.} \right) \quad . \tag{6.10b}$$

From (2b.16-18) we readily find $\partial A/\partial t$ which may be cast in the form,

$$\frac{\partial A}{\partial t} = \sum_{i,k} a_{ik} \xi_i^* \xi_k \quad , \tag{6.11}$$

where we set

$$\xi_1 = \phi_{12} \ , \quad \xi_2 = b_{\lambda 1} \ , \quad \xi_3 = d_1 \quad .$$

The matrix a is

$$a = \begin{pmatrix} -\Gamma_{12} \dfrac{d^{S^2}}{4\Omega_S^A} & \dfrac{2ig_\lambda^*}{n_\lambda^S} & \dfrac{ig_\lambda b_\lambda^{*S} d^{S^2}}{4\Omega_S^A} \\[4mm] -\dfrac{2ig_\lambda}{n_\lambda^S} & -\dfrac{4x_\lambda}{n_\lambda^S} & 0 \\[4mm] -\dfrac{ig_\lambda^* b_\lambda^S d^{S^2}}{4\Omega_S^A} & 0 & -\dfrac{1}{2T}\left(\dfrac{1}{4\Omega_S^A}\right) \end{pmatrix} \tag{6.12}$$

We readily observe that,

$$a_{ii} < 0 \quad . \tag{6.13}$$

Also,

$$\det(a) = \frac{-\Gamma_{12} d^{S^2} x_\lambda}{8Tn_\lambda^S \Omega_S^{A^2}} + \frac{|g_\lambda|^2}{2T\Omega_S^A |n_\lambda^S|^2} + \frac{|g_\lambda|^2 x_\lambda d^{S^\Delta}}{4\Omega_S^{A^2}} \quad . \tag{6.14}$$

Using the inequalities (5.17-18) we find, $\det(a) < 0$. Together with (6.13) this is a sufficient condition [23] to have $\partial A/\partial t < 0$. A similar calculation gives $\partial B/\partial t < 0$. Hence we arrive at the result,

$$(\delta^2 S)_S < 0$$

and

$$\frac{\partial}{\partial t} (\delta^2 S)_S > 0 \quad , \tag{6.15}$$

so that $(\delta^2 S)_S$ is a Liapounov function, the conditions (6.4a,b) being fulfilled. Our use of (6.3) as the fluctuation probability far from equilibrium indicates that the fluctuations are Gaussian, in accordance with the results of Risken [22] who uses quantum distribution functions to calculate the field correlations by treating the field as a classical random variable. Using (6.3) we find,

$$<|b_{\lambda 1}|^2> = K_B n_\lambda^S \quad , \tag{6.16}$$

$$<|\phi_{12}|^2> = 4\Omega_S^A K_B/d^{S^2} = d^{S^{-2}} <d_1^2> \quad , \tag{6.17}$$

and,

$$<b_{\lambda 1}^* b_{\lambda 1}^*> = 2K_B |n_\lambda^S|^2 (b_\lambda^S b_\lambda^S)^{-1} = <b_{\lambda 1} b_{\lambda 1}>^* \quad , \tag{6.18}$$

where it should be noted that in a fully quantum description, the last quantity would be zero because of gauge invariance requirements.

These results seem to indicate that the Einstein formula in the form (6.3) may be used far from equilibrium in conjunction with the definitions (4.1) and (4.12) for this model. It should be remarked, however, that in a fully quantum description, the photon mode is a pure Glauber state with $n = b^+b$, the photon number. It is well known [14] that the distribution above threshold is Poisson, but no information about this is available from the Einstein formula to this order. There is, thus, no reason to suppose that (6.3) holds far from equilibrium in general. This has been noticed recently in the case of chemical reactions wherein the Einstein formula gives non-Poissonian fluctuations away from equilibrium (contrary to experimental evidence), unless a local equilibrium assumption is made [24]. Such an assumption would, however, be meaningless for a monomode laser model.

References

1. Prigogine, I., George, C., Henin, F. and Rosenfeld, L., Chemica Scripta *4*, 5 (1973).
2. Bulsara, A.R., and Schieve, W.C., J. Chem. Phys. *65*, 2532 (1976) and in Int. Jl. of Quantum Chemistry, *S9*, 475 (1975).
3. Schieve, W.C., Int. Jl. of Quantum Chemistry, *S10*, 375 (1976).

4.   Henin, F., Physica *76*, 201 (1974).
5.   Hubert, D., J. Stat. Phys. *15*, 1 (1976).
6.   Sargent, M., Scully, M. and Lamb, W., *Laser Physics* (Addison-Wesley, Reading, Mass., 1974).
7.   Risken, H. and Nummedal, K., J. Appl. Phys. *39*, 4662 (1968).
8.   Kondepudi, D., Phys. Lett. *59A*, 17 (1976).
9.   Hofelich-Abate, E. and Hofelich, F., Z. Physik *211*, 142 (1968).
10. Walgraef, D., J. Stat. Phys. *14*, 399 (1976).
11. Glansdorff, P. and Prigogine, I., *Structure, Stability and Fluctuations* (J. Wiley, New York, 1971).
12. Vahey, D.W. and Yariv, A., Phys. Rev. *A10*, 1578 (1974).
13. Tang, C.L., J. Appl. Phys. *34*, 2935 (1963).
14. Haken, H., *Handbuch der Physik*, Vol. 25/2C (Springer-Verlag, Berlin, 1970) and in *Quantum Optics*, ed. Kay and Maitland (Academic Press, New York, 1970).
15. Bleistein, N. and Handelsman, R., *Asymptotic Expansions of Integrals* (Holt, Rhinehart and Winston, New York, 1975).
16. Weidlich, W. and Haake, F., Z. Physik *185*, 30 (1965).
17. Arzt, V., Haken, H., Risken, H., Sauermann, H., Schmid, Ch., and Weidlich, W., Z. Physik *197*, 207 (1966).
18. deGroot, S. and Mazur, P., *Non-Equilibrium Thermodynamics* (North-Holland, Amsterdam, 1969).
19. Klein, M. and Meijer, P., Phys. Rev. *96*, 250 (1954) and Meijer, P., in *Transport Processes in Statistical Mechanics* (IUPAP Symposium, Brussels, 1956, I. Prigogine, editor).
20. Callen, H., in *Transport Processes in Statistical Mechanics* (IUPAP Symposium, Brussels, 1956, I. Prigogine, editor).
21. Prigogine, I.P. and Nicolis, G., *Self-Organization in Non-Equilibrium Systems* (J. Wiley, New York, 1977).
22. Risken, H., Z. Physik *186*, 85 (1965).
23. Gantmacher, F.R., *Theory of Matrices* (Chelsea Publ, New York, 1960).
24. Nicolis, G., J. Stat. Phys. *6*, 195 (1972) and Gardiner, C.W., McNeil, K., Walls, D. and Matheson, I., J. Stat. Phys. *14*, 307 (1976).

THERMODYNAMIC ANALYSES OF LASER INSTABILITY AND SUPERRADIANCE

H. Hasegawa, S. Sawada and M. Mabuchi

Kyoto University, Kyoto, Japan

1. INTRODUCTION

The instability phenomena in nonlinear optics, laser and super-radiance, have received the interest of theoretical investigation from a macroscopic point of view. In the Third Rochester Confer-ence, Graham discussed[1] the problem of unifying the laser theory and the non-equilibrium thermodynamic theory of Glansdorff and Prigogine,[2] based on the entropy expression for the steady non-equilibrium state, $S = \ln P$, where P is the well-established steady-state solution of the laser Fokker-Planck equation.

The difficulty, which led Graham to say[1] that "the relation of the quantity $\ln P$ to the non-equilibrium thermodynamics remained still unclear" so that "a unified treatment of thermodynamics and fluctuations far from equilibrium is still missing", is concerned with the fact that the central role played as a thermodynamic potential in the Glansdorff-Prigogine theory is the entropy pro-duction rather than the entropy itself. Here we find a better answer to the problem by focusing our consideration on a rudi-mentary point of contact between the thermodynamic and fluctuation theories, on the basis of which results of our numerical analyses will be presented.

Our standpoint is that the thermodynamic framework can be statistically reformulated by an explicit use of generally time-dependent solutions of the Fokker-Planck equation, which removes the "local equilibrium" assumption in the thermodynamics and which also enables us to extend the framework so that it may apply to transient phenomena as well. This has been discussed fully in a series of papers by one of the authors.[3]

2. STATISTICAL DESCRIPTION OF THE THERMODYNAMICS OF A SYSTEM UNDER THE FOKKER-PLANCK EVOLUTION LAW

It is not an easy task to recapitulate the Glansdorff-Prigogine theory, but the following interpretative summary by means of the quantity $\mathcal{P}$, named entropy production, may be of assistance: (a) $\mathcal{P}$ is a non-negative quantity as a function of several variables, $\{x_\nu\}$ (state variables which characterize a thermodynamic state of the system evolving in time as $x_\nu(t)$), as well as of other time-independent parameters $\{\lambda_i\}$. (b) $\mathcal{P}$ attains its minimum value at the equilibrium (or, the steady non-equilibrium) state x_ν^0 towards which every state temporally approaches. (c) Evolution criterion $(d/dt)\mathcal{P} < 0$; stability against fluctuations $\delta^2\mathcal{P} > 0$, where $\delta^2\mathcal{P}$ implies the second increment of the functional $\mathcal{P}(x(t))$ with respect to the variation $\delta x_\nu(t)$ from the realizable time evolution of the system. (d) The macroscopic course $x_\nu(t)$ that is realized must be always accompanied with fluctuations. In the vicinity of the equilibrium state x_ν^0 the probability of occurrence of such fluctuations $\delta x_\nu(t)$, may be expressed in terms of the corresponding change in $\mathcal{P}$ through the Einstein formula $P \propto e^{\frac{1}{2}\delta^2 S}$ ($\delta^2\mathcal{P} = d/dt\,\delta^2 S$: the "excess entropy production").

The above aspects have been reviewed more fully by Nicolis,[4] who discussed the fluctuation property (d) to an extent by considering the Markoffian evolution law. Our view on this problem is that it is possible to complete Nicolis' argument so that essentially the entire aspect can be formulated on the Markoffian basis, in particular, on the Fokker-Planck evolution law.[3] We designate the space of the state variable $\{x_\nu\}$ (assumed to form a Cartesian coordinate system) as the "phase space", over which the probability distribution at time t, $\psi(x\ t)$, is defined. Let us take the form of equation satisfied by ψ as

$$\frac{\partial\psi}{\partial t} = -\frac{\partial}{\partial x_\mu}(K_\mu\psi) + \frac{1}{2}\frac{\partial^2}{\partial x_\mu \partial x_\nu}(K_{\mu\nu}\psi) \quad , \tag{1}$$

where, for simplicity the positive symmetric diffusion tensor $K_{\mu\nu}$ is assumed as constant and the drift velocity $K_\mu(x)$ (generally, a nonlinear function of x_ν's) is assumed to satisfy the potential condition, $\partial K_\mu/\partial x_\nu = \partial K_\nu/\partial x_\mu$. Then the steady-state solution $\psi_0(x)$ of Eq.(1) is obtainable from a direct integration

$$\psi_0(x) = \text{const.} \times \exp.\left[\int_{x0}^{x} 2K_{\mu\nu}^{-1} K_\nu(x')\, dx_\mu'\right] \tag{2}$$

which is identical with Graham's P mentioned in the Introduction.

Our plan to unify the Fokker-Planck and the Glansdorff-Prigogine theories begins with a firm variational principle which is equivalent to solving Eq.(1), because historically the thermodynamic

theory was initiated in connection with the variational principle
by Onsager[5] and expounded by Prigogine's school.[6] Two forms of
such principle are available:[3]

Onsager's variational principle

$$\int \mathcal{S} \frac{\partial \psi}{\partial t} \, dx - \frac{1}{2} \, \mathcal{P}\{ \mathcal{S}, \psi \} = \max.(\mathcal{S}) \quad , \tag{3}$$

where

$$\mathcal{P}\{ \mathcal{S}, \psi \} = \int \frac{1}{2} K_{\mu\nu} \frac{\partial \mathcal{S}}{\partial x_\mu} \frac{\partial \mathcal{S}}{\partial x_\nu} \, \psi \, dx \tag{3a}$$

with post-variation condition

$$\mathcal{S} = - \log (\psi/\psi_0) \qquad (\psi_0 \text{ defined by } (2)). \tag{3b}$$

Prigogine's variational principle (the local potential method)

$$\int \log\psi \left(\frac{\partial \psi}{\partial t}0 + \frac{\partial}{\partial x_\mu} (v_\mu \psi_0) \right) dx + \frac{1}{2} \mathcal{P}\{ \log\psi, \psi_0 \} = \min.(\log\psi) \quad ,\tag{4}$$

where

$$\mathcal{P}\{ \log\psi, \psi_0 \} = \int \frac{1}{2} K_{\mu\nu} \frac{\partial \log\psi}{\partial x_\mu} \frac{\partial \log\psi}{\partial x_\nu} \, \psi_0 \, dx \tag{4a}$$

with post-variation condition

$$\psi = \psi_0 \quad . \tag{4b}$$

Here ψ_0 is generally a time-dependent solution of Eq.(1) resulting
from the variational principle. The reason why the maximum principle (3) is attributed to Onsager is clearly that it is analogous to
the form:[5] [entropy production] - [dissipation function] = max.
by the association

$$\int \mathcal{S} \frac{\partial \psi}{\partial t} \, dx \quad \longrightarrow \quad \text{[entropy production]}$$

$$\mathcal{P}\{ \mathcal{S}, \psi \} \quad \longrightarrow \quad 2 \times \text{[dissipation function]} \quad .$$

It can be shown[3] that by the satisfaction of the variational principle (3) together with (3a,b) (hence for a solution ψ of Eq.(1)),
the following relations hold:

$$\int \mathcal{S} \frac{\partial \psi}{\partial t} \, dx = \mathcal{P}\{ \mathcal{S}, \psi \} = \frac{d}{dt} \int (\log(\psi_0/\psi))\psi \, dx \quad . \tag{5}$$

Therefore, the quantity $\mathcal{P}\{ \mathcal{S}, \psi \}$ (twice the "dissipation functional")
may be identified, under the Fokker-Planck evolution law resulting

from Onsager's variational principle, with the "entropy production"
- the total time derivative of a scalar quantity to be designated
as entropy (incidentally, as the very information entropy).

With the above understanding of the entropy production, it
can be seen intuitively that (b) $\mathcal{G}\{\mathcal{S},\psi\}$ tends to its minimum
value 0 as ψ tends to the steady-state distribution ψ_0 (when the
time goes to infinity), in the vicinity of which

(c)   the evolution criterion $\dfrac{d}{dt}\mathcal{G} \leq 0,$ (6)

and

            the stability against $\delta\mathcal{S}$ $\delta^2\mathcal{G} > 0$ (7)

hold. Furthermore, it can be shown that the stability condition
(7) is actually not restricted to the vicinity of the steady-state
but extends over all the time regions as far as the solution $\psi(x\ t)$
exists. As to the Einstein fluctuation formula for the non-equilibrium steady-state, we have

(d) $P(zt_2/yt_1) = \text{const.}\times\exp.\left[-\dfrac{1}{2}\int_{t_1}^{t_2} K_{\mu\nu}^{-1}\ \delta\dot{x}_\mu(t)\ \delta\dot{x}_\nu(t)\ dt\right]$ (8)

which can be derived from the general Onsager-Machlup formula[3]
based on a most-probable-path argument.

It is an interesting but difficult question whether the validity of the evolution criterion (6) holds unrestricted in the vicinity of the steady-state or not. In order to see the point of question in connection with Graham's argument,[1] let us reconsider the
evolution criterion in the ordinary framework with the entropy function $S(x) = \log\psi_0(x)(=\ln P(x))$ for which

$$\mathcal{G}(x(t)) \equiv \frac{dS}{dt} = \frac{\partial S}{\partial x_\mu}\ \dot{x}_\mu = 2K_{\mu\nu}^{-1}\ K_\mu K_\nu \ .$$ (9)

(We have used the potential condition and hence the expression (2)
and also the macroscopic evolution law, $\dot{x}_\mu = K_\mu\cdot$)

$$\frac{d}{dt}\mathcal{G} = 4K_{\mu\nu}^{-1}\ \dot{K}_\mu K_\nu = 4K_{\mu\nu}^{-1}\ \frac{\partial K_\mu}{\partial x_\nu}\ K_\lambda K_\nu = 2\ \frac{\partial^2 S}{\partial x_\lambda \partial x_\nu}\ K_\lambda K_\nu \ .$$ (10)

Therefore, the region of the phase space where the evolution criterion holds can be well identified with the one in which the curva-
ture tensor $\partial^2 S/\partial x_\mu \partial x_\nu$ is non-positive. This is certainly violated
in the vicinity of an unstable point (e.g., the unstable 0-photon
state above threshold of the laser action), as it should. Let us
now go over to the present new framework, for which we have derived
the formula[3]

$$\frac{d}{dt}\mathscr{P}\{\mathcal{S},\psi\} = \int 2 \frac{\partial^2 S}{\partial x_\mu \partial x_\nu} K_\mu K_\nu \, \psi \, dx - \int \frac{1}{2} K_{\mu\lambda} K_{\nu\sigma} \frac{\partial X_\lambda}{\partial x_\mu} \frac{\partial X_\sigma}{\partial x_\nu} \, \psi \, dx \quad ,$$

$$\tag{11}$$

where $\mathcal{S}$ is given by (3b) and

$$X_\lambda \equiv \frac{\partial}{\partial x_\lambda} \log(\psi_0/\psi) \quad . \tag{11a}$$

The first term on the r.h.s. of (11) is the same contribution as in (10) averaged over the distribution ψ, to which the second term is added. This additional contribution is always negative and is originated as an effect of fluctuations. Thus, one can say qualitatively that the effect of fluctuations associated with the macroscopic drift motion is such that they tend to make its unstable behavior mild so that the validity of the evolution criterion (6) may be widened much from the prediction (10). This will be subject to a direct quantitative test in Section 4.

3. SOLUTIONS OF THE LASER AND SUPERRADIANCE FOKKER-PLANCK EQUATIONS BY THE VARIATIONAL METHOD

The single-mode laser action, described by the Langevin equation for a complex Van der Pol oscillator corresponding to the active-mode amplitude, i.e.,

$$\dot{b} - \beta(d-b^*b)b = \Gamma(t), \quad \langle\Gamma(t)\Gamma^*(t)\rangle = Q\delta(t-t') \quad \text{etc.},$$

may have a counter description in terms of a probability distribution subject to the Fokker-Planck equation

$$\frac{\partial\psi}{\partial t} = -\beta\frac{\partial}{\partial b}\left((d-|b|^2)b\psi\right) + 4Q\frac{\partial^2\psi}{\partial b\partial b^*} \quad . \tag{12}$$

We restrict ourselves to the "uniform phase" case, and simplify (12) to

$$\frac{\partial\psi}{\partial t} = -2\frac{\partial}{\partial I}\left((a-I)I\psi\right) + 4\frac{\partial}{\partial I}\left(I\frac{\partial\psi}{\partial I}\right) \quad , \tag{12'}$$

changing the variable to $I \propto b^*b$ (photon number, i.e., the intensity of the active mode) and following Risken's scaling procedure.[7] Let us now apply the Onsager's variational principle (3)-(3a,b) to solve Eq.(12'), which may be rewritten as follows:

$$\int_0^\infty \mathcal{S}\frac{\partial\log\psi}{t}\,\psi\,dI - \int_0^\infty 2I\left(\frac{\partial\mathcal{S}}{\partial I}\right)^2\psi\,dI = \max. \tag{13}$$

with post-variation setting

$$\mathcal{S} = \log(\psi_0/\psi) \quad . \tag{13a}$$

A detail of the variational procedure has been described elsewhere[3] but may be summarized in a word by the Rayleigh-Ritz
method, viz.

$$\text{ansatz:} \quad \psi(I,t) = \exp.\times[u_0 + \frac{u_1}{2}I - \frac{u_2}{4}I^2 - \frac{u_3}{6}I^3 ----] \quad , \tag{14}$$

$$(u_n = u_n(t))$$

for which the coefficients $u_n(n\geq 1)$ can be determined from the variational principle, while u_0 can be determined from the normalization, $\int \psi \, dx = 1$. This condition may be re-expressed as

$$\int_0^\infty \frac{\partial \log\psi}{\partial t} \psi \, dI = \dot{u}_0 + \frac{<I>}{2} \dot{u}_1 - \frac{<I^2>}{4} \dot{u}_2 ---$$

$$= 0 \quad , \tag{15}$$

with

$$<I^n> \equiv \int_0^\infty I^n \psi \, dI \qquad \text{(instantaneous moments)}, \tag{16}$$

which eliminates $\dot{u}_0$ by means of other $\dot{u}_n$'s and the moments.

Truncating the power series in the exponent of ψ up to I^3, we
have derived a set of ordinary differential equations for u_n's. In
virtue of the known steady-state solution ψ_0 for which $u_1=a$ (the
pump parameter), $u_2=1$ and $u_n(n\geq 3)=0$, the equation may have the form

$$\frac{d}{dt} \begin{pmatrix} u_1 \\ u_2 \\ u_3 \end{pmatrix} = M_1^{-1} \cdot M_2 \begin{pmatrix} u_1-a \\ u_2-1 \\ u_3 \end{pmatrix} \quad , \tag{17}$$

where the two 3×3 matrices M_1 and M_2 are both symmetric given by

$$M_1 = \begin{bmatrix} \frac{-1}{4}<(I-<I>)^2> & \frac{1}{8}<(I-<I>)(I^2-<I^2>)> & \frac{1}{12}<(I-<I>)(I^3-<I^3>)> \\ & \frac{-1}{16}<(I^2-<I^2>)^2> & \frac{-1}{24}<(I^2-<I^2>)(I^3-<I^3>)> \\ * & & \frac{-1}{36}<(I^3-<I^3>)^2> \end{bmatrix}$$

$$M_2 = \begin{bmatrix} <I> & - & <I^2> & - & <I^3> \\ & & <I^3> & & <I^4> \\ * & & & & <I^5> \end{bmatrix}$$

(18)

A transient laser action can be simulated by the solution of Eq.(17) having the initial values

$$u_1 = -\infty \ , \quad u_2 = u_3 = --- = 0 \quad , \quad t=0$$

(19)

and evolving in time such that $u_1 \to a$, $u_2 \to 1$ and $u_n (n \geq 3) \to 0$, as t tends to infinity.

We have carried out the above task with computer, using the Runge-Kutta method (i.e., replacing the differential equation by successive difference equations), where the initial values of u_1 are chosen, which are negatively large. The results of the integration of u_n's for three values of the pump parameter, a=0, 4 and 8, are shown in Fig. 1. It has been assured that a different choice of the initial u_1-values yields a difference in the results that is insignificant after a first passage of duration (i.e., after the passage of t_0 at which a kink occurs for every u_2-curve in Fig. 1). Clearly, this duration corresponds to the period in which the system "falls" rapidly from the unstable equilibrium point. The strange behavior of the u_2-curve after this duration indicates that tremendous fluctuations are produced accompanying this sharp fall and then are reduced; this is more conspicuous for the larger pumping. An advantage of the present method is that various lower-order moments and cumulants are obtained simultaneously in the course of the Runge-Kutta integration process in a self-consistent manner with the distribution function. The simplest ones, viz. $<I>$ (intensity) and $<(I-<I>)^2>$ (intensity variance), are presented in Fig. 2. A remark should be made about the convergence problem, that is, the accuracy of the truncation up to n=3. An inquiry of this can be made by comparing the results with n=2 truncation and those with n=3, showing a fair result about u_n's and a satisfactory result about $<I>$ and $<(I-<I>)^2>$.

Let us consider, as a second example, the original simple model of superradiance (superfluorescence) due to Bonifacio et al.[8] expressed in the operator master equation

$$\frac{\partial W}{\partial t} = \frac{1}{2} I_1 \ ([R_-,WR_+] + [R_-W, R_+]).$$

The R_z-diagonal representation of this equation may be approximated by a Fokker-Planck equation, in accordance with the system-size

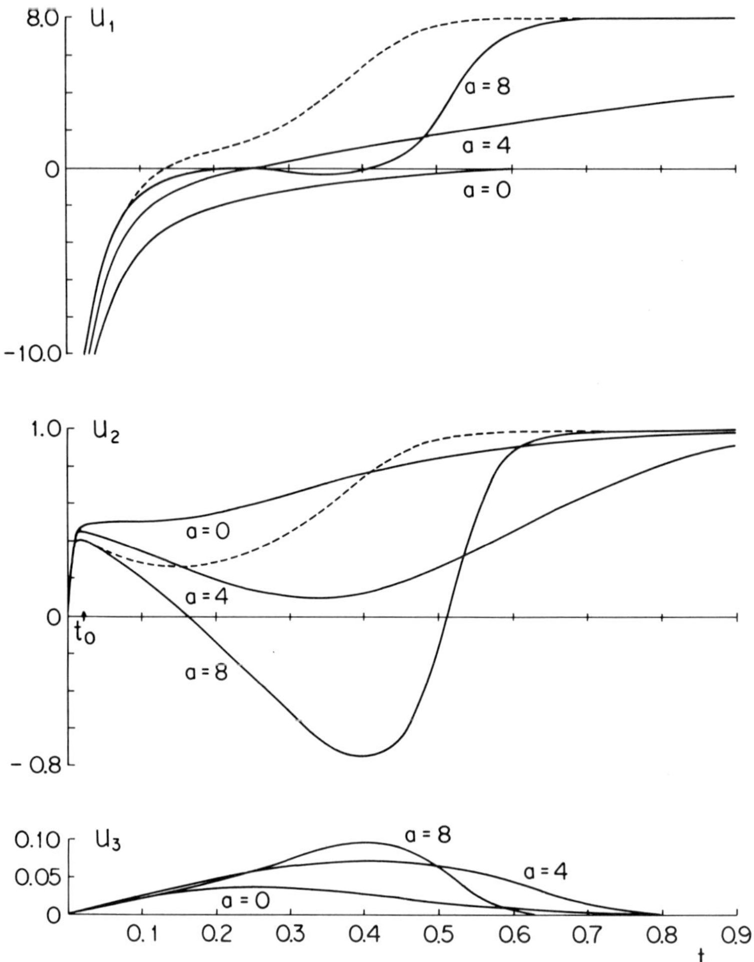

Fig. 1. The integrated three coefficients u_1, u_2 and u_3 in the exponent of the time-dependent laser distribution function for three values of the pump parameter a . The unit of abscissa is the scaled time in accordance with Risken's presentation.[7] For a comparison, the results for the truncation up to $n=2$ ($\dot{u}_3=u_3=0$ in Eq.(17)), a=8, are also shown (dotted lines).

expansion method,[9] as follows:

$$\frac{\partial \psi}{\partial t} = I_1 \, R\left(\frac{\partial}{\partial z} (1 + \varepsilon - z)\psi + \frac{\varepsilon}{2} \frac{\partial}{\partial z} (1-z^2)\frac{\partial \psi}{\partial z}\right) \tag{20}$$

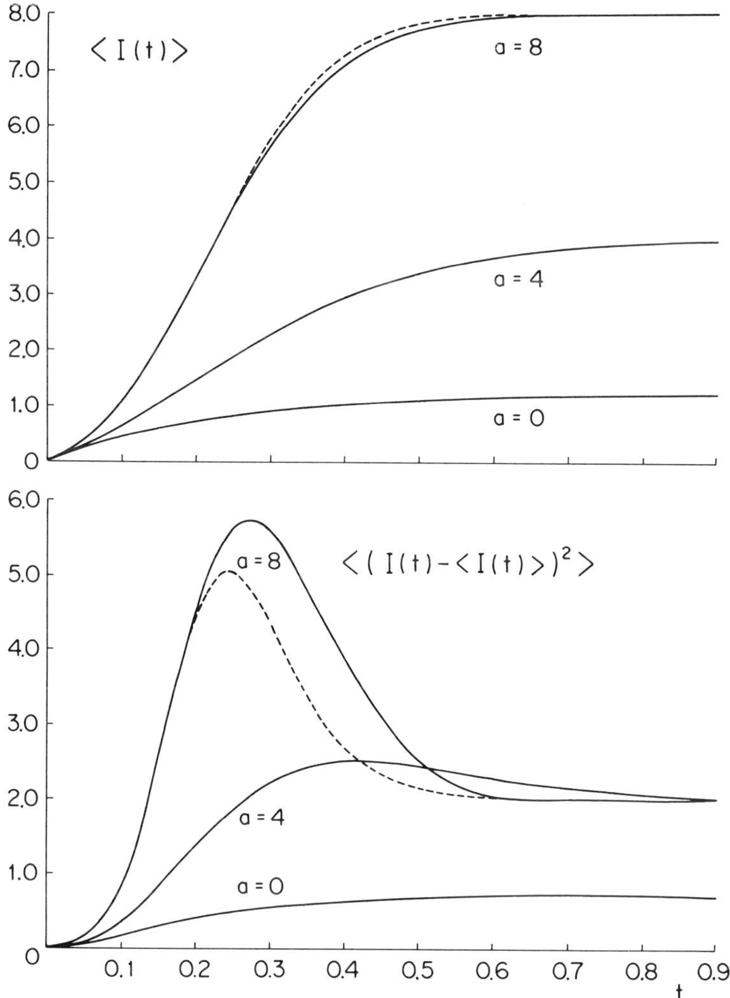

Fig. 2 The mean and the variance of the calculated intensity (in the same unit as Risken's) of the radiation for the laser. The n=2 truncation (dotted lines for a=8) disagrees with the corresponding curves due to Risken[7] (especially for the intensity variance between t=0.2 and 0.5). This is improved by taking into account the I^3-term (i.e., the n=3 truncation).

where $R=\varepsilon^{-1}$, the magnitude of the "spin", represents the system size (one-half of the number of active two-level atoms), and the variable z represents the cosine of the Bloch angle (measured from the north pole) so that

$$-1 \leq z \leq 1 \ . \tag{20a}$$

Here, the Prigogine's variational principle (4)-(4a,b) (the "local potential") is used to determine the approximate solution of Eq. (20) by taking a trial distribution function as

$$\psi(z) = const. \ exp[- \frac{R}{2\sigma} (z-m)^2] \ , \tag{21}$$

which also conforms to the spirit of the system-size expansion. By considering the mean m and the variance σ of z as two variation parameters, we have derived a set of differential equations for them as follows:

$$\frac{dm}{d\tau} = m^2 - 1 - \epsilon(m+1-\sigma) \tag{22a}$$

$$\frac{d\sigma}{d\tau} = 1 - m^2 + 4m\sigma - 3\epsilon\sigma \ . \tag{22b}$$

Here we have scaled time t such that $\tau=I_1Rt$, as usual.

The stationary points of Eqs.(22a,b) i.e., the zeroes of the r.h.s., can be easily located. In the physical region of the m-σ plane, $-1\leq m\leq 1$, $0\leq\sigma$, they are:

(i) $m = -1$, $\sigma = 0$ (stable equilibrium point)

(ii) $m = \frac{\epsilon}{2}$, $\sigma = \frac{1}{\epsilon} \gg 1$ (unstable "saddle point").

Thus, it is naturally expected that the system starting at any point inside the above physical region will tend to the stable equilibrium (i), as $\tau \rightarrow \infty$, i.e., to the "south pole" of the dipole pendulum around which the system becomes sharply distributed.

The nonlinear differential equations (22a,b) are easier to handle than Eq.(17) for the laser, some typical solutions of which are presented in Fig. 3. A significant point of discussion from the present result must be that our method yields the evolution equations for the mean and variance in order of the smallness param- eter $\epsilon(=1/R)$, one higher than that which has been formulated.[9] Thus, the familiar superradiance equation of the hyperbolic tangent type for the atomic inversion m is modified by the additional term which arises from the diffusion part of Eq.(20). Consequently, the present result yields a finite period in order of 0(log R) to attain the maximum intensity of the emitted radiation, starting from the complete inversion with a small, or even vanishing, dipole- fluctuation. Since the diffusion part of Eq.(20) is purely quantum in origin (as can be realized by checking its derivation and its existence in the case of vanishing thermal photons), we can say

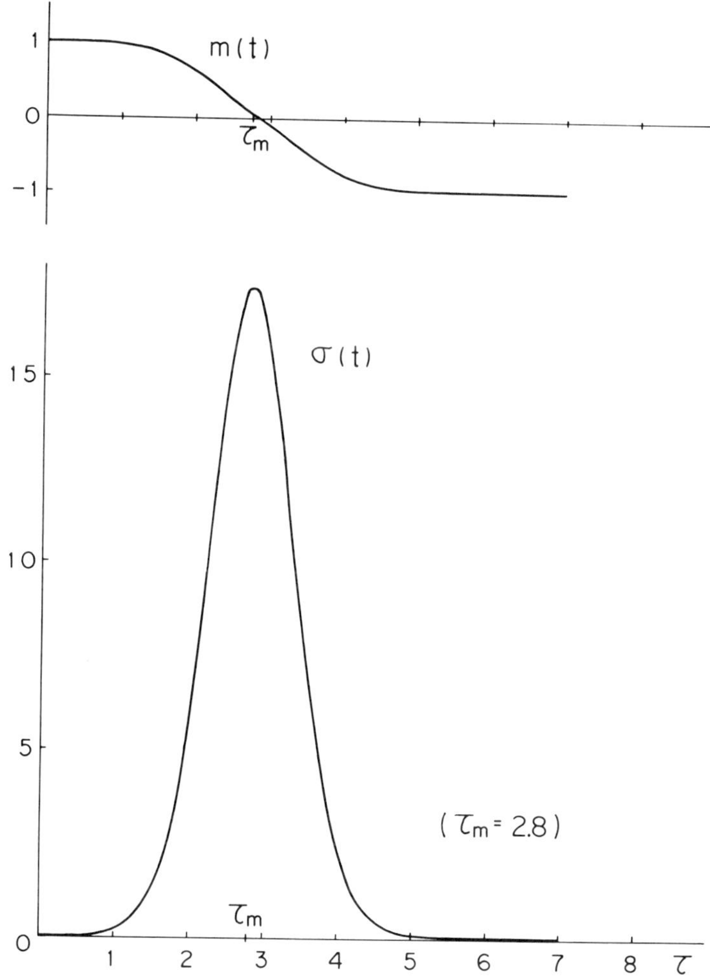

Fig. 3 The integrated mean m and variance σ of the normalized atomic inversion, $z=R_z/R$, for the superradiance with initial values: $m=\sigma=0$, at $\tau=0$ (R=100), where $\tau=RI_1t$.

that the quantum fluctuation assures us of this result.

4. A TEST OF THE GLANSDORFF-PRIGOGINE INEQUALITY AND OTHER REMARKS

We had conjectured that the evolution criterion (6) might be true in a much wider circumstance than restricted to near the steady state: in view of the idealized characteristic of entropy,[10] it might be true for any solution of the Fokker-Planck equation, so

far as it exists. The present computer results for the laser have
provided evidence that this is not the case, as shown in Fig. 4.
Thus, the Glansdorff-Prigogine inequality can be violated for stochastic systems with instability such as the laser.[11]

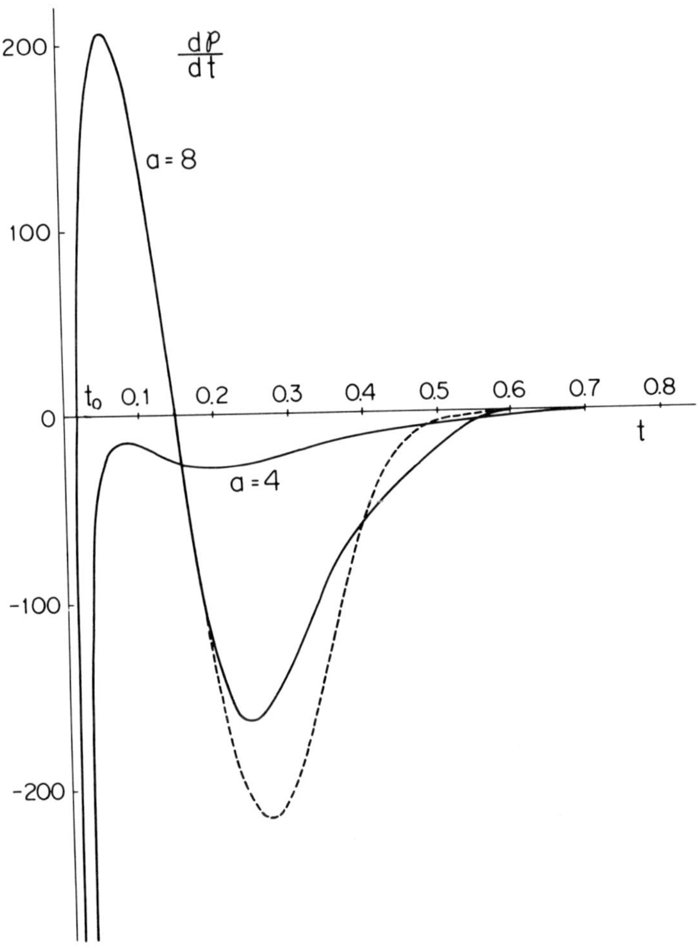

Fig. 4 The derivative in time of the calculated dissipation function (i.e., the entropy production rate) for the laser instability.
For a=4, the Glansdorff-Prigogine inequality is fulfilled all the
time, while for a=8 the curve crosses abscissa (which is unaltered
by going from the n=2 (dotted lines) to n=3 truncations), showing
that the instability action violates the said inequality.

We can show by a more analytic means that a similar situation occurs in superradiance when the atomic dipoles are set in initially at the point of full inversion with 0-fluctuations. It can be said that the entropy production decreases rapidly to attain a minimum value which is different from that corresponding to the final equilibrium (or, steady-) state: instead, it increases once and then decreases again, according to the real evolution criterion (see Fig. 5). The "anomalous fluctuation" produced in such instability action could be characterized by this entropy production increase or entropy overproduction. Hence, one might attribute the so-called "dissipative structure" formation to a kind of compensation effect for such an overproduction of fluctuations.

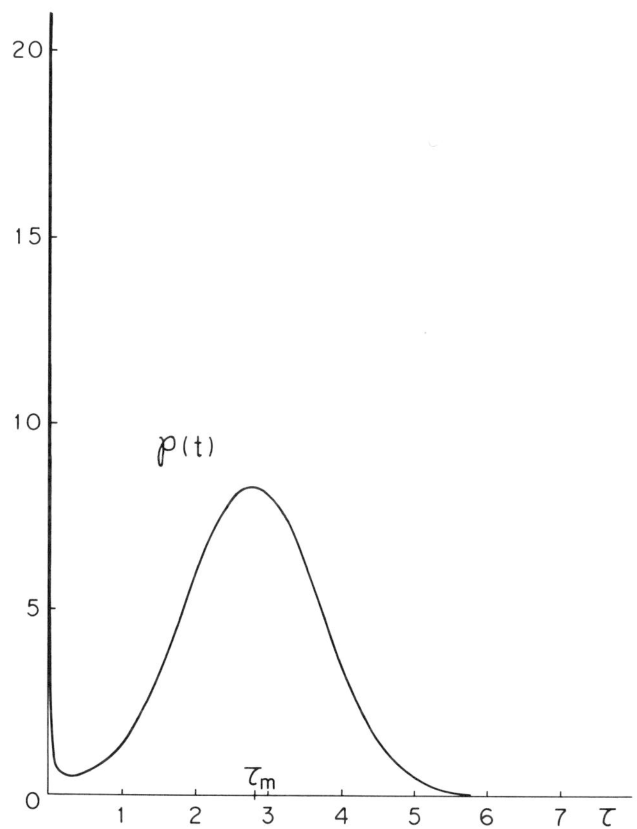

Fig. 5 The time behavior of the calculated dissipation function (i.e., the entropy production) for the superradiance showing a minimum point before τ_m (time of the maximum intensity), another "entropy production minimum".

In conclusion, the typical optical instabilities, the laser and superradiance, can be characterized from a thermodynamic point of view by saying that a unique principle governs the process, that is the least dissipation principle of Onsager.

Acknowledgements

We would like to thank Professor M. Suzuki and Mr. T. Armitsu for providing us with an account of their analyses before publication. Thanks are also due to Dr. K. Ikeda for a stimulating discussion.

References

1. R. Graham, in *Coherence and Quantum Optics*, edited by L. Mandel and E. Wolf (Plenum Press, New York, 1973) p. 851.
2. P. Glansdorff and I. Prigogine, *Thermodynamic Theory of Structure, Stabiltiy and Fluctuations* (Wiley-Interscience, London, 1971).
3. H. Hasegawa, Prog. Theor. Phys. *56*, 44 (1976); *57*, 1523 (1977); *58*, No. 1 (1977); see also Phys. Lett. *60A*, 171 (1977).
4. G. Nicolis, in *Advances in Chemical Physics, XIX*, edited by I. Prigogine and S.O. Rice (Wiley-Interscience, London, 1971) p. 209.
5. L. Onsager, Phys. Rev. *37*, 405 (1931).
6. *Non-Equilibrium Thermodynamics Variational Techniques and Stability*, edited by Donnelly, Hermann and Prigogine (The University of Chicago Press, Chicago and London, 1965).
7. H. Risken, in *Progress in Optics*, Vol. 8, edited by E. Wolf (North-Holland, Amsterdam, 1970) p. 239.
8. R. Bonifacio, P. Schwendimann and F. Haake, Phys. Rev. A *4*, 854 (1971).
9. R. Kubo, K. Matsuo and K. Kitahara, J. Stat. Phys. *9*, 51 (1973); M. Suzuki, Physica *84A*, 48 (1976); Prog. Theor. Phys. *57*, 380 (1977); T. Arimitsu and M. Suzuki, to be published (1977).
10. F. Hofflich, Z. Physik *226*, 395 (1969).
11. This statement deserves a careful account of its implication, which we point out here in order to avoid any misleading reference to the Glansdorff-Prigogine original expression,[2]:

$$J_\mu \dot{X}_\mu \leq 0 \ . \tag{A}$$

As noted by these authors, the inequality guarantees that $(d/dt)(J_\mu X_\mu) \leq 0$ in the near steady-state for which $J_\mu = L_{\mu\nu} X_\nu$ ($L_{\mu\nu} = L_{\nu\mu}$ and constant), but generally the l.h.s. of (A) cannot be identified with the total time-derivative of any scalar function of the potential nature. Therefore, one should look at the quantity $(d/dt)(J_\mu X_\mu)$, if at all concerned with the time behaviour of the "entropy production". (Note that the laser gives a typical example of the violation of (A) around the unstable point, when Graham's potential $\ln P$ is chosen as the entropy.) In the present context, the l.h.s. of (A) may be replaced by the average $\int J_\mu (\partial X_\mu / \partial t)\psi \, dx$, where $J_\mu = (1/2)K_{\mu\lambda}X_\lambda$ and $X_\mu = (\partial/\partial x_\mu)\log(\psi_0/\psi)$, showing that (A) is always valid.[3] However, what we consider here is the total derivative of $\mathcal{P} = \int J_\mu X_\mu \psi \, dx$.

OPERATOR ALGEBRAIC METHODS FOR LASER CAVITY MODES

David Stoler

Perkin-Elmer Corporation, Norwalk, Connecticut

1. INTRODUCTION

Wave dynamics in a source free region of space is governed by a homogeneous wave equation of the form:

$$\nabla^2 \psi - \frac{1}{c^2} \frac{\partial^2 \psi}{\partial t^2} = 0 \quad . \tag{1}$$

$\psi(x,y,z,t)$ is a complex scalar whose squared modulus is proportional to the field intensity. In the paraxial case where the energy flow is predominantly in a single direction which we take to be the positive z axis we may replace Eq.(1) with its paraxialized form:

$$\frac{\partial^2 W}{\partial x^2} + \frac{\partial^2 W}{\partial y^2} = 2ik \frac{\partial W}{\partial z} \quad ; \quad k = \frac{\omega}{c} \quad , \tag{2}$$

where, we have written $\psi(x,y,z,t) = W(x,y,z)\exp(i(\omega t - kz))$ and substituted this into Eq.(1). In deriving Eq.(2) we neglect $\partial^2 W/\partial z^2$ when compared to $k(\partial W/\partial z)$.

The solution to Eq.(2) can be written in the following useful way:

$$W(x,y,z) = \exp\left\{ \frac{-(z-z_0)i}{2k} \left(\frac{\partial^2}{\partial x^2} + \frac{\partial^2}{\partial y^2} \right) \right\} W(x,y,z_0) \quad . \tag{3}$$

We can propagate a wave front from z_0 to z using the evolution operator

$$U = \exp\left\{ \frac{-i(z-z_0)}{2k} \left(\frac{\partial^2}{\partial x^2} + \frac{\partial^2}{\partial y^2} \right) \right\} \ .$$

This approach to propagation is widely used in the quantum theory. In fact, the paraxial wave equation, Eq.(2), is formally identical to the Schroedinger equation for a free particle, of mass m, in the two dimensions. This can be seen by making the substitution

$$z = -\frac{\hbar k}{m}\, t \qquad \text{in Eq.(2)} \ .$$

The passage of a beam through an optical system is calculated by treating the system as a lumped circuit. The effects of extended bulk matter are introduced at a series of planes (stations) distributed throughout the optical train. These effects (gain, loss, thermal blooming, etc.) are included by multiplying the wave function by a transfer function that depends on the transverse coordinates and other relevant parameters. Reflections from mirrors and passage through apertures are handled in the same way with an appropriate transfer function. Between stations the wave propagates freely as governed by the paraxial wave equation. We will also use the lumped parameter approach to an optical system but will exploit the operator algebraic structure in a way that will lead to a powerful analytical and computational approach.

To begin, let us define the operators

$$p_x = -i\,\frac{\partial}{\partial x} \quad \text{and} \quad p_y = -i\,\frac{\partial}{\partial y}$$

in terms of which Eq.(3) becomes

$$W(x,y,z) = \exp\left\{ \frac{i(z-z_0)}{2k}\, \hat{p}^2 \right\} W(x,y,z_0) \ , \tag{4}$$

where

$$\hat{p}^2 = \hat{p}_x^2 + \hat{p}_y^2 = -\frac{\partial^2}{\partial x^2} - \frac{\partial^2}{\partial y^2} \ .$$

At this point it is convenient to switch to Dirac notation rather than the coordinate space notation we have been using.

The function $W(x,y,z)$ is represented by the ket vector and is related to it through the inner product $\langle x,y|W(z)\rangle$, i.e.,

$$W(x,y,z) = \langle x,y|W(z)\rangle \ .$$

In this notation Eq.(5) becomes

$$|W(x)> = \exp\left\{\frac{i(z-z_0)}{2k}\,\hat{p}^2\right\}|W(z_0)> \quad . \tag{5}$$

Note that in this equation the operator $\hat{p}^2 = \hat{p}_x^2 + \hat{p}_y^2$ is the abstract representative of the derivative operators defined in connection with Eq.(4) since it operates in the ket space and not in the space of functions of x and y.

We can use this method in two ways. The first is to calculate the progress of a wave through a portion of an optical system. Consider a system represented by N stations each of which is described by a transfer function $T_i(x,y)$. Let $|W(z_0)>$ represent the field at a plane $z = z_0$ before the first station (see Figure 1). The stations are located at

$$z = z_i \quad (i = 0, 1, 2, \ldots N) \quad .$$

We want to calculate the field at N-th station $|W(z_N)>$. The procedure is simply to propagate the field from the i-th station to the (i+1)-st station using the operator

$$\exp\left\{\frac{i(z_{i+1} - z_i)}{2k}\,\hat{p}^2\right\} \quad ,$$

multiply by the transfer function operator appropriate to the (i+1)-st station $T_{i+1}(x,y)$, and continue the process until the last station is reached. The result gives

$$|W(z_N)> = T_N(x,y) \ldots \hat{U}_{2,1}\,T_1(x,y)\,\hat{U}_{1,0}\,|W(z_0)> \quad , \tag{6}$$

where

$$U_{i+1,i} = \exp\left\{\frac{i\Delta_{i+1,i}}{2k}\,\hat{p}^2\right\} \quad ,$$

$$\Delta_{i+1,i} = \left|z_{i+1} - z_i\right| \quad .$$

We note here that this form of $U_{i+1,i}$ is valid for both directions of propagation. It is convenient to have all the stations equally spaced so that the propagation operator is always the same, i.e.,

$$U_{i+1,i} = \exp\left\{\frac{i\Delta}{2k}\,\hat{p}^2\right\} = U_\Delta \quad ,$$

where

$$\Delta_{i+1,i} = \Delta \quad .$$

Incorporating this into Eq.(6) and taking the inner product of both sides with the coordinate basis vector $|x,y\rangle$ we have the basic result:

$$\langle x,y|W(z_N)\rangle = \langle x,y|T_N(x,y)\ldots T_1(x,y)\hat{U}_\Delta|W(z_0)\rangle \quad . \tag{7}$$

There are several ways we can use this result. For example, suppose that we were able to multiply all of the operators in the sequence together in such a way that it became possible to evaluate the quantity $\langle x,y|T_N\hat{U}_\Delta\ldots T_1\hat{U}_\Delta|x',y'\rangle$. Then, knowing this quantity we could obtain our desired result by one double integration, i.e.,

$$\langle x,y|W(z_N)\rangle = \iint dx'dy'\langle x,y|T_N U_\Delta\ldots T_1 U_\Delta|x'y'\rangle\langle x'y'|W(z_0)\rangle \quad . \tag{8}$$

In deriving Eq.(8) we have used the completeness relation

$$\int dx'dy'|x',y'\rangle\langle x',y'| = 1 \quad .$$

In some cases this can be directly accomplished. In others the algebraic complexity would dictate use of approximation or intermediate integrations. To illustrate the point consider a situation for which $T_i(x) = \exp[i\alpha_i x^2]$. Then

$$T_i(x)U_\Delta = e^{i\alpha_i x^2} e^{i\beta\hat{p}^2} \quad ,$$

where for simplicity we consider only one transverse dimension and $(\beta = \Delta/2k)$. Now, using algebraic methods, one can easily show that

$$e^{i\alpha_i\hat{x}^2} e^{i\beta\hat{p}^2} = \exp[i\alpha_i x^2 + i\beta p^2 - i\alpha_i\beta(\hat{x}\hat{p} + \hat{p}\hat{x})] \quad .$$

Now it is also true that the product of any number of exponentials of the form $\exp[i\alpha x^2 + i\beta p^2 - i\alpha\beta(xp + px)]$ is equal to a single exponential whose exponent is a linear combination of x^2, p^2, and $(xp + px)$. Hence, all we need to do is evaluate an expression of the form

$$\langle x|\exp\{\alpha x^2 + \beta p^2 + i\gamma(\hat{x}\hat{p} + \hat{p}\hat{x})\}|x'\rangle \quad .$$

This expression can in fact be evaluated in closed form using a generating function for bilinear products of eigenfunctions (generalized Mehler's formula). This very simple example was chosen to indicate how utilization of the algebraic structure of the propagation formalism can reduce the amount of computer time required. We would require only a single twofold integration whereas the currently employed method would involve 2(N+1) integrations.

2. APPLICATION TO OPEN OPTICAL RESONATORS

The second way we shall use this technique is in the calculation of modes of open resonators. Let us consider a bare resonator consisting of two mirrors of focal lengths f_1 and f_2 separated by a distance L (see Figure 1). If $|W(0)>$ represents the field on mirror (1), and $T_1(x)$, $T_2(x)$ are the transfer functions corresponding to reflection at mirrors (1) and (2), the condition that the field replicate itself after one roundtrip is expressed by the equation

$$T_1(\hat{x}) \; \hat{U}_L \; T_2(\hat{x}) \; \hat{U}_L \; |W(0)> \; = \; \Lambda |W(0)> \quad . \tag{9}$$

For ease of writing we again only include a single transverse coordinate. Equation (9) is an operator eigenvalue equation for the resonator mode functions and eigenvalues. The propagation operator $\hat{U}_L$ is given by

$$\hat{U}_L \; = \; \exp \left\{ \frac{iL}{2k} \hat{p}^2 \right\} \; .$$

There are several ways we can use Eq.(9). We can derive an integral equation for the mode function $W(x,0) = <x|W(0)>$ by taking the inner product of Eq.(9) with $<x|$ and inserting the expression $1 = \int dx' |x'><x'|$ between $\hat{U}_L$ and $|W(0)>$, i.e.,

$$\int_{-\infty}^{\infty} <x|T_1(x) \; \hat{U}_L \; T_2(x) \; \hat{U}_L \; |x'> \; W(x',0) \; dx' \; = \; \Lambda W(x,0) \; . \tag{10}$$

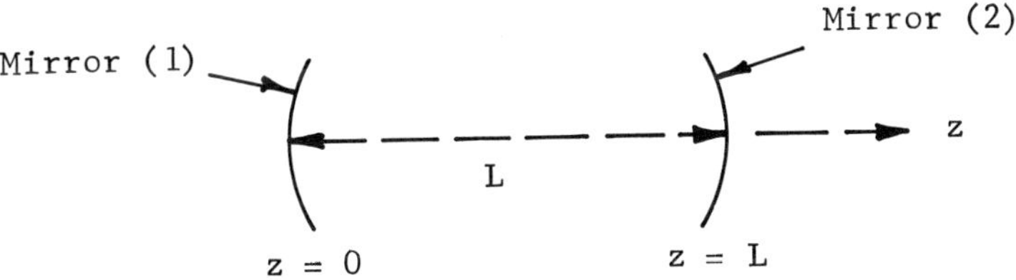

Figure 1. Resonator Cavity

Making use of the fact that $<p|W>$ is the Fourier transform of $<x|W>$, where $|p>$ is an eigenvector of $\hat{p}$, we have an integral equation for the Fourier transform of the mode function, i.e.,

$$\int_{-\infty}^{\infty} <p|T_1(x)\ \hat{U}_L\ T_2(x)\ \hat{U}_L|p'> W(p',0)dp' = \Lambda W(p,0) \quad . \tag{11}$$

This equation will prove useful when dealing with apertured resonators. This is because the aperture size occurs in the p-space kernel $<p|T_1\hat{U}_LT_2\hat{U}_L|p'>$ as an explicit parameter rather than as the range of a position space integral.

The algebraic approach widens the range of technique for calculating kernel functions and makes it possible to evaluate the effects on the kernel due to alterations in cavity configuration. In fact, it is possible to derive an expression for the change in the eigenvalue of a given mode by an alteration in the cavity, such as a mirror tilt, decentration, a change of aperture size, etc.

Rather then using Eq. (9) to yield a kernel and then solving an integral equation for the modes we can attempt to solve the operator eigenvalue equation directly. One approach to this is as follows.

The eigenvalue Eq. (9) contains a product of functions of the operators $\hat{x}$ and $\hat{p}$. The propagation operator $\hat{U}_L$ is an exponential of a very simple function of $\hat{x}$ and $\hat{p}$, namely just $\hat{p}^2$. If we write the entire product as a single exponential, say, $\exp(i\hat{S})$ where $\hat{S}$ is a relatively simple (e.g., algebraic) function of $\hat{x}$ and $\hat{p}$, Eq. (9) is replaced by the simpler eigenvalue problem

$$\hat{S}(\hat{p},\hat{x})\,|W> = \sigma|W> \quad , \tag{12}$$

and we have $\Lambda = \exp(i\sigma)$.

We illustrate this method for a strip mirror resonator with arbitrary focal lengths and mirror spacing. At this point we ignore the effects of finite apertures. We shall discuss aperturing later on. For the resonator shown in Figure 1 the transfer functions $T_j(x)$ are given by $T_j = \exp\{i\alpha_j\hat{x}^2\}$, where $\alpha_j = -(k/2f_j)$, f_j = focal length of j-th mirror, k = wave number, and $j = 1,2$. Denoting $L/2k$ by β in the evolution operator we find that Eq. (9) for this case becomes

$$e^{i\alpha_1\hat{x}^2}\ e^{i\beta\hat{p}^2}\ e^{i\alpha_2\hat{x}^2}\ e^{i\beta\hat{p}^2}\ |W_1> = \Lambda|W_1> \quad . \tag{13}$$

The ket $|W_1>$ represents the mode function on mirror number one. The task of writing the product of four exponentials as a single exponential is non-trivial because of the fact that $\hat{x}$ and $\hat{p}$ do not

commute. One cannot therefore multiply exponentials together simply
by adding their exponents. However, there exists a large body of
algebraic and group theoretic techniques for doing this sort of thing.
We have seen earlier that the product $e^{i\alpha\hat{x}^2} e^{i\beta\hat{p}^2}$ is equal to
$\exp\{i\alpha\hat{x}^2 + i\beta\hat{p}^2 - i\alpha\beta(\hat{x}\hat{p}+\hat{p}\hat{x})\}$. This result was obtained by recognizing
that the operators $\hat{x}^2$, $\hat{p}^2$, and $\hat{x}\hat{p}+\hat{p}\hat{x}$ together form the Lie algebra
$su(1,1)$ and by making use of the Baker-Hausdorff formula for the
associated group.

If we, for convenience, take mirror (2) to be a flat then $\alpha_2 = 0$,
and we have the half-symmetric case. Equation (12) becomes

$$[(\tfrac{L}{k})\hat{p}^2 - (\tfrac{k}{2f_1})\hat{x}^2 + (\tfrac{L}{2f_1})\ (\hat{x}\hat{p} + \hat{p}\hat{x})]|W_1> = \sigma|W_1> \quad . \tag{14}$$

This is the fundamental eigenvalue equation for modes of this
cavity. We can solve this equation for any values of f_1 and L.

We can conveniently study the properties of Eq.(14) using operator algebraic techniques. For example, the unitary transformation
generated by

$$\hat{V} = \exp \left\{ \frac{iL}{2k}\ \hat{p}^2 \right\}$$

will convert Eq.(14) into the equation

$$[\frac{L}{k}\ (1 + \frac{L}{2f_1})\ \hat{p}^2 - \frac{k}{2f_1}\ \hat{x}^2]\ |\phi_1> = \sigma|\phi_1> \quad , \tag{15}$$

where

$$|\phi_1> = \hat{V}|W_1> \quad .$$

When the parameters L, f_1 are such that the resonator is stable,
Eq.(15) has the form of the Schroedinger equation for the harmonic
oscillator for which the eigenfunctions and eigenvalues are well
known. When the resonator is unstable, Eq.(15) is the equation of
an inverted harmonic oscillator, i.e., with a repulsive parabolic
potential. The spectrum of Eq.(15) in this case is continuous and
unbounded. The eigenfunctions are related to the parabolic cylinder
functions.

3. EXACT EIGENVALUE EQUATION FOR MISALIGNED RESONATORS

The algebraic method leading to the derivation of an operator
eigenvalue equation for the resonator modes may be extended to the
case of misaligned resonators as well. For example, if the i-th

station is decentered it will be described by an operator transfer
function $\hat{T}_i' = \hat{R}\,\hat{T}_i\,\hat{R}^+$ where $\hat{R}$ is an operator generating the decen-
tering translation and $\hat{T}_i$ is the centered transfer function. $\hat{R}$ will
be of the form $\exp(i\hat{G})$ where $\hat{G}$ is the Hermitian generator of the
transformation. If $\hat{G}$ is a member of the Lie algebra of the centered
resonator, $\hat{T}_i'$ will be of the same general form as $\hat{T}_i$ but perhaps
having some additional members of the Lie algebra in its exponent.
The calculation proceeds as before with T_i' replacing T_i. If the
generator G is not a member of the original resonator Lie algebra,
the algebra must be extended via commutation to a larger algebra
that contains G. This approach permits the derivation of eigenvalue
equations for misaligned resonators that shall yield additional
insights into the effect of misalignments on the mode structure.
The misalignments are built into the equation for the mode and are
associated both with changes in value of parameters that occur in
the aligned case and also with the presence of new terms in the
mode equation.

4. EFFECTS OF PERTURBATIONS ON MODAL STRUCTURE

Another example of the general utility of the algebraic approach
is the derivation of a group of formulae that give the changes in
mode structure produced by an alteration in the configuration of a
cavity resonator. For example, if the transfer function T_1 describing an end mirror of a resonator is altered so that it becomes
$T_1 + \delta T_1$, the resulting change in the eigenvalue λ_n of the n-th
eigenmode can be seen to be given by

$$\frac{\delta\lambda_n}{\lambda_n} = \frac{\langle\phi_n^*\,|\,\delta T_1/T_1\,|\,\phi_n\rangle}{\langle\phi_n^*\,|\,\phi_n\rangle} \quad , \tag{16}$$

or in functional representation

$$\frac{\delta\lambda_n}{\lambda_n} = \frac{\displaystyle\int\phi_n^2(x)\,\frac{\delta T_1(x)}{T_1(x)}\,dx}{\displaystyle\int\phi_n^2(x)\,dx} \quad . \tag{17}$$

Equation (17) gives a quantitative evaluation of the effect of
the cavity distortion on the modes and shows how the modes themselves
influence that effect. It is interesting to note that the weight
factor in the integral multiplying $[\delta T_1(x)]/[T_1(x)]$ is the square of
the complex mode function and not the square modulus. This means

that the phase distribution as well as the intensity distribution
influences the effect of the cavity perturbation. We have derived
corresponding formulae for the mode coupling coefficients giving
the admixture of other unperturbed modes in a given perturbed mode.
We have also extended this approach to more complicated cavity configurations. Formula (17) has been tested in several cases where
approximate analytical solutions are known and has reproduced the
correct results.

We have thus far not dealt explicitly with the effects of
finite apertures in resonator caviities, although some results such
as Eq.(17) do apply to apertured resonators without any need for
modification.

There are a number of possible approaches to the treatment of
finite apertures. In the conventional treatment of resonators based
on Fox-Li integral equation the finite size of the aperture enters
into the calculation simply through the finite range of integration.
As we have seen in Eq.(11), one can formulate the integral for the
Fourier transform of the mode function in which case the aperture
size enters into the kernel as a parameter. This seems useful since
one can examine the dependence of the kernel on aperture size analytically and generate approximations to the kernel based on the
smallness or largeness of the aperture, etc.

Expressing the kernel functions as a matrix element of an operator product can lead to additional insights regarding the nature of
the solutions as well as further approximation techniques. For
example, there are many possibilities for approximating the kernel
by a degenerate kernel by inserting an appropriate complete set of
ket vectors between two operator factors and truncating the sum.
This can be done in a number of different ways, as indicated by the
following equations,

$$K(x,x') = \langle x|T_1(\hat{x})\hat{U}\,T_2(\hat{x})\,\hat{U}|x'\rangle$$

$$= \sum_n \langle x|T_1\,U|\phi_n\rangle\,\langle\phi_n|T_2\,U|x'\rangle \quad . \tag{18}$$

In this connection we should point out that the formalism being
developed can often greatly facilitate the derivation of useful general results such as the fact that the kernel of an open resonator
is normal. One can also demonstrate quite generally the fact that
the resonator kernel may always be made symmetric.

We do this as follows:

The integral equation is given by

$$\int K(x,x') \ W(x') \ dx' = \Lambda W(x) \tag{19}$$

where $K(x,x') = \langle x|\hat{T}_1\hat{U}\hat{T}_2\hat{U}|x'\rangle$ is not as it stands always symmetric. However, writing $K(x,x') = T_1(x)\langle x|\hat{U}T_2\hat{U}|x'\rangle$ and defining $f(x) = W(x)/\sqrt{T_1(x)}$, Eq.(19) becomes

$$\int [T_1(x)T_1(x')]^{\frac{1}{2}} \ \langle x|\hat{U}\ \hat{T}_2\ \hat{U}|x'\rangle \ f(x')dx' \quad . \tag{20}$$

The factor $\langle x|\hat{U}\hat{T}_2\hat{U}|x'\rangle$ is symmetric in x and x' as can be seen by inserting the relation $\int d\bar{x}|\bar{x}\rangle\langle\bar{x}|$ between $T_2(\hat{x})$ and $\hat{U}$, i.e.,

$$\langle x|\hat{U}\hat{T}_2\ \hat{U}|x'\rangle = \int d\bar{x}\ \langle x|\hat{U}\ \bar{x}\rangle\ T_2(\bar{x})\ \langle\bar{x}|\hat{U}|x'\rangle \quad . \tag{21}$$

Since $\langle x|\hat{U}|\bar{x}\rangle$ depends on x and $\bar{x}$ only through the combination $(x-\bar{x})^2$, we see that Eq.(21) defines a function symmetric in x and $\bar{x}$.

5. TREATMENT OF FINITE APERTURES

Methods for treating apertured resonators that capitalize as fully as possible on the power and flexibility of the operator algebraic approach, are currently being developed. Several approaches are possible here and we are pursuing each of them.

The direct extension of our derivation of an operator eigenvalue equation for resonator modes requires dealing with the fact that the transfer function of an apertured mirror is discontinuous. It is a constant or some given function on the mirror surface and it is zero off the mirror.

The method we have employed to derive the resonator equation required an explicit coordinate dependence for the transfer functions. The coordinate dependence of a step function cannot be globally defined in a simple way. One might define the step function as the limiting case of a sequence of continuous functions and do the operator algebraic manipulations first and then take the limit which gives the step function. Another approach is to use an integral representation of the discontinuous transfer function. This approach leads to a novel type of operator eigenvalue problem which is currently being investigated. We are also investigating the approach using the limiting sequence to represent the transfer function.

A more fundamental approach to apertures in this formalism should be based on the detailed properties of the operators involved, particularly their domains of definition and hermiticity. As is

usually the case in operator formalisms there is a need to pay
attention to the specific details of the abstract space in which
the operators act. It is important to define carefully the properties of the functions or ket vectors that comprise the space of
interest. For example, we note that the eigenvalue equation for a
two mirror resonator, Eq.(9), implies that the mode function
$W(x) = \langle x|W\rangle$ vanishes for values of x off the apertured mirror,
i.e., it vanishes wherever $T_1(x)$ does. This is an important point.
It is clear that there will be a non-zero field off the mirror due
to both geometric propagation as well as diffraction. The reason
for the vanishing off the mirror of the solution to Eq.(9) is that
Eq.(9) is gotten from the condition for self-replication of the
field which needs to be satisfied only by the field on the mirror.
We see that the approach being discussed here brings the differences
between the fields on the mirrors and the fields off the mirrors
(but in the mirror planes) into sharp relief. The field on the
mirror will have a standing wave component while the field off the
mirror clearly must be an outgoing traveling wave.

The proper domain of definition of the operators occurring in
Eq.(9) then is the class of functions that vanish off the mirror,
i.e., functions of compact support. This will have important consequences. For example, if we require the operator P to be defined
in this way, then the spectrum of its eigenvalues is discrete rather
than continuous, e.g., for a flat mirror of width 2a the eigenfunctions of P are the complex exponential functions $\exp[(\pm in\pi x)/(2a)]$
with n=1,2,... . The eigenfunctions of the propagation operator
$U = \exp(i\beta P^2)$ would be, under these circumstances, linear combinations of $\sin(n\pi x/2a)$ and $\cos(n\pi x/2a)$.

Using a $\hat{p}$-operator defined this way we have treated a flat
half-symmetric Fabry-Perot resonator by doing a simple analytical
iteration of Eq.(9). The resulting approximate solution has a form
closely resembling the solution obtained by Vainshtein for this
resonator. Numerical checks using this result agree well with the
computer calculations of Fox and Li.

It is clear from the above considerations that we ought to
examine the space of functions which are solutions to Eq.(9) and
which vanish off the mirror. This space, S, is important to us
even though the mode functions we seek are technically not in it.
We can calculate the field values off the mirror by a single propagation through the resonator starting from the mirror of interest.
The space S is still the natural object of study because it is the
natural domain of Hermiticity for operators like x and p. These
operators are Hermitian in the space S and therefore they need not
be Hermitian with respect to functions outside S. Hence an operator
like $\exp(i\beta P^2)$ acting on an eigenfunction outside of S could have
an eigenvalue whose modulus is less than unity.

THE MULTIPHOTON COHERENT HANLE EFFECT*

A. Flusberg, T. Mossberg, and S.R. Hartmann

Columbia University, New York, New York

After a lull of several years, a good deal of interest has
again been focussed on optical three-wave mixing in atomic vapors.
Six years ago Finn and Ward studied dc-electric-field induced
optical second-harmonic generation (SHG) in several inert gases[1].
This type of three-wave mixing depends on the mixing of atomic
states of opposite parity in a dc electric field. In the second
type of atomic three-wave mixing, which has only been reported
during the past year, no dc electric field is introduced; instead,
the generation occurs as a result of a higher-order (either E2 or
M1) multipole moment of an atomic two-photon excitation.

The recent interest in three-wave mixing in atomic vapors
began with the work of Bethune, Smith and Shen[2], who produced
E2 sum-frequency generation (SFG) by optically mixing two noncollinearly propagating laser beams in atomic Na vapor. Flusberg,
Mossberg and Hartmann produced M1 difference-frequency generation
(DFG) in atomic thallium vapor in an applied magnetic field[3];
later, they showed that magnetically induced three-wave mixing can
interfere either constructively of destructively with the mixing
due to noncollinear propagation[4]. Magnetically induced atomic
SHG was observed independently by Matsuoka et al.[5] and Flusberg
et al.[6,7] Finally, Bethune et al.[8] have shown that noncollinear
SFG can be used in conjunction with dc-electric-field induced SFG
to measure E2 matrix elements in terms of known electric-dipole
matrix elements[8].

The purpose of this paper is twofold. First, we will review
some of the characteristics of magnetically induced three-wave
mixing, giving special attention to some unexplained features of

SHG in atomic thallium vapor. Second, we will show that magnetically
induced three-wave mixing may be viewed as a type of multiphoton
forward-scattering ("coherent") Hanle effect. We will show that
it is completely analogous to the single-photon forward-scattering
Hanle effect which was first observed and explained several years
ago by Corney et al.[9] (The designation of the forward-scattering
and lateral-scattering Hanle effect as "coherent" and "incoherent,"
respectively, is ours, and will be explained below.)

Let us begin by explaining why collinear three-wave mixing
does not ordinarily occur in an atomic vapor. To begin with,
parity conservation rules out the mixing of three electromagnetic
quanta in the E1 approximation. However, it does not preclude
three-wave mixing in which an electric-quadrupole (E2) or magnetic-
dipole (M1) atomic matrix element is involved. We must examine
other conservation laws to understand why collinear three-wave
mixing is forbidden in an atomic vapor.

Consider three waves of frequency and wave vector $(\omega_1, \underline{k}_1)$,
$(\omega_2, \underline{k}_2)$, and $(\omega_3, \underline{k}_3)$, respectively, where $\omega_3 > \omega_2 \geq \omega_1$. The waves
are propagating collinearly in an atomic vapor. In order for mixing
to occur, the waves must satisfy certain relationships which may be
viewed as conservation laws. First, $\omega_3 = \omega_2 + \omega_1$ (photon energy
conservation). Second, the phase-matching conditions $\underline{k}_3 = \underline{k}_2 + \underline{k}_1$
("photon momentum" conservation) must be satisfied. Both laws
arise from the stipulation that, on the average, the atoms neither
absorb nor emit energy or momentum in the coherent transition
between photons of one kind and those of another. The second law
implies that if $\underline{k}_1$ and $\underline{k}_2$ are collinear, $\underline{k}_3$ must lie along them.
A third law which must be satisfied is associated with the conser-
vation of angular momentum among the three waves. For the case of
collinear propagation, we may quantize the angular momentum of the
photons along the direction of propagation $\hat{z}$. The atomic angular
momentum may be quantized along this direction as well, since the
atomic medium is isotropic. The fact that the atomic state is
unchanged in the coherent transition between the photons tells us
that the total angular momentum of the photons is conserved. The
transversality of an electromagnetic wave implies that each quantum
must carry ± 1 units of angular momentum along $\hat{z}$. This condition,
however, is incompatible with the conservation law $m_1 + m_2 = m_3$,
where m_i is the z-component of angular momentum of the ith photon.
Thus, we are forced to conclude that in the absence of a symmetry-
breaking field, three-wave mixing of collinear beams may not occur
in an atomic vapor.

The complete generality of this result, which was first stated
by Hänsch and Toschek[10], is striking. It is independent of the
atomic interaction (type of matrix element) which is responsible
for the mixing. Its validity does not depend on the electric-
dipole approximation.

In the remainder of this paper, we will restrict ourselves to three-wave mixing in which one of the waves is resonant or nearly resonant with an atomic M1 or E2 transition. Experimentally, this is the only situation which has so far been examined.

As demonstrated in Ref. 2, it is possible to obtain atomic three-wave mixing if the three waves are not collinear. One may use the dispersion of the medium to satisfy the energy and momentum conservation laws simultaneously. The angular-momentum conservation law may be satisfied, as well: if we quantize along $\underline{k}_3$, for example, we see that either quantum 2 or quantum 1 may have a nonzero probability amplitude of having m=0. Thus, either SFG or DFG may be obtained. In the SFG sodium experiment of Ref. 2, for example, the two incident waves were coupled to the atoms by the E1 interaction, while the third generated wave was coupled to the atoms by $3^2S_{1/2}$-4^2D E2 matrix element. Noncollinear three-wave mixing (DFG) has also been observed by Flusberg, Mossberg, and Hartmann on the $6^2P_{1/2}$-$6^2P_{3/2}$ M1 transition of atomic thallium[11].

If the two incident waves are collinear, three-wave mixing may be observed if a magnetic field $\underline{H}$ is applied perpendicular to the direction of propagation $\hat{z}$. This magnetically induced three-wave mixing, which has been described in Refs. 3-5, may be understood in terms of the conservation laws described above. Quantizing along $\hat{z}$, we see that a transverse magnetic field causes a precession of the atomic angular momentum about $\underline{H}$ and a change in the component of atomic angular momentum along $\hat{z}$; this angular momentum may be transferred to the generated light wave. Alternatively, we may say that each atom is not in a state of well-defined z-component of angular momentum, hence the angular momentum of the three quanta need not be conserved. Thus the mixing of three collinear optical waves is possible in the presence of a transverse magnetic field. (Alternatively, one may view this as four-wave mixing in which one "wave" is a dc magnetic field.)

To be explicit, let us specialize to the case of magnetically induced SFG in atomic Na, an effect which we have observed experimentally[4]. The incident optical waves are provided by two nitrogen-laser-pumped dye lasers producing 6-nsec-long pulses of frequencies ω_1 and ω_2 tunable near the 16 956-cm^{-1} $3^2S_{1/2}$-$3^2P_{1/2}$ and 17 594-cm^{-1} $3^2P_{1/2}$-4^2D Na absorption lines, respectively. The central laser frequencies are adjusted so that their sum ω_3 lies within the Doppler-broadened two-photon $3^2S_{1/2}$-4^2D absorption spectrum. At low Na density n (i.e. negligible dispersion) the two incident waves are phase matched with the SFG wave (of frequency ω_3) for collinear propagation. It is then found that the intensity I of the SFG wave is proportional to $n^2H^2E_1{}^2E_2{}^2$, where E_i is the electric field of wave i. The SFG wave electric $\underline{E}_H$ arises from a magnetically induced macroscopic E2 polarization radiating at ω_3. From symmetry considerations it may be shown that to lowest order in $\underline{E}_1$, $\underline{E}_2$ and $\underline{H}$, $\underline{E}_H$ can be written

$$\underline{E}_H = \mu_1 \hat{z} \cdot (\underline{E}_1 \times \underline{E}_2)\underline{H} + \mu_2[(\underline{E}_2 \times \underline{H}) \cdot \hat{z}\underline{E}_1 + (\underline{E}_1 \times \underline{H}) \cdot \hat{z}\underline{E}_2] \qquad (1)$$

where μ_1 and μ_2 are coefficients which depend upon the atomic parameters and the spectral profile of the excitation.

An interesting effect occurs if the Na density is increased enough so that phase matching occurs when the incident laser beams are noncollinear. In the absence of a magnetic field, this is simply the experimental situation of Ref. 2; a noncollinearly produced SFG wave of electric field $\underline{E}_{nc}$ ($\propto \underline{E}_1\underline{E}_2$) is generated. Now suppose a dc magnetic field H is applied. When $\underline{E}_1$ is parallel to $\underline{H}$ and $\underline{E}_2$ is approximately perpendicular to $\underline{H}$, the SFG intensity varies with H in the manner depicted in Fig. 1. The only difference between the two plots of Fig. 1 is a *reversal* of the magnetic-field direction. It will be seen that for $H \cong 30$ G a reversal of $\underline{H}$ changes the SFG intensity by an order of magnitude!

The explanation of this effect is straightforward. Since $\underline{E}_H$ reverses sign with respect to $\underline{E}_{nc}$ when $\underline{H}$ is reversed, the total SFG intensity I(H), which is proportional to $|\underline{E}_{nc} + \underline{E}_H|^2$, changes accordingly. According to this explanation, I(H) may be written as $I_0 + bH + cH^2$. As a check, we have used the data of Fig. 1 to plot $I_+ = I(H) + I(-H) - 2I_0$ and $I_- = I(H) - I(-H)$ against H (Fig. 2). As expected, $I_+ \propto H^2$, while $I_- \propto H$.

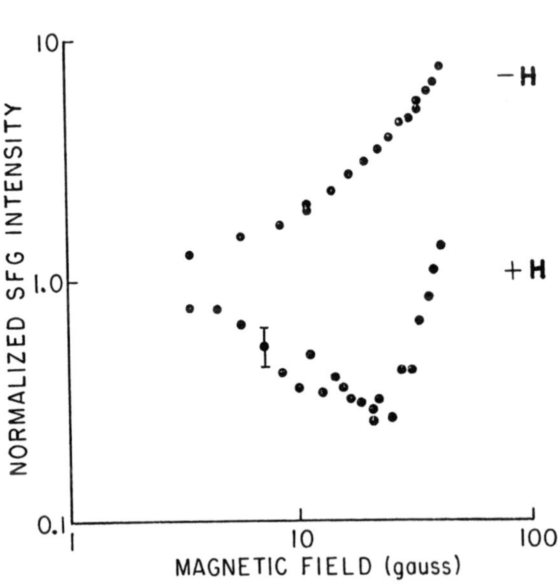

Fig. 1. Intensity of noncollinear SFG in atomic sodium vapor as a function of magnetic field H. The magnetic field direction for the data labelled -H is reversed with respect to its direction for the data labelled +H. The vertical scale represents the H-dependent intensity I(H) normalized by $I_0 = I(H=0)$ (Ref. 4).

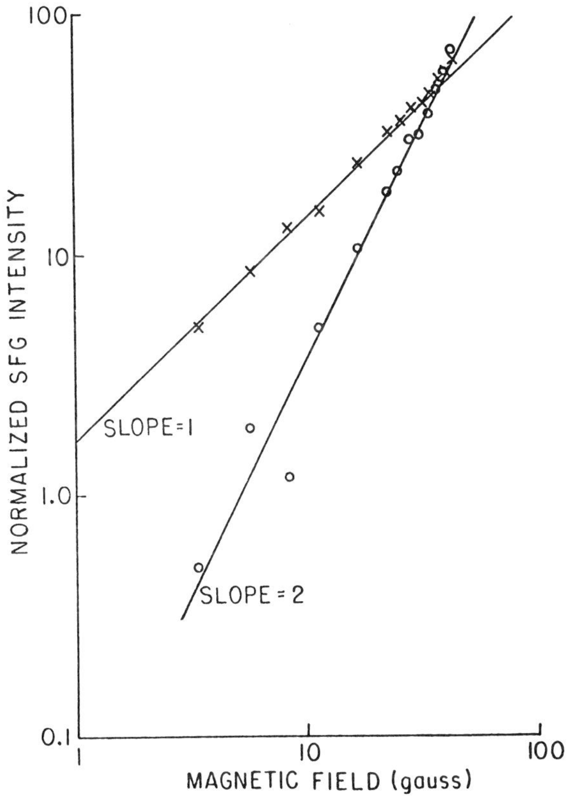

Fig. 2: Difference and sum of plots of Fig. 1. Crosses: abscissa is $[I(+H) + I(-H) - 2I_0]/I_0$; circles: abscissa is $[I(+H) - I(-H)]/I_0$. (Ref. 4).

In analogy with Eq. (1), we may also write down from symmetry considerations the expression for the electric vector $\underline{E}_H{}'$ of magnetically induced M1 SFG or DFG:

$$\underline{E}_H{}' = \underline{k}_3 \times \underline{B}_H \tag{2}$$

where

$$\underline{B}_H = \beta \underline{E}_1 (\underline{E}_2 \cdot \underline{H}) + \gamma \underline{E}_2 (\underline{E}_1 \cdot \underline{H}) + \delta \underline{H}(\underline{E}_1 \cdot \underline{E}_2) \tag{3}$$

(For DFG in which $\omega_3 = \omega_1 - \omega_2$, $\underline{E}_2$ must be replaced by $\underline{E}_2{}^*$ in Eq. (3).) A comparison of Eqs. (1) and (2) shows that M1 and E2 transitions may easily be distinguished from one another on the basis of their polarization dependence. In particular, E2 radiation vanishes when $\underline{E}_1$, $\underline{E}_2$ and $\underline{H}$ are parallel, while M1 radiation does not. The different properties of E2 and M1 three-wave mixing can, in principle, be used to determine the ratio between the E3 and M1 lifetime of a transition such as the $6^2P_{1/2}$-$6^2P_{3/2}$ transition of Tl.

When $\omega_1 = \omega_2$ in Eq. (1), we have the case of magnetically induced SHG. Our group has observed E2 SHG on the $6^2P_{1/2}$-$7^2P_{3/2}$ transition of Tl[6,7]. This effect has also been reported by Matsuoka, Nakatsuka, Uchiki and Mitsunaga on the $3^2S_{1/2}$-3^2D transition of Na and on the $4s^2$ $^1S_0$-$4s3d^1D_2$ transition of Ca[5].

At low atomic Tl density n the magnetically induced E2 SHG in Tl varies as $n^2H^2I^2$, where I is the incident laser intensity. As n is increased, the SHG intensity levels off and is modulated as a function of n[12]; the depth of the modulation is larger the smaller the intensity I. We believe this modulation is caused by a phase mismatch between the fundamental and second harmonic. The estimated dispersion at each wavelength is of the right order of magnitude to account for the modulation. We do not yet understand the intensity dependence of the modulation, however.

A puzzling effect which we do not yet understand is our observation of SHG on the $6^2P_{1/2}$-$7^2P_{1/2}$ transition of Tl. This SHG intensity is unpolarized; it varies as n^2I^2, and it is independent of H. It disappears if the fundamental laser frequency (of linewidth 1 Ghz) is mistuned by about 2 Ghz from either of the two Doppler-broadened two-photon resonances (there being two resonances because of the hyperfine splitting[13]). This indicates that the two-photon atomic resonance is necessary in order for the SHG to occur. The $6^2P_{1/2}$-$7^2P_{1/2}$ transition has no E2 moment and only a very small M1 moment. If the M1 moment were responsible for the effect, then one would expect an enhancement of the SHG in a transverse magnetic field due to mixing of adjacent hyperfine levels. However, no such enhancement is observed. Indeed, it is difficult to see how the SHG can occur at all in view of the conservation laws which must be satisfied. Yet the simple n^2 dependence and laser-frequency dependence point to an atomic effect, and the I^2 dependence to a second-order effect. Perhaps a proper account of the effect of $7^2S_{1/2}$-$7^2P_{1/2}$ stimulated emission, which occurs simultaneously with the SHG, will resolve this puzzle. We are looking into this possibility, as well as others.

We now show how magnetically induced three-wave mixing may be understood as a multiphoton coherent Hanle effect. We first review the derivation of the lateral-scattering ("incoherent") Hanle effect given by Franken[14], and then indicate what changes must be made in it to apply it to the forward-scattering ("coherent") Hanle effect. We then generalize the latter to an excitation which occurs via a multiphoton process.

In order to facilitate the analysis, we assume that the spectral density of the excitation wave (averaged over many experiments) is independent of frequency in the vicinity of the atomic resonances in question ("white-light" excitation). As a simple model of this type of excitation, the pump was is assumed to consist of a delta-function

in time, and the average energy radiated by the atoms (either
laterally or in the forward direction) after the passage of the
pump wave is calculated.

Consider an atom having a multiplet of ground states $|m\rangle$ and
excited states $|\mu\rangle$, connected by E1 matrix elements $\langle m|\underline{p}|\mu\rangle$. The
atom, initially in a particular state $|m\rangle$, is excited into a super-
position state $|\Psi\rangle \simeq a_m|m\rangle + \sum_\mu b_{m\mu}|\mu\rangle$ by a short burst of radiation
of polarization vector $\hat{f}$ at time $t=0$. For a weak enough excitation
(i.e. one in which the magnitude of a_m remains approximately equal
to 1), the amplitudes a_m and $b_{m\mu}$ at $t \geq 0$ are given by

$$a_m = e^{-i\Omega_m t}$$

$$b_{m\mu} = \text{constant} \times \langle\mu|\hat{f}\cdot\underline{p}|m\rangle e^{-\left((\Gamma/2)+i\Omega_\mu\right)t} \tag{4}$$

where $\underline{p}$ is the E1 operator, Γ is the inverse lifetime and $\hbar\Omega_m$ is the
energy of state $|m\rangle$. The probability that this atom will emit a
photon of polarization $\hat{g}$ and decay to a ground state $|m'\rangle$ is
proportional to $R(t) = |\langle m'|\hat{g}\cdot\underline{p}|\Psi\rangle|^2$. The ensemble average pro-
bability of the observation of a photon of polarization $\hat{g}$ is obtained
by summing $R(t)$ over m and m'; finally, the time-average probability
W is obtained by integrating $R(t)$ from $t=0$ to ∞. One obtains

$$W \propto \int_0^\infty dt \sum_{mm'} \left| \sum_\mu g_{m'\mu} f_{\mu m}\, e^{-\left(\frac{1}{2}\Gamma+i\Omega_\mu\right)t} \right|^2. \tag{5}$$

Here $f_{\mu m} = \langle\mu|\hat{f}\cdot\underline{p}|m\rangle$, $g_{\mu m} = \langle\mu|\hat{g}\cdot\underline{p}|m\rangle$. Note that the sum over μ is
inside the absolute-value signs, while that over m,m' is outside the
absolute-value signs, i.e., the sum over m,m' is *incoherent*.
Performing the integration one obtains

$$W \propto \sum_{\substack{mm'\\ \mu\mu'}} \frac{f_{\mu m}\, f_{m\mu'}\, g_{\mu'm'}\, g_{m'\mu}}{\Gamma - i\Omega_{\mu'\mu}} \tag{6}$$

which may be rewritten

$$W \propto \sum_{\substack{mm'\\ \mu\mu'}} \frac{\Gamma A_{\mu m\mu'm'}}{\Gamma^2 + \Omega_{\mu'\mu}^2} + \frac{B_{\mu m\mu'm'}\Omega_{\mu'\mu}}{\Gamma^2 + \Omega_{\mu'\mu}^2} \tag{7}$$

We now review the forward-scattering Hanle effect[9] (i.e., that
in which radiation at the excitation frequency is detected in the
direction of propagation of the incident light wave). For simplicity,

we assume all the atoms have the same spectrum (i.e. no inhomogeneous broadening). The short-pulse excitation of polarization $\hat{f}$ puts an atom which is initially in state $|m\rangle$ into the superposition-state $|\Psi\rangle$ having state amplitudes given by Eq. (4). The $\hat{g}$ component of the electric field of the radiation emitted by the atom is proportional to $\langle\Psi|\hat{g}\cdot\underline{p}|\Psi\rangle$. For an ensemble of atoms in different initial states $|m\rangle$, we obtain the emitted intensity of polarization $\hat{g}$ from

$$I = \left|\sum_m \langle\Psi|\hat{g}\cdot\underline{p}|\Psi\rangle\right|^2 = \left|\sum_{m\mu} g_{m\mu} f_{\mu m} e^{-(\frac{\Gamma}{2}+i\Omega_{\mu m})t} + c.c.\right|^2 \tag{8}$$

Note that the squaring is now carried out after the summation over the ground-state energy states $|m\rangle$, i.e. the summation over m is *coherent*. The reason is that the dipole moments created in the medium interfere coherently with one another to produce the forward scattering.

The average energy W' radiated is given by $\int_0^\infty I(t)dt$. We find

$$W' \propto \sum_{\substack{mm' \\ \mu\mu'}} \frac{g_{m\mu} f_{\mu m} g_{\mu'm'} f_{m'\mu'}}{\Gamma + i(\Omega_{\mu\mu'} - \Omega_{mm'})} \,. \tag{9}$$

W' may be written as

$$W' \propto \sum_{\substack{mm' \\ \mu\mu'}} \frac{\Gamma A'_{\mu m\mu'm'}}{\Gamma^2 + (\Omega_{\mu\mu'} - \Omega_{mm'})^2} + \frac{B'_{\mu m\mu'm'}(\Omega_{\mu\mu'} - \Omega_{mm'})}{\Gamma^2 + (\Omega_{\mu\mu'} - \Omega_{mm'})^2} \,. \tag{10}$$

W' differs from W, the expression for the incoherent radiation, in two respects: first, the coefficients A' and B' are different from A and B; and second, W' involves the ground-state and excited-state splittings in a symmetric way. In contrast, the ground-state splittings do not enter into the expression for W. Looking at I and R as functions of time, this difference is as pronounced. R, the incoherent radiation rate, contains quantum-beat modulation; as is well known, only the splitting in an excited atomic state leads to quantum beats. The modulation present in I, however, comes from interference between electric dipoles radiating coherently at several different frequencies; these frequencies are determined by the energy difference between the excited and ground states of the atom.

A more important difference between the coherent and incoherent Hanle effect is the role played by the Doppler broadening (ignored until now) in the former. Since the coherent Hanle effect is associated with interference between different atoms, rather than between a single atom and itself, the fact that there is a distribution of electric-dipole frequencies means that the relevant linewidth for the coherent effect includes the Doppler width Δ (FWHM) [9]. Qualitatively, one may replace Γ in Eq. (9) by the total linewidth of the medium, including Doppler broadening. Quantitatively, the lineshape factor $[\Gamma + i(\Omega_{\mu\mu'} - \Omega_{mm'})]^{-1}$ should be replaced by the proper lineshape factor $G(\Gamma,\Delta,\Omega_{\mu\mu'} - \Omega_{mm'})$ which is given by

$$G = \int_0^\omega dt \; e^{-\Delta^2 t^2/8\ln 2} \; e^{-(\Gamma+i[\Omega_{\mu\mu'} - \Omega_{mm'}])t} \tag{11}$$

and may be expressed in terms of the plasma dispersion function. In the strongly Doppler-broadened limit $\Delta \gg \Gamma$, for example, G may be written

$$G = \frac{\sqrt{2\pi\ln 2}}{\Delta} \; e^{-2(\ln 2)(\Omega/\Delta)^2} \; [1 + \mathrm{Erf}(i\sqrt{2\ln 2}\,\Omega/\Delta)] \tag{12}$$

where $\Omega = \Omega_{\mu\mu'} - \Omega_{mm'}$ and $\mathrm{Erf}(ix) = 2\pi^{-\frac{1}{2}} i\int_0^X e^{+t^2} dt$.

Before we generalize this coherent Hanle effect to the multiphoton case, we point out that we are actually using a model which may be inconsistent with the experimental situation, for we have calculated the average energy emitted by an ensemble of atoms prepared by a short-pulse excitation. For the coherent case, we are thus calculating the energy emitted during a free-induction decay. It is clear, however, that the results of the calculation would be the same so long as the excitation-pulse spectrum encompasses all the atomic transitions in question, i.e. so long as it may be treated as a "white-light" source. In the most general case, one must consider the source spectrum in detail; this leads to considerable complexity. We have therefore restricted the present analysis to the relatively simple case of white-light excitation.

Now consider a multiphoton Hanle effect, i.e. one in which the atomic superposition state $|\Psi\rangle$ is created by a multiphoton process. To be specific, assume an atomic ground state multiplet $|m\rangle$ and an excited state multiplet $|\mu\rangle$ which are connected by an M1 matrix element. (Formally, the same results obtain if the matrix element is E2). A *two-photon* pump pulse, i.e. a short light pulse travelling along $\hat{z}$ and containing two central frequencies ω_1 and ω_2, is incident on the atom. We assume the frequencies $\omega_{1,2}$ are far from any single-photon atomic resonance. If $\omega_1+\omega_2 \cong \Omega_{\mu m}$, the $|m\rangle$ - $|\mu\rangle$ superposition is created by a two-photon process. The superposition may radiate at $\omega_3 = \omega_2+\omega_1$ via the $|\mu\rangle$ - $|m\rangle$ magnetic-dipole moment $\mathfrak{m}$. We will consider only the coherent (phase-matched) radiation at ω_3. We must calculate the transverse component $M_x\hat{x} + M_y\hat{y}$ of $\underline{M} = \langle\Psi|\mathfrak{m}|\Psi\rangle$. In complete analogy with the single-photon coherent Hanle effect, we find

$$|\Psi\rangle = |m\rangle\, e^{-i\Omega_m t} + \sum_\mu |\mu\rangle\, f''_{\mu m}\, e^{-(\frac{\Gamma}{2}+i\Omega_\mu)t} \; , \tag{13}$$

where

$$f''_{\mu m} = \text{constant} \times \underset{=}{A}_{\mu m} : \underline{E}_1\underline{E}_2 \; . \tag{14}$$

Here $\underline{E}_{1,2}$ is the electric field amplitude at frequencies 1 and 2, respectively, and $\underset{=}{A}_{\mu m}$ is the two-photon polarizability tensor[13]. Defining $g''_{m\mu} = \langle m|\hat{g}\cdot\mathcal{m}|\mu\rangle$, we see immediately that the average energy W'' radiated coherently by the atoms at frequency ω_3 in the case of negligible Doppler broadening, is

$$W'' \propto \sum_{\hat{g}=\hat{x}\hat{y}} \sum_{\substack{mm' \\ \mu\mu'}} \frac{g''_{m\mu}\, f''_{\mu m}\, (g''_{m'\mu'})^*(f''_{\mu'm'})^*}{\Gamma + i(\Omega_{\mu\mu'} - \Omega_{mm'})} \; . \tag{15}$$

In the absence of a magnetic field, the energy differences $\hbar\Omega_{\mu\mu'}$, $\hbar\Omega_{mm'}$ are all zero for degenerate Zeeman levels. Thus for $H=0$ we have

$$W''(H=0) \propto \Big| \sum_{m,\mu} g''_{m\mu}\, f''_{\mu m} \Big|^2 \; . \tag{16}$$

It may be shown from symmetry considerations (angular momentum conservation) that if the incident pump beams are collinear, as we have assumed, then this sum [Eq. (16)] vanishes identically (i.e. there is no collinear three-wave mixing when H=0). For H≠0

$$W''(H) \propto \sum_{\hat{g}} \sum_{\substack{mm' \\ \mu\mu'}} \frac{g''_{m\mu}\, f''_{\mu m}(g''_{m'\mu'})^*(f''_{\mu'm'})^*\,[\Gamma-i(\Omega_{\mu\mu'}-\Omega_{mm'})]}{\Gamma^2 + (\Omega_{\mu\mu'} - \Omega_{mm'})^2} \; . \tag{17}$$

For the case of Doppler broadening ($\Gamma\ll\Delta$), one has only to replace $[\Gamma+i(\Omega_{\mu\mu'}-\Omega_{mm'})]^{-1}$ by the Doppler lineshape [Eq. (12)]. From the vanishing of Eq. (16) for collinear propagation it may be shown that the lowest-order contribution to W'' is quadratic in H, in agreement with experiment. However, if the incident pump beams are not collinear, then the sum $\sum g''_{m\mu}\, f''_{\mu m}$ does not vanish, i.e. three-wave mixing occurs in the absence of a magnetic field. The coherent wave at ω_3 then propagates along the phase-matching direction. From Eq. (17), it may easily be seen that $W''(H)$ will contain a term proportional to H. The same result will hold if Doppler broadening is properly taken into account. This explains

the graphs of Figs. 1 and 2[15]. In all cases, it is clear that to
lowest order W" is proportional to the product of the incident
laser intensities.

We have completed our goal of expressing magnetically induced
three-wave mixing in the language of the Hanle effect. The explicit
expressions for the energy W" of the generated wave may be evaluated
in terms of known atomic parameters. It should thus be possible to
make detailed comparison between theory and experiment for all values
of H. In particular, it should prove possible to measure the ratio
between the E2 and M1 radiation rates of transitions which have both
types of multipole moments. In the latter case the foregoing
equations apply if we replace g" by the μm component of
$g" = \hat{g} \cdot [\underline{M} + \hat{k} \times (\underline{P} + i\underline{k} \cdot \underline{Q})]$, where $\underline{Q}$ and $\underline{P}$ are the E2 and E1 moment
operators. The inclusion of $\underline{P}$ allows for the generation of an
induced polarization by a static electric field.

References

1.   R.S. Finn and F.J. Ward, Phys. Rev. Lett. *26*, 285 (1971).
2.   D.S. Bethune, R.W. Smith and Y.R. Shen, Phys. Rev. Lett. *37*,
      431 (1976).
3.   A. Flusberg, T. Mossberg and S.R. Hartmann, Phys. Rev. Lett.
      38, 59 (1977).
4.   A. Flusberg, T. Mossberg and S.R. Hartmann, Phys. Rev. Lett.
      38, 694 (1977).
5.   M. Matsuoka, N. Nakatsuka, H. Uchiki and M. Mitsunaga, Phys.
      Rev. Lett. *38*, 894 (1977).
6.   A. Flusberg, T. Mossberg and S.R. Hartmann, in *Cooperative
      Effects in Matter and Radiation*, C. Bowden, D.W. Howgate and
      H.R. Robl, eds. (Plenum Press, New York) to be published.
7.   A. Flusberg, T. Mossberg and S.R. Hartmann, Opt. Comm., to
      be published.
8.   Donald S. Bethune, Robert W. Smith and Y.R. Shen, Phys. Rev.
      Lett. *38*, 647 (1977).
9.   A. Corney, B.P. Kibble and G.W. Series, Proc. Roy. Soc.
      (London) *293A*, 70 (1966).
10.   T. Hänsch and P. Toschek, Z. Phys. *236*, 373 (1970).
11.   A. Flusberg, T. Mossberg and S.R. Hartmann, unpublished.
12.   There is also a hint of this effect in Fig. 3 of Reference 5.
13.   A. Flusberg, T. Mossberg and S.R. Hartmann, Phys. Rev. *A14*,
      2146 (1976).
14.   P.A. Franken, Phys. Rev. *121*, 508 (1961).
15.   The analysis of Ref. 4 is more general than the one presented
      here, since it does not make the "white-light"-pump assumption.
      In comparing the two, note that our Γ is twice that of Ref. 4.

APPLICATION OF COHERENT INTERACTIONS TO ISOTOPE SEPARATION

J.-C. Diels

University of Southern California

Los Angeles, California

The situation that will be considered here is that of a coherent pulse propagating in a vapor mixture of two isotopes. An exact analysis of coherent two-step photoionization is presented for the case of lithium vapor, showing that the production rate of selected ions can be one hundred times larger with coherent picosecond excitation than with conventional pulses, while the optical energy consumed to extract a unit quantity of the selected isotope is smaller in the case of coherent picosecond excitation.

To excite selectively a large volume of the rare element (in the forthcoming example: ^{6}Li with an abundance of 7.5%), a maximum transmission through the absorbing host of the abundant isotope (^{7}Li: 92.5%) is desired. A large transmission factor has been shown to be possible with near resonant "2π pulses",[1] or resonant "zero area pulses".[2] A study of pulse propagation stability in absorbing media [3] leads to the simple criterion for conservation of pulse frequency and energy with propagation distance: $\Delta W = W_\infty - W_0 = 0$ (W_0 and W_∞ are, respectively, the energy density in the medium before and after passage of the pulse). This condition is satisfied for a near resonant "2π pulse," as well as for a resonant "zero-area pulse". It was shown [4] that such "weakly interacting pulses" for one isotope can fully invert the other isotope. For instance, the same pulse can be simultaneously seen by ^{7}Li as a "2π pulse", and by ^{6}Li as a fully inverting "π pulse". For these pulses, ΔW is maximum and the criterion for stability in frequency and energy is not satisfied. Nevertheless, for the combination of both isotopes, one could hope for a stabilizing influence of the main line (^{7}Li) on the pulse propagation such as to achieve deep penetration of the selective excitation. This will indeed be seen to occur.

Let us consider the 2s-2p transition of Li at 6710 Å. At a temperature of 850°K, the Doppler width (FWHM in angular frequencies) of ^{6}Li is 24 GHz, and that of ^{7}Li is 22 GHz. The isotope shift has been measured [5] to be $\Delta\omega_0 \sim 66$ GHz, or roughly four Doppler widths. Excitation of the 2p level of ^{6}Li and ^{7}Li through the Gaussian pulse:

$$\mathcal{E}_{(t)} = \mathcal{E}_0 \exp \left\{-\left(\frac{t}{2\tau}\right)\right\}^2 \tag{1}$$

was calculated, solving numerically Bloch's equations.[1] In addition to the Doppler broadening, the contribution to the line-width from a phase relaxation time of $T_2 = 10$ ns and an energy relaxation time $T_1 = 20$ ns was taken into account. Figure 1 shows the variation of selectivity and excitation rate of ^{6}Li as a function of the pulse duration τ, for $\kappa \, \mathcal{E}_0 = 2\sqrt{\pi}/\tau$ in Eq. 1, ($\kappa = 2p/\hbar$ where p is the dipole moment of the 2s-2p transition). The separation factor should be expected to drop continuously with decreasing pulse duration (increasing pulse bandwidth). However, as the pulse duration becomes shorter than 10 ns, coherent interaction leads to an order of magnitude increase in selectivity. The latter ultimately decreases as the pulse spectrum covers both lines.

Figure 2 illustrates the excitation of ^{6}Li, and the separation factor (ratio of the fractional excitation of ^{6}Li to that of ^{7}Li) for a 75 ps [6] (FWHM of the intensity) pulse of Gaussian temporal shape applied at the resonance frequency of ^{6}Li, as a function of pulse amplitude. In contrast to the case of incoherent excitation (insert of Fig. 2), *both* the separation factor and the excitation rate of ^{6}Li are increasing with pulse amplitude. Both plots peak for the same value of $\kappa \mathcal{E}_0 T_2 \sim 275$, which corresponds to a Rabi precession frequency larger than the Doppler width of both lines. At the electric field amplitude corresponding to the peak of the pulse, the "power broadening" is such that both lines would be uniformly excited by incoherent excitation, as can be seen from the plot of the separation factor in the insert. In the case of coherent excitation, the peak in the separation factor is due to a reduced excitation of the off-resonance line (^{7}Li) as well as to a maximum excitation of ^{6}Li. The implication of the intensity dependence of the separation factor and excitation rate shown in Fig. 2 is that the full cross section of the beam can be used more efficiently in the case of short pulse coherent excitation than with incoherent excitation, as was demonstrated in Ref. 4.

Will a 75 ps FWHM Gaussian pulse of amplitude $\kappa \mathcal{E} T_2 = 275$ (yielding a maximum excitation rate of ^{6}Li in Fig. 2) exhibit the propagation characteristics of an unstable π pulse, or that of a 2π pulse? Figure 3 shows the Fourier amplitude of such a pulse as it propagates through the medium. As in the case of "self-induced transparency"[1] (S.I.T.) off resonance, the transmitted

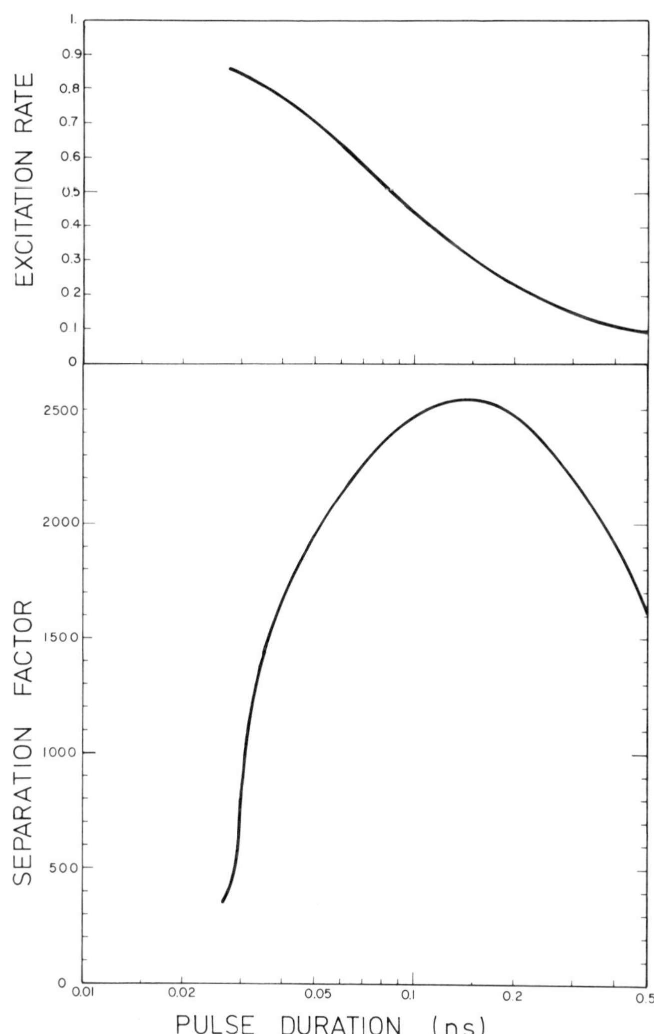

Fig. 1 Separation factor and fractional excitation of ^{6}Li by Gaussian pulses, as a function of the pulse duration τ (mean square deviation of the intensity). The pulse amplitude is that providing maximum (coherent) excitation in the middle of the Doppler profile ("π pulse excitation) of ^{6}Li. The separation factor is defined as the ratio of fractional excitation of ^{6}Li to that of ^{7}Li. The medium parameters are given in the text.

signal can be understood as a superposition of a spectrally narrow coherent pulse centered at the original pulse frequency (taken as reference in the plot of Fig. 3), and a phase modulated signal of increasing spectral width. In the first 15 linear absorption lengths, α_0^{-1}, the large absorption line of ^{7}Li is seen to deplete the overlapping spectral components of the pulse. The interpretation of this "hole burning" as an "incoherent interaction" between ^{7}Li and the radiation would be an oversimplification. After 15 linear absorption lengths, the spectral overlap of the absorption line (from ^{7}Li) increases as the pulse spectrum broadens

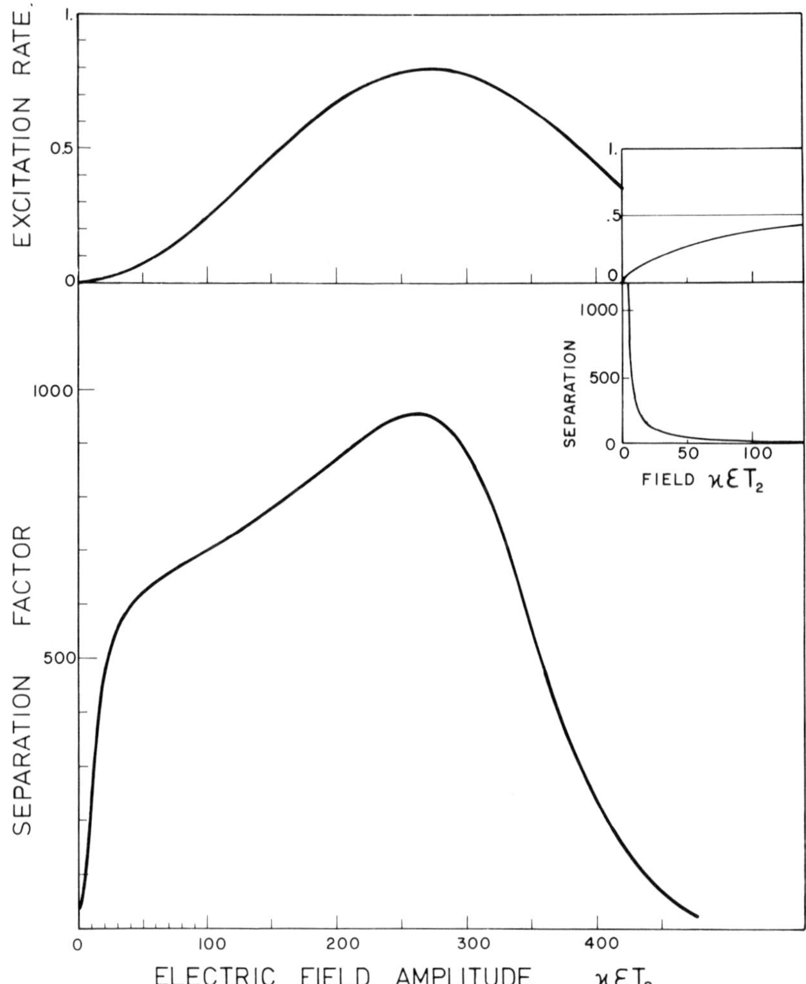

Fig. 2. Separation factor and fractional excitation of ^{6}Li as a
function of pulse amplitude. The pulse is of Gaussian temporal
shape and 75 ps FWHM (τ = 32 ps). The pulse amplitude is expressed
in Rabi frequency normalized to T_2^{-1}. For comparison, the separation factor and fractional excitation are given as a function of
electric field amplitude for CW or incoherent excitation in the
insert.

and frequency pushing effects [1] shift the "hole" away from the
main absorption (not shown in Fig. 3). The fractional excitation
of ^{7}Li qualitatively follows with distance the changes in spectral
overlap of the pulse spectrum with the ^{7}Li line, being reduced in

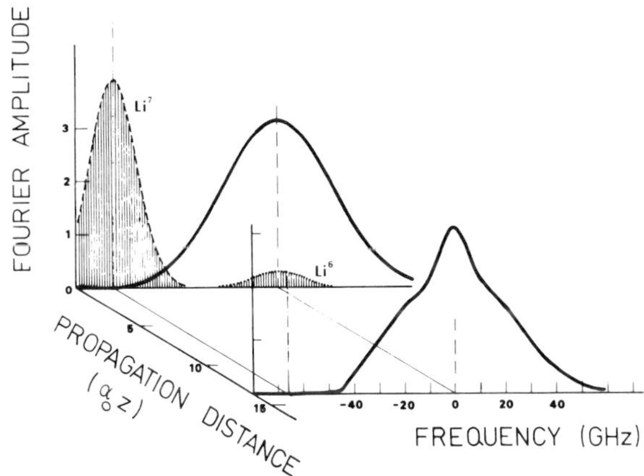

Fig. 3. Evolution with distance of the spectrum of an initially Gaussian pulse propagating through lithium vapor. The Doppler profiles of ^{6}Li and ^{7}Li are shown (in arbitrary units) together with the spectral amplitude of the pulse entering the medium. The carrier frequency of the radiation is chosen as reference frequency $\Omega = 0$. The pulse spectrum is the amplitude of the Fourier transform of the pulse envelope as defined in Diels and Hahn, Ref. 1, and its value at $\Omega = 0$ is the pulse "area"($=\pi$ for the initial pulse). The propagation distance is measured in units of linear absorption length [1] α_0^{-1}, where

$$\alpha_0 = \frac{4\pi\kappa^2 W_0}{c} \int \frac{g(\Omega T_2)d(\Omega T_2)}{1+ (\Omega T_2)^2}$$

$g(\Omega T_2)$ being the Doppler profile of ^{6}Li (ΩT_2 is the frequency normalized to the phase relaxation time T_2).

the first 13 absorption lengths, and enhanced thereafter. These changes in excitation of the abundant element result in an initially enhanced, thereafter decreasing, separation factor with distance (Fig. 4). The excitation of ^{6}Li, however, can be adequately described all along as a "π pulse excitation". The center of the Doppler profile of ^{6}Li is left excited by the pulse to more than 91% over 30 linear absorption lengths. However, a decreasing fraction of the full velocity distribution of the ^{6}Li atoms is excited as the "central coherent component" of the pulse spectrum narrows with propagation distance (Fig. 3). Comparison of pulse propagation in the presence (as for Fig. 2) or in the absence of the non-resonant transition of ^{7}Li clearly demonstrates the stabilizing influence of the latter. This is consistent with the analysis of Ref. 3, which shows "π pulse" propagation in an absorber to be unstable against frequency and energy fluctuations, and "2π pulse" propagation to be stable. Adding ^{7}Li to a vapor with a given number density of ^{6}Li is seen (in computer simulations) to result in an increased transmission at the peak of the pulse Fourier spectrum.

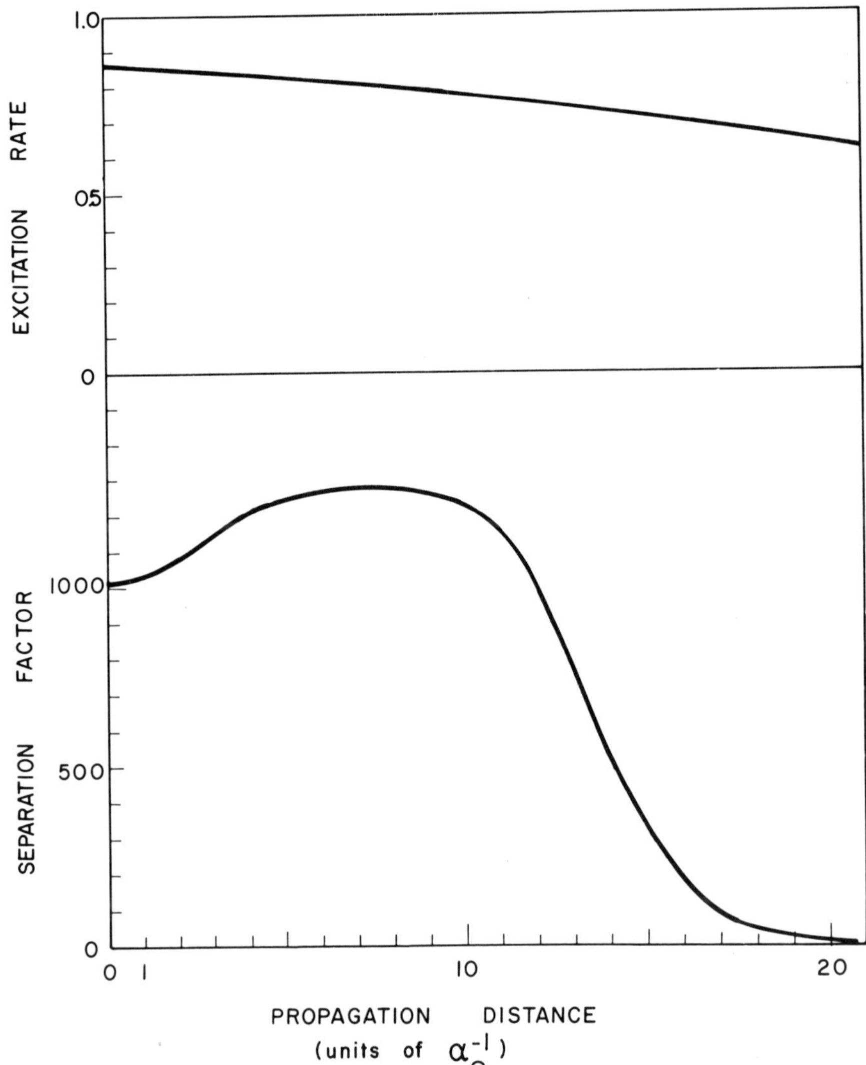

Fig. 4. Evolution with distance of the separation factor and fractional excitation of ^{6}Li as the initially Gaussian shaped pulse shown in Fig. 3 (75 ps FWHM of the intensity) propagates through lithium vapor.

Consequently, the center of the 2s-2p transition of ^{6}Li is excited to a higher rate (5% more after 13 absorption lengths) in the two component mixture (natural ratio of ^{7}Li to ^{6}Li) than in a vapor of pure ^{6}Li.

Isotope separation by two-step photoionization requires that the coherent excitation at ω_1 (frequency of the 2s-2p transition of lithium) be followed by a photoionizing pulse at a frequency ω_2. The energy of the ionizing pulse at ω_2 has to be sufficient to achieve complete depletion of the 2p states. For the coherent scheme proposed here, it is essential that the excitations be sequential. This is because the excitation of the unwanted element is not a monotonically increasing function of time, and even decreases in the tail of the coherent pulse. Two step photoionization is compared in Table 1 for coherent and incoherent excitation. For coherent two-step photoionization, the selecting pulse is a 75 ps FWHM Gaussian pulse of energy U_1. To ionize all the atoms excited in the 2p state, a strong 10 ns pulse of energy U_2 at 2500 Å (peak power of $\approx$ 300 W/cm^2) should be sent through the same interaction volume ($13\alpha_0^{-1}$ long) following selective picosecond excitation at 6712 Å. For comparison, a rate equation approach is used to calculate the energies required in conventional long pulse, simultaneous illumination at ω_1 and ω_2, to ionize 80% of the ^{6}Li in an interaction length of $13\alpha_0^{-1}$. A lower estimate is found by calculating the population of the 2p states from the equilibrium condition of the 2s-2p transition with ratiation at ω_1 only. More energy is required at both wavelengths for the incoherent scheme. More important, however, is that it takes at least one microsecond to ionize the quantity of lithium that is excited in less than 10 ns in the coherent excitation. By using a continuously mode-locked laser, the coherent process can be repeated at a frequency of 25 MHz. In the case of lithium, the absorption length is of the order of 1 μ. With an atomic flux of 10^4 cm/s across an interaction volume 2.5 cm deep and 1 cm diameter, the rate of photoionization is 10^{19} atoms/sec (at 850°K and an atomic density of 5.10^{14} at/cc). For the incoherent scheme A, the maximum rate is only 3.10^{16} atoms/sec or 300 times slower, while eight times more energy per atom is required at ω_2.

Table 1. Requirements to ionize 80% of the ^{6}Li in an interaction length of $13\alpha_0^{-1}$.

Requires:	Coherent	Incoherent		
		A	B	C
At ω_1 Energy	u_1 (75ps)	u_1	1.8 u_1	6u_1
At ω_2 Energy / Duration (μs)	u_2 / 0.01	8u_2 / 6	1.5u_2 / 1.2	1.25u_2 / 1
% Li^7 Ionized	0.07	0.03	0.05	0.2
MAX Rep Rate (MHz) (DUTY CYCLE 25%)	25	0.08	0.4	0.5

Can these results be applied to more complex elements than lithium? The isotope shift of medium heavy elements is not Doppler resolved. It was demonstrated [7] that the properties of selective coherent excitation of single photon transitions could be extended to two photon transitions, for which the Doppler broadening can be eliminated by counter-propagating excitation. The analysis of selective two-photon excitation is slightly complicated by the transient Stark shift, which is of the order of the Rabi precession frequency. However, by appropriately tuning the average pulse frequency, the Stark shift can be utilized to enhance the separation factor and excitation rate of the selected elements. For instance, in the case of the 2s-4s transition of Li, a "π pulse" Gaussian pulse of 105 ps FWHM [8] from a Rhodamine 6G laser at 5712Å, tuned to 70 GHz below resonance with the ^{6}Li transition, would achieve 50% excitation of ^{6}Li with a separation factor of 3000.

The isotope shift of heavy elements, as uranium, [9] is comparable to that of lithium. Further complications arise, however, because 1) degeneracy of the transitions and 2) broadening of the transition of isotopes of uneven atomic number due to the nuclear spin. Coherent propagation effects were initially thought to apply only to simple, non-degenerate transitions. [10] Zembrod and Gruhl [11] observed self-induced transparency in SF_6, a highly degenerate system. Gibbs, et al. [12] pointed out that, for a high j valued transition, although there are (2j + 1) values of the dipole moment ($p_m = p_j [(j + 1)^2 - m^2]^{\frac{1}{2}}$), the distribution of the dipole values p_m/p_j is strongly peaked around 1. It is, however, obvious that the selectivity of coherent excitation with Gaussian pulses will be affected by the degeneracy. The situation is different with zero-area pulses. On-resonance, since a zero-area pulse leaves the upper state of a transition unexcited independently of the value of the dipole moment, the degeneracy is expected to have little or no effect on the separation factor. This property will be confirmed by numerical analysis, after the discussion on selective zero-area pulse excitation of a simple 2s-2p transition which follows. Figure 5 (curve 1) shows the dependence of pulse duration of the separation factor and excitation rate for the 2s-2p transition of Li. The zero-area pulse is applied at resonance with the ^{7}Li transition, and is formed by the superposition of two Gaussian pulses 180° out of phase and delayed with respect to each other:

$$\mathcal{E}(t) = \mathcal{E}_1(t + \tfrac{\tau}{2}) - \mathcal{E}_1(t - \tfrac{\tau}{2}), \tag{2}$$

$\mathcal{E}_1(t)$ being the same Gaussian as in Eq. (1). The amplitude of the exciting pulses was chosen to be

$$\mathcal{E}_0 = \frac{2}{\kappa\tau} (2\pi)^{\frac{1}{2}} \tag{3}$$

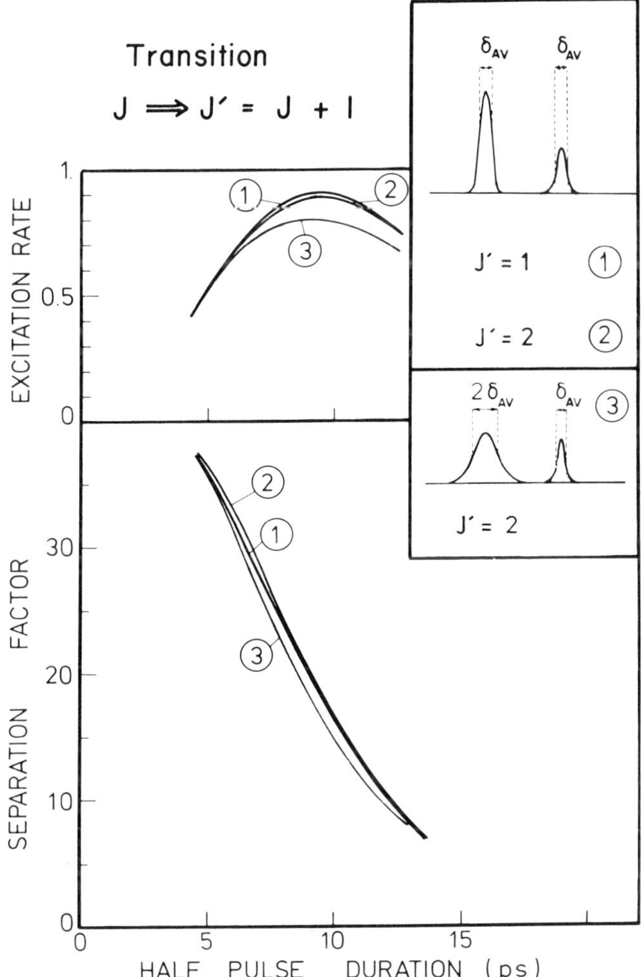

Fig. 5 Separation factor and fractional excitation of an off-resonance line, by the zero-area pulse given by Eqs.(2) and (3), as a function of pulse duration τ. Curve (1) pertains to the 2s-2p transition of lithium. Curve (2) pertains to a fictitious j=1→j'=2 transition that would have, except for the two-fold degeneracy, the same parameters (Doppler widths of ^{6}Li and ^{7}Li, phase and energy relaxation times, isotope shift) as the 2s-2p transition of lithium. The width of the line off-resonance (that is being excited) has been doubled in the case of curve (3) which otherwise pertains to the same set of parameters as curve (2).

for the calculations illustrated in Fig. 5. For long pulse duration, the zero-area pulse does not interact with ^{6}Li, and the separation factor is smaller than unity. As the pulse duration is decreased, more ^{7}Li is returned to the ground state by the zero-area pulse, and the off-resonance excitation of ^{6}Li reaches an optimum of 90%. This is to be contrasted with the case of Gaussian excitation illustrated in Fig. 2, where the separation factor as a function of pulse duration was seen to have a maximum. From a comparison of Fig. 1 and Fig. 5, one can also infer that zero-area pulse excitation is to be preferred over Gaussian pulse excitation if high densities of vapor

are to be treated. This is because, with increasing densities, the
longer Gaussian pulse will be more affected than the shorter zero-
area pulse by the decreasing phase relaxation time. Another advantage of the zero-area pulse is a uniform separation factor over the
cross-section of a finite beam (for τ = 10 ps, the separation factor
was calculated to be constant for $0 < \mathcal{E}_0 \leq 2/\kappa\tau \, (2\pi)^{\frac{1}{2}}$). Curves (2)
and (3) illustrate the effects of degeneracy and broadening of the
selected line. The same parameters are used for computing curve
(2) that are used in computing curve (1), except that the transition is assumed to be a j = 1 to j' = 2 transition (p-d). The
effect of degeneracy is clearly negligible. Curve (3) still refers
to a j = 1 to j' = 2 transition, but with the transitions of the
selected element spread over a line twice as broad as in the previous cases. The three sets of curves of Fig. 5 merge for short
pulse duration. With continuous radiation tuned to the selected
isotope (left line in the insert of Fig. 5), the excitation - hence
also the separation factor - would have been reduced by half in the
case (3) of the broader line. Zero-area pulse is the ideal form of
coherent excitation for uranium separation. Minimum interaction
with the unwanted element is obtained by tuning a picosecond zero-
area pulse to a narrow transition of ^{238}U. The same pulse will fully
invert *all the components* of the nuclear spin broadened transition[9]
of ^{235}U, at 60 GHz (angular frequency) below the corresponding narrow line of ^{238}U. This broadband but selective excitation is to
be contrasted with the very inefficient coupling provided by spectrally narrow ($\approx$ 1 GHz) CW radiation tuned to one of the 8 hyperfine components of the ^{235}U transition.

In summary, the analysis of *coherent* two-step photoionization
of lithium demonstrates that picosecond pulses rather than long pulses or CW radiation should be used for isotope selective excitation.
The properties of coherent selective excitation are extended to dipole forbidden two-photon transitions, and to degenerate transitions.
Selective excitation by zero-area pulses is shown to be particularly
insensitive to level degeneracy and broadening of the selected line.
Therefore, coherent two-step photoionization is a potentially valuable tool for the separation of isotopes in metal vapors.

References

1.   S.L. McCall and E.L. Hahn, Phys. Rev. *183*, 457 (1969); J.-C.
     Diels and E.L. Hahn, Phys. Rev. A *8*, 1084 (1973).
2.   H.P. Grieneisen, J. Goldhar, N.A. Kurnit and A. Javan, Appl.
     Phys. Lett. *21*, 559 (1972); J.-C. Diels and E.L. Hahn, Phys.
     Rev. A *10*, 2501 (1974).
3.   J.-C. Diels and E.L. Hahn, IEEE J. Quantum Electron. *QE-112*,
     411 (1976).

4. J.-C. Diels, Phys. Rev. A *13*, 1520 (1976).
5. D. A. Jackson and H. G. Kuhn, Proc. Roy. Soc. *173*, 278 (1939).
6. 75 ps FWHM corresponding to a mean square deviation (abscissa of Fig. 2) τ = 32 ps.
7. J.-C. Diels, Optical and Quantum Electronics *8*, 513 (1976).
8. As shown in Ref. 7, the two-photon coherent interaction of a 105 ps pulse compares to the single photon coherent interaction of a 75 ps pulse.
9. Laser Program, Annual Report 1974, Lawrence Livermore Laboratory, UCRL-50021-47, p. 492.
10. C. K. Rhodes, A. Szöke and A. Javan, Phys. Rev. Lett. *21*, 1151 (1968).
11. Z. Zembrod and T. Gruhl, Phys. Rev. Lett. *27*, 287 (1971).
12. H. M. Gibbs, S. L. McCall and G. J. Salamo, Phys. Rev. A *12*, 1032 (1975).

QUANTUM STATISTICS OF HOMODYNE AND HETERODYNE DETECTION

Horace P. Yuen and Jeffrey H. Shapiro

Massachusetts Institute of Technology

Cambridge, Massachusetts

1. INTRODUCTION

There are three basic techniques that may be employed to detect optical radiation: photon detection, homodyne detection, and heterodyne detection (see Fig. 1). From the photocounting theory of Kelley and Kleiner [1], one can readily show that photon detection, in the limit of unity quantum efficiency, can be interpreted as the quantum measurement of the total photon number operator N for any quantum state of a quasi-monochromatic input field. No similar quantum description is as yet available for homodyne or heterodyne detection. In this paper, we shall derive the abstract quantum description of homodyne detection. Because heterodyne detection involves the more complicated concept of simultaneous measurement of non-commuting observables, its quantum description will only be briefly discussed.

The quantum statistics of homodyne and heterodyne detection have significant ramifications. First of all, they permit us to theoretically evaluate the performance of optical information systems which employ such detection methods, as for example in optical communication. In addition, by relating the statistics of these detection schemes to the state of the input radiation, these methods become available as experimental techniques for probing the quantum characteristics of optical radiation. Finally, by showing that homodyne and heterodyne detection correspond to fully quantum-mechanical field measurements, we realize that, under appropriate circumstances, a quantum amplitude measurement can be realized with a quantum energy detector.

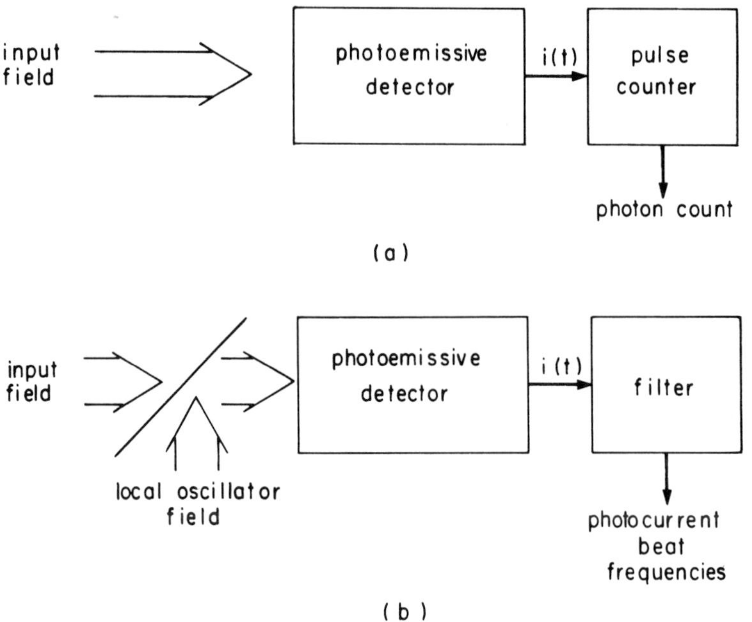

Fig. 1. Optical detection schemes: (a) photon detection;
(b) homodyne/heterodyne detection.

2. PHOTON DETECTION

We shall require, for deriving the statistics of homodyne detection, the quantum statistics of photon detection. For simplicity, our development both here and in the sequel will be restricted
to single-mode radiation fields. Analogous results hold in the
multimode case.

When a single mode quasi-monochromatic radiation field of
nominal frequency ω_o illuminates the photoemissive detector shown
in Fig. 1(a), the number of counts detected in a particular T second
interval, N, is a random variable whose statistics are determined
from the appropriate photon annihilation operator, a, for the field
mode in question by the following procedure. For a detector with
unity quantum efficiency, N comprises a measurement of $N = a^\dagger a$,
i.e., the probability that n counts are observed is

$$\Pr[N = n] = \langle n|\rho_a|n\rangle, \tag{1}$$

where ρ_a is the state of a and $\{|n\rangle\}$ are the eigenstates of the
number operator N [1]-[3]. Equation (1) is equivalent to the

generating function description

$$Q(\lambda) \equiv \sum_{n=0}^{\infty} (1-\lambda)^n \Pr[N=n], \tag{2}$$

$$= \sum_{n=0}^{\infty} (-\lambda)^n \langle a^{\dagger n} a^n \rangle / n!, \tag{3}$$

of the photocounting statistics, from which the counting probabilities may be recovered via

$$\Pr[N=n] = [(-1)^n/n\,!]\frac{d^n Q(\lambda)}{d\lambda^n}\bigg|_{\lambda=1}. \tag{4}$$

For a detector whose quantum efficiency, η, is less than one, the photocount generating function (2) is obtained by scaling the factorial moments in (3), viz [1]-[3]

$$Q(\lambda) = \sum_{n=0}^{\infty} (-\lambda\eta)^n \langle a^{\dagger n} a^n \rangle / n!, \tag{5}$$

whence

$$\Pr[N=n] = \sum_{m=n}^{\infty} \binom{m}{n} \eta^n (1-\eta)^{m-n} \langle m | \rho_a | m \rangle. \tag{6}$$

Note that Eqs. (5) and (6) are formally equivalent to unity quantum efficiency photon detection of the annihilation operator

$$a' = \eta^{\frac{1}{2}} a + (1-\eta)^{\frac{1}{2}} d, \tag{7}$$

where d is a photon operator that is in the vacuum state.

3. HOMODYNE AND HETERODYNE DETECTION

In homodyne detection of a quasi-monochromatic input field of nominal frequency ω_o, the field of a strong local oscillator laser of frequency ω_o is mixed with the input field on the active region of the photodetector. In heterodyne detection, a similar mixing occurs, except that the local oscillator frequency is offset by ω_{IF} from the nominal input frequency ω_o. In either case, the photocurrent i(t) is filtered to select the beat frequency components which are centered at 0 frequency for homodyne detection and the intermediate frequency ω_{IF} for heterodyne detection [4].

When the input field is in a coherent state, it has long been accepted (though not widely known) that homodyne and heterodyne detection may be interpreted, respectively, as the quantum measure-

ments of the electric field operator E and the positive-frequency
electric field operator $E^{(+)}$ [5,6]. There is no fully quantum-
mechanical proof of these assertions in the literature; they result
from comparing the statistics of the abstract measurements with
those derived by classical stochastic process analysis applied to
Poisson shot noise in the limit of unity quantum efficiency [5].
Thus, for input fields which do not possess well-behaved diagonal
P-representations the quantum statistics of homodyne and heterodyne detection remain to be found. Indeed, were one to assume that
for an arbitrary input state these detection schemes are quantum
measurements of E and $E^{(+)}$ when $\eta = 1$, the general results for
$\eta < 1$ would still have to be derived.

A. Homodyne Analysis

A single mode input field may be represented on a particular
T second interval by an appropriate photon annihilation operator a.
The abstract measurement of E on this time interval is equivalent
to the measurement of $a_1 \equiv (a+a^\dagger)/2$. The characteristic function
for the abstract a_1 measurement is, for real u, $\mathrm{Tr}(\rho_a \exp(iua_1))$
where ρ_a is the state of a. By use of the Baker-Hausdorff formula
[7] , this characteristic function can be related to the anti-
normally ordered characteristic function, $\chi_A^{\rho_a}$, for the state ρ_a,

$$\chi_A^{\rho_a}(\zeta_1,\zeta_2) \equiv \mathrm{Tr}(\rho_a \exp(-\zeta^* a)\exp(\zeta a^\dagger)), \tag{8}$$

where $\zeta \equiv \zeta_1 + i\zeta_2$, as follows

$$\mathrm{Tr}(\rho_a \exp(iua_1)) = \exp(u^2/8)\chi_A^{\rho_a}(0,u/2). \tag{9}$$

To show that homodyne detection measures a_1, it is sufficient to
verify that Eq. (9) also gives the characteristic function for the
homodyne system. The homodyne calculation is as follows.

Let b denote the annihilation operator for the appropriate
local oscillator mode on the T second interval of interest. For a
lossless beamsplitter of power transmission t, the annihilation
operator, c, which characterizes the mixed beam falling on the
photodetector in Fig. 1(b) satisfies

$$c = t^{\frac{1}{2}}a + (1-t)^{\frac{1}{2}}b. \tag{10}$$

The anti-normally ordered characteristic function, $\chi_A^{\rho_c}$, which determines the state ρ_c of the mixed beam is therefore [8]

$$\chi_A^{\rho_c}(\zeta_1,\zeta_2) = \chi_A^{\rho_a}(t^{\frac{1}{2}}\zeta_1,t^{\frac{1}{2}}\zeta_2)\chi_A^{\rho_b}((1-t)^{\frac{1}{2}}\zeta_1,(1-t)^{\frac{1}{2}}\zeta_2), \tag{11}$$

where $\rho_b = |K\rangle\langle K|$ for a coherent state $|K\rangle$ is the local oscillator state. For measurement of a_1 we take K to be real; for the measurement of $a_2 = (a-a_1)/i$ we take K to be purely imaginary. We shall treat only the a_1 measurement case; the a_2 measurement case is completely analogous.

In homodyne detection, the photocurrent is filtered to select the appropriate beat frequency components, bias terms are removed, the output is normalized to unit local oscillator strength while this strength is allowed to grow without bound [4]. In abstract language, this means that homodyne detection measures the limit as $K\to\infty$ of the operator

$$M \equiv (c^{\dagger}c - \eta t \, \mathrm{Tr}(\rho_a a^{\dagger}a) - \eta(1-t)K^2)/K. \tag{12}$$

The characteristic function for M measurement is, for real u and finite K,

$$\mathrm{Tr}(\rho_c \exp(iuM)) =$$
$$\sum_{n=0}^{\infty} \mathrm{Pr}[N=n] \exp[iu(n - \eta t\,\mathrm{Tr}(\rho_a a^{\dagger}a) - \eta(1-t)K^2)/K], \tag{13}$$

where N is the photocount random variable for photon detection with quantum efficiency η of the field c. From (6) we have

$$\mathrm{Tr}(\rho_c \exp(iuM)) = \mathrm{Tr}(\rho_c v^{c^{\dagger}c})$$
$$\times \exp[-iu\eta(t\,\mathrm{Tr}(\rho_a a^{\dagger}a) - (1-t)K^2)/K], \tag{14}$$

where

$$v \equiv \eta \exp(iu/K) + 1 - \eta. \tag{15}$$

The trace on the right side of (14) can be evaluated from

$$\rho_c = \int \chi_A^{\rho_c}(\zeta_1,\zeta_2)\exp(-\zeta c^{\dagger})\exp(\zeta^* c)d^2\zeta/\pi, \tag{16}$$

and the operator identity

$$\mathrm{Tr}(v^{c^{\dagger}c} \exp(-\zeta c^{\dagger})\exp(\zeta^* c)) = (1-v)^{-1}\exp[-v|\zeta|^2/(1-v)], \tag{17}$$

which is valid for any complex v. Thus, from (11), we find

$$\mathrm{Tr}(\rho_c \exp(iuM)) = (1-v)^{-1}\exp[-iu\eta(t\,\mathrm{Tr}(\rho_a a^{\dagger}a) + (1-t)K^2)/K]$$
$$\times \int \chi_A^{\rho_a}(t^{\frac{1}{2}}\zeta_1, t^{\frac{1}{2}}\zeta_2)\exp[i2\zeta_2(1-t)^{\frac{1}{2}}K - |\zeta|^2(1-t)]$$
$$\times \exp[-v|\zeta|^2/(1-v)]d^2\zeta/\pi. \tag{18}$$

By using the Fourier transform relation that exists between $\chi_A^{\rho_a}(\zeta_1,\zeta_2)$ and $\rho_a^{(n)}(\alpha) \equiv \langle\alpha|\rho_a|\alpha\rangle$ for coherent states $|\alpha\rangle$ we obtain

$$\mathrm{Tr}(\rho_c \exp(iuM)) = \exp\left[-iu\eta(t\mathrm{Tr}(\rho_a a^\dagger a) + (1-t)K^2)/K\right]$$

$$\times \int \rho_a^{(n)}(\alpha)\exp\left[-(1-v)|K(1-t)^{\frac{1}{2}} + \alpha t^{\frac{1}{2}}|^2/(1-t+vt)\right]d^2\alpha/\pi. \qquad (19)$$

Passing to the limit of infinite local oscillator strength we get

$$\lim_{K\to\infty} \mathrm{Tr}(\rho_c \exp(iuM)) = \exp[-u^2(1-t)\eta(1-2t\eta)/2]$$

$$\times \int \rho_a^{(n)}(\alpha)\exp(2iu\eta\alpha_1 t^{\frac{1}{2}}(1-t)^{\frac{1}{2}})d^2\alpha/\pi$$

$$= \exp[-u^2(1-t)\eta(1-2t\eta)/2]\chi_A^{\rho_a}(0,u\eta t^{\frac{1}{2}}(1-t)^{\frac{1}{2}}). \qquad (20)$$

Equation (20), which characterizes the statistics of homodyne detection, is to be compared with Eq. (9), which characterizes the statistics of the a_1 measurement. However, because $\langle M\rangle = 2\eta t^{\frac{1}{2}}(1-t)^{\frac{1}{2}}\langle a_1\rangle$, we should renormalize the homodyne detector to yield a measurement of

$$\lim_{K\to\infty} M' \text{ where } M' \equiv M/2\eta t^{\frac{1}{2}}(1-t)^{\frac{1}{2}}.$$

For M' we have

$$\lim_{K\to\infty} \mathrm{Tr}(\rho_c \exp(iuM')) =$$

$$\exp[-u^2(1-2t\eta)/8\eta t]\chi_A^{\rho_a}(0,u/2). \qquad (21)$$

We see, from comparison of Eqs. (9) and (21), that homodyne detection yields exactly the statistics for the quantum measurement of a_1 in the ideal limit $t\to 1$, $\eta\to 1$.

Interestingly, homodyne detection also yields an exact quantum measurement in the case $t\to 1$, $n<1$. Defining $M'' = M/2(\eta t)^{\frac{1}{2}}(1-t)^{\frac{1}{2}}$ we find for $t<1$

$$\lim_{K\to\infty} \mathrm{Tr}(\rho_c \exp(iuM'')) =$$

$$\exp[-u^2(1-2t\eta)/8t]\chi_A^{\rho_a}(0,\eta^{\frac{1}{2}}u/2)$$

$$\to \exp[(1-2\eta)u^2/8]\chi_A^{\rho_a}(0,\eta^{\frac{1}{2}}u/2) \qquad , \qquad (22)$$

when $t \to 1$. Using (7), (9), and (11), we find that (22) represents
the abstract measurement of $a'_1 \equiv \eta^{\frac{1}{2}} a_1 + (1-\eta)^{\frac{1}{2}} d_1$ where
$d \equiv d_1 + i d_2$ is a vacuum-state annihilation operator.

B. Examples

As a first example of the homodyne statistics, (9), for the
ideal case, we consider the quantum fluctuations in a two-photon
coherent state $|\beta; \mu, \nu\rangle$ [9]. Measurement of a_1 on this state re-
sults in a Gaussian probability density for the outcome with mean-
square fluctuation [9]

$$\langle \Delta a_1^2 \rangle = |\mu - \nu|^2/4 \quad , \tag{23}$$

where μ, ν are complex numbers which are arbitrary except for the
constraints that $|\mu|^2 - |\nu|^2 = 1$. Because this fluctuation can be
much smaller than $1/4$, which is the value obtained with a coherent-
state input, two-photon radiation has potential applications in
communication and other situations in which low noise levels are
desirable. For the more realistic case of $t \to 1$, $\eta < 1$, we find that
two-photon radiation yields a Gaussian probability density for the
measurement of a'_1 with

$$\langle \Delta a_1'^2 \rangle = \eta |\mu - \nu|^2/4 + (1-\eta)/4. \tag{24}$$

Equation (24) implies that the noise reduction factor attainable
with two-photon radiation will be limited (by finite detector quantum efficiency) to $1-\eta$; for $\eta = 0.9$ this is a 10dB reduction in
mean-square noise strength.

As a second example, we have directly calculated the characteristic function for homodyne detection (with $\eta < 1$, $t < 1$) of the number
state $|1\rangle$ using the Kelley-Kleiner photodetection model [1]. The
result agrees exactly with that obtained from (21), viz

$$\lim_{K \to \infty} \mathrm{Tr}(\rho_c \exp(iuM')) = (1-u^2/4)\exp(-u^2/8t\eta). \tag{25}$$

Note that the probability density obtained by Fourier transformation
of (25) is not Gaussian, in contrast to the semi-classical description of homodyne detection which always predicts Gaussian detection
statistics when the input field is not random [4].

C. Further Results

We shall briefly discuss a number of additional results that
we have obtained. First of all, the homodyne calculations just
presented readily generalize to the case of multimode input fields.

Thus, homodyne detection with a strong local oscillator yields
the abstract quantum measurement of the E field, or the magnetic
field operator if we alter the phase of the local oscillator with
respect to that of the input, for a multimode input field in an
arbitrary state ρ.

For heterodyne detection of a single-mode input field in state
ρ we have shown that the first and second moments (means and co-
variances) of the outcomes agree with the corresponding moments for
the abstract measurement of a.   The heterodyne detection moments
are obtained directly from photocounting theory via a stochastic
process manipulation using the configuration of Fig. 1(b) with $t \to 1$,
$n \to 1$.   The quantum measurement moments are obtained from the proba-
bility density $\langle \alpha | \rho | \alpha \rangle$ for the measurement of the photon annihila-
tion operator a [6].   These moment calculations can easily be ex-
tended to the multimode case.   The use of characteristic functionals
to complete the proof that heterodyne detection with a strong local
oscillator yields the abstract measurement of $E^{(+)}$ is being pursued.

4.   DISCUSSION

We conclude by elaborating a potential significance of homodyne and heterodyne detection as abstract quantum measurements of E
and $E^{(+)}$, respectively.   Photon detection is widely used to probe
the quantum properties of an input field, for example in photon
correlation experiments.   Recent theoretical work has shown that
certain novel quantum states, the two-photon coherent states (TCS)
treated in our first homodyne example, will result from stimulated
two-photon emission [9].   Now, although TCS counting statistics are
demonstrably non-Poisson (they can exhibit photon anti-bunching),
to verify the generation of such light by photon counting experiments may be perforce impossible for the case of multimode radiation
fields, i.e., if a single TCS mode is embedded in a multimode random
superposition of coherent states the photon detection experiment may
not be a sufficiently sensitive test of the field statistics.   On
the other hand, by means of post-detection filtering, both homodyne
and heterodyne detection can be made to respond to a single mode of
the input field.   Thus, that these procedures actually measure E and
$E^{(+)}$ means we have techniques for probing the quantum characteristics of a multimode field one mode at a time.   Indeed a further benefit accrues for TCS radiation when we can realize the E and $E^{(+)}$
measurements because the asymmetric quantum fluctuations in the
quadrature components of a TCS field [9], for example (23), become
directly observable.

*   Research supported in part by the Joint Services Electronics
    Program DAAB07-76-C-1400.

References

1. P. L. Kelley and W. H. Kleiner, Phys. Rev. *2A*, 136 (1964).
2. R. J. Glauber, in *Laser Handbook*, F. T. Arecchi and E. O. Schulz-Dubois ed. (Elsevier, New York, 1972), vol. 1, p. 1.
3. J. R. Klauder and E.C.G. Sudarshan, *Fundamentals of Quantum Optics* (Benjamin, New York, 1968).
4. R. M. Gagliardi and S. Karp, *Optical Communications* (Wiley, New York, 1976).
5. S. D. Personick, Tech. Rept. 477, M.I.T. Research Laboratory of Electronics (1970), Appendix A.
6. H. P. Yuen and M. Lax, IEEE Trans. on Info. Theory, vol. IT-19, 740, (1973).
7. G. Weiss and A. Maradudin, J. Math. Phys. *3*, 771 (1962).
8. H. P. Yuen, Tech. Rept. 482, M.I.T. Research Laboratory of Electronics (1971), part I.
9. H. P. Yuen, Phys. Rev. *A13*, 2226 (1976).

GENERAL THEORY FOR COLLISIONAL EFFECTS IN QUANTUM BEAT SPECTROSCOPY[†]

R.M. Herman and K.S. Meyer

The Pennsylvania State University, University Park, Pa.

The effects of collisions on quantum beats obtained following
sudden excitation of a multiplet of states which simultaneously decay
to a lower state or a lower multiplet are considered. The photobeats
are associated with single-photon decay, as opposed to two- or more-
photon correlations and, as such, represent the effects of intra-
atomic coherence[1,2].

The gas pressures are thought to be high enough that pressure
effects dominate Doppler effects. Accordingly, all velocity groups
participate in the radiative processes, and velocity changing
collisions are not of importance in the present study, for example.

The types of phenomena predicted in the present study are:

1. Collisionally-induced loss of coherence (i.e., broadening
of the quantum beat (QB) frequencies). This loss of coherence is
not closely related to ordinary spectral line broadening for the
transitions so that, in fact, even though the neighboring transitions
may be broadened so as to be totally unresolvable spectroscopically,
the photobeats may nevertheless be well defined.

2. Collisional shifts in the photobeat frequencies.

3. Coherent collisional transfer of excitation from a single
excited state to two or more other states which then may exhibit
photobeat amplitudes.

4. Collisional transfer from two or more coherent state
amplitudes to a single state in such a way as to give photobeat
amplitudes in the decay from the latter.

5.   In at least one case, decay rates inversely proportional to
pressure are predicted.

The expression for radiant intensity associated with a group
of N atoms excited by a sudden pulse (provided by a tuned laser,
for example) is given through the expression

$$I(t) \sim N\rho_{ij} \; \{<\psi_i(t)|\underline{p}^*|\phi_f(t)> \cdot <\phi_f(t)|\underline{p}|\psi_j(t)>\}_{Av} \tag{1}$$

where ρ_{ij} is the density matrix governing the excited multiplet
states at the moment of excitation; $\psi_i(t)$ and $\psi_j(t)$ are the state
functions as they evolve under the influence of collisions starting
from excited states i and j, respectively, as initially produced
by the sudden excitation; $\phi_f(t)$ is the final state(s) in the
Heisenberg representation (equivalently, one could express the final
states in any representation whatever, due to the unitarity of the
time-development operators for the final states); a sum over all
repeated indices is assumed wherever more than one state makes up
the multiplet, and finally $\{\}_{Av}$ denotes an ensemble average over
excited atoms.

Denoting the time-independent projection onto the final state(s),

$$\sum_f \underline{p}^*|\phi_f(t)> \cdot <\phi_f(t)|\underline{p} \quad , \text{ by the operator } \alpha^f \; ,$$

we obtain

$$I(t) \sim N\rho_{ij} \; \{<\psi_i(t)|\alpha^f|\psi_j(t)>\}_{Av} \; . \tag{2}$$

Let us now introduce the time development operator in the interaction
representation,

$$\psi_i(t) = U(t)\phi_i(t) \quad , \tag{3}$$

where $\phi_i(t)$, as mentioned, represents a Heisenberg state.
Accordingly,

$$U_{i'i}(t) = <\phi_{i'}(t)|\psi_i(t)> \quad \text{ is then valid.} \tag{4}$$

The states ψ_i and ϕ_i satisfy the equations

$$i\hbar\dot{\psi}_i(t) = (H_0 + V(t))\psi_i(t) \tag{5}$$

and

$$ih\dot{\phi}_i(t) = H_0\phi_i(t) \quad , \tag{6}$$

where H_0 is the unperturbed atomic Hamiltonian, which is time-independent, and $V(t)$ represents the interaction between the radiating atom in question and the perturbing atoms. The time-development equation for $U_{i'i}(t)$ is then, simply,

$$i\hbar\dot{U}_{i'i}(t) = \langle\phi_{i'}(t)|V(t)|\phi_{i''}(t)\rangle U_{i''i}(t) \quad , \tag{7}$$

subject to the initial condition

$$U_{i'i}(t=0) = \delta_{i'i} \quad . \tag{8}$$

With this definition of $U(t)$, Eq. (2) now becomes

$$I(t) \sim N\rho_{ij} \{U_{i'i}(t)^* \ U_{j'j}(t)\}_{Av} \ \alpha^f_{i'j'}(t) \quad . \tag{9}$$

Equation (9) is, indeed, a very beautiful expression, in view of the fact that the effects of state preparation, given by ρ_{ij}, and selection of outgoing radiation components (by polarization, observation geometry, etc.) denoted by one's choice of the α^f operator and consequently the matrix elements $\alpha^f_{ij}(t)$, are well separated from purely collisional effects, contained in the bracket $\{\}_{Av}$. Notice that the photobeat carrier frequencies are associated with $\alpha^f_{i'j'}(t)$, and depend only on unperturbed frequency differences in the initial, but not final, states.

Even though the states i', i, j', and j belong to the same (excited) multiplet, we may, at this point, *formally* consider $U_{i'i}$ and $U_{j'j}$ as matrices on operators in different spaces. Accordingly, one may use the doubled state notation for these matrix products, familiar in line broadening and multipole moment relaxation theory:

$$I(t) \sim N\rho_{ij} \{\langle\langle i'^* j'|U^{(i)}(t)^* U^{(j)}(t)|i^* j\rangle\rangle\}_{Av} \ \alpha^f_{i'j'}(t) \quad . \tag{10}$$

The method for solving for $\{\}_{Av}$ in the impact limit, for example, then relies upon one's finding eigenvalues and eigenvectors for the relaxation matrix under the effects of isolated collisions. Each matrix element is then found to relax with one or more complex exponential dependencies, yielding the phenomena mentioned above. Preliminary results of these solutions are presented, together with suggestions for future experiments.

Following methods which are standard in the impact limit of line broadening theory[3], we obtain, for the time dependence of $\{U^{(i)}(t)^* U^{(j)}(t)\}_{Av}$ the first order differential equation

$$\frac{d\{U^{(i)}(t)^* U^{(j)}(t)\}_{Av}}{dt} = \exp\left[\frac{i}{\hbar}(H_0^{(j)} - H_0^{(i)*})\right]t\ \Phi$$

$$\times \exp\left[-\frac{i}{\hbar}(H_0^{(j)} - H_0^{(i)*})\right]t\ \{U^{(i)}(t)^* U^{(j)}(t)\}_{Av} \tag{11}$$

where Φ is a collision operator having zero matrix elements between states not belonging to the excited multiplet, and having time-independent form (and having matrix elements formed with Schrödinger states)

$$\Phi = n\int vf(v)\,dv \int 2\pi b\,db \int \frac{d\Omega}{4\pi}\left[\left[1 + i\eta_2 P_2(\cos\theta_\Omega^{(i)})^* - \frac{1}{2}\left(\eta_2 P_2(\cos_\Omega^{(i)})^*\right)^2\right.\right.$$

$$\left.\left. + \ldots\right] \times \left\{1 - i\eta_2 P_2(\cos\theta_\Omega^{(j)}) - \frac{1}{2}\left(\eta_2 P_2(\cos\theta_\Omega^{(j)})\right)^2 + \ldots\right\} - 1\right] \tag{12}$$

with $H_0^{(i)*}$, for example, being the unperturbed atomic Hamiltonian (complex conjugate) operating in the space of the complex conjugate states i*, i'*, n being the number density of collision partners, and the integrals over v, b and Ω representing averages over velocity, impact parameter and the collision angles at the point of closest approach. The form of Eq. (12) is dependent on the assumed form of interatomic interaction,

$$V(t) = V_0(t) + V_2(t)\, P_2(\cos\theta_\Omega) , \tag{13}$$

where V_0 represents the isotropic portion of the interatomic interaction, and has no direct influence on the coherence problem, while the second term in Eq. (13), which leads to interesting collisional effects, arises from variation in the interatomic interaction energy depending upon the instantaneous position of the optically active electron relative to the interatomic axis at the peak of the collision[4]. (The form of V(t) is admittedly schematic, but provides an adequate representation of the atomic interaction in alkali-rare gas interactions, for example.) In Eq. (12), η_2 represents the integral of $V_2/\hbar$ over an entire collision,

$$\eta_2(b,v) = \frac{1}{\hbar}\left\{\int_{-\infty}^{\infty} V_2(t')dt'\right\} \qquad (14)$$

single collision at b,v ,

while $P_2^{(i)*}$ designates the second order Legendre polynomial operator in the i*, i'* space, etc.

Finally, using the definition

$$\chi = e^{-\frac{i}{\hbar}(H_0^{(j)}-H_0^{(i)*})t}\,\{U^{(i)}(t)*U^{(j)}(t)\}_{Av}\,e^{\frac{i}{\hbar}(H_0^{(j)}-H_0^{(i)*})t} \quad , \qquad (15)$$

Eq. (11) can be transformed into

$$\frac{d\chi}{dt} = \Phi\chi - \frac{i}{\hbar}\left[(H_0^{(j)}-H_0^{(i)*})\ ,\ \chi\right] \quad , \qquad (16)$$

while use of the addition theorem[5], and performance of the spherical average in Eq. (12) yields Φ in the following form,

$$\Phi = \frac{2\pi n}{5}\int vf(v)\,dv\int bdb\ \eta_2(v,b)^2$$

$$\times \sum_{M=-2}^{2}\left\{ Y_{2M}^{(i)*}\ Y_{2M}^{(j)} - \frac{1}{2}(-)^M\left(Y_{2M}^{(i)*}\ Y_{2-M}^{(i)*} + Y_{2M}^{(i)*}\ Y_{2-M}^{(j)}\right)\right\} . \qquad (17)$$

The solution to Eq. (16) now resides in finding the eigenvalues and eigenvectors of the resolvent operator

$$\Phi - \frac{i}{\hbar}\left[(H_0^{(j)}-H_0^{(i)*})\ ,\ \right] \quad ,$$

which appears to be straightforward, if somewhat complicated in practice. We shall comment on the application of Eqs. (10), (16), and (17) in two simpler cases of interest.

A. NORMAL ZEEMAN TRIPLET IN THE PRESENCE OF AN EXTERNAL FIELD

It can easily be shown that the commutator term in Eq. (16) leads to no effect, in the case of the normal Zeeman triplet $\ell = 1$ (or with very minor modification in theory, but no difference in results, $j = 1$ with $\ell \neq 0$). The presence of the magnetic field will,

of course, play a role in photobeat spectra from a normal Zeeman
triplet, but only through the time dependencies of the $\alpha^f_{ij}(t)$ in
Eq. (10). Accordingly, the solution to Eq. (16) resides in finding
the roots to the Φ-matrix which, in the present case, has the form

$$\langle\langle i'^* j' |\Phi| i^* j \rangle\rangle = \frac{1}{5}\, g
\begin{bmatrix}
-3 & & & & & & & & \\
& -4 & 1 & & & & & & \\
& 1 & -4 & & & & & & \\
& & & -3 & -1 & 2 & & & \\
& & & -1 & -2 & -1 & & & \\
& & & 2 & -1 & -3 & & & \\
& & & & & & -4 & 1 & \\
& & & & & & 1 & -4 & \\
& & & & & & & & -3
\end{bmatrix}$$

where the rows and columns are labeled respectively by the
magnetic quantum numbers of the doubled states, $(i^*,j) = (-1^*,1)$;
$(0^*,1)$; $(-1^*,0)$; $(1^*,1)$; $(0^*,0)$; $(-1^*,-1)$; $(1^*,0)$; $(0^*,-1)$; $(1^*,-1)$,
respectively. The roots, thereby, are:

      $- 0.6g$ (5-fold degenerate)

      $-\quad g$ (3-fold degenerate)

      0.0 (singlet)

with

$$g = \frac{3}{25}\, n \int v f(v)\, dv \int b\,db\; (\eta_2(v,b))^2 \quad .$$

These roots correspond to the relaxation, under the influence of
collisions, of spherical tensoral components T_{kq} (with $k = 2,1,0$,
respectively) which can be formed from $j = 1$ states inasmuch as,
indeed, each of the $\alpha^f_{i'j'}(t)$ corresponds to linear combinations
of spherical tensor matrix elements. Accordingly, in the absence
of an external magnetic field, upon production of an initial set
of amplitudes by the exciting laser beams, collisions alter the
various density matrix elements according to linear superpositions
of exponentials characterized by the above roots.

Upon application of an external magnetic field in the above problem, it can be seen that the following time dependencies could conceivably be observed in the various $\alpha^f_{i'j'}$, selected according to the analyzing geometry:

$\alpha_{1,1}$; $\alpha_{0,0}$; α_{-1-1} vary as linear combinations of 1, $e^{-0.6gt}$, e^{-gt}

$\alpha_{1,0}$; α_{0-1} vary as linear combinations of $e^{i\omega t}(e^{-0.6gt}$, $e^{-gt})$

$\alpha_{0,1}$; $\alpha_{-1,0}$ vary as $e^{-i\omega t}(e^{-0.6gt}$, $e^{-gt})$

$\alpha_{1,-1}$ varies as $e^{2i\omega t-0.6gt}$

$\alpha_{-1,1}$ varies as $e^{-2i\omega t-0.6gt}$

where ω is the Larmor frequency. Accordingly, in this simple case it is impossible to collisionally transfer quantum beats (as would be observed in σ fluorescence following simultaneous excitation of $m = 1$ and $m = -1$ substates) to any single level. (For example, the collisionally induced π radiational components would not possess photobeats.) Nonetheless, coherences between states, not all of which were necessarily populated by the exciting light pulse, can indeed by collisionally generated.

B.   FLUORESCENCE FROM A HYPERFINE DOUBLET, $j = 1$,
i (NUCLEAR SPIN) = $\frac{1}{2}$ IN THE ABSENCE OF EXTERNAL FIELDS.

In order to attempt to find collisional transfer of quantum beats into radiation components not initially present in the fluorescence, we have considered a system with hyperfine splitting ω. Space limitations preclude an extensive discussion of this problem; however, we wish to present some preliminary results showing that, in fact, such transfer does indeed occur. In this problem, as in the Zeeman problem, typical relaxation rates are of order g and 0. The presence of a hyperfine frequency modifies the time dependencies in χ in the following manner, for $g \gg \omega$, for example:

			1st order perturbation	2nd order perturbation
g	$\rightarrow$	g	$+\ i\omega$	$\pm\ 0\left(\dfrac{\omega^2}{g}\right) + \ .\ .\ .$
0	$\rightarrow$	0	$\pm\ i\omega$	$\pm\ \left(\dfrac{\omega^2}{g}\right) + \ .\ .\ .$

In addition, further $e^{\pm i\omega t}$ time dependencies may occur in the transformation of χ, Eq. (15) and in the $\alpha_{ij}^{f}(t)$.

While the second perturbation on the roots of χ is often negligible, this effect is decidedly not negligible when it constitutes the only real contribution to a root, since those components would then relax with rates *inversely proportional* to pressure - i.e., proportional to g^{-1}. To the authors' knowledge, such a slowly decaying component has not yet been observed experimentally. Pending a re-check of the present theoretical results, this would seem to be the most interesting aspect of the entire development, and it would be quite appropriate to attempt to observe this type of behavior in systems exhibiting hyperfine structure.

†Research supported by the U.S. Office of Naval Research.

References

1. W.W. Chow, M.O. Scully, and J.O. Stoner, Phys. Rev. *A11*, 1380 (1975).
2. R.M. Herman, H. Grotch, R. Kornblith, and J.H. Eberly, Phys. Rev. *A11*, 1389 (1975).
3. H.R. Griem, *Spectral Line Broadening by Plasmas* (Academic Press, New York, 1974).
4. K.S. Meyer and R.M. Herman, to be published.
5. M.E. Rose, *Elementary Theory of Angular Momentum* (John Wiley and Sons, Inc., New York, 1957).

GAS-KINETIC COOLING PROCESSES BY COHERENT RADIATION*

M. Garbuny

Westinghouse R&D Center, Pittsburgh, PA

1. ENERGY DEFECT EXCITATION

Recently we have proposed mechanisms by which coherent radiation can be converted, in principle, completely into mechanical work[1,2]. In simplest terms, the method uses: (1) a heat pump process in which radiation of frequency $h\nu_{12} = E_2 - E_1$ pumps molecules from an intermediate energy state E_1 to a long-lived upper state E_2 so that each absorbed photon withdraws an energy E_1 from the translational-rotational heat reservoir and stores an energy $h\nu_{12} + E_1$ in the upper state; and (2) as the second step, ultimate release of the energy E_2 to heat for the operation of a mechanical engine, the heat rejected being just compensated by that withdrawn in step (1). The process is therefore just the conversion of one type of energy of vanishing entropy into another, ideally, without net heat rejection.

The extent to which such a cycle can be realized depends on the amount of cooling that can be achieved and on the length of time over which it can be maintained in the presence of destructive thermal relaxation of the E_2 states. These two characteristics are of concern to the efficiency of radiative cooling in general. Of similar generality is the method of cooling by absorption of photons which have an energy less than that required for peak resonance. Figure 1 shows schematically the effect of such transitions for a 3-level system assuming equal statistical weights. E_1 represents any energy level which can thermalize fast with the translational heat reservoir of the gas. For example, cooling has been predicted for transitions from translational[3], rotational[1,2], or vibrational[4] levels. Short-lived temperature reductions of small amount caused by vibrational cooling have been observed[5,6] in molecular gas mixtures such as CO_2/N_2 pumped by CO_2 laser radiation.

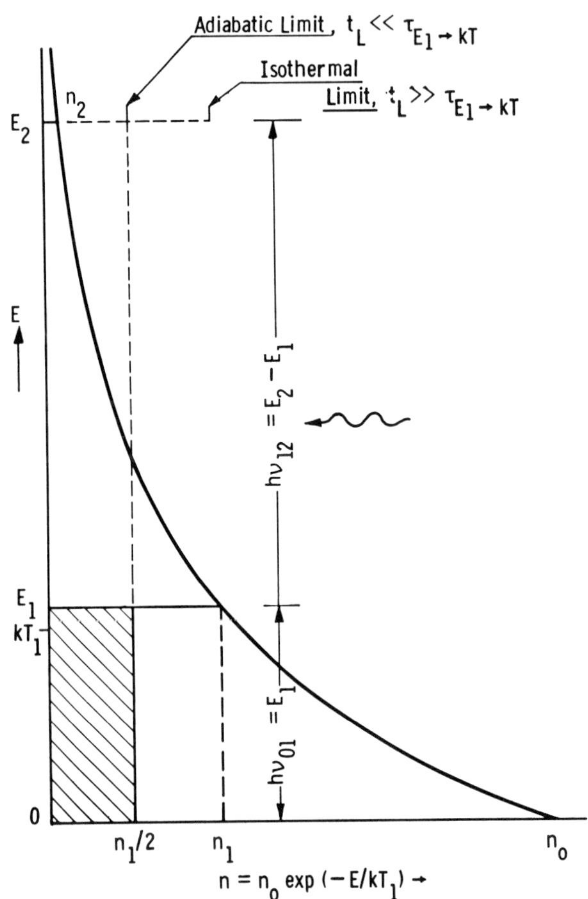

Fig. 1: Schematic presentation of radiative cooling by resonance defect absorption for a three-level system. Statistical weights $g_1 = g_2$.

In this study, we wish to determine the conditions under which radiative cooling of molecular gases is maximized. We will show that these effects can be quite large.

2.   REQUIREMENTS FOR RELAXATION TIMES

The rotational and translational (R/T) degrees of freedom of most molecular gases are tightly coupled, i.e., they form an ergodic system which reaches a thermal equilibrium after a few gas-kinetic collisions. If such a gas absorbs n_p photons of energy $h\nu_{12} = E_2 - E_1$, it loses an amount $n_p E_1$ as heat by transfer of energy from the R/T reservoir to the depleted levels E_1. This transfer is characterized

by a thermal relaxation time $\tau(E_1 RT)$, which is smallest when E_1
itself belongs to the R/T reservoir, but is considerably longer if
E_1 represents an excited vibrational state. To observe or utilize
the cooling effect, we must demand that the corresponding thermal
relaxation time of the level E_2 fulfills the condition
$\tau(E_2 RT) \gg \tau(E_1 RT)$. Any vibrational or electronic level $E_2 (>E_1)$
qualifies for this process provided its thermal relaxation time is
long enough so that heating is delayed. However, in the examples
later the choice of E_2 will be limited to the first excited vibrational level of the electronic ground state since the relevant data
are better known for them. If $\tau(E_2 RT)$ is sufficiently large, such
other relaxation processes as spontaneous emission or vibrational
energy transfer (VV) will reduce the lifetime of E_2 to a value $\tau(E_2)$.
We are particularly interested in approxiamtions for which the following inequalities are assumed:

$$\tau(E_1 RT) \ll \tau(E_2) \ll \tau(E_2 RT) \quad . \tag{1}$$

We assume that radiation will act on the gas during a time t_p which
is either the pulse length of the source or the time in which the
molecules of a flowing gas are exposed to the beam. The gas-kinetic
process depends decisively on the pulse duration t_p relative to the
relaxation times of condition (1).

3. RADIATION COOLING OF THREE-LEVEL SYSTEMS

A basic limit to the amount of heat $n_p E_1$ that can be pumped
from the R/T reservoir is the condition of saturation:

$$n_2/n_1 = g_2/g_1 \tag{2}$$

where n_1, g_1 and n_2, g_2 are the populations and statistical weights
(normalized to $g_0 = 1$ of the ground state) of E_1 and E_2, respectively.
The ratio E_2/E_1 will always be large enough so that the normal
thermal population n_2^T in E_2 is negligible.

(a) If $t_p \ll \tau(E_1 RT)$, the saturating number n_p of transitions
from E_1 to E_2 is that which distributes the original thermal populations n_1^T over the two levels so that Eq. (2) is fulfilled. For
$g_1 = g_2$, $n_1 = n_2 = \frac{1}{2}g_1 n_0 \exp(-E_1/kT)$ where n_0 is the population of
the ground state ($E_0 = 0$) and T is the gas temperature before the
onset of the radiation pulse. After a time $\tau(E_1 RT)$ the population
in E_1 termalizes, corresponding to a somewhat lower temperature.
The reduction of T results from the extraction of heat in the amount
of $n_p E_1$ (shown as the shaded rectangular area in Fig. 1). This
process, that may be termed *adiabatic fast* cooling (since zero

entropy is added to the irradiated gas column which is for a time
termally isolated), is in fact the radiative cooling effect
observed[4,5] by the lensing of CO_2 laser beams in the atmosphere.
In this case, E_1 corresponds to the vibrational level (10^00) and
E_2 (00^01) of atmospheric CO_2. Here $\tau(E_1 RT)$ is rather large and
condition (1) is not fulfilled so that magnitude and duration of
the cooling effects are small.

(b) Considerably larger amounts of heat can be withdrawn if
$\tau(E_1 RT) \ll t_p < \tau(E_2)$. This condition permits the maintenance of a
thermal equilibrium population n_1 (now omitting the superscript T)
in the level E_1 commensurate with the prevailing temperature, while
the radiative excitation is still underway. The rectangular area
of sides E_1 and n_2 $(=n_p)$ in Fig. 1 indicates the amount of heat
withdrawn from the R/T reservoir isothermally for $g_1 = g_2$. A third
type of operation (not shown) is adiabatic cooling at these longer
radiation pulses. In this case, the decreasing temperature reduces
the thermal equilibrium population n_1 so that, according to Eq. (2),
a smaller number $n_p = n_2$ is pumped to the bleaching point. The heat
withdrawn is then intermediate between the two operations discussed
before. The value for the heat ΔH extracted isothermally or
adiabatically in a 3-level system with statistical weights g_1 and
g_2 is given by

$$\Delta H = n_p E_1 = \frac{g_2 E_1 e^{-r}}{1 + (g_1 + g_2)e^{-r}} N \quad ; \quad r = E_1/kT \quad . \tag{3}$$

Equation (3) follows from Eq. (2) together with the thermal
equilibrium condition $n_1 = g_1 n_0 \exp(-r)$ and the distribution of the
total number N of molecules over n_0, n_1, and n_2. In adiabatic
operation, T represents the final temperature reached after
saturation. For a fixed T, there exists a maximum of ΔH for a
certain value E_{1m}. The latter is obtained from Eq. (3) in the
implicit form

$$(r_m - 1)\exp(r_m) = g_1 + g_2 \quad ; \quad r_m = E_{1m}/kT \quad , \tag{4}$$

so that

$$\Delta H_m = g_2 NkT\exp(-r_m) = N \frac{g_2}{g_1 + g_2} (E_{1m} - kT). \tag{5}$$

In adiabatic cooling from T' to T one has (if the specific heat
C_V is constant over that temperature range):

$$\Delta H = C_V(T' - T) \quad . \tag{6}$$

Thus Eqs. (4) and (5) define the conditions for the maximum temperature T' from which the gas can be cooled to T at saturation. Alternatively, we may define a merit factor M of adiabatic cooling:

$$M = (T' - T)/T \quad .\tag{7}$$

Since Eq. (4) is obtained also for the maximum of $\Delta H/T$, i.e., for

$$\left[\frac{\partial (\Delta H/T)}{\partial T} \right]_{E_1} = 0 \quad ,\tag{8}$$

it will be seen that Eq. (4) also determines the temperature T at fixed E_1 for which the merit factor is a maximum.

(c) The largest radiative cooling effects result, if $\tau(E_2) < t_p \ll \tau(E_2RT)$. This inequality implies the assumption that non-thermal relaxation processes exist which are much faster than the rate $\tau^{-1}(E_2RT)$. A simple example is radiative decay to the ground state as the dominant relaxation process with a time constant τ_r. If radiation continues while the population n_2 repeatedly returns to the ground state, the cooling process participates in this repetition. Thus isothermal cooling is enhanced by a factor approaching t_p/τ_r as the relevant rate equation will show. In adiabatic cooling, the effect becomes small for large ratios t_p/τ_r since the thermal equilibrium population n_1 will rapidly decrease as the temperature sinks below the optimum point of operation.

4. RADIATIVE COOLING OF DIATOMIC GASES

The theoretical and experimental work referred to[4-6] studied the cooling effects in polyatomic molecular systems in which E_1 and E_2 represented excited vibrational levels. The relaxation time conditions (1) and the optimum E_1/kT relationships were not fulfilled since these studies had been motivated by an interest in specific molecules rather than in a search for maximum cooling effects.

(a) The conditions derived here can be approached, however, for certain diatomic gases or rather their mixtures. Let, in the electronic ground state, E_1 be a rotational level J" of the vibrational ground state (v = 0) and E_2 a rotational level J' of the first excited vibrational level (v = 1). Here the thermal relaxation time × pressure P for E_1 is $\approx 10^{-10}$ sec-atm, except for molecules with very large rotational spacings such as HF. On the other hand, the thermal relaxation times of the vibrational energies, i.e. $\tau_{VRT} \equiv \tau(E_2RT)$, are usually many orders of magnitude larger. In particular, $\tau_{VRT}P \approx 1$ sec-atm for CO/CO or N_2/N_2 collisions[7]. Vibrational energy exchange occurs with time constants τ_{VV}

intermediate between τ_{RT} and τ_{VRT}. Finally, at low pressures the radiative relaxation time τ_r can be much smaller than τ_{VRT}. Thus condition (1) can be realized.

(b) A level (v,J) has the rotational energy $J(J + 1)B_v$ and the statistical weight 2J+1, where the rotational constants B_v differ only slightly for v=0 and v=1. A P-transition from $E_1(v=0,J'')$ to $E_2(v=1,J''-1)$ reduces the rotational energy from $J''(J''+1)B_0$ to $J'' \times (J''-1)B_1$. Thus in contrast to 3-level systems, here only a part of E_1 is extracted from the R/T reservoir. Exchange of the full amount of E_1 would require a transition to J'=0 which is forbidden by selection rules for J''>1.

Nevertheless, extraction of the entire energy $E_1=J''(J''+1)B_0$ is possible for a mixture of gases which have the term relationships shown (for the example of $^2HI/N_2$) in Fig. 2. Here a "dopant" gas I is excited in a P-transition to a level J' which is in resonance (within less than a rotational spacing) with the level (v=1, J=0)

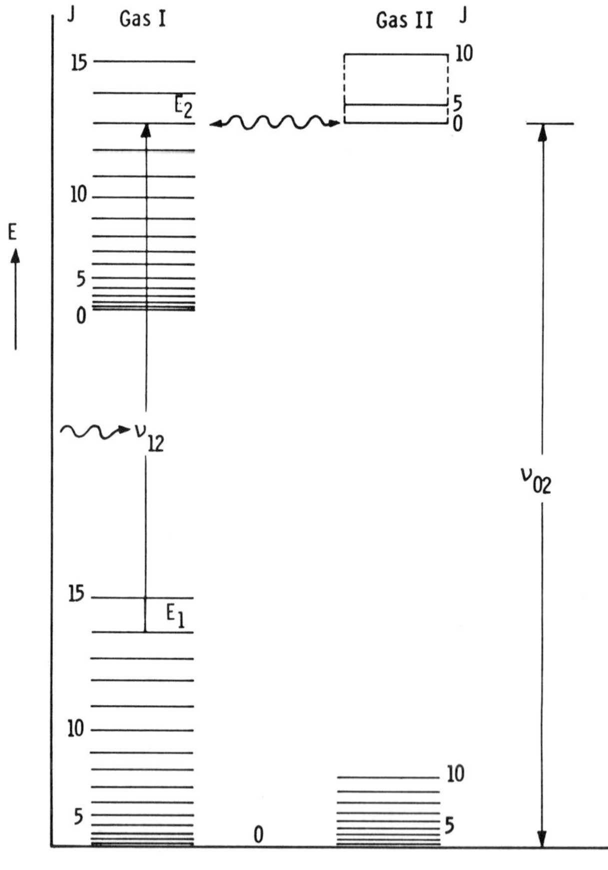

Fig. 2: Vibrational-rotational term schemes of two diatomic gases for the extraction of rotational energy $J(J+1)B_0$ from the R/T heat reservoir of the two gases by collisional resonance energy transfer. Term schemes are chosen for the example of $^2HI/N_2$.

of a host gas II. In general, collisional resonance energy transfer
is extremely fast. Thus the partial pressure of gas I can be as
low as required for the effective absorption of the radiation at
the frequency $h\nu_{12}$. The process important for the final energy
balance is therefore the vibrational excitation of the host gas II.
Again, each excitation event occurs with an energy defect E_1 at
the expense of the R/T reservoir with which the rotational and
translational energies of the molecules in all vibrational levels
are in equilibrium. The translational energies play an important
role in the excitation transfer between molecules of species I and
II. Typically a molecule I at energy $E_2(v=1,J')$ will transfer that
energy to a molecule II but return to a rotational level $J''=J'\pm1$
at the expense of translational energy.

(c) We assume now that the gas mixture is pumped to the limit
of saturating the population in the state $(v=1,J')$ while the latter
maintains an equilibrium with the rotational manifolds of both
gases at $v=1$. For the pulse length we assume the relation
$\tau_{RT} \ll t_p < \tau(E_2)$. By a process analogous to that for the
derivation of Eq. (3), we obtain for the heat extracted from the
R/T reservoir:

$$\Delta H = \frac{(2J'+1)(J'+1)(J'+2)B_o\exp[-(J'+1)(J'+2)B_o/kT]Q_{R1}}{Q_{RO} + (2J'+1)Q_{R1}\exp[-(J'+1)(J'+2)B_o/kT]} N \ . \quad (9)$$

Here J' and B_o refer to gas I. Q_{RO} and Q_{R1} are the rotational (and
nearly equal) partition functions at $v=0$ and $v=1$, of gas II, those
of the dopant gas I being ignored. The statistical weight factor
$(2J'+1)$ in nominator and denominator of the expression (9) results
from the assumption that the resonance energy transfer rate is
proportional to it. This assumption can be strictly correct only
for resonance transfer between simple (such as 3-level) systems.
However, it turns out that, after optimization, the effect of varying
that statistical weight factor from 1 to 101 is only to increase the
number of saturated transitions and hence ΔH by a factor of 3.3.

A maximum results for ΔH if $r=(J'+1)(J'+2)B_o/kT$ fulfills the
condition

$$(r_m-1)\exp(r_m) = (2J'+1)Q_{R1}/Q_{RO} \ . \quad (10)$$

Introduction of Eq. (10) into Eq. (9) yields for the maximum heat
withdrawn in the average per molecule

$$\Delta H_m/N = (r_m-1)kT \ , \quad (11)$$

where T is the temperature in isothermal operation or the final
temperature in adiabatic cooling. For the fraction of molecules

excited to v=1, one obtains the value

$$n_p/N = (r_m-1)/r_m \quad . \tag{12}$$

For example, with $J' = 10$, one obtains r_m=2.57, n_p/N=0.61, $\Delta H_m/N$ (isothermal) = 1.58 kT, and for adiabatic operation $M=(T'-T)/T = 0.63$. The last two values will be again considerably enhanced for pulse times t_p large enough for repetitions of the saturating process.

(d) In isothermal operation, quenching of vibrational energies by the walls has to be avoided since this would amount to a reduction of τ_{VRT}. It does not appear to be known at present whether wall materials can be developed and treated so that vibrational energies are not quenched and only the desired interaction between the R/T degrees of freedom of the gas and the phonon spectrum of the solid can interact. At any rate, there exist other methods. The gas, for example, may consist of a 1:10:100 mixture, respectively, of dopant, host, and a suitable heat transport gas such as helium. A suitable thermal transport gas will not reduce τ_{VRT} but serve as medium for coupling the R/T energies of a flowing gas to the walls of a fixed heat reservoir. An entirely different type of isothermal operation is radiative cooling of a gas during volume compression. This process, in which the heat of isothermal compression is just equal to that pumped by the radiation, may become a useful concept for energy converters of the type discussed in the beginning.

5. RADIATIVE COOLING OF CO ISOTOPE MIXTURES

Various heterogeneous gas mixtures, such as CO/N_2 or $^2HI/N_2$, are suitable media for enhanced radiative cooling. Alternatively, diatomic gases consisting of different isotopic species exhibit, because of the mass dependence of vibrational energies, term relationships as shown in Fig. 2 for $^2HI/N_2$. We choose as an example a mixture of $^{13}C^{18}O$ as dopant and $^{12}C^{16}O$ as host. The presence of other CO isotopes such as formed by isotope interchange does not alter the energy defect in the vibrational excitation of the host via absorption in $^{13}C^{18}O$.

A suitable pressure of the CO isotope mixture is about 100 Torr at which the thermal relaxation time[7] τ_{VRT} of the vibrationally excited molecules is in the order of 10 sec. In mixtures with helium at 77 K, $\tau_{VRT}P$ has been measured[8] as approximately 2 sec-atm. The relaxation time[7] $\tau(E_2)$ in CO of 10^{-7} sec (at 100 Torr) is almost completely determined by (VV) energy transfer in the collision processes 2CO (v=1) = CO(v=0) + CO (v=2) + 27 cm^{-1}. Escalation to yet higher levels follows quickly, so that the (VV) processes leave only a small fraction of CO molecules in the ground state. Nevertheless, spontaneous emission returns the molecules to their original distribution with a relaxation time of 33 msec. It is clear that the "renewal time", after which continuing radiation will repeat the saturation process, is not simply τ_r, but a value $\tau'(E_2)$, smaller than the radiative lifetime. Under the conditions described, we may assume that $t_p/\tau'(E_2)$ can be at least as large as 30 before destructive heating effects become important.

The vibrational-rotational energy levels and infrared transitions of various CO isotopes can be determined rather accurately from mass corrected Dunham coefficients[9]. The (v=1,J=0) state of $^{12}C^{16}O$ lies 99.58^{-1} above the corresponding state of $^{13}C^{18}O$, and the pertinent rotational constants of the latter are $B_0 = 1.75$ cm^{-1} and $B_1 = 1.73$ cm^{-1} so that the (v=1, J'=7) state of the dopant is in near resonance with the level (v=1, J=0) of the host gas. One must therefore excite the P(8) transition of $^{13}C^{18}O$ which starts from $E_1 = J''(J''+1)B_0 = 125.74$ cm^{-1}. Equation (10) yields $r_m = 2.39$ for maximum heat withdrawal corresponding to an optimal temperature of T = 76 K. From Eq. (12), one finds that 58% of the host gas has been excited to v=1 so that the radiative decay rate can be enhanced by stimulated emission. The average isothermally withdrawn heat equals [cf. Eq. (11)] 1.39 kT per molecule, i.e., 56% of its average heat content. These results apply of course to operation under the limits $\tau_{RT} \ll t_p < \tau(E_2)$. ΔH can yet be enhanced by the factor $t_p/\tau'(E_2)$.

For adiabatic operation with extended pulses, one can follow the decrease of temperature with time by an iteration procedure. The result is shown in Fig. 3 in the plot of temperature vs. $t/\tau'(E_2)$. Starting at room temperature, the gas mixture reaches 76 K after a

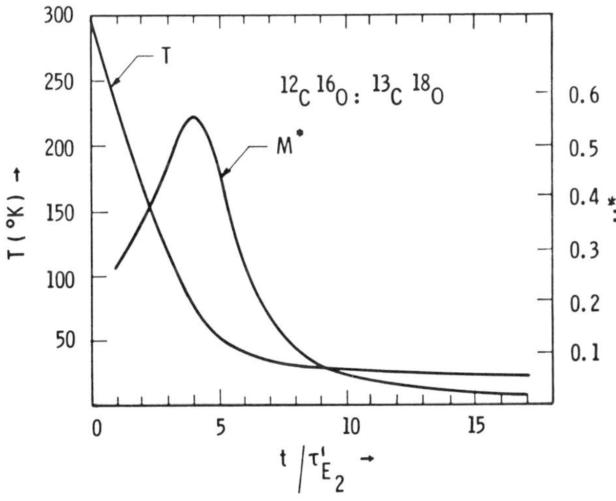

Fig. 3: Adiabatic temperature decrease of $^{13}C^{18}O/^{12}C^{16}O$ with time after onset of radiation pulse and merit factor $M^* = \Delta T/T$ for intervals of $\tau'(E_2)$.

few renewal times $\tau'(E_2)$ and then cools at a decreasing rate, approaching about 20 K asymptotically. Figure 3 also shows a merit factor M* which represents $M = \Delta T/T$ for intervals of $\tau'(E_2)$. M* reaches a maximum of 0.55 at T=76 K and vanishes for long pulse durations. The position of its peak with respect to the temperature varies with the molecular species and the transitions chosen. Since the optimum value of $r = E_1/kT$, as determined by Eqs. (4) and (10) is not far from 2.0, E_{1m} is in the order of 2kT. This simple relationship serves as a guide in the choice of molecular systems to achieve optimum operation at a desired temperature.

*Supported by NASA-Ames under Contract NAS2-9185.

References

1. M. Garbuny and M.J. Pechersky, Appl. Opt. *15* 1141 (1976).
2. M. Garbuny, Opt. Commun. *13*, 68 (1975).
3. T.W. Hänsch and A.L. Schawlow, Opt. Commun. *13*, 68 (1975).
4. A.D. Wood, M. Camac and E.T. Gerry, Appl. Opt. *10*, 1877 (1971).
5. F.G. Gebhardt and D.C. Smith, Appl. Phys. Lett. *20*, 129 (1972).
6. F.R. Grabiner, D.R. Siebert and G.W. Flynn, Chem. Phys. Lett. *17*, 189 (1972); D.R. Siebert, F.R. Grabiner and G.W. Flynn, J. Chem. Phys. *60*, 1564 (1974).
7. E. Weitz and G. Flynn, in Annual Review of Physical Chemistry *25*, 275 (1974), H. Eyring, ed., Annual Reviews Inc., Palo Alto, CA.
8. G. Karl, P. Kruus and J.C. Polanyi, J. Chem. Phys. *46*, 224 (1967); R.C. Millikan, J. Chem. Phys. *40*, 2594 (1964).
9. A.H.M. Ross, R.S. Eng and H. Kildal, Opt. Commun. *12*, 433 (1974).

ON THE LINEAR EQUATIONS OF SOLITON THEORY*

G.L. Lamb, Jr.

University of Arizona, Tucson, Arizona

1. INTRODUCTION

During the past few years there has been an increasing interest
in solitons. The soliton and its remarkable properties are well
known to those already familiar with the history of coherent optics
since one of the earliest examples of a soliton was the 2π pulse of
self-induced transparency [1]. In that instance a satisfactory
mathematical setting for the soliton was completely provided by the
coupled Maxwell and Schroedinger-Bloch equations. The situation is
not as satisfactory in regard to the equations that have more
recently been shown to exhibit soliton behavior, however. To solve
these equations, certain linear equations analogous to the equations
for the two-level atom have been introduced [2] in a somewhat *ad hoc*
fashion.

The purpose of this paper is to describe a simple geometric
context in which some of the more common soliton equations as well
as the above-mentioned linear equations are found to arise in a
rather natural way [3]. This method exploits the fact that the
Schroedinger-Bloch equations have a structure quite similar to one
component of the Serret-Frenet equations that arise in a consideration of twisted space curves. In fact, it was transformations
devised previously [4] for treating twisted space curves that have
already been used to obtain multisoliton solutions of the coupled
Maxwell and Schroedinger-Bloch equations [5]. All attempts by the
author to use the Serret-Frenet equations to obtain some understanding of the other soliton equations had met with failure until
a paper by Hasimoto [6] was pointed out [7].

It was shown by Hasimoto that the curvature and torsion of a thin vortex filament moving without stretching in an incompressible inviscid fluid leads quite directly to the nonlinear Shroedinger equation which is one of the standard soliton equations. One is then able to rely upon a general principle of soliton theory, namely, whatever applies for one equation applies for others as well. As expected, it is readily shown that other soliton equations can be associated with moving space curves [8].

The Serret-Frenet equations, which occur naturally in any consideration of twisted curves, then provide the linear equations in a very natural way. Application of these ideas to the sine-Gordon equation are outlined here. Consideration of the Hirota equation [9], which includes both the cubic Schroedinger equation and the modified Korteweg deVries equation as special cases, is included in [10].

2. EQUATIONS OF COHERENT OPTICAL PULSE PROPAGATION

In order to motivate the treatment of the Serret-Frenet equations we first recall the standard equations of coherent optical pulse propagation in slowly varying envelope approximation. Only the simplest situation of a carrier frequency at the center of a symmetrically broadened line will be considered here.

If the electric field is assumed to be a plane wave of the form

$$E(x,t) = \mathcal{E}(x,t) \cos(k_0 x - \omega_0 t) \, , \tag{2.1}$$

then the induced polarization per atom may be written

$$P(\zeta,x,t) = p[C(\zeta,x,t)\cos(k_0 x - \omega_0 t) + S(\zeta,x,t)\sin(k_0 x - \omega_0 t)], \tag{2.2}$$

where p is the dipole matrix element and $\zeta = \omega_{ab} - \omega_0$ is the detuning of the atom from the center frequency ω. The Maxwell equations then reduce to

$$\frac{\partial \mathcal{E}}{\partial t} + c \, \frac{\partial \mathcal{E}}{\partial x} = 2\pi\omega_0 p \int d\zeta \, g(\zeta) \, S(\zeta,x,t) \quad , \tag{2.3}$$

where c is the light velocity in the medium and $g(\zeta)$ is the spectrum of the detuning which is here assumed to be an even function of ζ.

Writing the wave function for the two-level atom as

$$\psi(\underline{r},t) = a(t) \, \psi_a(\underline{r}) + b(t) \, \psi_b(\underline{r}) \quad , \tag{2.4}$$

where a and b refer to the upper and lower levels respectively, one finds that standard time-dependent perturbation theory leads to the relations [11]

$$i\hbar\dot{a} = E_a a - pEb \; , \tag{2.5a}$$

$$i\hbar\dot{b} = E_b b - pEa \; , \tag{2.5b}$$

where the dot indicates a time derivative while E_a and E_b are the two energy levels. One may recast Eqs. (2.5) in a form analogous to the linear equations of soliton theory by first writing $a = u_1\exp(-i\omega_b t)$, $b = u_2\exp(-i\omega_a t)$ to obtain

$$\dot{u}_1 + i\omega_{ab}u_1 = -iVu_2\exp(-i\omega_{ab}t) \; , \tag{2.6a}$$

$$\dot{u}_2 - i\omega_{ab}u_2 = -iVu_1\exp(i\omega_{ab}t) \; , \tag{2.6b}$$

where $\omega_{ab} = \omega_a - \omega_b$ and $V = -pE/\hbar$. Setting $\omega_{ab} = \omega_0 + \zeta$ and writing

$$u_1 = w_1\exp\left(-i\omega_0 t - \frac{i}{2}\zeta t\right) \; , \tag{2.7a}$$

$$u_2 = w_2\exp\left(i\omega_0 t + \frac{i}{2}\zeta t\right) \; , \tag{2.7b}$$

one obtains

$$\dot{w}_1 + i\frac{\zeta}{2}w_1 = i\frac{p\mathcal{E}}{2\hbar}w_2 \; , \tag{2.8a}$$

$$\dot{w}_2 - i\frac{\zeta}{2}w_2 = i\frac{p\mathcal{E}}{2\hbar}w_1 \; . \tag{2.8b}$$

It is equations of this form that are introduced in solving the soliton equations [12] (if one further sets $w_1 = iv_1$ and $w_2 = v_2$).

When the polarization [11]

$$P = p(a^*b + ab^*) \tag{2.9}$$

is written in terms of w_1 and w_2 and compared with Eq. (2.2), one obtains

$$C = w_1^* w_2 + w_1 w_2^* \; , \tag{2.10a}$$

$$S = i(w_1 w_2^* - w_1^* w_2) \; . \tag{2.10b}$$

Differentiating with respect to time, using Eqs. (2.10) and noting that the population difference may be written as

$$n \equiv |a|^2 - |b|^2 = |w_1|^2 - |w_2|^2 \ , \tag{2.10c}$$

one recovers the Bloch equations in the form

$$\dot{S} = \frac{p\mathcal{E}}{\hbar}n + \zeta C \ , \tag{2.11a}$$

$$\dot{n} = -\frac{p\mathcal{E}}{\hbar}S \ , \tag{2.11b}$$

$$\dot{C} = -\zeta S \ . \tag{2.11c}$$

Use of linear equations such as Eqs. (2.8) and (2.11) to obtain multisoliton solutions has been considered elsewhere [5,13] and will not be repeated here. Instead, we consider how equations similar to these arise in an analysis of space curves.

3. SOME SOLITON EQUATIONS

Three of the most popular soliton equations are [14]:

$$u_{xt} = \sin u \qquad\qquad \text{sine-Gordon equation} \ ,$$

$$u_t + u^2 u_x + u_{xxx} = 0 \qquad\qquad \text{Modified Korteweg-deVries equation,}$$

$$iu_t + u_{xx} + |u|^2 u = 0 \qquad\qquad \text{Cubic Schroedinger equation.}$$

Pulse decomposition similar to that observed for $2n\pi$ pulses in self-induced transparency are exhibited by these equations. This is, of course, expected for the sine-Gordon equation since it arises as a limiting form of the coupled Maxwell-Bloch equations [15,16].

It has been found [2] that solutions of the above equations can be obtained by introducing linear equations of the form

$$v_{1x} + i\zeta v_1 = q v_2 \ , \tag{3.1a}$$

$$v_{2x} - i\zeta v_2 = r v_1 \ , \tag{3.1b}$$

where $r = -q = \frac{1}{2} u_x$ for the sine-Gordon equation, $r = -q = u$ for
the modified Korteweg deVries equation and $r = -q^* = -u^*$ for the
cubic Schroedinger equation. One also introduces

$$v_{1t} = Av_1 + Bv_2 \, , \tag{3.2a}$$

$$v_{2t} = Cv_1 + Dv_2 \, , \tag{3.2b}$$

where A, B, C and D are functions of u and, for some equations,
spatial derivatives of u as well. As an example, we consider the
sine-Gordon equation. Here one employs the linear equations [2]

$$v_{1x} + i\zeta v_1 = - \frac{1}{2} u_x v_2 \, , \tag{3.3a}$$

$$v_{2x} - i\zeta v_2 = \frac{1}{2} u_x v_1 \, , \tag{3.3b}$$

and

$$v_{1t} = \frac{i}{4\zeta} (\cos u) \, v_1 + \frac{i}{4\zeta} (\sin u) \, v_2 \, , \tag{3.4a}$$

$$v_{2t} = \frac{i}{4\zeta} (\sin u) \, v_1 - \frac{i}{4\zeta} (\cos u) \, v_2 \, . \tag{3.4b}$$

One readily obtains

$$v_{1xt} - v_{itx} = -i\zeta_t v_1 + \frac{1}{2} (\sin u - u_{xt}) \, v_2 \, , \tag{3.5a}$$

$$v_{2xt} - v_{2tx} = - \frac{1}{2} (\sin u - u_{xt}) \, v_1 + i\zeta_t v_2 \, . \tag{3.5b}$$

Hence the compatibility equations are satisfied provided ζ is a
constant and u satisfies the sine-Gordon equation. It is therefore
possible to obtain solutions to the nonlinear sine-Gordon equation
by solving the above equations that are linear in v_1 and v_2. The
solution techniques will not be considered here. Rather we take up
the question of obtaining linear equations of the sort listed above
from a consideration of twisted curves.

3. MOTION OF TWISTED CURVES

The shape of a twisted curve is completely specified by the
value of its curvature and torsion at each point along the curve.
The unit tangent $\hat{t}$, normal $\hat{n}$ and binormal $\hat{b}$ to the curve are related

by the Serret-Frenet equation [4]

$$\hat{t}_s = \kappa \hat{n} , \tag{4.1a}$$

$$\hat{b}_s = - \tau \hat{n} , \tag{4.1b}$$

$$\hat{n}_s = \tau \hat{b} - \kappa \hat{t} , \tag{4.1c}$$

where κ and τ are the curvature and torsion, respectively, while the subscript indicates differentiation with respect to arc length s. If, in addition, the curve distorts in time, then one may also express $\hat{t}_t$, $\hat{b}_t$ and $\hat{n}_t$ as linear combinations of $\hat{t}$, $\hat{b}$ and $\hat{n}$. Since these three vectors are orthogonal unit vectors, the time variation may be written

$$\hat{t}_t = \underline{\Omega} \times \hat{t} , \tag{4.2a}$$

$$\hat{b}_t = \underline{\Omega} \times \hat{b} , \tag{4.2b}$$

$$\hat{n}_t = \underline{\Omega} \times \hat{n} , \tag{4.2c}$$

where $\underline{\Omega} = \omega_1 \hat{t} + \omega_2 \hat{b} + \omega_3 \hat{n}$. (Equations (4.1) can be written in a similar fashion, i.e., $\hat{t}_s = \underline{d} \times \hat{t}$, etc., where $\underline{d}$ is the Darboux vector $\underline{d} = \tau \hat{t} + \kappa \hat{b}$.) Equating the mixed partial derivatives $\hat{t}_{ts} = \hat{t}_{st}$, $\hat{b}_{ts} = \hat{b}_{st}$ and $\hat{n}_{st} = \hat{n}_{ts}$, one obtains the three expressions

$$\frac{\partial \kappa}{\partial t} - \frac{\partial \omega_2}{\partial s} = \tau \omega_3 , \tag{4.3a}$$

$$\frac{\partial \tau}{\partial t} - \frac{\partial \omega_1}{\partial s} = -\kappa \omega_3 , \tag{4.3b}$$

$$\frac{\partial \omega_3}{\partial s} = \tau \omega_2 - \kappa \omega_1 . \tag{4.3c}$$

We now consider how Eqs. (4.3) lead to nonlinear evolution equations of soliton type. In particular, we shall concentrate on the sine-Gordon equation. A consideration of the other equations appears elsewhere [10].

We now list four ways of obtaining the sine-Gordon equations from Eqs. (4.3).

1. Set $\omega_1 = 0$ and $\kappa = $ const. $= \kappa_0$. Then

$$\frac{\partial \tau}{\partial t} = - \kappa_0 \omega_3 , \tag{4.4}$$

and Eqs. (4.3a) and (4.3b) can be combined and integrated. They yield $\omega_2^2 + \omega_3^2 = 1$ or equivalently, $\omega_2 = \cos u$, $\omega_3 = \sin u$ and hence $\tau = \partial u/\partial s$. Thus Eq. (4.4) becomes

$$\frac{\partial^2 u}{\partial s \partial t} = - \kappa_0 \sin u \ . \tag{4.5}$$

2. Set $\tau = \text{const.} = \tau_0$ and $\omega_2 = 0$. Then

$$\frac{\partial \kappa}{\partial t} = \tau_0 \omega_3 \ , \tag{4.6}$$

while $\omega_1 = \cos u$, $\omega_3 = \sin u$ and $\kappa = -\partial u/\partial s$. One obtains

$$\frac{\partial^2 u}{\partial s \partial t} = - \tau_0 \sin u \ . \tag{4.7}$$

3. Set $\omega_2 = 0$ and $\omega_1 = \text{const.} = \omega_0$. Then

$$\frac{\partial \omega_3}{\partial s} = - \omega_0 \kappa \ , \tag{4.8}$$

while $\kappa = \tau_0 \sin u$, $\tau = \tau_0 \cos u$ where τ_0 is the value of the torsion as $u \to 0$. (For the single soliton solution [5] $u = 4 \tan^{-1} \exp(as+t/a)$ this occurs as $s \to -\infty$.) Also $\omega_3 = \partial u/\partial t$ so that

$$\frac{\partial^2 u}{\partial s \partial t} = - \omega_0 \tau_0 \sin u \ . \tag{4.9}$$

4. Finally one may set $\omega_1 = 0$ and $\omega_2 = \text{const.} = \omega_0$. One then obtains

$$\frac{\partial^2 u}{\partial s \partial t} = - \omega_0 \kappa_0 \sin u \ , \tag{4.10}$$

with $\kappa = \kappa_0 \cos u$, $\tau = \kappa_0 \sin u$ and $\omega_3 = -\partial u/\partial t$.

We now consider the linear equations that may be associated with the sine-Gordon equation. We need merely follow a procedure that was devised long ago for relating the Serret-Frenet equations to a Riccati equation [4]. The Riccati equation is then replaced by a pair of linear first-order equations (instead of the customary single second-order equation). The resulting linear equations are just the linear equations postulated for solving the nonlinear evolution equations.

We begin by considering some one component of the vector Serret-Frenet equations, i.e.,

$$t_s = \kappa n \, , \tag{4.11a}$$

$$b_s = -\tau n \, , \tag{4.11b}$$

$$n_s = \tau b - \kappa t \, . \tag{4.11c}$$

These scalar equations have the first integral $t^2 + b^2 + n^2 = 1$ which may be rewritten as

$$\frac{n + it}{1-b} = \frac{1+b}{n - it} \equiv \phi \, . \tag{4.12}$$

Expressing n, b and t in terms of ϕ, one obtains [4,5]

$$n = \frac{2\,\mathrm{Re}\,\phi}{|\phi|^2 + 1} \, , \tag{4.13a}$$

$$b = \frac{|\phi|^2 - 1}{|\phi|^2 + 1} \, , \tag{4.13b}$$

$$t = \frac{2\,\mathrm{Im}\,\phi}{|\phi|^2 + 1} \, . \tag{4.13c}$$

When Eqs. (4.13) are expressed in terms of ϕ, they are found to be equivalent to the Riccati equation

$$\phi_s - i\kappa\phi + \frac{1}{2}\,\tau(\phi^2 + 1) = 0 \, . \tag{4.14}$$

Setting $\phi = v_1/v_2$, one finds that this Riccati equation may be replaced by the pair of linear equations

$$v_{1s} - i\,\frac{k}{2}\,v_1 = -\frac{1}{2}\,\tau v_2 \, , \tag{4.15a}$$

$$v_{2s} + i\,\frac{k}{2}\,v_2 = \frac{1}{2}\,\tau v_1 \, , \tag{4.15b}$$

which are of the form introduced to solve the sine-Gordon equation by inverse scattering methods [2]. Proceeding similarly with the equations for the time dependence, one obtains

$$\phi_t + i\omega_3\phi - \frac{1}{2}\,(\omega_1 + i\omega_3)\,\phi^2 - \frac{1}{2}\,(\omega_1 - i\omega_3) = 0 \, . \tag{4.16}$$

Again setting $\phi = v_1/v_2$, one obtains the linear equations

$$v_{1t} + \frac{i}{2}\,\omega_2 v_1 = \frac{1}{2}\,(\omega_1 - i\omega_3)\,v_2 \, , \tag{4.17a}$$

$$v_{2t} - \frac{i}{2} \omega_2 v_2 = - \frac{1}{2} (\omega_1 + i\omega_3) v_1 . \tag{4.17b}$$

Returning to the first method for obtaining the sine-Gordon equation that was outlined above, i.e., by setting $\kappa = \kappa_0$, $\tau = u_s$, $\omega_1 = 0$, $\omega_2 = \cos u$ and $\omega_3 = \sin u$, one obtains linear equations in agreement with Eqs. (3.3) and (3.4) when one sets $\kappa_0 = -2\zeta$ and $t' = \kappa_0 t$.

The twisted curve that corresponds to the single soliton solution, namely $u = 4 \tan^{-1}[\exp(as + t/a)]$, is shown in Fig. 1 for the case $a = 1$ and $\kappa_0 = 1$. For details on the determination of this curve shape, the reader is referred to Ref. 10.

By factoring the first integral of the Serret-Frenet equation in the form

$$\frac{b + it}{1-n} = \frac{1+n}{b - it} \equiv \phi \tag{4.18}$$

and proceeding along lines identical to those outlined above, one obtains linear equations for which the torsion plays the role of the eigenvalue parameter ζ. An example of the twisted curve that

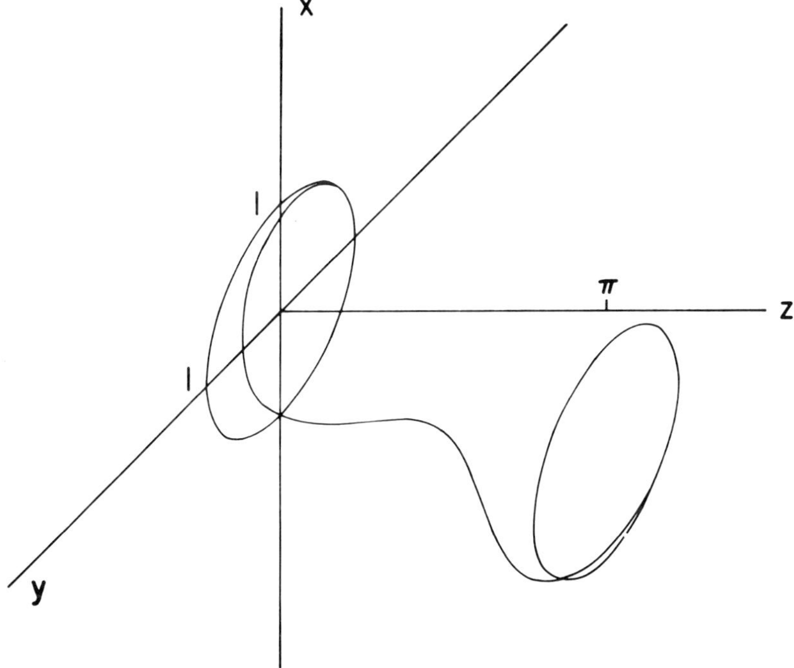

Figure 1.   Example of a single soliton curve of constant curvature, $\kappa=1$, $\tau=2$ sechs. (Reproduced from J. Math. Phys., Ref. 10, this paper.)

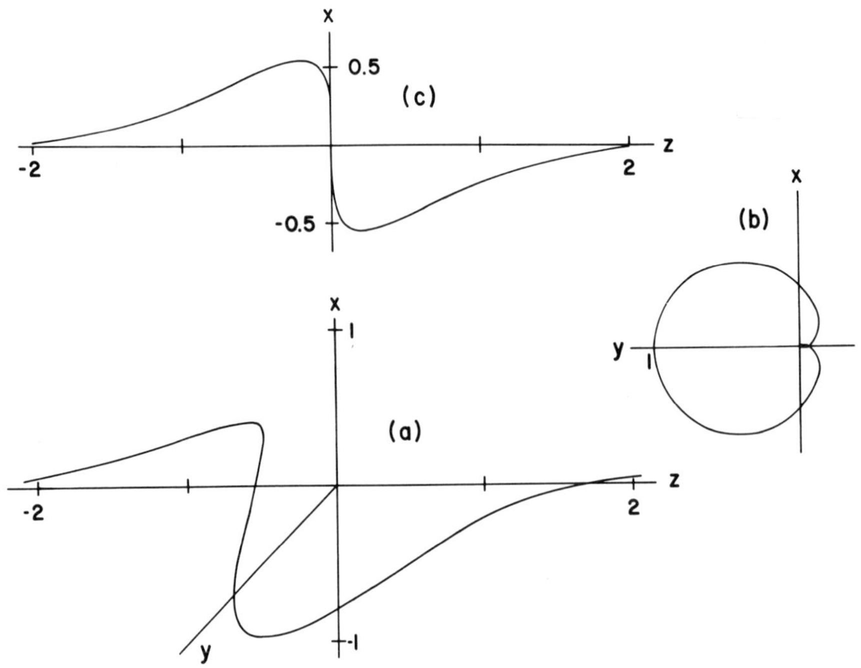

Figure 2.   (a) Example of a single soliton curve of constant torsion, τ=1, κ=2 sechs; (b) projection of curve on xy plane; (c) projection of curve on xz plane. (Reproduced from J. Math. Phys., Ref. 10, this paper.)

results is shown in Fig. 2. It is the same as one of the curves presented by Hasimoto [6] in his treatment of the cubic Shroedinger equation. That the shape of the curve is the same is to be expected since the linear equations for the spatial dependence, i.e., Eqs. (4.15) are the same for both equations. The equations governing the time dependence, and hence the motion of the twisted curve, are different for the two equations [10].

Determination of the shape of the curves requires the integration of a Riccati equation, Eq. (4.14), for the specified curvature and torsion. For cases 1 and 2 considered above, the solution is relatively straightforward and was considered long ago [17]. Integration of the Riccati equations associated with cases 3 and 4 does not seem to be feasible.

Application of the above considerations to the cubic Schroedinger equation and to the Hirota equation

$$\psi_t + 3A|\psi|^2\psi_x + iB|\psi|^2\psi + iC\psi_{xx} + D\psi_{xxx} = 0, \qquad AC = BD \qquad (4.19)$$

in which the dependent variable is complex is carried out by following more closely the work of Hasimoto in which the complex
quantity

$$\psi = \kappa \, \exp(i \int_0^s ds'\tau) \qquad\qquad (4.20)$$

is introduced. The details of this approach are carried out in
Ref. 10.

* This work was partly sponsored by the National Science Foundation
under Contract No. GP43070 with the University of Arizona.

References

1. S.L. McCall and E.L. Hahn, Phys. Rev. *183*, 457 (1969).
2. M.J. Ablowitz, D.J. Kaup, A.C. Newell and H. Segur, Studies in
   Appl. Math. *53*, 249 (1974).
3. For other geometric approaches to solitons see: H.D. Wahlquist
   and F.B. Estabrook, J. Math. Phys. *16*, 1 (1975); *17*, 1293 (1976);
   F. Lund and T. Regge, Phys. Rev. *D14*, 1524 (1976); R. Hermann,
   Phys. Rev. Letters *37*, 235 (1976); M. Lakshmanan, Phys. Lett.
   64A, 354 (1978).
4. L.P. Eisenhart, *A Treatise on the Differential Geometry of
   Curves and Surfaces* (Dover, New York, 1960), Sect. 13.
5. G.L. Lamb, Jr., Phys. Rev. *A9*, 422 (1974).
6. H. Hasimoto, J. Fluid Mech. *51*, 477 (1972).
7. The author is indebted to P.G. Saffman for bringing this work
   to his attention.
8. G.L. Lamb. Jr., Phys. Rev. Letters *37*, 235 (1976).
9. R. Hirota, J. Math. Phys. *14*, 805 (1974).
10. G.L. Lamb, Jr., J. Math. Phys. *18*, 1654 (1977).
11. W.E. Lamb, Jr., Phys. Rev. *134*, A1429 (1964).
12. A similar transformation is considered by D.W. McLaughlin and
    J. Corones, Phys. Rev. *A10*, 2051 (1974).
13. M.J. Ablowitz, D.J. Kaup, A.C. Newell and H. Segur, J. Math.
    Phys. *15*, 1852 (1974).
14. A.C. Scott, F.Y.F. Chu and D.W. McLaughlin, Proc. IEEE *61*, 1449
    (1973).
15. F.T. Arecchi and R. Bonifacio, IEEE J. Quantum Electron. *1*, 169
    (1965).
16. G.L. Lamb., Jr., Rev. Mod. Phys. *43*, 99 (1971).
17. R. Hoppe, J. f. Math. *60*, 182 (1862).

COHERENT OPTICAL PULSE PROPAGATION IN THICK RESONANT ABSORBERS

H.M. Gibbs

Bell Laboratories, Murray Hill, New Jersey

B. Bölger

Philips Research Laboratories, Eindhoven, The Netherlands

Recent [1] developments in self-induced transparency [2] (SIT) are summarized. The emphasis is placed on newly observed aspects of the on-resonance propagation of coherent optical pulses through simple absorbers.

Breakup of a 4π pulse into two completely separated sech [2] pulses has been seen. The input pulses were ten times shorter than the incoherent relaxation times. Large Fresnel numbers ($\approx 10^4$) were used to closely approximate a uniform plane wave in the center of the beam [3].

The destruction of this breakup at high absorptions is now attributed in part to spatial breakup of the beam into several self-focusing filaments. SIT transient on-resonance self-focusing [4] has been predicted by numerical solutions of coupled Maxwell-Bloch equations with radial and phase variations included; experimental confirmations [4] have been obtained for Fresnel numbers slightly larger than one and for absorption lengths of $\alpha L \approx 10$. Under these conditions the output pulse shapes differ greatly from those of uniform plane wave pulse propagation. The stronger focusing predicted for larger Fresnel numbers was not seen; instead the beam split into several small self-focusing beams. Presumably departures from the theoretically assumed equiphase uniform plane wave initiated these instabilities. It remains to be seen if SIT can be obtained in highly absorbing systems ($\alpha L > 20$) under uniform plane wave conditions by careful attention to smooth, equiphase input pulses. The importance of dynamic transverse effects in the evolution of inverted media with

Fresnel numbers close to one has not been assessed; these effects
may be significant in superfluorescence.

Degeneracies permit the study of pulse propagation in systems
without a unique dipole moment [5]. Degeneracies were found to have
surprisingly little effect on SIT on some transitions in Na [6].
With care one can see novel competition effects called *wobbling* and
leap frogging which are unique to systems of low degeneracy [7]. In
highly degenerate, non-overlapping systems, the clustering of dipole
values for the proper choice of light polarization leads to the prediction of near-ideal SIT propagation characteristics, i.e., peaking
and breakup [8]. Thus the near-ideal breakup [9] in SF_6 is expected
even without collisional equilibration. Faraday rotation of the
plane of polarization occurs when the frequency of linearly polarized
light is midway between the circularly polarized frequencies of a
degenerate system in the magnetic field. The SIT angle per unit
absorption length is the same as in the linear case, but SIT permits
many more absorption lengths with high transmission [10].

Coherent pulses of small area and energy propagate with anomalously low losses through narrowline absorbers as they reshape to
zero-area pulses [11,12]. The corresponding notch in the frequency
spectrum has also been seen [12].

Collisions of optical pulses can occur when they move with different velocities in the same direction (overtaking collisions) or
when they enter the sample from opposite directions (head-on collisions). Since the experimental collision work has not been published
elsewhere, the next section discusses this topic in more detail.

Studies have also been made of the stability of SIT solitons and
the evolution of near-resonance non-soliton input pulses, including
observations of precursors and phase modulations [13]. In general,
an on-resonance chirped pulse or an off-resonance chirped or unchirped pulse differing from 2π sech separates into a precursor with
strong phase modulation and a developing 2π sech pulse. Precursor
observation is facilitated if the carrier detuning absorption linewidth, and pulse spectral width are all approximately equal.

Finally, phonon echoes and SIT features have been observed in the
propagation of 0.7 GHz, 100 ns pulses through fused silica at 20 mK
$(T_1 \approx 200~\mu s,~T_2' \approx 15~\mu s)$ [14]. The observation and utilization of
coherent phenomena in an amorphous solid is indeed fascinating and
challenging.

OPTICAL SOLITON COLLISIONS

Many simulations of overtaking collisions show that soliton shapes are unchanged by the passage of a faster soliton through a slower one; only the pulse delays are slightly altered [15]. Pulse breakup may be thought of as half of an overtaking collision. Clearly, if the two 2π output pulses in 4π breakup were time reversed, they would coalesce into the original input pulse and would then breakup again as they traversed another sample length. From this point of view overtaking collisions have been studied to some extent. It would be interesting nonetheless to observe the complete collision process, varying the relative phase and angle between the pulses. In order to accomplish such overtaking collisions, one must be able to traverse an αL twice that for breakup. Attempts to do this have resulted in pulse deterioration, presumably from filament self-focusing. Care to avoid self-focusing may permit interesting studies of overtaking collisions.

Head-on collisions are difficult to treat theoretically because the electric field (and polarization) has components traveling in both directions. The coupled Maxwell-Bloch equations then yield an

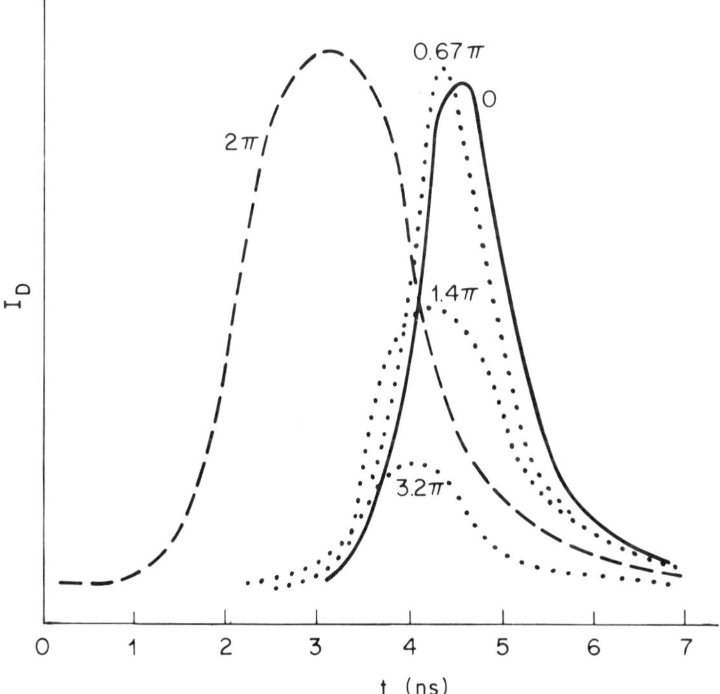

Figure 1. Head-on collisions of a 2π input pulse (dashed) with perturbing pulses with areas as labeled. The solid curve is the unperturbed output.

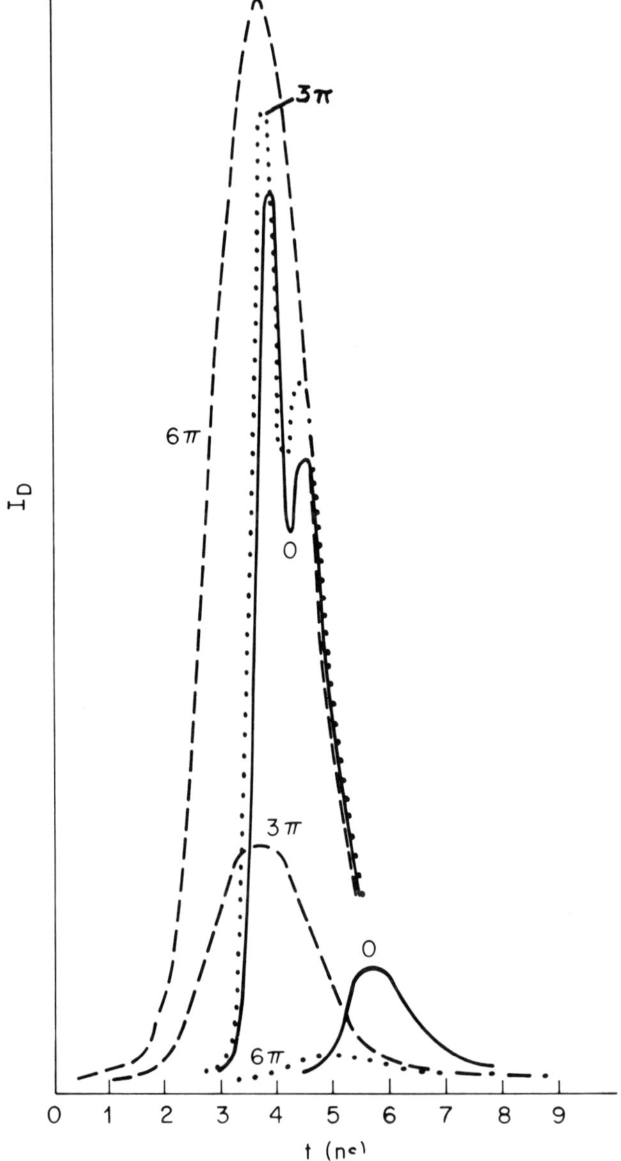

Figure 2. Head-on collisions of 6π and 3π pulses. The large pulses
are the 6π input (dashed) and outputs with no colliding pulse (solid)
and with a 3π colliding pulse (dotted). The small pulses are the
3π input (dashed) and outputs with no colliding pulse (solid) and
with a 6π colliding pulse (dotted). The high absorption makes the
outputs appear to have lower area than calibrated by 4π breakup with
$\alpha L \approx 5$.

infinite system of equations for the field and polarization envelopes and their spatial harmonics. These harmonics arise from the
reflection of the pulses from a population difference oscillating
spatially at e^{i2kz}. Morozov and Smolovich [16] estimate the reflection from this spatially periodic grating to be less than 20% and
neglect it in their calculations. A study of its importance would
be desirable. Morozov and Smolovich find, neglecting inhomogeneous
broadening, that for the cooperation time $\tau_c = (2T_2^*/\alpha c)^{\frac{1}{2}}$ greater
than the pulse duration τ_p, the colliding 2π pulses suffer only a
slight deformation and remain 2π pulses. But for $\tau_c < \tau_p$ the 2π
pulses are deformed in such a way that they become zero-area pulses.

The authors have observed head-on collisions in Na on the
D_2 F = 2 to F' = 1, 2, 3 transition. This transition exhibits non-
degenerate SIT for pulse spectra exceeding the excited state hyperfine structure [6]. Care was taken to ensure temporal synchroniza-
tion, spatial overlap, and collinearity of the colliding pulses.
Strong collisional effects were only seen when the laser frequency
was centered on the desired transition, so the oppositely directed
pulses interacted with the same atoms. The experimental parameters
yield $\tau_c \approx 0.04$ ns $<< \tau_p \approx 2$ ns, but no reshaping to zero-area
pulses was obvious. The principal conclusions are illustrated by
figures. A 2π pulse has its transmission increased by colliding
with a low-area ($<\pi$) pulse, but its transmission and delay decrease
rapidly as the area of the colliding pulse increases above π; see
Fig. 1. Similarly the transmission of a large area pulse is increased in collision with a pulse of much smaller area (Fig. 2).
Breakup, generally accepted as a signature of coherent propagation,
is destroyed when two equal large-area pulses collide (Fig. 3); the
transmitted pulse appears similar to a saturation output pulse.

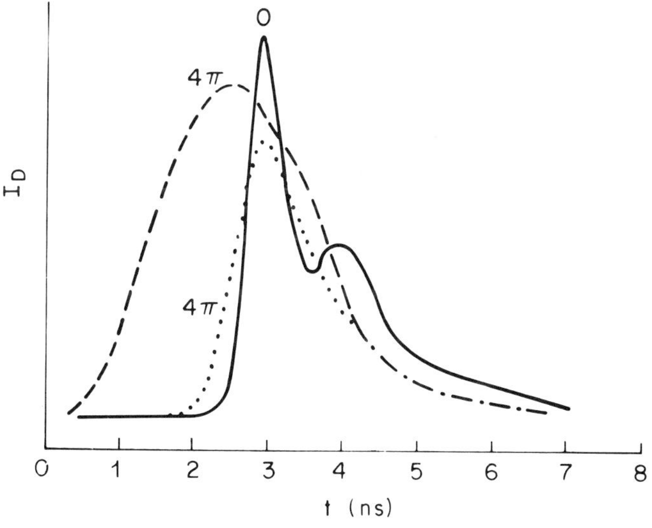

Figure 3. Head-on
collisions of 4π
pulses showing input
(dashed) and outputs
without (solid) and
with (dotted) the
colliding pulse pre-
sent.

Even though the sample length was not very much longer than the
spatial extent of one pulse in the sample, it is useful to divide
the sample into three regions as seen by the detected pulse: reshaping, colliding, and exiting. In the reshaping region the pulse
overlap is negligible and ordinary SIT reshaping occurs. In the
colliding region the coherence of a weak detected pulse can be destroyed by rapid oscillations of the population difference by the
strong colliding pulse, and the transmission of the weak pulse may
be greatly reduced or destroyed. Likewise a strong detected pulse
may be only slightly perturbed by a weak colliding pulse. In the
exiting region the detected pulse traverses atoms left partially
excited by the earlier passage of the colliding pulse. Thus the
absorption is reduced resulting in higher transmission and shorter
delays for the detected pulse. Clearly these data should be compared
with simulations with and without standing waves. Data should be
taken for higher αL under no self-focusing conditions so that the
pulse length can be made much shorter than the sample length.

Extension of this first collision experiment into optically
thicker samples will challenge both experimental and computational
techniques and elucidate both physical and mathematical phenomena.

AREAS FOR FURTHER STUDY

Achievement of true transparency with delay times equal to
several pulse widths would be satisfying and would permit the study
of overtaking collisions. It will require care to avoid local self-
focusing of the beam. Experimental and theoretical studies of coherent self-focusing should clarify the best procedures for maintaining
near-uniform wave fronts through thick absorbers. These dynamical
transverse effects should also be investigated for initially inverted
conditions, i.e., for superfluorescence. Collisions, both overtaking
and head-on, should be pursued and the importance of standing-wave
effects determined. The prediction that SIT in a highly degenerate
transition should always exhibit near-ideal breakup for the proper
choice of polarization (via dipole moment bunching) has never been
tested. SIT with subnanosecond pulses could reveal new interesting
effects. In particular, both intrinsic and applied chirps should be
studied. The distortionless propagation of pulse trains has never
been demonstrated [17]. Two-photon SIT offers an alternative to
pulse collisions for controlling one pulse with another [1c]. Much
work remains to be done on both one- and two-photon SIT in semiconductors in which delays and nonlinear transmission have been observed
[1c]. Perhaps eventually these phenomena will result in optoelec-
tronic discriminators, delay lines, double pulsers, compressors,
stretchers, oscillators, etc.

References

1. Primarily developments since the following reviews: (a) R.E.
    Slusher in *Progress in Optics XII*, E. Wolf, ed. (North-Holland,
    Amsterdam, 1974), 55-100; (b) L. Allen and J.H. Eberly, *Optical
    Resonance and Two-Level Atoms* (John Wiley, New York, 1975);
    (c) I.A. Poluektov, Yu. M. Popov and V.S. Roitberg, Sov. J.
    Quant. Electron. *4*, 423 and 719 (1974).
2. S.L. McCall and E.L. Hahn, Phys. Rev. Lett. *18*, 908 (1967) and
    Phys. Rev. *183*, 457 (1969).
3. B. Bölger, L. Baede and H.M. Gibbs, Opt. Commun. *18*, 67 (1976).
4. H.M. Gibbs, B. Bölger, F.P. Mattar, M.C. Newstein, G. Forster
    and P.E. Toschek, Phys. Rev. Lett. *37*, 1743 (1976) and refer-
    ences therein.
5. C.K. Rhodes, A. Szöke and A. Javan, Phys. Rev. Lett. *21*, 1151
    (1968); C.K. Rhodes and A. Szöke, Phys. Rev. *184*, 25 (1969);
    F.A. Hopf, C.K. Rhodes and A. Szöke, Phys. Rev. B *1*, 2833 (1970).
6. G.J. Salamo, H.M. Gibbs and G.G. Churchill, Phys. Rev. Lett. *33*,
    273 (1974); G.J. Salamo, Ph.D. Thesis, Brooklyn College of City
    University of New York (1973).
7. R.K. Bullough, P.J. Caudrey, J.D. Gibbon, S. Duckworth, H.M.
    Gibbs, B. Bölger and L. Baede, Opt. Commun. *18*, 200 (1976) and
    references therein. See also E.Y.C. Lu and L.E. Wood, Phys.
    Lett. *45A*, 373 (1973); Opt. Commun. *10*, 169 (1974); Lett. Nuovo
    Cimento *10*, 422 (1974).
8. H.M. Gibbs, S.L. McCall and G.J. Salamo, Phys. Rev. A *12*, 1032
    (1975).
9. A. Zembrod and T. Gruhl, Phys. Rev. Lett. *27*, 287 (1971).
10. E. Courtens, Phys. Rev. Lett. *21*, 3 (1968); N.S. Shiren, IEEE
    J. Quantum Electron. *QE-8*, 590 (1972); B.D. Silverman and J.C.
    Suits, Phys. Rev. A *6*, 847 (1972); H.M. Gibbs, G.G. Churchill
    and G.J. Salamo, Opt. Commun. *12*, 396 (1974).
11. M.D. Crisp, Phys. Rev. A *1*, 1604 (1970); Opt. Commun. *4*, 199
    (1971); Appl. Optics *11*, 1124 (1972); S.M. Hamadani, J. Goldhar,
    N.A. Kurnit and A. Javan, Appl. Phys. Lett. *25*, 110 (1974).
12. J.M. Friedman, H.M. Gibbs, T.N.C. Venkatesan, B. Bölger, D.
    Polder and M.F.H. Schuurmans, Opt. Commun. *20*, 183 (1977).
13. J.C. Diels and E.L. Hahn, IEEE J. Quant. Electron. *QE-12*, 411
    (1976); Phys. Rev. A *8*, 1084 (1973) and A *10*, 2501 (1974).
14. B. Golding and J.E. Graebner, Phys. Rev. Lett. *37*, 852 (1976)
    and to be published.
15. R.K. Bullough, P.J. Caudrey, J.C. Eilbeck and J.D. Gibbon,
    Opto-Electronics *6*, 121 (1974) and references therein.
16. V.N. Morozov and A.M. Smolovich, Sov. J. Quant. Electron. *4*,
    195 (1974). Head-on collisions are simulated, neglecting
    standing wave effects.
17. V. Nemec and L. Matulic, Opt. Commun. *13*, 380 (1975).

OPTICAL SOLITONS AND THEIR SPIN WAVE ANALOGUES IN ^{3}He

R. K. Bullough and P. J. Caudrey

The University of Manchester Institute of Science and

Technology, Manchester, U.K.

1. THE BACKGROUND IN DEGENERATE SIT

Solitons already have applications in statistical mechanics, low temperature physics, particle physics, plasma physics, and other areas. [1] Coherent non-linear optics has stimulated some of the remarkable advances in soliton theory made in the last few years. The sine-Gordon equation (s-G),

$$\phi_{xx} - \phi_{tt} = \sin \phi, \tag{1.1}$$

is a key equation. It is well known [2,3] that non-degenerate, strictly resonant, self-induced transparency (SIT) is governed by (1.1) in the sharp-line limit. Optical pulse break up, and in principle, pulse collisions are described by the multi-soliton solutions of this equation. [3]

In the case of hyperfine degeneracies with quantum number F = 2, equation (1.1) extends to the double s-G,

$$\phi_{xx} - \phi_{tt} = \pm (\sin \phi + \tfrac{1}{2} \sin \tfrac{1}{2} \phi) . \tag{1.2}$$

The selection rules are $\Delta F = 0$, $\Delta M_F = 0$ and the matrix elements are $\pm p, \pm\tfrac{1}{2}p$, 0 for the five possible transitions. This is a five-pseudo-spin problem, but the number 5 reduces to 2 because only the magnitudes of the matrix elements are significant.

There are eight equilibrium dielectrics associated with (1.2): they are the states at the four zeros of $\sin \phi + \tfrac{1}{2} \sin \tfrac{1}{2} \phi$: here there is no dipole moment and the system does not radiate. These

zeros may define attenuators or amplifiers. The ground state attenuator is the state $\phi = 0$ (or 4π) and corresponds to two psuedo
spins down: pulses propagating in this state are governed by (1.2)
with the +ve sign on the right side. Solutions are "wobbling" 4π
pulses which are bound pairs of 2π sech pulses. [4] They have
been seen in the $3s^2S_{\frac{1}{2}}(F=2) \to 3p^2P_{\frac{1}{2}}(F^{'} = 2)D_1$ transitions in Na
vapour [5] although complications from adjacent $F = 2 \to F^{'} = 1$
transitions 192 MHz off resonance change (1.2) to be a triple s-G.
[1,5]

This triple s-G proves to be

$$\phi_{xx} - \phi_{tt} = 3 \sin \phi + \sin \tfrac{1}{2} \phi + \sqrt{3} \sin \tfrac{1}{2} \sqrt{3} \phi. \qquad (1.3)$$

The additional matrix elements $\tfrac{1}{2}\sqrt{3}$ p on the $F = 2 \to F^{'} = 1, M_F =$
$\pm1 \to M_{F^{'}} = \pm 1$ transitions introduce the sine with irrational argument. The s-G (1.1) has the familiar 2π-sech electric field pulse
as single soliton solution. The double s-G has the 4π-wobbler
which has a soliton-like collision property [1,4] but is probably
not a true soliton. The triple s-G (1.3) is not periodic and the
Fig. 1(a) shows the remarkable form a solitary wave solution of
this equation would have to take. (The Fig. 1(b) shows the corresponding shape for D1 transitions $(F = 1,2 \to F^{'} = 1,2)$. The lower
F=1 level is 1772 MHz away from F=2 level and this F=1 level is
therefore just about resolvable for the Doppler broadening of 1700
MHz in the experiments [5]).

What is plotted in Figs. 1 (a) and 1(b) are trajectories in
a phase space - ϕ_t against ϕ essentially. But $\phi_t \propto \varepsilon$ the electric
field envelope in appropriate time units. [1-4] Thus the shapes of
the trajectories are roughly the shapes of the electric field
pulses. Since the solitary wave from Fig. 1(a) has infinite area,
it must adapt to a finite area input. In practice a pulse of area
close to 4π becomes a two-peaked pulse associated with the deep
minimum shown (at 4.2π) in Fig. 1(a). This is a zero of the right
side of (1.3) and so satisfies the area theorem [2] for this system.

The energy stored in the atoms for a pulse area θ is

$$W(\theta) = 7 - 3 \cos \theta - 2 \cos \tfrac{1}{2} \theta - 2 \cos \tfrac{1}{2} \sqrt{3} \theta. \qquad (1.4)$$

This is minimised at the zeros of the right side of (1.3). It
follows that the deep minimum at $\phi = 4.2\pi$ in Fig. (1a) is also a
deep minimum for energy stored in the atoms.

The experiments [5] show that the resulting double peaked
4.2π pulse behaves very much like the double peaked 4π pulse solution of (1.2), and agree with the computer simulations [6]. Figure
1(a) of Ref. 5 shows the form of wobbling intensity expected

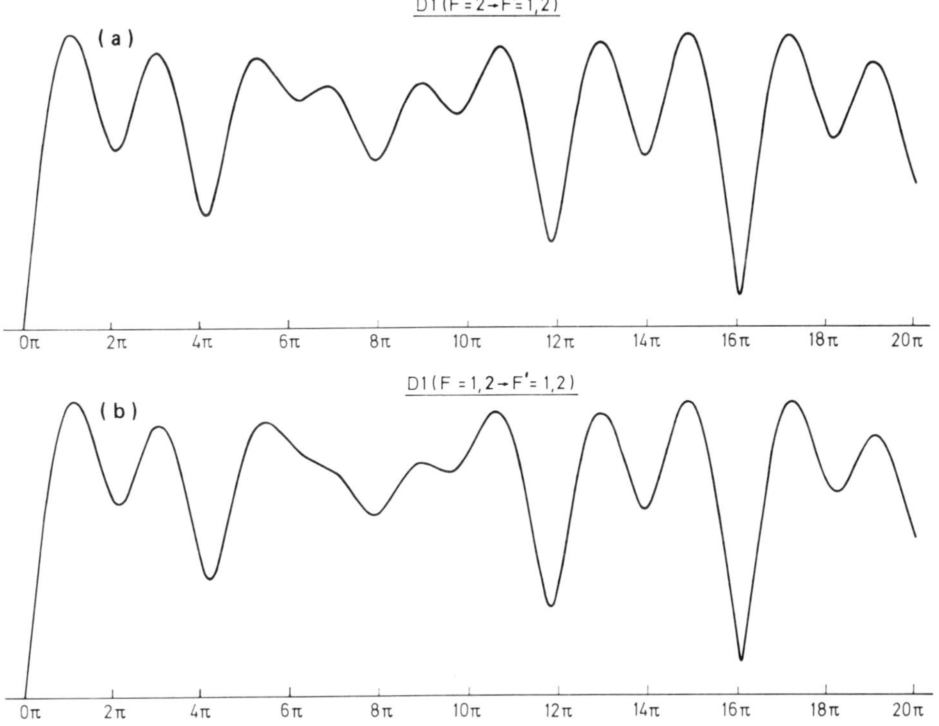

Fig. 1   Phase plane trajectories ϕ_x (or $-\phi_t$) against ϕ for
(a) D1(F=2→F´=1,2) and (b) D1(F=1,2→F´=1,2) transitions in Na vapour.

for 4π pulse solutions of (1.2) (the selection rules $\Delta F=0$, $\Delta M_F=0$
correspond to Q(2) symmetry). Figure 1(b) of Ref. 5 shows the wobbling degenerate 4.2π case whilst in Fig. 1(c) of Ref. 5, the hyperfine degeneracy is lifted by a magnetic field of 5k Gauss and pulses break up into two 2π-sech solutions of (1.1).

Whilst these optical behaviours seem remarkable enough we have
also looked theoretically at what appears to be a very different
problem. This is the problem of spin waves in liquid ^{3}He below
2.6mK. In the space available here we can only sketch our results
and we refer the reader to Leggett[7], Maki[8] and ourselves[9] for
further details.

2.   EQUATIONS OF MOTION FOR SPIN WAVES IN LIQUID ^{3}HE

Liquid ^{3}He is a Fermi liquid with an unpaired nuclear spin: it
undergoes a second order super-conducting phase transition at 2.6mK
and 34 atmospheres to the liquid A-phase [7]. At lower tem-

peratures and pressures it enters the B-phase which covers much
of the pressure-temperature phase diagram [7]. Both the A- and
B-phases are "anisotropic" liquids apparently carrying spin S=1 and
orbital angular momentum $\ell=1$. The finite angular momentum acts
to prevent hard core overlap between the two ^{3}He atoms forming a
Cooper pair (the binding forces are the weak van der Waals forces).

The order parameter is a 2x2 matrix (the "gap" matrix) with
elements $\Delta_{\uparrow\uparrow}$ $\Delta_{\downarrow\uparrow}$, etc. For $\ell=1$, $\Delta_{\uparrow\downarrow}=\Delta_{\downarrow\uparrow}$. In terms of Pauli
spin matrices $\underset{\sim}{\sigma}=(\sigma_u,\sigma_v,\sigma_w)$ $(u,v,w$ form an orthogonal system in
spin space) the gap matrix $\underset{\approx}{\Delta}$ is given by [10]

$$\underset{\approx}{\Delta}=\underset{\sim}{d}(\hat{n})\cdot\underset{\sim}{\sigma}\ \sigma_v=\begin{bmatrix} id_u + d_v - id_w & \\ & \\ -id_w & -id_u + d_v \end{bmatrix}. \qquad (2.1)$$

We are in momentum space and $\hat{n}$ is the normal to the Fermi surface.
The vector $\underset{\sim}{d}(\hat{n})$ in spin space is related to $\underset{\sim}{\hat{n}}$ through the $\hat{n}$-
independent dyadic $\underset{\approx}{D}$

$$\underset{\sim}{d}(\hat{n})=\underset{\approx}{D}\cdot\underset{\sim}{\hat{n}}, \qquad (2.2)$$

and $\underset{\approx}{D}$ is used to characterise the state. We refer the reader to
Refs. 7 or 10 where it is shown that $\underset{\approx}{\Delta}$ satisfies a BCS type integral
equation (the "gap" equation).

The dyadic $\underset{\approx}{D}$ has bivector character with one index $(u,v$ or $w)$
in spin space and one index $(x,y$ or $z)$ in orbital space. In BCS
approximation the states with spin and orbital axes, rotated arbitrarily with respect to each other, form a degenerate manifold.
The degeneracy is split by the weak spin dipole interactions. The
symmetric state with $S_z=0$ is apparently suppressed in the A-
phase [7] (and the matrix (2.1) is diagonal, $d_w=0$). The spin up
and spin down states coupled only by the dipole interaction are
formally equivalent to a Josephson junction. [1,7] It is well
known [1] that fluxons in the large area Josephson junction propagate as the 2π-kinks (solitons) of the s-G(1.1). We can therefore
expect that ^{3}He A propagates spin waves governed by (1.1) along
the field.

For P-wave states in the homogeneous case the dipole interaction takes the form [7]

$$H_D(\theta)=\text{constant}+g_D(T)\int 3\,|\underset{\sim}{\hat{n}}\cdot\underset{\approx}{\hat{D}}\cdot\underset{\sim}{\hat{n}}|^2 d\Omega/4\pi, \qquad (2.3)$$

and we suppose this applies also to the spatially inhomogeneous
case studied here. In this expression $\hat{n}(\Omega)$ is again the unit
vector normal to the Fermi surface and $\sim$

$$g_D(T) = \tfrac{1}{2}\pi\gamma^2\hbar\Delta_o^2 \ (T) \ \langle R^2\rangle_{Av} \quad ; \tag{2.4}$$

$\gamma = e(m_p c)^{-1}$ is the nuclear gyromagnetic ratio (for g-factor 2), $\langle R^2\rangle_{Av} \approx 1$ [7] and $\Delta_o(T)$ is the magnitude of the gap matrix; T is the temperature.

It is believed that the A-phase is described by the so-called ABM-like state [7] in which the gap matrix is determined by the dyadic $\underset{\sim}{D} = \sqrt{3} \ i \ \hat{\underset{\sim}{v}} \ \hat{\underset{\sim}{\ell}}^{(+)}$ in which $\hat{\underset{\sim}{v}}$ is a real unit vector in spin space and

$$\hat{\underset{\sim}{\ell}}^{(+)} \equiv i^{-\frac{1}{2}}(\hat{\underset{\sim}{\ell}}_x + i\hat{\underset{\sim}{\ell}}_y) \equiv i^{-\frac{1}{2}}(\hat{\underset{\sim}{\ell}}_1 + i \ \hat{\underset{\sim}{\ell}}_2)$$

is a spherical vector in the orbital (x,y)-plane with axis $\hat{\underset{\sim}{\ell}} \equiv \hat{\underset{\sim}{\ell}}_3 = i^{-1} \ \hat{\underset{\sim}{\ell}}^{(+)} \ x \ \hat{\underset{\sim}{\ell}}^{(-)}$ along z and normal to a constant magnetic field $\underset{\sim}{B}_o$ along y. From (2.3) it follows that $H_D(\theta)$ depends only on $(\underset{\sim}{v}\cdot\hat{\underset{\sim}{\ell}})^2 = \cos^2\theta$ and is stationary (actually minimum) with the spin vector $\hat{\underset{\sim}{v}}$ along the orbital axis $\hat{\underset{\sim}{\ell}}$. Precisely, if constants are neglected,

$$H_D(\theta) = - \frac{3}{5} \ g_D(T) \ \cos^2\theta \quad . \tag{2.5}$$

It is rotations of $\hat{\underset{\sim}{v}}$ through angles θ in the (x,z)-plane which advance as spin waves in the A-phase.

On the other hand the B-phase is thought to be in a BW-like state [7] in which the gap depends on the isotropic tensor

$$\sum_{i=1}^{3} \hat{\underset{\sim}{v}}_i \ \hat{\underset{\sim}{\ell}}_i \quad ;$$

the $\hat{\underset{\sim}{v}}_i$ are unit vectors defining the three orthogonal axes in spin space, the $\hat{\underset{\sim}{\ell}}_i$ in orbital space. Any rotation $R(\theta)$ (through angle θ about *some* axis) of the spin against the orbital space is also a BW-like state. The interaction (2.3) is now the projection of the outer product of two matrices $R(\theta)$ on a fourth rank tensor. Thus in this case, ignoring constants,

$$H_D(\theta) = \tfrac{1}{5} \ g_D(T) \ \{ (\mathrm{Tr} \ R(\theta))^2 + \mathrm{Tr} \ R^2(\theta)\}$$

$$= \tfrac{2}{5} \ g_D(T) \ \{ \ \mathrm{Tr} \ R(\theta) \ + \mathrm{Tr} \ R^2(\theta)\}$$

$$= \tfrac{2}{5} \ g_D(T) \ \{ 2 + 2\cos\theta + 2\cos 2\theta \ \} \quad . \tag{2.6}$$

This result suggests the double spin representation $(SU_2 \otimes SU_2)$ which underlies the double s-G(1.2).

A Hamiltonian density for both the A- and B- phases can be put in the form [9]

$$H(\underset{\sim}{x}) = \tfrac{1}{2}\gamma^2\chi^{-1}\sigma_w^2 - \gamma\sigma_w B_o + \tfrac{1}{2}\chi\gamma^{-2}\bar{c}^2\nabla\theta\cdot\underset{\approx}{\rho}\cdot\nabla\theta + H_D(\theta) \quad . \qquad (2.7)$$

χ is the static magnetic susceptibility scaled to a single atom by division by the superfluid number density ρ_s; B_o is the magnetic field along y; $\bar{c}$ is a velocity and ρ = diag (2,2,1) is the spin superfluid density tensor in the A-phase (the axis z is a special direction in the A-phase). In the B-phase the magnetic field is the special direction and $\underset{\approx}{\rho}$ = diag (2,1,2).(The factor ½ multiplying $\underset{\approx}{\rho}$ in (2.7) changes to ¼ in this case[9].) We omit a factor $\rho_s\hbar$ on H since it only changes canonical co-ordinates and not the spin wave equations of motion. Obvious commutation relations are

$$[\sigma_w(\underset{\sim}{x},t), \theta(\underset{\sim}{x}',t)] = -i\,\delta(\underset{\sim}{x}-\underset{\sim}{x}') \quad . \qquad (2.8)$$

The Hamiltonian (2.7) with the commutation relations (2.8) reduces to Leggett's "adiabatic Hamiltonian" system in the homogeneous case used by him to treat spin resonance.

From (2.7) and (2.8) we find equations of motion

$$\theta_t = \gamma^2\chi^{-1}\sigma_w - \gamma B_o \quad , \qquad (2.9a)$$

$$-\sigma_{w,t} = \frac{\delta H_D}{\delta\theta} - \chi\gamma^{-2}\bar{c}^2\nabla\cdot\underset{\approx}{\rho}\cdot\nabla\theta \quad . \qquad (2.9b)$$

From these we find

$$[\theta,\theta_t] = \gamma^2\chi^{-1}\,i\,\delta(\underset{\sim}{x}-\underset{\sim}{x}') \qquad (2.10)$$

and

$$\theta_{tt} = \bar{c}^2\nabla\cdot\underset{\approx}{\rho}\cdot\nabla\theta - \gamma^2\chi^{-1}(\delta H_D/\delta\theta) \quad . \qquad (2.11)$$

We consider a *single* space variable y and relabel it x. We set $\phi = -2\theta$ in the A-phase and find from (2.11) that

$$\phi_{tt} - \bar{c}_A^2\,\phi_{xx} = -\Omega_{\ell A}^2\sin\phi \quad . \qquad (2.12)$$

We set $\phi = +2\theta$ for the B-phase and find

$$\phi_{tt} - \bar{c}_B^2\,\phi_{xx} = \frac{16}{15}\Omega_{\ell B}^2[\sin\phi + \tfrac{1}{2}\sin\tfrac{1}{2}\phi] \quad . \qquad (2.13)$$

The commutation relations are

$$[\phi,\phi_t] = i\,\gamma_o\,\delta(\underset{\sim}{x}-\underset{\sim}{x}') \qquad (2.14)$$

where $\gamma_o \equiv 4\chi^{-1}\gamma^2\Omega^{-1}$ The frequency Ω_ℓ is the longitudinal nuclear magnetic resonance frequency: Ω_ℓ is $\Omega_{\ell A}$ in the A-phase and is $\sqrt{16/15}\,\Omega_{\ell B}$ in the B-phase ($\Omega_{\ell B}$ is the actual resonance frequency in this case and the $\sqrt{16/15}$ arises through the form of the dipole potential in the B-phase case). The pure number γ_o is large, $\sim 10^5$: it acts as a coupling constant in the quantised theory as (2.14) shows. Evidently we are concerned here with a strong quantum theory.

3. SPIN WAVES IN C-NUMBER FORM

We take the equations (2.12) and (2.13) in c-number form. The solutions of the s-G (2.12) are those of the s-G (1.1) and are well understood. The solutions of (2.13) are those of (1.2) with the *negative* sign. Thus they correspond to excited state solutions for the resonant degenerate optical pulse problem.

The zeros of the right side of (1.2) occur at $\phi=0$ (and 4π), $\phi=2\pi$, $\phi=\delta$ and $4\pi-\delta$ (where $\delta = 2\cos^{-1}(-\tfrac{1}{4})$). The states $\phi=\delta$ and $4\pi-\delta$ are the relevant excited states of the dielectric medium. They have zero dipole moment, so are in equilibrium in the absence of spontaneous emission. It might be possible to set up these states in the infra-red where the A-coefficients are small. In the ^{3}He spin wave problem these states are however the *ground* states of the dipole interaction energy (2.6).

The double s-G (1.2) with negative sign has the two kink (or antikink) solutions

$$\phi = \pm\, 4\, \tan^{-1}\left\{\sqrt{5/3}\ \tanh\left[\tfrac{1}{2}(\theta + \theta_o)\right]\right\}\ ,\tag{3.1}$$

$$\phi = \mp\, 4\, \tan^{-1}\left\{\sqrt{5/3}\ \coth\left[\tfrac{1}{2}(\theta + \theta_o)\right]\right\}\ ,\tag{3.2}$$

where $\theta \equiv kx-\omega t$ with $k^2-\omega^2 = (15/16)$. For the kinks (upper signs in (3.1) and (3.2)) we find: ϕ increases from $-\delta$ to $+\delta$ (mod 4π) as x runs from $-\infty$ to $+\infty$ in (3.1); ϕ increases from δ to $4\pi-\delta$ in (3.2). These are the 2δ and $(4\pi-2\delta)$ kinks, respectively; the lower signs are the corresponding antikinks. Derivatives of these kinks are bell shaped pulses of "areas" 2δ and $4\pi-2\delta$, respectively. These would be the electric field pulses in the optical case. Equation (2.9a) shows these derivatives are now pulses of magnetisation.

Although the states δ and $4\pi-\delta$ are of equal energy, the 2δ kink alone can take the state from $-\delta$(mod 4π) to $+\delta$; and the $4\pi-2\delta$ alone can take the state from $+\delta$ to $-\delta$(mod 4π). Thus, the state $-\delta$ transmits 2δ kinks; the state $+\delta$ transmits $4\pi-2\delta$ kinks and two kinks, 2δ and $(4\pi-2\delta)$ travelling in the state $-\delta$, *must retain their order*. Spin waves of this type therefore bump each other rather than pass through each other. Figure 2 shows the collision of a 2δ kink and

a (4π-2δ) kink with the 2δ to the left at t=0. The derivatives
rather than the kinks themselves are plotted. The pulses can be
identified by their areas and are found to maintain their order.

   The s-G's (1.1) and (1.2) are both Lorentz covariant. The
fundamental kink solutions have finite energies in their rest
frames: the mass of the 2π kink solution of the s-G (1.1) is 8
units; those of the 2δ and (4π-2δ) kink solutions of the double
s-G (1.2) are 11.3929 and 5.1097. The 2π kinks of the s-G pass
through each other as exact solitons (their derivatives are the 2π
sech pulses); the kinks of the double s-G (1.2) do not, as we have
seen. Further, in the state -δ a 2δ kink converts this to +δ and
a -2δ antikink converts this back to -δ. Likewise a -(4π-2δ) antikink converts the state -δ to +δ (mod 4π) whilst a (4π-2δ) kink
converts this to -δ (mod 4π). Thus through the state -δ the 2δ
kink-antikink pair can run; but also the -(4π-2δ) antikink-kink pair
can run (boundary conditions are ϕ = -δ at both ends).

   It is not possible to have a kink-antikink pair at rest so
the members of the pair collide or separate. If they collide they
can pass through each other if one type of pair changes to the
other type. The 2δ kink-antikink pair with mass 2x11.3929 can
always do this: it converts to a 4π-2δ pair with additional kinetic
energy (and also emits some "radiation") as Figure 3 shows. The 4π-
2δ pair with mass 2x5.1097 has an energy threshold for conversion,
and well below threshold the pair bounces. Figure 4 although just
above threshold, shows a similar bounce because of energy lost by
radiation. Figure 5, well above threshold, shows a (4π-2δ) kink-
antikink pair converting to a -2δ antikink-kink pair.

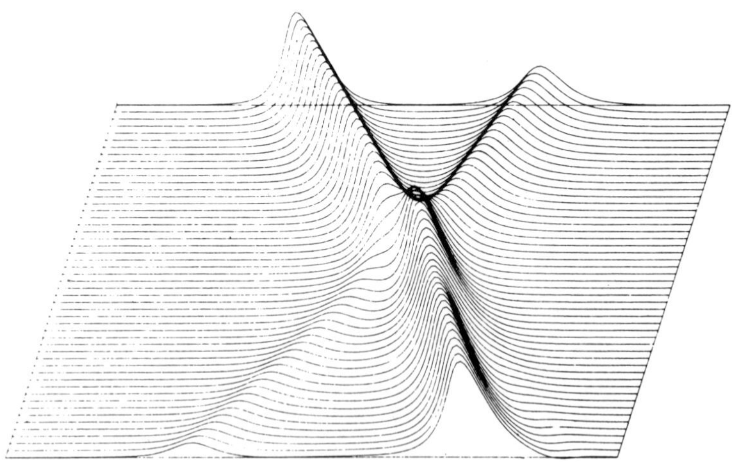

Fig. 2 Collision of a 2δ and a 4π-2δ pulse in liquid ³He B. The 2δ
is at the back initially and remains there. The kink derivatives are
plotted and the pulses can be identified by their areas.

Fig. 3. Collision of a 2δ kink-antikink pair creating a $-(4\pi-2\delta)$ antikink-kink pair. Boundary conditions are $\phi = -\delta$ for $|x| \to \infty$. Note the radiation emitted.

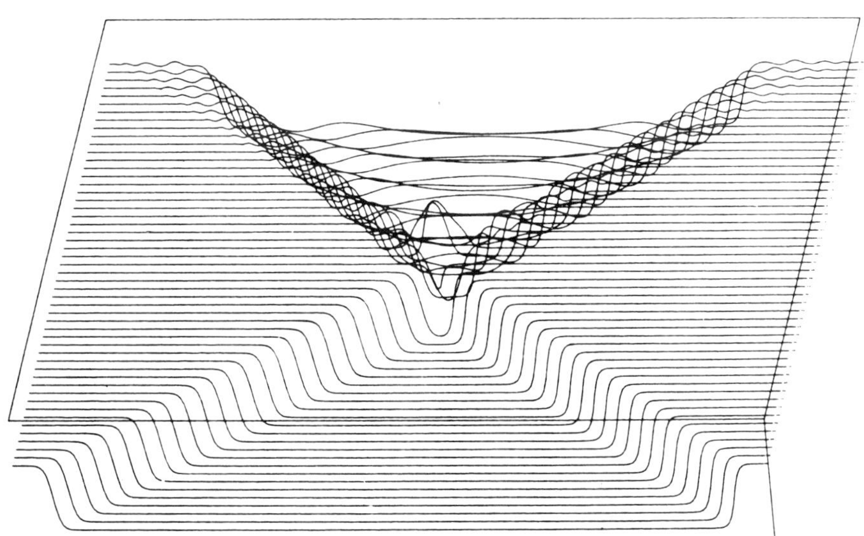

Fig. 4. Collision of a $(4\pi-2\delta)$ kink-antikink pair just above threshold. The pair still bounces because of the loss of energy by radiation. Note the short lived trough of the -2δ antikink-kink pair created initially.

Fig. 5 Collision of a (4π-2δ) kink-antikink pair well above
threshold. The -2δ antikink-kink pair is created but again note
the additional radiation.

Strictly speaking these behaviours are *not* soliton behaviours.
Solitons pass through each other with only a phase shift; [1] the
kinks of the s-G are true solitons in this sense. This means, in
practice, that we can actually solve the quantised problem (2.12)
in the A-phase. We have not solved the quantised problem (2.13) in
the B-phase.

4. QUANTISED EXCITATION SPECTRA IN THE A-PHASE

The sine-Gordon equation (1.1) is an exactly integrable system.
We take the s-G in the form

$$\phi_{xx} - \phi_{tt}\,\bar{c}_A^{-2} = \Omega_{\ell A}^2\,\bar{c}_A^{-2}\sin\phi \;, \tag{4.1}$$

and set $\hbar\Omega_{\ell A}\,\bar{c}_A^{-2} = m_A$, a "meson" mass. The Hamiltonian density
from which this equation of motion can be derived can be taken in
the form

$$\mathcal{H} = \gamma_0^{-1}\,[\tfrac{1}{2}\,\phi_x^{\;2} + \tfrac{1}{2}\,\phi_t^{\;2} - (1-\cos\phi)] \;, \tag{4.2}$$

where $\gamma_0 \equiv 4\gamma^2\chi^{-1}\,\Omega_{\ell A}^{-1}$ is a dimensionless coupling constant and we
work in units $\Omega_{\ell A}^{-1}$ for time and $\bar{c}_A\,\Omega_{\ell A}^{-1}$ for the space coordinate x.

By a deep canonical transformation we have transformed from the Hamiltonian $H = \int \mathcal{H}\, dx$ finally to the form[11]

$$
H = \sum_{i=1}^{K} \{64 m_A^2 \, \gamma_0^{-2} \, \bar{c}_A^4 + p_i^2 \bar{c}_A^2\}^{\frac{1}{2}}
$$

$$
+ \sum_{j=1}^{L} \{256 m_A^2 \gamma_0^{-2} \, \bar{c}_A^4 \, (\sin^2 \mu_j) + p_j^2 \bar{c}_A^2\}^{\frac{1}{2}}
$$

$$
+ \int_{-\infty}^{\infty} (\hbar^2 \, \Omega_{\ell A}^2 + \bar{c}_A^2 \xi^2)^{\frac{1}{2}} \, \rho(\xi) \, d\xi \tag{4.3}
$$

in which $\rho(\xi)$ is a density of occupied momentum states ξ.

This remarkable result (first obtained by Fadeev[12]) shows that the s-G, either as a resonant optical pulse problem or as a spin wave problem, is *exactly* equivalent to a collection of free massive relativistic particles: the particular form arises from K 2π kinks (K 2π sech pulses) of mass $8 m_A \, \gamma_0^{-1}$, L breathers (L 0π-pulses) of mass $16 m_A \, \gamma_0^{-1} \sin \mu_j$, and a set of background (radiation) modes of mass m_A per mode. Since μ_j is a free parameter $(0 \leq \mu_j \leq \pi/2)$ the breather spectrum forms a band of excitations over the mass range 0 to $16 m_A \, \gamma_0^{-1}$; at this point breathers break up into unbound kink-antikink pairs.

From the spin resonance results[7] $m \sim 10^{-26}$gm and becáuse γ_0 is so large ($\sim 10^5$), the kink masses are $\sim 10^{-32}$gm. There is a threshold here for kink excitations but this is so low in frequency ($\sim 10^{-1}$Hz) that it will be hard to see compared with the orbital wave excitations.[10]

The optical problem is quite different: $\gamma_0 \sim \sqrt{\alpha} \, \omega_s^{-1}$ where $\alpha = 2\pi n \, p^2 \, \omega_s \, \hbar^{-1}$, n is the number density, p is the dipole matrix element, ω_s is the non-degenerate two-level atom resonance frequency ($\sqrt{\alpha} = \tau_C^{-1}$ where τ_C is the maximum cooperation time - see our paper 'Theory of far infra-red superfluorescence', these Proceedings[13]). For $n \sim 10^{11}$ cm^{-3} and $\omega_s \sim 10^{15}$ Hz, $\sqrt{\alpha} \sim 10^9$ Hz and $\gamma_0 \sim 10^{-6}$. The kinks one sees are the 2π pulses on a nsec or shorter time scale.

The Hamiltonian (4.3) can be quantised[11,12]: the invariance of the Poisson brackets guarantees that the result is equivalent to quantising the sine-Gordon fields ϕ. One finds a single level for the kinks at $8 m_A \, \gamma_0^{-1}$. But because the phase space available to the breathers is compact, one finds that the internal momenta μ of the breathers take on discrete values μ_n:

$$
\mu_n = n\gamma_0/16 \quad . \tag{4.4}
$$

There is a renormalisation problem[11,12]: $\gamma_0 \to \gamma_0' = \gamma_0[1-(\gamma_0/8\pi)]^{-1}$ and the exact breather spectrum is

$$16m_A \; \gamma_0'^{\;-1} \sin \;(n\gamma_0'/16), \qquad n = 1,2,\ldots,N$$

where $N + 1 > 8\pi \; \gamma_0'^{\;-1} \geq N$. For $\gamma_0' > 8\pi$ there are no breather levels.

Since the *observed* value of γ_0 for ^{3}He A is $\gamma_0 \sim 10^5$, $\gamma_0' \sim 10^5$ and the breather spectrum would be eliminated if the s-G governing the spin waves in the A-phase is indeed quantised. To investigate this experimentally would seem to mean looking for a continuous breather band between mass zero and mass $16m_A \; \gamma_0'^{-1}$. As we have shown, this exists in the c-number case but not in the quantised case. Again this means looking in the frequency region where the orbital waves can be excited.

The SIT problem is also quantised. The proper choice for the coupling constant in this case is apparently the $\gamma_0 \sim \tau_c^{-1} \; \omega_s^{-1}$ introduced already. We indicate why.

Equations (3.4a) of Ref. 13 can be taken in the one way going form

$$P_t = \varepsilon N, \; N_t = -\varepsilon P \;\; ,$$

$$\varepsilon_t + c\varepsilon_x = \alpha P \;\; , \tag{4.5}$$

(we use α for $c\alpha$ compared with Ref. 13). These equations are to be read as operator equations and the commutation relations for the matter operators are $[P,N] = 2i \; Q$, etc. Q and P are the in and out of phase components of the dipole operator envelope and N is the inversion; in each case these operators are scaled to operators for a single atom. The matter-field commutation relations are $[N,\varepsilon] = [P,\varepsilon] = 0$ where ε is the operator field envelope since all field operators commute with all matter operators. Equations (4.6) become the s-G (1.1) by the substitutions $\varepsilon = \phi_t$, $P = -\sin\phi$, $N = -\cos\phi$ followed by the change of independent variables $\tau_c^{-1}t \to t, \tau_c^{-1}(t-2xc^{-1}) \to x$. When P,N and ϕ are operators, the substitution $P = -\sin\phi$, $N = -\cos\phi$ is not compatible with $[P,N] \neq 0$. To reach (1.1) in operator form we are therefore obliged to break the matter commutation relations. From $[\varepsilon,N] = [\varepsilon,P] = 0$ we then find $[\phi_t, -\cos\phi] = [\phi_t, -\sin\phi] = 0$ from which we infer $[\phi,\phi_t] = 0$. This result is not consistent with the Poisson brackets associated with the c-number form of the s-G (1.1).

Equation (1.1) works well for sharp line resonant non-degenerate SIT. At the same time we note that the many atom system should be described[14] by collective operators whose commutation relations are not necessarily the commutation relations for the individual

atoms since it is necessary, for example, to average the atomic
operators over all possible site occupations of the atoms. It seems
natural therefore to impose for the collective system the condition
$[\phi,\phi_t] = i\,\gamma_0\,\delta(x-x')$ where by scaling (4.2) for the optical problem
one sees that $\gamma_0 = 2\tau_C^{-1}\omega_s^{-1}$. This commutation relation means of
course that $[\varepsilon,N] \neq 0$ and $[\varepsilon,P] \neq 0$ and the suggestion is that collective spontaneous emission breaks the collective matter-field com-
mutation relations. We have discussed elsewhere [15] possible effects
of the reaction fields governing spontaneous emission and radiative
level shifts for single atoms on the atomic matter field commutation
relations.

If this argument for collective quantisation has any substance
the very *small* value of γ_0 now means that the "masses" $8m\gamma_0^{-1}$ of the
2π pulses are large compared with the fundamental masses m (these
are small however since τ_C^{-1} is large but c is also large compared
with the spin wave problem: m $\sim 10^{-39}$ gm). Because γ_0 is small the
quantised breather spectrum is also close to

$$nmc^2 = n\hbar\,\tau_C^{-1} , \qquad n = 1,2,\ldots\ldots$$

for small values of n. The level spacings are $\tau_C^{-1} \sim 10^9$ Hz, roughly
the radiative level shifts of individual atoms. These spacings are
so small that we can probably correctly infer that the effect of the
incoherent contribution of collective spontaneous emission on the
coherence of the transmitted pulses is also small. Nevertheless,
it would be interesting to know whether optical breather pulses (0π
pulses) have a discrete excitation spectrum or exhibit other quantised features. Unfortunately no true 0π pulses have yet been observed in SIT experiments as far as we are aware.

References

1. Bullough, R.K., Paper IAEA-SMR-20/51 in *Interaction of Radiation
    with Condensed Matter*, Vol. 1 (International Atomic Energy
    Authority, Vienna, 1977).
2. Lamb, G.L. Rev. Mod. Phys. *43*, 99 (1971).
3. Eilbeck, J.C., Gibbon, J.D., Caudrey, P.J., and Bullough, R.K.,
    J. Phys. A: Math. Nucl. Gen. *6*, 1337 (1973).
4. Duckworth, S., Bullough, R.K., Caudrey, P.J., and Gibbon, J.D.,
    Phys. Lett. *57A*, 19 (1976).
5. Bullough, R.K., Caudrey, P.J., Gibbon, J.D., Duckworth, S.,
    Gibbs, H.M., Bölger, G., and Baede, L., Opt. Commun. *18*, 200
    (1976).
6. Duckworth, S., Ph.D. Thesis, University of Manchester, 1976.
7. For a review and striking work (particularly on the NMR)
    referenced therein, see A.J. Leggett, Rev. Mod. Phys. *47*, 331
    (1975).

8.   Maki, K., Phys. Rev. B *11*, 4264 (1975).
9.   Bullough, R.K. and Caudrey, P.J., Preprint, May 1977.
10. Cross, M., Dissertation, Corpus Christi College, Cambridge, 1974.
11. Bullough, R.K. and Dodd, R.K., *Proceedings of the International Conference on Synergetics* (Schloss Elmau, May 1977) (to be published by Springer, 1977).
12. Fadeev, L.D., Lectures at Princeton, May 1975 (Copies from Institute for Advanced Study, Princeton, N.J. 08540.)
13. Bullough, R.K., Saunders, R. and Feuillade, C., "Theory of Far Infra-Red Superfluorescence", this Volume, p. 263.
14. Saunders, R. and Bullough, R.K., *Proceedings of the Cooperative Effects Meeting* (Redstone Arsenal, Ala., Dec. 1976) (to be published by Plenum Press, New York, 1977).
15. Saunders, R., Bullough, R.K. and Ahmad, F., J. Phys. A: Math. Gen. *8*, 759 (1975).

HOW MANY PHYSICALLY SIGNIFICANT SOLUTIONS ARE THERE TO THE SELF-
INDUCED TRANSPARENCY EQUATIONS?*

R.C. Harney

M.I.T. Lincoln Laboratory, Lexington, Massachusetts

The interaction of a coherent electromagnetic wave with a
collection of two-level atoms is of major theoretical importance
in the study of the interaction of light with matter, not only
because analytical solutions can be obtained but also because
experiments can be devised which closely approximate this ideal
case. In the past there has been some controversy concerning the
interpretation of the analytical solutions to this problem [1].
Although previously unpublished, the controversy is real and the
problem is of sufficient fundamental importance to demand its
resolution. Such is the purpose of the work described here.

In an inhomogeneously broadened two-level medium described by
a particle density n_0 and electric dipole matrix element p, an
incident electric field of the form

$$E(z,t) = (\hbar/p) \, \mathcal{E} \, (z,t) \, \cos(kz-\omega t)$$

will induce a macroscopic polarization of the form

$$P(z,t) = n_0 p \left[<C(\Delta\omega,z,t)>_{\Delta\omega} \cos(kz-\omega t) + <S(\Delta\omega,z,t)>_{\Delta\omega} \sin(kz-\omega t) \right]$$

where $\Delta\omega = \omega-\omega_0$ is the difference between the incident electric field
frequency and the resonant frequency of the two-level system and
$< >_{\Delta\omega}$ denotes an average over the inhomogeneous line profile.

The mutual interaction of the electric field and the induced
polarization can be exactly described using Maxwell's equations and
the Schroedinger equation for the medium. In the rotating wave
approximation, we obtain the following coupled nonlinear differential
equations

781

$$\frac{d\mathcal{E}}{dt} = \alpha <S>_{\Delta\omega} \quad ,$$

$$\frac{\partial C}{\partial t} = -\Delta\omega S \quad ,$$

$$\frac{\partial S}{\partial t} = \Delta\omega C + \eta\mathcal{E} \; ,$$

$$\frac{\partial \eta}{\partial t} = -S\mathcal{E} \quad ,$$

where $\alpha = n_0\omega_0 p^2/\hbar$ is the inverse Beer's length of the medium, η is the population inversion density, and C, S, and η obey the normalization condition $C^2 + S^2 + \eta^2 = 1$. These four equations are commonly referred to as the self-induced transparency equations.

Mishkin assumed that the steady state solutions to these equations would depend only on the retarded time [2]

$$u = \frac{1}{\tau} (t - \frac{z}{V}) \; ,$$

where τ is a characteristic time and V is the pulse propagation velocity in the medium. With this assumption

$$\mathcal{E} = \mathcal{E}_o \; e(u) \quad ,$$

$$C = \tau\Delta\omega \; A(\Delta\omega) \; e(u) \quad ,$$

$$S = -A(\Delta\omega) \; \frac{de(u)}{du} \; ,$$

$$\eta - \eta_o = (1/2)\mathcal{E}_o \; A(\Delta\omega) \; e^2(u) \quad ,$$

where $A(\Delta\omega)$ is a frequency response function and η_o is the population inversion which would exist in the medium in the absence of the electric field. Substituting these results into the normalization condition yields

$$\left(\frac{de}{du}\right) = \left[\left(\frac{\mathcal{E}_o\tau}{2}\right)^2 e^4 - \left[(\tau\Delta\omega)^2 + \frac{\eta_o\mathcal{E}_o\tau}{A(\Delta\omega)} \right] e^2 + \frac{1-\eta_o^2}{A^2(\Delta\omega)} \right]^{1/2}$$

The Jacobi elliptic functions [3] are solutions to this nonlinear differential equation.

Closer examination indicates that only four of the twelve Jacobi elliptic functions have real values of $\mathcal{E}_o$ and τ and are

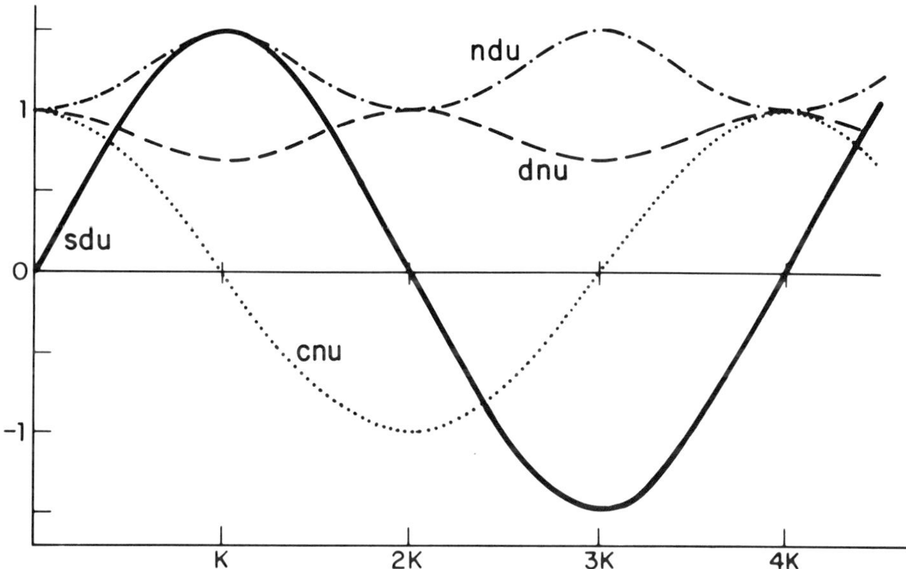

Fig. 1. The four real Jacobi elliptic function solutions to the
self-induced transparency equations.

therefore physically acceptable. The four acceptable solutions
(cnu, dnu, sdu, ndu) are shown in Fig. 1. The important physical
properties of these solutions have been derived by Mishkin [2] and
by the author [4] and are summarized in Table I.

The controversy concerning these solutions arises from the fact
that two of the solutions are related to the other two solutions by
a change of argument transformation

$$sd(u + K) = (1 - k^2)^{-1/2} \; cnu \; ,$$
$$nd(u + K) = (1 - k^2)^{-1/2} \; dnu \; ,$$

where k is the modulus of the solution (dependent on the physical
parameters of the system) and K is the elliptic integral of the
first kind. One contingent contends that the existence of this
transformation means that sdu is equivalent to cnu and that ndu is
equivalent to dnu [1]. Mishkin argues that the change of argument
is equivalent to a change of the physical situation and that all
four solutions are physically significant. Given this situation we
should fall back to the only true test of a scientific hypothesis,
experiment (or in this case, gedanken experiment). If an experiment
can be devised which can distinguish between two solutions, then
they are physically distinct. If no experiment can distinguish
between them, they are physically equivalent.

QUANTITY		cnu SOLUTION	dnu SOLUTION	ndu SOLUTION	sdu SOLUTION
Envelope of electric field	(E)	E_0 cnu	E_0 dnu	E_0 ndu	E_0 sdu
In-phase macroscopic polarization	(C)	$\tau \, \Delta\omega \, A\,(\Delta\omega)$ cnu	$\tau \, \Delta\omega \, A\,(\Delta\omega)$ dnu	$\tau \, \Delta\omega \, A\,(\Delta\omega)$ ndu	$\tau \, \Delta\omega \, A\,(\Delta\omega)$ sdu
Out-of-phase macroscopic polarization	(S)	$A\,(\Delta\omega)$ snu dnu	$k^2 A\,(\Delta\omega)$ snu cnu	$-k^2 A\,(\Delta\omega)$ sdu cdu	$-A\,(\Delta\omega)$ cdu ndu
Population inversion density	(η)	$\dfrac{A}{E_0\tau}\left[2\,\mathrm{dn}^2 u - 1 - (\tau\Delta\omega)^2\right]$	$\dfrac{A}{E_0\tau}\left[2\,\mathrm{dn}^2 u - 2 + k^2 - (\tau\Delta\omega)^2\right]$	$\dfrac{A}{E_0\tau}\left[k^2(1 - 2\,\mathrm{cd}^2 u) - (\tau\Delta\omega)^2\right]$	$\dfrac{A}{E_0\tau}\left[1 - 2k^2\,\mathrm{cd}^2 u - (\tau\Delta\omega)^2\right]$
Characteristic time relation	$(E_0\tau)$	$2k$	2	$2\,(1-k^2)^{1/2}$	$2k\,(1-k^2)^{1/2}$
Frequency response function	$(A\,(\Delta\omega))$	$\dfrac{2k}{\left[(1-(\tau\Delta\omega)^2)^2 + 4k^2\,(\tau\Delta\omega)^2\right]^{1/2}}$	$\dfrac{2}{\left[4\,(\tau\Delta\omega)^2 + (k^2-(\tau\Delta\omega)^2)^2\right]^{1/2}}$	$\dfrac{2\,(1-k^2)^{1/2}}{\left[(k^2-(\tau\Delta\omega)^2)^2 + 4\,(\tau\Delta\omega)^2\right]^{1/2}}$	$\dfrac{2k\,(1-k^2)^{1/2}}{\left[4k^2\,(\tau\Delta\omega)^2 + (1-(\tau\Delta\omega)^2)^2\right]^{1/2}}$
Period of oscillation	(T)	$4\tau\,K(k)$	$2\tau\,K(k)$	$2\tau\,K(k)$	$4\tau\,K(k)$
Integration constant	(η_0)	$\dfrac{1 - (\tau\Delta\omega)^2 - 2k^2}{\left[(1-(\tau\Delta\omega)^2)^2 + 4k^2\,(\tau\Delta\omega)^2\right]^{1/2}}$	$\dfrac{k^2 - 2 - (\tau\Delta\omega)^2}{\left[4\,(\tau\Delta\omega)^2 + (k^2-(\tau\Delta\omega)^2)^2\right]^{1/2}}$	$\dfrac{k^2 - 2 - (\tau\Delta\omega)^2}{\left[(k^2-(\tau\Delta\omega)^2)^2 + 4\,(\tau\Delta\omega)^2\right]^{1/2}}$	$\dfrac{1 - (\tau\Delta\omega)^2 - 2k^2}{\left[4k^2\,(\tau\Delta\omega)^2 + (1-(\tau\Delta\omega)^2)^2\right]^{1/2}}$
Ratio of light velocity to pulse velocity	$\left(\dfrac{C}{V}\right)$	$1 + \alpha\tau^2 \displaystyle\int_{-\infty}^{\infty} g\,(\Delta\omega)\left[(1-(\tau\Delta\omega)^2)^2 + 4k^2\,(\tau\Delta\omega)^2\right]^{-1/2} d\Delta\omega$	$1 + \alpha\tau^2 \displaystyle\int_{-\infty}^{\infty} g\,(\Delta\omega)\left[4\,(\tau\Delta\omega)^2 + (k^2-(\tau\Delta\omega)^2)^2\right]^{-1/2} d\Delta\omega$	$1 + \alpha\tau^2 \displaystyle\int_{-\infty}^{\infty} g\,(\Delta\omega)\left[(k^2-(\tau\Delta\omega)^2)^2 + 4\,(\tau\Delta\omega)^2\right]^{-1/2} d\Delta\omega$	$1 + \alpha\tau^2 \displaystyle\int_{-\infty}^{\infty} g\,(\Delta\omega)\left[4k^2\,(\tau\Delta\omega)^2 + (1-(\tau\Delta\omega)^2)^2\right]^{-1/2} d\Delta\omega$
Area per period		$0 \; \pi$	$2 \; \pi$	$2 \; \pi$	$0 \; \pi$
Area per half-period		$4\,\sin^{-1}(k)$	π	π	$4\,\sin^{-1}(k)$
Energy per period		$\dfrac{8 E_0}{k}\left[E\,(k) - (1-k^2)\,K\,(k)\right]$	$4 E_0\,E\,(k)$	$\dfrac{4 E_0}{(1-k^2)^{1/2}}\,E\,(k)$	$\dfrac{8 E_0}{k\,(1-k^2)^{1/2}}\left[E\,(k) - (1-k^2)\,K\,(k)\right]$

Table I.   Summary of results for the four real Jacobi elliptic
function solutions to the self-induced transparency equations.

Let us start by examining what parameters are subject to
experimental measurements. In principle the cnu and sdu solutions
can be distinguished from the dnu and ndu solutions by a time-
resolved intensity measurement as the former solutions have electric
field envelopes which pass through zero while the latter solutions
do not. The same measurement could also yield the period of oscillation T and the energy per period W carried by the pulse train. The
pulse propagation velocity V could be determined by observing the
propagation velocity of an infinitesimal perturbation to the field.
The initial population inversion density $\eta_0(\Delta\omega)$ which is characteristic of the medium could be measured by determining the gain or
attenuation of a weak frequency-tunable probe beam before the
electromagnetic wave was incident on the medium. The steady-state
population inversion density $\eta(\Delta\omega,t)$ could be determined in the
same manner after the steady state had been established. The
modulus k can be uniquely determined from the $\Delta\omega$-dependence of η_0
(see Table I). From a knowledge of T and k the characteristic time
can be unambiguously determined.

So far the only quantities which are not identical for the cnu
and sdu solutions or for the dnu and ndu solutions are the energy
per period, the intensity as a function of time, and the population
inversion as a function of time. Closer inspection of the expressions for these quantities indicates that $\mathcal{E}_0$ must be known in order
to be able to use them for distinguishing between the different
solutions. From a knowledge of τ and k, $\mathcal{E}_0$ can be predicted. A
different value of $\mathcal{E}_0$ is predicted for each solution, but the differences are such that using the predicted values leads to identical
values of the energy per period, etc. for the cnu and sdu solutions
and for the dnu and ndu solutions. Thus, to use $\mathcal{E}_0$ to discriminate
among the solutions it must be both measured and predicted and the
measurement compared with the predictions.

Unfortunately, $\mathcal{E}_0$ cannot be measured. The reason for this lies
in the fact that although $\mathcal{E}(t)$ can be inferred from a time-resolved
intensity measurement, $\mathcal{E}(u)$ cannot be precisely specified. The
problem that arises is one of where to place u = 0. Since a steady
state solution is one that has propagated for an infinite amount of
time through an infinite distance and is infinite in extent, an un-
arbitrary assignment of u = 0 implies a knowledge of one's absolute
position in space-time, which violates special relativity. The
same considerations prevent one from fixing the position of u = 0
within a single cycle (i.e. specifying the phase). Since the defi-
nition of $\mathcal{E}_0$ depends on knowing where u = 0 is, $\mathcal{E}_0$ cannot be measured.

One parameter which can be unambiguously determined is the maximum value of the electric field envelope $\mathcal{E}_{max}$. Comparing the four solutions we find

$$\mathcal{E}_{max}(cnu) = \mathcal{E}_{max}(dnu) = \mathcal{E}_o \; ,$$

and

$$\mathcal{E}_{max}(sdu) = \mathcal{E}_{max}(ndu) = \mathcal{E}_o / (1-k^2)^{1/2} \; .$$

Using the measurable quantity $\mathcal{E}_{max}$ we find

$$\mathcal{E}_{max}\tau = 2k \qquad \text{for both sdu and cnu} \; ,$$

$$\mathcal{E}_{max}\tau = 2 \qquad \text{for both ndu and dnu} \; .$$

Similar results obtain for all other quantities dependent on $\mathcal{E}_o$. In other words, we are unable to find any measurable quantity which can distinguish between cnu and sdu or between dnu and ndu.

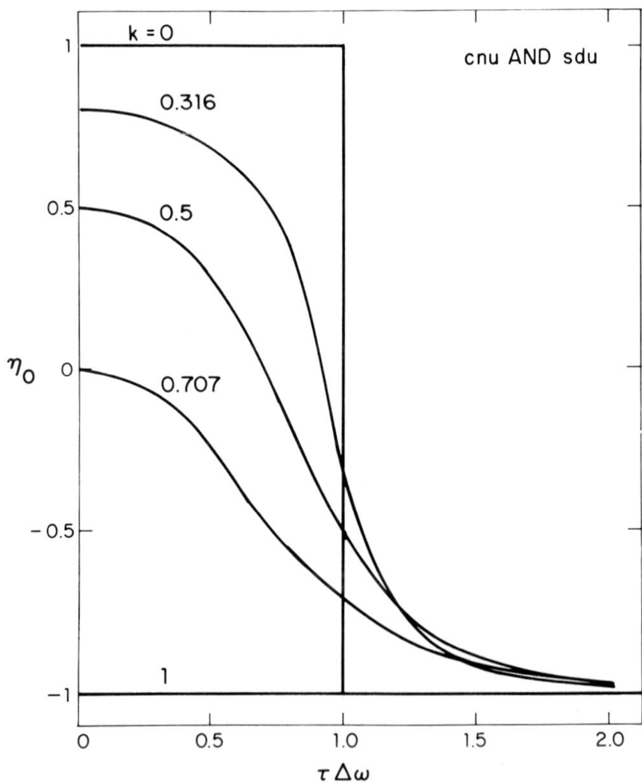

Fig. 2. The initial population inversion density η_0 as a function of $\tau\Delta\omega$ for cnu and sdu solutions. The curves correspond to values of $k = 1.0, .707, .5, .316,$ and 0.

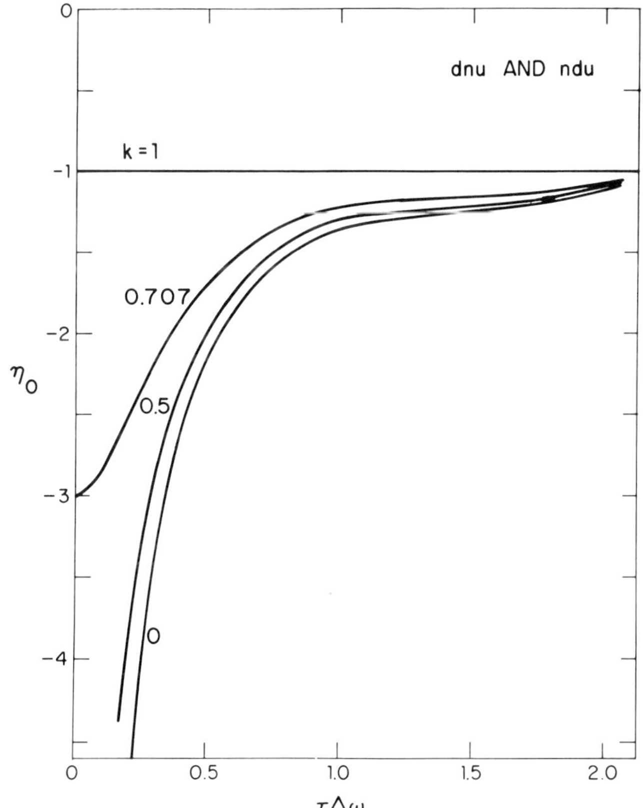

Fig. 3. The initial population inversion density η_0 as a function of $\tau\Delta\omega$ for the dnu and ndu solutions. The curves correspond to values of k = 1.0, .707, .5, and 0.

The preceding discussion sheds some light on the physical significance of the argument transformations. The transformation
$u \to u + K$ is just a displacement in space-time which would correspond to the addition of a constant phase factor. However, because
the initial phase cannot be specified, the physical situation is
unaltered. The modification of $\mathcal{E}_0$ by the factor $(1-k^2)^{-1/2}$ simply
expresses the fact that only $\mathcal{E}_{max}$ has meaning physically. Viewed
in this light, Mishkin's argument for four physically significant
solutions is seen to be in error.

Figures 2 and 3 show initial population inversion densities as
a function of $\tau\Delta\omega$ for several values of k required for the cnu (or
sdu) and dnu (or ndu) solutions to occur. It is immediately obvious
that η_0 assumes unphysical values ($\eta_0 < -1$) in the dnu (ndu) solution for every value of $k \neq 1$. At k = 1 dnu reduces to sechu just

as cnu does. However, k = 1 corresponds to $\eta_0(\Delta\omega) \equiv -1$ which implies
a temperature of absolute zero. This is also an unphysical situa-
tion. Consequently, the dnu and ndu solutions are not physically
significant.

In conclusion, we have examined the experimental distinguish-
ability of the four Jacobi elliptic function solutions to the self-
induced transparency equations. It was shown that sdu cannot be
experimentally distinguished from cnu and ndu cannot be experimentally distinguished from dnu. In addition, it was found that
the dnu and ndu solutions require unphysical values of the initial
population inversion density. We therefore conclude that cnu is the
only physically significant steady state solution to the self-
induced transparency equations.

*This work was sponsored by the Department of the Air Force.

References

1.   E.A. Mishkin, private communications and unpublished corre-
     spondence.
2.   E.A. Mishkin, Physica, *73*, 459 (1974).
3.   M. Abramowitz and I.A. Stegun, *Handbook of Mathematical
     Functions* (Dover Publications, New York, 1964) chapters 16
     and 17.
4.   R.C. Harney, submitted to Physica.

A SOLUTION OF THE BLOCH-MAXWELL'S EQUATIONS WITHOUT THE USE OF THE

"FACTORIZATION ASSUMPTION"

Ljubomir Matulic

St. John Fisher College, Rochester, New York

The dynamics of the interaction of an electromagnetic field
with an ensemble of two-level resonant atoms is usually described
by the coupled Maxwell-Schrödinger equations. When the electric
field is written as a carrier wave with slowly varying phase and
slowly varying amplitude, and if the medium through which this wave
propagates is inhomogeneously broadened, then the Maxwell-Schrödinger
equations for the steady state can be written as a set of five nonlinear coupled integro-differential equations in which the unknown
functions are the three components of the Bloch vector and the
envelope and phase of the field. A variety of solutions for this
system have been found [1], but all satisfy the so-called "factorization assumption", according to which the dependence of the absorption component of the Bloch vector on the detuning frequency is
factored out from its time dependence. In this note we investigate
the possibility of solving the above mentioned system of equations
without the benefit of the factorization assumption.

TRANSFORMATION OF EQUATIONS OF MOTION

We will adopt the notation of Matulic and Eberly [1] and write
our system of integro-differential equations in the following way:

$$\dot{u} = -(\gamma - \dot{\phi})\, v \,, \tag{1a}$$

$$\dot{v} = (\gamma - \dot{\phi})\, u + \kappa \varepsilon w \,, \tag{1b}$$

$$\dot{w} = -\,\kappa \varepsilon v \,, \tag{1c}$$

$$\dot{\varepsilon} = - \frac{\kappa}{m^2} \int_{-\infty}^{\infty} v\,(\xi,\gamma)\; g(\gamma)\; d\gamma \;, \tag{1d}$$

$$(\dot{\phi} + \frac{\Delta k}{\delta}) = \frac{\kappa}{m^2} \int_{-\infty}^{\infty} u(\xi,\gamma)\; g(\gamma)\; d\gamma \;. \tag{1e}$$

Many analytic solutions of the system (1) have been found [1,2], all satisfying the so-called "factorization assumption" [1,2]

$$v(\xi,\gamma) = F(\gamma)\; v_1(\xi) \;. \tag{2}$$

We have attempted to solve the system (1) without the simplifying condition of Eq. (2), by assuming that the components of the Bloch vector, u, v and w can be expanded in power series of γ/ω with time-varying coefficients:

$$u(\xi,\gamma) = \sum_{n=0}^{\infty} x_n\,(\xi)\,(\gamma/\omega)^n \;, \tag{3a}$$

$$v(\xi,\gamma) = \sum_{n=0}^{\infty} y_n\,(\xi)\,(\gamma/\omega)^n \;, \tag{3b}$$

$$w(\xi,\gamma) = \sum_{n=0}^{\infty} z_n\,(\xi)\,(\gamma/\omega)^n \;. \tag{3c}$$

We will solve the system (1) for the following initial conditions:

$$u(-\infty,\gamma) = v(-\infty,\gamma) = 0 \;, \tag{4a}$$

$$w(-\infty,\gamma) = -1 \;, \tag{4b}$$

$$\varepsilon(-\infty) = \dot{\phi}(-\infty) = 0 \;. \tag{4c}$$

Substituting Eqs. (3a) and (3b) into Eq. (1a) we get

$$\sum_{n=0}^{\infty} (\dot{x}_n - \dot{\phi} y_n)\,(\gamma/\omega)^n + \sum_{n=0}^{\infty} \omega y_n\,(\gamma/\omega)^{n+1} = 0 \;.$$

By changing the summation index in the second sum from n to n-1, we can write the above expression as

$$\dot{x}_o - \dot{\phi} y_o + \sum_{n=1}^{\infty} (\dot{x}_n - \dot{\phi} y_n + \omega y_{n-1})\,(\gamma/\omega)^n = 0 \;.$$

Since this is an identity in γ/ω, we deduce that

$$\dot{x}_n - \dot{\phi}y_n + \omega y_{n-1} = 0 , \qquad n = 0, 1, 2, \ldots$$

where we must define $y_{-1} = 0$.

In a similar way we can manipulate the other two Bloch equations (1b) and (1c) in order to arrive at

$$\dot{y}_n + \dot{\phi}x_n - \kappa\varepsilon z_n - \omega x_{n-1} = 0 , \qquad n = 0, 1, 2, \ldots$$

$$\dot{z}_n + \kappa\varepsilon y_n = 0 . \qquad\qquad n = 0, 1, 2, \ldots$$

We now substitute Eq. (3b) into the first Maxwell equation (1d) and obtain

$$\dot{\varepsilon} = - \frac{\kappa}{m^2} \sum_{n=0}^{\infty} I_n y_n ,$$

where

$$I_n = \int_{-\infty}^{\infty} (\gamma/\omega)^n g(\gamma) d\gamma . \tag{5}$$

We see that $\omega^n I_n$ are the moments of the atomic line-shape function $g(\gamma)$. A similar result is obtained when Eq. (3c) is substituted into the second Maxwell equation (1e):

$$(\dot{\phi} + \frac{\Delta k}{\delta}) = \frac{\kappa}{m^2} \sum_{n=0}^{\infty} I_n x_n .$$

Therefore we have transformed the system of five coupled *integro-differential* equations (1) for u, v, w, ε and ϕ into a set of *infinitely many coupled differential* equations for the time dependent coefficients x_n, y_n and z_n; and for the envelope ε and the chirp $\dot{\phi}$:

$$\dot{x}_n = \dot{\phi}y_n - \omega y_{n-1} , \qquad n = 0, 1, 2, \ldots \tag{6a}$$

$$\dot{y}_n = -\dot{\phi}x_n + \kappa\varepsilon z_n + \omega x_{n-1} , \qquad n = 0, 1, 2, \ldots \tag{6b}$$

$$\dot{z}_n = -\kappa\varepsilon y_n , \qquad n = 0, 1, 2, \ldots \tag{6c}$$

$$\dot{\varepsilon} = - \frac{\kappa}{m^2} \sum_{n=0}^{\infty} I_n y_n , \tag{6d}$$

$$(\dot{\phi} + \Delta k/\delta) = \frac{\kappa}{m^2} \sum_{n=0}^{\infty} I_n x_n . \tag{6e}$$

We must notice that the coefficients I_n given by Eq. (5) can become large, and even diverge for large n. We have always $I_0 = 1$, but the value of the remaining I_n's depends on $g(\gamma)$. For the pure Lorentzian, $I_n \to \infty$ for $n > 1$. However, the coefficients of x_n and y_n must be such that the sums in equations (5d) and (5e) remain finite. This is precisely what happens in the case of the hyperbolic secant solution.

It is, of course, doubtful if the system of differential equations (6a)-(6e) is any simpler than the original system (1). It seems, however, that system (6) is more amenable to successive approximations. One such scheme of approximation would be to solve the system in successive stages. In the first stage one could set n = 0 in the Bloch equations (6a)-(6c) and retain only the first terms on the right of the Maxwell equations (6d) and (6e). Next, one would put n = 1 in the Bloch equations and retain the two first terms in the Maxwell equations and then continue in this fashion to increase the number of Bloch equations and the terms in the Maxwell equations. We were able to solve directly only the case n = 0. It is also possible to show, as we will do later, that system (6) has a particular analytic solution which leads to the well known McCall-Hahn solution of the system (1).

However, before looking at these particular cases let us derive some general relations.

SOME GENERAL RESULTS

Multiplying (6a) by x_n, (6b) by y_n and (6c) by z_n, then adding the resulting equations together and integrating the sum with respect to time, we obtain

$$x_n^2 + y_n^2 + z_n^2 = - 2\omega \int_{-\infty}^{t} (x_n y_{n-1} - x_{n-1} y_n) \, dt' , \qquad (7)$$

for all n.

Combining (5c) and (5d) and integrating again with respect to time we get

$$\varepsilon^2 = \frac{2}{m^2} \sum_{n=0}^{\infty} I_n z_n + \bar{\varepsilon} ,$$

where $\bar{\varepsilon}$ is an integration constant that is determined by initial conditions, which yield

$$\varepsilon^2 = \frac{2}{m^2} \left(\sum_{n=0}^{\infty} I_n z_n + 1 \right) . \qquad (8)$$

This is an important first integral of the system (6).

Differentiation of Eq. (6e) yields

$$\varepsilon\ddot{\phi} + \dot{\varepsilon}\dot{\phi} + \frac{\Delta k}{\delta}\dot{\varepsilon} = \frac{\kappa}{m^2} \sum_{n=0}^{\infty} I_n \dot{x}_n \quad .$$

Using Eq. (6a) we eliminate x_n; then with the help of Eq. (6d) we can write:

$$\varepsilon\ddot{\phi} + 2\dot{\varepsilon}\dot{\phi} = - \frac{\Delta k}{\delta}\dot{\varepsilon} - \frac{\kappa\omega}{m^2} \sum_{n=0}^{\infty} I_n y_{n-1} \quad . \tag{9}$$

Equation (9) is a first-order differential equation in $\dot{\phi}$ whose integrating factor is ε^2. Straightforward integration yields:

$$\varepsilon^2\dot{\phi} = - \frac{\Delta k}{\delta}\varepsilon^2 + \frac{\omega}{m^2} \sum_{n=0}^{\infty} I_n z_{n-1} + \overline{\phi}$$

where $\overline{\phi}$ is an integration constant, which must be chosen so as to satisfy the initial conditions. As it turns out $\overline{\phi} = I_1/m^2$, so that the solution for $\dot{\phi}$ can be written in the following way

$$\dot{\phi} = - \frac{\Delta k}{2\delta} + \frac{\omega I_1}{m^2} \frac{z_o+1}{\varepsilon^2} + \frac{\omega}{m^2} \sum_{n=2}^{\infty} I_n \frac{z_{n-1}}{\varepsilon^2} \quad . \tag{10}$$

From the original Bloch equations (1a), (1b) and (1c) we derive

$$u^2 + v^2 + w^2 = 1 \quad . \tag{11}$$

Using the expansion (3) we can write Eq. (11) as:

$$[x_o + x_1(\gamma/\omega) + x_2(\gamma/\omega)^2 + ---]^2 +$$

$$[y_o + y_1(\gamma/\omega) + y_2(\gamma/\omega)^2 + ---]^2 +$$

$$[z_o + z_1(\gamma/\omega) + z_2(\gamma/\omega)^2 + ---]^2 = 1 \quad .$$

Collecting the coefficients of the successive powers of γ/ω we may also write:

$$\sum_{n=0}^{\infty} \left\{ \sum_{i=0}^{n} (x_i x_{n-1} + y_i y_{n-1} + z_i z_{n-1}) \right\} (\gamma/\omega)^n = 1 \quad .$$

Since this is an identity in (γ/ω) we must have

$$\sum_{i=0}^{n} (x_i x_{n-1} + y_i y_{n-i} + z_i z_{n-i}) = \delta_{on} \quad \text{for} \quad n = 0, 1, 2, \ldots \tag{12}$$

ZEROTH ORDER

We will solve the following system

$$\dot{x}_o = \dot{\phi} y_o \; , \tag{13a}$$

$$\dot{y}_o = -\dot{\phi} x_o + \kappa \epsilon z_o \; , \tag{13b}$$

$$\dot{z}_o = -\kappa \epsilon y_o \; , \tag{13c}$$

$$\dot{\epsilon} = -\frac{\kappa}{m^2} y_o \; , \tag{13d}$$

$$\dot{\phi} = -\frac{\Delta k}{\delta} + \frac{\kappa}{m^2} \frac{x_o}{\epsilon} \; . \tag{13e}$$

Conditions (4a) and (4b) imply through Eqs. (3) that

$$x_n(-\infty) = y_n(-\infty) = 0 \; , \tag{14a}$$

$$z_o(-\infty) = -1 \; , \tag{14b}$$

$$z_n(-\infty) = 0 \; , \quad n > 1 \; . \tag{14c}$$

The Chirp

An immediate consequence of the conditions (14) and equations (13a), (13b) and (13c) is that

$$x_o^2 + y_o^2 + z_o^2 = 1 \; , \tag{15}$$

which has already been established in Eq. (12).

The equations (13c) and (13d) lead to

$$\epsilon^2 = \frac{2}{m^2} (c_o + 1) \; , \tag{16}$$

which agrees with relation (8) when only the first term in the sum is taken.

In order to obtain an expression for $\dot{\phi}$ we multiply Eq. (13e) by ϵ, differentiate once and use Eqs. (13b) and (13d) in order to get

$$\epsilon \ddot{\phi} + 2\dot{\epsilon}\dot{\phi} = -\frac{\Delta k}{\delta} \dot{\epsilon} \; .$$

Integration of this equation yields at once

$$\dot{\phi} = - \frac{1}{2} \frac{\Delta k}{\delta} \quad ,$$

where the integration constant has been determined in accordance
with Eq. (4c). Using the same argument as in Ref. 1 following the
Eq. (3.4) we set this constant chirp equal to zero and obtain
$\Delta k = 0$, so that there is no dispersion due to the two level atoms
in zeroth order.

The Envelope

Differentiating (13d) and using Eqs. (13b) and (16) we obtain
a differential equation for ε:

$$\ddot{\varepsilon} = - \frac{\kappa^2}{2} \varepsilon^3 + \frac{\kappa^2}{m^2} \varepsilon \quad .$$

The integration of this equation is straightforward and leads to

$$\varepsilon = \frac{2}{\kappa\tau} \ \text{sech} \ \frac{\xi}{\tau} \quad , \tag{17}$$

where $\tau = m/\kappa$ is the pulse width.

Now Eq. (13a) gives $\dot{x}_0 = 0$, so that $x_0 = \text{const} = 0$, because of the
initial conditions. Equation (13d) gives

$$y_0 = 2 \ \text{sech} \ \frac{\xi}{\tau} \ \tanh \ \frac{\xi}{\tau} \quad , \tag{18}$$

while Eq. (16) yields

$$z_0 = -1 + 2 \ \text{sech}^2 \ \frac{\xi}{\tau} \quad . \tag{19}$$

The expression for pulse velocity turns out to be

$$V = (1/c_0 + \pi n \, \hbar \omega \kappa^2 \tau^2 / c \, \eta_0)^{-1} \quad , \tag{20}$$

as expected.

We notice at once that the solution to the zeroth order, Eqs.
(17)-(20), coincides with the "on resonance" solution of McCall
and Hahn. This must be so because Eqs. (13) are McCall-Hahn equations for $\gamma = 0$.

FACTORABLE SOLUTION

Up to this point we were unable to find any analytic solution of the system (5) when $\dot{\phi} \neq 0$. However, when $\dot{\phi} = 0$ it is a tedious, but straightforward, matter to verify that the system (5) admits the following solution:

$$x_{2m}(\xi) = 0 , \qquad\qquad m = 0, 1, 2, \ldots$$

$$x_{2m+1}(\xi) = (-1)^m (2)(\omega\tau)^{2m+1} \operatorname{sech} \frac{\xi}{\tau} , \qquad\qquad m = 0, 1, 2, \ldots$$

$$y_{2m}(\xi) = (-1)^m (2)(\omega\tau)^{2m} \operatorname{sech} \frac{\xi}{\tau} \tanh \frac{\xi}{\tau} , \qquad\qquad m = 0, 1, 2, \ldots$$

$$y_{2m+1}(\xi) = 0 , \qquad\qquad m = 0, 1, 2, \ldots$$

$$z_0(\xi) = -1 + 2 \operatorname{sech}^2 \frac{\xi}{\tau} ,$$

$$z_{2m}(\xi) = (-1)^m (2)(\omega\tau)^{2m} \operatorname{sech}^2 \frac{\xi}{\tau} , \qquad\qquad m = 1, 2, 3, \ldots$$

$$z_{2m+1}(\xi) = 0 , \qquad\qquad m = 0, 1, 2, \ldots$$

$$\epsilon(\xi) = \frac{2}{\kappa\tau} \operatorname{sech} \frac{\xi}{\tau} . \tag{21}$$

The dispersion Δk is given by

$$\Delta k = \delta \frac{\displaystyle\sum_{k=0}^{\infty} \omega^k \tau^{k-1} I_k}{\displaystyle\sum_{k=0}^{\infty} \omega^k \tau^k I_k} . \tag{22}$$

If the expressions for $x(\xi)$ in Eq. (21) are substituted into Eq. (3a) we obtain

$$\begin{aligned}
u(\xi,\gamma) &= \sum_{m=0}^{\infty} x_{2m+1}(\xi) \frac{\gamma^{2m+1}}{\omega^{2m+1}} \\
&= 2[\gamma\tau - (\gamma\tau)^3 + (\gamma\tau)^5 - \ldots] \operatorname{sech} \frac{\xi}{\tau} \\
&= \frac{2\gamma\tau}{1+(\gamma\tau)^2} \operatorname{sech} \frac{\xi}{\tau} .
\end{aligned} \tag{23}$$

Similarly we get:

$$v(\xi,\gamma) = \frac{2}{1+(\gamma\tau)^2} \text{ sech } \frac{\xi}{\tau} \text{ tanh } \frac{\xi}{\tau} , \qquad (24)$$

$$w(\xi,\gamma) = -1 + \frac{2}{1+(\gamma\tau)^2} \text{ sech}^2 \frac{\xi}{\tau} . \qquad (25)$$

Equations (23)-(25) are the factorable solutions of McCall and Hahn.

When the expressions for I_k are substituted into Eq. (22), we obtain

$$\Delta k = \delta \frac{\int_{-\infty}^{\infty} \gamma F(\gamma) g(\gamma) d\gamma}{\int_{-\infty}^{\infty} F(\gamma) g(\gamma) d\gamma} = \delta\overline{\gamma} ,$$

where $F(\gamma) = 1/(1 + \gamma^2\tau^2)$, in agreement with Eq. (3.5) of Ref. 1.

In conclusion, we have shown that the expansion (3) leads to the old factorable solution of McCall and Hahn for the system (1). In other words, the expansion (3) is equivalent to the factorization assumption (2) provided the solution of the system (6) is unique.

Acknowledgment

I would like to acknowledge the benefit of many discussions with Professor J.H. Eberly and Marek Konopnicki. I am also thankful to Darel Johnson for his helpful assistance during the early stages of the preparation of this paper.

References

1.   L. Matulic and J.H. Eberly, Phys. Rev. A*6*, 822 (1972) and ibid. *6*, 1258E, (1972).
2.   V. Nemec and L. Matulic, Opt. Commun. *13*, 380 (1975).
3.   S.L. McCall and E.L. Hahn, Phys. Rev. Letters *18*, 908 (1967); Phys. Rev. *183*, 457 (1969).

A LINK BETWEEN QUANTUM- AND SEMI-CLASSICAL THEORIES FOR SUPER-

RADIANCE VIA AN UNRESTRICTED MARKOVIAN MODEL

E. Ressayre and A. Tallet

Université Paris-Sud, Orsay, France

1. UNRESTRICTED MARKOVIAN MODEL

For the past few years many superfluorescence experiments
have been realized in the infrared (or near infrared) frequency
range for pencil-shaped samples [1-5]. In the two opposite directions the experiments exhibit the emission of an intense radiation
after a considerable delay with respect to the characteristic
superradiant time τ_R (about a hundred τ_R). These two radiations
consist, in general, of several successive pulses with decreasing
intensities, or ringings [1,3] (the single-pulse superfluorescence
in Cesium [5] will be discussed later).

A semi-classical description of superradiance [1,6,7] accounts
well for the observed delays and ringings, in spite of theoretical
reservations which can be raised (the problem of the beginning of
the cooperative emission, the problem of backscattering, etc.),
whereas the classical quantum one(two)-mode model [8] (mean-field
theory) fails generally [9] to interpret the superfluorescence
experiments. This breakdown of the mean-field theory was often
imputed to the Markov approximation and it was therefore concluded
that superfluorescence is not a Markovian process [7]. In fact
we will show here that the Markov approximation is not guilty of
the failure of the one-mode model. The one-mode model breaks down
basically because of its very conception: in the mean-field theory
any atom (or molecule) located at Z in the sample ($0 \leq Z \leq L$)
interacts with a forward electromagnetic field and a backward
electromagnetic field, each of these being the sum over the full
length of the microscopic polarizations. In particular, this means
that any atom located at Z "sees" the forward (backward) field
emitted by any atom located not only at $Z' \leq Z$ ($Z' \geq Z$) but also

at $Z' \geq Z$ ($Z' \leq Z$). Though such a model is convenient for atoms
located in a one-mode cavity, it appears to be unlikely for open
systems. As we will see later in detail, the Markov approximation,
when it is properly performed, removes the apparent lack of
causality: any atom located at Z will now see only the forward
(backward) field emitted by the atoms located at $Z' \leq Z$ ($Z' \geq Z$).
This implies the failure of the one (two)-mode model. A proper
treatment of the Markov approximation consists of adding the
damping term (the only one which is taken into consideration in
mean-field theory [8]), to the frequency shift term; this is
consistent with the need of causality for open superradiant systems.
And we will show that such an "unrestricted" Markovian model
accounts well for the observed signals with ringings, convenient
delays and widths as long as the retardation effects are negligible,
i.e., as long as the time L/c is not much greater than τ_R [7].
Thus, superfluorescence is basically a Markovian process.

When the discrepancy between the one-mode model and the
unrestricted Markovian model is really understood, it then becomes
obvious that this latter will lead to very large delays and ringings
since it individualizes the atoms of the sample. It follows that
much of the work concerning cooperative effects will have to be
reexamined in the light of the present result.

We briefly recall the hamiltonian of the full system atoms +
field.

$$H = \sum_{k,\sigma} \omega_k \, a^+_{k,\sigma} \, a_{k,\sigma} + \omega_0 \sum_j R_{3j} + \sum_{k,j,\sigma} \left[\alpha_{k,j} \, R^+_j \, a_{k,\sigma} + h.c. \right] \; ;$$

$$a_{kj} = \exp(i\underline{k} \cdot \underline{r}_j) \; \underline{\mu} \cdot \underline{\varepsilon}_k \; \sqrt{(2\Pi\hbar)/L^3\omega_k)} \quad . \tag{1}$$

R_{3j}, $R^{\pm}_j$ are the atomic operators for the two-level atom j, located
at $\underline{r}_j$ in the sample, (they are equivalent to the Pauli matrices),
and $\omega_0 = (2\Pi c)/\lambda$ and μ are the transition frequency and the dipole
moment of atom j, respectively. $a^{\pm}_{k,\sigma}$ are the annihilation and
creation operators for photons with wave vector $\underline{k}$ and polarization
$\underline{\sigma}$.

The Heisenberg equations for the atomic operators R_{3j}, $\tilde{R}^{\pm}_j$
expressed as $\exp(\pm i\omega_0 t) \, R^{\pm}_j$ are

$$\frac{d}{dt} R_{3j} = -\{\tilde{R}^+_j(t) \, \tilde{A}^+(t,\underline{x}_j) + h.c.\} \quad ,$$

$$\frac{d}{dt} \tilde{R}^-_j = + 2 \, R_{3j} \, \tilde{A}^+(\underline{x}_j,t) \quad . \tag{2}$$

$\tilde{A}^+(\underline{x}_j,t)$ stands for the scalar product of the positive frequency part of the potential vector operator expressed in the interaction picture with the atomic dipole moment $\underline{\mu}$. When it is expressed in terms of atomic operators, it is

$$\tilde{A}^+(\underline{x}_j,t) = \frac{\omega_o^2}{(2\Pi)^2\hbar} \sum_\ell \int \frac{d^3k}{c|\underline{k}|} \left[|\underline{\mu}|^2 - \frac{(\underline{\mu}\cdot\underline{k})^2}{|\underline{k}|^2}\right] \exp\left(i\underline{k}(\underline{x}_j-\underline{x}_\ell)\right)$$

$$\times \int_o^t dt' \exp\left(i(\omega_o-\omega)(t-t')\right) R_\ell^-(t') + \tilde{A}_o^+(t) , \qquad (3)$$

where $A_o^+(t)$ is the free potential-vector operator, the mean value of which will be taken equal to zero since we are dealing with spontaneous emission (we can verify that $A^+(\underline{x}_j,t)$ obeys the second-order Maxwell equation with sources).

For the treatment of the expression (3) we only take into account the electromagnetic field modes k lying inside two small opposite solid angles $\Delta\Omega_{\vec{oz}}$ and $\Delta\Omega_{\underset{oz}{\leftarrow}}$ about the oz axis. An estimate of the solid angles was given in a previous paper [10] for the resonant modes. This was deduced from the study of the real part $\Gamma_{j\ell}$ of the coupling between the atoms j and ℓ via the electromagnetic field, i.e. without taking into account the frequency shift term. The introduction of this latter term would not modify, at least qualitatively, our conclusions about the directivity of the emission and the field inhomogeneities. In this paper we will consider only the case for Fresnel numbers of order unity ($\mathcal{N} \simeq 1$) for which the effects of the field inhomogeneities are minimum. For $\mathcal{N} \simeq 1$, these effects can be considered as a perturbation with respect to the effect of the frequency shift terms and they will not be considered here.

With $\Delta\Omega_{\underset{oz}{\leftrightarrow}} = 2\Pi\Delta$, where Δ is the length of the integration interval of the direction cosines ($\Delta \simeq \lambda^2/S$ for $\mathcal{N} \simeq 1$, where S is the section), Eq. (3) becomes

$$\tilde{A}^+(\underline{x}_j,t) \simeq \frac{\mu^2\omega_o^2}{2\Pi c\hbar} \sum_{\ell\neq j} \int_o^t dt' \int_o^\infty dk\ k\ \exp\left(i(\omega_o-\omega)(t-t')\right) R_\ell^-(t')$$

$$\times \left[\int_{1-\Delta}^1 du + \int_{-1}^{-1+\Delta} du\right] J_o\left[k\sqrt{(1-u^2)}\ X_{ij}\right] \exp\left(iku(z_j-z_\ell)\right)$$

$$+ \frac{1}{2} \Gamma R_j^-(t) , \qquad (4)$$

where X_{ij} is the distance between the projections of $\underline{x_i}$ and $\underline{x_j}$ on a plane perpendicular to oz. In Eq. (4) we have neglected the dependence of Δ with respect to the wave vector $\underline{k}$ for $\underline{k}$ around $\underline{k_o}$. Then if we neglect the field inhomogeneities, Eq. (4) becomes approximately

$$\tilde{A}^+(x_j,t) \simeq \frac{\mu^2\omega_o^2\Delta}{2\Pi c^3\hbar} \sum_{\ell\neq j} \int_o^t dt' \, R_\ell^-(t')$$

$$\times \exp\left(ik_o(z_j-z_\ell)\right) \int_{-\infty}^{+\infty} dx \, \exp\left(ix(t-t'-z_j/c+z_\ell/c)\right) + \exp\left(-ik_o(z_j-z_\ell)\right)$$

$$\times \int_{-\infty}^{+\infty} dx \, \exp\left(-ix(t-t'+z_j/c-z_\ell/c)\right) + \frac{1}{2}\Gamma \, R_j^-(t) \tag{5}$$

$$= \frac{\mu^2\omega_o^2\Delta}{c^3\hbar} \sum_{\ell\neq j} \int_o^t dt'\left[\exp\left(ik_o(z_j-z_\ell)\right) \, \delta(t-t'-z_j/c+z_\ell/c)\right.$$

$$+ \exp\left(-ik_o(z_j-z_\ell)\right) \, \delta(t-t'+z_j/c-z_\ell/c)\left] R_\ell^-(t') \right. + \frac{1}{2}\Gamma \, R_j^-(t)$$

$$= \frac{\mu^2\omega_o^2\Delta}{c^3\hbar} \sum_{\ell\neq j} \left[\exp\left(ik_o(z_j-z_\ell)\right) \, U(z_j-z_\ell) \, U(t-z_j/c+z_\ell/c)\right.$$

$$\times R_\ell^-(t-z_j/c+z_\ell/c) + \exp\left(-ik_o(z_j-z_\ell)\right)U(z_\ell-z_j) \, U(t+z_j/c-z_\ell/c)$$

$$\left.\times R_\ell^-(t+z_j/c-z_\ell/c)\right] + \frac{1}{2}\Gamma \, R_j^-(t) \ . \tag{6}$$

(U(x) is the Heaviside function.) Then the coherent part of the radiation field at location z_j in the sample and at time t is the sum of a forward field, which is the superposition of the microscopic fields radiated by all the atoms located at $z_\ell \leq z_j$ and at retarded times $t-z_j/c+z_\ell/c$, plus a backward field which is the superposition of the microscopic fields radiated by the atoms located at $z_\ell \leq z_j$ and at retarded times $t+z_j/c-z_\ell/c$. Taking the Markovian limit of the expression (6) amounts to neglecting the retardation effects. Under these conditions we get

$$\tilde{A}^{+}(x_{j},t) = \Gamma\Delta \sum_{\ell\neq j} \left[\exp\left(ik_{o}(z_{j}-z_{\ell})\right) U(z_{j}-z_{\ell})\right.$$

$$\left. + \exp\left(-ik_{o}(z_{j}-z_{\ell})\right)U(z_{\ell}-z_{j})\right] R^{-}(t) + \frac{1}{2} \Gamma R_{\ell}^{-}(t). \qquad (7)$$

This is different from the standard $\tilde{A}^{+}(x_j,t)$ in mean-field theory, given by

$$\frac{1}{2} \Gamma\Delta \sum_{\ell\neq j} \left[\exp\left(ik_{o}(z_{j}-z_{\ell})\right) + \exp\left(-ik_{o}(z_{j}-z_{\ell})\right)\right] R_{\ell}^{-}(t) + \frac{1}{2} \Gamma R_{j}^{-}(t) .$$

$$(8)$$

Expression (7) can also be derived from Eq. (5) by taking the Markovian limit directly, i.e.

$$\int_{o}^{t} dt' \ \exp\left(i(\omega_{o}-\omega)(t-t')\right) R_{\ell}^{-}(t') \rightarrow R_{\ell}^{-}(t)\left[\Pi\delta(\omega-\omega_{o}) - \frac{i\mathcal{P}}{\omega-\omega_{o}}\right] \quad (9)$$

Then if we substitute the full expression (9) in Eq. (5) we easily find Eq. (7) again after using

$$-i\mathcal{P} \int dx \ \frac{\exp\left(-ix(z_{j}-z_{\ell})\right)}{x} = \Pi \ \text{sign} \ (z_{j}-z_{\ell}) \quad , \qquad (10)$$

while if we neglect the term $i\mathcal{P}/(\omega-\omega_{o})$, or frequency shift term, we get the classic expression (8) of the one (two)-mode model. (This cooperative frequency shift has been considered previously, especially by Banfi and Bonifacio [11], but it has always been treated separately.)

Finally, we put the unrestricted Markovian limit in the Heisenberg equations (2) and we take as usual the continuous limit

$$\sum_{j} \rightarrow \frac{N}{V} \int d^{3}x \quad ,$$

(where N and V are the total number of radiating atoms and the volume of the sample, respectively). Then the operators $W(z,t)$ and $R^{\pm}(z,t)$, defined by

$$\frac{N}{V} W(z,t) = \frac{1}{S} \int dS \sum_{j} \delta(\underline{x}-\underline{x}_{j}) R_{3j}$$

and

$$\frac{N}{V} R^{\pm}(z,t) = \frac{1}{S} \int dS \sum_{j} \delta(\underline{x}-\underline{x}_{j}) R_{j}^{\pm} \quad , \qquad (11)$$

obey the following equations of motion

$$\frac{d}{dt} W(z,t) = -\Gamma\left[W(z,t) + \frac{1}{2}\right] - \frac{1}{2\tau_R L}\left[\int_0^z dz'\ \exp\left(ik_o(z-z')\right)\right.$$

$$\times R^+(z,t)\ R^-(z',t) + \int_z^L dz'\ \exp\left(-ik_o(z-z')\right)$$

$$\left.\times R^+(z,t)\ R^-(z',t) + h.c.\right]$$

and

$$\frac{d}{dt} R^+(z,t) = -\frac{1}{2}\Gamma\ R^{\pm}(z,t) + \frac{1}{\tau_R L}\left[\int_0^z dz'\ \exp\left(-ik_o(z-z')\right)\right.$$

$$\left.\times R^+(z',t) + \int_z^L dz'\ \exp\left(ik_o(z-z')\right)R^+(z',t)\right] W(z,t)\ .$$

$$(12)$$

τ_R is the characteristic time for superradiance. It is defined, here, as $(2N\Gamma\Delta)^{-1}$.

The equations of motion for the expectation values $<W(z,t)>$ and $<R^+(z,t)\ R^-(z',t)>$ were solved by neglecting quantum correlations beyond the second-order and by neglecting the coupling between the forward and the backward fields in the equation of motion for any $<R^+(z,t)\ R^-(z',t)>$, i.e. by neglecting terms such as

$$\int_0 dz''\ \exp\left(ik_o(2z-z''-z')\right)\ <W(z,t)><R^+(z'',t)\ R^-(z',t)>\ .$$

Characteristic solutions for the forward (backward) radiation field at the exit of the sample $E_f(L,t)$ $[E_b(0,t)]$ are given in Fig. 1 as a function of the quantity $\Gamma\tau_R$, where the times Γ^{-1} and τ_R characterize the incoherent decay and the superradiant decay, respectively.

Figure 1 displays the variation of

$$|E_f(L,t)|^2 = |E_b(0,t)|^2 \propto \int_0^L dz'\ \int_0^L dz''\ \exp\left(\pm ik_o(z'-z'')\right)$$

$$\times <R^+(z')\ R^-(z'')>\qquad\qquad (13)$$

with respect to time for three different values of $\Gamma\tau_R$. Figure (1a) corresponds approximately to the experimental situation in HF[1] and Na[2], with a very small $\Gamma\tau_R$ of magnitude of order 10^{-6}; time is in

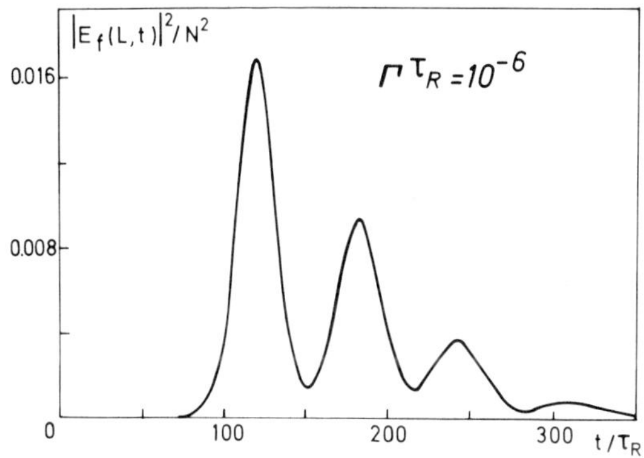

(a)

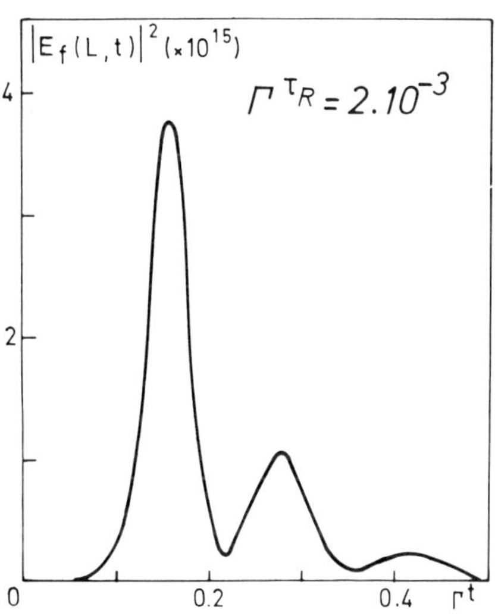

(b)

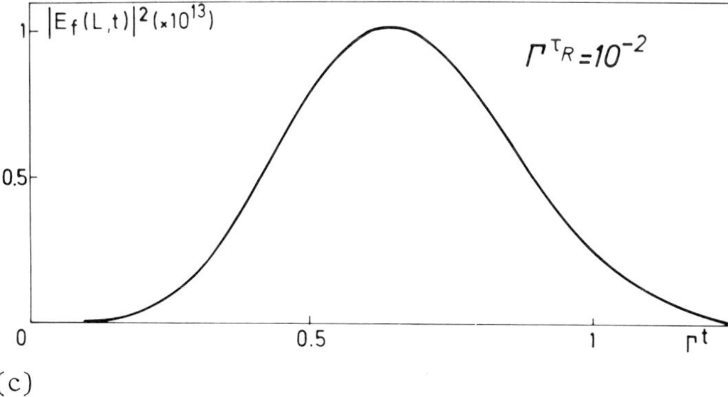

(c)

Figure 1. Intensities of the radiated field at the exit of the sample: (a) $\Gamma\tau_R = 10^{-6}$, $N = 10^{13}$, time is in units of τ_R; (b) $\Gamma\tau_R = 2\times10^{-3}$, $N = 5\times10^9$; (c) $\Gamma\tau_R = 10^{-2}$, $N = 10^9$. In the two latter cases time is in units of Γ^{-1}.

units of τ_R. We note the large delay of order 120 τ_R for the first
pulse and the regular ringings, as in the semi-classical approach,
which are in agreement with the observed signals. Figure (1b) and
Fig. (1c) correspond to $\Gamma\tau_R = 2\times10^{-3}$ and $\Gamma\tau_R = 10^{-2}$, (time is in
units of Γ^{-1}). In the three figures, Γ is constant and the total
number of atoms is 10^{13}, 5×10^9 and 10^9 in Figs. (1a), (1b) and (1c),
respectively. As the concentration decreases the delay t_d increases
and the intensities of the ringings decrease. The peak of the first
ringing decreases from 0.016 N^2 for $N = 10^{13}$ (Fig. (1a)) to $10^{-5}N^2$
for $N = 10^9$ (Fig. (1c)) and simultaneously t_d increases from
$1.2\times10^{-6}\Gamma^{-1}$ to $0.6\ \Gamma^{-1}$. Furthermore as $\Gamma\tau_R$ increases, the ringings
disappear. In Fig. (1a) the maximum of the second ringing is
approximately half the first maximum. In Fig. (1b) the second
maximum is only a quarter of the first maximum. Figure (1c) exhibits the single-pulse superfluorescence limit. So, as τ_R increases
with respect to Γ^{-1}, normal fluorescence prevents superfluorescence.

For $\Gamma\tau_R \gtrsim 10^{-2}$, the atomic energy $<W(z,t)>$ obeys the law
$[\exp(-\Gamma t)-\tfrac{1}{2}]$ of incoherent decay. If we solve the second equation
(12) with

$$<W(z,t)> = \exp(-\Gamma t) - \frac{1}{2} \ , \tag{14}$$

we get for the energy density of the forward (backward) pulse at
the exit of the sample

$$\left|E_f(L,t)\right|^2 = N \exp(-\Gamma t)\ \exp\left(-(t/2\tau_R)\right)\ \exp\left((1-e^{-\Gamma t})/\Gamma\tau_R\right) \ , \tag{15}$$

which reproduces Fig. (1c). For $\Gamma\tau_R$ smaller than 10^{-2}, $<W(z,t)>$
becomes a function of location z and ringings appear.

We have also performed numerical calculations with N and $\Gamma\tau_R$
as given by Vrehen [5] in the case L = 2 cm. The calculated signal
agrees well with the experimental one for the oscillatory regime
(the concentration n is 1.9×10^{11} cm^{-3}). But for the three cases
with smaller concentration the numerical delays are too large by a
factor of about 2, while the numerical widths are in agreement
with those observed. However if we adjust $\Gamma\tau_R$ in order to get
delays comparable to those observed, the numerical widths are too
narrow by a factor of about 2.

The above calculations rehabilitate the Markov approximation,
since it is proved without ambiguity that superfluorescence can be
a Markovian process. We point out that the Markov approximation
does not inhibit the ringings, i.e. the possibility that an atom
may be re-excited by the radiation field emitted by the others.
In mean-field theory [8], the ringings were inhibited because the
atomic system was considered as a single radiating supersystem
composed of N equivalent radiating emitters. Furthermore we show

that superfluorescence signals may evolve from signals with many ringings ($\Gamma\tau_R \ll 1$) to the single-pulse limit when $\Gamma\tau_R$ is of order 10^{-2} or larger.

2. THE SEMI-CLASSICAL LIMIT ($\Gamma\tau_R \lll 1$)

We assume that there exists a time t_ℓ, different from zero and uniform on the sample, from which the terms proportional to Γ in Eqs. (12) can be neglected. Then we find that $<W(z,t)>^2 + <R^+(z,t)\,R^-(z,t)>$ is a constant of the motion. With these conditions we may achieve the classical limit of the Heisenberg equations (12) by using the following correpondence between quantum operators and classical functions

$$W(z,t) \rightarrow -\frac{1}{2}\cos\,\theta(z,t)$$

$$R^{\pm}(z,t) \rightarrow \frac{1}{2}\sin\,\theta(z,t)\,\exp\big(\pm i\phi(z,t)\pm i\phi_0\big) \qquad (16)$$

where ϕ_0 is a uniform random phase such that any $<R^{\pm}(z,t)>^{\ell}$ and the field expectation value are zero. The random phase $\phi(z,t)$ occurs only via the two dimensional characteristic function

$$C(z,z',t) = <\exp\,\big(i\phi(z,t) - i\phi(z',t)\big)> \,, \qquad (17)$$

where the brackets mean that the average is taken over the phases with a given law of probability. As long as the cooperation between the atoms is negligible, $C(z,z',t)$ is constant and equal to $\exp(ik_0|z-z'|)$.

The closed equations of motion for the tipping angle $\theta(z,t)$ and the moments

$$C^{\pm}(z,z',t) = C(z,z',t)\,\exp\big(\pm ik_0(z-z')\big) \qquad (18)$$

can be easily derived by using the classical correspondence (16) together with the Heisenberg equations (12). For example, the equation of motion for $\theta(z,t)$ is (when retardation effects are not included)

$$\frac{d\theta}{dt}(z,t) = -\frac{1}{2\tau_R}\left[\frac{1}{L}\int_0^z dz'\,\sin\theta(z',t)\,C^+(z,z',t)\right.$$

$$\left. +\frac{1}{L}\int_z^L dz'\,\sin\theta(z',t)\,C^-(z,z',t)\right] \qquad (t \geq t_\ell).$$

$$(19)$$

We do not give here the equations of motion for $C^{\perp}(z,z',t)$. They can be easily deduced from Eqs. (12), (16) and (19). $C^{+}(z,z',t)$ and $C^{-}(z,z',t)$ are found to be directly coupled through their equations of motion and they satisfy the relation $C^{+}(z,z',t) = C^{-}(L-z,\ L-z',t)$.

Equation (19), together with the equation of motion for $C^{\pm}(z,z',t)$, was numerically solved with the following initial conditions

$$C^{\pm}(z,z',t_{\ell}) = 1 \tag{20}$$

$$\theta(z,t_{\ell}) = \text{arcos}\ (1-2\ e^{-\Gamma t_{\ell}})\ . \tag{21}$$

According to relation (16) solutions can be considered only for parameters corresponding to the case of superfluorescence with multiple ringings (cf. Fig. (1a)). Then we were led to choose the time t_{ℓ} such that the initial tipping angle is approximately equal to $\Pi-1/N^{\frac{1}{2}}$ in order to reproduce the figure (1a). If we choose an initial tipping angle which deviates further from Π, we get a signal with shorter delays and shorter widths for the ringings. We suspect that the order of magnitude of t_{ℓ} is determined by condition (20) for the atomic correlations. Indeed, it appears that the characteristic functions $C^{\pm}(z,z',t)$ vary very quickly as the atoms begin to decay even if the cooperative radiated power is still much smaller than the power radiated by normal fluorescence. Then t_{ℓ} has to be chosen very small compared to Γ^{-1} in order that $C^{\pm}(z,z',t_{\ell})$ may be taken equal to unity.

Numerical calculations show that, after a few tens of τ_{R}, $C^{\pm}(z,z',t)$ reach values approximately constant with respect to the time. As a first approximation these can be written as

$$C^{+}(z,z',t) \approx U(z - \frac{L}{2})\ U(z' - \frac{L}{2})\ ,$$

$$C^{-}(z,z',t) \approx U(\frac{L}{2} - z)\ U(\frac{L}{2} - z')\ . \tag{22}$$

Then if we substitute these expressions in the second member of Eq. (19), we get

$$\frac{d\theta}{dt}(z,t) = - \frac{1}{2\tau_{R}} \left[U(z - \frac{L}{2}) \frac{1}{L} \int_{0}^{z} dz'\ U(z' - \frac{L}{2})\ \sin\theta(z',t) \right.$$

$$\left. + U(\frac{L}{2} - z) \frac{1}{L} \int_{z}^{L} dz'\ U(\frac{L}{2} - z')\ \sin\theta(z',t) \right]\ . \tag{23}$$

This amounts to decoupling the forward and the backward radiations. Phenomenologically we can say that the system composed of N atoms behaves like two systems of $\frac{1}{2}$N atoms; one of them radiates a forward field (L/2 $<$ z $\leq$ L) with the classical equation of motion for $\theta_f(z,t)$

$$\frac{d}{dt}\,\theta_f(z,t) = -\frac{1}{\tau_R{}'}\,\frac{1}{L}\int_{L/2}^{z} dz'\;\sin\theta_f(z',t) \qquad (24a)$$

with

$$\left|E_f(L,t)\right|^2 = \left[\frac{1}{L}\int_{L/2}^{L} dz'\;\sin\theta(z',t)\right]^2 , \qquad (25)$$

where τ'_R is the characteristic superradiant time for $\frac{1}{2}$N atoms; the second one emits a backward radiation field with

$$\frac{d}{dt}\,\theta_b(z,t) = -\frac{1}{\tau_R{}'}\int_{z}^{L/2} dz'\;\sin\theta_b(z',t) \qquad (24b)$$

and

$$\left|E_b(0,t)\right|^2 = \left|E_f(L,t)\right|^2 .$$

These latter results explain why Feld et al. [1] were entitled to neglect the backscattering in order to reproduce their experimental signals.

In conclusion we point out that Eqs. (24) give a classical limit of the unrestricted Markovian model. It was achieved after we saw that the phase correlations between the microscopic radiated field "cut" the full sample (0 $\leq$ Z $\leq$ L) in half samples. The first one (L/2 $\leq$ Z $\leq$ L) radiates only in the forward direction and the second one (0 $\leq$ Z $\leq$ L/2) radiates in the backward direction. And for each of these two radiating systems, the phase of the electric field is well defined.

Retardation effects could be easily included if the phase correlations behave as previously. Then the time t ought to be replaced by t - z/c + z'/c and t + z/c - z'/c in the right-hand member of Eqs. (24a) and (24b), respectively. In these conditions the signals would be only shifted with respect to the time by L/c. Such a result would disagree with previous calculations of Pillet [7] who has found distortions of the superfluorescence signals as L/c becomes of order t_d. But the study of the phase correlations does not appear obvious if the retardation effects must be included and, on the other hand, the treatment of Pillet seems to us to be justified. An experimental study of the retardation effects would be illuminating if the variation of the ratio $(L/c\tau_R)$ (i.e. the variation of concentration, or length, or section, etc.) does not lead to another perturbation effect which disturbs the retardation effects.

ACKNOWLEDGEMENTS

We would like to acknowledge helpful discussions with G. Oliver. We thank Dr. M. Rousseau for helpful discussions.

References

1. N. Skribanowitz, I.P. Herman, J.C. MacGillivray and M.S. Feld, Phys. Rev. Lett. *30*, 309 (1973); I.P. Herman, J.C. MacGilli-vray, N. Skribanowitz and M.S. Feld, in *Laser Spectroscopy*, R.G. Brewer and A. Mooradian, Editors (Plenum, New York, 1974).
2. M. Gross, C. Fabre, P. Pillet and S. Haroche, Phys. Rev. Lett. *36*, 1035 (1976).
3. A. Flusberg, T. Mossberg and S.R. Hartmann, Phys. Lett. *58A*, 373 (1976).
4. Q.H.F. Vrehen, H.M.J. Hikspoors and H.M. Gibbs, Phys. Rev. Lett. *38*, 764 (1977).
5. Q.H.F. Vrehen, H.M.J. Hikspoors, H.M. Gibbs, this Vol., p. 543.
6. J.C. MacGillivray and M.S. Feld, Phys. Rev. A *14*, 1169 (1976).
7. P. Pillet, Thèse de 3ème cycle (Paris, 1977).
8. R. Bonifacio, P. Schwendimann and F. Haake, Phys. Ref. A *4*, 302 and 854 (1971); R. Bonifacio and L.A. Lugiato, Phys. Rev. A *11*, 1507 (1975) and *12*, 587 (1975).
9. R. Bonifacio and co-workers have recently shown that the mean-field theory may be valid in some limiting cases, particularly in the limiting case of single-pulse superfluorescence (See their communication in the proceedings of the Fourth Rochester Conference on Coherence and Quantum Optics, June 1977, this volume, p. **939**).
10. E. Ressayre and A. Tallet, Phys. Rev. Lett. *32*, 424 (1976); and Phys. Rev. A *15*, 2410 (1977).
11. G. Banfi and R. Bonifacio, Phys. Rev. Lett. *33*, 1259 (1974); and Phys. Rev. A *12*, 2068 (1974).

QUANTUM DISORDER IN MACROSCOPIC SYSTEMS OF INTERACTING ATOMS AND

RADIATION FIELDS

G. Compagno and F. Persico*

Università di Palermo, Palermo, Italy

1. INTRODUCTION

The linear interaction between a system of two-level atoms
and an electromagnetic field can be described as taking place
through a number of elementary acts in which photons are absorbed
or emitted, while atoms change their states. It is conceivable
that these processes tend to modify the original statistical properties characteristic of the atomic system and of the electromagnetic field at t = 0, when we assume that the interaction is
"turned on". The problem is of conceptual importance, and might
become of practical importance in connection with laser processes
in unusual ranges of frequency. In fact, it has recently received
increasing attention in the case of one-photon interactions [1]
and also for large times, where the interest has been focused
either mainly on the one-atom case [2-4] or on the many-atom case
[5,6]. In all cases in which at t = 0 the atomic system has been
taken to be in its unperturbed ground state and the electromagnetic
field in a coherent state [7], the z-component of the total dipole
atomic moment has been found to vanish at large times. As was
noticed early [2], in the absence of relaxation times of any sort,
this would seem to suggest caution in the use of the correspondence
principle for large N, since a classical angular momentum behaves
quite differently, exhibiting no damping in its z-component.
Moreover, the initial coherence of the field has been shown to
decrease with time during the interaction by calculating the appropriate correlation functions [4] or the variances of the field
amplitude [5,6].

The work we present here is divided as follows. In section 2
we explain briefly the method we have used. It is a short summary,

since the technique has been discussed elsewhere [6]. In the
third section we examine the quantum decay of the total dipole of
a system of N atoms interacting with a coherent field under rather
more general conditions than has been done hitherto, that is without assuming that all atoms are in the ground state at t = 0. In
section 4 we shall consider the consequences of the quantum effect
under consideration in a simple case of amplification of coherent
radiation. In section 5 we discuss the results obtained in connection with the limits in the efficiency of an ideal quantum
amplifier and with the usual interpretation of the correspondence
principle. Throughout this paper we shall consider only the case
of large fields, that is fields in which the average number n of
photons is larger than N, the number of two-level atoms.

2. THE STATE VECTOR

We take the system to be described by the resonant Dicke
Hamiltonian [8] for long-wavelength monochromatic radiation in the
rotating wave approximation

$$\mathcal{H} = \mathcal{H}_0 + V \; ; \quad \mathcal{H}_0 = \omega(S_z + \alpha^\dagger \alpha) \; ; \quad V = \frac{1}{2}\,\varepsilon(S_+\alpha + S_-\alpha^\dagger) , \qquad (2.1)$$

where the α's are Bose operators which refer to the electromagnetic
field mode and where

$$S_i = \sum_{\ell=1}^{N} S_i^\ell \qquad (i = +,-,z) ,$$

are collective spin operators describing the N two-level atoms,
S_i^ℓ being the single-atom spin operators represented by 2 × 2 Pauli
matrices. Since $|S|^2$ commutes with $\mathcal{H}$, we are entitled to work in
a space with fixed cooperation number S. We may then choose S to
be O(N) and constant.

We take the state of the field at t = 0 to be a Glauber
coherent state [7]

$$|a> = D|0> = \exp(a\alpha^\dagger - a^*\alpha)|0> , \qquad (2.2)$$

with $|a|^2$ = n, and the state of the atomic system to be a particular
Radcliffe state [9]

$$|\mu> = R_{\theta,\phi}|-S> = \exp(\zeta S_+ - \zeta^* S_-)|-S> \; ; \quad \zeta = \frac{\theta}{2}\,e^{-i\phi} , \qquad (2.3)$$

pointing along direction (θ,ϕ) on the Bloch sphere and obtainable
from the ground state of the atomic system with $S_z = -S$ by rotation
operator $R_{\theta,\phi}$ defined in Eq. (2.3). Thus the initial state of the

total system at t = 0 is

$$|a,\mu\rangle = DR_{\theta,\phi}|0,-S\rangle \quad . \tag{2.4}$$

Since $[\mathcal{H}_0,V] = 0$, state (2.4) at time t evolves into

$$|\psi\rangle = e^{-i\omega t a^{\dagger}\alpha}\, DR_z\, e^{-(i/2)\tau(X+Y/\sqrt{n})}\, R_{\theta,\phi}|0,-S\rangle \quad , \tag{2.5}$$

where

$$R_z = e^{-i\omega t S_z} \quad , \tag{2.6}$$

is a rotation operator about the z-axis with angular speed ω, $\tau = \varepsilon\sqrt{n}\,t$ is a reduced time and

$$X = S_+ + S_- \; ; \quad Y = \alpha S_+ + \alpha^{\dagger}S_- \quad . \tag{2.7}$$

We remark that in state (2.5) the "classical" features of the rotating electromagnetic field are represented by X, while the "quantum" fluctuations are included in Y. Since $[X,Y] \neq 0$, these two features cannot be immediately disentangled in (2.5), and we have to resort to approximations. Developing

$$\Phi(\tau) = e^{-(i/2)\tau(X+Y/\sqrt{n})} \tag{2.8}$$

in powers of $n^{-\frac{1}{2}}$, we obtain [6]

$$\Phi(\tau) = e^{-(i/2)\tau X}\,(1 + \sum_{j=1}^{\infty}\Phi_j') \; ;$$

$$\Phi_j' = \left(\frac{1}{i2\sqrt{n}}\right)^j \int_0^{\tau}\int_0^{\tau_1}\cdots\int_0^{\tau_{j-1}} Y'(\tau_1)Y'(\tau_2)\ldots Y'(\tau_j)\,d\tau_1 d\tau_2\ldots d\tau_j \; , \tag{2.9}$$

where

$$Y'(\tau) = e^{(i/2)\tau X}\,(\alpha S_+ + \alpha^{\dagger}S_-)\,e^{-(i/2)\tau X} \quad .$$

The integrations in Eq. (2.9) are easily worked out for j = 1,2 and it is found

$$\Phi_1' = \frac{1}{2\sqrt{n}}\,\{-(\alpha^{\dagger}-\alpha)S_z(\cos\tau-1) - i[(\alpha S_+ + \alpha^{\dagger}S_-)\tau$$

$$+ \frac{1}{2}(S_- - S_+)(\alpha^{\dagger}-\alpha)(\sin\tau-\tau)]\} \tag{2.10}$$

$$\Phi_2' = \frac{1}{4n} \left[i(\alpha S_+ + \alpha^\dagger S_-)(\alpha^\dagger - \alpha) S_z I_1 + (\alpha S_+ + \alpha^\dagger S_-)^2 I_2 \right.$$

$$+ \frac{1}{2}(\alpha S_+ + \alpha^\dagger S_-)(\alpha^\dagger - \alpha)(S_- - S_+) I_3 + \frac{i}{2}(\alpha^\dagger - \alpha)^2 (S_- - S_+) S_z I_4$$

$$+ \frac{1}{2}(\alpha^\dagger - \alpha)(S_- - S_+)(\alpha S_+ + \alpha^\dagger S_-) I_5 + \frac{1}{4}(\alpha^\dagger - \alpha)^2 (S_- - S_+)^2 I_6$$

$$+ (\alpha^\dagger - \alpha)^2 S_z^2 I_7 + i(\alpha^\dagger - \alpha) S_z (\alpha S_+ + \alpha^\dagger S_-) I_8$$

$$\left. + \frac{i}{2}(\alpha^\dagger - \alpha)^2 S_z (S_- - S_+) I_9 \right] \ , \tag{2.11}$$

where

$$I_1 = \sin\tau - \tau \ ; \quad I_2 = -\frac{1}{2}\tau^2 \ ; \quad I_3 = \cos\tau - 1 + \frac{1}{2}\tau^2 \ ;$$

$$I_4 = \frac{1}{2}\sin\tau\cos\tau - 2\sin\tau + \frac{3}{2}\tau \ ; \quad I_5 = -\tau\sin\tau - (\cos\tau - 1) + \frac{1}{2}\tau^2 \ ;$$

$$I_6 = -\frac{1}{2}(\sin\tau - \tau)^2 \ ; \quad I_7 = \frac{1}{2}(\cos\tau - 1)^2; \quad I_8 = \tau\cos\tau - \sin\tau \ ;$$

$$I_9 = \frac{1}{2}\sin\tau\cos\tau - \tau\cos\tau + \sin\tau - \frac{1}{2}\tau \ . \tag{2.12}$$

We are thus in the position to calculate explicitly the mean values of operators up to second-order terms in $n^{-\frac{1}{2}}$ by substituting Eqs. (2.9), (2.10) and (2.11) into Eq. (2.5). We have found it convenient, however, to move to a frame of reference which rotates about the z-axis at frequency ω, so that the direction of the electromagnetic field coincides with the new x-axis, and simultaneously rotates about this new x-axis with frequency $\varepsilon\sqrt{n}$. We call the frame of reference thus obtained the "doubly rotating frame" (DRF) and we note that in this frame, if the electromagnetic field were classical, the total atomic dipole $\bar{S}$ would stay at rest in the initial direction. The state vector in the DRF is

$$|\psi\rangle_R = e^{(i/2)\tau X} R_z^{-1} |\psi\rangle = e^{-i\omega t \alpha^\dagger \alpha} D\, e^{(i/2)\tau X} \Phi R_{\theta,\phi} |0,-S\rangle \ .$$

A further tip to the z-axis of the DRF to bring it along the negative direction of the atomic dipole moment takes us to a frame of reference where the state vector is

$$|\psi\rangle_{R_{\theta,\phi}} = R_{\theta,\phi}^{-1} |\psi\rangle_R = e^{-i\omega t \alpha^\dagger \alpha} D R_{\theta,\phi}^{-1} \left[1 + \sum_{j=1}^{\infty} \Phi_j'(\tau) \right] R_{\theta,\phi} |0,-S\rangle \ ,$$

$$\tag{2.13}$$

where we have used the expansion in Eq. (2.9). In this frame of reference, and in the absence of quantum fluctuations in the electromagnetic field, the total angular momentum should always remain aligned with the z-axis. In order to keep terms up to second order in $n^{-\frac{1}{2}}$ in Eq. (2.13), we have obtained explicit expressions for

$$R_{\theta,\phi}^{-1}\Phi_1'(\tau)R_{\theta,\phi}|0,-S\rangle = \frac{1}{2\sqrt{n}}(BS + CS_+)\alpha^\dagger |0,-S\rangle \; ;$$

$$R_{\theta,\phi}^{-1}\Phi_2'(\tau)R_{\theta,\phi}|0,-S\rangle = \frac{1}{4n}(P\alpha^{\dagger 2} + E + FS_+ + GS_+^2)|0,-S\rangle \; , \qquad (2.14)$$

where B, C, F and G are complicated functions of θ, ϕ and τ, while P is an operator containing the components of the total angular momentum. Here we need only to report the expression for C,

$$C = -\frac{1}{2}e^{-i\phi}\sin\theta(\cos\tau-1) + ie^{-i2\phi}\sin^2\frac{\theta}{2}\tau$$

$$+ \frac{i}{2}(\cos^2\frac{\theta}{2} + e^{-i2\phi}\sin^2\frac{\theta}{2})(\sin\tau-\tau) \; , \qquad (2.15)$$

since the companion quantities do not appear in a treatment which is limited to the z-component of $\underline{S}$.

3. QUANTUM DECAY OF S_z

We are now ready to calculate $\langle S_z\rangle_{R_{\theta,\phi}}$ as a function of time. We obtain from Eq. (2.13) and up to second order terms in $n^{-\frac{1}{2}}$

$$\langle (S_z + S)\rangle_{R_{\theta,\phi}} = \langle 0,-S| [(S_z + S) + R_{\theta,\phi}^{-1}(\Phi_1'^\dagger + \Phi_2'^\dagger)R_{\theta,\phi}(S_z + S)$$

$$+(S_z + S)R_{\theta,\phi}^{-1}(\Phi_1' + \Phi_2')R_{\theta,\phi} + R_{\theta,\phi}^{-1}\Phi_1'^\dagger R_{\theta,\phi}(S_z + S)R_{\theta,\phi}^{-1}\Phi_1' R_{\theta,\phi}]|0,-S\rangle \; .$$

$$(3.1)$$

Since $(S_z + S)|0,-S\rangle = \langle 0,-S|(S_z + S) = 0$, all the terms within square brackets in Eq. (3.1) vanish, except the last. Using Eq. (2.14) this gives

$$\frac{1}{4n}\langle 0,-S|(B^*S + C^*S_-)\alpha(S_z + S)(BS + CS_+)\alpha^\dagger|0,-S\rangle = \frac{S}{2n}|C|^2 \; ,$$

and

$$\langle S_z\rangle_{R_{\theta,\phi}} = -S + \frac{S}{2n}|C|^2 \; . \qquad (3.2)$$

From Eq. (2.15) we can easily obtain an expression for $|C|^2$ at large t which turns out to be of the form

$$|C(\tau \gg 1)|^2 \sim \frac{\tau^2}{4} (1 - \cos^2\phi \sin^2\theta) , \qquad (3.3)$$

and evidently shows a decrease of the average value of S_z which cannot be explained in semiclassical terms.

As for the variance of S_z, we proceed to calculate $\langle S_z^2 \rangle$ in the same way as for $\langle S_z \rangle$. We obtain, up to terms $0(n^{-1})$,

$$\langle (S_z^2 - S^2) \rangle_{R_{\theta,\phi}} = \langle 0, -S | R_{\theta,\phi}^{-1} \Phi_1^! {}^\dagger R_{\theta,\phi} (S_z^2 - S^2) R_{\theta,\phi}^{-1} \Phi_1^! R_{\theta,\phi} | 0, -S \rangle$$

$$= \frac{S(1-2S)}{2n} |C|^2 \qquad (3.4)$$

and

$$\langle S_z^2 \rangle_{R_{\theta,\phi}} = S^2 + \frac{S(1-2S)}{2n} |C|^2 . \qquad (3.5)$$

Consequently, within the same accuracy as before,

$$(\Lambda S_z)^2_{R_{\theta,\phi}} = \langle S_z^2 \rangle_{R_{\theta,\phi}} - \langle S_z \rangle^2_{R_{\theta,\phi}} = \frac{S}{2n} |C|^2 , \qquad (3.6)$$

and the variance of S_z is seen to increase with time. It is interesting to observe that Eq. (3.3) can be written as

$$|C(\tau \gg 1)|^2 \sim \frac{\tau^2}{4} (1 - \cos\gamma) , \qquad (3.7)$$

where γ is the initial angle between the total dipole and the x-axis in the laboratory frame, that is also the initial direction of the electromagnetic field. Thus γ is also the angle of the precession cone of S about the direction of the rotating electromagnetic field. Equation (3.7) together with expressions (3.2) and (3.6) shows that the larger is this angle (between 0 and $\pi/2$) the more efficiently the quantum fluctuations of the spin amplitude succeed in spreading the total dipole out of the initial coherent state. We may perhaps interpret this aspect of the decay in terms of fluctuations in the precession of S induced by the quantum fluctuations of the electromagnetic field. In fact when $\gamma = 0$, at large t the variance of S_z does not increase with time and the total dipole does not decay. On the contrary, the effect is maximum for $\gamma = \pi/2$, for example when all atoms are excited at $t = 0$, or when the atomic system is in its ground state.

4. COHERENCE DECAY IN AN AMPLIFICATION PROCESS

We wish now to consider in somewhat more detail a simple case of amplification of coherent radiation by our atomic system. We assume the atomic system to be initially in the fully excited state with $\mu = +S$, so that the state of the total system at $t = 0$ is $|a,+S\rangle$. We wish to study the total system at later times, particularly when most of the atomic energy has passed to the electromagnetic field. This is not a very general amplification process, since we are bound by our technique to the case $N \ll n$, but in spite of this, we shall be able to gather interesting information about the decay of coherence in the coupled system atoms + electromagnetic field.

We find from Eq. (2.5) with $\phi = 0$, $\theta = \pi$ and from development (2.9) the state vector

$$|\psi\rangle = e^{-i\omega t a^\dagger \alpha}\, DR_z e^{-(i/2)\tau X}(1 + \sum_{j=1}^{\infty} \Phi_j')|0,+S\rangle \quad . \qquad (4.1)$$

Terms (2.10) and (2.11) in (4.1) yield

$$\Phi_1'|0,+S\rangle = -\frac{1}{2\sqrt{n}}\,\alpha^\dagger[S(\cos\tau-1) + \frac{i}{2}(\sin\tau+\tau)S_-]|0,+S\rangle \quad ;$$

$$\Phi_2'|0,+S\rangle = \frac{1}{4n}\{[(2I_2 + I_3 + I_5 + \frac{1}{2}I_6)S - I_7 S^2]$$

$$+ [i(I_8 + \frac{1}{2}I_9) - i(I_8 + \frac{1}{2}I_4 + \frac{1}{2}I_9)S]S_-$$

$$+ [-i(I_8 + \frac{1}{2}I_9) + i(I_1 + I_8 + \frac{1}{2}I_4 + \frac{1}{2}I_9)S]\alpha^{\dagger 2}S_-$$

$$+ [-(I_5 + \frac{1}{2}I_6)S + I_7 S^2]\alpha^{\dagger 2} - \frac{1}{2}(I_5 + \frac{1}{2}I_6)S_-^2$$

$$+ (I_2 + \frac{1}{2}I_3 + \frac{1}{2}I_5 + \frac{1}{4}I_6)\alpha^{\dagger 2}S_-^2\}|0,+S\rangle \quad , \qquad (4.2)$$

where coefficients I_i have been defined in Eq. (2.12). The statistical properties of the field can be investigated directly in the laboratory frame of reference on the basis of Eq. (4.1). Since

$$D^{-1}e^{i\omega t a^\dagger \alpha}\,\alpha e^{-i\omega t a^\dagger \alpha}\,D = e^{-i\omega t}D^{-1}\alpha D = e^{-i\omega t}(\alpha + \sqrt{n}) \quad ,$$

we find immediately

$$\langle\psi|(\alpha - \sqrt{n}\,e^{-i\omega t})|\psi\rangle \equiv \langle(\alpha - \sqrt{n}\,e^{-i\omega t})\rangle$$

$$= \langle 0,+S| [\sum_{j,j'} (1 + \Phi_j'^{\dagger}) \alpha (1 + \Phi_{j'}')] |0,+S\rangle \, e^{-i\omega t} \qquad (4.3)$$

Clearly in (4.3)

$$\langle 0,+S| \, \alpha \, |0,+S\rangle = 0 \; ;$$

$$\langle 0,+S| (\Phi_1'^{\dagger}\alpha + \alpha\Phi_1') |0,+S\rangle = \langle 0,+S|\alpha\Phi_1'|0,+S\rangle$$

$$= -\frac{1}{2\sqrt{n}} \langle 0,+S| [S(\cos\tau-1)+\frac{i}{2}(\sin\tau+\tau)S_-]\alpha\alpha^{\dagger}|0,+S\rangle = -\frac{S}{2\sqrt{n}} (\cos\tau-1) \; ;$$

$$\langle 0,+S| (\Phi_1'^{\dagger}\alpha\Phi_1' + \Phi_2'^{\dagger}\alpha + \alpha\Phi_2') |0,+S\rangle = 0$$

where use has been made of expressions (4.2). Consequently, accurate up to terms $0(n^{-1})$, the expression for $\langle\alpha\rangle$ is

$$\alpha = \sqrt{n} \, [1 - \frac{S}{2n} (\cos\tau-1)] \, e^{-i\omega t} \; . \qquad (4.4)$$

Moreover, again neglecting terms of order higher than n^{-1}

$$(\alpha - \langle\alpha\rangle) \, |\psi\rangle =$$

$$= R_z e^{-(i/2)\tau X} [\alpha - e^{-i\omega t}\sqrt{n} + e^{-i\omega t}\frac{S}{2\sqrt{n}} (\cos\tau-1)]$$

$$\times \, e^{-i\omega t \alpha^{\dagger}\alpha} \, D(1 + \Phi_1' + \Phi_2') |0,+S\rangle$$

$$= R_z e^{-(i/2)\tau X} \, e^{-i\omega t \alpha^{\dagger}\alpha} \, D[\alpha + \frac{S}{2\sqrt{n}} (\cos\tau-1)]$$

$$\times \, (1 + \Phi_1' + \Phi_2') |0,+S\rangle \, e^{-i\omega t} \; . \qquad (4.5)$$

The variance of α is given by the length of vector (4.5)

$$(\Delta\alpha)^2 = ||\,(\alpha - \langle\alpha\rangle)\,|\psi\rangle\,||^2 \; ,$$

so that, using (4.2) again, we get

$$(\Delta\alpha)^2 = \frac{1}{16n} (\sin\tau+\tau)^2 \langle 0,+S|S_+S_-|0,+S\rangle = \frac{S}{8n} (\sin\tau+\tau)^2. \qquad (4.6)$$

Thus the variance of the field increases with time, indicating a progressive loss of coherence. It should be noted that no damping appears at $0(n^{-1})$ in the field amplitude (4.4), where the departure from $\sqrt{n}$ is only due to the periodic photon exchange with the atoms.

As for the atomic system, from Eq. (2.15) with $\theta = \pi$, $\phi = 0$ we obtain

$$C = \frac{i}{2} (\sin\tau + \tau) \,, \tag{4.7}$$

which can be substituted in Eq. (3.2) to give

$$\langle S_z \rangle_{R_{\pi,0}} = -S + \frac{S}{8n} (\sin\tau + \tau)^2 \,, \quad \text{or} \quad \langle S_z \rangle_R = S[1 - \frac{1}{8n} (\sin\tau + \tau)^2] \tag{4.8}$$

in the DRF. Also the variance of S_z can be obtained from Eqs. (3.6) and (4.7) as

$$(\Delta S_z)^2_R = \frac{S}{8n} (\sin\tau + \tau)^2 \,. \tag{4.9}$$

The decrease of $\langle S_z \rangle_R$ in Eq. (4.8) and the corresponding increase of the variance (4.9) are quantum effects due to the same mechanism as discussed in sec. 2. We now turn to the transverse dipole moment. In the DRF, state vector (4.1) is simply given by

$$|\psi\rangle_R = e^{-i\omega t a^\dagger a} D(1 + \sum_{j=1}^{\infty} \Phi'_j) |0,+S\rangle \,. \tag{4.10}$$

Thus, using (4.2) and neglecting terms with $j > 2$, after some algebra we find

$$\langle S_+ \rangle_R = \langle 0,+S | (\Phi_1'^\dagger S_+ \Phi_1' + S_+ \Phi_2') |0,+S\rangle = i[\frac{S^2}{2n} (\sin\tau - \tau) + \frac{S}{2n} (I_8 + \frac{1}{2} I_9)] \,. \tag{4.11}$$

The x- and y- components of $\underline{S}$ are readily found as

$$\langle S_x \rangle_R = \text{Re} \langle S_+ \rangle_R = 0 \;;$$

$$\langle S_y \rangle_R = |m\langle S_+ \rangle_R = \frac{S^2}{2n} (\sin\tau - \tau) + \frac{S}{4n} (\tfrac{1}{2}\sin\tau\cos\tau + \tau\cos\tau - \sin\tau - \tfrac{1}{2}\tau) \,, \tag{4.12}$$

where use has been made of Eq. (2.12). We see from Eq. (4.12) that the total dipole acquires a component along y, its x-component remaining constantly zero. For large S this component is proportional to S^2 and can be explained in terms of an advancement of the phase of the rotation of $\underline{S}$ about x, due to interaction with the photons emitted by the atoms during the "descending" half part of the cycle. The transverse variances can be calculated to be

$$(\Delta S_+)^2_R = \frac{S'^2}{4n} (\sin\tau+\tau)^2 \; ; \quad (\Delta S_-)^2_R = 2S - \frac{S(1-S)}{4n} (\sin\tau+\tau)^2 \; .$$

$$(4.13)$$

Comparing Eqs. (4.6) and (4.13) we have the relationship

$$(\Delta\alpha)^2 = \frac{1}{2S} (\Delta S_+)^2$$

between the field amplitude and one of the transverse atomic components variances which has been found in other cases also [5,6].

5. DISCUSSION

We are now ready to examine some implications of the results presented in the previous sections. We begin by considering the evolution of $\langle S_z \rangle$ from the fully excited state towards the ground state, in the presence of a large coherent field. As discussed in the previous section, we may think of this process as of an amplification process of a large coherent input signal. We observe that the atomic system does not seem to be able to reach the unperturbed ground state, since for $\tau \simeq \pi$ its energy is about $\pi^2\omega S/8n$ above the ground state. Admittedly, this cannot be a large quantity in ordinary circumstances, since it has to be $\leq\omega$. The fact remains however, that there is in principle some energy which is stored in the atomic system and which cannot be used for coherent amplification purposes. From the energetic point of view only, we may profitably apply this result to the operation of a machine made up of three-level atoms described in Fig. 1, where the population difference between levels 1 and 3 corresponds to a higher temperature than the population difference between levels 2 and 3, such as to yield a population inversion between levels 1 and 2. The efficiency of this machine has been discussed by Scovil and Schulz-DuBois [10] and more recently by Konyukhov and Prokhorov [11]. Here we wish to divide conceptually its operation in the following steps. First, a quantity of energy which we call Q_1 is absorbed from the high-temperature reservoir between levels 1 and 3. Secondly, a certain amount $Q_2 < Q_1$ of energy is released to the cold reservoir between levels 2 and 3. At the end of these two steps the machine is in the configuration shown in Fig. 1. The third step consists of isolating the three-level atoms from the reservoirs and of letting the system decay between levels 2 and 1 under the drive of a strong coherent signal ($n \gg N = n_2-n_1$) of frequency ω, which we wish to amplify. This last step can be discussed as in sec. 4, and ends when $\tau = \pi$. The cycle can then be repeated. The energy absorbed from the reservoirs in one cycle is $Q_1 - Q_2$, and we call E the energy extracted by the coherent signal during decay between levels

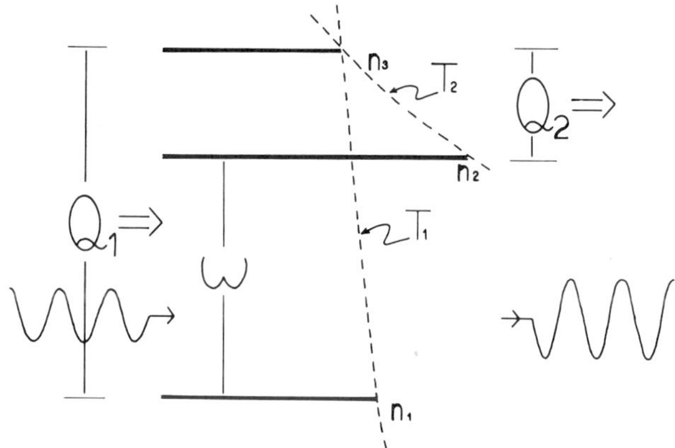

Figure 1. Ideal three-level machine working as an amplifier.
Q_1 is the energy pumped to levels 1-3 by thermal contact with a
source at temperature T_1. Q_2 is the energy given by levels 3-2
to a source at temperature $T_2 < T_1$. The system decays between
levels 2 and 1 under the drive of a strong coherent signal
$(n \gg n_2 - n_1)$ at frequency ω resonant with these levels.

2 and 1. We emphasize that this energy is given by

$$E = (n_2 - n_1)\, \omega \left(1 - \frac{\pi^2}{16n}\right) \, , \tag{5.1}$$

if the input signal is coherent, and not $(n_2 - n_1)\omega$ as one would
have expected. If the efficiency of the machine is defined as the
ratio $\eta = E/Q_1$, Eq. (5.1) shows that the quantum coherence effect
here discussed adds a term

$$\eta_c \sim -\frac{(n_2 - n_1)\pi^2}{16nQ_1}\, \omega$$

to the usual definition of efficiency which, however small in
ordinary circumstances, is conceptually of a fundamental nature and
should not be disregarded. Moreover at the end of the first amplification cycle the variance of the field amplitude is nonzero and
given by $\pi^2 S/8n$, so that the amplified field is no longer perfectly
coherent. As it is perhaps to be expected on physical grounds, if
the system is left undisturbed to evolve like in sec. 4, this loss
of coherence increases at later times, and we find that for $\tau \gg 1$,
$(\Delta\alpha)^2 \sim S\tau^2/8n$. Also the energy which is stored irreversibly in
the atomic system increases asymptotically as t^2.

The slow development of the $\langle S_y \rangle$ component in the DRF as given by Eq. (4.12) is made up of two contributions, one of which is proportional to S^2 and the other to S. We are interested in the case $S \gg 1$, so that the former predominates and for large τ can be written as $\langle S_y \rangle_R \sim -S^2 \tau/2n$. This result can be understood in terms of the average angular precession frequency of the total dipole about the electromagnetic field being slightly larger than $\varepsilon\sqrt{n}$, the angular rotation frequency of the DRF about the same direction. This happens because the precession takes place with average frequency

$$\varepsilon\sqrt{n+S} \sim \varepsilon\sqrt{n}\ (1 + S/2n) \ ,$$

due to the contribution of photons emitted by the atoms. This gives an increasing phase angle of $S\tau/2n$ which explains the S^2 term.

As far as the atomic system is concerned, the limits of validity of our approach at large times are given by the more restrictive of the two conditions

$$\tau^2/8n < 1 \quad ; \quad \tau S/2n < 1 \ . \tag{5.2}$$

The first of these can be obtained directly from Eq. (4.8), and the second can be deduced from the fact that for our development of Φ discussed in sec. 2 to be valid, the effects of terms with $j \geq 1$ in the series must be small, and among them the phase angle $S\tau/2n$ discussed above. Thus we obtain the two breakdown times

$$t^* = \begin{cases} (8n)^{\frac{1}{2}}(\varepsilon\sqrt{n})^{-1} & (1 < S < \sqrt{n}) \\[2ex] \dfrac{2n}{S}\ (\varepsilon\sqrt{n})^{-1} & (\sqrt{n} < S < n) \ , \end{cases} \tag{5.3}$$

as in a previous paper [6]. For $S < \sqrt{n}$ we note from Eq. (4.13) that at $t = t^*$ the spread in the transverse component of the total angular momentum is of $O(S)$; also from Eq. (4.6) we find that the variance of the field is of order $\sqrt{S}$ at the same time. On the other hand for $S > \sqrt{n}$, at $t = t^*$ the spread of the transverse dipole moment is $\sim\sqrt{n}$ while the variance of the field is of order $\sqrt{n/2S}$. In the two cases of (5.3), $(8n)^{\frac{1}{2}}/2\pi$ and $n/\pi S$ respectively represent the number of precessions of the total atomic dipole about the electromagnetic field that we can fairly reliably follow by the technique used here. This result shows the advantage of the present method over time-dependent perturbative approaches. If we extrapolate our results beyond t^*, we find from Eq. (4.6) that after $\sqrt{2}n/\pi\sqrt{S}$ precessions of S about the electromagnetic field, the coherence of the latter is entirely lost, the variance of α being $\sim\sqrt{n}$.

Finally we briefly discuss the persistence of the quantum decay of $<S_z>_R$ in the limit of large numbers. This is fairly obvious in the first of cases (5.3), where $t^* = 2\sqrt{2}/\varepsilon$ does not depend on n or S while $<S_z(t^*)>_R \simeq 0$ from (4.8). In the second case of (5.3) we find $<S_z(t^*)>_R \simeq S(1-n/2S^2)$. If we choose $\sqrt{n} = \beta S$ with $\beta < 1$, then $t^* = 2\beta/\varepsilon$ and $<S_z(t^*)>_R \simeq S(1-\beta^2/2)$ no matter how large are n and S. In other words, it seems that we can always find at least a sequence of values for n and S such that the quantum decay persists in the limit of large systems, contrary to the usual assumption that classical behavior is regained for large quantum numbers [12].

* Also partially supported by CRRNSM, Palermo, Sicily

References

1.   S. Carusotto, Phys. Rev. A *11*, 1629 (1975).
2.   I.R. Senitzky, Phys. Rev. A *6*, 1175 (1972).
3.   P. Meystre, E. Geneux, A. Quattropani, A. Faist, Nuovo Cim. B *25*, 521 (1975);
      H. Baltes, P. Meystre, A. Quattropani, Nuovo Cim. B *32*, 303 (1976).
4.   H.J. Kimble, L. Mandel, Phys. Rev. A *13*, 2123 (1976).
5.   F. Persico, G. Vetri, Phys. Rev. A *12*, 2083 (1975);
      G. Compagno, F. Persico, G. Vetri, Phys. Letters A *56*, 449 (1976).
6.   G. Compagno, F. Persico, Phys. Rev. A *15*, 2032 (1977).
7.   R.J. Glauber, Phys. Rev. *131*, 2766 (1963).
8.   R.H. Dicke, Phys. Rev. *93*, 99 (1954).
9.   J.M. Radcliffe, J. of Phys. A *4*, 313 (1971);
      F.T. Arecchi, E. Courtens, R. Gilmore, H. Thomas, Phys. Rev. A *6*, 2211 (1972).
10.   H.E.D. Scovil, E.O. Schulz-DuBois, Phys. Rev. Letters *2*, 262 (1959).
11.   V.K. Konyukhov, A.M. Prokhorov, Sov. Phys. Usp. *19*, 618 (1976).
12.   See e.g. A. Messiah, *Quantum Mechanics* (North-Holland, Amsterdam 1961), p. 29.

COLLECTIVE DECAY MODES OF MANY NONOVERLAPPING TWO-LEVEL ATOMS -

AN OUTLINE

V. Ernst

Universität München, München, Germany

We treat quantum mechanically the emission of N≥1 photons into
the continuum of modes, the photons being radiated by J≥N nonoverlapping, not necessarily identical, two-level atoms at given positions. Initially, at t = 0, there are no photons, but any N atoms
are excited. Only one ad hoc assumption and two well justified
standard approximations are needed to get in five essential steps
a "closed" solution of this rather complex problem:

(i) We start with "exact" equations of motion in the sense of
a certain variant of QED. (ii) We assume ad hoc that the atoms
are "two-level" atoms. This puts the theory into the framework of
Hilbertspace quantum mechanics. (iii) A variant of the rotating
wave approximation allows essential simplifications. Its proven
applicability shows that semiclassical approaches to many-photon
collective phenomena cannot be justified quantum mechanically.
(iv) All desired solutions of the simplified equations allow an
exact separation of the dependence on photon momentum angles and
spins from the dependence on time and frequencies. The latter is
controlled by given "radial" equations. (v) These equations can
be solved analytically if the distances between the atoms permit
within the natural lifetime of a single atom the exchange of
many "causal photon signals" between any two atoms.

This solution consists of a recursion which starts with given
expressions for the no-photon states at any t, and proceeds step
by step to the states with 1,2,...,N photons. It shows that the
dynamics of our emission process is governed by 2^J "collective
decay modes" corresponding to 2^J "collective decay constants."
These decisive quantities are defined collectively by all atoms and
the geometrical relations between them.

825

1. INTRODUCTION

It has been pointed out already in 1926 [1] that active atoms
in any source of light should not radiate independently. But only
in 1955 [2] it was realized that "macroscopic" many-photon collective phenomena could arise from the coupling of the active atoms
through the common radiation field. Inspired mainly by the
laser [3,4], much work [5] along these lines has been done since.

We give an outline [6] of investigations on the interaction
of the quantum field R of transverse photons over a continuum of
modes with active, nonoverlapping, not necessarily identical, two-
level atoms $A^1,...A^J$ at given positions $\underline{X}^1,...,\underline{X}^J$. The discussion
will be restricted finally to the case

$$|\underline{X}^j - \underline{X}^{j'}| \ll \Lambda \text{ for any } j,j' \varepsilon \{1,...,J\}. \tag{0}$$

Λ is typically of the order of magnitude of the coherence length of
the wavetrain emitted spontaneously by an isolated atom of the
same kind. This amounts to about $\Lambda \sim 1m$ so that (0) is not a serious
restriction in practice. It guarantees, however, that *many* photon-
signals *may* be exchanged between any two atoms during the natural
life time $\tau = \Lambda (\hbar = c = 1)$ of an isolated atom. Such a signal is a
"wavelet" which moves and spreads between the atoms like a retarded
solution of the D'Alambert equation.

The progress achieved so far is noted in the abstract. It
comprises in particular a reliable quantum mechanical "existence
theorem" for the drastic differences [2] to the case of independently radiating atoms which have been postulated up to now on a
rather formal base. In fact, most work on quantum optics has
been done outside or even in obvious contradiction to the accepted
principles of quantum mechanics [7] and causality. Risking an
outsider position we attempt an approach [8-10] within the domains
of these principles. The present work shows that this is possible
even in rather complicated situations. Its greater part, however,
is devoted to the discussion of the physical reliability of the
results obtainable from the formalism to be developed. Time has
not yet permitted many applications, but this is only a technical
problem now. See also Section 8.

2. THE QUANTUM ELECTRODYNAMICAL BACKGROUND

Let $\underline{A}(\underline{x},t)$ be the Heisenberg operator of the observable
"transverse part of the vectorpotential in the point $\underline{x}$" of the
system R, and denote by $\psi^j(\underline{x},t)$ the operator of the amplitude
of the fermion field whose quanta are the electrons (mass m_0) of
A^j. Let $V^j(\underline{x}-\underline{X}^j)$ be the "individual," real number valued, "Coulomb

type" potential of A^j, e the coupling constant, β, $\underline{\alpha} = (\alpha^1,\alpha^2,\alpha^3)$ the usual Dirac matrices. Consider the Heisenberg equations of motion

$$i \frac{d}{dt} \psi^j(\underline{x},t) = (-i\underline{\alpha}\nabla + \beta m_o + V^j(\underline{x}-\underline{X}^j) + e\,\underline{\alpha}A(\underline{x},t))\,\psi^j(\underline{x},t),$$

$$j = 1,\ldots,J, \tag{1a}$$

$$(\nabla^2 - \frac{\partial^2}{\partial t^2})\,\underline{A}(\underline{x},t) = e \sum_{j=1}^{J} [\psi^{+j}(\underline{x},t)\,\underline{\alpha}\psi^j(\underline{x},t)]_{tr} . \tag{1b}$$

$[\ ]_{tr}$ means the transverse part of $[\ ]$. Eqs. (1) amount for J=1 to a restriction of QED to transverse photons and a conserved number of electrons [8,9] which appears to be appropriate for applications in quantum optics. For J>1 they [9] reflect in an excellent way the postulated [1-5] coupling of the atoms through the common radiation field R. Each A^j reacts by (1a) on the observable $\underline{A}(\underline{x},t)$ of R which for A^j plays the role of an additional potential. By (1b) this potential has all atoms as "sources."

For good reasons [9] we prefer the Schrödinger picture for practical computations. Presently this allows us to restrict the discussion without any loss to one-electron atoms with complete sets of eigenstates $u = u_{a_j}^j(\underline{x}-\underline{X}^j)$ corresponding to eigenenergies $E_{a_j}^j$. a_j comprises all quantum numbers needed to specify one u of A^j. Let $K = (\underline{k},r)$ comprise wavevector $\underline{k} \in R^3$ and a polarization index $r\epsilon\{1,2,3\}$. It is convenient to formulate the Schrödinger theory in terms of complex-valued, transverse, cartesian tensors of rank n, $\alpha_n(t) = \alpha_n(t|a_1,\ldots,a_J|K_1;\ldots;K_n)$, with n = 0,1,2,$\ldots$ and each a_j varying over all u of A^j. The $\alpha_n(t)$ are symmetric in $K_1,\ldots,K_n$, square integrable in $\underline{k}_1,\ldots,\underline{k}_n$, and satisfy for any $n,t,a_1,\ldots,a_J$ and any $\nu\epsilon\{1,\ldots,n\}$ a transversality condition

$$\sum_{r^\nu=1}^{3} k_\nu^{r^\nu} \alpha_n(t|a_1,\ldots,a_J|\ldots;\underline{k}_\nu,r^\nu;\ldots) = 0 \text{ for any } \underline{k}_\nu \in R^3. \tag{2}$$

The $k_\nu^{r^\nu}$ are the cartesian components of $\underline{k}_\nu$. We use the $\alpha_n(t)$ in place of the usual Fock amplitudes [8-11] related to polarization vectors orthogonal to $\underline{k}$ because they allow [10,11] an unambiguous transition to the position space by Fourier transformation. The $\alpha_n(t)$ are of course the "components" of the state $|\alpha(t)>$ of the coupled system $R + A^1 + \ldots + A^J$ at the time t in the tensor product [7] of the Hilbert spaces of R and of $A^1,\ldots,A^J$. We are free to interpret a single $\alpha_n(t)$ as the instantaneous momentum space Schrödinger wave function of n photons at time t under the condition that at t the A^j are in the states u characterized by $a_1,\ldots,a_J$. Their spatial Fourier transforms can be interpreted correspondingly; they represent of course the same $|\alpha(t)>$.

The Schrödinger equation of the system $R + A^1 + \ldots + A^J$ reads with $k = |\underline{k}|$

$$i \frac{d}{dt} \alpha_n(t|a_1,\ldots,a_J|K_1;\ldots;K_n) =$$

$$(E^1_{a_1} + \ldots + E^J_{a_J} + k_1 + \ldots + k_n)\, \alpha_n(t|a_1,\ldots,a_J|K_1;\ldots;K_n)$$

$$+ V(n+1) \sum_{j=1}^{J} \sum_{b_j} \int dK\, \bar{M}^j(a_j,b_j;K)$$

$$\times \alpha_{n+1}(t|a_1,\ldots,a_{j-1},b_j,a_{j+1},\ldots,a_J|K;K_1;\ldots;K_n)$$

$$+ \frac{1}{\sqrt{n}} \sum_{\nu=1}^{n} \sum_{j=1}^{J} \sum_{b_j} M^j(b_j,a_j;K_\nu)$$

$$\times \alpha_{n-1}(t|a_1,\ldots,a_{j-1},b_j,a_{j+1},\ldots,a_J|K_1;\ldots;K_{\nu-1};K_{\nu+1};\ldots;K_n). \tag{3}$$

$\int dK\ldots$ comprises integration over $\underline{k}\varepsilon R^3$ and a summation over $r\varepsilon\{1,2,3\}$. The sums over b_j comprise all u of A^j. $M^j(a,b;K)$ is the complex conjugate of the transition element

$$\bar{M}^j(a,b;K) = \bar{M}^j(a,b;\underline{k},r) = e^{i\underline{k}\underline{X}^j}$$

$$\times \frac{e}{\sqrt{V(16\pi^3 k)}} \int d^3x\, e^{i\underline{k}\underline{x}}\, u_a^{j+}(\underline{x})\, \alpha^r u_b^j(\underline{x}). \tag{4}$$

$M^j(a,b;\underline{k},r)$ is square integrable in $\underline{k}$ for any couple u_a,u_b of eigenstates of the proper Dirac atom.[8]   We assume this to hold for all our present atoms A^j. Practically this excludes only the use of the dipole and related approximations which are responsible [12] for almost all "divergences" in the atom photon interaction as traded in the literature.

Eqs. (3) are still as exact as (1). They enforce that all photon wave packets move and spread like retarded solutions of the D'Alambert equation and in this sense meet the requirements of causality.[10]   It is clear that the Bose principle is satisfied from the outset.

3. THE TWO-LEVEL ASSUMPTION AND ITS DEEPER IMPLICATIONS

We must fear,[8,9] however, that no solutions exist. So we
must make some ad hoc assumption which allows at least some progress. In this sense we assume that only two levels of each A^j, say
$u_e^j(\underline{x})$, $u_g^j(\underline{x})$ with $E_e^j > E_g^j$, play an essential role for the effects we
are interested in.

We understand this traditional assumption to mean that each
a_j in (3) is restricted to two values, e,g, (for "excited" and
"ground") and that the sums over b_j cover only the terms b_j = e,g.
By defining a certain order in the systematic Weisskopf-Wigner
approach [4,8,9] this puts the theory into the framework of Hilbertspace quantum mechanics if infra-red problems are controlled appropriately.[8,9] The existence of a unique, finite, norm con-
serving solution of the "truncated" eqs. (3) is guaranteed for any
initial state compatible with this two-level assumption.

How can we justify this truncation? A unique solution exists
also if each A^j "has" 10 or 10^{10} levels, but not for infinitely
many. As atomic state diameters tend to infinity [13] if the
ionization limit is approached from below, we can infer that the
limits of the concept of "nonoverlapping" atoms are reached long
before any infinities can enter the theory. As we know for sure
that active atoms in lasers or superradiators do not overlap in
practice we must even "truncate" the eqs. (3). Otherwise we would
allow for transitions to and between states which provedly do not
exist in reality.

One might argue that the infinities of QED arise from "only
virtual" transitions to such states in "sums over intermediate
states." Such virtual transitions cannot occur in a theory equipped
with a unitary time evolution operator U(t). Let H be the hamiltonian and consider the equations

$$i \frac{d}{dt} \langle a|U(t)|b\rangle = \sum_z \langle a|H|z\rangle\langle z|U(t)|b\rangle \tag{5}$$

for the actual transition amplitudes $\langle a|U(t)|b\rangle$. Without any loss
we can assume that $|a\rangle$, $|b\rangle$ are elements of an orthonormal, com-
plete base B in the statespace of the system, and that the sum over
$|z\rangle$ covers the elements of B. We see that any intermediate state
$|z\rangle$ "counts" in the sum only with the "weight" $\langle z|U(t)|b\rangle$ with
which the transition $|b\rangle \to |z\rangle$ occurs really. As this "actual weight"
of an intermediate state is not known as long as U(t) is not known,
the formulae of perturbation theory contain necessarily some silent
assumptions on the weight of intermediate states. These assumptions
cannot be very realistic however, because in finite orders they lead

to infinite violations of the conservation of the sum of all transition probabilities. This occurs even in cases where $U(t)$ exists [12].

We feel therefore that the following compromise [9] with the infinities of the present version of QED is well justified from a positivistic point of view. We give up the highly questionable right to sum over "all states of A^j," and gain for this a unitary time evolution. Is it really hard to admit that the infinite space is never available for a single atom?

4. ROTATING WAVE APPROXIMATION AND ITS CONSEQUENCES

It was of course more realistic to include more than two states. We make the traditional two-level assumption for the trivial reason to get explicit solutions. But this is not yet sufficient. To get farther we use a variant of the traditional RWA. This is harmless [14,6] and neutral with respect to potential many photon collective phenomena [6] if and only if the transition $u_e^j(x) \leftrightarrow u_g^j(x)$ is favoured by the atomic selection rules in comparison with all reasonable competitors in A^j, and if [12,14] $\Delta^j = E_e^j - E_g^j \gg \varepsilon$ for any j. ε is of the order of magnitude of observed Lambshifts.[12] For $J = 1$ the RWA is equivalent to the omission of level shifts of this magnitude;[14] it has no influence on the dynamics of emission and absorption processes. In addition, it conserves unitarity for any J[9]. Technically it amounts to putting $M^j(e,e;K) = M^j(g,g;K) = M^j(g,e;K) = 0$ and retaining only $M^j(K) := M^j(e,g;K)$. This means: each A^j emits (absorbs) one photon if and only if it "jumps" from the upper (lower) to the lower (upper) level. It is clear that this consequence of the RWA is in best agreement with all experience of atomic physics and quantum optics.

We connect the simplification of (3) with a change of the notation. Instead by $a_1,\ldots,a_J$ with each a_j assuming the values e,g we can characterize [4] a state of J two-level atoms by noting the numbers $j^1,\ldots,j^m$ of the atoms $A^{j^1},\ldots,A^{j^m}$ which are in their upper states. In terms of the corresponding tensors $\alpha_n^m(t) = \alpha_n^m(t|j^1,\ldots,j^m|K_1,\ldots,K_n)$ with $n = 0,1,\ldots$ and $0 \leq m \leq J$ the simplified eqs. (3) read [6]

$$i \frac{d}{dt} \alpha_n^m(t|j^1,\ldots,j^m|K_1;\ldots K_n) =$$

$$(\Delta^{j^1} + \ldots + \Delta^{j^m} + k_1 + \ldots + k_n)\, \alpha_n^m(t|j^1,\ldots,j^m|K_1;\ldots;K_n)$$

$$+ V(n+1) \sum_{\mu=1}^{m} \int dK\, M^{j^\mu}(K) \alpha_{n+1}^{m-1}(t|j^1,\ldots,j^{\mu-1},j^{\mu+1},\ldots,j^m|K;K_1;\ldots;K_n)$$

$$+ \frac{1}{\sqrt{n}} \sum_{\nu=1}^{n} \sum_{\rho=1}^{J-m} M^{j_\rho}(K_\nu) \alpha_{n-1}^{m+1}(t|j^1,\ldots,j^m,j_\rho|K_1;\ldots;K_{\nu-1};K_{\nu+1};\ldots;K_n). \tag{6}$$

The sum over ρ covers the complement $j_1,\ldots,j_{J-m}$ of $j^1,\ldots,j^m$ relative to $1,\ldots,J$. It is thus a sum over the atoms in the ground-state. For convenience we have assumed $E_g^1+\ldots+E_g^J = 0$.

We note that eqs. (6) connect only amplitudes $\alpha_n^m(t)$ for which $n + m = N$ is constant, $N = 0,1,2,\ldots$ Technically this means that these equations can be solved separately for any value of N. This is indeed the clue to our further results. But this "decay" has deep-lying physical implications.

It means that the quantity N = (number of excited atoms plus number of existing photons) is conserved, with the possible eigen-values N = 0,1,2,... . Any "unsharpness" of the expectation values of N has therefore no influence on the dynamics of the interaction of photons and atoms. Conversely, this dynamics has no influence upon N, the sharpness or unsharpness of its expectation values included. As an unsharp number of photons is necessary, but not sufficient for a "classical wave" $\underline{E}(\underline{x},t),\underline{B}(\underline{x},t)$ to be associated to photons in the sense that $\underline{E}(\underline{x},t),\underline{B}(\underline{x},t)$ are the expectation values of the corresponding field operators in the state of these photons, the conservation of N deprives "semiclassical" theories of optical phenomena of any quantum mechanical justification. If, for example, precisely N atoms are excited initially (whatever else their state) and no photons impinge upon them, at t→∞ we must associate with the emitted photons the classical wave $\underline{E}(\underline{x},t) = \underline{B}(\underline{x},t) = 0$. This is hardly an acceptable classical result. If an unsharp number of atoms is excited initially, the same unsharp number of photons will exist finally. But as the time evolution for different N's takes place in orthogonal Hilbertspaces (this is why (6) decays), nonvanishing electromagnetic waves $\underline{E}(\underline{x},t),\underline{B}(\underline{x},t)$ remain an accident which has no influence on what goes on actually. This holds also if any "unsharp" number of photons impinges upon the atoms, e.g. photons in a coherent state.[15] So we must make a quantum theory, whatever its price.

5. A SEPARATION OF VARIABLES

In the general case this price is still very high. We restrict therefore the discussion to cases with no photons in the initial state. It is likely that these solutions are needed as "constituents" of the solutions with incident photons, as in the case J = 1.[16] In view of the preceding we can assume any sharp value for N with $0<N<J$, the case N = 0 being trivial. This amounts to an initial condition of (6) of the form

$$\alpha_n^m(0|j^1,\ldots,j^m|K_1;\ldots;K_n) = \chi(j^1,\ldots,j^m)\,\delta_{mN}\,\delta_{n0}\,. \tag{7}$$

$\chi(j^1,\ldots,j^m)$ is the amplitude for finding at $t = 0$ the atoms $Aj^1,\ldots,Aj^m$ excited, all other Aj deexcited. Note that the N "atom excitations" may be distributed arbitrarily over all J atoms.

The initial condition (7) allows the following separation of the dependence on the angular and spin variables of the photons from the dependence on t and the frequencies $k_1,\ldots,k_n$ (recall $k := |\underline{k}|$). We make an ansatz

$$\alpha_0^N(t|j^1,\ldots,j^N|\emptyset) = a_0^N(t|j^1,\ldots,j^N|\emptyset),$$

$$\alpha_n^m(t|j^1,\ldots,j^m|K_1;\ldots;K_n) =$$

$$\frac{(-i)^n}{\sqrt{(n!)}} \sum_{\iota_1=1}^{J}\ldots\sum_{\iota_n=1}^{J} M^{\iota_1}(K_1)\ldots M^{\iota_n}(K_n) a_n^m(t|j^1,\ldots,j^m|\iota_1,k_1;\ldots;\iota_n,k_n) \tag{8}$$

with unknown amplitudes $a_n^m(t)$ depending on the variables given explicitly in (8). With the abbreviation $h := (\iota,k)$ we find[6]: (8) yields the exact solution of (6) to the initial condition (7) if the $a(t|j^1,\ldots,j^m|h_1;\ldots;h_n)$ are symmetric in $h_1,\ldots,h_n$ and satisfy the "radial equations"

$$\frac{d}{dt}\,a_n^m(t|j^1,\ldots,j^m|h_1,\ldots,h_m) =$$

$$-i(\Delta^{j^1}+\ldots+\Delta^{j^m}+k_1+\ldots+k_n)\,a_n^m(t|j^1,\ldots,j^m|h_1;\ldots;h_n)$$

$$-\sum_{\mu=1}^{m}\sum_{\iota=1}^{J}\int_0^{\infty}dk\,m(j^\mu,\iota;k)\,a_{n+1}^{m-1}(t|j^1,\ldots,j^{\mu-1},j^{\mu+1},\ldots,j^m|\iota,k;h_1;\ldots;h_n)$$

$$+\sum_{\nu=1}^{n}\sum_{\rho=1}^{J-m}\delta_{\iota_\nu j_\rho}\,a_{n-1}^{m+1}(t|j^1,\ldots,j^m,j_\rho|h_1;\ldots;h_{\nu-1};h_{\nu+1};\ldots;h_n) \tag{9}$$

under the initial conditions

$$a_n^m(0|j^1,\ldots,j^m|h_1;\ldots;h_n) = \chi(j^1,\ldots,j^m)\,\delta_{mN}\delta_{n0}\,. \tag{10}$$

$m(j,j';k)$ is defined by

$$m(j,j';k) = k^2\int d^2\Omega\sum_{r=1}^{3}\overline{M}^j(k,\Omega,r)\,M^{j'}(K,\Omega,r)\,. \tag{11}$$

$\underline{k}$ is represented here by its polar coordinates k,Ω with $\Omega =(\theta ,\phi)$.

On a smaller scale, such a separation has been obtained [4,14, 10,17,18] previously. In all cases it proved to be the key not only for the formal calculation of the solution, but also for their physical interpretation. Its physical origin seems to be the iso- tropy of empty space relative to the motion and spread of influences transmitted by photons.

6. THE EXPONENTIAL APPROXIMATION

The integral equations (9) are still too complicated for a practical solution. To work out their essential content we apply now our second approximation which is also not new.[19,4] It is clear that the $a_n^m(t)$ should be appreciably different from zero only for k_ν around the mean Δ of the Δ^j for $\nu =1,\ldots,n$ and all n. The interaction picture shows indeed that the $a_n^m(t)$ are "built up" primarily by resonance denominators at Δ^j. In typical examples [12, 20] the function m(j,j;k) is in the optical region $k\backsim\Delta$ and far beyond it of the form m(j,j;k) = C k with constants C of the order of magnitude 10^{-8}. In agreement with this and for realistic u's the second factor in (4) remains nearly constant if the modulus k of some $\underline{k}$ varies within an interval of the order of magnitude of a typical natural line width $\Gamma =\Lambda^{-1}$. The factor $\exp(i\underline{k}(\underline{X}^{j'}-\underline{X}^j))$ in (11) varies weakly within the same limits if (0) is satisfied.

Under these conditions it should be allowed to "modify" eqs. (9) by the replacement [4]

$$\int_0^\infty dk\, m(j,j';k)\ldots \to \frac{\Gamma(j,j')}{\pi}\int_{-\infty}^{+\infty} dk\ldots := \frac{\Gamma(j,j')}{\pi}\lim_{K\to\infty}\int_{\Delta-K}^{\Delta+K} dk\ldots \quad (12)$$

with the definition $\Gamma(j,j') := \pi m(j,j';\Delta)$. The complex numbers $\Gamma(j,j')$ satisfy the hermitean symmetry relation $\bar{\Gamma}(j,j') = \Gamma(j',j)$. Hölder's inequality yields $\Gamma(j,j')\Gamma(j',j)\leq\Gamma(j,j)\Gamma(j',j')$. $\Gamma(j,j)$ is the constant in the exponential decay amplitude of A^j if A^j is isolated. The exponential form of this amplitude is a consequence of (12) which therefore is referred to as "exponential approxi- mation."[14] It amounts to the omission [12] of line shifts of the order of magnitude of observed Lambshifts. As level shifts of the same order have been omitted already, the use of (12) is actually more realistic than was the insistence upon the exact solutions of (9).

We combine the insertion of (12) into (9) with a transition to the equations for the Laplace transforms $A_n^m(s)$ of the $a_n^m(t)$. These read

$$(s+i(\Delta^{j^1}+\ldots+\Delta^{j^m}+k_1+\ldots+k_n)) \, A_n^m(s|j^1,\ldots,j^m|h_1,\ldots,h_n) =$$

$$\chi(j^1,\ldots,j^m) \, \delta_{mN} \, \delta_{n0}$$

$$-\sum_{\mu=1}^{m}\sum_{\imath=1}^{J} \frac{\Gamma(j^\mu,\imath)}{\pi} \int_{-\infty}^{\infty} dk \; A_{n+1}^{m-1}(s|j^1,\ldots,j^{\mu-1},j^{\mu+1},\ldots,j^m|\imath,k;h_1;\ldots;h_n)$$

$$+\sum_{\nu=1}^{n}\sum_{\rho=1}^{J-m} \delta_{\imath_\nu j_\rho} \; A_{n-1}^{m+1}(s|j^1,\ldots,j^m,j_\rho|h_1;\ldots;h_{\nu-1};h_{\nu+1};\ldots;h_n).$$

$$(13)$$

7.   INTRODUCTION OF THE COLLECTIVE DECAY MODES

These equations will be solved now.

   Consider for any m satisfying $1<m<J$ the linear vector space V^m of vectors X^m whose components $X^m(j^1,\ldots j^m)$ are labelled by the $\binom{J}{m}$ different choices without regard of order of m different numbers $j^1,\ldots,j^m$ out of $1,\ldots,J$. V^m has thus the dimension $\binom{J}{m}$. Consider on it the linear operator $X^m \rightarrow \Gamma^m X^m$ defined by

$$(\Gamma^m X^m)(j^1,\ldots,j^m) =$$

$$\sum_{\mu=1}^{m}\sum_{\imath=j^\mu,j_1,\ldots,j_{J-m}} \Gamma(j^\mu,\imath) X^m(j^1,\ldots,j^{\mu-1},\imath,j^{\mu+1},\ldots,j^m). \quad (14)$$

Γ^m is hermitean and nonnegative.[6] So it can be diagonalized, i.e. there are $\binom{J}{m}$ nonnegative eigenvalues γ_σ^m and $\binom{J}{m}$ mutually orthogonal, normalized eigenvectors U_σ^m with components $U_\sigma^m(j^1,\ldots,j^m)$ such that

$$(\Gamma^m U_\sigma^m)(j^1,\ldots,j^m) = \gamma_\sigma^m \, U_\sigma^m(j^1,\ldots,j^m) \qquad\qquad (15)$$

for any choice $j^1,\ldots,j^m$. For reasons to be seen below we call the U_σ^m the "collective decay modes of rank m" of our atoms which correspond to the "collective decay constants of rank m," $\gamma_\sigma^m \geq 0$. With their help we define the resolvents

$$R^m(s|j^1,\ldots,j^m|1^1,\ldots,1^m) = \sum_{\sigma=1}^{\binom{J}{m}} \frac{U_\sigma^m(j^1,\ldots,j^m)\bar{U}_\sigma^m(1^1,\ldots,1^m)}{s+\gamma_\sigma^m} \cdot \quad (16)$$

It will be convenient to define $U_1^0 = 1$, $\gamma_1^0 = 0$ so that (16)
holds also for $R^0(s|\emptyset|\emptyset) = 1/s$.

By means of these quantities we define

$$A_0^N(s|j^1,\ldots,j^N) =$$

$$\sum_q (R^N(s+i(\Delta^{j^1}+\ldots+\Delta^{j^N})|j^1,\ldots,j^N|q^1,\ldots,q^N)\chi(q^1,\ldots q^N). \quad (17)$$

The sum extends over the $\binom{J}{m}$ choices without regard of order of N
different numbers $q^1,\ldots,q^m$ out of $1,\ldots,J$. For $1\leq n\leq N$ we define
further on (with m short for N-n)

$$A_n^m(s|j^1,\ldots,j^m|h_1;\ldots;h_n) =$$

$$\sum_q \{R^m(s+i(\Delta^{j^1}+\ldots+\Delta^{j^m}+k_1+\ldots+k_n)|j^1,\ldots,j^m|q^1,\ldots,q^m)$$

$$\times \sum_{\nu=1}^{n} \sum_{\rho=1}^{J-m} \delta_{\ell_\nu q_\rho} A_{n-1}^{m+1}(s|q^1,\ldots,q^m,q_\rho|h_1;\ldots;h_{\nu-1};h_{\nu+1};\ldots;h_n)\}.$$

$$(18)$$

As above the first sum extends over the $\binom{J}{m}$ choices $q^1,\ldots,q^m$.
The sum over ρ covers the elements $q_1,\ldots,q_{J-m}$ of the complement
of $q^1,\ldots,q^m$ relative to $1,\ldots,J$. (17), (18) define a simple
recursion.

The point is that this recursion yields the exact solution [6]
of (13). The transition to the $a_n^m(t)$ by means of the convolution
theorem poses no problems. First hints on eigenvalue problems in
this connection are rather old. [4,17]

It could not be expected that the expressions are very simple
which describe the emission and the motion and spread in free
space of any number N of photons emitted by any number J>N of
nonoverlapping, not necessarily identical, two-level atoms in any
initial state $\chi(j^1,\ldots,j^N)$ at places $\underline{x}^j$ subject only to (0). The
problem was not so much to construct the solution technically, but
to gain the confidence that the results to be computed are trust-
worthy from the physical and mathematical point of view.

8. APPLICATIONS

The following applications have been worked out so far:

1. For $|\underline{x}^j - \underline{x}^{j'}| \gg \Lambda$ for any $j \neq j'$ the formalism leads still
to the correct result, namely the independent decay of the atoms.

It seems therefore that (0) need not be taken too seriously.

2. Within its proper domain the formalism leads to large and thus "interesting" differences to the case of independently radiating atoms. This follows from (15). The trace of Γ^m equals the sum of all γ_σ^m. As many nondiagonal terms of Γ^m are as large as the diagonal elements $\Gamma(j^1,j^1)+\ldots+\Gamma(j^m,j^m)$ which alone govern the independent decay of the atoms, some γ_σ^m must be appreciably larger, the other ones smaller than these diagonal elements. For example, we get the strong superradiance effects of Dicke [2] if $|\underline{x}^j - \underline{x}^{j'}|<<\Delta^{-1}$ for any j,j'.

3. In more detail: Eq. (17) shows that the no-photon amplitude disappears in accordance with a single exponential decay law if the initial state equals one of the collective decay modes U_σ^N. Depending on the corresponding decay constant γ_σ^N the "first photon" appears faster ("superradiance") or slowlier ("subradiance") than it would appear from independently radiating atoms. Eq. (18) shows in connection with (16) that also the creation of the second, third,... photon is governed by decay modes U, but of lower ranks.

4. If all atoms are excited initially, $N = J$, $\chi(1,\ldots,J) = 1$, we get $U_1^J = 1$, $\gamma_1^J = \Gamma(1,1)+\ldots+\Gamma(J,J)$. The no-photon amplitude assumes the form $\alpha_0(t|1,\ldots,J|\emptyset) = \exp\{-t(\Gamma(1,1)+\ldots+\Gamma(J,J))\}$ as suspected previously.[4] Surprisingly the first photon appears here as fast as it would appear from independently radiating atoms. It is clear that this is a consequence of the "initial symmetry" of this case, rather than an indication of the independence of the atoms. The second photon displays already their dependence.

5. Comparison with results [10] obtained for $N = 1$ without the use of (12) allows the conclusion [6] that the amplitudes $\alpha_n(t)$ are the result of the interference of many Schrödinger photon wavelets exchanged between the atoms in agreement with causality. Each wavelet corresponds to one of the ways n photons may have been emitted by J atoms, each photon being emitted by any A^j either spontaneously or following the potential absorption of a photon emitted previously by any other atoms. The amplitudes of all these possibilities occur in the present formalism. The point of (12) is that it approximates them in such a way that they can be summed up and thus yield the amplitudes $\alpha_n(t)$.

ACKNOWLEDGEMENTS

I am deeply indebted to Drs. E. Grimm, M. Stelzer, and J. Mühlstein who have patiently removed countless minor and some major obstacles on the long way to this work. Thanks are also due to Professor F. Bopp for his generous hospitality and the Deutsche Forschungsgemeinschaft for financial support.

References

1. P. A. M. Dirac, Proc. Roy. Soc. (London) A *112*, 661 (1926).
    See also V. Weisskopf, Ann. Phys. (Leipzig) *9*, 23 (1931).
    Dirac's arguments are quoted by C. S. Chang and P. Stehle on
    p. 739 of the first Ref. 5.
2. R. H. Dicke, Phys. Rev. *93*, 99 (1955).
3. F. Schwabl and W. Thirring, Erg. exakt. Naturw. *36*, 219 (1964).
4. V. Ernst and P. Stehle, Phys. Rev. *176*, 1456 (1968); V. Ernst,
    Z. Phys. *229*, 432 (1969).
5. Most work is referred to in: *Coherence and Quantum Optics,
    Proceedings of the Third Rochester Conference on Coherence and
    Quantum Optics* held at the University of Rochester, June 21-23,
    1972, Ed. by L. Mandel and E. Wolf (Plenum Press, New York -
    London, 1973). See also the references in: R. Jodoin and
    L. Mandel, Phys. Rev. A *9*, 873 (1974); M. Gronchi and L. A.
    Lugiato, Phys. Rev. A *13*, 830 (1976); J. C. MacGillivray and
    M. S. Feld, Phys. Rev. A *14*, 1169 (1976).
6. V. Ernst, to appear. We omit all proofs, but all quoted results
    have been proven.
7. J. M. Jauch, *Foundations of Quantum Mechanics,* (Addison-
    Wesley, Reading, Massachusetts, 1968).
8. E. Grimm and V. Ernst, J. Phys. A: Math. Nucl. Gen. *7*, 1664
    (1974); J. Phys. A: Math. Gen. (in print).
9. M. Stelzer, J. Mühlstein, and V. Ernst, J. Math. Phys. (in
    print).
10. J. Mühlstein and V. Ernst, Ann. Phys.(Leipzig) *34*, 105 (1977).
11. J. Mühlstein, Dissertation, Universität München (1977).
12. E. Grimm and V. Ernst, Z. Phys. A *274*, 293 (1975).
13. W. Weizel, *Lehrbuch der Theoretischen Physik*, Vol. II,
    (Springer Verlag, Berlin-Göttingen-Heidelberg, 1958) p. 871.
14. V. Ernst, Z. Phys. B *23*, 103, 113 (1976).
15. R. J. Glauber, Phys. Rev. *130*, 2529 (1963), *131*, 2766 (1963).
16. V. Ernst, J. Phys. A: Math Gen. *8*, 76 (1975).
17. V. Ernst, Z. Phys. *218*, 111 (1968).
18. M. Stelzer, Dissertation, Universität München (1976).
19. V. Weisskopf and E. Wigner, Z. Phys. *63*, 54 (1930).
20. E. Grimm, Dissertation, Universität München (1977).

INFLUENCE OF OMITTING THE P^2 TERM IN THE MULTIPOLE PHOTON-MATTER
HAMILTONIAN ON THE STABILITY AND PROPAGATION

M. Yamanoi

Meijo University, Tempaku-Ku, Nagoya, Japan

M. Takatsuji

Hitachi, Ltd., Kokubunji, Tokyo, Japan

1. INTRODUCTION

The second order phase transition in the Dicke system in the
state of thermal equilibrium, the so-called superradiant phase
transition (SPT), has been studied by many authors since the work
by Hepp and Lieb [1]. In the Dicke system it is assumed that many
two-level atoms interact with one another only through the transverse radiation field (the Maxwell field). In the mathematical
formulation for SPT, either the truncated hamiltonian without the
A^2 term in the Coulomb-gauge-vector-potential formulation (type-I)
[e.g., 1], or the one without the P^2 term in the multipole hamiltonian formulation (type-II) [e.g., 2,3] has been used.

Concerning the type-I formulation for SPT, however, Rzazewski
et al. [4] showed that SPT is forbidden by the TRK sum rule if the
hamiltonian with the A^2 term is used for the evaluation of the
thermodynamic free energy. In ref. 5, it is noted that the occurrence of the associated absolute instability in the Dicke system
originates entirely in the absence of the A^2 term. Concerning the
type-II formulation for SPT, on the other hand, Emeljanov and
Klimontovich [6] studied the effect of the $P^{\perp 2}$ term on SPT, treating
the $P^{\perp 2}$ term as arbitrarily variable one.

In the present paper we study from first principles the
linear instability (soft-mode instability) in the photon-matter
system S, which consists of radiation field and two-level atoms
with electric dipole-dipole Coulomb interaction V_{d-d}. In the

839

mathematical formulation for this problem, we shall use both the
type-I and type-II formulations to clarify the physical meaning of
the multipole hamiltonian with and without the P^2 term to get a
unified understanding of the problem. We study in sec. 2 the soft-
mode instability and the resultant ordered state in the system S
using the type-I formulation. It will be shown that the ordered
ground state is characterized by the appearance of transverse
polarization ($4\pi/3$ catastrophe) without the macroscopic vector
potential field. In sec. 3, we effect a canonical transformation
to get the multipole hamiltonian, and derive the instability conditions by evaluating the polariton eigenfrequencies for both cases
of omitting and taking into account the P^2 term. It will be shown
that the P^2 term is necessary in order to obtain the same result
as that in sec. 2. In sec. 4, we study a difference between the
implications of the photon operators in the type-I and type-II
formulations by evaluating the expectation values of the operator
for the electromagnetic momentum.

2. TYPE-I FORMULATION

The retarded interaction between atoms through radiation field
V_R, and the instantaneous Coulomb interaction between dipoles V_{d-d},
are separated clearly in the type-I formulation. Therefore, we
begin with the type-I formulation to see the contributions of V_R
and V_{d-d} to phenomena occurring in the system S. The hamiltonian
H_I for the system consisting of radiation field and one-electron
atoms is given in the electric dipole approximation by

$$H_I = \int \left[2\pi \underline{M}_I^2 + \frac{(\nabla \times A)^2}{8\pi} \right] d^3r + \sum_i \left[\frac{(\underline{P}_i - eA(\underline{R}_i))^2}{2m} + V(i) \right] + V_{d-d} \ ,$$

$$(2.1)$$

where $\underline{M}_I$ defined by

$$\underline{M}_I = \frac{\partial \mathscr{L}_I}{\partial \underline{\dot{A}}} = \frac{\dot{A}}{4\pi} = - \frac{E}{4\pi}$$

$$(2.2)$$

is the momentum density canonically conjugate to A, $\mathscr{L}_I$ the Lagrangian density in the type-I formulation, and V(i) the Coulomb-inter-
action energy between nucleus and electron within the i-th atom.
Throughout this paper we use units in which $\hbar = c = 1$. The third
term in Eq. (2.1) represents the interatomic Coulomb interaction
in the electric dipole approximation:

$$V_{d-d} = \frac{1}{2} \sum_{i \neq j} \frac{\underline{d}(i) \cdot \underline{d}(j) - 3\left(\hat{\underline{R}}_{ij} \cdot \underline{d}(i)\right)\left(\hat{\underline{R}}_{ij} \cdot \underline{d}(j)\right)}{R_{ij}^3} \ .$$

$$(2.3)$$

We assume N identical two-level atoms with energy separation ω_0, located at sites $\{R_i\}$ in the volume V, $(N/V = \rho)$. The quantum condition for the radiation field is given by [7]

$$[A_s(\underline{r}) , M_{Is'}(\underline{r}')] = i \, \delta^{\perp}_{ss'}(\underline{r}-\underline{r}') \; . \tag{2.4}$$

Making the mode expansion for A and M_I in terms of photon creation and annihilation operator $a_{\underline{k}}^{\dagger}$ and $a_{\underline{k}}$, we have for the hamiltonian

$$H_I = \sum_{\underline{k}} k a_{\underline{k}}^{\dagger} a_{\underline{k}} + i\omega_0 \sum_{\underline{k}} \sqrt{(2\pi\rho/k)} \, (\underline{d}_{12} \cdot \underline{e}^{\lambda}(\underline{k}))(b_{\underline{k}} - b_{-\underline{k}}^{\dagger})(a_{-\underline{k}} + a_{\underline{k}}^{\dagger})$$

$$+ (e^2/2m) \sum_{\underline{k}} (2\pi\rho/k)(a_{\underline{k}} a_{-\underline{k}} + a_{\underline{k}} a_{\underline{k}}^{\dagger} + a_{\underline{k}}^{\dagger} a_{\underline{k}} + a_{\underline{k}}^{\dagger} a_{-\underline{k}}^{\dagger})$$

$$+ \omega_0 \sum_{\underline{k}} b_{\underline{k}}^{\dagger} b_{\underline{k}} + \frac{1}{2} \sum_{\underline{k}} D(\underline{k})(b_{\underline{k}} + b_{-\underline{k}}^{\dagger})(b_{-\underline{k}} + b_{\underline{k}}^{\dagger}) \; , \tag{2.5}$$

where $b_{\underline{k}}$ is defined using the Pauli operator σ by

$$\sigma^-(i) = \frac{1}{\sqrt{N}} \sum_{\underline{k}} b_{\underline{k}} \, e^{-i\underline{k} \cdot \underline{R}_i} \; , \tag{2.6}$$

and $D(\underline{k})$ is defined by

$$D(\underline{k}) = \sum_i D(i,j) \, e^{-i\underline{k} \cdot (\underline{R}_i - \underline{R}_j)} \; ,$$

with

$$D(i,j) = \frac{\underline{d}_{12} \cdot \underline{d}_{12} - 3(\underline{d}_{12} \cdot \hat{\underline{R}}_{ij})(\underline{d}_{12} \cdot \hat{\underline{R}}_{ij})}{R_{ij}^3} \; .$$

The dipolar sum for a cubic lattice in the long wavelength limit leads to

$$D(\underline{k}) = -(4\pi/3) \, \rho \, [d_{12}^2 - 3(\underline{d}_{12} \cdot \hat{\underline{k}})^2] \; , \tag{2.7}$$

where $\hat{\underline{k}} = \underline{k}/k$. By regarding $b_{\underline{k}}^{\dagger}$ and $b_{\underline{k}}$ as Bose operators in Eq. (2.5), the polariton dispersion relation for the system S is obtained [8] in the form [cf. curve (a) in Fig. 1]:

$$k^2 = \omega^2 \, \varepsilon \, (\omega) \; , \tag{2.8}$$

with

$$\varepsilon(\omega) = (\omega_L^2 - \omega^2)/(\omega_T^2 - \omega^2) \; , \tag{2.9}$$

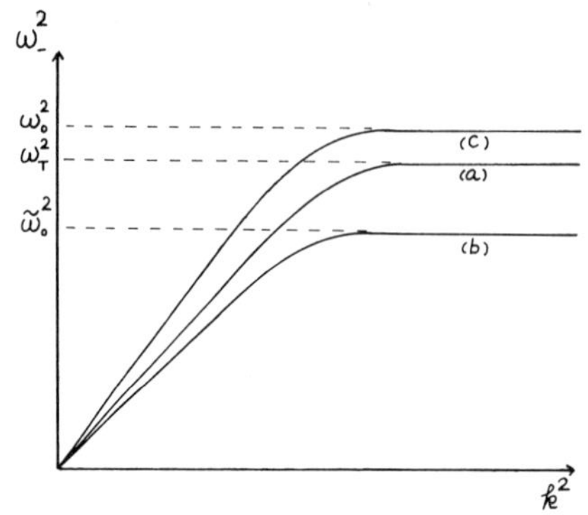

Fig. 1. Schematic representation of the lower branch of polariton dispersion relations derived from the full hamiltonian for the system S [curve (a)], from the Dicke hamiltonian [curve (b)], and from the Dicke hamiltonian plus $2\pi\int P^{\perp 2}d^3r$ [curve (c)]. The ω_T^2 and $\bar{\omega}_O^2$ are defined by Eqs. (2.10) and (3.12), respectively.

where

$$\omega_L^2 = \omega_O^2 + (16\pi/3)\omega_O\rho d^2 \ ,$$

$$\omega_T^2 = \omega_O^2 - (8\pi/3)\omega_O\rho d^2 \ . \tag{2.10}$$

The curve (a) in Fig. 1 shows that the lower branch of the polariton mode becomes unstable, $\omega_-^2 \leq 0$, if

$$(8\pi/3)\rho d^2 \geq \omega_O \ . \tag{2.11}$$

If we omit the radiation field in Eq. (2.5) and retain only the last two terms, we have a hamiltonian representing the matter system (matter-system hamiltonian). From this matter-system hamiltonian we again get the same Eq. (2.10) for the eigenfrequencies. Thus the softening of the transverse excitation in the system S originates entirely in the softening of the transverse excitation in the matter system with V_{d-d}.

The threshold condition (2.11) can be rewritten as

$$(4\pi/3)\rho\alpha(0) \geq 1 \ , \tag{2.12}$$

where $\alpha(0) = 2d^2/\omega_O$ is the low-frequency-limit atomic polarizability for the two-level atom. This is just the well-known condition for the $4\pi/3$ catastrophe [9]. Let us next study the ground state which will appear after this soft-mode instability. For this end we use the following hamiltonian, obtained by the renormalization of the A^2 term and single mode approximation of the new Bose field [10]:

$$H = \Omega\tilde{a}^{\dagger}\tilde{a} + \frac{\omega_O}{2}\sum_i \sigma^z(i) + \frac{\tilde{\lambda}}{\sqrt{N}}\sum_i \sigma^y(i)(\tilde{a}e^{i\underline{k}\cdot\underline{R}_i} + \tilde{a}^{\dagger}e^{-i\underline{k}\cdot\underline{R}_i})$$

$$+ \frac{1}{2}\sum_{i\neq j}\sum D(i,j)\sigma^x(i)\sigma^x(j) \tag{2.13}$$

where

$$\Omega = (k^2 + \omega_p^2)^{\frac{1}{2}} , \quad \tilde{\lambda} = \sqrt{(2\pi\rho/\Omega)} \; \omega_o (\underline{d}_{12} \cdot \underline{e}) .$$

In the mean field approximation and in the long wavelength limit, the ground state energy E_o of Eq. (2.13) is obtained as

$$E_o = \Omega\alpha^*\alpha - (\omega_o/2) \, N\cos\theta + \tilde{\lambda}\sqrt{N} \, \sin\theta \cdot \sin\phi(\alpha+\alpha^*)$$
$$+ (N/2)D(\underline{k})\sin^2\theta \cdot \cos^2\phi , \tag{2.14}$$

where α is the c-number amplitude for the transformed annihilation operator $\tilde{a}$, and θ, ϕ are polar and azimuthal angles of the Bloch vector, respectively; $D(k)$ is given by Eq. (2.7). The stationary conditions, $\partial E_o/\partial\alpha = \partial E_o/\partial\alpha^* = \partial E_o/\partial\theta = \partial E_o/\partial\phi = 0$, lead to

$$\Omega\alpha^* + \tilde{\lambda}\sqrt{N} \, \sin\theta \, \sin\phi = 0 ,$$

$$\Omega\alpha + \tilde{\lambda}\sqrt{N} \, \sin\theta \, \sin\phi = 0 ,$$

$$\frac{\omega_o}{2} N \, \sin\theta + \tilde{\lambda}\sqrt{N} \, \cos\theta \, \sin\phi(\alpha+\alpha^*) + N \, D(k) \, \sin\theta \, \cos\theta \, \cos^2\phi = 0 ,$$

$$\tilde{\lambda}\sqrt{N} \, \sin\theta \, \cos\phi \, (\alpha+\alpha^*) - N \, D(k) \, \sin^2\theta \, \cos\phi \, \sin\phi = 0 .$$

From these equations we see that the non-trivial solution with $\alpha+\alpha^* \neq 0$ exists if

$$(4\tilde{\lambda}^2/\Omega) > \omega_o . \tag{2.15}$$

This condition (2.15) is just the same as that derived in ref. 4 and forbidden by the TRK sum rule. In the case of $\alpha+\alpha^* = 0$, however, a non-trivial solution with $\langle\sigma^x\rangle \neq 0$ is allowed, for the inversionless atomic state and the transverse wave, if

$$(8\pi/3) \, \rho d^2 \geq \omega_o ,$$

which is just the condition (2.11).

We see from these results that, after the soft-mode instability in the system S, the only possible type of ordering in the ground state is the appearance of the transverse polarization without the macroscopic vector potential field ($\alpha+\alpha^* = 0$) in the long wavelength limit ($4\pi/3$ catastrophe).

3. LINEAR INSTABILITY IN THE TYPE-II FORMULATION

The canonical transformation from the type-I to the type-II formulation is essentially shifting both the momentum of atomic electrons and the momentum density of radiation field. In the electric dipole approximation, this shift is performed by changing the Lagrangian [7]:

$$L_{II} = L_I - \frac{d}{dt} \sum_i e\underline{r}_i \cdot \underline{A}(\underline{R}_i) = L_I - \frac{d}{dt} \int \underline{P}^{\perp}(\underline{r}) \cdot \underline{A}(\underline{r}) \, d^3r \, ,$$

where

$$\underline{P}(\underline{r}) = \sum_i e \, \underline{r}_i \, \delta(\underline{r}-\underline{R}_i) = \sum_i \underline{d}(i) \, \delta(\underline{r}-\underline{R}_i) \quad . \tag{3.1}$$

Performing this shift, we get the multipole hamiltonian H_{II} from Eq. (2.1) in the form:

$$H_{II} = \int \left[2\pi(M_{II} + \underline{P}^{\perp})^2 + \frac{(\nabla \times A)^2}{8\pi} \right] d^3r + \sum_i \left[\frac{P_i^2}{2m} + V(i) \right] + V_{d-d} \, , \tag{3.2}$$

where

$$M_{II} = \frac{\partial \mathcal{L}_{II}}{\partial \dot{\underline{A}}} = \frac{\dot{\underline{A}}}{4\pi} - \underline{P}^{\perp} = - \frac{1}{4\pi} (\underline{E} + 4\pi\underline{P}^{\perp}) \quad . \tag{3.3}$$

The quantum condition for canonical variables $\underline{A}$ and $\underline{M}_{II}$ is taken as

$$[A_s(\underline{r}) \, , \, M_{IIs'}(\underline{r}')] = i \, \delta^{\perp}_{ss'}(\underline{r}-\underline{r}') \quad . \tag{3.4}$$

The mode expansion satisfying Eq. (3.4) is usually taken as the same form as that in the type-I formulation, regarding $\underline{M}_{II}$ as pure field operator [11]:

$$\underline{A}(\underline{r}) = \sum_{k\lambda} \sqrt{(2\pi/kV)} \, \underline{e}^{\lambda}(\underline{k}) \, [a_{k\lambda} \, e^{i\underline{k}\cdot\underline{r}} + a^{\dagger}_{k\lambda} e^{-i\underline{k}\cdot\underline{r}}] \, , \tag{3.5}$$

$$\underline{M}_{II}(\underline{r}) = -(1/4\pi) \sum_{k\lambda} i\sqrt{(2\pi k/V)} \, \underline{e}^{\lambda}(\underline{k}) \, [a_{k\lambda} e^{i\underline{k}\cdot\underline{r}} - a^{\dagger}_{k\lambda} e^{-i\underline{k}\cdot\underline{r}}] \quad .$$

We first show that the term $2\pi\int \underline{P}^{\perp 2} d^3r + V_{d-d}$ in Eq. (3.2) may be expressed as

$$2\pi \int \underline{P}^{\perp 2} \, d^3r + V_{d-d} = \frac{4\pi}{3} \int \underline{P}^2 \, d^3r \quad . \tag{3.6}$$

Using the expression for $\delta^{\perp}_{ss'}(\underline{r})$ of the form [7]:

$$\delta^{\perp}_{ss'}(\underline{r}) = (2/3)\,\delta_{ss'}\,\delta(\underline{r}) - (1/4\pi r^3)(\delta_{ss'} - 3\,\hat{r}_s\,\hat{r}_{s'})\ ,$$

and the definition of polarization density $\underline{P}(\underline{r})$ given by Eq. (3.1), we get the expression for $\underline{P}^{\perp}(\underline{r})$:

$$\underline{P}^{\perp}(\underline{r}) = \frac{2}{3}\,\underline{P}(\underline{r}) - \sum_i \frac{\underline{d}(i) - 3\hat{\underline{x}}_i\left(\hat{\underline{x}}_i \cdot \underline{d}(i)\right)}{4\pi\left|\underline{r} - \underline{R}_i\right|^3}\ ,$$

where

$$\hat{\underline{x}}_i = (\underline{r} - \underline{R}_i)/\left|\underline{r} - \underline{R}_i\right|\ .$$

Therefore it follows that

$$\int \underline{P}^{\perp} \cdot \underline{P}^{\perp}\, d^3 r = \int \underline{P} \cdot \underline{P}^{\perp}\, d^3 r$$

$$= \frac{2}{3}\int \underline{P}^2 d^3 r - \sum_i \sum_j \frac{\underline{d}(i) \cdot \underline{d}(j) - 3\left(\hat{\underline{R}}_{ij} \cdot \underline{d}(i)\right)\left(\hat{\underline{R}}_{ij} \cdot \underline{d}(j)\right)}{4\pi R_{ij}^3}\ .$$

If we omit the i=j terms (the self energy term) in the double summation, we obtain the desired result (3.6).(Q.E.D.) Thus, under the two-level atom approximation, H_{II} in Eq. (3.2) takes the following form:

$$H_{II} = \int\left[2\pi\underline{M}_{II}^2 + \frac{(\nabla\times A)^2}{8\pi}\right]d^3 r + \sum_i \frac{\omega_o}{2}\,\sigma^z(i) + 4\pi\int\underline{M}_{II} \cdot \underline{P}^{\perp}\, d^3 r$$

$$+ \frac{4\pi}{3}\int\underline{P}^2\, d^3 r\ , \tag{3.7}$$

where

$$\underline{P}(\underline{r}) = \sum_i \underline{d}_{12}(i)\,\sigma^x(i)\,\delta(\underline{r} - \underline{R}_i)\ . \tag{3.8}$$

If the last term in Eq. (3.7) is omitted, the remaining term is the commonly used hamiltonian H_{II}^D for the study of the Dicke system:

$$H_{II}^D = \sum_{\underline{k}} k a_{\underline{k}}^{\dagger} a_{\underline{k}} + \omega_o \sum_{\underline{k}} b_{\underline{k}}^{\dagger} b_{\underline{k}} - i\sum_{\underline{k}}\sqrt{2\pi k\rho}\,(\underline{d}_{12} \cdot \underline{e})(a_{-\underline{k}} - a_{\underline{k}}^{\dagger})(b_{-\underline{k}}^{\dagger} + b_{\underline{k}})$$

$$\tag{3.9}$$

where we have made use of Eqs. (2.6), (3.5) and (3.8).

We first examine the instability property of H_{II}^D. By regarding $b_{\underline{k}}^{\dagger}$ and $b_{\underline{k}}$ as Bose operators, the polariton dispersion relation is derived in the form [cf. curve (b) in Fig. 1]:

$$k^2 = \varepsilon(\omega)\,\omega^2\ , \tag{3.10}$$

with

$$\varepsilon(\omega) = (\omega_o^2 - \omega^2)/(\tilde{\omega}_o^2 - \omega^2) \ , \tag{3.11}$$

where

$$\tilde{\omega}_o^2 = \omega_o^2 - 8\pi\omega_o\rho d^2 \ . \tag{3.12}$$

From these results we see that the lower branch of the polariton mode becomes unstable if the condition

$$8\pi\rho d^2 \geq \omega_o$$

is satisfied, which is just the strong coupling condition [e.g.,2] for SPT to occur in the system described by the Dicke hamiltonian H_{II}^{D}. However, Eq. (3.11) does not satisfy the Lorentz-Lorenz formula, which Eq. (2.9) satisfies, and Eq. (3.12) does not represent the correct Lorentz shift given by Eq. (2.10) in the absence of the factor 1/3. These shortcomings are shown to be removed by adding the term $(4\pi/3)\int \underline{P}^2 d^3r$. For this end, we examine the instability property of H_{II} in Eq. (3.7), by expressing the P^2 term in terms of $b_k^{\dagger}$ and b_k defined by Eq. (2.6). From the definition (3.1) of $\underline{P}(\underline{r})$ we have

$$\int \underline{P}^2 d^3r = \sum_i \sum_j \underline{d}(i)\cdot\underline{d}(j) \ \delta(\underline{R}_i - \underline{R}_j) \ . \tag{3.13}$$

Using the Fourier expansion form for $\delta(\underline{R}_i - \underline{R}_j)$:

$$\delta(\underline{R}_i - \underline{R}_j) = \frac{1}{V} \sum_k e^{i\underline{k}\cdot(\underline{R}_i - \underline{R}_j)} \ ,$$

and writing the dipole moment operator as

$$\underline{d}(i) = \underline{d}_{12}\left(\sigma^+(i) + \sigma^-(i)\right) \ ,$$

we get

$$(4\pi/3) \int \underline{P}^2 d^3r = \frac{4\pi}{3} \rho d_{12}^2 \sum_k (b_k + b_{-k}^{\dagger})(b_{-k} + b_k^{\dagger}) \ . \tag{3.14}$$

Adding the term (3.14) to H_{II}^{D} in Eq. (3.9) we have the full hamiltonian H_{II} for the system S. Regarding $b_k^{\dagger}$ and b_k as Bose operators as before, we get the polariton dispersion relation which is the same as Eq. (2.8).

We note that if V_{d-d} is not taken into account, the term $2\pi\int \underline{P}^{\perp 2}d^3r$ should be added to H_{II}^{D}, which eliminates the occurrence of the soft-mode instability [cf. Eq. (4.3) and curve (c) in Fig.1].

Therefore, from the results of sections 2 and 3, it is seen that the inclusion of V_{d-d} is essential to produce the soft-mode instability in our treatment, and the revised condition of SPT is the same as that for the $4\pi/3$ catastrophe.

4. MOMENTUM AND ENERGY FLUX OF ELECTROMAGENTIC WAVES

IN TYPE-I AND TYPE-II FORMULATIONS

In the previous section we studied some properties of multipole hamiltonians from the viewpoint of polariton dispersion relations, and achieved a unified understanding of linear instability in both formulations. However, to see the characteristic of the multipole hamiltonian, there remains the point to be observed that the different conjugate field variables $(\underline{A},\underline{M}_I)$ in Eq. (2.4) and $(\underline{A},\underline{M}_{II})$ in Eq. (3.5) are expanded in terms of the same Bose operators.

As a consequence of these different definitions of $a_{\underline{k}}^{\dagger}$ and $a_{\underline{k}}$, we shall show below that if we evaluate the expectation value of the operator $\sum_{\underline{k}} \underline{k} a_{\underline{k}}^{\dagger} a_{\underline{k}}$, in the type-I formulation, we obtain Abraham's result, whereas if we work in the type-II formulation, we get Minkowski's result for the momentum of electromagnetic waves. To evaluate the thermal average of the operator $a_{\underline{k}}^{\dagger} a_{\underline{k}}$, we use the retarded two-time photon Green's function defined by [12]

$$G_{\underline{k}}(t) = -i\theta(t)<[a_{\underline{k}}(t),\ a_{\underline{k}}^{\dagger}(o)]> \ .$$

The thermal average $<a_{\underline{k}}^{\dagger} a_{\underline{k}}>$ is given by the spectral theorem:

$$<a_{\underline{k}}^{\dagger} a_{\underline{k}}> = \int_{-\infty}^{\infty} \frac{-\frac{1}{\pi} I_m G_{\underline{k}}(\omega)}{e^{\beta\omega} - 1}\ d\omega \ ,$$

where the Fourier transform $G_{\underline{k}}(\omega) = \ll a_{\underline{k}} : a_{\underline{k}}^{\dagger} \gg_{\omega}$ is to be evaluated by the equation of motion:

$$\omega \ll a_{\underline{k}} : a_{\underline{k}}^{\dagger} \gg_{\omega} = <[a_{\underline{k}},a_{\underline{k}}^{\dagger}]> + \ll [a_{\underline{k}},H] : a_{\underline{k}}^{\dagger} \gg_{\omega} \ .$$

We shall evaluate $G_k(\omega)$ for the photon-matter system consisting of a radiation field and N two-level atoms without V_{d-d} under the linear approximation. The hamiltonian in the type-I formulation is given by Eq. (2.5) without the last term. The corresponding Green's function $G_{\underline{k}}^{I}(\omega)$ is obtained as

$$G_{\underline{k}}^{I}(\omega) = \frac{(\omega^2-\omega_0^2)\left(\omega+k+(4\pi\omega_0\rho d^2/k)\right)+(4\pi\omega_0^3\rho d^2/k)}{\omega^4-\omega^2[\omega_0^2+8\pi\omega_0\rho d^2+k^2]+\omega_0^2 k^2} \quad . \tag{4.1}$$

The hamiltonian in the type-II formulation is given by Eq. (3.9)
with $2\pi\int P^{\perp 2}d^3r$. The corresponding Green's function $G_{\underline{k}}^{II}(\omega)$ is then
obtained as

$$G_{\underline{k}}^{II}(\omega) = \frac{(\omega^2-\omega_0^2-8\pi\omega_0\rho d^2)(\omega+k)+4\pi k\omega_0\rho d^2}{\omega^4-\omega^2(\omega_0^2+8\pi\omega_0\rho d^2+k^2)+\omega_0^2 k^2} \quad . \tag{4.2}$$

We note that both Green's functions have identical denominators
leading to the same polariton dispersion relation given by [cf.
curve (c) in Fig. 1]

$$k^2 = \frac{\omega_0^2 + 8\pi\omega_0\rho d^2 - \omega^2}{\omega_0^2 - \omega^2} \omega^2 \quad . \tag{4.3}$$

However, the numerators are different from each other. The consequence of this appears as the difference in Z-factors in Eq. (4.4)
below. For both Green's functions, the spectral function $D_{\underline{k}}(\omega)$
defined by

$$D_{\underline{k}}(\omega) = -(1/\pi)\ \mathrm{Im}\ G_{\underline{k}}(\omega)$$

is expressed in the form:

$$D_{\underline{k}}(\omega) = Z_1\delta(\omega-E_1) + Z_1'\delta(\omega+E_1) + Z_2\delta(\omega-E_2) + Z_2'\delta(\omega+E_2) \quad , \tag{4.4}$$

where E_1 and E_2 are the eigenfrequencies for the lower and upper
branches of the polariton mode, respectively, and the factors Z
are the residues at each pole.

In evaluating the momentum or energy flux $\langle S_{\underline{k}}\rangle$ accompanying
to the plane wave state $\cos(\underline{k}\cdot\underline{r} - \omega t)$, we must take into account
the contributions not only of the positive frequency part of the
state $\underline{k}$ but also of the negative frequency part of the state $-\underline{k}$:

$$\langle S_{\underline{k}}\rangle = \underline{k}\ \langle a_{\underline{k}}^{\dagger}a_{\underline{k}} + 1/2\rangle_{+} - \underline{k}\ \langle a_{-\underline{k}}^{\dagger}a_{-\underline{k}} + 1/2\rangle_{-} \tag{4.5}$$

where

$$\langle a_{\underline{k}}^{\dagger}a_{\underline{k}} + 1/2\rangle_{+} = \frac{1}{2}\int_0^{\infty} \coth(\beta\omega/2)\ D_{\underline{k}}(\omega)\,d\omega \quad .$$

$$\langle a_{-\underline{k}}^{\dagger}a_{-\underline{k}} + 1/2\rangle_{-} = \frac{1}{2}\int_{-\infty}^{0} \coth(\beta\omega/2)\ D_{-\underline{k}}(\omega)\,d\omega \quad .$$

In the above expression we included the zero-point fluctuation for the sake of convenience which makes no contribution to $\sum_k <S_{-k}>$. Noting that $D_k(\omega) = D_{-k}(\omega)$, we rewrite Eq. (4.5) as

$$<S_{-k}> = \frac{1}{2} \hat{k} \int_0^\infty \coth(\beta\omega/2) k \, [D_k(\omega) + D_k(-\omega)] \, d\omega \ . \qquad (4.6)$$

With the help of Eq. (4.4), the function $D_k(\omega) + D_k(-\omega)$ is given in the form:

$$D_k(\omega) + D_k(-\omega) = (Z_1 + Z_1') \delta(\omega - E_1) + (Z_2 + Z_2') \delta(\omega - E_2) \ , \qquad (4.7)$$

where terms which make no contribution to the integral in Eq. (4.6) are omitted.

For $G_k^I(\omega)$ given by Eq. (4.1), it follows that

$$k(Z_1 + Z_1') = E_1 \, (\partial E_1/\partial k) \ .$$

The spectral function in Eq. (4.6) is thus obtained in the form

$$k[D_k(\omega) + D_k(-\omega)] = E_1 (\partial E_1/\partial k) \delta(\omega - E_1) + \begin{array}{l}\text{contribution} \\ \text{from branch 2}\end{array} \qquad (4.8)$$

Equation (4.8) shows that the momentum is given by group velocity $\partial E/\partial k$ multiplied by energy E. This correponds to Abraham's result for the momentum of electromagnetic waves (true-momentum) [13,14]. For $G_k^{II}(\omega)$ given by Eq. (4.2), we have

$$k(Z_1 + Z_1') = \varepsilon(E_1) \, E_1 \, (\partial E_1/\partial k) \ , \qquad (4.9)$$

where

$$\varepsilon(\omega) = 1 + \frac{8\pi\omega_0\rho d^2}{\omega_0^2 - \omega^2}$$

is the dielectric function for the medium consisting of two-level atoms without V_{d-d}. Therefore in the type-II formulation we have

$$k[D_k(\omega) + D_k(-\omega)] = \varepsilon(\omega) E_1 (\partial E_1/\partial k) \delta(\omega - E_1) + \begin{array}{l}\text{contribution} \\ \text{from branch 2}\end{array} \qquad (4.10)$$

Equation (4.10) shows that the momentum is equal to Abraham's result multiplied by the dielectric constant.

This result corresponds to Minkowski's result for the momentum of electromagnetic waves (pseudomomentum) [13,14]. This pseudomomentum is obtained also in the type-I formulation if we consider the operator

$$(1/4\pi)\int (\underline{E} + 4\pi\underline{P}^{\perp}) \times \underline{B}\ d^3r = \sum_{\underline{k}} \underline{k}\ a_{\underline{k}}^{\dagger}a_{\underline{k}} + \sum_{\underline{k}} \underline{k}\ B_{\underline{k}}A_{\underline{k}}$$

where

$$B_{\underline{k}} = -i\sqrt{(2\pi\rho d^2/Vk)}\ (b_{\underline{k}} + b_{-\underline{k}}^{\dagger})\ ,\qquad A_{\underline{k}} = a_{\underline{k}}^{\dagger} + a_{-\underline{k}}\ .$$

Evaluating the Green's function $\ll B_{\underline{k}} : A_{\underline{k}}\gg_{\omega}$, we get the corresponding spectral function in the form:

$$k[D_{\underline{k}}(\omega) + D_{\underline{k}}(-\omega)] = 4\pi\chi(E_1)E_1(\partial E_1/\partial k)\delta(\omega - E_1) + \begin{array}{l}\text{contribution}\\ \text{from branch 2}\end{array}\ ,$$

$$(4.11)$$

where

$$4\pi\chi(\omega) = \frac{8\pi\omega_0\rho d^2}{\omega_0^2 - \omega^2}\ .$$

The addition of Eq. (4.11) to Eq. (4.8) leads to the total spectral-function which is just equal to Eq. (4.10).

References

1.   K. Hepp and E.H. Lieb, Ann. Phys. (New York) *76*, 360 (1973).
2.   M. Takatsuji, Phys. Rev. A*10*, 1437 (1974).
3.   W.R. Mallory, Phys. Rev. A*11*, 1088 (1975).
4.   K. Rzazewski, K. Wodkiewicz and W. Zakowicz, Phys. Rev. Letters *35*, 432 (1975).
5.   M. Yamanoi, Phys. Letters *58A*, 437 (1976).
6.   V.I. Emeljanov and Yu. L. Klimontovich, Phys. Letters *59A*, 366 (1976).
7.   E.A. Power, *Introductory Quantum Electrodynamics* (Longmans, Green and Co. LTD, 1964).
8.   J.J. Hopfield, Phys. Rev. *112*, 1555 (1958).
9.   C. Kittel, *Introduction to Solid State Physics*, 4th edition (John Wiley & Sons, Inc.), Chap. 14.
10. S. Takeno and M. Nagashima, Progr. Theor. Phys. *57*, (1977).
11. E.A. Power and S. Zienau, Phil. Trans. Roy. Soc. *251*, 427 (1959).
12. D.N. Zubarev, Usp. Fiz. Nauk *71*, 71 (1960); [Soviet Phys. - Uspekhi *3*, 320 (1960)].
13. J.P. Gordon, Phys. Rev. A*8*, 14 (1973).
14. R. Peierls, Proc. R. Soc. Lond. A*347*, 475 (1976).

THE CRITICAL PROPERTIES OF TWO-LEVEL MODELS OF THE FUNDAMENTAL

HAMILTONIANS*

K. Wódkiewicz[†] and J.H. Eberly

University of Rochester, Rochester, New York

It has been shown [1] recently that the well-advertised "phase-transition" implied by the so-called Dicke Hamiltonian fails to occur if the two-level model of the minimal-coupling atomic Hamiltonian includes the A^2 terms in the interaction, and if the Thomas-Reiche-Kuhn sum rule is invoked. However, the question arises: since the "phase transition" is due to the properties of the $\underline{p} \cdot \underline{A}$ term of the minimal coupling Hamiltonian, and since those properties are shared by the $\underline{d} \cdot \underline{E}$ term in the dipole Hamiltonian, can we not recover the "phase transition" simply by making the unitary transformation from minimal coupling to dipole Hamiltonian at the outset, and then working entirely with $\underline{d} \cdot \underline{E}$ thereafter? The answer is that we cannot, and the reason why not is connected with a much-neglected term in the dipole Hamiltonian, and with the nature of two-level models.

To see this clearly we have to study carefully how a two-level model of a fundamental theory is obtained. If the starting point is the usual one: the minimal coupling theory with the assumption that all the atoms are contained within a volume of dimensions much smaller than the wavelength of the radiation field (the dipole approximation), then we have to deal with the following fundamental Hamiltonian:

$$H = \frac{1}{8\pi} \int d^3x (\underline{E}^2 + \underline{B}^2) + \sum_{i=1}^{N} \left[\frac{(\underline{P}_i - \frac{e}{c} \underline{A})^2}{2m} + V(\underline{r}_i) \right] \tag{1}$$

---

† permanent address: Warsaw University, Warsaw, Poland

where the notation has its standard meaning. According to well-known procedures [2] we can obtain a 2 × 2 truncation of the fundamental Hamiltonian H. If only one mode of the quantized electromagnetic field is taken into account, we obtain the following two-level model Hamiltonian of the minimal coupling theory:

$$H_2 = \hbar\omega a^\dagger a + \hbar\omega_o S_3 + \lambda(a+a^\dagger)(S+S^\dagger) + N\kappa(a+a^\dagger)^2 \tag{2}$$

where

$$S_3 = \sum_{i=1}^{N} \sigma_3^i \quad , \quad S = \sum_{i=1}^{N} \sigma^i \quad , \quad \text{and} \quad S^\dagger = (S)^\dagger$$

are atomic operators built out of individual Pauli matrices representing a single two-level atom (spin 1/2) algebra. The coupling constants λ and κ are given by the following formulas:

$$\lambda = \omega_o d \left(\frac{2\pi\hbar}{\omega V}\right)^{\frac{1}{2}} \quad , \tag{3a}$$

$$\kappa = \frac{e^2 \pi\hbar}{m\omega V} \quad , \tag{3b}$$

where all the constants can be obtained from the fundamental Hamiltonian [1] and d is the dipole moment between the two atomic levels. Another form of the fundamental Hamiltonian can be obtained from the minimal coupling Hamiltonian H by means of the following unitary transformation [3]

$$\mathscr{H} = U^\dagger H U \tag{4a}$$

$$U = \exp\left[-\frac{ie}{\hbar c} \sum_{i=1}^{N} \underline{r}_i \cdot \underline{A}\right] . \tag{4b}$$

If we denote all of the transformed variables as well as the transformed Hamiltonian by script letters (e.g., $\mathscr{E} = U^\dagger E U$), the dipole Hamiltonian can be written in the following form [3]:

$$= \frac{1}{8\pi} \int d^3x (\underline{\mathscr{E}}^2 + \underline{\mathscr{B}}^2) + \sum_{i=1}^{N} \left[\frac{\underline{\pi}_i^2}{2m} + V(r_i)\right] - e\sum_{i=1}^{N} \underline{r}_i \cdot \underline{\mathscr{E}} + \sum_{ij=1}^{N} V_{ij} \tag{5}$$

where

$$V_{ij} = 2\pi \int d^3x \, |\underline{P}^i \cdot \underline{P}^j| \tag{6a}$$

and

$$P^i(\underline{x},t) = e \int \frac{d^3k}{(2\pi)^3} \left[\underline{r}^i - \frac{k(k \cdot \underline{r}^i)}{k^2} \right] e^{i\underline{k} \cdot \underline{x}} \quad . \tag{6b}$$

The vector P^i given by Eq. (6b) is usually called the transverse part of the polarization vector. The contribution of the polarization vector of the fundamental Hamiltonian $\mathcal{H}$ we will call, for simplicity, the polarization part or the P^2 term.

We can proceed now in the standard way to obtain the two-level truncation of the fundamental Hamiltonian $\mathcal{H}$. In a single-mode approximation of the electromagnetic field we obtain the following two-level model Hamiltonian of the dipole theory [4]

$$\mathcal{H}_2 = \hbar\omega a^\dagger a + \hbar\omega_o S_3 + \tilde{\lambda}(a+a^\dagger)(S+S^\dagger) + \tilde{\kappa}(S+S^\dagger)^2 \tag{7}$$

where the coupling constants are given by the following formulas

$$\tilde{\lambda} = d \left(\frac{2\pi\hbar\omega}{V} \right)^{\frac{1}{2}} , \tag{8a}$$

$$\tilde{\kappa} = \frac{2\pi d^2}{V} \quad . \tag{8b}$$

The last term in expression (7) represents the two-level truncation of the P^2 term from the Hamiltonian $\mathcal{H}$. If we adopt the rotating wave approximation (RWA), thus eliminating in the Hamiltonian the terms which in first order of perturbation theory do not conserve energy, we obtain from H_2 and $\mathcal{H}_2$ the following Hamiltonians:

$$H_2^{RWA} = \hbar\omega a^\dagger a + \hbar\omega_o S_3 + \lambda(aS^\dagger+Sa^\dagger) + N\kappa(aa^\dagger+a^\dagger a) , \tag{9}$$

$$\mathcal{H}_2^{RWA} = \hbar\omega a^\dagger a + \hbar\omega_o S_3 + \tilde{\lambda}(aS^\dagger+Sa^\dagger) + \tilde{\kappa}(S^\dagger S+SS^\dagger) \quad . \tag{10}$$

It is easy to check that both truncated Hamiltonians given by Eqs. (9) and (10) have the following constants of motion

$$S^2 = S_1^2 + S_2^2 + S_3^2 \quad ; \qquad C = S_3 + a^\dagger a \quad . \tag{11}$$

These two constants of motion allow us to decompose the entire Hilbert space of the system into irreducible representations of the group SU(2) with spin $S^2 = S(S+1)$ and excitation number C:

$$\mathcal{F} = \overset{N/2}{\underset{S \geq 0}{\oplus}} r(N,S) \overset{\infty}{\underset{C=-S}{\oplus}} \mathcal{F}(N,S,C) \quad ,$$

and now within each $\mathcal{F}(N,S,C,)$ space the Hamiltonian $\mathcal{H}_2^{RWA}$ and H_2^{RWA} can be diagonalized exactly [4]. It is possible to show that there exists a state with energy lower than the ground state of the system $E_O = - N\hbar\omega_O/2$ if

$$f^{RWA} = \frac{\hbar\omega_O(\hbar\omega+2\kappa N)}{N\lambda^2} < 1 \quad \text{for} \quad H_2^{RWA} \quad , \tag{12a}$$

and

$$\tilde{f}^{RWA} = \frac{\hbar\omega\left(\hbar\omega_O+2\tilde{\kappa}(N-1)\right)}{N\tilde{\lambda}^2} < 1 \quad \text{for} \quad \mathcal{H}_2^{RWA} \quad . \tag{12b}$$

Condition (12a) is possible to satisfy if the A^2 term is not taken into account. If $\kappa = 0$, $f^{RWA} > 1$ for $\lambda > \lambda_{crit}$. The same thing happens for the dipole theory if $\tilde{\kappa} = 0$ or $N = 1$, $\tilde{f}^{RWA} > 1$ for $\lambda > \tilde{\lambda}_{crit}$ (and $\tilde{\lambda}_{crit} = (\omega_O/\omega)\lambda_{crit}$). If the A^2 term is taken into account condition (12a) is impossible to satisfy if the Thomas-Reiche-Kuhn (TRK) sum rule is recalled and used to relate the constants λ and κ:

$$\frac{\hbar e^2}{md^2\omega_O} < 2 \quad . \tag{13}$$

Condition (12b) with $\tilde{\kappa} \neq 0$ and $N \neq 1$ is impossible to satisfy and this can be shown without any use of the atomic sum rule. In both cases (12a) and (12b) it is possible to show that min $f^{RWA} = 2$ and min $\tilde{f}^{RWA} = 2$.

We then see that in both theories the instability of the ground state predicted by Narducci et al [5] for the Hamiltonian H_2 with $\kappa = 0$ is impossible if all the terms are taken into account. The same prediction is true for the dipole theory, and the crucial role is now played by the P^2 term.

However we must also remark that, apart from the collapsing properties of the ground state, the two Hamiltonians H_2^{RWA} and $\mathcal{H}_2^{RWA}$ have different energy spectra, i.e., are not unitarily equivalent. This point is discussed in detail elsewhere [4].

If we do not adopt the RWA, the operators C and S^2 are no longer constants of motion and an exact diagonalization of the Hamiltonians is impossible. But it is possible to use the variational method to establish the energy of the ground state for H_2 and $\mathcal{H}_2$. For the trial variational eigenfunction we will choose the following state:

$$\psi = |\alpha> \otimes |\theta,\rho> \quad , \tag{14}$$

where $|\alpha\rangle$ is the coherent state of the photon operator and $|\theta,\phi\rangle$ is the coherent Bloch state. We obtain the energy of the ground state by a proper minimalization of the following expectation values:

$$W = \langle\psi|H_2|\psi\rangle \quad , \quad \mathcal{W} = \langle\psi|\mathcal{H}_2|\psi\rangle \quad . \tag{15}$$

Again we can check that the energy of H_2 and $\mathcal{H}_2$ is different but the collapse of the ground state is impossible if the A^2 term and the P^2 term are taken into account in the Hamiltonians [4]. The inclusion of the counter rotating terms lowers the lower bound of the functions f and $\tilde{f}$: min f = 1 and min $\tilde{f}$ = 1.

In the thermodynamic limit the Hamiltonian H_2 without the A^2 term was discussed by Hepp and Lieb [6]. The crucial role of the A^2 term in the phase transition problem was already shown in the literature [1]. The partition function for the theory represented by the Hamiltonian H_2 was computed with the help of the Wang-Hioe method [1]. This method cannot be applied to the Hamiltonian $\mathcal{H}_2$. The term coming from the truncation of the P^2 term is the source of troubles and a simple evaluation of the trace over the atomic states is impossible. In order to compute the partition function

$$\mathcal{Z}_N = \mathrm{Tr}\, \exp(-\beta\mathcal{H}_2) \quad , \tag{16}$$

we modify the Wang-Hioe method in the following way:

i) compute the trace over the atoms, treating them "classically" (replace the atom operators by coherent Bloch amplitudes);
ii) compute the trace over the field exactly;
iii) use the steepest descent method to evaluate the bifurcation properties of the canonical partition function.

According to powerful theorems of Lieb [7] this procedure is rigorous in the thermodynamic limit. After these calculations, we obtain the following condition for the second order phase transition [4]

$$\frac{\hbar\omega(\hbar\omega_0 + 4\tilde{\kappa}N)}{4N\tilde{\lambda}^2} < 1 \tag{17}$$

with the values of $\tilde{\kappa}$ and $\tilde{\lambda}$ given by Eqs. (8a) and (8b). It is easy to check that this condition is never satisfied and due to the presence of the P^2 term the phase transition is impossible. This condition has been obtained without using the TRK sum rule.

In conclusion we have shown that a system of N two-level atoms coupled to a single mode of a quantized radiation field may have different energy spectra depending upon which two-level truncation of the fundamental Hamiltonian is taken into consideration. We have

seen that the two-level truncation of the fundamental theories is
not a unitarity-preserving procedure. The two-level models H_2 and
$\mathcal{H}_2$ fail to be unitarily equivalent. We have shown that the predictions of the instability of the ground state given by Narducci
et al [5] are no longer valid if the A^2 term is taken into the
minimal coupling Hamiltonian and the P^2 term into the dipole Hamiltonian. Thus this general property of stability is independent of
the particular form of the two-level Hamiltonian used in the calculations. The same kind of argument is also valid for the second
order phase transition problem. The general conclusion of this
paper can be summarized by the noncommutativity of the following
diagram:

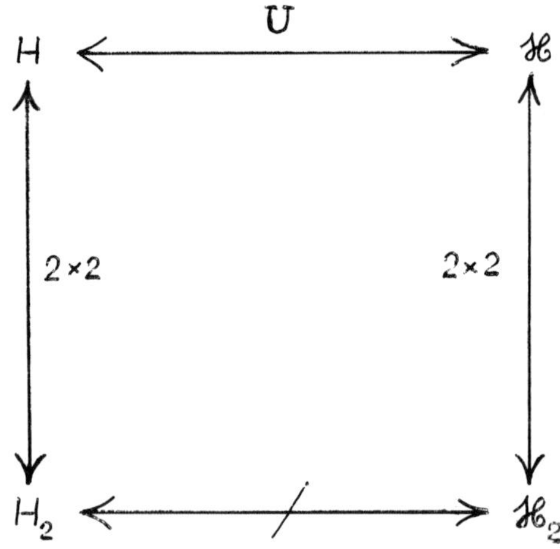

Acknowledgements

     We thank Prof. L. Narducci for his letter to one of us (JHE)
in which he raised the basic question (mentioned in the opening
paragraph) at an early stage of our interest in the subject.

* Research partially supported by the US Energy Research and
Development Administration.

References

1.   K. Rzążewski, K. Wódkiewicz and W. Żakowicz, Phys. Rev. Lett.
     35, 432 (1975).
2.   L. Allen and J.H. Eberly, *Optical Resonance and Two-Level
     Atoms* (Wiley, New York, 1975).

3.   E.A. Power and S. Zienau, Phil. Trans. R. Soc. A*251*, 427 (1959).
4.   K. Wódkiewicz, Ph.D. Thesis (University of Rochester, 1976).
5.   L.M. Narducci, M. Orszag and R.A. Tuft, Phys. Rev. A*8*, 1892,
     (1973). L.M. Narducci, M. Orszag and R.A. Tuft, Collective
     Phenomena 1, 113 (1973).
6.   K. Hepp and E.H. Lieb, Ann. Phys. (New York) *76*, 360 (1973).
7.   E.H. Lieb, Comm. Math. Phys. *31*, 377 (1973).

CONTRIBUTIONS OF THE A^2 AND COUNTER-ROTATING TERMS TO THE PHASE TRANSITION IN THE DICKE HAMILTONIAN

R. Gilmore

University of South Florida, Tampa, Florida

C. M. Bowden

Redstone Arsenal, Alabama

Recently, Rzążewski, Wódkiewicz, and Żakowicz [1] showed that the phase transiton disappears when the A^2 term is included in the Dicke model. They also claim that these results do not depend on the presence of the counter-rotating terms.

The purpose of this paper is to show that the phase transition will persist in the presence of the A^2 terms if the counter-rotating terms are neglected. In this case, the gap equation reduces to the expression first derived by Hepp and Lieb. We study the Dicke model Hamiltonian including the counter-rotating and A^2 terms

$$H = a^+a + \varepsilon \sum_{k=1}^{N} \frac{1}{2}\sigma_k^z + \frac{\lambda}{\sqrt{N}} \sum_{k=1}^{N} (a^+\sigma_k^- + a\sigma_k^+ + r^*a^+\sigma_k^+ + ra\sigma_k^-)$$

$$+ (\gamma^*a^+ + \gamma a)(\gamma^*a^+ + \gamma a) . \tag{1}$$

The relative phases of the ground and excited state wave-functions may be so chosen that the coupling constant for the resonant terms ($a^+\sigma_k^-$, $a\sigma_k^+$) is real. Then, the coupling constants λr, λr^* for the counter-rotating terms are, in general, complex. The phase of r and γ are not independent, since $r^*\gamma^2$ has constant phase. The constants appearing in (1) are

$$\lambda^2 = \varepsilon^2 d^2 \left(\frac{2\pi\hbar}{w}\right)\frac{N}{V} \quad ,$$

$$\gamma^*\gamma = \frac{e^2}{2m}\left(\frac{2\pi\hbar}{w}\right)\frac{N}{V} = k \quad .$$

(2)

The thermodynamic properties of (1) can be determined using the method introduced by Wang and Hioe.[2] However, an even simpler method [3,4] is preferred because this new method leads to a much more transparent understanding of the extent to which the phase transition depends on the presence or absence of the counter-rotating terms. This new method was suggested by Gibberd's treatment [5] of the Dicke model using a "thermodynamically equivalent Hamiltonian." All three methods lead to the same gap equation and critical behavior.

In this alternative treatment, the Hamiltonian is linearized by expanding each shift operator with respect to a disposable c-number parameter: $a = (a - \mu) + \mu$; $\sigma_k^- = (\sigma_k^- - \nu_k) + \nu_k$. These disposable parameters are then chosen to minimize the difference between the free energy associated with (1) and the free energy associated with the linearized version of (1). The choices that minimize this difference and yield a thermodynamically equivalent Hamiltonian are $\mu = <a>_1$, $\nu_k = <\sigma_k^->_2$, where the expectation values are taken with respect to $H_1(\nu_k)$ and $H_2(\mu)$, respectively. The Hamiltonian $H_1(\nu_k)$ is obtained by replacing $\sigma_k^- \to <\sigma_k^-> = \nu_k$ in (1); $H_2(\mu)$ is obtained similarly. Since μ and ν_k are expectation values of shift operators a, σ_k^-, they are interpreted as order parameters for the field and atomic subsystems, respectively. This procedure gives a system of coupled nonlinear order parameter equations.

The order parameter $\mu = <a>$, $\nu_k = <\sigma_k^-> = \nu$ are zero in the disordered state and nonzero in the ordered state. An alternative procedure for computing a system of coupled nonlinear order parameter equations involves computing the equation of motion $d<a>/dt = (i/h)<[H,a]>$. In thermodynamic equilibrium the commutator is zero. In the thermodynamic limit the expectation value factors into the product of expectation values for the field and atomic operators with respect to $H_1(\nu_k)$ and $H_2(\mu)$, respectively. The coupled nonlinear equations obtained are

$$\begin{bmatrix} 1 + 2\gamma^*\gamma & 2\gamma^{*2} \\ 2\gamma^2 & 1 + 2\gamma^*\gamma \end{bmatrix}\begin{bmatrix} \mu \\ \mu^* \end{bmatrix} = -\lambda\sqrt{N}\begin{bmatrix} 1 & r^* \\ r & 1 \end{bmatrix}\begin{bmatrix} \nu \\ \nu^* \end{bmatrix}$$

(3)

and

$$\varepsilon \begin{bmatrix} \nu \\ \\ \nu^* \end{bmatrix} = \frac{\lambda \langle \sigma^z_k \rangle}{\sqrt{N}} \begin{bmatrix} 1 & r^* \\ \\ r & 1 \end{bmatrix} \begin{bmatrix} \mu \\ \\ \mu^* \end{bmatrix} \ . \tag{4}$$

In the neighborhood of the critical temperature, if there is a critical temperature, the nonzero-order parameters approach zero as $|T - T_c| \to 0$. To locate the critical temperature, linearize (3) and (4) at T_c. This involves taking the limit $\langle \sigma^z \rangle \to -\tanh \tfrac{1}{2} \beta_c \varepsilon$. The resulting coupled linear equations can then be solved by standard eigen-value methods. There will be a nontrivial solution if, and only if,

$$\det \left| \ I_2 \ - \ \frac{\lambda^2 \tanh \tfrac{1}{2} \beta_c \varepsilon}{\varepsilon} \ A^{-1} B^2 \ \right| \ = \ 0 \ . \tag{5}$$

In this expression, A is the matrix on the left in (3), and B is the matrix on the right. If ζ_1, ζ_2 are the two roots of $A^{-1} B^2$, the two gap equations are

$$\zeta_i \ \frac{\lambda^2}{\varepsilon} \ \tanh \tfrac{1}{2} \beta_c \varepsilon \ = \ 1 \ . \tag{6}$$

For r and γ real, $0 \le r \le 1$, the roots ζ_i are $\zeta_1 = (1 + r)^2 / 1 + 4k$, $\zeta_2 = (1 - r)^2$.

The gap equation for ζ_1 is

$$\tanh \tfrac{1}{2} \beta_c \varepsilon \ = \ \left\{ \frac{\varepsilon}{4\lambda^2} \ + \ \frac{\varepsilon k}{\lambda^2} \right\} \left(\frac{2}{1 + r} \right)^2 \ . \tag{7}$$

Since $(2/1 + r)^2 \ge 1$ for $0 \le r \le 1$, $\varepsilon/4\lambda^2 > 0$, and $\varepsilon k/\lambda^2 > 1$, by [1] the Thomas-Reiche-Kuhn sum rule $e^2 h/2m > \varepsilon d^2$, the root ζ_1 will never lead to a critical temperature.

The gap equation for ζ_2 is

$$\frac{\lambda^2}{\varepsilon} \ (1 - r)^2 \ \tanh \tfrac{1}{2} \beta_c \varepsilon \ = \ 1 \ . \tag{8}$$

In the absence of the counter-rotating terms ($|r| = 0$), (8) reduces to the gap equation for the Dicke Hamiltonian without A^2 terms first derived by Hepp and Lieb.

The decisive contribution made by the counter-rotating terms on the behavior of the critical temperature is apparent from (7) and (8). In the absence of the A^2 terms, the counter-rotating terms increase the critical temperature. In the presence of the A^2 term, the counter-rotating terms decrease the critical temperature. Under the assumptions γ and r real, the phase transition disappears for $r > 1 - (\sqrt{\epsilon}/\lambda)$.

We have shown that the location of the critical temperature for the Hamiltonian (1) depends critically on the extent to which the counter-rotating terms are present, contrary to the claim made in Ref. (1). This erroneous claim was made because of a misapplication, by one of the authors, of Laplace's method for estimating the asymptotic form of an integral, at a saddle point rather than at a minimum [6]. The mistake was first noted in Ref. (7), and corrected in Ref. (8).

The extreme detuning sensitivity of gain in a CW laser is well known. If we assume that the detuning sensitivity has a Lorentzian shape of the form $|\lambda(\omega)|^2 = |\lambda(\omega_0)/(1 + i\tau(\omega-\omega_0))|^2$ around the resonance ω_0, with $\lambda(\omega_0)$ the coupling constant at the resonant frequency and τ a characteristic bath relaxation time, then typically $\omega_0 \sim 10^{15}$ sec^{-1}, $\tau \sim 10^{-8}$ sec. The ratio of the contributions of the counterrotating terms to the resonant interaction terms is typically $r \sim 10^{-7} \simeq 0$. In view of the recently discovered isomorphism [9] between the Dicke model subject to nonequilibrium steady-state boundary conditions (laser model) and the Dicke model subject to equilibrium boundary conditions (Hepp-Lieb), the claims by Rzążewski, Wódkiewicz, and Żakowicz and r = 1 are unsupportable.

References

1. K. Rzążewski, K. Wódkiewicz, and W. Żakowicz, Phys. Rev. Lett. *35*, 432 (1975).
2. Y. K. Wang and F. T. Hioe, Phys. Rev. A*7*, 831 (1973).
3. R. Gilmore and C. M. Bowden, Phys. Rev. A*13*, 1898 (1976).
4. R. Gilmore and C. M. Bowden, J. Math. Phys. *17*, 1617 (1976).
5. R. W. Gibberd, Aust. J. Phys. *27*, 241 (1974).
6. K. Wódkiewicz, private communication. K. Rzążewski and W. Żakowicz, private communication.
7. R. Gilmore, Phys. Lett. A*55*, 459 (1976).
8. R. Gilmore and C. M. Bodwen, unpublished.
9. R. Gilmore and L. M. Narducci, "Laser as Catastrophe", this Volume, p. 81.

TIME-DEPENDENT PROJECTION-OPERATOR TECHNIQUES IN QUANTUM OPTICS

R.H. Picard

Rome Air Development Center, Hanscom Air Force Base, Mass.

C.R. Willis

Boston University, Boston, Mass.

1. INTRODUCTION

We wish to describe the applications in quantum optics of a
systematic approach to the analysis of coupled systems. This
approach enables one to treat the coupled systems on an equal
footing without assuming that any part of the system is large and
reservoir-like. In the Schrödinger picture this approach leads to
a description of the system in terms of exact generalized master
equations (ME's) which are obtained by operating on the quantum
Liouville equation with a time-dependent projection operator (TDPO).

The projection-operator (PO) techniques developed by
Nakajima [1] and Zwanzig [2] for deriving exact ME's are well-known
and in recent years have found considerable use in the formulation
of quantum-optical problems.[3] In the Nakajima-Zwanzig approach,
one partitions the density operator (DO) $F(t)$ into relevant and
irrelevant parts, $F_R(t)$ and $F_I(t)$, by a projection operator P. By
eliminating $F_I(t)$ from the Liouville equation, one can obtain a
closed exact equation for $\dot{F}_R(t)$, the time rate of change of the
relevant part of the DO, alone. Robertson [4] and Kawasaki and
Gunton [5] introduced generalized PO's which are time-dependent
$P=P(t)$ for application to the description of the dynamics of macroscopic system variables in the hydrodynamic regime. More recently,
we have used a different type of TDPO to derive ME's for a variety
of problems in the kinetic theory of fluids and in quantum
optics.[6-9]

Since the exact ME's are no more easily solvable than the
starting Liouville equation, the advantage of a particular PO lies

in the ease with which it allows one to obtain approximate ME's
adequately describing the phenomena being investigated. TDPO
techniques provide alternative starting points for the systematic
derivation of approximate ME's. Their use allows one to obtain
especially powerful descriptions of coupled systems in a low level
of approximation. In this paper we will summarize the application
of TDPO techniques to quantum optics by citing examples which we
hope will illustrate the logical simplicity, power, and versa-
tility of TDPO techniques.

2. COMPARISON OF PO AND TDPO APPROACHES TO DERIVATION OF ME'S

    In both the ordinary PO and the TDPO approaches, the starting
point is the quantum Liouville equation

$$\dot{F}(t) = LF(t), \tag{1}$$

where F is the system density operator and L is the Liouvillian
operator related to the Hamiltonian H by $L \equiv (-i/\hbar)[H,\ldots]$. We
then decompose F into relevant and irrelevant parts,

$$F(t) = F_R(t) + F_I(t), \tag{2}$$

by use of a PO.

    In the ordinary PO approach, we have

$$F_R(t) \equiv P\ F(t) \tag{3}$$

and

$$F_I(t) = (1-P)\ F(t), \tag{4}$$

where

$$P^2 = P. \tag{5}$$

The exact closed ME for $\dot{F}_R$ is then obtained by operating on Eq. (1)
with P to obtain an equation for $\dot{F}_R$ and with (1-P) to obtain an
equation for $\dot{F}_I$, formally integrating the F_I equation, and using
the result to eliminate F_I from the $\dot{F}_R$ equation. The result is

$$\dot{F}_R(t) - PLF_R(t) = PLG(t-t_o)F_I(t_o)$$

$$+ \int_{t_o}^{t} dt'\ PLG(t-t')(1-P)\ LF_R(t'), \tag{6}$$

where

$$G(\tau) \equiv \exp[(1-P)L\tau] . \tag{7}$$

The left-hand side describes the time-development in the relevant
subspace, while the right-hand side describes the development in
the irrelevant subspace expressed in terms of $F_R(t)$. The DO $F_I(t)$
appears only in terms of its value at the initial instant t_o. For
most choices of F_R, this term vanishes, since

$$F_I(t_o) = 0 \tag{8}$$

In the TDPO approach one proceeds in a similar fashion.
Equations (3) and (4) are replaced simply by

$$F_R(t) = P(t)F(t) \tag{9}$$

$$F_I(t) = [1-P(t)]F(t), \tag{10}$$

while the PO satisfies a more general relation than Eq. (5),

$$P(t)P(t') = P(t) \tag{11}$$

for any t and t'. In fact, one should consider Eq. (11) to be
the defining property of a TDPO. The exact closed ME for $\dot{F}_R$ is
obtained by operating on Eq. (1) with $P(t)$. The resulting equations, replacing Eqs. (6) and (7), are

$$\dot{F}_R(t) - PLF_R(t) = P(t)LG(t,t_o)F_I(t_o)$$

$$+ \int_{t_o}^{t} dt' \; P(t)LG(t,t')[1-P(t')] \; LF_R(t') \; , \tag{12}$$

$$G(t,t') \equiv T \exp\{\int_{t'}^{t} dt''[1-P(t'')]L\} \; . \tag{13}$$

Notice that G is now no longer a function of a time difference; T
is the Dyson time-ordering operator. The interpretation of the
terms in Eq. (12) is much the same as that for Eq. (6), and we
assume that Eq. (8) still applies. In order to derive and use
Eq. (12), we must impose a third condition on the TDPO $P(t)$, in
addition to those contained in Eqs. (9) and (11), brought about by
the fact that P no longer commutes with time differentiation as it
did in the case of a time-independent PO. We require
$[P(t), d/dt] F(t) = 0$, otherwise stated as

$$\dot{F}_R(t) = P(t)\dot{F}(t). \tag{14}$$

3. THE QUANTUM-OPTICAL SYSTEM

We consider a radiation field (degrees of freedom denoted f)
coupled to N atoms (degrees of freedom denoted $1,\ldots,\alpha,\ldots,N$),

referred to as the system. The coupling constant is denoted by μ.
The system is open, since each of the atoms and the field are
coupled to independent reservoirs which account for pumping, decay,
and dephasing.

The system DO is a function of all system degrees of freedom,

$$F(t) \equiv F(1,\ldots,\alpha,\ldots,N,f;t). \tag{15}$$

We will usually be able to suppress the dependence of operators on
f without loss of clarity. Several reduced DO's will be important
in our work; these are the DO's for N atoms

$$\rho_N(t) \equiv \mathrm{Tr}_f \, F(t), \tag{16}$$

for the αth atom and the field

$$\chi(\alpha,t) \equiv \underset{\substack{\beta \\ (\beta \neq \alpha)}}{\Pi'} \; \mathrm{Tr}_\beta \, F(t), \tag{17}$$

for the αth atom

$$\rho_1(\alpha,t) \equiv \mathrm{Tr}_f \, \chi(\alpha,t), \tag{18}$$

and for the field

$$R(t) \equiv \mathrm{Tr}_\alpha \, \chi(\alpha,t). \tag{19}$$

The system Hamiltonian H is the usual quantum-optical Hamiltonian, defined in Eqs. (4.3) - (4.7) of Ref. 6, for example. We
will not use the explicit form below, but note only that H is a
sum of matter, field, and interaction Hamiltonians

$$H \equiv H^O + H' \equiv H_m + H_f + H' \tag{20}$$

and that H_m and H' are sums of single-atom operators,

$$H_m = \sum_{\alpha=1}^{N} H_\alpha^O \, , \qquad H' = \sum_{\alpha=1}^{N} H_\alpha' \, . \tag{21}$$

The corresponding Liouville operators L^O, L', and so forth, satisfy
relations similar to Eqs. (20) and (21).

The reservoirs will be characterized by simple relaxation
rates and treated in the simplest possible fashion. Under these
conditions, we can easily include the effects of reservoirs in the

ME's at a later time rather than incorporating them from the beginning. We will quote exact ME's below without incorporating reservoirs, to simplify the notation, while approximate ME's will include their effects.

We will consider three distinct choices for $F_R(t)$ below as examples of applications of TDPO techniques to quantum optics. These three cases differ according to whether or not F_R contains atom-atom (AA) and atom-field (AF) correlations.

4.   CASE 1:   AA CORRELATIONS; NO AF CORRELATIONS

In the first case the relevant part of the system DO is chosen to be a product of the reduced field and N-atom DO's,

$$F_R(t) = R(t)\ \rho_N(t) \ . \tag{22}$$

The TDPO which yields this F_R is

$$P(t) = R(t)\ Tr_f + \rho_N(t)\ Tr_m - R(t)\ \rho_N(t)\ Tr, \tag{23}$$

where Tr_m is a trace over all matter variables and $Tr \equiv Tr_m Tr_f$. It can be checked easily that $P(t)$ given by Eq. (23) satisfies the three required conditions, Eqs. (9), (11), and (14). This TDPO should be contrasted with the usual time-independent PO's designed for treating the relaxation of a material system or of a field in the presence of a reservoir, [10,3] $P = R_{eq}\ Tr_f$ and $P = \rho_{eq}\ Tr_m$, respectively, R_{eq} and ρ_{eq} being equilibrium DO's. Whereas the TDPO treats the coupled systems symmetrically, the usual PO's do not, but are set up to deal with situations where one of the coupled systems is near equilibrium.

The exact ME for $\dot{F}_R$, Eq. (12), can be transformed into *exact* closed coupled ME's for R and ρ_N by tracing over matter and field, respectively,

$$\dot{R}(t) - (L_f + <L'>_{m,t})\ R(t)$$
$$= \int_{t_o}^{t} dt'\ Tr_m L' G(t,t')\ \Delta_{t'} L' R(t')\ \rho_N(t'), \tag{24a}$$

$$\dot{\rho}_N(t) - (L_m + <L'>_{f,t})\ \rho_N(t)$$
$$= \int_{t_o}^{t} dt'\ Tr_f\ L' G(t,t')\ \Delta_{t'} L' R(t')\ \rho_N(t') \ , \tag{24b}$$

where

$$\langle L'\rangle_{m,t} \equiv \mathrm{Tr}_m L' \rho_N(t), \quad \langle L'\rangle_{f,t} \equiv \mathrm{Tr}_f L' R(t) \tag{25a}$$

are partial averages with respect to matter or field and

$$\Delta_t L' \equiv L' - \langle L'\rangle_{m,t} - \langle L'\rangle_{f,t}. \tag{25b}$$

In the case in which there is no average field or polarization, Eqs. (24) simplify, since $\langle L'\rangle_m = \langle L'\rangle_f = 0$ and $\Delta_t L' = L'$.

From Eqs. (24) one may derive approximate coupled ME's for $\dot{R}$ and $\dot{\rho}_N$ in the first Born approximation (FBA). The FBA is obtained by setting $L' = 0$ in G in Eqs. (24), in which case the right-hand sides of the ME's are $\sim \mu^2$. The equation for $\dot{\rho}_N$ becomes

$$\dot{\rho}_N(t) - (L_m + \langle L'\rangle_{f,t} - K_m)\,\rho_N(t)$$

$$= \int_{t_o}^{t} d\tau e^{-\nu\tau}\,\mathrm{Tr}_f\, L'\tilde{\Delta}_{t-\tau} L'(-\tau)\,\tilde{R}(t-\tau)\,\tilde{\rho}_N(t-\tau), \tag{26}$$

where $L'(-\tau)$, $\tilde{R}$, and $\tilde{\rho}_N$ are interaction-picture operators, coinciding with Schrödinger-picture operators at time t and $\tilde{\Delta}_t L'$ is obtained from $\Delta_t L'$ by replacing R, ρ_N by $\tilde{R}$, $\tilde{\rho}_N$. We have incorporated the influence of reservoirs through the linear operator K_m and the factor $e^{-\nu\tau}$, where ν is the fastest reservoir relaxation rate. Equation (26) and a completely analogous equation for $\dot{R}(t)$ are a non-Markovian generalization of the coupled superradiance ME's which we introduced at the last Rochester Conference.[11,12] These equations with average fields set equal to zero allow for stimulated processes, non-Markovian behavior including retardation, and a multimode radiation field, all of which appear to be necessary to describe superradiance adequately in some regimes.[13]

5.   CASE 2:   NO AA OR AF CORRELATIONS

In the second case we destroy the AA correlations in the F_R of case 1 and consider that F_R is a product of N+1 reduced DO's, one for the field and one for each atom,

$$F_R(t) = R(t)\,\prod_\alpha \rho_1(\alpha,t). \tag{27}$$

The TDPO yielding this form of F_R is

$$P(t) = R(t) \; \mathrm{Tr}_f \sum_\alpha \prod_\beta{}' \; \rho_1(\beta,t) \; \mathrm{Tr}_\beta$$

$$+ \prod_\alpha \rho_1(\alpha,t) \; \mathrm{Tr}_m - NR(t) \prod_\alpha \rho_1(\alpha,t) \; \mathrm{Tr}, \qquad (28)$$

where a prime on a product (or sum) indicates a previously occurring index is omitted (in this case $\beta=\alpha$). The ME for $\dot{F}_R$ is equivalent to *exact* coupled ME's for $\dot{R}$ and $\dot{\rho}_1$, where the $\dot{R}$ equation is identical to Eq. (24a) and the $\dot{\rho}_1$ equation is

$$\dot{\rho}_1(\alpha,t) - (L_m + <L'>_{f,t}) \; \rho_1(\alpha,t)$$

$$= \int_{t_0}^{t} dt' \mathrm{Tr}_f L'_\alpha \prod_\beta{}' \; \mathrm{Tr}_\beta \, G(t,t') \Delta_{t'} L' R(t') \prod_\gamma \rho_1(\gamma,t'). \qquad (29)$$

The exact coupled ME's, Eqs. (24a) and (29), can be used as a starting point to derive ME's for a single-mode laser, including effects higher order in N. In order to obtain coherent behavior one must go beyond the FBA, that is to infinite order in μ^2. If only terms of order N are kept in G and one makes the Markov approximation and sets average-field terms equal to zero, one has the physical content of most laser theories. The approach is probably closest in spirit to the laser Boltzmann equations derived by one of us.[14] In addition, one can obtain correction terms of order N^2, that is two-atom contributions. We will not pursue this approach any further, however, since the evaluation of two-atom contributions will be simpler with the F_R of case 3.

6. CASE 3: AF CORRELATIONS; NO AA CORRELATIONS

In the third case [8] we relax the condition assumed until now that F_R contains no AF correlations. This will allow us to obtain a good description of coherent laser behavior already to lowest order in μ^2, by contrast with case 2. First, however we will discuss exact results for this case. The relevant part of the DO is

$$F_R(t) = \sum_\alpha \chi(\alpha,t) \prod_\beta{}' \; \rho_1(\beta,t) - (N-1)R(t) \prod_\alpha \rho_1(\alpha,t), \qquad (30)$$

where we have allowed one atom to be correlated with the field at a time through the presence of χ. As a check, we note that replacing χ by $R\rho_1$ leads to the F_R of case 2 [Eq. (27)]. For example, in the special case N = 2, F_R leads to the transparent form

$$F_{2,R}(1,2,t) = \chi(1,t)\rho_1(2,t) + \chi(2,t)\rho_1(1,t)$$

$$- R(t)\rho_1(1,t)\rho_1(2,t). \qquad (31)$$

The TDPO associated with Eq. (30) is

$$P(t) = \sum_{\alpha \beta} \Pi' \, \rho_1(\beta,t) \mathrm{Tr}_\beta + \sum_\alpha \chi(\alpha,t) \mathrm{Tr}_\alpha \mathrm{Tr}_f {\sum_{\beta \gamma}}' \Pi'' \, \rho_1(\gamma,t) \mathrm{Tr}_\gamma$$

$$- (N-1) \sum_\alpha \chi(\alpha,t) \, \Pi' \, \rho_1(\beta,t) \mathrm{Tr} - (N-1) \, P_{nc}(t), \qquad (32)$$

where $P_{nc}(t)$ is the completely uncorrelated PO of case 2 given by Eq. (28).

The exact ME for $\dot{F}_R$ leads to *exact* coupled closed ME's for $\dot{\chi}$, $\dot{R}$, and $\dot{\rho}_1$, namely

$$\dot{\chi}(\alpha,t) - (L_f + L_\alpha{}^o + L_\alpha') \, \chi(\alpha,t) - {\sum_{\substack{\beta \\ (\beta \neq \alpha)}}}' \mathrm{Tr}_\beta L'{}_\beta F_{2,R}(\alpha,\beta,t)$$

$$= \sum_{\substack{\beta \\ (\beta \neq \alpha)}} \int_{t_o}^t dt' \mathrm{Tr}_\beta L'{}_\beta \, \Pi'' \mathrm{Tr}_\gamma G(t,t') \sum_\delta \Delta'{}_{t'} L'{}_\delta$$

$$\times {\sum_{\substack{\varepsilon \\ (\varepsilon \neq \delta)}}}' \Delta\chi(\varepsilon,t') \quad \Pi' \, \rho_1(\zeta,t'), \qquad (32a)$$

$$\dot{R}(t) - L_f R(t) = \sum_\beta \mathrm{Tr}_\beta L'{}_\beta \chi(\beta,t), \qquad (32b)$$

and

$$\dot{\rho}_1(\alpha,t) - L_\alpha{}^o \rho_1(\alpha,t) = \mathrm{Tr}_f L_\alpha' \chi(\alpha,t), \qquad (32c)$$

where

$$\Delta'{}_t L'{}_\alpha \equiv L'{}_\alpha - \langle L'{}_\alpha \rangle_{\alpha,t} - \langle L'{}_\alpha \rangle_{f|\alpha,t}. \qquad (33)$$

$\langle L'{}_\alpha \rangle_{\alpha,t}$ is defined analogously to Eqs. (25a) and vanishes when there is no average field,

$$\langle L'{}_\alpha \rangle_{f|\alpha,t} \equiv \mathrm{Tr}_f L'{}_\alpha \, \chi(f|\alpha,t) \qquad (34)$$

is a conditional average, relative to a conditional DO defined by

$$\chi(\alpha,t) \equiv \chi(f|\alpha,t) \, \rho_1(\alpha,t) \equiv \chi(\alpha|f,t) \, R(t), \qquad (35)$$

and $\Delta\chi \equiv \chi - R\rho_1$. Equations (32b,c) are just the first equations of the BBGKY hierarchy [15] relating $\dot{R}$ and $\dot{\rho}_1$ to χ whereas Eq. (32a) for $\dot{\chi}$ replaces exactly all of the succeeding equations in the hierarchy and contains all of the interesting physics.

One can obtain an improvement over the usual laser theories by treating Eq. (32a) in the FBA and the Markov approximation. Assuming no average field and, for simplicity, zero detuning, the Born-Markov ME obtained for χ is

$$\dot{\chi}(\alpha,t) - (L_f + L_\alpha{}^0 + L'_\alpha - K_{\alpha f})\chi(\alpha,t)$$

$$\sideset{}{'}\sum_{\substack{\beta \\ (\beta \neq \alpha)}} \langle L'_\beta \rangle_{\beta | f,t}\, R(t)\, \rho_1(\alpha,t)$$

$$= T_2 \sideset{}{'}\sum_{\substack{\beta \\ (\beta \neq \alpha)}} Tr_\beta L'_\beta [\Delta'_t\, L'_\beta\, \rho_1\,(\beta,t)\, \Delta\chi\,(\alpha,t) + (\alpha \leftrightarrow \beta)], \qquad (36)$$

where the reservoir has been introduced through $K_{\alpha f}$ (the reservoir decay operator) and T_2 (the polarization dephasing time, which is the shortest reservoir damping time); $\langle L'_\alpha \rangle_{\alpha | f}$ is a conditional average completely analogous to that of Eq. (34). The first line of Eq. (36) together with Eqs. (32b, c), contains usual laser theory including all single-atom terms. The second and third lines of Eq. (36) contain two types of two-atom correlation terms whose magnitude we have been able to calculate explicitly. The third line consists of irreducible two-atom correlation terms; their magnitude relative to one-atom terms is $\sim T_2/2T_f$, where T_f is the cavity photon loss rate, or $\sim 10^{-2} - 10^{-6}$ for many typical lasers. The second line of Eq. (36) contains reducible two-atom correlation terms. It can be shown that the latter terms yield corrections of the order of ordinary spontaneous emission, that is $a^\dagger a + 1$ is replaced by $a^\dagger a + \langle n^- \rangle$, a and $a^\dagger$ being the photon annihilation and creation operators and $\langle n^- \rangle$ the mean occupation number per atom of the terminal laser state.

Other possible applications of Eqs. (32) can be envisioned in addition to the one just discussed. Fruitful areas of application should be to the theory of amplified spontaneous emission and to the theory of superradiance.

7. COUPLED-SYSTEMS APPROACH IN THE HEISENBERG PICTURE

As one might expect, there exists an exact counterpart in the Heisenberg picture of the TDPO approach to ME's for coupled systems. We will illustrate the Heisenberg counterpart for the F_R of case 1

above. In the Schrödinger picture we start from the Liouville
equation, Eq. (1), and break up the system DO according to Eq. (2).
Equations (22) and (23) yield $F_R(t) = P(t) F(t)$. One noteworthy
property of F_R pertains to its effect on expectation values of
products of operators. If we assume an operator corresponding to an
observable which is a product of matter and field operators,

$$\mathcal{O} = \mathcal{O}_m \mathcal{O}_f, \tag{37}$$

then

$$\langle \mathcal{O} \rangle_R \equiv \mathrm{Tr}\mathcal{O}F_R = \langle \mathcal{O}_m \rangle \langle \mathcal{O}_f \rangle; \tag{38}$$

that is the average relative to F_R of the product is the product
of the averages. Similarly, we have

$$\langle \mathcal{O} \rangle_I \equiv \mathrm{Tr}\mathcal{O}F_I = \langle \Delta\mathcal{O}_m \Delta\mathcal{O}_f \rangle, \tag{39}$$

where $\Delta\mathcal{O}_m \equiv \mathcal{O}_m - \langle \mathcal{O}_m \rangle$, and so forth.

In the Heisenberg counterpart of the TDPO approach, we wish
to preserve properties (38) and (39) of the expectation values.
We start with the Heisenberg equation of motion

$$\dot{\mathcal{O}}(t) = - L\mathcal{O}(t), \tag{40}$$

and break up dynamical variables rather than DO's into relevant
and irrelevant parts,

$$\mathcal{O}(t) = \mathcal{O}_R(t) + \mathcal{O}_I(t). \tag{41}$$

The relevant part is obtained from $\mathcal{O}$ by means of a PO,

$$\mathcal{O}_R(t) = P\mathcal{O}(t). \tag{42}$$

If we assume that $\mathcal{O}$ is the product operator of Eq. (37), then we
demand that

$$\langle \mathcal{O}_R \rangle = \langle \mathcal{O}_m \rangle \langle \mathcal{O}_f \rangle, \quad \langle \mathcal{O}_I \rangle = \langle \Delta\mathcal{O}_m \Delta\mathcal{O}_f \rangle . \tag{43}$$

The PO resulting in these properties is easily shown to be

$$P = \mathrm{Tr}_f R + \mathrm{Tr}_m \rho_N - \mathrm{Tr}R\rho_N . \tag{44}$$

Notice that $P^2 = P$ and that P is *time-independent*.

Just as we obtained a closed ME for $\dot{F}_R$ above by eliminating F_I, we now obtain a closed *exact* equation of motion for $\mathcal{O}_R(t)$ by eliminating $\mathcal{O}_I(t)$,

$$\dot{\mathcal{O}}_R(t) + PL\mathcal{O}_R(t) - \int_{t_0}^{t} dt'PL\ G(t-t')(1-P)L\mathcal{O}_R(t')$$

$$= -\ PL\ G(t_0-t)\ \mathcal{O}_I(t_0). \tag{45}$$

This equation is an exact generalized Langevin equation similar to those obtained by the Mori formalism.[16] The integral on the left side represents a generalized damping term while the right side is a generalized fluctuating force term.

Further exploration of this approach should be of particular interest because of the considerable amount of recent activity in the Heisenberg approach to quantum-optical problems.[17] For now we note only that a Heisenberg counterpart can be formulated for any $F_R(t)$, for example the F_R of case 2 or of case 3. Finally, it should be pointed out that an interaction-picture counterpart can also be formulated.

8. CONCLUSION

We have extended the number of possible exact ME's available as starting points for deriving approximate ME's by picking new forms for the relevant part $F_R(t)$ of the density operator. These new forms are the basis of a new approach to coupled-systems analysis, which is adapted to dealing with two or more coupled systems, neither of which is reservoir-like, and treats the coupled systems symmetrically. The implementation of this approach requires that one generalize the Nakajima-Zwanzig PO techniques for deriving ME's to allow TDPO's. The TDPO techniques are particularly well-suited to problems involving strongly coupled dynamical systems, both of which may be far from equilibrium. We have shown how they allow one to obtain powerful descriptions of quantum-optical systems in a low order of approximation. Finally, we discussed the Heisenberg-picture counterpart of the TDPO approach to coupled-systems analysis.

References

1. S. Nakajima, Prog. Theor. Phys. *20*, 948 (1958).
2. R. W. Zwanzig, J. Chem. Phys. *33*, 1338 (1960); *Lectures in Theoretical Physics*, Vol. III, edited by W. E. Brittin, B. W. Downs, and J. Downs (Interscience, New York, 1961), p. 106.

3. See, for example, the excellent reviews by G. S. Agarwal, *Progress in Optics*, Vol. XI, edited by E. Wolf (North-Holland, Amsterdam, 1973), p. 1, and by F. Haake, Springer Tracts in Mod. Phys. *66*, 98 (1973).

4. B. Robertson, Phys. Rev. *144*, 151 (1966); *160*, 175 (1967); *166*, 206 (1968).

5. K. Kawasaki and J. D. Gunton, Phys. Rev. A *8*, 2048 (1973).

6. C. R. Willis and R. H. Picard, Phys. Rev. A *9*, 1343 (1974).

7. C. R. Willis, Gen. Rel. Grav. *7*, 69 (1976).

8. R. H. Picard and C. R. Willis, Bull. Am. Phys. Soc. *21*, 114 (1976); Phys. Rev. A *16*, 1625 (1977).

9. For other recent applications of TDPO's, see H. Grabert and W. Weidlich, Z. Phys. *268*, 139 (1974) and H. Grabert, Z. Phys. B *27*, 95 (1977).

10. P. N. Argyres and P. L. Kelley, Phys. Rev. *134*, A98 (1964).

11. C. R. Willis and R. H. Picard, *Coherence and Quantum Optics*, edited by L. Mandel and E. Wolf (Plenum Press, New York, 1973), p. 675.

12. R. H. Picard and C. R. Willis, Phys. Rev. A *8*, 1536 (1973).

13. See, for instance, S. Haroche, in this Volume, p. 539; and Q.H.F. Vrehen, H.M.J. Hikspoors and H.M. Gibbs, in this Volume, p. 543.

14. C. R. Willis, *Kinetic Equations*, edited by R. L. Liboff and N. Rostoker (Gordon and Breach, New York, 1971), p. 235.

15. See, for example, T. - Y. Wu, *Kinetic Equations of Gases and Plasmas* (Addison-Wesley, Reading, Massachusetts, 1966), p. 62.

16. H. Mori, Prog. Theor. Phys. *33*, 423 (1965).

17. For example, K. Wódkiewicz and J. H. Eberly, Ann. Phys. (N.Y.) *101*, 574 (1976), and references therein.

PHASE STATES IN QUANTUM OPTICS

F. Rocca

Université de Nice, Nice Cedex, France

I. INTRODUCTION

The coherent states of bosons, among a lot of applications in
different fields of Physics, are frequently used in the description
of systems with "phase properties": lasers in Quantum Optics,
superfluids and superconductors (in terms of Cooper pairs) in
Statistical Mechanics... More or less explicitly, the basic idea
in such approaches is the following: the coherent state (for a
single mode) $|\alpha\rangle = ||\alpha|e^{i\theta}\rangle$ is the state of the system with mean
number of particles $\langle N\rangle = |\alpha|^2$ and well-defined phase θ.

At this point, from general quantum principles, we would need
to define an observable phase as the canonical conjugate variable
of the particle number, i.e. a selfadjoint operator Φ such that

$$[N,\Phi] = i \tag{1}$$

and leading consequently to the uncertainty relation

$$\Delta N.\Delta\Phi \geq 1/2 \quad . \tag{2}$$

It is now well-known that such an operator Φ cannot exist.
More precisely, its existence is mathematically incompatible with
the property that the spectrum of N is lower bounded.[1,2]

The introduction of operators "cosine" C and "sine" S of the
phase [1] which are well-defined but don't commute together, is a
trick which permits some successful applications, but which don't
solve the theoretical problem.

Restricting ourselves to the quantum optics domain, it is very enlightening to consider simply a plane wave. In classical electromagnetic theory, we shall write a plane wave

$$g(\vec{x},t) = a \, \frac{e^{i(\vec{k}\vec{x}-|\vec{k}|t+\phi)}}{\sqrt{2|\vec{k}|}} \, \varepsilon \tag{3}$$

(ε: polarization vector)

and there is no mystery about the classical phase ϕ, which is a quantity physically measurable and well defined with respect to some origin. The total energy of this plane wave is proportional to the norm of g and is consequently infinite. There is no trouble from this point: an infinitely extended plane wave is evidently only a theoretical tool in the theory. But this tool is very useful and so it is logical to try to define its quantum analogue. In this latter, to keep at least the same physical contents, we must find an observable phase and an infinite energy, i.e. a strictly infinite number of photons. This implies that the usual Fock representation of Canonical Commutation Relations (C.C.R.) is completely unsuitable (since the probability to have $N = \infty$ vanishes in Fock space). In particular, the usual coherent states are not the quantum plane wave states.

However, one can give an explicit construction of quantum plane wave states, starting from coherent states. It is the matter of the next part.

II. QUANTUM PLANE WAVE STATES

Let us recall some generalities of the usual Fock representation in second quantization for a photon system.

Let L be the linear space of one-particle wave functions, i.e., the four-dimensional functions of $x = (\vec{x}, t)$ satisfying

i) $\Box f(x) = 0$

ii) $\partial^\mu f_\mu(x) = 0$ (Lorentz condition)

iii) $(f,g) = \int \tilde{f}^\mu(\vec{k})^* \tilde{g}_\mu(\vec{k}) \, \dfrac{d^3\vec{k}}{2|\vec{k}|}$ $\tag{4}$

is a positive (semi-definite) scalar product.

Let $A_\mu(x) = A_\mu^+(x) + A_\mu^-(x)$ the field operator of the photon field. We shall work essentially with its smeared out form $A_F(f) = A_F^+(f) + A_F^-(f)$, $f \varepsilon L$ defined by

$$A_F^{-}(f) = [A_F^{+}(f)]^* = (f, A_F^{-}) = \int \tilde{f}^{\mu}(\vec{k})^* \tilde{A}_{\mu}(\vec{k}) \, \frac{d^3\vec{k}}{2|\vec{k}|} \tag{5}$$

and satisfying

$$[A_F^{\pm}(f), \, A_F^{\pm}(g)] = 0$$

$$[A_F^{-}(f), \, A_F^{+}(g)] = (f,g) \tag{6}$$

The Fock space $\mathcal{H}_F$ can be considered as generated from the vacuum $|\Omega_F\rangle$ by application of operators $A_F(f)$, for all $f \epsilon L$ ($|\Omega_F\rangle$ is cyclic in $\mathcal{H}_F$ for $A_F(f)$).

The coherent states are then defined, for each $g \epsilon L$, by:

$$|\Omega_g\rangle = e^{-iA_F(g)}|\Omega_F\rangle \tag{7}$$

They satisfy in particular:

$$A_F^{-}(f)|\Omega_g\rangle = (f,g)|\Omega_g\rangle \tag{8}$$

which is the second-quantization form of the one-mode relation

$$a|\alpha\rangle = \alpha|\alpha\rangle$$

We start now our construction in considering a photon system described by the coherent gauge invariant density matrix $\rho_V = \int_o^{2\pi}|\Omega_{g_V}\rangle\langle\Omega_{g_V}|d\phi/2\pi$ where g_V is the function of L which is given at $t = 0$ by

$$g_V(\vec{x},0) = \chi_V(\vec{x}) \; a \; \frac{e^{i(\vec{k}\vec{x}+\phi)}}{\sqrt{2|\vec{k}|}} \; \varepsilon \tag{9}$$

$\chi_V(\vec{x})$ is the conveniently regularized characteristic function of the volume $V \subset \mathbb{R}^3$.

Note that g_V describes a photon of polarization ε, localized within the volume V at the time $t = 0$. In the limit $V \to \infty$, this function becomes the plane wave (2) and so describes a photon of definite momentum $\vec{k}$. The mean value of the particle number operator N_F is:

$$\langle N_F\rangle = \mathrm{Tr}(\rho_V N_F) = \int_o^{2\pi}\langle\Omega_{g_V}|N_F|\Omega_{g_V}\rangle \, \frac{d\phi}{2\pi}$$

$$= (g_V, g_V) \int_o^{2\pi}\frac{d\phi}{2\pi} = \|g_V\|^2. \tag{10}$$

It is clear that $\|g_V\|^2 \to \infty$ as V^{-1} when $V \to \infty$, but the mean density $\bar{d} = <N_F>/V = \|g_V\|^2 V^{-1}$ remains constant in this limit. In the following, we look precisely at the limit state in this limit $V \to \infty$. Roughly speaking, one expects by this procedure to define generalized coherent states whose mean number of particles is strictly infinite, but with finite mean density.[3] At this point, it is convenient to introduce the functional

$$E_{g_V}(f) = \mathrm{Tr}(\rho_V \, e^{iA_F(f)}) \tag{11}$$

which is easily calculated with (5) and (6):

$$E_{g_V}(f) = \int_0^{2\pi} <\Omega_{g_V}| e^{iA_F(f)} |\Omega_{g_V}> \frac{d\phi}{2\pi}$$

$$= \int_0^{2\pi} e^{-1/2(f,f)+2i\,\mathrm{Im}(f,g_V)} \frac{d\phi}{2\pi} \, . \tag{12}$$

The function E_{g_V} satisfies the relations

i) $E_{g_V}(0) = 1$

ii) $\displaystyle\sum_{j,k=1}^{n} C_j{}^* C_k E_{g_V}(f_k - f_j)\, e^{i\,\mathrm{Im}(f_j,f_k)} \geq 0$

$$\forall C_j \in \mathbb{C}$$

$$\forall f_j \in L \, , \quad j = 1,2,\ldots,n; \forall_n \, .$$

iii) $\lambda \in \mathbb{R} \to E(f+\lambda g)$ is continuous $\forall f, g \in L$

which are fundamental for a photon (and more generally a boson) system. Indeed, it is known in algebraic field theory, from the so-called Gelfand-Naimark-Segal (G.N.S.) theorem that the definition of any functional E on L satisfying the previous relations (i - iii) is equivalent to the definition of a (unique) triplet $(\mathcal{H}, A(f), |\Omega>)$:

- $\mathcal{H}$, Hilbert space,

- $A(f)$ field operator on $\mathcal{H}$ satisfying the commutation relations (4) and (5),

- $|\Omega> \in \mathcal{H}$ cyclic vector in $\mathcal{H}$ for $A(f)$, with the relation $E(f) = \langle\Omega|e^{iA(f)}|\Omega\rangle$.

We perform now the limit $V \to \infty$ on the functional E_{g_V} . In this limit we have, from (3) and (9):

$$(f,g_V) \to (2\pi)^{3/2} \frac{a}{\sqrt{2|\vec{k}|}} \tilde{f}^{\mu}(\vec{k})^* e^{i\phi}\varepsilon_{\mu} \tag{13}$$

and so, introducing the notation $\tilde{f}_\varepsilon(\vec{k}) = \tilde{f}^{\mu}(\vec{k})\varepsilon_{\mu}$, one gets:

$$E_\lambda(f) = \lim_{V\to\infty} E_{g_V}(f) = \frac{1}{2\pi} e^{-1/2(f,f)}\int_0^{2\pi}d\phi\, e^{i\lambda\mathrm{Im}[\tilde{f}_\varepsilon(\vec{k})^* e^{i\phi}]}$$

$$= \frac{1}{2\pi} e^{-1/2(f,f)}\int_0^{2\pi}d\phi\, e^{i\lambda[-\mathrm{Im}\tilde{f}_\varepsilon(\vec{k})\cos\phi+\mathrm{Re}\tilde{f}_\varepsilon(\vec{k})\sin\phi]}$$

$$= e^{-1/2(f,f)} J_0(\lambda|\tilde{f}_\varepsilon(\vec{k})|) \tag{14}$$

with $\lambda = (2\pi)^{3/2} a \sqrt{\dfrac{2}{|\vec{k}|}}$.

The G.N.S. triplet corresponding to this limit is:

$$\mathcal{H} = \mathcal{H}_F \otimes M \tag{15}$$

$$A^-(f) = [A^+(f)]^* = A_F^-(f) \otimes \mathbb{1}_M + \mathbb{1}_{\mathcal{H}_F} \otimes \frac{1}{2}\lambda\tilde{f}_\varepsilon(\vec{k})\Theta \tag{16}$$

$$|\Omega\rangle = |\Omega_F\rangle \otimes \chi_0 , \tag{17}$$

where:

- M is the Hilbert space of square integrable functions on the unit circle,

- $\chi_0(\alpha) = 1 \quad \forall\alpha\in[0,2\pi]$

- $\{\Theta\chi\}(\alpha) = e^{i\alpha}\chi(\alpha) \quad \chi\in M$.

Indeed, it is easy to verify directly, from (14) - (17), the relation

$$E_\lambda(f) = \langle\Omega|e^{iA(f)}|\Omega\rangle . \tag{18}$$

In this representation, the particle number N (in its general definition of infinitesimal generator of gauge transformations of the first kind: $e^{iN\theta}A(f)e^{-iN\theta}=A(fe^{i\theta})$) is given by:

$$N = N_F \otimes \mathbb{1}_M + \mathbb{1}_{\mathcal{H}_F} \otimes i\frac{d}{d\alpha} \tag{19}$$

and there exists a phase operator Φ defined by

$$e^{i\Phi} = \mathbb{1}_{\mathcal{H}_F} \otimes \Theta \tag{20}$$

with

$$[N,\Phi] = i \ , \tag{21}$$

It is important to realize that the spectrum of N given by (19) is the whole set $\mathbb{Z}$ of integers. Thus, the particle number is not observable, in the usual sense, in this state, where, in compensation, there exists a phase operator. Recalling that conversely a phase operator doesn't exist in the Fock representation, where the particle number is observable, we obtain the true meaning of the duality between N and Φ which is intuitively expressed by the uncertainty relation (2).[2]

We give now more properties of the state defined by the functional E_λ.

Using (4), one can deduce from (16) the expression of the field operator components in a normalized Minkowsky basis (e_μ), $\mu = 0, 1, 2, 3$:[4]

$$A_\mu^-(x) = (A_\mu^+(x))^* = A_{F,\mu}^-(x) \otimes \mathbb{1}_M + a\,\frac{e^{ikx}}{\sqrt{2k_o}}\,(\varepsilon.e_\mu)\,e^{i\Phi} \tag{22}$$

$$(k_o = |\vec{k}|)$$

and calculate the different correlation functions:

$$\langle\Omega|\Omega\rangle = 1 \tag{23}$$

$$\langle\Omega|A_\mu^+(x)|\Omega\rangle = \langle\Omega|A_\mu^-(x)|\Omega\rangle = 0 \tag{24}$$

$$\langle\Omega|\ \prod_{i=1}^{n} A^{+}_{\mu_i}(x_i)\ \prod_{j=1}^{m} A^{-}_{\mu_j}(x_j)|\Omega\rangle = 0 \qquad \text{if } n \neq m \tag{25}$$

$$= \prod_{i=1}^{n} A^{*}_{\mu_i}(x_i)\ \prod_{j=1}^{n} A_{\mu_j}(x_j) \qquad \text{if } n = m$$

where

$$A_{\mu}(x) = a\ \frac{e^{ikx}}{\sqrt{2k_o}}\ (\varepsilon.e_{\mu}) \qquad (k_o = |\vec{k}|) \tag{26}$$

is precisely the four potential corresponding to a plane electro-magnetic wave of momentum $\vec{k}$ and polarization ε.

In conclusion, we have exhibited a state in which

· the even correlation functions factorize; this has to be connected with the conditions of complete coherence for the electromagnetic field,

· the modulus and the phase of the field are both observables.

These properties are characteristic, in classical theory, of plane waves, as discussed in the introduction and for example in [5]. This enables us to consider that the state defined by E_{λ} is actually the quantum analogue of plane waves.

III. APPLICATION TO THE DICKE MODEL WITH CLASSICAL SPINS

We shall give now an explicit example where the previously constructed phase state appears in a concrete situation. By the way, we shall see explicitly that the existence of a phase operator, and consequently of specific phase properties for the system is linked to the condensation properties appearing below the critical temperature of a thermodynamical phase transition. This seems to be a (unproved) general result in Statistical Mechanics.

We consider the system of N two levels atoms (spins) with positions $\vec{x}_i$ (i = 1, ..., N) interacting with one mode of electromagnetic field in a box of volume V, in the rotating wave approximation.[6] The hamiltonian of the system reads:

$$H_{N,V} = \sum_{\vec{k},\sigma} |\vec{k}|\, a^{+}_{\vec{k},\sigma} a_{\vec{k},\sigma} + \varepsilon \sum_{i=1}^{N} S^{z}_i + \frac{\lambda}{\sqrt{N}} \sum_{i=1}^{N} (a^{+}_{\vec{k}_o,\sigma_o} S^{-}_i e^{i\vec{k}_o \cdot \vec{x}_i} + h.c) \tag{27}$$

We suppose furthermore that the spins are classical vectors of modulus one in $\mathbb{R}^3$ ($|S_i^\pm|^2 + (S_i^z)^2 = 1$). This condition, which is technically very simplifying, will not modify qualitatively the results.

At a given temperature T, the photon field is entirely defined by the (Gibbs equilibrium) functional:

$$E_{N,V}(f) = \frac{\mathrm{Tr}\{\exp(-\beta H_{N,V})\, e^{iA_F(f)}\}}{\mathrm{Tr}\{\exp(-\beta H_{N,V})\}} \tag{28}$$

This quantity can be, after easy calculations, put in the form:

$$E_{N,V}(f) = E_{N,V}^B(f)\, E_{N,V}^S(f) \tag{29}$$

$E_{N,V}^B$ is the functional describing a free photon field in a box:

$$E_{N,V}^B(f) = \prod_{\vec{k},\sigma} \exp\left\{- \frac{(2\pi)^3}{V}\, \frac{1}{2|\vec{k}|}\left(\frac{1}{2}+\rho(\vec{k})\right)|\tilde{f}_\sigma(\vec{k})|^2\right\} , \tag{30}$$

where

$$\rho(\vec{k}) = \left(e^{\beta|\vec{k}|} - 1\right)^{-1} . \tag{31}$$

$E_{N,V}^S$ is a functional with respect to the spin variables:

$$E_{N,V}^S = \frac{\displaystyle\sum_{\{S\}}\left[\exp(-\beta H_{N,V}^{\{S\}})\exp\left\{-i\,\frac{(2\pi)^{3/2}}{\sqrt{2|\vec{k}_o|}}\,\frac{\lambda}{V|\vec{k}_o|}(\tilde{f}_\sigma(\vec{k}_o)^* \sum_{i=1}^{N} S_i^- + h.c)\right\}\right]}{\displaystyle\sum_{\{S\}}\exp(-\beta H_{N,V}^{\{S\}})} \tag{32}$$

with

$$H_{N,V}^{\{S\}} = \varepsilon \sum_{i=1}^{N} S_i^z - \frac{\lambda^2}{|\vec{k}_o|V}\left|\sum_{i=1}^{N} S_i^+\, e^{i\vec{k}_o\cdot\vec{x}_i}\right|^2 . \tag{33}$$

Taking now the limit $V \to \infty$, $N \to \infty$, keeping constant the spin density $d = N/V$ (thermodynamical limit), we get:

$$E^B_{N,V}(f) \rightarrow E^B(f) = e^{-1/2(f,f)} \, e^{-(f,\rho f)}$$

(34)

where $\quad \{\widetilde{\rho f}\}(\vec{k}) = \rho(\vec{k})\widetilde{f}(\vec{k})$,

This is nothing but the state of an infinite free system of photons in thermal equilibrium.[7]   The thermodynamical limit of $E^S_{N,V}(f)$ is performed in noting that, for classical spins:

$$\sum_{\{S\}} \cdot = \prod_{i=1}^{N} \, [\frac{1}{4\pi} \int \cdot \delta(|\vec{S}_i|-1) \, dS_i^{\,x} dS_i^{\,y} dS_i^{\,z} \, ,$$

(35)

and using a two-dimensional steepest descent method.[8]   It appears a critical temperature $1/\beta_c$ given by the gap equation:

$$\coth(\beta_c \varepsilon) - \frac{1}{\beta_c \varepsilon} = \frac{|\vec{k}_o|\varepsilon}{2\lambda^2 d}$$

(36)

such that

    - for $\quad \beta < \beta_c \quad E^S_{N,V}(f) \rightarrow 1$

       and so $\quad E_{N,V}(f) \rightarrow E(f) = E^B(f)$

(37)

    - for $\quad \beta > \beta_c \quad E^S_{N,V}(f) \rightarrow J_o(\eta|\widetilde{f}_{\sigma_o}(\vec{k}_o)|)$

       and so $\quad E_{N,V}(f) \rightarrow E(f) = E^B(f)J_o(\eta|\widetilde{f}_{\sigma_o}(\vec{k}_o)|)$

(38)

(the constant η is completely, but not very easily, determined from the initial parameters).

Our model presents a second order phase transition at the temperature $1/\beta_c$.

For $\beta > \beta_c$, one can write:

$$e^{1/2(f,f)} \, E(f) = e^{-(f,\rho f)} \, J_o(\eta|\widetilde{f}_{\sigma_o}(\vec{k}_o)|)$$

(39)

$$= e^{1/2(f,f)} \, E^B(f) \, e^{1/2(f,f)} \, E_\eta(f)$$

where E_η is the functional corresponding to the quantum plane wave constructed in part II (see 14).   This formula enables us to extend

to this infinite mode case a notion which is well known for the
finite mode case. For example, for one mode, if the density matrix
has a diagonal coherent state representation (P-representation)

$$\rho = \int P(\alpha) |\alpha><\alpha| d^2\alpha \tag{40}$$

the corresponding functional E_ρ is such that

$$E_\rho(f) = \text{Tr}\{\rho e^{i(a^+f+af^*)}\}$$
$$= e^{-1/2|f|^2} \int P(\alpha)\, e^{i(\alpha^*f+\alpha f^*)} d^2\alpha \tag{41}$$

or equivalently

$$e^{1/2|f|^2} E_\rho(f) = \tilde{P}(f) \ . \tag{42}$$

Thus, a formula of the type (39) will imply in the one mode case
that the P-distribution of the state E is the convolution product
of the P-distributions of states E^B and E_η, i.e. the field described
by E is the coherent superposition of the fields described respectively by E^B and E_η. [9]

Extending this notion to the general case, one can finally
conclude that the infinite Dicke model at the thermodynamical equilibrium presents a phase transition; we have a thermal light above
the critical temperature and a coherent superposition of a thermal
light and a quantum plane wave below the critical temperature.

References

1. See for example: P. Carruthers, and M. M. Nieto, Rev. Mod.
    Phys. *40*, 411 (1968).
2. F. Rocca and M. Sirugue, Commun. Math. Phys. *34*, 111 (1973).
3. J. P. Provost, F. Rocca and G. Vallee, Ann. of Phys. *94*, 307
    (1975).
4. J. P. Provost, F. Rocca, G. Vallee and M. Sirugue, J. Math.
    Phys. *15*, 2079 (1974).
5. C. L. Mehta, E. Wolf and A. P. Balachandran, J. Math. Phys. *7*,
    133 (1966).
6. For a study of the Dicke model in Statistical Mechanics, see
    K. Hepp and E. H. Lieb, Ann. of Phys. *76*, 360 (1973); Y. K.
    Wang and F. T. Hioe, Phys. Rev. A*7*, 831 (1973).
7. H. Araki and E. J. Woods, J. Math. Phys. *4*, 637 (1963).
8. J. P. Provost, F. Rocca and G. Vallee, Ann. of Phys. (to appear).
9. R. J. Glauber, Phys. Rev. *131*, 2766 (1963).

QUANTUM POINT PROCESS MODEL FOR PHOTODETECTION

M.D. Srinivas

University of Madras, Madras, India

1.   INTRODUCTION

In classical physics, phenomena which involve a random
sequence of events in time have been successfully investigated
by means of the theory of classical point processes (CPP)[1].
Hence, it is not surprising that most of the theoretical approaches
[2-6] to the photon-counting problem are also based on the theory
of CPP. The central objective of these investigations is to derive an expression (referred to as the counting formula) for the
probability $p\big((t,t+T],m\big)$ that m counts are observed in the time
interval (t,t+T]. This naturally leads to the study of a situation where the detector performs continuous measurements on the
electromagnetic field in the interval (t,t+T].

When we attempt to formulate a quantum theory of counting experiments along similar lines, we will have to take into account
the fact that the statistics of successive observations in quantum
theory are not describable in terms of the classical theory of probability as systematised by Kolmogorov. This is best done by employing the recently developed[7,8] framework of quantum probability theory, which has been shown to be well-suited to the analysis of correlations between successive measurements performed on a
quantum system. It has also been proposed[9] that counting experiments in quantum physics should be studied in terms of point processes which are now defined in the framework of quantum probability theory. A general mathematical analysis of these quantum
point processes (QPP)[10,11] has also led to the establishment of
a quantum counting formula for a large class of these processes.

In this article we shall indicate how the theory of QPP can be applied to the photon-counting problem. We first show that most of the so-called quantum theoretical approaches[12] to this problem are not 'truly quantum mechanical', in the sense that statistical quantities like $p\big((t,t+T],m\big)$ are, as a rule, calculated on the basis of classical probability theory. In order to obtain a 'truly quantum mechanical' theory of photodetection, we shall employ the theory of QPP, where a quantum counting formula has been derived[11] in terms of quantum theoretical joint probabilities of an appropriate sequence of observations. Based on this QPP approach, we first show that, in general, the photon-counting statistics are completely determined once we specify the nature of the measurement performed by the detector, and the evolution of the electromagnetic field. We shall also enumerate the conditions under which the well-known Glauber-Mandel (G-M) formula (cf. Eq.(2.10) below), can be established. An important conclusion which emerges from our analysis (which is not in any way restricted to a consideration of free electromagnetic fields only), is that, in general, the counting statistics depend not only on the initial state of the electromagnetic field, but also on its dynamics.

2.   THE CLASSICAL COINCIDENCE APPROACH TO PHOTODETECTION

We begin with a brief review of the coincidence approach to the theory of CPP, which has been systematised only recently by Macchi[13]. A CPP is essentially a collection $\{\mathcal{N}(t,t+\tau]\}$ of integer-valued random variables indexed by the intervals $(t,t+\tau]$ such that each realisation of the process is also an integer-valued measure on the real line. For the study of most of the counting phenomena encountered in physics, it is sufficient to consider only the so-called 'regular CPP'[13], which can be characterised by a set of 'coincidence probability densities' (CPD) and 'exclusion probability densities' (EPD) defined as follows: if the times $\{t_i | 1 \le i \le r\}$ are such that $t < t_1 < t_2 < \ldots < t_r < t+T$, then the CPD $h_r(t_1,t_2,\ldots,t_r)$ is the joint probability density that one count is observed around each of the instants t_i ; it is important to note that in this definition nothing is specified about the rest of the duration of the experiment. The EPD $\tilde{p}_r(t_1,t_2,\ldots,t_r)$ is defined as the joint probability density that one count is observed around each of the instants t_i and no count is observed in the rest of the duration of the interval $(t,t+T]$. Macchi[13] has shown that if the CPP is 'completely regular', then for each positive integer r, both h_r and $\tilde{p}_r$ exist and satisfy the following relations:

$$h_r(t_1,t_2,\ldots,t_r) = \sum_{j=0}^{\infty} \frac{1}{j!} \int_{-\infty}^{\infty} .. \int_{-\infty}^{\infty} d\theta_1 \, .. \, d\theta_j$$

$$\times \, \tilde{p}_{r+j}(t_1,\ldots,t_r,\theta_1,\ldots\theta_j) \quad ; \qquad (2.1)$$

$$\tilde{p}_r(t_1, t_2, \ldots, t_r) = \sum_{j=0}^{\infty} \frac{(-1)^j}{j!} \int_{-\infty}^{\infty} \cdots \int_{-\infty}^{\infty} d\theta_1 \cdots d\theta_j$$

$$\times\, h_{r+j}(t_1, \ldots, t_r, \theta_1, \ldots, \theta_j) \quad . \tag{2.2}$$

The relations (2.1) and (2.2) play a very important role in the derivations of the counting formula and will be referred to as the 'CPP relations'.

From the definition of the EPD given above, it is clear that the probability that m counts are observed in the interval $(t, t+T]$ can be expressed in terms of the EPD as follows:

$$p\big((t,t+T],m\big) = \frac{1}{m!} \int_t^{t+T}\cdots\int_t^{t+T} dt_1 \cdots dt_m$$

$$\times\, \tilde{p}_m(t_1, t_2, \ldots, t_m) \quad . \tag{2.3}$$

It now follows from (2.2) and (2.3) that the counting statistics of any given counting experiment (which can be described in terms of a completely regular CPP), is completely determined once we specify the CPD on the basis of the physical model under consideration.

In the classical theory of photodetection the basic random quantity is the fluctuating electromagnetic field considered as a random analytic signal $V(\underline{r},t)$. The fundamental physical assumption can be most conveniently expressed in the form

$$h_r(t_1, \ldots, t_r) = \alpha^r \langle I(t_1) \ldots I(t_r) \rangle \quad , \tag{2.4}$$

which essentially states that the CPD are proportional (α is an efficiency factor) to the correlation functions of the fluctuating field intensity I(t) given by

$$I(t) = \iiint_D d^3r\, V^*(\underline{r},t)\, V(\underline{r},t) \quad , \tag{2.5}$$

where D is the volume of the detector[14]. From (2.2)-(2.4), we obtain the well-known Mandel formula[2]

$$p\big((t,t+T],m\big) = \left\langle \frac{W^m}{m!} e^{-W} \right\rangle \quad , \tag{2.6}$$

where

$$W = \alpha \int_t^{t+T} I(t')dt' \quad . \tag{2.7}$$

In most of the so-called quantum theoretical investigations of
the photodetection process[12], the central objective has been to
arrive at the following expression for the CPD:

$$h_r(t_1,t_2, \ldots ,t_r) = \alpha^r \, \mathrm{Tr}\left[\hat{\rho} \, T_N\{\hat{I}(t_1) \, \hat{I}(t_2)\ldots\hat{I}(t_r)\}\right] \qquad (2.8)$$

In Eq.(2.8) $\hat{\rho}$ is the density operator characterising the state (in
the Heisenberg picture) of the electromagnetic field (and the sources, if any), T_N indicates that the operators enclosed by the braces
$\{\ \}$ should be placed in the normal order and in the apex time-
sequence (i.e., the $\hat{V}^-(\underline{r},t)$ are ordered chronologically, while the
$\hat{V}^+(\underline{r},t)$ are ordered antichronologically from left), and $\hat{I}(t)$ is
given by [14]

$$\hat{I}(t) = \iiint_D \hat{V}^-(\underline{r},t) \, \hat{V}^+(\underline{r},t) \, d^3r \quad , \qquad (2.9)$$

where $\hat{V}^-(\underline{r},t)$ and $\hat{V}^+(\underline{r},t)$ are respectively the negative and positive
frequency parts of the field operator. From (2.2), (2.3) and (2.8)
we can easily derive the G-M formula

$$p\big((t,t+T],m\big) = \mathrm{Tr}\left[\hat{\rho} \, T_N\{\frac{\hat{W}^m}{m!} \, e^{-\hat{W}}\}\right] \quad , \qquad (2.10)$$

where

$$\hat{W} = \alpha \int_t^{t+T} \hat{I}(t') \, dt' \quad . \qquad (2.11)$$

From our discussion it should be clear that the two important
steps in the derivation of the counting formulae (2.6) and (2.10)
are (i) the specification of the CPD as given by (2.4) and (2.8),
and (ii) the calculation of the EPD from the CPD by means of the
CPP relation (2.2). The above procedure is completely justified
as far as the classical theory of photodetection is concerned,
(since the basic framework is that of a classical stochastic theory
of fluctuating electromagnetic fields), provided that the intensity
correlation functions on the r.h.s. of (2.4) satisfy all the properties required of the CPD of a completely regular CPP. On the contrary, quantum mechanics is a theory with its own probability prescriptions and therefore joint probabilities like the CPD and EPD
have to be calculated from the basic algorithms of quantum theory;
moreover, the relations between these CPD and EPD will now have to
be deduced from the statistical prescriptions of quantum theory and
not from any extraneous considerations based on the framework of
classical probability theory. In fact, it is very important to
realise that the statistical prescriptions of quantum theory are
mainly in the form of joint probabilities for successive observations performed on a system and these joint probabilities satisfy

a probability calculus quite different from the (more or less uni-
versally used) classical probability calculus.

We shall illustrate the basic nonclassical feature of quantum
theoretical joint probabilities by the following example. Consider
a system in state $\hat{\rho}$ to be subjected to the following sequence of
experiments - the observable $\hat{A}(t_1)$ is measured at time t_1 and the
observable $\hat{B}(t_2)$ is measured at time t_2 ($t_1 < t_2$), where $\hat{A}(t_1)$ and
$\hat{B}(t_2)$ are self-adjoint operators with a purely discrete spectrum
and the following spectral resolution:

$$\hat{A}(t_1) = \sum_{i=1}^{n} \lambda_i \, \hat{P}_i(t_1) \quad , \tag{2.12}$$

$$\hat{B}(t_2) = \sum_{j=1}^{m} \mu_j \, \hat{Q}_j(t_2) \quad , \tag{2.13}$$

where $\hat{P}_i(t_1)$ and $\hat{Q}_j(t_2)$ are projection operators. From the basic
principles of quantum theory we can draw the following conclusions:
(i) the probability of observing the value of $\hat{A}(t_1)$ to be λ_i is
$\text{Tr}[\hat{\rho} \, \hat{P}_i(t_i)]$; (ii) if the value of $\hat{A}(t_1)$ is observed to be λ_i, then
the state $\hat{\rho}$ of the system is transformed into the (unnormalised)
state $\hat{\rho}'$ via the transformation

$$\hat{\rho} \to \hat{\rho}' = \hat{P}_i(t_1) \, \hat{\rho} \, \hat{P}_i(t_1) \quad ; \tag{2.14}$$

(iii) the joint probability for observing the value of $\hat{A}(t_1)$ to be
λ_i and that of $\hat{B}(t_2)$ to be μ_j is given by the expression

$$\text{Tr}[\hat{Q}_j(t_2) \, \hat{P}_i(t_1) \, \hat{\rho} \, \hat{P}_i(t_1) \hat{Q}_j(t_2)] \quad . \tag{2.15}$$

A basic nonclassical feature of quantum theoretical probabilities is expressed by the following inequality

$$\sum_{i=1}^{n} \text{Tr}[\hat{Q}_j(t_2) \, \hat{P}_i(t_1) \, \hat{\rho} \, \hat{P}_i(t_1) \, \hat{Q}_j(t_2)] \neq \text{Tr}[\hat{Q}_j(t_2) \, \hat{\rho} \, \hat{Q}_j(t_2)] \quad , \tag{2.16}$$

which states that the probability for observing the value of $\hat{B}(t_2)$
to be μ_j given only that $\hat{A}(t_1)$ has been measured earlier (l.h.s. of
(2.16)), is in general different from the probability for the same
event when no measurement of $\hat{A}(t_1)$ was carried out at t_1 (r.h.s. of
(2.16)). More generally, it can be shown[15] that joint and conditional probabilities for a set of events in quantum theory depend
also on the sequence of experiments performed on the system. This
has sometimes been referred to as the 'quantum interference of

probabilities' (cf. ref. [8], [15] and references cited therein).

We will now show that, because of the quantum interference of probabilities, the CPD and EPD in quantum theory are not, in general, related to each other via the CPP relations (2.1), (2.2). To start, we should clarify the exact sense in which the correlation function in the r.h.s. of Eq.(2.8) is interpretable as a joint probability in quantum theory. For this, we rewrite (2.8) in the following form:

$$h_r(t_1,t_2, \ldots ,t_r) = \iiint\limits_{D} d^3r_1 \cdots \iiint\limits_{D} d^3r_r$$

$$\times \, \mathrm{Tr}\left[\hat{V}^+(\underline{r}_r,t_r) \cdots \hat{V}^+(\underline{r}_1,t_1) \, \hat{\rho} \, \hat{V}^-(\underline{r}_1,t_1) \cdots \hat{V}^-(\underline{r}_r,t_r)\right]$$

$$(2.17)$$

Let us now postulate that the nature of the measurement performed by the photodetector is such that the event that one count is observed in the infinitesimal interval $(t,t+\Delta t]$ corresponds to the following measurement transformation (in the same way as (2.14) corresponds to the event that the value of $\hat{A}(t_1)$ is observed to be λ_i):

$$\hat{\rho} \to \alpha\left[\iiint\limits_{D} \hat{V}^+(\underline{r},t) \, \hat{\rho} \, \hat{V}^-(\underline{r},t) \, d^3r\right] \Delta t \quad . \qquad\qquad (2.18)$$

Then, for $t<t_1<t_2<..<t_r<t+T$, the quantity $h_r(t_1,t_2..,t_r)$ as given by (2.8) can be interpreted as the joint probability density that one count is observed around each of the instants t_i, when measurements are carried out only during the intervals $(t_i,t_i+\Delta t_i]$ $(1\leq i\leq r)$. In other words, the CPD as given by (2.8) refers to a situation where the detector is active only during the intervals $(t_i,t_i+\Delta t_i]$ and carries out no observations in the rest of the duration of the interval $(t,t+T]$ [16]. Therefore, the CPD for different sets of $\{t_i\}$ refers to different sequences of experiments performed on the system, and none of them corresponds to the actual situation of the counting experiment where the detector performs continuous observations on the electromagnetic field.

Let us now proceed to the definition of the EPD. In quantum theory, the EPD $\tilde{p}_r(t_1,t_2,\ldots,t_r)$ can only be defined as the joint probability density of the following sequence of observations: no counts are observed in the interval $(t,t_1]$, one count is observed in the interval $(t_1,t_1+\Delta t_1]$, no counts are observed in the interval $(t_1+\Delta t_1,t_2]$, and so on. Clearly all the EPD, when defined this way, correspond to a situation where the detector performs continuous observations in the interval $(t,t+T]$.

We have thus shown that in quantum theory the CPD (as defined, for example, by (2.8)) and the EPD correspond to totally different experimental situations. However, relations like (2.16) show that, in general, the quantum theoretical joint probabilities are *not* related to each other as in classical probability theory, whenever they correspond to *different* sequences of experiments being performed on the system. Hence, we conclude[17] that the quantum mechanical CPD and EPD are not related to each other via the CPP relations (2.1) and (2.2) - relations, which are crucially dependent on the framework of classical probability theory. It is for this reason that we shall employ the framework of quantum probability theory (which formalises the statistics of successive observations in quantum theory) for deriving the quantum photon-counting formula.

3. QUANTUM PHOTON-COUNTING FORMULA

We shall present only a brief sketch of certain aspects of the theory of QPP, as the details can be found in references [10],[11]. Since a QPP is after all a point process defined in the framework of quantum probability theory, we will first introduce some of its basic notions. The starting point, of course, is the recognition [8] that, in quantum theory, to each event (that the outcome of some experiment is found to lie in a certain domain of a value space), there is associated a measurement transformation(cf. Eqs. (2.14), (2.18))

$$\hat{\rho} \rightarrow \zeta \, \hat{\rho} \ , \tag{3.1}$$

where the 'operation' ζ is a linear positive mapping of the space V of self-adjoint trace-class operators into itself, which also satisfies the condition

$$\mathrm{Tr}\left[\zeta \, \hat{\rho}\right] \leq 1 \ . \tag{3.2}$$

Also $\mathrm{Tr}\left[\zeta \, \hat{\rho}\right]$ is the probability that the event (represented by) ζ is observed when the system is in state $\hat{\rho}$. The joint probability of observing the event ζ_1 followed by another event ζ_2 (when a system in state $\hat{\rho}$ is subjected to the above sequence of experiments), is given by $\mathrm{Tr}\left[\zeta_1\zeta_2\hat{\rho}\right]$ (cf. as in Eqs.(2.15), (2.17)). In this framework, an observable (or random variable) X is characterised by specifying all the operations $\{X(E)\}$, where for each domain E of the value space, X(E) corresponds to the event that the value of X is found to lie in E. This definition of an observable includes also the more conventional characterisation of observables as self-adjoint operators (cf. Eq.(2.12)), provided we restrict ourselves to those cases where the conventional theory also specifies the corresponding measurement transformations (cf. Eq.(2.14)).

In analogy with the notion of CPP, a QPP can also be defined as a collection $\{\mathcal{N}_{(t,t+\tau]}\}$ of observables (random variables), which satisfies certain consistency conditions[11]. By definition, the operation $\mathcal{N}_{(t,t+\tau]}(\{m\})$ corresponds to the event that m counts are observed in the interval $(t,t+\tau]$; therefore, if $\hat{\rho}$ represents the initial state of the system, then we have

$$p\big((t,t+\tau],m\big) = \mathrm{Tr}\big[\mathcal{N}_{(t,t+\tau]}(\{m\})\hat{\rho}\big] \ . \tag{3.3}$$

The basic consistency relation among the various $\mathcal{N}_{(t,t']}$ arises from the following fact. For arbitrary t' such that $t<t'<t''$, the event 'm counts are observed in $(t,t'']$' occurs if, and only if, for some $m_1,m_2>0$ such that $m_1 + m_2 = m$, the succession of events 'm_1 counts are observed in $(t,t']$' and 'm_2 counts are observed in $(t',t'']$' is found to occur. Thus we have the relation

$$\mathcal{N}_{(t,t'']}(\{m\}) = \sum_{m_1+m_2=m} \mathcal{N}_{(t',t'']}(\{m_2\}) \, \mathcal{N}_{(t,t']}(\{m_1\}) \ , \tag{3.4}$$

for all $t<t'<t''$. If we now make the identification

$$\mathcal{N}_{(t,t']}(\{0\}) = S_{t,t'} \ , \tag{3.5}$$

then it follows from (3.4) that $\{S_{t,t'}\}$ satisfies the following semigroup relation:

$$S_{t,t''} = S_{t',t''} \, S_{t,t'} \ , \tag{3.6}$$

for all $t<t'<t''$. If $\tilde{J}_t$ is the generator of this semigroup, then we can formally write

$$S_{t_1,t_2} = T \exp\Big[\int_{t_1}^{t_2} \tilde{J}_t \, dt\Big] \ , \tag{3.7}$$

where T is the time ordering operator which orders the various $\tilde{J}_t$ chronologically from right to left.

If the QPP is also regular (by which we mean that the $\{\mathcal{N}_{(t,t+\tau]}\}$ satisfy certain additional conditions which essentially rule out of the occurence of multiple counts), and if (for each t) $\tilde{J}_t$ is a densely defined operator on V, then the following conclusions can be drawn[11]:

I. For each t,

$$\lim_{\tau \to 0} \frac{\mathcal{N}_{(t,t+\tau]}(\{1\})}{\tau} = J_t \tag{3.8}$$

exists and is a bounded positive operator on V and satisfies

$$\mathrm{Tr}\left[J_t \hat{\rho}\right] = -\mathrm{Tr}\left[\tilde{J}_t \hat{\rho}\right] \tag{3.9}$$

for all density operators $\hat{\rho}$ in the domain of $\tilde{J}_t$.

II. The CPD given by

$$h_r(t_1, t_2, \ldots, t_r) = \mathrm{Tr}\left[J_{t_r} \cdots J_{t_2} J_{t_1} \hat{\rho}\right] \tag{3.10}$$

represents the joint probability density that one count is observed
in each of the intervals $(t_i, t_i + \Delta t_i]$ $(1 \leq i \leq r)$ when measurement is
carried out only during these intervals. The EPD given by

$$\tilde{p}_r(t_1, t_2, \ldots, t_r) = \mathrm{Tr}\left[S_{t_r, t+T}\, J_{t_r} S_{t_{r-1}, t_r} \cdots J_{t_1} S_{t, t_1}\, \hat{\rho}\right] \tag{3.11}$$

represents the joint probability density that one count is observed
in each of the intervals $(t_i, t_i + \Delta t_i]$ $(1 \leq i \leq r)$ and no counts are observed during the rest of the duration of the interval $(t, t+T]$. The
EPD (3.11) and the CPD (3.10) are not, in general, related to each
other via the CPP relations (2.1), (2.2).

III. For each non-negative integer m, and T>0

$$\mathcal{N}_{(t,t+T]}(\{m\}) = \mathrm{Tr}\left[T\left\{\frac{\left(\int_t^{t+T} J_{t'}\, dt'\right)^m}{m!}\right.\right.$$

$$\left.\left. \times \exp \int_t^{t+T} \tilde{J}_{t'}\, dt'\right\} \hat{\rho}\right] . \tag{3.12}$$

From (3.12) and (3.1), we obtain the *quantum counting formula*

$$p\big((t,t+T], m\big) = \mathrm{Tr}\left[T\left\{\frac{\left(\int_t^{t+T} J_{t'}\, dt'\right)^m}{m!}\right.\right.$$

$$\left.\left. \times \exp \int_t^{t+T} \tilde{J}_{t'}\, dt'\right\} \hat{\rho}\right] . \tag{3.13}$$

In addition to I, II and III the following result can also be established:

IV.   The linear operators J_t and $\tilde{J}_t$ (on V) completely characterise a regular QPP via Eq.(3.12), provided that for each t,

a) J_t is a bounded positive operator on V;

b) $\tilde{J}_t$ is densely defined and generates a semigroup of operations;

c) J_t and $\tilde{J}_t$ satisfy Eq.(3.9).

We now proceed to show that the theory of QPP outlined above provides the appropriate probabilistic basis for the derivation of the photon-counting formula in quantum theory. We shall first demonstrate that the well known G-M formula (2.10) can be derived as a particular case of (3.13), provided certain assumptions are made regarding the nature of the measurement performed by the photo-detector. From our discussion in Section 2, it follows that the conventional approaches to the theory of photodetection correspond to the stipulation that J_t is given by the relation[18]

$$J_t \hat{\rho} = \alpha \iiint_D \hat{V}^+(\underline{r},t)\ \hat{\rho}\ \hat{V}^-(\underline{r},t)\ d^3r \quad . \qquad (3.14)$$

By substituting (3.14) in (3.10) we obtain the CPD which also coincides with the CPD as given by (2.8) in Section 2; moreover, the physical interpretation of these CPD as given in II above, turns out to be the same as the one we were led to in our discussion in Section 2.

However, in order to obtain a counting formula, it is not just sufficient to specify J_t alone. The measurement performed by the detector is completely characterised only when *both* J_t and $\tilde{J}_t$ are specified in such a way that they also satisfy the conditions of IV. In physical terms this essentially means that we should also specify the measurement transformations $\{S_{t,t'}\}$ associated with events of the form, (cf. Eq.(3.5)) 'no counts are observed in the interval $(t,t']$'. Without a knowledge of the $S_{t,t'}$ (or, equivalently, of the $\tilde{J}_t$), it is not possible even to compute the EPD, as is evident from (3.11). However (as we saw in Section 2), in the conventional approach, an attempt is made at completely bypassing this problem by reverting to the framework of classical probability theory and employing the CPP relation.

In order to show that a specification of J_t alone in the form given by (3.16) is not by itself sufficient to arrive at the G-M formula, we consider a class of QPP investigated by Davies[10]. In the Davies' approach with each J_t is associated a particular $\tilde{J}_t$ (among all those which, together with J_t, satisfy the conditions of IV); for the particular J_t given by (3.16), Davies' prescription leads to the following expression for $\tilde{J}_t$:

$$\tilde{J}_t \hat{\rho} = \frac{1}{2} \left[\hat{I}(t) \hat{\rho} + \hat{\rho} \hat{I}(t) \right] \ , \tag{3.15}$$

where $\hat{I}(t)$ is as in Eq.(2.9). If we now substitute (3.14) and (3.15) in (3.13), we obtain a photon-counting formula, which does not bear any resemblance whatsoever to the G-M formula (2.10), even for the case of free fields. Davies' model is therefore a clear example of a situation where (the measurement performed by the photo-detector is such that), the CPD and EPD do not satisfy the CPP relation (2.2) even for free fields.

What are, then, the conditions under which the highly success-ful G-M formula (2.10) can be deduced from (3.13)? If we first assume that J_t is given by Eq.(3.14), then it is quite easy to see that Eq.(3.13) leads to Eq.(2.10) provided it is possible for us to set $\tilde{J}_t = -J_t$ (which is not very unreasonable as Eq.(3.9) is auto-matically satisfied). Hence, we can conclude that the counting sta-tistics is given by the G-M formula (2.10) whenever the measurement performed by the photodetector is such that J_t is given by (3.14) and $\tilde{J}_t = -J_t$ and, in addition, J_t and $\tilde{J}_t$ satisfy all the conditions of IV.

Let us briefly note some of the other advantages of the QPP approach, apart from the fact that it provides the appropriate probabilistic basis for the derivation of the quantum counting formula. First of all, it is now possible to determine the *complete counting statistics*, in the sense that we can now calculate quanti-ties like $p\big((t_1,t_1'],m_1;(t_2,t_2'],m_2\big)$ which is the joint probability that 'm_1 counts are observed in $(t_1,t_1']$' and 'm_2 counts are observed in $(t_2,t_2']$', when the detector performs observations only during the intervals $(t_1,t_1']$ and $(t_2,t_2']$ (with $t_1<t_1'<t_2<t_2'$). While in the con-ventional approach there is no prescription for calculating such joint probabilities, the QPP formulation leads to the expression

$$p\big((t_1,t_1'],m_1;(t_2,t_2'],m_2\big) = \mathrm{Tr}\left[\mathcal{N}_{(t_2,t_2']}(\{m_2\}) \, \mathcal{N}_{(t_1,t_1']}(\{m_1\}) \, \hat{\rho} \right] \tag{3.16}$$

which can be computed once J_t and $\tilde{J}_t$ are specified.

The QPP approach is general enough to be applicable to various forms of counting experiments. For example, for the case of the counting experiments performed with the so-called 'quantum counter', the counting formula derived by Mandel[19] can be obtained from

(3.13), if we stipulate

$$J_t \hat{\rho} = \alpha \iiint_D \hat{V}^-(\underline{r},t) \; \hat{\rho} \; \hat{V}^+(\underline{r},t) \; d^3r \quad , \tag{3.17}$$

and also assume that J_t, along with $\tilde{J}_t = -J_t$, satisfies all the conditions of IV. Finally, it must be emphasized that the QPP approach is equally applicable to the analysis of counting experiments with the free field as well as fields in interaction with sources and reservoirs. In fact, our considerations clearly show that the counting statistics depend not only on the initial state of the field but also on its dynamics.

References

1. See for example, *Stochastic Point Processes: Statistical Analysis, Theory and Applications*, ed. P.A.W. Lewis (Wiley, New York, 1972).
2. L. Mandel, Proc. Phys. Soc. (London) *72*, 1037 (1958); see also L. Mandel in *Progress in Optics, Vol. II*, ed. E. Wolf (North-Holland, Amsterdam, 1963).
3. R.J. Glauber, Phys. Rev. Lett. *10*, 84 (1963); see also R.J. Glauber in *Quantum Optics and Electronics*, ed. C. deWitt, A. Blandin and C. Cohen-Tannoudji (Gordon and Breach, New York, 1965).
4. L. Mandel, E.C.G. Sudarshan and E. Wolf, Proc. Phys. Soc. (London) *84*, 435 (1964).
5. P.L. Kelley and W.H. Kleiner, Phys. Rev. *136A*, 316 (1964).
6. M. Lax and M. Zwanziger, Phys. Rev. A *7*, 750 (1973); see also Appendix A of a preliminary version of this paper.
7. E.B. Davies and J.T. Lewis, Commun. Math. Phys. *17*, 239 (1970).
8. M.D. Srinivas, Jour. Math. Phys. *16*, 1672 (1975)(to be reprinted in *The Logic-Algebraic Approach to Quantum Theory, Vol. II*, ed. C.A. Hooker (Dordrecht, Netherlands, 1977)).
9. In fact the study of QPP (and much of the quantum probability framework) originated from the consideration of photon-counting experiments by Davies [see Ref. 10]). The so-called 'quantum stochastic processes', introduced by Davies for this purpose, are nothing but a particular class of QPP.
10. E.B. Davies, (a)Commun. Math. Phys. *15*, 277 (1969); (b)ibid. *19*, 83 (1970); (c)ibid. *22*, 51 (1971). See also E.B. Davies, *Quantum Theory of Open Systems* (Academic Press, New York, 1976).
11. M.D. Srinivas, Jour. Math. Phys. *18*, 2138 (1977).
12. See for example, Refs. 5 and 6. Unfortunately, it has been the case with several other investigations of this problem, that there is no clear statement of the basic probability relations employed.
13. O. Macchi, Adv. Appl. Prob. *1*, 83 (1975).

14. For the sake of simplicity, we have suppressed all the vector
     indices and we have also restricted ourselves to the case of
     a single detector only. Another generalisation, which can
     also be easily carried out, involves replacing (2.5) by the
     relation

$$I(t) = \iiint_D d^3r \iiint_D d^3r' \; K(\underline{r},\underline{r}') \; V^*(\underline{r},t) \; V(\underline{r}',t) \quad ,$$

     where $K(\underline{r},\underline{r}')$ is sometimes interpreted as a correlation func-
     tion characterizing the detector. A similar change can also
     be effected in Eqs.(2.9) and (2.18).
15. M.D. Srinivas, "Conditional Probabilities and Statistical
     Independence in Quantum Theory", Jour. Math. Phys. (in press).
16. This is in marked contrast with the situation in classical
     theory where the CPD are the same, irrespective of whether
     or not any experiments are performed during the intervening
     periods.
17. Our arguments can also be presented in more formal terms on
     the basis of the formalism outlined in Section 3.
18. It should be noted that, since J_t as given by (3.14) is not a
     bounded operator on V, our model does not strictly correspond
     to a regular QPP as per conditions of IV. However, Davies[10]
     has indicated how the QPP framework can be suitably generalised
     in the case of such unbounded J_t.
19. L. Mandel, Phys. Rev. *152*, 438 (1966).

COMPARISON OF $\underline{d} \cdot \underline{E}$ VERSUS $\underline{p} \cdot \underline{A}$ INTERACTIONS FOR A CHARGED HARMONIC OSCILLATOR

Kazimierz Rzążewski

Warsaw University, Warsaw, Poland

1. INTRODUCTION

There are two competing interaction Hamiltonians used to describe the coupling of nonrelativistic charged particles with an electromagnetic field. One is obtained through the prescription of minimal coupling:

$$\underline{p} \rightarrow \underline{p} - e\underline{A}(\underline{x}) \tag{1.1}$$

where $\underline{p}$ is the canonical momentum of a particle, e is its charge, and $\underline{A}(\underline{x})$ is the vector potential of the electromagnetic field evaluated at the position $\underline{x}$ of the particle. The other is expressed entirely in terms of electric and magnetic field strengths $\underline{E}$ and $\underline{B}$, coupled to different multipoles of the system. For a single charged particle in the dipole approximation the two Hamiltonians read:

$$H_1 = -\frac{e}{m}\, \underline{p} \cdot \underline{A} + \frac{e^2}{2m}\, (\underline{A})^2 \tag{1.2}$$

$$H_2 = -e\, \underline{x} \, \underline{E} + 2\pi \int |\underline{P}^\perp|^2 \, d^3r \quad . \tag{1.3}$$

The Coulomb gauge is assumed in Eq. (1.2), $\underline{E}$ denotes the electric field, and $\underline{P}^\perp$ is the transverse part of the polarization vector [3], which in the dipole approximation becomes:

$$\underline{P}^\perp(\underline{r}) = \frac{e}{(2\pi)^3} \int d^3k \left[\underline{x} - \frac{\underline{k} \cdot (\underline{k} \cdot \underline{x})}{k^2} \right] \exp(i\underline{k} \cdot \underline{r}) \quad . \tag{1.4}$$

Due to the dipole approximation, fields A and E are evaluated at the fixed mean position of the particle, which is assumed to be the origin of the coordinate system.

The connection between Hamiltonians (1.2) and (1.3) has been studied by many authors: Göppert-Mayer [1], Richards [2], Power and Zienau [3] and recently Davidovich [4]. On the one hand, a unitary relation between the Hamiltonians has been established; on the other hand possible observable differences between predictions concerning the spontaneous emission lineshape have been discussed by Lamb [5].

The aim of the present paper is to illustrate the relation between the two Hamiltonians in a simple exactly solvable model of a charged harmonic oscillator interacting with the electromagnetic field. The solution to this model in the framework of the $p \cdot A$ interaction has been recently discussed by Żakowicz and Rzążewski [6] and Rzążewski and Żakowicz [7].

In Section 2, we solve the model with the $d \cdot E$ interaction. In Section 3 the comparison is made of the solutions. To this end the unitary relation between the dynamical variables in the two cases is exploited.

2. SOLVING THE MODEL WITH $\underline{d} \cdot \underline{E}$ INTERACTIONS

In this section we will explicitly solve the Heisenberg equations of motion for a charged harmonic oscillator interacting with a transverse electromagnetic field via the coupling of the dipole moment of the system to the transverse part of the electric field. The total Hamiltonian reads:

$$H = \frac{\pi^2}{2m} + \frac{1}{2}\,m\omega_o^2\,\underline{x}^2 + \frac{1}{8\pi}\int d^3r\ \left[\left(\underline{E}(\underline{r})\right)^2 + \left(\underline{B}(\underline{r})\right)^2\right] +$$

$$- e\,\underline{x} \cdot \underline{E}(0) + 2\pi \int |\underline{P}^{\perp}|^2\,d^3r\ , \tag{2.1}$$

where m is mass of oscillating particles and ω_o is the frequency of its free oscillations. In addition, we denote by x the actual position of the particle, π its kinetic momentum, $\underline{E}$ the electric field, $\underline{B}$ the magnetic field, and $\underline{P}^{\perp}$ the transverse part of the polarization vector. By introducing the creation and annihilation operators $c_{\underline{k}\mu}^{\dagger}$ and $c_{\underline{k}\mu}$ of photons with definite wave vector k and polarization $\mu(\mu=1,2)$, we can write the standard plane wave decomposition of the transverse electric field:

$$\underline{E}(\underline{r}) = \frac{i}{2\pi} \sum_{\mu} \int d^3k \sqrt{k} \; \underline{e}_{\underline{k}\mu} \left(c_{\underline{k}\mu} \exp(i\underline{k}\cdot\underline{r}) - c_{\underline{k}\mu}^{\dagger} \exp(-i\underline{k}\cdot\underline{r}) \right) \qquad (2.2)$$

where $\underline{e}_{\underline{k}\mu}$ denotes the polarization of the photon. The creation and annihilation operators $c_{\underline{k}'\mu}^{\dagger}$ and $c_{\underline{k}'\mu}$ satisfy commutation relations of the following form:

$$[c_{\underline{k}'\mu'} , c_{\underline{k}\mu}^{\dagger}] = \delta_{\mu\mu'} \, \delta_{(3)}(\underline{k}-\underline{k}') \; ; \quad [c_{\underline{k}\mu}, c_{\underline{k}'\mu'}] = 0 \; . \qquad (2.3)$$

The particle canonical variables $\underline{\pi}$ and $\underline{x}$ commute with all field operators. Exactly as in [7] we will temper the singularity of the dipole approximation by inserting in the Fourier decompositions a convergence factor:

$$\frac{1}{\sqrt{k}} \rightarrow g(k) = \frac{1}{\sqrt{k}} \cdot \frac{\beta}{\sqrt{k^2+\beta^2}} \quad , \qquad (2.4)$$

which corresponds to a spatial extension of the source.

After the regularization has been performed, the Hamiltonian (2.1) takes the form:

$$H = \frac{\pi^2}{2m} + \frac{1}{2} m\omega_o^2 \, \underline{x}^2 + \sum_{\mu} \int d^3k \; k c_{\underline{k}\mu}^{\dagger} c_{\underline{k}\mu} - i \, \frac{e}{2\pi} \, \underline{x} \int d^3k g(k) k \underline{e}_{\underline{k}\mu} (c_{\underline{k}\mu} - c_{\underline{k}\mu}^{\dagger})$$

$$+ \frac{1}{2} m \left[\frac{2e^2}{3\pi m} \int_{-\infty}^{+\infty} k^3 g^2(k)\, dk \right] \underline{x}^2 \; . \qquad (2.5)$$

The last term in this Hamiltonian is the transverse polarization vector density integrated over whole space. In spite of regularization the integral over dk is linearly divergent. We shall see, however, that so is the $\underline{d} \cdot \underline{E}$ term, and the sum does make mathematical sense. Note that, in the case of a harmonic oscillator, the $2\pi \int |P\perp|^2 d^3r$ term has the meaning of a renormalization counterterm, because the divergent integral can be regarded as a correction to the bare frequency ω_o (compare [4]).

The Hamiltonian (2.5) yields the following equations of motion:

$$\frac{d\underline{x}}{dt} = \frac{\underline{\pi}}{m}$$

$$\frac{d\underline{\pi}}{dt} = -m \left(\omega_o^2 + \frac{2e^2}{3\pi m} \int_{-\infty}^{+\infty} k^3 g^2(k)\, dk \right) \underline{x} + i \frac{e}{2\pi} \int d^3k \; k g(k) \underline{e}_{\underline{k}\mu} (c_{\underline{k}\mu} - c_{\underline{k}\mu}^{\dagger})$$

$$\frac{dc_{k\mu}^{\dagger}}{dt} = i k c_{k\mu}^{\dagger} + \frac{e}{2\pi} (\underline{x} \cdot \underline{e}_{-k\mu}) kg(k) \ . \tag{2.6}$$

Equations (2.6) are similar to the equation (2.7) of [7]. Using the convenient Laplace transform method, we find the solution to equations (2.6) in the form:

$$\underline{x}(t) = \frac{\underline{\pi}(0)}{m} \int_{\Gamma} \frac{e^{zt}}{H(z)} \frac{dz}{2\pi i} + \underline{x}(0) \int_{\Gamma} \frac{ze^{zt}}{H(z)} \frac{dz}{2\pi i}$$

$$- \frac{ie}{2\pi m} \int_{\Gamma} \frac{e^{zt}}{H(z)} \sum_{\mu} \int kg(k) \underline{e}_{-k\mu} \left[\frac{c_{k\mu}^{\dagger}(0)}{z-ik} - \frac{c_{k\mu}(0)}{z+ik} \right] d^3k \frac{dz}{2\pi i}$$

$$\tag{2.7}$$

$$c_{k\mu}^{\dagger}(t) = e^{ikt} c_{k\mu}^{\dagger}(0) + \frac{e}{2\pi} kg(k) \int_{\Gamma} \frac{e^{zt}}{H(z)} \frac{\underline{e}_{k\mu}}{z-ik} \left(\frac{\underline{\pi}(0)}{m} + z\underline{x}(0) \right) \frac{dz}{2\pi i}$$

$$+ \frac{ie^2}{4\pi^2 m} kg(k) \int_{\Gamma} \frac{e^{zt}}{H(z)(z-ik)} \sum_{\nu} \int d^3p \frac{pg(p)}{H(z)} (\underline{e}_{p\nu} \cdot \underline{e}_{k\mu})$$

$$\times \left[\frac{c_{p\nu}(0)}{z+ip} - \frac{c_{p\nu}^{\dagger}(0)}{z-ip} \right] \frac{dz}{2\pi i} \ , \tag{2.8}$$

where the contour Γ in the inverse Laplace transform should lie to the right of the singularities of the integrands. The basic function $H(z)$ is equal to:

$$H(z) = z^2 + \omega_o^2 + \frac{2e^2}{3\pi m} z^2 \int_{-\infty}^{+\infty} \frac{k^3 g^2(k)}{z^2+k^2} dk \ , \tag{2.9}$$

and is identical with its counterpart in [7]. Note that if the term $2\pi \int |\underline{p}^{\perp}|^2 d^3r$ were missing we would have the very different function

$$H'(z) = z^2 + \omega_o^2 - \frac{2e^2}{3\pi m} \int_{-\infty}^{+\infty} \frac{k^5 g^2(k)}{z^2+k^2} dk \tag{2.10}$$

instead, which is ill-defined even with regularization (2.4). The analytic function $H(z)$ given by Eq. (2.9) has two branches, and the one serving for the evolution for the future can be evaluated with the help of Eq. (2.4) in the form:

$$H(z) = z^2 + \omega_o^2 + \frac{2e^2}{3m} z^2 \beta^2/(z+\beta) \quad .$$

(2.11)

For the discussion of the zeroes of this function see [7]. We quote here approximate values of these zeroes:

$$z_{1,2} = \pm i\omega_o \frac{1}{\left(1+(2e^2\beta/3m)\right)^{\frac{1}{2}}} - \frac{e^2\omega_o^2}{3m} \frac{1}{\left(1+(2e^2\beta/3m)\right)^2}$$

(2.12a)

$$z_3 = -\beta \left(1 + \frac{2e^2\beta}{3m}\right) \quad .$$

(2.12b)

We can now perform a space-time reconstruction of the solutions, expressing $x(t)$, $\pi(t)$, $E(r,t)$, $B(r,t)$ in terms of the initial conditions $\underline{x(0)}$, $\underline{\pi(0)}$, $\underline{E}''(\underline{r},\overline{0})$, $B(\overline{r},\overline{0})$. To this end we explore the relation:

$$\sum_\nu \int d^3p\sqrt{p} \; e_{\underline{p}\nu} \left[\frac{c_{\underline{p}\nu}(0)}{z+ip} - \frac{c_{\underline{p}\nu}^\dagger(0)}{z-ip}\right] = -\frac{i}{2} \int d^3r' \frac{e^{-zr'}}{r'} \left[z\underline{E}(\underline{r}',0)\right.$$

$$\left. + \text{curl} \; \underline{B}(\underline{r}',0)\right] \quad .$$

(2.13)

The motion of the attenuating charge, driven by incoming radiation, is given by:

$$\underline{x}(t) = \frac{\underline{\pi}(0)}{m} \int_\Gamma \frac{e^{zt}}{H(z)} \frac{dz}{2\pi i} + \underline{x}(0) \int_\Gamma \frac{ze^{zt}}{H(z)} \frac{dz}{2\pi i}$$

$$+ \frac{e}{4\pi m} \int_\Gamma \int \frac{e^{z(t-r')}}{H(z)r'} \left[z\underline{E}(\underline{r}',0) + \text{curl} \; \underline{B}(\underline{r}',0)\right] \frac{dz}{2\pi i} d^3r' \quad .$$

(2.14)

Notice that only the term proportional to $\underline{x}(0)$ is different from the corresponding term in Eq. (3.16) in ref. [7]. Similarly, as in [7], the electric and magnetic fields can be written as a sum of three terms: free field, source field and scattered field. For the sake of our discussion it is enough to quote here the expressions for $\underline{E}^{source}$ and $\underline{E}^{scatt}$. For $\underline{r} \neq 0$ we get:

$$
E_i^{source}(\underline{r},t) = e \int_\Gamma \frac{\exp[z(t-r)]}{H(z)} \frac{z^2}{r} \left(\delta_{ij} - \frac{r_i r_j}{r^2}\right) \left(\frac{\pi_j(0)}{m} + z\, x_j(0)\right)
$$

$$
+ e \int_\Gamma \frac{\exp[z(t-r)]}{H(z)} \frac{z}{r^2} \left(\delta_{ij} - 3\frac{r_i r_j}{r^2}\right) \left(\frac{\pi_j(0)}{m} + z\, x_j(0)\right)
$$

$$
+ e \int_\Gamma \frac{\exp[z(t-r)]}{H(z)} \frac{1}{r^3} \left(\delta_{ij} - 3\frac{r_i r_j}{r^2}\right) \left(\frac{\pi_j(0)}{m} + z\, x_j(0)\right) .
$$

$$(2.15)$$

The expression above is to be compared with Eq. (3.14b) of Ref. [7].
The source field is different in the following respects:

1. In deriving Eq. (2.15) we encounter integrals of the form:

$$
\int_0^\infty \frac{k^3 \sin kr}{k^2 + z^2} \, dk = \frac{1}{2} \frac{d}{dr} \delta(r) - \frac{\pi}{2} z^2 e^{-zr} .
$$

Hence Eq. (2.8) contains a part which builds up a singularity at the
origin only.

2. The expression (2.15) represents a causal outgoing wave.
Contrary to Eq.(3.14) of Ref.[7] it does not have a constant static
dipole field tail outside the sphere of radius t. This is because Eq.
(2.15) satisfies the initial condition $E(\underline{r},0) = 0$. For the scattering part of the electric field one easily gets (dropping singular
delta function terms):

$$
\underline{E}^{scatt}(\underline{r},t) = - \frac{e^2}{4\pi m} \int d^3r' \int \frac{dz}{2\pi i} \frac{\exp[z(t-r-r')]}{H(z)} \frac{1}{r'}
$$

$$
\times \left[\frac{z^2}{r} \left(\hat{I} - \frac{\underline{r} \otimes \underline{r}}{r^2}\right) + \left(\frac{z}{r^2} + \frac{1}{r^3}\right) \left(\hat{I} - 3\frac{\underline{r} \otimes \underline{r}}{r^2}\right) \right]
$$

$$
\times [z\, \underline{E}(\underline{r}',0) + \mathrm{curl}\, \underline{B}(\underline{r}',0)] .
$$

$$(2.16)$$

Although the scattered part of $c_{k\mu}^\dagger(t)$ is different from the scattered
part of $a_{k\mu}^\dagger(t)$ in [7], the space-time dependent field in Eq. (2.16)
is identical with Eq. (3.14c) of [7]. In other words, the difference
in momentum space amounts to a singular term with support at the
origin only.

3. CONCLUSIONS

Unitary relation between Hamiltonians:

$$H_I = [\left(\underline{p} - e\underline{A}(0) \right)^2 /2m] + V(\underline{x}) + \frac{1}{8\pi} \int [(\underline{E}_T)^2 + (\underline{B})^2]d^3r \qquad (3.1)$$

and

$$H_{II} = (\pi^2/2m) + V(\underline{x}) + \frac{1}{8\pi} \int [(\underline{E})^2 + (\underline{B})^2] - e\underline{x} \cdot \underline{E} + 2\pi \int |P\bot|^2 d^3r$$

$$(3.2)$$

found by Power and Zienau [3] can be stated in the Heisenberg picture, suitable for our discussion, as follows. The unitary transformation:

$$U = \exp[ie\ \underline{x} \cdot \underline{A}(0)] \quad , \qquad\qquad (3.3)$$

when applied to variables describing the system in Eq. (3.1) gives:

$$U^\dagger \underline{x}\ U = \underline{x} \quad , \qquad\qquad (3.4a)$$

$$U^\dagger \underline{p}\ U = \underline{p} - e\ \underline{A}(0) = \pi \quad , \qquad\qquad (3.4b)$$

$$U^\dagger a_{\underline{k}\mu}^\dagger\ U = a_{\underline{k}\mu}^\dagger + \frac{ie\underline{x}}{2\pi} g(k)\ \underline{e}_{\underline{k}\mu} \overset{df}{=\!=} c_{\underline{k}\mu}^\dagger \quad , \qquad\qquad (3.4c)$$

$$U^\dagger a_{\underline{k}\mu}\ U = a_{\underline{k}\mu} - \frac{ie\underline{x}}{2\pi} g(k)\ \underline{e}_{\underline{k}\mu} \overset{df}{=\!=} c_{\underline{k}\mu} \quad . \qquad\qquad (3.4d)$$

Substituting $\pi(p,a,a\dagger)$, $c\dagger(x,a\dagger)$ and $c(\underline{x},a)$ into H_{II} one finds H_I. In agreement with the last statement one can easily verify by direct calculation that solutions (2.7), (2.8), after substituting Eq. (3.4), coincide with solutions (3.11) of [7]. Nevertheless, they are not physically equivalent. The oscillator is damped differently in the two cases: poles of the integrand in the inverse Laplace transform are located in exactly the same positions (and this is not a statement depending on perturbation theory). The residues are different. The damping process, in both cases consists of two parts: the initial phase, controlled by zero z_3 of $H(z)$ in Eq. (2.12b) and the main part controlled by z_1 and z_2 in Eq. (2.12a). The amount of energy lost in the initial phase is always small for realistic atomic physics parameters. Still this amount is greater in the $\underline{d} \cdot \underline{E}$ case.

The relevant ratio of residues is $\sim \beta/\omega_0 \gg 1$. The explanation of this fact has two aspects:

(1) The notion of the photon (radiation part of the field) is different in the two cases (3.4c,d); therefore, in speaking of the preparation of the initial excited state of the mechanical system

with no photons we speak of different states depending on the interaction used. It is therefore understandable that the evolution is
not the same.

(2) It is not possible to discriminate between one interaction
and the other. They are rather distinguished by different preparation mechanisms used to produce the initial state. To see this we
should observe that in the case of $x(0) \neq 0$, the total electric
field in both cases satisfies essentially different initial condi-
tions. In the $\underline{p} \cdot \underline{A}$ case $E(0)$ is a static Coulomb field produced by
a dipole $\underline{x}(0)$ on the whole space, but in the $\underline{d} \cdot \underline{E}$ case $E(0) = 0$
(outside point $r = 0$). This means that $\underline{p} \cdot \underline{A}$ is suitable for the
situations in which preparation of the initial state has been done
adiabatically, so that at t=0 all the wave fields produced during
the excitation process escaped to infinity, and then at t=0 the
interaction is switched on. On the other hand $\underline{d} \cdot \underline{E}$ can serve for
the case of instantaneous excitation at t=0. Neither of these two
extreme situations seems to be very realistic in atomic physics.
Atomic physics experiments are usually of the scattering type, e.g.
the atom is usually in a ground state at the initial moment. But
the scattering part of the field is exactly the same in the two
cases. We have shown this explicitly in our model (2.16) It can
be understood from relations (3.4c,d) as the expectation value of
x tends to zero in the remote future and the remote past if the
total energy of the system is kept finite. This means that photons
in the asymptotic spaces (in-and-out-photons) are the same in both
cases. Finally it has been shown by Davidovich [4] that for a quite
general potential the scattering matrix is the same in the $\underline{p} \cdot \underline{A}$ and
$\underline{d} \cdot \underline{E}$ cases.

Having in mind that $E(t=0) = 0$ for $\underline{r} \neq 0$ in the $\underline{d} \cdot \underline{E}$ case one
can ask how the Maxwell equation

$$\mathrm{div}\underline{E} = 4\pi\rho \tag{3.5}$$

can be satisfied. The answer is, that for a dipole,

$$\rho(\underline{r}) = e(\underline{x} \cdot \underline{\nabla})\,\delta(\underline{r}) \;, \tag{3.6}$$

there exists a distribution-like solution of Eq. (3.5):

$$\underline{E}_0 = 4\pi e\,\underline{x}\,\delta(\underline{r}) \;\;, \tag{3.7}$$

with support in $r = 0$ only, besides the more familiar Coulomb
longitudinal field

$$\underline{E}_L = \frac{e}{r^3}\left(\hat{1} - 3\,\frac{\underline{r} \otimes \underline{r}}{r^2}\right) \cdot \underline{x} \;\;. \tag{3.8}$$

The difference between $\underline{p} \cdot A$ and $\underline{d} \cdot \underline{E}$ lies in the different splitting of the general solution of Eq. (3.5) into its special solution and the general solution of the homogeneous vacuum equation (the latter being usually interpreted as the wave field). To summarize, using the simple model of a charged harmonic oscillator, we have demonstrated that a difference between physically measurable predictions from the $\underline{p} \cdot A$ and $\underline{d} \cdot \underline{E}$ interactions appears only for quantities involving the mechanical system excited in the initial state, and is due to different splitting of the total electric field between the part depending directly on the excitation and on the remaining radiation field. This difference only affects the near field. The $\underline{p} \cdot A$ versus $\underline{d} \cdot \underline{E}$ problem has nothing to do with causality or with quantum theory, as is believed by some authors [3].

I am indebted to Professor J.H. Eberly for helpful conversations and for his invitation and hospitality extended to me at Rochester, where part of this work was done. I would like to thank J. Ackerhalt, L. Davidovich and W. Żakowicz for valuable discussions.

References

1.   M. Göppert-Mayer, Ann. Phys. Lpz., *9*, 273 (1931).
2.   P. Richards, J. Phys. Rev. *73*, 254 (1948).
3.   E.A. Power and S. Zienau, Phys. Trans. R. Soc. A. *254*, 427 (1959).
4.   L. Davidovich, Ph.D. thesis, University of Rochester (1975).
5.   W.W. Lamb, Phys. Rev. *85*, 259 (1952).
6.   W. Żakowicz and K. Rzążewski, J. Phys. A. Nucl. Gen. *7*, 869 (1974).
7.   K. Rzążewski and W. Żakowicz, J. Phys. A. Nucl. Gen. *9*, 1159 (1976).

ORDERING OF AN EXPONENTIATED MULTIMODE QUADRATIC OPERATOR

G. P. Agrawal[*]

The City College of CUNY, New York, N.Y.

C. L. Mehta

Indian Institute of Technology, New Delhi, India

1. INTRODUCTION

In quantum optics one often encounters a hamiltonian which is quadratic in boson annihilation and creation operators. One of the specific examples is a harmonic oscillator interacting with a reservoir. [1,2] The mutual exchange between the quanta of the harmonic oscillator and the reservoir leads to the quadratic coupling. The system hamiltonian can be written as [3]

$$\hat{H} = \hbar\omega_0 \hat{a}^\dagger \hat{a} + \hbar \sum_{j=1}^{N} \omega_j \hat{b}_j^\dagger \hat{b}_j + \hbar \sum_{j=1}^{N} (g_j \hat{b}_j^\dagger \hat{a} + g_j^* \hat{b}_j \hat{a}^\dagger) + \hbar \sum_{i=j} \nu_{ij} \hat{b}_i \hat{b}_j^\dagger, \qquad (1)$$

where the symbols have their usual meaning. The coupling between the oscillator and N-mode reservoir leads to the well-known Lamb shift in the frequency and the exponential decay of the oscillator excitation energy [2].

Another system of considerable interest is an assembly of two-level atoms interacting with an external electromagnetic field. The system hamiltonian is now of the form [4]

$$\hat{H} = \hbar\omega_0 \hat{a}^\dagger \hat{a} + \hbar\omega \hat{J}_z + \hbar(g\hat{a}^\dagger \hat{J}_- + g^* \hat{a}\hat{J}_+), \qquad (2)$$

where the incident field is assumed to be monochromatic of frequency ω_0, $\hbar\omega$ is the energy gap between two atomic levels and $\hat{J}_z$, $\hat{J}_\pm$ are

the angular momentum operators representing the atomic system
collectively. If we make use of Schwinger's representation [5]
of the angular momentum operators in terms of two harmonic oscil-
lator operators $\hat{b}$, $\hat{b}^\dagger$ and $\hat{c}$, $\hat{c}^\dagger$, we can rewrite Eq. (2) in the
form

$$\hat{H} = \hbar\omega_0\hat{a}^\dagger\hat{a} + \frac{\hbar\omega}{2}(\hat{c}^\dagger\hat{c} - \hat{b}^\dagger\hat{b}) + \hbar(g\hat{a}^\dagger\hat{b}^\dagger\hat{c} + g^*\hat{a}\hat{b}\hat{c}^\dagger). \qquad (3)$$

The interaction part of the hamiltonian (3) is now trilinear in
the boson operators. However, under certain very reasonable conditions, one of the three modes is highly excited and could be
assumed to behave classically [6] . In this case the hamiltonian
(3) reduces to the bilinear form. Such a reduction is, for example,
possible when the incident field is either very intense or when the
interaction is such that most of the atoms are either in the ground
state or in the excited state throughout the interaction.

The dynamics of any such system is governed through the time-
evolution operator

$$\hat{V} = \exp(i\hat{H}t/\hbar). \qquad (4)$$

While evaluating the quantum-mechanical average of the system
operators of interest, it is often convenient to employ the phase
space description. This requires the association of a classical
function with $\hat{V}$ and to that end it is necessary to put $\hat{V}$ in
some well-ordered form. For instance if the initial density
operator is expressed in the diagonal coherent state representation
[7,8], the normal ordered form of $\hat{V}$ is useful. Similarly, if we
are interested in thermodynamics of the system, the evaluations of
the partition function $z = Tr(e^{-\hat{H}/kT})$ could be simplified if the
operator $\exp(-\hat{H}/kT)$ is suitably ordered. The above discussion suggests that the ordered form of an exponentiated multimode quadratic
hamiltonian may be quite useful in various problems of interest.

Although various orderings of an operator have been discussed
in the literature [9-11], the Weyl, the normal and the antinormal
orderings are the most useful ones. An operator is said to be
in Weyl ordered form if it is completely symmetric in the ordering
of $\hat{a}$ and $\hat{a}^\dagger$. Similarly, if all powers of the creation operator $\hat{a}^\dagger$
occur to the left (right) of all powers of the annihilation operator
$\hat{a}$, the operator is said to be in the normal (antinormal) ordered
form. We shall use the notation $\hat{G}_\mu$ where $\mu = W$, N or A to represent
the Weyl, the normal and the antinormal ordered forms of the
operator $\hat{G}$. On the other hand $\{\hat{G}\}_\mu$ denotes an operator obtained
from $\hat{G}$ by rewriting it in μ-ordered form without making use of the
commutation relations. Thus in particular we note that $\hat{G}$ is, in
general, not necessarily equal to $\{\hat{G}\}_\mu$, but we always have

$$\hat{G} = \hat{G}_\mu = \{\hat{G}_\mu\}_\mu \ . \tag{5}$$

It may further be remarked that $\{\hat{G}\}_N$ and $\{\hat{G}\}_A$ are identical to $:\hat{G}:$ and $"\hat{G}"$ respectively, a notation used quite often in the literature [10].

In the present article we consider the Weyl, the normal and the antinormal ordered forms of the operator $\hat{G} = e^{\hat{P}}$ where $\hat{P}$ is the most general quadratic in boson operators

$$\hat{P} = \sum_{i,j=1}^{n} \{\alpha_{ij}\hat{a}_i\hat{a}_j + \beta_{ij}\hat{a}_i^\dagger\hat{a}_j^\dagger + \gamma_{ij}(\hat{a}_i^\dagger\hat{a}_j + \hat{a}_j\hat{a}_i^\dagger)\}$$

$$+ \ 2\sum_{i=1}^{n}(\delta_i\hat{a}_i + \varepsilon_i\hat{a}_i^\dagger) \ . \tag{6}$$

For the sake of generality we have included the linear terms as well and have not assumed $\hat{P}$ to be hermitian. Terms of the form $\hat{a}_i\hat{a}_j$ occur in the physical problems when the rotating wave approximation is not made and/or the interaction energy proportional to the square of the vector potential is also included. In order to illustrate our method, we consider in Sec. 2 the single mode case [12] ($i = j = 1$) and consider the generalization to n-modes [13] in Sec. 3. In Sec. 4, we discuss some applications.

2. SINGLE MODE CASE

Consider the operator $\hat{G} = \exp(\hat{P})$ where $\hat{P}$ is now given by

$$P = \alpha\hat{a}^2 + \beta\hat{a}^{\dagger 2} + \gamma(\hat{a}^\dagger\hat{a} + \hat{a}\hat{a}^\dagger) + 2\delta\hat{a} + 2\varepsilon\hat{a}^\dagger \ . \tag{7}$$

The first step consists of eliminating the linear terms in (7). We rewrite (7) in the form

$$P = \alpha\hat{u}^2 + \beta\hat{v}^2 + \gamma(\hat{u}\hat{v} + \hat{v}\hat{u}) + \theta, \tag{8}$$

where

$$\hat{u} = \hat{a} + (\varepsilon\gamma - \beta\delta)/\lambda^2 \ , \tag{9a}$$

$$\hat{v} = \hat{a}^\dagger + (\gamma\delta - \alpha\varepsilon)/\lambda^2 \ , \tag{9b}$$

$$\theta = (\alpha\varepsilon^2 + \beta\delta^2 - 2\gamma\varepsilon\delta)/\lambda^2 \ , \tag{9c}$$

$$\lambda = (\gamma^2 - \alpha\beta)^{1/2} \ . \tag{9d}$$

Note that although $\hat{u}$ and $\hat{v}$ are not hermitian adjoints of each other, they satisfy the commutation relation $[\hat{u},\hat{v}] = 1$. We define two new operators

$$\hat{d} = S_{11}\hat{u} + S_{12}\hat{v} \; ; \qquad \hat{c} = S_{21}\hat{u} + S_{22}\hat{v} \; , \tag{10}$$

as linear combinations of $\hat{u}$ and $\hat{v}$ and require that $[\hat{d},\hat{c}] = 1$, and that expression (8) now becomes of the form

$$\hat{P} = \lambda(\hat{c}\hat{d} + \hat{d}\hat{c}) + \theta \; . \tag{11}$$

When $\lambda \neq 0$, we may always determine S_{ij}, though the choice may not be unique. A possible choice is given by

$$S_{11} = (\gamma + \lambda)/2\lambda, \; S_{12} = \beta/2\lambda, \; S_{21} = \alpha/(\gamma + \lambda) \text{ and } S_{22} = 1. \tag{12}$$

It is worth pointing out that the matrix

$$S = \begin{pmatrix} S_{11} & S_{12} \\ S_{21} & S_{22} \end{pmatrix} \tag{13}$$

is symplectic, i.e., it leaves the matrix

$$z = \begin{pmatrix} 0 & 1 \\ -1 & 0 \end{pmatrix} \tag{14}$$

invariant under the transformation

$$Sz\tilde{S} = z, \tag{15}$$

where $\tilde{S}$ denotes the transpose of S.

The reason for making such a transformation and writing $\hat{P}$ in the form of Eq. (11) is that the Weyl ordered form of $\exp[\lambda(\hat{c}\hat{d} + \hat{d}\hat{c})]$ may readily be derived as shown in the Appendix. Since the transformations (9) and (10) are linear, any symmetric combination in $\hat{c}$ and $\hat{d}$ will automatically be symmetric in $\hat{a}$ and $\hat{a}^{\dagger}$ as well. Hence the Weyl ordered form of $\exp(\hat{P})$ in $\hat{c}$ and $\hat{d}$ is also Weyl ordered in $\hat{a}$ and $\hat{a}^{\dagger}$. Note that such a statement is not valid, in general, for other ordered forms.

Using Eq. (A8) of the Appendix and Eq. (11), we obtain

$$\hat{G} = \exp(\hat{P}) = e^{\theta}\mathrm{sech}\lambda\{\exp(2\hat{c}\hat{d}\, \tanh\lambda)\}_W \; . \tag{16}$$

From Eqs. (8), (10) and (13), we finally obtain

$$\hat{G}_W = e^{\theta}\,\mathrm{sech}\lambda\left\{\exp\left[\frac{\tanh\lambda}{\lambda}\,(\hat{P} - \theta)\right]\right\}_W \ , \tag{17}$$

where θ and λ are defined in Eqs. (9).

Once we have obtained the Weyl ordered form of $\hat{G}$, it could be used to determine the normal and the antinormal ordered forms by using the well-known relations among the Fourier transforms Γ_μ of $\hat{G}_\mu$;

$$\Gamma_\mu(X,X^*) = \int G_\mu(v,v^*)e^{(Xv^* - X^*v)}d^2v. \tag{18}$$

Here $G_\mu(v,v^*)$ is the c-number function obtained on replacing $\hat{a}$ and $\hat{a}^\dagger$ in $\hat{G}_\mu$ by v and v^* respectively. This function is actually the classical function associated with $\hat{G}$ in μ-ordering rule of association. These relations among the Fourier transforms are given below [10]

$$\Gamma_W(X,X^*) = e^{|X|^2/2}\Gamma_N(X,X^*) = e^{-|X|^2/2}\Gamma_A(X,X^*) \ . \tag{19}$$

From Eqs. (18) and (19), we obtain

$$G_N(v,v^*) = \frac{2}{\pi}\int G_W(v',v^{*\prime})\ e^{-2|v-v'|^2}d^2v' \ . \tag{20}$$

The normal ordered form may now be obtained by substituting Eq. (17) in (20) and carrying out the integration. Similarly from Eq. (17), (18) and (19) we may obtain the antinormal ordered form as well. The resulting normal and antinormal ordered forms are given below

$$\hat{G}_{\substack{N\\A}} = e^{\theta}\left(\cosh 2\lambda \mp \frac{\gamma}{\lambda}\,\sinh 2\lambda\right)^{-1/2} x$$

$$\left\{\exp\left[\frac{\hat{P} \mp 2\lambda(\hat{a}^\dagger + y)(\hat{a} + x)\tanh\lambda - \theta}{2(\lambda\coth 2\lambda \mp \gamma)}\right]\right\}_{\substack{N\\A}} \ , \tag{21}$$

where

$$x = (\varepsilon\gamma - \beta\delta)/\lambda^2; \qquad y = (\gamma\delta - \alpha\varepsilon)/\lambda^2 \ , \tag{22}$$

and λ and θ are given by Eqs (9).

3.   MULTIMODE CASE

Consider the operator $\hat{G} = e^{\hat{P}}$ where $\hat{P}$ is now given by Eq. (6).
We rewrite (6) in the matrix form

$$\hat{P} = \hat{\tilde{A}}\xi\hat{A} + 2\tilde{\eta}\hat{A} \tag{23}$$

where the coefficient matrix ξ is the $(2n\times2n)$ matrix

$$\xi = \begin{pmatrix} \alpha & \gamma \\ \tilde{\gamma} & \beta \end{pmatrix}$$

$\hat{A}$ and η are 2n-dimensional column vectors

$$\hat{A} = \{\hat{a}_1,\ldots,\ \hat{a}_n;\ \hat{a}_1^\dagger,\ldots,\ \hat{a}_n^\dagger\} \tag{25a}$$

$$\eta = \{\delta_1,\ldots,\ \delta_n;\ \varepsilon_1,\ldots,\ \varepsilon_n\}\ . \tag{25b}$$

The $(n\times n)$ matrices α and β can be taken to be symmetric without any
loss of generality. Matrix ξ is therefore symmetric. The commutation
relations

$$[\hat{a}_i,\hat{a}_j^\dagger] = \delta_{ij}\ ,\qquad [\hat{a}_i,\hat{a}_j] = [\hat{a}_i^\dagger,\hat{a}_j^\dagger] = 0\qquad (i,j = 1,\ldots,\ n) \tag{27}$$

could be written in the matrix form

$$\hat{A}\hat{\tilde{A}} - \widetilde{\hat{A}\hat{\tilde{A}}} = z\ , \tag{28}$$

where z is the $(2n\times2n)$ antisymmetric matrix

$$z = \begin{pmatrix} 0 & 1 \\ -1 & 0 \end{pmatrix}\ . \tag{29}$$

Here 0 and 1 are $(n\times n)$ null and unit matrices, respectively,

As before, we eliminate the linear terms in Eq. (23) by rewriting
it in the form

$$\hat{P} = \hat{\tilde{U}}\xi\hat{U} - \tilde{\eta}\xi^{-1}\eta\ , \tag{30}$$

where

$$\hat{U} = \hat{A} + \xi^{-1}\eta\ . \tag{31}$$

The components of $\hat{u} = \{\hat{u}_1,\ldots, \hat{u}_n; \hat{v}_1,\ldots, \hat{v}_n\}$ satisfy the same commutation relations as those of $\hat{A}$.

Let us make a linear symplectic transformation

$$\hat{B} = \{\hat{d}_1,\ldots, \hat{d}_n; \hat{c}_1,\ldots,\hat{c}_n\} = S\hat{U} \ , \tag{32}$$

which reduces $\hat{P}$ to a simpler form

$$\hat{P} = \tilde{\hat{B}}\Lambda\hat{B} - \tilde{\eta}\xi^{-1}\eta \tag{33}$$

with

$$\Lambda = \begin{pmatrix} 0 & \lambda \\ \lambda & 0 \end{pmatrix} \tag{34}$$

and the ($n\times n$) matrix λ is diagonal, $\lambda_{ij} = \lambda_i\delta_{ij}$. From Eqs. (30), (32) and (33) we find that ξ and Λ are related according as

$$\xi = \tilde{S}\Lambda S \ . \tag{35}$$

The symplectic transformation S defined in Eq. (15) is a commutation-relations preserving transformation, i.e., the components of $\hat{B}$ satisfy the same commutation relations as those of $\hat{U}$ (or of $\hat{A}$). The existence of such a transformation can be established, since ξ is symmetric, although it may not be unique. However, as it turns out we shall not require the explicit form of S.

From Eqs. (30)-(35), we obtain

$$\hat{G} = e^{\hat{P}} = e^{-\tilde{\eta}\xi^{-1}\eta} \prod_{i=1}^{n} \exp[\lambda_i(\hat{c}_i\hat{d}_i + \hat{d}_i\hat{c}_i)]. \tag{36}$$

Since the Weyl order form of $\exp[\lambda_i(\hat{c}_i\hat{d}_i + \hat{d}_i\hat{c}_i)]$ is known (cf. Eq. A8) and different $\hat{c}_i$ and $\hat{d}_i$ commute, we may readily write the Weyl ordered form of $\hat{G}$. Writing the resulting expression in the matrix form, we obtain after some simplifications

$$\hat{G}_W = e^{-\tilde{\eta}\xi^{-1}\eta}|\mathrm{sech}z\Lambda|^{1/2}\{\exp(\tilde{\hat{B}}z \tanh z\Lambda\hat{B})\}_W \tag{37}$$

where $|R|$ denotes the determinant of matrix R.

The final step consists of eliminating Λ and rewriting (37) in terms of the original matrices ξ and $\hat{A}$. For this purpose we observe that

$$z\xi = S^{-1}z\Lambda S \, , \tag{38}$$

which readily follows from Eqs. (15) and (35). In fact any function of $z\xi$ is related to the same function of $z\Lambda$ as $f(z\xi) = S^{-1}f(z\Lambda)S$, or inversely

$$f(z\Lambda) = Sf(z\xi)S^{-1} \, . \tag{39}$$

Using Eqs. (15), (35), (37) and (39) and the relation $|S| = 1$, we obtain the required Weyl ordered form of $\hat{G}$:

$$\hat{G}_W = |\mathrm{sech}z\xi|^{1/2}\, e^{\tilde{\eta}(T_W-1)\xi^{-1}\eta}\, \{\exp(\hat{\tilde{A}}\xi T_W\hat{A} + 2\tilde{\eta}T_W\hat{A})\}_W \tag{40}$$

where the matrix T_W depends on ξ only and is given by

$$T_W \equiv T_W(\xi) = \tanh(z\xi)/z\xi \, . \tag{41}$$

The Weyl ordered form (40) could be used to obtain the normal and the antinormal ordered forms of $\hat{G}$. Equation (20), valid for the single mode case, now generalizes to

$$G_N(V) = \left(\frac{2}{\pi}\right)^n \int\ldots\int G_W(V)\, \exp\{-(V^\dagger-V'^\dagger)(V-V')\}d^{2n}V' \, , \tag{42}$$

where $V = \{v_1,\ldots, v_n;\ v_1^*,\ldots, v_n^*\}$ and the **row** vector $V^\dagger$ is its hermitian adjoint, $V^\dagger = (v_1^*,\ldots, v_n^*;\ v_1,\ldots, v_n)$. The c-number function $G_W(V)$ is obtained from $\hat{G}$ by replacing $\hat{a}_i$ by v_i and $\hat{a}_i^\dagger$ by v_i^* for $i = 1,\ldots, n$. The integration is carried out to obtain the normal ordered form of $\hat{G}$. An analogous procedure is adopted to obtain the antinormal ordered form. Our results for all three orderings could be expressed in the following convenient form:

$$\hat{G} \equiv \hat{G}_\mu = K_\mu\{\exp(\hat{\tilde{A}}\xi T_\mu\hat{A} + 2\tilde{\eta}T_\mu\hat{A})\}_\mu \, , \tag{43}$$

where

$$K_\mu = \left|\frac{\sinh z\xi}{z\xi}\right|^{-1/2} |T_\mu|^{1/2}\, e^{\tilde{\eta}(T_\mu-1)\xi^{-1}\eta} \, , \tag{44}$$

$$T_\mu = \left(\frac{\sinh z\xi}{z\xi}\right)(\cosh z\xi + k_\mu yz\, \sinh z\xi)^{-1} \, . \tag{45}$$

Here z is the antisymmetric matrix defined in (29), y is the (2n×2n) matrix

$$y = \begin{pmatrix} 0 & 1 \\ 1 & 0 \end{pmatrix} \tag{46}$$

and

$$k_W = 0, \quad k_N = 1 \text{ and } k_A = -1 . \tag{47}$$

It may readily be verified that Eq. (43) agrees with the previously known special cases. For the case when $\alpha = \beta = 0$, the normal and the antinormal ordered forms agree with those obtained in Ref. 10. Using a method involving group theory and parametric differentiation, Berezin [14] has obtained the normal ordered form of $e^{\hat{P}}$ when $\hat{P}$ is hermitian (i.e., when $\alpha = \beta^*$, $\gamma = \varepsilon^*$ and $\tilde{\gamma} = \gamma^*$). Our result in this special case is in agreement with his result.

It is worth remarking that the Weyl ordered form of $e^{\hat{P}}$ always exists. However, the normal ordered form does not exist whenever

$$\left| \cosh z \xi' + yz \sinh z \xi \right| = 0. \tag{48}$$

Similarly the antinormal ordered form does not exist if

$$\left| \cosh z \xi - yz \sinh z \xi \right| = 0. \tag{49}$$

4. APPLICATIONS

Having obtained the antinormal ordered form of $\hat{G}$. One may readily write the diagonal coherent state representation [7,8]:

$$\hat{G} = \int \ldots \int \phi(\{v\},\{v^*\}) \, |\{v\}><\{v\}| \, d^{2n}\{v\} , \tag{50}$$

where $|\{v\}>$ is the multimode coherent state. From the relation

$$\phi(\{v\},\{v^*\}) = \frac{1}{\pi^n} G_A(\{v\},\{v^*\}) , \tag{51}$$

and Eq. (43), we obtain

$$\hat{G} = \frac{K_A}{\pi^n} \int \ldots \int \exp\{\tilde{V}\xi T_A V + 2\tilde{\eta} T_A V\} \, |\{v\}><\{v\}| \, d^{2n}\{v\} , \tag{52}$$

where

$$V = \{v_1, \ldots, v_n, v_1^*, \ldots, v_n^*\} . \tag{53}$$

The expression (52) may be used to obtain the trace of $\hat{G}$. We find that

$$\mathrm{Tr}\hat{G} = \frac{K_A}{\pi^n} \int \ldots \int \exp\{\tilde{V}\xi T_A V + 2\tilde{\eta}T_A V\} \, d^{2n}\{v\} . \tag{54}$$

Carrying out the required integration we obtain

$$\mathrm{Tr}\hat{G} = \frac{1}{2^n} \left| y\sinh z\xi \right|^{-1/2} \exp(-\tilde{\eta}\xi^{-1}\eta) . \tag{55}$$

As an application of Eq. (55) we derive the partition function of a harmonic oscillator interacting with a reservoir whose hamiltonian is given by Eq. (1). In matrix notation the partitions function Z is given by

$$Z = e^{\frac{\hbar}{2\kappa T}\mathrm{Tr}\gamma} \, \mathrm{Tr}[\exp(\hat{\tilde{A}}\xi\hat{A})] \tag{56}$$

where

$$\hat{A} = \{\hat{a}, \hat{b}_1, \ldots, \hat{b}_N; \hat{a}^\dagger, \hat{b}_1^\dagger, \ldots, \hat{b}_N^\dagger\} \tag{57}$$

and

$$\xi = \frac{-\hbar}{2\kappa T} \begin{pmatrix} 0 & \gamma \\ \tilde{\gamma} & 0 \end{pmatrix} \tag{58}$$

Here 0 is a $(N+1)\times(N+1)$ null matrix and γ is given by

$$\gamma = \begin{pmatrix}
\omega_0 & g_1 & g_2 & \cdots & g_N \\
g_1^* & \omega_1 & v_{12} & \cdots & v_{1N} \\
g_2^* & v_{21} & \omega_2 & \cdots & \cdot \\
\cdot & \cdot & \cdot & & \cdot \\
\cdot & \cdot & \cdot & & \cdot \\
\cdot & \cdot & \cdot & & \cdot \\
g_N^* & v_{N1} & \cdots\cdots & & \omega_N
\end{pmatrix} \tag{59}$$

Using Eqs. (29), (46), (55), (56) and (58), we obtain

$$Z = \frac{1}{2^n} \exp(\frac{\hbar}{2\kappa T} \, \mathrm{Tr}\gamma) \left| \sinh\frac{\hbar}{2\beta T}\gamma \right|^{-1} . \tag{60}$$

On further simplification Eq. (60) could be rewritten in the following form:

$$Z = \prod_{\ell=0}^{N} \left[\exp(\frac{\hbar}{kT} \, \lambda_i) - 1 \right] , \tag{61}$$

where λ_0, $\lambda_1,\ldots,$ λ_N are the eigenvalues of the matrix γ given by Eq. (59). The partition function could be used to evaluate various thermodynamical quantities of interest.

Appendix: Weyl Ordered Form of $\exp\{\gamma(\hat{a}^\dagger\hat{a} + \hat{a}\hat{a}^\dagger)\}$

In this Appendix we obtain the Weyl ordered form of the operator $\exp\{\gamma(\hat{a}^\dagger\hat{a} + \hat{a}\hat{a}^\dagger)\}$. For this purpose we first obtain its normal ordered form. We find that

$$\langle v| \exp\{\gamma(\hat{a}^\dagger\hat{a} + \hat{a}\hat{a}^\dagger)\}|v\rangle$$

$$= \sum_{n=0}^{\infty} \langle v| \exp\{\gamma(\hat{a}^\dagger\hat{a} + \hat{a}\hat{a}^\dagger)\}|n\rangle\langle n|v\rangle , \tag{A1}$$

where $|v\rangle$ is the coherent state [8]

$$|v\rangle = \sum_{n=0}^{\infty} e^{-|v|^2/2} \frac{v^n}{\sqrt{n!}} |n\rangle \tag{A2}$$

and $|n\rangle$ is the number state. Using (A2) we rewrite (A1) as

$$\langle v| \exp\{\gamma(\hat{a}^\dagger\hat{a} + \hat{a}\hat{a}^\dagger)\}|v\rangle = e^{\gamma}e^{-|v|^2(1-e^{2\gamma})} . \tag{A3}$$

Hence the normal ordered form of $\exp\{\gamma(\hat{a}^\dagger\hat{a} + \hat{a}\hat{a}^\dagger)\}$ is given by

$$\exp\{\gamma(\hat{a}^\dagger\hat{a} + \hat{a}\hat{a}^\dagger) = e^{\gamma}\{e^{-\hat{a}^\dagger\hat{a}(1-e^{2\gamma})}\}_N . \tag{A4}$$

From Eq. (18) we thus obtain the Fourier transform of the normal ordered form

$$\Gamma_N(\chi,\chi^*) = \frac{\pi e^{\gamma}}{1-e^{2\gamma}} \exp\{-|\chi|^2(1-e^{2\gamma})^{-1}\} \quad . \tag{A5}$$

Using Eqs. (19) and (A5) we then find the Fourier transform of the Weyl ordered form

$$\Gamma_w(\chi,\chi^*) = \frac{\pi e^{\gamma}}{1-e^{2\gamma}} \exp\left\{-|\chi|^2\left(\frac{1+e^{2\gamma}}{1-e^{2\gamma}}\right)\right\}. \tag{A6}$$

Again from (18) we obtain, on taking the inverse Fourier transform the following expression for the Weyl ordered form:

$$\exp\{\gamma(\hat{a}^{\dagger}\hat{a} + \hat{a}\hat{a}^{\dagger})\} = \mathrm{sech}\gamma\{\exp(2\hat{a}^{\dagger}\hat{a}\,\tanh\gamma)\}_w \quad . \tag{A7}$$

Finally we observe that the result (A7) is a consequence of rearranging powers of $\hat{a}$ and $\hat{a}^{\dagger}$ which depends only on their commutation relation and not on the fact that $\hat{a}^{\dagger}$ is hermitian adjoint of $\hat{a}$. Hence a result similar to (A7) viz.

$$\exp\{\gamma(\hat{c}\hat{d} + \hat{d}\hat{c})\} = \mathrm{sech}\gamma\{\exp(2\hat{c}\hat{d}\,\tanh\gamma)\}_w \tag{A8}$$

is valid for any operators $\hat{d}$ and $\hat{c}$ which satisfy

$$[\hat{d},\hat{c}] = 1 \quad . \tag{A9}$$

References and Footnotes

*Supported in part by ARO
1.   W. H. Louisell, Quantum Statistical Properties of Radiation (John Wiley & Sons, New York, 1973).
2.   E. Braun and S. V. Godoy, Physica *86A*, 337 (1977).
3.   We put a circumflex over all quantities which denote operators.
4.   R. H. Dicke, Phys. Rev. *93*, 99 (1954).
5.   J. Schwinger, Quantum Theory of Angular Momentum, ed. by L. C. Biedenharn and H. Van Dam (Adademic, New York, 1965).
6.   S. Kumar and C. L. Mehta, presented at the International Conference on Multiphoton Processes, University of Rochester, Rochester, New York, June 6-9, 1977.
7.   E.C.G. Sudarshan, Phys. Rev. Lett. *10*, 277 (1963).
8.   R.J. Glauber, Phys. Rev. *131*, 2766 (1963).
9.   L. Cohen, J. Math. Phys. *7*, 781 (1965).

10. C.L. Mehta, J. Phys. A (Proc. Phys. Soc. London) *1*, 385 (1968).
11. G.S. Agarwal and E. Wolf, Phys. Rev. *D2*, 2161 (1970).
12. C.L. Mehta, J. Math. Phys. *18*, 404 (1977).
13. G.P. Agrawal and C.L. Mehta, J. Math. Phys. *18*, 408 (1977).
14. F.A. Berezin, *The Method of Second Quantization* (Academic, New York, 1966) p. 143; see also R.M. Wilcox, J. Math. Phys. *8*, 962 (1967); R. Balian and E. Brezin, Nuovo Cimento *B64*, 37, (1969).

SWEPT-GAIN SUPERRADIANCE IN CO_2-PUMPED CH_3F

J.J. Ehrlich, C.M. Bowden, D.W. Howgate, S.H. Lehnigk

Redstone Arsenal, Alabama

A.T. Rosenberger and T.A. DeTemple

University of Illinois-Champaign, Urbana, Illinois

1. INTRODUCTION

Since the first experiment by Feld et al.[1,2] further experimental studies of FIR superradiance have been carried out by Rosenberger, Petuchowski and DeTemple,[3] this time in the homogeneously broadened pressure regime of CO_2 pumped CH_3F. At about the same time, experimental observations of superradiance and superfluorescence were reported in the near IR spectral region in metal vapors by a number of investigators working independently.[4-6]

A phenomenon closely related to Dicke superradiance (and which may compete with it depending upon the material geometry, density and the method of excitation), was recently predicted by Bonifacio, Hopf, Meystre and Scully[7] by means of amplifier theory, and has come to be known as swept-gain superradiance. A fully quantum mechanical treatment of this phenomenon was given in the small signal regime by Hopf and Meystre[8] and by Hopf, Meystre and McLaughlin.[9] These theories predict that under certain conditions, intense short solitary pulses of the superradiant type can be generated provided that the gain-to-loss ratio is sufficiently high. Due to the aspect of the swept excitation, there exists no limitation in terms of cooperation length. Although this phenomenon is closely associated with Dicke superradiance in its very nature, no experimental studies have been reported to date. It is worth noting that all experimental studies of superradiance and superfluorescence have used longitudinal (swept) excitation as opposed to transverse excitation. Furthermore, swept-gain superradiance is a potentially very useful way to generate short, intense coherent pulses.

We report in this paper what we believe to be the first studies of the transition of a given system from the Dicke superradiant regime through the swept-gain superradiant regime as a function of pressure. In the next section, we present the experimental results. For a complete discussion of the experimental arrangement the reader is referred to reference 10. The essential difference between the present case and that of reference 10 is that the CH_3F cell is 6m in length in the case discussed here. Section 3 is devoted to a theoretical development which attempts to account for the qualitative aspects of the effects of the pump pulse on the evolution of the swept-gain superradiance pulse. It is shown that the theory is in qualitative agreement with the experimental results.

A summary of the results and interpretation is given in Section 4. Also in this section we outline further extensions of our work which we plan to accomplish in the near future.

2.   EXPERIMENTAL RESULTS

The measured FIR intensity versus the square of the pressure of the CH_3F in the cell is presented in Fig. 1(a) for the pressure range from 0.025 Torr to 0.20 Torr. Each data point in this and all other figures which follow is the average of three scope traces taken in succession. The straight line fitted to these points is consistent with the results previously reported in reference 3 over the same pressure range but with a 3.5m length cell. This result is presented on a different scale in Fig. 1(b) together with measurements of FIR intensity over the higher pressure range from 0.25 Torr to 0.60 Torr.

It is seen from Fig. 1(b) that there are two distinct pressure regimes for the variation of the FIR pulse intensity, both apparently varying linearly as the square of the pressure but having slopes which differ significantly by more than a factor of 2, in fact, by about 3.25. Also plotted in Fig. 1(b) is the ratio of the forward to backward FIR pulse intensity as a function of the pressure squared. This curve is taken from Fig. 1(c). The curve clearly shows a deviation of the forward to backward pulse intensity ratio from unity below about 0.2 Torr to a linear variation with the square of the pressure and with slope greater than zero above 0.2 Torr up to about 0.6 Torr. If we call the swept-gain regime that pressure region where the forward FIR pulse intensity is greater than the backward one, then Fig. 1(b) indicates a clear breakpoint at about 0.2 Torr. The pressure regime P < 0.2 Torr can be considered the Dicke regime and P > 0.2 Torr can be considered the swept-excitation regime.

For pressures below 0.3 Torr, the FIR and the IR pulses are clearly separated in time as shown in Fig. 3(a). However, above this pressure the FIR pulse begins to move into the IR; the situation is shown in the bottom right trace in Fig. 3(a). The FIR pulses corresponding to the three points plotted in Fig. 1(b) at the higher pressures clearly show, in their scope traces, overlap with the IR pump pulse, yet the intensities appear to lie along the same curve with the other measurements above 0.2 Torr. The traces on the scope also show IR pump depletion at the three higher pressures where the IR and FIR overlap.

The pulse width versus inverse pressure is presented in Fig. 2(a). The line through the data points at the lower pressures corresponds to the Doppler broadened regime whereas the data points lying along the straight line of greater slope correspond to the homogeneously broadened regime. The demarcation from Doppler to homogeneous broadening on the graph occurs at about 0.09 Torr and the calculated value[3] is about 0.08 Torr. Again, the linear dependence of pulse width with inverse pressure over the pressure range from 0.10 Torr to 0.2 Torr is consistent with the results of reference 3.

In Fig. 2(b) is shown the pulse delay versus inverse pressure. A straight line variation is indicated with a vertical intercept at about minus 35 nsec, indicating that the initiation of the FIR pulse evolution does not begin at the pump cut-off. All delays were measured from the pump pulse cut-off to the FIR pulse peak (see Fig. 3(a)). The slope of the line through the points is again consistent with the results reported in reference 3. It is to be noted that the data points corresponding to pressures lower than about 0.07 Torr seem to be along a different slope. For this reason, these data points were not used to fit the line shown in the figure, since they belong to the Doppler broadened region which affects the pulse width variation shown in Fig. 2(a).

Actual traces from the IR pump and FIR pulse scope photographs are shown in Fig. 3(a) for pressures from 0.06 Torr to 0.70 Torr. The FIR pulse delays and widths are clearly seen to diminish with increasing pressure while the intensities increase nonlinearly, even when the FIR and the IR overlap in time. Furthermore, the pump intensity remains perceptively the same over the full pressure range until the temporal overlap region occurs. For the example shown at 0.70 Torr, not more than 10% of the pump pulse remains.

3. THEORY OF PUMP EFFECTS ON PULSE EVOLUTION

It is clear from Fig. 2(b) that the evolution of the FIR pulse does not begin at the pump cut-off, but actually starts to evolve

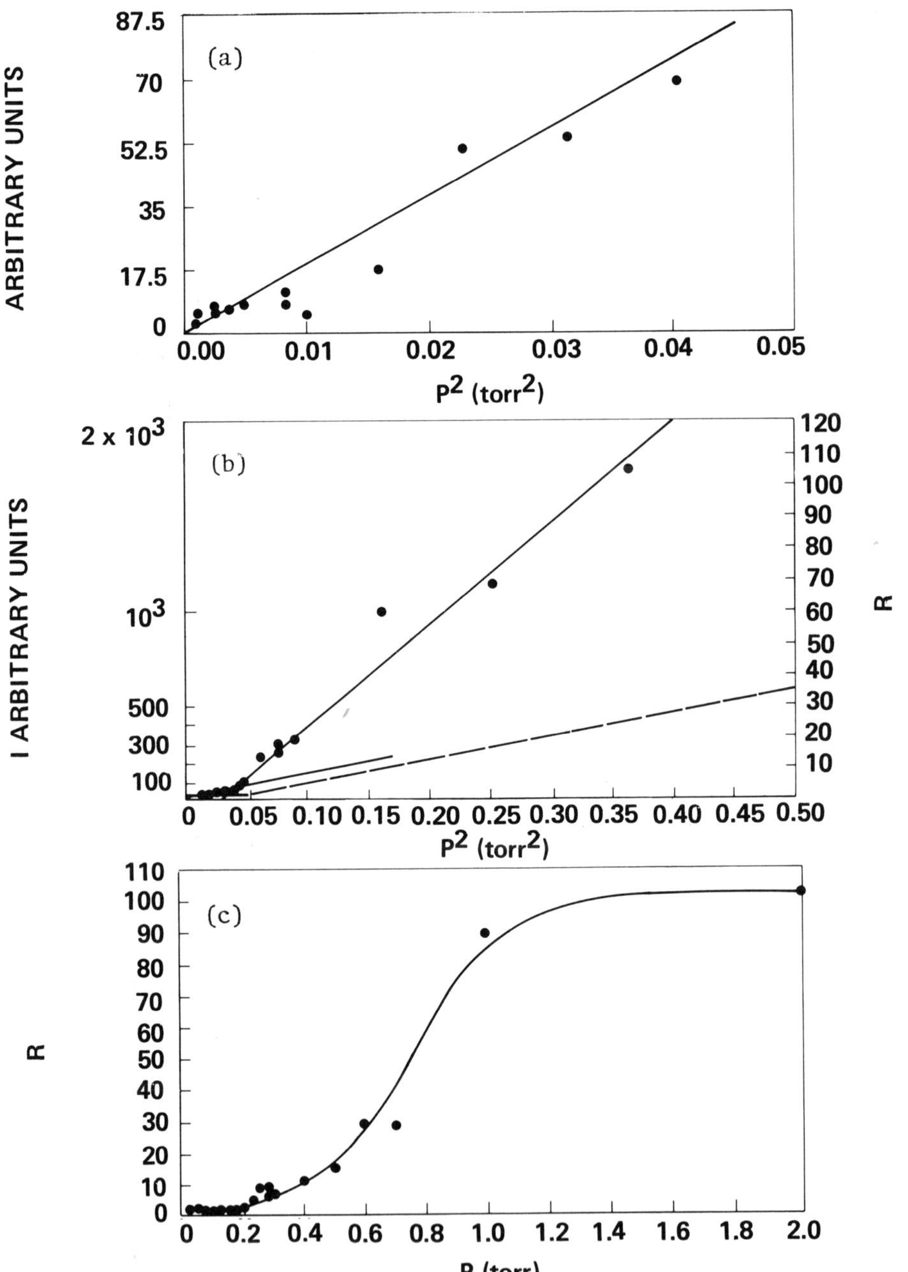

Fig. 1 Relative intensity measured with low temperature phosphorous-doped silicon detector vs. pressure squared. (a) Pressure range from 0.025 Torr through 0.20 Torr. Each solid circle is the average of three shots taken in succession. The straight line is the best fit to the points. (b) Pressure range from 0.025 Torr through 0.60 Torr. Each solid circle is the average of three shots taken in succession. The solid straight lines are the best fit to the points. The solid straight line of smaller slope is the best fit to the data points for pressures below 0.20 Torr and is taken from Fig. 1(a), whereas the solid straight line of the larger slope is the best fit to the data points for pressures greater than 0.2 Torr. The broken straight line is the ratio R (right-hand scale) of the forward to backward FIR pulse intensity as a function of the square of the pressure. This curve was taken from the best fit to the data points of Fig. 1(c). (c) Ratio R of the measured forward to backward FIR intensity vs. pressure for cell pressures over the range from 0.025 Torr through 2.0 Torr.

---

somewhere inside the pumping pulse. Therefore, we would expect the pump duration time and perhaps its shape as well as intensity, to have a significant effect upon the FIR pulse evolution, especially at the higher pressures where the delays are short. In order to determine what the effect might be, we consider a three-level model which includes the pump field as well as the FIR field and the associated principle transitions.

This model is depicted in Fig. 4. Consistent with the rotational selection rules for the experimental situation under consideration, we assume that level ε_1 and level ε_0 are not radiatively coupled.

We calculate the hierarchy of rate equations represented by the level scheme shown in Fig. 4 and couple these with Maxwell's equations for the fields $E_0(z,t)$, $E_1(z,t)$. The rotating wave and slowly varying envelope approximations are imposed to generate a set of coupled Bloch-Maxwell equations for this model.

In order to simplify the calculation, and consistent with our initial assumption that levels ε_1 and ε_0 are not radiatively coupled, we set equal to zero the off-diagonal elements in the rate equations and their derivatives which connect these levels. This ignores AC Stark effects and stimulated Raman, but results in a partial decoupling of the set of Bloch equations for the $(0 \longleftrightarrow 2)$ manifold from the set of Bloch equations for the $(2 \longleftrightarrow 1)$ manifold.

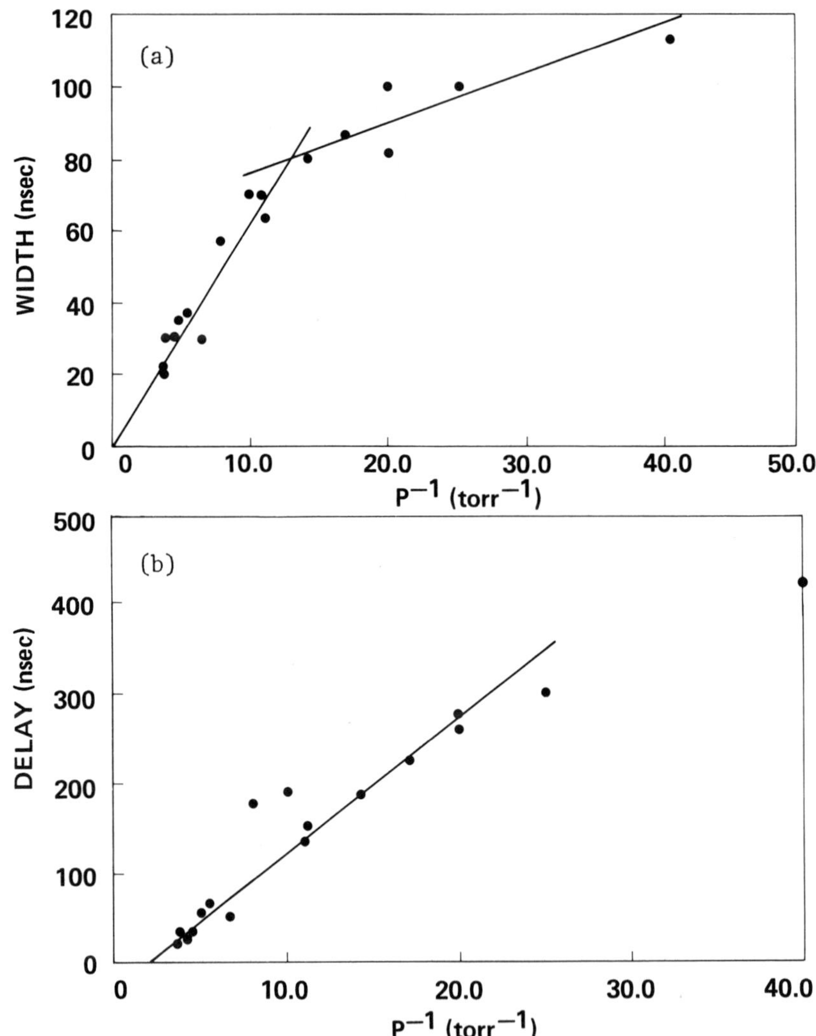

Fig. 2(a) Width of the FIR pulses in nsec vs. inverse pressure
over the pressure range from 0.025 Torr through 0.30 Torr. Each
solid circle corresponds to the average of three shots taken in
succession. The FIR pulse measurements were made using a low
temperature phosphorous-doped silicon detector. The solid straight
line of larger slope is the best fit to the data points corresponding to pressures above 0.09 Torr and the solid straight line of
smaller slope is the best fit to the data points corresponding to
pressures below 0.09 Torr. (b) Delay in nsec of peak of the FIR
pulse from the IR pulse cut-off vs. inverse pressure over the pressure range from 0.30 Torr to 0.025 Torr. The solid straight line
is the best fit to the data points corresponding to pressures below 0.09 Torr. The FIR pulse measurements were made with a low
temperature phosphorous-doped silicon detector.

We are interested in the development of the FIR field envelope.
It is noted from the experimental observations that, consistent
with the fact that for the particular pump transition chosen only
about 6×10^{-3} fraction of the total population is pumped,[3] the
cell medium is nearly transparent to the IR pump as long as the IR
and FIR pulses are temporally separate. This is very nearly true
even in most of the regime where the pulses overlap. Consistent
with this observation, we take the IR pump envelope Λ_I as independent of η, i.e., we assume that $\Lambda_I \neq \Lambda_I(\eta)$, where $\eta = z/c$ and
$t = \tau-\eta$ in the retarded time frame. This assumption allows us
to completely decouple the Bloch equations of the $(2 \longleftrightarrow 1)$ manifold from the corresponding set of equations for the $(2 \longleftrightarrow 0)$
transition. We look for asymptotic (i.e., $\eta \to \infty$) solutions to the
resulting Bloch-Maxwell equations for the $(2 \longleftrightarrow 1)$ transition. The
result is the set of equations in the asymptotic region,

$$\dot{r}_3 = - g \ (r_1^2 - \Gamma^2/g^2) - \gamma r_3 \ , \tag{1}$$

$$\dot{r}_1 = g \ r_1 r_3 - \gamma r_1 \ , \tag{2}$$

and

$$\Lambda_F = g \ r_1 \ , \tag{3}$$

where $\gamma = \dfrac{1}{T_1} = \dfrac{1}{T_2}$

is the phenomenological damping rate and

$$\Gamma(\tau) \equiv \frac{\Lambda_I(\tau)}{\sqrt{2}}$$

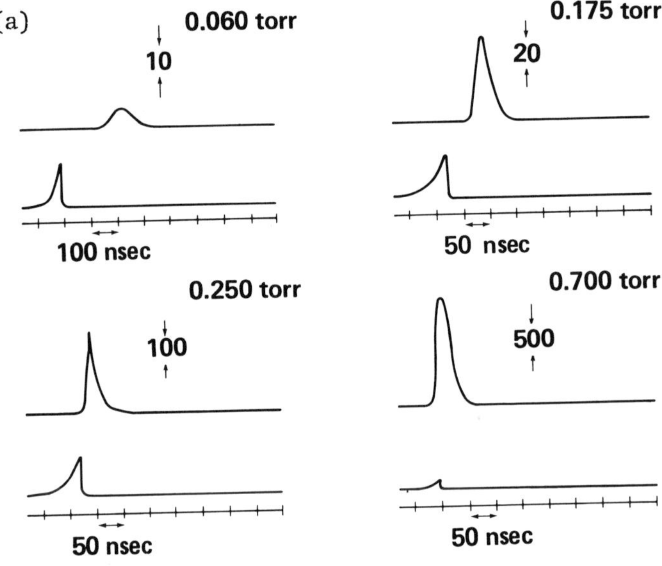

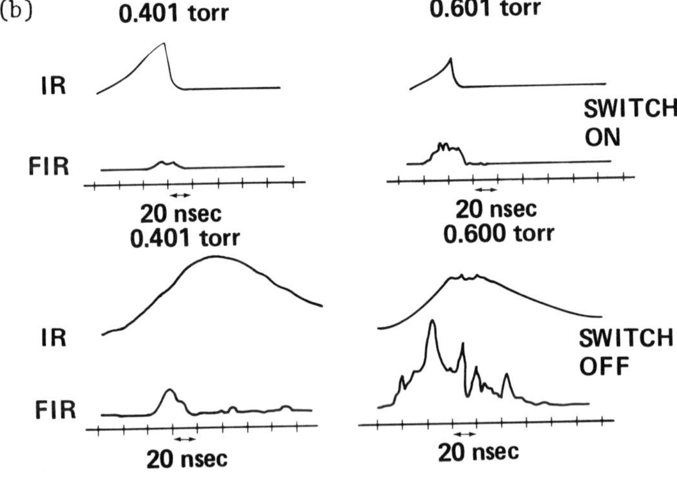

Fig. 3(a) Traces of the oscilloscope photographs of the FIR and
pump IR pulses in the same time frame for four different cell
pressures of CH_3F. The vertical scale is in mv and pertains to
the FIR pulse which is the upper curve in each case. The vertical
scale for the IR pulse, lower curve, is the same in each case. The
horizontal scale is in nsec. The FIR pulses were monitored using
a low temperature phosphorous-doped silicon detector. (b) Traces
of the oscilloscopic photographs of the IR and FIR pulses in the
same time frame for two different pressures. The FIR pulses were
monitored with a MOM diode detector. The upper traces correspond
to UV gas breakdown switch on, which chops the IR pulse. The
lower traces correspond to the UV gas breakdown switch turned off,
i.e., the IR pulse is not cut off.

---

is determined by the effective pump Rabi rate Λ_I which, as stated
above, is assumed independent of η . Here, r_3 and r_1 are the energy
and inquadrature components of the Bloch vector $\underline{F}$, respectively, and
g is the gain-to-loss ratio. In Eq.(3) Λ_F is the FIR field envelope.

The equations (1)-(3) are identical with the asymptotic equations for the two-level model with the impulse excitation of reference 7, except for the source term Γ^2/g in (1). If $\Gamma^2/g \ll 1$ for all
$\tau > 0$, then Eqs. (1)-(3) lead to solitary wave solutions in the intensity saturated regime as in reference 7.

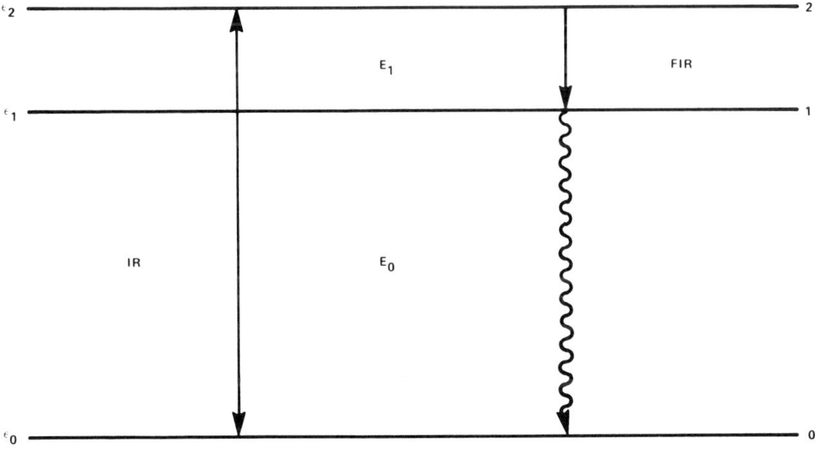

Fig. 4 Energy level diagram for the three-level model with energy
values ϵ_0, ϵ_1 and ϵ_2. The electric fields E_0 and E_1 couple the
transitions $(0 \longleftrightarrow 2)$ and $(1 \longleftrightarrow 2)$, respectively. The levels 0
and 1 are considered not radiatively coupled.

Since the manifold $(2 \longleftrightarrow 1)$ is essentially unpopulated before the pump $\Gamma(\tau)$ is turned on at time $\tau=0$, the length of the Bloch vector F is not normalizable and evolves from zero from the pump turn-on time. Thus, $F = F(\tau)$, and this leads us to impose the ansatz

$$r_3(\tau) = F(\tau) \, \cos\phi(\tau) \quad , \tag{4a}$$

$$r_1(\tau) = F(\tau) \, \sin\phi(\tau) \quad . \tag{4b}$$

The functions $F(\tau)$ and $\phi(\tau)$ are obviously to be interpreted as the instantaneous Bloch vector and tipping angle, respectively.

The ansatz (4), together with (1) and (2), leads to the equations of motion

$$\dot{\phi} = \Lambda \left[1 - \frac{\Gamma^2}{\Lambda^2} \, \sin^2\phi \right] \tag{5}$$

and

$$\dot{F} = \frac{\Gamma^2}{g} \, \cos\phi - \gamma F \quad , \tag{6}$$

where all subscripts have now been dropped. It is convenient to express in (6) the Bloch vector length F in terms of the FIR field envelope Λ . This is done by eliminating $F(\tau)$ from (6) by using the asymptotic condition (3), together with the ansatz (4b) in Eq.(6); the resulting equation of motion for $\Lambda(\tau)$ is

$$\dot{\Lambda} = \Lambda^2 \cot\phi - \gamma\Lambda \quad . \tag{7}$$

The desired set of coupled equations is therefore (5) and (7).

To see how the pulse envelope varies in time in the IR-FIR temporal overlap regime, numerical results for Λ as a function of τ are presented in Figs. 5(a) and 5(b) for an arbitrary set of values of the parameters. Here Γ is taken as a step function such that $\Gamma = 0$, $\tau < 0$, and $\Gamma \neq 0$, $\tau > 0$.

For large Γ/γ, Fig. 5(a) shows a series of pulses of diminishing intensity and increasing widths, whereas a smaller value of Γ/γ results in a larger initial "delay" and "ringing". The delay in Figs. 5(a) and 5(b) is a result of the fact that the FIR manifold in Fig. 4 is initially empty of population and when the pump is turned on at $\tau = 0$, the r_3 component of the Bloch vector begins to grow. As it does, the tipping angle ϕ also begins to grow, but $\dot{\phi}$ is smaller than it would be for no pumping (and for the same instantaneous value for the Bloch vector F and tipping angle ϕ) because

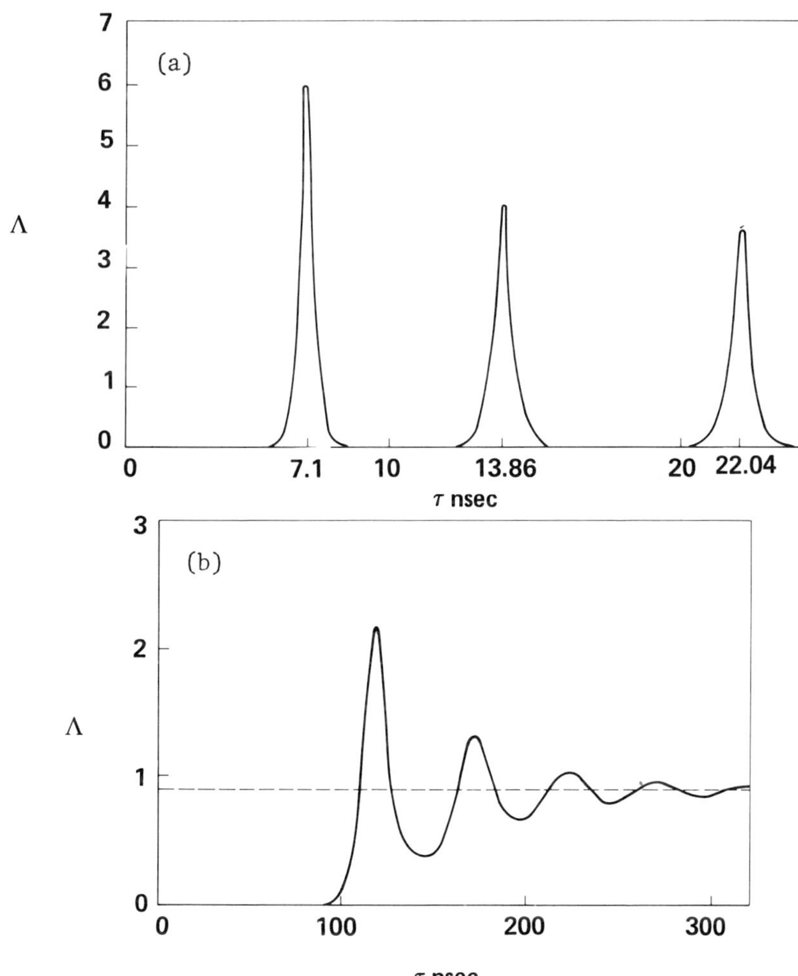

Fig. 5(a) Λ in units of Γ vs. time in nsec for values of the parameters $\gamma/\Gamma = 4\times10^{-2}$, $\phi_i = 4\times10^{-8}$, $\Lambda_i = 1.6$; (b) Λ in units of Γ vs. time in nsec for values of the parameters $\gamma/\Gamma = 4\times10^{-1}$, $\phi_i = 4\times10^{-8}$, $\Lambda_i = 1.6$.

934   EHRLICH et al.

the pumping continues to build the Bloch vector by increasing r_3.
The result is that ϕ remains small for a longer time until the
Bloch vector, hence the population of inversion in the upper level,
becomes appreciable; then ϕ and hence Λ become large quickly.

These results are qualitatively borne out in the observed FIR
pulse behavior in the high pressure region where the IR and the FIR
overlap temporally. This is seen in Fig. 3(b) which shows two cases
each with the pump switch on so the IR pump is chopped, and also
with the switch off. In these observations the pump certainly does
not correspond to a step function, but the qualitative features of
the analysis are evident. In the upper left-hand traces, the double
"hump" FIR pulse evolution is evident, consistent with the analysis
of the effects of the pump and its turn-off from Eqs.(5) and (7).
The multiple pulse generation and "ringing" are evident from the
lower traces corresponding to a longer pump duration. Due to absorption the average or effective pump intensity is larger for the
lower pressures so the lower left-hand trace corresponds to the
relatively large Γ/γ whereas the lower right-hand trace corresponds
to the smaller value of this ratio in the qualitative interpretation given above in connection with Figs. 5(a) and 5(b). It
should be emphasized that the observed FIR pulses shown in Fig.3(b)
were taken with a MOM diode which has a faster response than the
low temperature phosphorous-doped silicon detector used to obtain
the data of Fig. 3(a), but which gives wide variations in intensity from shot to shot. Consequently, large intensity variations
can occur from one shot to the next, but the main structure of the
traces shown in Fig. 3(b) are quite reproducible. One further point
should be noted in connection with the lower left-hand trace of the
FIR pulse in Fig. 3(b). The spacing between successive pulses in
this trace is, indeed, uncomfortably close to 40 nsec, and since
the CH_3F cell is 6m in length, spurious effects cannot be completely
discounted in this case. However, reasonable steps were taken to
determine the existence of feedback and spurious effects with completely negative results. The effect is entirely reproducible.

It is perhaps useful to note that in the case where the pump
is suddenly turned off at some time $\tau_0 > 0$ during the pulse evolution, we have the following area theorem: we let $\Gamma = 0$ in Eq.(5)
which gives $\Lambda = \dot{\phi}$. If we use this in Eq.(7),

$$\ddot{\phi} = \dot{\phi}^2 \cot\phi - \gamma\dot{\phi} \ . \tag{8}$$

In complete analogy with the results of reference 7, this leads
to the equation for the pulse area $\theta \left(\theta = \phi(\tau)\big|_{\tau\to\infty} \right)$,

$$\tan \tfrac{1}{2}\theta = e^{1/\gamma \, \tau_s} \tan \tfrac{1}{2}\phi_0 \ , \tag{9}$$

where, in this case, the characteristic time $\tau_s = (\Lambda_0/\phi_0)^{-1}$ and $\phi_0 = \phi(\tau_o)$, $\Lambda_0 = \partial\phi/\partial\tau|_{\tau = \tau_o}$. Noting that $1/\gamma\,\tau_s = \Lambda_0/\phi_0\gamma$ or $\tau_s = \phi_0/\Lambda_0$, in analogy with the appendix of reference 7, we define the effective "cooperation number", N_c , as

$$\tau_s = \frac{\tau_{SP}}{N_c} \quad . \tag{10}$$

Thus

$$N_c = \frac{\tau_{SP}\Lambda_0}{\phi_0} \quad , \tag{11}$$

where τ_{SP} is the spontaneous relaxation time from the upper to the intermediate level in Fig. 4. Thus, the "cooperation number", and hence the peak intensity of the subsequent pulse, very much depend upon the values of Λ and ϕ when the pump is turned off, i.e., Λ_0 and ϕ_0 . A large population in the upper level of Fig. 4 corresponding to a small ϕ_0 gives a large ratio Λ_0/ϕ_0 and hence large N_c . As Λ_0/ϕ_0 becomes large, Eq.(9) shows that the free pulse area θ approaches π . It should be noted that although we eliminated the explicit dependence of the equations of motion on the gain g through the transformation which led to (5) and (7), it remains implicit and effectively reappears in the initial conditions for the free evolution of the pulse, i.e., Λ_0 and ϕ_0 . The values for Λ_0 and ϕ_0 are explicitly determined by the pump intensity and duration as well as its shape. These do not strongly depend upon the corresponding values which must be inserted as initial conditions when the pump is turned on as long as these initial values are sufficiently small.

Since the free pulse evolution, its delay, width and intensity are very much dependent on the values Λ_0 and ϕ_0, at the pump termination and these values, in turn, are determined by the pump intensity, duration and shape, the effect of the pump characteristics is expected to prevail significantly even in the regime where the pump pulse and the FIR pulse do not appreciably temporally overlap. The implications of the role of the pump characteristics on pulse evolution are expected to be rather strong even in the superradiant regime.

4. CONCLUSIONS

We have studied experimentally the transition from the homogeneously broadened Dicke superradiant regime for CH_3F reported earlier[3] through the swept excitation regime, which is probably better referred to as the log-swept regime in terms of predictions of the model of reference 7. We have also studied the Doppler broadened Dicke superradiance regime for CH_3F which has not been

previously reported. Although the pressure region from 0.2 Torr
to 0.3 Torr corresponds to swept excitation superradiance in that
the forward to backward intensity ratio is greater than unity, the
pulse is certainly not in the asymptotic regime due to the short-
ness of the cell length. Further studies will involve extending
the cell length, increasing the pump intensity and reducing its
time duration to about 10 to 15 nsec. We shall study also the
details of the Dicke to swept transition region.

The theoretical model which incorporates the effects of the
pump and the FIR pulse generation given in the last section is expected to give good quantitative as well as qualitative results
for coherent pump excitation as long as the two pulses do not overlap appreciably in time and the FIR intensities are small. When
the intensities become large enough such that Stark and Raman
effects become important, then the assumption of the decoupling
of levels ε_1 and ε_0 which leads to the partially decoupled sets
of Bloch equations would not be valid (stimulated Raman effects
have been observed[12] in D_2O under similar experimental conditions).
For the experiment under consideration, the examples shown in Figs.
5(a) and 5(b) can be taken only as highly qualitative. Future work
will include the relaxing of the assumptions made in the model which
precludes the incorporation of Stark and Raman contributions, and
detuning and level degeneracy will also be incorporated. Account
will also be taken that the pump pulse is Doppler broadened. The
area theorem given in the last section demonstrates the pump effects
on the free pulse evolution explicitly in terms of the values Λ_0
and ϕ_0 at the pump cut-off which, in turn, depend upon the development of these quantities during the build-up when the pump is on.
Thus, the pump characteristics can play an important role in the
free evolution of the FIR pulse even when the pump cut-off and FIR
peak intensity are well separated. There are strong implications
for the role of the pump characteristics even for the superradiant
regime.

The final results of this work will be reported in a forthcoming publication.[11]

References

1.   N. Skribanowitz, J.P. Herman, J.C. MacGillivray and M.S. Feld,
     Phys. Rev. Lett. *30*, 309 (1973).
2.   J.C. MacGillivray and M.S. Feld, Phys. Rev. A *14*, 1169 (1976).
3.   A.T. Rosenberger, S.J. Petuchowski and T.A. DeTemple, in
     Cooperative Effects in Matter and Radiation, eds. C.M. Bowden,
     D.W. Howgate and H.R. Robl (Plenum, New York, 1977) p. 15.
4.   S. Haroche, J.A. Paisner and A.L. Schawlow, Phys. Rev. Lett.
     30, 948 (1973).

5.   H.M. Gibbs, in *Cooperative Effects in Matter and Radiation*, eds. C.M. Bowden, D.W. Howgate and H.R. Robl (Plenum, New York, 1977) p. 61; also this Volume, p. 543.

6.   Q.H.F. Vrehen, in *Cooperative Effects in Matter and Radiation*, eds. C.M. Bowden, D.W. Howgate and H.R. Robl (Plenum, New York, 1977) p. 79; also this Volume, p. 543.

7.   R. Bonifacio, F.A. Hopf, P. Meystre and M.O. Scully, Phys. Rev. A *12*, 2568 (1975).

8.   F.A. Hopf and P. Meystre, Phys. Rev. A *12*, 2534 (1975).

9.   F.A. Hopf, P. Meystre and D.W. McLaughlin, Phys. Rev. A *13*, 777 (1976).

10.   A.T. Rosenberger, S.J. Petuchowski and T.A. DeTemple, this Volume, p. 555.

11.   J.J. Ehrlich, C.M. Bowden, D.W. Howgate, S.H. Lehnigk, A.T. Rosenberger and T.A. DeTemple, to be published.

12.   S.J. Petuchowski, A.T. Rosenberger and T.A. DeTemple, IEEE J. Quantum Electronics *QE 13*, 476 (1977).

SUPERFLUORESCENCE: MAXWELL-BLOCH EQUATIONS, MEAN-FIELD APPROACH

AND CS EXPERIMENT[*]

R. Bonifacio,[†] M. Gronchi,[†] and L.A. Lugiato

Istituto di Fisica dell-Università, Milano, Italy

A.M. Ricca

C.I.S.E., Milano, Italy

1. MEAN-FIELD THEORY

Let us first summarize the predictions of Mean-Field Theory (MFT) [1,2] for a "pencil-shaped" atomic volume of length L, section A and Fresnel number $\mathcal{F} \simeq 1$. There are three Superfluorescence (SF) regimes depending on the value of L/L_c, where L_c is the cooperation length [3] defined below.

1) $L/L_c \ll 1$. Pure SF: single pulse, sech^2 shape, time duration ruled by $\tau_R = 8\pi T_1/3N\mu \propto N^{-1}$ and intensity $I \propto 1/\tau_R^2 \propto N^2$. Here T_1 is the spontaneous lifetime and μ is the Rehler-Eberly shape factor [4].

2) $L/L_c \simeq 1$. Weak oscillatory SF: few pulses, time duration and intensity as before.

3) $L/L_c \gg 1$. Strong oscillatory SF: many pulses, time duration as $\tau_c = L_c/c = \sqrt{\tau_R L/c}$ and intensity $I \propto \sqrt{N}$.

In the Cs-vapour experiment [5] pure SF has been observed for $L/L_c \ll 1$ and the transition to oscillatory SF took place at $L/L_c \simeq 0.7$. There was also some evidence of the change of the time scale from τ_R to $\sqrt{\tau_R}$ in the sense that going from pure SF to oscillatory SF, the time duration experimentally did scale as $1/\sqrt{N}$.

---

† Also Istituto Nazionale di Ottica, Firenze, Italy.

Hence, we state that there is an excellent qualitative agreement between MFT and the Cs-vapour experiment. In what follows we show that quantitative agreement for $L/L_c < 1$ is within a factor 1.5.

In regard to the time scale, MFT gives for the delay time τ_D, the time width of the pulse τ_W and their ratio [6]

$$\tau_D = \tau_R \left(2 + \frac{L}{L_c}\right) |ln\theta_0| \quad , \quad \tau_W = \tau_R \, 1.76 \left(2 + \frac{L}{L_c}\right) \quad , \qquad (1)$$

$$\frac{\tau_D}{\tau_W} = \frac{|ln\theta_0|}{1.76} \quad . \qquad (2)$$

For the quantity θ_0 two different cases have been discussed: $\theta_0 = 1/\sqrt{N}$ [2] and $\theta_0 = 1/\sqrt{N\mu}$ [4]. We stress that, whatever it is, θ_0 contributes to τ_D only via a *numerical factor* which depends logarithmically on the number of atoms so that it is a scale factor which is practically constant in each experimental situation. It can be changed only by changing the experiment. Therefore, we do not give much importance to its theoretical definition [7]. We prefer to determine it by fitting numerical and experimental data.

2. MAXWELL-BLOCH EQUATIONS

We solve Maxwell-Bloch Equations (MBE) simulating the spontaneous emission noise which starts SF by tipping the Bloch vector initially through an angle θ_0 constant along the sample. The same approach has been followed by Saunders and Bullough [8], whereas for MacGillivray and Feld [9] θ_0 is the value of an "equivalent input field". The "input field" and the "tipping angle" approaches are the same in the case of one-directional propagation. We prefer the second approach because we introduce two directions of propagation and the coherent coupling between them (standing-wave effect). The equations are written in the Appendix. Two-directional and one-directional propagation problems, however, give the same results for very small values of θ_0 because the two waves do not interact, but do not give the same results for larger values of θ_0 for which back scattering is important in decreasing the ringing (see later on).

In Ref. 9 one-directional MBE have been solved suggesting the following choice for θ_0

$$\theta_0 \simeq \left[\sqrt{2\pi} \, N_0 \, (\alpha L)^{3/2}\right]^{-1/2} \quad , \qquad (3)$$

where N_O is the total number of initially excited atoms and αL is
the initial amplitude gain at the line center. Equation (3) leads
to the value $\theta_O \simeq 10^{-8}$ which, as the authors point out, is appropriate for their HF experiment. However, the observed ringing in
this experiment is much smaller than the ringing predicted by MBE
with $\theta_O \simeq 10^{-8}$. Hence, in order to fit the experimental results,
they must introduce strong linear diffraction losses which according
to the formulas require $\mathcal{F} \simeq 0.1$ whereas they found that in their
experiment $\mathcal{F} \simeq 1$. This is a contradiction which has to be clarified.

On the basis of their numerical data, the authors of Ref. 9
give the following analytical formulas for τ_D, τ_W and for the number
of pulses $\mathcal{N}$

$$\tau_D = 4B^2\tau_R \ , \qquad \tau_W = 4B\tau_R \ , \qquad \mathcal{N} = B \equiv \frac{1}{4}\left|\ell n\left(\frac{\theta_O}{2\pi}\right)\right| \quad . \qquad (4)$$

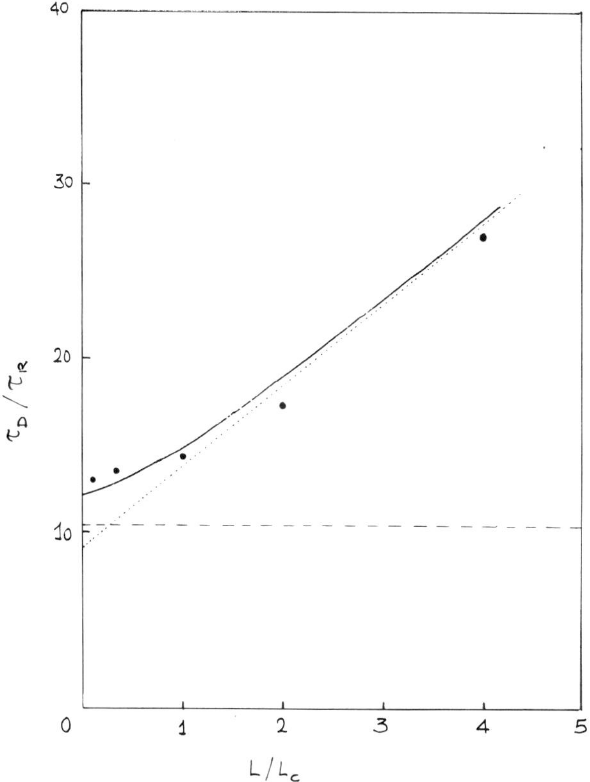

Fig. 1. Delay time in units τ_R vs. L/L_c for $\theta_O = 10^{-2}$: comparison
between the predictions of Ref. 9 (broken line), the MFT predictions
(dotted line), the best fit formula [6] (solid line) and the numerical solutions of MBE (dots).

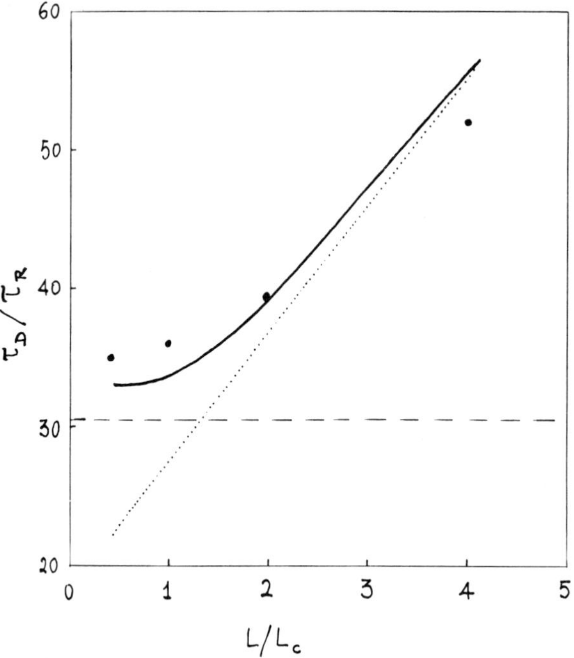

Fig. 2.   Delay time in units τ_R vs. L/L_C for $\theta_0 = 10^{-4}$: comparison between the predictions of Ref. 9 (broken line), the MFT predictions (dotted line), the best fit formula [6] (solid line) and the numerical solutions of MBE (dots).

Note that this implies

$$\frac{\tau_D}{\tau_W} = \mathcal{N} \; . \tag{5}$$

These formulas lead to qualitative conclusions and quantitative conclusions completely different from those of MFT. In what follows we show that these formulas fit numerical solutions of MBE if, and only if, $L/L_c \ll 1$ *and* θ_0 is very small. On the contrary, if $\theta_0 \gtrsim 10^{-2}$ *or* $L/L_C > 1$, the MFT predictions are in very good agreement with MBE and Eqs. (4) and (5) do not apply.

We have solved MBE with θ_0 ranging from 10^{-1} to 10^{-4} and for different values of L/L_C. All our numerical data for τ_D are fitted by the following best-fit formula (see Figs. 1 and 2)

$$\tau_D = \tau_R \left\{ 2 + \frac{L}{L_C} + \left[e^{-L/L_C} (B-1) \right] \right\} \left| \ell n \; \theta_0 \right| \; . \tag{6}$$

The term in the square brackets represents the correction to the MFT formula due to propagation effects. As one sees, the correction is always qualitatively irrelevant because its maximum value is (B-1) which depends only logarithmically on the number of atoms. The propagation correction becomes quantitatively relevant only if $L/L_c \ll 1$ *and* $B \gg 1$ ($\theta_0 \ll 10^{-2}$). In this limiting case Eq. (6) reduces to Eq. (4) for τ_D.

On the contrary, if $\theta_0 \gtrsim 10^{-2}$ the propagation correction becomes even quantitatively irrelevant and Eq. (6) reduces to Eq. (2) for τ_D for all values of L/L_c. We stress that for $L/L_c \gg 1$ the MFT formulas are correct for all values of θ_0, predicting correctly the change of time scale from τ_R to $\tau_R L/L_c = \tau_c \propto 1/\sqrt{N}$. In conclusion, regarding the time scale let us say that the MFT formulas have lost the logarithmic dependence on the number of atoms for τ_D and τ_W, whereas Eqs. (4) and (5) have lost the linear dependence on L/L_c.

In Figs. 1 and 2 we give τ_D/τ_R as a function of L/L_c for $\theta_0 = 10^{-2}$ and 10^{-4}, respectively. The broken line represents Eq. (4), the dotted line represents Eq. (2), the solid line represents Eq. (6) and the dots are our numerical data. We find that in the region $10^{-4} \leq \theta_0 \leq 10^{-1}$, τ_D/τ_W shows a linear behaviour as a function of $|\ell n\theta_0|$ (see Fig. 3). In particular, in the case $\theta_0 = 10^{-2}$ one obtains

$$\frac{\tau_D}{\tau_W} \simeq 2 \; ,$$

which is very close to the MFT evaluation

$$\frac{\tau_D}{\tau_W} = \frac{|\ell n\theta_0|}{1.76} \quad .$$

For $\theta_0 = 10^{-4}$ MBE give

$$\frac{\tau_D}{\tau_W} \simeq 3.6 \; ,$$

(see Fig. 3). This difference in the ratio τ_D/τ_W going from 10^{-2} to 10^{-4} is relevant for the forthcoming discussion of the Cs-vapour experiment.

Let us now discuss the ringing. From our calculations and the calculations in Refs. 8 and 9 we conclude that MBE give two kinds of ringing: propagation ringing, due to field disuniformity along the sample which is not described by MFT; ringing due to stimulated processes, which is described by MFT. The appearance or not of the

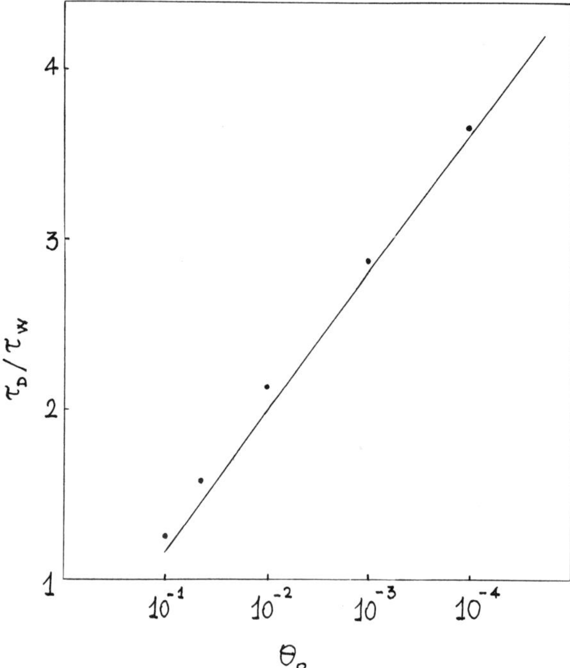

Fig. 3. Ratio of time delay to time width vs. θ_0 for $L/L_C \simeq 0.3$.
The dots are the numerical solutions of MBE whereas the solid line
is an interpolation.

first type of ringing depends on θ_0, whereas the second type of
ringing is ruled by L/L_C. More precisely, the number of pulses is
ruled by the following expression

$$\mathcal{N} \simeq \frac{\tau_D}{\tau_W} \simeq \frac{L}{L_C} + e^{-L/L_C} \, (B\text{-}1) \, , \tag{7}$$

where the first term is the MFT contribution and the second term is
the propagation contribution. If $L/L_C \gg 1$, only the MFT ringing
survives. If $L/L_C < 1$, MFT ringing disappears and if $B \gg 1$, one
gets Eqs. (4) and (5) and a strong propagation ringing. We stress
that the appearance or not of the propagation ringing depends
logarithmically on the number of atoms. Hence, propagation ringing,
unlike MFT ringing, can hardly be eliminated by changing the density.
In other words, there are experiments, such as the HF experiment,
with propagation ringing and experiments, such as the Cs-vapour
experiment, *without* ringing because these two experiments are

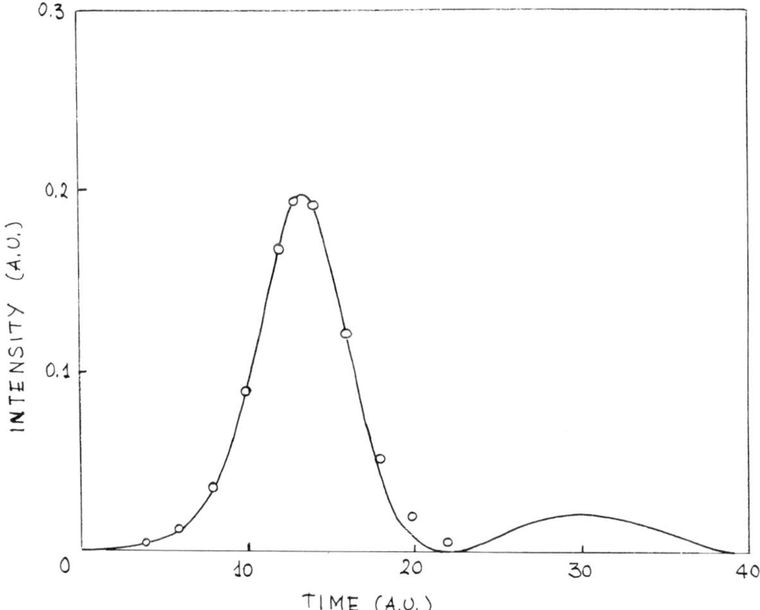

Fig. 4. Comparison between the solution of MBE with standing-wave effect for $L/L_c \simeq 0.3$ and $\theta_0 = 10^{-2}$ (solid line) and a sech^2 shape with the same height and width (circles). The ringing ratio is about 0.1.

characterized by different values of the quantity B. If this quantity is larger than 1, MFT does not apply. On the contrary, if $B \lesssim 1$ ($\theta_0 \gtrsim 10^{-2}$) MFT applies since propagation ringing is always very small and negligible.

In Figs. 4 and 5 we give the time evolution of the intensity for $\theta_0 = 10^{-2}$ and $L/L_c \simeq 0.3$ and for $\theta_0 = 4.5 \times 10^{-2}$ and $L/L_c \simeq 0.6$, respectively. Defining the ringing ratio as

$$R = \frac{\text{intensity of the 1-st pulse}}{\text{intensity of the 2-nd pulse}} ,$$

one obtains from Figs. 4 and 5 $R \simeq 0.1$ and 0.06, respectively. In Ref. 8 the ringing ratio is about twice larger because back scattering has not been taken into account. Furthermore, we have tested the MFT predictions about the sech^2 shape. As shown in Figs. 4 and 5 for $\theta_0 \gtrsim 10^{-2}$ the sech^2 shape is a very good approximation to the pulse shape given by MBE.

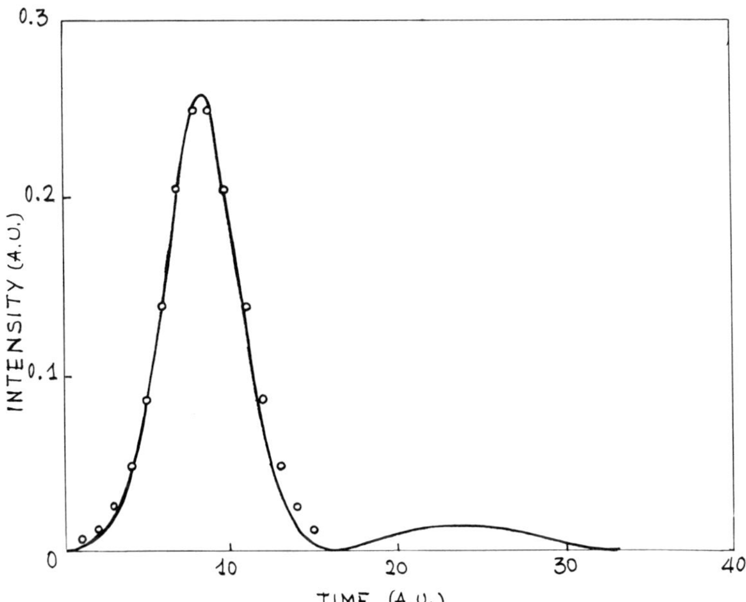

Fig. 5. Comparison between the solution of MBE with standing-wave effect for $L/L_c \simeq 0.6$ and $\theta_0 = 4.5 \times 10^{-2}$ (solid line) and a $sech^2$ shape with the same height and width (circles). The ringing ratio R is about 0.06.

In conclusion, all the MFT predictions are in excellent qualitative and quantitative agreement with MBE for $\theta_0 \gtrsim 10^{-2}$. For $\theta_0 < 10^{-2}$ the logarithmic correction due to propagation becomes quantitatively relevant and propagation ringing appears.

3. COMPARISON WITH Cs-VAPOUR EXPERIMENT (SINGLE PULSE)

The first thing one must understand and compare with theories are the qualitative features of an experiment and one can worry about numbers only if there is qualitative agreement. In fact, an agreement between numbers is meaningless if not supported by qualitative agreement.

Let us now recall the main features of the Cs-vapour experiment [5] which are the following:

i) existence of pure SF in which the shape of the single pulse is nearly a $sech^2$ for $L/L_c < 0.7$, and the appearance of oscillations for $L/L_c > 0.7$;

ii) fixed ratio between the observed τ_D and τ_W: $\tau_D/\tau_W \simeq 2$.

Using the analytical criteria (3), (4) and (5) one would predict $\theta_0 \simeq 5\times10^{-6}$ leading to strong ringing when a single pulse has been observed and to $\tau_D/\tau_W = 3.5 = \mathcal{N}$. Hence, these analytical criteria are in contradiction with the main features of the Cs-vapour experiment. Further comparisons in our opinion would be meaningless for the reason we have given earlier. On the contrary, we stress the excellent qualitative agreement with MFT.

In order to understand i) and ii) with lossless MBE one must abandon criterium (4) for θ_0 and go to higher θ_0. Choosing $\theta_0 \simeq 1/\sqrt{N} \simeq 10^{-4}$ the ringing and the ratio τ_D/τ_W are always too high as can be seen from Table 1. More precisely, the analytical criteria (4) and (5) would give $\tau_D \simeq 30\tau_R$, $\tau_W \simeq 10\tau_R$ leading to $\tau_D/\tau_W \simeq 3$. Our numerical solution of MBE gives for this choice of θ_0, $\tau_D \simeq 35\tau_R$, $\tau_W \simeq 9.6\ \tau_R$ and $\tau_D/\tau_W \simeq 3.6$. Furthermore, let us note that our numerical solutions show that adding linear losses one suppresses the ringing and increases τ_D and τ_W (see Fig. 6), but the ratio τ_D/τ_W remains practically constant. Hence, the only possible correct way to obtain the observed ratio τ_D/τ_W from MBE without looking for other physical effects is to choose $\theta_0 \simeq 10^{-2}$ or a little larger, as can be seen from Fig. 3. Let us note that this value of θ_0 in the case of the Cs-vapour experiment corresponds to the choice $\theta_0 \simeq 1/\sqrt{N\mu}$.

TABLE 1

	RINGING	τ_D/τ_W	τ_D	τ_W
Cs-vapour experiment: single pulse (Ref. 5)	none	~2	12.5 nsec	~6 nsec
MBE with $\theta_0 \simeq 5\times10^{-6}$ (Ref. 9)	strong	$3.5=B$	$49\tau_R=4B^2$	$14\tau_R=4B$
MBE with $\theta_0 = 10^{-4}$ (numerical solution)	strong	3.6	$35\tau_R$	$9.6\tau_R$
MBE with $\theta_0 = 10^{-2}$ (numerical solution)	very small	2.14	$13.4\tau_R$	$6.3\tau_R$
MFT for $L/L_c \simeq 0.3$	none	2.6	$10.6\tau_R$	$\sim4\tau_R$

Experimental data are in agreement with MBE with $\theta_0 = 10^{-2}$ and with MFT if $\tau_R \simeq 1$ nsec. There is no choice of τ_R which gives agreement between experiment and MBE with $\theta_0 \simeq 5\times10^{-6}$ or 10^{-4}.

As one can see from Table 1, in this region there is good qualitative agreement between MBE with standing-wave effect and MFT, and a
quantitative agreement within a factor 1.3 for τ_D and 1.5 for τ_W.
We stress that these numerical factors are still smaller for
$\theta_0 > 10^{-2}$ and come from a logarithmic dependence on the number of
atoms due to propagation effects.

Let us now come to the quantitative comparison with the experiment. As one sees from Table 1, the experimental values of τ_D and
τ_W can be easily understood if τ_R is about 1 nsec. The value of
τ_R evaluated through the classical formula

$$\tau_R = T_1/N\mu_0 \quad ,$$

with

$$\mu_0 = 3\lambda^2/8\pi A \quad ,$$

is about 0.7 nsec. However, this expression for μ holds only for
$\mathcal{F} \gg 1$. If $\mathcal{F} \simeq 1$, the numerical evaluation of the shape factor
gives $\mu = 0.78\ \mu_0$ [5]. This leads to $\tau_R \simeq 0.9$ nsec and then to a
good quantitative agreement between MBE and the single pulse

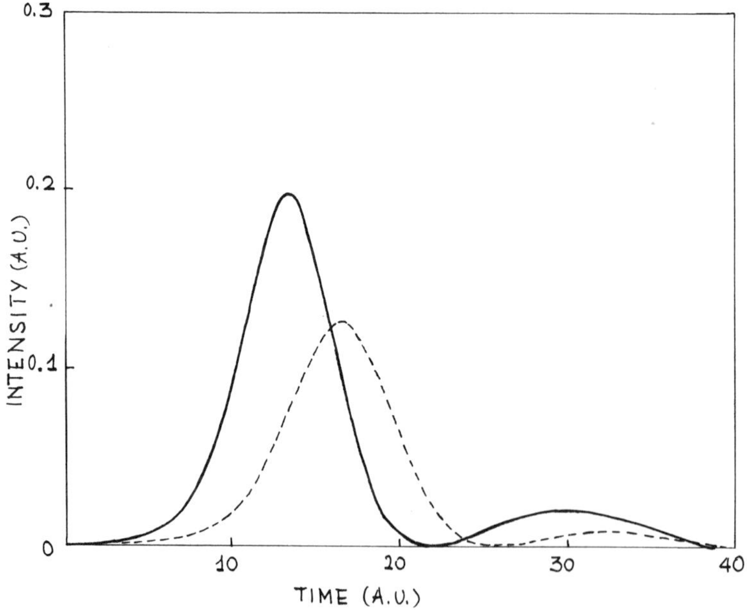

Fig. 6. Numerical solution of MBE with standing-wave effect for
$L/L_c \simeq 0.3$ and $\theta_0 = 10^{-2}$ with (broken line) and without (solid line)
linear damping. Note that even without damping the ringing ratio
R is only about 0.1.

experiment. In any case, due to the large uncertainties for the density ($\pm$ 40%) and for the Fresnel number (0.5-2), numerical comparisons which involve the evaluation of τ_R should not be taken too strictly. On the contrary, what should be strictly respected in fitting the experiment are the ratio τ_D/τ_W and the ringing. In this sense, as shown in Table 1, the choices $\theta_o \simeq 5\times10^{-6}$ and 10^{-4} do not fit the experiment whereas $\theta_o \simeq 10^{-2}$ does fit. To be more clear, even adjusting τ_R so that there is perfect agreement for τ_D or τ_W with the smallest values of θ_o, the ratio between them can not be adjusted since it is independent of the specific value of τ_R. On the contrary, with the choice $\tau_R \simeq 1$ nsec, one can fit the experimental results both qualitatively and quantitatively with $\theta_o \simeq 10^{-2}$.

4. CONCLUSIONS

MFT and MBE for $\theta_o \gtrsim 10^{-2}$ are in excellent qualitative and quantitative agreement and give a good description of single pulse SF in the Cs-vapour experiment. The advantage of MFT is that it is an analytically solvable quantum mechanical model able to calculate quantum fluctuations [1] which still must be experimentally tested.

APPENDIX

The MBE for an atomic system, confined in a "pencil-shaped" geometry along the z-direction with the origin in its middle, and two coherently coupled opposite waves can be written in the following way

$$\frac{\partial}{\partial t} P(z,t) = \varepsilon(z,t) \, \Delta_o(z,t) - \varepsilon(-z,t) \, \Delta_1(z,t) \ ,$$

$$\frac{\partial}{\partial t} \Delta_o(z,t) = -[\varepsilon(z,t) \, P(z,t) + \varepsilon(-z,t) \, P(-z,t)] \ ,$$

$$\frac{\partial}{\partial t} \Delta_1(z,t) = \frac{1}{2} [\varepsilon(z,t) \, P(-z,t) + \varepsilon(-z,t) \, P(z,t)] \ ,$$

$$\left(\frac{\partial}{\partial t} + c \, \frac{\partial}{\partial z}\right) \varepsilon(z,t) = \frac{1}{\tau_c^2} P(z,t) \ ,$$

where: i) $\varepsilon(z,t) = p \, E(z,t)/\hbar$ is the Rabi frequency and p is the dipole moment of the transition; ii) $\tau_c = \sqrt{\hbar\lambda/4\pi^2 \, p^2 n}$ is the cooperation time [3] which is also related to the SF time τ_R through $\tau_c^2 = \tau_R \, L/c$, n is the atomic density and λ is the wavelength associated with the atomic transition; iii) $P(z,t)$ is the macroscopic

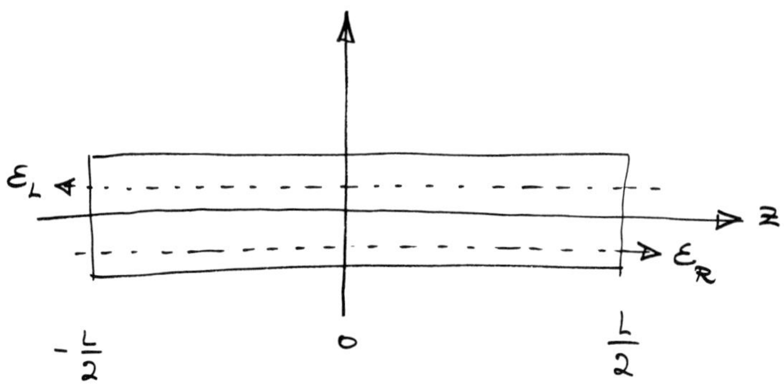

Fig. 7. Atomic geometry with the two opposite waves traveling
through it.

atomic polarization; and iv) $\Delta_0(z,t)$ and $\Delta_1(z,t)$ are two components
of the population difference which arises from the **following** position

$$\Delta(z,t) = \Delta_0(z,t) - 2\Delta_1(z,t) \cos 2kz ,$$

where $k = 2\pi/\lambda$. Further on, we have assumed that no space harmonics
of order higher than the second are in play and that

$$\varepsilon_L(z,t) = \varepsilon_R(-z,t) = \varepsilon(-z,t) ,$$

$$P_L(z,t) = P_R(-z,t) = P(-z,t) ,$$

where the subscripts R and L label electric field and atomic polarization for the two waves, as shown in Fig. 7. A detailed derivation
of these equations, which can be obtained on the line of Ref. 10,
will be given elsewhere. Finally, let us state that these equations
in the Mean Field approximation lead to the damped pendulum equation
as described in Ref. 6 without standing-wave effect.

* Work partially supported by ERO Grant No. DA-ERO-77-G-054.

References

1.   R. Bonifacio, P. Schwendimann and F. Haake, Phys. Rev. A *4*,
     302 and 854 (1971).
2.   R. Bonifacio and L.A. Lugiato, Phys. Rev. A *11*, 1507 (1975)
     and *12*, 587 (1975).
3.   F.T. Arecchi and E. Courtens, Phys. Rev. A *2*, 1730 (1970).

4.   N.E. Rehler and J.H. Eberly, Phys. Rev. A *3*, 1735 (1971).
5.   H.M. Gibbs, Q.H.F. Vrehen and H.M.J. Hikspoors, Phys. Rev.
      Lett. *39*, 547 (1977); Q.H.F. Vrehen, in *Cooperative Effects in
      Matter and Radiation*, eds. C.M. Bowden, D.W. Howgate and H.
      Robl (Plenum Publ. Co., New York, 1977) p. 79.
6.   R. Bonifacio, M. Gronchi, L.A. Lugiato and A.M. Ricca, in
      Cooperative Effects in Matter and Radiation, eds. C.M. Bowden,
      D.W. Howgate and H. Robl (Plenum Publ. Co., New York, 1977)
      p. 193.
7.   This point has been extensively discussed by E. Ressayre and
      A. Tallet attempting to construct a quantum theory which de-
      scribes propagation ringing. See their paper in these Pro-
      ceedings, p. 799.
8.   R. Saunders and R.K. Bullough, in *Cooperative Effects in Matter
      and Radiation*, eds. C.M. Bowden, D.W. Howgate and H. Robl
      (Plenum Publ. Co., New York, 1977) p. 209.
9.   J.C. MacGillivray and M.S. Feld, Phys. Rev. A *14*, 1169 (1976).
10.   J.A. Fleck, Jr., Appl. Phys. Lett. *13*, 365 (1968).

THEORY OF NATURAL LINE SHAPE

Luiz Davidovich*

Institut für Theoretische Physik, ETH, Zürich, Switzerland

H.M. Nussenzveig[†]

Universidade de São Paulo, São Paulo, Brazil

1. INTRODUCTION

The treatment of the emission of light by an atom in quantum electrodynamics has been strongly influenced by Weisskopf and Wigner's early contribution [1] to this subject. While their work was highly successful in accounting for the observed line shape, several disturbing questions about the underlying assumptions remained unsettled:

(i) The initial state assumed for the system (atom in excited state with no photons present) seems quite unphysical. The state preparation and the dependence of the decay on the excitation should be discussed.

(ii) The Weisskopf-Wigner exponential decay "Ansatz" cannot be valid for all times. Deviations from it are expected to be extremely small for long-lived decaying states such as the atomic ones. However, the range of validity of the exponential decay law should be determined.

(iii) The state space was restricted to a two-level atom and to the vacuum and one-photon sectors, without any indication of how to proceed in order to improve the approximation. For such a basic problem, a clear-cut formulation should be provided, as well as a systematic procedure for deriving corrections to the Weisskopf-Wigner approximation.

Measurements of the Lamb shift in hydrogen yielded a sensitive experimental test of the predicted line shape. It was remarked by Lamb [2] that the Weisskopf-Wigner line shape disagrees with experiment in this case if the usual minimal-coupling Hamiltonian is employed, and that one must use instead the interaction $-e\underline{r}\cdot\underline{E}^{\perp}$, where $\underline{E}^{\perp}$ is the transverse electric field. As will be seen below, the discrepancy is orders of magnitude larger than the present accuracy in the measurement. Since the minimal-coupling Hamiltonian is widely employed in quantum electrodynamics, this discrepancy should be resolved.

An extensive study of the line shape problem was made around 1950 by Heitler, Arnous and collaborators [3-5]. They applied a dressing transformation in order to go over from "bare" states of the system to physical states. They began [3] by imposing unphysical constraints on the dressing transformation; later [4,5], these constraints were removed, but they employed a cumbersome formalism, and focussed attention on the evaluation of radiative corrections to the resonant term in a transition between two atomic states. The effect of nonresonant terms was not considered, and there was no discussion of the time development of the system. Similar remarks apply to Low's [6] covariant $\underline{S}$-matrix treatment.

The relationship between time evolution and the analytic properties of the S-matrix as a function of energy has been clarified in nonrelativistic potential scattering, where decaying states can be described in terms of propagators associated with complex poles on unphysical sheets [7]. This has also been verified in some field-theoretical models of unstable particles.

The present paper is a survey of results obtained in a new treatment [8] of the line shape problem for the nonrelativistic hydrogen atom. A more detailed version [8] will appear elsewhere.

2. THE MODEL

We consider a nonrelativistic hydrogen atom (recoil is neglected) interacting with the quantized radiation field. The Hamiltonian of the system is (we take $\hbar = c = 1$ throughout)

$$H = H_A + H_F + H_{I,1} + H_{I,2} \ , \tag{2.1}$$

$$H_A = \underline{p}^2/2m - e^2/r \ , \tag{2.2}$$

$$H_F = \frac{1}{8\pi} \int d^3r \ [(\underline{E}^{\perp})^2 + (\underline{\nabla}\times\underline{A})^2] \ , \tag{2.3}$$

where $\underline{A}$ is the vector potential in the Coulomb gauge, and $\underline{E}^{\perp}$ is the transverse electric field, and

$$H_{I,1} = - \frac{e}{m}\, \underline{p} \cdot \underline{A}(\underline{r}) \; , \tag{2.4}$$

$$H_{I,2} = \frac{e^2}{2m}\, \underline{A}^2(\underline{r}) \; . \tag{2.5}$$

We rewrite H_A in terms of the H-atom stationary states $|n\rangle$ (where n stands for a complete set of quantum numbers) and the corresponding energies E_{0n}, as

$$H_A = \sum_n E_{0n}\, |n\rangle\langle n| \; , \tag{2.6}$$

where the summation is to be understood as integration for eigenstates in the continuum. However, we will be concerned mainly with the discrete spectrum.

We employ the multipole expansion

$$\underline{A}(\underline{r}) = 2 \sum_{\tau=0}^{1} \sum_{J=1}^{\infty} \sum_{M=-J}^{J} \int_0^{\infty} dk\sqrt{k}\; a_{JM\tau}(k)\; \underline{A}_{JM\tau}(k,\underline{r}) + \text{h.c.} \; , \tag{2.7}$$

in terms of the usual basis [9], where $\underline{A}_{JM\tau}$ represents an electric ($\tau=0$) or magnetic ($\tau=1$) multipole field of order 2^J. The operator $a_{JM\tau}(k)$ annihilates photons of frequency k with the set of quantum numbers $\beta = (JM\tau)$, so that

$$[a_{\beta}(k),\, a_{\beta'}^{\dagger}(k')] = \delta_{\beta\beta'}\delta(k-k') \; . \tag{2.8}$$

After zero-point energy subtraction, H_F becomes

$$H_F = \sum_{\beta} \int_0^{\infty} dk\, k\, a_{\beta}^{\dagger}(k)\, a_{\beta}(k) \; . \tag{2.9}$$

The exact matrix elements $\langle n|\underline{p}\cdot\underline{A}_{\beta}|m\rangle$ for the nonrelativistic H-atom (including retardation) have been evaluated by Moses [10]. To express $H_{I,1}$ in terms of them, it is convenient to adopt "atomic units" for the frequency k, measuring it in units of the inverse Bohr radius a_B^{-1}, i.e., setting

$$a_B^{-1} \equiv \alpha m = 1 \; , \tag{2.10}$$

where α is the fine-structure constant. We get

$$H_{I,1} = \sqrt{\lambda} \sum_{nm\beta} \int_0^\infty dk\sqrt{k}\ f_{nm\beta}(k)\ a_\beta(k)\ |n><m| + h.c.\ , \qquad (2.11)$$

where we have defined a coupling constant (in the units (2.10))

$$\lambda = (e/m)^2 = \alpha^3 \qquad (\approx 4 \times 10^{-7})\ . \qquad (2.12)$$

Let $n = (N_n, j_n, M_n)$, where N_n, j_n and M_n are the principal quantum number, angular momentum and magnetic quantum number of the state n, respectively. The matrix elements $f_{nm\beta}$ vanish unless they fulfill the following exact selection rules, which follow from angular momentum and parity conservation:

$$|j_n - j_m| \leq J \leq j_n + j_m\ ; \qquad M = M_n - M_m\ ;$$

$$J + j_n + j_m \equiv \tau\ (\text{mod. } 2)\ . \qquad (2.13)$$

Under these conditions, it can be shown [8] that $|f_{nm\beta}(k)|^2$ is a rational function of k^2, of the following remarkably simple form:

$$|f_{nm\beta}(k)|^2 = (k^2 + K_{nm}^2)^{-p_{nm}}\ P_{nm\beta}(k)\ , \qquad (2.14)$$

where $P_{nm\beta}(k)$ is a polynomial (in k^2) with real coefficients, that are $\mathcal{O}(1)$ in the units (2.10), $p_{nm} \geq 4$ is an integer, $|f_{nm\beta}(k)|^2 = \mathcal{O}(k^{-8})$ as $k \to \infty$, and $K_{nm} = \mathcal{O}(1)$ is an inverse transition radius of the order of a_B^{-1}. The functions (2.14) play the role of exact atomic form factors, introducing a natural cutoff at distances of the order of the Bohr radius. The extremely simple analytic properties of (2.14) play an important role in a soluble model discussed below.

3. THE RESOLVENT OPERATOR

Our treatment is based on the resolvent operator method, as developed by Van Hove [11]. We briefly recall its main features.

The resolvent $\mathcal{G}(z)$ associated with the Hamiltonian $H = H_0 + H_I$ is an operator-valued function of the complex variable z defined, for Im $z \neq 0$ (where it is expected to be holomorphic), by

$$\mathcal{G}(z) = (z - H)^{-1}\ . \qquad (3.1)$$

The unperturbed resolvent $\mathcal{G}_0(z)$ is obtained by replacing H by H_0.

The operator $\mathcal{G}(z)$ can be split into a diagonal part with respect to H_0, $\mathcal{D}(z)$, and a nondiagonal part $\mathcal{N}(z)$:

$$\mathcal{G}(z) = \mathcal{D}(z) + \mathcal{N}(z) \; , \tag{3.2}$$

where, for every eigenstate $|\alpha\rangle$ of H_0, $H_0|\alpha\rangle = E_{0\alpha}|\alpha\rangle$, we have

$$\mathcal{D}(z)|\alpha\rangle = \mathcal{D}_\alpha(z)|\alpha\rangle \; . \tag{3.3}$$

The eigenvalue $\mathcal{D}_\alpha(z)$ determines the persistence amplitude of the state $|\alpha\rangle$, i.e., if we start out from the initial state $|\alpha\rangle$ at $t = 0$, the probability amplitude to find the system in the state $|\alpha\rangle$ at time t is given by

$$\langle\alpha|\exp(-iHt)|\alpha\rangle = -\frac{1}{2\pi i}\int_C \exp(-izt)\,\mathcal{D}_\alpha(z)\;dz \; , \tag{3.4}$$

where C is a straight line parallel to the real axis and taken slightly above it. The amplitude (3.4) therefore depends crucially on the analytic properties of $\mathcal{D}_\alpha(z)$.

The analogue $\Sigma(z)$ of Dyson's mass operator [12] satisfies

$$\mathcal{D}(z) = \mathcal{G}_0(z) + \mathcal{D}(z)\,\Sigma(z)\,\mathcal{G}_0(z) \; , \tag{3.5}$$

and it can be pictured diagrammatically by

$$\Sigma(z) = [H_I + H_I\,\mathcal{D}(z)\,H_I + \ldots]_{\text{i.d.}} \; , \tag{3.6}$$

where the index "i.d." stands for a summation over all irreducible diagonal [11] diagrams. As z approaches a point E on the real axis, we have [11]

$$\lim_{z\to E\pm i0}\Sigma(z) = \Delta(E)\mp\frac{i}{2}\,\Gamma(E) \; , \tag{3.7}$$

where $\Gamma(E)$ is a positive semidefinite operator, $\Gamma(E) \geq 0$. Thus, denoting by an index α the eigenvalues of diagonal operators in the state $|\alpha\rangle$,

$$\lim_{z\to E\pm i0}\mathcal{D}_\alpha(z) = [E - E_{0\alpha} - \Delta_\alpha(E) \pm \frac{i}{2}\,\Gamma_\alpha(E)]^{-1} \; . \tag{3.8}$$

If $\Gamma_\alpha(E) \neq 0$, we see that E lies on a cut of $\mathcal{D}_\alpha(z)$, which is holomorphic for $\text{Im } z \neq 0$ (physical sheet). The Weisskopf-Wigner approximation would correspond to taking $\mathcal{D}_\alpha(z) = [z - E_\alpha + (i/2)\Gamma_\alpha]^{-1}$,

which would violate the analyticity on the physical sheet. We see
also that $\Delta(E)$ and $\Gamma(E)$ play the roles of level shift and level
width operators, respectively.

Let us assume that the equation (cf. (3.8))

$$E - E_{0\alpha} - \Delta_\alpha(E) = 0 \tag{3.9}$$

has one and only one real root, $E = E_\alpha$. Then, there are only three
possibilities:

(i) $\Gamma_\alpha(E_\alpha) \neq 0$: In this case, by (3.8), $E = E_\alpha$ lies on a cut
of $\mathcal{O}_\alpha(z)$. Van Hove calls this a *dissipative* state; typically, it
decays in the presence of the interaction, and $\Delta_\alpha(E_\alpha)$ and $\Gamma_\alpha(E_\alpha)$
represent the energy shift and linewidth due to the interaction.

(ii) $\Gamma_\alpha(E) = 0$ for all real E: In this case, $\mathcal{O}_\alpha(z)$ has a
simple pole at $E = E_\alpha$ and no cut. The state $|\alpha\rangle$ is classified by
Van Hove as *asymptotically stationary*, because its asymptotic
evolution is unaffected by the interaction, which produces only
transient effects, as in scattering from a short-range potential.

(iii) $\Gamma_\alpha(E_\alpha) = 0$, but $\Gamma_\alpha(E) \neq 0$ for some real E: In this
case, $|\alpha\rangle$ is not asymptotically stationary. The interaction pro-
duces *persistent perturbation effects*, including self-energy and
"dressing" ("cloud") effects.

The nondiagonal part $\mathcal{N}(z)$ in (3.2) may be written as

$$\mathcal{N}(z) = \mathcal{O}(z)\,\mathcal{U}(z)\,\mathcal{O}(z) \; , \tag{3.10}$$

where the transition operator $\mathcal{U}(z)$ has a representation similar to
Eq. (3.6), summed over irreducible nondiagonal diagrams [11].

If $|\alpha\rangle$ and $|\alpha'\rangle$ are asymptotically stationary, the S-matrix
element $\langle\alpha|S|\alpha'\rangle$ between these states can be shown to exist (with
no need of adopting the unphysical procedure of adiabatic switching
of the interaction). It is then given by

$$\langle\alpha|S|\alpha'\rangle = \delta(\alpha-\alpha') - 2i\pi\delta(E_\alpha - E_{\alpha'}) \langle\alpha|\mathcal{U}|\alpha'\rangle \; , \tag{3.11}$$

and the corresponding differential cross-section, for one-photon
incidence, is given by

$$\sigma_{\alpha\alpha'} = (2\pi)^4 |\langle\alpha|\mathcal{U}|\alpha'\rangle|^2 \, \rho_\alpha(E_\alpha) \; , \tag{3.12}$$

where ρ_α is the density of final states.

4. POSSIBLE APPROXIMATIONS

Our approach is based on successive approximations, retaining first only a part (as large as we can make it) of the Hamiltonian such that the problem can be solved exactly, and then adding back the omitted pieces of the Hamiltonian and investigating how they affect the solution. A related approach for Weisskopf-Wigner-type theories has been advocated by Grimm and Ernst [13].

The following simplifications lead to an exactly soluble model, to be treated in Section 5:

(I) Finite number of atomic states: The spectrum of H_A is cut off beyond some discrete level N, so that, in (2.6) and (2.11), $n \leq N$, $m \leq N$. The two-level atom is a particular case.

(II) Generalized RWA ("Rotating-Wave Approximation"): We consider a specific (fixed) excited state $|r>$ ("resonant"), which will be selectively singled out by resonant processes, so that we keep only those terms in (2.11) where one of the levels involved in the transition is r. The generalized RWA corresponds to the following choice of interaction Hamiltonian:

$$H_I = \sqrt{\lambda} \sum_{n\beta} \int_0^\infty dk\sqrt{k}\ f_{rn\beta}(k)\ a_\beta(k)\ |r><n| + \text{h.c.}\ ,\qquad (4.1)$$

where the "counter-rotating" terms, which differ from those in Eq. (4.1) by the interchange of r and n in the summation, are neglected. This is an extended version of the usual RWA, to which it reduces for a two-level atom. It couples the *absorption* of a photon only with transitions that *end* in r, and *emission* of a photon only with those that *start* from r, whether the transitions are to levels above or below r.

If we call "resonant atomic excitation" the occupation of level r, the generalized RWA leads to an additional conservation law, corresponding to the conservation of the number of photons plus the resonant atomic excitation, represented by the operator

$$\mathcal{E} = \frac{1}{2}\ |r><r| - \frac{1}{2} \sum_{n\neq r} |n><n| + \sum_\beta \int_0^\infty dk\ a_\beta^\dagger(k)\ a_\beta(k)\ ,\qquad (4.2)$$

which commutes with $H = H_A + H_F + H_I$, i.e., $[\mathcal{E},H] = 0$. This leads to a splitting of Hilbert space into sectors, allowing an exact solution in the sector of interest (cf. Section 5). This is also related with the fact that the generalized RWA excludes persistent perturbation effects, as will be seen below.

The following two well-known approximations will *not* be employed in the soluble model of Section 5:

(III) Dipole Approximation: This amounts to substituting, in Eqs. (2.4) and (2.5), $\underline{A}(\underline{r}) = \underline{A}(\underline{0})$. As a consequence of this, only electric dipole waves remain coupled to the atom, i.e., the photon index β ranges only over the values

$$J = 1 ; \quad M = 0, \pm 1 ; \quad \tau = 0 \quad . \tag{4.3}$$

Since this corresponds to neglecting retardation over atomic dimensions, the electric dipole form factors in Eq. (2.14) go over into their $k \to 0$ limit, which is constant. This would yield divergent integrals, so that this approximation is often coupled with approximation (IV).

(IV) Sharp Cutoff: This substitutes the effect of the atomic form factors by a sharp cutoff,

$$\rho(k) = \theta(K - k) \quad , \tag{4.4}$$

where θ is the Heaviside step function, and the cutoff parameter K is of the order of the inverse Bohr radius, $K \sim a_B^{-1}$.

The combination of approximations (III) and (IV), although it has often been employed in conjunction with two-level atomic models, leads to spurious effects. It gives rise to non-decaying, non-ergodic contributions to the time evolution for a Weisskopf-Wigner initial state [8]. This does not happen for a smooth form factor, such as Eq. (2.14).

5. EXACTLY SOLUBLE MULTILEVEL MODEL

This model is defined by the Hamiltonian

$$H = \sum_n E_{0n} |n\rangle\langle n| + \sum_\beta \int_0^\infty dk \; k \; a_\beta^\dagger(k) \; a_\beta(k)$$

$$+ \sqrt{\lambda} \sum_{n\beta} \int_0^\infty dk\sqrt{k} \; [f_{rn\beta}(k) \; a_\beta(k) |r\rangle\langle n| + h.c.] \quad , \tag{5.1}$$

where $n \leq N$, and β also ranges only over a finite set of values, due to the selection rules (2.13). The coefficients $f_{rn\beta}(k)$ are taken to be the exact H-atom matrix elements. This corresponds to making approximations (I) and (II) of Section 4 and neglecting the quadratic interaction Hamiltonian (2.5).

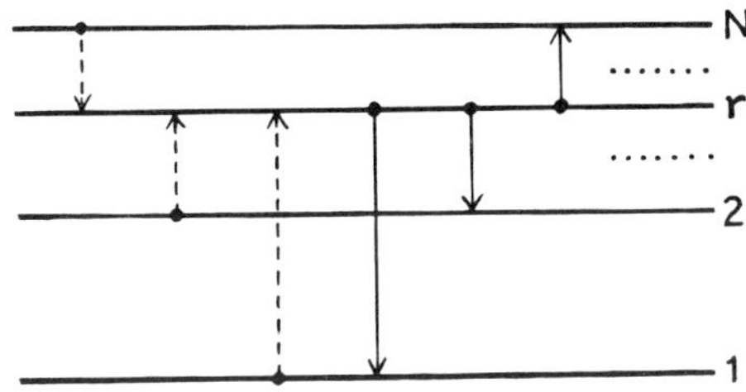

Fig. 1. Transitions taken into account by the soluble N-level model. r = resonant level;

$-----\rightarrow$ absorption;

$\longrightarrow$ emission.

The transitions allowed by the interaction Hamiltonian in Eq. (5.1) are schematically represented in Fig. 1. In contrast with Davies' multilevel model [14], the present one allows for transitions (within the generalized RWA) between the resonant level and other excited levels, but it does not take into account the decay of the other levels.

We restrict our discussion to the Hilbert space sector associated with the eigenvalue $\mathcal{E} = 1/2$ of the operator (4.2). Any state vector in this sector is of the form

$$|\Psi> = u \ |r;0> + \ |\Phi> \ , \qquad <\Psi|\Psi> = 1 \ , \tag{5.2}$$

where

$$|\Phi> = \sum_{n \neq r, \beta} \int_0^\infty dk \ \phi_{n\beta}(k) \ |n;1\beta,k> \ , \tag{5.3}$$

$$|r;0> = |r>|0) \ , \qquad |n;1\beta,k> = a_\beta^\dagger(k) |n>|0) \tag{5.4}$$

are direct products of atomic state vectors and photon state vectors, the latter being denoted by round brackets. Cascade and multiphoton processes are excluded in this model, but the effects of transitions to other levels on the excitation and decay of the resonant level are partially taken into account.

The condition

$$\frac{\lambda}{E_{0r}} \sum_{n \neq r, \beta} \int_0^\infty \frac{|f_{rn\beta}(k)|^2}{k + E_{0n}} \ k \ dk < 1 \ , \tag{5.5}$$

which follows from the smallness of the coupling constant (2.12), ensures that H has only a continuous spectrum [15], ranging from the unperturbed ground-state energy E_{01} (which we take as the zero level of energy, $E_{01} = 0$) to infinity. The unperturbed ground state is also the interacting ground state (no persistent effects). The resolvent has a cut along the real axis from 0 to ∞.

The matrix elements of the resolvent in this sector are [8]

$$\langle r;0|\mathcal{G}(z)|r;0\rangle = \langle r;0|\mathcal{O}(z)|r;0\rangle = \mathcal{O}_{r;0}(z) = [D(z)]^{-1} \, , \quad (5.6)$$

$$\langle r;0|\mathcal{G}(z)|\Phi\rangle = \frac{\sqrt{\lambda}}{D(z)} \sum_{n\neq r,\beta} \int_0^\infty \frac{\sqrt{k}\, f^*_{rn\beta}(k)\phi_{n\beta}(k)}{z - k - E_{0n}} \, dk \, , \quad (5.7)$$

and a similar expression for $\langle\Phi|\mathcal{G}(z)|\Phi'\rangle$, where

$$D(z) = z - E_{0r} - \frac{1}{2\pi} \sum_{n\neq r,\beta} \int_0^\infty \frac{\Gamma_{rn\beta}(k)}{z-k-E_{0n}} \, dk \, , \quad (5.8)$$

$$\Gamma_{rn\beta}(k) = 2\pi\lambda k \, |f_{rn\beta}(k)|^2 \, . \quad (5.9)$$

The analytic properties of $\mathcal{O}_{r;0}(z)$ and of the other matrix elements of the resolvent are determined by those of the denominator function $D(z)$. This function can be explicitly computed from Eq. (2.14). It is given by [8]

$$D(z) = z - E_{0r} - \frac{1}{2\pi} \sum_{n\neq r,\beta} \Gamma_{rn\beta}(z-E_{0n}) \, \ln[(E_{0n}-z)/K_{rn}]$$

$$+ \lambda \sum_{n\neq r,\beta} \left[(z-E_{0n})^2 + K_{rn}^2\right]^{-p_{rn}} Q_{rn\beta}(z-E_{0n}) \, , \quad (5.10)$$

where, on the physical sheet,

$$\ln[(E_{0n}-E \mp i0)/K_{rn}] = \ln|(E_{0n}-E)/K_{rn}| \mp i\pi\theta(E-E_{0n}) \, , \quad (5.11)$$

θ being the Heaviside step function, and $Q_{rn\beta}(z)$ is a polynomial with real coefficients, that are $\mathcal{O}(1)$ in the units (2.10), with degree $Q_{rn\beta} \leq 2p_{rn}-1$. Thus,

$$D(z) = \mathcal{O}(z) \qquad (|z| \to \infty) \, . \quad (5.12)$$

The function $D(z)$ is holomorphic and zero-free on the physical sheet. The coefficients of the polynomial $Q_{rn\beta}(z)$ follow from Eqs. (2.14), (5.9), (5.10) and the condition that $D(z)$ has no singularities at $z = E_{0n}\pm iK_{rn}$ on the physical sheet. According to Eq. (5.10), $D(z)$ has a logarithmic branch point at each unperturbed bound state energy E_{0n}, except E_{0r}. The logarithmic terms, while reminiscent of the "Bethe-log" contributions to the Lamb shift, are exact within this model.

6. APPLICATIONS TO DECAY AND RESONANCE SCATTERING

We consider these problems within the framework of the exactly soluble model of Section 5.

(a) Decay of a Weisskopf-Wigner Initial State

The initial state $|r;0\rangle$ defined by Eq. (5.4) is a dissipative state in the sense of Van Hove (Section 3). According to Eqs. (3.4) and (5.6),

$$\langle r;0|\exp(-iHt)|r;0\rangle = -\frac{1}{2\pi i}\int_C \frac{\exp(-izt)}{D(z)}\,dz\ . \qquad (6.1)$$

Let us deform the path C in the way indicated in Fig. 2, into a series of paths C_1, C_2, ..., C_N directed parallel to the bisector of the fourth quadrant, such that C_j winds around the branch point E_{0j}. Thus, the l.h.s. of C_j and the r.h.s. of C_j are on different Riemann sheets, but the r.h.s. of each path is on the same sheet as the l.h.s. of the following one. These deformations are allowed by Eq. (5.12). Each time the path winds around a branch point E_{0j}, a new term $+i\sum_\beta \Gamma_{rj\beta}(z-E_{0j})$ is added to the determination of $D(z)$ in the corresponding Riemann sheet (cf. (5.10)-(5.11)). On each sheet, the integrand has a finite number of poles, the positions of which can be determined [8] from Eq. (5.10). With the above choice of contour, it may be shown [8] that only one pole z_r

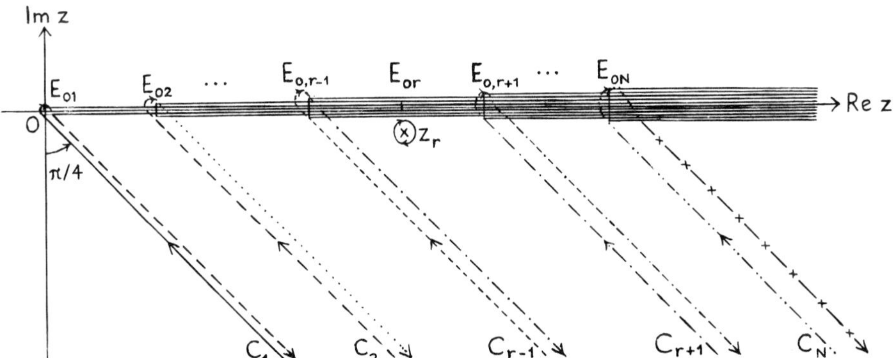

Fig. 2. Equivalent path of integration for (6.1). The positive real axis is covered by a series of superimposed branch cuts associated with the threshold branch points, and the path winds around successive branch points; portions belonging to the same Riemann sheet are similarly represented. The pole z_r of (6.2) is indicated.

(Fig. 2) will be crossed, with

$$z_r = E_r - \frac{i}{2} \Gamma_r = E_{0r} + \frac{1}{2\pi} \sum_{n \neq r, \beta} P \int_0^\infty \frac{\Gamma_{rn\beta}(k)}{E_{0r} - k - E_{0n}} \, dk$$

$$- \frac{i}{2} \sum_{\beta, n=1}^{r-1} \Gamma_{rn\beta}(E_{0r} - E_{0n}) + \mathcal{O}(\lambda) , \qquad (6.2)$$

where only those thresholds crossed up to E_{0r} contribute to the imaginary part, and P denotes Cauchy's principal value.

For times much longer than the optical period ($E_{0r}t \gg 1$), the asymptotic behavior of Eq. (6.1) is given by [8]

$$<r;0|\exp(-iHt)|r;0> \approx [1 + \mathcal{O}(\lambda^2)] \exp(-iE_r t - \frac{1}{2} \Gamma_r t)$$

$$+ \sum_{n \neq r, \beta} A_{rn\beta} \, t^{-2\gamma_{rn\beta}} \exp(-iE_{0n}t) , \quad (6.3)$$

where $A_{rn\beta}$ are constants and $\gamma_{rn\beta} = J_{rn\beta} + \tau_{rn\beta}$, where $J_{rn\beta}$ and $\tau_{rn\beta}$ are associated with the multipolarity of the transition; for electric dipole transitions, $\gamma_{rn\beta} = 1$.

The first term in Eq. (6.3) corresponds to the Weisskopf-Wigner "Ansatz", with the level shift E_r-E_{0r} given by the principal-value integral in Eq. (6.2) (including, as we have seen, "Bethe-log"-like contributions). The half-width Γ_r, according to Eq. (6.2), is the sum of the partial widths for transitions to all *lower* levels, as it should be. Both the level shift and the half-width include contributions that were not taken into account in previous exactly solved models (in particular, those involving two-level atoms).

The corrections to exponential decay in Eq. (6.3) arise from the integrals around the cuts. A rigorous estimate [8] shows that, for electric dipole transitions, such corrections are negligible up to times t_0 such that $t_0^2 \exp(-\Gamma_r t_0/2) \sim (2e^2/3\pi) <r|r^2|r>$. For the Lyman-$\alpha$ line, this yields $t_0 \sim 96$ lifetimes, so that the corrections are extremely small in this case. Note that they behave as t^{-2} for $t \to \infty$, just like a free-photon wave packet when account is taken of causal propagation [16]. Analogous features have been found for massive particles [7], suggesting that limitations in the validity of the exponential decay law are to be traced to limitations of the particle concept itself when applied to unstable particles. To discuss the observability of such deviations, however, would require a theoretical analysis of the measurement process. Furthermore, the corrections are associated with threshold (infrared) effects, and multiphoton processes are not taken into account by the present model.

(b) Resonance Scattering of a Wave Packet

The dependence of the decay on the excitation can be investigated by discussing the scattering of a photon wave packet by the atom. In order for the model to remain reasonably realistic, we must assume that both the detuning from resonance and the width γ of the incident wave packet are much smaller than the level spacing, so that we are considering resonance fluorescence.

As a consequence of the generalized RWA, the state $|1;1\beta,k\rangle$ defined by Eq. (5.4) is asymptotically stationary in the sense of Van Hove, so that the incident wave packet is of the form (5.3) with n=1 (atom in the ground state). The excitation amplitude of the atom is given by (cf. Eq. (5.7))

$$\langle r;0|\exp(-iHt)|\Phi\rangle = -\frac{1}{2\pi i}\int_C \exp(-izt)\langle r;0|G(z)|\Phi\rangle\,dz. \qquad (6.4)$$

By suitable choice of the wave packet (e.g., a Lorentzian), this can be treated [8] similarly to Eq. (6.1). Typically, the excitation probability is characterized [5,7] by a rise time $T_{rise} \sim \min(\gamma^{-1},\Gamma_r^{-1})$ and by a decay time $T_{decay} \sim \max(\gamma^{-1},\Gamma_r^{-1})$. For excitation by a broad line ($\gamma \gg \Gamma_r$), the excitation and the decay may be regarded to some extent as independent processes; for a Lorentzian wave packet, the Weisskopf-Wigner state is excited with a probability $\sim(\Gamma_r/\gamma)^2$ and it decays with the natural line shape.

The scattered wave packet and the emitted line shape can also be obtained [8]. The asymptotic behavior for large times is found to display a marked dependence on the excitation.

7. PERSISTENT EFFECTS AND DRESSING TRANSFORMATION

When the "counter-rotating" terms that were omitted from the Hamiltonian in Section 5 are reintroduced, they lead to virtual transitions from the unperturbed ground state to excited states with the emission of a photon, followed by the reverse transition in which the photon is absorbed. This produces persistent perturbation effects in the sense of Van Hove, i.e., both self-energy and cloud effects. The state $|1;0\rangle$ is no longer the interacting ground state (it "gets dressed by a photon cloud"), and the states $|1;1\beta,k\rangle$ are no longer asymptotically stationary, so that they cannot be employed as a basis to build up wave packets in the treatment of resonance scattering.

In order to deal with this situation, we first apply a dressing transformation in order to get rid of the persistent effects. After this is done, the resolvent operator method can, in principle, be employed as before, with the transformed Hamiltonian now generating the dynamics of "dressed" states of the system. However, exact results are no longer obtained: the transformation is defined order-by-order in perturbation theory; we discuss the results obtained to second order. Mass renormalization must also be performed.

Our procedure is based on a generalization of a dressing transformation proposed by Faddeev [17]. We transform $H = H_0 + \lambda V$, where λ is the small perturbation parameter, into

$$H' = U H U^{-1} = H_0' + V' \quad ; \quad U = \exp(iW) \ , \quad W = W^\dagger, \qquad (7.1)$$

through the unitary transformation U, where

$$V' = \sum_{n \geq 1} \lambda^n V_n' \ , \qquad W = \sum_{n \geq 1} \lambda^n W_n \ . \qquad (7.2)$$

Equating the coefficients of equal powers of λ, we get

$$V_1' = V + i[W_1, H_0] \ , \qquad (7.3)$$

$$V_2' = i[W_1, V] + \frac{i^2}{2!} [W_1, [W_1, H_0]] + i[W_2, H_0] \ , \qquad (7.4)$$

and so on.

We now choose W_1 so that $i[W_1, H_0]$ cancels out, in Eq. (7.3), those terms in V that lead to persistent effects, thus eliminating these effects to first order in λ. With this choice for W_1, we next choose W_2 in Eq. (7.4) so as to eliminate persistent effects from the nondiagonal part of V_2' (the diagonal part is incorporated into H_0'). In principle, this procedure can be continued up to arbitrarily high order in λ.

The terms associated with persistent perturbation effects (cf. Section 3) are those that lead to $\mathrm{Im}\Sigma_\alpha(E) \neq 0$ for non-dissipative states. They are readily recognized in the present problem as those terms that give rise to transitions from the ground state.

The choice of W_1, W_2, ... is clearly not unique. This is so because two dressing transformations whose effects differ by an arbitrary admixture of transient (non-persistent) terms are equivalent by the above definition. In practice, our choice of W_1, W_2,... is guided by simplicity: their form is determined by that of the

terms we wish to cancel out, with the insertion of suitable energy
denominators to result in cancellation when the commutator with H_0
is taken.

8. DRESSING TRANSFORMATION FOR THE HYDROGEN ATOM

For the sake of simplicity, and in order to facilitate the
comparison with the usual treatment, we restrict our discussion to
the dipole approximation (cf. Section 4). The Hamiltonian of
Section 2 becomes

$$H = H_0 + H_{I,1} + H_{I,2} + H_{ren} \quad , \qquad (8.1)$$

$$H_0 = \sum_{n \geq 1} E_{0n} |n><n| + \sum_{M=-1}^{1} \int_0^\infty k\, a_M^\dagger(k)\, a_M(k)\, dk \quad , \qquad (8.2)$$

$$H_{I,1} = \sum_{ijM} \lambda_{ijM} |i><j| \int_0^\infty \rho(k)\, [a_M(k) + a_M^\dagger(k)]\, \sqrt{k}\, dk \quad , \qquad (8.3)$$

where the cutoff factor $\rho(k)$ is usually taken in the form (4.4);

$$\lambda_{ijM} = -\frac{1}{m}\, (2\alpha/3\pi)^{\frac{1}{2}} <i|p_M|j> \quad , \qquad (8.4)$$

p_M being the spherical components [9] of $\underline{p}$;

$$H_{I,2} = \lambda' \sum_{M=-1}^{1} (-)^M \int_0^\infty dk\, \sqrt{k}\, \rho(k) \int_0^\infty dk'\, \sqrt{k'}\, \rho(k')$$

$$\times [a_M(k)\, a_{-M}(k') + (-)^M a_M^\dagger(k)\, a_M(k') + h.c.] \quad , \qquad (8.5)$$

$$\lambda' = \alpha/3\pi m \quad . \qquad (8.6)$$

We have already written $H_{I,2}$ in normally-ordered form. In order
that m be the experimental electron mass, the usual nonrelativistic
mass-renormalization counterterm H_{ren} is added. To second order
(to which we limit our calculation), it is given by

$$H_{ren,2} = (\Delta m/m)\, \underline{p}^2/2m \quad , \qquad \Delta m/m = 4\lambda' \int_0^\infty \rho(k)\, dk \quad . \qquad (8.7)$$

In order to apply the dressing transformation, some slight
modifications are required in Eq. (7.4), owing to the fact that,

besides terms linear in the coupling constant, the Hamiltonian also contains quadratic terms. The transformed Hamiltonian, to second order, is

$$H' = H_0' + H_1' + H_2' \quad , \tag{8.8}$$

where H_1' is given by Eq. (8.3) with the counter-rotating terms connected with the ground state subtracted out, and H_2' will not be written out here [8]; H_0', which incorporates the diagonal contributions found in second order, differs from Eq. (8.2) by the replacement $E_{0n} \rightarrow E_{0n}'$, where

$$E_{0n}' = E_{0n} + \langle n|H_{ren,2}|n\rangle \quad , \qquad n \neq 1 \quad , \tag{8.9}$$

$$E_{01}' = E_{01} + \mathcal{L}_1 \quad , \tag{8.10}$$

$$\mathcal{L}_1 = \sum_{n \neq 1,M} \lambda_{n1M}^2 \, (E_{0n} - E_{01}) \int_0^\infty \frac{\rho(k) \; dk}{k + E_{0n} - E_{01}} \quad . \tag{8.11}$$

If $\rho(k)$ is replaced by Eq. (4.4), $\mathcal{L}_1$ becomes identical with the nonrelativisitic contribution to the ground-state Lamb shift, as computed by Bethe [18]. Thus, for the ground-state energy, the dressing transformation cancels out the contribution from the mass-renormalization counterterm, and it adds the Lamb-shift correction. Furthermore, the new interaction terms H_1' and H_2' produce no persistent effects to second order. We can therefore apply the resolvent operator method, with H' as generator of the time evolution, to discuss decay and resonance fluorescence by procedures similar to those of Section 6. We confine ourselves to a description of the main results [8].

The time evolution of an initial excited state $|n;0\rangle$ is determined, as in Section 6(a), by the analytic properties of $\mathcal{O}_{n;0}(z)$, the associated eigenvalue of the diagonal part of the resolvent. We find that, in analogy with Eq. (6.2), the dominant exponentially-decaying term arises from an unphysical-sheet pole

$$z_n = E_n - \frac{i}{2} \Gamma_n = E_{0n} + \mathcal{L}_n - \frac{i}{2} \sum_{j<n,M} \Gamma_{njM} \quad , \tag{8.12}$$

where $\mathcal{L}_n$ is the nonrelativistic contribution to the Lamb shift for state n, and Γ_{njM} are the partial widths for the transitions to all lower levels. Thus, the analogue of the energy correction (8.10) for excited states (cancellation of the mass-renormalization counterterm in Eq. (8.9) and Lamb-shift correction) appears, as it should, in the poles of the analytic continuation of the resolvent, and not through the dressing transformation, which must lead to the correct energy only for the ground state. This essential difference between ground and excited states was obscured in previous treatments where dressing transformations were attempted.

The discussion of resonance fluorescence also proceeds in analogy with Section 6(b). We define the line shape in terms of the differential cross-section for scattering of a photon with momentum k and circular polarization λ from an atom in the ground state, leading to a photon with momentum k' and circular polarization λ'. The result [8], obtained with the help of Eq. (3.12), is given by a modified Kramers-Heisenberg dispersion formula:

$$\frac{d\sigma}{d\Omega} = r_0^2 \left| \hat{\underline{\varepsilon}}_{k\lambda} \cdot \hat{\underline{\varepsilon}}_{k'\lambda'}^* + \frac{1}{m} \sum_{n \neq 1} \frac{\langle 1 | \underline{p} \cdot \hat{\underline{\varepsilon}}_{k\lambda} | n \rangle \langle n | \underline{p} \cdot \hat{\underline{\varepsilon}}_{k'\lambda'}^* | 1 \rangle}{E_{01} - E_{0n} - k} \right.$$

$$\left. + \frac{1}{m} \sum_{n \neq 1} \frac{\langle 1 | \underline{p} \cdot \hat{\underline{\varepsilon}}_{k'\lambda'}^* | n \rangle \langle n | \underline{p} \cdot \hat{\underline{\varepsilon}}_{k\lambda} | 1 \rangle}{E_{01}' + k - [E_{0n}' + \Delta_n(E_{01} + k)] + \frac{i}{2}\Gamma_n(E_{01} + k)} \right|^2 , \quad (8.13)$$

where r_0 is the classical electron radius, $\hat{\underline{\varepsilon}}_{k\lambda}$ and $\hat{\underline{\varepsilon}}_{k'\lambda'}$ are polarization vectors of the incident and scattered photons, and Δ_n and Γ_n are defined in terms of the resolvent by Eq. (3.8). The first line of Eq. (8.13) contains the effects of Thomson scattering and of antiresonant terms. If the energy of the incident photon approaches a resonance associated with a given level n, the corrections Δ_n and Γ_n to the Kramers-Heisenberg formula in the corresponding term of the second line of Eq. (8.13) become important. In this case, $E_{0n} + \Delta_n - (i/2)\Gamma_n$ is close to z_n (cf. Eq. (8.12)), so that the Lamb-shift correction and the linewidth are also properly taken into account in the resonance scattering cross-section.

9. INTERACTION HAMILTONIAN AND LINE SHAPE

We can now discuss Lamb's remark, mentioned in Section 1, by comparing Eq. (8.13), obtained from the minimal-coupling Hamiltonian in dipole approximation, with the corresponding result obtained with the interaction Hamiltonian $\overline{H}_I = -e\underline{r} \cdot \underline{E}^\perp(0)$. The relation between the two Hamiltonians for a quantized field has been discussed by Power and Zienau [19] and by Woolley [20]. In Woolley's treatment, $\underline{E}^\perp$ is indeed the transverse electric field, while in Power and Zienau's treatment it is replaced by $\underline{D} = (\underline{E} + 4\pi\underline{P})^\perp$, where $\underline{P} = e\underline{r}\delta(\underline{q})$ is the polarization operator in dipole approximation.

Both interpretations are possible, and the difference between them simply corresponds to regarding the transformation connecting the two Hamiltonians from the active or from the passive point of view [8].   In the active point of view [20], which will be adopted here, the new Hamiltonian $\overline{H}$ is connected with H (as given by Eq. (2.1) in dipole approximation) by

$$\overline{H} = \exp(-i\Sigma)\, H\, \exp(i\Sigma) \quad , \qquad \Sigma = e\underline{r}\cdot\underline{A}(\underline{0}) \quad , \tag{9.1}$$

leading to $\overline{H} = H_A + H_F + \overline{H}_I$, where H_A and H_F are given by Eq. (2.2) and Eq. (2.3), respectively, and

$$\overline{H}_I = -e\underline{r}\cdot\underline{E}^{\perp}(\underline{0}) + 2\pi \int [\underline{P}^{\perp}(\underline{q})]^2\, d^3q \quad , \tag{9.2}$$

where the last term contributes to second-order (or higher-order) calculations, e.g., in the evaluation of the Lamb shift.   Here, $\overline{H}$ is regarded as a new Hamiltonian, expressed in terms of the old canonical variables, whereas, in the passive point of view [19], we get the old Hamiltonian expressed in terms of new canonical variables.

Since the two Hamiltonians are connected by a unitary transformation [21], they lead to equivalent results, provided that the state vectors are correspondingly transformed.   If one tries to define the natural line shape in terms of the decay of a specified Weisskopf-Wigner initial state vector without taking into account the transformation of this state vector (it can be shown [8] that this correponds to adopting different definitions of the photon vacuum), the two Hamiltonians lead to different results, but one cannot say a priori which (if any) is to be preferred, because this depends on how realistic it is to regard the associated Weisskopf-Wigner state as being produced by the excitation process, which must be taken into account for an unambiguous definition of the line shape.

If we adopt a definition in terms of resonance fluorescence, we must discuss the effect of the transformation (9.1) on the S-matrix.   The exact S-matrix elements for transitions between corresponding physical states are indeed identical for H and $\overline{H}$. However, when they are defined in terms of the usual adiabatic hypothesis, this is true only after wave function renormalization [8], and the corresponding renormalization constants Z and $\overline{Z}$ (which represent the probability of finding the unperturbed ground state in the interacting ground state) are different.   Since Z and $\overline{Z}$ differ from unity only by terms of order e^2, the Kramers-Heisenberg matrix element is the same for both Hamiltonians, as is well-known [22].   However, this need not apply to Eq. (8.13), which already includes partial summations over higher-order terms, embodied in Δ_n and Γ_n.   It can be shown [8], nevertheless, that, for photon energies k within the resonance width associated with each resonance denominator, these corrections also agree, up to order e^2.

Let r be the resonant level, R the corresponding resonant term (the term n=r in the second line of Eq. (8.13)), and B the background, i.e., the sum of all remaining terms in Eq. (8.13). The result just stated then implies that, to order e^2,

$$d\sigma/d\Omega = r_0^2 \left| R + B \right|^2 = r_0^2 \left| \overline{R} + \overline{B} \right|^2 \quad . \tag{9.3}$$

However, we have $R \neq \overline{R}$, $B \neq \overline{B}$. If, as is often done, one approximates the result by retaining only the resonant term, the results are indeed different; in order to find out which is the better approximation, we estimate the effect of the background.

Let us do this first for the Lyman-α line, for which

$$\overline{R}/R = k^2/(E_{02} - E_{01})^2 \quad , \tag{9.4}$$

where E_{02} is the energy of the 2p level. To compare the line shapes, we characterize them by two parameters: the photon energy k_0 at the peak of the curve and the asymmetry δ, defined by

$$\delta = \left[\left(\frac{d\sigma}{d\Omega}\right)_{k_0 + \frac{\Gamma}{2}} - \left(\frac{d\sigma}{d\Omega}\right)_{k_0 - \frac{\Gamma}{2}} \right] \Big/ \left(\frac{d\sigma}{d\Omega}\right)_{k_0} \quad , \tag{9.5}$$

where Γ is the linewidth associated with the 2p level.

In terms of these parameters, we find that

$$(\overline{k}_0 - k_0)/\Gamma = \frac{1}{2}\,\Gamma/k_0 \quad , \tag{9.6}$$

and that R is symmetric ($\delta=0$), whereas, for $\overline{R}$,

$$\overline{\delta} = 2\Gamma/k_0 \quad . \tag{9.7}$$

Since $\Gamma/k_0 \approx 4\times10^{-8}$, both deviations are very small in this case. We may still ask, however, which one is a better approximation.

In order to find out, we must compute B, the "background" contribution. This can be done in closed form, with the help of the Coulomb Green's function, which has been employed by Gavrila [23] to compute the Kramers-Heisenberg matrix element. The result [8] for the photon energy k_0' at the peak when both R and B are included is

$$(k_0' - k_0)/\Gamma \approx -0.22\Gamma/k_0 \quad , \tag{9.8}$$

and the corresponding asymmetry parameter is

$$\delta' \approx -1.8\Gamma/k_0 \quad . \tag{9.9}$$

Comparing these results with Eqs. (9.6) and (9.7), we see that the corrections associated with $\bar{R}$ have the wrong sign, so that, in this case, it is R which represents a better approximation.

As a final example, let us discuss the line shape for the Lamb-shift transition, i.e., for the induced decay from the meta-stable $2s_{\frac{1}{2}}$ state to the $2p_{\frac{1}{2}}$ state, in the presence of near-resonant microwave photons of frequency k_0 (followed by a Lyman-α transition to the ground state). This corresponds to a more recent version [24] of Lamb's experiment, which does not employ magnetic-field tuning of the $2s_{\frac{1}{2}} - 2p_{\frac{1}{2}}$ energy difference.

We make use of the fact that, up to second order, $\mathrm{Im}\Sigma_{2s} = 0$ (cf. Eq. (3.7)), so that the metastable $2s_{\frac{1}{2}}$ state may be treated as stable, to this order. However, to the same order, $\mathrm{Re}\Sigma_{2s} \neq 0$, so that one must correct the energy of the $2s_{\frac{1}{2}}$ state in order for it to behave as an asymptotically stationary state. This is achieved by adding a term to the unperturbed Hamiltonian (8.2) and then subtracting the same term from the interaction Hamiltonian. This term should be nondiagonal, so as to remove the degeneracy between the $2s_{\frac{1}{2}}$ and $2p_{\frac{1}{2}}$ states [25], and it must not affect the ground-state energy. A suitable choice is [8]

$$H_L = 8\pi\varepsilon_0 \, a_B^3 \, \delta(\underline{r}) - 8\varepsilon_0 |1><1| \quad ,\tag{9.10}$$

where a_B is the Bohr radius and ε_0 is the (unrenormalized) nonrelativistic contribution to the Lamb shift of the $2s_{\frac{1}{2}}$ state.

This leads to the replacement

$$E_{02s} \rightarrow E_{02s} + \varepsilon \tag{9.11}$$

in Eq. (8.9), where ε is the renormalized nonrelativistic contribution to the Lamb shift of the $2s_{\frac{1}{2}}$ state, whereas Eq. (8.10) remains unaffected. Corresponding modifications must be made in the dressing transformation to ensure that the $2s_{\frac{1}{2}}$ state (as well as the ground state) remains asymptotically stationary to second order.

A contact term similar to that in Eq. (9.10) appears in the usual treatment [26] of the nonrelativistic contribution to the Lamb shift. It has also been employed by Fried [27] as a phenomenological term in a treatment of the problem based upon a semiclassical Hamiltonian.

Taking the above modifications into account, we find expressions for the differential cross-section to second order [8] that can be analysed in terms of resonant and background contributions, as in Eq. (9.3). We find that R is peaked at a photon energy

$$k_0 \approx \varepsilon - \Gamma^2/8\varepsilon \quad , \tag{9.12}$$

whereas $\overline{R}$ is peaked at

$$\overline{k}_0 \approx \varepsilon + \Gamma^2/8\varepsilon \quad , \tag{9.13}$$

while the corresponding asymmetries (9.5) are given by

$$\delta \approx -\Gamma/2\varepsilon \quad , \qquad\qquad \overline{\delta} \approx \Gamma/2\varepsilon \quad . \tag{9.14}$$

Since $\Gamma/\varepsilon \approx 10^{-1}$, we have $(\overline{k}_0 - k_0)/\varepsilon \approx 2.5\times10^{-3}$, which is about three orders of magnitude larger than the present accuracy in the measurement of the Lamb shift. Thus, the differences between R and $\overline{R}$ are easily detectable in this case.

The background contribution B can again be evaluated in closed form [28], with the help of Coulomb Green's function. The results for k_0' and δ' when both R and B are included [8] agree with $\overline{k}_0$ and $\overline{\delta}$ of (9.13) and (9.14) up to terms of the order of $\varepsilon(\Gamma/\varepsilon)^4$ and $(\Gamma/\varepsilon)^2$, respectively. This remains valid when fine and hyperfine structure contributions are included.

Thus, in agreement with Lamb's remark [2], the resonant term derived from Eq. (9.2) indeed yields better results for the line shape in the Lamb-shift transition. However, this does not hold true in other cases, as shown by our discussion of the Lyman-α line. Similar conclusions were reached by Fried [27].

The line shape in the Lamb-shift transition, defined in terms of the usual minimal-coupling Hamiltonian of quantum electrodynamics, illustrates the need to go beyond the Weisskopf-Wigner approximation. In order to obtain agreement with experiment, one must take into account the effect of (virtual) transitions to all nonresonant levels. These background corrections to the line shape can be systematically evaluated by employing the Coulomb Green's function.

As the accuracy of Lamb-shift measurements increases, the evaluation of background corrections to the line shape becomes of comparable or possibly greater significance than that of higher-order radiative corrections to the resonant term, since the result of the measurement is directly affected by the line shape, which arises from the interference between resonant and background contributions.

* Supported by Swiss National Fund.

† Work partially supported by the National Research Council of Brazil.

References

1. V. Weisskopf and E.P. Wigner, Z. Phys. *63*, 54 (1930).
2. W. Lamb, Phys. Rev. *85*, 259 (1952).
3. W. Heitler and S.T. Ma, Proc. Roy. Ir. Acad. *52*, 109 (1949);
    E. Arnous and S. Zienau, Helv. Phys. Acta *24*, 279 (1951);
    E. Arnous and K. Bleuler, Helv. Phys. Acta *25*, 581, 631 (1952).
4. E. Arnous and W. Heitler, Proc. Roy. Soc. (London) A *220*,
    290 (1953).
5. W. Heitler, *The Quantum Theory of Radiation*, 3rd ed. (Oxford
    University Press, London, 1954).
6. F.E. Low, Phys. Rev. *88*, 53 (1952).
7. H.M. Nussenzveig, *Causality and Dispersion Relations* (Academic
    Press, New York, 1972), Chapter 4.
8. L. Davidovich, Ph.D. thesis, University of Rochester (1975);
    L. Davidovich and H.M. Nussenzveig, to be published.
9. M.E. Rose, *Elementary Theory of Angular Momentum* (Wiley, New
    York, 1957).
10. H.E. Moses, Phys. Rev. A *8*, 1710 (1973).
11. L. Van Hove, Physica *21*, 901 (1955).
12. F.J. Dyson, Phys. Rev. *75*, 486, 1736 (1949).
13. E. Grimm and V. Ernst, J. Phys. A *7*, 1664 (1974); Z. Phys. A
    274, 293 (1975).
14. E.B. Davies, J. Math. Phys. *15*, 2036 (1974).
15. Cf. E. Grimm and V. Ernst, Ref. 13, for an example where this
    condition is not fulfilled.
16. E.C.G. Stueckelberg and D. Rivier, Helv. Phys. Acta *23*, 215
    (1950); M. Fierz, Helv. Phys. Acta *23*, 731 (1950).
17. L.D. Faddeev, Sov. Phys. Doklady *8*, 881 (1964).
18. H.A. Bethe, Phys. Rev. *72*, 339 (1947).
19. E.A. Power and S. Zienau, Nuovo Cimento *6*, 7 (1957), Phil.
    Trans. Roy. Soc. (London) A *251*, 427 (1959); M. Babiker,
    E.A. Power and T. Thirunamachandran, Proc. Roy. Soc. (London)
    A *338*, 235 (1974).
20. R.G. Woolley, Molec. Phys. *22*, 1013 (1971).
21. A cutoff in momentum space is required in order for (9.1) to
    define a proper unitary transformation. One can adopt the
    same cutoff as in (8.3).
22. P.A.M. Dirac, *The Principles of Quantum Mechanics*, 4th ed.
    (Oxford University Press, London, 1958).
23. M. Gavrila, Phys. Rev. *163*, 147 (1967).
24. S.R. Lundeen and F.M. Pipkin, Phys. Rev. Lett. *34*, 1368 (1975).
25. If this degeneracy is not removed, the first-order contribution
    of the minimal-coupling interaction to the $2s_{1/2} \to 2p_{1/2}$ + one
    photon transition vanishes, and one must compute higher-order
    contributions, leading essentially to the same results (cf.
    E.J. Kelsey, Phys. Rev. A *15*, 647 (1977)).
26. J.J. Sakurai, *Advanced Quantum Mechanics* (Addison-Wesley,
    Reading, Mass., 1967).
27. Z. Fried, Phys. Rev. A *8*, 2835 (1973).
28. S. Klarsfeld, Lett. N. Cim. *1*, 682 (1969).

QUADRATIC NONLINEARITIES AND TURBULENCE IN A LASER SYSTEM

F.T. Arecchi

Istituto Nazionale d'Ottica, Firenze, Italy

A.M. Ricca

CISE, Milano, Italy

1. INTRODUCTION

As is well known, the cubic nonlinearity in a many mode laser system is responsible for mode competition and for the concentration of the emitted power into one or a few excited modes. On the other hand, wide classes of nonlinear, nonequilibrium systems are known where the existence of nonlinearities with different symmetry (e.g., the Navier-Stokes nonlinearity in a driven fluid) gives rise to the spread of the energy initially fed into one mode over a wide spectrum of excited modes. These phenomena go under the general name of turbulence. When they occur in a driven fluid, the nonlinearity is the Navier-Stokes nonlinearity $\underline{V} \cdot \underline{\nabla} V$ where $\underline{V}$ is the velocity field. By Fourier-transforming this nonlinearity $\underline{V} \cdot \underline{\nabla} V$ within a finite volume, a mode-mode coupling as $\sum_{K'} K' \, V_{K-K'} V_{K'}$ arises which is quadratic in the mode amplitudes V_K. An example of amplitude equations with a cubic as well as a quadratic nonlinearity is given in the Fourier expansion of the Benard instability. Here, the quadratic interaction takes the form[1]

$$-\delta \sum_{K_1, K_2} A^*_{K_1} A^*_{K_2} \, \delta_{K_1 + K_2 + K, 0} \quad , \tag{1.1}$$

to be added in the dynamic equation for the K mode amplitude A_K. Notice that hydrodynamics allows for a very wide excitation spectrum (indeed, energy conservation implies $\omega_{K'} + \omega_{K''} = \omega_K$) as well as for three-dimensional structures, so that the momentum closure condition implied by the δ-function in Eq.(1.1) gives rise to the well-known hexagonal structures of the Benard instability.

Having this in mind we describe a physical interaction whereby the amplitudes of a laser oscillator can be perturbed by a quadratic nonlinearity, thus giving rise to turbulence-like effects. If we want to introduce terms like (1.1) in the mode amplitude equations of a laser[2], we must remember that (i) in a laser the gain line is narrow ($\Delta\omega \ll \omega_{osc.}$); (ii) the laser is usually a one-dimensional device. The first statement means that a physical, quadratic inter-action cannot couple an unbroken sequence of modes over a wide spec-tral range, as in hydrodynamics. However, such an effect can stem from an effective Hamiltonian of the form

$$H' = \hbar\, g \sum_{KK'q} (a_K^+ a_{K'}^+ a_q + a_K a_{K'} a_q^+)\, \delta_{K+K'-q,0} \quad , \tag{1.2}$$

which couples the modes K, K' within the gain line with the second harmonic modes q at frequency 2ω. This interplay between two rather narrow lines around ω and 2ω allows for a momentum closure relation

$$\underline{K} + \underline{K}' = \underline{q} \tag{1.3}$$

in one dimension, as imposed from statement (ii). On the other hand, the effective Hamiltonian (1.2) introduces terms as[3]

$$\dot{a}_K \propto \sum_{K'q} a_{K'}^+ a_q \quad , \tag{1.4}$$

which are quadratic in the amplitudes, and hence change the symme-try of the usual amplitude equations. These considerations have led us to the physical model described in the next section.

2.   THE PHYSICAL SYSTEM

A suitable way to introduce a quadratic nonlinearity in the laser rate equations is shown in Fig. 1. We consider a Fabry-Perot cavity with mirrors highly reflecting both at ω and 2ω, an active medium with inverted population on a homogeneously broadened transition at frequency ω. Also, within the cavity there is a non-linear crystal where parametric up- and down-conversion processes occur, yielding 2ω photons by second harmonic generation (SHG) and also photons by subharmonic generation. The population rate equa-tions for the modes K under the gain of the active medium and for the modes q obtained by SHG are:

$$\dot{n}_K = (\gamma_K\, n_K - \beta_{KK}\, n_K^2 - \sum_{K' \neq K} \theta_{KK'}\, n_{K'}\, n_K) + G_K - L_K \quad ,$$

$$\dot{n}_q = -\gamma_q\, n_q + G_q - L_q \quad . \left.\rule{0pt}{40pt}\right\} \tag{2.1}$$

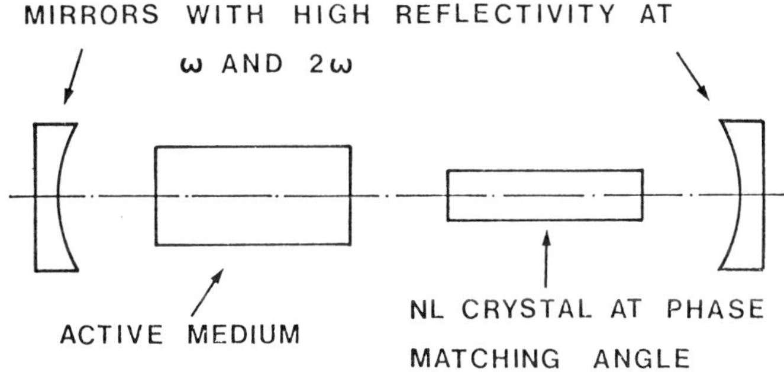

Fig. 1

The terms in the parenthesis represent the usual rate equation
terms for a many mode gas laser for which phase relations among
the modes have been neglected[2]. γ_K represents the difference
between linear gain and linear loss for the mode K; β_{KK} is usually
referred to as the cubic (in the field) nonlinearity loss coefficient for the mode K. $\theta_{KK'}$ are the cubic coefficients giving rise
to mode-mode competition. Also they refer to a cubic nonlinearity
in the amplitude equations. L and G account, respectively, for the
gain and the loss due to three photon processes caused by the nonlinear crystal. By virtue of the Fermi golden rule L_K and G_K are
given by

$$
\left.
\begin{aligned}
L_K &\propto \sum_{K'} \left| \left\langle \cdots \right\rangle \right|^2 + \left| \left\langle \cdots \right\rangle \right|^2 \;, \\[2em]
G_K &\propto \sum_{K'} \left| \left\langle \cdots \right\rangle \right|^2_{q=K+K'} + \left| \left\langle \cdots \right\rangle \right|^2_{q=2K} \;,
\end{aligned}
\right\}
\tag{2.2}
$$

where the diagrams within brackets show the matrix elements envis-
aged for the process. Here we assume a Hamiltonian given by (1.2).
Performing the calculations we have (see Appendix A)

$$
\dot{n}_K = \left(\gamma_K n_K - \beta_{KK} n_K^2 - \sum_{\substack{K' \\ K' \neq K}} \theta_{KK'} n_K n_{K'} \right) + S \sum_{\substack{K' \\ K' \neq K}} \{ n_{K+K'} n_K + n_{K+K'} n_{K'}
$$

$$
- n_K n_{K'} + n_{K+K'} \} + S \{ 4 n_K n_{2K} - n_K^2 + n_K + 2 n_{2K} \}
\tag{2.3}
$$

and

$$\dot{n}_q = -\gamma_q n_q - S \sum_{\substack{K' \\ 2K' \neq q}} \{n_q n_{q-K'} + n_q n_{K'} - n_{q-K'} n_{K'} + n_q\}$$

$$- S\{4n_K n_q - n_K^2 + n_K + 2n_q\} \, \delta(2K-q) \quad . \tag{2.4}$$

Here S is a constant to be determined, whose value is proportional to the coupling g and to the density of states. To give a numerical estimate we illustrate the process in the very special case of only one mode. Denoting by n_0 and $\tilde{n}_0$ the photon numbers at ω and 2ω , we have the following two rate equations:

$$\dot{n}_0 = \gamma n_0 - \beta n_0^2 + S(4n_0 \tilde{n}_0 + n_0 + 2\tilde{n}_0 - n_0^2) \tag{2.5}$$

and

$$\dot{\tilde{n}}_0 = -\tilde{\gamma} \tilde{n}_0 - S(4n_0 \tilde{n}_0 + n_0 + 2\tilde{n}_0 - n_0^2) \quad . \tag{2.6}$$

If we look at the equilibrium values we have

$$\tilde{n}_0 \simeq \frac{S n_0^2}{\tilde{\gamma} + 4 S n_0} \quad . \tag{2.7}$$

This equation can be considered in two limits. When $(S n_0/\tilde{\gamma} \ll 1)$, it yields the quadratic relation

$$\tilde{n}_0 \simeq \frac{S}{\tilde{\gamma}} n_0^2 \quad . \tag{2.8}$$

Comparing (2.8) with a large set of experimental data for SHG we find that $S/\tilde{\gamma}$ can be of the order of 10^{-14} (see Appendix B) for $Ba_2 Na Nb_5 O_{15}$. Hence for a cavity 4m long (to have many modes under the gain line) and with 1% mirror transmission ($\tilde{\gamma} \sim 10^6 sec^{-1}$), we obtain

$$S \sim 10^{-8} \, sec^{-1} \quad . \tag{2.9}$$

On the other hand, when $S n_0/\tilde{\gamma} \gg 1$, Eq.(2.7) saturates, giving:

$$\tilde{n}_0 \simeq \frac{n_0}{4} \quad . \tag{2.10}$$

This is a very important relation, to our knowledge given here for the first time. In fact, there are claims of $\sim 100\%$ SHG conversion [4], but they refer to a cavity whose output mirror is 99% reflecting at ω and is transparent at 2ω. Hence a power yield at 2ω equal to the power available at ω means that the photon number inside the cavity at 2ω and at ω, respectively, are in the ratio 1/100, still far away from the saturation above indicated.

To simplify the mathematical treatment to the utmost we shall assume from now on that the profile of the gain line is flat, see Fig. 2, extending around ω_0 in order to allow many modes to oscillate. Also we assume equal cubic coefficients $\beta_{KK} = \theta_{KK'} = \beta$. This implies a homogeneously broadened gain line. With these assumptions Eqs.(2.3) and (2.4) become

$$\dot{n}_K = \gamma_K n_K - \tilde{\beta} \sum_{K'} n_K n_{K'} + S \sum_{K' \neq K} \left\{ n_{K+K'} (n_K + n_{K'}) + n_{K+K'} \right\}$$

$$+ S \left\{ 4 n_K n_{2K} + 2 n_{2K} \right\} \quad , \tag{2.11}$$

and

$$\dot{n}_q = \tilde{\gamma} n_q - S \sum_{\substack{K' \\ K' \neq q/2}} \left\{ n_q n_{q-K'} + n_q n_{K'} - n_{q-K'} n_{K'} + n_q \right\}$$

$$- S \left\{ 4 n_K n_q - n_K^2 + n_K + 2 n_q \right\} \delta_{2K,q} \quad . \tag{2.12}$$

In deducing (2.11) and (2.12) we have assumed $\gamma_K = \gamma$, $\gamma_q = \tilde{\gamma}$ and $\tilde{\beta} = \beta + S$ with $\beta \gg S$.

If we consider the overall number of photons N_0 of the ω band and the corresponding number $\tilde{N}_0$ of photons of the 2ω band, from Eqs.(2.11) and (2.12), we have

$$\frac{d}{dt} (N_0 + \tilde{N}_0) = \gamma N_0 - \beta N_0^2 - \tilde{\gamma} \tilde{N}_0 \quad . \tag{2.13}$$

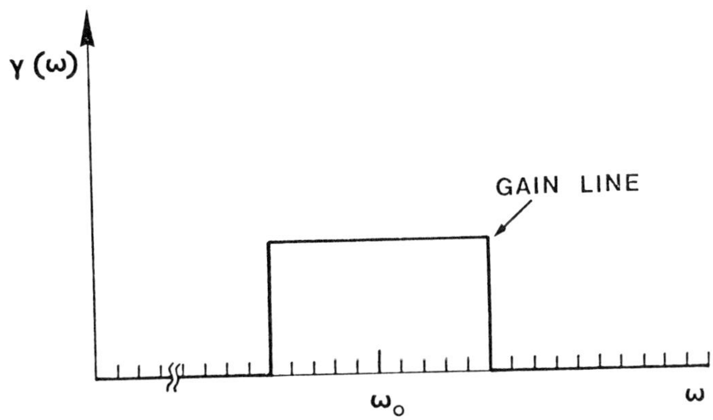

Fig. 2

Since $N_0 \gtrsim \tilde{N}_0$ and, as we shall see later, $\gamma \gg \tilde{\gamma}$, the equilibrium value $N_0^{eq.}$ of the photon number of the ω band is

$$N_0^{eq.} \simeq \gamma/\beta \quad ; \tag{2.14}$$

that is, the overall photon number is approximately a constant for any possible distribution of photons among the modes. This is a straight consequence of the assumed flatness of the gain line. In this case there is a strong mode-mode competition and, as a result, only one mode oscillates[5]. Let us indicate by n_0 the number of photons pertaining to this preferred mode, and by $\tilde{n}_0$ the number of photons of the mode whose frequency is twice the frequency of the preferred mode. In addition, let us indicate by n and $\tilde{n}$ the number of photons of the other modes under the gain line and of the SHG modes. Taking the number M of modes within the ω band as very large and neglecting the edge effects coming from the finite size of the band, one can deduce (see Appendix C) a set of coupled equations for n_0, n, $\tilde{n}_0$, $\tilde{n}$. However instead of n and $\tilde{n}$, it is more useful to introduce the quantities

$$n_1 = M\,n \tag{2.15}$$

and

$$\tilde{n}_1 = 2M\,\tilde{n} \quad , \tag{2.16}$$

which represent the overall photon populations of modes, besides the privileged ones, respectively at ω and 2ω. The equations of motion are then (see Appendix C)

$$\left.\begin{aligned}
\dot{n}_0 &= (\gamma - \tilde{\beta}\,n_0 - \tilde{\beta}\,n_1 + \tfrac{1}{2}\,S\,\tilde{n}_1 + 4\,S\,\tilde{n}_0)\,n_0 \;, \\[4pt]
\dot{n}_1 &= (\gamma - \tilde{\beta}\,n_0 - \tilde{\beta}\,n_1 + 2\,S\,\tilde{n}_0 + S\,\tilde{n}_1)n_2 + \tfrac{1}{2}\,S\,\tilde{n}_1 n_0 \;, \\[4pt]
\dot{\tilde{n}}_0 &= -(\tilde{\gamma} + 2S\,n_1 + 4S\,n_0)\,\tilde{n}_0 + S\,n_0^2 \;, \\[4pt]
\dot{\tilde{n}}_1 &= -(\tilde{\gamma} + S\,n_1 + S\,n_0)\,\tilde{n}_1 + 2S\,n_0 n_1 + S\,n_1^2 \;.
\end{aligned}\right\} \tag{2.17}$$

The above equations describe the interplay among the preferred laser mode and all the others within the gain line under the simultaneous presence of a nonlinearity cubic in the field amplitudes related to the laser process, and a quadratic nonlinearity related to a parametric interaction within the cavity.

3. THE ONSET OF TURBULENCE: STABILITY ANALYSIS

In this section we show that Eqs.(2.17) describe two different regimes between which a phase transition occurs. In the first regime, terms coming from the cubic nonlinearity of the laser prevail on the quadratic nonlinearity coming from the Hamiltonian (1.2), and as a result, there is competition with only one emerging mode for each band. In the second regime the quadratic nonlinearity overcomes the cubic nonlinearity and as a result, all modes are equally populated. Since turbulence is an effect in which energy is spread over many modes, we shall call the second regime a turbulent-like regime. Let us introduce the dimensional parameter α defined as

$$\alpha \equiv \frac{S\,\gamma}{\beta\,\tilde{\gamma}} \qquad\qquad (3.1)$$

If $\alpha \ll 1$, it can be easily seen that a steady-state solution of Eqs.(2.11) is

$$P_1 \begin{cases} n_0 = d \\ n_1 = 0 \\ \tilde{n}_0 = S\,d^2/\tilde{\gamma} \\ \tilde{n}_1 = 0 \end{cases}, \qquad\qquad (3.2)$$

where we have indicated by d the ratio γ/β. Also, it can be shown (see Appendix D) that this solution is stable. The above solution means that all energy pertaining to the ω band, is stored in one mode ω_0 and that the 2ω band is empty but the $2\omega_0$ mode is slightly excited.

If $\alpha \gg 1$, it can be shown that a stable (see Appendix D) steady-state solution is

$$P_2 \begin{cases} n_0 = 0 \\ n_1 = d \\ \tilde{n}_0 = 0 \\ \tilde{n}_1 = d \end{cases}. \qquad\qquad (3.3)$$

The above solution illustrates a situation in which all modes are equally populated in the two bands. Moreover, the two bands have the same number of photons. Of course, for mode number $M \gg 1$, it is practically irrelevant that the once-privileged model is empty. In terms of energy this means that if $\alpha \ll 1$, the energy stored within the cavity is $\hbar\omega d$. If, however, $\alpha \gg 1$, the energy stored is $3\,\hbar\omega d$. This can be understood in the following way: when only

one mode oscillates, it feeds the mode at double frequency, which
cannot grow too much because of its tendency to dissociate. However, when the phase transition has occurred, the 2ω band is filled
but each mode has a few photons and therefore the tendency to decay
is reduced.

4. CONCLUSION

By insertion of an SHG crystal into a laser cavity reflecting
at both the laser frequency ω as well as at 2ω, we have shown that
as the excitation overcomes a critical threshold, a phase transition occurs, in the sense that modes usually killed by the cubic
competition become populated. In the ideal case of a flat homogeneous gain line, the laser goes from a single, excited mode to
all available modes being equally excited. The approach to the
critical point is like a first-order phase transition, even though
the concept of one slow variable (order parameter) becomes questionable near this threshold. The passage from one stable regime
to the other is characterised by a dimensionless number

$$\alpha = \frac{S\,\gamma}{\beta\,\tilde{\gamma}} \quad , \tag{4.1}$$

being of the order of 1. Here γ is the difference between linear
gain and loss and β is the cubic coefficient of the laser; $\tilde{\gamma}$ is
the cavity loss rate at 2ω and S is the SHG coupling coefficient.
In a more realistic approach that we have recently completed[6],
by considering a non-flat gain line, we show that the onset of the
lateral modes is not simultaneous, but sequential, through a series
of bifurcation points which makes the phenomenon resemble hydrodynamic turbulence[1].

In order to justify the phase averaging, it seems important to
have incommensurable mode frequencies. In a laser medium the regularity in the sequence of the Fabry-Perot modes on the two sides of
the central mode is already destroyed by linear pulling, provided
the central mode is not located exactly at the center of the atomic
line. When the modes are regularly spaced, phased contributions
introduce new cubic, as well as quadratic terms and thus the SHG
nonlinearity may bring about mode-locking rather than turbulence.

APPENDIX A

Evaluation of Matrix Element in Eq.(2.2)

For $K' \neq K$ we have by virtue of (1.2)

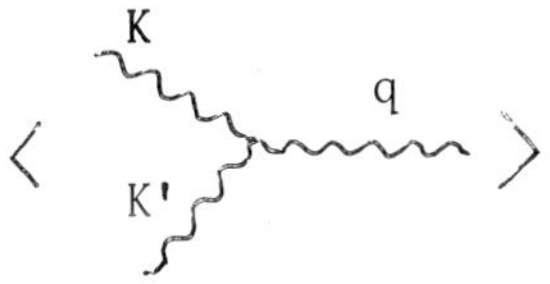

$$\propto \langle ..,n_K - 1,..,n_{K'} - 1,..,n_q + 1,.. | a_q^+ a_K a_{K'} | ..,n_K,..n_{K'},..n_q \rangle$$

$$= \sqrt{n_K n_{K'} (n_q + 1)} \quad . \tag{A.1}$$

For $K' = K$

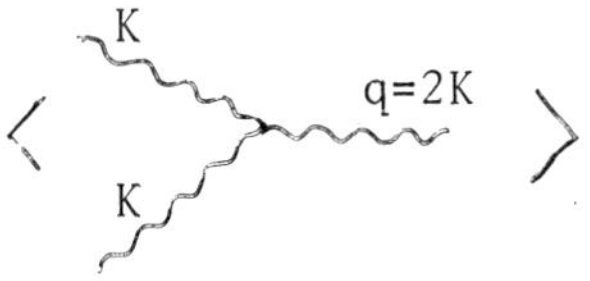

$$\propto \langle ..n_q + 1,..n_K - 2,.. | a_q^+ (a_K)^2 | ..,n_q,..n_K,.. \rangle$$

$$= \sqrt{(n_q + 1) n_K (n_K - 1)} \quad . \tag{A.2}$$

For $K' \neq K$ and $q = K+K'$

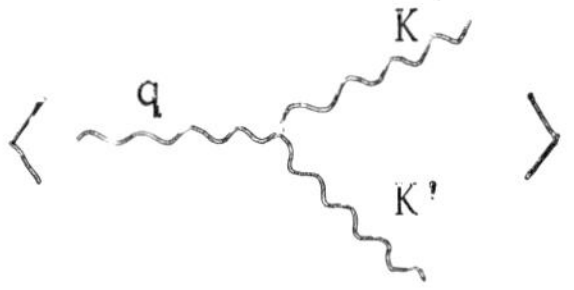

$$\propto \langle ..n_q - 1,..,n_K + 1,..n_{K'} + 1,.. | a_q a_K^+ a_{K'}^+ | ..n_q,..n_K,..n_{K'},.. \rangle$$

$$= \sqrt{n_q (n_K + 1)(n_{K'} + 1)} \quad . \tag{A.3}$$

For K=K' and q = 2K

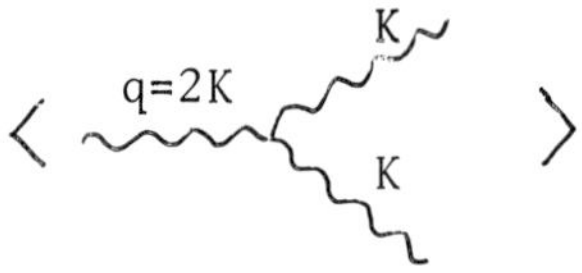

$$\propto \left\langle ..n_q - 1,..,n_K + 2,.. \left| a_q (a_K^+)^2 \right| ..n_q,..n_K,.. \right\rangle$$

$$= \sqrt{n_q (n_K + 1) (n_K + 2)} \quad . \tag{A.4}$$

APPENDIX B

Numerical Evaluation of SHG Coefficient

The SHG conversion efficiency parameter q, defined as

$$q = \frac{P^{2\omega}}{(P^{\omega})^2} \quad , \tag{B.1}$$

where $P^{2\omega}$ and P^{ω} are the radiation powers at 2ω and ω, can be as high as 5×10^{-3} W^{-1} for the crystal $Ba_2 Na Nb_5 O_{15}$. By virtue of (B.1) and (2.8) we have

$$\frac{S}{\tilde{\gamma}} = \frac{\hbar\omega q\ c}{2L} \quad , \tag{B.2}$$

L being the length of the cavity. If we consider a cavity whose length is of the order of 4m, having many modes under the gain line and a transition in the visible, we have

$$\frac{S}{\tilde{\gamma}} \sim 10^{-14} \quad . \tag{B.3}$$

$\tilde{\gamma}$ can be roughly determined neglecting diffraction losses as

$$\tilde{\gamma} \simeq \frac{c}{L} (1 - R) \quad , \tag{B.4}$$

R being the reflectivity of the mirrors. Hence we have for R = 99%

$$S \sim 10^{-8} \ sec^{-1} \quad . \tag{B.5}$$

APPENDIX C

Derivation of the Coupled Photon Equations for a Flat Gain Line

From Eqs. (2.11) and (2.12) one has with straightforward calculations:

$$\dot{n}_0 = (\gamma - \tilde{\beta}n_0 - \tilde{\beta}Mn + SM\tilde{n} + 4S\tilde{n}_0)n_0 + SM\tilde{n}n + SM\tilde{n} + 2S\tilde{n}_0 \quad , \tag{C.1}$$

$$\dot{n} = (\gamma - \tilde{\beta}n_0 - \tilde{\beta}Mn + 2S\tilde{n}_0 + 2SM\tilde{n})n + S\tilde{n}n_0 + S\tilde{n}_0 + Sm\tilde{n} \quad , \tag{C.2}$$

$$\dot{\tilde{n}}_0 = -(\tilde{\gamma} + 2SMn + SM + 4Sn_0)\tilde{n}_0 + SMn^2 + Sn_0^2 - Sn_0 \quad . \tag{C.3}$$

The rate equation for $\tilde{n}$ is more troublesome to deduce, namely, the number of channels (K, K') in which a mode q can decay depends on the position of q in the 2ω band in contrast with the assumed flat photon population profile both at ω and 2ω. To overcome this difficulty and to have a certain degree of self-consistency, we proceed in the way hereafter described. We neglect small order terms in Eqs. (C.1)-(C.3) and rewrite the above equations in terms of n_0, n_1, $\tilde{n}_0$, $\tilde{n}_1$ only. We have

$$\dot{n}_0 = (\gamma - \tilde{\beta}n_0 - \tilde{\beta}n_1 + \frac{1}{2}S\tilde{n}_1 + 4S\tilde{n}_0)n_0 \quad , \tag{C.4}$$

$$\dot{n}_1 = (\gamma - \tilde{\beta}n_0 - \tilde{\beta}n_1 + 2S\tilde{n}_0 + S\tilde{n}_1)n_1 + \frac{1}{2}S\tilde{n}_1 n_0 \quad , \tag{C.5}$$

$$\dot{\tilde{n}}_0 = -(\tilde{\gamma} + 2Sn_1 + 4Sn_0)\tilde{n}_0 + Sn_0^2 \quad . \tag{C.6}$$

On the other hand, we know from Eq. (2.12) that the motion equation for $\tilde{n}_1$ must be of the form

$$\dot{\tilde{n}}_1 = -(\tilde{\gamma} + \alpha Sn_1 + S\mu n_0)\tilde{n}_1 + \eta Sn_0 n_1 + S\varepsilon n_1^2 \quad . \tag{C.7}$$

The coefficients α, μ, η, ε are chosen so as to satisfy Eq. (2.13). Therefore we easily obtain

$$\dot{\tilde{n}}_1 = -(\tilde{\gamma} + Sn_1 + Sn_0)\tilde{n}_1 + 2Sn_0 n_1 + Sn_1^2 \quad . \tag{C.8}$$

$$\textit{APPENDIX D}$$

Stability Analysis of Eqs.(2.17) around the Equilibrium Points

With reference to the solution P_1 we set for $\alpha \ll 1$

$$
\left.
\begin{aligned}
n_0 &= d + \Delta_0 \quad, \\[4pt]
n_1 &= \Delta_1 \quad, \\[4pt]
\tilde{n}_0 &= Sd^2/\tilde{\gamma} + \tilde{\Delta}_0 \quad, \\[4pt]
\tilde{n}_1 &= \tilde{\Delta}_1 \quad,
\end{aligned}
\right\} \qquad (D.1)
$$

where Δ_0, Δ_1, $\tilde{\Delta}_0$, $\tilde{\Delta}_1$ are deviations assumed to be small from equilibrium values. From Eqs.(2.17), neglecting second-order terms and remembering that $\alpha \ll 1$, we have

$$
\left.
\begin{aligned}
\dot{\Delta}_0 &= -(\beta d)\Delta_0 - (\beta d)\Delta_1 + \frac{1}{2}(Sd)\Delta_1 \quad, \\[10pt]
\dot{\Delta}_1 &= -\frac{2S^2 d^2}{\tilde{\gamma}}\,\Delta_1 + \frac{1}{2}(Sd)\,\tilde{\Delta}_1 \quad, \\[10pt]
\dot{\tilde{\Delta}}_0 &= -\tilde{\gamma}\,\Delta_0 \quad, \\[10pt]
\tilde{\Delta}_1 &= -\tilde{\gamma}\,\Delta_1 \quad.
\end{aligned}
\right\} \qquad (D.2)
$$

It is easily seen from Eqs.(D.2) that the solution P_1 is stable since the deviations from equilibrium values go to zero as t goes to infinity.

Analogously with reference to the solution P_2 we set

$$
\left.
\begin{aligned}
n_0 &= \Delta_0 \quad, \\[4pt]
n_1 &= d + \Delta_1 \quad, \\[4pt]
\tilde{n}_0 &= \tilde{\Delta}_0 \quad, \\[4pt]
\tilde{n}_1 &= d + \tilde{\Delta}_1 \quad,
\end{aligned}
\right\} \qquad (D.3)
$$

for $\alpha \gg 1$. By virtue of Eqs.(2.17) and neglecting second-order terms, we have

$$
\begin{aligned}
\dot{\Delta}_0 &= -\tfrac{1}{2}(Sd)\Delta_0 \quad , \\[4pt]
\dot{\Delta}_1 &= -(\beta d)\Delta_1 - (\beta d)\Delta_0 \quad , \\[4pt]
\dot{\tilde{\Delta}}_0 &= -(2Sd)\tilde{\Delta}_0 \quad , \\[4pt]
\dot{\tilde{\Delta}}_1 &= -(Sd)\tilde{\Delta}_1 + (Sd)\Delta_1 + (Sd)\Delta_0 \quad .
\end{aligned}
\right\} \quad (D.4)
$$

From a direct inspection of Eqs.(D.4) we see that the solution P_2 is stable since the deviations from equilibrium values go to zero as t goes to infinity.

References

1. a) L. Landau and E. Lifshitz, *Fluid Mechanics* (Pergamon Press, London, 1959); b) R. Graham, Phys. Rev. Lett. *31*, 479 (1973); c) H. Haken, Rev. Mod. Phys. *47*, 67 (1975); d) see e.g. papers by H. Haken and P.C. Martin delivered at the Conference on Statistical Physics, Budapest, 25-29 August, 1975.
2. W.E. Lamb, Jr., Phys. Rev. *134*, A1429 (1964); see also M. Sargent, III, M.O. Scully and W.E. Lamb, *Laser Physics* (Addison-Wesley, Reading, Mass., 1974).
3. The introduction of a Hamiltonian and the derivation of a dynamic equation as (1.4) from it, is a purely formal device introduced for the sake of clarifying the terminology. The photon rate equations written later are a combination of the semi-classical laser equations, plus second-order terms evaluated by a perturbative approach.
4. J.E. Geusic et al., Appl. Phys. Lett. *12*, 306 (1968).
5. Any slight deviation from a flat gain line is sufficient to break the symmetry, thus giving rise to the privileged mode.
6. F.T. Arecchi, A.M. Ricca, to be published.
7. F.T. Arecchi and E. Schulz-Du Bois, *Laser Handbook* (North-Holland, Amsterdam, 1972).
8. See for instance, A. Yariv, *Quantum Electronics*, 2nd ed. (Wiley, New York, 1975) p. 430.

ON SOLUTIONS DESCRIBING SELF-INDUCED TRANSPARENCY OF ULTRA-SHORT

PULSES[†]

A. Kujawski* and J.H. Eberly[#]

University of Rochester, Rochester, New York

1. INTRODUCTION

Recent theoretical investigations [1-5] into systems of equations describing self-induced transparency (SIT) have revealed several interesting points concerning propagation of ultra-short pulses. The main theoretical reason for which these pulses are of interest appears in the fact that the standard theory of McCall and Hahn [6] is based on the slowly varying envelope assumption (SVEA) which cannot comprise a very good approximation for sufficiently short pulses. In the theory of ultra-short pulses in absorbers, attempts to get results better than the ones based on the original McCall-Hahn equations can be divided into two groups. The first group [4,5] offers systematic perturbation approaches in which $\rho = \kappa\mathcal{E}/\omega_0$, or $1/\omega_0\tau$, is a small parameter. Here τ denotes the pulse length, ω_0 is the resonant transition frequency for two-level atoms, κ is the atomic dipole matrix element p in frequency units, i.e. $p = \frac{1}{2}\hbar\kappa$, and $\mathcal{E}$ is the envelope of the electric field. In another group [1-3] of papers one derives a system of equations which may be treated as an improved version of McCall and Hahn's equations. These new equations can be solved in an analytic way yielding corrections to the known SIT solutions. There are a few conclusions which follow from both approaches and which agree between themselves confirming new features of solutions to the coupled Maxwell-Bloch equations.

In this paper we shall discuss solutions to the system of equations in which an approximate wave equation similar to the one proposed for the first time by Courtens [1] is taken into account. We shall take into account dispersion and we shall discuss in detail the role of detuning. We neglect non-resonant losses and consider

989

pulses much shorter than the relaxation times of the atomic system. It is found that different stationary solutions exist for different values of the detuning parameter.

2. WAVE EQUATION

As is well known, through using the variables $p = t - z/c$, $q = t + z/c$ the wave equation can be written down in the form

$$\frac{\partial^2 E}{\partial q \partial p} = -\Pi \left[\frac{\partial^2 P}{\partial p^2} + 2 \frac{\partial^2 P}{\partial q \partial p} + \frac{\partial^2 P}{\partial q^2} \right] \quad , \tag{1}$$

where P is the polarization that drives the electric field E. Since we consider steady-state solutions we assume that all envelope functions depend on $\zeta = t - z/V$ with V the pulse speed. The total phase ϕ_{tot} will be assumed to have one of the most general forms: $\phi_{tot}(z,t) = \omega_0 t - k_0 z + \phi(z,t) = [\omega_0 + (\partial \phi_L/\partial t)]t - [(\omega_0/c) + (\partial \phi_L/\partial z)]z + \phi_{NL}(\zeta) = \omega t - Kz + \phi_{NL}(\zeta)$ where $k_0 = \omega_0/c$, ω is a carrier frequency, K is an unknown wave vector and $\phi = \phi_L + \phi_{NL}$. In this way ϕ_L depends linearly on z and t, and the nonlinear part ϕ_{NL} depends on ζ only. Using this notation we have

$$E(z,t) = \mathcal{E}(\zeta) \, [\exp(i\phi_{tot}) + c.c.] \quad , \tag{2a}$$

$$P(z,t) = \tilde{\mathcal{P}}(\zeta) \, \exp(i\phi_{tot}) + c.c. = \mathcal{P}(\zeta) \, \exp[i(\omega t - Kz)] + c.c. \tag{2b}$$

where $\mathcal{P} = \tilde{\mathcal{P}} \exp(i\phi_{NL})$. After changing the variables z and t into p and q and denoting $\varepsilon = c/V - 1$, $\mu = K/k - 1$ where $k = \omega/c$, one gets for each right-hand term of (1)

$$\frac{\partial^2 P}{\partial p^2} = \frac{1}{4} (\varepsilon+2)^2 \ddot{\mathcal{P}} e^{i\psi} + \frac{i}{2} \omega (\varepsilon+2)(\mu+2) \dot{\mathcal{P}} e^{i\psi} - \frac{1}{4} \omega^2 (\mu+2)^2 \mathcal{P} e^{i\psi} + c.c., \tag{3a}$$

$$\frac{\partial^2 P}{\partial q^2} = \frac{1}{4} \varepsilon^2 \ddot{\mathcal{P}} e^{i\psi} - \frac{i}{2} \omega \varepsilon \mu \dot{\mathcal{P}} e^{i\psi} - \frac{1}{4} \omega^2 \mu^2 \mathcal{P} e^{i\psi} + c.c. \quad , \tag{3b}$$

$$\frac{\partial^2 P}{\partial q \partial p} = \frac{1}{4} \varepsilon(\varepsilon+2) \ddot{\mathcal{P}} e^{i\psi} - \frac{i}{4} \omega(\varepsilon+2)\mu \dot{\mathcal{P}} e^{i\psi} - \frac{i}{4} \omega \varepsilon (\mu+2) \dot{\mathcal{P}} e^{i\psi}$$

$$- \frac{1}{4} \omega^2 \mu(\mu+2) \mathcal{P} e^{i\psi} + c.c. \quad . \tag{3c}$$

In (3) the dot denotes differentiation with respect to ζ, and

ψ = ωt-Kz. Since we look for solutions for which V $\simeq$ c, ε is a small parameter. Moreover we assume K $\simeq$ k so that μ is a small parameter too. Later, after solving the complete system of equations, it will be seen that the last condition is satisfied. Under this assumption (3b) is smaller than (3a) and (3c) and after neglecting the last term on the right-hand side of (1) we obtain

$$\frac{\partial^2 E}{\partial p \partial q} = -\Pi \left(\frac{\partial^2 P}{\partial p^2} + 2 \frac{\partial^2 P}{\partial q \partial p} \right) \quad . \tag{4}$$

Integration of (4) from $-\infty$ to p together with the condition P=0, E=0 for p = $-\infty$ gives:

$$\frac{\partial E}{\partial q} = -\Pi \left(\frac{\partial P}{\partial p} + 2 \frac{\partial P}{\partial q} \right) \quad . \tag{5}$$

The linear wave equation equivalent to (5) has been derived for the first time by Courtens [1] using different arguments associated with the SVEA. The derivation presented above doesn't make any explicit use of the SVEA arguments. Although we dropped (3b), Eq. (1) still contains terms with the factors ε^2, μ^2, $\varepsilon\mu$. In further calculations they will be neglected if they appear together with the ε and μ terms.

3. SYSTEM OF MAXWELL-BLOCH EQUATIONS

We solve (5) together with the optical Bloch equations making use of the phase form ϕ_{tot} = $\omega_0 t - k_0 z + \phi(z,t)$ without any assumption about the dependence of ϕ on z and t. One gets the same results by putting ϕ_{tot} = $\omega t - K z + \phi_{NL}(\zeta)$ and keeping ω and K as independent parameters whose possible values will be specified by conditions for the existence of solutions possessing a clear physical interpretation. We should stress that the assumption $\phi(z,t) = \phi(\zeta)$ leads to a narrower class of solutions. This fact is easily seen when one writes $\phi(\zeta)$ = $\phi_L(\zeta) + \phi_{NL}(\zeta)$ = $-\gamma\zeta + \phi_{NL}(\zeta)$, where γ is the detuning parameter, because in this case K = $k_0 - \gamma/V$ and there is no dispersion effect for the exact resonant case γ=0.

Introducing the in- and out-of-phase components u and v of the polarization, $\hat{P}$ = $\frac{1}{2}$ N$\hbar$K(u + iv), the cooperative time τ_c of the resonant medium, τ_c^{-2} = $\frac{1}{2}$ πN$\hbar\omega$,K^2, and denoting ρ = $K \mathcal{E}/\omega_0$, we use the optical Bloch equations

$$\dot{u} = \frac{\partial \phi}{\partial t} v \quad , \tag{6a}$$

$$\dot{v} = - \frac{\partial \phi}{\partial t} u + \rho\omega_0 w \quad , \tag{6b}$$

$$\dot{w} = -\rho \omega_o v \quad , \tag{6c}$$

to obtain from (2) and (5) the two equations:

$$\varepsilon \dot{\rho} = - \frac{2}{\omega_o \tau_c^2} v - \frac{c}{\omega_o^2 \tau_o^2} \left(\frac{1}{V} \frac{\partial \phi}{\partial t} + \frac{\partial \phi}{\partial z} \right) v \quad , \tag{7a}$$

$$\rho \left(c \frac{\partial \phi}{\partial z} + \frac{\partial \phi}{\partial t} \right) = - \frac{2}{\omega_o \tau_o^2} u - \frac{c}{\omega_o^2 \tau_o^2} \left(\frac{1}{V} \frac{\partial \phi}{\partial t} + \frac{\partial \phi}{\partial z} \right) u - \frac{\rho w}{\omega_o^2 \tau_o^2} (\varepsilon - 2) \quad . \tag{7b}$$

Here we do not consider inhomogeneous broadening and the detuning $\gamma = \omega_0 - \omega$ is the same for all atoms. Its value specifies the carrier frequency ω. It is instructive to recall that the approximate first order wave equations based on the SVEA approach are of the form of (7), having on the right-hand sides the terms $-2v/\omega_0^2\tau_c^2$, $-2u/\omega_0^2\tau_c^2$ only. Courtens [1] and Lee [2] derived equations similar to (7) making use of $\varepsilon/\omega\tau$ as a small parameter. Here Eqs. (7) are derived in a different way without any explicit use of $1/\omega\tau$ under the assumptions that ε and μ are small parameters only. We shall briefly sketch how stationary solutions to (6) and (7) can be found. From (6a) we calculate $\partial\phi/\partial t$, and $\partial\phi/\partial z$ from (7a). After differentiating $\partial\phi/\partial t$ with respect to z and $\partial\phi/\partial z$ with respect to t, together with the fact that u, v, w, ρ depend on ζ only, and after making equal the second order mixed derivative, one obtains:

$$\varepsilon \dot{\rho} = - \frac{C\omega_o}{s} v \quad , \tag{8}$$

$$\frac{1}{V} \frac{\partial \phi}{\partial t} + \frac{\partial \phi}{\partial z} = \frac{\omega_o}{c} (C-2) \quad . \tag{9}$$

where C is an arbitrary constant and where we denote $s = \omega_o^2 \tau_c^2$ to make the notation simpler. Equation (9) gives an interpretation of C if one remembers that the nonlinear part ϕ_{NL} depends on ζ and the left-hand side of (9) specified for ϕ_{NL} vanishes. In a similar way using $\partial\phi/\partial t$ from (6a) and $\partial\phi/\partial z$ from (7b), or using (9), we get

$$\varepsilon \frac{\dot{u}}{v} - \frac{C\omega_o}{s} \frac{u}{\rho} + \frac{w\omega_o}{s} (\varepsilon - 2) - \omega_o (C-2) = 0 \quad . \tag{10}$$

Since, from (6c), (7a), (9) and the condition w=-1 for ρ=0, one has w= $(1/2C)\rho^2 \varepsilon$ s-1, after inserting w and v calculated from (8) into (10), a solution for u may be found

$$u = \frac{(\varepsilon-2)\varepsilon s}{8C^2} \rho^3 - \frac{\rho}{2C}[\varepsilon-2+s(C-2)] \quad . \tag{11}$$

From (6a), (9) and (11) we obtain

$$\frac{\partial\phi}{\partial t} = - \frac{3(\varepsilon-2)\omega_0}{8C}\rho^2 - \gamma \quad , \tag{12}$$

$$\frac{\partial\phi}{\partial z} = \frac{3(\varepsilon-2)\omega_0}{8CV}\rho^2 - \frac{\gamma}{c}(\varepsilon-1) - \frac{\omega_0(\varepsilon-2)}{cs} \quad , \tag{13}$$

$$\frac{K}{k} - 1 = \frac{\gamma\varepsilon/\omega_0+(\varepsilon-2)/s}{1-\gamma/\omega_0} \quad . \tag{14}$$

The constant C depends on γ in the following way

$$\frac{\gamma}{\omega_0} = - \frac{\varepsilon-2}{2s\varepsilon} - \frac{C-2}{2\varepsilon} \quad . \tag{15}$$

One can easily check that (12), (13) and (15) satisfy (9). Using the relation $u^2+v^2+w^2 = 1$, one finds v and, next, from (8):

$$(\dot{\rho}^2) = \pm\rho^2(a\rho^4 + b\rho^2 + d)^{\frac{1}{2}} \quad , \tag{16}$$

where

$$a = - \frac{\omega_0^2(\varepsilon-2)^2}{16C^2} \quad , \qquad b = \omega_0^2\left[(\varepsilon-2)\frac{\varepsilon-2+s(C-2)}{2Cs\varepsilon} - 1\right] \quad ,$$

$$d = \frac{4\omega_0^2}{s\varepsilon}\left[C - \frac{[\varepsilon-2+s(C-2)]^2}{4\varepsilon s}\right] \quad . \tag{17}$$

In (17) for simplicity of notation we use C but by (15) one can express all formulas in terms of γ only. Moreover from (16), (17) and (15) it is seen that solutions to (16) also depend on the detuning parameter γ. Equation (12) also shows that chirping is present.

4. CLOSE-TO-RESONANT SOLUTIONS

Solutions to (15) can easily be found and classified according to values of d. For $d > 0$ together with the condition that for $\zeta \rightarrow \pm\infty$ one has $\rho = 0$, one gets

$$\rho^2 = \frac{d}{\sqrt{\Delta}\,\cosh^2(\zeta/\tau) - \frac{1}{2}(\sqrt{\Delta} + b)} \quad , \tag{18}$$

where $\Delta = b^2 - 4ad$ and the pulse duration is $\tau = 2/\sqrt{d}$. The values of γ specified by $d > 0$ and (15) satisfy the following inequality

$$-\frac{1}{s} + \frac{1}{s} \sqrt{\frac{2(1+s)}{\varepsilon}} > \frac{\gamma}{\omega_0} > -\frac{1}{s} - \frac{1}{s} \sqrt{\frac{2(1+s)}{\varepsilon}} \quad , \tag{19}$$

but, within these limits can be chosen arbitrarily, yielding the pulses given by Eq. (18). Although in our considerations of stationary solutions γ is an arbitrary parameter some special cases are interesting. For exact resonance, $\gamma = 0$, one has $C = 2 - (\varepsilon - 2)/s$. Thus, for stationary exact resonant pulses with different velocities V, one has to use different values of C. However when $C = 2(1+1/s)$ is independent of V, the detuning is fixed at $\gamma = -\omega_0/2s$ for each stationary pulse. When $\gamma = 0$ one obtains

$$\frac{\partial \phi}{\partial t} = -\frac{3(\varepsilon - 2)\omega_0}{8[2 - (\varepsilon - 2)/s]} \rho^2 \quad , \tag{20}$$

$$\frac{cK}{\omega_0} - 1 = \frac{\varepsilon - 2}{s} \quad , \tag{21}$$

$$\varepsilon = \frac{2\tau^2(1+1/s)}{\tau_c^2(1 - \tau^2/s\tau_c^2)} \quad . \tag{22}$$

Up to now we haven't used the fact that $1/s = (\omega_0\tau_c)^{-2}$ takes small values, because for typical situations $\tau_c \simeq 10^{-10}$ sec and for optical frequencies $\omega_0\tau_c \simeq 10^5$. If one makes use of the relations $1/s, \varepsilon \ll 1$ in (20)-(22) and makes the additional approximation $\varepsilon - 2 \simeq -2$, one reproduces almost all the corrections obtained by various perturbation approaches [4,5]. Our result given by (22) with $1/s \ll 1$ demonstrates that the basic formula of SIT, $\varepsilon = 2\tau^2/\tau_c^2$, is valid for arbitrarily short pulses to an extremely good approximation. It is valid equally well for other γ satisfying (19). Moreover Eqs. (20) and (21) have a remarkable property. Their left-hand sides vanish for $\varepsilon = 2$, corresponding to $c/V = 3$ or $\tau = \tau_c$. Of course, this result is beyond our approximation $\varepsilon \ll 1$. However, it has been found previously [1,4,7] in a few different ways and this fact suggests that Eqs. (7) are valid not only in the region of small values of ε. Courtens' [1] results show it in a different way. Finally from (14) and (21) one finds that our assumption $\mu \ll 1$ made in the beginning is well satisfied. Inequality (19) may be expressed in another way if one puts $1+s \simeq s$ and inserts $\varepsilon = 2\tau^2/\tau_c^2$

$$-\frac{1}{s} + \frac{1}{\omega_0\tau} > \frac{\gamma}{\omega_0} > -\frac{1}{s} - \frac{1}{\omega_0\tau} \quad . \tag{23}$$

Although we don't find any mathematical restriction on values of τ from the physical point of view we cannot consider pulses for which $\omega_0\tau$ would be equal to or smaller than one. From (23) it then follows that for an extremely short pulse for which $\omega_0\tau \gtrsim 1$, the detuning γ may reach a value close to the resonant frequency ω_0. Moreover it means that possible values of ε are given by $\varepsilon \gtrsim 2/s$. When C=2 we have $\gamma/\omega_0 = -(\varepsilon-2)/2s\varepsilon \simeq 1/s\varepsilon$, the result obtained by Lee [3]. In this case γ is completely specified by the pulse velocity V and cannot be put equal to zero. Our considerations treat the general case and show that other solutions for which γ satisfies (23) are also possible.

5. OFF-RESONANT PULSES

Assuming d=0 in (16) we get a new type of pulse shape. In this case the integration of (16) yields

$$\rho^2 = \frac{1}{b\zeta^2/4 - a/b} \quad . \tag{25}$$

In (25) a and b are defined for such values of C, or γ if (15) is used, which we got from d=0

$$\frac{\gamma}{\omega_0} = -\frac{1}{s} \pm \frac{1}{s}\sqrt{\frac{2(1+s)}{\varepsilon}} \quad . \tag{26}$$

Taking into account the fact that $1/s$ and ε are small parameters, and neglecting terms of the order of $1/s$, ε, ε/s and keeping the zero order terms, from (17) one gets

$$b = \omega_0^2 \left(-1 + \sqrt{(2/s\varepsilon)}\right) \quad , \tag{27}$$

where the plus sign in (26) has been considered. The minus sign in (26) leads to b < 0 and (25) doesn't represent a solution with a clear physical meaning. Since $b = 4/\tau^2 > 0$, we get $\varepsilon < 2/s$. Because $1/s$ is a small parameter a possible solution like (25) describes pulses with velocity V very close to c. The last condition $\varepsilon < 2/s$ is satisfied for arbitrary values of τ as can be easily seen from

$$\varepsilon = \frac{2}{s(4/\omega_0^2\tau^2+1)^2} \tag{28}$$

As in the previous section we consider as physical solutions those for which $\omega_0\tau \gtrsim 1$. It sets the lower limit of ε at $\varepsilon \simeq 2/25s$. In this way Lorentzian shape solutions (25) have propagation velocities in a narrow range very close to the value c. The condition $\varepsilon < 2/s$, together with (26), leads to the conclusion that the solutions described by (25) are always off-resonant ones. When C = 2 and $\varepsilon-2 \simeq -2$ then from d = 0 one gets $\varepsilon = \frac{1}{2}s$. This is the case discussed by Lee [2] and is a special case of the off-resonant class of solutions discussed here.

For d < 0 one obtains off-resonant solutions of the oscillating type. We don't discuss them here.

6. SUMMARY

We have discussed propagation of the SIT ultra-short stationary pulses that may have their carrier frequency different from the resonant transition frequency of the two level atoms. We found that all near-resonant solutions are of the form similar to the sech type. Since our considerations concern the stationary pulses, we are not able to fix the value of the carrier frequency. Using the inequality (19) we illustrate qualitatively our results in Fig. 1. The condition $\varepsilon \gtrsim 2/s$ corresponds to solutions for which $\tau \gtrsim 1/\omega_0$ when γ/ω_0 lies between two curves (a) and (b). These are the close-to-resonant solutions. When $2/25s \lesssim \varepsilon \lesssim 2/s$ and γ/ω_0 takes values on the curve (a), one gets pulses of Lorentzian shape. The heavy dashed curve in Fig. 1 corresponds to possible solutions when the selection C = 2 is made, the case discussed earlier [3].

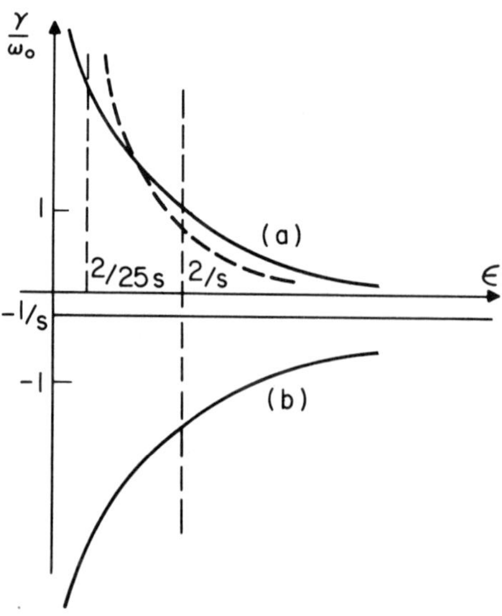

Figure 1

In our discussion of non-resonant pulses one should realize, however, that there exists another restriction of possible values of γ. The Bloch equations are derived under the "rotating wave" approximation and one has to satisfy conditions stronger than $|\gamma/\omega_0| \lesssim 1$, say $|\gamma/\omega_0| \lesssim 1/10$. It immediately leads to $\omega_0\tau \gtrsim 10$, a stronger limit on the validity of the ultra-short solutions discussed above. The existence of this RWA limit of validity has been overlooked in some of the previous work on the subject [1,3].

ACKNOWLEDGEMENTS

One of us (A.K.) wishes to thank Dr. Z. Białynicka-Birula and Dr. R. Trautman from the Institute of Physics, Polish Academy of Science for helpful discussions. Also he is grateful to the Quantum Optics group at the University of Rochester for their hospitality during his visits in 1973 and 1977.

[†]Research partially supported by U.S. National Science Foundation and U.S. Energy Research and Development Administration.

[*]Exchange scholar, U.S. National Academy of Science - Polish Academy of Sciences, 1977. Permanent address: Institute of Physics, Polish Academy of Sciences, Al. Lotnikow 32, 02-668 Warsaw, Poland.

[#]Visiting Fellow, JILA, University of Colorado and the National Bureau of Standards, Boulder, CO, during the period August 1977 - August 1978.

References

1.   E. Courtens, paper presented at the Sixth International Quantum Electronics Conference, Abstracts, p. 298, Kyoto, Japan, 1970.
2.   C.T. Lee, Opt. Commun. *9*, 1 (1973).
3.   C.T. Lee, Opt. Commun. *10*, 111 (1974).
4.   R.A. Marth, D.A. Holmes and J.H. Eberly, Phys. Rev. A*9*, 2733 (1974).
5.   Z. Białynicka-Birula, Phys. Rev. A*10*, 999 (1974).
6.   S.L. McCall and E.H. Hahn, Phys. Rev. *183*, 457 (1969).
7.   L. Davidovich and J.H. Eberly, Opt. Commun. *3*, 32 (1971).

AUTHOR INDEX